ANNALS OF
THE NEW YORK ACADEMY
OF SCIENCES

Volume 912

EDITORIAL STAFF

Executive Editor
BARBARA M. GOLDMAN

Managing Editor
JUSTINE CULLINAN

Associate Editors
JOHN W. KENNEDY
STEFAN MALMOLI

The New York Academy of Sciences
2 East 63rd Street
New York, New York 10021

THE NEW YORK ACADEMY OF SCIENCES
(Founded in 1817)

BOARD OF GOVERNORS, September 1999–September 2000

BILL GREEN, *Chairman of the Board*
TORSTEN WIESEL, *Vice Chairman of the Board*
RODNEY W. NICHOLS, *President and CEO* [ex officio]

Honorary Life Governors
WILLIAM T. GOLDEN   JOSHUA LEDERBERG

JOHN T. MORGAN, *Treasurer*

Governors

D. ALLAN BROMLEY	LAWRENCE B. BUTTENWIESER	PRAVEEN CHAUDHARI
JOHN H. GIBBONS	RONALD L. GRAHAM	HENRY M. GREENBERG
ROBERT G. LAHITA	MARTIN L. LEIBOWITZ	JACQUELINE LEO
WILLIAM J. McDONOUGH	KATHLEEN P. MULLINIX	JOHN F. NIBLACK
SANDRA PANEM	RICHARD RAVITCH	RICHARD A. RIFKIND
	SARA LEE SCHUPF	JAMES H. SIMONS

ELEANOR BAUM, *Past Chairman of the Board*

HELENE L. KAPLAN, *Counsel* [ex officio]   PETER H. KOHN, *V.P. & Secretary* [ex officio]

GAS HYDRATES

CHALLENGES FOR THE FUTURE

ANNALS OF THE NEW YORK ACADEMY OF SCIENCES
Volume 912

GAS HYDRATES

CHALLENGES FOR THE FUTURE

Edited by
Gerald D. Holder and P. R. Bishnoi

The New York Academy of Sciences
New York, New York
2000

Copyright © 2000 by the New York Academy of Sciences. All rights reserved. Under the provisions of the United States Copyright Act of 1976, individual readers of the Annals *are permitted to make fair use of the material in them for teaching and research. Permission is granted to quote from the* Annals *provided that the customary acknowledgment is made of the source. Material in the* Annals *may be republished only by permission of the Academy. Address inquiries to the Permissions Department (editorial@nyas.org) at the New York Academy of Sciences.*

Copying fees: *For each copy of an article made beyond the free copying permitted under Section 107 or 108 of the 1976 Copyright Act, a fee should be paid through the Copyright Clearance Center, Inc., 222 Rosewood Drive, Danvers, MA 01923 (www.copyright.com).*

♾ *The paper used in this publication meets the minimum requirements of American National Standard for Information Sciences—Permanence of Paper for Printed Library Materials. ANSI Z39.48-1984.*

Library of Congress Cataloging-in-Publication Data

Gas hydrates: challenges for the future / edited by Gerald D. Holder and P.R. Bishnoi.
    p. cm. — (Annals of the New York Academy of Sciences; v. 912).
    Includes bibliographical references and index.
    ISBN 1-57331-242-8 (cloth : alk. paper) — ISBN 1-57331-243-6 (pbk : alk. paper)
    1. Natural gas—Hydrates. I. Holder, Gerald D. II. Bishnoi, P.R. III. Series
Q11.N5   v. 912
[TN884]
500 s—dc21
[665.7]

                                                00-029223
                                                    CIP

K-M Research/PCP
Printed in the United States of America
ISBN 1-57331-242-8 (cloth)
ISBN 1-57331-243-6 (paper)
ISSN 0077-8923

ANNALS OF THE NEW YORK ACADEMY OF SCIENCES

Volume 912

GAS HYDRATES

CHALLENGES FOR THE FUTURE[a]

Editors
GERALD D. HOLDER AND P.R. BISHNOI

Coeditors
CHARLES W. BYRER, JEAN PIERRE MONFORT,
YASUHIKO H. MORI, PHILLIP K. NOTZ, CHARLES K. PAULL,
JOHN RIPMEESTER, E. DENDY SLOAN, JR.,
P. SOM SOMASUNDARAN, M.H. YOUSIF, AND ROBERT WARZINSKI

CONTENTS

Preface. *By* GERALD D. HOLDER xiii

Part I. Reviews

Hydrate Research—From Correlations to a Knowledge-based Discipline:
The Importance of Structure. *By* JOHN A. RIPMEESTER 1

Gas Hydrate and Humans. *By* KEITH A. KVENVOLDEN 17

Part II. Resource Characterization

Comparisons of *In Situ* and Core Gas Measurements in ODP Leg 164 Bore Holes.
By C.K. PAULL, T.D. LORENSON, G. DICKENS, W.S. BOROWSKI,
W. USSLER III, AND K. KVENVOLDEN 23

An *In Situ* Experiment of Methane Sequestration as Gas Hydrate, Authigenic
Carbonate, and Loss to the Water Column and/or Atmosphere.
By MIRIAM KASTNER, BOBB CARSON, DOUGLAS B. BARTLETT,
JOHN JAEGER, KELLY BIDLE, AND HANS JANNASCH 32

Methane Hydrate Estimates from the Chloride and Oxygen Isotopic Anomalies:
Examples from the Blake Ridge and Nankai Trough Sediments.
By RYO MATSUMOTO 39

Reservoir Characterization of Marine and Permafrost Associated Gas
Hydrate Accumulations with Downhole Well Logs.
By TIMOTHY S. COLLETT AND MYUNG W. LEE 51

[a]This volume is the result of the **Third International Conference on Gas Hydrates**, which was held on July 18–22, 1999 in Salt Lake City, Utah, USA.

North Cascadia Deep Sea Gas Hydrates. *By* G.D. SPENCE, R.D. HYNDMAN,
  N.R. CHAPMAN, R. WALIA, J. GETTRUST, AND R.N. EDWARDS 65

The Nature of Gas Hydrates on the Nigerian Continental Slope.
  By JAMES M. BROOKS, WILLIAM R. BRYANT, BERNIE B. BERNARD,
  AND NICK R. CAMERON 76

Relation between Gas Hydrate and Physical Properties at the Mallik 2L-38
  Research Well in the Mackenzie Delta. *By* W.J. WINTERS, S.R. DALLIMORE,
  T.S. COLLETT, K.A. JENNER, J.T. KATSUBE, R.E. CRANSTON, J.F. WRIGHT,
  F.M. NIXON, AND T. UCHIDA 94

Gas Hydrates of Siberia. *By* F.A. KUZNETSOV 101

The First Discovery of the Gas Hydrates in the Sediments of the Lake Baikal.
  By M.I. KUZMIN, V.F. GELETIY, G. KALMYCHKOV, F.A. KUZNETSOV,
  E.G. LARIONOV, A.YU. MANAKOV, YU.I. MIRONOV, B.S. SMOLJAKOV,
  YU.A. DYADIN, A.D. DUCHKOV, N.M. BAZIN, AND G.M. MAHOV 112

Rock Physics Characterization for Gas Hydrate Reservoirs: Elastic Properties.
  By MICHAEL B. HELGERUD, JACK DVORKIN, AND AMOS NUR 116

Part III. Geophysics

A Double Gas-Hydrate Related Bottom Simulating Reflector at the
  Norwegian Continental Margin. *By* KARIN ANDREASSEN, JÜRGEN MIENERT,
  PETTER BRYN, AND SATISH C. SINGH 126

Methane Hydrate Accumulation along the Western Nankai Trough.
  By YUTAKA AOKI, SHOSHIRO SHIMIZU, TERUMASA YAMANE,
  TOMOYUKI TANAKA, KAZUO NAKAYAMA, TSUTOMU HAYASHI,
  AND YOSHIHISA OKUDA 136

Resource Evaluation of Marine Gas Hydrate Deposits Using Seafloor
  Compliance Methods. *By* ELEANOR C. WILLOUGHBY,
  KONSTANTIN LATYCHEV, R. NIGEL EDWARDS, AND GEORGE MIHAJLOVIC .. 146

Electromagnetic Modeling and *In Situ* Measurement of Gas Hydrate in
  Natural Marine Environments. *By* RÉMI BOISSONNAS, DAVE GOLDBERG,
  AND SANEATSU SAITO 159

Relationship between Electrical and Thermal Conductivities for Evaluating
  Thermal Regime of Gas Hydrate Bearing Sedimentary Layers.
  By OSAMU MATSUBAYASHI AND R. NIGEL EDWARDS 167

Messoyakh Gas Field (W. Siberia): A Model for Development of the Methane
  Hydrate Deposits of Mackenzie Delta. *By* JAN KRASON 173

Microscopic Character of Marine Sediment Containing Disseminated
  Gas Hydrate: Examples from the Blake Ridge and the Middle
  America Trench. *By* THOMAS D. LORENSON 189

Part IV. Global Climate Change

Gas Hydrates and Global Climate Change. *By* PETER G. BREWER 195

Changes of the Hydrate Stability Zone of the Norwegian Margin from Glacial to Interglacial Times. *By* JÜRGEN MIENERT, KARIN ANDREASSEN, JÖRG POSEWANG, AND DIRK LUKAS 200

A Calculation Model for Liquid CO_2 Injection into Shallow Sub-Seabed Aquifer. *By* KYURO SASAKI AND SATOSHI AKIBAYASHI 211

The Impact of CO_2 Clathrate Hydrate on Deep Ocean Sequestration of CO_2: Experimental Observations and Modeling Results. *By* ROBERT P. WARZINSKI, RONALD J. LYNN, AND GERALD D. HOLDER 226

CO_2 Hydrate Formation in Various Hydrodynamic Conditions. *By* A. YAMASAKI, H. TENG, M. WAKATSUKI, Y. YANAGISAWA, AND K. YAMADA 235

MRI Measurement of Hydrate Growth and an Application to Advanced CO_2 Sequestration Technology. *By* SHUICHIRO HIRAI, YUTAKA TABE, KUNIHIRO KUWANO, KUNIYASU OGAWA, AND KEN OKAZAKI 246

Strength of CO_2 Hydrate Membrane in Sea Water at 40 MPa. *By* K. YAMANE, I. AYA, S. NAMIE, AND H. NARIAI 254

The Effect of CO_2–Air Mixture Compositions on the Formation and Dissociation of CO_2 Hydrate. *By* HIRONORI HANEDA, TAKESHI KOMAI, AND YOSHITAKA YAMAMOTO 261

Dynamics of Reformation and Replacement of CO_2 and CH_4 Gas Hydrates. *By* TAKESHI KOMAI, YOSHITAKA YAMAMOTO, AND KOTARO OHGA 272

Part V. Transportation and Offshore Hydrate Engineering

A New Class of Kinetic Hydrate Inhibitor. *By* MALCOLM A. KELLAND, THOR M. SVARTAAS, JORUNN ØVSTHUS, AND TAKASHI NAMBA 281

Hydrate Plug Properties: Formation and Removal of Plugs. *By* TORSTEIN AUSTVIK, XIAOYUN LI, AND LARS HENRIK GJERTSEN 294

Hydrate Dissociation in Pipelines by Two-Sided Depressurization: Experiment and Model. *By* DAVID PETERS, M. SAMI SELIM, AND E. DENDY SLOAN, JR.. 304

Comparison of Laboratory Results on Hydrate Induction Rates in a THF Rig, High-Pressure Rocking Cell, Miniloop, and Large Flowloop. *By* LARRY D. TALLEY, GARRICK F. MITCHELL, AND RUSSELL H. OELFKE .. 314

Flow Properties of Hydrate-in-Water Slurries. *By* VIBEKE ANDERSSON AND JÓN STEINAR GUDMUNDSSON 322

Flow Loop Experiments Determine Hydrate Plugging Tendencies in the Field. *By* PATRICK N. MATTHEWS, PHIL K. NOTZ, MARK W. WIDENER, AND GABRIEL PRUKOP 330

Pilot Loop Tests of a Threshold Hydrate Inhibitor. *By* T. PALERMO AND S.P. GOODWIN 339

Hydrate Plug Computer Tool. *By* GUNNAR BORTHNE AND BEN BLOYS 350

Flow Loop Tests on a Novel Hydrate Inhibitor to be Deployed in the
North Sea ETAP Field. *By* TH. PALERMO, C.B. ARGO, S.P. GOODWIN,
AND A. HENDERSON 355

Hydrate Challenges in Deep Water Production and Operation. *By* AJAY MEHTA,
JOHN WALSH, AND SUSAN LORIMER. 366

Can We Estimate the Amount of Gas Hydrates by Seismic Methods?
By AKIO SAKAI. 374

A Consistent Thermodynamic Model for Predicting Combined Wax-Hydrate in
Petroleum Reservoir Fluids. *By* A.R. TABATABAEI, A. DANESH, B. TOHIDI,
AND A.C. TODD 392

Part VI. Production

Hydrate Technology for Capturing Stranded Gas. *By* J.S. GUDMUNDSSON,
V. ANDERSSON, O.I. LEVIK, AND M. MORK. 403

Gas Hydrates and Offshore Drilling: Predicting the Hydrate Free Zone.
By K.K. ØSTERGAARD, B. TOHIDI, A. DANESH, AND A.C. TODD 411

A Simple Model for Natural Gas Production from Hydrate Decomposition.
By GOODARZ AHMADI, CHUANG JI, AND DUANE H. SMITH 420

Mathematical Models of Gas Hydrates Dissociation in Porous Media.
By GEORGE G. TSYPKIN 428

Part VII. General Papers

Department of Energy Methane Hydrate Research and Development Program:
An Update. *By* EDITH ALLISON. 437

Growth and Inhibition of Ethylene Oxide Clathrate Hydrate. *By* ROAR LARSEN,
CHARLES A. KNIGHT, KEVIN T. RIDER, AND E. DENDY SLOAN, JR. 441

Hydrogen and Oxygen Isotope Fractionation in Water during Gas Hydrate
Formation. *By* TATSUO MAEKAWA AND NOBORU IMAI. 452

Methane Hydrate Fuel Storage for All-Electric Ships: An Opportunity
for Technological Innovation. *By* M.D. MAX, V.T. JOHN,
AND R.E. PELLENBARG. 460

Methane Hydrate: Melting and Memory. *By* P. MARK RODGER. 474

Eutectic Freeze Crystallization Using CO_2 Clathrates. *By* R.J.C. VAESSEN,
F. VAN DER HAM, AND G.J. WITKAMP 483

A Unified Nucleation Theory for the Kinetics of Hydrate Formation.
By BJØRN KVAMME. 496

The Occurrence of Methane Hydrate in Ternary and Quaternary Systems
of Methane, Water, Certain Organics, and Sodium Chloride.
By M.M. MOOIJER-VAN DEN HEUVEL, M. REUVERS, R.M. DE DEUGD,
C.J. PETERS, AND J. DE SWAAN ARONS 502

Control of Gas Hydrate Formation Using Surfactant Systems: Underlying Concepts and New Applications. *By* GLEN IRVIN, SICHU LI, BLAKE SIMMONS, VIJAY JOHN, GARY MCPHERSON, MICHAEL MAX, AND ROBERT PELLENBARG 515

A Remote Station to Monitor Gas Hydrate Outcrops in the Gulf of Mexico. *By* THOMAS M. MCGEE AND J. ROBERT WOOLSEY 527

Part VIII. Hydrate Kinetics

Growth Kinetics of Single Crystal sII Hydrates: Elimination of Mass and Heat Transfer Effects. *By* P. BOLLAVARAM, S. DEVARAKONDA, M.S. SELIM, AND E.D. SLOAN, JR. 533

Methane Hydrate Dissociation Rates at 0.1 MPa and Temperatures above 272 K. *By* SUSAN CIRCONE, LAURA A. STERN, STEPHEN H. KIRBY, JOHN C. PINKSTON, AND WILLIAM B. DURHAM 544

Determination of the Intrinsic Rate of Gas Hydrate Decomposition Using Particle Size Analysis. *By* MATTHEW CLARKE AND P.R. BISHNOI 556

Mechanisms of Methane Hydrate Crystallization in a Semibatch Reactor: Influence of a Kinetic Inhibitor—Polyvinylpyrrolidone. *By* J.-S. PIC, J.-M. HERRI, AND M. COURNIL 564

Kinetics of Ethane Hydrate Growth on Latex Spheres Measured by a Light Scattering Technique. *By* PHILLIP SERVIO, PETER ENGLEZOS, AND P. RAJ BISHNOI 576

Microscopic Measurements and Modeling of Hydrate Formation Kinetics. *By* S. SUBRAMANIAN AND E.D. SLOAN, JR. 583

In Situ Observations of Methane Hydrate Formation Mechanisms by Raman Spectroscopy. *By* T. UCHIDA, R. OKABE, S. MAE, T. EBINUMA, AND H. NARITA 593

Calorimetry to Study Metastability of Natural Gas Hydrate at Atmospheric Pressure and Temperatures below 0°C. *By* O.I. LEVIK AND J.S. GUDMUNDSSON 602

Part IX. Thermodynamic and Mass Transfer in Hydrates

A Generalized Model for Calculating Equilibrium States of Gas Hydrates: Part II. *By* S.-Y. LEE AND G.D. HOLDER 614

Mechanisms for Methane Gas Accumulation under Hydrate Deposits in Sediments. *By* KEVIN L. GERING, ROBERT S. CHERRY, AND DAVID M. WEINBERG 623

Modeling of Simultaneous Heat and Mass Transfer to/from and across a Hydrate Film. *By* YASUHIKO H. MORI AND TAKAAKI MOCHIZUKI 633

Numerical Simulation of Transient Heat and Mass Transfer Controlling the Growth of a Hydrate Film. *By* TAKAAKI MOCHIZUKI AND YASUHIKO H. MORI 642

An Engineering Approach to Kinetic Inhibitor Design Using Molecular
  Dynamics Simulations. *By* E.M. FREER AND E.D. SLOAN, JR..   651

Configuration-Biased Monte Carlo Simulations of Poly(vinylpyrrolidone)
  at a Gas Hydrate Crystal Surface. *By* TIM J. CARVER,
  MICHAEL G.B. DREW, AND P. MARK RODGER.   658

A Molecular Dynamics Study of the Mechanism of Kinetic Inhibition.
  By MARK T. STORR AND P. MARK RODGER.   669

Molecular Dynamics Simulation of Dissociation Process for Methane Hydrate.
  By KENJI YASUOKA AND SUGURU MURAKOSHI   678

Molecular Orbital Calculations for Polyhedral Water Clusters Including Gas
  Molecules. *By* AKIRA HORI AND TAKEO HONDOH   685

Molecular Dynamics Simulations of Clathrate Hydrate: Intramolecular Vibrations
  of Methane. *By* HIDENOSUKE ITOH AND KATSUYUKI KAWAMURA   693

Optimizing Thermodynamic Parameters to Match Methane and Ethane
  Structural Transition in Natural Gas Hydrate Equilibria.
  By ADAM L. BALLARD AND E.D. SLOAN, JR.   702

Application and Extension of Aasberg-Petersen Model for Prediction of
  Gas Hydrate Formation Conditions in Mixtures of Aqueous Electrolyte
  Solutions and Alcohol. *By* JAFAR JAVANMARDI, MAHMOOD MOSHFEGHIAN,
  AND ROBERT N. MADDOX.   713

Measurements and Predictions of Hydrate Equilibrium Conditions.
  By LARS HENRIK GJERTSEN AND FINN HALLSTEIN FADNES   722

Part X. Hydrate Inhibition and Control

Effect of Surfactants on Hydrate Formation Rate. *By* UGUR KARAASLAN
  AND MAHMUT PARLAKTUNA.   735

Experiments Related to the Performance of Gas Hydrate Kinetic Inhibitors.
  By T.M. SVARTAAS, M.A. KELLAND, AND L. DYBVIK.   744

Kinetics of Gas Hydrates Formation and Tests of Efficiency of Kinetic Inhibitors:
  Experimental and Theoretical Approaches. *By* J.P. MONFORT,
  L. JUSSAUME, T. EL HAFAIA, AND J.P. CANSELIER   753

Study of Methane Hydrate Inhibition Mechanisms Using Copolymers.
  By B. CINGOTTI, A. SINQUIN, J.P. DURAND, AND T. PALERMO   766

Kinetics and Mechanisms of Gas Hydrate Formation and Dissociation with
  Inhibitors. *By* Y.F. MAKOGON, T.Y. MAKOGON, AND S.A. HOLDITCH   777

Effect of Inhibitor Methanol on the Microscopic Structure of Aqueous Solution.
  By YOSHITAKA YAMAMOTO, KAZUSHIGE NAGASHIMA, TAKESHI KORNAI,
  AND AKIHIRO WAKISAKA   797

Calculation of Gas Hydrate Equilibrium in Presence of Aqueous Salt
  Solutions Using a New Predictive Activity Model. *By* ASLE JØSSANG
  AND ELLEN STANGE   807

Thermodynamic Inhibitors for Hydrate Plug Melting. *By* XIAOYUN LI,
LARS HENRIK GJERTSEN, AND TORSTEIN AUSTVIK 822

Part XI. Hydrate Technologies

A Novel Approach for Oil and Gas Separation by Using Gas Hydrate Technology.
By K.K. ØSTERGAARD, B. TOHIDI, A. DANESH, R.W. BURGASS, A.C. TODD,
AND T. BAXTER 832

Feasibility of Storing Natural Gas in Hydrates Commercially.
By R.E. ROGERS AND YU ZHONG.................................. 843

Laboratory for Continuous Production of Natural Gas Hydrates.
By JON STEINAR GUDMUNDSSON, MAHMUT PARLAKTUNA,
ODD IVAR LEVIK, AND VIBEKE ANDERSSON........................ 851

Part XII. Properties of Hydrates

The Application of Raman Spectroscopy to the Study of Gas Hydrates.
By C.A. TULK, J.A. RIPMEESTER, AND D.D. KLUG 859

Structural Transition Studies in Methane + Ethane Hydrates Using Raman and
NMR. *By* S. SUBRAMANIAN, R.A. KINI, S.F. DEC, AND E.D. SLOAN, JR. .. 873

Formation of Natural Gas Hydrates in Marine Sediments: Gas Hydrate Growth
and Stability Conditioned by Host Sediment Properties.
By M. BEN CLENNELL, PIERRE HENRY, MARTIN HOVLAND,
JAMES S. BOOTH, WILLIAM J. WINTERS, AND MICHEL THOMAS 887

NMR Imaging Study of Hydrates in Sediments. *By* MARIT MORK,
GRETHE SCHEI, AND ROAR LARSEN 897

Rheological Characterization of Hydrate Suspensions in Oil Dominated Systems.
By R. CAMARGO, T. PALERMO, A. SINQUIN, AND P. GLENAT 906

A Model for Systems with Soluble Hydrate Formers. *By* M.D. JAGER,
R.M. DE DEUGD, C.J. PETERS, J. DE SWAAN ARONS, AND E.D. SLOAN 917

Improving the Accuracy of Gas Hydrate Dissociation Point Measurements.
By B. TOHIDI, R.W. BURGASS, A. DANESH, K.K. ØSTERGAARD,
AND A.C. TODD 924

Equilibrium Conditions of Methane and Ethane Hydrates in Aqueous
Electrolyte Solutions. *By* TATSUO MAEKAWA AND NOBORU IMAI 932

Crystal Growth, Structure Characterization, and Schemes for Economic
Transport: An Integrated Approach to the Study of Natural Gas Hydrates.
By D. MAHAJAN, T.F. KOETZLE, W.T. KLOOSTER, L. BRAMMER,
R.K. MCMULLAN, AND A.N. GOLAND.............................. 940

Natural Gas Storage Properties of Structure H Hydrate. *By* A.A. KHOKHAR,
E.D. SLOAN, AND J.S. GUDMUNDSSON 950

Mechanical Properties of Water/Hydrate-Former Phase Boundaries
and Phase-Separating Hydrate Films. *By* R. OHMURA,
T. SHIGETOMI, AND Y.H. MORI 958

Double Gas Hydrates at High Pressures: The Highest Decomposition
  Temperatures. *By* E.G. LARIONOV, A.YU. MANAKOV,
  YU.A. DYADIN, AND F.V. ZHURKO 967

In Situ Observation of CO_2 Hydrate by X-ray Diffraction.
  By SATOSHI TAKEYA, TAKEO HONDOH, AND TSUTOMU UCHIDA 973

High-Pressure Optical Cell for Hydrate Measurements Using Raman
  Spectroscopy. *By* V. THIEU, S. SUBRAMANIAN, S.O. COLGATE,
  AND E.D. SLOAN, JR. 983

Mechanical Stability of Gas Hydrates under Pressure. *By* V.R. BELOSLUDOV,
  V.P. SHPAKOV, J.S. TSE, R.V. BELOSLUDOV, AND Y. KAWAZOE 993

Laboratory Measurements of Compressional and Shear Wave Speeds through
  Methane Hydrate. *By* WILLIAM F. WAITE, MICHAEL B. HELGERUD,
  AMOS NUR, JOHN C. PINKSTON, LAURA A. STERN,
  STEPHEN H. KIRBY, AND WILLIAM B. DURHAM. 1003

Dissociation of Natural Gas Hydrates Observed by X-ray CT Scanner.
  By JUN MIKAMI, YOSHIHIRO MASUDA, TAKASHI UCHIDA,
  TOHRU SATOH, AND HIDEAKI TAKEDA 1011

Occurrences of Natural Gas Hydrates beneath the Permafrost Zone in Mackenzie
  Delta: Visual and X-ray CT Imagery. *By* TAKASHI UCHIDA,
  SCOTT DALLIMORE, AND JUN MIKAMI 1021

Hydrate Phase Composition for Multicomponent Gas Mixtures.
  By HENG-JOO NG 1034

Index of Contributors 1041

The New York Academy of Sciences believes it has a responsibility to provide an open forum for discussion of scientific questions. The positions taken by the participants in the reported conferences are their own and not necessarily those of the Academy. The Academy has no intent to influence legislation by providing such forums.

Preface

The Engineering Foundation Conference on Gas Hydrates, *Gas Hydrates and Global Warming,* turned out to be an enormous success. Although we expected between 100–125 participants, more than 250 people from around the world ultimately attended. Although this led to some crowding of the schedule, the intellectual, scholarly, and educational content of the conference was at the very highest levels.

This volume represents the heart of the conference. All of the papers included here have been reviewed by two reviewers prior to acceptance and they cover a rich variety of topics related to gas hydrates. The sections are designed to cover all of the areas where hydrates might be important to chemical engineers and chemists, geophysicists and oceanographers, and petroleum engineers. The subjects are broadly divided into resource characterization and geophysics, global climate change including storage of carbon dioxide as a hydrate, offshore hydrate engineering, production of gas from hydrates, hydrate kinetics, hydrate thermodynamics, mass transfer in gas hydrates, inhibition of hydrate formation in transportation lines, control of hydrates to make them more easily transportable, and properties of hydrates. Thus, this publication can answer how gas and oil production is affected by hydrate formation and prevention—a problem of immediate and critical interest for current energy suppliers. It also addresses how hydrates might serve as a clean fuel—a critical issue for future energy suppliers.

One of the more important papers of this volume is the paper by Edith Allison, which outlines a major Department of Energy strategy for supporting research in gas hydrates for the next decade. Such research is essential to the continued advancement of our knowledge of gas hydrates and their role in the world's energy cycle.

The advances made in characterizing hydrate resource have been tremendous. A companion volume, Bulletin 544, *Scientific Results from JAPEX1JNOC/GSC Mallik 2L-38 Gas Hydrate Research Well, Mackenzie Delta, Northwest Territories, Canada (1999),* prepared by the Geological Survey of Canada and edited by S.R. Dallimore, T. Uchida, and T.S. Collett, provides a great deal of breadth to the section on resource characterization and is recommended reading.

I would like to thank each of the members of the conference organizing committee: P.R. Bishnoi, University of Calgary; Charles W. Byrer, U.S. Department of Energy; Jean Pierre Monfort, ENSIGC; Yasuhiko H. Mori, Keio University; Phillip K. Notz, Texaco Group, Inc.; Charles K. Paull, University of North Carolina at Chapel Hill; John Ripmeester, National Research Council of Canada; E. Dendy Sloan, Jr., Colorado School of Mines; P. Som Somasundaran, Columbia University; M.H. Yousif, Shell E&P Technology Company; and Robert Warzinski, U.S. Department of Energy; and the staff of the Engineering Foundation including Barbara Hickernell, Rosa Landinez, and Antoinette Chartier, and my Administrative Assistant Linda Iams for making the Conference and this publication successful. I want to also thank those organizations that contributed financially to the conference including BP-Amoco, the National Science Foundation, and the U.S. Department of Energy.

Finally I want to thank superb editorial staff of the New York Academy of Sciences and of K-M Research, particularly Dr. John W. Kennedy and Mahmuda Huq, for seeing this *Annals* volume through the press.

Gerald D. Holder
University of Pittsburgh

Hydrate Research—From Correlations to a Knowledge-based Discipline

The Importance of Structure

JOHN A. RIPMEESTER[a]

Steacie Institute for Molecular Sciences, National Research Council of Canada, Ottawa, Ontario, Canada K1A 0R6

ABSTRACT: This contribution gives a short historical perspective on the development of fundamental knowledge about gas hydrates, and how such fundamental knowledge is important for a variety of problems related to hydrate prevention (pipelines), hydrate formation (CO_2 sequestration), or hydrate decomposition (permafrost or marine natural gas hydrates). It is shown that early correlations derived from measurements on hydrates, many of which were made as early as the 1800s, could not be understood properly until around 1950 when X-ray diffraction measurements gave a structural understanding of the materials we now know as clathrates, or host–guest materials. In turn, this led to a statistical mechanical description and a thermodynamic model that has considerable predictive power. Recently, it has become apparent that there is considerable complexity and subtlety in the relationship between structure and the size of the hydrate formers once more than a single species is present. This emphasizes the need to obtain structural information together with thermodynamic measurements, especially in multicomponent systems. Such developments have encouraged the adoption of additional structural techniques such as NMR and vibrational spectroscopy. Current interest in hydrate formation and decomposition requires the adaptation and development of new approaches that allow the acquisition of time-resolved structural information. Furthermore, an understanding of the morphology of hydrate crystals and how to modify this morphology, as well as understanding of interfacial properties, are all key to learning how to control hydrates. Although a great deal of progress has been made, much remains to be learned.

INTRODUCTION

Hydrate researchers today face an impressive number of practical challenges. Generally speaking, these can be classified as problems in *control* and *prediction*: the control of hydrate formation, decomposition, or hydrate morphology, be it in connection with flow assurance in pipelines, the sequestration of carbon dioxide, or the recovery of gas from natural formations, and, the prediction of the location and quantity of gas in natural formations in connection with planetary energy reserves and the potential for global climate change. To achieve such ends researchers study phase equilibria, the kinetics of hydrate formation, as well as a number of fundamental

[a]Telecommunication. Voice: 613-993-2011; fax: 613-998-7833.
jar@ned1.sims.nrc.ca

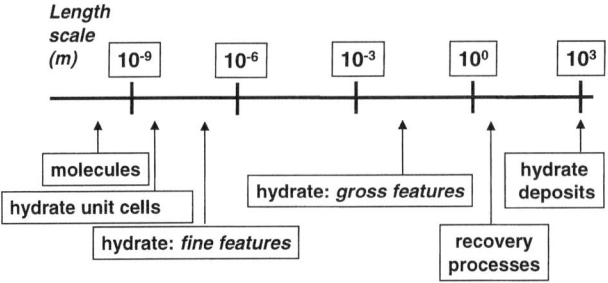

FIGURE 1. Scope of hydrate research, illustrating the length scale on which hydrate structure needs to be addressed.

properties, for example, the speed of sound, the elastic constants, and the thermal conductivity. Structure, on various length scales, is the underlying property of fundamental importance and is central to the development of a knowledge base that will lead to the understanding required for control and prediction. FIGURE 1 illustrates that hydrate research, as a multidisciplinary activity, needs to be carried out on a variety of length scales. We need to work (1) on the scale of *molecular structure*, that of guest and host molecules as probed by spectroscopic and other local order techniques in the study of solution structure and hydrate precursors; (2) on a nanometer scale, that of the *crystal unit cells,* with techniques such as diffraction and scattering; (3) on a scale of the *morphological structure* using light scattering, optical microscopy, and direct visual observation to study hydrates up to the millimeter size range; and (4) ultimately we reach the scale of proposed hydrate recovery on a scale of tens of meters, and the *geological structure of the hydrate deposits* themselves at scales that may range up to kilometers in length, where well logging and other geophysical tools need to be employed. My experience is very much with the small end of the scale, and in this paper I hope to convince you that molecular-scale knowledge is fundamental to developing an understanding of hydrates at all levels.

EARLY CORRELATIONS

The history of hydrate research is extensive; the first report on hydrate research is nearly two hundred years old. In many ways, hydrate research played a leading role in the development of modern science, as we can see from the involvement of well known researchers of the day. It is worthwhile to examine some of this early work on hydrates to see how correlations were established that were either confirmed or discarded, and how the transition to predictions based on understanding required the development of a structural basis.

Following reports on chlorine[1] and bromine[2] hydrate, the nineteenth century saw some 40 different hydrate forming species.[3] It was recognized that small, especially hydrophobic species, were able to form solid hydrates. This early correlation still teaches us that in order to take advantage of the vast store of hydrate information we should not confine ourselves to the literature on hydrocarbon hydrates since these form only a small subgroup of all hydrates.

In the late nineteenth century, a correlation known as Villard's rule[4] postulated that the hydration number should be six for all hydrates, but eventually this was shown to be incorrect. However, the large amount of activity in determining hydrate compositions did serve to show that direct determination is extremely difficult. Furthermore, the extensive phase equilibrium studies on hydrates led to the elaboration of the phase rule by Bakhuis Roozeboom.[5] This had an impact that went far beyond direct applications to hydrates.

The next, and entirely distinct, stream of input came from the engineering community starting in the 1930s. Hammerschmidt[6] proposed that hydrates were more likely to block pipelines than ice, thus leading to a large number of studies on the phase equilibria of hydrocarbon hydrates and their inhibition.[7] Katz and coworkers summarized an impressive amount of work (see, for example, Ref. 8), and in these complex systems they were often able to use the density of natural gas as a correlating parameter. By 1950 quite a large database on hydrates was available, yet what was still lacking was a knowledge of the structure of hydrates and the factors that kept the water and the other components together in a single compound. Von Stackelberg, active in hydrate research during the 1940s and 1950s, summarized the

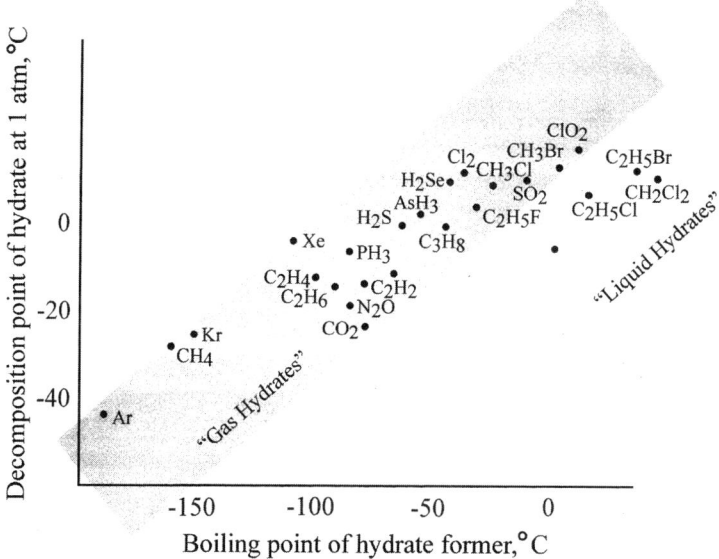

FIGURE 2. Von Stackelberg's classification of hydrates as *gas* hydrates and *liquid* hydrates. Note that the gas hydrates are structure I, with the exception of Ar, Kr, and propane hydrates. (Source: Von Stackelberg.[9])

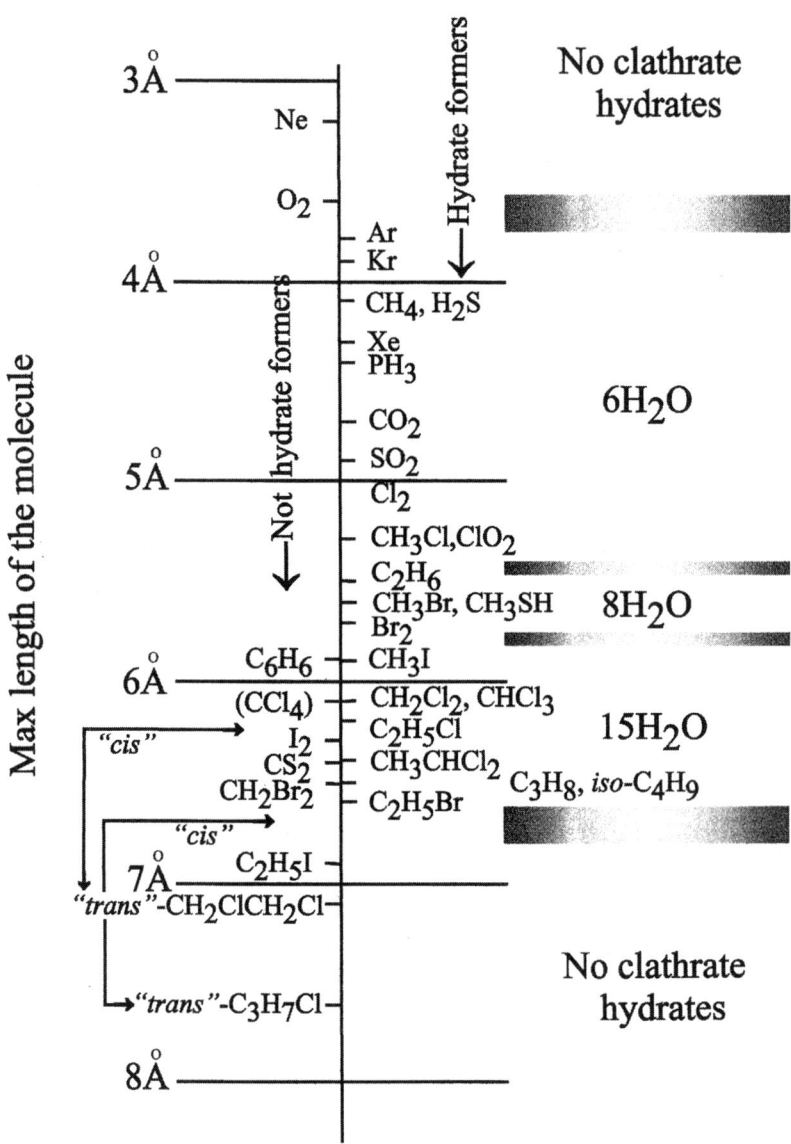

FIGURE 3. Von Stackelberg's correlation of molecular size with hydration number. Note that O_2 and CCl_4, listed as not forming hydrates are actually hydrate formers. The fact that the size ranges of molecules that are listed as hydrate formers and nonhydrate formers overlaps is anomalous, since size is the key property for hydrate formation.

state of hydrate research at the time.[9] He made the distinction between *gas* hydrates and *liquid* hydrates, correlating the decomposition temperature at 0°C with the boiling point of the guest (see FIGURE 2). We know now that the gas hydrates include both structures I and II hydrates, so that this correlation is not very useful if we want to classify hydrate structure. Von Stackelberg also produced a correlation between molecular size and hydration number (see FIGURE 3). This recognized three sets of hydrate groups with hydration numbers 6, 8, and 15. In both correlations there are anomalies that we recognize today — for instance, the listing of O_2 and CCl_4 as not forming hydrates, and the overlap of size ranges of hydrate formers and nonformers.

THE ADVENT OF CLATHRATE SCIENCE

The recognition of clathrates as a structurally unique class of materials and the coining of the word "clathrate" by Powell,[10] can be considered the advent of a new level of understanding. The observation that one material, the guest, could be trapped within another, the host, without obvious directional bonding between components, that was discovered for quinol clathrate, applied equally well to the clathrate hydrates. Within a few years of Powell's discovery, Claussen,[11], von Stackelberg and coworkers,[12] and Pauling and Marsh[13] had identified the hydrates as clathrates. This gave us the families of hydrate structures well-known as cubic structures I and II, and the three constituent cages, $D-5^{12}$, $T-5^{12}6^2$ and $H-5^{12}6^4$. The crystal structures of hydrates were explored at great length by Jeffrey and coworkers[14] in the 1960s and 1970s, who categorized and classified a large number of actual and hypothetical structures.

The next major development was the presentation of a model that links stability criteria to composition and the crystal structure of clathrates. Van der Waals and Platteeuw applied the ideal solid solution model first to the quinols and later to hydrate lattices,[15] and this elegant formulation still is fundamental to an understanding of gas hydrates. A key equation, linking $\Delta\mu$, the difference in free energy between the hypothetical empty hydrate lattice and ice, the minimum cage occupancies for hydrate lattice stability θ_i and the hydration number N is[1]

$$\Delta\mu = kT/N\Sigma_i n_i \ln(1-\theta_i). \tag{1}$$

An important feature of this equation is that for a hydrate with a single guest type, any two of the variables $\Delta\mu$, N, or θ_i/θ_j define the third. Together with the assumptions that the hydrate structure is independent of guest type and cage occupancy, this equation shows that it is possible to measure hydrate compositions from a knowledge of $\Delta\mu$ for the structure and the cage occupancy ratio, θ_i/θ_j. Thus, an instrumental method suitable for measuring cage occupancy ratios should be able to give hydrate compositions even in the presence of excess water, ice, or other impurities. As stated previously, measuring hydrate compositions directly is notoriously difficult, as attested by the numerous indirect methods that have been proposed. Application of the solid solution model also depends on the evaluation of the interaction energy between the guest and the host cage, leading to the calculation of Langmuir constants and the vapor pressure over the hydrate under equilibrium conditions. This procedure was improved and implemented by a number of groups, first using the Lennard-Jones and Devonshire spherical cell models, and later by means of the Kihara potential and explicit atomic positions for host lattice atoms.[16]

A true test of any modeling procedure is the prediction of previously unknown structures or properties. The prediction by Holder and Manganiello[17] that structure II hydrate might in fact be more stable than structure I for small guests such as Ar and Kr was borne out by experimental measurements of the crystal unit cell parameters for these hydrates,[18] as well as those of O_2 and N_2. This event was significant in that it inspired confidence that predictive modeling had come of age. One remaining challenge would seem to be the observation that carbon monoxide forms structure I hydrate:[19] CO is the same size as N_2, a structure II former, and also is isoelectronic with the latter molecule.

The structural work reported above essentially completes the molecular size—structure type—hydration number relationship that we still accept today (see FIGURE 4).

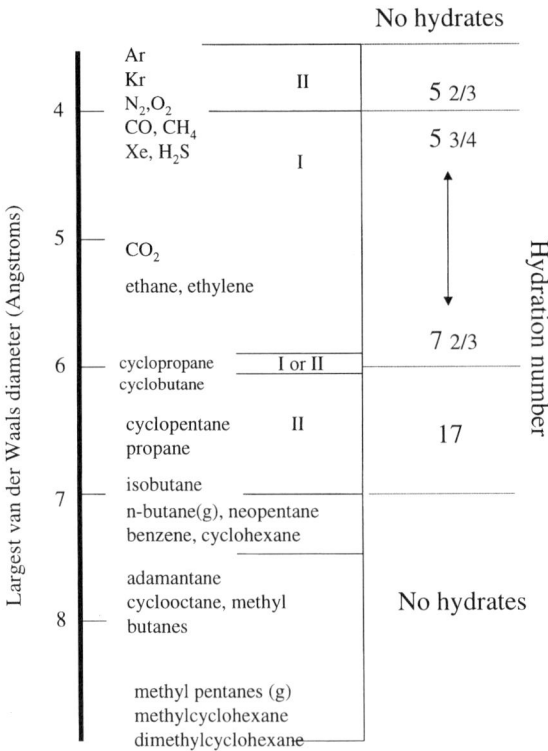

FIGURE 4. Guest size–structure–hydration number correlation accepted today for hydrates of single guests (note that there are many more guests than appear on this list).

NEW INSTRUMENTAL METHODS

In order to use equation (1) and to bring new experimental procedures to bear upon hydrate problems, NMR spectroscopic work using the isotope ^{129}Xe was implemented around 1980. Xe is a hydrate former similar in size to guests such as CH_4 and H_2S, and it forms hydrates of the same structure and similar composition. From the point of view of NMR spectroscopy, the large number of electrons around a xenon atom suggest it to be a highly polarizable species, one that is likely to show a strong sensitivity to local order in its chemical shift spectrum. The initial efforts on xenon hydrate were richly rewarded in that the chemical shift for xenon in different cages gave isotropic chemical shifts that reflected the cage size, and an anisotropic chemical shift that reflected the cage shape.[20] The combination of observed chemical shifts for a xenon-containing hydrate gives a unique signature linked directly to the crystal structure of the hydrate.[21] Such measurements also allows a completely experimental measurement of $\Delta\mu$ for structure I hydrate based on θ_i/θ_j and N measurements for a pure Xe hydrate sample.[22]

Vibrational spectroscopy also has been used to give site-specific information on clathrate hydrates. Much of the early work on infrared spectroscopy was carried out by Bertie,[23] and later, Devlin, who observed cage-dependent shifts for H_2S[24] and CO_2 trapped in hydrates. Similarly, Raman spectroscopy has been used by a number of Japanese researchers[25] in the study of air hydrates, and by Sloan and coworker[26] for methane and CO_2 hydrate. It is quite clear that encagement affects the bond lengths of molecules such as CH_4, giving rise to cage-dependent shifts of the vibrational frequencies. It should be added that the line intensities depend on the derived polarizability tensors that also vary from cage to cage, thus making quantification of the cage occupancy ratios rather less than straightforward.[27] Crosscalibration with NMR spectroscopy, and a calculation of the derived polarizability tensors will put Raman spectroscopy on firmer footing as a quantitative technique.

STRUCTURE H HYDRATE

The Xe NMR tool described above was instrumental in the characterization of a new hydrate structure now known as structure H.[28] This work was initiated in an attempt to test the upper size limit of the guest molecules that might fit into the structure II large cage. It became clear that it was relatively easy to incorporate guests that were far too large for this cage. For such samples, Xe NMR spectroscopy revealed that in fact there were two distinct small cages in addition to a large nonspherical cage as shown by ^{2}H NMR. Powder diffraction measurements showed a unique Laue symmetry, and comparison with structurally related clathrasils suggested the structure of clathrasil dodecasil-1H. This assignment was confirmed only recently by single crystal diffraction measurements.[29] The significance of the new structure can be summarized as follows: first new hydrate family structure identified since the 1950s, extended size of largest possible guests ($5^{12}6^8$ cage), and the first observation of square faces in a hydrate cage ($4^35^66^3$ cage). Subsequent studies of phase equilibria for structure H[30] resulted in a thermodynamic model[31] and the incorporation of larger guests into the prediction packages for hydrate formation. A prediction made in our original publication that structure H hydrate should occur naturally, was confirmed for hydrates found in the Gulf of Mexico in 1995.[32]

FURTHER STRUCTURAL COMPLEXITIES

Double hydrates, that is, hydrates with two guest types have been known for a long time and continue to be a source of structural complexity. De Forcrand published a list of double hydrates, based on halogenated hydrocarbons and H_2S, in about 1883.[33] Close examination shows that he prepared double hydrates with isobutyl chloride or bromide as the large guest. A comparison of guest sizes shows that these two are similar in size to isopentane—now a known structure H hydrate former. Therefore, it is likely that de Forcrand was the first to prepare structure H hydrate,

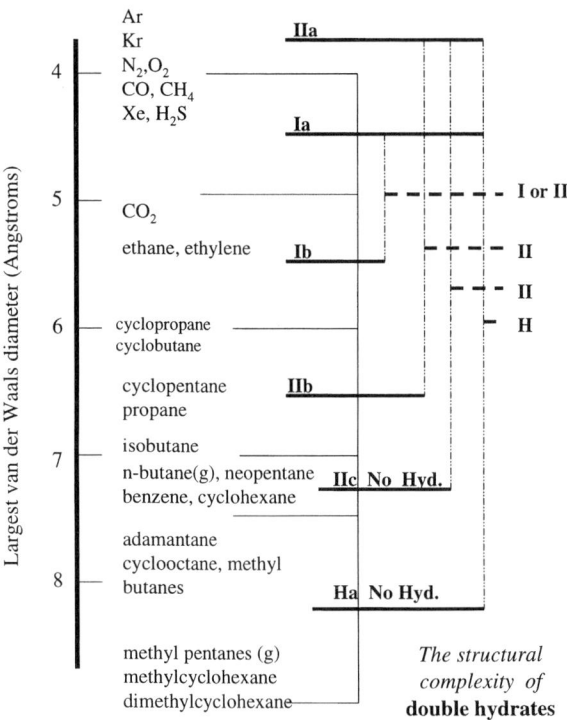

FIGURE 5. Complexity in guest size–structure correlations for hydrates with two guest species. Molecules are classified by how they would behave as single guests (i.e., structure II guests come in three size ranges, a, b, and c. The following combinations are known, although some of the boundaries may not have been firmly established: Ia + Ib → structures I or II; IIa or Ia + Ib → structure II; Ia or IIa +IIc → structure II; and Ia or IIa +Ha → structure H.

but without the tools to determine or confirm structure the discovery lay dormant for about 100 years.

Of course, hydrates based on natural gas almost always have at least two components, so these also are likely to offer a fertile ground for the discovery of further structural complexities. A case of recent interest is the prediction by Hendriks *et al.*[34] that certain combinations of methane and ethane, both structure I hydrate formers, are likely to form structure II hydrate under certain conditions. As presented in this volume by Sloan *et al.*,[35] modern instrumental approaches confirm structural transitions in methane/ethane hydrates. However, since the basic property of hydrate formers is molecular size, relevant data can be found much earlier. Von Stackelberg's list of single and double hydrates from the mid-fifties shows[36] that guest combinations of H_2S with CH_3Br, COS, and CH_3CHF_2—all structure I hydrate formers individually—give structure II lattice parameters when H_2S is combined with any of the others. This suggests that all small structure I guests (those that occupy the 5^{12} cage to a significant extent) when combined with large structure I guests (those that do not occupy the 5^{12} cage) may give structure II hydrate under certain circumstances. We may well ask, does the actual structure of the hydrate really matter? We see an answer when we examine the calculations of Holder and Hand[37] who compare their results with the earlier measurements of Deaton and Frost on methane-ethane hydrates. Although there are good fits for most compositions shown, there is considerable disagreement for the sample with about 90% methane, probably because the hydrate is structure II rather than structure I, the assumed structure in the calculation. Taking these observations together with those on other double hydrates, gives the correlation shown in FIGURE 5. For hydrates with two types of guest, the structure–size relationship is one of considerable complexity.

Further complexities may be inferred from the structure we published recently.[38] The structure is remarkable in that its *c* dimension is about 90Å and consists of

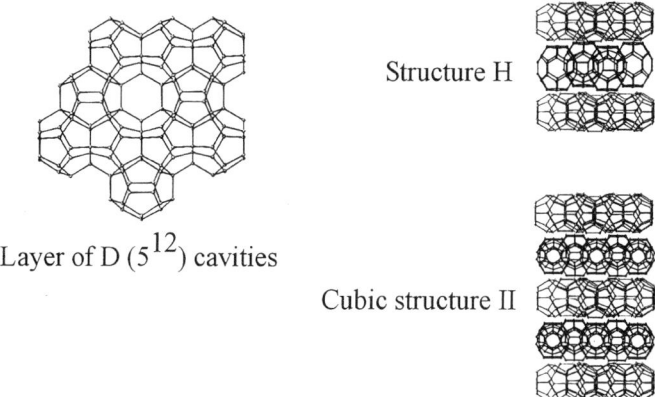

FIGURE 6. Layers of D (5^{12}) cavities is the same for structure H and for cubic structure II. Hydrate structure based on layers of pentagonal dodecahedra: structure II and structure H.

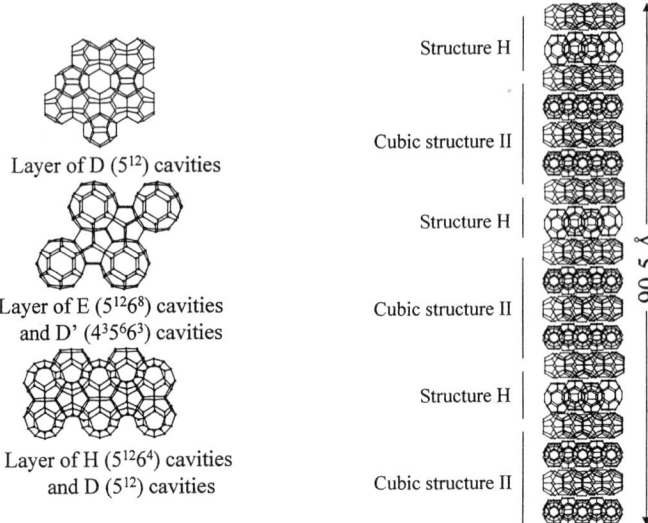

FIGURE 7. New complex hydrate structure reported recently also based on three layers of pentagonal dodecahedra.

alternating sequences of layers of structure II and structure H hydrate. The ideal structure contains four of the known hydrate cages (D, D´, H, and E), and although it has not been prepared with hydrocarbon guests it has reasonable prospects for existence as a stable material. It should be noted that this structure, just like structures II and H, derives from layers of 5^{12} cages stacked in a variety of ways (see FIGURES 6 and 7) (A*A*, structure H; ABC*ABC*, cubic structure II; CABBCAABC*CABBCAABC*, new structure). Naturally occurring hydrates that are based on polytypes or layers with occasional intergrowths of layers of large cages may very well occur, thus posing a challenge both for structural characterization and an appropriate thermodynamic description.

KINETIC STUDIES

Kinetic studies also have achieved considerable popularity over the last few decades. Thus far, the studies have been macroscopic in nature, the main observable being the amount of gas taken up as a function of time, either on ice or in aqueous solution. Work carried out by Bishnoi and coworkers[39] has defined many of the issues that need to be addressed, including the presence of induction times, memory effects, the determinate versus random nature of nucleation processes, and so forth. Of course, the random side of the nucleation phenomena cannot be addressed effectively. Lekvam and Ruoff[40] have reported a set of kinetic models that satisfactorily account for the hysteresis in hydrate formation—decomposition cycles; thus, it addresses the determinate part of the process. We note that an essential assumption of the model is that water and gas, after dissolution, form an oligomeric precursor. One may well ask, what is the evidence for such a precursor phase, and if it exists, what might its structure be? In order to obtain answers to such questions, suitable experimental approaches must be found. Techniques discussed previously as useful for identifying hydrate structures must now also incorporate time resolution. Diffraction, as the primary structural technique, depends on long-range order and as such may have limitations when applied to materials where long-range order does not exist. Techniques that give information about the time development of the radial distribution function in guest-water systems have promise. Examples are methods such as EXAFS,[41] which usually require a high intensity source to provide the necessary high signal to noise ratio for time resolution. Techniques considered previously as showing sensitivity to local order, such as NMR and vibrational spectroscopies, lack the sensitivity to obtain time-resolved data unless specialized approaches are employed. For Raman spectroscopy the sensitivity can be improved by employing a CCD detector to shorten the data acquisition times.[42] In the case of NMR, the field has benefited from the recent development of the technology to produce hyperpolarized (HP) xenon, resulting in an increase in sensitivity in spectroscopic measurements by as much as a factor of 10^4. An early example of the latter was provided a few years ago,[43] and additional refinements have improved the prospects yet further.[44] The sorption of HP xenon on ice allows the collection of kinetic data with site resolution. The entire experiment follows the formation of a hydrate layer some 1000Å thick on the surface of ice, and this is done easily within the lifetime of HP xenon, as witnessed by the slow decay of the xenon signals. The xenon data show analogies to gas uptake results in that an induction time is observed before the stage of rapid hydrate growth commences (see FIGURE 8, bottom). The local structural feature that can be measured is the cage occupancy ratio which, after the induction time, approaches the value 4, much as is determined for the equilibrium crystal. During the induction time this ratio drops to values of 1 (with some uncertainty) at the earliest times, indicating that something is made with a structure very different from the equilibrium crystal as there are as many occupied small as large cages (see FIGURE 8, top). These results can be taken as the first real evidence for a precursor phase. It is hoped that this development will spur the pursuit of realistic models for hydrate nucleation and growth, both from the experimental and modeling sides.

Structure on a larger length scale becomes important when one attempts to understand the action of kinetic inhibitors. The latter are thought to adsorb on the surface of hydrate crystals and to delay the growth of certain crystal faces.[45] The observation and quantitative description of morphological changes under different growth conditions or in the presence of inhibitors, is still very much in its infancy. Some other examples of the varied hydrate morphologies can be found in Makogon's book,[46] or in a recent publication on bromine hydrate.[47] We may expect that new techniques will be brought to bear upon the emerging subdiscipline of hydrate morphological studies, including the study of crystal surfaces and interfaces.

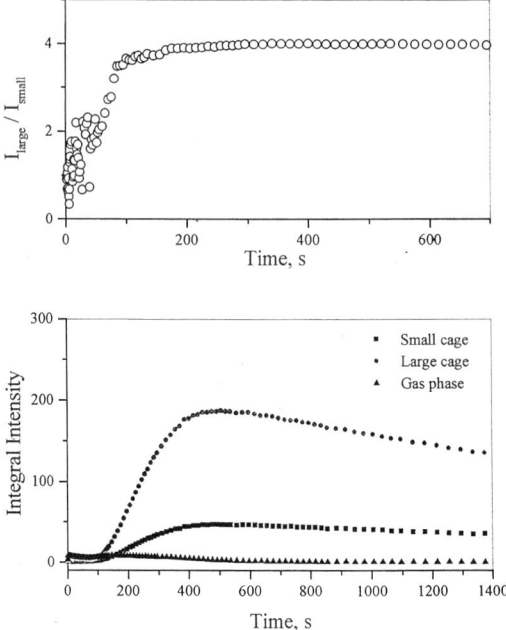

FIGURE 8. Kinetics of formation of hydrate on the surface of ice with hyperpolarized xenon; *bottom,* the integrated intensities of the lines for xenon in small and large cages and the gas recorded as a function of time; *top,* the cage occupancy ratio derived from intensities as a function of time. During the induction time, the ratio is very small, indicative of the presence of a precursor phase. After the induction time the ratio is equal to four, characteristic of the equilibrium crystal.

CONTRIBUTIONS TO HYDRATE RESEARCH BY DON DAVIDSON

Before continuing the discussion on fundamental studies of recovered natural hydrate samples, I would like to pay tribute to Don Davidson's contributions to hydrate research. During many years of basic research on hydrates[48] he pioneered the application of a variety of laboratory methods that played a large role in furthering the state of knowledge on hydrates, including recovered natural samples. During the late seventies Don was able to take advantage of funding opportunities to expand the hydrate program that he had established as early as about 1960. Initially this was strictly a curiosity-oriented research program, developing, first of all, dielectric spectroscopic, and later, broadline NMR spectroscopic applications to hydrate studies to a high degree of sophistication. By about 1972, pulsed NMR spectroscopy had been added, followed by Fourier transform NMR spectroscopy in 1978, with information becoming available on dynamics of the guest molecules, host lattice dynamics, and the development of clathrate-specific spectral signatures. New techniques added in the early eighties included powder X-ray diffraction (PXRD), thermodynamic measurements, and computational modeling. Additional personnel (Tse, Handa, and Ratcliffe) joined the existing group (Davidson, Garg, Gough, and Ripmeester) to initiate and develop these various methods. Achievements of the expanded hydrate group include: the identification of small structure II guests by powder X-ray diffraction and the identification of new hydrate guests;[49] the measurement of thermal expansivities;[50] the synthesis of essentially pure hydrate samples and the measurement of many thermodynamic properties for such samples;[51] the measurement of $\Delta\mu$ for a pure xenon hydrate sample; the initiation of molecular dynamics and density of states calculations;[52] measurements of thermal conductivity of hydrate.[53] By way of collaboration, isotope ratios were measured with R. Hesse,[54] and the refractive index of THF hydrate with R. O'Brien.[55] Before the group could really develop to its full capacity, funding cuts took away a number of positions in 1984, and Don's illness and death in 1986 more or less terminated the fully supported hydrate program at NRC.

NATURAL GAS HYDRATES FROM THE BLAKE RIDGE AND THE GULF OF MEXICO

Perhaps the best illustration of the full capacity for hydrate work of the expanded group was a study[56] of natural gas hydrates, after samples recovered from the Blake Ridge and the Gulf of Mexico became available. Samples were subjected to analysis by powder X-ray diffraction, dielectric analysis, ^{13}C and ^{1}H solid state NMR and calorimetry. As well as identifying the Gulf of Mexico sample as structure II hydrate, the application of magic angle spinning (MAS) NMR clearly showed the potential for this technique for hydrate analysis. The observation of a weak signal for methane in the structure II large cage led to a more detailed study with isotopically labeled material confirming that a cage-dependent shift was likely for many different guests.[57] Calorimetry revealed the presence of the self-preservation effect by showing gas release well above the normal range of decomposition temperatures at 1 atm.

NATURAL GAS HYDRATE FROM THE MALLIK WELL

This leads us to consider the more recently recovered hydrates from the Mallik well.[58] More than 10 years have passed since the Gulf of Mexico samples were analyzed, and after some lean years there is again a group at NRC able to carry out analysis of hydrates with a variety of techniques. This time, the hydrate samples were considerably more difficult to handle, and they first required a separation of the hydrate from the mineral components by crushing the recovered composites and floating the hydrate on top of the mineral fraction. Gas release showed small but significant differences in gas composition, some samples containing essentially pure methane, others containing 1–3% of heavier gases identified as CO_2 and propane. The pure methane hydrates showed the PXRD and Raman signatures of essentially pure methane hydrate, whereas the other sample showed a somewhat ambiguous PXRD pattern and an intensity distribution in the Raman spectrum that is not easily rationalized. These rather subtle differences do lead to distinct decomposition behavior in DSC studies, illustrating that there is a connection between macroscopic behavior, the composition, and the, as yet unresolved, structural subtleties of the Mallik hydrates.

SUMMARY

Over the many years of hydrate research, a number of distinguishing features can be used to describe this fascinating discipline. Hydrates are remarkably subtle materials, as sometimes even minute changes in composition or synthetic conditions may induce structural changes that are not necessarily easy to gauge. They have provided a continuing challenge to prevailing views. For instance, as the chemists of the day were finding that most hydrated materials had integer hydration numbers, the fractional hydration numbers for the gas hydrates proved difficult to accept, and even as late as 1950, experimentally determined hydration numbers for gas hydrates were sometimes rounded off. As clathrates, the gas hydrates were part of a class of materials that defied description for many years since they did not fit the standard classifications of covalent, ionic, or coordination compounds. Hydrate research continues to have considerable impact on problems that are of great fundamental interest today, such as the nature of hydrophobic hydration, and the anomalous low thermal conductivity of hydrates as compared to that of ice.

Finally, note that since the basic property of hydrate formers is the size of the molecule, hydrate research from the near and distant past continues to be of relevance today, even if the vast majority of hydrate researchers are interested primarily in hydrates of hydrocarbons or CO_2. The quotation: "If I have seen further it is because I have stood on the shoulders of giants" (Newton to Hooke, 1671) is something that all hydrate researchers should bear in mind when contemplating the progress made in recent years. As to the actual state of knowledge required for the control and prediction of hydrates—all researchers can judge for themselves—but I expect that in spite of the impressive amount of effort and progress made, most would agree that this is very much a work in progress.

ACKNOWLEDGMENTS

The author would like to thank his current hydrate coworkers at NRC, C.I. Ratcliffe, K.A. Udachin, G.D. Enright, C.A. Tulk, I.L. Moudrakovski, and A.Sanchez, for their many valued contributions. The author also thanks E.D. Sloan, Jr. for his continuing interest in and support of hydrate research at NRC. T. Collett, S. Dallimore, F. Wright, T. Uchida, and S. Okada are acknowledged for their kind collaboration on recovered natural gas hydrate samples.

REFERENCES

1. DAVY, H. 1811. Phil. Trans. Roy. Soc. London **101:** 1. FARADAY, M. 1823. Quart. J. Sci. **15:** 71.
2. LÖWIG, C. 1828. Mag. Pharm. **23:** 12. 1829. Ann. Chim. Phys. **42**(Series 2): 113.
3. SCHROEDER, W. 1926. Sammlung Chem. Chem.-Tech. Vortrage **29:** 1.

4. VILLARD, P. 1897. Ann. de Chim. Et de Phys. **11:** 289.
5. BAKHUIS ROOZEBOOM, H.W. 1918. "Die Heterogenen Gleichgewichte von Standpunkte der Phasenlehre", vol. II part 2, p191, Vieweg and Sohn, Braunschweig.
6. HAMMERSCHMIDT, E.G. 1934. Ind. Eng. Chem. **2.** 851.
7. DEATON, W.M. & E.M. FROST, JR. 1946. U. S. Bur. Mines Monograph 8.
8. KATZ, D.L., D. CORNELL, R. KOBAYASHI, F.H. POETMANN, J.A. VARY, J.R. ELLENBAAS & C.F. WEINAUG. 1959. Handbook of Natural Gas Engineering, McGraw-Hill Book Company, Inc., New York.
9. VON STACKELBERG, M. 1949. Naturwissenschaften **11:** 327.
10. POWELL, H.M. 1948. J. Chem. Soc. London 61.
11. CLAUSSEN, W.F. 1951. J. Chem. Phys. **19:** 259, 662. 1951. J. Chem. Phys. **19:** 1425.
12. VON STACKELBERG, M. & H.R. MULLER. 1951. J. Chem. Phys. **19:** 1319. 1951. Naturwissenschaften **38:** 456. MULLER, H. & M. VON STACKELBERG. 1952. Naturwissenschaften **39:** 20.
13. PAULING, L. & R.E. MARSH. 1952. Proc. Nat. Acad. Sci. U.S. **38:** 112.
14. JEFFREY, G.A. & R.K. MCMULLAN. 1967, Progr. Inorg. Chem. **8:** 43. JEFFREY, G.A. 1984. Inclusion Compounds, Vol. 1, J.A. ATWOOD, J.E.D. DAVIES & D.D. MACNICOL, Eds. Academic Press, London.
15. VAN DER WAALS, J.H. & J.C. PLATTEEUW. 1959. Advan. Chem. Phys. **2:** 1.
16. PARRISH, R.P. & J.M. PRAUSNITZ. 1972. Ind. Eng. Chem. Proc. Des. Dev. **11:** 26. NG, H.-J. & D.B. ROBINSON. 1976. Ind. Eng. Chem., Fundam. **15:** 293. JOHN, V.T. & G.D. HOLDER. 1985. J. Phys. Chem. **89:** 3279.
17. HOLDER, G.D. & D.J. MANGANIELLO. 1982. Chem. Eng. Sci. **37:** 9.
18. DAVIDSON, D.W., Y.P. HANDA, C.I. RATCLIFFE, J.S. TSE & B.M. POWELL. 1984. Nature **311:** 142.
19. DAVIDSON, D.W., M.A. DESANDO, S.R. GOUGH, Y.P. HANDA, C.I. RATCLIFFE, J.A. RIPMEESTER & J.S. TSE. 1987. Nature **328:** 418.
20. RIPMEESTER, J.A. 1982. J. Am. Chem. Soc. **104:** 289.
21. RIPMEESTER, J.A., C.I. RATCLIFFRE & J.S. TSE. 1988. J. Chem. Soc. Farad. I **84:** 3761.
22. DAVIDSON, D.W., Y.P. HANDA & J.A. RIPMEESTER. 1986. J. Phys. Chem. **90:** 6549.
23. BERTIE, J.E. & D.A. OTHEN. 1972. Can. J. Chem. **50:** 3443. BERTIE, J.E. & S.M. JACOBS. 1977. Can. J. Chem. **55:** 1777. BERTIE, J.E. & S.M. JACOBS. 1978. J. Chem. Phys. **69:** 4105.
24. RICHARDSON, H.H., P.J. WOOLDRIDGE & J.P. DEVLIN. 1985. J. Chem. Phys. **83:** 4387.
25. NAKAHARA, J., Y. SIGESATO, A. HIGASHI, T. HONDOH & C.C. LANGWAY, JR. 1988. Phil. Mag. **57:** 421.
26. SUM, A.K., R.C. BURRUSS & E.D. SLOAN, JR. 1997. J. Phys. Chem. B. **101:** 7371.
27. TULK, C.A., D.D. KLUG & J.A. RIPMEESTER. Ann. N.Y. Acad. Sci. **912:** this volume.
28. RIPMEESTER, J.A., J.S. TSE, C.I. RATCLIFFE & B.M. POWELL. 1987. Nature **325:** 135. RIPMEESTER, J.A. & C.I. RATCLIFFE. 1990. J. Phys. Chem. **94:** 8773.
29. UDACHIN, K.A., C.I. RATCLIFFE, G.D. ENRIGHT & J.A. RIPMEESTER. 1997. Supramol. Chem. **8:** 173.
30. LEDERHOS, J.P., A.P. MEHTA, G.B. NYBERG, K.J. WARN & E.D. SLOAN, JR. 1992. AIChE J. **38** 1045: MEHTA, A.P. & E.D. SLOAN, JR. 1994. J. Chem. Eng. Data **39:** 887. DANESH, A., B. TOHIDI, R.W. BURGASS & A.C. TODD. 1994. Chem. Eng. Res. Des. **72**(A2): 197.
31. MEHTA, A.P. & E.D. SLOAN, JR. 1994. AIChE J. **40:** 312.
32. SASSEN, R. & I.R. MACDONALD, 1994, Org. Geochem. **22:** 1029.
33. DE FORCRAND, R. 1883. Ann. Chim. Phys. **28:** 5.
34. HENDRIKS, E.M., B. EDMONDS, R.A., MOORWOOD, R. SZCZEPANSKI. 1996. Fluid Phase Equilib. **117:** 193.
35. SUBRAMANIAN, S.R., A. KINI, S. DEC & E.D. SLOAN, JR. Ann. N.Y. Acad. Sci. **912:** this volume.
36. VON STACKELBERG, M. & W. JAHNS. 1954. Z. Elektrochem **58:** 162.
37. HOLDER, G.D. & J.H. HAND. 1982. AIChE J. **28:** 440.
38. UDACHIN, K.A. & J.A. RIPMEESTER. 1999. Nature **397:** 420.

39. VYSNIAUSKAS, A. & P.R. BISHNOI. 1983. Chem. Eng. Sci. **38:** 1061. 1985. *ibid.* **40:** 299. ENGLEZOS, P., N. KALOGERAKIS, P.D. DHOLOBAI, P.R. BISHNOI. 1987. Chem. Eng. Sci. **42:** 2647.
40. LEKVAM, K. & P. RUOFF. 1997. J. Cryst. Growth **179:** 618.
41. BOWRON, D.T., A. FILIPPONI, M.A. ROBERTS & J.L. FINNEY. Phys. 1998. Rev. Lett. **81:** 4164.
42. SUBRAMANIAN, S., E.D. SLOAN, JR. Ann. N.Y. Acad. Sci. **912:** this volume.
43. PIETRASS, T, H.C. GAEDE, A. BIFONE, A. PINES & J.A. RIPMEESTER. 1995, J. Am. Chem. Soc. **117:** 7520.
44. MOUDRAKOVSKI, I.L., A. SANCHEZ, C.I. RATCLIFFE & J.A. RIPMEESTER. Unpublished results.
45. LARSEN, R., C.A. KNIGHT & E.D. SLOAN, JR. 1998. Fluid Phase Equil. **150:** 353. LARSEN, R. 1997. Clathrate Hydrate Single Crystals: Growth and Inhibition. Dissertation. Norwegian University of Science and Technology.
46. MAKOGON, Y.F. 1997 Hydrates of hydrocarbons, PennWell Books, Tulsa
47. UDACHIN, K.A., G.D. ENRIGHT, C.I. RATCLIFFE & J.A. RIPMEESTER. 1997. J. Am. Chem. Soc. **119:** 11481.
48. DAVIDSON, D.W. 1973. *In* Water. A Comprehensive Treatise, Vol. 2, Ed. F. Franks, Plenum, N. Y. Ch. 2. DAVIDSON, D.W. & J.A. RIPMEESTER. 1984. Inclusion Compounds, Vol. 3. Ed. J.A. ATWOOD, J.E.D. DAVIES & D.D. MACNICOL. Academic Press, London.
49. DAVIDSON, D.W., Y.P. HANDA, C.I. RATCLIFFE, J.A. RIPMEESTER, J.S. TSE. J.R. DAHN, F. LEE & L. CALVERT. 1986. Mol. Cryst. Liq. Cryst. **141:** 141.
50. TSE, J.S., W.R. MCKINNON & M. MARCHI, 1987. J. Phys. Chem. **91:** 4188.
51. HANDA, Y.P. 1986. J. Chem. Thermodyn. **18:** 891, 915.
52. TSE, J.S., M.L. KLEIN & I.R. MACDONALD. 1983. J. Phys. Chem. **87:** 4198. TSE, J.S. & M.L. KLEIN. 1987. J. Phys. Chem. **91:** 5789.
53. HANDA, Y.P. & J.G. COOK. 1987. J. Phys. Chem. **91:** 6327.
54. DAVIDSON, D.W., D.G. LEAST & R. HESSE. 1983. Geochim. Cosmochim Acta **47:** 2293.
55. DAVIDSON, D.W., R.N. O'BRIEN, P. SAVILLE & S. VISSAISOUK. 1986. J. Opt. Soc. Am. B **3:** 864.
56. DAVIDSON, D.W., S.K. GARG, S.R. GOUGH, Y.P. HANDA, C.I. RATCLIFFE, J.A. RIPMEESTER, J.S. TSE & W.F. LAWSON. 1986. Geochim. Cosmochim. Acta **50:** 619.
57. RIPMEESTER, J.A. & C.I. RATCLIFFE. 1988. J. Phys. Chem. **97:** 337.
58. TULK, C.A., C.I. RATCLIFFE & J.A. RIPMEESTER. 1999. Bull. Geological Survey of Canada **544:** 251–263.

Gas Hydrate and Humans

KEITH A. KVENVOLDEN[a]

U.S. Geological Survey, 345 Middlefield Road, MS 999, Menlo Park, California 94025, USA

ABSTRACT: The potential effects of naturally occurring gas hydrate on humans are not understood with certainty, but enough information has been acquired over the past 30 years to make preliminary assessments possible. Three major issues are gas hydrate as (1) a potential energy resource, (2) a factor in global climate change, and (3) a submarine geohazard. The methane content is estimated to be between 10^{15} to 10^{17} m^3 at STP and the worldwide distribution in outer continental margins of oceans and in polar regions are significant features of gas hydrate. However, its immediate development as an energy resource is not likely because there are various geological constraints and difficult technological problems that must be solved before economic recovery of methane from hydrate can be achieved. The role of gas hydrate in global climate change is uncertain. For hydrate methane to be an effective greenhouse gas, it must reach the atmosphere. Yet there are many obstacles to the transfer of methane from hydrate to the atmosphere. Rates of gas hydrate dissociation and the integrated rates of release and destruction of the methane in the geo/hydro/atmosphere are not adequately understood. Gas hydrate as a submarine geohazard, however, is of immediate and increasing importance to humans as our industrial society moves to exploit seabed resources at ever-greater depths in the waters of our coastal oceans. Human activities and installations in regions of gas-hydrate occurrence must take into account the presence of gas hydrate and deal with the consequences of its presence.

INTRODUCTION

Interest by humans in gas hydrate began in the early part of the nineteenth century when chemists discovered clathrate hydrates of various gases. Engineers became aware of gas hydrate in the 1930s when hydrate formation was discovered to be the cause of pipeline blockage during the transmission of natural gas. This early history of gas hydrate was summarized in detail by Sloan,[1] including reference to the first observation of naturally occurring gas hydrate in permafrost regions by Makogon in 1965. In the 1970s, naturally occurring gas hydrate was found and exploited in the Siberian Messoyakha gas field,[2] and geophysicists[3] and geochemists[4] recognized that gas hydrate occurs naturally, not only in polar continental regions, but also in shallow sediment under deep water of oceanic outer continental margins.

Interest in the relationships between naturally occurring gas hydrate and humans has continued to increase during the last 30 years. First, there was recognition of the potentially enormous amount of methane in the hydrate reservoir.[5,6] Second, there is the idea that gas hydrate can directly affect humans in at least three important ways:

---

[a]Telecommunication. Voice: 650-329-4196; fax: 650-329-5441.
kk@octopus.wr.usgs.gov

(1) as a potential energy resource, (2) as a factor in global climate change, and (3) as a submarine geohazard.[7]

GAS HYDRATE AS A POTENTIAL RESOURCE

The enormous amount of methane, currently estimated to be between (1 and 50) $\times 10^{15}$ m^3 at STP,[8] apparently sequestered in gas hydrate at shallow sediment depths within 2000 m of the surface of the Earth, and the widespread geographic distribution (see FIGURE 1) of gas hydrate have generated interest in its role as a potential energy resource. However, there is not yet consensus that methane can be extracted economically from gas hydrate deposits.

One of the problems for gas extraction stems from a geologic consequence of the shallow occurrence of oceanic gas hydrate. Much of the sediment that gas hydrate can occupy is unconsolidated or only semiconsolidated. When gas hydrate dissociates, free methane gas and water are produced. A real challenge for the production engineer is how to move the free methane to collectors through an unconsolidated mix of sediment and water.[8]

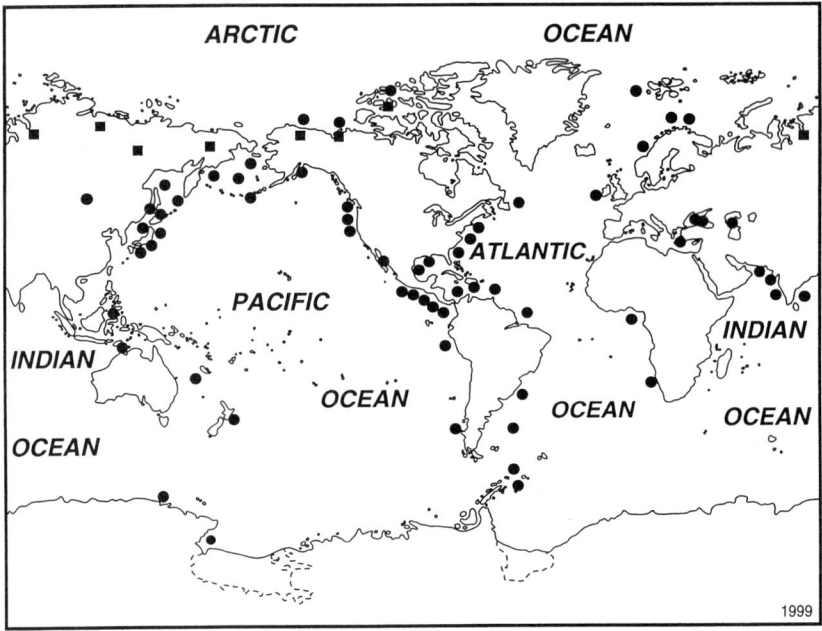

FIGURE 1. Map showing worldwide locations of known and inferred gas-hydrate deposits in oceanic (aquatic) sediments of outer continental margins (*solid circles*) and in polar (permafrost) regions (*solid squares*). Modified from Ref. 7.

A general lack of industrial interest over the past 30 years in the extraction of methane from gas hydrate reflects the low cost of other available energy resources. At present, however, interest is increasing. Certainly nations such as Japan and India, with immediate energy needs and a lack of alternative energy sources, have undertaken major efforts to investigate gas hydrate as an energy source.[9] Successful recovery of methane from gas hydrate requires the attention and infrastructure of the gas industry. It is likely that the location of the infrastructure and markets will dictate where, if ever, gas hydrate is to be first produced commercially.

Although naturally occurring gas hydrate was recognized in the 1960s, the gas industry has been slow to develop methodologies to recover methane from gas hydrate due in part to a generally abundant conventional gas supply and a lack of economic incentives. Based on economic arguments, Rogner[10] concluded that "in the foreseeable future, there will be little need for the development of gas hydrate" deposits, although the amount of methane in these deposits "is remarkable and should be acknowledged".

Development of the Messoyakaha gas field in Western Sibera during the past 30 years has been cited as an example of successful gas-hydrate exploitation.[11] Even this single example is now questioned,[12] because the observed gas hydrate may be secondary, a result of the production of conventional gas from this field. Thus more work must be done to demonstrate gas hydrate as a potential energy resource for humans. Attempts at full-scale production of methane for energy from gas hydrate will likely happen in the twentyfirst century.[1]

GLOBAL CLIMATE CHANGE

Methane is an important trace component of the atmosphere and is radiatively active. The Earth's atmosphere has a wide variety of sources and sinks for methane,[13] including methane hydrate, which exists in metastable equilibrium and is affected by changes in pressure and temperature that occur mainly due to factors associated with changes in sea level. A substantial instantaneous release of methane from gas hydrate could have an impact on atmospheric composition and thus on the radiative properties of the atmosphere that affect global climate.[14]

There may be obstacles, however, that limit the amount of hydrate methane that reaches the atmosphere. Whether or not hydrate methane gets into the atmosphere depends on the rates of (a) hydrate dissociation, (b) gas migration and trapping in sediments, (c) gas venting into the water column, and (d) methane oxidation resulting in carbon dioxide—also a radiatively active gas when in the atmosphere. The net effect of all of these processes is not known with certainty and it undoubtedly varies with the geological situation. Regarding methane oxidation, a recent study in the Atlantic Ocean by Rehder *et al.*[15] concluded that "the oxidation rate of methane in the water column is one of the unknown parameters in the climate scenario of methane emissions from gas hydrates to the atmosphere as the cause for past global warming". Methane that does reach the atmosphere reacts with hydroxyl radicals[13] in about 10 years (one half-life), unless the supply of radicals is overwhelmed. Thus, the chemical processes in both the ocean and atmosphere tend to limit the effects that hydrate methane may have on global climate change.

The potential behavior of gas hydrate in the present climate regime has been evaluated by Kvenvolden,[16] who suggested that gas-hydrate deposits of the polar continental shelves are presently most vulnerable to global change. This idea was tested by measuring methane concentration in water overlying the Beaufort Sea continental shelf of Alaska when ice was present and absent during 1990–1995. Results (summarized in Ref. 8) showed that the average difference in methane concentrations between ice-covered and ice free conditions is about 15 nanomoles; however, carbon isotopic determinations indicated that most of the methane came from coastal microbial processes and not from gas hydrate. One ephemeral submarine seep did release methane that may have been from gas hydate, but the results could not be confirmed. Nevertheless, the Beaufort Sea continental shelf of Alaska appears to be a seasonal source of a very minor contribution of methane to the atmosphere. Although this test failed to show significant release of methane from assumed gas-hydrate occurrence, the test remains inconclusive until similar research is carried out on the more extensive continental shelves of the Siberian Arctic Ocean.

Large negative excursions in the carbon isotopic record of carbonate in oceanic sediment from the latest Paleocene thermal maximum have been attributed to massive dissociation of gas hydrate.[17] Although this idea is attractive in explaining the marine isotopic record, the results do not necessarily require that hydrate methane actually reached the atmosphere to affect global climate change. However, excursions in atmospheric concentrations of methane during the Holocene have been noted in ice cores from Antarctica.[18] Blunier et al.[19] considered the role that gas hydrate might have played in producing these excursions and concluded, based on ice cores from both Antarctica and Greenland, that gas hydrate was a relatively small source of methane during at least the later Holocene.

Thus the role of gas hydrate in global climate change is uncertain. With respect to humans, it is probably not of immediate importance because the time scales involved in gas-hydrate–induced climate change may be very long. To be on the safe side, however, it would be best for humans not to deliberately perturb natural gas hydrate stability by continuing societal practices that enhance global warming, and thus ocean warming, resulting in gas-hydrate dissociation.

GEOLOGIC HAZARD

Of the three major gas-hydrate issues, resource, climate change, and geohazard, only the geohazard issue is likely of immediate importance to humans on both regional and local scales.[7,8] Regionally, there are many examples of the possible connection between gas hydrate and submarine sediment failures—surficial slides and slumps on the continental slope and rise of West Africa; slumps and collapse features on the U.S. Atlantic continental slope; large submarine slides on the Norwegian continental margin; sediment blocks on the sea floor in fjords of British Columbia; and massive bedding-plane slides and rotational slumps on the Alaskan Beaufort Sea continental margin (summarized in Refs. 7 and 8).

These submarine disruptions of the seafloor, caused by gas-hydrate dissociation, have an impact on humans if engineering structures are located in regions of potential failure. Since human interest expands to uses of the seafloor at increasing water depths, such as the active search for new conventional oil and gas deposits, stability of the sea-floor becomes increasingly important for any engineering structures. On the local scale, gas-hydrate dissociation in oceanic sediments, caused by heat transfer during petroleum production can lead to sediment failure and collapse of engineering structures.[20] Risks to drilling and production through gas hydrate-bearing sediment include blowouts and casing failures as experienced in Arctic regions.[21] These same concerns, such as casing collapse, gas leakage outside of the conductor casing, and gas blowouts, are applicable to gas hydrate of deep oceanic regions, only the problems will likely be more severe.[22] Thus, gas hydrate is a significant geohazard of immediate and increasing importance as our industrial society moves to exploit ever-deeper seabed resources in the waters of our coastal oceans. It should be remembered that the geohazard aspects of gas hydrate provide an additional constraint on exploiting oceanic gas hydrate as a future energy resource.

CONCLUSIONS

The amount of methane in natural gas hydrate is undoubtedly very large, probably greater than 10^{15} m^3, but considerably less than 10^{17} m^3 at STP. How this methane can or will affect us is not yet clear. There is much current interest in gas hydrate as a potential (1) energy resource, (2) factor in global climate change, and (3) submarine geohazard. Of these three issues, only submarine geohazards are considered to be of immediate importance, because as industry moves to exploit resources in the seabed at ever-greater water depths, gas hydrate becomes an increasing problem. Activities and engineering installations in regions of gas hydrate occurrence, must take into account the presence of gas hydrate and deal with the consequences of gas-hydrate dissociation, or means must be devised to prevent its dissociation.

It is of current importance, however, to evaluate gas hydrate as a potential energy resource, but because of complicated geological and technological problems, large-scale economic exploitation of methane for energy purposes from gas hydrate is not likely to be undertaken immediately. Likewise, evaluation of the role of gas hydrate in global climate change is of current interest, but the long-term effects on humans are still very uncertain.

REFERENCES

1. SLOAN, E.D. 1998. Clathrate Hydrates of Natural Gas, 2nd edit. Marcel Dekker, New York.
2. MAKOGON, Y.F. et al. 1971. Detection of a pool of natural gas in a solid (hydrate) state. Doklady Academii Nauk SSSR **196:** 197–206.
3. STOLL, R.D. et al. 1971. Anomalous wave velocities in sediments containing gas hydrates. J. Geophys. Res. **76:** 2090–2094.
4. CLAYPOOL, G.E. & I.R. KAPLAN. 1974. The origin and distribution of methane in marine sediments. In Natural Gases in Marine Sediments. I.R. KAPLAN, Ed.: 99–139. Plenum. New York.
5. CHERSKII, N.V. & Y.F. MAKOGON. 1970. Solid gas world reserves are enormous. Oil and Gas International **10:** 82–84.
6. KVENVOLDEN, K.A. 1988. Methane hydrate—a major reservoir of carbon in the shallow geosphere? Chem. Geol. **71:** 41–51.

7. KVENVOLDEN, K.A. 1993. Gas hydrates—geological perspective and global change. Rev. Geophys. **31:** 173–187.
8. KVENVOLDEN, K.A. 1999. Potential effects of gas hydrate on human welfare: Proc. Natl. Acad. Sci. USA **96:** 3420–3426.
9. COLLETT, T.S. & V.A. KUUSKRAA. 1998. Hydrates contain vast store of world gas resources. Oil & Gas J. **96**(19): 90–95.
10. ROGNER, H.-H. 1997. An assessment of world hydrocarbon resources. Annu. Rev. Energy Environ. **22:** 217–262.
11. COLLETT, T.S. 1993. Natural gas production from Arctic gas hydrates. U.S. Geol. Surv. Prof. Pap. **1570:** 299–311.
12. COLLETT, T.S. & G.D. GINSBURG. 1998. Gas hydrates in the Messoyakha gas field of the West Siberian Basin—a re-examination of the geological evidence. Int. J. Offshore and Polar Engineering **8**(1): 22–29.
13. CICERONE, R.G. & R.S. OREMLAND. 1988. Biogeochemical aspects of atmospheric methane. Global Biogeochem. Cycles **2:** 299–327.
14. MACDONALD, G.T. 1990. Role of methane clathrates in past and future climates. Clim. Change **16:** 247–281.
15. REHDER, G. *et al.* 1999. Methane in the northern Atlantic controlled by microbial oxidation and atmospheric history. Geophy. Res. Letters **26:** 587–590.
16. KVENVOLDEN, K.A. 1988. Methane hydrates and global climate. Global Biogeochem. Cycles **2:** 221–229.
17. DICKENS, G.R. *et al.* 1997. A blast of gas in the latest Paleocene: simulating first order effects of massive dissociation of oceanic methane hydrate. Geology **25:** 259–262.
18. CHAPPELLAZ, J. *et al.* 1990. Synchronous changes in atmospheric CH_4 and Greenland climate between 40 and 8 kyr BP. Nature **366:** 443–445.
19. BLUNIER, T. *et al.* 1995. Variations in atmospheric methane concentrations during the Holocene epoch. Nature **374:** 46–49.
20. CHAOUCH, A. & J.L. BRIAUD. 1997. Post melting behavior of gas hydrates in soft ocean sediments. Offshore Technology Conference **1**(OTC 8298): 217–224.
21. YAKUSHEV, V.S. & T.S. COLLETT. 1992. Gas hydrates in Arctic regions: Risk to drilling and production. Proc. 2nd Internal. Offshore and Polar Engr. Conf. **1:** 669–673.
22. BAGIROV, E. & I. LERCHE. 1997. Hydrates represent gas source, drilling hazard. Oil & Gas J. **95**(48): 99–104.

Comparisons of *In Situ* and Core Gas Measurements in ODP Leg 164 Bore Holes

C.K. PAULL,[a] T.D. LORENSON,[b] G. DICKENS,[c] W.S. BOROWSKI,[d] W. USSLER III,[a] AND K. KVENVOLDEN[b]

[a]*Monterey Bay Aquarium Research Institute, Moss Landing, California 95039-0628, USA*

[b]*U. S. Geological Survey, Menlo Park, California 94025, USA*

[c]*James Cook University, Townsville, Queensland 4811, Australia*

[d]*EXXON Exploration Co., Houston, Texas 77210-4778, USA*

ABSTRACT: During Ocean Drilling Program Leg 164, an unprecedented effort was made to determine the amounts of gas and gas hydrate in the sediments from Sites 994, 995, and 997. For the first time in the history of academic drilling, a pressure core sampler (PCS) worked well enough to generate an independent stratigraphy of *in situ* gas concentrations and compositions with depth. Here, gas concentrations and composition data produced by routine shipboard gas sampling techniques are compared with PCS data.

INTRODUCTION

Methane (CH_4) in marine sediments can occur dissolved in pore water, trapped in gas hydrates, and as free gas bubbles. Current understanding of the distribution and amounts of interstitial gases stored in marine sediments is rudimentary. The global gas hydrate reservoir is estimated to be about 10^4 gigatons CH_4 carbon.[1] Similar estimates of the size of the other CH_4 phases are not available. However, CH_4 concentrations in other phases may exceed the size of the gas hydrate reservoir. Ignorance about *in situ* CH_4 concentrations within continental margin sediment persists because much of the interstitial gas is lost before cores are available for sampling.

Most of the published information on gases in marine sediments comes from the Deep Sea Drilling Project (DSDP) and the Ocean Drilling Program (ODP). Evidence for out-gassing is commonly reported in freshly recovered sediment cores including: visual observations of bubbles moving through the core liners, occurrence of gas voids within core liners, and extrusion of core material from core liners. Presence of gas bubbles and expansion of core material indicate that gas saturation was achieved within cores during ascent to the surface. However, establishing the original gas phase and concentration before the coring process disturbs the sediments is not routinely possible.

Most of the gas measurements made during the DSDP were from free gas (FG) samples taken from the voids within the core liner. Free gas measurements provide ratios of the relative abundance of CH_4 to higher hydrocarbon gases (usually ethane and propane) and carbon dioxide (CO_2). During ODP headspace (HS) gas measurements, which relate the amount of gas evolved to the volume of sediment from which it is derived, became common and reduced the reliance on sampling free gas. Headspace gas (HS) measurements are usually reported as PPM CH_4 or as $\mu L_{CH_4}/L$ wet sediment. This is a concentration per volume of sediment; corrections for sediment porosity, which would provide gas concentrations per volume of pore fluid, have not been applied.

Core gas measurements have a characteristic trend relative to the geochemical layers and depth into the sediments. Typical HS gas concentrations remain very low until the level at which sulfate depletion occurs. Below this depth, HS concentrations rise to a subsurface maximum, then commonly show significant scatter around the same concentration or decrease with increasing depth as shown in FIGURE 1. Patterns seen in gas concentration data commonly vary with lithologic changes and have influenced interpretations about *in situ* gas occurrence.

Determination of *in situ* gas concentrations in DSDP/ODP bore holes is an inherently difficult task because of the one- to two- orders of magnitude change in the CH_4 saturation level caused by changes in pressure and temperature during sediment recovery. The primary variable affecting CH_4 saturation in core samples is the extreme pressure decrease during core retrieval. For example, at pressures associated with 3 km water depths, approximately 190 mM of CH_4 are required to saturate the pore waters,[2] whereas at STP conditions, pore waters are saturated with CH_4 at about 1.1 mM. Because significant amounts of degassing may occur during core recovery,

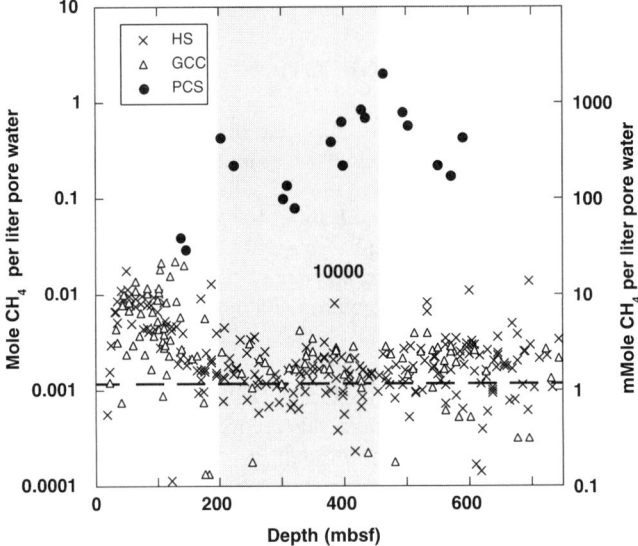

FIGURE 1. Plot showing methane concentration data versus depth using both mole CH_4 and mM CH_4 per liter of pore water from ODP Sites 994, 995, and 997 as measured with headspace (HS), gas collection tube (GCC), and pressure core sampler (PCS). *Dashed line* indicates CH_4 saturation under surface laboratory conditions (1 ATM., 25°C, and 35‰ salinity[14]). CH_4 is the sedimentary gas with the lowest saturation value under surface conditions. Shipboard core gas sampling techniques (HS and GCC) grossly underestimate *in situ* gas concentrations as indicated by PCS data. The main gas-hydrate–bearing zone is *shaded*.

it is unclear how measurements of the gas concentration in freshly recovered deep-water sediment cores can be related to the true *in situ* gas concentration.

Most gas-charged samples that have been collected during DSDP and ODP drilling legs are largely composed of CH_4 with secondary amounts of CO_2. Gas isotope measurements and methane to ethane (C_1/C_2) ratios indicate that gases of a microbial origin predominate.[3,4]

Microbial CH_4 generation in marine sediments proceeds along predictable pathways: (1) Microbial sulfate reduction must first consume the interstitial sulfate before CH_4 can be produced and accumulate.[5] (2) Microbial CH_4 production commences near the base of the sulfate reduction zone. Interstitial CH_4 concentrations should increase below this level. As the quality of organic matter deteriorates with burial, inhibiting methanogens, the production rate should also decrease with depth. Ultimately, CH_4 concentrations reflect both production and losses through absorption onto sedimentary particles and migration. (3) After threshold concentrations are achieved, any additional CH_4 production will result in the CH_4 being sequestered in gas hydrate if the pressure and temperature conditions are appropriate.[6] Below the base of gas hydrate stability, CH_4 saturation must be exceeded before CH_4 bubbles can form.[7]

During ODP Leg 164, Sites 994, 995, and 997 were drilled to 700–750 mbsf along a 9.6 km long transect across the Blake Ridge.[8] A distinct gas-hydrate–bearing zone occurs between 200–450 mbsf. At ODP Sites 995 and 997, well log and vertical seismic profiler data show low velocities at and below 450 mbsf that indicates the occurrence of gas bubbles.

Measurements of *in situ* gas concentrations were made using a *pressure core sampler* (PCS) and provided the first profiles of *in situ* gas concentrations.[9] The PCS is a down hole sampling tool that seals a small core *in situ* allowing recovery at nearly *in situ* pressures.[10] Because PCS samples do not degas during core recovery, the original gases can be recovered during depressurization under controlled shipboard conditions. PCS samples contained CH_4 and other gases that evolved from pore water or gas hydrate. The objective of this paper is to understand the significance of traditional core gas measurements,[11] with respect to the existing PCS measurements.

METHODS

Gas samples were obtained with four techniques during ODP Leg 164: (1) FG samples from core voids, (2) HS gas, (3) gas collection chambers (GCC), and (4) PCS. Detailed CH_4 concentration profiles are available for a few meters below the base of sulfate reduction at Sites 994B and 995B. Methane, CO_2, ethane, and higher molecular weight hydrocarbon concentrations were measured by gas chromatography at these sites.[8]

Free gas samples were extracted minutes after core recovery by inserting a piercing tool through the core liner into core voids. At shallow depths core gas was extracted from gas voids by opening a valve connected to an evacuated container. Below about 50 mbsf, the gas within voids escaped under its own pressure into 60-cc syringes.

Headspace gas concentrations were measured by extracting a calibrated volume of sediment (usually 5 cc) from the core with a cork borer shortly after the core was retrieved. Each sample was placed into a 21.5-cm^3 glass serum vial, sealed with a septum and a metal crimp cap, and heated at 60°C for 30 minutes to evolve gas for analysis. Headspace gas measurements were reported in units of μL/kg of wet sediment.[8] However, for comparison to other shipboard gas data, HS measurements given here are converted into units of moles CH_4 per liter of pore water by using measured porosity and wet bulk density data[8] using the ideal gas law. CO_2 measurements were not made on the HS samples.

Gas collection chambers (GCC) were used to collect the gases that evolved from core sections as the cores warmed within the core laboratory. Cores were cut into normal 1.5-meter sections and placed in 1.54-meter GCC tubes that were slightly larger in diameter than the core liners. The GCC were constructed from standard PVC schedule-40 pipe sections (approximately 3.5 inches i.d.) with o-ring-fitted end caps that could be quickly attached to make a gas tight seals. Evolved gas flowed out through a port in the sealed end caps of the GCC into tygon tubing. The tubing passed into inverted 1-liter graduated cylinders that floated in baths filled with NaCl saturated water. The cylinders filled with the gas as the gas exited the tubing. Because the inverted graduated cylinders were initially water-filled, the volume of water displaced by the gas could be measured directly. The tops of the cylinders were fitted with a stopcock valve that allowed samples to be extracted into syringes. For comparison with other shipboard gas data, GCC measurements are converted into units of moles CH_4 per liter of pore water using the porosity and wet bulk density data.[8] Since the gases were typically at least 99% CH_4 the volume estimates assume that all the gas captured by the GCC was CH_4. Because the GCC samples were not heated, only the evolved gas in excess of surface saturation was measured.

Detailed sampling for pore water and gas measurements were conducted near the base of the sulfate reduction zone at Sites 994B and 995B. Sections of core were sealed in 1.5-m long PVC tubes and flushed with nitrogen to protect the core from oxidation. After the base of the sulfate reduction zone was identified from pore water measurements, the sealed cores were reopened for microbial incubation studies. Subsampling of the preserved cores occurred one and two weeks after initial core recovery at Sites 994B and 995B respectively. At the time, additional sub-samples were taken for CH_4 concentration measurement. Gas concentrations in these samples were calculated using shipboard porosity and wet bulk density and are reported as moles CH_4 per liter of pore water for Site 994B[8] and Site 995B.[12]

The PCS recovered core samples at *in situ* pressure. Upon recovery, the gases within the sealed vessel were vented[9] and reported in units of moles CH_4 per liter of pore water.

RESULTS

Gas samples collected during ODP Leg 164 were composed primarily of CH_4, with secondary amounts of CO_2, tertiary amounts of ethane, and traces of higher hydrocarbon gases (C_3–C_7). No significant difference was observed in the gas composition profiles between Sites 994, 995, and 997.[8] However, significant apparent differences in the concentrations and relative composition occur among the sampling techniques.

All CH_4 concentration profiles from Sites 994, 995, and 997 show a distinct increase at about 20 mbsf, which coincides with the level of sulfate depletion.[13] However, the initial gradients vary with the technique (see FIGURES 1 and 2). Detailed CH_4 concentration measurements from Sites 994B and 995B provide the greatest resolution at the top of the CH_4 production zone. Linear fits to these data show that CH_4 concentrations increase between 20.05–26.50 mbsf at Site 994B at a rate of about 0.11 mM per meter, and between 20.25–22.83 mbsf at Site 995B at about 0.19 mM per meter (FIGURE 2, inset). The CH_4 concentration data collected by the HS and GCC techniques are at a lower sampling density and show greater

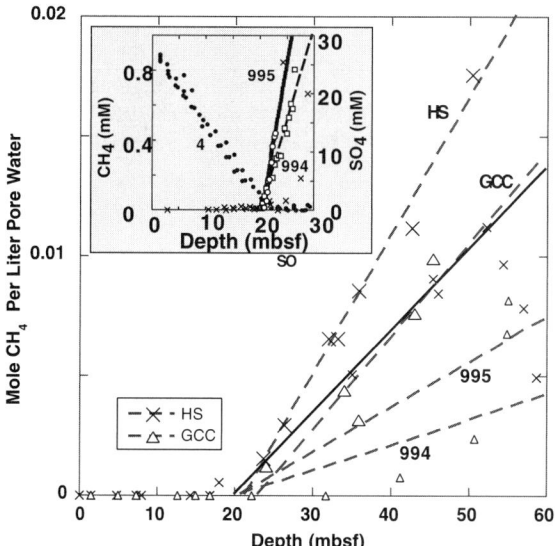

FIGURE 2. The main figure shows plots of methane concentration data measured on core samples spanning the upper 60 mbsf at ODP Sites 994, 995, and 997 from routine headspace (HS, indicated with *crosses*), gas collection chambers (GCC, indicated with *triangles*), and detailed HS sampling from Sites 994B and 995B. Best linear fits to higher values selected to represent the onset of higher CH_4 concentration data are indicated with *larger symbols*. Best fit line for the HS gas samples show the steepest slope of the onset of significant CH_4 accumulations (*HGS;* $Y = -0.012 + 0.00058X$, $R^2 = 0.99$), followed by the gas tube samples (*GCC;* $Y = -0.00089 + 0.00039X$, $R^2 = 0.89$). *Solid line* in main figure connects between zero at 20 mbsf and 0.40 mole/l CH_4 at 136.70 mbsf as measured by the PCS. *Inset shaded figure* shows data from closely spaced HS samples used to define the sulfate-methane interface at Sites 994B and 995B. Data used for Sites 994 (*squares;* $Y = -0.0039 + 0.00019X$, $R^2 = 0.92$) and 995 (*circles;* $Y = -0.0022 + 0.00011X$, $R^2 = 0.95$) fits in main plot are given in the *inset*. Small crosses in *inset* are other 994 data. The beginning of CH_4 accumulation begins at about 20 mbsf. Sulfate concentrations for 994, 995, and 997 are indicated with smaller filled symbols.

scatter. However, lines fitted to the higher values below the base of the sulfate reduction zone (20–80 mbsf), indicate initial concentration increases at a rate of 0.58 mM per meter in the HS data and 0.39 mM per meter in the gas tube data (FIGURE 2, main figure).

Maximum HS CH_4 concentrations occur between 40–80 mbsf, where the 13 samples have a mean value of 9.2 ± 3.3 mM CH_4. Thus, the samples within this zone are up to 800% saturated with CH_4 at surface conditions (approximately 1.1 mM). Although there is considerable scatter and some erratic higher values, the HS CH_4 concentrations below 200 mbsf show a mean value of 1.9 ± 1.7 mM ($n = 168$), which is about 170% of surface CH_4 saturation.

The GCC concentration profiles are generally similar in shape to the HS data (FIGS. 1 and 2). Samples evolved gas below 20 mbsf and a broad maximum occurred at 110 to 145 mbsf where the mean value is 7.8 ± 7.1 mM ($n = 13$) per liter pore water. The highest individual value was at 130 mbsf where the samples evolved 22.2 mM of CH_4 gas. The mean value below 200 mbsf is 1.6 ± 1.2 mM ($n = 68$). Both HS and GCC data indicated a subsurface maximum; however, the positions of the maximum GCC values were displaced downhole. Of the 18 successful PCS deployments, only two were at depths shallower than 200 mbsf. The shallowest successful PCS samples (at 136.7 mbsf and 146.9 mbsf) showed CH_4 concentrations of 400 mM and 300 mM respectively. These samples are more than 55 m below the zone of maximum HS measurements, and roughly correspond with the maximum gas concentration indicated by the GCC, but contain 40 to 50 times more CH_4 than

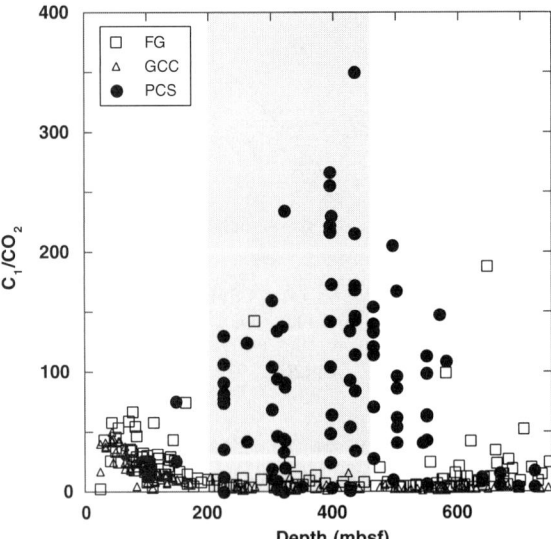

FIGURE 3. Plot comparing values of CH_4 volume to carbon dioxide ratios (C_1/CO_2) measured on FG, GCC, and PSC samples collected from ODP Sites 994, 995, and 997. The main gas-hydrate–bearing zone is *shaded*.

the maximum GCC sample. The PCS concentrations show further increase in CH_4 with depth to a maximum value of 2,020 mM at 462.2 mbsf, at Site 997.

Methane to carbon dioxide molar ratios (C_1/CO_2) show considerable scatter with all techniques (see FIGURE 3). Values of C_1/CO_2 from FG (17 ± 16, $n = 154$) and GCC (13 ± 12, $n = 138$) samples are less than the PCS samples (90 ± 73, $n = 93$). The FG and GCC data show similar stratigraphic trends in C_1/CO_2 values with the highest values occurring in the upper 80 mbsf (where the mean C_1/CO_2 values are 38 ± 14 [$n = 14$] for FG) and 33 ± 10 ($n = 20$) for GCC. Below 80 mbsf the values diminish. Below 200 mbsf the mean values are 9 ± 9 ($n = 92$) for FG and 2.4 ± 2.7 ($n = 20$) for GCC techniques. Conversely, the PCS samples have C_1/CO_2 values that range up to 350, with a maximum value at 462 mbsf, and slowly diminishing values above and below this depth.

Methane to ethane concentration ratios (C_1/C_2) show the same trend in all holes (see FIGURE 4). The highest values are in the upper section of the core and values decrease with depth. However, average values vary depending on the sampling technique used. In the upper 50–80 mbsf depth range, mean values of C_1/C_2 were 26,000 ± 3,500 ($n = 12$) for FG samples, 23,000 ± 6,200 ($n = 15$) for GCC samples, and 15,300 ± 4,300 ($n = 10$) for HS gas samples. No PCS samples are available at shallow depths. No abrupt change in the C_1/C_2 ratios was observed across the base of the gas hydrates at these sites (about 450 mbsf). Mean C_1/C_2 values below 450 mbsf are 1,900 ± 530 ($n = 38$) for the PCS, 1,300 ± 400 ($n = 53$) for FG, 1,000 ± 330 ($n = 31$) for GCC, and 380 ± 180 ($n = 92$) HS samples.

DISCUSSION AND CONCLUSIONS

The four sampling techniques used to estimate the amount of gas within the core samples during ODP Leg 164 produce dramatically different results from the *in situ* concentration data provided by the PCS. All core gas measurements of CH_4

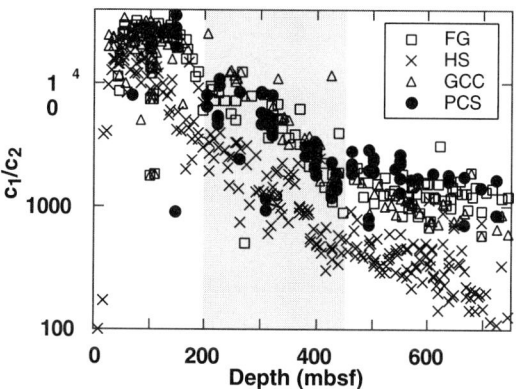

FIGURE 4. Plot comparing values of methane volumes to ethane ratios (C_1/C_2) measured FG, HS, GCC, and PSC samples collected from ODP Sites 994, 995, and 997. The main gas-hydrate–bearing zone is *shaded*.

concentration indicate an initial downward increase from low concentrations toward the surface saturation values. These gradients suggest that concentrations exceeding surface saturation (approximately 1.1 mM) would be achieved at less than 10 meters below the base of sulfate reduction. Thus, core samples from below about 30 mbsf should evolve gaseous CH_4 during core recovery resulting in CH_4 loss. Core CH_4 concentrations peak at about eight times surface saturation between 80–140 mbsf. At greater depths (more than 200 mbsf) the mean core CH_4 concentration data decreases to values that are less than twice surface saturation. In contrast, the PCS data indicate continuing CH_4 increases with depth and the average CH_4 concentration below 450 mbsf was more than 400 times over surface saturation values. In these samples more than 99% of the original CH_4 escaped before shipboard core sampling. Thus, the FG, HS, and GCC gas profiles do not reflect *in situ* gas abundance. The only proper use of HS and GCC CH_4 concentration data is to establish the rate of CH_4 accumulation between the base of sulfate depletion and the depth at which surface saturation concentrations are achieved. However, this has to be done soon after coring because even preserved cores de-gas.

Peaks in CH_4 concentration observed with HS and GCC data clearly have little to do with *in situ* gas volumes. We suggest that the head space and gas chamber measurements reflect aspects of sediment strength. The amount of interstitial gas increases until sediment failure occurs as a consequence of the internal pressure generated by evolving interstitial gas bubbles. Apparently failure usually occurs when the amount of dissolved gas exceeds 8 to 10 times surface saturation.

Both the C_1/C_2 and C_1/CO_2 values are considerably different in the PCS data than in the core gas measurement. These discrepancies indicate that CH_4 escapes preferentially during the outgassing of cores that exceed surface saturation. Thus, gas composition ratios measured on core gases collected from sediment that contained CH_4 at concentrations that exceed surface saturation are strongly influenced by degassing. This bias will artificially decrease the C_1/C_2 values and make pure microbial gases appear to contain a component of thermal gas.

REFERENCES

1. KVENVOLDEN, K. 1988. Methane hydrate—a major reservoir of carbon in the shallow geosphere? Chem. Geol. **71**: 41–51.
2. DUAN, Z., N. MØLLER, J. GREENBERG & J.H. WEARE. 1992. The prediction of methane solubility in natural waters to high ionic strength from 0 to 250°C and from 0 to 1600 bar. Geochim. Cosmochim. Acta **56**: 1451–1460.
3. BERNARD, B.B., J.M. BROOKS & W.M. SACKETT. 1978. Light hydrocarbons in recent Texas continental shelf and slope sediments, J. Geophys. Res. **83**: 4053–4061.
4. WHITICAR, M.J., E. FABER & M. SCHOELL. 1986. Biogenic methane formation in marine and freshwater environment, CO_2 reduction versus acetate fermentation, isotopic evidence. Geochim. Cosmochim. Acta **50**: 693–709.
5. MARTENS, C.S. & R.A. BERNER. 1974. Methane production in the interstitial waters of sulfate-depleted marine sediments. Science **185**: 1167–1169.
6. SLOAN, D.E., JR. 1990. Clathrates Hydrates of Natural Gases. Marcel Dekker. New York.
7. PAULL, C.K., W. USSLER, III & W. BOROWSKI. 1994. Sources of biogenic methane to form marine gas-hydrates: *in situ* production or upward migration? Ann. N.Y. Acad. Sci. **715**: 392–409.

8. PAULL, C.K., R. MATSUMOTO, P. WALLACE et al. 1996. Proceedings ODP, Initial Reports 164. College Station, TX (Ocean Drilling Program).
9. DICKENS, G.R., C.K. PAULL, & P. WALLACE. 1996. Direct measurement of in situ methane quantities in a large gas-hydrate reservoir. Nature **385:** 426–428.
10. PETTIGREW, T.L. 1992. Design and operation of a wireline pressure core sampler (PCS). Ocean Drilling Program Technical Note 17, College Station, TX.
11. KVENVOLDEN, K. & T. LORENSON. 2000. Methane and other hydrocarbon gases in sediments from the southeastern North American Continental Margin, ODP Scientific Results 164. College Station, TX (Ocean Drilling Program). 29–36.
12. HOELHER, T.M. et al. 2000. Model, stable isotope, and radiotracer characterization of anaerobic methane oxidation in gas hydrate-bearing sediments of the Blake Ridge, ODP Scientific Results, 164. College Station, TX (Ocean Drilling Program). 79–86.
13. BOROWSKI, W.S. et al. 2000. Significance of anaerobic methane oxidation in methane-rich sediments overlying the Blake Ridge Gas Hydrates, ODP Scientific Results, 164. College Station, TX (Ocean Drilling Program). 87–100.
14. YAMAMOTO, S., J.B. ALCAUSKAS & T.E. CROXIER. 1976. Solubility of methane in distilled water and seawater. J. Chem. Eng. Data **21:** 78–80.

An *In Situ* Experiment of Methane Sequestration as Gas Hydrate, Authigenic Carbonate, and Loss to the Water Column and/or Atmosphere

MIRIAM KASTNER,[a,b] BOBB CARSON,[c] DOUGLAS B. BARTLETT,[b]
JOHN JAEGER,[c] KELLY BIDLE,[b] AND HANS JANNASCH[d]

[b]*Scripps Institution of Oceanography, University of California,
San Diego, La Jolla, California 92093-0212, USA*

[c]*Lehigh University, Department of Earth & Environmental Sciences,
Bethlehem, Pennsylvania 18015, USA*

[d]*Monterey Bay Aquarium Research Institute, Moss Landing, California 95039, USA*

ABSTRACT: A one year *in situ* experiment to quantify the flux of carbon, primarily as methane, from an overpressured thrust fault at the Cascadia convergent margin (ODP Site 892B) is described. Most of the expelled carbon is sequestered in two solid phases, methane hydrate and authigenic carbonates and an unknown portion is lost in the water column and/or atmosphere. ^{14}C ages of clam shells from the vicinity of the site suggest active methane venting for at least 21–24 kyrs. The water column chemistry provides information on the potential effects of global warming on rapid massive gas hydrate dissociation and on the effects of microbial oxidation of the released methane on the local oceanic oxygen and CO_2 contents. The local and global implications of these processes for the oceanic carbon cycle are being assessed.

INTRODUCTION

Oceanic Distribution and Importance of Oceanic Gas Hydrates and Authigenic Carbonates

In ocean margins, particularly in convergent margins, gas hydrates and authigenic carbonates are widespread and may be massive at or near the seafloor. The gas hydrates only extend to the base of the gas hydrate stability field, often characterized by a bottom simulating reflector (BSR) that parallels the seafloor. The carbonates and gas hydrates are especially well developed and abundant in the vicinity of fluid-active thrust faults and fractures that breach the seafloor, common at tectonic ridges. For the most part, dissolved and gaseous methane are discharged; low concentrations of higher polymerized hydrocarbons are not uncommon.[1–3] Although much of the expelled methane is sequestered in the solid phases, authigenic Ca-Mg carbonates and gas hydrates, some escapes to the water column and/or to the atmosphere. The

---

[a]Telecommunication. Voice: 858-534-2065; fax: 858-534-0784.
mkastner@ucsd.edu

expelled methane-rich fluids are mixtures of passively incorporated interstitial sediment pore water (seawater) with diagenetically derived fresher water released by dehydration and transformation reactions at depths exceeding 1 km.[4–7] The associated gases are derived from microbial fermentation and thermal maturation of the sediment organic matter, with methane as the most abundant gas.[8,9] The carbon isotope values of the methane in the authigenic carbonates and gas hydrates indicate that it is mostly microbially produced. It probably, however, includes some thermogenic methane, with an origin at greater depths where temperatures are, or exceed, 80–90°C. For example, in the Gulf Coast, known for its abundant hydrocarbon reservoirs, the methane that breaches the seafloor and forms gas hydrates and authigenic

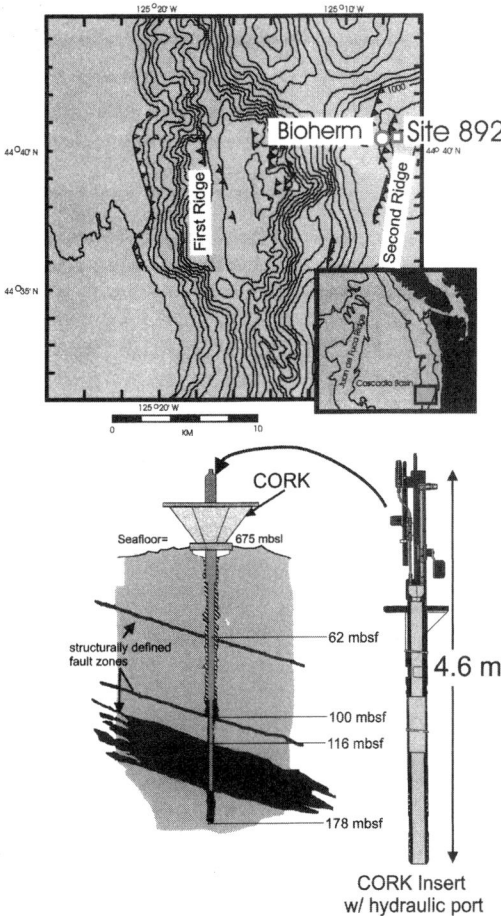

FIGURE 1. Location map of Cascadia margin ODP Site 892B on the central Oregon margin. The CORK insert with the hydraulic port to which the year long experiment is connected is shown in the *lower half* of the figure.

carbonates has a thermogenic origin. If the ascending methane-rich fluids, that are tectonically and/or diagenetically driven, are warmer than the surrounding gas hydrate bearing sediments they destabilize the gas hydrates within the stability field above the BSR. Hence, these fluids become further enriched in methane and more buoyant from the fresh water released from the dissociating gas hydrate. Because of the high sulfate concentration in seawater (about 29 mM) in the sulfate reduction zone, some of the advecting methane is anaerobically oxidized by bacteria that are directly or indirectly linked to sulfate reduction. The hydrogen sulfide produced may be incorporated in the subseafloor gas hydrate and stabilize it. For example, close to 10% of the cages in the methane hydrate recovered at ODP Site 892B (see FIGURE 1) are occupied by H_2S.[10]

The best characterized convergent margin with massive authigenic carbonates (*chemoherm*) and gas hydrate is the Hydrate Ridge in the Cascadia convergent margin off central Oregon, where ODP Site 892B is situated (FIG. 1).[1,3,6,11–13] Similar authigenic carbonates and gas hydrates have been observed at other convergent margins—Barbados, Peru, Costa Rica, and Northern California.[14–17] The global importance of these deposits is, however, as yet unknown.

Oceanic methane hydrates are estimated to contain a very large amount of carbon on the order of 10^{19} grams methane C,[18] which exceeds the other known combustible hydrocarbon reserves. The potential effects of global warming on the stability of this enormous quantity of methane hydrates and the implications of rapid release of large quantities of methane for ocean and atmospheric chemistry and biology, are key unknowns that must be pursued. Because methane gas is an important contributor to the atmospheric radiation balance, any flux of methane into the atmosphere in addition to the present annual growth rate would enhance global warming.

No similar estimate of the amount of carbon sequestered in oceanic authigenic carbonates is available, but it also must be enormous. For example, Carson *et al.*[3] used GLORIA imagery to establish that the Oregon "chemoherm" was just the most obvious surficial expression of an extensive carbonate deposit, related to a fluid-active fault, that covers about 17 km^2 of the Oregon prism. Assuming the deposit averages just 0.5 m in thickness, it represents approximately 2.2×10^7 metric tons of carbonate deposited on one 20 km portion of the Oregon margin, probably since the Plio-Pleistocene era. ^{14}C ages of two clams cemented by authigenic aragonite from the bioherm indicates active methane venting and authigenic carbonate formation near Site 892B for at least 21 to 24×10^3 years, and most probably longer.

Considering the volume of the oceanic gas hydrate and authigenic carbonate deposits and their probable importance for the oceanic carbon cycle, global warming and continental slope destabilization, it is timely to determine the rates and processes involved in their formation or the rate of methane hydrate dissociation, in response to environmental perturbations.

OBJECTIVES AND THE ONE YEAR *IN SITU* EXPERIMENT

In order to quantify the carbon flux from the Cascadia accretionary prism, the rates of carbon sequestration in authigenic carbonates and gas hydrates, and the balance lost to the ocean and/or atmosphere, we have developed the following one year *in situ* experiment. The instrumental package consists of two newly designed experimental chambers, one aerobic for microbially mediated carbonate deposition and one anaerobic for gas hydrate formation. The package has been recently deployed and connected through a valve to the borehole seal (CORK) at ODP Site 892B (FIG. 1); this borehole fluid is naturally overpressured and delivers 2–3 liters/day. ODP Site 892 is located about 16 km east of the Cascadia trench, off central Oregon, at approximately 675 m water depth. The bottom water temperature is 4.3°C, thus the seafloor is within the stability field of pure methane hydrate and the BSR is at 68 mbsf;[19] a CH_4-H_2S gas hydrate was recovered at 2–19 mbsf.[10] The borehole intersects a hydrologically active out-of-sequence landward-dipping thrust fault. Its surface trace, at 0.5 km west of the site, is associated with massive authigenic carbonate deposits (the chemoherm in FIG. 1) and sustains prolific benthic communities. In addition to the two chambers, the experimental package also includes: (1) two OsmoSamplers for continuous sampling of borehole water and gases, respectively; (2) three systems for monitoring long-term variabilities of fluid composition and flow rates, and hence, of the fluid and chemical fluxes that enter into the two experimental chambers—one of the flow rate systems utilizes two independent geochemical tracers; and (3) an *in situ* pressure monitor that once per hour records

FIGURE 2. The one year experimental package ready to be deployed is shown. The grey box on the far left side contains the oxic carbonate chamber; in the lower right corner the upright semicircular feature wrapped in black tape contains the anoxic gas hydrate chambers. The two horizontally lying white cylindrical tubes, in line with the anoxic chambers, are the OsmoSamplers, and the fluid flow and pressure monitoring systems are located in the center of the experimental package.

the borehole versus hydrostatic pressure. The complete experimental package ready for its deployment is shown in FIGURE 2. In addition to monitoring the fluid flux, the chemical and physical conditions of authigenic carbonate and gas hydrate formation and how much of each of these solid phases has precipitated in the corresponding experimental chambers, and the microbial conditions and processes and the specific microbial groups found in the borehole and experimental chamber and their role in the formation of the carbonates and gas hydrates will also be characterized. In order to understand the role of microorganisms in methane hydrate formation and dissociation, molecular phylogenetic studies of microbial communities found in the ODP 892B borehole fluid and in nearby sediment pore waters are being performed.

THE FATE OF METHANE IN THE OCEAN WATER COLUMN

To quantify the amount of methane lost to the water column and characterize the effect of microbial methane oxidation on the local seawater chemistry, in particular the O_2 and CO_2 contents at seven sites in the vicinity of ODP Site 892B, the water column was sampled and the chemistry and methane oxidation rates determined. The objective was to assess the potential global effects of seafloor methane seepages in active convergent margins on seawater, and possibly also on atmospheric chemistry.

In addition to the previously described small near-surface methane concentration maximum, produced biologically in the oceans,[20] we identified two regional more intense methane concentration maxima (greater than 35 nM) in the area of Site 892B, one at 450–550, the second at about 150–200 m depth. These methane concentrations are 10–20 times higher than its solubility at the *in situ* pressure–temperature conditions.[21]

The seawater chemistry, particularly the pH, alkalinity, and dissolved inorganic carbon (DIC) concentrations, indicate that in seawater methane is oxidized by two microbially mediated reactions: (1) methane oxidation to CO_2 and depletion in O_2 content, but no change in alkalinity

$$CH_4 + 2O_2 \rightarrow CO_2 + 2H_2O \tag{1}$$

$$\Delta Alk = 0,$$

and (2) methane oxidation to CO_2 and reduction in sulfate content, with an increase in alkalinity

$$CH_4 + SO_4^{-2} + 2H^+ \rightarrow CO_2 + 2H_2O + H_2S \tag{2}$$

$$\Delta Alk = 2.$$

The water column chlorofluorocarbon (CFC) concentration and ratio measurements indicate that bottom water at ODP Site 892B is about 30 years old. This "apparent age" helps to assess the penetration rate of heat from global warming to the depth of the seafloor gas hydrates. The manner and rate of gas hydrate dissociation and the release to seawater of the enormous reservoir of methane stored in the oceanic gas hydrates have important implications for the ocean carbon cycle, ocean DIC—thus carbonate isotopic composition, and global warming.

ACKNOWLEDGMENTS

We thank the crew members and pilots of R/V Atlantis and Alvin for their support and the students from UCSD and Lehigh University for their enthusiastic shipboard help. We greatly appreciate the shipboard and shore-based help by Drs. A. Paytan, B. Ransom, and Y. Weinstein, and by G. Robertson and J. Plant. We also thank M. Vollmer for the water column CFC analyses. The work was supported by National Science Foundation Grant #OCE 97-12135 to M. Kastner and D. Bartlett, Grant #OCE 97-12174 to B. Carson, and Grant #OCE-9817612 to M. Kastner and A. Payton.

REFERENCES

1. KULM, L.D., E. SUESS, J.C. MOORE, B. CARSON, B.T.R. LEWIS, S.D. RITGER, D.C. KADKO, T.M. THORNBERG, R.W. EMBLEY, W.D. RUGH, G.J. MASSOTH, M.R. LANGSETH, G.R. COCHRANE & R. SCAMMAN. 1986. Oregon subduction zone: venting, fauna, and carbonates. Science **231**: 561–566.
2. CARSON, B., E. SUESS & J.C. STRASSER. 1990. Fluid flow and mass flux determinations at vent sites on the Cascadia Margin accretionary prism. J. Geophys. Res. **95**: 8891–8897.
3. CARSON, B., E. SEKE, V. PASKEVICH & M.L. HOLMES. 1994. Fluid expulsion sites on the Cascadia accretionary prism: Mapping diagenetic deposits with processed GLORIA imagery. J. Geophys. Res. **99**: 11,959–11,969.
4. SUESS, E., R. VON HUENE et al. 1988. Ocean Drilling Program Leg 112, Peru continental margin: Part 2, sedimentary history and diagenesis in a coastal upwelling environment. Geology **16**: 939–943.
5. KASTNER, M., H. ELDERFIELD & J.B. MARTIN. 1991. Fluids in convergent margins: What do we know about their composition, origin, role in diagenesis, and importance for oceanic chemical fluxes? Philosophical Trans. Royal Soc. London **A335**: 243–259.
6. KASTNER, M., K.A. KVENVOLDEN, M.J. WHITICAR, A. CAMERLENGHI, & T.D. LORENSON. 1995. Relation between pore fluid chemistry and gas hydrates associated with bottom-simulating reflectors at the Cascadia margin, sites 889 and 892. *In* Proc. Ocean Drill. Program Sci. Res. B. Carson, G.K. Westbrook, R.J. Musgrave & E. Suess, Eds. **146**: 175–187.
7. GIESKES, J.M., P. VROLIJK & G. BLANC. 1990. Hydrogeochemistry of the northern Barbados accretionary complex transect: Ocean Drilling Project Leg 110. J. Geophys. Res. **95**: 8809–8818.
8. CLAYPOOL, G.E. & KAPLAN, I.R. 1974. The origin and distribution of methane in marine sediment. *In* Natural Gases in Marine Sediments, I.R. Kaplan, Ed.: 99–139. Plenum, New York.
9. WHITICAR, M.J., M. HOVLAND, M. KASTNER & J.C. SAMPLE. 1995. Organic geochemistry of gases, fluids, and hydrates at the Cascadia accretionary margin. Proc. ODP Sci. Results, Leg 146. B. Carson, G.K. Westbrook, R.J. Musgrave, E. Suess, Eds.: 385–397. Ocean Drilling Program, College Station, TX.
10. KASTNER, M., K.A. KVENVOLDEN & T. LORENSON. 1998. Chemistry, isotopic composition, and origin of methane hydrogen sulfide hydrate at the Cascadia subduction zone. Earth Planet Sci. Lett. **156**: 173–183.
11. RITGER, S., B. CARSON & E. SUESS. 1987. Methane-derived authigenic carbonates formed by subduction-induced pore-water expulsion along the Oregon/Washington margin. Geol. Soc. Amer. Bull. **98**: 147–156.

12. CARSON, B. & G.K. WESTBROOK. 1995. Modern fluid flow in the Cascadia accretionary wedge: a synthesis. *In* Proc. Ocean Drill. Program Sci. Res. G.K. Westbrook, R.J. Musgrave, & E. Suess, Eds. **146:** 413–424.
13. SUESS, E., M.E. TORRES, G. BOHRMANN, R.W. COLLIER, J. GREINERT, P. LINKE, G. REHDER, A. TREHU, K. WALLMANN, G. WINCKLER & E. ZULEGER. 1999. Gas hydrate destabilization: enhanced dewatering, benthic material turnover, and large methane plumes at the Cascadia Convergent margin. Earth Planet. Sci. Lett. **170:** 1–15.
14. THORNBURG, T.M. & E. SUESS. 1990. Carbonate cementation of granular and fracture porosity: implications for the Cenozoic hydrologic development of the Peru continental margin. Proc. ODP Sci. Results, Leg 112. E. Suess & R. von Huene, Eds.: 95–109. Ocean Drilling Program, College Station, TX.
15. MARTIN, J.B., M. KASTNER, P. HENRY, X. LE PICHON & S. LALLEMENT. 1996. Chemical and isotopic evidence for sources of fluids in a mud volcano field seaward of the Barbados accretionary wedge. J. Geophys. Res. **101:** 20,325–20,345.
16. MCADOO, B.G., D.L. ORANGE, E.A. SILVER, K. MCINTOSH, L. ABOTT, J. GALEWSKY, L. KAHN & M. PROTTI. 1996. Seafloor structural observations, Costa Rica accretionary prism. Geophys. Res. Lett. **23:** 883–886.
17. BROOKS, J.M., M.E. FIELD & M.C. KENNICUTT. 1991. Observations of gas hydrates in marine sediment, offshore northern California. Mar. Geol. **96:** 103–109.
18. KVENVOLDEN, K.A. 1988. Methane hydrate—a major reservoir of carbon in the shallow geosphere? Chem. Geology **71:** 41–51.
19. DICKENS, G.R. & M.S. QUINBY-HUNT. 1994. Methane hydrate stability in seawater. Geophys. Res. Lett. **19:** 2115–2118.
20. SCRANTON, M.I. & P.G. BREWER. 1997. Occurrence of methane in the near-surface waters of the western subtropical North-Atlantic, Deep-Sea Res. **24:** 127–138.
21. YAMAMOTO, S., J.B. ALCAUSKAS & T.E. CROZIER. 1976. Solubility of methane in distilled water and sea water. J. Chem. Eng. Data **21:** 78–81.

Methane Hydrate Estimates from the Chloride and Oxygen Isotopic Anomalies

Examples from the Blake Ridge and Nankai Trough Sediments

RYO MATSUMOTO[a]

Department of Earth and Planetary Science, School of Science, University of Tokyo, 7-3-1 Hongo, Tokyo 113-0033, Japan

ABSTRACT: Oxygen isotopic fractionation between gas hydrate and ambient water is determined as α_{GH-IW} = 1.0037 at 12–16°C and 31 Mpa, on the basis of direct measurements of gas hydrate-derived waters and ambient pore waters recovered from the Blake Ridge during ODP Leg 164. Oxygen isotopic anomalies give us the amount of gas hydrate of 7 to 9% (pore filling), which is almost twice as much as estimates from chloride anomalies. The difference is probably due to uncertainties in determining base-line profiles of the *in situ* pristine pore waters, and partially due to the effects of selective filtration/adsorption during pore water extraction. Two 250 meter-deep holes were drilled in the eastern Nankai trough off central Japan at a water depth of 950 m, where strong BSRs occur at about 300 mbsf. Massive hydrates were not recovered during this drilling but a number of soupy horizons suggest the existence of subsurface gas hydrate. Chloride concentration and $\delta^{18}O$ of interstitial waters are observed to vary in a remarkable zigzag pattern with spiky anomalies, reflecting hydrate dissociation during core-recovery and water extraction. The concentration of gas hydrate in sediments is estimated to be about 3–7% with a spiky maximum value of 30% from chloride anomalies and between 5 and 30% from $\delta^{18}O$ anomalies. Significant difference in vertical distribution between nearby two holes in Nankai Trough probably reflect heterogeneous fluid migration through particular conduits in an accretionary wedge.

INTRODUCTION

The most important question in the study of methane hydrate is: How much methane hydrate is stored in marine sediments? The environmental impact of gas hydrate depends primarily on the total amount of methane trapped in gas hydrate of a shallow geosphere. Resource assessment of gas hydrates requires refinement of the distribution, amounts, occurrence, and reserves of gas hydrate deposits. The amount of subsurface gas hydrate is given by the areal distribution and concentration of gas hydrate within the host sediments. The areal extent of gas hydrate distribution was thought to be nearly identical to the distribution of BSRs. Ocean Drilling Program (ODP) Leg164 has revealed that gas hydrate can occur in sediments without BSRs,[1] but

[a]Telecommunication. Voice: +81-3-5841-4522; fax +81-3-5841-4569.
ryo@eps.s.u-tokyo.ac.jp

BSRs are still useful tools for estimating the minimum areal extent of subsurface gas hydrate. On the other hand, the concentration of gas hydrate in sediments are not readily obtained, requiring sophisticated remote sensing techniques, such as VSP and well-logging or direct measurement of sediment cores. Because of the ephemeral nature of gas hydrate, several proxy analyses have been employed in estimating the amount of gas hydrate in core samples; among which the chloride anomaly technique provides the most reliable proxy of methane hydrate amounts. However, recently possible effects of selective filtration/adsorption during mechanical squeezing has been discussed,[2] requiring reassessment of the chloride anomaly technique.

Formation and dissociation of gas hydrate in marine sediments modify the oxygen isotopic composition of ambient waters as well as the chloride concentration, because gas hydrate concentrates isotopically heavier oxygen (^{18}O) in its cage structures.[3,4] Matsumoto[5] has identified isotopically heavy oxygen containing siderites in the Blake Ridge sediments, suggesting a close relationship between the formation of siderites and the dissociation of gas hydrate during burial diagenesis.

The interstitial waters squeezed from gas hydrate-bearing sediments are variably enriched in ^{18}O, depending on the amounts of gas hydrate contained in the sediments. Given the isotopic fractionation between gas hydrate and ambient water (α_{GH-IW}), pore saturation of gas hydrate is easily estimated from the ^{18}O anomaly.[6] ODP Leg164 recovered a number of massive gas hydrate samples and collected hundreds of interstitial waters from Blake Ridge sediments. Matsumoto[7] measured $\delta^{18}O$ values of gas hydrate and a number of interstitial waters in an attempt to determine the fractionation factor (α_{GH-IW}) between methane hydrate and ambient waters. In this paper, a short summary of $\delta^{18}O$ variations on Blake ridge and isotopic fractionation between gas hydrate and water are presented. This information is applied to determine the amount of subsurface gas hydrate in Blake Ridge and Nankai Trough sediments.

GEOLOGIC SETTINGS

The Blake Ridge, located about 150 km off east coast of north America, is a spit-like submarine rise with a relief of about 3 km (see FIGURE 1). The Ridge is characterized by rapidly accumulated, moderately calcareous hemipelagic sediment deposited by strong contour currents. *B*ottom *s*imulating *r*eflectors (BSRs) are well developed in the crest of the Ridge, covering about 26000 km^2 of the area. In 1995, ODP Leg 164 drilled 750–800 meter-deep holes at Sites 994, 995, and 997 at a water depth of 2800 meters and penetrated BSRs on the crestal area of the Ridge.

The Nankai Trough lies in a subduction zone between the Eurasian and Philippine Sea Plate, extending about 700 km SSW–NNE offshore SE Japan (see FIGURE 2). The accretionary prism is largely composed of siliciclastic hemipelagic sediments with abundant sandstone turbidites. Strong BSRs have been recognized within this sequence, extending about 35000 km^2. In 1997, 250 meter-deep test holes, BH-1 and BH-2, were drilled by Japan National Oil Corporation (JNOC)–Japan Petroleum Exploration (JAPEX) in the Eastern Nankai trough. The two holes, about 100 m apart, were drilled on a flat-topped, deep-sea terrace. Experimental data concerning hydrate stability, *in situ* temperature measurements, and seismic survey data indicate that BSR occurs at around 290 ± 10 mbsf.

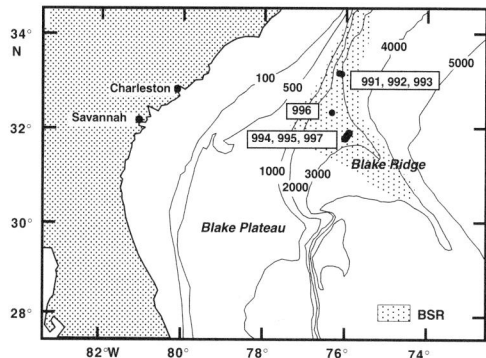

FIGURE 1. Location of Sites 991 to 997 of ODP Leg164 on the Blake Ridge off East coast of North America.

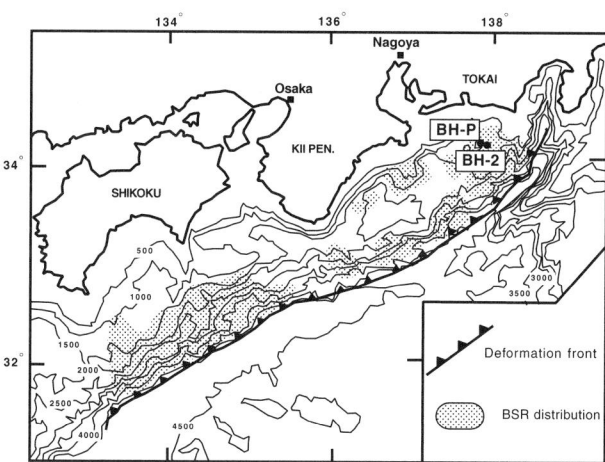

FIGURE 2. Location of drill holes BH-1 and BH-2 in the eastern Nankai Trough off central Japan.

ISOTOPIC FRACTIONATION BETWEEN GAS HYDRATE AND WATER

Experimental Methods

Each gas hydrate sample was dissociated in a Teflon-coated dissociation chamber at room temperature. Gas pressure within the chamber steadily increased and reached a stable value in approximately 10 minutes. Gas was transferred to a gas collection tube for later gas analysis. The volume of residual water was measured and stored in sealed glass tubes for isotopic analysis. Chloride concentration of the residual water was measured to estimate the mixing ratio of gas hydrate water and the pristine pore water. An accurately measured volume (1.0–1.5 ml; 28 to 42 mmol of O_2) of the water was placed into a small flask, stirred, and mixed with 0.10 mmol of CO_2 of known isotopic composition in a water bath at 25.0°C for 15–20 hours to attain isotopic equilibration between water and gas. Oxygen isotopic ratio $^{18}O/^{16}O$ of the equilibrated CO_2 gas was determined using a Finnigan Delta S mass spectrometer. The results are represented as delta per mil notation ($\delta^{18}O‰$) relative to the SMOW standard. The standard deviation (2σ) of independent analysis was 0.01–0.05‰ and the reproducibility of the measurements was about 0.10‰. Chloride concentrations of pore waters were measured by ion chromatography (Model ICA-5000, TOA Dempa Company).

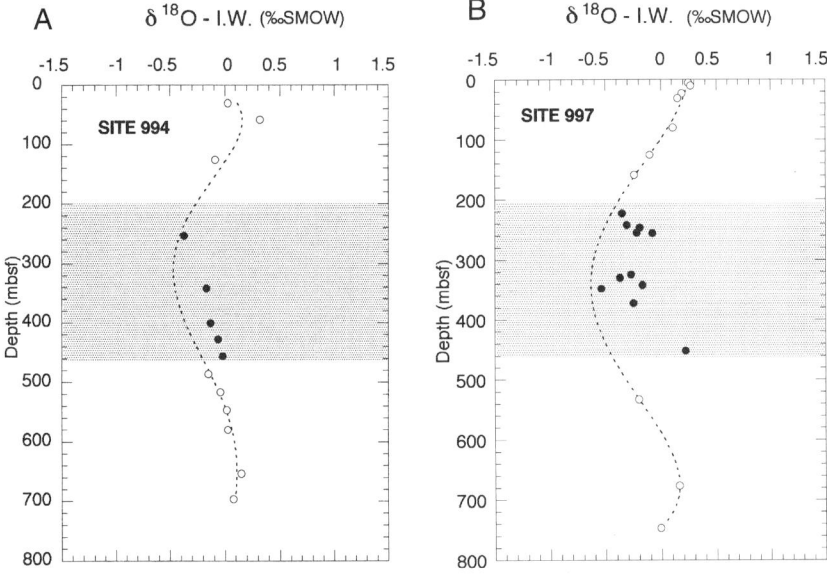

FIGURE 3. The $\delta^{18}O$ (‰ SMOW) of interstitial waters at (*A*) Site 994 and (*B*) Site 997. *Dotted lines* represent baselines for the interstitial waters seemingly unaffected by gas hydrate dissociation. Baseline for Site 994 is, $Y = 0.886 - 1.20 \times 10^{-02}X + 4.23 \times 10^{-05}X^2 - 8.06 \times 10^{-08}X^3 + 1.01 \times 10^{-10}X^4 - 5.85 \times 10^{-14}X^5$, ($r^2 = 0.9906$), and for Site 997 is, $Y = 0.245 - 7.43 \times 10^{-04}X - 2.44 \times 10^{-05}X^2 + 7.29 \times 10^{-08}X^3 - 4.82 \times 10^{-11}X^4 - 6.34 \times 10^{-15}X^5$, ($r^2 = 0.9890$), where Y is $\delta^{18}O_{H_2O}$ (‰ SMOW) and X is depth (mbsf).

$\delta^{18}O$ of Gas Hydrates and Waters from the Blake Ridge

The $\delta^{18}O$ values of a gas hydrate sample recovered from 259.9 mbsf (994C-31X-7) at Site 994 is 2.67‰ SMOW and four samples from 330.15 mbsf (997A-42X-3) at Site 997 were 2.82–3.51‰ SMOW. In comparison, the $\delta^{18}O$ of 14 water samples from Site 994 were between 0.32‰ and −0.17‰ SMOW and 21 water samples from Site 997 range from 0.25‰ to −0.54‰ SMOW (see FIGURE 3). The depth profiles at both sites separate into an upper smooth zone (0–200 mbsf), a middle zigzag zone, which corresponds to gas hydrate zone (200–450 mbsf), and a lower smooth zone (450–750 mbsf). The zigzag pattern of the middle zone is likely to have been caused by heavy oxygen containing water from gas hydrate. Regression curves to represent the depth profile of the interstitial waters of gas hydrate free sediments were prepared from the data of the upper and lower zones.

Oxygen Isotopic Fractionation

The difference in $\delta^{18}O$ values between gas hydrate and the ambient water ($\Delta\delta^{18}O_{GH-IW}$) is 3.1‰ at Site 994 and 3.3–3.8‰ (mean, 3.6‰) at Site 997 (see FIGURE 4). Assuming that the gas hydrate samples were in isotopic equilibrium with the ambient waters (T = 12–16°C and P = 31 MPa), the equilibrium isotopic

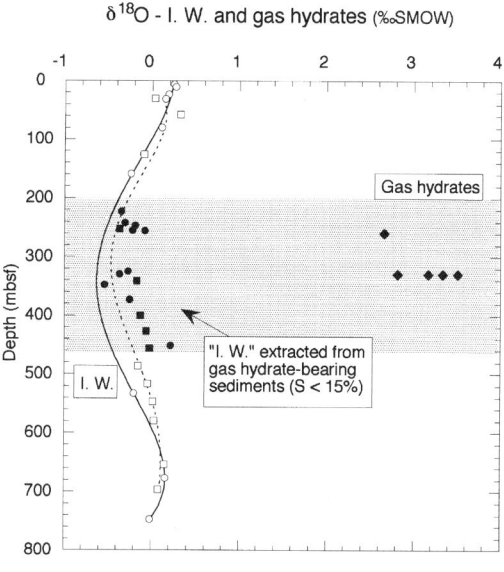

FIGURE 4. Diagram showing the $\delta^{18}O$ of gas hydrates, $\delta^{18}O$ of the interstitial waters and baselines at Sites 994 (*open squares* and *broken line*) and at Site 997 (*open circles* and *thin line*). The difference in $\delta^{18}O$ values ($\Delta\delta^{18}O_{GH-IW}$) is 3.1‰ at Site 994 and 3.3‰ to 3.8‰ at Site 997. *Filled symbols* show positive excursions due to dissociation of gas hydrate.

fractionation factor (α_{GH-IW}) is calculated to be 1.0034–1.0040 (mean, 1.0037). For comparison, the oxygen isotopic fractionation between ice and water is 1.0027–1.0035[8–10] and that of THF hydrate–water is 1.00268 ± 0.00003 at 0–4°C.[11]

GAS HYDRATE AT BLAKE RIDGE

Assuming a stoichiometry of the Blake Ridge gas hydrate ($CH_4 \cdot 5.75H_2O$), gas hydrate density of 0.97g/cm^3, and gas hydrate–water fractionation ($\Delta\delta^{18}O_{GH-IW}$) value of 3.1‰ at Site 994 and 3.6‰ at Site 997, the pore filling fraction (X) of gas hydrate is given by

$$X = (\delta^{18}O_M - \delta^{18}O_P)/(2.8 - 0.11\delta^{18}O_P) \quad (1)$$

For Site 997, equation (1) is

$$X = (\delta^{18}O_M - \delta^{18}O_P)/(3.2 - 0.11\delta^{18}O_P) \quad (2)$$

M and P denote measured $\delta^{18}O$ values of extracted waters and $\delta^{18}O$ values of the pristine pore waters, respectively, estimated from FIGURE 1. Gas hydrate amounts (% pore filling) are given in FIGURE 5 along with gas hydrate estimates from chloride anomalies.[1]

At Site 994, four samples in the gas hydrate zone yield a pore filling of 7.5 ± 1%, and the average amount at Site 997 is about 9%, both of which are almost twice the

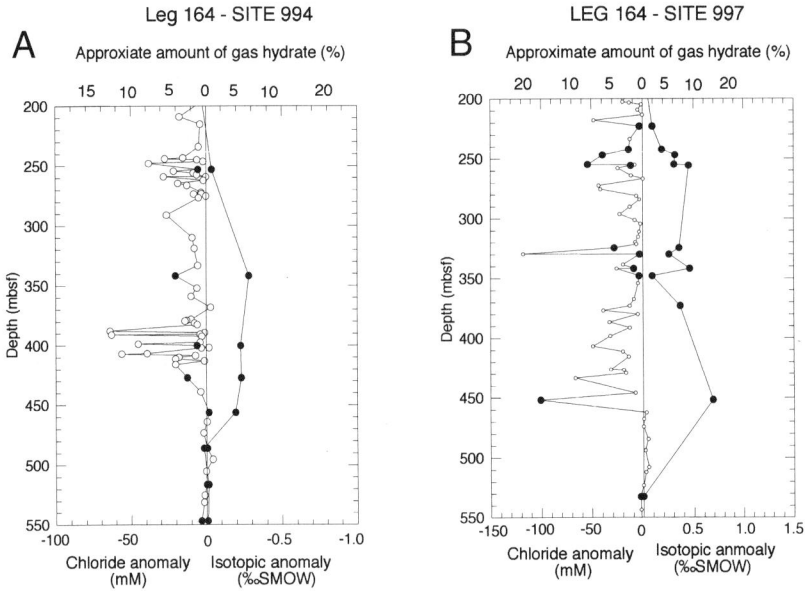

FIGURE 5. Magnitude of chloride and isotopic anomalies (*lower axis*) and approximate amount of gas hydrate (*upper axis*) at Sites 994 and 997. *Filled symbols* for chloride anomaly data correspond to isotopic anomaly data on the right.

estimates from chloride anomalies. At Site 997, gas hydrate tends to concentrate in the upper and lower part of gas hydrate zone (200–450 mbsf) with minimum value in the middle (330–370 mbsf) of the zone. This pattern is identical to that of the chloride anomaly.

GAS HYDRATE IN NANKAI TROUGH

Sediment Lithology

A number of spot cores were collected at both holes and 41 m-long and 68 m-long sediment cores were recovered from BH-1 and BH-2, respectively. Significant differences were not observed between the two holes. The upper part of the holes is dominated by clay to clayey silt to siltstone with occasional thin ash layers, whereas the deeper part contains significant amounts of sand and sandy beds of possible turbidite origin. Core recovery rate dramatically dropped downhole; empty core-barrels with thin sandy films or sand grains inside the barrel suggest that unrecovered, lost intervals were dominantly sand and sandstone beds. We did not recover solid hydrate, but the existence of massive hydrate horizons were indicated by occasional soupy horizons and anomalies in the interstitial water chemistry.

$\delta^{18}O$ and Chloride Concentration of Waters

About forty interstitial water samples were measured for oxygen isotopic composition and chloride concentration at each site. The $\delta^{18}O$ of the interstitial water is about $-0.5‰$ SMOW at the top of the holes, tends to decrease downward with a strong zigzag pattern, then reaches to -1.0 to $-1.5‰$ at about 250 mbsf (see FIGURE 6). In BH-2, the amplitude of zigzag fluctuation is observed to increase in

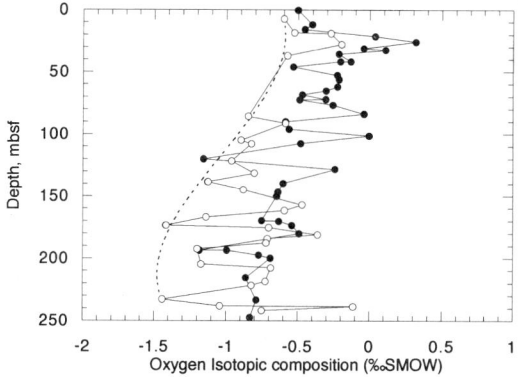

FIGURE 6. $\delta^{18}O$ of pore waters from drillholes BH-1 (*dots*) and BH-2 (*open circles*) in the Eastern Nankai Trough. Hypothetical baseline for pristine pore waters is indicated by the *thin broken line*. Isotopic anomalies are determined as the difference between the base line values and measured values.

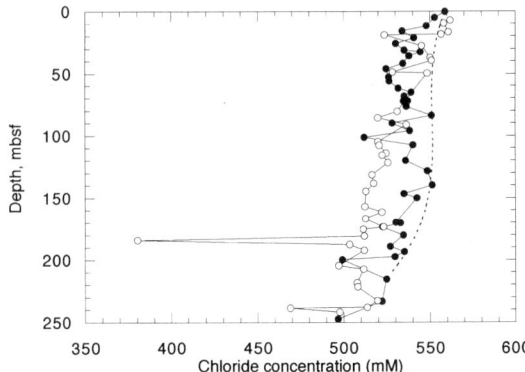

FIGURE 7. Chloride concentration of pore waters at BH-1 (*dots*) and BH-2 (*open circles*) in the Eastern Nankai Trough. The hypothetical baseline for the pristine pore water is shown by a *thin broken line*. Chloride anomalies are determined as the difference between the baseline values and measured values.

the lower sandstone-rich horizons (FIG. 6). Early diagenetic processes such as free convection, fluid migration, and ionic diffusion within highly porous sediments should produce a smooth depth profile. Thus, the observed zigzag profiles are thought to be an artifact caused by dissociation of gas hydrate during core-recovery and water extraction. This implies that the heavier shifts and spikes reflects gas hydrate dissociation, suggesting that the lighter values represent depth profile of the pristine pore waters. A broken line connecting the lighter values is taken as a hypothetical baseline for isotopic variations.

The chloride concentration tends to decrease downhole from about 560 mM at the core-top, toward about 500 mM at the bottom of the holes (see FIGURE 7), but the depth profiles show irregular zigzag patterns as in case of isotopic variation. Negative spikes and excursions reflect dissociation of gas hydrate. A strong negative spike at 180 mbsf of BH-2 corresponds to approximately 30% pore filling by gas hydrate.

Amounts of Gas Hydrates

FIGURE 8 depicts the distribution and amounts of gas hydrate at Site BH-1 and BH-2 from the Eastern Nankai Trough. The distance between the sites is only 100 m and the lithologies are quite similar to each other. However, BH-1 tends to have more gas hydrate in the upper part, whereas BH-2 has more gas hydrate in the middle and lower parts. This pattern is also reflected by oxygen isotopic profiles that depict 30–40 meter-thick hydrate bearing horizons in the interval 100–250 mbsf (see FIGURE 9).

Pore filling of gas hydrate calculated from chloride anomalies are mostly around 3 to 7% with a spiky maximum value of 30%, whereas the estimates from $\delta^{18}O$ anomalies are between 5 and 25% (FIG. 9). These values are somewhat similar to, or a bit larger than, those of Blake Ridge sediments.

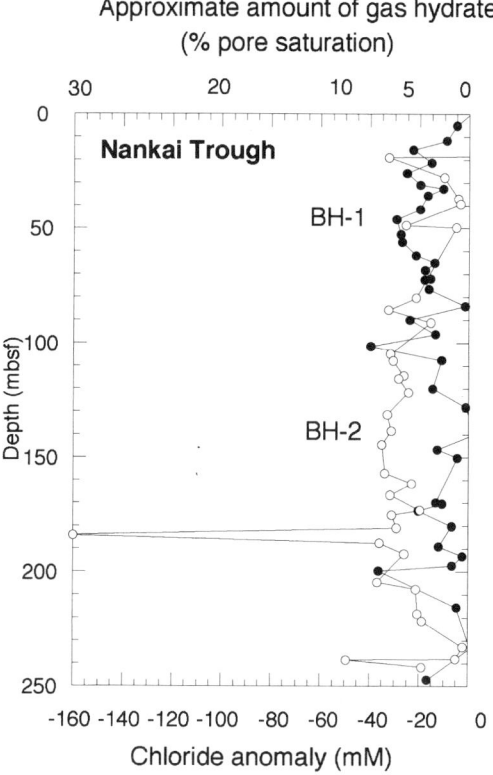

FIGURE 8. Depth profiles of the chloride anomalies at BH-1 (*dots*) and BH-2 (*open circles*). Chloride anomaly values are converted to the approximate amount of gas hydrate (percent pore filling) on the upper axis.

DISCUSSION

Chloride and Isotopic Anomalies

The observed discrepancy between the two estimates of gas hydrate pore filling is not large but it is significant. The possible causes of the discrepancy are: (1) uncertainty in the determination of baselines (*in situ* trends) of both $\delta^{18}O$ and chloride concentration (especially at the Blake Ridge, where sampling density is low); (2) differential chemical effects between ^{18}O and chloride concentration during mechanical squeezing; and (3) sample deterioration, which may have occurred during storage of water samples. Among these, the baseline problem is most critical. As for chloride, pristine waters may have changed only slightly or may not have changed throughout the depth interval. If this is the case, hydrate estimates would be increased by the factor of 1.5–2.0. Furthermore, the sampling density may partially explain the apparent discrepancy for Blake Ridge data. Chloride measurements were

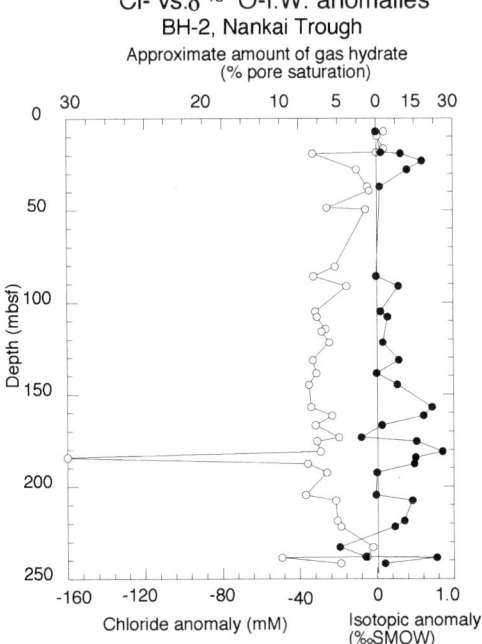

FIGURE 9. Comparison between chloride anomaly (*open circles*) and isotopic anomaly (*dots*) at BH-2. Approximate amount of gas hydrate is indicated on the *upper axis*. Note that gas hydrate tends to concentrate in the upper horizon (20–40 mbsf) and lower horizon (150–250 mbsf).

more densely spaced than those for isotope analyses. Interstitial water samples were preferentially taken from undisturbed sediment cores, which are usually poor in or free of gas hydrate.

Blake Ridge and Nankai Trough

Leg 164 has revealed two distinct modes of occurrence of gas hydrate in marine sediments: (1) massive, megascopic aggregates of pure hydrate, and (2) fine crystalline hydrate disseminated in fine grained sediments. The first type occurs as nodules or veins filling open fractures and cavities. We did not recover massive aggregates of gas hydrates from BH-1 and BH-2 of the Eastern Nankai trough, probably because the holes do not penetrate fractures or faults in the area. As for the second type of gas hydrate, we conclude that the concentration in fine-grained sediments is less than 15% (pore filling) in both the passive and active margin sites. A notable difference between the Blake Ridge and Eastern Nankai is the vertical distribution. In the former, zone of gas hydrate is substantially limited between 200 and 450 mbsf, however in the Eastern Nankai, significant amount of gas hydrate accumulate even in shallow subsurface at about 20 mbsf. This may be related to active gas venting and fluid seeps that are characteristic phenomena of the Eastern Nankai.[12] Active fluid migration through particular conduits in the accretionary prism may also account for the difference in vertical distribution pattern of gas hydrates at BH-1 and BH-2.

CONCLUSIONS

1. Oxygen isotopic fractionation between gas hydrate and ambient water is about 3.2‰ at 12–16°C and 31 Mpa, based on field samples.

2. The amounts of gas hydrate in Blake Ridge is a few percent of pore saturation (chloride anomalies) to about 9% (isotopic anomalies), whereas in Nankai trough the amounts range between 3 and 30% (chloride anomalies) and from 5 to 25% (isotopic anomalies).

3. The discrepancy between the chlorine and $\delta^{18}O$ anomaly estimates is probably due to uncertainties in determining the baseline profiles for the pristine interstitial waters, and at least partially due to filtration/adsorption effects during water extraction.

ACKNOWLEDGMENTS

Drilling and coring of BH-1 and BH-2 were performed by Fugro Japan, being conducted by JNOC and JAPEX. Onsite observation, description, and sampling of sediment cores and squeezing of the interstitial waters were performed by the Hydrate Research Group of Drs. Y. Hiroki (Osaka Kyoiku University), Lu Hailong (Tokyo University), T. Fujii (JNOC-TRC), T. Uchida, A. Waseda, K. Baba, M. Yagi (JAPEX), and RM.

REFERENCES

1. PAULL, C.K., R. MATSUMOTO & P.J. WALLACE. 1996. Proc. ODP, Init. Repts. 164. Ocean Drilling Program, College Station, TX.
2. CAVE, M., L. GRIFFAULT & S. REEDER. 1998. The extraction and characterization of pore-water from lower permeability argillaceous rock samples. 23rd General Assembly of EGS, Nice. Abst., SE406.
3. HESSE, R. & W.E. HARRISON. 1981. Gas hydrates (clathrate) causing pore-water freshening and oxygen isotope fractionation in deep-water sedimentary sections of terrigenous continental margins. Earth and Planet. Sci. Lett. **55:** 453–462.
4. HARRISON, W.E. & J.A. CURIALE. 1982, Gas hydrates in sediments of Holes 497 and 498A, Deep Sea Drilling Project Leg 67. *In* Init. Repts., DSDP 67. J. Aubouin & R. von Huene *et al.*,Eds.: 591–595. U. S. Govt. Printing Office. Washington.
5. MATSUMOTO, R. 1989. Isotopically heavy oxygen-containing siderite derive from the decomposition of methane hydrate. Geology **17:** 707–710.
6. USSLER, W. III & C.K. PAULL. 1995. Effects of ion exclusion and isotopic fraction on pore water geochemistry during gas hydrate formation and decomposition. GeoMarine Lett. **15:** 37–44.

7. MATSUMOTO, R. & W.S. BOROWSKI. 2000. Gas hydrate estimates from newly determined oxygenisotopic fractionation (α_{GH-IW}) and $\delta^{18}O$ anomalies of the interstitial waters. ODP Leg 164, Blake Ridge. Proc. ODP, Sci. Results, 164. Ocean Drilling Program, College Station, TX.
8. O'NEIL, J.R. 1968. Hydrogen and oxygen isotope fractionation between ice and water. J. Phys. Chem. **72**: 3682–3684.
9. CRAIG, H. & B. HOM. 1968. Relationship of deuterium, oxygen-18, and chlorinity in the formation of sea ice. Trans. Am. Geophys. Union **49**: 216–217.
10. JAKLI, G. & D. STASCHEWSKI. 1977, Vapor pressure of H_2O ice (–50 to 0°C) and H_2O water (0–170°C). J. Chem. Soc. Faraday Trans. **I 73**: 1505–1509.
11. DAVIDSON, C.W., D.G. LEAIST & R. HESSE. 1983. Oxygen-18 enrichment in the water of a clathrate hydrate. Geochim. Cosmochim. Acta **47**: 2293–2295.
12. LE PICHON, X., K. KOBAYASHI & KAIKO-NANKAI SCIENTIFIC CREW. 1992. Fluid venting activity within the Eastern Nankai trough accretionary wedge: a summary of the 1989 Kaiko-Nankai results. Earth Planet. Sci. Lett. **109**: 303–318.

Reservoir Characterization of Marine and Permafrost Associated Gas Hydrate Accumulations with Downhole Well Logs

TIMOTHY S. COLLETT[a] AND MYUNG W. LEE[b]

U.S. Geological Survey, Denver Federal Center, Box 25046, MS-939, Denver, Colorado 80225 USA

ABSTRACT: Gas volumes that may be attributed to a gas hydrate accumulation depend on a number of reservoir parameters, one of which, gas-hydrate saturation, can be assessed with data obtained from downhole well-logging devices. This study demonstrates that electrical resistivity and acoustic transit-time downhole log data can be used to quantify the amount of gas hydrate in a sedimentary section. Two unique forms of the Archie relation (standard and quick look relations) have been used in this study to calculate water saturations (S_w) [gas-hydrate saturation (S_h) is equal to ($1.0 - S_w$)] from the electrical resistivity log data in four gas hydrate accumulations. These accumulations are located on (1) the Blake Ridge along the Southeastern continental margin of the United States, (2) the Cascadia continental margin off the Pacific coast of Canada, (3) the North Slope of Alaska, and (4) the Mackenzie River Delta of Canada. Compressional wave acoustic log data have also been used in conjunction with the Timur, modified Wood, and the Lee weighted average acoustic equations to calculate gas-hydrate saturations in all four areas assessed.

INTRODUCTION

The research focus of this study followed two general paths: (1) gas hydrate well log response modeling and (2) field data characterization and verification. The well log response modeling phase included the assessment of existing and newly developed gas hydrate well log evaluation techniques. In the field verification phase of this study, the gas hydrate well log evaluation techniques developed in the response modeling phase were tested and used to calculate gas-hydrate saturations with the log data from four known gas hydrate accumulations (see FIGURE 1): (1) the Blake Ridge along the Southeastern continental margin of the United States, (2) the Cascadia continental margin off the Pacific coast of Canada, (3) the North Slope of Alaska, and (4) the Mackenzie River Delta of Canada.

Electronic mail: [a]tcollett@usgs.gov [b]mlee@usgs.gov

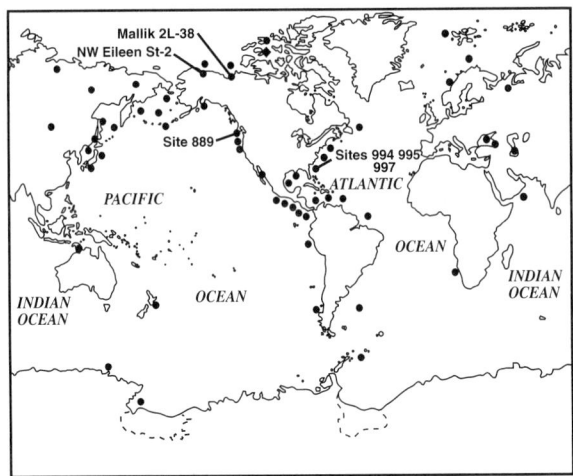

FIGURE 1. Location of known and inferred gas hydrate occurrences. Also shown are the locations of logged wells with confirmed gas hydrate occurrences.

GAS HYDRATE WELL LOG RESPONSE MODELING

Electrical Resistivity Logs

Presented in this section of the report and utilized in the field data application section, are two forms of the Archie relation that appear to yield useful gas-hydrate saturation data.[1] The Archie equation is an empirical relationship between water content (saturation) and the resistivity of water-saturated sediments:

$$S_w = \left(\frac{aR_w}{\emptyset^m R_t}\right)^{1/n}, \quad (1)$$

where R_t is the formation resistivity (from log), ohm-m; a, an empirically derived parameter; R_w, the resistivity of formation water, ohm-m; $\emptyset$, porosity, decimal percent; m, an empirically derived parameter; S_w, water saturation [gas-hydrate saturation (S_h) is equal to $(1.0 - S_w)$], decimal percent; and n is an empirically derived parameter.

Another resistivity approach for assessing gas-hydrate saturations is based on the so-called quick look log analysis technique that compares the resistivity of a water saturated sediment (R_o) to the resistivity of a hydrocarbon-bearing sediment (R_t) using the following modified Archie relation:

$$S_w = \left(\frac{R_o}{R_t}\right)^{1/n} \quad (2)$$

Acoustic Transit-Time Logs

Timur developed the first three component time-average equation (modified Wyllie time-average equation) that can be used to directly calculate the volume of gas hydrate within a rock interval.[2] Pearson and others[3] were first to apply the Timur equation to hydrate-bearing rocks, and they concluded that it adequately predicted the acoustic properties of gas hydrates in consolidated rock media. They used the following three component Timur time-average equation:

$$\frac{1}{V_b} = \frac{\varnothing(1-S_h)}{V_w} + \frac{\varnothing S_h}{V_h} + \frac{1-\varnothing}{V_m}, \qquad (3)$$

where V_b is the well log measured velocity, km/sec; V_w, the water compressional velocity, km/sec; V_h, the gas hydrate compressional velocity, km/sec; and V_m is the matrix compressional velocity, km/sec.

Several workers have shown that the observed velocity behavior of sedimentary rocks are not always consistent with the predictions of the Timur time-average model. It has been proposed that a modified version of the Wood equation, which is approximately valid for particles in suspension, could be used to overcome problems with the time-average equation.[2] Like the three component time-average equation, the modified Wood equation for gas-hydrate–bearing sediments can be written as:

$$\frac{1}{\rho_b V_b^2} = \frac{\varnothing(1-S_h)}{\rho_w V_w^2} + \frac{\varnothing S_h}{\rho_h V_h^2} + \frac{1-\varnothing}{\rho_m V_m^2}, \qquad (4)$$

where ρ_b is the bulk-density, g/cm³ [$\rho_b = (1-\varnothing)\rho_m + (1-S_h)\varnothing\rho_w + S_h \varnothing \rho_h$]; ρ_w is the water density, g/cm³; ρ_h is the hydrate density, g/cm³; and ρ_m is the matrix density, g/cm³.

Similar to the Timur equation, the modified Wood equation does not always accurately predict observed velocity–porosity relations in marine sediments. Lee and others[2] have proposed an acoustic equation for interval velocities within marine gas-hydrate–bearing sediments that uses weighted means of the Timur and modified Wood equations, that is:

$$\frac{1}{V_b} = \left[\frac{W\varnothing(1-S_h)^r}{V_{Wood}}\right] + \left[\frac{1-W\varnothing(1-S_h)^r}{V_{Timur}}\right], \qquad (5)$$

where r is the gas hydrate cementation constant; W, the weighting factor (for most cases this is 1); V_{Wood}, the results of the Wood equation, km/sec; and V_{Timur} is the results of the Timur equation, km/sec.

FIELD APPLICATION

In this section of the report, the quantitative gas hydrate well-log evaluation techniques described in the response modeling phase of this study have been tested and used to calculate gas-hydrate saturations within four known and logged gas hydrate occurrences, see FIGURES 1, 2a, and 2b and TABLE 1.

For both the "standard" (**1**) and "quick look" (**2**) derivation, the Archie relations have been used to calculate water saturations (S_w) [gas-hydrate saturation (S_h) is equal to $(1.0 - S_w)$] from the available downhole electrical resistivity logs in wells

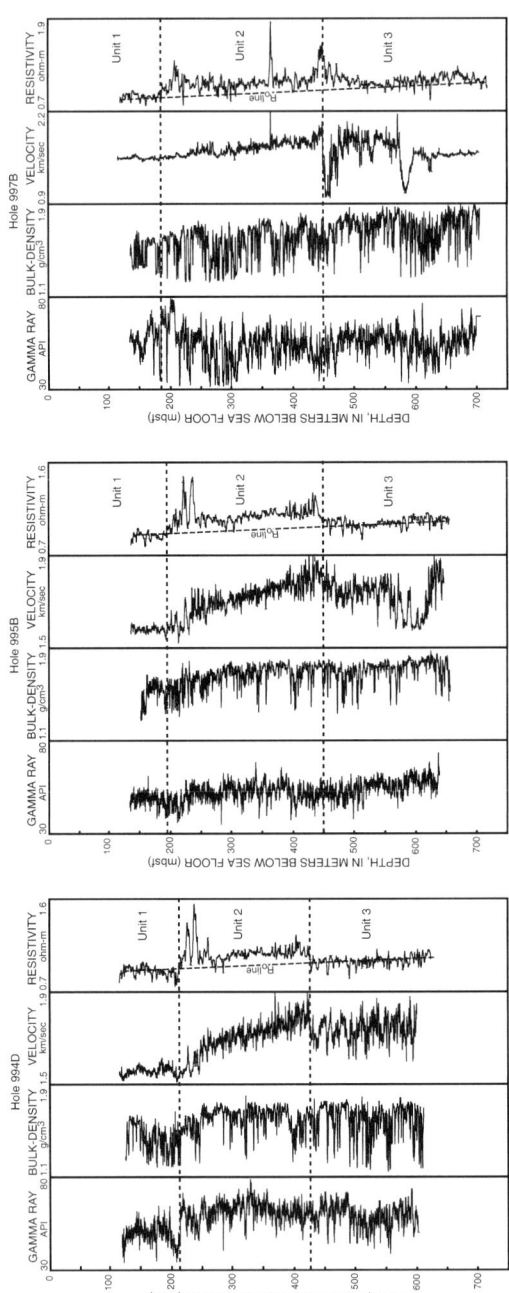

FIGURE 2a. Downhole log data from ODP Holes 994D, 995B, and 997B.

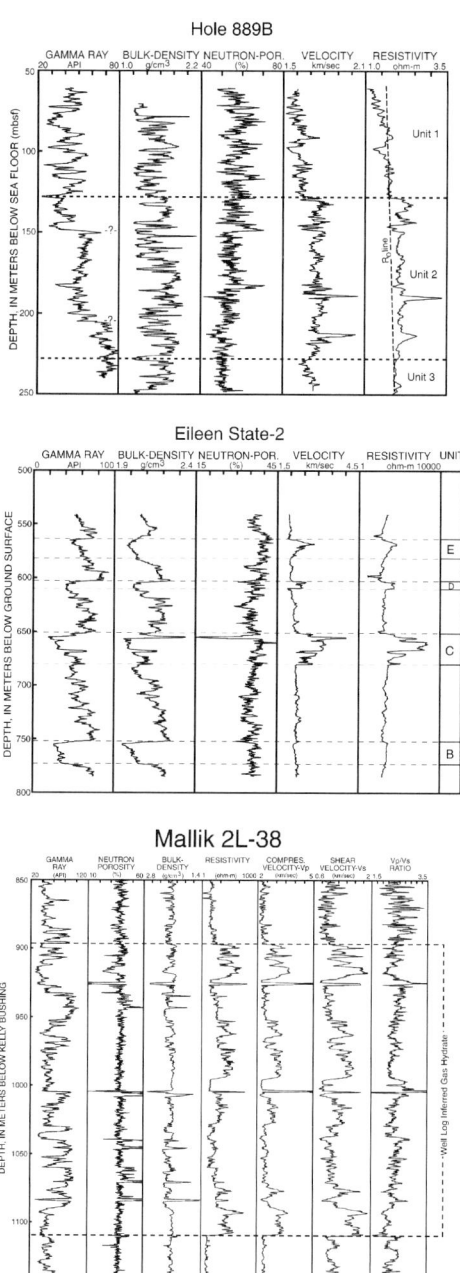

FIGURE 2b. Downhole log data from ODP Hole 889B, Northeast Eileen State-2 well, and the Mallik 2L-38 well.

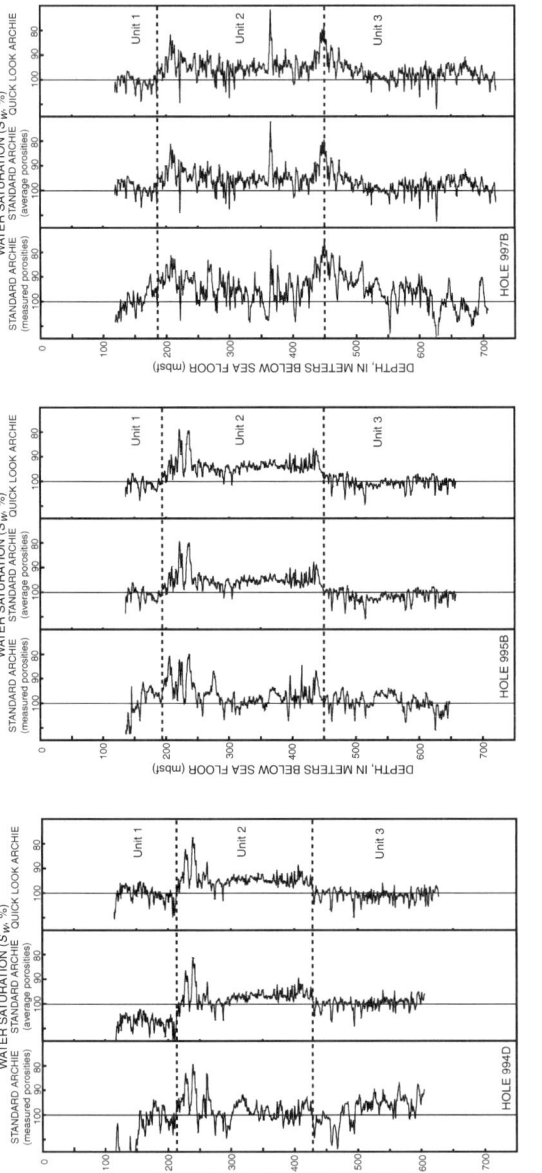

FIGURE 3a. Standard and quick look Archie derived water saturations in ODP Holes 994D, 995B, and 997B.

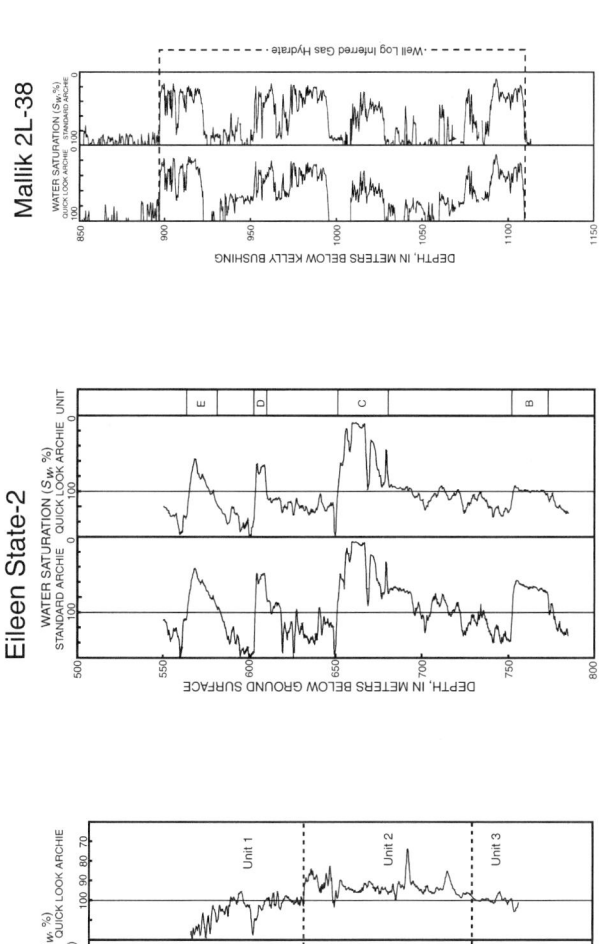

FIGURE 3b. Standard and quick look Archie derived water saturations in ODP Hole 889B, Northwest Eileen State-2 well, and the Mallik 2L-38 well.

TABLE 1. Resistivity and acoustic constants used to calculate water and gas-hydrate saturations within the downhole log-inferred gas hydrate occurrences at ODP sites 889, 994, 995, and 997, and in the Northwest Eileen State-2 and Mallik 2L-38 wells

Site/well identification	Archie a	Archie m	R_w	R_o	Porosity	Lee r	Lee W
Site 889	0.97	2.81	Ref. 1	Fig. 2b	Neutron log Av. core. por.	1.0	1.2
Site 994	1.05	2.56	Ref. 4	Fig. 2a	Meas. core por. Av. core por.	1.0	1.1
Site 995	1.03	2.53	Ref. 4	Fig. 2a	Meas. core por. Av. core por.	1.0	1.1
Site 997	1.07	2.59	Ref. 4	Fig. 2a	Meas. core por. Av. core por.	1.0	1.1
Eileen St-2	0.62	2.15	Ref. 1	Ref. 1	Density log	1.0	1.0
Mallik 2L-38	0.62	2.15	Ref. 6	Ref. 6	Density log	1.0	1.6

that have been drilled into the four gas hydrate accumulations assessed in this study (FIGURE 1). In the first computation, the "standard" Archie equation (1) has been used with sediment porosities calculated from log and core derived physical property data. Formation water resistivities (R_w) calculated from the recovered core water samples or waters collected during formation production test were used in the "standard" Archie calculation along with assumed or log-core values calculated for the a and m Archie constants. In all cases the value of the empirical constant n was assumed to be 1.9386, as determined by Pearson and others.[3] In FIGURES 3a and 3b, the results of the standard Archie calculations are shown as water saturation (S_w) log traces for all of the gas hydrate accumulations assessed in this study.

The next resistivity log approach used to assess gas-hydrate saturations is based on the modified quick look, Equation (2), Archie log analysis technique. In order to determine R_o, we used the measured, deep resistivity log data from the non-gas–bearing portions of each borehole to project R_o trend-lines for the gas-hydrate–bearing intervals within each of the accumulations assessed (FIGS. 2a and 2b). Knowing R_t (well log measurement), R_o, and n, it is possible to use the modified quick look Archie relationship to estimate water saturations. Displayed in FIGURES 3a and b, along with the standard Archie derived water saturations, are the water saturations calculated by using the quick look Archie method.

Compressional-wave acoustic log data from the drill-holes in the four gas hydrate accumulations assessed in this study have been used along with the Timur equation (3), modified Wood equation (4), and Lee weighted average acoustic equations (5) to calculate gas-hydrate saturations. In this set of acoustic gas-hydrate saturation calculations (S_h), the sediment porosities were obtained from available core and well log data. The remaining variables in the Timur and Wood equations were assigned constant values: we assumed a water velocity (V_w) of 1.5 km/sec, marine sediment matrix velocity (V_m) of 4.37 km/sec, terrestrial sediment matrix velocity (V_m) of 4.65

km/sec, gas hydrate velocity (V_h) of 3.35 km/sec, water density (ρ_w) of 1.0 g/cm^3, marine sediment matrix density (ρ_m) of 2.7 g/cm^3, terrestrial sediment matrix density (ρ_m) of 2.65 g/cm^3, and a gas hydrate density (ρ_h) of 0.9 g/cm^3. The bulk compressional-wave velocity of the formation (V_b) is obtained directly from the transit-time well logs. The remaining two variables needed before conducting the acoustic gas-hydrate saturation calculations are the weight factor (W) and the gas-hydrate cementation exponent (r) in the Lee weighted average equation, both of which were determined from available core and log data.[2] In FIGURES 4a and 4b, the results of the Timur, Wood, and Lee acoustic velocity calculations are shown as gas-hydrate saturation log traces for all of the gas hydrate accumulations assessed in this study. The results of the Timur acoustic saturation calculations have not been included in FIGURES 4a and 4b for the marine gas hydrate accumulations, because the Timur equation failed to indicate the presence of gas hydrates in these settings. Our calculations of the amount of gas within the known gas hydrate accumulations assessed in this study are listed in TABLE 2.

Blake Ridge—Atlantic Ocean. Leg 164 of the Ocean Drilling Program (ODP)[4] was designed to investigate the occurrence of gas hydrate in the sedimentary section

TABLE 2. Volume of natural gas within the downhole log inferred gas hydrate occurrences at ODP Sites 889, 994, 995, and 997, and in the Northwest Eileen State-2 and Mallik 2L-38 wells. The average gas-hydrate saturations listed in this table were calculated by the standard Archie method

Site/Well identification	Depth of log inferred gas hydrates (m)	Thickness of hydrate-bearing zone (m)	Sediment porosity (%)	Average gas-hydrate saturation (%)	Volume of gas within hydrate per square km (m^3)[a]
Ocean Drilling Program Drill-Sites:					
Site 889	127.6–228.4	100.8	51.8	5.4	466,635,705
Site 994	212.0–428.8	216.8	57.0	3.3	699,970,673
Site 995	193.0–450.0	257.0	58.0	5.2	1,267,941,673
Site 997	186.4–450.9	264.5	58.1	5.8	1,449,746,073
Northwest Eileen State-2 Drill Site:					
Unit C	651.5–680.5	29.0	35.6	60.9	1,030,904,796
Unit D	602.7–609.4	6.7	35.8	33.9	133,382,462
Unit E	564.0–580.8	16.8	38.6	32.6	346,928,811
			Total for Northwest Eileen State-2		1,511,216,069
Mallik 2L-38 Drill Site:					
Hydrate Unit	897.3–1,109.5	212.2	36.0	33.2	4,150,000,000

[a]In this assessment we have assumed a hydrate number of 6.325 (90% gas fill clathrate), which corresponds to a gas yield of 164 m^3 of methane (at STP) for every cubic meter of gas hydrate.

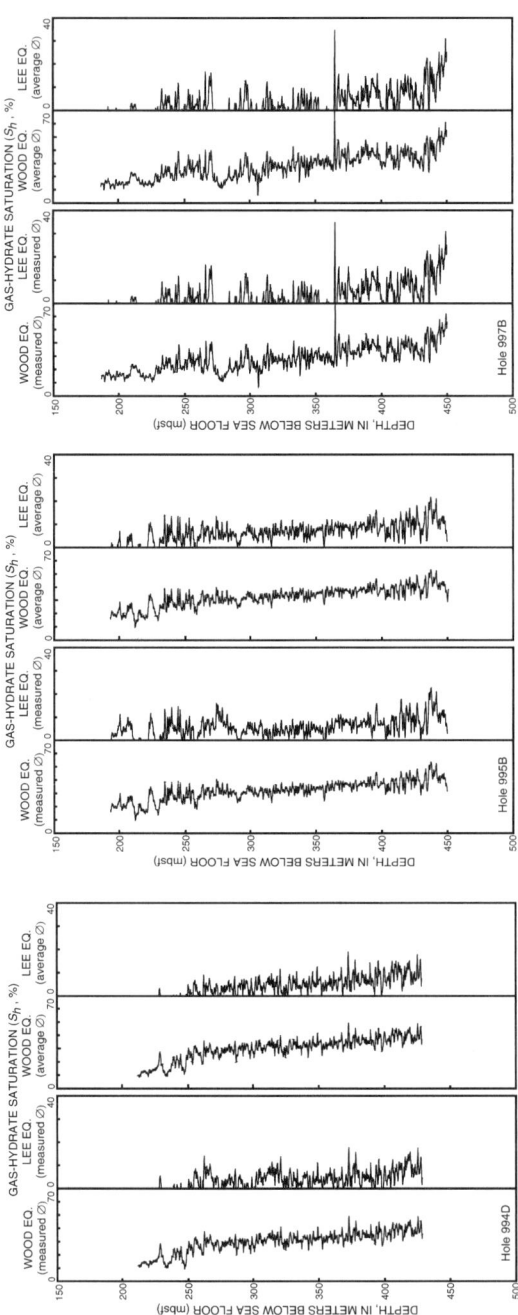

FIGURE 4a. Acoustic velocity derived gas-hydrate saturations in ODP Holes 994D, 995B, and 997B.

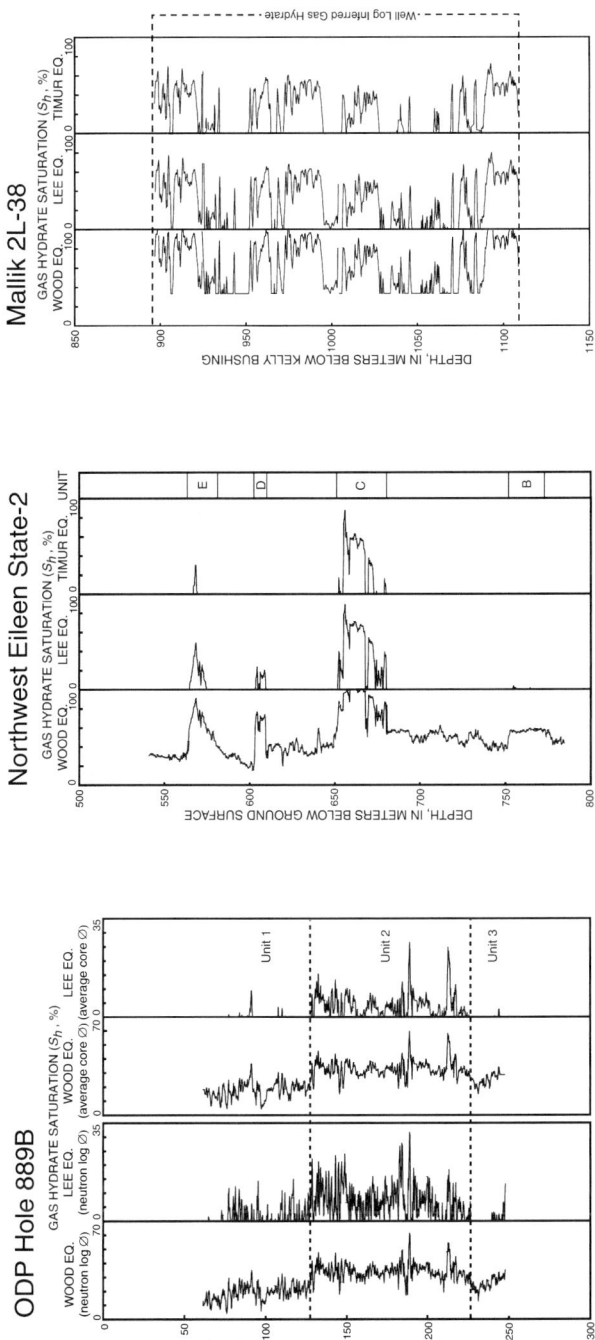

FIGURE 4b. Acoustic velocity derived gas-hydrate saturations in ODP Hole 889B, Northwest Eileen State-2 well, and the Mallik 2L-38 well.

beneath the Blake Ridge (FIG. 1). The logged interval in the ODP Leg 164 boreholes at Sites 994, 995, and 997 were separated into three *logging units* and it was determined that disseminated gas hydrates occur throughout logging unit 2 at all three sites.[4]

The standard Archie relation yielded water saturations (S_w) ranging from about 100% to a minimum of about 80% in the Blake Ridge boreholes (FIG. 3a). The quick look calculated water-saturations are very similar to the water saturations calculated from the standard Archie relation. For the Blake Ridge boreholes, the Wood equation yielded gas-hydrate saturations (S_h) ranging from a minimum of about 20% to a maximum near 60%, which is significantly higher than the gas-hydrate saturations calculated by other methods. However, for the most part, the Lee equation yielded gas-hydrate saturations ranging from 0% to a maximum of about 20% and these are more compatible with the gas-hydrate saturations calculated from interstitial-water chloride-freshening trends and electrical resistivity log data.

Cascadia Continental Margin—Pacific Ocean. Leg 146 of the Ocean Drilling Program[5] was designed in part to examine the distinct bottom-simulating reflectors (BSRs) on the Cascadia Contiental margin (FIG. 1). ODP Leg 146 Site 889, located off the west coast of Vancouver Island, yielded a wealth of data pertaining to the *in situ* nature of gas hydrates on the Cascadia margin. The logged interval in Hole 889B was separated into three *logging units* on the basis of obvious changes in the acoustic velocity and electrical resistivity measurements (FIG. 2b). At Site 889, disseminated gas hydrates are inferred to occur in Logging Unit 2 and may occur in Logging Unit 1.

In Hole 889B, the standard Archie relation yielded water saturations ranging from about 100% to a minimum of about 70% (FIG. 3b). Also displayed in FIGURE 3b, along with the standard Archie derived water saturations, are the water saturations obtained by the quick look Archie method. The quick look calculated water-saturations are very similar to the water saturations calculated by the standard Archie relation. In FIGURE 4b, the results of the Wood and Lee acoustic velocity calculations are shown as gas-hydrate saturation log traces for Hole 889B. In Logging Unit 2 of Hole 889B, the Wood equation yielded gas-hydrate saturations ranging from a minimum of about 20% to a maximum near 60%, which is significantly higher than the gas-hydrate saturations calculated by other methods at Site 889. However, for the most part, the Lee equation yielded gas-hydrate saturations in Logging Unit 2 of Hole 889B ranging from 0% to a maximum of about 30%. These are more compatible with the gas-hydrate saturations calculated from the electrical resistivity log data.

North Slope—Alaska. The Northwest Eileen State-2 well drilled five (FIG. 2b; Units B–F) of the six Prudhoe Bay–Kuparuk River gas-hydrate–bearing stratigraphic units described by Collett.[1] Units C, D, and E were determined to contain gas hydrates in the Northwest Eileen State-2 well, whereas Unit B was determined to contain both water and free-gas (no gas hydrate).

The quick look Archie approach yielded average water saturations for the three gas-hydrate–bearing stratigraphic units (Units C, D, and E) in the Northwest Eileen State-2 well ranging from 47 to 88% (FIG. 3b). The average water saturations calculated using the standard Archie relation within the three gas-hydrate–bearing units (Units C, D, and E) in the Northwest Eileen State-2 well range from about 39 to 67%. Furthermore, the standard Archie relation yielded an average free-gas saturation for

Unit B of about 32%. As is shown in FIGURE 4b, the Timur equation yielded relatively low gas-hydrate saturations for most of the gas-hydrate–bearing stratigraphic units (Units C, D, and E) in the Northwest Eileen State-2 well. The Lee equation yielded, for the most part, reasonable gas-hydrate saturations among the three gas-hydrate–bearing stratigraphic units (Units C, D, and E) ranging from 10 to about 40%. These are compatible with, but slightly lower than, the gas-hydrate saturations calculated from the electrical resistivity log data in the Northwest Eileen State-2 well. The Wood equation yielded for the most part unreasonable gas-hydrate saturations.

Mackenzie River Delta—Canada. The JAPEX/JNOC/GSC Mallik 2L-38 gas hydrate research well was designed to investigate the occurrence of *in situ* natural gas hydrates in the Mallik area of the Mackenzie River Delta of Canada (FIG. 1).[6] Downhole electrical resistivity and acoustic transit-time logs from the Mallik 2L-38 well confirm the occurrence of *in situ* gas hydrates at the Mallik drill-site within the subsurface depth interval between 897.25 and 1,109.5 m (FIG. 2b).

The average water saturations calculated from the standard Archie relation within the gas-hydrate–bearing units (897.25 and 1,109.5 m) of the Mallik 2L-38 well range from near zero to about 90% (FIG. 3b). In FIGURE 3b, the results of the quick look Archie calculations are also shown as a water saturation log trace for the subpermafrost portion of the Mallik 2L-38 well. As shown in FIGURE 4b, the Timur equation yielded relatively low gas-hydrate saturations for most of the gas-hydrate–bearing stratigraphic units, in the Mallik 2L-38 well. The Lee equation yielded, for the most part, reasonable gas-hydrate saturations within the gas-hydrate–bearing stratigraphic units, ranging from near zero to about 80%, values that are compatible with, but slightly lower than the gas-hydrate saturations calculated from the electrical resistivity log data in the Mallik 2L-38 well. The Wood equation yielded unreasonable gas-hydrate saturations.

SUMMARY

The standard Archie relation (**1**) and quick look Archie method (**2**) yielded, for the most part, gas-hydrate saturations that compare favorably with gas-hydrate concentrations calculated by other means. The Lee weighted average acoustic equation (**5**) yielded gas-hydrate saturations that compared favorably with saturations calculated by other means. The Timur equation (**3**), which adequately predicts the acoustic properties of gas hydrates in consolidated rock media, yielded reasonable gas-hydrate saturations with the acoustic data from only the Northwest Eileen State-2 and Mallik 2L-38 wells. The Wood equation (**4**) yielded unreasonable gas-hydrate saturations in all of the gas-hydrate accumulations assessed.

REFERENCES

1. COLLETT, T.S. 1998. Well log evaluation of gas hydrate saturations. Trans. Soc. Professional Well Log Analysts. Thirty-Ninth Annual Logging Symposium, May 26–29, 1998, Keystone, Colorado, USA, Paper MM.
2. LEE, M.W., D.R. HUTHINSON, W.P DILLON, J.J. MILLER, W.F. AGENA & B.A. SWIFT. 1993. Method of estimating the amount of *in situ* gas hydrates in deep marine sediments. Marine Petrol. Geol. **10**: 493–506.

3. PEARSON, C.F., P.M. HALLECK, P.L. MCGUIRE, R. HERMES & M. MATHEWS. 1983. Natural gas hydrate deposit: a review of *in situ* properties. J. Phys. Chem. **87:** 4180–4185.
4. SHIPBOARD SCIENTIFIC PARTY. 1996. Sites 994, 995, & 997 (Leg 164). *In* Proceedings, Ocean Drilling Program, Initial Reports. C.K. Paull *et al.* **164**: 99–623. College Station, Texas.
5. SHIPBOARD SCIENTIFIC PARTY. 1994. Sites 892 & 889 (Leg 146), *In* Proceedings, Ocean Drilling Program, Initial Reports. G.K. Westbrook *et al.* **146:** 301–396. College Station, Texas.
6. COLLETT, T.S., R.E. LEWIS, S.R. DALLIMORE, M.W. LEE, T.H. MROZ & T. UCHIDA. 1999. Detailed evaluation of gas hydrate reservoir properties using JAPEX/JNOC/GSC Mallik 2L-38 gas hydrate research well downhole well-log displays. *In* Scientific Results from JAPEX/JNOC/GSC Mallik 2L-38 Gas Hydrate Research Well, Mackenzie Delta, Northwest Territories, Canada. S.R. Dallimore, T. Uchida & T.S. Collett, Eds. Geological Survey of Canada Bulletin **544**: 295–312.

North Cascadia Deep Sea Gas Hydrates

G.D. SPENCE,[a] R.D. HYNDMAN,[a,b] N.R. CHAPMAN,[a] R. WALIA,[a]
J. GETTRUST,[c] AND R.N. EDWARDS[d]

[a]School of Earth and Ocean Sciences, University of Victoria,
Victoria, British Columbia, Canada

[b]Pacific Geoscience Centre, Geological Survey of Canada,
Sidney, British Columbia, Canada

[c]Naval Research Laboratory, Stennis Space Center, Mississippi, USA

[d]Department of Physics, University of Toronto, Toronto, Ontario, Canada

ABSTRACT: The Cascadia accretionary margin off Vancouver Island is one of the best studied margins world-wide for the determination of *in situ* properties of marine gas hydrates. Most quantitative information has come from cores and downhole logs of the Ocean Drilling Program (ODP) Leg 146 and from extensive seismic and other geophysical surveys. As part of the ODP site surveys, large-offset multichannel seismic lines outlined the regional distribution of hydrate, and seismic velocity analyses coupled with full waveform inversion provided estimates of the vertical distribution of hydrate and gas. High resolution single-channel seismic surveys indicate correlations between hydrate/gas concentrations and topographic highs, and between geothermal flux and topography; these correlations provide insight into fluid and methane flow through the sediments. A deep-towed multichannel seismic survey (DTAGS), together with other data from 20–600 Hz, provide constraints on gradients at the base of the hydrate and gas layer. Extensive heat flow measurements, from probes and from depths of the bottom simulating reflector (BSR), have been modelled numerically, including the regional effects of sediment thickening and advective fluid flow in the accretionary prism. Other geophysical surveys include several seafloor electrical sounding experiments and seafloor compliance measurements of hydrate. ODP drilling has provided valuable downhole log data, core physical property data, and detailed pore fluid chemistry and isotopes. Several semi-independent estimates of hydrate and gas concentrations have been obtained; all are dependent on reference sediment properties for which no hydrate and no gas is present. From a vertical seismic profile and other seismic data, the velocity increase in a hydrated region is consistent with hydrate concentrations of 20–30% of the pore space in a 100-m interval above the BSR. Similar values are determined from log resistivity data and from core pore fluid chlorinity.

INTRODUCTION

The Northern Cascadia subduction zone accretionary prism has received the most detailed studies of any convergent margin for examining the properties of marine gas hydrate. Key information for understanding the properties and formation processes of hydrates has been derived from drill holes of the Ocean Drilling Program (ODP) Leg 146, which have provided estimates of hydrate concentration through downhole seismic, resistivity, and chlorinity measurements. Off Vancouver Island, hydrate on the continental slope has been the target for an exceptionally broad range of geophysical studies. These have included regional multichannel seismic surveys, finely-gridded single channel seismic programs, deep-towed seismic measurements with a high-frequency source, heat flow measurements and numerical models, seafloor electrical sounding experiments, seafloor compliance measurements, and physical property analyses from seafloor piston cores. The objective of this paper is to review briefly the principal geophysical methods used offshore Vancouver Island, and to provide examples of what we have learned about hydrates in this environment, both their broad regional distribution and their fine vertical structure. The primary focus is on seismic methods and results.

REGIONAL MULTICHANNEL SEISMIC LINES AND HYDRATE DISTRIBUTION

Lateral Distribution of Hydrate

Hydrates within the Cascadia accretionary prism were first recognized in 1985 through the characteristic bottom-simulating reflector (BSR) in conventional multi-channel seismic lines.[1,2] Additional multichannel lines collected in 1989 as part of the ODP site survey provided new information on the regional distribution of gas hydrate (see FIGURE 1). A hydrate BSR occurs in a 20–30 km wide band along much of the 250 km length of the continental slope off Vancouver Island. FIGURE 2 shows

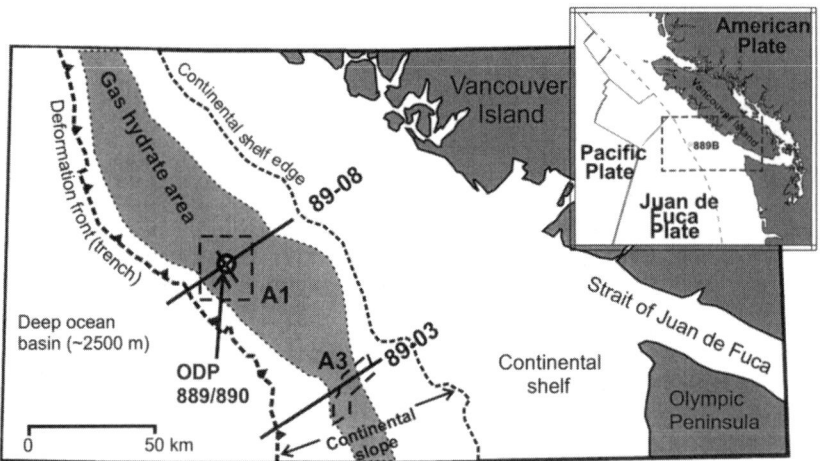

FIGURE 1. The Vancouver Island continental margin, showing the portion of the continental slope where gas hydrates are found. Detailed grids of seismic lines were collected in regions A1 and A3, indicated by the *dashed-line polygons*. Track lines for seismic surveys in region A1 are shown in FIGURE 4. Multichannel lines 89-03, 89-08, and 89-10 (perpendicular to 89-08) were collected as part of a site survey for ODP Leg 146.

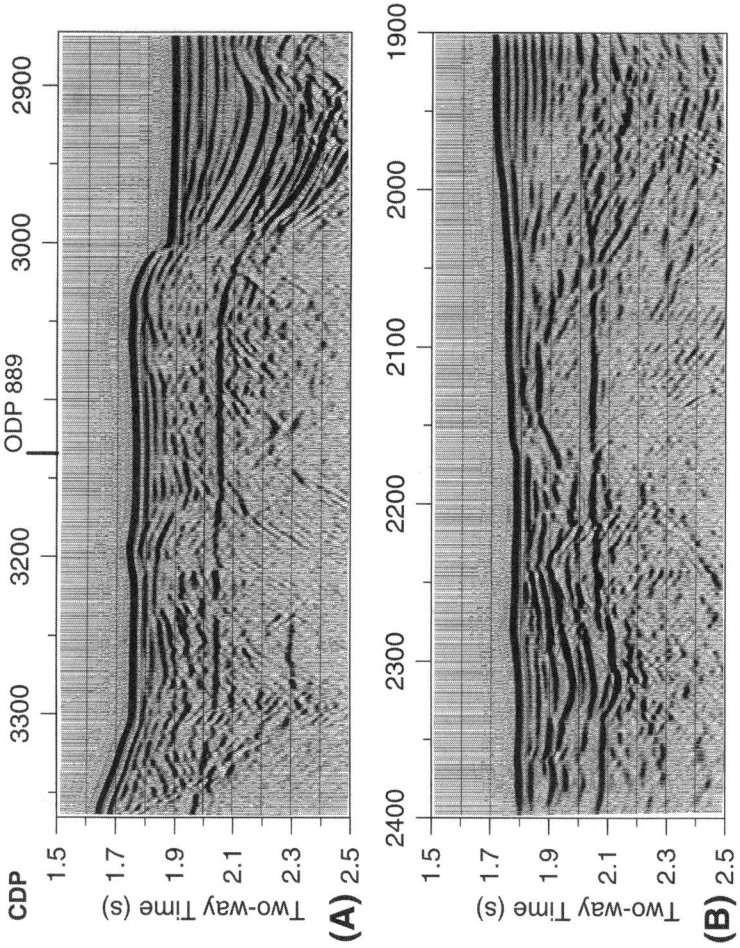

FIGURE 2. Reflection seismic sections 89-08 (**A**) and 89-10 (**B**) near ODP Site 889 on the lower slope region. The BSR, representing the base of the hydrate stability zone, is particularly strong in this region.

portions of two orthogonal reflection lines 89-08 and 89-10 across ODP Sites 889/890 (FIG. 1) where the BSR is continuous and particularly strong.

Landward of the deformation front, the accretionary prism is formed as coarse elastic sediments are scraped off the downgoing ocean plate. The accretionary wedge thickens landwards and the sediments are increasingly deformed and compacted. Most sediment stratigraphy is lost and clear sediment horizons are typically observed only in recent slope basins (e.g., CDP 2900-3000 in FIG. 2A). Sediment compaction results in the expulsion of pore fluids that rise upwards and carry methane with them to be concentrated as hydrate when the fluids enter the hydrate stability field above the BSR.[3] A BSR is generally not observed in the well-bedded slope basin sediments (FIG. 2A), but interpretations vary as to whether or not hydrate is present. Hydrate may not be formed because the well-bedded sediments reduce permeability, thus inhibiting vertical fluid and methane flow;[4] alternatively, hydrate may be present but tectonic subsidence in the basin results in downward movement of the base of the stability field, so that the gas layer is transformed to hydrate and the BSR is much weakened.[5]

Velocity–Depth Information from Multichannel Seismic and ODP Drillhole Data

Several independent methods for estimating the velocity–depth profile above and below the BSR are available on the Cascadia margin. Downhole sonic logs from ODP Site 889[6] provide detailed velocity information from about 50 metres below the seafloor (mbsf) to the BSR at 225 mbsf (see FIGURE 3). The best constraint for low velocities below the BSR, due to the presence of small quantities of gas (probably less than 1%), comes from a vertical seismic profile (VSP) at Site 889.[7] Unfortunately, these measurements are limited to a depth of only about 30 m below the BSR and thus no conclusions can be made about the thickness of the gas layer.

The multichannel data provide complementary and additional velocity information. Careful semblance velocity analyses could be carried out on sporadic reflectors down to depths of about 2000 mbsf.[8] Most importantly, the trend established by the deep velocities, when extrapolated to the surface, represent the only estimate available for the reference velocity of sediments unaffected by either hydrate or free gas. Above the BSR, semblance velocities agree very well with both the sonic log and VSP data. Velocities just above the BSR are more than 200 m/s greater than the reference trend (FIG. 3).

Additional information on velocity structure is contained in the phase and amplitude variations, with offset for a given common midpoint (CMP) gather. A sophisticated inversion scheme of the full seismic waveform has been applied to selected CMP gathers along lines 89-08 and 89-10 (FIG. 1), in order to investigate the detailed velocity variations at and near the BSR.[9] The inversion is very dependent on the starting model, which is derived from the sonic, VSP, and semblance velocity profiles. The results indicate that velocities as low as 1500 m/s occur in a 25–50 m thick layer below the BSR (FIG. 3). Based on the reference velocity, estimated 1650 m/s at the BSR, we conclude that the BSR in this area appears to come from both a velocity increase due to hydrate above the BSR and a velocity decrease due to free gas below the BSR.

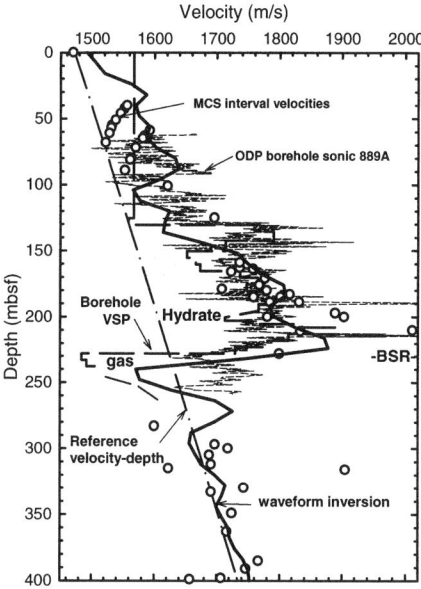

FIGURE 3. Velocity-depth estimates on the Cascadia continental slope where the BSR is located at 225 mbsf. Sonic log and VSP velocities are from ODP Site 889. Above the BSR, MCS velocities from semblance analyses are in excellent agreement with the *in situ* measurements and with the results from the full waveform inversion (*solid heavy curve*). Below the BSR, the VSP and inversion results indicate a low velocity zone, produced by a small quantity of free gas. A reference velocity–depth profile is provided by the MCS velocities (*dotted dash line*), showing significant velocity enhancement due to hydrate above the BSR.

DETAILED SEISMIC SURVEYS

Extensive seismic surveys and related studies off Vancouver Island include (see FIGURE 4):
1. High resolution closely spaced single channel and short-offset multichannel surveys.

In 1993, single channel seismic (SCS) data were recorded over a tight grid of lines near ODP Site 889/890.[10,11] Maximum BSR reflection coefficients of 0.15–0.18 were observed beneath topographic highs, whereas maximum seafloor reflection coefficients of up to 0.5–0.6 were found on the flanks of the highs. Furthermore, the seafloor reflection in these areas typically showed a larger amplitude for data recorded with the 75 Hz airgun source, relative to amplitudes from single-fold 30-Hz data (extracted from coincident MCS lines collected in 1989). We interpreted this seafloor amplitude character as evidence for a thin high-velocity layer, with the velocity and density of carbonate; amplitudes for the lower frequency data are reduced because the layer thickness (about 2 m) is significantly less than the 30-Hz quarter-wavelength.

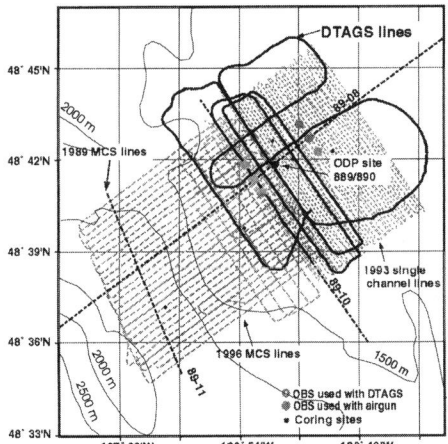

FIGURE 4. Seismic lines in the vicinity of ODP Site 889/890. A 120 in^3 airgun was used as the source for the grid of single channel lines in 1993 and short-offset multichannel lines in 1996. The source was also recorded by arrays of OBSs in both surveys. The *solid lines* indicate a survey collected in 1997 using a deep-towed source (DTAGS).

In 1996, several grids of short-offset multichannel seismic data were acquired in the continental slope region. Piston coring was also carried out at 18 sites and physical property analyses (including velocity, density, resistivity, and porosity) allowed ground-truth calibration of the seismically-derived seafloor reflection coefficients.[12] Over a grid of seismic lines southeast of Site 889 (FIG. 1), heat flow was calculated from the depth of the BSR. The regional variation of heat flow, decreasing landwards from about 80 to 65 mW/m^2, was consistent with tectonic thickening of accretionary wedge sediments.[13] Significant local variations in heat flow were also observed, notably low heat flow values over topographic highs and high heat flow values over the flanks of the highs (see FIGURE 5). Although much of this variation may be produced by topographic effects, a component may result from dynamic effects, including the upward migration of fluids along faults and the displacement of isotherms by thrust faulting.

2. Ocean bottom seismometer (OBS) surveys.

Several deployments of OBSs were carried out in 1993 and 1997, recording a 120 in^3 airgun source with a dominant frequency of 75 Hz. Simultaneous travel time inversion of data from four instruments and from coincident normal-incidence lines gave velocities of 1.83–1.95 km/s just above the BSR, slightly higher than VSP and sonic log data from ODP Site 889.[14]

3. High resolution deep-towed multichannel seismic survey.

With the deep-tow acoustics/geophysics system (DTAGS), a survey was carried out in 1997 in the area of ODP Site 889.[15–17] The frequency band of DTAGS was

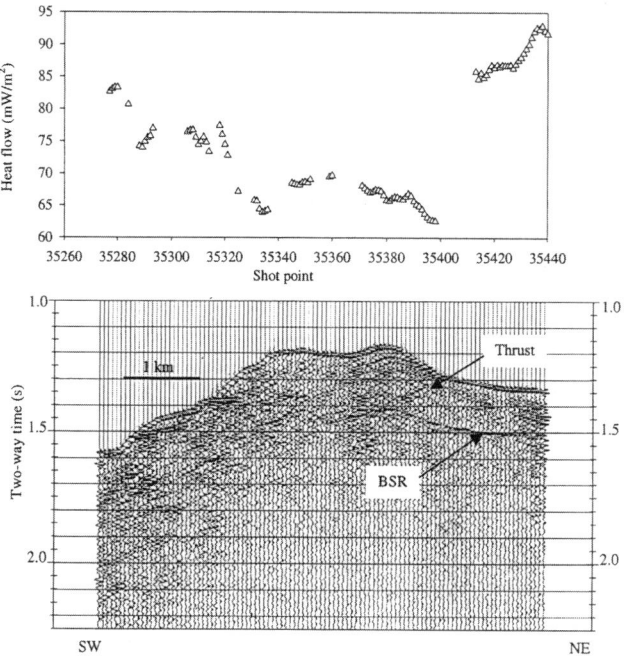

FIGURE 5. Part of a migrated seismic section from short-offset multichannel seismic line 27, located in region A3 (FIG. 1) about 6.5 km southeast of line 89-03. Heat flow was calculated from the depth of the BSR using a simple conductive model, in which temperature at the BSR was determined from the pressure–temperature stability curve for hydrate. A heat flow minimum is located over the topographic high, whereas high heat flow is located on the flanks of the topographic high. The large increase just east of SP 35400 is partially associated with a possible thrust fault.

250–650 Hz, providing a vertical and horizontal resolution of 2 m and 25 m, respectively. Combined with five previous surveys in the area, this enables us to determine the frequency dependence of hydrate-related reflectors over the broad band from 20–650 Hz (see FIGURE 6). The BSR is very clear for low frequency data but its amplitudes becomes significantly smaller for the higher frequencies and it is barely discernible at the peak frequency of the DTAGS source. This behavior suggests that the BSR is produced, not by a sharp impedance contrast, but rather by a negative velocity gradient. The vertical scale of the gradient is much smaller than the wavelength of the low frequency data, but greater than the wavelength of the DTAGS data. By modelling this frequency dependence of amplitude using synthetic seismograms (see FIGURE 7), we infer that the velocity gradient at the BSR, where the velocity decreases by 250 m/s, has a thickness of about 10 m.

72 ANNALS NEW YORK ACADEMY OF SCIENCES

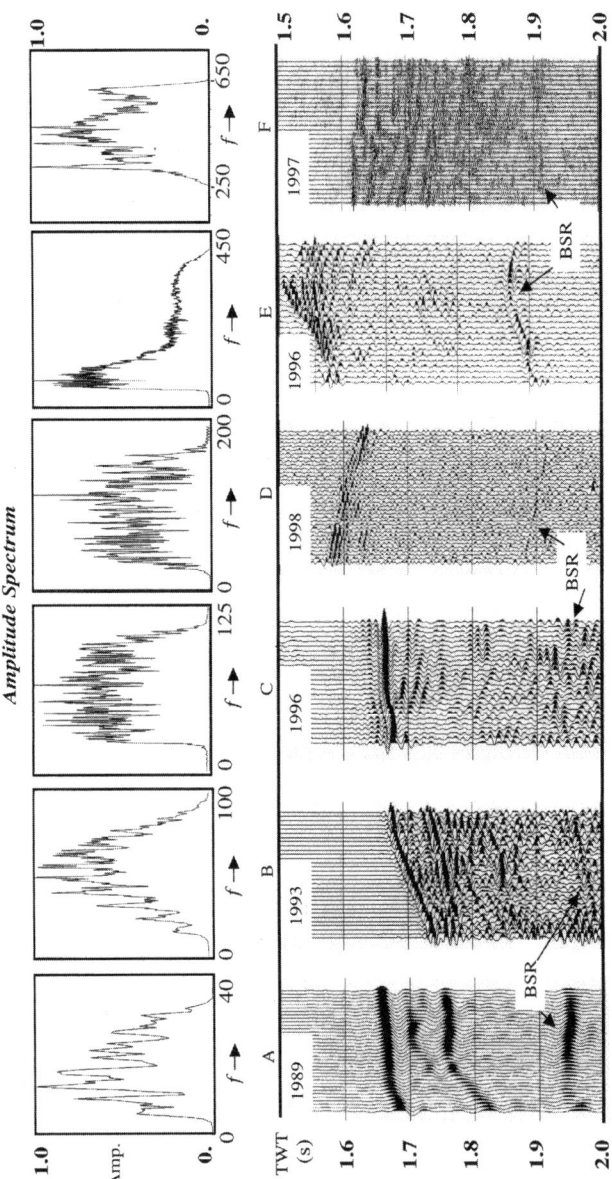

FIGURE 6. Samples of single channel or stacked multichannel seismic data from six recent surveys using sources with different frequency bandwidths. The sections are all NW–SE and are located within a few kilometers of each other near ODP Site 889. 1989, multichannel survey using large airgun array with total volume 120 L; 1993, single channel survey, 1.97-L airgun source; 1996, short-offset multichannel survey, 1.97-L airgun source; 1998, single channel survey, 0.65-L airgun source; 1996, single channel survey, 0.7/1.7-L GI-Gun source; 1997, single channel acoustic streamer survey, with deep-towed DTAGS Helmholtz resonator source.

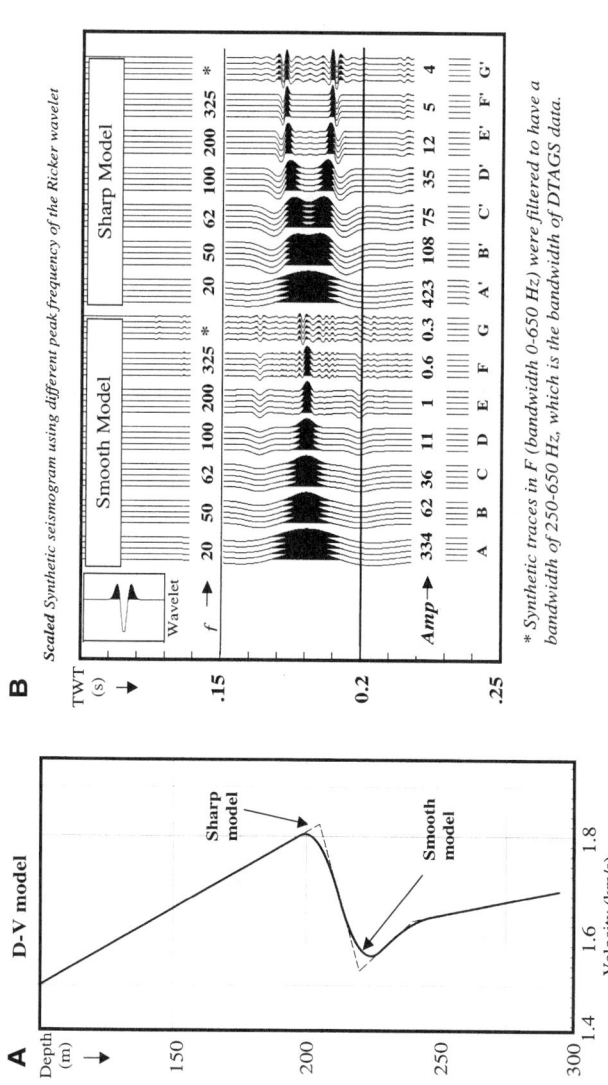

FIGURE 7. (A) Velocity-depth model for the BSR based on downhole, sonic, and multichannel velocities (FIG. 3). The BSR is modelled as a zone with a negative velocity gradient, not as a sharp impedance contrast. (B) Scaled synthetic seismograms for the "sharp" and "smooth" model BSRs with a range of peak source frequencies. The peak frequencies correspond to the data of FIGURE 6 (corresponding letters). The traces were scaled to give each a peak amplitude of one. The unscaled amplitudes shown (Amp) vary by a factor of 1000. Amplitude decreases with increasing frequency, since the gradient zone appears smooth to the shorter wavelengths of the higher frequency sources.

SEAFLOOR ELECTRICAL PROFILING AND SEAFLOOR COMPLIANCE MEASUREMENTS

Electrical resistivity is an alternative parameter that is very sensitive to gas hydrate, because electrically resistive hydrate replaces conductive seawater in the pore spaces. A seafloor electrical system has been developed recently to map deep sea hydrate, based on the time for an electrical signal to diffuse through sediment from a seafloor source to a seafloor receiver.[18] In several surveys near Site 889, resistivities near 2 ohm-m have been obtained down to depths of abut 300 m, consistent with resistivity logs from the drillhole.

Another independent method for estimating hydrate concentration is the measurement of seafloor compliance, since hydrate cementation increases the elastic moduli, especially the rigidity. Compliance is obtained by using a seafloor gravimeter and a precision seafloor pressure gauge to measure the seafloor deformation due to ocean surface gravity waves and longer period internal waves. Successful data have been obtained from six sites near ODP Site 889.[19]

CALCULATION OF HYDRATE AND FREE GAS CONCENTRATIONS

The primary estimators for hydrate concentration used to date in this area are seismic velocity, electrical resistivity, and core pore fluid chlorinity. All methods are dependent on reference sediment properties for which no hydrate and no gas is present. One method of calculating hydrate concentration from velocity (see FIGURE 8A) assumes that the observed velocity increase relative to the reference (FIG. 3) is due simply to a reduction in porosity as the hydrate fills the pore spaces.[8] Resistivity estimates from downhole logs and recovered core samples have average

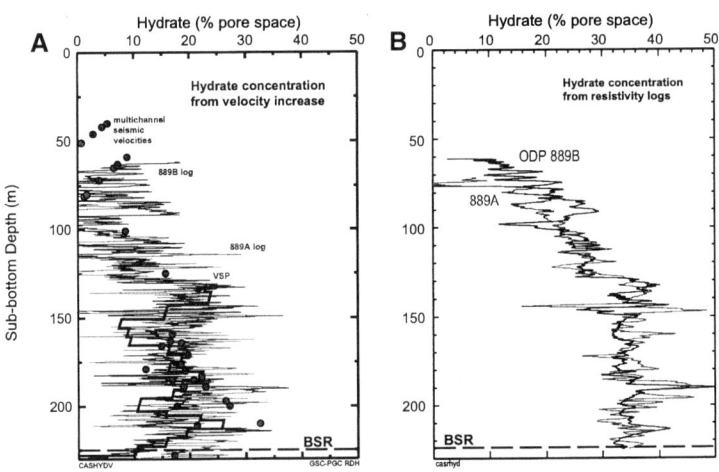

FIGURE 8. Hydrate concentration at ODP Site 889/890 estimated from (**A**) velocity data from sonic logs, VSP and MCS velocities, and (**B**) log resistivities and core salinities.

values near 2 ohm-m at Sites 889/890 where hydrate was found, compared to 1.0 ohm-m at a nearby reference site.[20] Calculation of hydrate concentrations from resistivity (FIGURE 8B) requires allowance for the effect of reduced salinity of the in situ pore fluid. The velocity and resistivity data give concentration estimates of 20–30%of the pore space in a 100 m interval above the BSR.[8,20] Similar values are determined from core pore fluids chlorinity data.[8]

REFERENCES

1. DAVIS, E.E. & R.D. HYNDMAN. 1989. Accretion and recent deformation of sediments along the northern Cascadia subduction zone. Geol. Soc. Am. Bull. **101**: 1465–1480.
2. DAVIS, E.E. *et al.* 1990. Rates of fluid expulsion across the northern Cascadia accretionary prism: constraints from new heat flow and multichannel seismic reflection data. J. Geophys. Res. **95**: 8869–8889.
3. HYNDMAN, R.D. & E.E. DAVIS. 1992. A mechanism for the formation of methane hydrate and seafloor bottom-simulating reflectors by vertical fluid expulsion. J. Geophys. Res. **97**: 7025–7041.
4. ZUEHLSDORFF, L. *et al.* 2000. BSR occurrence, near surface reflectivity anomalies and small scale tectonism imaged in a multi-frequency seismic data set from the Cascadia accretionary prism. Geol. Runds. **88**: 655–667.
5. VON HUENE, R. & I. PECHER. 1999. Vertical tectonics and the origins of BSRs along the Peru margin, Earth Planet. Sci. Lett. **166**: 47–55.
6. WESTBROOK, G.K. *et al.* 1994. Proc. Ocean Drill. Program, Init. Reports. 146.
7. MACKAY, M.E. *et al.* 1994. Origin of bottom simulating reflectors: Geophysical evidence from the Cascadia accretionary prism. Geology **22**: 459–462.
8. YUAN, T. *et al.* 1996. Seismic velocity increase and deep-sea gas hydrate concentration above a bottom-simulating reflector on the northern Cascadia continental slope. J. Geophys. Res. **101**(13): 655–671.
9. YUAN, T. *et al.* 1999. Seismic velocity studies of a gas hydrate bottom-simulating reflector on the northern Cascadia continental margin: Amplitude modelling and full waveform inversion. J. Geophys. Res. **104**: 1179–1191.
10. SPENCE, G.D. *et al.* 1995. Seismic structure of methane gas hydrate, offshore Vancouver Island. Proc. Ocean Drill. Program, Sci. Results, **146**: 163–174.
11. FINK, C.R. & G.D. SPENCE. 1999. Hydrate distribution off Vancouver Island from multifrequency single-channel seismic reflection data. J. Geophys. Res. **104**: 2909–2922.
12. MI, Y. 1998. Seafloor Sediment Coring and Multichannel Seismic Studies of Gas Hydrate, Offshore Vancouver Island. M.Sc. Thesis. University of Victoria, Victoria, Canada.
13. GANGULY, N. *et al.* 2000. Heat flow variation from bottom simulating reflectors on the Cascadia margin. Mar. Geol. **164**: 53–68.
14. HOBRO, J.W. *et al.* 1998. Tomographic seismic studies of the methane hydrate stability zone in the Cascadia margin, Geol. Soc. Lond. Spec. Publ. **137**: 133–140.
15. CHAPMAN, N.R. *et al.* 1999. High resolution deep-towed seismic survey of deep sea gas hydrates off western Canada. Geophysics. Submitted for publication.
16. GETTRUST, J. *et al.* 1999. High resolution seismic studies of deep sea gas hydrate using the DTAGS deep towed multichannel system. EOS Trans. AGV **80**: 439–440.
17. WALIA, R.K. & D. HANNAY. 1999. Source and receiver geometry corrections for deep towed multichannel seismic data. Geophys. Res. Lett. **26**: 1993–1996.
18. EDWARDS, R.N. 1997. On the resource evaluation of marine gas hydrate deposits using a seafloor transient electric dipole-dipole method. Geophys. **62**: 63–74.
19. WILLOUGHBY, E.C. & R.N. EDWARDS. 1997. On the resource evaluation of marine gas hydrate deposits using seafloor compliance methods. Geophys. J. Int. **131**: 151–166.
20. HYNDMAN, R.D. *et al.* 1999. The concentration of deep sea gas hydrates from downhole resistivity logs and laboratory data. Earth Planet. Sci. Lett. **172**: 167–177.

The Nature of Gas Hydrates on the Nigerian Continental Slope

JAMES M. BROOKS,[a] WILLIAM R. BRYANT,[a,b]
BERNIE B. BERNARD,[a] AND NICK R. CAMERON[a]

[a]*TDI-Brooks Int'l Inc., 1902 Pinon, College Station, Texas 77845, USA*

[b]*Department of Oceanography, Texas A&M University,
College Station, Texas 77845, USA*

ABSTRACT: Gas hydrates were collected in six-meter piston cores during surface geochemical exploration (SGE) surveys in the deep and ultra deepwaters of Nigeria during 1991, 1996, and 1998. To date, gas hydrates have been collected in about 21 cores out of the more than 800 core collections on the Nigerian margin. This represents a 2.5% recovery ratio of gas hydrated cores on this margin at sites that are potential conduits for the upward migration of hydrocarbons (i.e., the core locations are sited based on two- and three-dimensional seismic overfaults, mounds, acoustic wipe-outs, etc.). Unlike the Northern Gulf of Mexico where the authors have retrieved a significant percentage of thermogenic hydrates in piston cores, all the gas hydrate collections offshore Nigeria to date are primarily biogenic in nature (methane more than 99% of the hydrocarbon gases; $\delta^{13}C$ generally light, -60 to -117‰). A few of these gas hydrated sites do contain a mixed thermogenic gas component (ethane to butane gases up to a few hundred ppm of total hydrocarbon gas) but even at these sites the primary gas in the hydrates is methane.

INTRODUCTION

Although gas hydrates have been known to exist in upper continental slope sediments for many years, they have not often been collected. The global distribution of gas hydrates has been deduced primarily from bottom simulating reflectors (BSRs) and the occasional collection, hundreds of meters deep in the subsurface during deep-sea drilling (i.e., DSDP and ODP) cores. Brooks and coworkers[1-6] have documented the occurrence of gas hydrates in shallow subsurface marine sediments overlying several of the hydrocarbon generative basins throughout the world (e.g., Gulf of Mexico, Northern California, and offshore Nigeria). The gas hydrates have generally been collected from the upper five meters of piston cores taken at water depths greater than 400 m. These gas hydrates occur in close proximity to faults and other conduits for gas migration. In the Gulf of Mexico, from submersibles, hydrates containing gases from thermogenic and biogenic sources have been observed to outcrop at the seafloor.[5,7] Observations of gas hydrates at the seafloor at water depths near their upper stability zone suggests that slight changes in bottom water temperature or pressure could cause the hydrates to disassociate and, thereby, dramatically increase the release of gas to the ocean. It is not clear to what degree shallow hydrates act as barriers to the seepage of gas from the seafloor, because bubbling gas seeps are common in areas containing extensive shallow hydrates.[3,8]

The Niger Delta occupies the central region of West Africa's Gulf of Guinea. With a land area of some 75,000 km², it forms the largest delta system in Africa.[9] The delta owes its size to the focus provided by the Benue arm of the Niger Triple Junction for sediment delivery from interior Africa to the Atlantic Ocean. The modern delta began its growth in the late Eocene.[10,11] Since that time the delta top, as defined by the 200 meter isobath, has prograded South and South-Westward from the Cretaceous shelf-edge hinge line by some 300 km, across previously deepwater settings. The distal edge of the delta lies about 80 to 170 km further seaward. The continental slope forms the intermediate region and has been the focus of surface geochemical exploration (SGE) cores containing the hydrates reported here.

SEA FLOOR GAS HYDRATE COLLECTIONS

The initial hydrate discoveries in the Gulf of Mexico, offshore West Africa, Northern California, and elsewhere have resulted from piston cores acquired for the purpose of surface SGE. SGE studies are used to define the aerial distribution of oil, condensate, and gas seepage on the continental margin. These study high grade areas and prospects by defining areas of active oil migration and charge. Active migration acts to charge accompanying reservoirs in the same geological system. From many such studies, especially in Tertiary delta systems in West Africa, the Gulf of Mexico, and elsewhere, we know that there is considerable macroseepage of *live* oil and gas into seafloor sediments throughout broad regions from the shelf–slope break extending to ultra deep waters (greater than 1,500 m).

Core locations for SGE studies are chosen from both two- and three-dimensional seismic data where there are possibly deep conduits (i.e., faults and fractures) for the upward migration of hydrocarbons. The optimum targets are deep cutting faults that link the source succession to the seabed. These are best developed where there is ongoing tectonism, for example in clay diapir or salt tectonic provinces. However, even in tectonically quiet regions breaks are usually present, especially where the section is thick and/or where there has been differential movement and reactivation across basement features such the Benue and Charcot Fracture Zones in Nigeria. The ideal faults are those associated with: (1) amplitude anomalies (*flags*) and/or BSRs; (2) seabed constructional features, such as carbonate accumulations and mud-gas mounds; (3) gas vent pits; and (4) gas chimneys. Thus, the sites chosen for SGE studies are focused to optimize the chance for retrieving upward migrated gaseous and liquid hydrocarbons.

Gas hydrates are most often recognized visually in the cores retrieved on deck as white ice-like nodules or lenses in the core. They are also inferred from large gas expansion pockets in cores retrieved. If large gas nodules are present, the hydrate is sometimes placed in a 23-cc Parr bomb to collect the hydrate decomposition gas in a high pressure cylinder.[6]

GEOGRAPHIC DISTRIBUTION

TABLE 1 documents the sites at which the authors have collected cores for SGE programs and the estimated number of gas-hydrated cores obtained. The table shows that at water depths exceeding 500 meters the chance of obtaining a gas-hydrated core in Nigeria is 2.5%. Chances are greater at sites targeted that are based on three-dimensional seismic data. Targeting core locations based on three-dimensional seismic data increases ones ability to select the best locations for hitting upward migrated hydrocarbons in shallow sediments using deep fault extensions into shallow sediments along with amplitude anomalies and edge maps.

TABLES 1 and 2 indicates that 21 gas hydrated cores were acquired in three surveys consisting of more than 800 cores at water depths exceeding 500 meters

TABLE 1. Gas hydrate recovery rates in offshore continental slope regions (more than 500 meters water depth) collected from SGE piston coring programs

	Number of Cores	Hydrate Cores	Percent
Northern Gulf of Mexico[a]			
Central/Eastern Gulf (1997–1999)	425	28	6.6
Western Gulf (1997–1999)	361	8	2.2
Total Northern Gulf	**786**	**36**	**4.6**
West Africa			
1994 Nigerian deep water[b]	310	6	1.9
1996 Nigerian deep and ultra deep	186	6	3.2
1998 Nigerian deep and ultra deep	330	9	2.7
Total Nigeria	**826**	**21**	**2.5**
Gabon (1994–1998)	307	0	0.0
Congo (1997–1998)	16	0	0.0
Angolan (1994–1998)	1,330	0	0.0
Namibia (1994)	90	0	0.0
Total Non-Nigerian	**1,743**	**0**	**0.0**
Northern California[c]			
Eel River	74	7	9.5
Point Arena	90	0	0.0

[a]These Gulf of Mexico core numbers represent only those cores obtained by the author since 1996 but do not include several thousand additional SGE cores obtained prior to 1996.
[b]After Brooks *et al.*, Ref. 3.
[c]After Brooks *et al.*, Ref. 4.

TABLE 2. Locations of gas hydrates offshore Nigeria

Sample ID	Latitude	Longitude	Water depth (m)	Comment
N-074C3	3° 33.7′ N	6° 31.8′ E	677	Ref. 3
N-074C4	3° 33.7′ N	6° 31.8′ E	675	Ref. 3
N-082C3	3° 31.4′ N	6° 20.9′ E	770	Ref. 3
N-138C2	3° 57.6′ N	5° 16.6′ E	560	Ref. 3
N-138C3	3° 57.6′ N	5° 16.6′ E	560	Ref. 3
N-138C6	3° 57.6′ N	5° 16.6′ E	563	Ref. 3
PEF005	3° 40.9′ N	7° 25.3′ E	549	Nodules 2.0–2.6 m subbottom; largest nodule was 4–7 cm thick; H_2S present
PEF013	3° 40.9′ N	7° 45.9′ E	440	Present near bottom of 4.6 m core; inferred site; H_2S present
PT028a	3° 17.1′ N	6° 01.0′ E	1,528	H_2S present
PT028b	3° 17.1′ N	6° 01.0′ E	1,528	White massive hydrate; H_2S present
PEX005a	5° 31.5′ N	4° 15.2′ E	1,176	Massive white hydrate in bottom at 2.2 m (TD), H_2S present
PEX05d2	5° 31.5′ N	4° 15.2′ E	1,172	Small white hydrates, H_2S present
NGC102	3° 15.0′ N	6° 42.4′ E	1,147	Inferred based on large gas voids; H_2S present
NGC103	3° 14.1′ N	6° 42.0′ E	1,185	10-cm of solid white hydrate in core catcher; H_2S present
NGC226	4° 56.9′ N	4° 19.2′ E	1,341	White hydrates present; in depression, H_2S present
PCO005	3° 29.7′ N	6° 54.8′ E	738	Hydrates throughout 0.4 m core; in depression; H_2S present
PTX004	3° 28.7′ N	5° 34.1′ E	1,378	Small white hydrate nodules; in small depression; H_2S present
PTX017	3° 34.6′ N	5° 24.6′ E	1,333	Hydrates in bottom of 2.2 m core; In abrupt depression; H_2S present
PTX026	3° 28.2′ N	5° 33.6′ E	1,405	Abundant hydrates; H_2S present
PAG008	4° 52.2′ N	4° 41.8′ E	569	Hydrate present; H_2S present
PAG013	4° 48.2′ N	4° 29.3′ E	971	Hydrates in 0.2 to 0.4 m subbottom; H_2S present

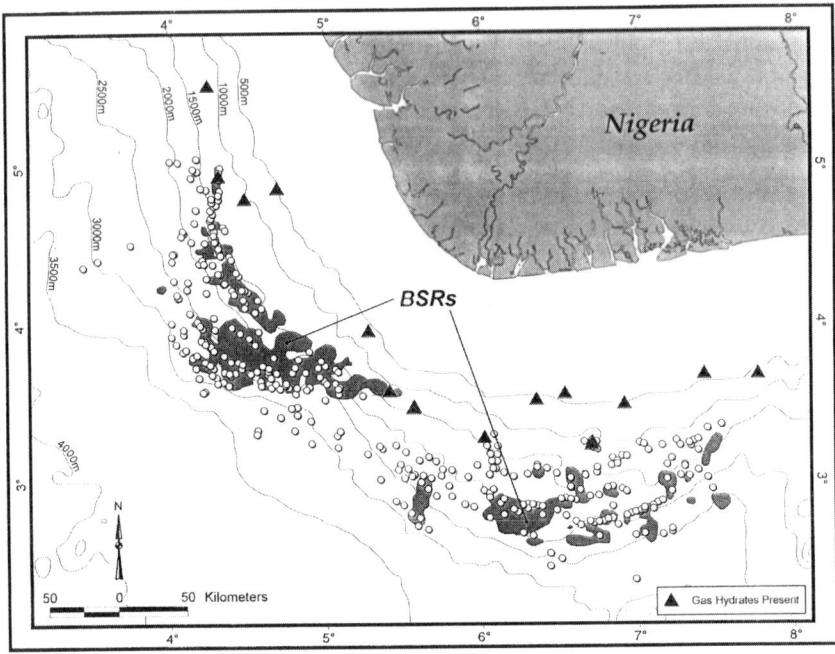

FIGURE 1. Locations of gas hydrates on the Nigerian continental margin. BSRs mapped by Cunningham and Lindholm[13] are also noted.

offshore Nigeria. FIGURE 1 shows the locations of the Nigerian gas hydrate collections. The sites range in water depths from 440 to 1,528 meters. Although most of the cores had small, dispersed, gas hydrates either throughout the core or in the bottom of the cores, several cores bottomed into a massive hydrate 10 to 15 cm in thickness that came up plugging the end of the core. All Nigerian gas hydrates were white, contained mostly methane, and were found predominately in clay-rich sediment. All the hydrated cores contained hydrogen sulifde gas indicating anoxic conditions. Since most sediments on the slope are not anoxic in the top 3–4 meters subbottom, the presence of H_2S in the hydrated cores indicates the occurrence of active bacterial sulfate reduction, possibly using the gaseous hydrocarbons as the substrate.

HYDRATE ORIGIN AND GAS ON THE NIGERIAN MARGIN

The nature of the hydrate gas offshore Nigeria can be inferred from an examination of headspace gases obtained from the shallow piston cores. TABLE 3 shows the headspace gas concentration in the cores containing the gas hydrates. Unless noted otherwise, the values are the average of three measurements in the bottom half of each core. The $C_1/(C_2 + C_3)$ ratios indicate that the molecular compositions contain mostly biogenic gas,[12] although small thermogenic components might be present at

TABLE 3. Headspace gas concentrations in gas hydrated cores on the Nigerian continental slope[a]

Sample ID	Methane (ppm)	Ethane (ppm)	Propane (ppm)	i-Butane (ppm)	n-Butane (ppm)	$\dfrac{C_1}{C_2+C_3}$
N-074C3	6,250	108	8.7	2.7	1.4	54
N-074C4	35,700	116	6.0	0.7	0.2	292
N-082C3	29,600	12	3.4	0.3	0.3	1,920
N-138C2	75,100	11	0.4	0.0	0.4	6,590
N-138C3	69,800	5.6	0.5	0.0	1.5	11,400
N-138C6	77,000	6.6	0.4	0.0	0.0	11,000
PEF005	37,600	17.2	1.2	0.3	0.2	1,990
PEF013	36,000	94.3	23.8	1.5	0.3	300
PT028a	16,400	5.2	0.9	0.7	0.5	2,250
PT028b	27,100	10.2	4.5	1.7	1.0	1,560
PEX005a[b]	5,470	41.6	4.4	0.9	0.2	116
NCG102[b]	44,500	23.3	1.6	0.4	0.5	1,720
NGC103	106,000	79.1	2.1	0.9	0.4	1,280
NGC226[b]	81,500	13.6	3.6	0.9	0.4	4,440
PCO005	423,000	101	3.1	2.1	2.1	3,910
PTX004	50,200	68.3	2.3	0.1	0.1	709
PTX017	62,800	19.3	2.4	0.2	0.2	2,840
PTX026[c]	1,240,000	3,340	2,080	738	125	198
PAG008	35,900	21.8	37.0	9.9	7.1	474
PAG013	59,700	55.7	3.3	5.2	0.9	917

[a]Unless otherwise noted, concentrations are the average of three headspace cans distributed in the bottom half of the core, generally (1) at the bottom, (2) at the bottom minus 1-meter, and (3) near the middle of the core.

[b]Concentration represents the can from the bottom of the core.

[c]The reason the headspace volume is over 100% methane is that the can was over pressureized. This one sample is not averaged, but the concentration in the bottom can taken from this core.

locations for $C_1/(C_2+C_3)$ ratios less than 1,000. With one exception, methane makes up greater than 99% of the hydrocarbon gases. This is consistent with other headspace gas carbon isotopic ratios from high gas containing cans from these same Nigerian SGE surveys (TABLE 4). TABLE 4 lists the carbon isotope values reported as $\delta^{13}C_{PDB}$ (‰) measured in alkane gases of concentration greater than 500 ppmV in the headspace of the selected cans from the 1998 program. The data in TABLE 4 with values more negative (lighter) than −100‰ represent cores that contain only biogenic gas. Although thermogenic gas is typically represented by $\delta^{13}C_{PDB}$ of

TABLE 4. Carbon isotope ratios of selected headspace gases offshore Nigeria (1998 Program)

Core number	Section	Methane	Ethane
NGC124	22	−77.5	
NGC128	21	−117.1	
NGC151	25	−116.2	
NGC158	26	−106.1	
NGC190	25	−71.5	
NGC206	22	−85.0	
NGC219	19	−73.0	
NGC224	19	−62.3	
NGC226	18	−67.6	
NGC230	09	−53.5	−34.3

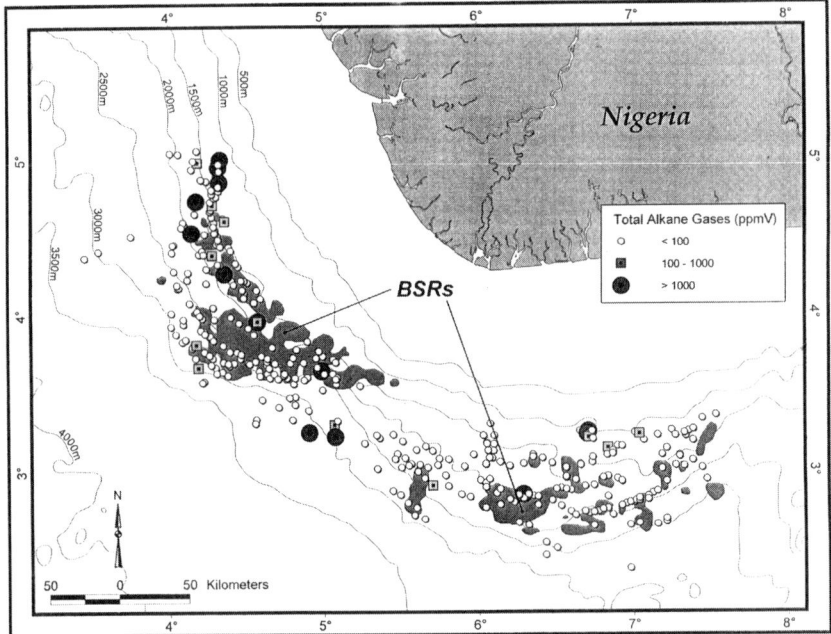

FIGURE 2. Distribution of high alkane gas (more than 100 ppmV) containing cores relative to the location of the BSRs mapped by Cunningham and Lindholm.[13]

methane from −40 to −50‰, values between −50 to −85‰ are routinely observed in sediment gases with higher than biogenic levels of C_{2+} alkane gases. We interpret these sites as having some component of thermogenic gas mixed with predominately biogenic gas. This small component of thermogenic gas does not change the basic biogenic nature of the gas hydrated cores.

FIGURES 2 and 3 show the distribution of C_{2+} alkanes greater than 100 ppmV and the presence of hydrogen sulfide in the core bottom relative to the BSRs mapped by Cunningham and Lindholm.[13] Most of the high gas containing cores are outside of the areas of mapped BSRs, possibly indicating that the hydrate can form a partial barrier to the upward migration of gas. However, an examination of the seismic data at sites where macroseepage of liquid hydrocarbons exist in the ultra deep water, show that BSRs are generally present. Thus, we do not believe that the BSRs act as a significant barrier for the upward migration of liquid and, therefore, gaseous hydrocarbons along deep cutting faults on the slope. No gas hydrated cores (TABLE 2) contained visible liquid hydrocarbons, although several contained significant amounts of liquid hydrocarbon microseepage (i.e., liquid hydrocarbons were only detected analytically). In general, the presence of gas hydrated cores on the Nigerian margin is decoupled from the seepage of liquid hydrocarbons to the seafloor, which is consistent with the biogenic nature of the gas hydrates. The presence of reducing conditions in the cores, as indicated by the presence of hydrogen sulfide

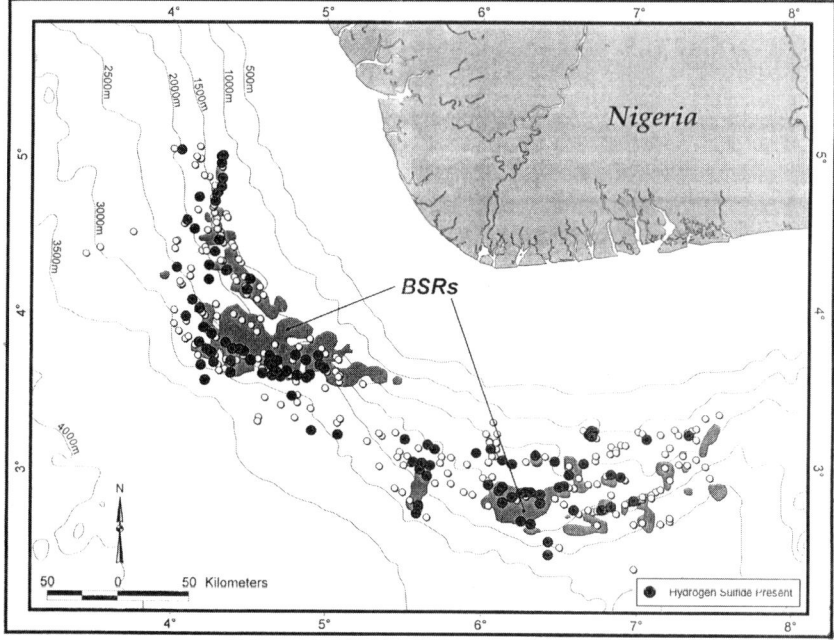

FIGURE 3. Distribution of cores with hydrogen sulfide in the bottom of the core relative to the location of the BSRs mapped by Cunningham and Lindholm.[13]

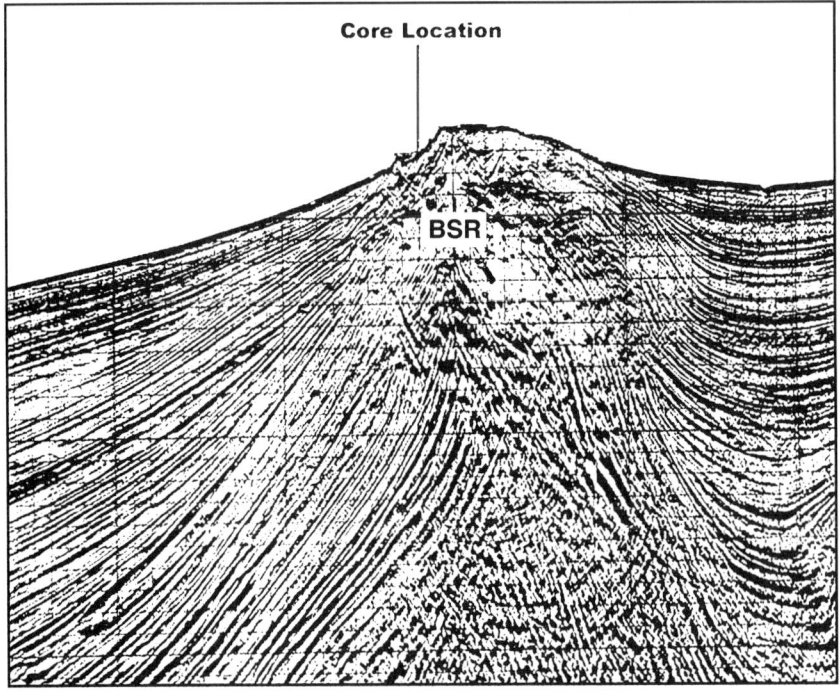

FIGURE 4. Example of a BSR associated with a Nigerian core site that contains visual oil-staining in the core.

in the bottoms of the cores, did not show any coupling with the presence of the BSRs (FIG. 3)

BOTTOM SIMULATING REFLECTORS

FIGURE 4 shows an example of a BSR over a macroseepage core site offshore Nigeria. Cunningham and Lindholm[13] mapped the BSRs using two-dimensional regional seismic data offshore West Africa. They reported that BSRs are extensive on the continental margin off the Niger and Congo River Delta but absent elsewhere in the Nigerian to Angolan corridor of West Africa. This corresponds to the collection of gas hydrates reported here offshore Nigeria in shallow cores, but the complete lack of any shallow gas hydrate collections offshore Angola in more than 1,300 cores collected using the same techniques and core settings as used in Nigeria (TABLE 2). Cunningham and Lindholm[13] report that the cumulative surface area of BSR areas offshore Nigeria and Congo are 11,000 and 4,000 km^2, respectively. FIGURE 1 shows the sites of gas hydrate core collections offshore Nigeria and their correspondence with the BSRs. An amazing observation is that most of the seafloor collections of gas

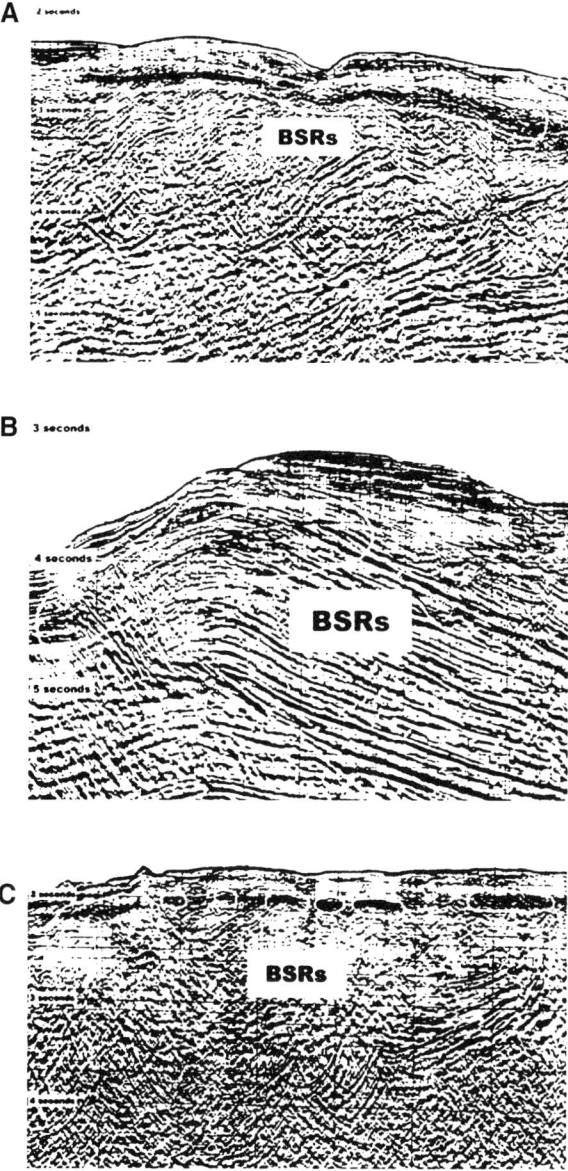

FIGURE 5. Examples of (**A**) the relationship between an extensive BSR to a large thrust fault, (**B**) an extensive BSR in water depths of 1,350 meters at approximately 270 meters below the seafloor, and (**C**) a well-defined BSR in water depths of 1,800 meters and 175 milliseconds twtt below the seafloor.

hydrates are shoreward of the major BSR trends, despite the fact that many cores were obtained over faults in the BSR regions.

In Nigeria, the BSRs are generally associated with complex structural types that are contractional in origin (i.e., imbricated and fault-related folds) at water depths greater than 1,200 meters.[8,13] Hovland et al.[8] from his studies in the OPL-213/215 area reports that the BSRs cover 8.5% of the study area and tend to follow the up-dipping strata formed by the anticlinal compressional ramp structures, where the mud volcanoes tend to form at the summit of these ramps. Our examination of the BSRs along the Nigerian margin generally correspond to those areas identified by Cunningham and Lindholm[13] and by Hovland et al.[8] The BSRs are common along the distal portions of the prodelta, where large thrust faults create bathymetric highs adjacent to the flat Atlantic seafloor. FIGURE 5A illustrates the relationship between an extensive BSR and a large thrust fault. The BSR occurs at approximately 500 meters below the seafloor and at a water depth of 2,400 meters. There is little blanking above the BSR at this location. FIGURE 5B is a two-dimensional seismic profile that illustrates the nature of occurrence of an extensive BSR at water depths of 1,350 meters. The BSR is 300 milliseconds two-way travel-time (twtt) below the seafloor, which is approximately 270 meters at a sediment velocity of 1,800 m/sec. The BSR is above and on the flanks of a diapiric structure. There is extensive blanking above and below the BSR. FIGURE 5C illustrates the development of a well-defined BSR at water depths of 1800 meters and 175 milliseconds twtt below the seafloor. The BSR extends below a seafloor depression and there is extensive blanking above the BSR.

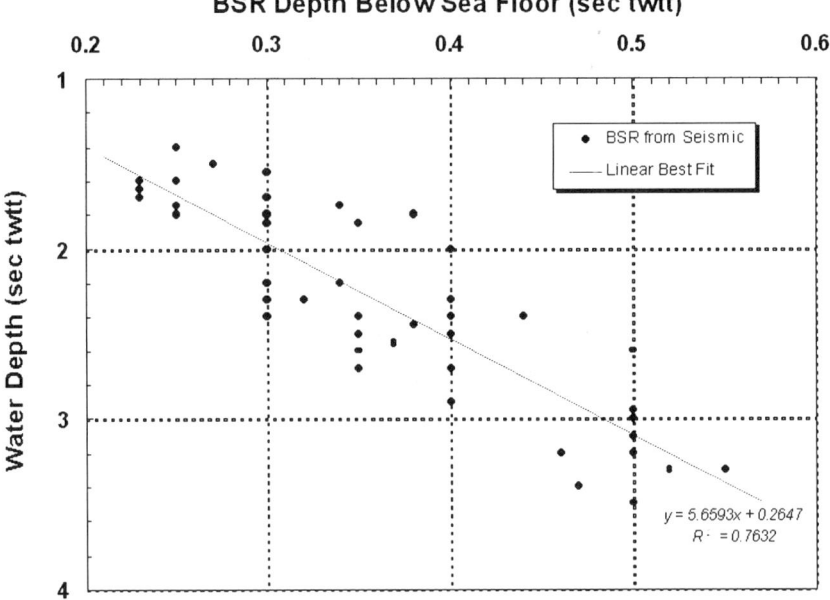

FIGURE 6. Cross-plot of water depth versus BSR depth below the seafloor.

The water depth and the depth below the seafloor of BSRs in the offshore portions of Nigeria, determined from an extensive two-dimensional regional seismic survey, are presented in FIGURE 6. The equation of the best fit line was determined to be:

Water Depth (sec twtt) = 0.2647 + 5.6593 BSR Depth Below Seafloor (sec twtt)

This equation has an R^2 value of 0.76. The BSRs depths shown in FIGURE 6 are at similar geographic locations to those presented by Cunningham and Lindholm[13] as shown in FIGURE 1.

Hovland et al.[8] concluded that the mean maximum amount of gas hydrates and free gas residing in sediments above and below the Nigerian BSRs is 1–3% and 1–5% by volume, respectively. They argue that BSRs and natural gas hydrates form at locations where there is a relatively high flux of methane to shallow sediments due to fluid migration, and, that the Nigerian margin is an area of active fluid flux. Our studies support these arguments since:

- most of the sites of known liquid oil macroseepage to the surface are associated with BSRs thus at these sites vertical fluid migration must be occurring or have occurred in the recent past;
- all our gas hydrate-containing sites were over conduits (i.e., faults, mud mounds, and depressions) for the upward migration of fluids and hydrocarbons; and
- gas hydrates as well as macroseepage of oil and gas are common on the Nigerian margin.

There is the general assumption that the deeper BSRs in the complex structural zones are fed by upward migrating thermal gas from the active petroleum systems that exist on the Nigerian continental margin. Our analyses indicate that whereas thermogenic gas components are sometimes present, thermal gas is a minor component. A few of the hydrate sites actually have significant levels of thermogenic liquid hydrocarbon microseepage, but even these sites are predominately methane (more than 99%) with light isotopes ($\delta^{13}C$ −50 to −70‰). Cunningham and Lindholm[13] reports that one of the gas hydrated sites had a $\delta^{13}C$ of −54‰, indicating a thermogenic component even though it was more than 99% methane. Our conclusion is that although there are thermogenic gas components in some of the hydrates as evidenced by the presence of small amounts of C_{2+} gases and carbon isotopes of methane in the −50 to −70‰ range, hydrate gas is predominately being supplied by biogenic process, presumably in the shallow (upper few hundred meters) subsurface.

REGIONAL HEAT FLOW AND GEOTHERMAL GRADIENT ON THE NIGERIAN MARGIN

Regional heat flow measurements were conducted on the continental margin offshore Nigeria in June 1998 (see FIGURE 7) for the primary purpose of determining basal heat flow for thermal maturation studies of the petroleum systems. Heat flow measurements were acquired at 112 sites at water depths between 500 and 3,400 meters, using the Dalhousie heat flow probe from Dalhousie University, which measures the geothermal gradient at eight depths in the first five meters of sediment, and

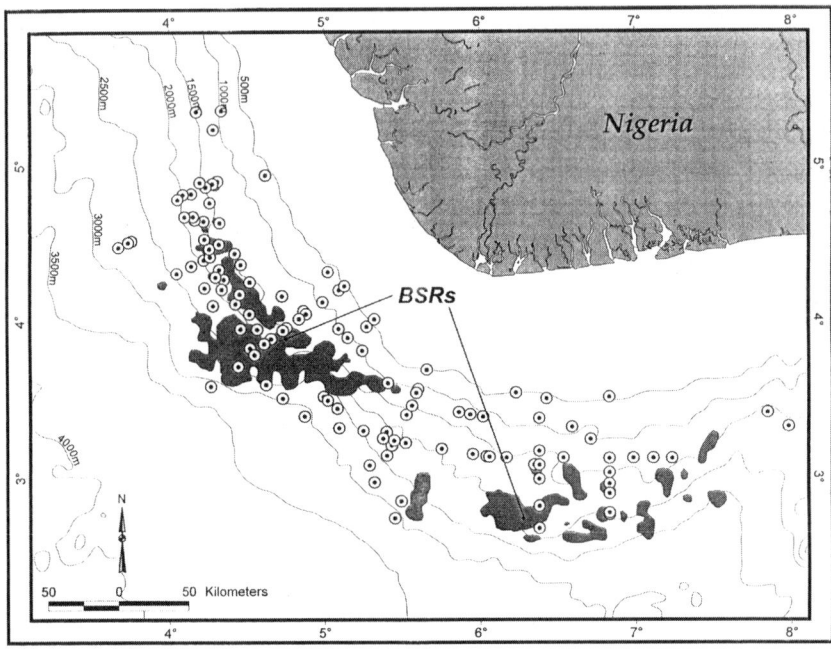

FIGURE 7. Locations of heat flow sites collected on the Nigerian continental margin.

the *in situ* thermal conductivity at the corresponding intervals. Details of the instrument and measurements can be found in Hutchison and Owen.[14] Sites were chosen, away from conduits for fluid migration (i.e., faults) and in quiescent zones, to best reflect the regional heat flow of the area. Heat flow in the study area ranged from 18.8 to 123.7 mW/m^2, with an average of 58.2 mW/m^2. FIGURE 8 is a bar chart of the distribution of heat flows for the 112 sites. The chart shows a great predominance of heat flows between 40 and 70 mW/m^2. FIGURE 9 shows the bottom water temperature at each site as a function of water depth, as well as the dissolution boundary for methane hydrates from literature values.[15,16] as a function of water depth. This figure illustrates that the sediment surface at every site except the most shallow one (500 m) is at a temperature and pressure regime within the stability zone for methane hydrates. Thus, the thickness and the bottom of the hydrate stability zone for each of these sites depends on the temperature/pressure regime within the sediment.

The bottom of the stability zone can be predicted (to first order) if the composition of the hydrate, bottom water temperature, water depth, and geothermal gradient are known. FIGURE 10 shows the geothermal gradient (in milliKelvins per meter) versus the water depth measured at each of the 112 heat flow sites. The figure illustrates that there is no distinct trend in geothermal gradient with increasing water depth at the stations measured. Because of this, an average thermal gradient cannot be assumed to predict the bottom of the hydrate stability zone in the region. However, we have

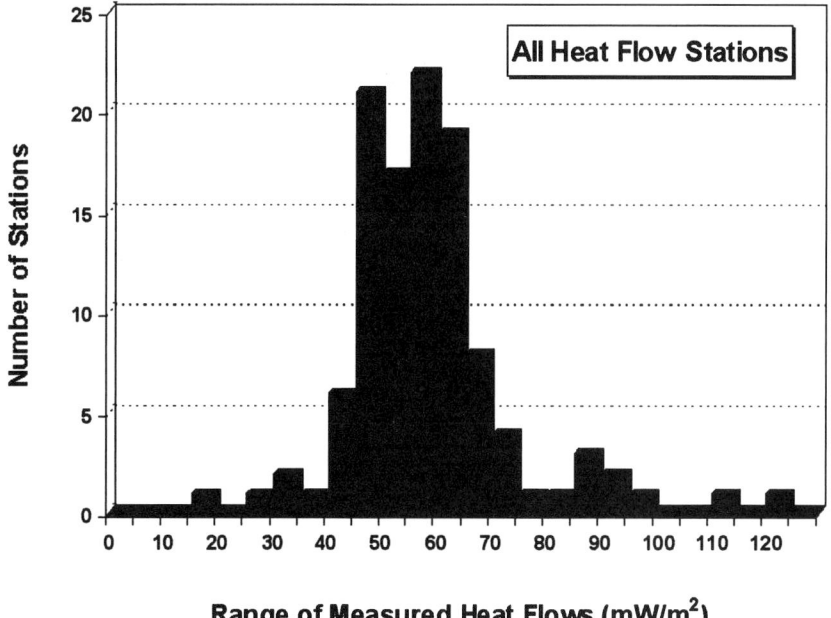

FIGURE 8. Histogram of heat-flows measurements for 122 sites.

measured the bottom water temperature, the geothermal gradient, and the water depth at each heat flow site, as well as the composition of gas in hydrates recovered at a few sites. Based on these measurements, the predicted bottom of the methane hydrate stability zone can be calculated for each site.

FIGURE 11 shows the calculated bottom of the methane hydrate stability zone versus water depth for each site. The figure also shows the calculated zone-bottom based on using the average geothermal gradient measured for the region. The plotted values generally follow the trend shown by the average gradient, but deviations from this trend line illustrate the effect of higher or lower than average heat flows at the various sites. FIGURE 11 also shows the depth values measured on the seismic records, interpreted as BSRs. The BSR depth values are generally somewhat deeper than would be predicted from the calculations of the bottom of the methane hydrate stability zone. At water depths greater than 1,200 meters, the linear best fit BSR line is increasing deeper than the similar line from the calculated base of the methane hydrate stability zone. This difference can be easily explained by the variability in heat flow for the region (i.e., the values predicted for the methane hydrate stability zone overlaps the observed BSR depth considering the range of measured heat flow

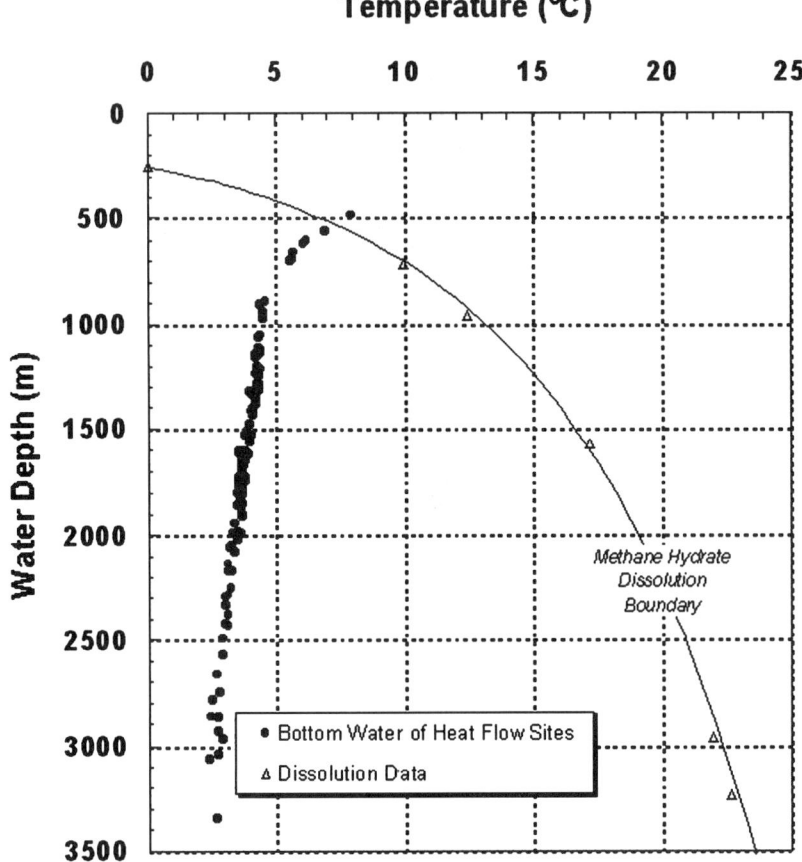

FIGURE 9. Plot of bottom water temperature (measured by the heat-flow probe) and the methane hydrate dissolution temperature boundary versus water depth.

on the slope) or that the average thermal gradient in the upper six-meters is not representative of the upper 200 to 600 meters. Other explanations for the difference include (1) the inclusion of non-methane gases that shift the hydrate stability to increasing depths; and (2) the estimate of 1,800 m/sec for the sound speed of all sediments is high. One could argue that the BSR is deeper than predicted for a base of the methane hydrate stability zone because of the inclusion of other thermal hydrocarbons (C_2–C_5). Although there is a general coincidence of the calculated gas hydrate stability zone from the thermal data and the observed depth of the BSR, we suggest that for the reasons noted above it may be unreliable to use the depth of the BSR as a means of predicting regional heat flow for the region.

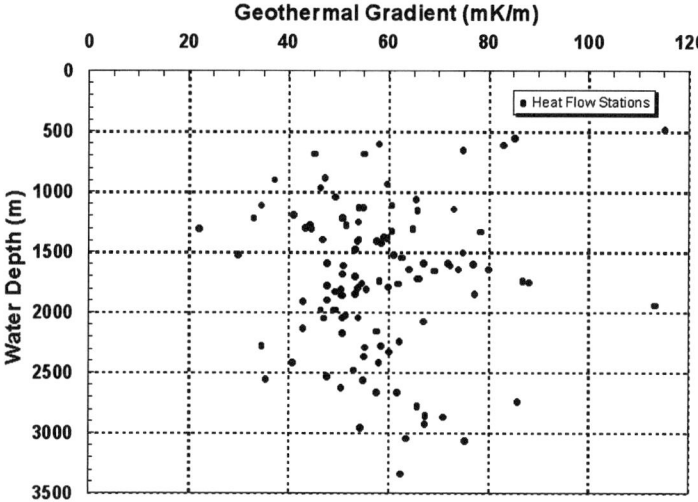

FIGURE 10. Cross-plot of the measured geothermal gradient versus water depth for all heat flow sites.

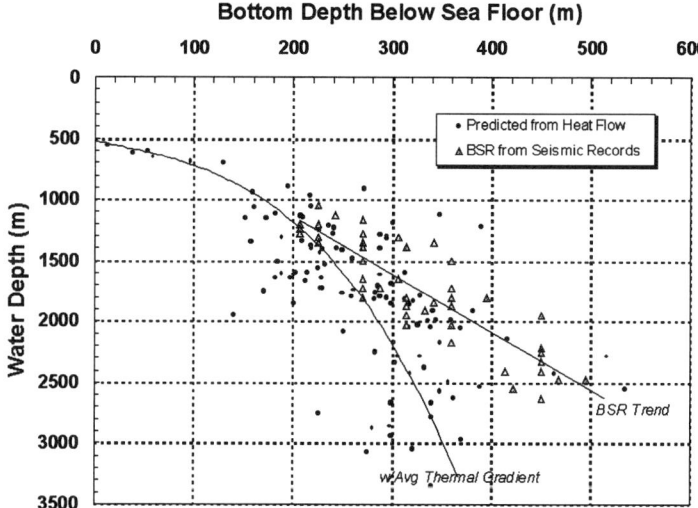

FIGURE 11. Comparison of the calculated bottom of methane hydrate stability zone to actual measured depth of the BSR from an examination of seismic records. The linear best fit for the BSR is indicated ($y = 5.6593x + 0.2647$); $R^2 = 0.7632$.

SUMMARY AND CONCLUSIONS

Gas hydrates were encountered in about 2.5% of the cores in the deep and ultradeep waters offshore Nigeria. The hydrates are of biogenic nature with some containing a small component of gas of thermogenic nature. The collection of gas hydrated cores is generally independent of the macroseepage of liquid hydrocarbon core sites. BSRs are often associated with the oil/gas seepage core sites offshore Nigeria. BSRs are common on the seismic records of the Nigerian continental slope. When gas hydrates are collected in cores they often consist of disseminated nodules a few centimeters in diameter within the mud matrix at a few meters subbottom, or they are massive (5 to 10+ cm thick) and emerge as the bottom of the core. The depth of the BSRs are generally similar or are at depths shallower than the calculated base of the methane hydrate stability zone using known bottom water temperatures and thermal gradients for the region. The average heat flow for the Nigerian continental margin is 58.2 mW/m^2 with a range from 18.8 to 123 m/Wm^2.

ACKNOWLEDGMENTS

The two-dimensional seismic data from offshore Nigeria was provided by Mabon Ltd. We wish to thank Dr. Keith Louden of Dalhousie University for supervision of the heat flow collections and the analysis of the heat flow data. We thank Neil Summers and others of the TDI-Brooks field staff that participated in the collection of the hydrated cores offshore Nigeria. We thank Bob Cunningham at Exxon for providing an expanded figure of his BSR map from offshore Nigeria. Dr. Yuri Makogon provided useful instruction for our models of the hydrate stability zone using the thermal gradient data obtained from the heat flow measurements. We especially want to thank our oil company consortium that funded the collection of these Nigerian cores.

REFERENCES

1. BROOKS, J.M., H.B. COX, W.R. BRYANT, M.C. KENNICUTT II, R.G. MANN & T.J. MCDONALD. 1986. Association of gas hydrates and oil seepage in the Gulf of Mexico. Org. Geochem. **10:** 221–234.
2. BROOKS, J.M., M.C. KENNICUTT II, R.A. FAY, T.J. MCDONALD & R. SASSEN. 1984. Thermogenic gas hydrates in the Gulf of Mexico. Science **225:** 409–411.
3. BROOKS, J.M., A.L. ANDERSON, R. SASSEN, I.R. MACDONALD, M.C. KENNICUTT II & N.L. GUINASSO, JR. 1994. Hydrate occurrences in shallow subsurface cores from continental slope sediments. Ann. N.Y. Acad. Sci. **715:** 381–391.
4. BROOKS, J.M., M.E. FIELDS & M.C. KENNICUTT II. 1991. Observations of gas hydrates in marine sediments, offshore Northern California. Mar. Geol. **96:** 103–109.
5. MACDONALD, I.R., N.L. GUINASSO, JR., R. SASSEN, J.M. BROOKS, L. LEE & K.T. SCOTT. 1994. Gas hydrates that breaches the sea floor on the continental slope of the Gulf of Mexico. Geol. **22:** 699–702.
6. KENNICUTT, M.C. II, J.M. BROOKS & H.B. COX. 1993. The origin and distribution of gas hydrates in marine sediments. *In* Organic Geochemistry. M.H. Engel & S.A. Macko, Eds.: 535–544. Plenum Press, N.Y.

7. SASSEN, R., I.R. MACDONALD, N.L. GUINASSO, JR., S. JOYE, A.G. REQUEJO, S.T. SWEET, J. ALCALA-HERRERA, D.A. DEFRITAS & D.R. SCHINK. 1998. Bacterial methane oxidation in sea-floor gas hydrate: significance to life in extreme environment. Geol. **26**(9): 851–854.
8. HOVLAND, M., J.W. GALLAGHER, M.B. CLENNELL & K. KEKVAM. 1997. Gas hydrate and free gas volumes in marine sediments: example from the Niger Delta front. Marine Petroleum Geol. **14**(3): 245–255.
9. BURKE, K. 1996. The African Plate. South African J. Geol. **99**(4): 341–410.
10. BURKE, K. 1972. Longshore drift, submarine canyons, and submarine fans in development of Niger Delta. Bull. Am. Assoc. Petrol. Geol. **56**(10): 1975–1983.
11. DOUST, H. & E. OMATSOLA. 1990. Niger Delta. Am. Assoc. Petrol. Geol. Memoir No. **48:** 201–238.
12. BERNARD, B.B., J.M. BROOKS & W.M. SACKETT. 1977. A geochemical model for characterization of hydrocarbon gas sources in marine sediments. Proc. 9th Annual Offshore Technology Conference, Houston. **2934:** 435–438
13. CUNNINGHAM, R. & R.M. LINDHOLM. 1997. Seismic evidence for widespread gas hydrate formation, offshore West Africa. Petroleum Systems of the South Atlantic Margin, Rio de Janeiro, 16–19 November. Extended abstract.
14. HUTCHISON, I. & T. OWEN. 1989. In Handbook of Seafloor Heat Flow. J.A. Wright & K.E. Louden, Eds.: 71–89. CRC Press, Boca Raton.
15. KAPLAN, I.R. 1974. Natural Gases in Marine Sediments. Plenum Press, New York.
16. MAKOGON, Y.F. 1977. Hydrates of Hydrocarbons. Pennwell Publishing Company, Tulsa.

Relation between Gas Hydrate and Physical Properties at the Mallik 2L-38 Research Well in the Mackenzie Delta

W.J. WINTERS,[a,b] S.R. DALLIMORE,[c] T.S. COLLETT,[d] K.A. JENNER,[e]
J.T. KATSUBE,[c] R.E. CRANSTON,[e] J.F. WRIGHT,[c] F.M. NIXON,[c] AND T. UCHIDA[f]

[b]*U.S. Geological Survey, 384 Woods Hole Road, Woods Hole, Massachusetts 02543, USA*

[c]*Geological Survey of Canada, Ottawa, Ontario K1A 0E8, Canada*

[d]*U.S. Geological Survey, Denver, Colorado 25046, USA*

[e]*Geological Survey of Canada, Dartmouth, Nova Scotia B2Y 4A2, Canada*

[f]*Japan Petroleum Exploration Company, Mihama-ku Chiba, Japan*

ABSTRACT: As part of an interdisciplinary field program, a 1150-m deep well was drilled in the Canadian Arctic to determine, among other goals, the location, characteristics, and properties of gas hydrate. Numerous physical properties of the host sediment were measured in the laboratory and are presented in relation to the lithology and quantity of *in situ* gas hydrate. Profiles of measured and derived properties presented from that investigation include: sediment wet bulk density, water content, porosity, grain density, salinity, gas hydrate content (percent occupancy of non-sediment grain void space), grain size, porosity, and post-recovery core temperature. The greatest concentration of gas hydrate is located within sand and gravel deposits between 897 and 922 m. Silty sediment between 926 and 952 m contained substantially less, or no, gas hydrate perhaps because of smaller pore size.

INTRODUCTION

The Japan Petroleum Exploration Co. Ltd./Japan National Oil Corporation/ Geological Survey of Canada Mallik 2L-38 gas hydrate research well was drilled in February and March 1998, at latitude 69° 27′ 40.71″N, longitude 134° 39′ 30.37″W, on the Northeastern edge of the Mackenzie Delta, Northwest Territories, Canada. A primary goal of the 1,150-m deep well was the collection of core samples to document an occurrence of Arctic gas-hydrate.[1] The sediment physical properties determined from the Mallik well were used to verify well log data and to relate the location of gas hydrate to the physical nature of the host material. These data also complement sedimentology studies,[2] petrophysical analyses,[3] and a variety of modeling investigations.

Water content, grain density, and Atterberg limits (water contents that separate different remolded sediment consistency states) were measured from core subsamples collected between 109 and 175 m and between 886 and 952 m, from widely

[a]Telecommunication. Voice: 508-457-2358; fax: 508-457-2310.
bwinters@usgs.gov

differing sediments. Other parameters, such as porosity, wet and dry bulk density, void ratio, and total unit weight were calculated from the index properties. The physical property measurements discussed in this report are supplemented by other data[4] and by Winters et al.[5]

GAS HYDRATE LOCATION AND GEOLOGICAL SETTING

Unconsolidated and lithified sediments from three stratigraphic sequences, which range from Pleistocene to Oligocene (Iperk Sequence 0–346 m, Mackenzie Bay Sequence 346–926.5 m, and Kugmallit Sequence 926.5–1,150 m),[2] were penetrated during drilling. Permafrost is present from slightly below ground surface to a depth of 640 m (relative to the kelly bushing on the drill rig, which was 8.31 m above ground surface). In total, 37.3 m of core were obtained between the depths of 886 and 952 m to document a thick gas hydrate zone that had been identified at a nearby site (Mallik L-38) during exploration drilling by Imperial Oil Ltd. in 1972.[6]

Gas hydrate typically either filled sediment voids or coated grains within coarse sand and gravel deposits in samples obtained at depths between 897 and 922 m.[1,7] Nodules less than 1 cm in diameter and thin veins less than 1 mm thick were also observed. In addition, the presence of hydrate indicated by resistivity well logs suggest that hydrates exist below the level of sample recovery to a depth of about 1,110 m.[7] The "standard" Archie relation used for interpretation of electrical resistivity measurements indicates that hydrates occupy an average of 47% of the void space. Well logs from the nearby Mallik L-38 well indicate that gas hydrate occupies more void space (about 67%) than is present at the 2L-38 well.[8]

METHODS

A number of different coring systems and techniques were used during the project. A wireline retrievable system produced 50 mm diameter cores in permafrost. A conventional core barrel was typically used in the gas hydrate interval to obtain larger (133 and 89 mm) diameter cores. Pore water near gas hydrate deposits occasionally froze during recovery due to endothermic cooling caused by gas-hydrate dissociation. This process probably stabilized the sandy sections. Considerable care was taken at the drill site to characterize the physical properties of core samples with emphasis on time-critical observations such as bonding or strength, gas hydrate or permafrost characteristics, and core temperatures. Subsamples were taken directly at the drill site for water content, pore-water geochemistry, petrophysical studies, and a variety of other investigations. Four 26-cm long, large-diameter samples containing gas hydrate were also preserved for specialized physical-property studies.[9] Detailed physical property measurements and geochemical analyses were conducted at a field laboratory in Inuvik, located approximately 170 km south of the drill site where additional subsamples were obtained.

Core subsamples, used for physical-property measurements, were obtained using two methods. A plunger or syringe, which produced a cylindrical plug, provided a uniform specimen from which volume was calculated using measured sample dimensions. A larger number of irregularly shaped specimens were also collected. Volumes of components in these samples were corrected for pore water salinity and were back calculated using density values, assuming 100% saturation of the pore voids by either ice or by a combination of water and gas hydrate according to interpretation of the resistivity log.

The specimens were dried, typically after gas hydrate dissociation, at a temperature of 90°C for at least 24 hours in order to determine the amount of fresh water and solids present. The dried specimen was then broken into granule-size pieces, and the volume of dried solids was determined with an automatic gas pycnometer, using helium as the purge and expansion gas.[10]

Sediment pores are filled with a combination of water or gas hydrate *in situ* (and ice in the permafrost zone). The physical property calculations were adjusted to account for the *in situ* gas hydrate quantities interpreted by Collett *et al.*[7] from resistivity well log data. Grain sizes were determined using two methods. Dry sieving was used to measure the size of coarse material and a Brinkmann particle size analyzer using time of transition theory was used for the fine fraction.

RESULTS AND DISCUSSION

Measurements of water content and porosity are important to understanding host sediment conditions. These characteristics indicate the quantity of water and void space available to form hydrate, provided that gas concentration is not a limiting factor. Water contents (mass of pore water/mass of sediment grains, M_w/M_s) in the zone beneath permafrost vary from 35.5% to about 1% in a thin cemented sandstone encountered at 926 m. Porosities change from 50% to 3% over the same interval. Most of the samples from the deeper section exhibit porosities less than 44%.[4]

Physical property index measurements on sediment cores and derived geotechnical parameters are summarized for the gas hydrate interval in FIGURE 1 and are plotted with profiles of well-log-derived bulk density, porosity, and gas hydrate concentration. Petrophysical determinations of porosity and bulk density using a helium porosimeter are also presented. In general, within the gas-hydrate interval from 897 m to 922 m, the core-measured values and the values derived from well logs are strongly correlated. For some samples, a difference exists between the index-derived values and the measured helium porosimetry values. Good agreement between all techniques is observed in a thin layer at 926 m, which has porosity values between 2% and 4% (dolomite cemented sandstone).

Salinity determinations on pore water extracted from core samples beneath permafrost reveal pronounced changes, ranging from 60 to 4 ppt. Within the gas hydrate interval, many sediment specimens containing high hydrate concentrations had values less than 10 ppt and usually less than 5 ppt. However, other core samples within this interval had salinity values from 26 ppt to over 50 ppt. These samples generally correlate with apparent interbeds having lower gas hydrate concentrations and may imply solute migration associated with gas hydrate formation. Salinity in finer grained sediments above and below this interval had more consistent salinity values in the range of 25 to 35 ppt.

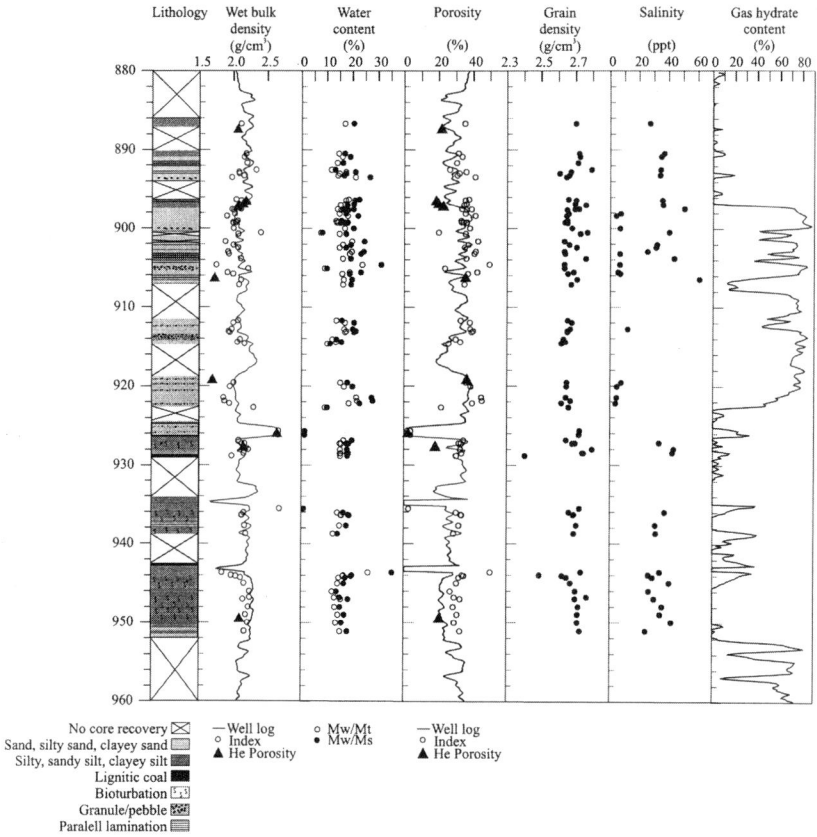

FIGURE 1. Core measurements of wet bulk density, water content, porosity, grain density, porewater salinity and gas hydrate concentration (percent occupancy of non-sediment grain void space). Water content (M_w/M_t represents mass water/mass total; M_w/M_s represents mass water/mass sediment grains), grain density, and pore-water salinity were measured from core subsamples and were used to calculate wet bulk density and porosity (assuming full water saturation and corrected for pore water salinity and presence of gas hydrate). Additional porosity and wet bulk density measurements by the helium porosimetry technique are also provided for petrophysical samples. Well log values are shown for wet bulk density, porosity, and hydrate saturation of pore voids from Collett et al.[7] A complete listing of physical properties is provided in Dallimore et al.[4]

FIGURE 2 shows grain size, core temperatures (measured immediately after sample extrusion from the core barrel), and visual qualitative estimates of gas hydrate content made by scientists at the drill site. The main gas hydrate interval from 897 to 922 m is dominated by sands, although some silty sands and low plastic silts are present in interbeds with lower gas hydrate concentrations. Sediments above this interval have similar grain size characteristics, however sediments below consist of low plastic silts with trace amounts of sand and clay.

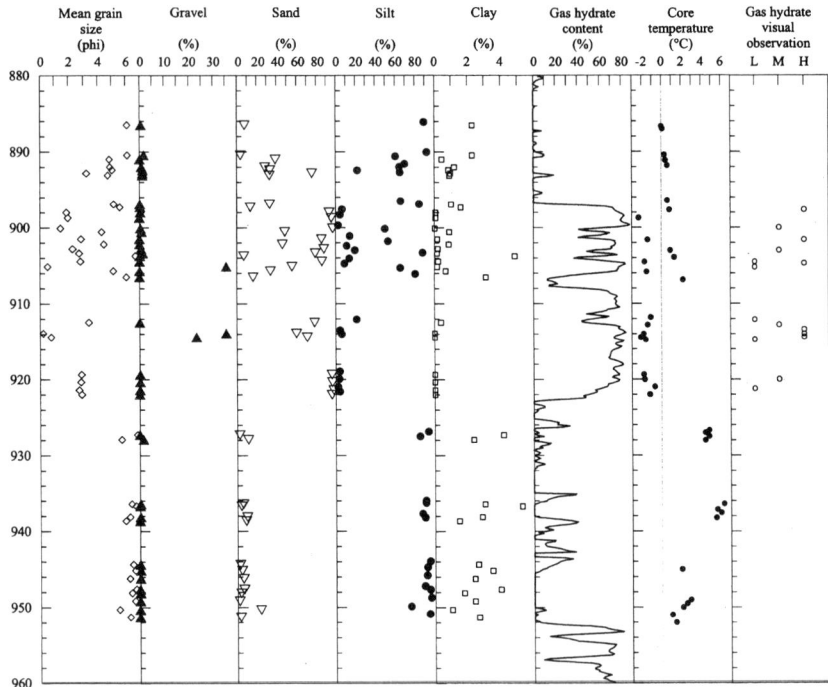

FIGURE 2. Grain size data, temperature measurements (made immediately after core extrusion), and qualitative estimates of gas hydrate concentration based on observations at the drill site. Well log gas-hydrate saturation values are from Collett et al.[7]

As discussed by Dallimore et al.,[1] core temperature measurements can provide a proxy to the presence of gas hydrate. Given that *in situ* temperatures are almost certainly above 8°C and mud temperatures were above 2°C, measurements of core temperatures below 2°C are interpreted to result from the endothermic cooling effect occurring because of gas hydrate dissociation. Furthermore, sediment cemented by gas hydrate and ice was observed at depths where temperatures below 0°C were measured. Quantitative visual estimates of gas hydrate concentration were typically made by immersing representative subsamples in water and estimating the intensity and volume of gas hydrate dissociation. Good correlation is observed between visual observations and negative core temperatures.

Gas hydrate was typically recovered in sandy material with a porosity of 32 to 45% and in gravel with a lower porosity of 23 to 29% (corrected for the presence of gas hydrate).[4] Silts and silty sands that are present above, within, and below the different gas hydrate layers (but do not contain gas hydrates) have similar or slightly lower porosities (26–41%) compared to the sands. This suggests that porosity alone is not a primary control on the distribution of gas hydrate and that gas hydrate may preferentially form in the larger void spaces of sand and gravels, whereas formation is inhibited within the smaller voids of the finer-grained sediment. Also, the lower permeability of the finer silts and sandstone could influence gas migration and hydrate formation. Localized elevated pore water salinity, which may have resulted from salt exclusion caused by initial gas hydrate formation in surrounding sediments, may also inhibit hydrate formation.[11]

CONCLUSIONS

Gas hydrate was typically recovered in sandy sediments and in layers with a high gravel content. Silts and silty sands that are present above, within, and below the gas hydrate layers, but contain lesser amounts of hydrate, have similar porosities to the sands but they exhibit higher pore-water salinities. This suggests that porosity may not be a primary control on the distribution of gas hydrate and that gas hydrate may preferentially form in the larger void spaces of sand and gravels, whereas formation may be somewhat inhibited within the smaller voids of the finer-grained and more saline sediments. Perhaps the lower permeability of the finer silts and sandstone also influences gas migration and hydrate formation. Other considerations such as the supply of gas may also be a factor.

ACKNOWLEDGMENTS

John Bratton and Monica Relle provided helpful reviews. Barbara Medioli constructed the figures. The drillers and staff at the drill site are thanked for their assistance.

REFERENCES

1. DALLIMORE, S.R., T. UCHIDA & T.S. COLLETT. 1999. Overview of the scientific program for the JAPEX/JNOC/GSC Mallik 2L-38 research well. Geological Survey of Canada Bulletin 544.
2. JENNER, K.A., S.R. DALLIMORE, F.M. NIXON, W.J. WINTERS & T. UCHIDA. 1999. Sedimentology of methane hydrate host strata from JAPEX/JNOC/GSC Mallik 2L-38. Geological Survey of Canada Bulletin 544.
3. KATSUBE, J.T., S.R. DALLIMORE, T. UCHIDA, K.A. JENNER, T.S. COLLETT & S. CONNELL. 1999. Petrophysical environment of sediments hosting gas hydrate, JAPEX/JNOC/GSC Mallik 2L-38 gas hydrate research well. Geological Survey of Canada Bulletin 544.
4. DALLIMORE, S.R., R. LAFRAMBOISE & M. FOTIOU, Eds. 1999. JAPEX/JNOC/GSC Mallik 2L-38 research well. Multi-media CD ROM, Geological Survey of Canada Open File Report.
5. WINTERS, W.J., S.R. DALLIMORE, T.S. COLLETT, J.T. KATSUBE, K.A. JENNER, R.E. CRANSTON, J.F. WRIGHT, F.M. NIXON & T. UCHIDA. 1999. Physical properties of sediments from the JAPEX/JNOC/GSC Mallik 2L-38 gas hydrate research well. Geological Survey of Canada Bulletin 544.
6. BILY, C. & J.W.L. DICK, 1974. Naturally occurring gas hydrates in the Mackenzie Delta, N.W.T. Bull. of Can. Pet. Geol. 22(3): 340–352.

7. COLLETT, T.S., R. LEWIS, S.R. DALLIMORE, M.W. LEE, T.H. MROZ & T. Uchida. 1999. Mallik 2L-38 downhole well log displays—detailed evaluation of gas hydrate reservoir properties. Geological Survey of Canada Bulletin 544.
8. COLLETT, T.S. & S.R. DALLIMORE. 1998. Quantitative assessment of gas hydrates in the Mallik L-38 well, Mackenzie Delta, N.W.T., Canada. *In* Proceedings of the 7th International Conference on Permafrost. Yellowknife, Canada. 189–194.
9. WINTERS, W.J., I.A. PECHER, J.S. BOOTH, D.H. MASON, M.K. RELLE & W.P. DILLON. 1999. Properties of samples containing natural gas hydrate from the Mallik 2L-38 research well, MackenzieDelta, NWT determined using Gas Hydrate And Sediment Test Laboratory Instrument (GHASTLI). Geological Survey of Canada Bulletin 544.
10. AMERICAN SOCIETY FOR TESTING AND MATERIALS. 1997. Standard test method for specific gravity of soil solids by gas pycnometer D 5550-94. *In* American Society for Testing and Materials, Annual Book of ASTM Standards, 04.09, Soil and Rock. West Conshohocken, PA. 380–383.
11. WRIGHT, J.F., S.R. DALLIMORE & F.M. NIXON. 1999. Influences of grain size and salinity on pressure-temperature thresholds for methane hydrate stability in Mallik sediments. Geological Survey of Canada Bulletin 544.

Gas Hydrates of Siberia

F.A. KUZNETSOV[a]

Institute of Inorganic Chemistry, Siberian Branch, Russian Academy of Sciences, Lavrentiev Avenue 3, 630090 Novosibirsk, Russia

ABSTRACT: Natural gas hydrates were first observed on 21 July 1961 (the discovery was officially registered on 24 December 1969).[1] It soon became clear that these compounds are not just chemical curiosities, or a nuisance for gas transportation in pipe lines, but rather that they are related significant natural phenomena. It is possible that in the past natural gas hydrates have manifested themselves in marine and land incidents. In future gas hydrates may offer an inexhaustible source of energy and chemical raw material. However, if not accessed with sufficient knowledge, gas hydrates may be destabilized to create many local problems and possibly global problems as well. In the USSR and more recently in Russia, a number of groups are traditionally involved in investigation of different aspects of gas hydrates problems. These groups have made significant contributions to present knowledge about the nature and properties of the gas hydrates. The Siberian branch of the Russian Academy of Sciences recently started an interdisciplinary program to comprehensively investigate all major aspects of gas hydrates, with special emphasis on natural gas hydrates and gas hydrates deposits in Siberia and its adjacent regions. The program consists of three components: (1) Geology of gas hydrates in the cryolito-zone of Siberia and in the bottom deposits of the Arctic and East seas, and of Lake Baikal. (2) Physicochemical study of hydrates of different gases. (3) Ecological monitoring. In this paper we offer a short summary of the main results of these three components of the program. We emphasize the second component, since it is close to the scientific interests of the author.

GEOLOGY OF GAS HYDRATES IN CRYOLITO-ZONE OF SIBERIA, IN BOTTOM DEPOSITS OF ARCTIC SEAS AND OF LAKE BAIKAL

Gas-hydrate stability zones have been outlined (see FIGURE 1) based on analysis of data collected to date. The data include geothermal conditions, pressure, mineralization of underground waters, and presence of gas (required for formation of hydrates). The scheme outlines the most favorable areas for the search for gas hydrate accumulations that formed from free gas that existed in these areas prior to rock and sediment freezing, or from gas that migrated to the hydrate-formation zone from lower layers. The analysis shows that early conclusions need to be reconsidered.

More detailed mapping of the gas-hydrate stability zones (GHSZ) in Northern Siberian regions was performed with use of computer technology. The aim was to determine the upper and lower boundaries of the zone. FIGURE 2 shows the results

---

[a]Telecommunication. Voice: (7 3832) 34 44 88; fax (7 3832) 34 44 89.
FK@che.nsk.su

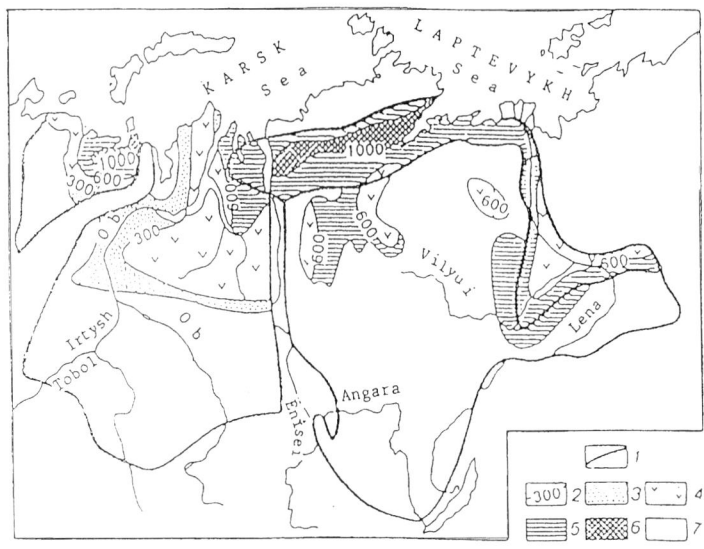

FIGURE 1. Map of GHSZ in boundaries of oil-gas fields in Northern Siberia. *1*, boundaries of oil-gas fields; *2*, isolines of thickness of the zones, with possible presence of gas hydrates; area with different thickness of the zones in meters; *3*, up to 300; *4*, 300–600; *5*, more then 1000; *7*, absence of hydrates.

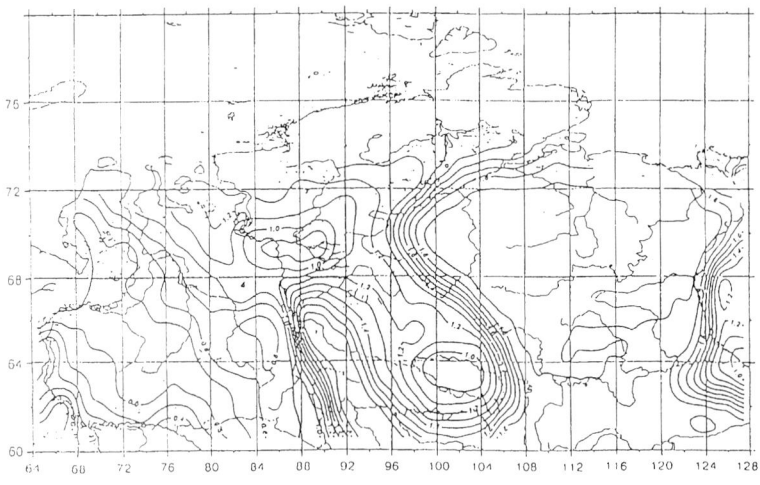

FIGURE 2. Estimated thickness of gas-hydrate stability zones.

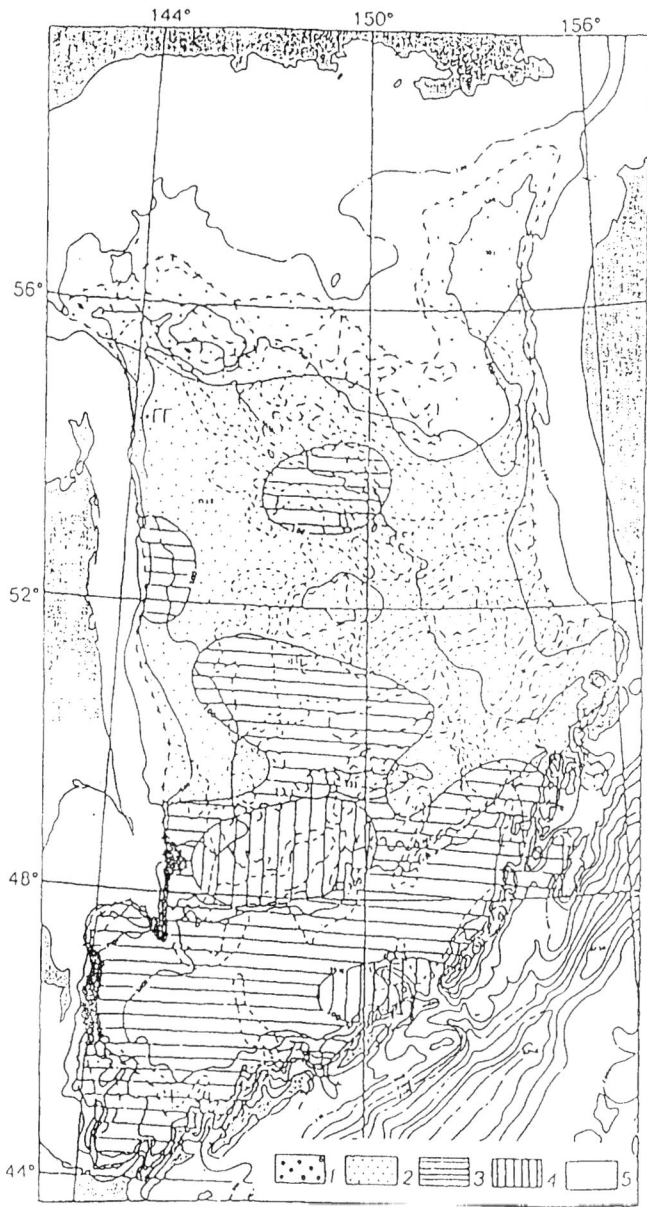

FIGURE 3. GHSZ in Okhotsk Sea. ГГ indicates point of direct visual observations of hydrates in bottom probes. Thickness of hydrates deposits in meters: *1*, less then 100; *2*, 100–200; *3*, 200–300; *4*, 300-400; *5*, areas with no hydrates.

from these calculations. The upper boundary of the GHSZ over almost all Siberia is at the depth of 100 meters. The depth of the lower boundary varies. In West Siberia it occurs at depths around 1,000 meters; in East Siberia, which is the coolest part of the region, the lower boundary is down to 2,000 meters. These data demonstrate a greater extent of the GHSZ then was previously accepted.

New data about the concentration of helium in natural gas has been obtained for the Myasoyakh gas field. An increase in He concentration is attributed to the formation of hydrates because the formation of gas hydrate would, in effect, scavenge methane and other hydrocarbon gases, leaving helium. It follows from variation in He concentration that the central part of the Myasoyuakh gas field most probably does not contain hydrates, whereas peripheral areas (characterized by higher concentration of He) may contain hydrate accumulations.

FIGURE 3 shows GHSZ at the bottom of Okhotsk Sea Derived from work of Prof. G. Ginsburg.[2,3] Finally, concerning hydrates in Lake Baikal bottom deposits; for a number of years a team from the South Branch of Institute of Oceonology RAS has performed seismic investigation of the Lake Baikal bottom.[4] As a result of this investigation a map of positions of lower boundary of gas hydrates was constructed (see FIGURE 4).

In March 1997 international expedition working on the *Baikal-drilling* program took a bottom probe in the center of the depression south of the Lake Baikal.[5] On heating this probe, an intensive evolution of gas from the sample was observed. Volumetric measurements and chemical and X-ray analysis proved the presence of methane hydrate in the sample. On the basis of isotope analysis, it was concluded that methane in the hydrate had a biogenic origin.

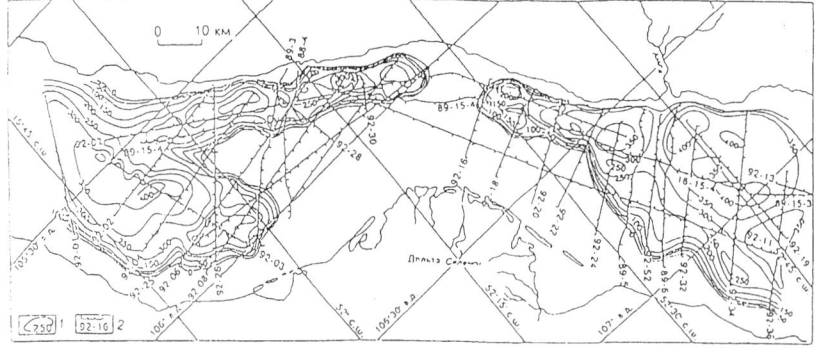

FIGURE 4. Map of the lower boundary for gas-hydrate deposits in Lake Baikal obtained from seismic data (depths of BSR). *1*, position of the lower boundary in meters; *2*, year and number of the profile.

PHYSICO-CHEMICAL STUDY OF HYDRATES OF DIFFERENT GASES

Phase Equilibrium in Systems with Gas Hydrate Formation

We report on new results from a program of systematic investigations of clathrate compounds (including clathrate hydrates), obtained by a team headed by Prof. Yu. A. Dyadin.

All the hydrophobic gases can be divided into two groups relative to formation of clathrate hydrates.[6,7] The first group includes three gases: hydrogen, helium, and neon with molecular diameter less then 3.5 Å. For these molecules classical clathrate hydrates are not known. However, due to the small size of the molecules they can be included within small voids in ice-Ih, -Ic, and -II (only these modifications of ice have voids within which such small molecules can be included, with formation of solid solutions or clathrate hydrates[8–11]).

The second group includes gases with molecules that are 3.8–9.2 Å in size. These gases do not dissolve in ice, but individually or in mixtures they can form classical hydrates: cubic-I and -II (CS-I and CS-II), hexagonal-III (HS-III or "Structure H"). It is one example of tetragonal structure ($Br_2 \cdot 8 \cdot 6H_2O$) formation.

Our recent investigations show that argon and krypton at sufficiently high pressure can form hydrates based on ice-II (analagous to hydrogen or neon). Experimental study of the phase equilibrium under pressures up to 15 kbar was performed using the technique described previously.[14]

Results for noble-gas hydrate-equilibrium are shown in FIGURE 5.[12–15] For the case of heavy noble gases, it can clearly be seen that decomposition curves of classical hydrate structures have a dome-like form. Stability of the hydrates decreases with decreasing guest molecule size. The low pressure section of the diagrams for hydrogen and neon shows the same feature (dome-like shape). This indicates, that in this pressure interval, hydrogen and neon form hydrates in which a pair of molecules occupies large voids. For the case of nitrogen, inclusion of two molecules in the large voids of CS-I and CS-II structures was reported by Kuhs and others.[16] The same process apparently operates for even smaller hydrogen molecules. This conclusion is also supported by the experimental observation that hydrogen and neon establishes equilibrium much slower then other gases for which the voids are occupied by only one guest molecule.

As it can be seen from FIGURE 5, to the right of the dome-like section, the higher pressure parts of the curves show that decomposition temperature sharply increases with pressure (in all the systems except xenon). It was shown, by using X-ray diffraction, that for hydrogen systems this part of the diagram corresponds to decomposition of a hydrate phase based on the ice-II crystal lattice. The similar character of the high pressure part of the curves for other systems leads to the conclusion that the hydrates based on the of ice-II crystal lattice are also formed in systems with neon, argon and krypton at pressures of 4 kbars, 9.6 kbars, and 13.4 kbars, respectively.

Thus, we conclude that in systems with neon, hydrogen, argon, krypton, and xenon classical clathrate hydrates are formed. Their stability decreases in direction from xenon to neon. For all the gases, except xenon, at high pressure hydrates based on the ice-II crystal lattice are formed. The pressure required for formation of this phase increases with increasing molecule size. For the case of xenon, the gas

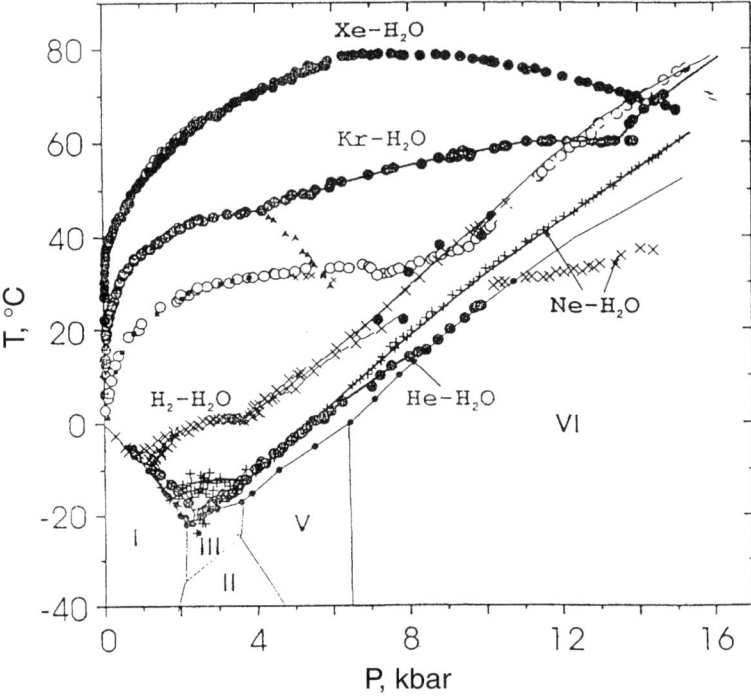

FIGURE 5. Decomposition of of noble gas and hydrogen hydrates. The *thin lines* show field of crystallization of different modifications of ice.

molecule is obviously too large for inclusion in voids of the ice-II crystal lattice and hydrate CS-I is stable at least for pressures up to 15 kbar.

The same experimental technique was used to investigate of phase diagrams of the binary systems methane–water and propane–water, and a section of the ternary system C_3H_8-CH_4-H_2O. The results from this study are shown in FIGURE 6.

In the system with methane (described previously[17]) two hydrate phases are formed: at pressures less than 6.2 kbars, hydrate with CS-I structure is stable. At higher pressure another hydrate phase is formed having higher density. The structure of this phase is not known yet. In the water-propane system three different hydrate phases are formed. At low pressure a hydrate with CS-II structure forms. Stability of this phase decreases with pressure. At a pressure of 1.45 kbars this structure is transformed into hydrate with higher density. Stability of this phase increases with pressure. At 6.45 kbars and 31.5°C a new hydrate phase with even higher density is formed.

In the propane system nothing but clathrate structure can be expected. Thus, taking into consideration what is know about the water clathrate structure, we propose the following sequence of transformations in this system at elevated pressures:

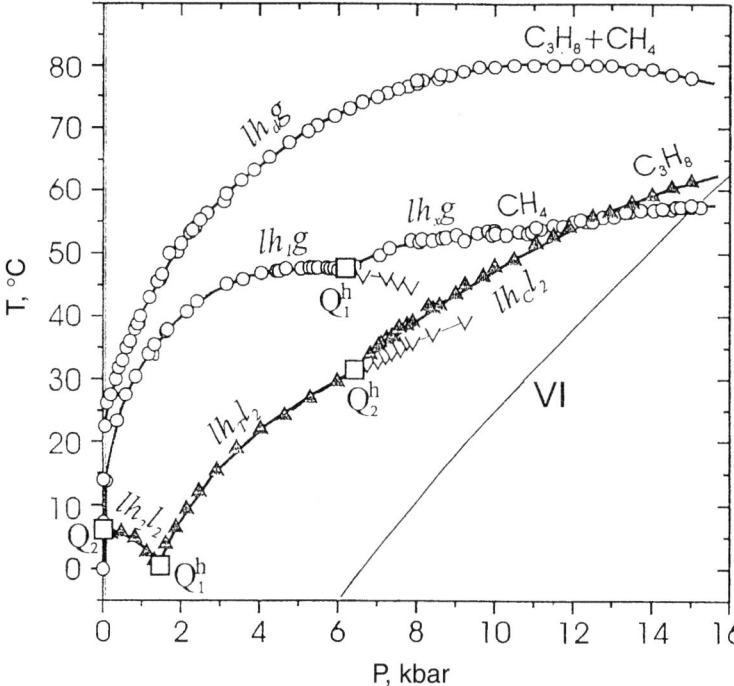

FIGURE 6. Decomposition of hydrates of methane, propane, and binary hydrate $C_3H_8CH_4 \cdot 17H_2O$ l and l_2, liquid water and propane phases, respectively; h_1, h_2, h_d, h_T, h_C, hydrates of cubic I and II, binary hydrate, tetragonal, and cubic I phases, respectively.

Hydrate CS-II → Hydrate TS-I → Hydrate CS-I

where TS-I is tetragonal structure I.

In methane–propane mixtures, even at low concentrations of propane hydrate, CS-II with composition $C_3H_8 \cdot CH_4 \cdot 17H_2O$ is formed.[18] As can be seen from FIGURE 6 there are no phase transitions in the system within the range of pressure values used (up to 15 kbars). This results from the fact that the structure is well packed from the beginning and thus phase transformation is not required for additional densification.

Physical Aspects of Structure and Interactions in Gas Hydrates

The theoretical part of the project aims at the construction of a quantitative theory of clathrate compounds. Work in this direction was started by van der Waals and Platteuw[18] and continues to attract the attention of researchers.[19–21] Problems, that we try to solve within the framework of the project, are related to microscopic mechanisms of phase transformations in clathrate hydrates and in different modifications of ice. Using the method of crystal network dynamics, we demonstrate the significant role played by guest molecules in the stability of hydrate structures. Another problem of interest to us is the relation between various forms of ice and hydrates, and the structure and mechanical instability of these compounds. Despite of the similarity of short-range order in hydrates and ice the hydrates demonstrate very peculiar mechanical and thermodynamic properties. Clathrate hydrate transforms at high pressure into a high-density amorphous phase. The transformation is reversible—at reduced pressures the initial structure returns. This is known as structure memory effect.[22] In the case of ice amorphisation under high pressure is irreversible.[23,24]

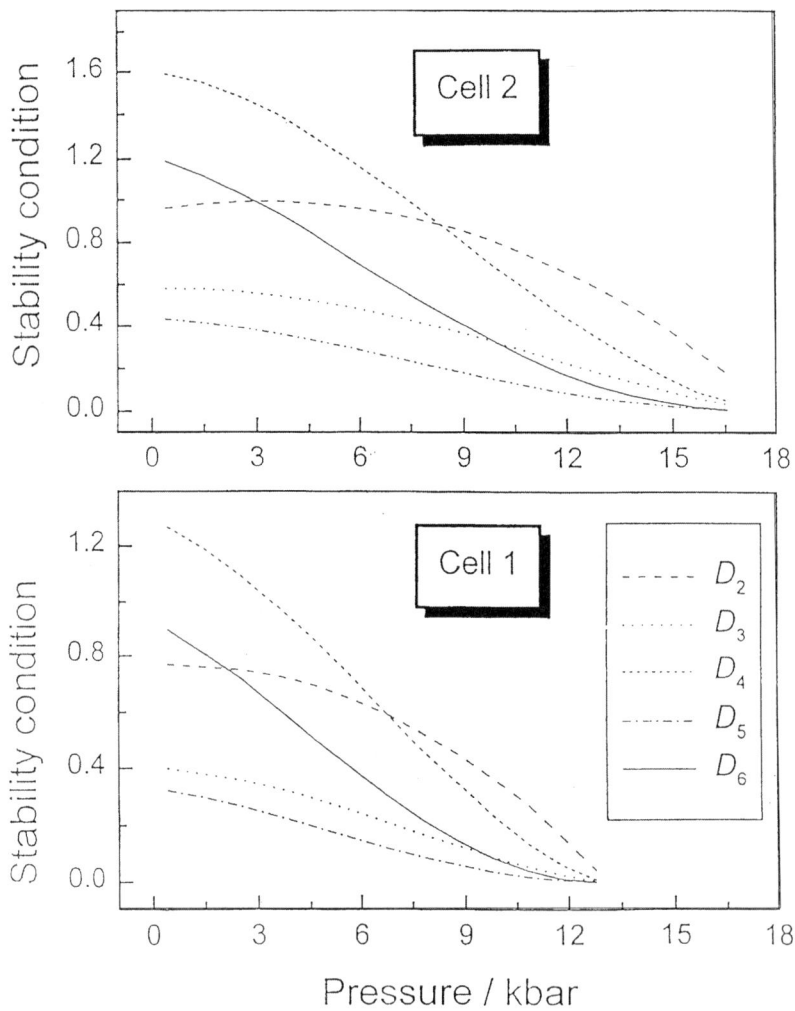

FIGURE 7. Conditions of stability of methane hydrate.

The methods of lattice dynamics permit investigation of the dynamic properties of both guest molecules and the network formed by host molecules. Thus, the role of the guest molecules in the stability of the clathrate structure can be studied. The method can also be used to calculate dynamic and thermodynamic properties of the clathrates and to find their boundary for mechanical stability.

Previously, the molecular dynamics approach was used to analyze various properties of the clathrate hydrates:

(a) The effect of filling of host lattice voids with guest molecules on the stability of the structure was investigated and a phase diagram for the H_2O-Xe system was calculated.[25,26]

(b) A strong interaction between the acoustic and optical branches, associated with motion of the guest molecules, was revealed.[27]

(c) The pressure dependence of thermal expansion and the stability boundaries for hydrates and different modifications of ice were found.[28,29]

New results have now been obtained that support the hypothesis of a relation between amorphization of clathrates with mechanical instability of crystal structures at high pressures and low temperatures. For methane and xenon hydrates and for an empty network of hydrate with CS-I structure, the elastic modules, $C_{ij}(P)$ in Voigh's notation ($i,j = 1, 2, ..., 6$) for $T = 10$ K were calculated. The mechanical stability boundaries for the structures were determined from the pressure dependence of main minors D_k ($k = 2, 3, ..., 6$) of the matrix $C_{ij}(P)$. According to Born criteria all values of D_k must be positive in the range of stability.[30] It was shown that the stability of hydrates strongly depends on presence of a guest molecule. The effect of stabilization also significantly depends on the type of proton ordering in the host crystal network.

FIGURE 7 shows results obtained for pressure dependence of principal minors for two different versions of proton ordering for methane hydrate. For the first version the hydrate becomes unstable at pressure of 13.0 kbar, for the second version at $P = 16.5$ kbar. Respective values for xenon hydrate are 16.5 kbar and 24.5 kbar. In future we plan to make an X-ray study of the gas hydrates at high pressure. This will permit us to demonstrate phenomena that have been theoretically predicted.

ECOLOGICAL MONITORING

In this paper we report on our results for modeling methane emission to the atmosphere as a result of marine gas hydrate decomposition caused by global warming. In our model, hydrate deposits were assumed to be located at Arctic and Antarctic coastal lines. It was accepted that decomposition starts when the temperature exceeds the thermodynamic decomposition temperature by 0.1 to 2.0°C. At that moment, the concentration of methane is assumed to increase to 100 times higher that of the background concentration of methane in ocean water (50 ppb). Three-dimensional diffusion in the ocean was calculated. Two different warming scenarios were considered: (a) instantaneous increase of ocean surface temperature by 3°C and (b) gradual warming by 0.08°C a year. Distribution of methane concentration in time at different depths was obtained by calculation. An example of these results is shown in FIGURE 8.

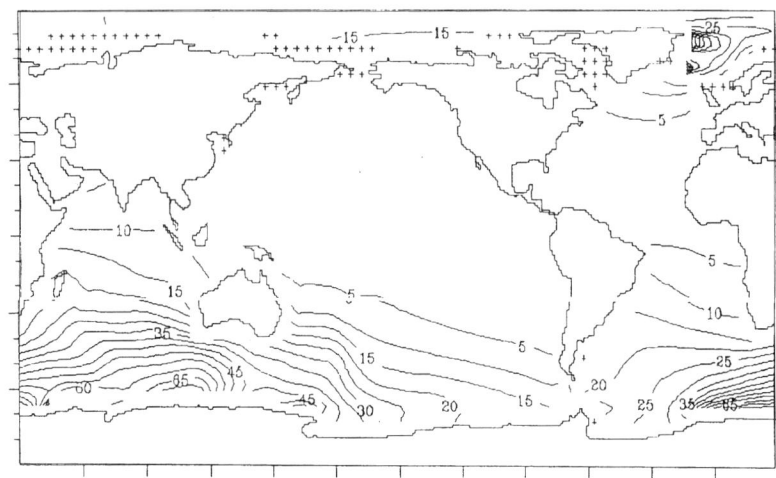

FIGURE 8. Iso-concentration lines for dissolved methane at a depth of 690 meters after 50 years of warming.

For these two warming scenarios emission of methane was also calculated. For the case of 3°C instantaneous warming and a decomposition threshold of 0.1°C, over a period of four years the total flow of methane into the atmosphere exceeds 1 terragram (TG) and reaches 14 TG in 50 years. In the case of gradual warming by 0.08°C/year the methane flow reaches a value of 1 TG in 17 years and 3.8 TG in 50 years. More details of these calculations can be found in.[31–33]

REFERENCES

1. Vasiliev, V.G., Yu.F. Makagon, F.A. Trebin, A.A. Trofimuk & N.V. Cherski. 1970. Ability of natural gases to exist in the Earth crust in the solid state. USSR Register of discoveries (Otkrytiia v SSSR 1968–1969). Moscow Nauka **10:** 3–5 (In Russian).
2. Ginzburg, G.D & V.A. Soloviev. 1994. Submarine gas hydrates. VNIIOceangeology, St. Petersburg. (In Russian).
3. Kremlev, A.N., G.D. Ginzburg & V.A. Soloviev. 1997. Investigations of submarine gas hydrates by seismic methods. *In* Abstracts of Seminars of Gas Hydrates in the Earth's Eco-system. Novosibirsk. (In Russian).
4. Golmshtok, A.Yu., A.D. Duchkov & D.R. Hutchinson. 1997. Estimation of heat flow in lake Baikal by seismic data on the lower boundary of gas hydrate layer. Russ. Geol. Geophys. **38:** 1677–1728.
5. Kuzmin, M.I. *et al.* 1998. The first finding of gas hydrates in bottom deposits of Lake Baikal. *In* Proceeding of RAS-Doklady Academii Nauk. **362:** 541–544. (In Russian).
6. Istomin, V.A. & V.S. Yakushev. 1992. Gas Hydrates in Natural Conditions. Nedra. Moskva. (In Russian).
7. Kuznetsov, F.A, Yu.A. Dyadin & T.V. Rodionova. 1997. Gas hydrates inexhaustible source of hydrocarbons. Russ. Chem. J. **41:** 28–34.
8. Namiot, A.Yu. & E.B. Bukhalter. 1965. Clatrates in ice. J. Struct. Chem. (Russ.) **6:** 911–913.

9. ARNOLD, G.P. et al. 1971. Neutron diffraction study of the ice polymorphs under heleum pressure. J. Chem. Phys. **55:** 589–595.
10. LONDONO, D., W.F. KUHS & J.L. FINNEY. 1992. Formation, stability and structure of helium hydrate at high pressure. J. Chem. Phys. **97:** 547–553.
11. VOS, W.L. et al. 1993. Novel H_2–H_2O clathrates at high pressures. Phys. Rev. Lett. **71:** 3150–3153.
12. DYADIN, YU.A. et al. 1996. Hydrate formation in the krypton–water and xenon–water systems up to 10 kbar. Proceedings of the 2nd International Conference on Natural Gas Hydrates. Toulouse, France. 59–66.
13. DYADIN, YU.A. 1996. Gas hydrates: thermodynamics and structural aspects. In Proceedings of the 2nd International Conference on Natural Gas Hydrates, Toulouse, France. 625–632.
14. DYADIN, YU.A. et al. 1997. Clathrate formation in the Ar–H_2O system under pressures up to 15000 bar. Mendeleev Comm. 32–33.
15. DYADIN, YU.A. et al. 1997. Clathrate formation in the Kr–H_2O and Xe–H_2O systems under pressures up to 15 kbar. Mendeleev Comm. 74–76.
16. KUHS, W.F. et al. J. 1996. Raman spectroscopic and neutron diffraction studies on natural and synthetic clathrates of air and nitrogen. Proceedings of the 2nd International Conference on Natural Gas Hydrates, Toulouse, France. 9–16.
17. DYADIN, YU.A. & E.YA. ALADKO. 1996. Decomposition of the methane hydrate up to 10 kbar. Proceedings of the 2nd International Conference on Natural Gas Hydrates, Toulouse, France. 67–70.
18. VAN DER WAALS, J.H. & J.C. PLATTEEUW. 1959. Clathrate Solution. Adv. Chem. Phys. **2:** 1–57.
19. BELOSLUDOV, V.R., M.YU. LAVRENTIEV & YU. A. DYADIN. 1991. Theory of clathrates. J. Inc. Phenom. Molec. Recog. Chem. **10:** 399–422.
20. ZUBKUS, V.E., E.E. TORNAU & V.R. BELOSLUDOV. 1992. Theoretic physicochemical problems of clathrate compounds. Adv. Chem. Phys. **lLXXXI:** 269–359.
21. TSE, J. 1994. Dynamical properties and stability of clathrate hydrates. Ann. N.Y. Acad. Sci. **715:** 187–206.
22. HANDA, Y.P., J.S. TSE, D.D. KLUG & E. WHALLEY. 1991. Pressure-induced phase transitions in clathrate hydrates. J. Chem. Phys. **94:** 623–627.
23. MISHIMA, O., L.D. CALVERT & E. WHALLEY. 1984. Melting ice I at 77 K and 10 kbar: a new method of making amorphous solids. Nature **310:** 393–395.
24. MISHIMA, O. 1996. Relationship between melting and amorphization of ice. Nature **384:** 546–549.
25. BELOSLUDOV, V.R., M.YU. LAVRENTIEV & S.A. SYSKIN. 1988. Dynamical properties of the molecular crystals with electrostatic interaction taken into account. Low Pressure Ice Phases (I_h and I_c) Phys. Stat. Sol. **B149:** 133–142.
26. BELOSLUDOV, V.R. et al. 1990. Dynamic and thermodynamic properties of clathrate hydrates. J. Inc. Phenom. Molec. Recog. Chem. **8:** 59–69.
27. TSE, J.S., V.P. SHPAKOV, V.V. MURASHOV & V.R. BELOSLUDOV. 1997. The low frequency vibrations in clathrate hydrates. J. Chem. Phys. **107:** 9271–9274.
28. SHPAKOV, V.P., J.S. TSE, V.R BELOSLUDOV & R.V. BELOSLUDOV. 1997. Elastic moduli and instability in molecular crystals. J. Phys. Cond. Mat. **9:** 5853–5865.
29. SHPAKOV V.P., J.S. TSE, C. TULK et al. 1998. Elastic moduli calculation and instability in structure I methane clathrate hydrate. Chem. Phys. Lett. **282:** 107–114.
30. BORN, M. & H. HUANG. 1954. Dynamical Theory of Crystal Lattices. Clarendon Press, Oxford.
31. SHERBAKOV, A.V., V.V. MALAKHOVA & E.N. ANTSIS. 1997. Numerical model of the World Ocean considering influence of Arctic Ocean. Preprint RAS Siberian Branch. Inst. Of Numerical Matematical Geophysics 1106, Novosibirsk.
32. OBJIGOV, A.I. 1993. Gas-geochemical fields of bottom layers of oceans and seas. Nauka, Moscow.
33. SCHERBAKOV, A.V. & V.V. MALAKHOVA. 1996. Numerical model of methane transport by the ocean currents.. Bull Nov. Comp. Center, Numerical Modeling in Atmosphere. **4:** 67–80.

The First Discovery of the Gas Hydrates in the Sediments of the Lake Baikal

M.I. KUZMIN,[a] V.F. GELETIY,[a] G. KALMYCHKOV,[a] F.A. KUZNETSOV,[b]
E.G. LARIONOV,[b] A.YU. MANAKOV,[b] YU.I. MIRONOV,[b] B.S. SMOLJAKOV,[b]
YU.A. DYADIN,[b,c] A.D. DUCHKOV,[d] N.M. BAZIN,[e] AND G.M. MAHOV[e]

[a]*Institute of Geochemistry,*
Siberian Branch of the Russian Academy of Sciences, 664033, Irkutsk, Russia

[b]*Institute of Inorganic Chemistry,*
Siberian Branch of the Russian Academy of Sciences, 630090, Novosibirsk, Russia

[d]*United Institute of Geology, Geophysics and Mineralogy,*
Siberian Branch of the Russian Academy of Sciences, 630090, Novosibirsk, Russia

[e]*Institute of Chemical Kinetics and Combustion,*
Siberian Branch of the Russian Academy of Sciences, 630090, Novosibirsk, Russia

ABSTRACT: In 1997 an underwater borehole was drilled in the South part of Lake Baikal (water depth, 1,433 m; borehole depth, 225 m; the temperature at these depths is about 10°C). Samples of sediments containing methane hydrate were collected at depths 121 and 161 m below bottom. At present, this is the only example of the existence of gas hydrates in a fresh-water basin under non-permafrost conditions.

SAMPLE COLLECTION

The existence of gas hydrates in the sediments of Lake Baikal was predicted in the course of geophysical investigations during 1989 and 1992. In 1997 samples of frozen sand-aleuritic materials (mineral grain size about 0.2 mm) were collected at depths 121 and 161 m below bottom in the BDP-97 borehole (Baikal Drilling International Program) at latitude 51°47′51″ North and longitude 105°29′14″ East. The water depth at this point is 1,433 m, the borehole depth is 225 m, the temperature at the depth of sample collection is about 10°C. Unusually high emission of a gas from these samples was observed accidentally immediately after they were lifted onto a ship. This observation served as the basis for a hypothesis about the presence of the gas hydrates in this region. The samples were kept in the usual refrigerator for two days, and only after that were they frozen in liquid nitrogen.

INVESTIGATION OF THE SAMPLES

First the dependence of emission of gas from the samples on sample temperature was studied. The following stages were established with a rise in temperature:

[c]Telecommunication. Voice: 7-3832-391346; fax: 7-3832-344489.
CLAT@che.nsk.su

1. violent gas emission at low temperatures; we interpret this stage as evaporation of liquid nitrogen from the pores of the sample and, possibly, decomposition of nitrogen hydrate formed while keeping the sample in liquid nitrogen;
2. absence of gas emission at low temperature;
3. slow gas emission;
4. rapid gas emission; and
5. absence of gas emission from the thawed sample.

We interpreted stages 3 and 4 as decomposition of gas hydrate present in the sample. The volume of gas emitted at stages 3 and 4 was 6.0 ml per gram of the rock.

More detailed study of gas emission at stages 3 and 4 led us to interesting results (see FIGURE 1). The following experimental procedure was used. About 1 gram of the sample (2–3 mm pieces or carefully pounded at the temperature of liquid nitrogen up to the size of the mineral particles) was placed into the isolated vessel at the temperature of liquid nitrogen. The vessel was rapidly heated to about −180°C and stored in the temperature interval −180 to −150°C for about 10 minutes to complete evaporation of the liquid nitrogen (or, probably, to complete decomposition of the nitrogen hydrate—decomposition stage 1). Subsequently, the vessel was slowly heated (0.5–1 deg/min.) and the volume of the gas in the vessel was measured at atmospheric pressure. Taking into account the low hydrate content in the sample and the small heating rates used in our experiments, we conclude that experimental conditions do not affect the clathrate decomposition temperature measured. The temperature of the sample was measured by the means of a thermocouple immersed in the sample. Small pieces and powder from the frozen sample were used for *Experiments 1* and *2*, respectively. We interpret the results from *Experiment 1* as an effect of self-conservation[1] of thermodynamically unstable methane hydrate in the pieces of the sample. In the case of powdered sample (*Experiment 2*) the main gas emission occurred at temperatures from −70 to −80°C—the decomposition temperature of methane hydrate falls in this interval. A weak endothermic effect was detected in the sample at the same temperature by means of a differential-thermal analysis method. Under the conditions in which this sample was collected and kept, the partial preservation of the gas hydrates was possible only as a result of the self-conservation effect.

Chromatographic analysis of the composition of gaseous mixture, emitted by the sample, gave following results (in milliliters per gram of rock): methane, 5.9; nitrogen, 44.7; oxygen + argon, 4.5; and carbon dioxide, 0.5. Methane was the only hydrocarbon in the gas mixture. Therefore, the data on the composition of the gas mixture accord with our interpretation of stages 3 and 4 as the decomposition of methane hydrate, preserved in the sample. The amount of hydrate can be estimated as 3.3 mass percent (assuming that composition of the hydrate $CH_4 \cdot 6H_2O$). In our opinion, impurities of oxygen and carbon dioxide originated from air when the sample was immersed in liquid nitrogen. Taking into consideration the fact that the sample contains 17 mass percent of water and that the main part of the hydrate decomposed at the sample collection stage, the maximum initial concentration of gas hydrate in the rock was about 20 mass percent.

X-ray powder diffraction patterns were recorded for original samples and for the same samples while being thawed. Comparison of these patterns shows that ice and at least one additional phase disappear in the course of thawing the samples. We

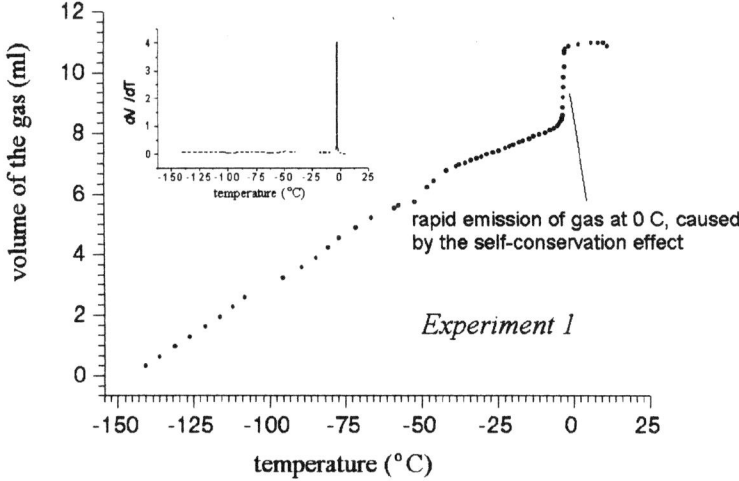

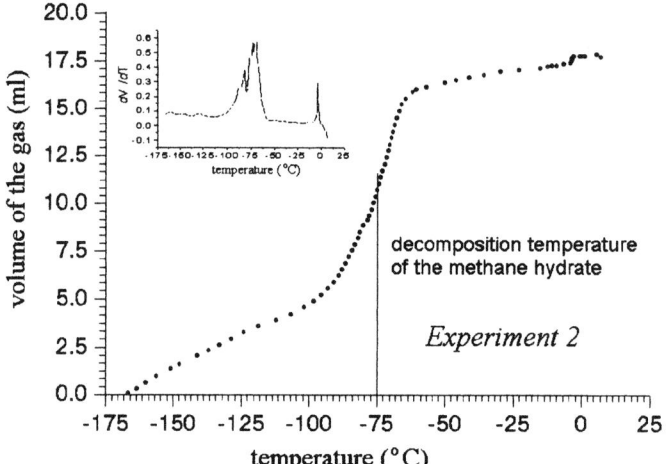

FIGURE 1. The dependence of emission of gas from samples on temperature. Small pieces of the sample were taken for *Experiment 1*, a powdered sample was taken for *Experiment 2*. Small figures illustrate differential variants of the curves shown in the main figures.

consider this phase to be methane hydrate. Taking into account the low content of gas hydrate in the sample (below the sensitivity of this method) we were only able to observe corresponding reflexes only because ice and gas hydrate cover the surface of mineral particles rather than being uniformly distributed throughout the sample. Unfortunately only poor quality diffraction patterns were obtained and these did not allow us to reach any conclusions about the structure of the hydrate in the sample

TABLE 1. Content of different ions (in micrograms per gram of rock) and pH of the water extract of the sample

pH	NH_4^+	$Ca^{2+}+Mg^{2+}$	Na^+	K^+	Cat^-	HCO_3^-	F^-	Cl^-	NO_3^-	SO_4^{2-}	An^+	Σ
6.28	48	18	12	12	90	151	6	17	2.5	7	184	274

(the strong low-angle reflection that is characteristic of CS-II was absent from our patterns).

According to the determination of ^{13}C content in emitted methane gas,[2] a biogenic origin of the methane should be assumed. The content of different ions and the pH of the water extract of the sample were examined (TABLE 1). The predominant ions in the water extract are ammonium and hydrocarbonate indicative of the reductive conditions at the point of sample collection. This fact supports the hypothesis of a biogenic origin for the methane. The Cl^- ion content in the extract is lower than that of the data for the BDP-93 borehole,[2] the latter being characteristic of hydrate-containing rock.

CONCLUSIONS

The main result of this work is that it provides the first experimental evidence of the existence of gas hydrates in the sediments of a freshwater reservoir (Lake Baikal) in the absence of permafrost. A self-conservation effect (a specific feature of gas hydrates) is the main fact confirming presence of gas hydrate in the samples studied. Taking into consideration the data on the composition of the gas mixture (methane is the only hydrocarbon), pressure–temperature conditions at the point of sample collection, and the decomposition temperature of the hydrate in the sample, we conclude that the samples studied contain pure methane hydrate.

ACKNOWLEDGMENTS

This work was supported by Presidium SD RAS Grant No. 97-18.

REFERENCES

1. HANDA, Y.P. 1988. Calorimetric study of naturally occurring gas hydrates. Ing. Eng. Chem. Res. **V27**(5): 872–874.
2. KUZMIN, M.I. *et al.* 1998. Pervaja nahodka gazogidratov v osadochnoj tolshe ozera Bajkal. (The First Discovery of the Gas Hydrates in the Sediments of the Lake Baikal.) Doklady Akademii Nauk (C.R. Russian Academy of Sciences) **362**(4): 541–543. (In Russian.)

Rock Physics Characterization for Gas Hydrate Reservoirs

Elastic Properties

MICHAEL B. HELGERUD,[a] JACK DVORKIN,[b] AND AMOS NUR[c]

Stanford Rock Physics Laboratory, Geophysics Department,
Stanford University, Stanford, California 94305-2215, USA

ABSTRACT: We offer a first-principle-based, effective medium model for elastic-wave velocity in unconsolidated, high porosity, ocean bottom sediments containing gas hydrate or free gas. The dry sediment frame elastic constants depend on porosity, elastic moduli of the solid phase, and effective pressure. Elastic moduli of saturated sediment are calculated from those of the dry frame using Gassmann's equation. To model the effect of gas hydrate on sediment elastic moduli we use two separate assumptions: (a) hydrate modifies the pore fluid elastic properties without affecting the frame and (b) hydrate becomes a component of the solid phase, reducing porosity and modifying the elasticity of the frame. The goal of the model is to predict the amount of hydrate in sediments from sonic or seismic velocity data. We apply the model to sonic and VSP data from ODP Hole 995 and obtain hydrate concentration estimates from assumption (b) that are consistent with estimates obtained from resistivity, chlorinity, and evolved gas data.

INTRODUCTION

Natural gas hydrates are nonstoichiometric crystalline solids comprised of a hydrogen-bonded water lattice and entrapped "guest" molecules, usually predominantly methane.[1] Gas hydrate with methane as the guest species is stable at the pressure and temperature conditions present in the sediments beneath most of the world's continental margins and deep inland seas and also in arctic sediments below the permafrost layer. Enormous amounts of methane are believed to be trapped by hydrates,[2] both in the hydrate crystal structure itself and also in sediments beneath hydrate deposits.[3,4] However, all hydrate-related methane estimates are very rough because accurate estimates of the amount of methane hydrate *in situ* are not currently available on a regional or site specific basis. A remote sensing technique that can accurately assess the amount and distribution of hydrate in natural deposits is needed before these hydrate-related methane estimates can be improved.

The best technique for remotely probing sediments several hundred meters below the surface or beneath deep bodies of water is seismic reflection profiling. Interpreting

Telecommunication.
[a]Voice: 650-723-7910; fax: 650-725-7344. helgerud@pangea.stanford.edu
[b]Voice: 650-725-9296. jack@pangea.stanford.edu
[c]Voice: 650-723-9526. nur@pangea.stanford.edu

seismic data to deduce the amount of gas hydrate in place requires a relation between the hydrate fraction in the sediments and the elastic properties of the hydrate-sediment composite. Unfortunately, very little is known about the elastic properties of gas hydrate and sediment-hydrate composites. Elsewhere we report on our collaboration with the USGS in Menlo Park, CA, in which we successfully measured the compressional and shear wave speeds of pure methane hydrate.[5] In this paper, we present some effective medium models for the elastic properties of sediments containing methane gas hydrate.

BACKGROUND

Several recent attempts have been made to construct a relation between hydrate fraction and compressional velocity in ocean bottom sediments. The most commonly used relations are based on modifications of Wyllie's time average[6] or weighted combinations of Wyllie's time average and Wood's[7] relation.[3,8–15] These authors generally achieve a good fit between the data and their model by fine-tuning the input and weighting parameters contained in their equations. A problem with using this technique to model marine sediments is that Wyllie's original time average equation is strictly empirical and was derived from a consolidated rock database. It is not based on first-principle physics,[16] and combining Wyllie's time average with Wood's relation in a weighting scheme provides little physical insight. As a result, it is hard to establish a rational pattern for adapting *free* parameters to site-specific conditions.

A different, physically intuitive approach was taken by Hyndman and Spence.[17] They constructed an empirical relation between porosity and velocity for sediments without gas hydrate and approximated the effect of hydrate formation on sediment velocity by a simple reduction in porosity. By doing so they effectively assumed that hydrate becomes part of the frame without altering the frame's elastic properties.

Here, instead of using Wyllie's time average or an empirical relationship to estimate velocity, we introduce a physics-based model for hydrate-bearing sediments. We start with the rock physics model of Dvorkin *et al.*,[18] for the elastic properties of ocean-bottom sediments without gas hydrate. This model relates the stiffness of the dry frame to porosity, mineralogy, and effective pressure. The effect of water saturation is modeled by Gassmann's equation.[19] We further develop this model to account for the effect of hydrate formation by separately applying two assumptions: (a) hydrate modifies the pore fluid elastic properties without affecting the frame; (b) hydrate becomes a component of the load bearing solid phase, reducing the porosity and modifying the compressibility of the frame. We apply the theory and assumptions to sonic and VSP data from ODP site 995. The results show that using assumption (b) in conjunction with this model produces accurate estimates of *in situ* hydrate concentrations.

THEORY

The porosity at which a granular composite ceases to be a suspension and becomes grain supported is called the critical porosity (ϕ_c). For a dense random packing of nearly identical spheres, ϕ_c is approximately 0.36–0.40.[20] A number of laboratory experiments have shown that the elastic properties of porous materials are best modeled as mixtures of solid and critical porosity material instead of solid material and void space.[20] Our 100% water saturated baseline model for ocean bottom sediments uses the effective moduli of a dense random packing of nearly identical spheres as its starting point.

We calculate the effective bulk (K_{HM}) and shear (G_{HM}) moduli of the dry rock frame at ϕ_c using the Hertz-Mindlin contact theory:[21]

$$K_{HM} = \left[\frac{n^2(1-\phi_c)^2 G^2}{18\pi^2(1-\nu)^2}P\right]^{1/3}, \quad G_{HM} = \frac{5-4\nu}{5(2-\nu)}\left[\frac{3n^2(1-\phi_c)^2 G^2}{2\pi^2(1-\nu)^2}P\right]^{1/3}, \quad (1)$$

where P is the effective pressure; G and ν are, respectively, the shear modulus and Poisson's ratio of the solid phase; and n is the average number of contacts per grain in the sphere pack (about 8–9).[22] Effective pressure is calculated as the difference between the lithostatic and hydrostatic pressure: $P = (\rho_b - \rho_w)gD$, where ρ_b is sediment bulk density, ρ_w is water density, g is the acceleration due to gravity, and D is depth below sea floor. ν can be calculated from G and K (the solid phase bulk modulus) using:

$$\nu = \frac{1}{2}\frac{K-(2/3)G}{K+(1/3)G} \quad (2)$$

For porosity (ϕ) less than ϕ_c the bulk (K_{Dry}) and shear (G_{Dry}) moduli of the dry frame are calculated from the modified lower Hashin-Shtrikman (H-S) bound:[22]

$$K_{Dry} = \left[\frac{\phi/\phi_c}{K_{HM}+\frac{4}{3}G_{HM}} + \frac{1-\phi/\phi_c}{K+\frac{4}{3}G_{HM}}\right]^{-1} - \frac{4}{3}G_{HM},$$

$$G_{Dry} = \left[\frac{\phi/\phi_c}{G_{HM}+Z} + \frac{1-\phi/\phi_c}{G+Z}\right]^{-1} - Z, \quad Z = \frac{G_{HM}}{6}\left(\frac{9K_{HM}+8G_{HM}}{K_{HM}+2G_{HM}}\right), \quad (3)$$

which represents the weakest possible combination of solid and critical porosity material (see FIGURE 1). For porosity above critical, K_{Dry} and G_{Dry} are calculated via the modified upper H-S bound:

$$K_{Dry} = \left[\frac{(1-\phi)/(1-\phi_c)}{K_{HM}+\frac{4}{3}G_{HM}} + \frac{(\phi-\phi_c)/(1-\phi_c)}{\frac{4}{3}G_{HM}}\right]^{-1} - \frac{4}{3}G_{HM},$$

$$G_{Dry} = \left[\frac{(1-\phi)/(1-\phi_c)}{G_{HM}+Z} + \frac{(\phi-\phi_c)/(1-\phi_c)}{Z}\right]^{-1} - Z, \quad (4)$$

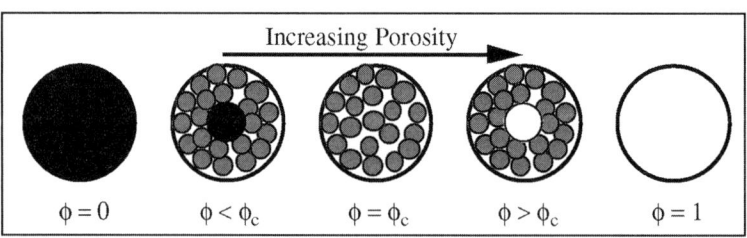

FIGURE 1. Hashin-Shtrikman arrangements of sphere pack, solid, and void.

where Z is defined as in (3). Equation (4) represents the strongest possible combination of critical porosity material and void space (FIG. 1).

If the sediment is saturated with pore fluid of bulk modulus K_f, the shear modulus G_{Sat} is the same as that of the dry frame, and the bulk modulus K_{Sat} is calculated from Gassmann's equation[19] as

$$K_{Sat} = K \frac{\phi K_{Dry} - (1+\phi) K_f K_{Dry}/K + K_f}{(1-\phi) K_f + \phi K - K_f K_{Dry}/K}. \quad (5)$$

Once the elastic moduli are known, the elastic wave velocities are calculated from

$$V_p = \sqrt{(K_{Sat} + (4/3) G_{Sat})/\rho_B}, \quad V_s = \sqrt{G_{Sat}/\rho_B}, \quad (6)$$

where ρ_B is bulk density.

In a common case of mixed mineralogy, the effective elastic constants of the solid phase can be calculated from those of the individual mineral constituents using Hill's average formula:[23]

$$K = \frac{1}{2}\left[\sum_{i=1}^{m} f_i K_i + \left(\sum_{i=1}^{m} f_i/K_i\right)^{-1}\right], \quad G = \frac{1}{2}\left[\sum_{i=1}^{m} f_i G_i + \left(\sum_{i=1}^{m} f_i/G_i\right)^{-1}\right], \quad (7)$$

where m is the number of mineral constituents, f_i is the volumetric fraction of the ith constituent in the solid phase, and K_i and G_i are the bulk and shear moduli of the ith constituent, respectively. The solid phase density is calculated as

$$\rho = \sum_{i=1}^{m} f_i \rho_i, \quad (8)$$

where ρ_i is the density of the ith constituent.

We can account for the presence of gas hydrate by assuming that either (a) hydrate is part of the pore fluid and does not affect the stiffness of the dry frame; or (b) hydrate is a component of the dry frame. In the latter case hydrate formation acts to reduce porosity and alter the solid phase elastic properties.

In case (a), the volumetric concentration of gas hydrate in the pore space is given by $S_h = C_h/\phi$, where C_h is the volumetric concentration of hydrate in the rock. If we assume the gas hydrate and water are homogeneously distributed throughout the pore space, then the effective bulk modulus of the water/hydrate pore fluid mixture is the Reuss average[24] of the water and gas hydrate bulk moduli:

$$\bar{K}_f = \left[\frac{S_h}{K_h} + \frac{1-S_h}{K_f}\right]^{-1}, \quad (9)$$

where K_h is the bulk modulus of gas hydrate. $\bar{K}_f$ should now be used in Gassmann's equation (5) instead of K_f, where the effective elastic moduli of the dry frame are given by Equations (3) and (4) and those of the solid phase are given by Equation (7).

In case (b), hydrate is modeled as a load bearing component of the frame. It reduces the original porosity ϕ to $\bar{\phi} = \phi - C_h$ and changes the effective mineral modulus as calculated in (7) where f_i should now be replaced by:

$$\bar{f}_i = f_i(1-\phi)/(1-\bar{\phi}), \quad (10)$$

and gas hydrate has to be treated as an extra mineral component with fraction $\bar{f}_h$ given by

$$\bar{f}_h = C_h/(1-\bar{\phi}). \tag{11}$$

If instead of hydrate we have gas in the pore space, we can model the effect on velocity by modifying the water saturation calculation represented by Equation (5). The particular modification we apply depends on how we assume the gas is distributed in the pore space. If we assume the gas is homogeneously distributed (i.e., each pore has the same well mixed amounts of gas and water), then we need to modify the fluid bulk modulus by taking the Reuss (isostress) average of the water (K_w) and gas (K_g) bulk moduli:

$$\bar{K}_f = \left[\frac{S_w}{K_w} + \frac{1-S_w}{K_g}\right]^{-1}. \tag{12}$$

S_w in (12) is the water saturation of the pore space. Using $\bar{K}_f$ for K_f in Equation (5) accounts for the effect of homogeneously distributed gas on the saturated sediment bulk modulus. The shear modulus is unaffected ($G_{Sat} = G_{Dry}$), but the bulk density changes according to:

$$\rho_B = \phi(S_w \rho_w + (1-S_w)\rho_g) + (1-\phi)\rho_{solid}. \tag{13}$$

The new velocities are calculated from (6).

If we instead assume that gas is located in patches where gas bubbles much larger than the average pore size are fully surrounded by water-saturated sediment, then we need to calculate a saturated sediment effective bulk modulus (K_{Sat}) using the relation:[25]

$$\frac{1}{K_{Sat} + 4G_{Sat}/3} = \frac{S_w}{K_{SatW} + 4G_{Sat}/3} + \frac{1-S_w}{K_{SatG} + 4G_{Sat}/3} \tag{14}$$

where K_{SatW} and K_{SatG} are the bulk moduli of the sediment fully saturated with water and gas, respectively:

$$\begin{aligned}K_{SatW} &= K\frac{\phi K_{Dry} - (1+\phi)K_w K_{Dry}/K + K_w}{(1-\phi)K_w + \phi K - K_w K_{Dry}/K}, \\ K_{SatG} &= K\frac{\phi K_{Dry} - (1+\phi)K_g K_{Dry}/K + K_g}{(1-\phi)K_g + \phi K - K_g K_{Dry}/K}.\end{aligned} \tag{15}$$

As in the homogeneous gas saturation case, the shear modulus is unchanged and the bulk density changes according to (13), where S_w is the average water saturation of the sediment.

Velocities calculated using the homogeneous gas saturation assumption are extremely sensitive to very small amounts of gas. A tiny amount of gas causes a large decrease in velocity because the easily compressed gas dominates the compressibility of the gas-water composite fluid. The patchy saturation model leads to much smaller velocity variation with gas saturation and higher inferred amounts of gas for a given decrease in compressional velocity. The homogenous and patchy saturation models are two endmember cases of potential gas distributions in nature. In this study, both models are applied, allowing upper and lower estimates of gas saturation to be made.

RESULTS

We apply our theory with assumptions (a) and (b) to an analysis of compressional-wave velocity data from ODP Leg 164, Site 995. The site is located off the coast of the Southeastern United States on the Blake-Bahama Ridge in a water depth of 2800 m. Available data include well logs, core analyses, and vertical seismic profiles (VSP) as well as independent hydrate concentration estimates from resistivity logs, chloride anomalies, and methane gas volumes evolved from a pressure core sampler (PCS).

The mineralogy at the site was determined from smear slides and XRD data and is reported in the ODP Preliminary Report Volume for Leg 164.[26] The interval of interest extends from 190 meters below the sea floor (mbsf) to the bottom of the hole (about 700 mbsf). This interval is lithologically uniform and is predominantly comprised of clay, calcite, and quartz.[26] For the purpose of calculating the effective moduli of the solid component we take volume percentages of clay, calcite, and quartz to be 60, 35, and 5%, respectively, consistent with the mineralogy data. The elastic properties and densities used in this study for clay, calcite, and quartz[27] are shown in TABLE 1.

The bulk modulus and density of seawater in the formation were calculated versus depth as a function of temperature and pressure using the method of Batzle and Wang.[28] The average bulk modulus was 2.5 GPa. The calculated density was practically constant in the interval at 1.032 g/cm^3. The bulk modulus and density of methane gas *in situ* were calculated from Sychev *et al.*[29] Effective pressure was calculated as the difference between the overburden and hydrostatic pressure. Porosity was taken from shipboard core measurements. Two other input parameters needed in our model, coordination number (n) and critical porosity (ϕ_c), were taken as 9 and 0.37, respectively, the theoretical values for a dense packing of identical spheres.[30] The model results are not sensitive to the values chosen for n and ϕ_c provided physically reasonable values are used.

The properties of pure methane hydrate used in this study are given in TABLE 1. Density was calculated from the physical dimensions of the methane hydrate unit cell by assuming about 92% cage occupancy. Elastic properties were determined from ongoing laboratory velocity measurements of compacted, synthetic polycrys-

TABLE 1. Properties of sediment constituents

Constituent	Bulk volume[a] (%)	Bulk modulus (GPa)	Shear modulus (GPa)	density (g/cm^3)
clay	$60(1-\phi)$	20.9	6.85	2.58
calcite	$35(1-\phi)$	76.8	32.0	2.71
quartz	$5(1-\phi)$	36.6	45.0	2.65
methane hydrate	variable	7.9	3.3	0.90
water	variable	2.4–2.6	0	1.032
methane gas	variable	0.10–0.12	0	0.23

[a]Bulk volume for clay, calcite, and quartz include factor $(1-\phi)$ to signify that they are 60, 35, and 5% of the solid phase, respectively.

talline methane hydrate samples.[5,31] In these experiments, samples with porosities below about 2% yield compressional and shear wave speeds of 3,700 ± 70 and 1,920 ± 50 m/s, respectively.

In FIGURE 2A, we plot sonic log data (smoothed with a 61-point median filter) together with the calculated velocity profiles for 100% water saturation, as well as 2, 4, and 8% bulk hydrate and 0.5 and 1% bulk gas. The gas or hydrate are assumed to be homogeneously distributed in the pore space. FIGURE 2B shows modeling results using a patchy distribution for gas (1, 2, and 4%) and assumption (b) for the effects of hydrate. Chlorinity data and PCS gas volumes suggest that hydrate is located in the interval from 190 to 450 mbsf.[4,26] The 100% water saturation line in FIGURE 2 intersects the velocity curve at approximately these same depths, confirming the validity of our baseline model and input parameters. The constant methane gas and gas hydrate saturation lines in FIGURES 2A and 2B are level curves for estimating bulk methane and hydrate percent from well log sonic velocity. Comparing FIGURES 2A and 2B we see that the hydrate in the fluid model predicts considerably more hydrate *in situ*. For reference, FIGURE 2C compares hydrate concentration estimates derived from the (smoothed) resistivity log at the depths where core porosity values were measured to the inversion of the hydrate in frame model results shown in FIGURE 2B. The resistivity based estimates are derived from the *quick look* Archie approximation.[15,26] The velocity based hydrate distribution estimate from the hydrate in the frame model shown in FIGURE 2C compares well with independent estimates from resistivity (FIG. 2C) and pore water chlorinity (FIG. 3C). It is also consistent with values derived from PCS evolved gas data which suggest that 0–9% (by volume) of the pore space contains gas hydrate.[4]

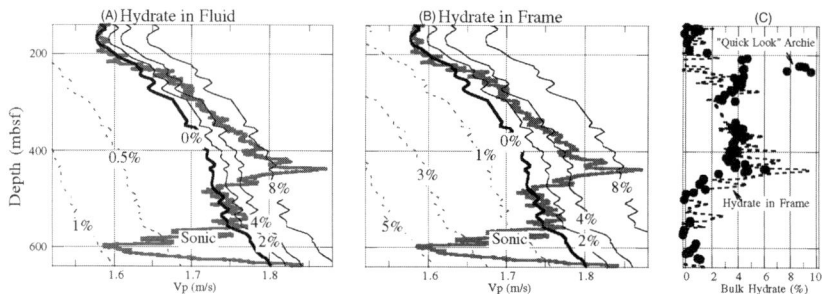

FIGURE 2. Hydrate concentration from compressional wave velocity and resistivity logs. **(A)** Comparison of well log V_p with model results assuming hydrate (*solid lines*) or homogeneously distributed gas (*dashed lines*) are part of the pore fluid. **(B)** Comparison with model results assuming hydrate is part of the sediment frame or gas patchily distributed in the pore space. For both **(A)** and **(B)**, model values calculated at core depths and results fit with smoothed curves. **(C)** Comparison of hydrate concentration estimates from the resistivity log and results from the elastic model assuming that the hydrate becomes a load bearing component of the sediment.

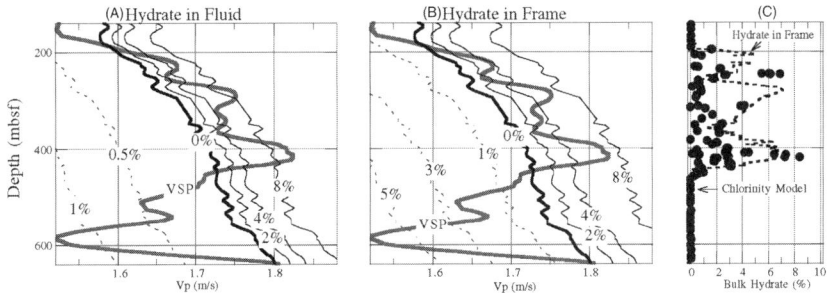

FIGURE 3. Hydrate concentration from VSP and chlorinity data. (**A**) Comparison of VSP velocity with model results assuming hydrate (*solid lines*) or homogeneously distributed gas (*dashed lines*) are part of the pore fluid. (**B**) Comparison with model results assuming hydrate is part of the sediment frame or gas to be patchily distributed in the pore space. For both (**A**) and (**B**), model values calculated at core depths and results fit with smoothed curves. (**C**) Comparison of hydrate concentration estimates from pore water chlorinity data and results from the elastic model assuming hydrate becomes a load bearing component of the sediment.

In FIGURE 3 we compare compressional wave velocity from VSP data[3] to calculations made using assumptions (a) and (b) in our model (FIGS. 3A and 3B, respectively). Again we see that the hydrate in fluid model predicts significantly more hydrate. In FIGURE 3C we compare the inversion of the hydrate in frame model results shown in FIGURE 3B to gas hydrate concentration estimates derived from chlorinity data.[26] The model results shown in FIGURES 3A and 3B are the same as those in FIGURES 2A and 2B. Differences in estimated hydrate concentration and distribution between FIGURES 2 and 3 are due solely to the difference in velocity detected by the two (sonic and VSP) measurement techniques.

Although there are differences in distribution and absolute amount of hydrate predicted by the hydrate in frame model based on the different velocity data sets (sonic and VSP), the results are of the same magnitude and general distribution. Additionally, the hydrate in frame modeling results compares favorably with the concentration estimates derived from independent data (i.e., resistivity, chlorinity, and evolved gas). Therefore, we conclude that the hydrate present at ODP site 995 acts as a load bearing component of the sediment frame and comprises on average 2–4% of the sediment by volume, with peak concentrations up to about 9%.

CONCLUSIONS

We present a new physics-based technique for calculating the elastic moduli of shallow unconsolidated marine sediments containing gas hydrate or free gas. The theory is based on first-principle physics and uses only physics-based input parameters. The model can be used to estimate hydrate and/or gas concentrations in ocean bottom sediments from velocity data, provided accurate porosity, solid phase density and solid phase elastic moduli can be found.[18] Applied to porosity, mineralogy and velocity data obtained at ODP Site 995, this model predicts hydrate acts as a load bearing component of the sediment frame from about 190–450 mbsf, comprising on average 2–4% of the sediment by volume with peak concentrations of 8–9%. Bulk gas concentrations potentially as high as 7% are indicated in the sediments beneath the hydrate.

REFERENCES

1. SLOAN, E.D., JR. 1998. Clathrate Hydrates of Natural Gases, 2nd edit. Marcel Dekker, Inc., New York. 705.
2. KVENVOLDEN, K.A. 1993. Gas hydrates as a potential energy resource—a review of their methane content. USGS Prof. Paper 1570, 555–561.
3. HOLBROOK, W.S. et al. 1996. Methane hydrate and free gas on the Blake Ridge from vertical seismic profiling. Science **273**: 1840–1843.
4. DICKENS, G.D. et al. 1997. Direct measurement of in situ methane quantities in a large gas hydrate reservoir. Nature **385**: 426–428.
5. WAITE, W.F. et al. 1999. Laboratory measurements of compressional and shear wave speeds through methane hydrate. Ann. N.Y. Acad. Sci. **912**: this volume.
6. WYLLIE, M.R.J. et al. 1956. Elastic wave velocities in heterogeneous and porous media. Geophysics **21**: 41–70.
7. WOOD, A.B. 1941. A Textbook of Sound. G. Bell and Sons, Ltd., London.
8. PEARSON, C. et al. 1986. Acoustic and resistivity measurements on rock samples containing tetrahydrofuran hydrates: laboratory analogues to natural gas hydrate deposits. JGR **91**: 14132–14138.
9. MILLER, J.J. et al. 1991. An analysis of a seismic reflection from the base of a gas hydrate zone, offshore Peru. AAPG Bull. **75**: 910–924.
10. BANGS, N.L. et al. 1993. Free gas at the base of the gas hydrate zone in the vicinity of the Chile triple junction. Geology **21**: 905–908.
11. SCHOLL, D.W. & P.E. HART. 1993. Velocity and amplitude structures on seismic-reflection profiles—possible massive gas-hydrate deposits and underlying gas accumulations in the Bering Sea Basin. USGS Prof. Paper 1570. 331–351.
12. MINSHULL, T.A. et al. 1994. Seismic velocity structure at a gas hydrate reflector, offshore western Colombia, from full waveform inversion. JGR **99**: 4715–4734.
13. WOOD, W.T. et al. 1994. Quantitative detection of methane hydrate through high-resolution seismic velocity analysis. JGR **99**: 9681–9695.
14. LEE, M.W. et al. 1996. Seismic velocities for hydrate-bearing sediments using weighted equation. JGR **101**: 20,347–20,358.
15. COLLETT, T.S. 1998. Well log evaluation of gas hydrate saturations. Transactions of the Society of Professional Well Log Analysts, 39th Annual Logging Symposium, Paper MM.
16. DVORKIN, J. & A. NUR. 1998. Time-average equation revisited. Geophysics **63**: 460–464.
17. HYNDMAN, R.D. & G.D. SPENCE. 1992. A seismic study of methane hydrate marine bottom simulating reflectors. JGR **97**: 6683–6698.
18. DVORKIN, J. et al. 1999. Elasticity of marine sediments: Rock physics modeling. Geophys. Res. Lett. **76**: 1781–1784.
19. GASSMANN, F. 1951. Elasticity of porous media: Uber die elastizitat poroser medien. Vierteljahrsschrift der Naturforschenden. Gesselschaft **96**: 1–23.
20. NUR, A. et al. 1998. Critical Porosity: A key to relating physical properties to porosity in rocks. The Leading Edge **17**: 357–362.
21. MINDLIN, R.D. 1949. Compliance of elastic bodies in contact. Trans. SDME **71**: A-259.
22. DVORKIN, J. & A. NUR. 1996. Elasticity of high-porosity sandstones: theory for two North Sea datasets. Geophysics **61**: 1363–1370.

23. HILL, R. 1952. The elastic behavior of crystalline aggregate. Proc. Phys. Soc. (Lond.) **A65**: 349–354.
24. REUSS, A. 1929. Berechnung der Fliessgrenzev von Mischkristallen auf Grund der Plastizitätsbedingung für Einkristalle. Zeitschrift für Angewandte Mathematik und Mechanik **9**: 49–58.
25. DVORKIN, J. & A. NUR. 1998. Acoustic signatures of patchy saturation, Int. J. Solids Structures. **35**: 4803–4810.
26. PAULL, C.K., R. MATSUMOTO, P.J. WALLACE *et al.* 1996. Proc. ODP, Init. Repts. 164. College Station, TX (ODP).
27. MAVKO, G. *et al.* 1998. The Rock Physics Handbook: Tools for Seismic Analysis in Porous Media, Cambridge University Press, New York.
28. BATZLE, M. & WANG, Z. 1992. Seismic properties of pore fluids. Geophysics **57**: 1396–1408.
29. SYCHEV, V.V. *et al.* 1987. Thermodynamic Properties of Methane. Hemisphere Publishing Corp, Washington.
30. MURPHY, W.F. III. 1982. Effects of Microstructure and Pore Fluids on the Acoustic Properties of Granular Sedimentary Materials. Ph.D. Dissertation, Stanford University.
31. HELGERUD, M.B. *et al.* 1999. Laboratory measurement of compressional and shear wave speeds through methane hydrate. EOS Trans. Suppl. **80**: T51A-03.

A Double Gas-Hydrate Related Bottom Simulating Reflector at the Norwegian Continental Margin

KARIN ANDREASSEN,[a,b] JÜRGEN MIENERT,[b] PETTER BRYN,[c] AND SATISH C. SINGH[d]

[b]*University of Tromsø, Department of Geology, N-9037 Tromsø, Norway.*

[c]*U&P Division, Norsk Hydro, N-0246 Oslo, Norway*

[d]*Bullard Laboratories, University of Cambridge, Cambridge CB3 0EZ, United Kingdom*

ABSTRACT: An unusual pattern of two bottom simulating reflections (BSRs) has been observed on seismic profiles from the continental margin offshore Western Norway. One of these reflections (BSR1) extends over large areas and has the characteristics of the classical BSR, that is a phase-reversed reflection from the base of the gas-hydrate stability zone. The second BSR (BSR0) occurs at approximately 70 ms two-way travel time beneath BSR1 and is here called a double BSR. The distribution of BSR0 is more local than that of BSR1 and it does not show the phase-reversal relative to the sea floor reflection that is characteristic for a BSR at the gas hydrate–free gas boundary. Results from an industrial borehole, from full waveform inversion of multichannel seismic data, from high-frequency ocean bottom hydrophones, and interpretation of seismic profiles, clearly indicate that BSR1 is reflected from the base of the methane hydrate equilibrium field. Results from full waveform inversion indicate that BSR0 corresponds to a 16–20 m zone where the velocity drops from about 1.8 km/s to a minimum of 1.4 km/s and then increases again. The low velocity of 1.4 km/s suggests the presence of free gas. The results support the hypothesis that BSR0 is a reflection from the base of gas hydrates containing hydrocarbons with a heavier molecular weight in addition to methane gas. Interference of reflections from the top and base of the low-velocity zone associated with BSR0 explain why BSR0 is not phase-reversed.

INTRODUCTION

A bottom simulating reflection (BSR) at a depth corresponding approximately to the base of the methane hydrate equilibrium field is the most widely used indicator of the presence of gas-hydrate bearing sediments beneath the sea floor. BSRs are observed worldwide on reflection seismic profiles from continental margins.[1] The most recent Ocean Drilling Program (ODP) drillings penetrating gas hydrate related BSRs,[2–4] and recent seismic investigations studying BSRs[5–7] have indicated that the BSR is mainly due to small amounts of free gas beneath the hydrate stability, and to a lesser degree, to gas hydrate in the sediments above the BSR. The total amount of

[a]Telecommunication. Voice: +47 77 644420; fax: +47 77 645600.
karina@ibg.uit.no

bulk gas hydrate to be expected within the gas hydrate stability zone seems to be about 1–3%, combined with about 1–5% of sediment bulk volume of free gas below the BSR.[8]

Hydrate related BSRs are characterised by reversed polarity relative to the sea floor reflection, indicating a decrease in acoustic impedance and, hence, most probably in velocity, since density is not expected to be affected significantly by the presence of hydrocarbons or gas in the sediments. The seismic image of the BSR depends greatly on the frequencies used. The use of a very high-frequency (250–650 Hz) deep-towed reflection system (DTAGS) on the Blake Ridge revealed, for example, that the BSR was not the continuous reflection that it appeared to be from more conventional reflection seismic data (10–240 Hz). Instead, it appeared as an abrupt change in reflection amplitude of dipping reflections, from anomalously high amplitude below to very low amplitude above the BSR.[9]

The presence of gas hydrates and free gas in sediments beneath the sea floor of the continental margin of Western Norway has been documented from scientific and industrial data of various frequencies and reflection angles.[10–12] The first documentation of a double hydrate-related BSR pattern was from this area (see FIGURE 1), recorded by Posewang and Mienert,[12] who concluded that the lower BSR segment may represent a relict of former changes of the hydrate stability field (from glacial to interglacial times), or the base of gas hydrates with a gas composition including heavier hydrocarbons than methane. A similar double BSR pattern has recently been observed offshore Japan on high-resolution reflection seismic lines from the Nankai Trough.

A geotechnical industry drilling in 1997 through the main upper BSR (FIG. 1B; BSR1) provided, by correlation with seismic data, new information about the gas-hydrate related BSR that is presented here. The nature of the lower, double BSR (FIG. 1; BSR0), which was not penetrated by the drilling, was studied by applying the full waveform inversion method of Singh and coworkers, a method has proven to be reliable for investigating the velocity structure of hydrate related BSRs.[5,13,14]

GAS HYDRATE AND ASSOCIATED FREE GAS INDICATED FROM ACOUSTIC DATA

The strong bottom-simulating reflection (FIG. 1; BSR1) has, throughout a grid of seismic profiles, been identified over large areas of the Mid-Norwegian continental margin. BSR1 is subparallel to the sea floor reflection, it cuts across reflections from bedding planes, and it exhibits reversed polarity relative to the sea floor reflection (see FIGURE 2). These are characteristics of a seismic reflection from the base of the gas hydrate stability zone. P-wave velocity profiles from high-frequency ocean bottom hydrophone (HF-OBH)-data from the nothern rim of the Storegga Slide[11] is shown in FIGURE 3.

BSR0 occurs as a reflection segment 0.06–0.07 s TWT (approximately 45 m) below BSR1, and is here called a *double BSR*. Since this reflection is observed on seismic data recorded with different acquisition systems it seems unlikely that it is an artefact,[12] and it does not seem, from its distribution and from geometrical considerations, to be a multiple reflection. The BSR0 segment crosses reflections from

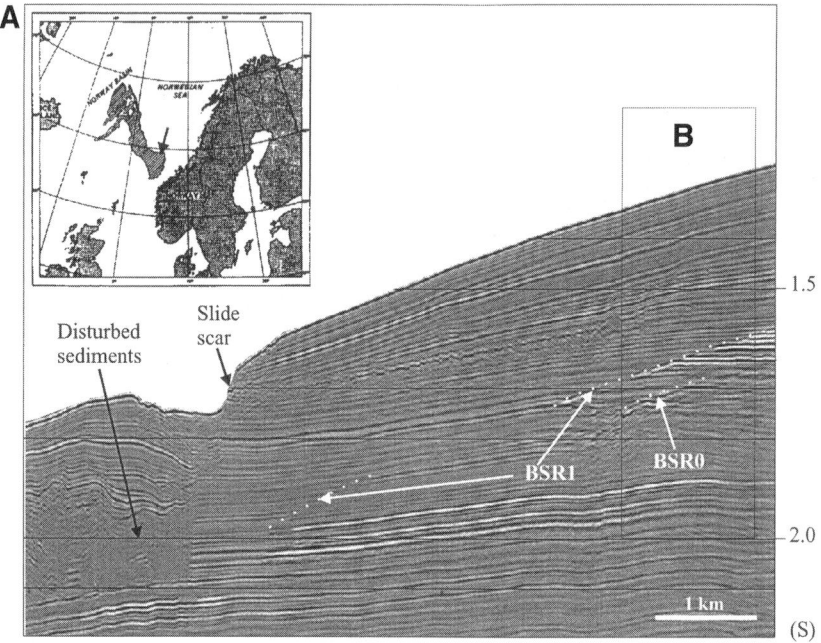

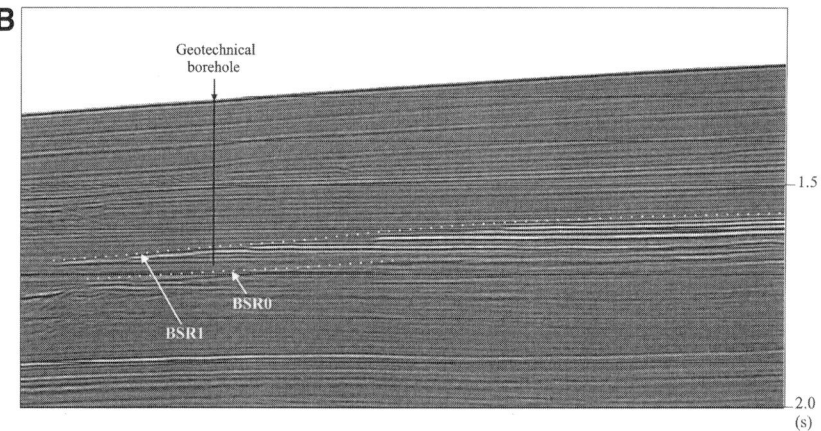

FIGURE 1. A. Stack section of multichannel seismic profile NH9651-202. Location of the section is indicated by the *arrow* on the small map, which also shows the Storegga Slide (*shaded area* offshore of Norway). Acquisition parameters for this line are shown in TABLE 1. **B.** Stack section of multichannel seismic line NH9651-202. Location of the geotechnical borehole 6404/5-GB1A is indicated on the profile. Location of the section is indicated in **A**.

sedimentary strata, and is subparallel to the sea floor reflection and to BSR1. BSR0 seems to have the same reflection polarity as the sea floor reflection (FIG. 2).

DRILLING RESULTS

An industrial geotechnical bore made in 1997 penetrated the western flank of BSR1 (FIG. 1B). A pressure core sampler was not used. Gas hydrates were not recovered. Gas dissolved in the sediment pore space was indicated at several depths. The P-wave velocity trend obtained from vertical seismic profiling (VSP) results from the borehole is shown in FIGURE 4. Water content measured in the borehole sediments is indicated in FIGURE 5.

FULL-WAVEFORM INVERSION

The full-waveform inversion technique developed by Singh and co-workers[5,13,14] was used to study a supergather of four common midpoint gathers (CMP) containing the geotechnical borehole of line NH9651-202 (FIG. 2). This location is suitable for

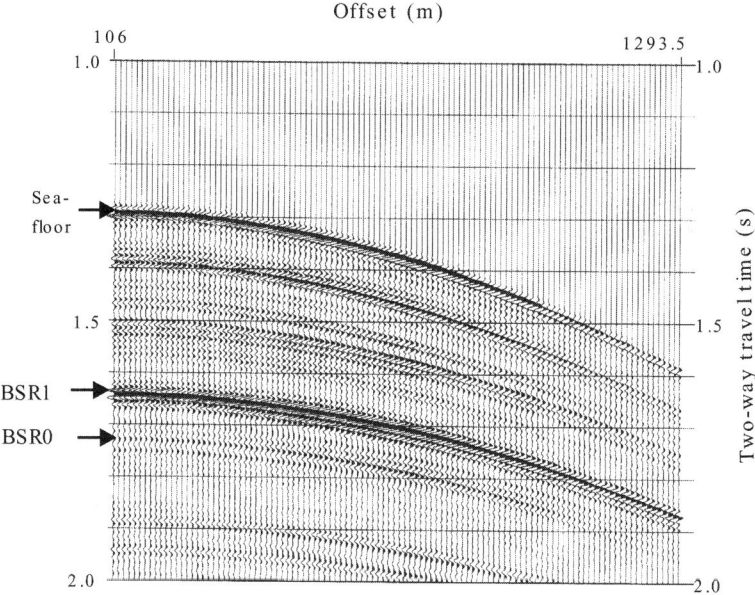

FIGURE 2. Supergather of four CMPs covering the geotechnical orehole 6404/5-GB1A (location of the borehole is indicated in FIGURE 1B) displaying the seismic reflections from the seabed, BSR1, and the underlying BSR0. The sea floor reflection and BSR0 exhibit the same reflection polarity, whereas BSR1 shows reversed polarity.

applying the one-dimensional full-waveform inversion technique, since both the sea floor and the two BSRs are almost horizontal (about 1.5 degree slope) and a one-dimensional structure can be assumed. In this study the approach was to use the filtered long-wavelength velocity-trend obtained from zero-offset VSP studies in the well (FIG. 4) and densities obtained from the well logging as input for the inversion. S-wave velocities were calculated from the relation of Castagna *et al.*[5] Seismic quality factors Q, describing attenuation were set to typical values for marine sediments (Q_p = 200 and Q_s = 100). Below the base of the borehole the velocity was increased uniformly from the value of 1.628 km/s, measured at the borehole base (1.263 km), to 1.8 km/s at 1.5 km, since the VSP report provided no velocity estimation for this depth.

The inversion procedure was used to minimise the misfit between the data and the calculated wave field in the frequency-slowness domain, sample by sample, using a local iterative search algorithm (conjugate gradient) method.[16] The final best-fit velocity-profile from the full waveform inversion is shown in FIGURES 5B and 6A.

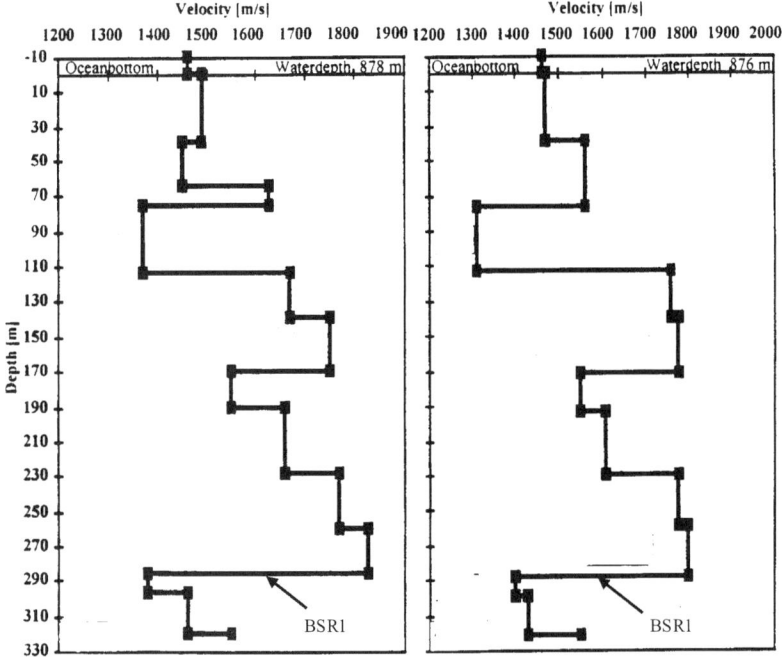

FIGURE 3. V_p-depth profiles obtained from the two high-frequency ocean bottom hydrophone stations HF-OBH1 and HF-OBH2 from the Northern rim of the Storegga Slide.[11]

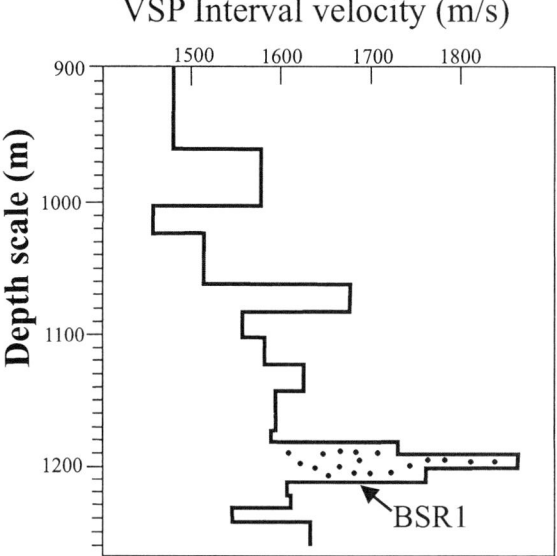

FIGURE 4. P-wave velocity-trend obtained from vertical seismic profiling (VSP) results from borehole 6404/5-GB1A (Statoil Report SP22-RE-01R-00000-98). The *dotted areas* indicate zones with relatively high velocity.

DISCUSSION AND CONCLUSIONS

BSR1

BSR1 shows all the seismic characteristics of a reflection from the interface between sediments partly saturated with gas hydrates overlying sediments containing free gas. Pressure and temperature conditions place BSR1 at the approximate base of the stability zone for a methane hydrate and sea water (see FIGURE 7). The velocity profile from the full waveform inversion (FIG. 6A) indicates that the BSR is produced by an abrupt drop in velocity to 1.3–1.4 km/s, which is most likely due to the presence of free gas.[17] This observation is in good accordance with velocities obtained from the HF-OBH- (FIG. 3) and the VSP-data (FIG. 4); additionally, the inversion provides better resolution both in velocity and time/depth. The low-velocity zone beneath BSR1 is at the drill site location (FIG. 1B), estimated to be thin (8–10 m) both by the VSP velocities (FIG. 4) and by the inversion velocities (FIG. 6A). The relatively high velocity of the two zones above BSR1 (FIG. 5B; *dotted areas*) is probably at least partly due to gas hydrate in the sediment pore space. This conclusion is supported by the low water content borehole measurements in these zones (FIG. 5A). Pore-water freshening is assumed to be a useful signal for the presence of gas hydrates.[18] Low water content has been measured in gas hydrate bearing sediments from several locations (e.g., the Mallik 2L-38 well, Mackenzie Delta,

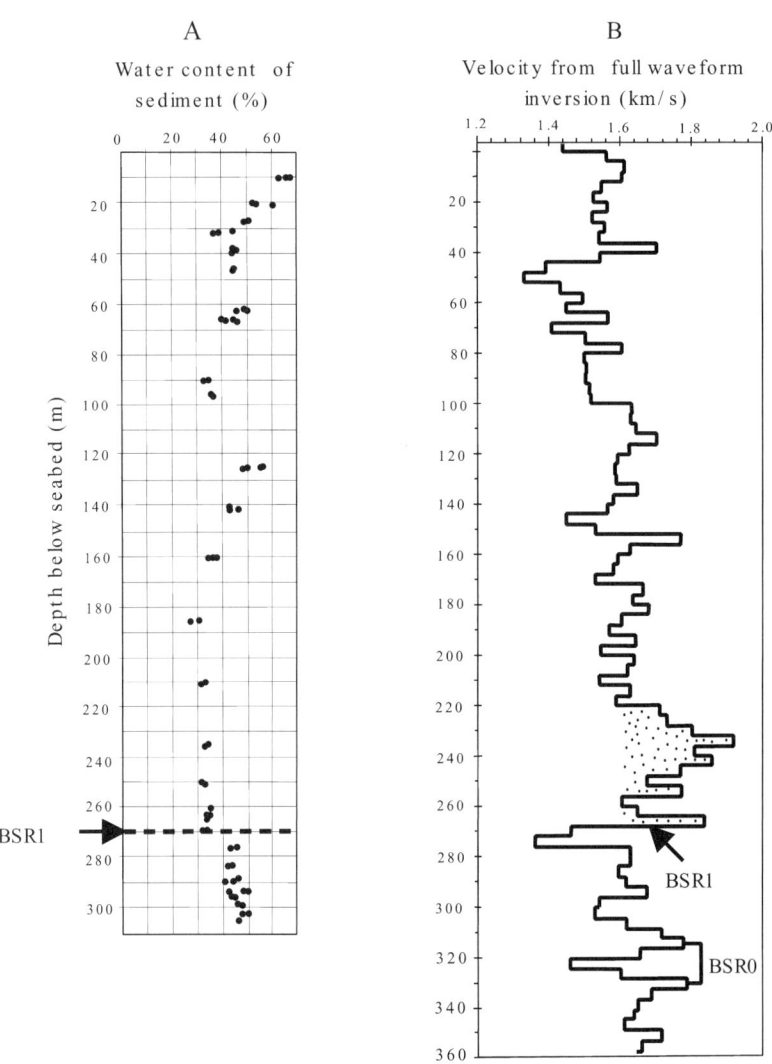

FIGURE 5. A. Measurements of water content from borehole 6404/5-GB1A. **B.** Final interval velocity trend from best fit inversion plotted against depth. *Dotted areas* indicate zones with relatively high velocities.

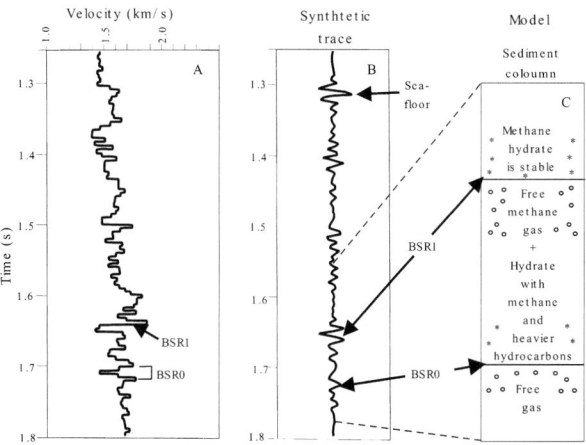

FIGURE 6. A. Final interval velocity trend from best-fit inversion plotted against two-way travel time. **B.** Synthetic seismogram (slowness 0.05 s/km) constructed using the final velocity trend of **A**. **C.** Suggested geological model.

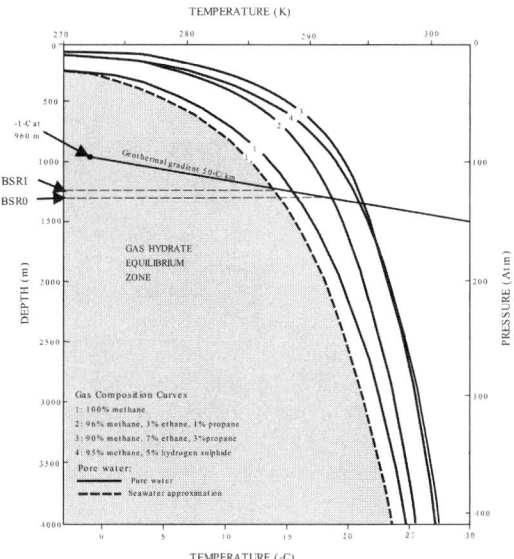

FIGURE 7. Equilibrium diagram for methane hydrate (*shaded*) and for hydrates of varing composition (*curves* 2–4), obtained from compilation of Sloan.[21] The sea floor temperature of –1°C and geothermal gradient of 50°C/km obtained from borehole 6464/5-GB1A.

Canada[19]), and experimental studies of gas hydrate bearing sediments have showed a decrease in water content of sand samples after formation of gas hydrate.[20]

The Double Bottom-Simulating Reflection, BSR0

BSR0 is, like BSR1, subparallel to the sea floor reflection, cutting across reflections from bedding planes. It is assumed to be a real reflection and not an acoustic artefact. Three potential explanations for BSR0 have been evaluated:

1. It is a paleo gas-hydrate BSR, a reflection from a diagenetic relict of a former set of pressure/temperature-conditions.

2. It is a reflection from the base of sediments partially saturated with gas hydrates that have a gas composition different from that of the hydrate associated with the overlying BSR1. Small amounts of heavier hydrocarbons, in addition to methane, shift the base of the gas hydrate equilibrium zone to higher temperatures and hence deeper in the sediments (FIG. 7).

3. It is a reflection from a diagenetic surface that is not associated with gas hydrates (e.g., quartz diagenesis).

The supergather of FIGURE 2 and from the synthetic seismogram of FIGURE 6B clearly indicate that BSR0 has the same reflection polarity as the sea floor reflection, suggesting that BSR0 represents a positive acoustic impedance contrast. The inverted velocity trend of FIGURE 6A, combined with the synthetic seismogram of FIGURE 6B indicate, however, that BSR0 is not a reflection from one single interface, but corresponds to a 16–20 m zone where the velocity drops from about 1.8 km/s to a minimum of 1.4 km/s before it increases again. The low velocity of 1.4 km/s suggests that free gas is also present in the sediments at the depth of BSR0. The existence of free gas beneath BSR0 fits well with the model for hypothesis (2), but not so easily with the models for hypotheses (1) and (3). Hypothesis (2), where BSR0 is a reflection from the base of sediments partially saturated with gas hydrates that have a gas composition different to that of the hydrate associated with the overlying BSR1 (FIG. 6C), seems, therefore, to be the most likely explanation for BSR0.

ACKNOWLEDGMENTS

The University of Tromsø acknowledges support of this research by Landmark Graphics Corporation via the Landmark University Grant Program. The seismic inversion was carried out at University of Cambridge, where the first author spent a sabbatical, supported by a grant from the Norwegian Research Council for Science and Humanities (NFR) and by the Seabed Project of oil companies working in deep water offshore of Western Norway. We offer our sincere thanks.

REFERENCES

1. KVENVOLDEN, K.A. *et al.* 1993. Worldwide distribution of subaquatic gas hydrates. Geo-Marine Lett. **13**: 32–40.
2. BANGS, N.L.B. *et al.* 1993. Free gas at the base of the gas hydrate zone in the vicinity of the Chile Triple Junction. Geol. **21**(10): 905–908.

3. MACKAY, M. *et al.* 1994. Origin of bottom simulating reflectors: Geophysical evidence from the Cascada accretionary prism. Geol. **22:** 459–462
4. HOLBROOK, W.S. *et al.* 1996. Methane hydrate and free gas on the blake ridge from vertical seismic profiling. Sci. **273:** 1840–1843.
5. SINGH, S.C. *et al.* 1993. Velocity structure of a gas hydrate reflector. Sci. **260:** 204–207.
6. PECHER, I., C.R. RANERO, R. VON HUENE, T.A. MINSHULL & S. SINGH. 1998. The nature and distribution of bottom simulating reflectors at the Costa Rica convergent margin. Geophys. J. Int. **133:** 219–229.
7. TINIVELLA, U.E. *et al.* 1998. Seismic tomography strudy of a bottom simulating reflector off the south Shetland Islands (Antarctica). *In* Gas Hydrates: Relevance to World Margin Stability and Climate Change. J.-P. Henriet & J. Mienert, Eds. Geol. Soc., London, Spec. Publ. **137:** 141–151.
8. HOVLAND, M. *et al.* 1997. Gas hydrate and free gas volumes in marine sediments: example from the Niger Delta front. Marine Petroleum Geol. **14**(3): 245–255.
9. WOOD, W.T. & J.F. GETTRUST. 1998. Evidence for concentration of methane hydrate from fluid flux along faults. Fifth International Conference on Gas in Marine sediments. Bologna Italy, September 9-12, 1998. Abstracts and Guide Book 56–59.
10. BUGGE, T. *et al.* 1987. A giant three-stage submarine slide off Norway. Geo-Marine Letters. **7:** 191–198.
11. MIENERT, J. *et al.* 1998. Gas hydrates along the northeastern Atlantic margin: possible hydrate-bound margin instabilities and possible release of methane. *In* Gas Hydrates: Relevance to World Margin Stability and Climate Change. J.-P. Henriet & J. Mienert, Eds. Geol. Soc., London, Spec. Publ. **137:** 275–291.
12. POSEWANG, J. & J. MIENERT. 1999. The enigma of double BSRs: Indicators for changes in the hydrate stability field? Geo-Marine Lett. **19:** 150–156,
13. SINGH, S.C. & T.A. MINSHULL. 1994. Velocity structure of a gas hydrate reflector at Ocean Drilling Program Site 887 from a global seismic waveform inversion. J. Geeophys. Res. **99**(B12): 24221–24233.
14. MINSHULL, T.A. *et al.* 1994. Seismic velocity structure at a gas hydrate reflector from full waveform inversion. J. Geophys. Res. **99:** 4715–4734.
15. CASTAGNA, J.P. *et al.* 1985. Relationship between compressional- and shear-wave velocities in clastic silicate rocks. Geophys. **50:** 571–581.
16. KORMENDI, F. & M. DIETRICH. 1991. Non-linear waveform inversion of plane-wave seismograms in stratified elastic media. Geophys. **56:** 664–674.
17. DOMENICO, S.N. 1977. Elastic properties of unconsolidated porous sand reservoirs. Geophys. **50:** 571–581.
18. KVENVOLDEN, K.A. 1993. Gas hydrates—geological perspective and global change. Rev. Geophys. **31**(2): 173–187.
19. COLLETT, T.S., T. UCHIDA, S.R. DALLIMORE, T.H. MROZ & M.W. LEE. 1998. Downhole well log evaluation of gas hydrates in the Mallik 2L-38 well, Mackenzie Delta, N.W.T., Canada. International Symposium on Methane Hydrates. Resources in the Near Future? JNOC-TRC, Chiba City, Japan, October 29–22, 1998. Proceedings 349–358.
20. CHUVILIN, E.M. & V.S. YAKUSHEV. 1998. Structure and some properties of frozen hydrate-containing soils. International Symposium on Methane Hydrates. Resources in the Near Future? JNOC-TRC, Chiba City, Japan, October 29–22, 1998. Proceedings 239–252.
21. SLOAN, E.D. 1990. Clathrate Hydrates of Natural Gases, 1st edit. Marcel Dekker, Inc. New York and Basel.

Methane Hydrate Accumulation along the Western Nankai Trough

YUTAKA AOKI,[a] SHOSHIRO SHIMIZU,[a] TERUMASA YAMANE,[a]
TOMOYUKI TANAKA,[a] KAZUO NAKAYAMA,[a] TSUTOMU HAYASHI,[a]
AND YOSHIHISA OKUDA[b]

[a]*JGI, Inc., 1-5-21, Otsuka, Bunkyo-ku, Tokyo, Japan*

[b]*Geological Survey of Japan, 1-1-3, Higashi, Tsukuba-shi, Ibaraki, Japan*

ABSTRACT: A one-dimensional gas hydrate accumulation model is developed to simulate the existence of hydrate in the Nankai Trough margin, where the total organic carbon content is extremely low. In order to study the accumulation of gas hydrate along a seismic profile, our model assumes an initial 1000 m thick turbidite sedimentation at the trench axis, and then the sediment experiences tectonic uplift to form the Nankai accretionary prism. Methane generated by microbial processes during the sedimentation is partly trapped in the pore space and partly migrated upwards due to compaction and subsequent porosity decrease at deeper levels. Upward methane gas flux from the deeper part (below 1000 m) is predicted by the model because the maximum thickness of the accreted sediment exceeds ten kilometers in the Nankai Trough accretionary prism. The basic geological parameters input to the model are those obtained from the ODP site 808, which was drilled at the lower inner trench slope. The methane hydrate stability zone is essentially controlled by bottom water temperature, regional heat flow and hydrostatic pressure. The simulation shows that methane hydrate will not accumulate in the Nankai Trough margin if only microbial methane generation is considered because the organic carbon content in the Nankai Trough area is as low as 0.75%. However, if we assume an upward flux of 5 kg of methane per square meters per 10,000 years, it will cause approximately 10 to 25% hydrate saturation under current physical conditions. The origin of the methane may be thermogenic and/or microbial. The resultant distribution of methane hydrate shows very good correlation with the distribution of the BSR.

INTRODUCTION

The Nankai Trough is formed at the convergent margin between the Philippine and Eurasian plates (see FIGURE 1), where the Philippine plate is subducting at an angle of approximately 10 degrees and a speed of about 4 cm per year.[1] This plate boundary is characterized by a well developed accretionary prism, seen on multi-channel seismic sections, in particular in the Western Nankai Trough.[2,3] This region is one of the most extensively studied convergent margins in the world and it has been established that a thick pile of trench fill of turbidite origin and hemipelagic Shikoku basin sediment are accreted to form the prism. A well developed decollement and a series of thrust faults also characterize this margin. ODP site 808 was drilled through the lower trench slope and reached the basaltic oceanic basement.[4] The total penetration was more than 1200 meters below the sea bottom. Unlike other

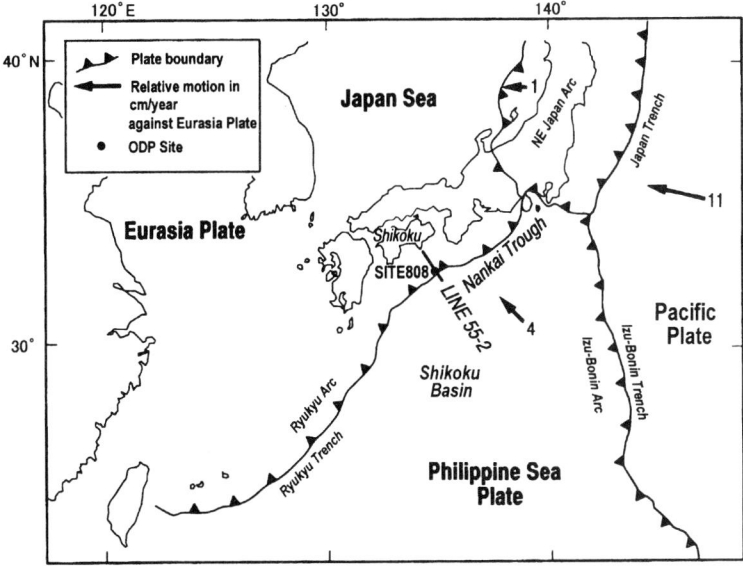

FIGURE 1. Index map of location (modified from Leg131 Shipboard scientific party, 1991).

trench systems in the Northwestern Pacific, such as the Japan and Ryukyu trenches, a wide spread occurrence of gas hydrate has been inferred in the inner trench slope by the appearance of a bottom simulating reflector (BSR). The ODP drilling results indicated actual occurrence of gas hydrates.[4] It is believed that the Pacific side of the Japanese island arc is not prospective from the petroleum point of view because organic carbon content is generally very low and thermal conditions are not favorable to maturation of the source materials. Indeed, the major Japanese oil and gas fields, although small in size, are located along the Sea of Japan side of the island and Hokkaido, the Northernmost island of Japan. Furthermore, total organic carbon in the sediment of the Nankai Trough is reported to be less than one percent indicating a very small possibility of hydrocarbon generation.[5] Nonetheless, the BSR is most significant both in intensity and extent in the Nankai Trough area when compared to other regions. This is generally accounted for by the process of sediment accretion and the associated migration of gas from the deeper part. A similar mechanism has been reported from other accretion type convergent margins including Cascadia.[6] Here we simulate the accumulation of gas hydrate with a one-dimensional model.

METHOD

In order to evaluate the accumulation of methane hydrate, a basin simulation technique as described by Nakayama[7] was modified and applied. In the original

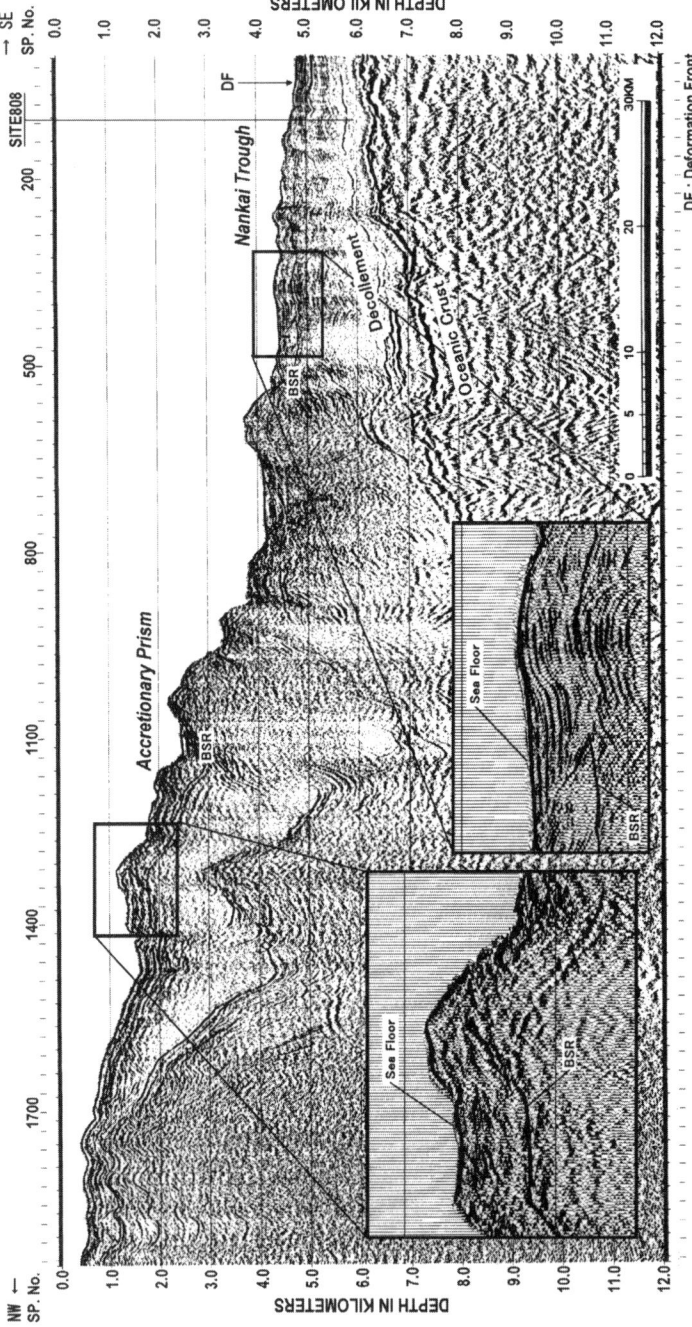

FIGURE 2. Seismic profile of LINE 55-2 with depth (vertical magnification 400%). ODP Site 808, which is offset a few hundreds meters from the section, is projected.

model, the geologic section is represented by a system of cells. The burial of each cell and its associated parameters are calculated for a certain time interval defined by the geological age assigned to each individual cell. Geologic and physical properties are assumed to be uniform within a cell. In the modified model, microbial methane generation and migration in the shallow sediment below sea bottom are taken into account, as well as the migration of thermogenic methane. It is assumed that methane hydrate is formed only when the amount of methane exceeds the solubility in the pore water. The locations to be studied by simulation were chosen every 5 km (100 shot points interval) on the seismic section, LINE 55-2, which crosses the Nankai Trough almost perpendicularly from Southeast to Northwest (FIG.1). This section is illustrated in FIGURE 2. All 19 locations were studied, from SP106 to SP1906. SP106 is located near the trench axis or deformation front (DF) and SP1906 is located on the upper trench slope. The BSR is developed on most of the inner trench slope but it cannot be identified at the ODP Site 808. The upper boundary of the subducting oceanic plate or basaltic basement is traceable on the seismic section up to 90 km from the trench axis. Most of the parameter values necessary for this simulation are assumed to be the same as the well data obtained at ODP Site 808. This well is located a few hundred meters to the west of SP106 of LINE 55-2 and detailed geological data is available including, formation density, grain size of matrix, TOC, porosity, sea bottom water temperature, and geothermal gradient. The numerical values of some input parameters were decided a priori because of no information was available. The cell size of the simulation was 1 m × 1 m × 25 m (height) in this particular study. It was found that the cell size is not a critical parameter for the simulation, but other parameters proved to be very important. The calculation was limited to the interval between the sea floor and 1000 m below sea floor because the BSR depth and microbial methane generation were limited to this range. Therefore, the total number of cells was 40. To simulate the accretion process, it was assumed that each point has the same burial history as the ODP Site 808 and then experienced continuous uplifting to the present position. Major parameters and input values for the simulation are described below.

Porosity. To define a compaction coefficient, five porosity versus depth values were input for the location SP106, based upon the results from ODP Site 808. They range from 0.6 to 0.26. The same data was assumed for the other locations (from SP206 to SP1906).

Present water temperature of sea bottom. The data is based on the measurement at Site 808 (1.7°C). For shallower locations, statistical data around Japan was used.

Total organic carbon (TOC). Total organic carbon at the time of sedimentation was used. A value of 0.0075 (0.75%), which is close to the average at the ODP Site 808, was used for all locations.

Utilized organic carbon. This parameter refers to the fraction of organic carbon that can become methane. A value of 0.15 was adopted for geochemical reasons.[8]

Methane generation ratio. This parameter defines the cumulative amount (%) of carbon that can change into methane during burial to the corresponding depth. We assume that microbial methane generation is most active in the shallowest part (see FIGURE 3). The methane generation rate has the numerical value between 0% and 100%, where 0 means that organic carbon does not change into the methane at all, and 100 means that all carbon changes into methane at the given depth.

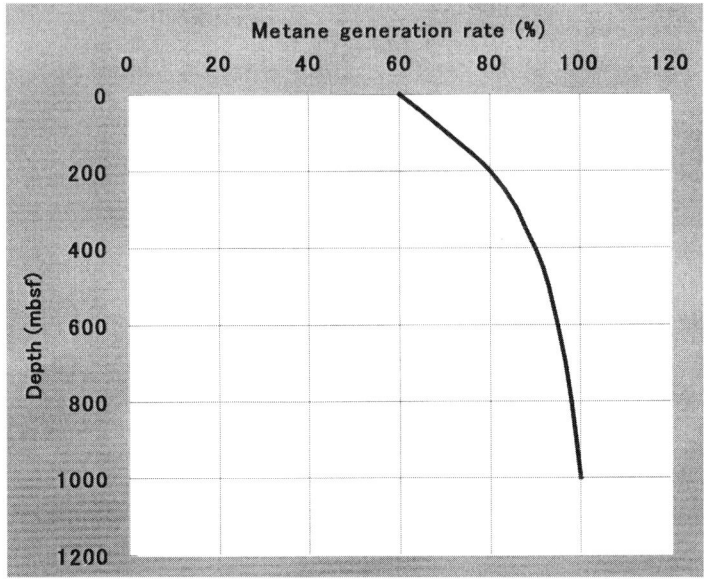

FIGURE 3. Assumed microbial methane generation curve (integration) versus depth.

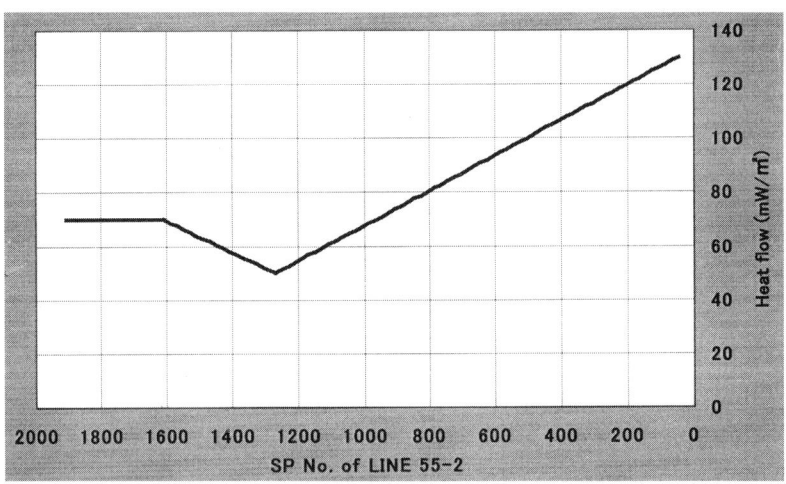

FIGURE 4. Heat flow along seismic line 55-2.

Methane flux from the deep zone. The amount of methane flow per unit area (1 m × 1 m) from the deeper zone. The origin of this methane is not necessarily thermogenic. In the accretionary prism, both trapped microbial methane and thermogenic methane may also migrate upwards. We assumed this flux to be 5 kg per 10,000 years for all locations.

Solubility of methane. The solubility of methane in the pore water was specified as a function of temperature and pressure.

Methane volume coefficient. The volume ratio of decomposed methane to hydrate. A value of 155 was assumed at standard temperature and pressure.

Irreducible water value. The percentage of irreducible water. This parameter was used to calculate capillary pressure that affects gas migration.

Gas hydrate phase boundary. Pure methane and sea water are assumed.

Methane hydrate generation speed. In order to explain the vertical distribution of the hydrate, the concept of methane hydrate generation speed was introduced. This parameter was described in terms of the fraction of methane transformed into hydrate per 10,000 years. Without this parameter, hydrate formed only in the vicinity of the base of hydrate stability zone (BHSZ). Alternatively, the parameter could be replaced by the permeability of the hydrate, but in the simulation methane hydrate was presumed to be completely impermeable. The following parameters are used for all locations.

Depth (mbsf) (m)	Fraction
0	0.002
250	0.0002
500	0.00002
750	0.000002
1,000	0.0000002

Geothermal Gradient. The geothermal gradient, or temperature as a function of depth, must be defined. In the current simulation, the geothermal gradient based on the heat flow data near the Nankai Trough (see FIGURE 4, Yamano and Kinoshita, 1989) was used together with the conductivity at Site 808. The calculated geothermal gradient at SP106 is 13.3°C/100 m at the sea bottom and 6.9°C/100 m at 1000 m below sea floor.

Density. Values from 2.43 to 1.77 were used based on the Site 808.

Grain size of matrix. This parameter was used to calculate capillary pressure. A value of 0.002 cm was assumed, based on the data at Site 808.

Age. The geological age for each cell must be defined in order to calculate the sedimentation rate. The age for location SP106 was determined from the ODP Site 808, where the geological age is well known from core analysis. For other locations, it was assumed that there is movement out towards the upper trench slope at a speed of 0.5 cm per year. This speed is derived from the age of Nabae formation, which outcrops in the Muroto peninsular, Shikoku island, and is known to be a part of an accretionary prism formed about 20 million years ago.[9]

Seal function. Upward migration of methane was inferred from the presence of methane hydrate. In order to simulate this effect, a linear seal function was defined. For example, the hydrated layer will be completely impermeable if the hydrate

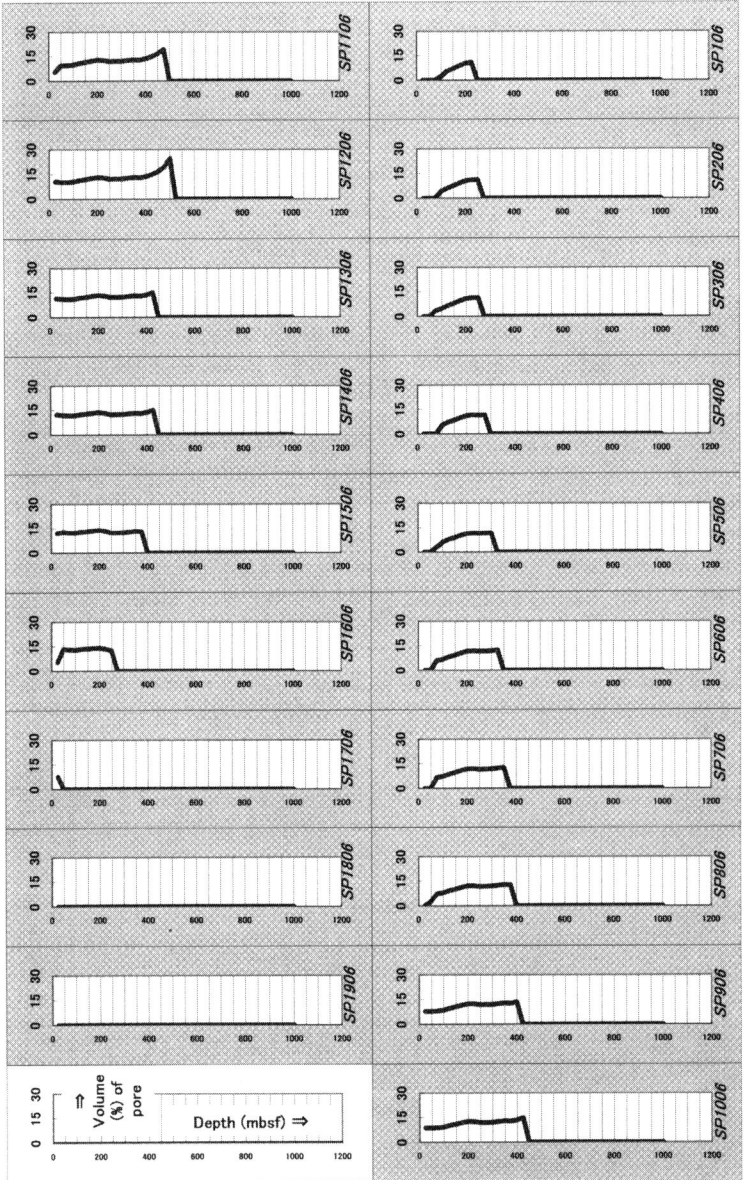

FIGURE 5. Hydrate volume (%) of pore space (saturation) obtained for the locations from SP106 to SP1906.

saturation is 100%, and 50% of methane will migrate during one step of simulation if the hydrate saturation is 50%.

RESULTS AND DISCUSSION

The volume percent (%) of the methane hydrate in the pore space (hydrate saturation) is shown in FIGURE 5. According to the simulation, the methane hydrate extends over almost the entire range of the inner trench slope up to around SP1806 where water depth is no longer sufficient to stabilize the hydrate. A small amount of hydrate should exist around 200-m below the sea floor at SP106. Here no BSR was identified on the seismic section. However, hydrate-cemented sand was recovered between 90–140 m below seafloor at the ODP Site 808.[4] A broader depth distribution of methane hydrate can be seen for the more landward locations. The lower boundary of the methane hydrate distribution corresponds to the BSR depth. The depth of the boundary from the sea floor is not constant but varies in a U-shaped distribution as shown in FIGURE 6. In the figure, the BSR depth identified on the seismic section is also shown. If the water bottom temperature and the geothermal gradient are constant, the BSR depth should become shallower as the water depth decreases. However, this is not the case in the Western Nankai Trough. The main reason for this phenomenon is believed to be the heat flow pattern shown in FIGURE 4. It is known that heat flow values show a local minima on the inner trench slope.[10] The discrepancy between the depth of the methane hydrate base from the simulation and the BSR depth may be due to the uncertainty of the seismic interval velocity used for the depth conversion of the seismic section. LINE 55-2 is not depth migrated but is converted to a depth section by using filtered stacking velocities. The concentration of

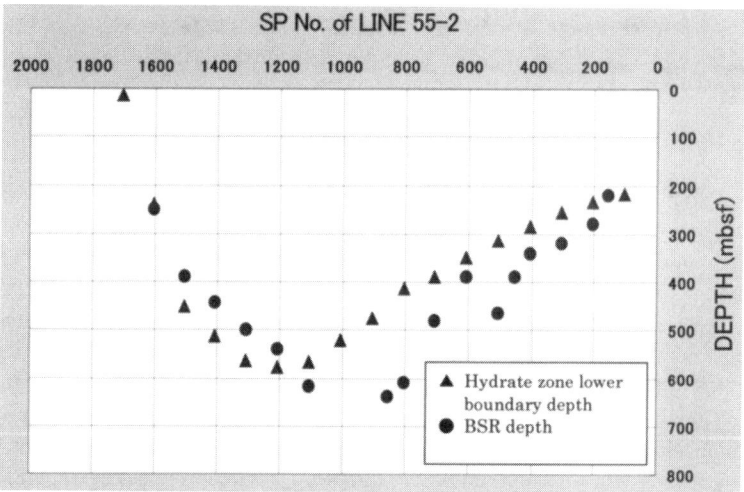

FIGURE 6. BSR depth and lower boundary of hydrate occurrence from the simulation, along seismic line 55-2.

methane hydrate in the stability zone, where it exists, is approximately 10–15% at most locations. However, locations from SP1006 through SP1406 have a very small peak just above the base of hydrate stability zone. This is explained by the upward flux of methane from the deeper zone and the porosity decrease with depth. In FIGURE 5, the maximum hydrate saturation just above the base of hydrate stability zone is approximately 25%. It should be noted that these values are still speculative because we assumed values for a number of important parameters. However, an important point is that methane hydrate will not form in the Nankai Trough margin if only *in situ* methane formation is taken into account. This is clear because of the low TOC content and is demonstrated by the simulation. In the current study we assumed an upward supply of methane of 5 kg/10,000 years from the deeper column. No quantitative data has been observed to enable us to estimate the amount of this flow, but the assumed value of 5 kg/10,000 years per square meters may be considered a minimum. If the flux is smaller than this value, hydrate accumulation or the existence of BSR in the lower trench slope will not occur. This is also verified by the simulation. The hydrate saturation resulting from the simulation is only 10 to 25%. This figure also varies depending on the amount of the upward methane flow. If a larger flux exists, more methane hydrate will concentrate. We do not include the effect of gas charged fluid flow along fractures or faults that are frequently observed as methane rich cold seepages in the Nankai Trough margin. This type of flow is localized and will accelerate the migration of methane in the deeper part but it will reduce the concentration of methane in the shallower zone. If we knew the exact rate of seepage, it would be possible to include this effect in the simulation. As for the vertical distribution of methane hydrate, there is an ambiguity. As already mentioned, we introduced the concept of methane hydrate generation speed. This is quite artificial and values such as 0.1% per 10,000 years are quite slow compared with formation rates measured in the laboratory.[11]

CONCLUSIONS

The simulation shows that methane hydrate accumulation is possible along the Nankai Trough margin despite an organic carbon content as low as 0.75%, provided that the upward migration of methane is sufficient. If we assume a flux of 5 kg of methane per square meters per 10,000 years, this causes approximately 10 to 25% hydrate saturation under current physical conditions. This upward migration of methane may be accelerated by growth of the Nankai accretionary prism, which has a thickness of more than ten kilometers near Shikoku island. Free gas is also trapped beneath the base of hydrate stability zone. The current simulation was based on a one-dimensional burial and uplifted model and we need to assume values for some important parameters because no observations are available yet. However, this study shows that a quantitative study on hydrate accumulation is possible by a proper simulation. Increased sophistication of the model and proper choice of physical parameters will help very much in understanding the occurrence of the natural gas hydrate.

ACKNOWLEDGMENTS

We would like to express appreciation to our colleagues at GSJ and JGI for their valuable discussion and suggestions. We would also like to thank to P. Ward for his critical reading of this manuscript. This study is conducted as a part of the methane hydrate research project by New Energy and Industrial Technology Development Organization, Japan.

REFERENCES

1. KIMURA, S. & K. OKANO. 1998. Source region of the 1946 Nankai earthquake inferred from its aftershock distribution. J. Seis. Soc. Japan. **50:** 461–470.
2. AOKI, Y., T. TAMANO & S. KATO. 1983. Detailed structure of Nankai Trough from migrated seismic sections. AAPG. Mem. **34:** 309–322.
3. MOORE, G.F. et al. 1991. Structural framework of the ODP leg131 area, Nankai Trough. In Proceedings of the Ocean Drilling Program, Initial Reports, 131.
4. TAIRA, A., I. HILL, J. FIRTH et al. 1991. Site 808 Proceedings of the Ocean Drilling Program, Initial Reports, 131.
5. KAGAMI, H., D.E. KARIG, C.S. BARY et al. 1986. Initial Report of Deep Sea Drilling Project. 87A.
6. HYNDMAN, R.D. & E.E. DAVIS. 1992. A mechanism for the formation of methane hydrate and seafloor bottom-simulating reflectors by vertical fluid expulsion. J. Geophys. Res. **97:** 7025–7041.
7. NAKAYAMA, K. 1987. Hydrocarbon-expulsion model and its application to the Niigata Area, Japan. AAPG Bull. **71:** 810–821.
8. WASEDA, A. 1998. Organic carbon content, bacterial methanogenesis, and accumulation processes of gas hydrates in marine sediments. Geochem. J. **32:** 143–157.
9. TAIRA, A. 1981. Process of formation Shimanto belt. Kagaku. **51:** 516–523. (In Japanese).
10. KINOSHITA, H. & M. YAMANO. 1986. The heat flow anomaly in the Nankai Trough area. In H. Kagami et al., Eds. Initial Reports DSDP **87:** 737–743.
11. SLOAN, E.D. 1998. Clathrate Hydrates of Natural Gases. Marcel Dekker, New York.

Resource Evaluation of Marine Gas Hydrate Deposits Using Seafloor Compliance Methods

ELEANOR C. WILLOUGHBY,[a] KONSTANTIN LATYCHEV,[a]
R. NIGEL EDWARDS,[a] AND GEORGE MIHAJLOVIC[b]

[a]*Department of Physics, University of Toronto, Toronto, Ontario, Canada, M5S 1A7*

[b]*Scintrex Ltd., 222 Snidercroft Rd., Concord, Ontario, Canada, L4K 1B5*

ABSTRACT: We introduce the theory and practice of the compliance method: a new tool for assessing offshore methane hydrate deposits. Compliance is defined as the transfer function between the vertical displacement of the seafloor and the corresponding pressure expressed as a function of frequency. It is sensitive to the elastic parameters of the underlying sediments, particularly the shear modulus. We have measured normalized compliance from 0.001 to 0.049 Hz, using ocean surface gravity waves as a source, at sites in Cascadia near the Ocean Drilling Program (ODP) hole 889B. A differential pressure gauge, datalogger, and self-leveling gravimeter were lowered to the seafloor, and each site was occupied for eight hours. The compliance estimates are reproducible and are consistent with other available data and simple models of sediment physical properties. Shear strength is increased from a normal profile in the uppermost few hundred meters, possibly an effect of grain cementation within a known hydrate layer. The magnitude of the increase may be associated with the total mass of hydrate present irrespective of the details of its distribution.

INTRODUCTION

Methane hydrate deposits are expected to become a very important natural energy resource. They are estimated to account for 53% of the total organic carbon in the Earth, twice as much carbon as all the other fossil fuels—coal, oil and existing sources of natural gas—combined. The recovery process is also a possible threat to humans, since methane is a powerful greenhouse gas. The release of as little as 0.065% of the Earth's methane deposits into the atmosphere would be equivalent to doubling the atmospheric CO_2 and would lead to global warming. The temperature increase could cause hydrates to dissociate, creating a positive feedback loop, with potentially catastrophic results. It is clear that the detection and evaluation of the extent of subseafloor methane hydrate deposits is of great importance, not only to determine the potential for resource recovery but also for natural hazard assessment.

PROBLEM

The seismic reflection method is commonly used to identify the existence of a hydrate layer. The depth of the base of the hydrate layer is clearly marked by the bottom simulating reflector (BSR), but the diffuse upper boundary of the hydrated sediment is not clearly defined and thus the total volume of hydrate and its value as a resource cannot be obtained directly from these data. Pure methane hydrate, like ice in permafrost, does change the physical properties of the material in which it is found. There is a small decrease in density, a small increase in the bulk modulus, but a significant increase in the shear modulus. The shear modulus is a sensitive indicator of skeletal structure and strength of sediments, and hence the relative amounts of fluid and hydrate in the pore space, provided the hydrate forms on grain contacts and not simply in pore space. The direct measurement of shear properties requires a source vibrator that shakes the seafloor laterally, a process difficult to implement in soft sediments. P-S converted shear wave velocities can be measured in soft sediments, but few such measurements exist since they are difficult to make.

SOLUTION: SEAFLOOR COMPLIANCE

Our alternative solution is to monitor the palpitations of the seafloor caused by the natural motion of sea surface waves. Wind excited gravity, and coastline excited infragravity surface waves palpitate the seafloor causing a corresponding vertical displacement and acceleration, a method developed originally by Yamamoto and Torii[1] for shallow water and extended by Crawford and others[2] to full ocean depths. The term compliance is given to the transfer function between seafloor deformation and pressure as a function of frequency. The displacement of the seafloor depends on the oceanic crustal density and elastic parameters, particularly the shear properties. Compliance measurements made at specific frequencies are tuned to structure at specific depths. They may be inverted using standard methods to a shear velocity profile as a function of depth, which can in turn be related to the hydrate content, assuming hydrate has formed on grain contacts. The accuracy of the measured data is related to instrument quality, the availability of suitable signals, and the ambient noise level. The surface gravity wave dispersion relation is $\omega^2 = gk\tanh(kH)$, where ω is the angular frequency, k is the wave number, g is the acceleration due to gravity, and H is the water depth. The maximum frequency waves to exert pressure on the seafloor have wavelengths around H (between 0.5 and 2 times the water depth depending on the wave amplitude). Therefore, the cutoff frequency is $f_c = \sqrt{g/2\pi nH}$, $0.5 < n < 2$, and the maximum period is approximately $T = \sqrt{2\pi H/g}$, which, for H equal to 100 m and 1,000 m, is 8 and 25 seconds, respectively.

The principal source of natural noise on all measurements is due to microseismic activity. In the deep ocean, the activity is confined to periods shorter than 10 s. The infragravity wave signal and microseismic noise are therefore in completely separate bands. In water shallower than 100 m, the signal and noise bands start to overlap. The water depth in the region we are studying is typically of the order of 1 km, thus there is no concern about overlap between the microseismic and infragravity frequency bands.

We have computed the theoretical compliance responses of numerical models of hydrated sediments and established the resolution of the method for hydrate assessment and the optimum frequency spectrum.[3] In FIGURE 1 three models of seafloor sediments, with no hydrate (M0), up to 20% of pore space filled by hydrate (M1), and up to 40% of pore space filled by hydrate (M2), and their associated compliance

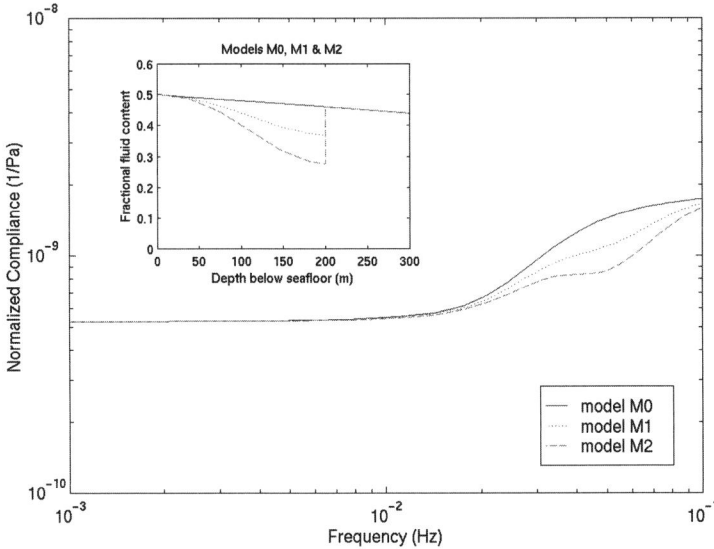

FIGURE 1. Calculated compliance response as a function of frequency for models M0 (no hydrate), M1 (up to 20% of pore space filled with hydrate), and M2 (up to 40% of pore space filled with hydrate). The models are shown *inset*.

response are illustrated. The addition of hydrate stiffens the sediments, increasing the shear modulus and decreasing the compliance function at frequencies related to the associated depth below the seafloor.

We have been asked repeatedly if the three-dimensional, heterogeneous nature of hydrate deposits invalidates any conclusions based on our layered earth modeling. Using finite difference methods, we investigated the response of several three dimensional structures, including a finite sized plate shown in FIGURE 2. The uppermost panel shows the response of a 500 m^2 plate of hydrate in a *numerical box* with a 1 km^2 surface area and periodic boundary conditions on the sides. The concentration of hydrate increases with depth as in layered model M2. For clarity, the stiffness, or the inverse of compliance is plotted, at a frequency of 0.056 Hz, which corresponds to the maximum response to the one dimensional models. Somewhat to our relief, over the centre of the plate the response is comparable to that calculated for layered earth models. At the edge of the plate the response returns smoothly to the background value, much like a gravity anomaly over a massive body. The bottom two panels of FIGURE 2 show the effect of breaking the plate into four pieces and varying the underlap. When the four pieces are separated by 30 m there is just a hint of this separation in the response, but the total amplitude is decreased by 10%. At 100 m separation, the four peaks in stiffness are well separated and the maximum is decreased by a further 10%. This suggests that the compliance method would not be hindered by discontinuities in slabs of hydrate provided that sufficient drop sites are used to sample adequately any major discontinuities.

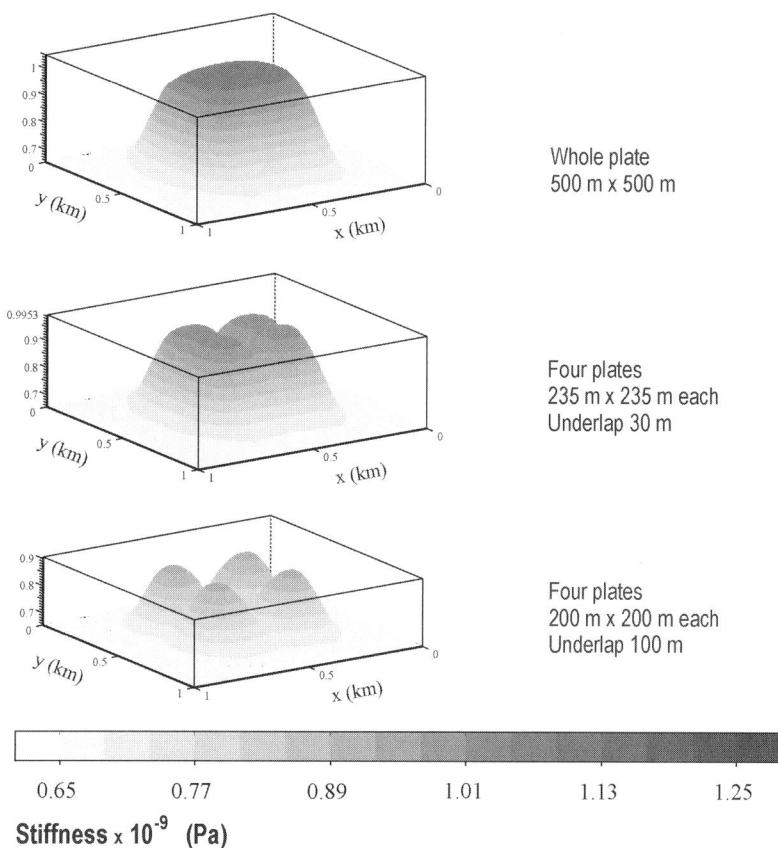

FIGURE 2. Divided plate model (wavelength, 500 m; frequency, 0.056 Hz). Stiffness, or inverse of compliance, response at 0.056 Hz over a finite plate or plates of hydrate in seafloor sediments. The *uppermost panel* depicts a 500 m² plate within 1 km², and the *lower two panels* depict the same plate in four quarters with 30 and 100 m separation, respectively.

We have also considered the response of clumped hydrate deposits built from different shaped hydrate bearing blocks. The blocks are randomly located with the *numerical boxes,* shown in FIGURE 3. Thus the connectivity of the blocks is randomly determined. The concentration of hydrate within the shaded regions increases with depth as in the layered model M2. The hydrate bearing blocks occupy 50% of the available volume. On average over the 1 km² surface area, the two different structures generate almost the same response. The response is illustrated for gravity waves of both 500 and 1000 m wavelength. The lateral variations clearly correlate with the extent of the building blocks.

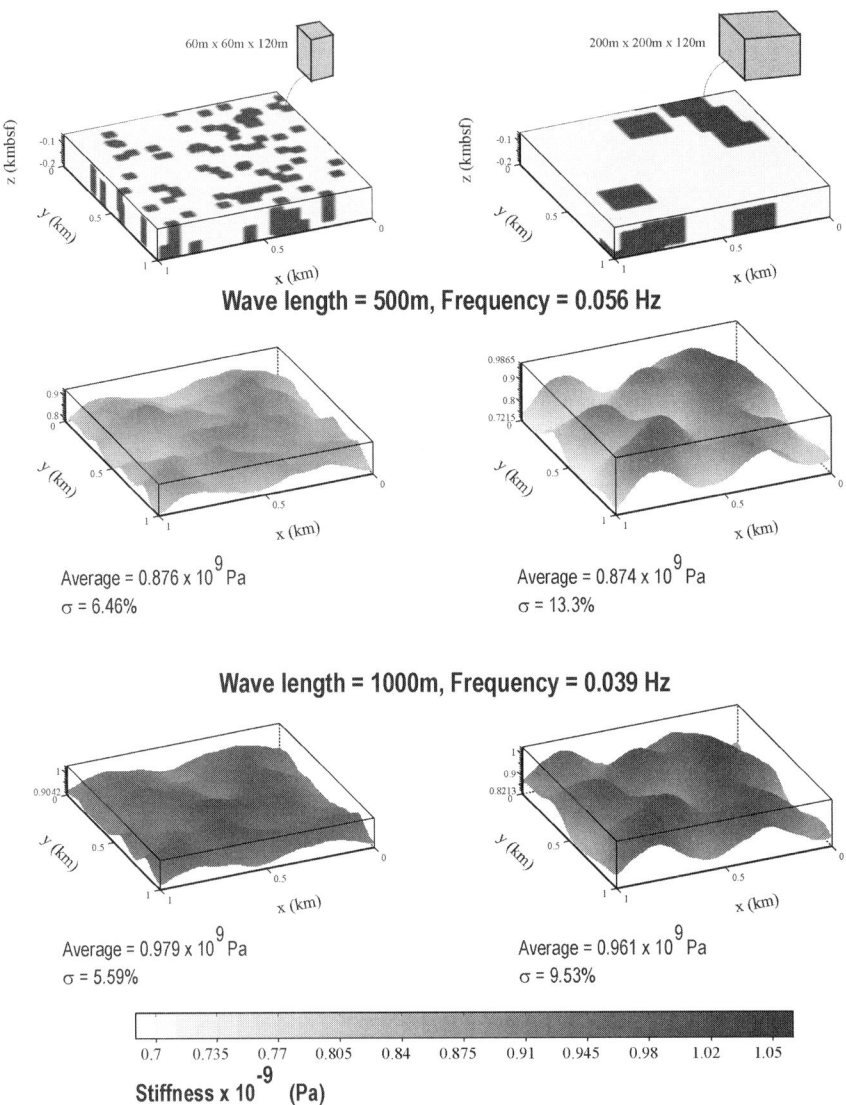

FIGURE 3. Random medium, effect of clumping. Stiffness response, for both 500 and 1000 m wavelength gravity waves, of two different random distributions of clumps of hydrate that fill 50% of the available volume.

In FIGURE 4 we show the effect of varying the amount of hydrate bearing volume for two different distributions, at 0.056 Hz or 500 m wavelength. The fractional change in compliance appears to be a linear function of the total fractional volume

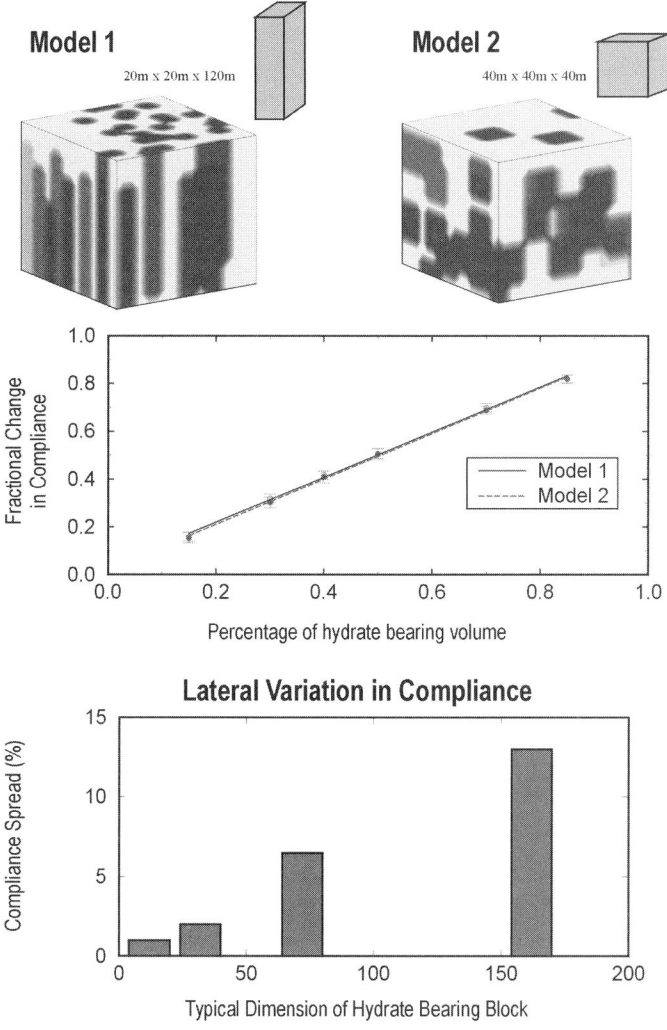

FIGURE 4. Effect of varying the hydrate bearing volume in a random medium two different distributions, at 0.056 Hz or 500 m wavelength. The *lower panel* shows the fractional change in compliance versus the total fractional volume occupied by hydrates.

occupied by the hydrates, regardless of the connectivity pattern. However, the degree of lateral variation in compliance is a function of the typical size of the anomaly. For example, we see from the bar chart that there is a 15% variation in compliance for lateral variations that are comparable to the thickness of the hydrate layer.

EXPERIMENTAL APPARATUS

The measurement of pressure to the required degree of accuracy is relatively straightforward using a differential pressure gauge. The measurement of the vertical displacement or its analogue, the vertical acceleration, is a more difficult problem technically, particularly if a single instrument is required to cover the whole of the relevant spectrum. The pressure field at the seafloor, induced by surface gravity and infragravity waves is measured with a Cox/Webb differential pressure gauge. The deformation or acceleration of the seafloor under this load requires more elaborate instrumentation. For their shallow water experiments, Yamamoto et al.[4] used a Teledyne-Geotech Model S-750 seismometer to measure seafloor movement. The instrument has a flat frequency response between 0.01 and 100 s. Whereas pressure measurements could be obtained over a period band from 7 to 200 s, displacement/acceleration measurements were limited by instrument noise to periods less than 30 s. Typical values of the coherent seafloor acceleration spectrum were 10 $\mu m/s^2/\sqrt{Hz}$ at 10 s. Crawford et al.[2] used a LaCoste-Romberg underwater gravimeter.[5,6] The useful frequency range of this instrument is two decades lower than is typical for ocean-bottom seismometers. Typical values of the observed coherent ocean floor acceleration spectrum were 0.03 $\mu m/s^2/\sqrt{Hz}$ at 80 s and they were recovered from 45 minutes of recorded data.

We used a self-leveling SeaGrav gravimeter built by Scintrex Ltd. The inherent frequency response of the unit makes it a possible replacement for both the S-750 type seismometer and the LaCoste underwater gravimeter for compliance studies. The gravimeter was mounted in an existing pressure vessel and seafloor stand, developed by the marine geophysics group at the University of Toronto. The system self-levels automatically in less than 10 s after landing, to an accuracy of a few arcseconds, accommodates tilts of up to 36° between the plane of the stand and the true horizontal, and measures gravity to a resolution better than 0.01 $\mu m/s^2$. The basic Scintrex gravimeter needed to be modified for our use. The software as originally configured determined the mean value of gravity by sampling at 1 s intervals and averaging the samples over 30 s or more to reduce the statistical noise. On the seafloor, we are interested in the temporal variations of gravity, the noise that was filtered out by the original software. The analogue acceleration signals, the tilt signals, the pressure signals from a Cox/Webb sensor, were interfaced to existing multichannel dataloggers for seafloor data collection.

FIELDWORK

The marine geophysics group was granted ship time on the CCGS TULLY to conduct seven days of field trials in the summer of 1997 and ten days in the summer of 1998 to examine the physical properties of the methane hydrate deposits off Canada's West coast. The Cascadia region is the site of much research, including ODP holes dedicated to hydrate exploration. FIGURE 5 is a bathymetric map of the region studied showing drop sites. We obtained many hours of pressure and gravity data. There is a very high level of coherence between the two channels in the 1998

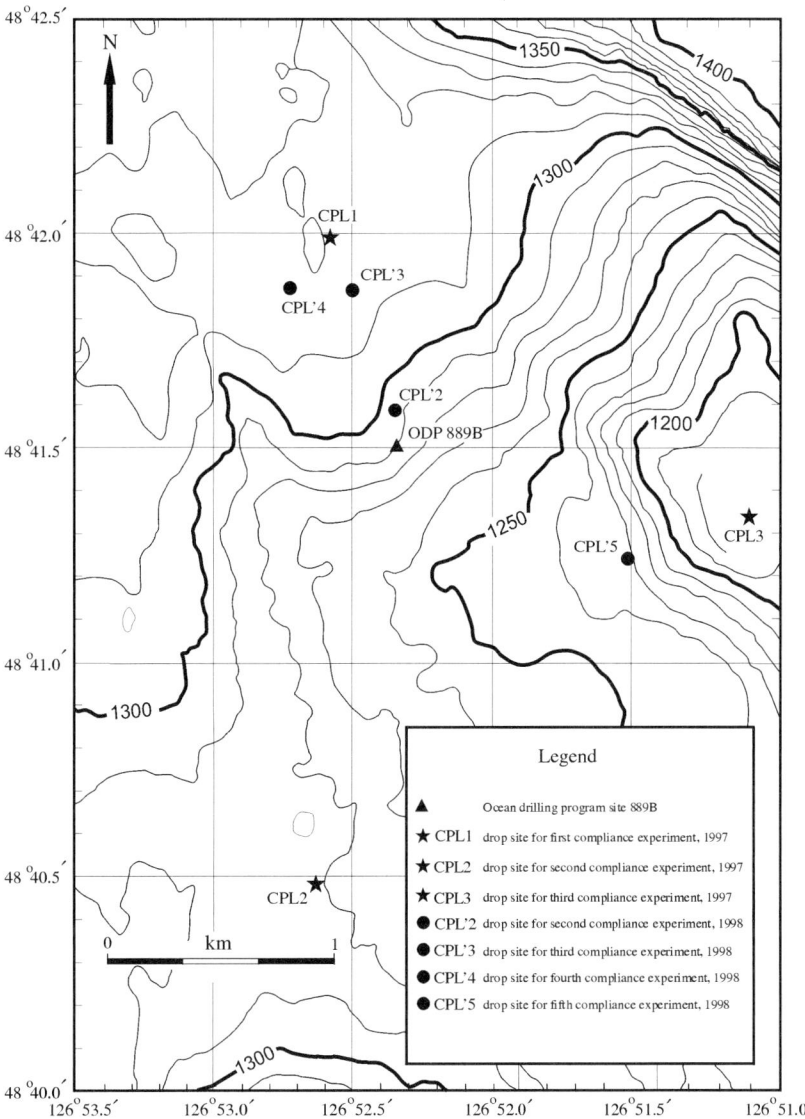

FIGURE 5. Bathymetric map showing sites where compliance experiments were performed.

data, which indicates that the shear modulus profile as a function of depth is well constrained.

FIGURE 6 shows actual field data from Site 5, 1998. The raw data in the gravity wave band clearly shows a correlation between the two channels. The power spectral density, shown in FIGURE 7, also reveals this correlation. Note that the power is significantly reduced at the end of the gravity wave band (around the cut-off frequency of 0.05 Hz) and increases again in the microseismic band (around 0.1 Hz). The noise in the gravity channel can be seen in this cut-off region to be less than 0.1 μgal.

FIGURE 8 shows the upper and lower bounds on the transfer function between gravity and pressure data; the compliance, collected at the fifth drop site during our 1998 cruise. As the bounds approach one another, the uncertainty in the compliance function decreases and the accuracy to which we can invert the data for the shear modulus or hydrate content as a function of depth increases. Note that the bounds in the compliance band, where the energy source is known to be infragravity surface waves, are quite close, meaning that the compliance function is well constrained. Unsurprisingly, in the region of the cutoff frequency, the maximum frequency of waves to exert pressure on the seafloor, the compliance is not well constrained. The compliance is well constrained again in the microseismic band. FIGURE 9 shows coherence as a function of frequency between these gravity and pressure data. High coherence is an indication of high quality data that constrains the shear modulus profile at this site. Note that the coherence in the compliance band is remarkably high. In the region of the cutoff frequency the coherence is greatly reduced, as expected.

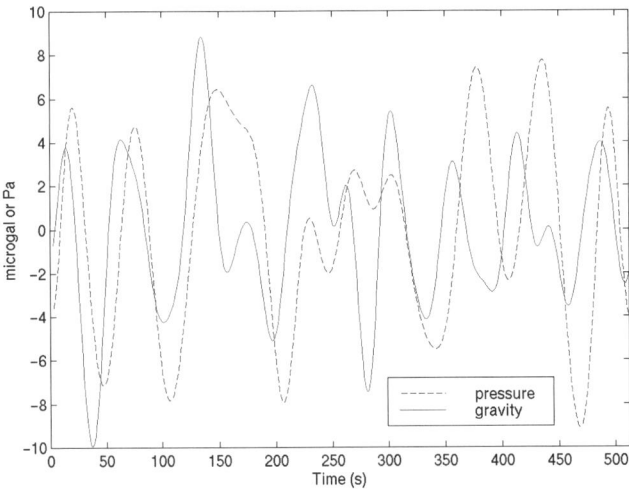

FIGURE 6. Filtered gravity and pressure field data from site CPL'5, in the gravity wave band. The data were processed to remove frequencies above and below the gravity wave band, using analog and digital filters, respectively. Acceleration of 1 μgal corresponds to a displacement of 2.5 microns at 0.01 Hz.

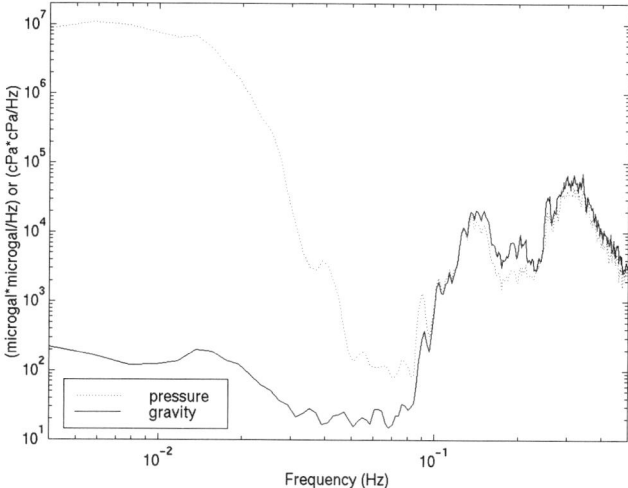

FIGURE 7. Calculated power spectral density from site CPL'5.

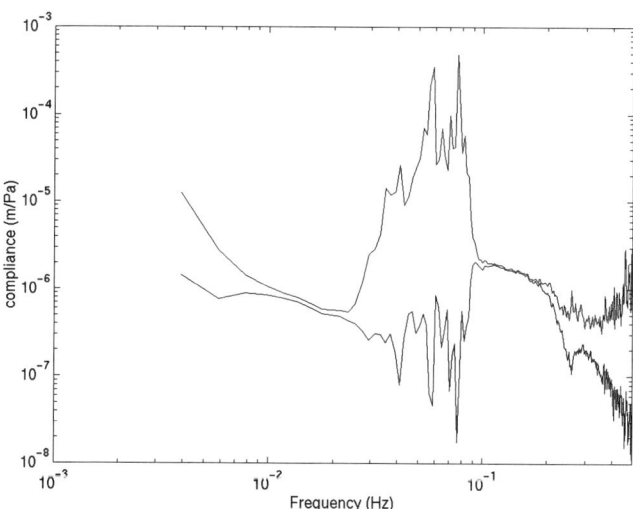

FIGURE 8. Upper and lower bounds on the transfer function between displacement and pressure, from site CPL'5. The bounds are quite close in the gravity wave band, diverge at the cutoff frequency and approach each other again in the microseismic band.

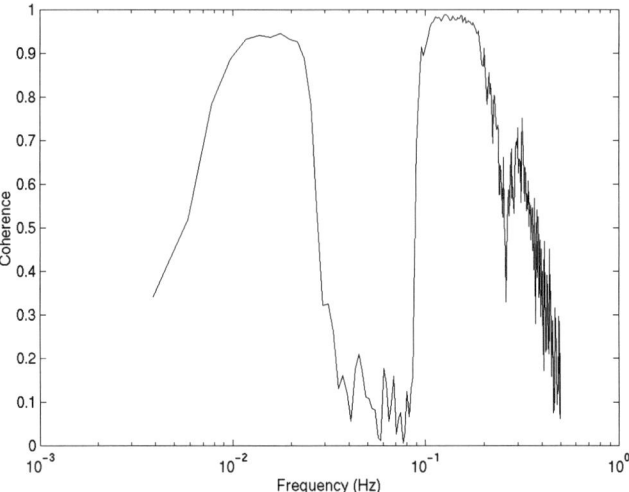

FIGURE 9. Coherence between pressure and gravity data form site CPL'5 is quite high in the gravity and microseismic bands, but low in between.

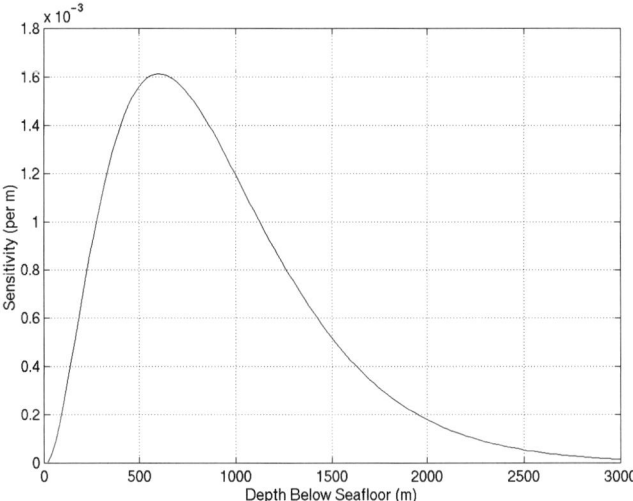

FIGURE 10. The sensitivity function at 0.02 Hz showing the fractional change in compliance due to a given fractional change in V_s or V_p in a 1-m layer at a given depth below the seafloor.

The coherence increases again in the microseismic band. Note also that the compliance and microseismic bands are distinctly separate.

The sensitivity function, shown in FIGURE 10, expresses the fraction change in compliance for a given fractional change in shear velocity or compressional velocity in a layer 1 m thick at a specified depth in the half space, at 0.02 Hz slightly below the cutoff frequency for 1300 m water depth. It is evident from this figure that compliance data are significantly more sensitive to shear than compressional velocities. The datum at this frequency is most sensitive to a change in shear velocity around 600 mbsf, much lower than the hydrate layer. If the water depth was 1000 m, the datum at the associated cutoff frequency would be most sensitive to changes in shear velocity near 300 mbsf, which would be more appropriate to studying the hydrated sediments. Thus we intend to repeat these experiments in shallower water.

FIGURE 11 shows the calculated normalized compliance (the transfer function between vertical displacement and vertical stress multiplied by the wave number of the source) at the best three sites. The data was reproducible from station to station over a one-nautical mile radius. We have included a very simple model based on what is known about the density, compressional and shear velocity depth profiles in this region from ODP data, including the effect of a hydrate and an underlying free gas layer, shown in the inset. The calculated compliance response, indicated by the solid line, of the model fits the data nicely. This model is certainly not unique, and can not be considered an inversion of the data. The vertical resolution of the structure of the uppermost few hundred meters of sediment is limited, because the cutoff frequency is too low to resolve accurately the hydrate layer.

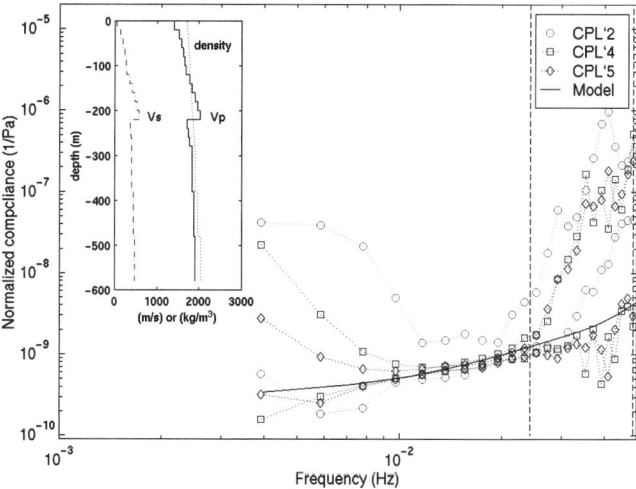

FIGURE 11. The upper and lower bounds of normalized compliance in the gravity wave band for our three best sites and a possible model of sedimentary structure (*inset*) along with its calculated compliance response (*solid line*). The cutoff frequency must lie between the *dashed vertical lines*.

CONCLUSIONS

The paper presents an introduction to our work on the use of compliance for marine gas hydrate evaluation. We have managed to show that the experimental data support the theory and are reproducible from station to station. The data fits a model with reasonable values of sediment compressional and shear velocity and density. We have shown modeling in one dimensional is an appropriate way of interpreting the data. Future work needs to concentrate on surveys over hydrate in shallower water to improve the resolution of the upper 400 meters.

ACKNOWLEDGMENTS

The marine geophysics group is supported by grants and student scholarships from the Natural Sciences and Engineering Research Council and the Centre for Research in Earth and Space Technology, an Ontario Centre of Excellence.

REFERENCES

1. YAMAMOTO, T. & T. TORII. 1986. Seabed shear modulus profile inversions using surface gravity (water) wave-induced bottom waves. Geophys. J. R. Astron. Soc. **85:** 413–431.
2. CRAWFORD, W.C., S.C. WEBB & J.A. HILDEBRAND. 1991. Seafloor compliance by long-period pressure and displacement measurements. J. Geophys. Res. **96:** 16151–16160.
3. WILLOUGHBY, E.C. & R.N. EDWARDS. 1997. On the resource evaluation of marine gas-hydrate deposits using seafloor compliance methods. Geophys. J. Int. **131:** 751–766.
4. YAMAMOTO, T., M.V. TREVORROW, M. BADIEY & A. TURGUT. 1989. Determination of the seabed porosity and shear modulus profiles using a gravity wave inversion. Geophys. J. Int. **98:** 173–182.
5. LACOSTE, L.J.B. 1967. Measurement of gravity at sea and in the air. Rev. Geophys. **5:** 477–526.
6. HILDEBRAND, J.A., J.M. STEVENSON, P.T.C. HAMMER, M.A. ZUMBERGE, R.L. PARKER, C.G. FOX & P.J. MEIS. 1990. A seafloor and sea surface gravity survey of Axial Volcano. J. Geophys. Res. **95:** 12751–12763.
7. CASTAGNA, J.P., M.L. BATZLE & R.L. EASTWOOD. 1985. Relationships between compressional-wave and shear-wave velocities in clastic silicate rocks. Geophys. **50:** 571–581.

Electromagnetic Modeling and *In Situ* Measurement of Gas Hydrate in Natural Marine Environments

RÉMI BOISSONNAS,[a] DAVE GOLDBERG,[b] AND SANEATSU SAITO[c]

Lamont-Doherty Earth Observatory, Borehole Research Group, Route 9W, Palisades, New York 10964, USA

ABSTRACT: During Ocean Drilling Program Leg 170, electromagnetic logs were acquired at three sites on the Costa Rica continental margin. Gas hydrate was observed in sediment cores from nearby holes. These data provide the first opportunity to study and quantify hydrate concentration using logging-while-drilling, a technique used to record logs before *in situ* conditions change significantly. We develop a mixture model that allows for separation of the sediment, pore fluid, and hydrate volume fractions from dielectric permittivity and conductivity measurements. However, the relatively low frequency (2 MHz) of these measurements and the high formation conductivity in this setting do not allow for quantitative estimates of hydrate concentration using the dielectric effect. Pseudoporosity logs can nevertheless be derived from the mixture model and compare well with core measurements. We recommend that future *in situ* electromagnetic measurements be made at frequencies greater than about 20 MHz in order to increase the influence of the dielectric permittivity on such measurements.

INTRODUCTION

The occurrence of gas hydrate on shallow continental margins has been well established by numerous methods and in many studies, but our knowledge of their *in situ* properties still remains poor.[1] Downhole logging at Ocean Drilling Program (ODP) sites has provided some of the most representative subsurface profiles of the properties and distribution of gas hydrate to date. Because of the ephemeral behavior of hydrate outside their limited pressure–temperature window, logging-while-drilling (LWD) measurements made within minutes of drilling may best represent the undisturbed physical and chemical state of their *in situ* environment.[2] This paper presents the first LWD data acquired in hydrate-bearing sediments.

ODP Leg 170 drilled several holes into the sedimentary wedge on the Costa Rican margin near to Deep Sea Drilling Project (DSDP) Site 565.[3,4] This wedge consists of an apron of Plio-Pleistocene age hemipelagic sediments that are mostly derived from the Central American peninsula. The wedge was cored at Sites 1040–1043 and

---

[a]Present address: 52 rue de Richelieu, 75001 Paris, France.
[b]Telecommunication: D. Goldberg. Voice: 914-365-8674; fax: 914-365-3182. goldberg@ldeo.columbia.edu
[c]Present address: Ocean Research Institute, University of Tokyo, 1-15-1 Minamidai, Nakano-ku, Tokyo 164-8639, Japan.

logged at Sites 1040, 1042, and 1043 using LWD.[3] All three sites are located on the apron, upslope from the trench (see FIGURE 1). Hydrate was observed in fractures, cemented ashes, and diffuse pore space in the sediments recovered from Sites 1040, 1041, and 1043.[3] No hydrate was observed at Site 1042, a likely consequence of low core recovery.[3]

Schlumberger's Compensated Dual Resistivity™ (CDR) tool was used in Holes 1040D, 1040E, 1042C, and 1043B. The CDR tool is a conductivity-seeking device that emits a 2 MHz electromagnetic wave 50–150 cm into the formation. The phase shift and attenuation of the electromagnetic wave propagating between two receivers are negligibly affected by modest borehole irregularities and can be transformed into two apparent resistivities.[5] The phase shift resistivity (PSR) produces a shallow measurement with better vertical resolution than the attenuation resistivity (ATR). Both measurements penetrate deeper as formation resistivity increases and their vertical resolution improves as formation conductivity increases.[6] Our objective is to model the effect of gas hydrate on the propagation of electromagnetic waves and assess their occurrence and distribution in these holes. We also compute pseudoporosity logs from the model and the CDR data.

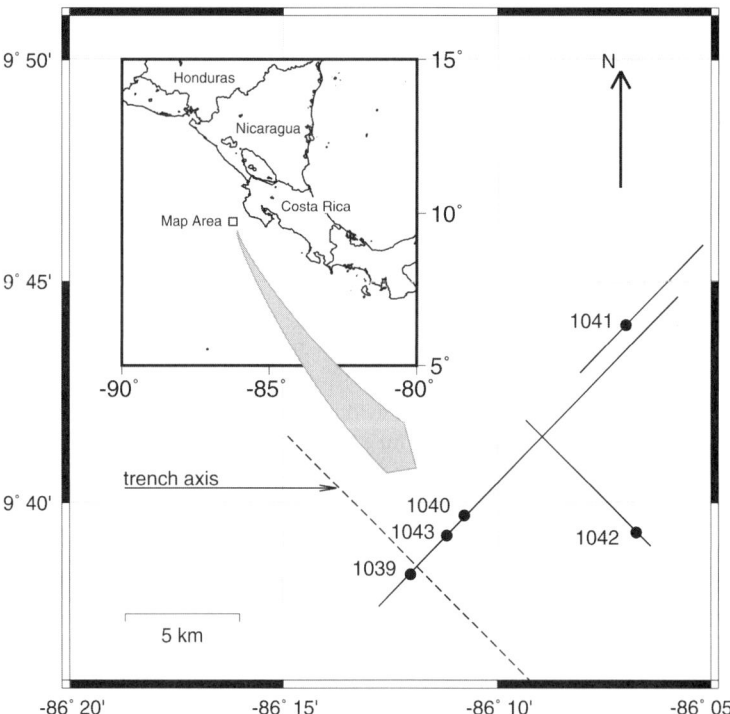

FIGURE 1. Map of ODP Leg 170 drilling Sites 1039–1042 and intersecting seismic lines near the Costa Rica trench. Gas hydrate was recovered in cores from sites 1040, 1041, and 1043 upslope from the trench on the sediment apron.

MODELING

Our mixture model allows for the distinction of gas hydrate from electromagnetic data by exploiting the differences in dielectric properties among hydrate, pore fluid, and sediment. For example, the relative dielectric permittivity, ε_r, is typically less than 10 in sediments.[7] The relative permittivity of massive gas hydrate, $\varepsilon_r = 58$, is much greater and about three-quarters that of fresh water ($\varepsilon_r = 78$). The resistivity of massive hydrate is expected to be about 50 times greater than that of fresh water.[8] Compared to brine-saturated marine sediment, the resistivity of hydrate is therefore quite high and can be detected *in situ* using well logs.[2] In summary, gas hydrate behaves like fresh water with respect to the dielectric permittivity and like sediment with respect to conductivity. FIGURE 2 illustrates these differences.

In this application of the mixture model, we assume that the resistivity and dielectric permittivity both have assessable effects on these *in situ* measurements and that the mixture is homogeneous. The propagation of an electromagnetic wave may then be described by a complex wave vector $k = k_r + jk_i$. We model this by simple weight averaging of the three complex components of the mixture:

$$k = \varphi_W k_W + \varphi_{GH} k_{GH} + \varphi_{MA} k_{MA} \tag{1}$$

where φ_W, φ_{GH}, φ_{MA} and k_W, k_{GH}, k_{MA} are the volume fractions and the complex wave vectors of the pore fluid, hydrate, and matrix (sediment), respectively. When no hydrate is present ($\varphi_{GH} = 0$), Equation (1) reduces to $k = \phi k_W + (1 - \phi) k_{MA}$ where ϕ is the formation porosity. This relationship is commonly used for fluid-filled sedimentary rocks and allows the pseudoporosity to be estimated because all values of k fall between k_{MA} and k_W end-members of the model.[7]

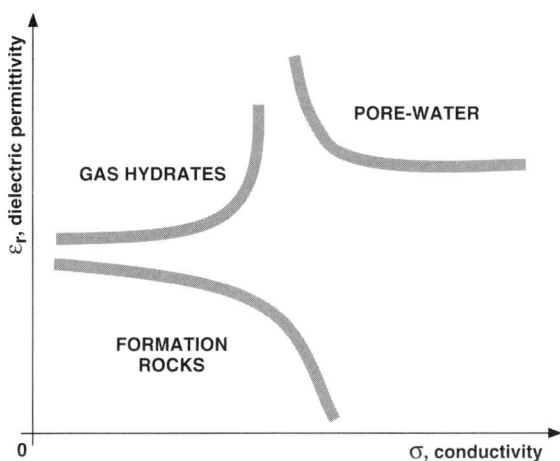

FIGURE 2. Theoretical domains of the electromagnetic properties of gas hydrate, formation rock, and pore water. Sediments typically have low conductivity and low dielectric permittivity (if clay is not present). Pore fluid (water) has high conductivity and high dielectric permittivity. Gas hydrate is predicted to have high dielectric permittivity and low conductivity.

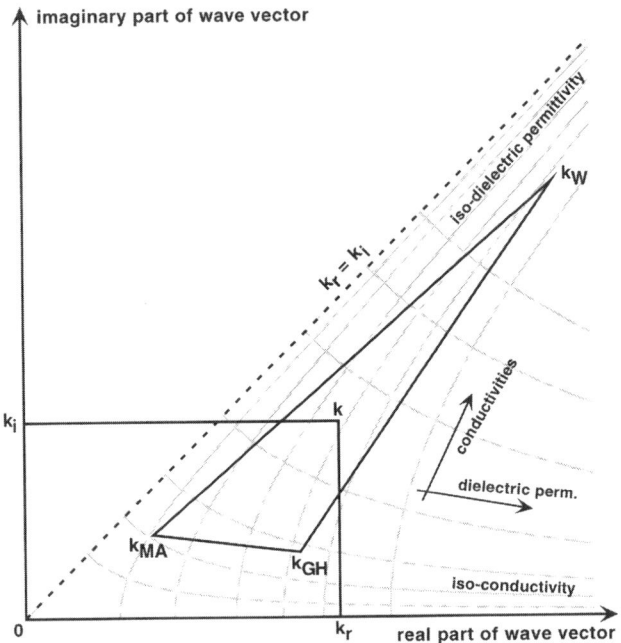

FIGURE 3. Representation of the mixture model in the complex plane k_r, k_i. Contours illustrate isoconductivity and isodielectric permittivity (see Eq. **4**) with arrows indicating the directions of increasing value. The model requires $k_r - k_i > 0$ so that all points fall below the dashed line. The set of potential $k = k_r + jk_i$ values for the weighted-average mixture of the formation matrix (k_{MA}), pore fluid (k_W), and gas hydrate (k_{GH}) is shown by the area enclosed within the k_{MA}-k_W-k_{GH} triangle.

The system described by (1) is also constrained by $\varphi_W + \varphi_{GH} + \varphi_{MA} = 1$ and $\varphi_W > 0$, $\varphi_{GH} > 0$, and $\varphi_{MA} > 0$. Separating the complex wave vectors into real and imaginary parts, the following linear system yields one solution (φ_W, φ_{GH}, φ_{MA}) at each depth:

$$\varphi_W k_{rW} + \varphi_{GH} k_{rGH} + \varphi_{MA} k_{rMA} = k_r$$
$$\varphi_W k_{iW} + \varphi_{GH} k_{iGH} + \varphi_{MA} k_{iMA} = k_i \quad (2)$$
$$\varphi_W + \varphi_{GH} + \varphi_{MA} = 1.$$

FIGURE 3 illustrates this system in the complex wave number plane. Each wave vector $k = k_r + jk_i$ is a weighted barycenter of $k_W = k_{rW} + jk_{iW}$, $k_{GH} = k_{rGH} + jk_{iGH}$, $k_{MA} = k_{rMA} + jk_{iMA}$. The values assigned to k_W, k_{GH}, and k_{MA} are critical because any datum falling outside the three end-member system will yield negative, and therefore unrealistic, values for the volume fraction.

The dielectric permittivity ε_r and the conductivity σ are mixed functions of k_r and k_i and may be described by:

$$k_r = \frac{\omega}{c}\left(\frac{\mu_r\varepsilon_r}{2}\right)^{1/2}\left\{\left(1+\left(\frac{\sigma}{\omega\varepsilon_0\varepsilon_r}\right)^2\right)^{1/2}+1\right\}^{1/2}$$

$$k_i = \frac{\omega}{c}\left(\frac{\mu_r\varepsilon_r}{2}\right)^{1/2}\left\{\left(1+\left(\frac{\sigma}{\omega\varepsilon_0\varepsilon_r}\right)^2\right)^{1/2}-1\right\}^{1/2}$$

(3)

where μ_0, μ_r and ε_0, ε_r are the magnetic permeability and the dielectric permittivity of the void space and of the rock, respectively, and ω/c is the ratio of the angular velocity of the wave to the speed of light.[5] Using $\mu = 1$, $\mu_0\varepsilon_0 c^2 = 1$, $\omega = 4\pi 10^6$ rad·s^{-1}, $\varepsilon_0 = 8.85 \cdot 10^{-12}$ F·m^{-1}, Equation (3) gives the solution to Maxwell equations for a one-dimensional harmonic wave propagating through a conductive body.[9] If we assume $\sigma/(\omega\varepsilon_0\varepsilon_r) \gg 1$ because of the high conductivity (and/or low hydrate concentration) in these marine sediments, Equation (3) can be simplified to $k_r = k_i$ and $\sigma_a = (2/(\omega\mu))k_r^2$ where $\mu = \mu_0\mu_r$. This allows the pseudoporosity to be computed.

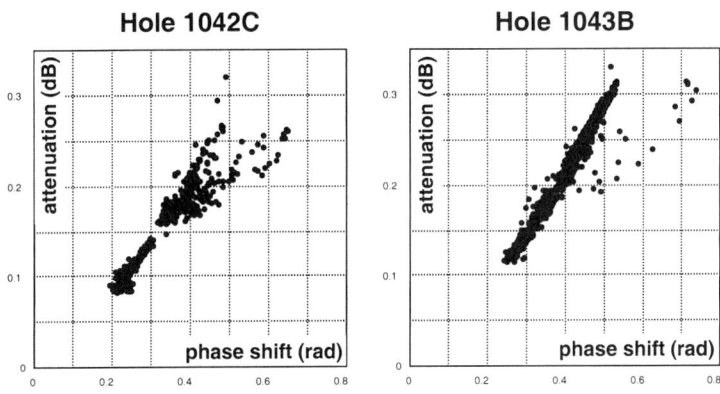

FIGURE 4. Crossplots of attenuation versus phase shift of CDR data in ODP Holes 1042C (0–297 in depth) and 1043B (0–485 in depth). Scatter from the linear trend of the data in both examples is attributed to hole conditions, geometric effects, and the possible occurrence of distributed hydrate in sediment pores. In Hole 1043B, the large excursion corresponds to a caliper anomaly at about 74 mbsf, possibly due to a fault or fluid conduit intersecting the borehole.

PSEUDOPOROSITY RESULTS

The phase shift and attenuation of the electromagnetic wave are computed using inverse resistivity transforms[5] of PSR and ATR in Holes 1040D, 1040E, 1042C, and 1043B. In FIGURE 4, the crossplot of phase shift and attenuation measurements show linear trends in Holes 1042C and 1043B. This suggests the presence of a simple fluid-sediment mixture with no gas hydrate ($\varphi_{GH} = 0$). Equation (2) may then be reduced to

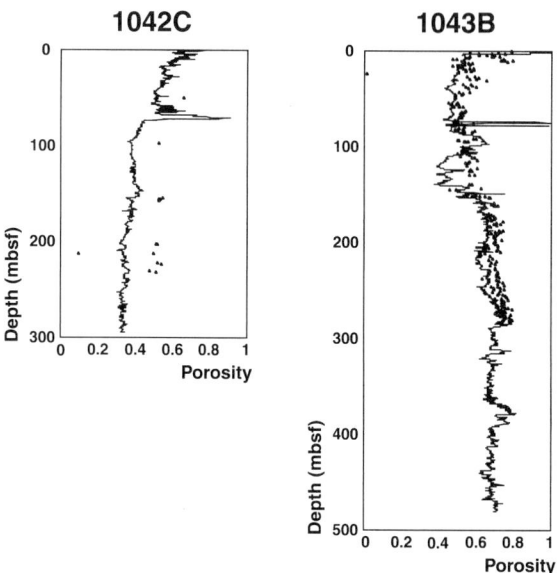

FIGURE 5. Computed pseudoporosity logs (*solid line*) in ODP Holes 1042C and 1043B (see text) and porosity measurements on core samples (*dots*). Offset between core and log data is attributed to expansion of the samples under ambient laboratory conditions. The overall agreement of these trends with depth gives confidence in the model parameters.

$$\phi = \frac{\sigma_a^{1/2} - \sigma_{MA}^{1/2}}{\sigma_W^{1/2} - \sigma_{MA}^{1/2}}, \qquad (4)$$

where ϕ is porosity, σ_a is apparent conductivity; and σ_{MA} and σ_W are matrix and fluid conductivity, respectively.

FIGURE 5 shows pseudoporosity logs in Hole 1042C and 1043B computed using mean values for σ_{MA} = 0.01 S/m and σ_W = 2.5 S/m in **(4)**. The fluid conductivity is computed from interstitial water measurements on core samples at corrected salinity and temperature from Site 1042.[3] Depth-dependent rebound due to expansion of the core samples at ambient surface conditions (approximately 5–10%) was not corrected in the porosity data.[10] Few core data are available at Site 1042, so porosity variation with depth and between sites was sparsely sampled. At Site 1043, the pseudoporosity log correlates well with the trend of the core measurements, giving confidence in the estimates of σ_{MA} and σ_W as well as the method.

DISCUSSION

Although gas hydrate was recovered at these sites, the predicted three end member mixture is not evident in FIGURE 4. A likely explanation is that the frequency response of the CDR tool is inadequate for measuring the dielectric permittivity in these low-resistivity formations. At 2 MHz frequency, the phase shift and attenuation depend strongly on the resistivity and weakly on the dielectric constant because the dielectric term $\omega\varepsilon_0\varepsilon_r$ in Equation (3) is relatively small. The formation is too conductive at this propagating frequency for the dielectric permittivity to have a significant effect on the electromagnetic wave. Although this allows for the high-conductivity approximation ($\sigma \gg \omega\varepsilon_0\varepsilon_r$) to be made and accurate pseudoporosity logs to be inferred, a higher working frequency would increase the dielectric effect. Assuming $\varepsilon_r \approx 10$, the σ and $\omega\varepsilon_0\varepsilon_r$ terms in high conductivity formations become nearly equal at frequencies approaching 1 GHz. Some logging tools use electromagnetic frequencies of 25 MHz up to 1.1 GHz, but their depth of investigation into the formation is considerably less than that of the CDR tool.[7] Unfortunately, these tools are deployed by a wireline and not connected to the drillstring, thus acquiring data in unstable marine formations may be more difficult than using the CDR.

The local distribution of hydrate in the formation may also influence the model. In FIGURE 4, we observe several excursions and some broadening around the linear trend of the data. Massive, layered, nodular, and disseminated hydrate all possibly have different electromagnetic properties. If the effect of disseminated hydrate on the propagation of an electromagnetic wave is small or close to the behavior of a simple sediment–fluid mixture, then broadening may partly be explained by the presence of distributed hydrate in pores. Partial dissociation of hydrate due to the drilling process may also reduce their local concentrations. It is not easy to distinguish these effects from differences between the vertical resolution of the PSR and ATR logs, all of which could produce broadening around the trend, or from data scatter due to localized borehole conditions. For example, a zone of borehole enlargement (washout) at approximately 74 m depth in Hole 1043B is identified by the excursion of several data points with anomalous phase from the linear trend (see FIG. 4). Shoulder effects near beds with large conductivity differences, bedding dip, formation anisotropy, and fluid invasion may also generate similar scatter of the data, but these effects are not likely to be significant in the sediments on the Costa Rica margin.

The effects of clay on these measurements must also be considered. Clay has a dielectric constant varying between approximately 5 and 25, which lies between the dielectric constants for sediment, hydrate, and water.[7] The conductivity of clay is greater than most marine sediments and generally less than their pore fluid. The clay effect could potentially be included in the model as a matrix correction or as a fourth "$\varphi_{CLAY} k_{CLAY}$" end-member. In future, independent laboratory studies on the electromagnetic behavior of hydrate over a broad frequency range and *in situ* logs in a variety of hydrate-bearing sediment would aid considerably in distinguishing the effects of clay from hydrate using this model.

CONCLUSIONS

We present a mixture model to represent the effect of hydrate on electromagnetic logs. The main supposition in this model is that hydrate has electromagnetic properties that differ from both the pore fluid and matrix in marine sediments. We apply the model to data recorded using an electromagnetic LWD tool on the Costa Rica margin. Despite the evidence of hydrate in recovered cores, we were not able to quantify hydrate concentration from the electromagnetic logs. Because of the influence of frequency and conductivity on the electromagnetic measurements, it is most likely that the dielectric effect of the hydrate in these data is small. The hydrate may also be diffusely distributed, rapidly dissociated during drilling, and/or masked by unaccounted for effects, such as clay, in the model. Pseudoporosity logs can be derived from a simplified mixture model. These results agree well with porosity measurements made on core samples.

Future research should aim at extending laboratory measurements to 2 MHz and higher in order to estimate the apparent conductivity and dielectric permittivity of gas hydrate and their host sediments. Independent measurement of porosity, pore fluid chemistry, pressurized core samples, and an understanding of the effects of clay are needed to fully evaluate this dielectric model. We recommend that electromagnetic tools using higher frequencies be deployed in the future.

REFERENCES

1. KVENVOLDEN, K. 1993. Gas hydrates—geological perspective and global change. Rev. Geophys. **31**(2): 173–187.
2. GOLDBERG, D. 1997. The role of downhole measurements in marine geology and geophysics. Rev. Geophys. **35**(3): 315–342.
3. KIMURA, G., E. SILVER, P. BLUM *et al.* 1997. Proceedings of the Ocean Drilling Program, Initial Reports, **170**. Ocean Drilling Program, College Station, TX.
4. VON HUENE, R., J. AUBOUIN *et al.* 1985. Initial Reports of the Deep Sea Drilling Project, **84**. U.S. Government Printing Office, Washington, D.C.
5. SCHLUMBERGER, 1995. Logging While Drilling. Educational Services, Houston, TX.
6. CLARK, B. *et al.* 1998. Electromagnetic propagation logging while drilling: theory and experiment. Society of Petroleum Engineering, paper 18117.
7. SCHLUMBERGER, 1989. Log Interpretation Principles. Educational Services, Houston, TX.
8. SLOAN, E.D. 1990. Clathrate Hydrates of Natural Gas. M. Dekker, New York, NY.
9. GORBECHEV, Y.I. 1995. Well Logging: Fundamentals of Methods. J. Wiley, Chichester, U.K.
10. HAMILTON, E.L. 1976. Variations of density and porosity with depth in deep-sea sediments. J. Sedimentary Petrology **46**(2): 280–300.

Relationship between Electrical and Thermal Conductivities for Evaluating Thermal Regime of Gas Hydrate Bearing Sedimentary Layers

OSAMU MATSUBAYASHI[a,b] AND R. NIGEL EDWARDS[c]

[b]*Geological Survey of Japan, 1-1-3 Higashi, Tsukuba, Ibaraki, 305-8567 Japan*
[c]*Department of Physics, University of Toronto, Toronto, Canada*

ABSTRACT: In an attempt to explore the potential use of electrical resistivity data combined with seafloor heat flow measurements as an alternative predictor of bottom of gas hydrate stability (BGHS) in marine gas hydrate-bearing geological environments, a model for conversion of electrical conductivity to thermal conductivity has been formulated. It was found that a thermal conductivity versus electrical conductivity model assuming hydrate-coated-grains gives lower thermal conductivity than other models, because gas hydrates characteristically have a low value of thermal conductivity compared with sediment grains. The authors consider the hydrate-coated-grain model more realistic than other models. Once we can establish a relationship between electrical resistivity and thermal resistivity of sedimentary formations like that proposed in this paper, it will be possible to provide an independent source of information about the depth of BGHSs even with an absence of seismic BSR data.

INTRODUCTION

The purpose of this work is to develop a new method for predicting the accurate thermal regime of offshore sedimentary layers that contain gas hydrates, with a goal to estimate the depth of the bottom of gas hydrate stability (BGHS) even without seismically observed BSRs. If we reverse the logical steps adopted by Yamano *et al.*[1] in evaluating terrestrial heat flow from the estimated depth of the BGHS, we are able to use the field measurements of heat flow on the seafloor as a proxy for BSR records.[2] In cases where BSR signatures on seismic profiles are available, our approach should give an independent data source to check the physical nature of those seismically estimated BGHSs. We show that electrical conductivity observations for the area of our interests can be used as a predictor for the thermal conductivity distribution with depth, based on the theoretical relationship between thermal conductivity and electrical conductivity.

The thermal conduction equation in steady state, $Q = \int k(z) dT/dz$, can be rewritten as a formula to give $T(z)$, based on the seafloor temperature and the depth integral of thermal resistivity ($1/k$). The BGHS is the depth at which the $T(z)$ curve crosses

[a]Telecommunication. Voice: +81-298-61-3675; fax: +81-298-61-3666.
matsu@gsj.go.jp

the phase boundary curve for methane hydrates. Observed data required as input for this method are:

1. Seafloor water temperature $T(0)$;
2. Heat flow probe measurements $Q(0)$ on the seafloor, for which development of a reliable field technique to remove the thermal effects of seasonal variations in bottom water temperature are still necessary;
3. Estimation of the accurate thermal resistance as function of depth, converted from the electrical resistance, and assuming a consistent physical model for both thermal and electrical conduction, as detailed below (hydrate-coated-grain model). In the conversion procedure realistic values for other parameters, such as porosity variation with depth need to be assumed. A family of standard curves showing the relationship between thermal conductivity and electrical conductivity for gas hydrate containing sediments will be produced. We discuss the possible differences among different models as well as the reason for our preferred model over others.

When the relationship between thermal and electrical conductivities is based on firm physical grounds, and a reliable heat flow measurement on the seafloor is given, our standard curves are more useful in defining a better $T(z)$ profile at the site than the profile conventional assumed using a *linear* temperature–depth relationship to predict the BGHS.[3] In the present work, various time-dependent processes are not taken into account.

THE THERMAL CONDUCTIVITY MODELS FOR GAS HYDRATED SEDIMENT

Composite Thermal Conductivity Model for Grains Completely Coated with Gas Hydrates

As one end-member, we commence by modeling the thermal conductivity of bulk marine sediments using a random mixture of solid phase (sediment grains completely coated with gas hydrates) and liquid water phase. The thermal conductivity of the solid phase is calculated as follows:

$$\kappa_c = (1 + 2R)/(1 - R)\kappa_f \tag{1}$$

$$R = \frac{R_{hf} + \lambda R_{sh}(1 + R_{hf})}{1 + 2\lambda R_{sh} R_{hf}} \tag{2}$$

where $R_{hf} = (\kappa_h - \kappa_f)/(\kappa_h + 2\kappa_f)$ and $R_{sh} = (\kappa_s - \kappa_h)/(\kappa_s + 2\kappa_h)$, with subscripts s, h, and f corresponding to grain, hydrate and fluid, respectively; and $\lambda = (1 - \phi)/(1 - S\phi)$, where ϕ and S stand for porosity and fluid saturation in the pore space.

The composite thermal conductivity can be calculated as the geometric mean of the solid phase and liquid phase conductivities[4]

$$\langle \kappa \rangle = \kappa_f^{(1-\phi)} \cdot \kappa_c^{(1-S\phi)}. \tag{3}$$

The overall model as obtained from (3) is called the *hydrate-coated-grain* model. This model incorporates compatibility with the simple electrical conductivity model, assuming that both grains and hydrates are electrical insulators, hence the well-known expression from Archie's law is valid:

$$\langle\sigma\rangle = (S\phi)^2 \cdot \sigma_f \tag{4}$$

where $\langle\sigma\rangle$ is the composite electrical conductivity, and σ_f stands for the electrical conductivity of the fluid phase. In other words, it is implied that the mode of thermal contact among the three components (grain, hydrate, and water) is similar to that of electrical conductivity. The relationship between thermal conductivity and electrical conductivity is shown in the form of standard curves (see FIGURE 1A). This is our preferred model. FIGURE 1A shows that higher thermal conductivity correlates with lower electrical conductivity, but the range of thermal conductivity variation is much smaller than that of electrical conductivity.

Comparison with Other Composite Thermal Conductivity Models

A large difference is obvious among different thermal conductivity mixing models. A mixing relationship widely used for composite thermal conductivity is a model that assumes a three-phase mixture with all the components having equal weights (extended *geometric mean* formula originally proposed by Woodside and Messmer.[4] This is defined as follows:

$$\langle\kappa\rangle = \kappa_f^{S\phi} \cdot \kappa_h^{(1-S)\phi} \cdot \kappa_s^{(1-\phi)}. \tag{5}$$

FIGURE 1B shows the relationship between thermal and electrical conductivities for this model.

The hydrate-coated-grain model as discussed above might correspond to an extreme case in nature, giving a lower bound on the thermal conductivity, but we believe that the model represents actual situations rather well. It is interesting that the three-phase geometric mean mixing model (5) predicts significantly higher composite thermal conductivity values than the first model for the same porosity (ϕ) and gas hydrate saturation (1-S), assuming the set of parameters listed in TABLE 1 (FIG. 1). The difference between the two models amounts to a factor of 1.7 for cases with electrical conductivity values higher than 0.1 S/m. Such a large difference may be due to the fact that the effect of highly conductive grains (assumption: $\kappa_s = 2.0$) become minimum in the large-scale conductive thermal regime, if the sediment grain surfaces are totally coated with gas hydrates (hydrates $\phi_h = 0.45$ are thermally less conductive than liquid water). On the other hand, the geometric mean model corresponds to a random mixture of hydrates, water, and grains, where grains are modeled to have large surface areas in direct thermal contact with one another. For that reason, even the same amount of hydrate present in the random mixture may contribute less to the overall thermal insulation, and therefore result in a higher composite thermal conductivity value in comparison with the hydrate-coated-grain model. This is a significant difference for assessment of the thermal field $T(z)$ around gas hydrate bearing formations.

The reason that the hydrate-coated-grain model is considered the most suitable is that the gas hydrate forming process in sediments requires a nucleus to enable the initial growth of a hydrate phase. The surfaces of sediment grains act as such nuclei. Therefore, it is very likely that grains are coated with gas hydrate phase much more easily than the growth of isolated hydrate phase at random within the fluid phase.

We can formulate other mixing models for composite thermal conductivity, such as one that assumes a dispersed and isolated gas hydrate phase in very small quantities.

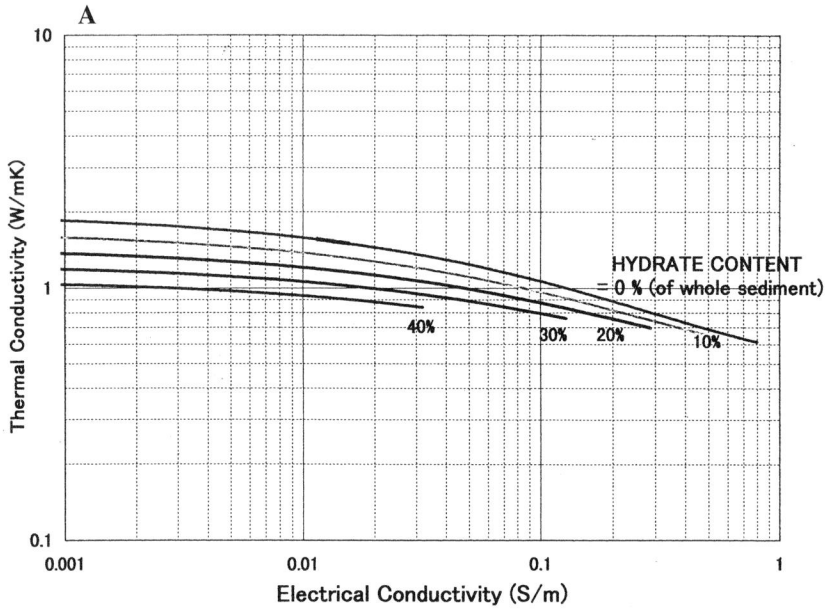

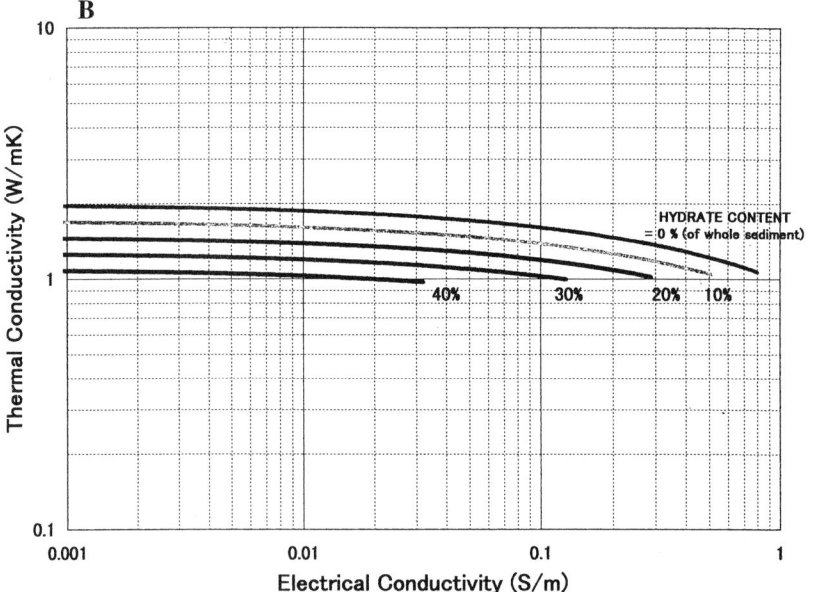

FIGURE 1. Thermal conductivity of sediments containing gas hydrate as function of electrical conductivity based on **(A)** the hydrate-coated-grain model and **(B)** the geometric mean model. Assumed properties of the components are shown in TABLE 1.

TABLE 1. Properties assumed in the standard curves of composite thermal conductivity as function of electrical conductivity (FIG. 1)

Electrical conductivity		
	water	3.2 S/m
Thermal conductivities		
	water	0.57 W/mK
	pure methane hydrate	0.45 W/mK
	rock matrix	2.0 W/mK

In this opposite extreme case, the contribution of gas hydrates in reducing the composite thermal conductivity is minimized, when assuming the same volume fraction of gas-hydrate phase. The latter model is not considered to have a large effect on thermal conductivity in natural situations.

PRACTICAL CONSIDERATION

We indicated previously that the final goal of this study is to better predict BGHS depths from the electrical survey data.[5] Resistivity data from well logs would provide better information for this purpose in cases where drillholes penetrate our target geological formations. However, the range of error in converting electrical resistivity data to thermal resistivity, and the estimated BGHS thus obtained, depends on many uncertain factors, such as poorly known thermal conductivity values for sediment grains having different lithology (κ_s in this paper) and also the assumption of Archie's formula (4). Furthermore, at present we can only derive the variation of thermal conductivity with depth, over an interval of at least 20 m. Improving the accuracy of thermal conductivity as well as the depth resolution necessary for meaningful conversion from electrical to thermal resistance is a matter to be investigated by refining our model and establishing a better thermal conductivity data base by using laboratory measurements.

CONCLUSIONS

Assuming a three-component (grain, hydrate, and fluid) thermal conductivity mixing law, a physically consistent relationship between electrical conductivity and thermal conductivity for gas hydrate bearing sediments can be modeled, so that field observations of electrical resistance as a function of depth may be converted to thermal resistance as a function of depth. The latter then provides information about how the temperature versus depth curve behaves due to the occurrence of gas hydrates, if it is supplemented with sediment surface heat flow and bottom seawater temperature data. We have discussed the magnitude of differences in thermal conductivity predictions according to different assumptions about the thermal conductivity mixing law, as well as the physical reasons for such differences.

REFERENCES

1. YAMANO, M. *et al.* 1982. Estimates of heat flow derived from gas hydrates. Geology **10:** 339–343.
2. MATSUBAYASHI, O. 1998. Heat flow measurement as an exploration tool for subbottom methane hydrates. Bull. Geol. Surv. Japan **49:** 541–549.
3. RUPPEL, C. 1998. Anomalously cold temperatures observed at the base of the gas hydrate stability zone on the U.S. Atlantic passive margin. Geology **25:** 699–702.
4. WOODSIDE, W. & J.H. MESSMER. 1961. Thermal conductivity of porous media. J. Appl. Phys. **32:** 1688–1706.
5. EDWARDS, R.N. 1997. On the resource evaluation of marine gas hydrate deposits using sea-floor transient electric dipole-dipole methods. Geophys. **62:** 63–74.

Messoyakh Gas Field (W. Siberia)
A Model for Development of the Methane Hydrate Deposits of Mackenzie Delta

JAN KRASON[a]

Geoexplorers International, Inc., 5701 East Evans Avenue, Suite 22, Denver, Colorado 80222, USA

ABSTRACT: The Messoyakh gas field of Northwestern Siberia was discovered in 1967. This field has been thoroughly studied and described in many publications, including References 1–9. At least one-third and, most likely, two-thirds of the Messoyakh reservoir, which for 13 years has been in commercial production, occurs in the form of natural gas hydrates.

INTRODUCTION

The Messoyakh gas field is located within an arctic climatic zone underlaid by 400 to 450 m of permafrost and a subjacent 200 to 250 m interval in which gas hydrates are stable. The anticlinal structure (measuring 13 by 20 km), within the flank of the Tanam arch, traps the Messoyakh gas field. Structural closure of 84 m occurs at the top of the Cenomanian age reservoir formation.[5] The trap at the Messoyakh gas field covers approximately 165 km^2. Krason and Finley[9] calculated and concluded that the Messoyakh reservoir originally contained 62 bn m^3 (2.2 tcf) of gas in the form of both hydrate and free gas.

The reservoir formation is composed of sands and poorly indurated sandstone. Interbeds of shale constitute up to 30% of the unit. The sandstone ranges in porosity from 16 to 30%, averaging 25%.[5] Permeability ranges from 10 to 1000 md.[6] Makogon et al.,[1] reported that the reservoir temperature ranges from 8 to 12°C.

Because of the presence of gas hydrates, the Messoyakh gas field differs from the typical, conventional gas deposit. Among other features, the resistivity values in hydrate zone indicate an anomalously low permeability, greater than those of water-filled sand but less than those of a free-gas-bearing unit. The caliper and microresistivity curves show substantial caving and mudecake above an elevation of −755 m, suggesting that the base of the hydrate stability zone is located at −755 m. However, the negative shift in SP and positive resistivity and neutron kick at −782 m are characteristic of the transition from a hydrate-bearing zone to an underlying free-gas zone.[9]

The natural gas produced from Messoyakh averages 98.5% methane and about 0.1% heavier hydrocarbons. The lack of heavier hydrocarbons suggests a biogenic source of gas. However, the isotopic signature of the Messoyakh gas indicates a thermogenic source.

[a] geoexpl@geoexplorers.com

At least 50 production wells were drilled within the Messoyakh structure, with spacing of 500 to 1000 m. Gas production commenced in 1970 and continued until 1977 at an average of 3 million m^3/day (110 mmcf/day). Following a four-year shut-in period, production was reestablished at rates of 0.2 to 0.5 million m^3/day (7 to 18 mmcf/day; Refs. 3 and 4). Wells completed in the hydrate zone constantly produced less than nearby wells completed in the deeper, free-gas part of the reservoir. Because of difficulties in the efficiency of production of gas from the hydrated reservoir, two methods were applied to maximize production at the gas field. The wells were preferentially perforated beneath the gas hydrate stability zone and stimulation of hydrate dissociation was attempted in wells completed in the hydrate zone. Methanol and $CaCl_2$ were used both alone and mixtures. Methanol was more effective at increasing gas production but $CaCl_2$ cost less. The most economic treatments were mixtures of the two compounds. The most effective treatment involved injection of about 3 to 4 m^3 of methanol over a period of 24 to 26 hours under a pressure of 100 to 150 atm.

Messoyakh records, especially those including information on production methods and encountered problems, teach highly valuable lessons. Certainly, these lessons prompted the decision by Japan National Oil Corp. (JNOC), in cooperation with other members of the consortium, to drill an exploratory well, and selected for this well the Mackenzie Delta, Northwest Territories, Canada. The presence of gas hydrates in the Mackenzie Delta had been already known for some time. Numerous previously-drilled wells penetrated the several hundred meters thick permafrost, which is underlaid by gas hydrates. The exploratory well known as Mallik 2L-38 was completed in March, 1998. The main objective for this drilling had been exploration and in depth research on many aspects and characteristics of gas hydrates. Although such investigations are necessary and should be carried out, the results of drilling and preliminary investigations indicate the discovery of a giant gas field (see the Proceedings of the International Symposium on Methane Hydrates, Resources in the Near Future?; JNOC, Chiba City, Japan, October 20–22, 1998). It is evident that there are many similarities between these two gas fields. Therefore, it appears prudent and very encouraging to use the Messoyakh gas field at least as a temporary "model" for production tests and feasibility studies, with consideration for near future commercial development of the methane hydrate and free gas deposit trapped in the Mallik anticlinal structure in the Mackenzie Delta.

RATIONALE

During the last two decades our knowledge of occurrence, formation, stability, and potential resources of gas associated with natural hydrates has advanced more rapidly than during the period, since their discovery by Davy in 1810. It is now well known that gas hydrates composed mostly of methane occur in numerous locations on continental margins. They are common in offshore deep sea waters.[10–13] The presence of methane hydrates in onshore environments, especially in the regions underlying permafrost has been confirmed and well documented.[9,14–16] Many authors assess the methane hydrates resource to be very large.[13,16–18] Moreover, JNOC seriously considers methane hydrates to be an energy resource for the near future.[19,20] Therefore, in the context of present knowledge of the formation-controlling factors mentioned above it is not surprising that the theme of this volume focusses on gas hydrates and challenges for the future.

The Messoyakh in Western Siberia is the only field in which up to one third of natural gas was trapped in methane hydrates that has been commercially produced. There are many publications pertaining to this gas field's geology, geophysical characteristics, chemical composition of gas (including production records), applied technology, and inter-related problems.[9] There is also a lot of information pertaining to gas hydrates in the Beaufort Sea and the Mackenzie Delta, Northwestern Territories, Canada.[11,21-23] Therefore, considering all readily available information, including data from drilling results of the Mallik 2L-38 well, its stratigraphy, sedimentology, structural setting, and log interpretation published by JNOC-TRC in the Proceedings of the International Symposium on Methane Hydrates Resources in the near Future (held in Chiba City, Japan, October 20–22, 1998), certainly one may be encouraged to compare them both. Furthermore, thorough scientific research, it is recommended that the Messoyakh gas field be considered as at least a tentative model for production tests and the feasibility study of the methane hydrates of Mackenzie Delta for commercial development. The Messoyakh gas field may serve as a case study of the problems encountered in exploration and production of shallow gas fields in arctic regions.

GEOGRAPHIC ANALOGY

The Messoyakh gas field and Mallik 2L-38 well at the Mackenzie Delta are located in two different continents, far apart from each other although the latitude of both areas are almost the same (see FIGURES 1 and 2). Both regions are confined to arctic climatic conditions underlain by a thick zone of permafrost (400–450 m and 650 m, respectively) and they both coincide with major hydrocarbon-bearing provinces. Because of the latter coincidence, both fields are highly favorable for formation, stability, and both contain large resources of gas hydrates at a relatively shallow depth.

In Western Siberia geomorphologic methods have been useful in the mapping of the location and extent of Quaternary tectonic features. At the Messoyakh gas field, 400 m of Quaternary uplift has been detected and reported by Kulakov and Mokhatina.[25]

STRUCTURAL SETTING AND TRAPS

The Messoyakh gas field is located within West Siberian cratonic basin containing up to 10,000 m of Jurassic to Quaternary sediments deposited on folded basement. The basement of underlying Messoyakh is probably composed of altered miogeosynclinal rocks.[8]

The Messoyakh gas field occupies a part of the Southeast flank of the larger arch trending about N600E. The arch was uplifted 1,500 m in Late Jurassic to Early Cretaceous, producing an unconformity in the Middle to Late Jurassic section of the

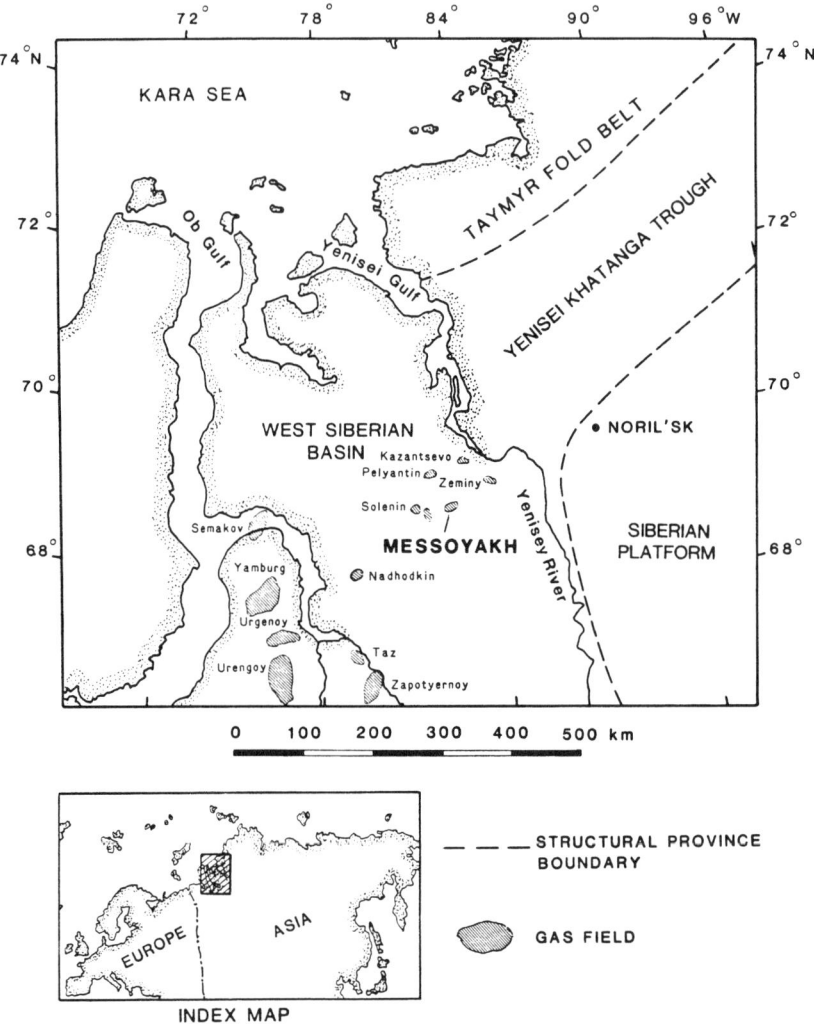

FIGURE 1. Map showing the Western Siberian petroleum-bearing province, the Messoyakh gas field, other fields, and major structural units (after Krason and Finley, 1992).

arch. The arch was reactivated in the Tertiary, with about 1,300 m of uplift occurring between Oligocene and Pliocene.

The Messoyakh gas field is confined to an anticlinal structure measuring 20 by 13 km, covering approximately 165 km^2, with 84 m of closure. Diminished permeability caused by the formation of natural gas hydrates in the upper part of the Dolgan Formation or the bottom-most Dorozhkov Formation have contributed to gas trapping.[9]

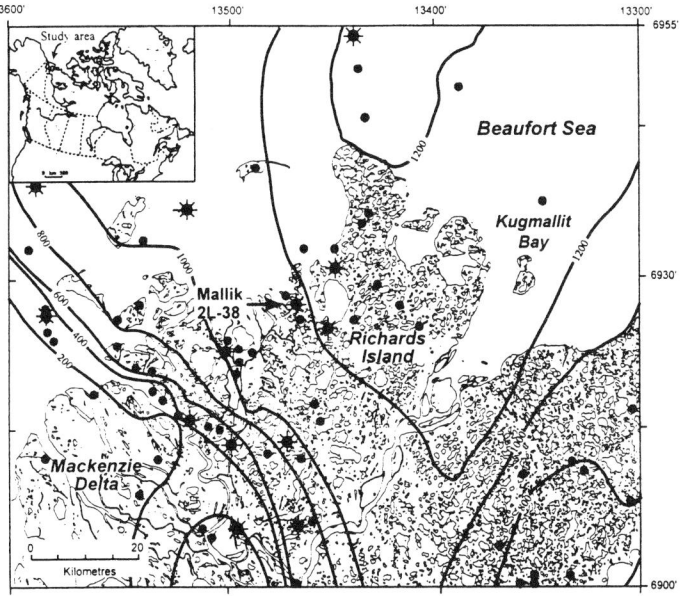

FIGURE 2. Location map of the Mackenzie Delta region showing JAPEX/JNOC/GSC Mallik 2L-38 drill site. The base map also shows the calculated thickness of the methane hydrate stability zone based on ground temperature data and those exploration wells with well log interpreted gas hydrate occurrences (see Ref. 24; modified from Judge and Majerowicz, 1992; Judge et al., 1988). Key: ✷, gas hydrate; ●, well; ⌒, contours of base methane hydrate (m).

Rifting and sea floor spreading during the Lower Cretaceous created the present Canadian Basin and the passive continental margin of the Beaufort Sea. Uplift of the Brooks Range by thrusting in the Cretaceous and early Tertiary provided vast quantities of sediments that were deposited over the ancestral rifts and progressed seaward to construct the sedimentary prism underlying the present shelf, slope, and rise in the Canadian and Alaskan Beaufort Sea offshore.[11]

Dallimore et al.,[24] report that for the selection of the drill site of the subsequent JAPEX/JNOC/GSC Mallik 2L-38 Research Well primary consideration included the thickness and amount of the gas hydrates, geologic setting, availability of background data required for safe execution of the program, and site logistics. During the course of the site selection process, 25 onshore wells with previously identified gas hydrates in the Mackenzie Delta area were reexamined.[23] Four hydrocarbon wells with Mallik prefixes were drilled by Imperial Oil Limited in the early 1970s. These wells were drilled within 15 km of each other along a Northwest-Southeast trend, roughly coincident with a broad faulted anticline feature. Each well was apparently targeted to examine conventional oil and gas potential of mid-Tertiary and older sediments below 1,500 m depth. This anticline, delineated by 13 wells, covers an area approximately 10 × 20 km. Dallimore et al.[24] report after Dixon et al.[21] that of the above mentioned four wells, "only one Mallik L-38 was ranked as a discovered resource with an assigned pool size of natural gas of between 0.3 to 2.8 billion m^3 (10 to 100 bcf)". Reinterpretation of older data indicated that Mallik L-38 encountered about 110 m of gas-hydrate-bearing strata in distinct layers between 819 m and 1111 m. Mallik 2L-38 was drilled at the same location as the Mallik L-38 well, drilled by Imperial Oil Limited in 1972.

According to Collett et al.,[15] a relatively thick gas-hydrate–bearing stratigraphic section within the depth interval from 897.25 to 1,109.5 m in the Mallik 2L-38 well has been confirmed. Downhole log data also confirm the occurrence of a free-

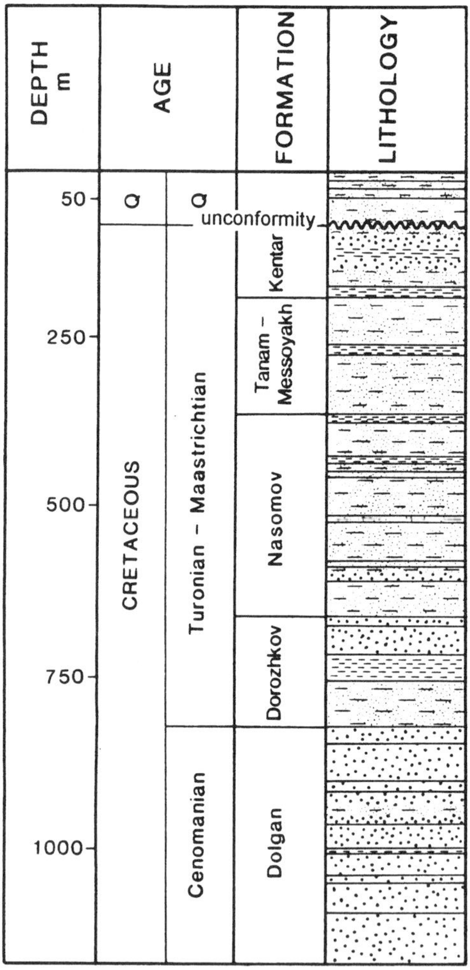

FIGURE 3. Schematic stratigraphic column, Messoyakh gas field (see Ref. 1). Key: ⊡, siltstone; ⊟, mudstone; ⊡, sandstone.

gas–bearing unit at the predicted base of gas hydrate stability zone in the Mallik 2L-38 well. Sediment porosities for the gas-hydrate-bearing units in the Mallik 2L-38 well average 28%. Gas-hydrate saturations within the gas-hydrate–bearing units of the Mallik 2L-38 well, calculated from the "standard" Archie relation, average 47%.

STRATIGRAPHY

The Messoyakh methane hydrates are hosted in a 74 m-thick Dolgan Formation of Cenomanian age (see FIGURES 3 and 4). This Formation occurs between depths of 850 to 780 m (the net thickness of production zone is 78 m) and it is composed of sands and poorly indurated sandstone; with up to 30% interbeds of shale. The sandstone porosity ranges from 16 to 38%, averaging 25%. Permeability ranges from 10 to 1,000 md, with an average of 125 md. Temperature ranges from 10 to 12°C.

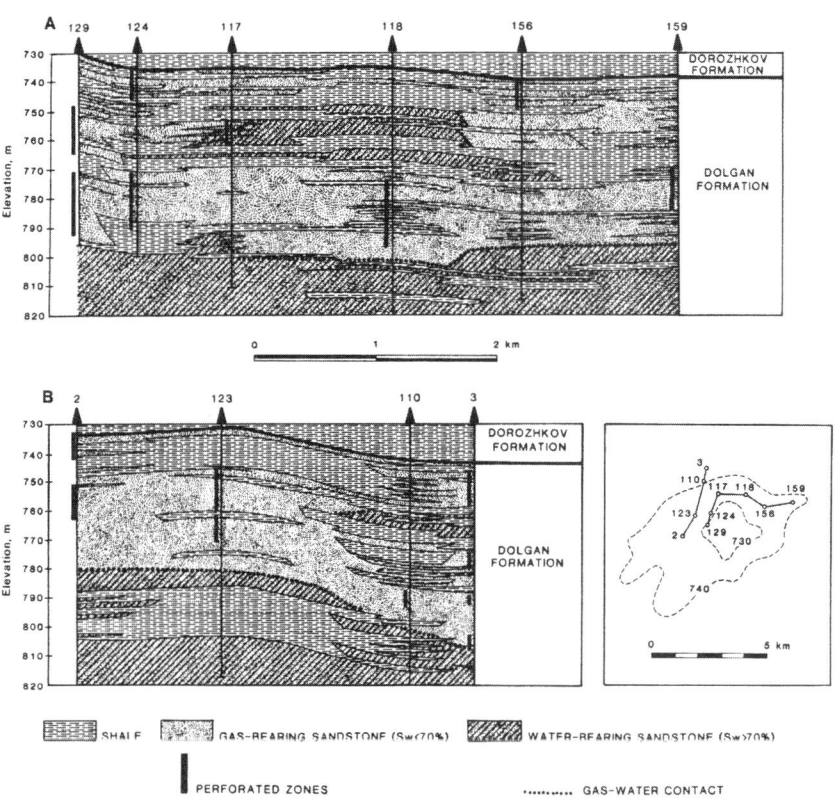

FIGURE 4. Cross sections of the Messoyakh gas field. Depths are sub-sea. **A.** *Dark vertical lines* represent permafrost zones (see Ref. 9; adapted from Ref. 5).

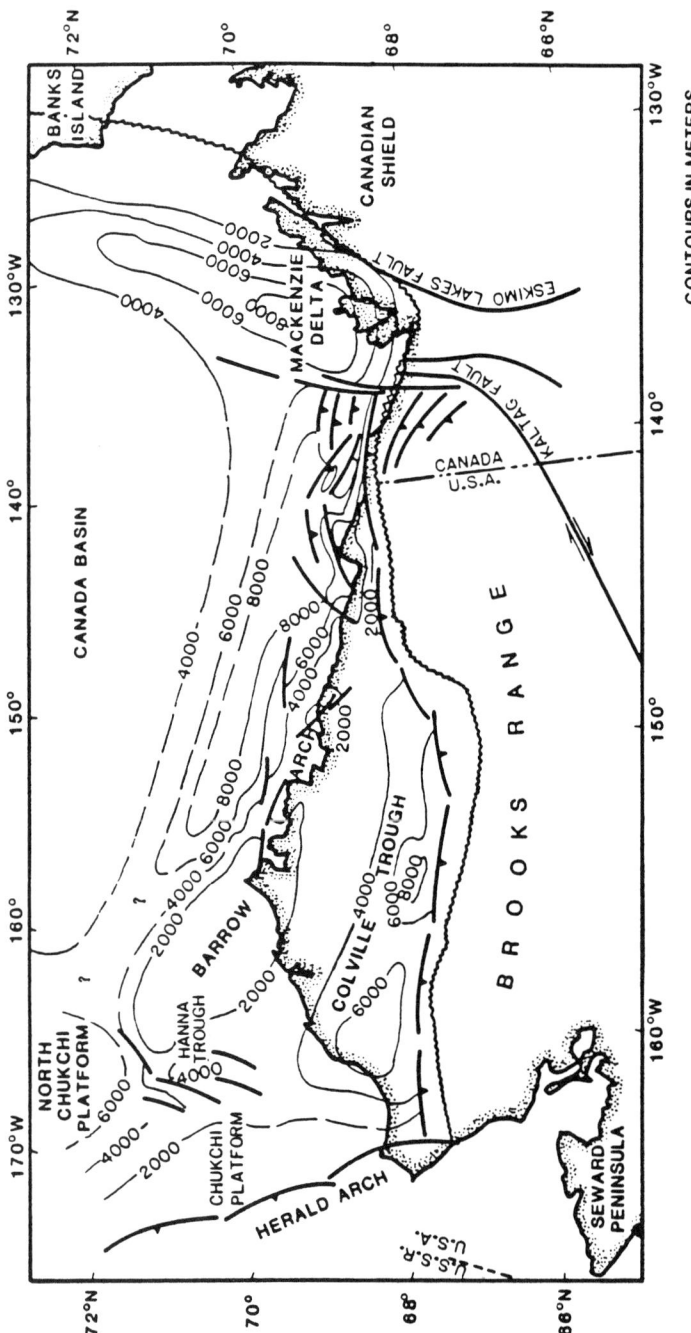

FIGURE 5. Total sediment thickness, Beaufort Sea and Mackenzie delta regions (modified from Ref. 11; adapted from Hubbard *et al.*, 1987).

Most of the sediments underlying the Beaufort Sea, including the Mackenzie Delta and its offshore region, also extensively studied by Finley and Krason,[11] are terrigeneous clastic detritus derived from the uplift of the Brooks Range and associated structures (see FIGURE 5). The continental slope beneath the Beaufort Sea offshore of Canada is dominated by sediment accumulation from the Mackenzie River. The Mackenzie Delta has prograded steadily seaward throughout the Tertiary (Willumsen and Cote, 1982). The slope offshore of the Mackenzie Delta is composed of rapidly deposited delta-front sediments and turbidite deposits (see FIGURE 6).

Although in the Mackenzie Delta, especially at Mallik 2L-38, methane hydrates occur in much younger stratigraphic units (Oligocene through Pliocene), the host rocks lithology is generally similar to that in the Messoyakh gas field. According to Jenner *et al.*,[26] thirty-seven meters of core were recovered between 886 m and 952 m to characterize gas hydrate-bearing host strata. The sediments are part of the early Miocene to late Oligocene lower Mackenzie Bay and upper Kugmallit. Sequences that regionally comprise cemented sandstone are interbedded with sand and silt.

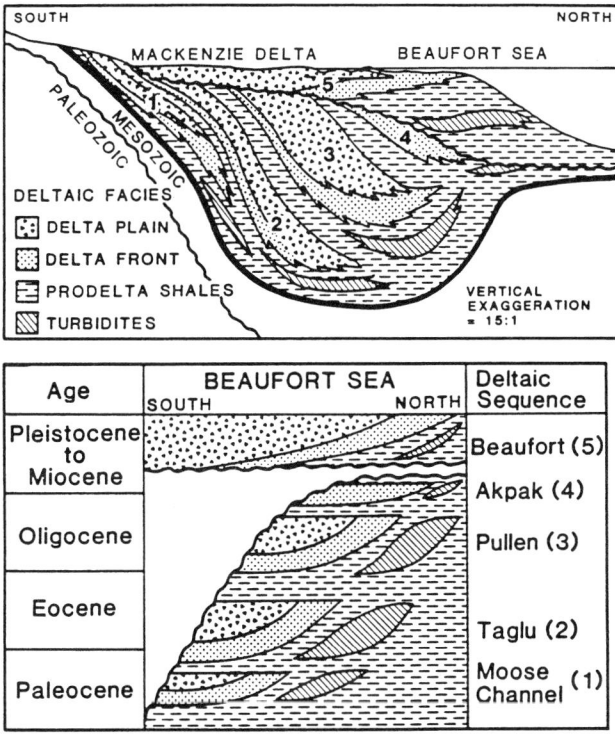

FIGURE 6. Cross section of the Tertiary Delta sequences, Mackenzie Delta and Beaufort Sea, offshore Canada (after Finley and Krason, 1988, adapted from Willumsen and Cote, 1982).

Three sedimentary facies with distinct sedimentary structures and textures were weakly bioturbated clayey silt with thin interbeds of low rank fissile coal and silty sand. The overlying facies, between 926.5 m and 908.5 m are comprised of sand and interbedded gravel and is the predominant host strata for gas hydrate observed in the core. Fining upward succession of basal matrix-supported gravel to pebbly sand and fine sand occur in the succession. A dolomite-cemented sandstone forms a distinct subfacies at the base of this succession between 926.5 m and 925.0 m; the basal contact at 926.5 m is tentatively interpreted from sedimentologic data as the boundary between the Kugmallit and Mackenzie Bay Sequences. The uppermost facies from between 908.5 and 886.2 m are composed of a fining upward succession of basal fine to medium sand and interbedded gravel into fine sand with an agradational increase in silt content.

Visible, pore-space methane hydrate was identified from core samples primarily within the sand and gravel facies between 952.2 m to 926.5 m. Gas hydrate content was variable and is interpreted to be partly governed by lithology. Within the pebbly sand, white, opaque gas hydrate was observed as pore fillings, particle coatings, and along nodules less than 2 cm. Infrequently, clasts are supported by massive hydrate, forming an estimated 20% to 30% gas hydrate by volume. Pore-space hydrate predominates in the well sorted, clean sand; rare megascopic hydrate occurs as particle coatings or veins less than 1 mm thick. Visible gas hydrate was rarely observed in silt-rich sediment.

GAS RESOURCES

Gas reserves for the Messoyakh gas field have been reported by numerous authors and they differ considerably, ranging from 37 to 400 bn m^3 (1.3 to 14 tcf). Finley and Krason (1992), after thorough analysis and critical evaluation of all available information, also assessed gas resources in the Messoyakh gas field. As well as considering geological and interrelated factors, they used production data reported by Makogon,[3,4] especially his projected pressure curve (see FIGURE 7), which shows that the free-gas reservoir that would decline from 7.8 to 4 Mpa to produce 8 bn m^3 of free gas in the absence of hydrates. Finley and Krason explain that "applying the gas state equation ($PV = ZnRT$) to that pressure drop indicates that in-place free gas originally totaled 15 bn m^3 (530 bcf). At 78 atm and 10° to 12°C, the free gas originally present in the reservoir occupied 1.6×10^8 m^3 (5.6×10^9 ft^3). At a mean S_W of 40%, the free gas would occupy 2.5×10^8 m^3 of pore space or 1×10^9 m^3 of 25% porosity rock.

Based on the derived volume of the free-gas reservoir, Finley and Krason calculated the amount of gas in the hydrate portion of the reservoir. Assuming the reservoir at Messoyakh is two-thirds hydrate and one-third free gas, the hydrate portion of the reservoir occupies about 5×10^8 m^3 of pore space. Assuming that 40% of the reservoir pore space is occupied by hydrates, then each m^3 of pore space contains 72 m^3 of gas from the hydrate and 55 m^3 of gas in the remaining pore space for a total 127 m^3. Based on these assumptions, Finley and Krason calculated that the 2×10^9 m^3 hydrate portion of the reservoir originally contained 62 bcf (2.2 tcf) of gas in place in both hydrate and free gas.

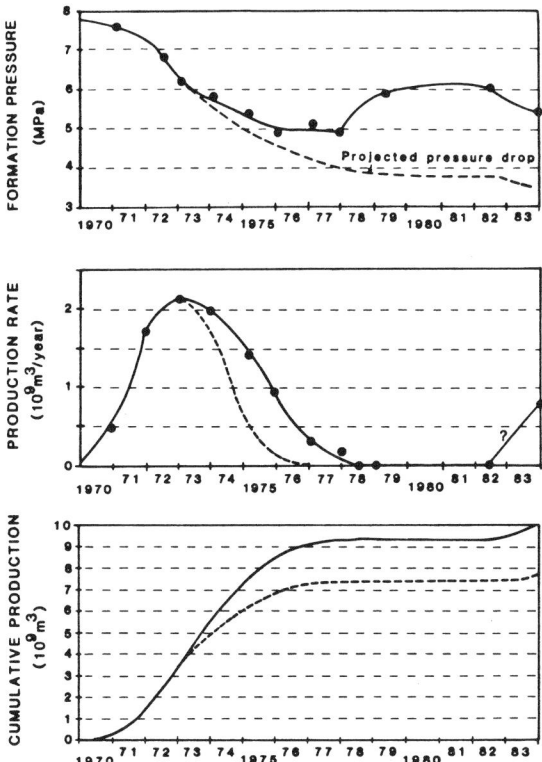

FIGURE 7. Production history of the Messoyakh gas field. Dashed lines indicate projected formation pressure, production rate, and cumulative production from the deeper, gas-free zones plus contributions from the hydrate gas zone above. Shut-in period during which reservoir pressure increased, extended from mid-1978 to mid-1981. The pressure buildup and latter increase in production rate resulted from the dissociation of hydrates to free gas (from Ref. 9; adapted from Refs. 3 and 4).

Although gas resources for the anticlinal structure in which the JAPEX/JNOC/ GSC Mallik 2L-38 Research Well is located and drilled have not been revealed, certainly, it is evident that this well confirmed the presence of a very impressive amount of gas in a relatively thick (212.25 m) gas-hydrate- bearing stratigraphic section, at a depth interval from 897.25 to 1,109.5 m. This well also confirmed the presence of free gas below the deepest downhole log inferred gas-hydrate.[27] Moreover, prior to drilling the Mallik 2L-38 well, Collet and Dallimore[22] made "a simplified resource assessment of the volume of natural gas contained within the gas hydrate sequence in Mallik L-38" and suggested that "about 4.3 bn m^3 (about 150 bcf) of gas would occur if the gas hydrate layers were continuous over a one square kilometer area surrounding the drill site". Then, if considering "the corrected density log derived sediment porosities for the gas-hydrate-bearing units in the Mallik 2L-38 well average

28%. Gas-hydrate saturations within the gas-hydrate-bearing units of the Mallik 2L-38, calculated from [the] standard Archie relation, average 47%",[28] it is possible to estimate that potential gas reserves within the Mallik anticlinal structure may be an order of magnitude larger. Finley and Krason[11] estimated that the continental shelf offshore of Canada, adjacent to the Mackenzie Delta, contains 1×10^{12} m^3 (140 tcf) of gas in hydrate form.

PRODUCTION MODEL

Although the primary objective for drilling of the JAPEX/JNOC/GSC Mallik 2L-38 well in the Mackenzie Delta was "a comprehensive scientific research program to study the geology, geochemistry, geophysics, and engineering properties of an arctic gas accumulation",[24] the information and data with its preliminary evaluation, included in the Proceedings of the Symposium, held in Chiba City, Japan, mentioned above are encouraging for production tests and a feasibility study for development of an evidently giant gas field. Certainly this idea and recommendation is not a prior one, especially for JAPEX and JNOC. However, it has to be realized that in spite of the importance and necessity of thorough research, scientists often prefer to expand their investigation instead of swift use of the research results for practical purposes and much quicker benefits.

Since the Messoyakh is the only gas field where natural gas has been produced commercially from hydrates and free gas, among various technological production methods that could be considered for production tests in Mallik anticlinal structure at Mackenzie Delta, priority should be directed to the model already tested.

With reference to production methods applied to and rates obtained at the Messoyakh gas field, among numerous other authors, Finley and Krason (1992) report the following: in view of the difficulties in efficiently producing gas from hydrated reservoirs, two approaches were taken to maximize production at Messoyakh. Since the deeper portions of the Messoyakh reservoir in which free gas is present are similar to conventional gas reservoirs, wells were preferentially perforated beneath the gas hydrate stability zone. Stimulation of hydrate dissociation was attempted in wells completed in the hydrate zone.

Compounds that had been found effective in preventing hydrate formation in pipelines were injected into a number of wells at Messoyakh.[6,29] Methanol and $CaCl_2$ were used alone and in mixtures. Methanol was found to be more effective at increasing gas production, but $CaCl_2$ cost less. The most economic treatments were mixtures of the two compounds.

Makogon et al.[1] reported on the methanol treatment of Wells 133 and 142. Similar production levels were obtained before and after methanol application, but higher pressures could be maintained at a given production level subsequent to the treatment. Production figures reported by Makogon et al.[1] are normalized to account for the difference in well-head pressure and are presented in FIGURE 8.

One limit on hydrate production rates is retrograde hydrate formation in some holes. The dissociation of hydrates by depressurization lowers formation temperatures to the point that hydrates may reform at and clog formation pores. Sumets[29]

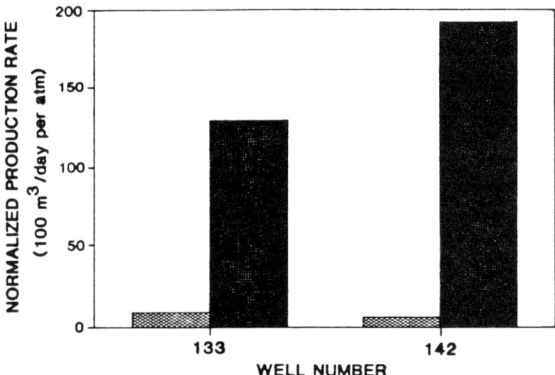

FIGURE 8. Effect of methanol injection on gas production rates at Messoyakh gas field (from Ref. 9; adapted from Ref. 1). Key: ▨, no treatment; ■, methanol treatment.

discussed methanol stimulation treatment methods that minimized retrograde hydrate formation. The most effective treatment involved injection of about 3 to 4 m^3 of methanol over a period of 24 to 26 hours under a pressure of 100 to 150 atm. Subsequently, air was forced into the perforated zone to drive the methanol deep into the formation. The methanol dissociated the hydrates and prevented hydrates from reforming in the reservoir during production.

Makogon[3,4] and Trofimuk et al.[30] published production rates and formation pressure data from 13 years of production at Messoyakh (FIG. 7). Production increased from 1970 when the field went on-stream to reach a peak rate of about 2.1 bn m^3/yr (74 bcf/yr) in 1972. Production diminished until 1978, when the field was shut down. Production was recommenced in 1982, although the three principal sources for the production data[3,4,30] give conflicting estimates of rates for the more recent production cycle.

The contribution of hydrate gas to the total production from the Messoyakh field is shown in FIGURE 7. The dashed line on the formation pressure curve in FIGURE 7 indicates the pressure decline originally estimated for Messoyakh based on initial declines and calculated reservoir volume. The estimated decline shown by the dashed line assumes that the reservoir was occupied by free gas and water, with no hydrates present. The actual pressure curve diverges from the estimated pressure curve at a pressure of about 5.8 MPa when gas from dissociating hydrates began contributing substantially to the formation pressure. During this time, hydrates continued to dissociate, releasing free gas into the reservoir and increasing the formation pressure to 6.2 Mpa. Subsequent production has been accompanied by decreasing formation pressure as withdrawal rate exceeds the dissociation rate of the hydrates.

Relatively high production rates may be possible from the field in the future. Low production rates have been reported for the production period subsequent to the shut down period. Makogon[3] and Trofimuk et al.[30] reported rates of 0.3 to 0.7 bcm/year (10 to 25 bcf/year) of gas, whereas Makogon[4] indicated a gas production rate of about 0.2 bcm/year (7 bcf/year). During the shutdown period formation pressure increased to about 6.2 MPa. The annual gas production rate of about 2 bcm/year (7 bcf/year) was attained in 1973, when the formation pressure was in the range of 5.8 to 6.2 MPa. Although other factors may dictate the low production rates of the most recent production phase, the reservoir pressure attained during the shut down period is sufficient to permit much higher rates.

CONCLUSIONS

It is evident that the Messoyakh of the Western Siberia gas field and Mallik anticlinal structure in the Mackenzie Delta, Northwest Territories of Canada may be considered textbook examples of conventional type gas deposits. The anticlinal traps, with methane hydrates and free gas underneath, at relatively shallow depth, are located within larger uplifts. Both deposits are located within major hydrocarbon provinces, in arctic climatic conditions, with a thick permafrost zone overlying a thick sequence, of mostly elastic sediments, ranging in age from Jurassic through Quaternary. Both hydrocarbon provinces have been extensively explored, methane hydrates appear to be widely spread and trap an enormous volume of natural gas. Their presence, previously interpreted in at least 25 wells in the Mackenzie Delta and in numerous wells offshore, have been successfully confirmed by JAPEX/JNOC/GSC Mallik 2L-38 Research Well. The thick sequence and physical characteristics of the methane-hydrate-bearing strata, as preliminarily determined and reported by numerous Japanese, Canadian, and U.S. scientists, indicate the discovery of a giant gas field in the Mackenzie Delta.[19]

Considering the latter context, it therefore appears to be prudent, already at this stage of the research results, to compare them with the abundance of information pertaining to the production methods, rates, and interrelated difficulties applied and experienced at the Messoyakh gas field.

Since this author thoroughly studied petroleum geology, formation, and stability of gas hydrates of both regions and many others,[9,11–13,31] he concludes that the challenge of using methane hydrates as an energy resource, especially from the Mackenzie Delta, is already foreseeable in the near future.

REFERENCES

1. MAKOGON, Y.F., F.A. TREBIN, A.A. TROFIMUK & V.P. CHERSKII. 1971. Obnaruzheniye zalezhi prirodnogo gaza v tverdom (gazogidratnom) sostoyanii (Detection of a pool of natural gas in a solid [hydrated state]). Doklady Akademii Nauk USSR **196**(1): 197–200.

2. MAKOGON, Y.F. 1974. Gidrati prirodnykh gazov (Hydrates of natural gas; trans. W.J. Cieslewicz, 1978). Geoexplorers Associates, Inc., Denver.

3. MAKOGON, Y.F. 1984. Razrabotka gazogidratnoy zalezhi. (Production from natural gas hydrate deposits.) Gazovaya Promishliennost **10**: 24–26.

4. MAKOGON, Y.F. 1988. Natural gas hydrates-the state of study in the USSR and perspectives for its using. Third Chemical Congress of North America, Toronto, Ontario, Canada, June 1988.

5. SAPIR, M.H., E.N. KHRAMENKOV, I.D. YEFREMOV, G.D. GINZBURG, A.E. BENIAMINOVICH, S.M. LENDA & V.L. KISLOVA. 1973. Geologicheskie i promislovo-geofizicheskie osobennosti gazogidratnoi zalezhi Messoiakhskogo gazovogo mestorozhdenia. (Geologic and geophysical features of the gas hydrate deposits in the Messoyakh field.) Geologiia Nefti i Gaza **6**: 26–34.
6. SHESHUKOV, N.L., A.F. BEZNOSIKOV, Y.H. KRAMENKOV & I.D. YEFREMOV. 1972. O zaleganii gaza v gidratnom sostoyanii na Mesoyakhskom mestorozhdenii. (Occurrence of gas in the hydrate state in Messoyakh field.) Gazovoe Delo **6**: 8–10.
7. SHESHUKOV, N.L. 1973. Priznaki zalezhei gaza, soderzhazchikh gidrati. (Features of gas bearing strata with the hydrates.) Geologiia Nefti i Gaza **6**: 20–26.
8. MEYERHOFF, A.A. 1980. Petroleum basins of the Soviet Arctic. Geological Magazine **117**(2): 101–210.
9. KRASON, J. & P. FINLEY. 1992. Messoyakh gas field—Russia, West Siberian basin; Treaties of Petroleum Geology Atlas of Oil and Gas Fields, Structural Traps VII, Am. Assoc. Petrol. Geol. 197–220.
10. KVENVOLDEN, K.A & L.A. BARNARD. 1983. Hydrates of natural gas in continental margins. *In* Studies in Continental Margin Geology. J.S. Watkins &C.L. Drake, Eds.: 631. Am. Assoc. Petroleum Geologists Mem. 34.
11. FINLEY, P. & J. KRASON. 1988. Basin analysis, formation and stability of gas hydrates of the Beaufort Sea; geological evolution and analysis of confirmed or suspected gas hydrate localities. US Department of Energy publication, DOE/MC/21181-1950. **12**: 212.
12. FINLEY, P. & J. KRASON. 1989. Basin analysis, formation and stability of gas hydrates; geological evolution and analysis of confirmed or suspected gas hydrate localities. Summary Report, US Department of Energy publication, DOE/MC/21181-1950, **15**: 1.
13. KRASON, J. 1994. Study of 21 marine basins indicates wide prevalence of hydrates. Offshore, Penn Well Publ, 34–35.
14. MAKOGON, Y.F. 1978. Hydrates of natural gas; Geoexplorers Associates, Inc. Denver, Colorado.
15. COLLETT, T.S., K.J. BIRD, K.A. KVENVOLDEN & L.B. MAGOON. 1988. Geologic interrelations relative to gas hydrates within the North Slope of Alaska. US Geological Survey, Open-File Report 88–389.
16. COLLETT, T.S. 1993. Natural gas hydrates of the Prudhoe Bay and Kuparuk River area, North Slope, Alaska. Am. Assoc. Petrol. Geol. Bull. **77**(5): 793–812.
17. KVENVOLDEN, K.A. 1988. Methane hydrate—a major reservoir of carbon in the shallow geosphere? Chemical Geology **71**: 41–51.
18. COLLETT, T.S. & V.A. KUUSKRAA. 1998. Hydrates contain vast store of world gas resources; Oil Gas J. May 11, 90–95.
19. JNOC-TRC. 1998. Proceedings of International Symposium on Methane Hydrates Resources in the Near Future? JNOC-TRC, Chiba City, Japan, October 20–22, 1998, 399.
20. KRASON, J. 1999. Methane hydrates impetus for research and exploration; Offshore, **59**: 76–79.
21. DIXON, J., G.R. MORRELL, J.R. DITRICH, G.C. TAYLOR, R.M. PROCTOR, R.F. CONN, S.M. DALLIMORE & J.A. CHRISTIE. 1994. Petroleum resources of the Mackenzie Delta and Beaufort Sea; Geological Survey of Canada, Bull. 474.
22. COLLETT, T.S. & S.R. DALLIMORE. 1998. Quantitative assessment of gas hydrates in Mallik L-38 Well, Mackenzie Delta, N.W.T. Proc. 7th Int. Confer. Permafrost, Yellowknife, Canada, June 1998.
23. DALLIMORE, S.R. & T.S. COLLETT. 1998. Gas hydrates associated with deep permafrost in the Mackenzie Delta, N. W. T., Canada; Regional overview; Proceedings of the 7th Int. Conf. of Permafrost, Yellowknife, Canada, June 1998.
24. DALLIMORE, S.R., T. UCHIDA & T.S. COLLETT. 1998. JAPEX/JNOC/GSC Mallik L2-38 gas hydrate research well: overview of science program. Proc. Int. Sympos. Methane Hydrates: Resources in the Near Future? JNOC- TRC, Japan, Oct. 20–22, 1998, 311–318.
25. KULAKOV, Y.N. & G.P. MAKHOTINA. 1985. Recent tectonics of the Yenisey-Khatanga regional downwarp. Petroleum Geology **23**: 77–85.

26. JENNER, K.J., S.R. DALLIMORE, F.M. NIXON, W.J. WINTERS & T. UCHIDA. 1998. Sedimentology and methane hydrate host strata from JAPEX/JNOC/GSC Mallik 2L-38. Proc. Int. Sympos. Methane Hydrates: Resource in the Near Future? JNOC-TRC, Japan, Oct. 20–22, 1998, 319–326.
27. COLLETT, T.S. & G.D. GINSBURG. 1998. Gas hydrates in the Messoyakha gas field of the western Siberian basin—re-examination of the geologic evidence; Int. J. Offshore Polar Eng. **8**(1): 22–29.
28. COLLETT, T.S. 1998. Methane hydrate: an unlimited energy resource? *In* Methane Hydrates: Resource in the Near Future? Proc. Int. Sympos. JNOC-TRC, Japan, Oct. 20–22. 1–13.
29. SUMETS, V.I. 1974. Predotvrashchenie gydratobrazovaniya v prizaboynoy zone. (Prevention of hydrate formation in gas wells zone.) Gazovaya Promishlennost **2**: 24–26.
30. TROFIMUK, A.A., Y.F. MAKOGON, M.V. TOLKACHEV & N.V. CHERSKII. 1984. Some distinctive features of the discovery, prospecting and exploitation of gas hydrate deposits. Geologiia i Geofizika **25**(9): 1–7.
31. KRASON, J. 1998, Gas hydrates in the context of basin analysis. Proc. Int. Sympos. Methane Hydrates: Resource in the Near Future?; JNOC-TRC, Japan, Oct. 20–22, 1998, 27–40.

Microscopic Character of Marine Sediment Containing Disseminated Gas Hydrate

Examples from the Blake Ridge and the Middle America Trench

THOMAS D. LORENSON[a]

United States Geological Survey, Menlo Park, California 94025, USA

ABSTRACT: The presence of disseminated gas hydrate was inferred based on pore fluid geochemistry and downhole logging data, but was rarely observed at Ocean Drilling Program (ODP) Leg 164 (Blake Ridge), and Leg 170 (Middle America Trench, offshore from Costa Rica) drilling sites. Gas hydrate nucleation is likely to occur first in larger voids rather than in constricted pore space, where capillary forces depress the temperature-pressure stability field for gas hydrate formation. Traditional macroscopic descriptions of sediment fail to detect the microscopic character of primary and secondary porosity in sediment hosting disseminated gas hydrate. Light transmission and scanning electron microscopy of sediments within and below the depth of gas hydrate occurrences reveal at least four general types of primary and secondary porosity: (1) microfossils (diatoms, foraminifera, and spicules) void of infilling sediment, but commonly containing small masses of pyrite framboids; (2) infauna burrows filled with unconsolidated sand and or microfossil debris; (3) irregularly shaped pods of nonconsolidated framboidial pyrite; and (4) nonlithified volcanic ash.

INTRODUCTION

Downhole logging of Ocean Drilling Program (ODP) holes at Leg 164 (Blake Ridge) and Leg 170 (offshore Costa Rica) sites (see FIGURE 1)[1,2] demonstrates that the majority of gas hydrate deposits occur as disseminated deposits. However, disseminated gas hydrate samples were rarely recovered and preserved, nor could the samples be examined microscopically with *in situ* gas hydrate still intact. Thus, the *in situ* distribution of these disseminated gas hydrate accumulations, inferred to occur worldwide, remains poorly understood.[1] Gas hydrate accumulations in Blake Ridge sediment typically comprise no more than about 15% of the bulk volume of the host sediment and are known to be concentrated in coarser-grained sediment relative to finer-grained sediment.[3] It is suggested that gas hydrate nucleation and growth are much more likely to occur in open pore space or voids rather than in more confined areas between sediment grains.[4,5] Capillary pressure forces within pore

---

[a]Address for correspondence: Tom Lorenson, U.S. Geological Survey, Western Coastal and Marine Geology Team, 345 Middlefield Road, MS-999, Menlo Park, CA 94025, USA. Voice: 650-329-4186; fax: 650-329-5441.

tlorenson@usgs.gov

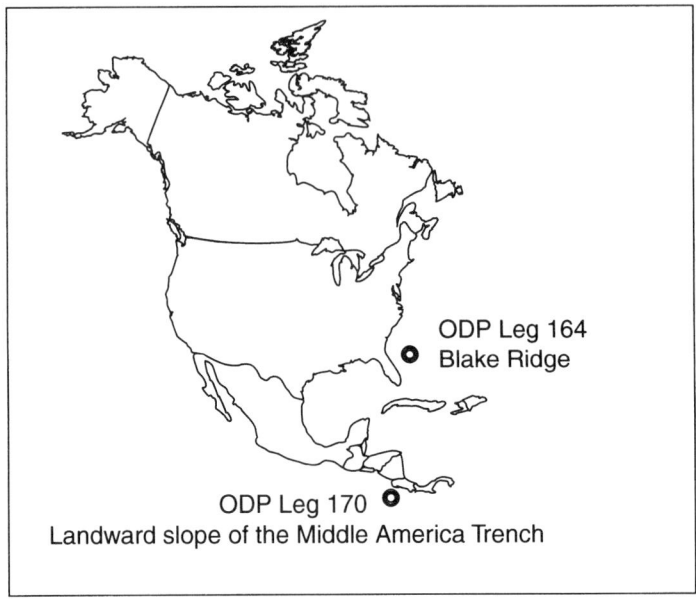

FIGURE 1. Location of ODP drilling sites referred to in text.

spaces and pore throats of fine-grained sediment depress the temperature for gas hydrate formation by as much as 4°C at a given pressure.[6–8] Thus, gas hydrate formation should be expected to occur first in areas of high porosity, coarser-grained sediment, rather than in much smaller, constricted, pore spaces between clay grains within finer-grained sediment. Closer examination of gas-hydrated sediment from two ODP sites on the sedimentary drift deposits of the Blake Ridge and from the Middle America Trench, an active margin offshore from Costa Rica, reveals that the zones of high primary and secondary porosity available for gas hydrate formation is quite varied.

RESULTS AND DISCUSSION

Sediment recovered during Leg 164 was examined with light transmission microscopy immediately after core recovery and photographed at magnifications usually less than 100×. For these observations a small piece of core was removed and broken open to reveal a fresh unoxidized surface. Scanning electron microscopy (SEM) was completed on Legs 164 and 170 core material frozen in liquid nitrogen. These samples were thawed, broken open, finely coated with gold and/or carbon, and then viewed by SEM.

Large-void primary and secondary porosity (visible porosity, not including small scale, less than about 10 nanometer porosity between clay particles) in Blake Ridge sediments consists of foraminifera, diatoms, accumulations of spicules, sand-like

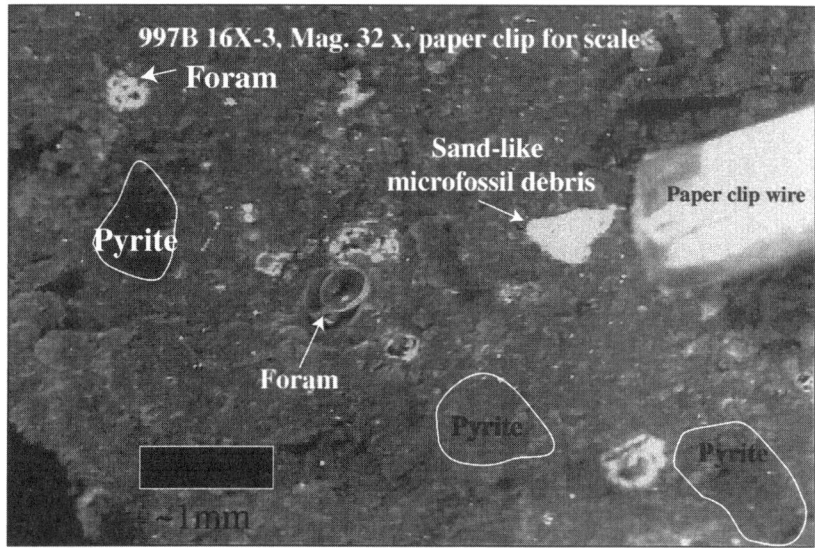

FIGURE 2. Microphotograph of the principal forms of primary and secondary porosity as indicated in ODP leg 164 sediment. Note the close spatial relationship. Sediment is from 997B 16X3, 505 meters below the sea floor (mbsf).

microfossil debris, and pods or lenses of framboidal pyrite seen collectively in FIGURE 2. These forms of porosity occurred throughout the drilled section (750 m) and appeared to be more frequent in stratigraphic intervals known to contain more disseminated gas hydrate. Such areas were identified primarily by pore water chlorinity anomalies and downhole resistivity and acoustic logging.[1] Each of these areas of enhanced porosities are in contrast with the clay-rich sediment matrix, which offers less room for gas hydrate crystals to grow.

FIGURE 3 shows a suite of microfossils commonly found in Blake Ridge sediments. White, sand-like accumulations of spicules or microfossil debris were found in sheets, stringers, and pods and they appear to represent infilling of former infaunal burrows or paleoburrows. A variety of foraminifera and diatom tests are also commonly present, each one void of sediment, with the exception of minor amounts of secondary framboidal pyrite that grew within the test.

Framboidal pyrite is common in Blake Ridge sediment (FIGURE 4). Iron monosulfide precursors of framboidal pyrite are in Blake Ridge sediment[1] and are known to be associated with organic-rich sediment or microenvironments supersaturated with iron monosulfide mineraloids.[9] However, the mechanism of framboidal pyrite formation and its direct link with organic matter in this regard is still unknown.[9] In Blake Ridge sediment, the nondisseminated occurrence of framboidal pyrite and association with microfossils (possible organic matter source) likely links its origin with former concentrations of organic matter. Regardless of the genesis of framboidal pyrite, its frequent occurrence and high porosity in Blake Ridge sediment makes framboidal pyrite locations ideal places for gas hydrate nucleation and growth.

Disseminated gas hydrate was recovered from the landward slope of the Middle America Trench as a pore filling cement in beds of volcanic ash[2] (see FIGURE 5). Within the depth range of 234 to 280 meters below the sea floor, the ash is mainly composed of unaltered, vesicular siliceous shards lacking significant compaction, thus retaining much of its original primary porosity. Visual inspection of an ash sample revealed no visible gas hydrate, yet when a sample was allowed to warm in a syringe, unusually vigorous gas expansion followed, confirming the presence of gas hydrate. The ash remaining after gas hydrate dissociation was completely non-consolidated, thus disseminated gas hydrate in these ash deposits is likely to occur as pore-filling cement. In contrast to gas-hydrated sediments of the Blake Ridge sediments, gas-hydrated sediments of the Middle America Trench appear to be largely devoid of framboidal pyrite, spicules, or microfossils.

FIGURE 3. SEM photographs of primary porosity in OPD Leg 164 sediment within microfossils. The **upper left** photograph shows white sand-like accumulations of spicules or microfossil debris. **Center** shows a foraminifer with minor amounts of framboidal pyrite inside. The photograph on the **upper right** shows part of a test from a diatom. The primary porosity available in microfossils and fossil debris are likely to provide ideal loci for gas hydrate nucleation and growth. Photographs are from ODP Leg 164 (997A 69X2, 510 mbsf).

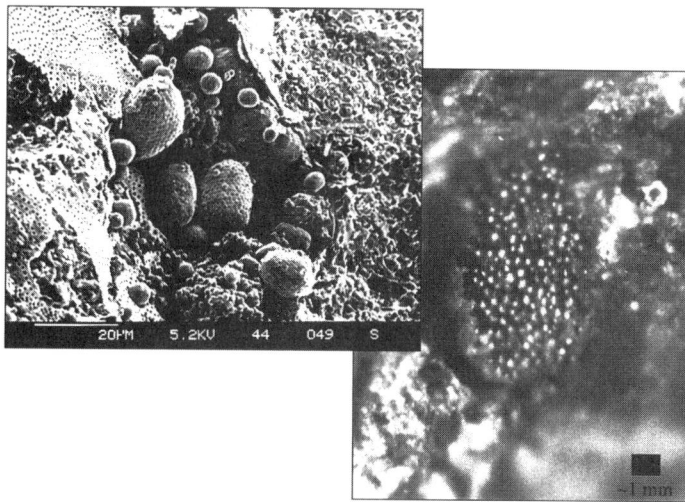

FIGURE 4. SEM photograph (**upper left**) of framboidal pyrite associated with diatom tests. Light transmission microphotograph of a typical pod or lens of framboidal pyrite. Photographs are from ODP Leg 164 (997A 69X2, 510 mbsf).

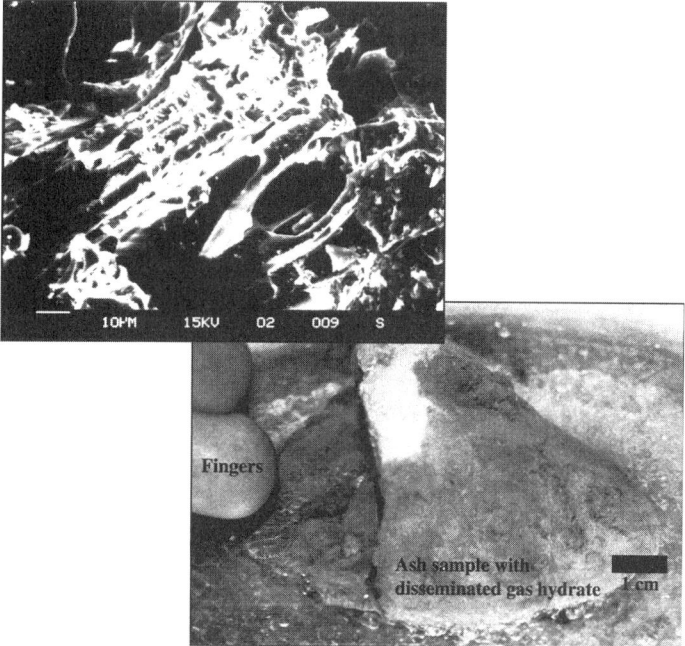

FIGURE 5. SEM photograph (**upper left**) and photograph of gas hydrate-cemented volcanic ash from ODP Leg 170 Hole 1041B 11R5, 256 mbsf. Disseminated gas hydrate is not visible in this ash sample shown in liquid nitrogen awaiting dissociation and gas analyses. The highly porous, vesicular nature of the volcanic ash is evident from the SEM photograph.

CONCLUSIONS

Observations of gas-hydrate-bearing sediment at microscopic resolution reveal a wide variety primary and secondary porosity, including the interiors of foraminifera tests, diatom tests, paleoburrows filled with sand-like microfossil debris, framboidal pyrite accumulations, and vesicular volcanic ash. Because disseminated gas hydrate accumulations preferentially occur in coarse-grained sediment, it is suggested that disseminated gas hydrate is more likely to occur in sediment with higher primary and secondary porosity that contain more microfossils, paleoburrows, framboidal pyrite accumulations, and vesicular volcanic ash rather than sediment lacking these large-aperture pore spaces.

REFERENCES

1. PAULL, C.K. *et al.* 1996. Proceedings of the Ocean Drilling Program, Initial Reports, **164**. College Station, TX.
2. KUMURA, G. *et al.* 1997. Proceedings of the Ocean Drilling Program, Initial Reports, **170**. College Station, TX.
3. GINSBURG, G.D. *et al.* 2000. Sediment grain-size control on gas hydrate presence, ODP Sites 994, 995, and 997. Proceedings of the Ocean Drilling Program, Scientific Results, **164**. College Station, TX. 237–246.
4. GINSBURG, G.D. *et al.* 1997. Methane migration within the submarine gas-hydrate stability zone under deep-water conditions. Mar. Geol. **137**: 49–57.
5. LORENSON, T.D. & T.S. COLLETT. 2000. Gas content and composition of gas hydrate from sediments of the southeastern North American continental margin. Proceedings of the Ocean Drilling Program, Scientific Results, **164**. College Station, TX. 37–46.
6. CLENNELL, M.B. *et al.* 1995. Role of capillary forces, coupled flows and sediment-water depletion in the habitat of gas hydrate (abs.). EOS Trans. AGU **76**: S164–S165.
7. HANDA, Y.P. & D. STUPIN. 1992. Thermodynamic properties and dissociation characteristics of methane and propane hydrates in the 70-angstrom-radius silica gel pores. J. Phys. Chem. **96**: 8599–8603.
8. RUPPEL, C. 1997. Anomalously cold temperatures observed at the base of the gas hydrate stability zone on the U.S. Atlantic passive margin. Geol. **25**: 669–702.
9. WILKEN, R.T., & H.L. BARNES. 1997. Formation processes of framboildal pyrite. Geochim. Cosmochim. Acta **61**: 323–339.

Gas Hydrates and Global Climate Change

PETER G. BREWER[a]

Monterey Bay Aquarium Research Institute, P.O. Box 628,
Moss Landing, California 95039, USA

> ABSTRACT: The gas hydrate phenomenon intersects with societal concerns about global climate change in at least two distinct ways: the possibility of destruction of gas hydrates exposed on or near the sea floor by global warming; and the possibility of avoiding this and other problems by taking advantage of CO_2 hydrate formation during active disposal of fossil fuel CO_2 in the deep ocean. This extraordinary picture of contemporary discussion of the possible destruction by mankind of one hydrate, and the creation of another, both in vast quantities in the ocean, is a tribute to the unusual nature and power of this chemical cage.

GAS HYDRATES AND GLOBAL WARMING

High on the list of questions surrounding the enormous quantities of gas hydrates in the shallow geosphere is their stability in the face of global warming, and the environmental consequences of significant methane release. Gas hydrates were not recognized in natural systems until discoveries in Russia during the 1960s, and concerns about their stability soon followed. The first formal analysis of the problem was given by Revelle,[1] following earlier discussion by MacDonald.[2] Since that time we have learned a great deal about both the quantity and distribution of gas hydrates in nature[3] and about the course of global warming,[4] although it has rarely been possible uniquely to put the two together in the same observational and computational framework. New evidence of the possible role of gas hydrates in global changes of the geologic past has been uncovered.[5] In this review I discuss the present state of knowledge and ask whether the imminent threat is real and if it is possible to devise important research programs that can shed further light on this topic to provide objective assessments on which policy might be based.

The interest in gas hydrates and global change has so far derived almost solely from the atmospheric connection; that is that methane gas released from the ocean floor escapes to the atmosphere. Yet from my perspective this is a narrow view. Very substantial and societally important oceanic effects must also be considered in addition to the potential atmospheric signal.

There is no doubt that methane gas makes a very substantial contribution to the atmospheric radiation balance. A recent IPCC report[6] estimates that the adjustment time for a pulse of methane added to the atmosphere is 12 (± 3) years, the present concentration is about 1720 ppbv, and the present growth rate is about 0.6% per year, resulting from an estimated net flux into the atmosphere of about 37 Tg methane per

---

[a]Telecommunication. Voice: 831-775-1706; fax: 831-775-1620.
brpe@mbari.org

year. Therefore, any flux of methane into the atmosphere of approximately this scale would exert a significant effect in enhancing the present trend of global warming.[7]

For purely oceanic effects we must also consider the possibility of continental slope destabilization, the loss of a potential economic resource, and the large reduction in local oxygen content from the microbial oxidation of the released methane in sea water.[8,9]

Unfortunately there is a gap between those scientists concerned with the acoustic detection of hydrates and their presence deep in the sediment column, and those scientists concerned with the properties of the fluid ocean and the penetration of the industrial signature of the twentieth century into the upper ocean. Therefore, reliable ocean water column data are rarely available at sites where hydrates occur and there is an urgent need bridge this gap.

For a specific example consider a recent study[10] of the gas hydrate exposure off the Eel River Basin, Northern California[11] in August 1997, during the strong El Nino warming event. At this location the observed temperature profile in the water overlying the shallowest observed hydrate exposure, can be compared with the P–T phase boundary curve for methane hydrate.[12] It is reported that the shallowest depth (510 m) at which hydrates have been reported[11] by other researchers was, on the day the temperature profile was taken, above the hydrate formation point. Clearly this system is poised almost exactly at the phase boundary and thus any small thermal perturbation can cause release of gas. A significant gas vent at what would "normally" be below the hydrate formation boundary was indeed found.

Is this system unique? Hardly, and in fact it may represent the general case. MacDonald *et al.*[13] have documented the response of a hydrate system in the Gulf of Mexico to the passage of warm eddies and have shown clearly that gas releases occur in response to even small transient warming events. Notably, although most discussion takes a pure methane hydrate as the model case, the system studied by MacDonald *et al.*[13] is a mixed gas hydrate of 88–99% methane and other trace hydrocarbons. Nature has, over short geologic time scales, poised this system at its own boundary, different in P–T space to pure methane, but nonetheless in balance and responsive to change. Whether this equilibrium is static, or dynamic with constant balance between supply and dissolution, is very much a matter for current research.

We do not have direct evidence that the ocean is measurably warming today, but we do observe that the atmospheric warming trend and the growth of the greenhouse gases are well correlated. We can observe the penetration of the greenhouse gas tracer signals, of CFCs, and fossil fuel CO_2 into the ocean at depths well beyond the upper limit of the methane hydrate phase boundary. Thus, we may infer that a warming trend at these depths is already taking place.

GAS HYDRATES AND OCEAN CO_2 DISPOSAL

Direct ocean disposal of CO_2 from fossil fuel was first suggested by Marchetti[14] and, with the adoption of the Kyoto protocols to the United Nations Framework Convention on Climate Change, this idea has now received much greater attention. One important and exciting scientific development is that over the last three years this topic has moved from the province of policy, laboratory, and modeling studies to active fieldwork on an experimental scale. Recent papers from our laboratory showed first that detailed work on controlled methane releases at oceanic depths within the hydrate regime could be accomplished[15] and this was soon extended to the more difficult problem of CO_2 releases.[16] Rapid advances in technique then lead to the first successful controlled experimental transfer of CO_2 to depths below 2700 m where the release becomes gravitationally stable.[17]

The results were extraordinary, showing the rapid reaction of water and liquid CO_2 to form a massive hydrate within only a few hours. Under the experimental conditions at the chosen site (3600 m depth, 1.6°C) the density contrasts within the sea water–hydrate–liquid CO_2 system are such that the hydrate skin falls into the liquid CO_2, creating a solid mass at the base of the reaction container. The resulting volume changes are unexpectedly large, with the result that remaining unreacted liquid is pushed upward above the hydrate mass, and in our case expelled freely onto the ocean floor (see FIGURE 1). It had been expected that the freely released CO_2 would react with carbonate sediments, but, although we could make no direct measurement of this, the visual appearance of the reaction is that the hydrate skin wrapping the CO_2 globules effectively prevented contact and inhibited any reaction. This experiment indicates that hydrate formation from CO_2 in the deep sea can occur on very short time scales, yielding a strongly gravitationally stable hydrate mass and greatly assisting the sequestration time scale from the atmosphere.

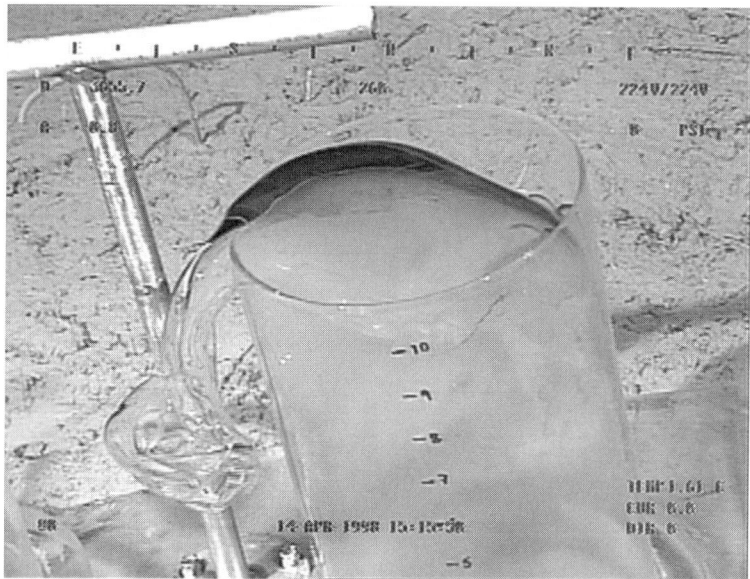

FIGURE 1. Image of liquid CO_2 being expelled onto the sea floor at a depth of 3600 m. The liquid is contained within a glass tube (approximately 30 cm high, 17.5 cm diameter) inserted into the sediments. Rapid reaction of the liquid CO_2 with surrounding sea water to form a dense hydrate, which sinks to the bottom, has taken place, with a resulting large change in volume. This causes overflow of the remaining un-reacted liquid.

These early field experiments have opened an important window on the behavior of real fluids and hydrates in part of the P–T phase space not normally available to ocean chemists, and into direct investigation of biological responses, plume dynamics, and carbonate dissolution. Many fundamental scientific issues remain unanswered.

On a larger scale it is clear that some form of advanced CO_2 sequestration is required in order "to achieve...stabilization of greenhouse gas concentrations in the atmosphere at a level which would prevent dangerous anthropogenic interference with the climate system" as the goal adopted by the IPCC. Analyses of this goal in comparison with the widely accepted "business as usual" scenario show that massive displacements of CO_2 from its present atmospheric trajectory are required. For instance, the widely cited paper by Wigley et al.[18] indicates that approximately 1 GT C/yr by 2025, and 4 GT C/yr by 2050 need to be sequestered in order to meet the goal of stabilizing the atmosphere at about 550 ppm CO_2.

These are truly enormous numbers that have the capacity to profoundly change the ocean chemistry signals we see today, even if only a part of the required sequestration is oceanic. Little practical scientific knowledge yet exists on the geochemical and biological consequences of such actions, although many modeling scenarios have been published.

In summary, two major research themes emerge: an urgent need to provide a direct formal link between the poise of hydrates in the shallow geosphere and the rapidly evolving tracer/warming signal; and the provision of the science base, through innovative experiments, needed for society to make wise judgements about oceanic CO_2 sequestration. I hope this paper demonstrates that both themes are directly addressable today and that the results are exciting and useful.

REFERENCES

1. REVELLE, R.R. 1983. Methane hydrates in continental slope sediments and increasing atmospheric carbon dioxide. In Changing Climate: Report of the Carbon Dioxide Assessment Committee. 252–261. National Academy Press.
2. MACDONALD, G.J. 1982. The Long-term Impacts of Increasing Atmospheric Carbon Dioxide Levels. Ballinger. Cambridge, MA.
3. KVENVOLDEN, K.A. 1993. Gas hydrates–geological perspective and global change. Rev. Geophys. **31:** 173–187.
4. HANSEN, J., R. RUEDDY & M. SATO. 1996. Global surface air temperature in 1995: return to pre-Pinatubo level. Geophys. Res. Lett. **23:** 1665–1668.
5. DICKENS, G.R., M.M. CASTILLO & J.G. WALKER. 1997. A blast of gas in the latest Paleocene: Simulating first-order effects of massive dissociation of oceanic methane hydrate. Geology **25:** 259–262.
6. IPCC. 1996. Technical Summary. In Climate Change 1995, The Science of Climate Change. 12–49. Cambridge University Press.
7. GORNITZ, V. & I. FUNG. 1994. Potential distribution of methane hydrates in the world's oceans. Global Biogeochem. Cycles **8:** 335–347.
8. SCRANTON, M.I. & P.G. BREWER. 1978. Consumption of dissolved methane in the deep ocean. Limnol. Oceanogr. **23:** 1207–1213.
9. REHDER, G., R.S. KEIR, M. RHEIN & E. SUESS. 1999. Methane in the ocean controlled by microbial oxidation and atmospheric history. Geophys. Res. Lett. In press.
10. BREWER, P.G. et al. 1997. Gas hydrates and global change: a preliminary case study offshore Northern California. Eos. Trans. AGU **78:** (Supplement), F340.

11. BROOKS, J.M., M.E. FIELD & M.C. KENNICUTT II. 1991. Observations of gas hydrates in marine sediments, offshore Northern California. Marine Geology **96:** 103–109.
12. PELTZER, E.T. & P.G. BREWER. 1999. Accurate prediction of methane hydrate phase boundaries: a comparison of models and data. Eos. Trans. AGU. In press.
13. MACDONALD, I.R., N.L. GUINASSO, JR., R. SASSEN, J.M. BROOKS, L. LEE & K.T. SCOTT. 1994. Gas hydrate that breaches the sea floor on the continental slope of the Gulf of Mexico. Geology **22:** 699–702.
14. MARCHETTI, E. 1977. On geoengineering and the CO_2 problem. Climate Change **1:** 59–68.
15. BREWER, P.G., F.M. ORR, JR., G. FRIEDERICH, K.A. KVENVOLDEN, D.L. ORANGE, J. MCFARLANE & W. KIRKWOOD. 1997. Deep-ocean field test of methane hydrate formation from a remotely operated vehicle. Geology **25:** 407–410.
16. BREWER, P.G., F.M. ORR, JR., G. FRIEDERICH, K.A. KVENVOLDEN & D.L. ORANGE. 1998. Gas hydrate formation in the deep sea: *In situ* experiments with controlled release of methane, natural gas, and carbon dioxide. Energy and Fuels **12:** 183–188.
17. BREWER, P.G., G. FRIEDERICH, E.T. PELTZER & F.M. ORR, JR. 1999. Direct experiments on the ocean disposal of fossil fuel CO_2. Science **284:** 943–945.
18. WIGLEY, T.M.L., R. RICHELS & J.A. EDMONDS. 1996. Economic and environmental choices in the stabilization of atmospheric CO_2 concentrations. Nature **379:** 240–243.

Changes of the Hydrate Stability Zone of the Norwegian Margin from Glacial to Interglacial Times

JÜRGEN MIENERT,[a,b] KARIN ANDREASSEN,[b]
JÖRG POSEWANG,[c] AND DIRK LUKAS[d]

[b]*Institute of Geology, University of Tromsø, Dramsveien 201, N-9037 Tromsø, Norway*

[c]*SFB 313 of University of Kiel, D-24118 Kiel, Germany*

[d]*GEOMAR, D-24148 Kiel, Germany*

ABSTRACT: During the last two decades, the detection and quantitative determination of gas hydrate deposits, as well as the study of clathrate formation and dissociation kinetics, have become central fields of marine research. The presence of gas hydrates in oceanic sediments along the Norwegian continental margin is documented in high-frequency near-vertical and wide-angle seismic reflection data. The base of the hydrate stability zone (HSZ) is detected in reflection seismic sections by the occurence of a strong bottom simulating reflector (BSR). The BSR mimics the shape of the sea floor, crosses the sedimentary strata, and is characterized by a strong phase-reversed event. Below the BSR a low velocity layer is interpreted as a gas-bearing zone. The inferred thickness of the hydrate existence zone (HEZ) allows for the estimation of the total amount of carbon trapped in gas hydrates at the present time. Modeling the HSZ as a function of temperature and pressure shows a distinct decrease of the HSZ at the Norwegian margin from the last glacial maximum (LGM) to the present time. This highly dynamic HSZ system may provide a complex seismic expression of gas hydrate occurrences, a phenomenon that is far from being well understood.

INTRODUCTION

Oceanic gas hydrates are known to occur in sea floor sediments at numerous regions around the world, seemingly not confined to certain latitudes. They are very efficient methane collectors: dissociation of 1 m^3 gas hydrate may release about 164 m^3 methane under standard temperature and pressure (STP) conditions.[1] For these reasons clathrates are considered to be capable of having an impact on the global climate system through the release of the greenhouse gas methane. However, it is not yet known what portion of methane from hydrates can be released from the ocean to the atmosphere.

In this context, gas hydrates might cause severe problems because global warming may increase water temperatures down to the intermediate water depth and, after some time, the temperature signal will reach the sea floor and the sediments where

[a]Telecommunication. Voice: 0047-7764-4446; fax: 0047-7764-5600.
juergen.mienert@ibg.uit.no

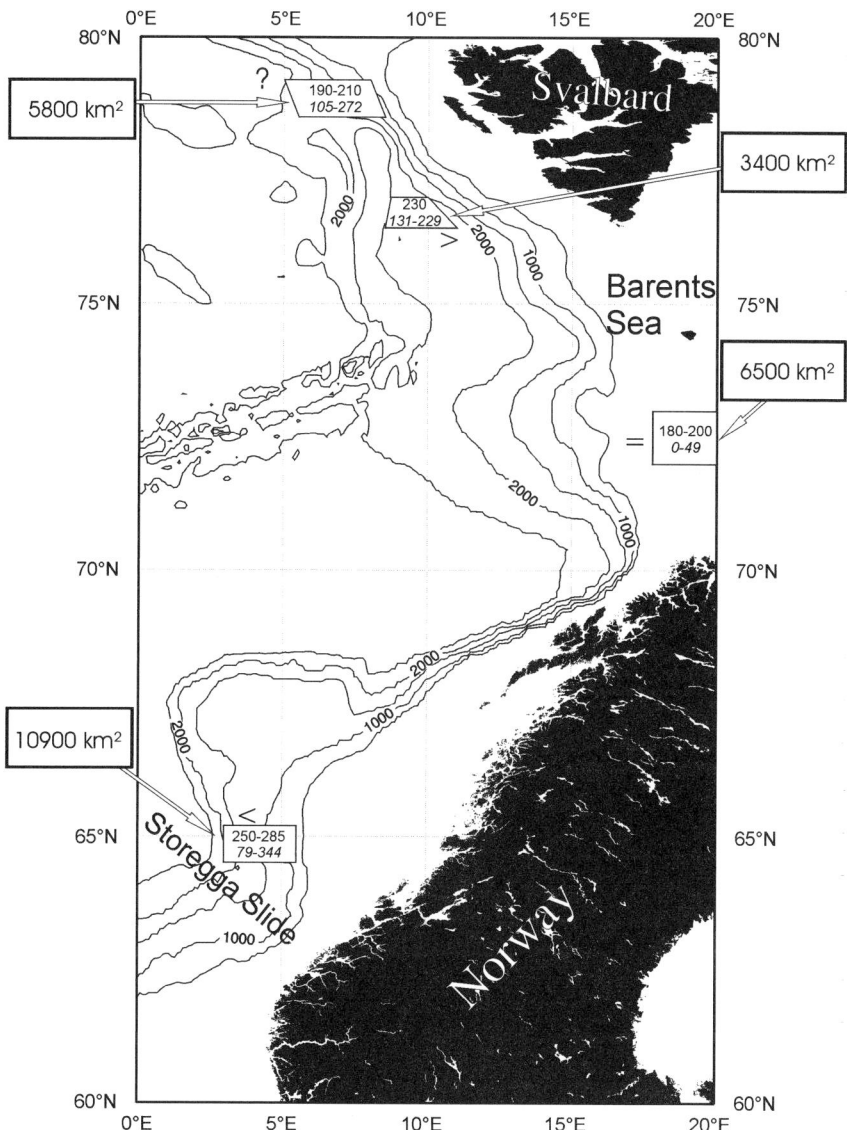

FIGURE 1. Map showing the four working areas with clathrate fields. The numbers in the boxes denote ranges of present BSR depths (*upper numbers* inferred from seismic, *lower numbers* calculated by the model of this work).

hydrates reside. Since clathrate stability depends most critically on temperature,[2] partial hydrate dissociation will release considerable amounts of methane that are expected to break their way through the sediments and to enter the ocean. Such localized pathways or gas channels have been clearly identified by seismic measurements on the Norwegian Continental Margin and are associated with "pockmarks"

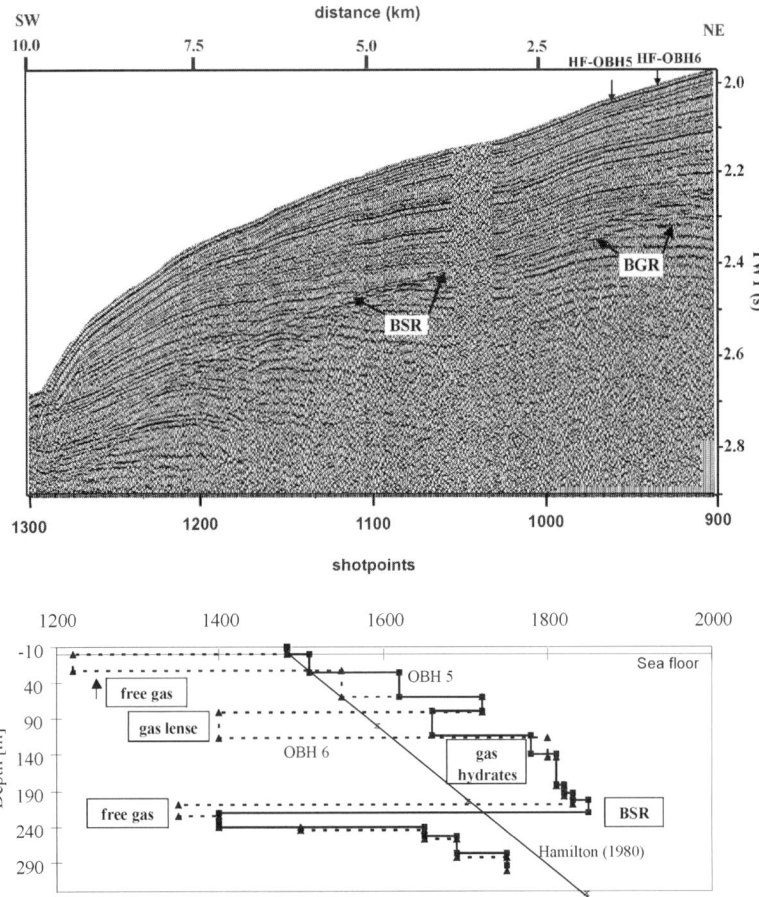

FIGURE 2. Section of a seismic reflection profile from working area 4 (FIGS. 1 and 3) based on a 2-liter airgun and a 6-channel streamer. The *arrows* mark the locations of the HF-OBH stations 5 and 6 at the sea floor. A strong BSR occurs at a depth of 0.25 s TWT bsf. The BSR parallels the sea floor, crosses the discordant sedimentary strata and is characterized by high amplitudes. Below the BSR the base of gas-bearing sediments is marked by the BGR. The velocity depth models below are based on HF-OBH data that show several zones of alternating high and low velocities. The anomalous high-velocity values up to 1840 m/s are an indicator for gas hydrated sediments, whereas values below 1500 m/s are characteristic of gas bearing sediments. For comparison the standard velocity profile by Hamilton for unhydrated sediments is shown.

visible during side-scan sonar surveys.[3] These are excellent candidates for relicts of past hydrate dissociation events that might be part of the subseafloor response to palaeoclimatic changes, especially since the last glacial maximum (LGM). Indeed, a number of presently known hydrate deposits are accompanied by gas escape features.[4]

The purpose of this paper is to estimate the changes in the hydrate stability zone since the last glacial maximum (LGM), and the associated changes in volume of methane bound in hydrates over a limited study area: the Norwegian continental margin (see FIGURE 1). This hydrocarbon-rich region is one of the prototype areas in which seismic-based clathrate investigations were successful. Several locations reveal extended (and partially strong) bottom simulating reflectors (BSRs) as well as pockmark fields, these are the region West of Svalbard,[5] the Storegga Slide region,[3,6] and the Barents Sea.[7] Additionally, deployments of high-frequency ocean bottom hydrophones (HF-OBHs) allowed for the determination of high-resolution velocity-depth models (see FIGURE 2), which clearly confirm the ideas (1) that hydrate-cemented sediments are characterized by enhanced P-wave velocities, and (2) that the measurable depth of the BSR marks the lower boundary of the hydrate stability zone (HSZ).[5] Moreover, calculated predictions of the depth of the lower HSZ boundary from the hydrate stability curve and from the geothermal gradient (linear temperature field) have been shown to be a good approximation, especially at Storegga where evidence stems from an industrial drill site.[8]

We aim at the large-scale estimation of the total HSZ volume and the methane bound in hydrates. Relying on seismic-based clathrate investigations (FIG. 2), bathymetry, measured bottom water temperatures (see FIGURE 3), mapped heat flow data and an experimentally confirmed methane–sea water hydrate stability curve, we present a simple model calculation for the extension of the present HSZ. Taking into account referenced palaeoceanic temperature shifts, the calculation is repeated for the LGM, and the position change of the lower HSZ boundary (measured in mbsf).

METHOD OF CALCULATION

Ocean bottom temperatures have been measured (CTD) by other workers during several expeditions in the area of our interest. The data are available in electronic form from the Bundesamt für Seeschiffahrt und Hydrographie (BSH), Germany. We interpolated and extrapolated these data numerically to obtain a set of bottom water temperatures for the Norwegian Continental Margin (FIG. 3). The calculation was performed using readily implemented standard algorithms provided by the generic mapping tools (GMT) software package. The procedure avoids spatial aliasing by calculating averages of measured values on a number of nodes of a regularly-spaced two-dimensional grid. Each node is associated with an angle step width of 6 minutes × 6 minutes, or an area element of 36 nm^2 (nm—nautical mile) multiplied by the cosine of its latitude.

To estimate of the LGM temperature field, calculated paleotemperatures were used.[9,10] One data set[9] covers sea surface temperature changes dependent on latitude within the time frame between 11–13 ka and the present day. The other data sets[10] indicate temperature changes with water depth (more than about 800 m) between

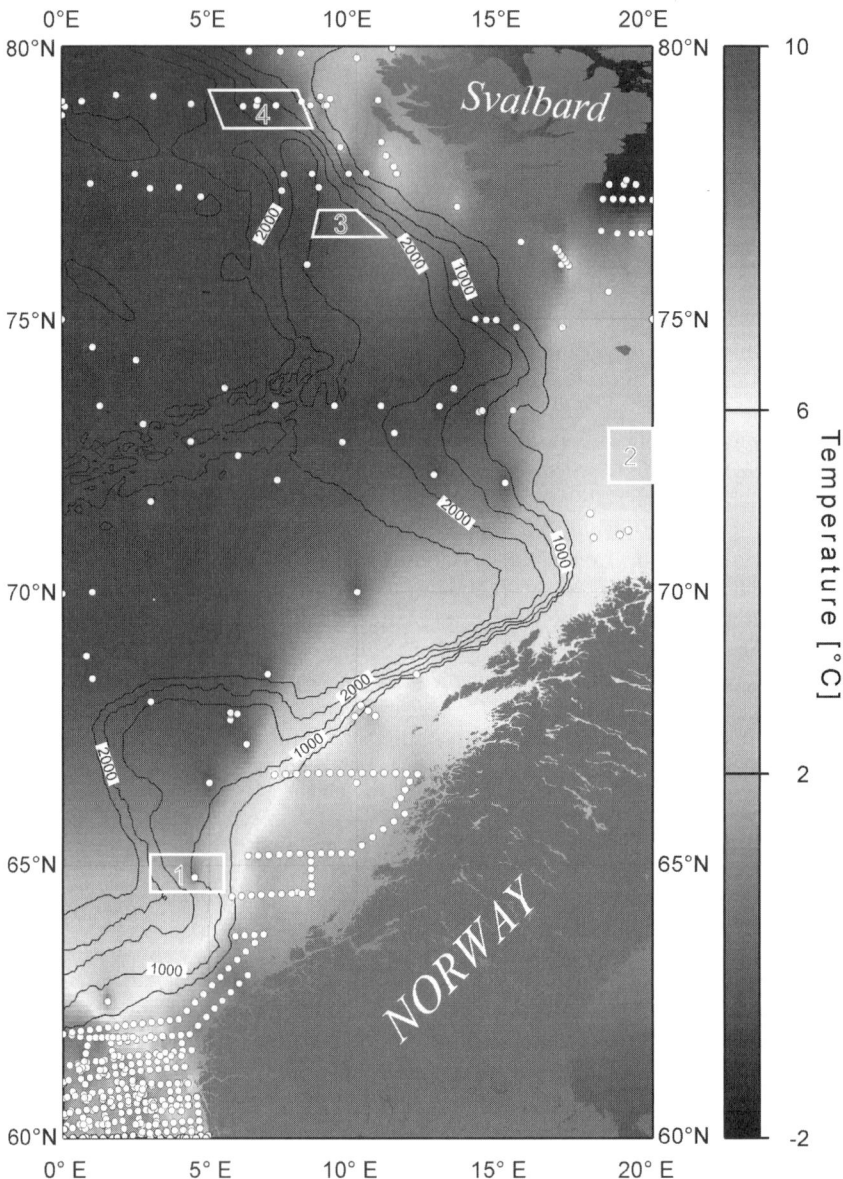

FIGURE 3. Present day *in situ* bottom water temperatures on the Norwegian Margin interpolated from CTD station data (*white dots*). The *white boxes* are numbered and indicate the working areas for seismic clathrate investigations.

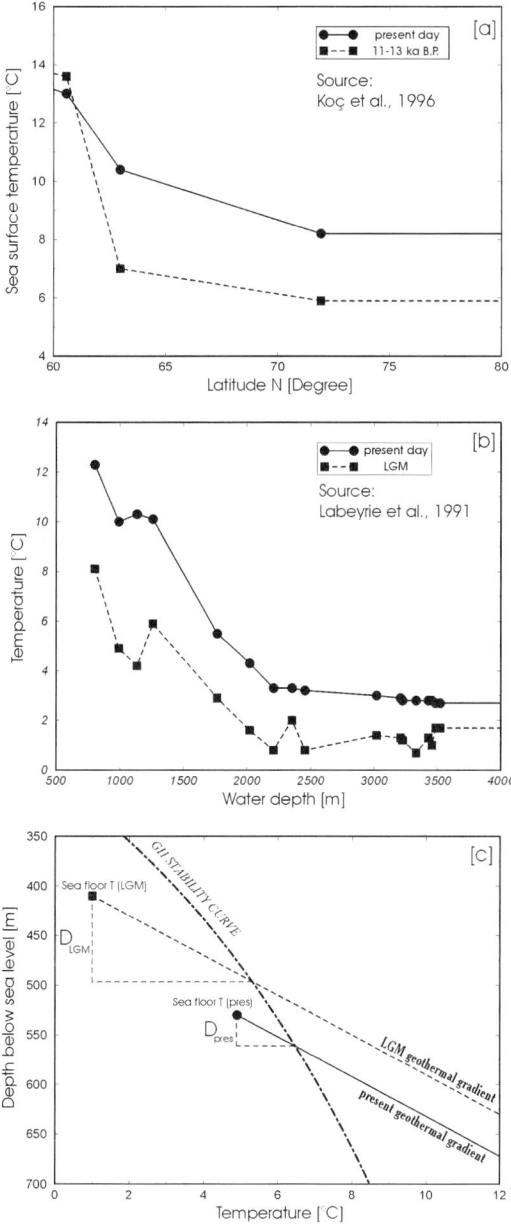

FIGURE 4. (a) Surface water temperature changes since 11–13 ka, and (b) temperature changes with depth since the LGM. The data in (b) are based on core data from the region 35–55°N, 2–32°W and are considered generic North Atlantic. *Lines* in (a) and (b) indicate linear interpolation and constant extrapolation used in this work. (c) Schematic calculation of the HSZ thickness change, which is identified with the difference $D_{pres} - D_{LGM}$.

LGM and today. Referenced data points are presented in FIGURE 4, and we interpolate linearly as suggested by the drawn lines. If the range of given data is left we approximate the palaeotemperature difference as a constant equal to the last referenced value. However, since the ocean water was not frozen down to the sea floor, we prohibit bottom water temperatures below −1.9°C.

Hydrostatic pressure at the sea floor is taken from the bathymetry of the GMT database, multiplying water depth by 0.01 MPa/m. The pressure is reduced by 1.2 MPa for the LGM to take into account an overall sea level decrease of roughly 120 m.[11]

The geothermal gradient G at each grid node is estimated on the basis of mapped heat flow data.[12] Under steady state conditions the (one-dimensional) heat flow f is related to G by

$$f = -k\frac{dT}{dx} = -kG \quad (1)$$

where x denotes subbottom depth and k is the sediment thermal conductivity. The minus sign indicates that heat flow is positive in the direction of decreasing temperature. Referring to conductivity measurements at various sites of ODP Leg 162 in the North Atlantic region,[13] $k = 1$ W/(K*m) seems to be a reasonable approximation for a typical k value. Consequently, G varies between 50 and 200°C/km in accord with f being reported essentially between 50 and 200 mW/m^2.[12] The BSR depths inferred from seismic and calculated by our model show one distinct discrepancy that is in the Barents Sea (FIG. 1). This can be explained by the heat flow gradient (50°C/km) used in this model, which appears to be too high for this region. The higher heat flow gradient caused a shallower BSR in our model.

We take into account the gas hydrate stability curve for a methane-sea water system.[14] The authors proposed a simple interpolation formula based on their measurements

$$\frac{1}{T} = 3.79 \times 10^{-3} - 2.83 \times 10^{-4} \log p, \quad (2)$$

which can be slightly improved by introducing a third parameter:

$$\frac{1}{T} = 3.724 \times 10^{-3} - 9.7 \times 10^{-5} \ln(p - 1.032). \quad (3)$$

Numerical values for T and p should be entered in K and MPa units, respectively. Moreover, the third parameter is expected to be useful when searching corresponding formulae for various gas compositions involving higher hydrocarbons. Note that log in Equation (2) denotes the decadic logarithm whereas Equation (3) uses the natural logarithm.

The presently well accepted picture of the hydrate stability zone (HSZ) is that of a subseafloor region, extending from the ocean bottom down to a depth (associated with hydrostatic pressure) at which the subbottom temperature exceeds the maximum value for hydrate stability. Hence, concerning the pT-diagram, the depth of the lower HSZ boundary—usually in good agreement with the seismically detectable depth of the bottom simulating reflector (BSR)—is located at the intersection of the linear temperature curve and hydrate stability curve. Since *linear* temperature dependence on subbottom depth is, in general, well supported by drilled findings, we also assume linearity for the LGM. Hence, since both temperature curves must not

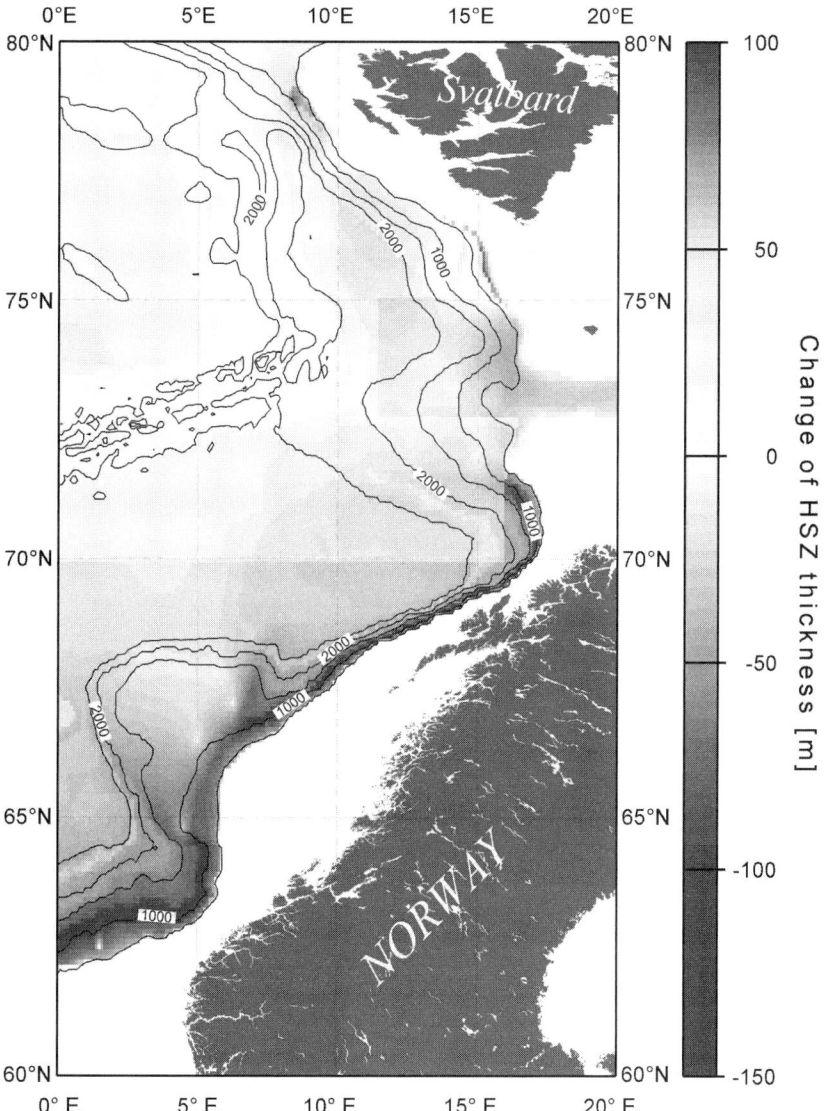

FIGURE 5. Map of estimated decrease in HSZ thickness changes. *Grey* and *white* reflect the local HSZ thickness changes $D_{pres} - D_{LGM}$ in meters as indicated by the (grey) scale on the *right*.

intersect at any finite depth, they are expected to be parallel. In other words, we suppose that the geothermal gradient has not changed since the LGM.

There are two major uncertainties involved in estimating the HSZ volume of the Norwegian margin. First, the geothermal gradient was extracted from a readily compiled heat flow map by assuming that thermal conductivity is 1 W/mK everywhere. This is a reasonable first approximation, but it is well known that conductivity can differ locally. Consequently, the geothermal gradient bears some uncertainty. However, since the gradient is also a quantity of high lateral variability, the chosen model values, between 50 and 250°C/km, are in an acceptable range for regional averages.

Second, when modelling the HSZ of the LGM, we assume that the geothermal gradient remains unchanged. From the viewpoint of physics, one would expect the gradient to change since the LGM because bottom water temperatures have increased. Quantifying the magnitude of the probable gradient change requires (1) more detailed information about thermal conductivities, and (2) more detailed bottom water temperature evolution functions for the study area. The temperature evolution could serve as a boundary condition for a diffusive heat transfer problem. For a first approximation, we regard the error associated with neglecting the gradient change to be small compared with the uncertainties in heat flow.

In summary, the geothermal gradient, as well as the sea floor temperature and pressure for the present and the LGM, are approximated for every node of the working area grid. The corresponding temperature curves (straight lines) are given analytically, and the depths of intersection with the hydrate stability curve, see Equation (3), can be calculated. Usually, for each node two distinct HSZ thicknesses, D_{pres} and D_{LGM} (see FIGURE 5), are obtained and then integration of their differences over all nodes yields an estimate for the total HSZ volume change. The calculated HSZ volume change since the LGM—which turns out to be decrease at almost every node—is considered a worst case scenario with a maximum release of methane.

RESULTS AND CONCLUSIONS

In order to determine the total HSZ volume and its change since the LGM, contributions of nodes with present ocean bottom depths between 300 and 4000 m have been integrated. The volume changes are reflected in a gray scale map, where the shades represent HSZ thickness changes in meters. It is obvious that the HSZ has decreased since the LGM almost everywhere except for a small zone south of Svalbard. This exception stems from the bottom water temperatures below $-1°C$ that have been measured there and that are quite close to our assumed minimum of $-1.9°C$ for the LGM. On the other hand, since water depth is generally small in that area, hydrate stability depends more significantly on pressure than for deep waters, and the sea level rise since the LGM effectively stabilized the clathrate reservoir.

In contrast, the slope of the hydrate stability curve with respect to pressure is quite small and the influence of temperature is dominant for water depths between about 500 to 1,500 m. Seemingly, hydrated sediments in intermediate water depths react more sensitively to changes in bottom water temperature.

As a second step, we estimate the total volume of methane (STP, standard temperature and pressure conditions) and the corresponding carbon mass of the hydrate reservoirs (today and LGM) on the following basis. The typical porosity for the topmost sediment layers below the sea floor is commonly observed to be about 50–60%.[13] The clathrate content of pore space C is based on using empirical approaches for quantitatively estimating of gas hydrate concentration.[15] The concentration of gas hydrates is derived by comparing theoretical velocity curves in the absence of clathrates with the actual velocity curves obtained from HF-OBH measurements. We applied our methods to results from ODP Leg 164,[16] where $C = 2$–7%. Absolute values for percentage of hydrates in the pore space agree, to within a few percent, with the values suggested by pore water chemistry.[15] Finally, employing the well-known fact that 1 m^3 methane (STP) contains 536 g carbon (mole volume, 22.4 l; mole weight carbon, 12 g), we obtain a minimum value of 190 GT and a maximum value of 798 GT for the LGM, and a minimum value of 174 GT and a maximum value of 731 GT for the present. The minimum and maximum values in GT are based on $C = 2\%$ and porosity equal to 50%, and $C = 7\%$ and porosity 60%, respectively. Compared to global estimates from other authors, our findings are of reasonable order of magnitude. However, if our results were taken as representative and projected to a global scale, they would exceed present estimates[17] by a factor 2–3.

To conclude, considerable decreases in the HSZ have taken place since the LGM, and considerable amounts of methane (2.9–12.4×10^{13} m^3) and carbon may have since left the submarine hydrate deposits of the Norwegian continental margin area. A reasonable estimate for the released carbon mass lies between 16 and 67 gigatons.

ACKNOWLEDGMENTS

The results presented were obtained as part of the SFB 313 of the University of Kiel funded by the Deutsche Forschungsgemeinschaft and the European North Atlantic Margin (ENAM II) project (MAS3- CT9-0003) funded by the Marine Science and Technology (MAST III) programme of the European Commission in Brussels.

REFERENCES

1. KVENVOLDEN, K.A. 1993. Gas hydrates—geological perspective and global change. Rev. Geophys. **31**(2): 173–187.
2. SLOAN, E.D. 1990. Clathrate Hydrates of Natural Gas. Marcel Dekker, New York.
3. MIENERT, J., J. POSEWANG & M. BAUMANN. 1998. Gas hydrates along the Northeastern Atlantic Margin: possible hydrate-bound margin instabilities and possible release of methane. *In* Gas Hydrates: Relevance to World Margin Stability and Climatic Change. J.P. Henriet & J. Mienert, Eds. Geological Society, London, Special Publications **137**: 275–291.
4. VOGT, P.R., K. CRANE, E. SUNDVOR, M.D. MAX & S.L. PFIRMAN. 1994. Methane-generated(?) pockmarks on young, thickly sedimented oceanic crust in the Arctic: Vestnesa ridge, Fram strait. Geology **22**: 255–258.
5. POSEWANG, J. & J. MIENERT. 1999. High-resolution seismic studies of gas hydrates west of Svalbard. Geo-Marine Lett. **19**: 150–156.

6. BUGGE, T., R.H. BELDERSON & N.H. KENYON. 1988. The Storegga Slide. *In* Philos. Trans. Roy. Soc. (Lond.) **A325:** 357–388.
7. ANDREASSEN, K. 1995. Seismic reflections associated with submarine gas hydrates. Dr. Scient. Thesis, University of Tromsø, Norway. Unpublished.
8. MIENERT, J. & J. BRYN. 1997. Gas Hydrate Drilling conducted on the European Margin. EOS Trans. Amer. Geophys. Union **78:** 49, 567–571.
9. KOC, N., E. JANSEN, M. HALD & L. LABEYRIE. 1996. Late glacial-Holocene sea surface temperatures and gradients between the North Atlantic and the Norwegian Sea: implications for the Nordic heat pump. Palaeoceanography of the North Atlantic Margins **111:** 177–185.
10. LABEYRIE, L.D., A. JUILLET-LECLERC, N. KALLEL & P.-L. BLANC. 1991. Sea level and oceanic thermohaline circulation: changes over a glacial/interglacial cycle. Klimageschichtliche Probleme der letzten 130000 Jahre. *In* B. Frenzel, Ed.: 197–214. Akademie der Wissenschaften und der Literatur, Mainz, Germany, Gustav Fischer Verlag, Stuttgart, New York.
11. BARD, E., R. FAIRBANKS, M. ARNOLD, P. MAURICE, J. DUPRAT, J. MOYES & J.-C. DUPLESSY. 1989. Sea-level estimates during the last deglaciation based on $\delta^{18}O$ and accelerator mass spectrometry ^{14}C ages measured in *Globigerina bulloides*. Quatern. Res. **31:** 381–391.
12. CRANE, K. & A. SOLHEIM. 1995. Seafloor atlas of the Northern Norwegian-Greenland Sea. Norsk Polarinstitutt Medd. 138, Oslo.
13. JANSEN, E., M.E. RAYMO & P. BLUM. 1996. Initial reports. Proceedings of the Ocean Drilling Program College Station, TX (Ocean Drilling Program). 162.
14. DICKENS, G.R. & M.S. QUINBY-HUNT. 1994. Methane hydrate stability in seawater. Geophys. Res. Lett. **21**(19): 2115–2118.
15. TINIVELLA, U., D. LUKAS, E. LODOLO, J. POSEWANG, A. CAMERLANGHI & J. MIENERT. 1999. Two models for the quantitative estimation of gas hydrates concentrations based on borehole data: application to ODP Leg 164 results. EUG 10, J. Conf. Abs. **4:** 250.
16. PAULL, C.K., R. MATSUMOTO, P.J. WALLACE *et al.* 1996. Proceedings of the Ocean Drilling Program. Initial Reports: Texas A&M Univ., College Station, TX, 164.
17. GORNITZ, V. & I. FUNG. 1994. Potential distribution of methane hydrates in world's oceans. Global Biochem. Cycles **8**(3): 335–347.

A Calculation Model for Liquid CO_2 Injection into Shallow Sub-Seabed Aquifer

KYURO SASAKI[a] AND SATOSHI AKIBAYASHI

Department of Earth Science and Technology, Akita University, 1-1, Tegata-Gakuencho, Akita, 010-0852, Japan

ABSTRACT: This study provided a model for calculating the aquifer transmissibility, the CO_2 injection rate, the inner diameter of the injection well, and the number of wells for liquid CO_2 disposal in the aquifer. The possibility of disposing liquid CO_2 in an aquifer just beneath the sea floor was shown, based on the equilibrium lines in the pressure and temperature map. Our study focused on the feasibility of liquid CO_2 disposal below the critical temperature because CO_2 can be denser in the low-pressure range (below the critical temperature) than above the critical temperature. An aquifer about 200 m under the sea floor, at a water depth of around 500 m (700 m below the sea surface), will serve for liquid CO_2 disposal. In the aquifer the absolute pressure is approximately 7.3 MPa, sea-floor temperature is about 4–6°C, and aquifer temperature is about 15–20°C. Therefore, it can be assumed that CO_2 dissolves in the aquifer water, and liquid CO_2 replaces the water. This means that under the previous conditions, more CO_2 can be injected into the aquifer compared to supercritical conditions. Furthermore, by forming a cap of CO_2 hydrates, the sediment between the sea floor and the aquifer, prevents CO_2 leakage to the sea. Even without the cap, liquid CO_2 and CO_2 hydrates form at the sea floor, so the CO_2 exerts no large environmental impact.

INTRODUCTION

Japan has made a commitment to the reduction of green house gases emissions to six percent below the 1990 level by the year 2012 in the Kyoto Agreement (COP3) (December, 1997).[1] Annual CO_2 production in Japan has been estimated to be equivalent to 3×10^8 tons in carbon; hopefully, Japan can reduce its current fossil fuel combustion rate by 16% from the present level.[1] On the other hand, fixation, disposal in forests, oceans, or underground can control CO_2 emissions into the atmosphere.

Investigation by The Engineering Advancement Association of Japan (ENAA-Japan) (Tanaka *et al.*[2,3]) used an economic feasibility study and a numerical simulation of CO_2 disposal into offshore aquifers and underground sequestration. The maximum CO_2 fixation into underground aquifers is about 8.9×10^{10} tons in CO_2.[2] Their investigations mainly concerned CO_2 disposal in deep sub-seabed aquifers under supercritical conditions. They investigated increasing CO_2 solubility in aquifer water using the latest petroleum numerical simulator.

[a]Telecommunication. Voice: +81-18-889-2395; fax: +81-18-837-0401.
sasaki@uws47.mine.akita-u.ac.jp

Koide et al.[4,5] explored CO_2 disposal in subseabed aquifers. They discussed CO_2 trapping by the CO_2 hydrates layer in the sea bottom sediment near the upper side of the aquifer.

The report by Law et al.[6] reported simulation results for CO_2 disposal into the Alberta-Basin, the biggest land region aquifer in the world. The Alberta-Basin includes oil and natural gas layers, and all the CO_2 separated from the extracted natural gas is returned to the basin for fixation. Their analysis used a current numerical simulator for oil fields. They suggested that CO_2 fixation is a mineral trapping chemical reaction between CO_2 and basic aluminosilicate and that this is important for simulating injectivity.

The report by Goldverg[7] detailed two possible aquifer locations for CO_2 injection (Sites 995A and 395A), previously drilled and studied by the Deep-Sea Drilling Project. The report provided long-term *in situ* data for evaluating CO_2 injection systems and conditions. The study also mentioned a methane hydrates layer providing an impermeable seal for CO_2 sequestration.

There are two ways to transfer CO_2 from an electrical power plant to a liquid injection site. The first is a pipeline on the sea bottom,[1] and the second is a tanker for liquid CO_2 (see FIGURE 1). When an aquifer has a short lifetime for its capacity the tanker system is preferable to the fixed pipeline system. Our study assumed a CO_2 injection system with tankers, an injection platform, and vertical injection well.

The objective of our study was to investigate the injectivity of liquid CO_2 into a relatively shallow subseabed aquifer with a simple flow model. An advantage to a shallow subseabed aquifer is that the aquifer temperature close to the seafloor temperature, 4–7°C, is usually lower than that of onshore aquifers. If liquid CO_2 can be kept in the aquifer at a relatively low pressure, then both the operating cost for the injection and initial costs for the required facilities decrease. Our calculation model focuses on the absolute pressure in the injection wells and in the aquifer. Therefore, our model allows us to calculate the injection rate based on transmissibility of the aquifer, combinations of different inner diameters of injection wells, and the number of injection wells drilled into a continuous aquifer.

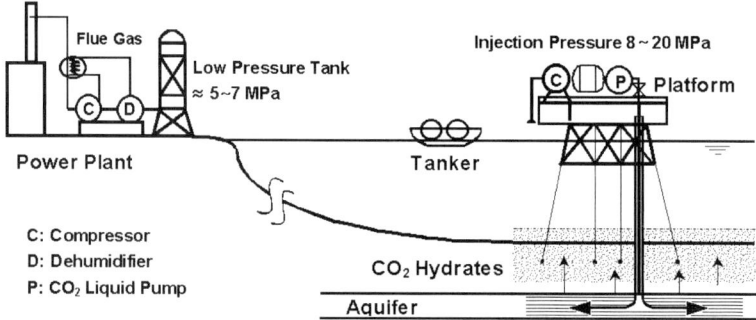

FIGURE 1. Schematic system for CO_2 disposal into aquifer under sea-floor.[1,2]

CALCULATION MODEL FOR CO_2 INJECTION

Equilibrium Conditions

The critical temperature and pressure for CO_2 are 31.2°C and 7.39 MPa, respectively. The relationship between absolute pressure P_l (Pa) and temperature t_l (°C) in CO_2 liquid–gas equilibrium can be approximated by the following equation over the range $t_l = 0$–30°C:[8]

$$\frac{P_l}{10^6} = \exp\left(10.826 - \frac{3497.5}{365.1 + t_l}\right) \quad (1)$$

CO_2 hydrate formation equilibrium lines for pure water and seawater were formulated by Ohgaki et al.[9]

$$\text{Pure water: } \frac{P_h}{10^6} = \exp\left(4.8596 - \frac{8885.95}{t_h + 273.2} + 4.9764 \cdot \ln(t_h + 273.2)\right) \quad (2)$$

$$\text{Sea water: } \frac{P_h}{10^6} = \exp\left(4.8372 - \frac{8891.81}{t_h + 273.2} + 5.0125 \cdot \ln(t_h + 273.2)\right) \quad (3)$$

where P_h (Pa) is the formation pressure and t_h (°C) is the formation temperature. The experimental results[9] are almost the same as those of Robinson and Mehta,[10] and Ng and Robinson.[11] The CO_2 hydrates become stable at lower temperature and higher pressure. In seawater, the hydrate formation temperature is about 2°C lower than that of pure water.

FIGURE 2 shows CO_2 density contour lines (liquid CO_2 density; $\rho = 100$–1000 kg/m³) with the equilibrium lines for CO_2 gas-liquid and the hydrate formations for pure water and seawater. If CO_2 disposal is carried out with injection flow in liquid phase, the CO_2 hydrates formations in water must be avoided to prevent plugging. The liquid phase region at relatively high density is shown in FIGURE 2 around the contour line of density $\rho = 900$ kg/m³. The region may show the targeted temperature and pressure conditions for economical CO_2 disposal since higher density and lower pressure decrease the operating costs. This shows that more CO_2 can be injected compared with the supercritical condition for the same pressure or sea level.

Targeted Condition for CO_2 Disposal

As shown in FIGURE 3, the closed area (a-b-c-d) indicates the temperature and the absolute pressure conditions in the targeted aquifers. It can be assumed not only that the CO_2 spreads into the aquifer water with the solution, but also that the liquid CO_2 replaces the water. Furthermore, when the sediment between the sea floor and the aquifer is under the conditions for the closed area A-B-C-D in FIGURE 3, the sediment blocks CO_2 leakage from the aquifer toward the sea floor, because CO_2 hydrates form with either pure water or seawater in the sediment. Even if the sediment is unable to prevent leakage, there will be no large environmental impact, since the conditions on the sea floor allow liquid CO_2 formation, and CO_2 hydrate generation with seawater.

For the above physical conditions (FIGS. 2 and 3), the range of absolute pressure in the injection well or the aquifer, $P(x)$, can be determined as follows:

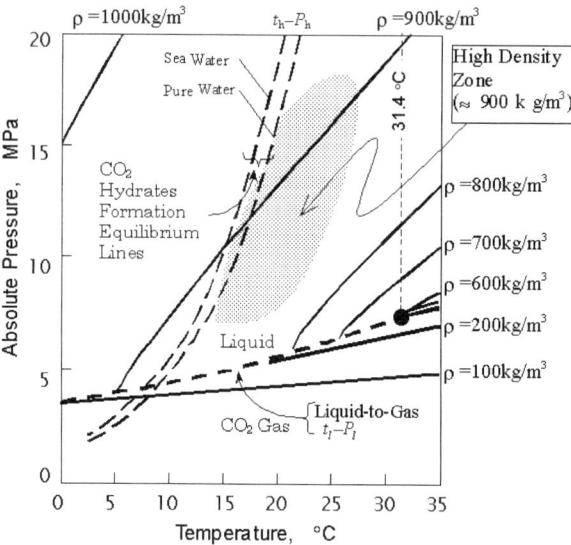

FIGURE 2. CO_2 density contour lines. The hydrate formation lines are based on Ohgaki et al.[6]

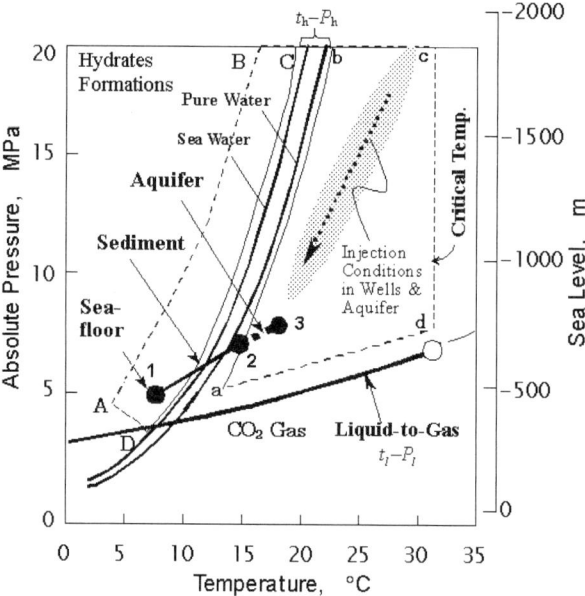

FIGURE 3. Aquifer conditions and equilibrium lines.

$$8 \text{ MPa} < P(x) < 20 \text{ MPa}. \tag{4}$$

These bounds were derived by considering the reliability of the injection well for liquid CO_2, and knowing that hydrate formation temperature is close to the critical temperature for pressures above 20 MPa. The temperature in the injection well is limited to 20 to 27°C, however it is relatively easy to maintain the CO_2 liquid temperature in the well in an ocean by means of absolute pressure control.

Aquifer Condition

The initial absolute pressure in the sediment and the aquifer. P_{aq} (Pa) was converted using the depth from the sea surface, x (m); acceleration due to gravity, g; and the seawater density, ρ_{SW} (kg/m^3) as shown in the following equation (see FIGURE 4).

$$P_{aq}(x) = \rho_{SW} g(x + 10.3). \tag{5}$$

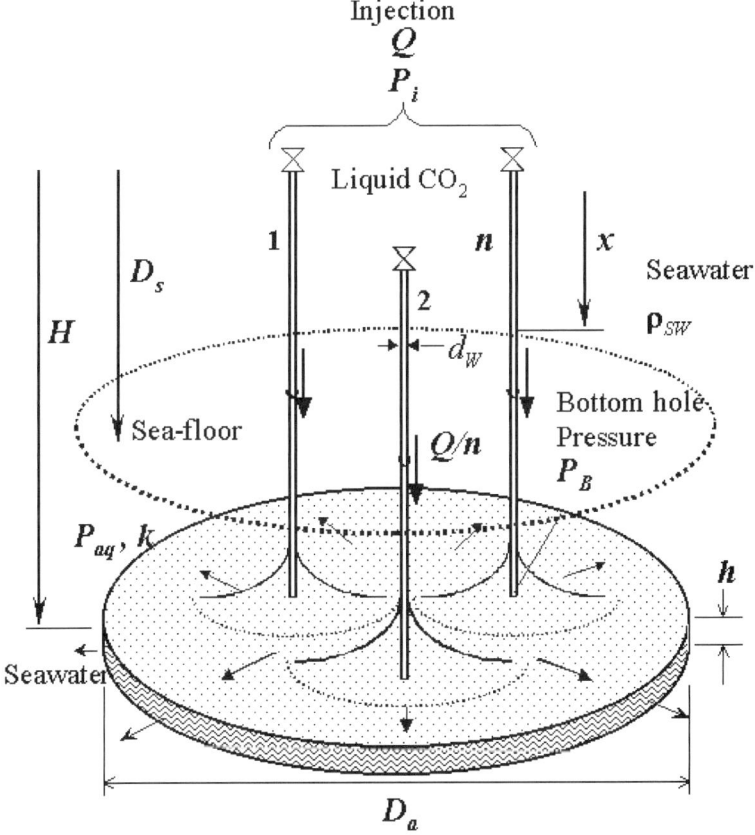

FIGURE 4. Calculation flow model for CO_2 liquid injection into subseabed aquifer.

The aquifer temperature, t_{aq} (°C), is estimated as a function of the geothermal gradient, q_E/λ_E (°C/m) (q_E geothermal flow; λ_E thermal conductivity), sea-floor temperature t_{SB}, depth x, and sea depth as D_s (m):

$$t_{aq} = t_{SB} + \frac{q_E}{\lambda_E}(x - D_S). \tag{6}$$

The lines 1-2 and 2-3 in FIGURE 3 show the pressure-temperature changes modeled for the sediment and the aquifer, respectively; where $t_{SB} = 7°C$; $q_E/\lambda_E = 0.035°C/m$; depth of the aquifer, $H = 700$ m; depth of sea-floor $D_s = 450$ m; and aquifer thickness, $h = 50$ m (see FIG. 4). The line indicates an expected absolute pressure of approximately $P_{aq} = 7.3$ MPa, and an aquifer temperature $t_{aq} = 15–18°C$.

An interesting physical property is that the equilibrium lines for gas-liquid and hydrates cross each other in the temperature interval of $t = 8–10°C$, and at an absolute pressure of approximately 4 MPa. In the region where the absolute pressure is lower than where the equilibrium lines cross, the hydrate formation temperature is higher than the gas-liquid equilibrium temperature. This shows that even if liquid CO_2 changes to gaseous CO_2 in the water, the mixture of gas and water can still form CO_2 hydrates.

Targeted Range for Injection Flow Rate and Life of the Aquifer

Assuming a 250 to 500 MW electrical power station using fossil fuel, where the CO_2 generation rate is roughly estimated at 50 kg/s (about 4,300 ton/day), then the total injection mass rate into a continuous aquifer in the model was about 1–200 kg/s (0.03–6.1 Mton/year). The life of the aquifer, N_a (year), is defined as the time to fill up the aquifer porous-space with liquid CO_2, and can be estimated from the following equation,

$$N_a = \frac{\pi \cdot D_a^2 \cdot h \cdot \phi}{4Q} \cdot \frac{1}{3600 \times 24 \times 365}, \tag{7}$$

where D_a is the diameter, h is the thickness, ϕ is the porosity of the aquifer, and Q is the volumetric injection rate. FIGURE 4 shows N_a versus the injection rate, ρQ (kg/s). For example, assuming $D_a = 1500$ m, $h = 50$ m, and $Q = 50$ kg/s, the total effective pore volume in an aquifer, $\phi h \pi D_a^2/4$, must exceed 2×10^7 m^3, or the porosity, ϕ, must exceed 0.22, to extend the life of the aquifer by 10 years.

Estimation of Injection Pressure

The present flow model was applied by assuming that the fluid in the injection well is liquid CO_2 and the aquifer contains seawater. In practice, the injection fluid may be changed gradually from seawater to liquid CO_2.

Based on the numerical simulation results due to Law[6] using the simulator STARS™ for the Nisku aquifer in the Alberta-Basin, the cumulative CO_2 injection increases with elapsed time over 30 years under a constant pressure of either 30.12 or 25.15 MPa. Thus, the injection rate is almost constant and independent of the injection time. Therefore, in the present study, the steady state flow model was assumed to simulate the effects of aquifer transmissibility, inner diameter, number of wells, and injection rate versus the absolute pressure in both the aquifer and the pipe. The calculation model for the aquifer used in this study is a steady state two-dimensional, radial-flow, and multiwell system. The aquifer with diameter D_a, effective permeability k, and effective thickness h, is open at its outer boundary. Muskat[12] presented a theory of the total water flow rate of a multi-well system, under various conditions, and where there is interaction of fluid flow between the wells (see FIGURE 5). However, this work includes variables that cannot be determined in the present feasibility study. Therefore, a simple approximation, using index power (σ) was employed to express the interaction, the number of wells (n), and to estimate pressure drop in the aquifer against volumetric flow rate (Q). The index used in our study was approximated by $\sigma = 0.6$, to give an almost equal pressure drop calculated with Muskat theory for $n = 2$. Darcy's law and the multi-well system give the absolute pressure at the bottom of the well as

$$P_B = P_{aq} + \frac{\mu_w \cdot \ln(D_a/d_w)}{2\pi \cdot kh} \frac{Q}{n^\sigma}, \tag{8}$$

where μ_w is seawater viscosity, and P_{aq} is the initial pressure in the aquifer. Seawater viscosity is much larger than that for liquid CO_2. Thus, the calculation using the seawater viscosity shows the pressure drop in the aquifer at the initial injection period.

Transmissibility, as usually defined, includes the viscosity and the density of fluid. Since the system includes two kinds of fluids in the model, transmissibility, T, is defined by $T = hk$ (m^3).[13]

On the other hand, the pressure drop, ΔP_L, during the flow in the pipe, from its well head to the location, x, is expressed by Darcy-Weisbach's equation[14]

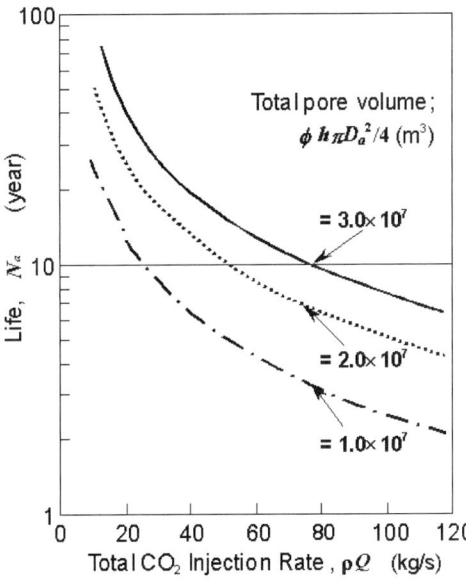

FIGURE 5. Life of aquifer to carry out CO_2 disposal against total injection rate.

$$\Delta P_L(x) = \frac{8}{\pi^2} \cdot \rho \cdot f \frac{x}{d_w^5} \left(\frac{Q}{n}\right)^2, \tag{9}$$

where f is a friction factor for the well internal surface, which is almost constant for the Reynolds number range, 10^6 to 10^8. For our calculations, f was set to 0.01 or 0.001 from the Moody diagram, by assuming the relative roughness order for commercial steal pipe to be 0.00001.[14]

The absolute pressure in the injection well, consisting of pressure drops and the initial pressure in the aquifer, and the liquid CO_2 head, is expressed by

$$P(x) = P_{aq} + \frac{\mu_w \cdot \ln(D_a/d_w)}{2\pi h k} \frac{Q}{n^{0.6}} + \frac{8}{\pi^2} \cdot \rho \cdot f \frac{H-x}{d_w^5} \left(\frac{Q}{n}\right)^2 - \rho g(H-x). \tag{10}$$

The injection pressure at the sea surface expressed as P_i, is given by

$$P_i = P(0). \tag{11}$$

RESULTS AND DISCUSSION

Bottom Hole Pressure

FIGURE 6 shows calculated examples of the relationship between transmissibility, T, of the aquifer, and the total injection rate satisfying the absolute pressure at the bottom hole, P_B, in the range of 8–20 MPa. The total injection rate increases with the transmissibility (FIG. 6).

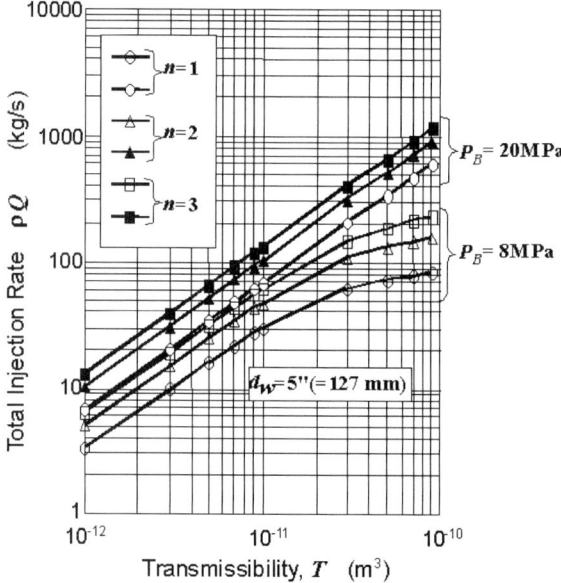

FIGURE 6. Calculation results for total injection rate versus transmissibility.

The calculated examples of the relationship between the inner diameter, d_w, and transmissibility satisfying the bottom hole pressure for the previous pressure range, are shown in FIGURE 7. For a constant value for the bottom hole pressure, the aquifer transmissibility is not sensitive to d_w.

Thus, the maximum and minimum injection rates are calculated for $n = 1$, $d_w = 0.14$ m (5.5 inch), and $P_B = 20$ MPa, and for $n = 1$, $d_w = 0.0254$ m (1 inch), and $P_B = 8$ MPa, respectively, satisfying the pressure range given by Equation (**4**). If the range of the injection rate is $\rho Q = 1-200$ kg/s, then the transmissibility range is

$$T = 6.0 \times 10^{-13} \text{ to } 2.2 \times 10^{-11} \text{ (m}^3\text{)}. \tag{12}$$

Thus, the transmissibility measured by the field injection test will be in the range given by Equation (**12**) for the injection rate range.

Pressure Gradient in Injection Well

FIGURE 9 shows calculations for the pressure drop in the well. The number of wells, n, was changed to $n = 2$ at $\rho Q = 53$ kg/s, and $n = 3$ at $\rho Q = 88$ kg/s, considering power consumption. The effects of the number of wells and their inner diameters are clear. The pressure drop, ΔP_L, must be less than 20 MPa to satisfy the absolute pressure range given by (**4**) as shown in FIGURE 8. The total energy consumption was calculated to be about 2 MW at injection, $\rho Q = 50$ kg/s, by $P_i = 20$ MPa, and efficiencies of compressors and pumps are 0.6. The net power consumption for liquid injection is roughly estimated at 12 kWh/CO_2ton. The multiwell system shows an advantage in reducing the injection energy consumption. However, the initial cost to drill multiple wells increases with n.

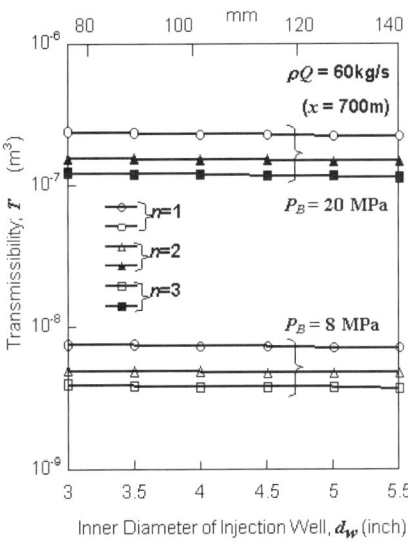

FIGURE 7. Calculation results for transmissibility versus inner diameter of well.

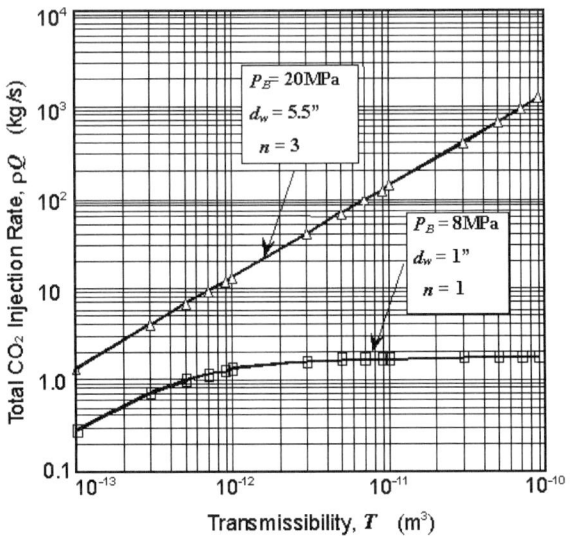

FIGURE 8. Relationship between aquifer transmissibility and total injection rate.

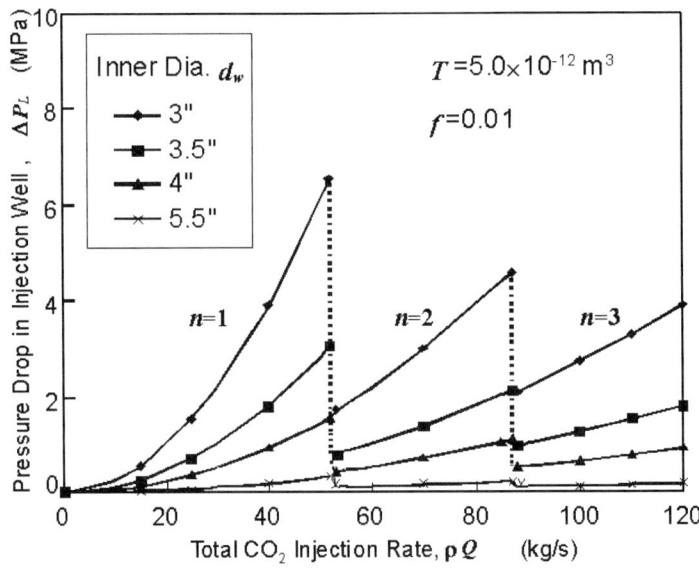

FIGURE 9. Calculation results for pressure drop in injection wells.

FIGURES 10, 11, and 12 show calculations of absolute pressure distribution in the injection pipe; though there are many combinations of variables d_w, n, and ρQ. Furthermore, the absolute pressure gradient along the injection well, $\partial P/\partial x$, changes widely from negative to positive values against the variable combinations. In order to satisfy **(4)**, the maximum permissible pressure gradient along the injection well is roughly estimated at 12 MPa/700 m (17 kPa/m) for the present model. Calculations of the relationship between the pressure gradient in the well, $\partial P/\partial x$, and the injection rate, ρQ, to satisfy **(4)**, are shown in FIGURE 13 for various inner diameters, d_w. The figure shows that the injection rate gradually increases with an increasing pressure gradient. The inner diameter increases with the injection rate. The results for $\partial P/\partial x = 0$ give almost average injection rate against the diameter. Thus, it is relatively easy to satisfy **(4)** when the pressure gradient is nearly zero;

$$\frac{\partial P(x)}{\partial x} \cong 0 \qquad (13)$$

This equation provides not only an initial design scheme for liquid injection at constant absolute pressure, but also the engineering design basis for other facilities.

Some combinations of d_w and n versus injection rate, ρQ, for different friction factors $f = 0.001$ and 0.01 are plotted in FIGURE 14. The diameter range for the smoother pipe with $f = 0.001$ decreases to about half compared with that for $f = 0.01$. The economic evaluation for each combination must be made, so the combination with the minimum total cost and system risk may be selected.

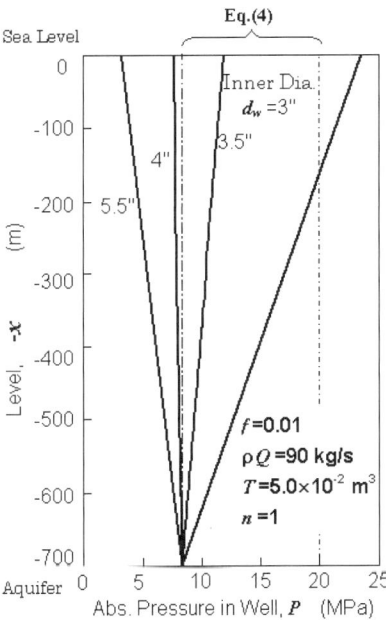

FIGURE 10. Effects of the injection well diameter on the absolute pressure distribution along the injection well.

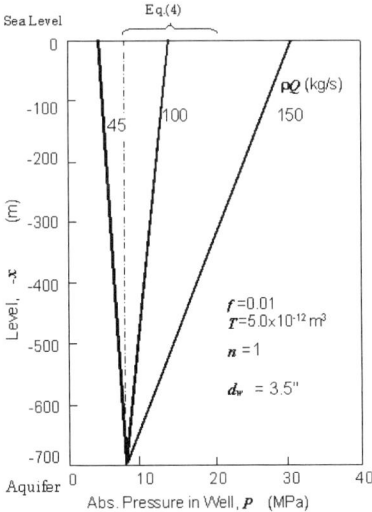

FIGURE 11. Effects of CO_2 injection rate on the absolute pressure distribution along the injection well.

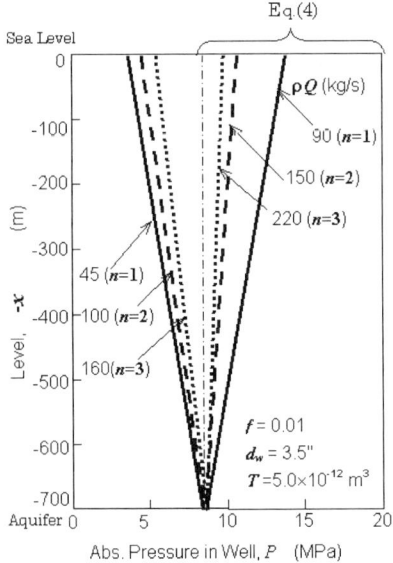

FIGURE 12. Effects of the number of wells and CO_2 injection rate on the absolute pressure distribution along injection well.

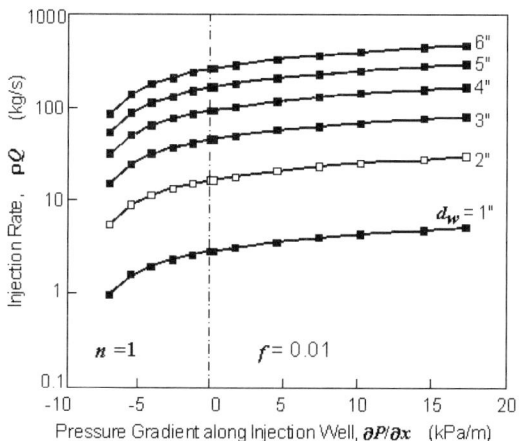

FIGURE 13. Effects of pressure gradient in the injection wells on the CO_2 injection rate.

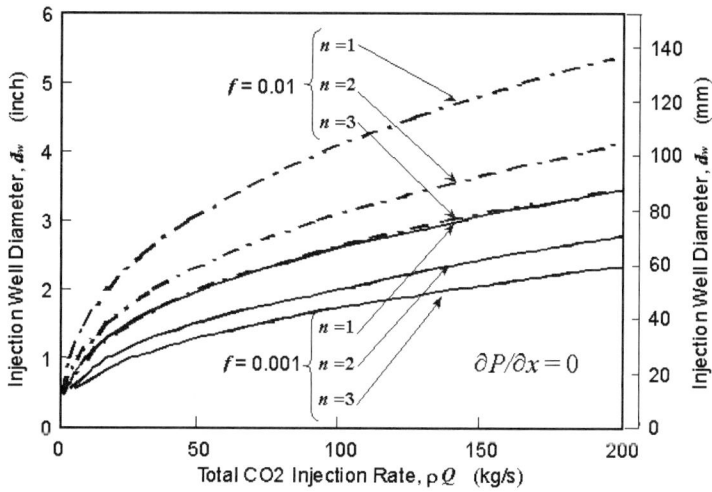

FIGURE 14. Relationship between total injection rate and inner diameter of well for $\partial P/\partial x = 0$.

CO₂ Injection System Design

Since there are many parameters involved in designing an injection system, we propose the following simple design scheme:
(a) With *in situ* water injection measurements, obtain a target aquifer with transmissibility T in the 6.0×10^{-13}–2.2×10^{-11} (m³) range, and the life of aquifer N_a, over the period assumed in the disposal scenario.
(b) Obtain the injection rate range, ρQ, using FIGURE 8.
(c) Determine several combinations of n and d_w versus the injection rate range from FIGURE 14.
(d) Perform economic and system risk evaluations for each combination by estimating total costs, and then select the best combination.
(e) Determine values in the injection system design.

For example, suppose an aquifer located at $x = 700$ m, with $T = 5.0 \times 10^{-12}$ (m³) and $D_a = 1500$ m. The injection range is selected as $\rho Q = 2$–67 (kg/s) from FIGURE 8. Therefore, some combinations of ρQ at 2–67 kg/s, d_w with standard commercial pipes and n can be found in FIGURE 14.

CONCLUSION

The basic design scheme for disposing liquid CO_2 into an aquifer just under the sea floor was studied using an injection flow model. Based on the equilibrium lines in the pressure and temperature map, an aquifer located about 200 m under the sea floor, or about 700 m below the sea surface, can be targeted as a zone for liquid CO_2 disposal. This shows that more CO_2 can be injected compared to the supercritical condition, since the CO_2 has a higher density. Furthermore, the sediment between the sea floor and the aquifer makes an impermeable sealing cap against CO_2 leakage to the sea floor, by forming CO_2 hydrates with either pure water or seawater. Our numerical study focused on the transmissibility range in the targeted aquifer, a design scheme for deciding the total CO_2 injection rate, and the number and inner diameters of injection wells to satisfy the absolute pressure range, as determined by the CO_2 equilibrium conditions. Our disposal system shows an advantage in reducing the targeted area, because liquid CO_2 density is larger in the low-pressure range, than that above the critical temperature. Our study focused on the design procedure for liquid CO_2 injection below the critical temperature (31.2°C).

Our results can be summarized as follows:
1. An aquifer about 200 m below the bottom of the sea (500 m below the sea surface) can be injected with liquefied CO_2. CO_2 leakage to the sediment over the aquifer will be blocked by the hydrates generated.
2. The range of transmissibility, defined by effective permeability multiplied by the height of the aquifer, should be 6.0×10^{-13}–2.2×10^{-11} m³ against the CO_2 injection rate 1–200 kg/s. This keeps the bottom hole pressure range at 8–20 MPa, which satisfies the conditions for liquid CO_2 in the aquifer.
3. To maintain the absolute pressure in the injection well within the absolute pressure range, the gradient along the well is held to near zero for the basic design scheme. This scheme proposes a map that shows the combination of the total

CO_2 injection rate ($\rho Q = 1$–200 kg/s), the number of injection wells ($n = 1$–3), and the inner diameters of the wells ($d_w = 1$–6 inch).

ACKNOWLEDGMENTS

Our study was supported by a Grant-in-Aid for Scientific Research from the Ministry of Education, Science, Sport and Culture of Japan (International Scientific Research Program #10044123 and Research [B] #10555348). The authors would like to express their thanks to Dr. D. Goldverg (Columbia University) and Dr. P.R. Bishnoi (University of Calgary).

REFERENCES

1. ENVIRONMENT AGENCY OF JAPAN. 1997. Environment White Paper 1997 edit. General Comments. Environment Agency of Japan.
2. TANAKA, S. et al. 1996. Investigation on disposal of CO_2 in aquifers-rep. of 1995–. ENAA-Rep. for MITI, ENAA (in Japanese).
3. TANAKA, S. et al. 1997. Investigation on disposal of CO_2 in aquifers-rep. of 1996–. ENAA-Rep. for MITI, ENAA (in Japanese).
4. KOIDE, H. et al. 1992. Subterranean containment and long-term storage of carbon dioxide in unused aquifers and in depleted natural gas reservoirs, Energy Convers. Mgmt. **33**: 619–626.
5. KOIDE, H. et al. 1997. Hydrate formation in sediments in the sub-seabed disposal of CO_2. Energy **22**(2/3): 279–283.
6. LAW, D. 1996. In Aquifer Disposal of Carbon Dioxide. B. Hitchen, Ed.: 59–92. Geo-Sci. Pub. Ltd., Canada.
7. GOLDVERG, D. 1999. CO_2 sequestration beneath sea-floor: evaluating the in situ properties of natural hydrate-bearing sediments and oceanic basalt crust. Int. J. Soc. Materials Eng. Resources **7**(1): 11–16.
8. FOGG, P.G.T. & W. GERRARD. 1991. Solubility of Gasses in Liquids. John Wiley & Sons 241–264.
9. OHGAKI, K. et al. 1993. Formation of CO_2 hydrate in pure and sea waters. J. Chem. Eng. (Japan) **26**(5): 558–564.
10. NG, H.J. & D.B. ROBINSON. 1983. Equilibrium phase compositions and hydrating conditions in systems containing methanol, light hydrocarbons, carbon dioxide and hydrogen sulfide. Research Report, RR-66. Gas Processors Association. Tulsa.
11. ROBINSON, D.B. & B.R. MEHTA. 1971. Hydrates in the propane carbon dioxide-water system. J. Canad. Petrol. Technol. **10**(1): 33–35.
12. MUSKAT, M. 1937. The Flow of Homogeneous Fluids through Porous Media. McGraw-Hill, New York.
13. RAUDKIVI, A.J. & R.A. CALLANDER. 1976. Analysis of Groundwater Flow, 130–140. Arnold.
14. BOURGOYNE, A.T., JR. et al. 1991. Applied Drilling Engineering. SPE Textbook Series, **2**: 33–35. Society of Petroleum Engineers.

The Impact of CO_2 Clathrate Hydrate on Deep Ocean Sequestration of CO_2

Experimental Observations and Modeling Results

ROBERT P. WARZINSKI,[a,b] RONALD J. LYNN,[b] AND GERALD D. HOLDER[c]

[b]*United States Department of Energy, National Energy Technology Laboratory, Pittsburgh, Pennsylvania 15236, USA*

[c]*University of Pittsburgh, School of Engineering, Pittsburgh, Pennsylvania 15261, USA*

ABSTRACT: CO_2 clathrate hydrate is a crystalline compound that can form under temperature and pressure conditions associated with the injection and storage of CO_2 in the deep ocean (below 500 m). At depths being considered for injection of CO_2 (between 1,000 and 1,500 m), in the absence of hydrate formation, the buoyant CO_2 would simply rise as it dissolved in the seawater. If, however, the hydrate phase forms, it will affect this process. The impact could be positive or negative, depending on how hydrate forms and whether it is associated with undissolved CO_2. This paper summarizes experimental and theoretical information relating to formation conditions for the hydrate, the relative density of the hydrate, the formation of a hydrate shell on drops of liquid CO_2, and the impact that a hydrate shell has on dissolution of CO_2. The future direction of the work is also briefly described.

INTRODUCTION

Processes for sequestering CO_2 in the deep ocean must take into account the formation of CO_2 clathrate hydrate ($CO_2 \cdot nH_2O$; $6 < n < 8$). Recent reports provide new insights on the ease of formation of hydrates in the deep ocean. In one case, several liters of liquid CO_2 were introduced into open containers located at 3,627 m depth in Monterey Bay.[1] Hydrate formed rapidly at the seawater/CO_2 interface and sank through the pool of liquid CO_2, which resulted in expansion of the pool beyond the confinement of the containers. In another case, Ohmura and Mori recently published an analysis of the mechanical forces associated with a hydrate film at a seawater/CO_2 interface at similar depths.[2] Critical conditions, beyond which such a film would not remain stable at the interface, were described. These observations and calculations show that a submerged CO_2 lake could be significantly disturbed by the formation of hydrate rather than being a quiescent pool slowly dissolving into the deep ocean.

Injection of CO_2 into the ocean at shallower depths (below 2,700 m), where CO_2 is less dense than seawater, could also be affected by the formation of hydrate. Theoretically, pure hydrate particles should sink in the ocean.[3] This would facilitate sequestration by transporting CO_2 to even greater depths than used for injection.

[a]Telecommunication. Voice: 412-386-5863; fax: 412-386-4806.
warzinsk@netl.doe.gov

However, our prior work has shown that various scenarios are possible at anticipated injection depths (1,000 m to 1,500 m) depending on the conditions under which hydrates are formed. In one case, hydrate structures that initially formed from a two-phase system (seawater and CO_2) floated in the seawater phase, likely due to CO_2 trapped within the hydrate particles but not incorporated into the hydrate lattice.[4] Similar observations were made in the recent experiments in the ocean.[1] If such floating hydrates form, sequestration would be adversely impacted because the CO_2 would ultimately end up at shallower depths than planned or even in the atmosphere. On the other hand, a more dense, sinking hydrate was formed when CO_2 was first dissolved in the seawater prior to reaching hydrate-forming conditions.[4] Another possible occurrence of hydrate is as a thin shell on CO_2 drops. At 1,000 m to 1,500 m the shells would not be thick enough to cause any but the smallest drops to sink.[3] In addition, shells retard the dissolution of CO_2 into the seawater and, therefore, frustrate sequestration by allowing the hydrate-encased CO_2 drops to rise to shallower depths.[4]

Understanding hydrate formation occurrences and processes is therefore critical to successful deployment of strategies for introducing CO_2 into the deep ocean in a manner that leads to long-term storage of CO_2 in the ocean. The importance of developing models and scenarios consistent with experimental observations was recently pointed out in a review of state-of-the-art hydrate film modeling.[5] This paper describes observations made in our laboratories concerning hydrate formation in seawater. These observations are compared to the predictions from mathematical models we have also developed.[6] Finally, the future direction of our experimental work is briefly described.

EXPERIMENTAL

Experimental observations were made using a high-pressure, variable-volume viewcell (HVVC) of 10 cm^3 to 40 cm^3 capacity. The HVVC was enclosed in a chamber where the temperature could be maintained in the region of interest (0°C to 10°C). Agitation in the HVVC was provided by a glass- encased magnetic stirring bar. More complete descriptions of the HVVC and the basic procedures have been published.[3,4] General purpose seawater, salinity of 35, was obtained from Ocean Scientific International, Ltd., Petersfield, Hampshire, U.K. SFC purity (above 99.99%) CO_2 was used.

RESULTS

Hydrate Formation from Dissolved CO_2

The bulk density of CO_2 hydrate is affected by the mode of its formation.[4] Formation from a single-phase solution of CO_2 and seawater results in the formation of transparent, sinking hydrates. Several experiments were performed to investigate the amount of dissolved CO_2 required to form hydrate at conditions similar to those anticipated for ocean injection at approximately 1,500 m. In an experiment with an initial dissolved CO_2 concentration of 59 mg/g seawater, hydrate formed readily with agitation at 4°C and 15.0 MPa. In contrast, in another experiment at a lower initial dissolved CO_2 level (44 mg/g seawater) hydrate did not form at similar conditions. However, hydrate was formed in this experiment when conditions were changed to simulate deeper ocean depths (2°C, 27.2 MPa). It was also noted in this experiment that at higher pressures the hydrate mass became more difficult to detect visually. At 31 MPa the hydrate mass could not be distinguished from the seawater phase. Brewer *et al.*, made the same observation in experiments at 3,650 m depth in the ocean.[1]

Hydrate Shell Formation on CO_2 Drops

A model was developed in our previous work to estimate the initial and steady-state thicknesses of hydrate shells on CO_2 drops.[6] The model assumed that the initial thickness was determined by the degree of oversaturation of the water surrounding the drop with CO_2 relative to the equilibrium saturation concentration, C_H, at the hydrate equilibrium pressure at a given temperature. The excess CO_2 could accumulate in the water during the induction period commonly observed in hydrate studies. Since the amount of dissolved CO_2 is a function of pressure, C_H would be expected to decrease with temperature in the hydrate formation region, since lower hydrate equilibrium pressures are associated with decreasing temperature. This has been observed experimentally at 30 MPa.[7]

After the initial formation, our model assumed that the thickness of the shell was governed by the rates of diffusion of the CO_2 through the hydrate shell and diffusion or convection of dissolved CO_2 away from the hydrate-covered particle. Based on these assumptions, the model predicts that the initial hydrate shell forming around drops of CO_2 injected into the deep ocean would be thin (less than 0.1 cm thick). Over time, in water unsaturated relative to hydrate forming conditions, the model also predicts that a stable hydrate shell thickness on the order of 10^{-2} to 10^{-4} times the radius of the drop will eventually result. Thus, based on this model, for drop sizes anticipated for ocean injection (up to about 1-cm radius) the initially formed hydrate shell should become thinner over time.

With the HVVC it is possible to inject drops (typically 0.5 cm to 1.0 cm diameter) of CO_2 into pressurized seawater using a high-pressure syringe pump. Such experiments have shown that drops of CO_2 only formed a hydrate shell if sufficient dissolved CO_2 is present. In the experiment described above with an initial CO_2 concentration of 44 mg/g seawater, drops of CO_2 were added and their behavior observed. The first several drops introduced into the seawater phase at 2°C, 14.5 MPa dissolved without forming a hydrate shell. Each added drop increased the dissolved CO_2 concentration. When the concentration reached 51 mg CO_2/g seawater, subsequent drops formed hydrate shells within several seconds of injection at 2°C, 17.5 MPa. An illustration of this phenomenon is shown in FIGURE 1.

FIGURE 1A shows two drops of liquid CO_2 resting at the top of the HVVC immediately prior to hydrate shell formation. FIGURE 1B shows the hydrate shell beginning to form on the lower left corner of the drop. Shell formation began at a single point and rapidly grew to completely envelop the drops, which coalesced during the process, in 1 to 2 seconds. Others have assumed that hydrate shell formation occurs uniformly across the surface of the drop, growing from hydrate clusters or

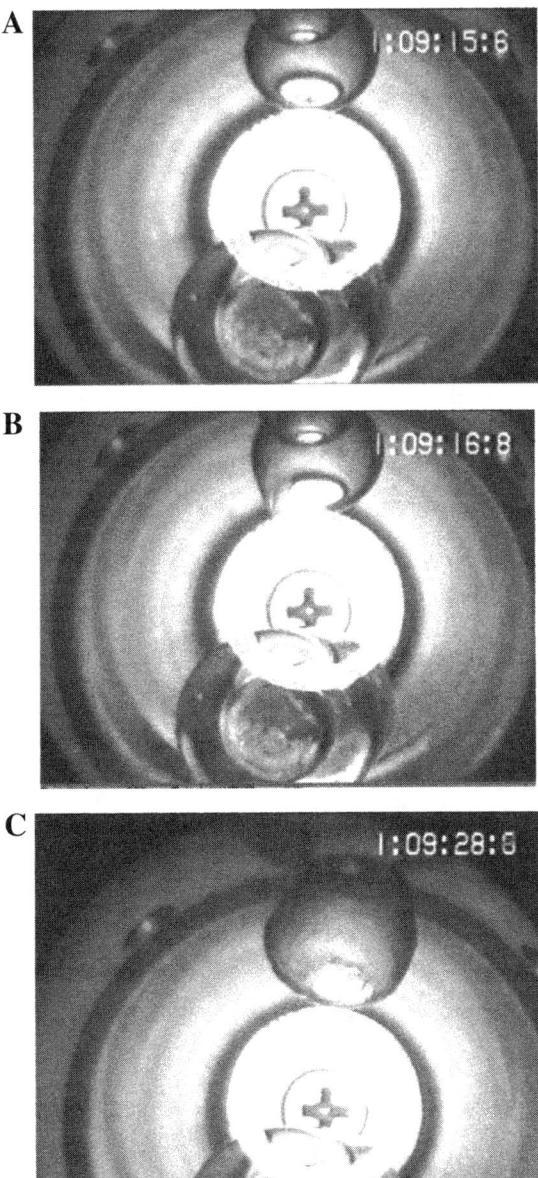

FIGURE 1. Images showing the formation of a hydrate shell on CO_2 drops in seawater. The cell is full of seawater at 2°C, 17.5 MPa. The object in the *lower part* of the images is an end-on view of a glass-encased magnetic stir bar that has a diameter of 0.01 m. In the *upper right corner* of each image is an elapsed time indication reading out to 0.1 s.

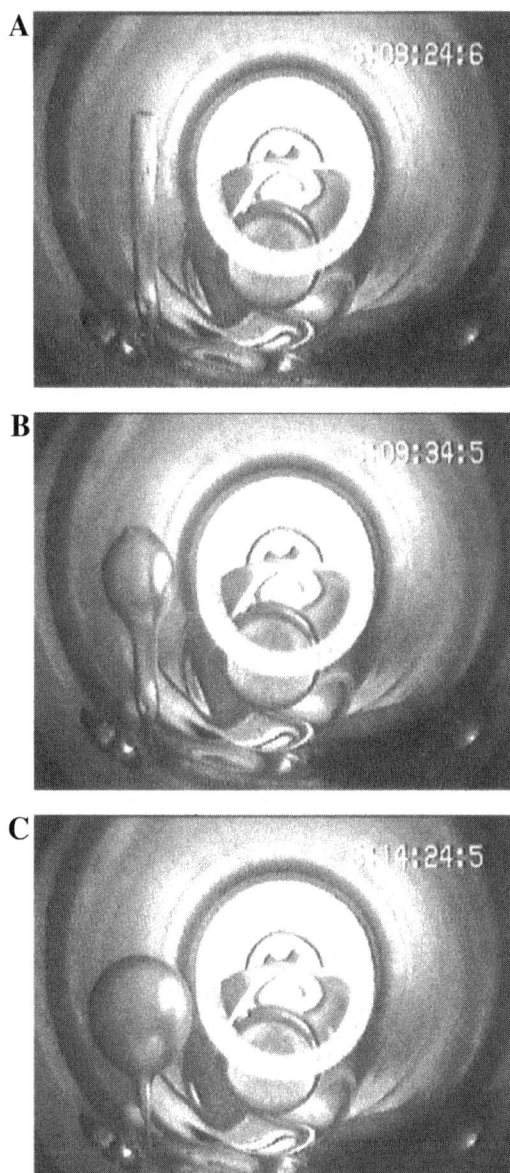

FIGURE 2. Images showing the apparent thinning of a hydrate shell. (See FIG. 1 caption for other information.)

crystallites;[5] however, our observations on hydrate-forming systems that have just attained sufficient dissolved CO_2 do not show this to be the case. Hydrate shell formation starts at a point, usually where contact to a foreign object occurs (but not always as in FIG. 1) and then rapidly advances to envelop the entire drop. A 0.5-cm to 1.0-cm diameter drop is typically enveloped in 1 s to 2 s. FIGURE 1C shows the drop covered with a rough-textured hydrate shell. Over time the texture became smoother and the drop assumed a nearly spherical shape, indicating that the shell was thinning with time, as predicted by our model.

Another more dramatic example of a hydrate shell thinning with time after initial formation is shown in the images contained in FIGURE 2. In this experiment, the HVVC was at 2.4°C and 15.8 MPa. The dissolved CO_2 concentration was approximately 60 mg/g seawater at the time the drop was injected. The hydrate shell formed immediately and was thick enough so that its strength permitted the formation of the long tubelike structure shown in FIGURE 2A. Within seconds, however, the structure began to expand to a more spherical shape as shown in FIGURE 2B, likely due to the thinning of the shell as it approached an equilibrium thickness. FIGURE 2C shows this process has continued over a five-minute interval. This phenomenon was predicted by the modeling work summarized above.

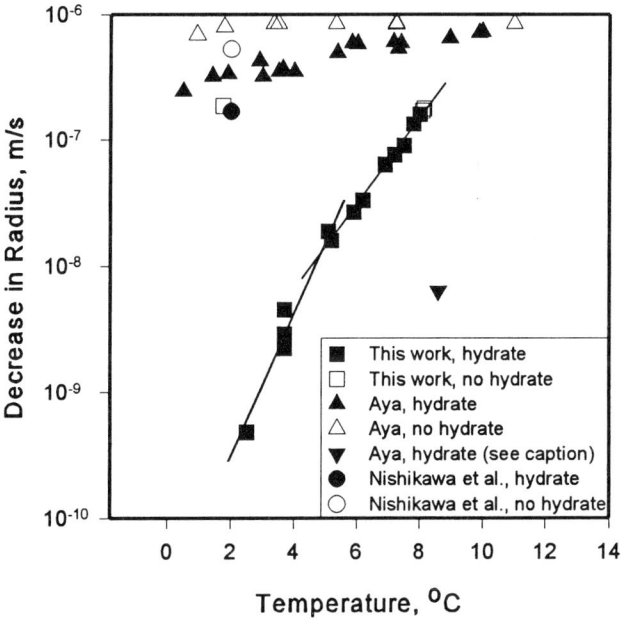

FIGURE 3. CO_2 drop dissolution data. Contains our work in seawater, Aya's work in fresh water, and the work of Nishikawa *et al.* in seawater. Aya's observation at approximately 7 wt% dissolved CO_2 is indicated by the symbol ▼.

Effect of a Hydrate Shell on CO_2 Drop Dissolution

The dissolution of CO_2 drops in water and seawater, with and without hydrate shells, expressed as the rate of radial decrease as a function of temperature, are shown in FIGURE 3. Our data were obtained during the experiment with an initial dissolved CO_2 concentration of 44 mg/g seawater. Measurements of radial shrinkage were made from recorded video images. The pressure was maintained near 17 MPa during these observations. The first several drops were injected at 1.7°C and dissolved without forming a hydrate shell. The dissolution of these drops increased the dissolved CO_2 concentration to 51 mg/g seawater. The next drops were introduced at 2°C and formed a hydrate shell within several seconds of injection. The formation of the shell caused a decline in the rate of radial shrinkage of the drops of nearly three orders in magnitude when compared to dissolution in the absence of hydrate. Additional measurements were made on the same hydrate-covered drops over a period of five days at 2.5°C to 3.7°C as they slowly dissolved. Measurements were then made over the next two days as the temperature was incrementally increased to 8.0°C, at which point these drops completely dissolved. Two final drops were then

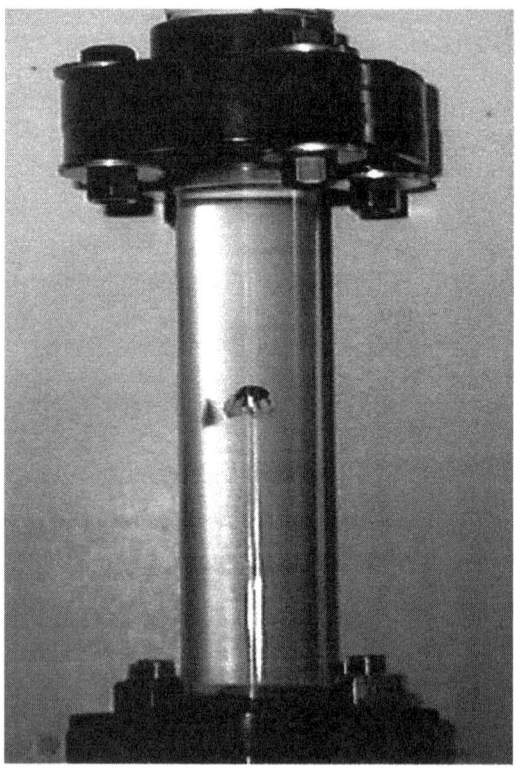

FIGURE 4. Test section of the LWTF with an air bubble stabilized in a downward flow of water.

introduced at 8.1°C that dissolved without forming a hydrate shell. The data in FIGURE 3 show that the rate of shrinkage increased with temperature, likely due to thinning of the hydrate shell that was apparent in the visual observations. The hydrate shells at the lower temperatures were rough textured for extended periods of time indicating a greater shell thickness, whereas at higher temperatures the shells became smooth. The different slopes in our data also suggest that the shells may not have thinned to a steady state thickness during the observation periods at the lower temperatures.

Data from other researchers are also shown in FIGURE 3. The data of Aya were obtained in a 32 L vessel at 30 MPa in fresh water.[8] Nishikawa et al., obtained their data in a 16.7 L vessel at 30 MPa in artificial seawater recirculated at different flow rates.[9] The two data points of Nishikawa et al. shown in FIGURE 3 were obtained by extrapolation to zero flow. Neither investigator mentions dissolved CO_2 content, except that Aya notes one observation in water with about 7 wt% CO_2.

Previous investigators did not always indicate the levels, if any, of dissolved CO_2 present in their studies. Based on the results reported here, this information should be provided by investigators in future. The data in FIGURE 3 show dramatic differences, not only in the rates of radial decrease, but also in the relative differences

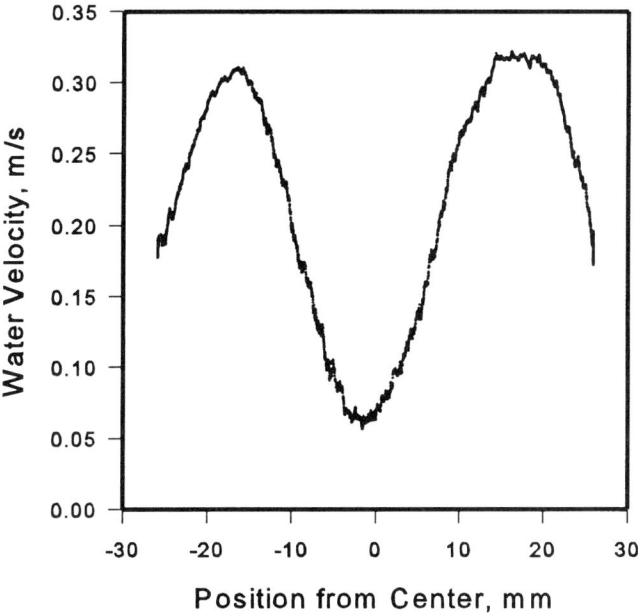

FIGURE 5. Example of the measured velocity profile across the viewing section of the LWTF. The measurements were made with an S-type pitot tube constructed of 1/16″ dia stainless steel tubing and housed within a 3/16″ tube. The pitot tube was translated across the bore using a stepper motor under control of a computer which also collected 5,000 data points during the transverse. Calibration of the probe was achieved by timed positive displacement measurements.

between drops with and without hydrate, especially at the lower temperatures. The different pressures, drop sizes, and vessel configurations undoubtably are involved. Our vessel had a much smaller volume than the others, which would permit dissolved CO_2 levels to more rapidly equilibrate.

Development of a Water Tunnel Facility

To more accurately simulate CO_2 in the oceanic water column, a high-pressure water tunnel facility is being built in our laboratory. The device will permit stabilization of rising or sinking CO_2 drops in a visual observation area for extended periods of time using a countercurrent flow of water or seawater following design principles established in the literature.[10,11] In this device, as in the ocean, a drop will not touch a containment structure but will encounter only the fluid phase while being subjected to changing conditions of temperature, pressure, and dissolved CO_2 content, giving a more accurate simulation of the drop's journey in the ocean water column.

A low-pressure water tunnel facility (LWTF) has been constructed for the purpose of refining design geometries of the viewing section and flow conditioning elements, to permit stabilization of the drop in both the radial and axial directions. In initial experiments, air bubbles and plastic spheres of varying density were successfully stabilized in a countercurrent flow of water. FIGURE 4 shows an air bubble that has been stabilized in the LWTF. Such bubbles have been stabilized and observed for several hours. FIGURE 5 shows an example of the velocity profile measured across the center of the viewing section. The velocity minimum in the center of the viewing section provides for radial stabilization of the object. Further refinements in the design are being made to optimize stabilization at the slower flow rates needed for objects that would have densities closer to that of compressed liquid CO_2.

REFERENCES

1. BREWER, P.G., G. FRIEDERICH, E.T. PELTZER & F.M. ORR, JR. 1999. Science **284:** 943–945.
2. OHMURA, R. & Y.H. MORI. 1998. Environ. Sci. Technol. **32:** 1120–1127.
3. HOLDER, G.D., A.V. CUGINI & R.P. WARZINSKI. 1995. Environ. Sci. Tech. **29:** 276–278.
4. WARZINSKI, R.P. & G.D. HOLDER. 1997. In Proceedings of the 9th International Conference on Coal Science 1879–1882.
5. MORI, Y.H. 1998. Energy Convers. Mgmt. **39:** 1537–1557.
6. HOLDER, G.D. & R.P. WARZINSKI. 1996. In Preprints of Papers, American Chemical Society, Division of Fuel Chemistry. **41:** 1452–1457.
7. AYA, I., K. YAMANE & H. NARIAI. 1997. Energy (Oxford) **22:** 263–271.
8. AYA, I. 1995. In Direct Ocean Disposal of Carbon Dioxide. N. Handa & T. Ohsumi, Eds.: 233–238. Terra Scientific Publishing Co., Tokyo.
9. NISHIKAWA, N., M. ISHIBASHI, H. OHTA, N. AKUTSU, M. TAJIKA, T. SUGITANI, R. HIRAOKA, H. KIMURO & T. SHIOTA. 1995. Energy Convers. Mgmt. **36:** 489–492.
10. MAINI, B.B. & P.R. BISHNOI. 1981. Chem. Eng. Sci. **36:** 183–189.
11. MOO-YOUNG, M., G. FULFORD & I. CHEYNE. 1971. Ind. Eng. Chem. Fundam. **10:** 157–160.

CO_2 Hydrate Formation in Various Hydrodynamic Conditions

A. YAMASAKI,[a,b] H. TENG,[c] M. WAKATSUKI,[b]
Y. YANAGISAWA,[b] AND K. YAMADA[b]

[b]*Department of Chemical System Engineering, University of Tokyo, 7-3-1 Hongo, Bunkyo-ku, Tokyo 113, Japan*

[c]*AVL Powertrain Engineering, Plymouth, Michigan 48170, USA*

ABSTRACT: The influence of hydrodynamics on formation of CO_2 hydrate in a column-type reactor and a stirred-vessel reactor was investigated experimentally. In the column-type reactor, the buoyant motion of the CO_2 drops released into the reactor was balanced by the counterflow of the water in the reactor and, therefore, the CO_2 drops were suspended stably in the test column. The drop Reynolds number affected hydrate formation significantly. For drops with small Reynolds numbers, hydrate formed almost immediately on the drops. For drops with large Reynolds numbers, it took up to 20 minutes for hydrate to cover the drops. In cases where the Reynolds number reached a certain critical value, a continuous process of hydrate formation on the drops and hydrate shedding from the drops was observed. In the stirred-vessel reactor, the kinetics of hydrate formation was governed by the strength of agitation. In cases without agitation, hydrate formed only at the liquid CO_2–water interface. In cases with a weak agitation, hydrate was shed from the interface; however, the interface was still the only location for hydrate formation. In cases with a strong agitation, the liquid CO_2–water interface broke, liquid CO_2 dispersed throughout the water phase, and hydrate particles of about 0.1 mm in size formed in water simultaneously. The hydrate particles and dispersed CO_2 drops coagulated, forming a cluster with a dimension of a few centimeters when stirring was stopped. The bulk density of the cluster varied with the duration of agitation. At early stages, the bulk density of the cluster was less than that of water; however, it increased with time under a continuous agitation. The critical time of duration, at which buoyancy of the cluster changed from positive to negative, decreased dramatically with increase in the rate of agitation.

INTRODUCTION

Various scenarios of ocean disposal of anthropogenic CO_2 have been proposed as a means to mitigate global warming.[1] The majority of these scenarios are concerned with discharging liquid CO_2 into the ocean below the mixed layer (typically deeper than 500 m) and sequestering the CO_2 in the ocean via dissolution. At these disposal depths, the conditions for formation of CO_2 hydrate (i.e., pressure $p \geq 44.5$ bar and

[a]Telecommunication. Voice: +81-3-5841-7311; fax: +81-3-5684-3298.
akihiro@chemsys.t.u-tokyo.ac.jp

temperature $T \leq 283$ K)[2] are generally satisfied. Therefore, in modeling the behavior of CO_2 disposed of in the ocean, hydrate must be taken into consideration.

Several investigators have reported laboratory studies of CO_2 hydrate formation under pressures and temperatures simulating the deep ocean waters. Aya et al.[3] and Fujioka et al.[4] investigated hydrate formation on a single CO_2 drop in high-pressure and low-temperature water. They reported that hydrate covered the drop immediately when the drop was introduced into the water, and that the hydrate film formed on the surface of the drop was stable, although it did not stop drop dissolution in water. Since the effect of drop buoyancy was limited in their experiments, the influence of hydrodynamics of the ambient water on hydrate formation was not included in the results of Aya et al. and Fujioka et al. Saji et al.[5] studied the formation of CO_2 hydrate in a stirred-vessel reactor. They found that the hydrodynamics of the water in the reactor affected hydrate formation significantly. The experiment of Saji et al. suggests that the results of Aya et al. and Fujioka et al. may reflect only part of the behavior of a CO_2 drop in the ocean because the hydrodynamic effect on hydrate formation was barely included in their experiments. In reality, buoyant motion of a CO_2 drop in the ocean influences the CO_2 concentration around the drop and the surface stability of the drop, which, in turn, affects the chemical reaction at the drop surface considerably.

Being a solid crystal, hydrate changes the behavior of CO_2 in the ocean significantly. To properly model the behavior of the CO_2 disposed of in the ocean, it is important to understand how the hydrodynamics of the ambient seawater affects the hydrate formation. To date, no systematic study of the hydrodynamic effect on hydrate formation has been reported. To fill this gap, we investigated the hydrate formation in a column-type reactor and a stirred-vessel reactor under various hydrodynamic conditions. The results of our investigation are reported below. We believe that the phenomena found in our experiments will be of a great help in understanding of the behavior of the CO_2 disposed of in the ocean.

EXPERIMENTAL SYSTEMS

Our investigation of the hydrodynamic effect on hydrate formation was conduced in two different kinds of reactors: a column-type reactor and a stirred-vessel reactor. FIGURE 1 is a schematic diagram of the column-type reactor. The observatory or test column of the reactor was a tapered polycarbonate tube of 40 mm (at the top) and 60 mm (at the bottom) inner diameters, and 200 mm length. The reactor could stand pressures up to 300 bar that corresponds to ocean depths down to 3,000 m. At these pressures and ocean temperatures, liquid CO_2 has positive buoyancy. In order to suspend the CO_2 drops in the test column, the water in the reactor was downwardly circulated. A high-pressure pump drove the water circulation. The water flow rate could be controlled in the range of 0–6.4 L/min that corresponds to 0–8 cm/s in linear velocity at the top of the test column. A flow stabilizer section filled with drinking straws of different lengths was connected to the test section in order to regulate the flow and suspend the drops stably. The pressure of the reactor was controlled to an accuracy of ±0.3 bar by a piston pump with a PID controller. The temperature of the reactor was controlled to an accuracy of ±0.3 K by a heat exchanger where the water

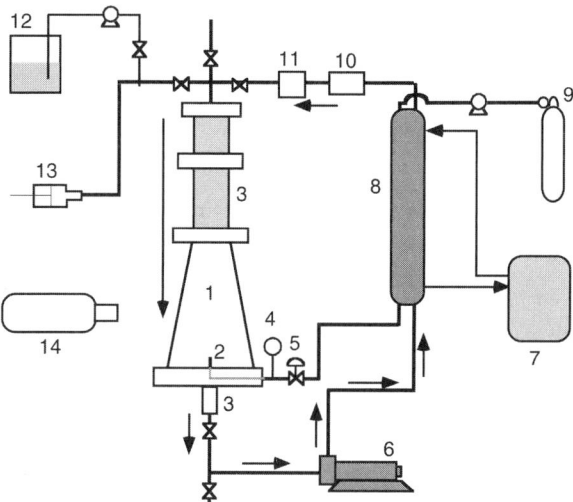

FIGURE 1. Schematic diagram for the column-type reactor. *1*, Polycarbonate tube; *2*, CO_2 injection nozzle; *3*, flow stabilizer; *4*, pressure gauge; *5*, needle valve; *6*, high-pressure pump; *7*, thermal bath; *8*, heat exchanger; *9*, CO_2 cylinder; *10*, thermometer; *11*, flow rate transducer; *12*, water tank; *13*, piston pump with pressure controller; *14*, CCD camera.

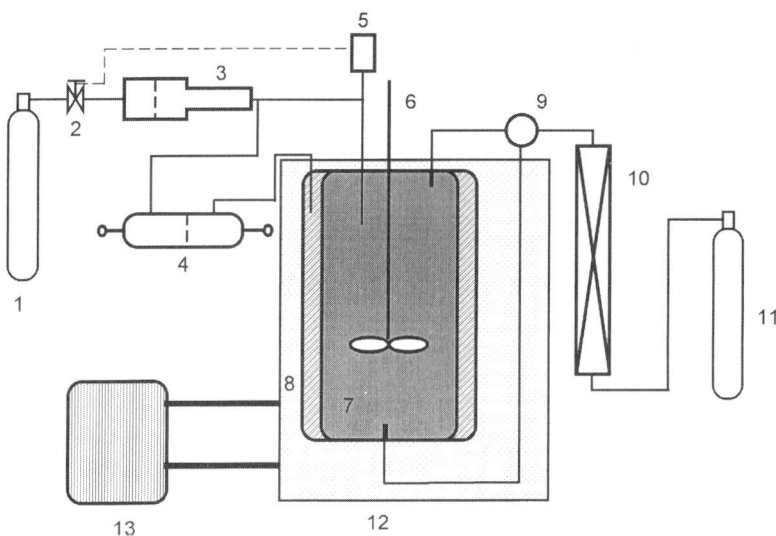

FIGURE 2. Schematic drawing for the stirred-vessel reactor. *1*, N_2 cylinder (for pressure controller); *2*, solenoidal valve; *3*, pressure controller; *4*, pressure equilibrator; *5*, pressure sensor; *6*, agitator; *7*, Pyrex glass tube; *8*, polycarbonate tube; *9*, three-way valve; *10*, heat exchanger (cooling unit); *11*, CO_2 cylinder; *12*, water bath; *13*, cooling unit.

in the system exchanged heat with a thermal bath. Liquid CO_2 from a CO_2 cylinder was introduced into the reactor through a 2-mm orifice from the bottom of the test column. By properly controlling the pressure, temperature, and water flow rate, CO_2 drops introduced into the reactor could be stably suspended in the test column. Because the polycarbonate tube was transparent, the process of hydrate formation in the reactor could be recorded with a CCD camera.

FIGURE 2 shows a schematic diagram of the stirred-vessel reactor. The main part of this reactor was a coaxial-structured vessel with an inner Pyrex-glass tube of 100 mm inner diameter, 5 mm wall thickness, and 250 mm length and an outer polycarbonate tube of 120 mm inner diameter, 60 mm thickness, and 250 mm length. The reactor could stand pressures up to 200 bar. Units similar to those in the column-type reactor controlled the pressure and temperature of the reactor, to accuracy of ±0.5 bar for pressure and ±0.1 K for temperature. In this reactor, liquid CO_2 was also introduced into the reactor through a 2-mm orifice from the bottom of the reactor. The agitation was performed by a stirrer, whose speed varied in the range of 0–750 rpm. The process of hydrate formation in the reactor could be observed directly and video-recorded with a CCD camera.

EXPERIMENTAL RESULTS

Hydrate Formation in a Column-Type Reactor

CO_2 drops of various sizes and shapes, produced by controlling the injection rate and the injection pressure, were tested in the column-type reactor. These drops could be suspended in the test column by means of proper control of the rate for the downward-flow water. For a CO_2 drop with a dimension reaching that of the column, its shape became a bullet-like liquid plug. For such a big drop or liquid plug, it took about 20 minutes for hydrate formation to start. Hydrate was formed first at the rear of the plug and then spread throughout the entire plug surface in a few tens of seconds. However, hydrate formation was found to be limited only on the surface of the plug (see FIGURE 3). For CO_2 drops of sizes falling in a range 5–10 cm, hydrate shedding was observed from the rear of the drops, which reduced the sizes of the drops rapidly by removing the mass of CO_2 from the drops (see FIGURE 4). The phenomenon of hydrate shedding from a CO_2 drop was highly related to its size and shape. For a large drop, hydrate shedding was a continuous process; however, it became intermittent for smaller drop sizes. The frequency of hydrate shedding was reduced with decreasing drop size. When the drop size reached a critical value, the phenomenon of hydrate shedding stopped completely. Such drops were usually spherical in shape. For the sizes of CO_2 drops in the range 0.5–1.5 cm, a thin hydrate film formed immediately on the surfaces of the drops upon them in contact with water. These drops were generally spherical in shape and no observable hydrodynamic effect on hydrate formation was detected. The drops covered with a hydrate film were easily agglomerated to form a grape-like cluster when interacting with each other (see FIGURE 5). The drop cluster was apparently stable and no drop coalescence was observed at least over a period of one hour.

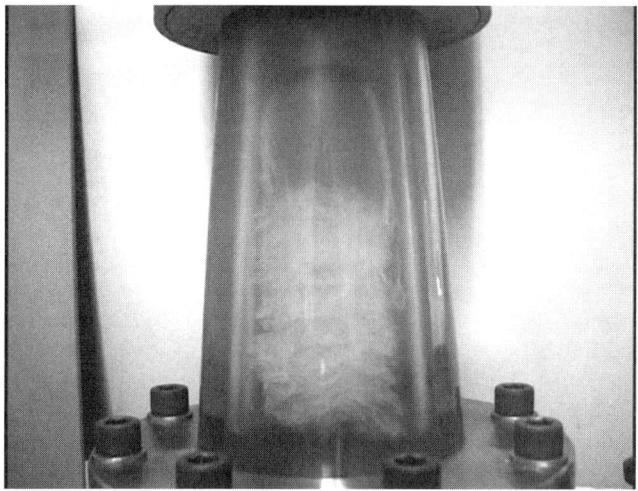

FIGURE 3. Hydrate formation on a liquid CO_2 plug. Temperature, 276 K; pressure, 50 bar; water flow rate, 6 L/min.

Hydrate Formation in Stirred-Vessel Reactor

In the stirred-vessel reactor tests, the reactor was first filled with a certain amount of water and then liquid CO_2 was introduced into the reactor until a desired pressure in the reactor was obtained. A hydrate film formed at the liquid CO_2–water interface

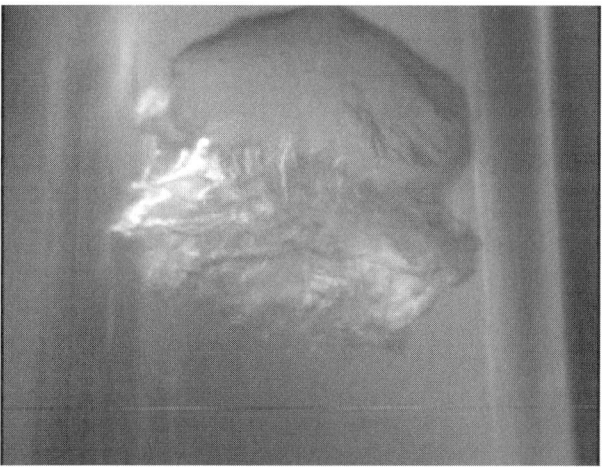

FIGURE 4. Hydrate shedding from liquid CO_2 drop. Temperature, 276 K; pressure, 50 bar; water flow rate, 6 L/min.

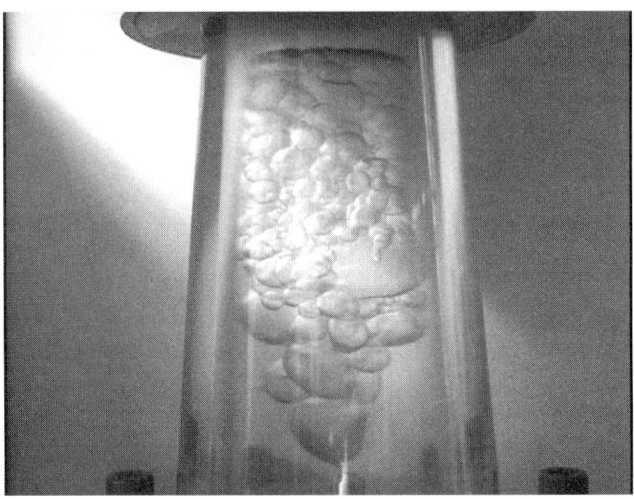

FIGURE 5. Grape-like cluster of liquid CO_2 drops covered with hydrate. Temperature, 276 K; pressure, 50 bar; water flow rate, 6 L/min.

less than a half minute after the CO_2 entered the reactor. The stability of the hydrate film was highly influenced by agitation. Without agitation, the hydrate film was flat and remained stably at the interface. Under weak agitation (stirrer speed less than 100 rpm), the film became wavy but was basically stable. When the stirring rate reached 120 rpm, shedding of hydrate pieces from the interface into the water was observed (see FIGURE 6). The hydrate shedding was a continuous process as long as the agitation continued. The hydrate pieces shed from the interface stacked in the neighborhood of the liquid CO_2–water interface and their sizes increased continuously. When the stirrer speed increased to 150 rpm, the wavy liquid CO_2–water interface broke and CO_2 drops of a few millimeters in diameter were produced. The

FIGURE 6. Shedding of hydrate from liquid CO_2 phase. Temperature, 279 K; pressure, 50 bar; stirring rate, 120 rpm.

FIGURE 7. Grape-like cluster of CO_2 drops covered with hydrate film. Temperature, 279 K; pressure, 50 bar; stirring rate, 150 rpm.

FIGURE 8. Hydrate and liquid CO_2 drop cluster with positive buoyancy. Temperature, 279 K; pressure, 50 bar; stirring 5 min at 740 rpm.

stronger the agitation (i.e., the higher the stirrer speed), the smaller the drop sizes. These small drops agglomerated forming a grape-like cluster (see FIGURE 7). When the stirrer speed became higher than 350 rpm, the liquid CO_2–water interface lost its stability completely and, as a result, the liquid CO_2 dispersed into water forming many dispersed drops in water. The dispersed CO_2 drops were covered with a thin hydrate film. Small hydrate particles also were noticed to exist as a dispersed phase in water. Sizes of the small dispersed CO_2 drops decreased with increase in the stirring speed and became about 0.1 mm in diameter at a stirring rate about 500 rpm. However, further increase in the stirring rate did not result in further decrease in the drop sizes, and the drop sizes were basically unchanged when the stirring rate was increased from 500 to 740 rpm. When the agitation was stopped, the CO_2 drops and hydrate particles in water agglomerated forming a cluster of a few centimeters in dimension (see FIGURE 8). The cluster had positive buoyancy in water if the duration of agitation was short. When the agitation was continued for long enough, buoyancy of the cluster reversed and the cluster sank to the bottom of the reactor. The critical stirring time for the buoyancy reversal decreased with increase in the stirring rate dramatically (see FIGURE 9). The pressure of the reactor was influenced by agitation only at the early stages: the pressure of the reactor dropped rapidly when the agitation was just started. However, after about one minute, the changes in pressure became small. No obvious pressure variation was detected at the critical stirring time for the buoyancy reversal of the cluster. FIGURE 10 shows the variations in the reactor pressure with the stirring time or duration of agitation.

Discussion

Experimental results from our study for hydrate formation in two different types of reactors demonstrated that hydrodynamics of the ambient water significantly

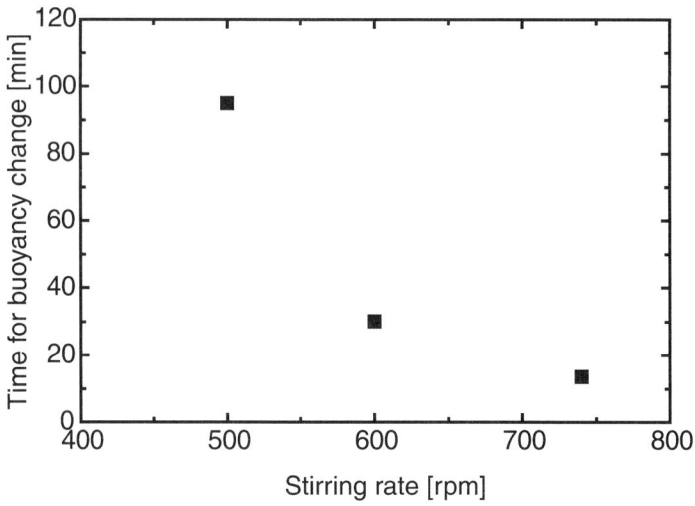

FIGURE 9. Effect of stirring rate on the critical stirring rate for the buoyancy reverse.

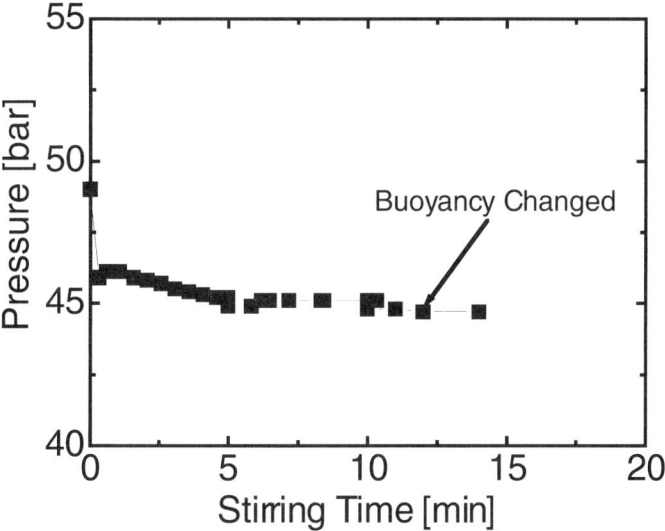

FIGURE 10. Pressure change for the stirred-vessel reactor. Temperature, 279 K; pressure, 50 bar; stirring rate, 740 rpm.

affects the process of hydrate formation. For the cases with or without a weak hydrodynamic disturbance, a thin hydrate film formed rapidly but only at the liquid CO_2–water interface, and it was apparently stable. Mass transfer from the liquid CO_2 into water was reduced significantly by this solid hydrate interphase. Hydrate formation on small CO_2 drops in the column-type reactor and at the liquid CO_2–water interface in the stirred-vessel reactor with weak agitation (stirrer speed less than 100 rpm) were such cases. For cases that the system underwent a strong hydrodynamic disturbance, the hydrate film formed at the liquid CO_2–water interface became unstable and hydrate pieces were shed from the interface into the water. These cases were encountered for hydrate formation on large drops (5–10 cm) in the column-type reactor and at the interface in the stirred-vessel reactor when the stirrer speed was at least 120 rpm. Interestingly, the corresponding Reynolds numbers (drop diameter based) for the cases where hydrate shedding was observed were of the same order of magnitude for the two different types of reactors: about 3,000–6,000 for the column reactor and 5,000 for the stirred-vessel reactor. This indicates that hydrate shedding takes place in a similar range of Reynolds numbers for both reactors.

We believe that, in both types of reactors, hydrate shedding is due to the fact that the liquid CO_2–water interface (where hydrate is formed) loses its stability and breaks up, and that the instability and breakup of the interface is induced largely by the shear forces at the interface. Under this shedding mechanism, because the shear forces are due to the difference in velocities crossing the liquid CO_2–water interface, hydrate shedding in both reactors is influenced strongly by the hydrodynamics in the reactors. In the column-type reactor, the rate of hydrate shedding decreased with decrease in the drop size. Since increase in the drop dimension leads to a higher drop velocity, the Reynolds number increases considerably with the drop size. This suggests that a larger Reynolds number results in a faster shedding rate. In the stirred-vessel reactor, under a strong hydrodynamic disturbance (i.e., stirrer speed at least 150 rpm), the liquid CO_2–water interface became unstable and broke up, resulting in small drops. These small drops stayed in the neighborhood of the interface, or more accurately, they formed a new type of the interphase between the liquid CO_2 and water phases. Because the interphase formed by the CO_2 drops covered with a hydrate film was more stable than a thin hydrate interface, the shedding was stopped. The interphase formed by smaller drops should be more stable than for larger drops and, therefore, it has greater resistance to the hydrodynamic disturbance. Note that the CO_2–drop interphase was formed because of the positive buoyancy of the drops. The smaller the drop size, the smaller the ratio of buoyant to hydrodynamic forces. The hydrodynamic forces increase with further increase in the stirring rate. When the stirring rate was so high that the hydrodynamic forces became much larger than the buoyant force of the CO_2 drops, the CO_2 drops disrupted to form a dispersed phase in the water. This also led to the entire liquid CO_2 phase forming a dispersed phase in water due to the loss of the interphase. Tiny hydrate particles in water formed either from the breakup hydrate pieces (shed from the liquid CO_2–water interface previously) or were produced in water due to a considerably increased CO_2 concentration in water from the contribution of the dispersed drops. Because hydrate surfaces are very reactive, the hydrate particles and the CO_2 drops covered with hydrate formed clusters when they interacted with each other. The bulk density of a cluster was dependent upon the proportion of hydrate it contained. This increased with stirring time. When the portion of hydrate in the cluster was so large that the bulk density of the cluster exceeded that of water, the buoyancy of the cluster reversed. The time for the buoyancy reversal was a function of both the stirring rate and duration of the agitation. The stirring rate indicates the strength of the hydrodynamics in the reactor, which may be characterized by the Reynolds number; whereas the duration of the agitation reflects the overall kinetics for hydrate formation under given hydrodynamic conditions.

CONCLUSIONS

CO_2 hydrate formation in two different types of reactors were investigated by observation. It was found that the kinetics for hydrate formation were significantly affected by the hydrodynamics in both types of reactors. With or without a weak hydrodynamic disturbance, a thin hydrate film formed rapidly at the liquid CO_2–water interface and the hydrate film was apparently stable. Increase in the strength of the hydrodynamic disturbance increased the shear forces at the interface causing the interface to become unstable. The instability and breakup of the interface led to a phenomenon of hydrate shedding from the interface. The rate of hydrate shedding increased with increase in the Reynolds number. Further increase in the strength of the hydrodynamic disturbance resulted in a different pattern for the interface breakup: the interface broke up into smaller CO_2 drops covered with a hydrate film. These drops formed grape-like clusters when they interacted with each other. The stronger the hydrodynamic disturbance, the smaller the drops from the interface breakup. Small hydrate particles were also formed in the water-rich phase when the reactor underwent a strong hydrodynamic disturbance. These small CO_2 drops and hydrate particles agglomerated forming a cluster under interaction. At early stages, the bulk density of the cluster was less than that of water and, hence, had positive buoyancy. However, with continued agitation, the bulk density of the cluster increased with time and a buoyancy reversal occurred at a critical time under the given hydrodynamic condition.

REFERENCES

1. HALMANN, M.M. & M. STEINBERG. 1999. Greenhouse gas carbon dioxide mitigation. Lewis Publishers. Boca Raton, FL.
2. SONG, K.Y. & R. KOBAYASHI. 1987. Water content of CO_2 in equilibrium with liquid water and/or hydrates. SPE Formation Evaluation **1987:** 500–508.
3. AYA, I., K. YAMANE & N. YAMADA. 1992. Stability of clathrate-hydrate of carbon dioxide in highly pressurized water. HTD, ASME **215:** 17–22.
4. FUJIOKA, Y., Y. SHINDO, K. TAKEUCHI & H. KOMIYAMA. 1994. Shrinkage of liquid CO_2 droplets in water. Int. J. Energy Res. **18:** 765–769.
5. SAJI, A., H. NODA, Y. TAKAMURA, T. TANII, T. TANAKA, H. KITAMURA & T. KAMATA. 1995. Dissolution and sedimentation behavior of carbon dioxide clathrate. Energy Conv. Mgmt. **36:** 493–496.

MRI Measurement of Hydrate Growth and an Application to Advanced CO_2 Sequestration Technology

SHUICHIRO HIRAI,[a,b] YUTAKA TABE,[b] KUNIHIRO KUWANO,[b]
KUNIYASU OGAWA,[b] AND KEN OKAZAKI[c]

[b]*Research Center for Carbon Recycling and Utilization, Tokyo Institute of Technology, 2-12-1, O-okayama, Meguro-ku, Tokyo 152-8552, Japan*

[c]*Department of Mechanical Engineering and Science, Tokyo Institute of Technology, 2-12-1, O-okayama, Meguro-ku, Tokyo 152-8552, Japan*

> ABSTRACT: MRI measurements of hydrate thickness growth have been measured and this phenomenon applied to advanced CO_2 ocean dissolution technology. CO_2 droplets dissolve during the process of sinking from their release point into deep ocean, by forming fine hydrate particles inside CO_2 droplets before the droplets are released from a towed pipe on a moving ship. This results in a sufficiently long sequestration period from the atmosphere and further reduces biological impact. The increasing rate of hydrate film thickness in forming hydrate particles was measured by an ultrahigh-pressure magnetic resonance imaging (MRI) technique.

INTRODUCTION

To combat global warming it is required to take measures to restrict the increase of atmospheric CO_2 concentration. It is predicted that CO_2 concentration, now at 360 ppm, would become 500 ppm in the year 2100, if we continue to burn fossil fuels at the present rate.[1] Increasing the storage of CO_2 in the ocean artificially is one measure that was first proposed by Marchetti.[2] Injection into ocean of CO_2 removed from flue gas at fossil-fuel-fired power plants is considered to be efficient, since ocean has a huge storage capacity for CO_2. Two problems, however, are anticipated as a result of injecting large quantities of CO_2 in the ocean.

The first is that acidification of the seawater caused by CO_2 injection will influence marine life in the ocean. This effect is strongly dependent on the release methods employed. Release of CO_2 droplets from a pipe on a moving ship[3] is one of the more advanced methods, since CO_2 dissolved in sea water is then diluted due to motion of the release point. This method alone, however, is insufficient because of the second problem, the length of time CO_2 is sequestrated in ocean. CO_2 released droplets from a towed pipe on a moving ship, with the tip of the pipe at a maximum

[a]*Address for correspondence: Shuichiro Hirai, Research Center for Carbon Recycling and Utilization, Tokyo Institute of Technology, 2-12-1, O-okayama, Meguro-ku, Tokyo 152-8552, Japan. Voice: 81-3-5734-3336; fax: 81-3-5734-3554.*
    hirai@mes.titech.ac.jp

depth of about 1,500 m, entails dissolution while the CO_2 droplets are rising due to their buoyancy in sea water. Wong and Matear have noted that disposal of CO_2 into intermediate water of the North Pacific is not useful because of the short ventilation time, 20 to 50 years.[4] Simulation using a general ocean circulation model due to Nakashiki and Ohsumi suggests that artificially injected CO_2 at a depth of 950 m will reach the sea surface in 20 years.[5]

Injection of liquid CO_2 in seawater is accompanied by CO_2 clathrate-hydrate formation at the liquid CO_2–seawater interface under pressure higher than 4.45 MPa and temperature lower than 283.4 K.[6] CO_2 clathrate-hydrate is an ice-like solid in which water molecules are linked by hydrogen bonds creating a lattice structure with cavities in which carbon dioxide molecules can be observed. NMR studies on CO_2 clathrate-hydrate have been conducted by Ratcliffe and Ripmeester.[7]

In this paper, magnetic resonance imaging (MRI) of CO_2 clathrate-hydrate growth is presented and the phenomenon is applied to an advanced CO_2 ocean dissolution technology. The growth rate of the thickness of CO_2 clathrate-hydrate

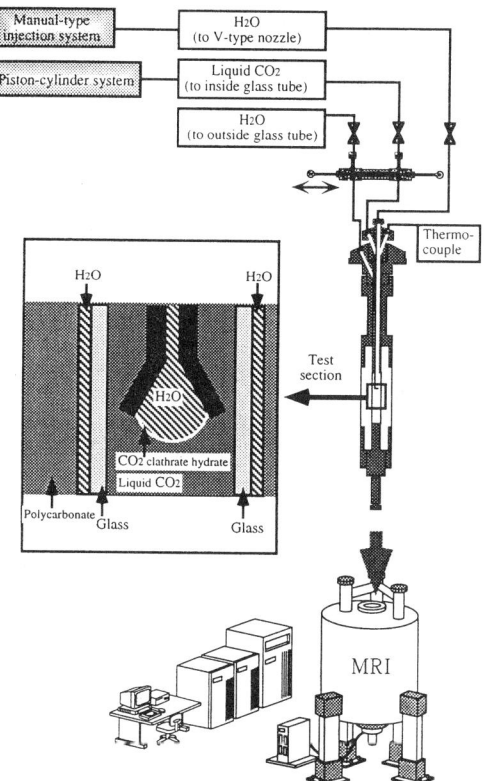

FIGURE 1. High pressure vessel and magnetic resonance imaging (MRI) apparatus. The V-type nozzle is water filled. CO_2 clathrate-hydrate forms at the interface between liquid CO_2 and water.

formed at the surface of water droplet inside liquid CO_2 was measured, fundamentally by a MRI system, using a high pressure vessel designed and constructed for this purpose. The phenomenon was applied to a new concept of CO_2 sequestration in the ocean. CO_2 clathrate-hydrate forms fine particles inside CO_2 droplets as they sink from the release point into deep ocean.[8] This process results in a sufficiently long sequestration period from the atmosphere. Our paper also provides numerical simulation results for the change of pH due to CO_2 dissolution, indicating that the decrease in pH is less than that of previous schemes.

MRI MEASUREMENT OF CO_2 CLATHRATE-HYDRATE

The thickness of the CO_2 clathrate-hydrate depends on the external conditions. For a liquid CO_2 droplet in water, thin CO_2 clathrate-hydrate film forms at the droplet surface as CO_2 dissolves into water and the droplet shrinks. For a water droplet inside liquid CO_2, CO_2 clathrate-hydrate film, first formed at the droplet surface, extends into the water droplet.[9] The increase in CO_2 clathrate-hydrate film thickness results from the fact that water solubility in liquid CO_2 is very low.

A high pressure vessel for the measurement of CO_2 clathrate-hydrate growth was designed for insertion inside a commercial rf MR imaging system probe (VARIAN; UNITY INOVA 300/150 SWB) depicted in FIGURE 1. The high pressure vessel was connected to the piston-cylinder system placed at the outside of the MRI room to elevate and maintain constant pressure. The high pressure vessel was designed to operate safely to a pressure of 40 MPa. The outer part of the test section was made of polycarbonate, 57 mm o.d. and 16 mm i.d. Since polycarbonate is one of the plastics used, a new technique to avoid contact between liquid CO_2 and the polycarbonate tube was developed. A thin glass tube was inserted inside the polycarbonate tube. The pressure of liquid CO_2 inside the glass and water filling the gap between glass tube and polycarbonate tube were elevated simultaneously, maintaining similar pressures. For details on the design of the pressure vessel see Reference 10. Water was injected inside the V-type nozzle placed in the liquid CO_2 using another manual type injection system. Images were obtained by using a spin-echo sequence with the following parameters: $T_E = 17$ ms; $T_R = 3000$ ms; field of view (FOV), 20 mm × 20

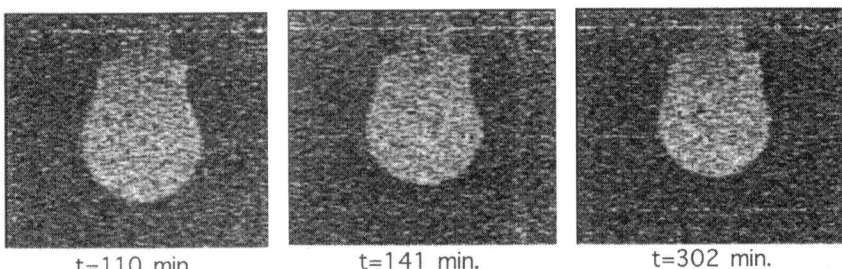

t=110 min.   t=141 min.   t=302 min.

FIGURE 2. As CO_2 molecules diffuse through the hydrate layer and form CO_2 hydrate at the water side, the area of water decreases with time.

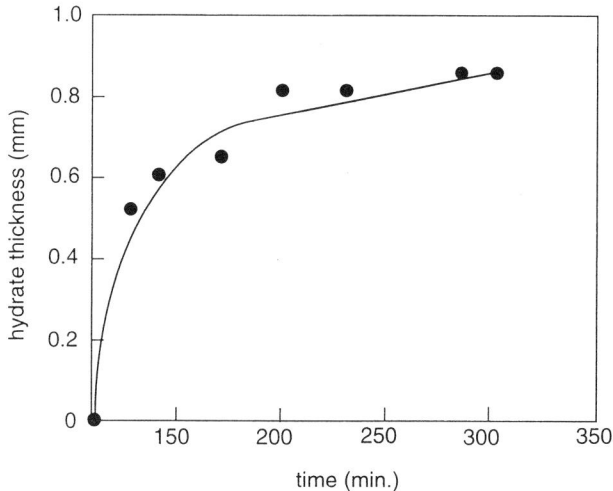

FIGURE 3. Hydrate thickness versus time.

mm; image matrix, 512 × 128; and slice thickness, 0.5 mm. The spatial resolution was 0.039 × 0.156 × 0.5 mm^3. Pressure and temperature were fixed at 10 MPa and 277.2 K, respectively. The temperature was adjusted using data from by Caulfield.[15]

FIGURE 2 shows our images 110, 141, and 302 min after injecting water into the nozzle. The figure shows an area of 0.984 × 0.859 cm^2. Three phases coexist inside the pressure chamber, water, liquid CO_2, and CO_2 clathrate-hydrate. CO_2 molecules diffuse through the CO_2 clathrate-hydrate layer and form CO_2 clathrate-hydrate at the water side. Therefore, the thickness of CO_2 clathrate-hydrate increases. Since CO_2 clathrate-hydrate is an ice-like solid, MRI can detect only the water phase. A decrease of water area corresponds to an increase of CO_2 clathrate-hydrate thickness. FIGURE 3 depicts the relation between CO_2 clathrate-hydrate thickness and time. The CO_2 clathrate-hydrate thickness data are obtained from the decrease of water image area with time, as depicted in FIGURE 2. CO_2 clathrate-hydrate thickness increases to 200 min and the rate of increase is low thereafter. This characteristic increase is thought to be related to the diffusion of CO_2 molecules through the CO_2 clathrate-hydrate layer, and the formation rate of CO_2 clathrate-hydrate.

NEW METHOD FOR CO_2 DISSOLUTION

Since the CO_2 clathrate-hydrate thickness increases when water is contained inside liquid CO_2, this phenomenon can be employed to force liquid CO_2 to sink in seawater. When small water droplets, formed by a spray nozzle, are injected inside the CO_2 droplets before their release from a towed pipe on a moving ship, water droplets become CO_2 clathrate-hydrate particles (see FIGURE 4b). CO_2 clathrate-hydrate thickness increases by 500 μm in 17 minutes (FIG. 3) when water droplets of 40 µm diameter are injected into a CO_2 droplet. Water droplets covered with the CO_2 clathrate-hydrate film needs to grow by 20 µm to form CO_2 clathrate-hydrate particles; thus, this takes 40 seconds. Since the velocity of a CO_2 droplet rising or sinking is about 10 cm/s, a vertical displacement of 4 m is required. CO_2 clathrate-hydrate film covering the CO_2 droplet surface retains CO_2 clathrate-hydrate particles inside the CO_2 droplet. The density of CO_2 clathrate-hydrate is approximately 1.1 g/cm^3.[9] When more than 40% of the total volume of CO_2 clathrate-hydrate particles is inside CO_2 droplet at a pressure 15 MPa (1,500 m ocean depth), CO_2 droplets are heavier than ambient seawater and hence they sink from the release point. Heat released due to CO_2 clathrate-hydrate formation causes the temperature to rise. We measured the temperature inside liquid CO_2 droplets while forming CO_2 clathrate-hydrate particles. The temperature increased to 281.5 K and then decreased gradually. It did not raise beyond the dissociation temperature of CO_2 clathrate-hydrate. We note also that input of energy required to spray water is low compared with the total energy consumption for CO_2 disposal in the ocean.

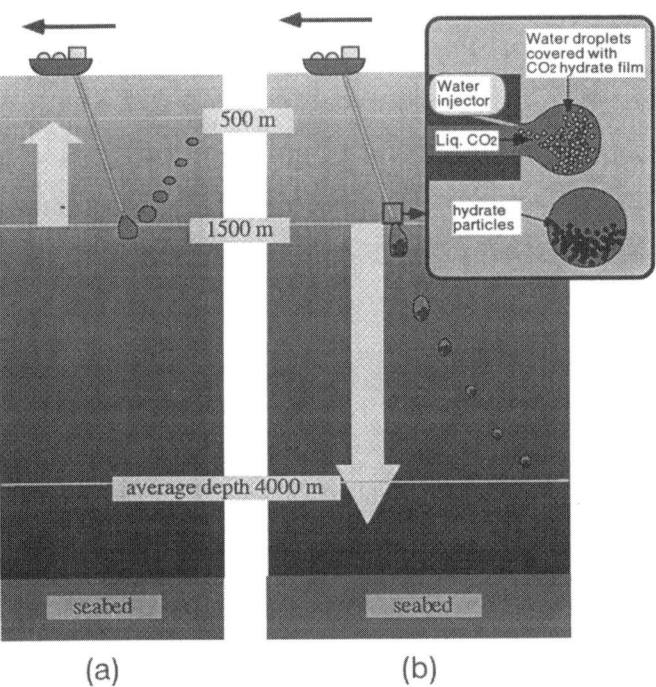

FIGURE 4. (a) A previous scheme for dissolution of liquid CO_2 during rising motion, and (b) a new concept for liquid CO_2 dissolution while sinking. Fine hydrate particles formed inside the liquid CO_2 droplets with a simple system for injecting water droplets before releasing liquid CO_2.

ACIDIFICATION OF SEAWATER BY CO_2 DISSOLUTION

In this section we discuss quantitative estimation of the acidification due to CO_2 dissolution for both cases, CO_2 droplets sinking and rising. The effect of CO_2 dissolution on acidification can be calculated from the amount of dissolved CO_2 (see Refs. 11 and 12 for details). About 16,000 ton/day (185 kg/s) of CO_2 is exhausted from a 1 GW coal-fired power plant. We simulated the case where 185 kg/s of liquid CO_2 is released from a towed pipe on a moving ship both cases of CO_2 droplets rising and sinking. The simulation described here deals with pH perturbations due to CO_2 dissolution. Assuming the ship to move at a speed of 6 knot (3.09 m/s), our calculation deals with the case where CO_2 is dissolved homogeneously in a path 3 m wide in the direction of movement of the ship. The release diameter was set to be 1.5 cm for the rising case, where CO_2 dissolution is completed at a sea depth of 500 m, before liquid CO_2 gasification occurs. It was set to 2.5 cm for the sinking case, where CO_2 dissolution is completed at a sea depth of about 4,000 m. The maximum dilution was obtained for the rising case, since then the travel distance used for dissolution is a maximum. Dissolved CO_2 follows the reactions between carbonic acid (H_2CO_3), bicarbonate ion (HCO_3^-) and carbonate ion (CO_3^{2-}) with the hydrogen ions (H^+) in seawater. The pH was calculated under these conditions.

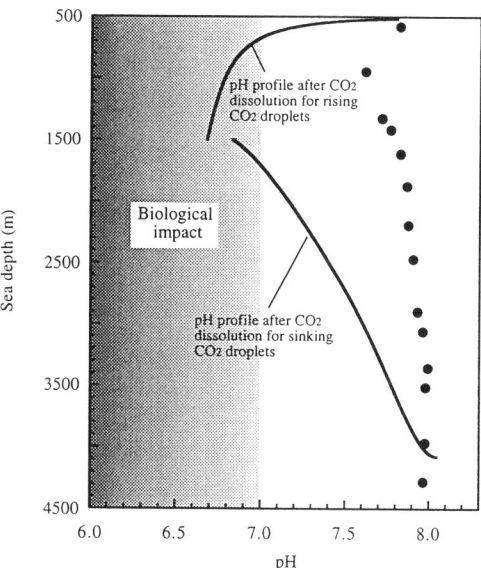

FIGURE 5. Effect of CO_2 dissolution on pH reduction. The initial diameters of the released droplets are 1.5 cm and 2.5 cm for rising and sinking motion, respectively. This gives the minimum change in pH due to CO_2 dissolution. Biological impact is a function, not only of pH, but also of the exposure time.[13] Here we adopt pH = 7.0, where mortality of marine life is zero, even at an exposure time of about 1,000 hours.[13] *Plotted points* are measured pH profiles by Millero.[14]

$$CO_2 + H_2O \leftrightarrow H_2CO_3 \quad (1)$$

$$H_2CO_3 \leftrightarrow H^+ + HCO_3^- \quad (2)$$

$$HCO_3^- \leftrightarrow H^+ + CO_3^{2-} \quad (3)$$

The result is shown in the FIGURE 5. This technology exhibits a decrease in pH to 7.0, whereas the previous scheme decreased the pH to 6.5. It is estimated that influences on marine life begin to appear when pH is lower than 7.0.[13] The present method, using sinking CO_2 droplets, offering a large travel distance of 3,000 m, is an effective method for avoiding biological impact in comparison with the case of rising CO_2 droplets.

SEQUESTRATION PERIOD

Our advanced scheme enables the CO_2 sequestration period to be longer than that for the rising droplet case. To delay the return of CO_2 to the atmosphere, discharge at greater depths is necessary. Return of the CO_2 content of deeper water is controlled by oceanographic processes such as advection and diffusion transport. The deep waters of the North Pacific Ocean are relatively poorly known. It is suggested, however, that for CO_2 dispersed below 3,000 m could achieve 3% of the disposed CO_2 in 200 years and 15% in 1,000 years.[4] The present scheme possesses a high potential for dissolved CO_2 not to be returned to the atmosphere for a longer period than that when dissolving CO_2 droplets in the rising droplet case.

CONCLUDING REMARKS

Although releasing CO_2 droplets from a moving ship reduces the biological impact by dilution of dissolved CO_2, this alone is insufficient, since CO_2 would out gas to the atmosphere in 20–100 years. This paper reports a new concept for CO_2 sequestration in the ocean that results in CO_2 dissolution at sea depths below the release point (i.e., 1,500 m) and this further reduces the biological impact. The fundamental CO_2 clathrate-hydrate growth rate data were measured by magnetic resonance imaging. This can be employed in the formation of CO_2 clathrate-hydrate fine particles inside liquid CO_2 droplets before releasing CO_2 droplets. CO_2 droplets including fine CO_2 clathrate-hydrate particles sink from release point into deep ocean, which lengthens the sequestration period. A long travel distance for CO_2 droplet dissolution also offers an advantage in that CO_2 is further diluted than in the rising case.

ACKNOWLEDGMENTS

The authors would like to thank Prof. K. Hijikata for useful discussion and assistance on this CO_2 ocean disposal project. The authors are also grateful to Mr. H. Takao (Nitto Koatsu Co.) for assistance with the experiments.

REFERENCES

1. WONG, C.S. & S. HIRAI. 1997. Ocean Storage of CO_2; A Review of Oceanic Carbonate and CO_2 Hydrate Chemistry. IEA Greenhouse Gas R&D Programme.
2. MARCHETTI, C. 1977. On geoengineering and the CO_2 problem. Climate Change **1:** 59–68.
3. NAKASHIKI, N. et al. 1995. Technical view on CO_2 transportation onto the deep ocean floor and dispersion at intermediate depths. In Direct Ocean Disposal of Carbon Dioxide. N. Handa & T. Ohsumi, Eds.: 183–194. Terrapub, Tokyo.
4. WONG, C.S. & R.J. MATEAR. 1997. Ocean disposal of CO_2 in the North Pacific Ocean: assessment of CO_2 chemistry and circulation on storage and return to the atmosphere. Waste Mgmt. **17**(5/6): 329–355.
5. NAKASHIKI, N. & T. OHSUMI. 1997. Dispersion of CO_2 injected into the ocean at the intermediate depth. Energy Convers. Mgmt. **38:** S355–S360.
6. UCHIDA, T. et al. 1995. Physical data of CO_2 hydrate. In Direct Ocean Disposal of Carbon Dioxide. N. Handa & T. Ohsumi, Eds.: 45–61. Terrapub, Tokyo.
7. RATCLIFFE, C.I. & J.A. RIPMEESTER. 1986. ^1H and ^{13}C NMR studies on carbon dioxide hydrate. J. Phys. Chem. **90:** 1259–1263.
8. AYA, I. et al. 1997. Solubility of CO_2 and density of CO_2 hydrate at 30 MPa. Energy **22**(2/3): 263–271.
9. UCHIDA, T. & J. KAWABATA. 1995. Observations of water droplets in liquid carbon dioxide. 906–910. International Conference on Technologies for Marine Environment Preservation, Tokyo, Japan.
10. HIRAI, S. et al. 2000. High-pressure magnetic resonance imaging up to 40 MPa. Magnetic Resonance Imaging. To be published.
11. HIRAI, S. et al. 1997. Dissolution rate of liquid CO_2 in pressurized water flow and effect of clathrate film. Energy Int. J. **22**(2/3): 285–293.
12. HIRAI, S. et al. 1997. Numerical simulation for dissolution of liquid CO_2 droplets covered with clathrate film in intermediate depth of ocean. Energy Convers. Mgmt. **38:** S313–S318.
13. CAULFIELD, J.A. et al. 1997. Near field impacts of reduced pH from ocean CO_2 disposal. Energy Convers. Mgmt. **38:** S343–S348.
14. MILLERO, F.J. 1996. Chemical Oceanography. 237–279. CRC Press, Boca, Florida.
15. CAULFIELD, J.A. et al. 1996. Environmental Impacts of Carbon Dioxide Ocean Disposal: Plume Predictions and Time Dependent Organism. Master Thesis, Massachusetts Institute of Technology.

Strength of CO_2 Hydrate Membrane in Sea Water at 40 MPa

K. YAMANE,[a] I. AYA,[a,b] S. NAMIE,[c] AND H. NARIAI[d]

[a]*Osaka Branch, Ship Research Institute, 3-5-10 Amanogahara, Katano, Osaka 576-0034, Japan*

[c]*Power and Energy Division, Ship Research Institute, Mitaka, Tokyo 181-0004, Japan*

[d]*Institute of Engineering Mechanics, University of Tsukuba, Tsukuba, Ibaraki 305-8573, Japan*

ABSTRACT: The strength of the CO_2 hydrate membrane that forms at the interface between liquid CO_2 and artificial sea water at 40–45 MPa was measured with Du-Nouy type surface tension meter. At low temperatures with a subcooling greater than 5 K, the membrane strength, initially abut 0.1 N/m, decreased with increasing temperature. However, it increased sharply and reached a peak of about 0.9 N/m just below the dissociation temperature and abruptly drops to zero at the dissociation temperature. This abnormal tendency of the membrane strength was previously observed by the authors in an experiment with fresh water. The temperature of the abnormality, however, shifts to lower temperatures and the peak decreases with increasing salinity. This new phenomenon could exert major influences on the various CO_2 ocean sequestration methods that have been proposed to mitigate global warming. It can be explained in terms of a model in which the dissociation process of hydrate that occurs near the dissociation temperature enhances the diffusion of water molecules in the hydrate membrane and makes the membrane thicker.

INTRODUCTION

The technology of CO_2 ocean sequestration is thought to offer a promising measure to mitigate global warming. However, its influences on the ocean environment need to be evaluated appropriately, before applying this idea. At the interface between sea water and liquid CO_2 in ocean deeper than 500 (North Pacific) to 900 meters (North Atlantic), CO_2 hydrate appears as a thin film or membrane. Therefore, it is important in its evaluation to clarify the behavior of the CO_2 hydrate membrane, especially of the mechanical strength.

In 1992, the authors attempted to measure the membrane strength at 5 MPa in fresh water and obtained an approximate value of 1.3 N/m,[1] which is 17 times the surface tension of water at the same temperature. In 1995, Uchida *et al.*[2] tried to measure the strength by the pressure difference applied to the membrane. The authors' second trial, using the same method of Uchida *et al.*, revealed that the strength at 5 MPa was scattered between 0.1 to 3.0 N/m, but that the data showed a

[b]Telecommunication. Voice: +81-720-91-6273; fax: +81-720-91-6274.
aya@srimot.go.jp

tendency to increase as the CO_2 concentration of the bulk water increased.[3] In a third trial, the membrane strength in fresh water was accurately measured by using a newly built facility that can simulate the pressure and temperature of the ocean down to 4,500 meters depth. From the data,[4] it was found that the membrane strength gradually decreased with increasing of the temperature below 8°C, but sharply increased and reached a very high peak at about 12°C. Its peak is about 10 times that observed at low temperatures. Finally, the strength abruptly drops to almost zero at the dissociation temperature of 12.7°C.

Recently, the authors conducted a fourth series of measurements with the purpose of confirming the strength abnormality in sea water, and to make clear the influences of salinity on the temperature of the abnormality and on the maximum strength observed.

METHOD OF MEASUREMENT

FIGURE 1 shows the loop type facility used in the third and fourth experiments. A propeller fixed at the left lower corner circulates water clockwise or counter clockwise to ensure a good average temperature and CO_2 concentration in the entire loop. Precise temperature control (±0.1°C) is achieved by means of a heat exchanger that covers almost all of the loop. A Du-Nouy type surface tension meter is installed in the observation chamber (center of lower side). The loop was filled with saline

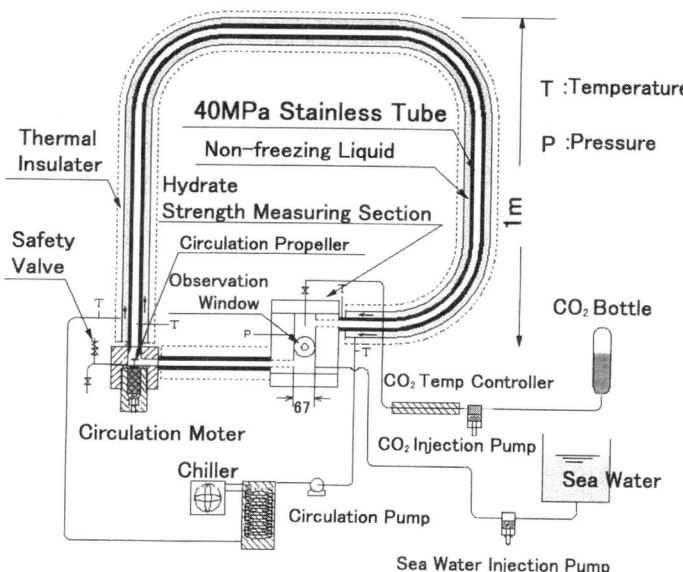

FIGURE 1. Circulation type simulation facility for use in ocean 4,500 meters deep.

water in which the salinity was adjusted by changing the ratio of ion-exchanged fresh water and artificial sea salts, *Jamarin*. After pressurizing the loop to 45 MPa, a beaker with diameter and height 30 mm and installed in the measuring section was filled with liquid CO_2 that was heavier than saline water under these conditions.

FIGURE 2 shows the principle for measuring the membrane strength. The twisting force exerting on the wire is transferred through a lever to a platinum ring that pushes down on the membrane. The twisting of the wire was started 10 minutes after formation of hydrate membrane. The twisting angle, θ, and membrane deformation angle, $\Delta\theta$, just before the rupture of the membrane were measured from images recorded through two windows by a small CCD camera. The membrane strength, σ [N/m], can be calculated from the following equation:

$$\sigma = K(\theta - \Delta\theta), \tag{1}$$

where K is a kind of spring constant, determined by the material, the length and the diameter of the wire, the length of lever, and the diameter of the ring. In this experiment $K = 0.0167$ N/m/deg. The twisting rate of the wire was fixed at a constant rotation speed (1.48 deg/sec) by a servomoter. This twisting rate corresponds to a free ring downward velocity of 1.65 mm/sec.

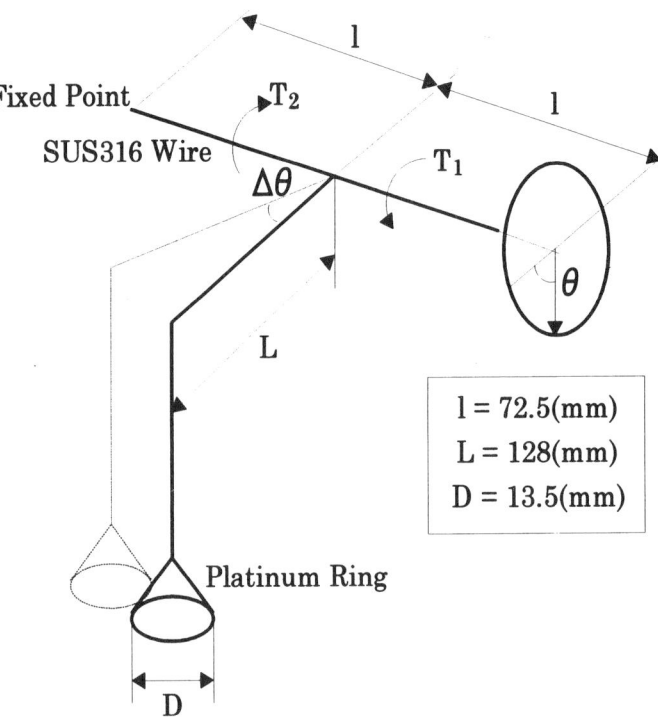

FIGURE 2. Principle of hydrate membrane strength measurement.

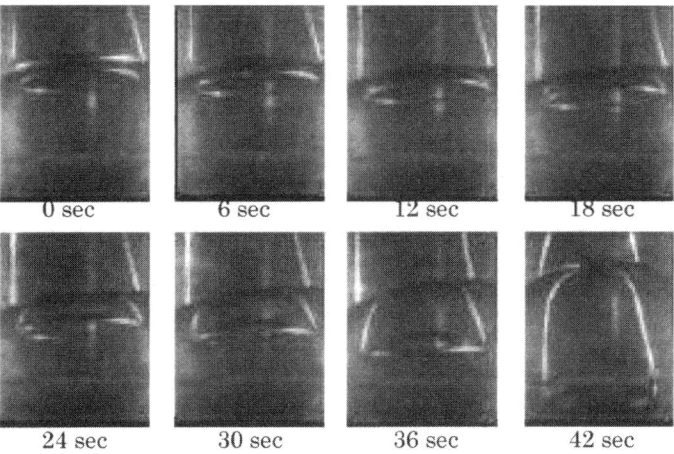

FIGURE 3. Deformation and rupture of CO_2 hydrate membrane in saline water.

EXPERIMENTAL RESULTS

FIGURE 3 shows typical examples of the membrane deformation process as the platinum ring is pushed down the membrane formed in average sea water at 10°C. The horizontal line seen in every frame is the upper edge of the beaker. The round curve seen around the ring is the CO_2 hydrate membrane appearing at the interface

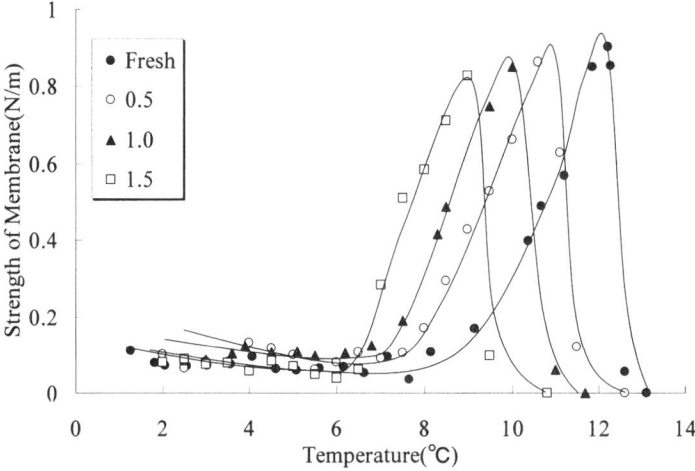

FIGURE 4. Strength of CO_2 hydrate membrane in various saline waters. (Average for seawater, 1.0.)

between saline water and liquid CO_2. The location of the ring changes noncontinuously between frames 7 and 8, which means that the membrane breaks between 36 and 38 seconds. If the membrane has no resistance against its deformation, the location of the ring at 36 second should be far lower than shown in the frame. This means the membrane has a considerable resistance against its deformation. From 30 photos recorded each second, a more precise rupture time that corresponds to θ was determined. The deformation angle Δθ was evaluated from the location of ring just before the membrane ruptured. The membrane strength in FIGURE 3 was calculated at 0.87 N/m using Equation (1).

FIGURE 4 shows the membrane strength evaluated by above method for 0.5, 1.0, and 1.5 times the salinity of average sea water. For these data, the temperature dependence of membrane strength is very similar to the case of fresh water but the peak shifts to lower temperatures and the maximum value to decreases with increasing salinity.

DISCUSSION

There is no theory to explain the new finding, that the strength of the CO_2 hydrate membrane increases abnormally just below the dissociation temperature. The authors have proposed a model[4] that can qualitatively explain the strength abnormality in fresh water. In that model, the dissociation of the hydrogen bond between the water molecules in CO_2 hydrate plays a large role in making the membrane thicker. That is, the dissociation probability of the hydrogen bond is not negligible compared with the association probability just below the dissociation temperature. The mutual diffusion between CO_2 and water molecules would be enhanced at this point, and the thickness of the membrane could increase just below the dissociation temperature. The dissociation probability for lower temperatures, however, is negligible, and the usual diffusion phenomena in a membrane prevails. In the latter case, the thickness of membrane is determined by the balance between the diffusion process in the membrane and the dissolution rate of the membrane into water. Because the dissolution rate of the hydrate membrane is controlled only by the diffusion process in water[5], and the solubility of CO_2 coexisting with hydrate increases with temperature[6], the membrane thickness decreases with temperature. This is the reason why the membrane strength of CO_2 hydrate in fresh water decreases with temperature below 8°C, where the dissociation probability is negligible. The fact that the temperature dependence of membrane strength in saline waters is the same as that in fresh water suggests that the above model can also be applied to the case of saline water.

The following equation explains this idea:

$$\frac{CP_D^n(P_A - P_D) + \lambda_D(P_A - P_D)}{\delta} = u, \quad (2)$$

where C is a multiplier; u the dissociation rate of hydrate; n the dissociation effect exponent (≥1); P_A and P_D the association and dissociation probability of a hydrogen bond, respectively; λ_D the diffusivity of water molecule in hydrate membrane; and δ is the thickness of membrane. This equation can qualitatively explain the abnormality of hydrate strength observed just below the dissociation temperature.

As to the reason why the peak and dissociation temperatures shift to lower values with increasing salinity, the authors think that the salts dissolved in water make the chemical activities of the solution weak. That is, a drop in the chemical potential due to the existence of salts compensated for by a shift in the dissociation temperature to lower values. In general, an inhibitor (in this case various salts dissolved in artificial sea water) makes the dissociation temperature lower.[7] The temperature shift for each salinity used in this experiment can be estimated from following equation:

$$\Delta T = \frac{2335W}{100M - MW}, \qquad (3)$$

where ΔT is the downward shift in temperature; W is the weight percent of inhibitor; and M is the molecular weight of inhibitor. FIGURE 5 compares the measured temperature shifts of dissociation temperature with the estimated values, where the effects of Na, Cl, Mg, S, Ca, and K in saline water are evaluated from (3). It can be seen that (3) provides a good estimate for the shift of dissociation temperature. Therefore, the shift of peak temperature seen in FIGURE 4 is also explained as an effect of salinity. In other words, the good agreement between measured and estimated salinity effect guarantees the high accuracy of the data obtained this time. The salinity may also have an effect on the membrane strength, however, it remains to explain theoretically why the maximum membrane strength is lowered by about 0.06 N/m for the salinity of average sea water.

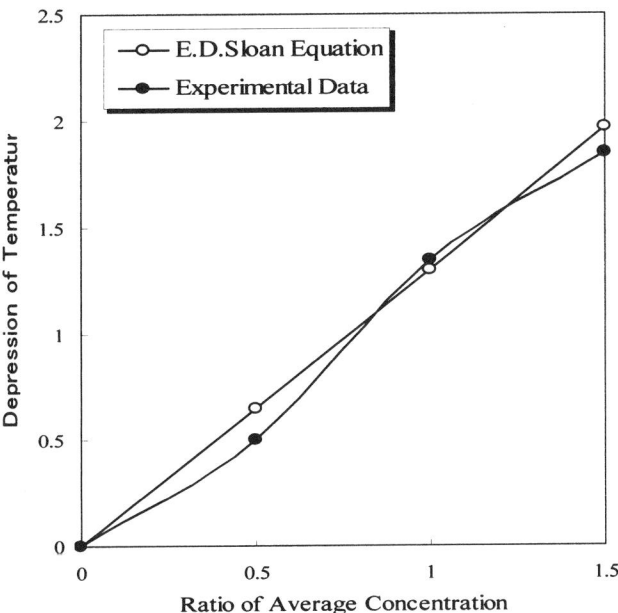

FIGURE 5. Relation between the temperature depression at peak strength and salinity level. (Average for seawater, 1.0.)

CONCLUSIONS

The measurement of membrane strength of CO_2 hydrate in artificial saline water reveals that a similar high strength peak appearing just below the dissociation temperature in fresh water was also observed in saline water. The dissociation and strength peak temperatures decrease by 1.7 K for the salinity of average sea water. These temperature shifts were well explained by the published equation that estimates the effect of inhibitor. A model to consider the effect of dissociation temperature can qualitatively explain the appearance of abnormally high peak strength.

REFERENCES

1. AYA, I. *et al.* 1992. Stability of clathrate-hydrate of carbon dioxide in highly pressurized water. ASME HTD-Vol.215. Fundamentals of Phase Changes: Freezing, Melting and Sublimation. 17-22. Anaheim, CA.
2. UCHIDA, T. *et al.* 1997. Measurements of mechanical properties of the liquid CO_2–water–CO_2 hydrate system. Energy Int. **22**(2/3): 357–361.
3. SUZUKI, T. *et al.* 1997. Strength measurement of CO_2 hydrate membrane. 34th National Heat Transfer Symposium of Japan. F162. 329–330. Sendai, JP.
4. YAMANE, K. *et al.* 1998. Strength abnormality of CO_2 hydrate membrane just below dissociation temperature. 4th Int. Conf. on Greenhouse Gas Control Technologies. Interlaken, Switzerland.
5. AYA, I. *et al.* 1997. Solubility of CO_2 and density of CO_2 hydrate at 30 MPa. Energy Int. **22**(2/3): 263–271.
6. YAMANE, K. *et al.* 1997. Temperature dependence of CO_2 solubility in hydrate region. 2nd ISOPE Ocean Mining Symposium. Seoul, Korea. 154–158.
7. SLOAN, E.D., JR.1990. Clathrate Hydrate of Natural Gases. Marcel Dekker. New York.

The Effect of CO_2-Air Mixture Compositions on the Formation and Dissociation of CO_2 Hydrate

HIRONORI HANEDA,[a] TAKESHI KOMAI, AND YOSHITAKA YAMAMOTO

National Institute for Resources and Environment, AIST, MITI, Sapporo, Japan

ABSTRACT: The disposal of carbon dioxide to the marine and sea bed sediments as CO_2 gas hydrate is an innovative technique for solving the global environment issue. Experiments on the formation and dissociation of gas hydrate have been carried out using a pressure vessel to investigate the effect of carbon dioxide concentration in the gas phase. From the experiment results, the following are clarified: (1) There is a strong relationship between the partial pressure of carbon dioxide, concentration, and the temperature of formation and dissociation of gas hydrate. Therefore, the use of this relation enables the estimation of equilibrium conditions of the gas mixture. (2) The initial formation rate varies from 0.1 to 0. 5 ml/(min·g). In terms of average values, the initial formation rate increases as the carbon dioxide concentration of the initial gas mixture increases. (3) From the analysis of component gas of gas hydrate and space gas, it can be assumed that nitrogen and oxygen are also incorporated into the hydrate structure cage as guest molecules. Moreover, it can be seen that the carbon dioxide concentration in the initial space gas is higher than that in the space gas at the time of gas hydrate formation. Therefore, this hydrate technology applies to the concentration of carbon dioxide. In future, we will attempt to carry out tests on the formation and dissociation of CO_2 hydrate under a low concentration of CO_2. Furthermore, we will analyze the structure of gas hydrate using Raman spectroscopy to clarify that nitrogen and oxygen are incorporated into the gas hydrate cage as guest molecules.

INTRODUCTION

According to the Kyoto protocol of the Third Conference of the Parties (COP3) of the United Nations Framework Convention on Climate Change held in December 1997, Japan's goal is to reduce the total emission of greenhouse gases by 6% in reference to that in the year 1990 by the target period, 2008–2012.[1] As one of the measures that is being considered to accomplish this goal is the development of a sequestration technique in which carbon dioxide (CO_2), the major substance that causes the greenhouse effect, can be converted to CO_2 hydrate, and fixed to the ocean bed.

---

[a]Address for correspondence: Hironori Haneda, Hokkaido Coal Mine Research Center, National Institute for Resources and Environment, Kita 1-25, Heiwadori 3-chome, Shiroishi-ku, Sapporo, 003-0029, Japan. Voice: +81-11-861-2191; fax: +81-11-864-3469.
haneda@nire.go.jp

In the area of CO_2 hydrate research, Ohgaki et al.[2] proposed a method to produce methane hydrate using the heat released during CO_2 hydrate formation as the heat source, considering the relationship between methane hydrate and CO_2. Ohgaki et al.[3] also reported the phase equilibrium of the hydrate of a CO_2-methane gas mixture. John et al.[4] applied a thermodynamic model analysis to gas mixtures such as CO_2/CH_4 and CO_2/propane. Parrish et al.[5] performed phase equilibrium calculations using a physical model and demonstrated a correlation between their data and conventional experimental data. Gas mixture experiments were carried out using systems such as methane–nitrogen–water and methane–propane–water. Ripmeester et al.[6] pointed out that CO_2 is occupied in small cages of structure II in systems such as CO_2 propane and CO_2 neohexane.

We aim to stabilize the upper seam of a hydrate reservoir by injecting CO_2 into the porous medium, generated by using methane hydrate, to induce the formation of CO_2 hydrate. The CO_2 gas injected into the porous medium consists of a mixture of CO_2 released from industries, and air. We evaluate the characteristics of the hydrate formed.

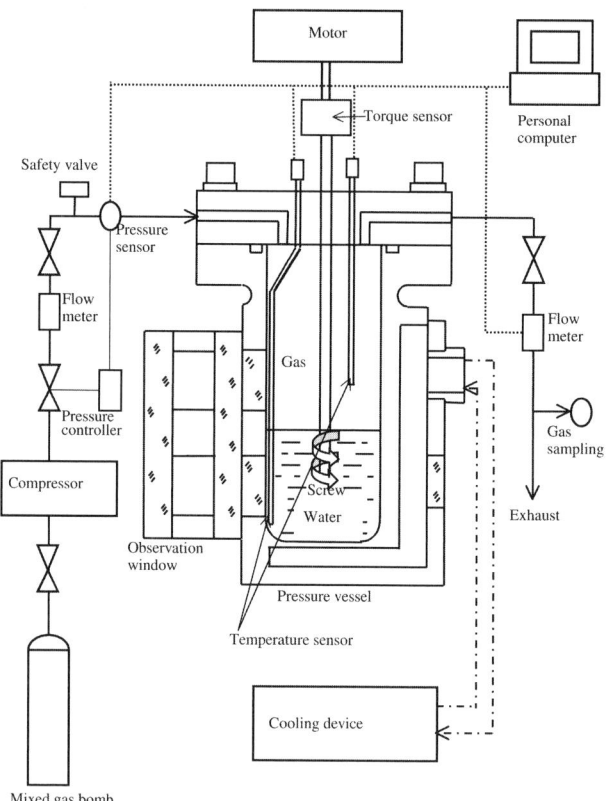

FIGURE 1. Experimental apparatus.

The current study examined the changes in the equilibrium conditions of CO_2–air hydrate formation/dissociation and the initial formation rate, observed by varying the CO_2 concentration in a system of CO_2–air–water. We also studied the changes in the gas compositions during the formation and dissociation of CO_2–air hydrate, the amounts and the compositions of dissociated gas hydrate, and the possibility of gas concentration using the phenomenon of gas hydrate formation.

EXPERIMENTAL APPARATUS AND EXPERIMENTAL METHOD

Experimental Apparatus

FIGURE 1 shows a schematic representation of the gas hydrate formation/dissociation apparatus used in the experiment. The apparatus consists of a pressure cell, a gas control system, and a data measurement system. The main unit, a pressure cell, is made of stainless steel. It is capable of sustaining a maximum operating pressure of 10 MPa, and has a volume of 390 ml. The main unit also includes a jacket for temperature control; a coolant is circulated from the external cooling system to control the temperature inside the cell. An observation window is installed on the side of the cell to allow observation into the cell. A rotary axis, temperature sensor, and gas inlet/outlet are installed on the top of the pressure cell. The rotary axis is rotated using an external motor for stirring the contents of the cell.

The CO_2-air mixture is injected into the pressure cell using a compressor. A pressure sensor, pressure controller, safety valve, flow meter, and high-pressure valve are installed along the injection line. A gas sampling tube, flow meter, and high-pressure valve are incorporated along the gas exhaust line.

The data on pressure and temperature in the pressure cell and the flow volume in each gas line were measured using sensors, and the data were recorded on a personal computer.

Formation and Dissociation Experiment

First, we studied the changes in formation/dissociation conditions for CO_2 concentrations in the CO_2–air gas mixture of 60, 80, and 100%. In the experiment, the phase transition from water and CO_2 (gas) to CO_2 hydrate + CO_2 (gas) was defined as formation, and the opposite process was defined as dissociation.

Prior to the study, CO_2-air gas mixtures with 60, 80, and 100% CO_2 were prepared, the gas mixture was transferred to the cell, and the cell was sealed. Pure water (50 ml) was introduced into a pressure cell, and the air inside was replaced by the gas mixture. Complete replacement of the air by the gas mixture was confirmed by analyzing a sample via the gas sampling tube, shown in FIGURE 1, by gas chromatography. Thereafter, the gas mixture was introduced into the pressure cell using a compressor to reach the desired pressure.

The liquid phase in the pressure cell was stirred at 400 rpm using a mixer and the temperature was lowered to approximately 0°C at a constant rate of 0.1°C/min. The near zero temperature was either maintained or the temperature was gradually reduced to facilitate the formation of the gas hydrate. Thereafter, the low temperature was maintained for the continued formation of gas hydrate. Finally, the cell temperature was increased and the gas hydrate was dissociated.

During the course of the above process, the changes of states inside the pressure cell were observed directly or recorded on video. At the same time, the changes in pressure were monitored using pressure sensors installed along the gas lines, and the changes in the temperature of the gas phase, liquid phase, and gas hydrate were measured using the temperature sensors installed inside the pressure cell.

The pure water used in the experiment was replaced with new pure water for each experiment. The pressure of the gas mixture was changed in each experiment.

Gas Analysis Study

In the process of formation of the gas hydrate from a gas mixture, the ratio of the gas molecules in the gas hydrate and the ratio of the spatial gases not forming gas hydrate are considered to be different from the corresponding ratios in the initial gas mixture. In order to study the composition ratios at different stages of hydrate formation, gas analysis was performed using the following procedure. First, after replacing the air in the pressure cell with the gas mixture, the composition of the initial gas mixture was determined by analyzing a gas sample obtained from the gas sampling tube shown in FIGURE 1. Next, after the formation of the gas hydrate using the procedure described in FORMATION AND DISSOCIATION EXPERIMENT, to prevent the dissociation of the gas hydrate, the temperature in the pressure cell was reduced to approximately −7°C, the spatial gas not forming the hydrate was sampled via the sampling tube, and the gas composition analyzed.

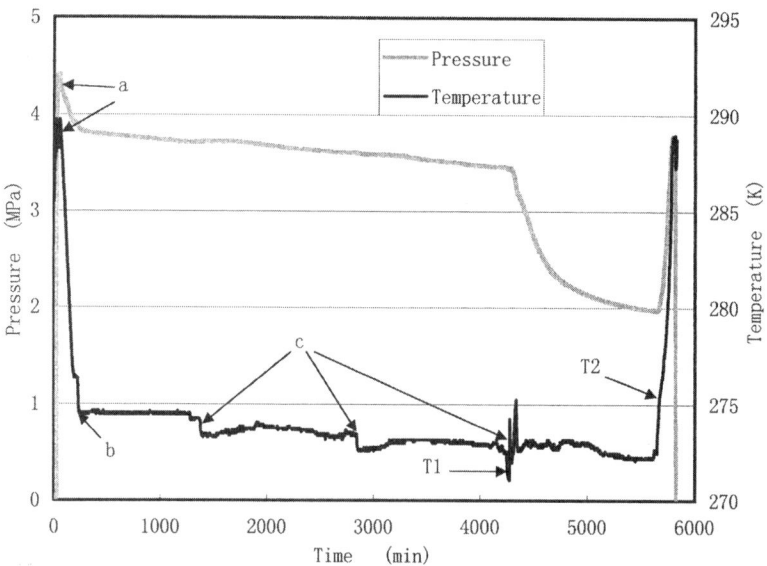

FIGURE 2. Pressure and temperature curves for CO_2 hydrate formation and dissociation.

The spatial gas was recovered until the inside of the pressure cell returned to atmospheric pressure. The cell was then sealed and the temperature was increased to facilitate the dissociation of gas hydrate. After complete dissociation, all spatial gas in the pressure cell was collected in a Tedlar bag, and its composition and volume measured.

By subtracting the volume of the spatial gas present prior to dissociation from that present after dissociation and adding the result to the volume of the gas dissolved in water, the gas composition and volume of the hydrate gas can be calculated.

The instrument used for the analysis was a microprocessor-type gas chromatograph. The column used for the detection of CO_2 was a Porapak, and molecular sieves were used for nitrogen and oxygen measurements. The temperature of each column was set to 40°C and 60°C, respectively. A dry gas meter was used in the volume measurements.

EXPERIMENTAL RESULTS AND DISCUSSION

Formation and Dissociation Points

FIGURE 2 shows an example of the changes in pressure in the cell and temperature of the liquid phase obtained in the formation and dissociation experiments. As shown by point "a" in FIGURE 2, the initial pressure in the cell was set to 4.5 MPa and the temperature to 15°C, and these conditions were maintained for 30 min. Thereafter, the temperature was reduced at a constant rate (0.1°C/min) to point "b" and maintained. However, no gas hydrate was formed. Therefore, the temperature was further

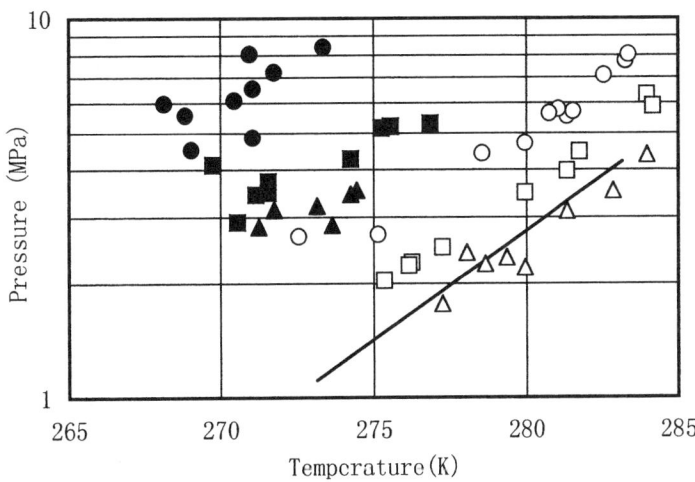

FIGURE 3. Relation between pressure and temperature in the formation and dissociation of CO_2 hydrate. ▲, pure CO_2, formation; ■, 80 vol% CO_2, formation; ●, 60 vol% CO_2, formation; △, pure CO_2, dissociation; □, 80 vol% CO_2, dissociation; ○, 60 vol% CO_2, dissociation; ——, Sloan's result.

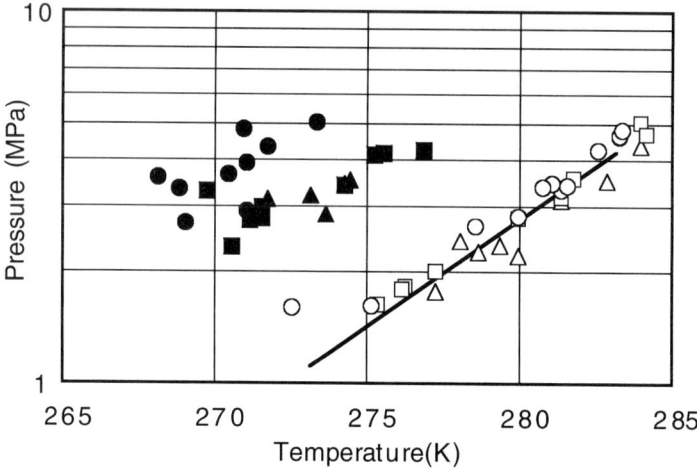

FIGURE 4. Formation and dissociation conditions corrected for partial pressure. ▲, pure CO_2, formation; △, pure CO_2, dissociation; ■, 80 vol% CO_2, formation; □, 80 vol% CO_2, dissociation; ●, 60 vol% CO_2, formation; ○, 60 vol% CO_2, dissociation; ———, Sloan's result.

lowered to point "c". After approximately 4,300 min, when the temperature was lowered for the third time, heat generation was observed (see FIG. 2, T1). Simultaneously, a drop in the pressure of the gas phase was observed. At this point, the liquid phase abruptly became opaque, indicating the formation of the gas hydrate.

After formation of gas hydrate, the cell temperature was increased at a constant rate (0.1°C/min) for 5,620 min. After approximately 5670 min, a decrease in the rate of increase of temperature (FIG. 2, T2) and an increase in the gas-phase pressure were observed. At this point, generation of water droplets were observed from the observation window and the dissociation of the gas hydrate was confirmed.

It has been found that a significant temperature change accompanies the phase transition of gas hydrate. Thus, the points at which the temperature changes were observed were designated as the formation point (FIG. 2, T1) and the dissociation point (FIG. 2, T2).

CO_2 Concentration in the Gas Mixture and the Formation/Dissociation Conditions

FIGURE 3 shows the formation and dissociation points observed for different CO_2 concentrations. In FIGURE 3, as the CO_2 concentration decreases from 100% to 80% and then to 60%, the formation and the dissociation conditions for gas hydrate shifted to lower temperatures and higher pressures.

In FIGURE 3, the pressure on the Y axis (ordinate) represents pressures in the pressure cell (total pressure). FIGURE 4 shows the formation and dissociation conditions (pressures) corrected by partial pressure of CO_2 based on their initial concentration in gas mixtures, as a function of the temperatures.

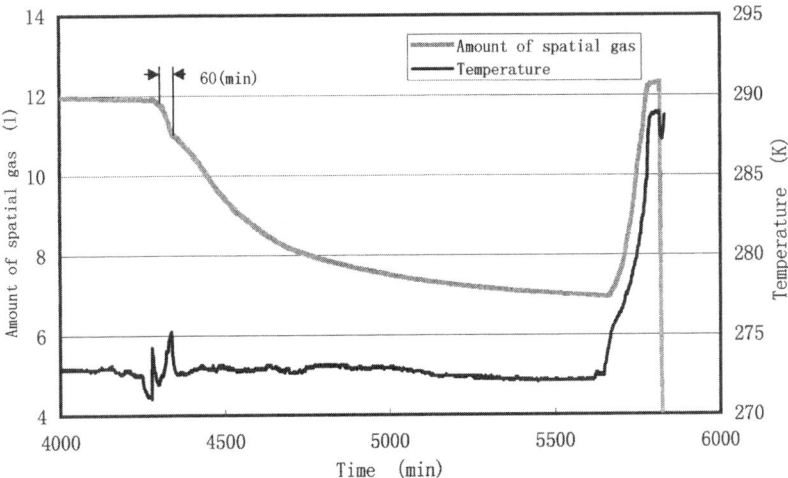

FIGURE 5. The amount of spatial gas and temperature obtained during CO_2 hydrate formation and dissociation.

When the data are represented by partial pressures based on the initial concentrations, all dissociation conditions essentially agreed with the values obtained by Sloan.[7] However, the data for the formation conditions exhibited large variability compared with those for the dissociation conditions. This suggests a supercooling

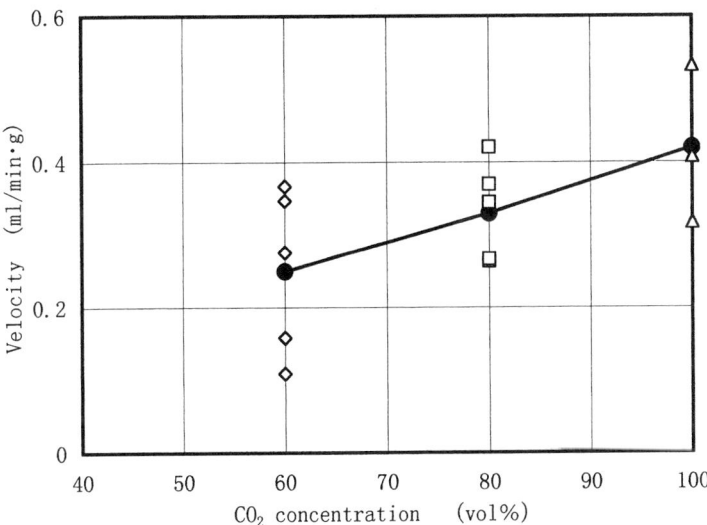

FIGURE 6. Relation between CO_2 concentration and initial rate of formation.

phenomenon during formation resulting in an unstable condition; however, a superheating phenomenon was not observed during dissociation, and, the system was considered to be stable.[6]

Our results confirm that the formation of gas hydrate decreases, whereas the dissociation of gas hydrate occurs more easily with decreasing CO_2 concentration. Furthermore, the equilibrium conditions for CO_2 hydrate depend on the concentration of CO_2 in the gas mixture.

Initial Formation Rate

The amount of spatial gas at the time of the formation and dissociation of CO_2 hydrate in the pressure cell was calculated from the cell pressure and temperature. FIGURE 5 shows the change in volume and temperature. At approximately 4,300 min, simultaneously with the release of heat and formation of CO_2 hydrate, the spatial gas volume in the pressure cell decreased sharply. The amount of spatial gas in the pressure cell increased simultaneously with the endothermic reaction on CO_2 hydrate dissociation, at approximately 5,670 min. It also increased with increasing temperature.

FIGURE 6 shows the formation rate calculated for the initial period of gas hydrate formation, in which region the gas volume decreased linearly (60 min). FIGURE 6 shows that the initial CO_2 hydrate formation rate was 0.1–0.5 ml/(min·g). The gas volume incorporated into one gram of water per minute was converted into the standard state value; there was, however, significant variability in the data. The average initial formation rate decreased with decreasing CO_2 concentration.

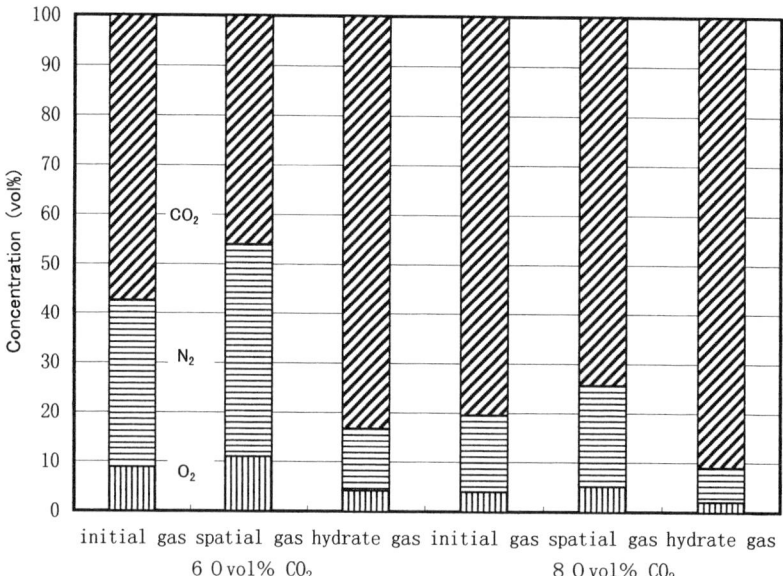

FIGURE 7. Gas component ratio at each stage.

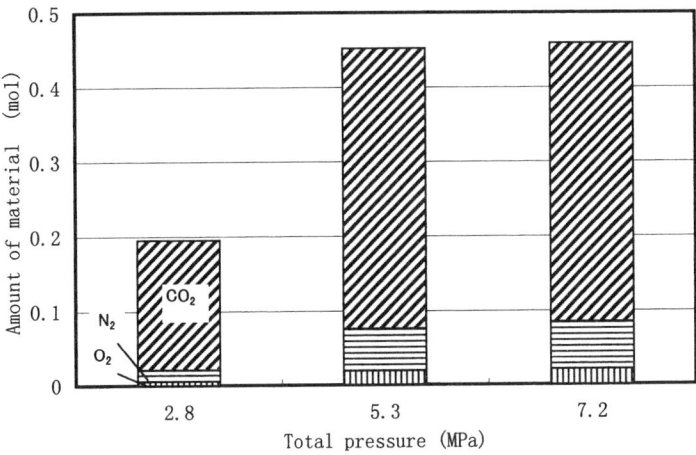

FIGURE 8. Amount of guest-molecule material at each total pressure.

Gas Composition Analysis

FIGURE 7 shows the composition of the initial gases, the composition of spatial gases that were not incorporated into the gas hydrate and the gas composition of hydrates formed from gas mixtures with 60, 80, and 100% CO_2.

The CO_2 concentrations of spatial gases decreased in comparison with the initial CO_2 concentration for both 60 and 80% CO_2 in the gas mixture. On the other hand, the CO_2 concentration in the hydrate was higher than that during the initial stages of the experiment. This result indicates that CO_2 preferentially formed the gas hydrate. This phenomenon can be applied to CO_2 concentration technology using gas hydrate.

FIGURE 8 shows the relationship between the pressure in the cell after the formation of the gas hydrate and the content in the gas hydrate for a CO_2 concentration of 60% in the gas mixture. According to Kuustraa *et al.*,[8] the hardness of nitrogen and oxygen hydrate formation is 10 times greater than that for CO_2.

When we began this study, we expected that CO_2 alone would form gas hydrate, however, in reality, when hydrate-forming experiments were performed at 60% CO_2, at 5.3 MPa, the gas hydrate also contained 0.0560 mol nitrogen and 0.0190 mol oxygen. The amounts of nitrogen and oxygen dissolved in water (50 ml) before the

TABLE 1. Gas content of 60% CO_2/air hydrate (ml/ml)

Pressure after hydrate completely formed (MPa)	CO_2	N_2	O_2	Total
2.8	78.24	6.82	2.40	87.46
5.3	168.74	25.48	8.48	202.70
7.2	167.86	28.68	9.22	205.76

formation of the gas hydrate (0°C, 6.82 MPa) were 0.00119 mol (nitrogen) and 0.00653 mol (oxygen). These values are very small compared with the volume of nitrogen and oxygen gases actually contained in the gas hydrate. According to Maeno,[9] air does not fit in ice crystal lattices. Therefore, even if ice were formed under these conditions, it would release the dissolved gas or incorporate the dissolved gas as bubbles. Thus, it is very unlikely that ice incorporates any significant amount of gas dissolved in water. The results suggest that it is very likely that the nitrogen and oxygen contained in the gas hydrate were in the form of guest molecules of the gas hydrate.

According to Sloan,[7] when the gas hydrate adopts structure I, in which two small cages and six large cages form one unit, and if all of these were occupied, then the hydration number of the guest molecules would be 5.75 (216 ml/ml). If only the large cages were occupied, the hydration number would be 7.73 (161 ml/ml). In the case of structure II, where 16 small cages and eight large cages form a unit, and if all of these were occupied, then the hydration number would be 5.67 (219 ml/ml). CO_2 is taken up mostly by the large cages in structure I, but some is also taken up by small cages.

TABLE 1 lists the gas content for each structure shown in FIGURE 8, expressed by conversion to values per 1 ml of water. The volume of CO_2 contained in the hydrate is approximately 168 ml/ml at the pressures of 5.3 and 7.2 MPa; this is similar to the situation in which guest molecules occupied only the large cages of structure I. The ratios of the total volumes of the nitrogen and oxygen to CO_2 were 0.12 at 2.8 MPa, 0.2 at 5.3 MPa, and 0.22 at 7.2 MPa. If the large cages of structure I were occupied with CO_2, and all the small cages were filled with nitrogen and oxygen molecules, the ratio of the gas volumes would be 0.25, a value close to that accommodated by the small cages.

From these discussions, it is assumed that gas hydrate of air adopt structure II; however, the mixed gas hydrate for 60% CO_2 and 40% air adopted structure I.

CONCLUSIONS

The equilibrium condition for CO_2 hydrate formation/ dissociation bears a close relationship to the CO_2 partial pressure in the system, and the equilibrium conditions can be predicted on the basis of the composition of the gas mixture (up to 60% of CO_2).

The initial rate of formation of CO_2 hydrate decreases with decreasing CO_2 concentration in the gas mixture. Under the conditions used in this study, it was approximately 0.1–0.5 ml/min·g.

The CO_2 concentrations in the gas hydrate were high compared with the initial CO_2 concentration in the gas mixture. Thus, it is possible to apply the gas hydrate formation phenomenon to CO_2 concentration technology.

Judging from the results of the gas analyses in the current study, nitrogen and oxygen molecules were also taken up simultaneously with CO_2 as guest molecules during the formation of the gas hydrate. If the concentration of CO_2 is low and the percentage of air increased, the gas hydrate adopts structure II.

The results of this study are considered to be useful for application to the formation control of gas hydrates in terms of fixing a gas mixture of CO_2–air to the ocean bed.

In future, we propose to perform further gas hydrate formation and dissociation experiments using lower concentrations of CO_2 than those used in this study. Furthermore, we will perform structural analyses of the gas hydrates formed using analytical equipment to clarify their structure in detail.

REFERENCES

1. MITI MATERIAL INVESTIGATION ASSOCIATION. 1998. Environmental Surveying 1999. The Ministry of International Trade and Industry Environment Location Bureau.
2. OHGAKI, K., K. TAKANO, H. SANGAWA, T. MATSUBARA & S. NAKANO. 1996. J. Chem. Eng. Jpn. **29**: 478–483.
3. OHGAKI, K., K. TAKANO & M. MORITOKI. 1994. Kagakukogaku Ronbunshu **20**: 121–123.
4. JOHN, V.T., K.D. PAPADOPOULOS & G.D. HOLDER. 1985. AIChE J. **31**: 252–259.
5. PARRISH, W.R. & J.M. PRAUSNITZ. 1972. Ind. Chem. Process Des. Develop. **11**: 26–35.
6. RIPMEESTER, J.A. & C.I. RATCLIFFE. 1998. Energy Fuels **12**: 197–200.
7. SLOAN, E.D., JR. 1998. Clathrate Hydrates of National Gases. 2nd edit. 586–601. Marcel-Dekker, New York.
8. KUUSTRAA, A. 1983. Handbook of Gas Hydrate Properties and Occurrence. Lewin and Associates.
9. MAENO, N. 1984. Science of Ice. Hokkaido University Books Publication Association.

Dynamics of Reformation and Replacement of CO_2 and CH_4 Gas Hydrates

TAKESHI KOMAI,[a,b] YOSHITAKA YAMAMOTO,[b] AND KOTARO OHGA[c]

[b]National Institute for Resources and Environment, Safety Engineering Department, 16-3 Onogawa, Tsukuba, Ibaraki, 305-8569 Japan

[c]Faculty of Engineering, Hokkaido University, Kita-ku, Sapporo, 060 Japan

ABSTRACT: The dynamics of reformation and replacement of gas hydrates were studied under nonequilibrium conditions. It was found that the reformation of gas hydrates is largely affected by the state of the gas–water system and the restarting temperature. This suggests that the effects are caused by changes in the structure of the aqueous solution at a molecular level. Pure samples of CH_4 gas hydrate were synthesized from ice crystals and the dissociated solution using the reformation method. Replacement of CO_2 gas hydrate is achieved within a short duration in the solid-phase sample of CH_4 gas hydrate, if the pressure and temperature is precisely controlled in a pressure vessel.

INTRODUCTION

Recently, the properties of gas hydrates were clarified for equilibrium conditions.[1] However, the dynamic process of their formation and dissociation under nonequilibrium conditions are not yet clearly understood, particularly from the viewpoint of nucleation and crystal growth of gas hydrates.[2] We have performed experiments on the reformation of gas hydrates and the replacement of guests, using specially designed pressure cells. In this paper, we present experimental results on the hysteresis process of gas hydrate formation and dissociation, and discuss the reformation mechanism of nucleation of hydrate crystals in the liquid phase. Experimental results on the replacement of guests in solid-phase gas hydrates are reported to aid in understanding the dynamics of gas hydrate reformation.

EXPERIMENTAL

Reformation Experiment

The formation and dissociation processes of gas hydrates and the reformation behavior were observed and measured, using a high-pressure cell with an observation system. FIGURE 1 shows the experimental setup and the measurement system. The apparatus consists of a pressure cell, a high-pressure pump, and a control system. The cell with an internal volume of 90 ml is also equipped with magnetic mixing

---

[a]Telecommunication. Voice: +81-298-61-8795; fax: +81-298-61-8796.
koma@nire.go.jp

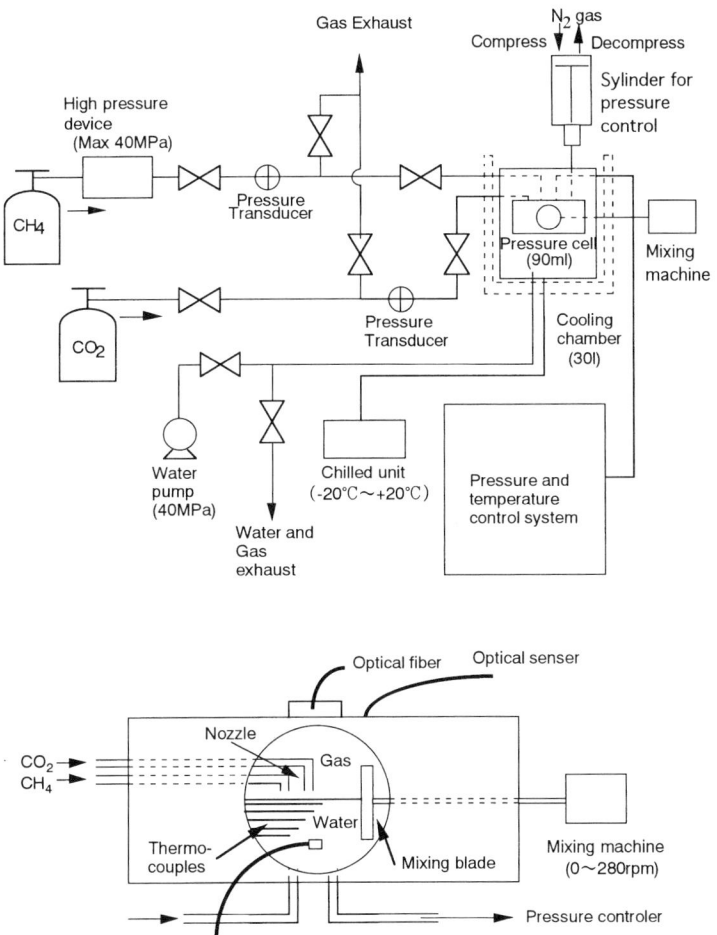

FIGURE 1. Experimental design for gas hydrate reformation and the measurement system.

equipment and three glass windows for the observation of gas hydrate formation. Pure water and CH_4 (or CO_2) gas were introduced into the cell using a high-pressure pump system. Sufficient gas transport into the solution was achieved by mixing the interface. The temperature was decreased gradually until the gas hydrate formed. After hydrate growth in the solution, the temperature was increased to observe the dissociation. The rate of temperature decrease and increase was controlled at 0.16 K/min. The mixing mechanism was operated continuously during the entire formation and dissociation processes. The cell was maintained as a closed system, without introducing pressure control. Temperatures were measured using five thermocouples installed inside the cell to represent the average temperature in the system. Three to five continuous runs of the process were conducted.

Replacement Experiment

Solid-phase samples of CH_4 gas hydrate were obtained for the replacement experiment, using the special method from ice crystals and the dissociated solution with CH_4. After the solid sample of CH_4 gas hydrate was installed in the cell, pressurized CO_2 was introduced for the replacement of guests in gas hydrates. The temperature and pressure were controlled by means of the procedure described below to maintain the solid phase of CH_4 hydrate prior to conducting the experiment and to prevent the formation of solid-phase CO_2 in the cell. In the first stage of the experiment, the temperature and pressure were 193 K and 0.1 MPa in the presence of pure CH_4 hydrate. Then, carbon dioxide gas was gradually introduced to set the temperature and pressure to 253 K and 3.5 MPa, respectively, conditions under which both CH_4 and CO_2 hydrates are stable. In the final stage of the experiment, the temperature and pressure were maintained between values for the equilibria of CH_4 and CO_2 hydrates. Changes in the gas atmosphere components were analyzed by gas chromatography as functions of time after CO_2 introduction and environmental conditions. Raman spectroscopy was also applied for *in situ* analysis of components and the structure of solid gas hydrates.

RESULTS AND DISCUSSION

Dynamics of Gas Hydrate Reformation

FIGURE 2 shows a typical result from the process for CH_4 hydrate reformation in which three continuous experimental runs were conducted without any changes in the components of the liquid and gas. The starting point for each run was 293 K and 10.0 MPa. The gas hydrate formation started in the liquid during the period of temperature decrease. During the mixing procedure, numerous fine fragments of CH_4 hydrate crystals appeared in the liquid phase and accumulated in the solution. A rapid increase of temperature due to the heat of formation and a decrease of pressure were observed at the formation point. The pressure gradually decreased as the temperature dropped as a result of the changing solubility of the liquid and the ideal gas law. Dissociation of the gas hydrate was observed during the period of temperature increase. After the dissociation point was reached, the rate of temperature increase declined slightly and then began to rise again. During the course of these processes it was found that the temperature of the reformation point gradually increased with the number of continuous runs. These critical temperatures under nonequilibrium conditions are important for understanding the phase behavior of gas hydrates.[3]

Hysteresis Process of Gas Hydrate Formation/Dissociation

FIGURE 3 illustrates the relationship between pressure and temperature in the experiment with 100% CH_4 hydrate. The formation process follows an oscillating curve between a1 and d, and the dissociation process follows the curve between e and f. The figure shows that there were large differences in temperature and pressure

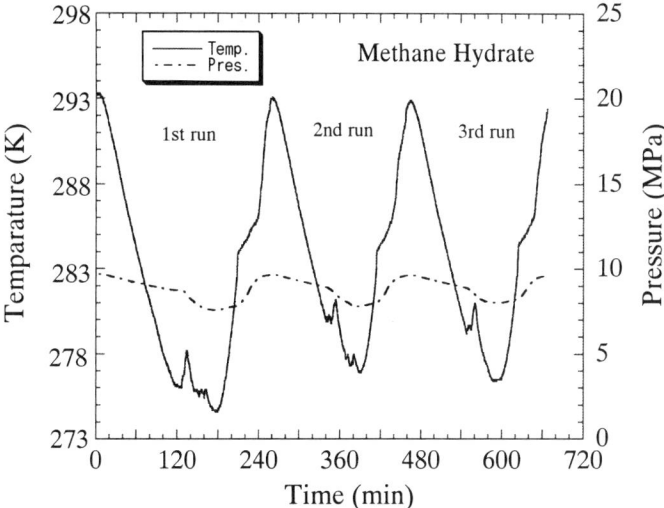

FIGURE 2. Trend curves for pressure and temperature obtained from the hysteresis experiment.

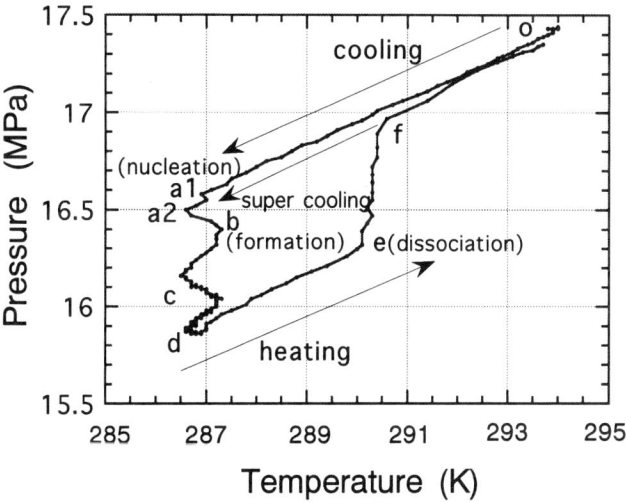

FIGURE 3. Pressure and temperature history curve for gas hydrate formation and dissociation.

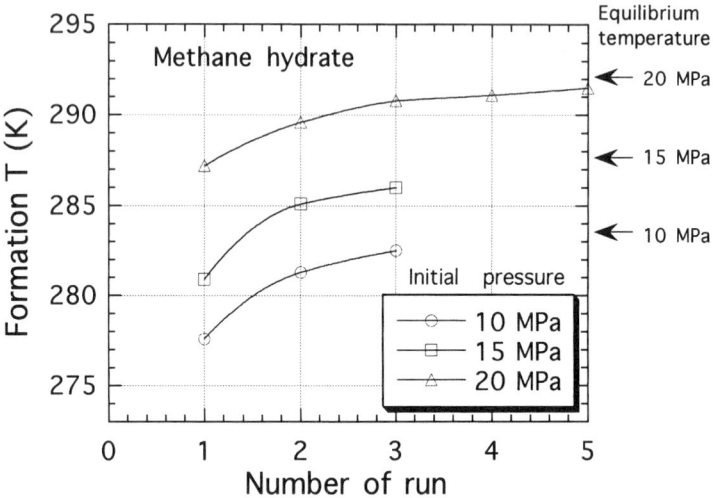

FIGURE 4. Hysteresis data for formation temperature.

during formation and dissociation of the gas hydrate. The temperature difference in the formation and dissociation equilibrium is usually regarded as being due to a supercooling effect. In the case of CH_4 hydrate, the temperature difference at supercooling and in equilibrium states was in the range of 2.5 to 4.5 K. FIGURE 4 gives the hysteresis data for the nonequilibrium formation temperature T_f and the pressure P obtained from continuous runs of the formation/dissociation processes of the gas hydrate. It is apparent from these data that the formation temperature tends to increase with the number of repetitions, and that it approaches the equilibrium temperature. This hysteresis process suggests that less activation energy is required to form the hydrate a second time, due to the presence of hydrogen bonds that remain in the solution.[4]

Effect of Restarting Conditions on Reformation

FIGURE 5 shows the effect of restart temperature on the formation temperature during the second run for CH_4 hydrate. In this experiment, the holding period of the controlled temperature between each run was set to 1 hour. It is clear that the formation of hydrate is strongly affected by the restart conditions. This phenomenon is generally considered to be due to the memory effect. In another experiment, the second run showed that the memory effect was retained if the holding period of the restart temperature ranged from 1 hour to 12 hours. The effect can be explained as being due to the following fundamental mechanism. If the restart temperature is close to the equilibrium value, most molecular bonds remain in the solution, even above the dissociation temperature. In other words, a significant change in the nucleation process of gas hydrates is caused by the structure of clusters in the solution. It was reported that the process of nucleation of gas hydrates was affected by the structure of water molecules and heterogeneous conditions in the solution.[5] To

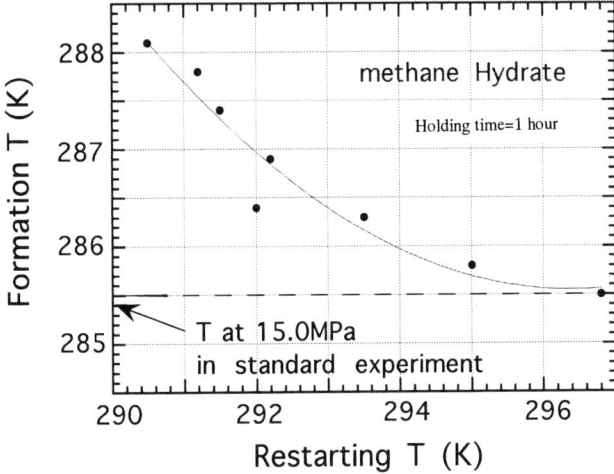

FIGURE 5. Effect of restart temperature on formation temperature.

clarify this phenomenon, it is necessary to conduct further research on the structure of water and the relevant changes caused by the presence of hydrogen bonds at the molecular level.

Phase Behavior of CH_4 and CO_2 Gas Hydrates

FIGURE 6 shows the formation and dissociation relations obtained during the first run for CH_4 and CO_2 gas hydrates. The upper points (●, CH_4; ▲, CO_2) represent the relationship between nonequilibrium formation temperatures T_f and pressures P_f, and the lower points (○, CH_4; △, CO_2) the relation between dissociation temperature T_d and pressure P_d. The middle curve corresponds to the phase equilibrium estimated using a theoretical thermodynamic approach. It can be seen that the relation between P_f, (P_d) and T_f, (T_d) is approximately linear in a semilog plot. In addition, it was found that the nonequilibrium formation relation differs greatly from the theoretical data for phase equilibria. However, the dissociation relation agrees well with the theoretical equilibrium data.[1] The differential temperatures between the nonequilibrium formation and dissociation for CO_2 hydrate were a slightly larger than those for CH_4 hydrate. These properties may reveal interesting phenomena concerning the mechanism of gas hydrate nucleation under nonequilibrium conditions.

Solid-Phase Formation of Methane Hydrate Sample

Pure CH_4 hydrate samples were synthesized as the solid phase for the replacement experiment, using fine grains of ice crystals (average diameter: 0.1 mm) and the solution for which CH_4 gas hydrate was dissociated at a temperature slightly higher than the equilibrium value. It took a relatively short time to reform the gas hydrate when the dissociated solution was used. When the temperature was set 2.0 K

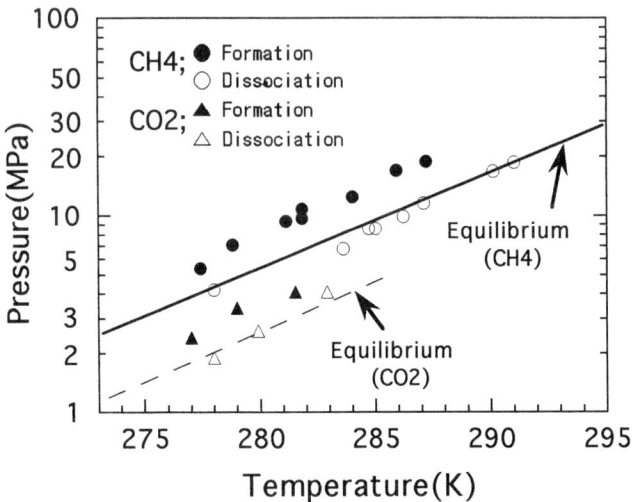

FIGURE 6. The relations of formation/dissociation pressure and temperature for CO_2 and CH_4 hydrates.

lower than the equilibrium value, the reformation of CH_4 hydrate could be achieved for a period of 24 hours. The hydration number of the samples obtained was in the range of 6.0 to 6.4; that is, the ratio of occupancy of small cages can be estimated to be approximately 85–95 percent.

Replacement of CO_2 and CH_4 Gas Hydrates

The replacement experiment for CO_2 and CH_4 gas hydrates was carried out using a bulk-scale testing apparatus. The sample of CH_4 gas hydrate was synthesized as a solid phase from ice crystals and the dissociated solution. The preliminary process was carried out under the conditions discussed in the experimental procedure. During the process of replacement, the pressure of CO_2 was 3.5 MPa and the temperature was 276 ± 0.5 K. FIGURE 7 illustrates the Raman spectra obtained for the replaced hydrate sample, observed one hour after CO_2 introduction. Two typical peaks appeared in the Raman spectra, 2905/2915 cm^{-1} for CH_4 hydrate and around 1275/1380 cm^{-1} for CO_2 hydrate. This implies that both CH_4 and CO_2 gas hydrate coexist in the structure of the replaced hydrate sample.[6] According to the qualitative analysis of components, the ratios of occupancy for CO_2 and CH_4 gas hydrates in the sample were 40 and 45 percent, respectively.[7] The portion of CO_2 gas hydrate increased with time. FIGURE 8 shows the trend curves for components in gas atmosphere and in solid gas hydrate after the introduction of CO_2, as a function of time. This result indicates that the replacement of guests can be achieved in solid gas hydrate for a short period, such as 12 hours, if the pressure and temperature are precisely controlled in a pressure cell. It is considered that a period of 12 hours is required only for the nucleation of CO_2 hydrate. Because rapid transport of CO_2

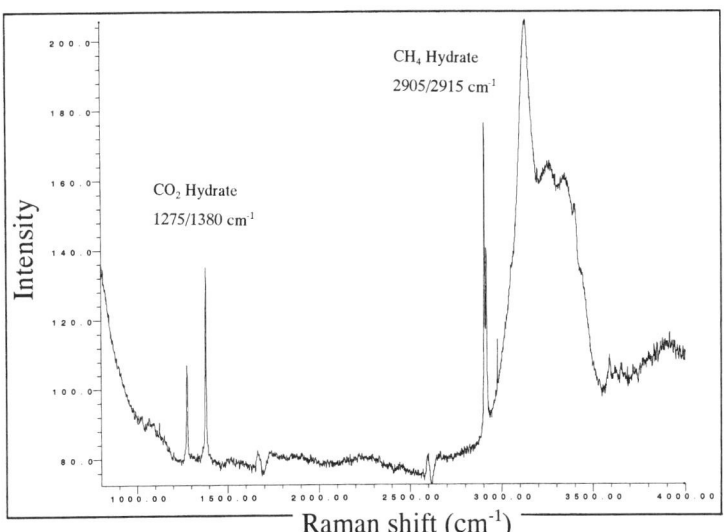

FIGURE 7. Raman spectral data for gas hydrate sample replaced by CO_2.

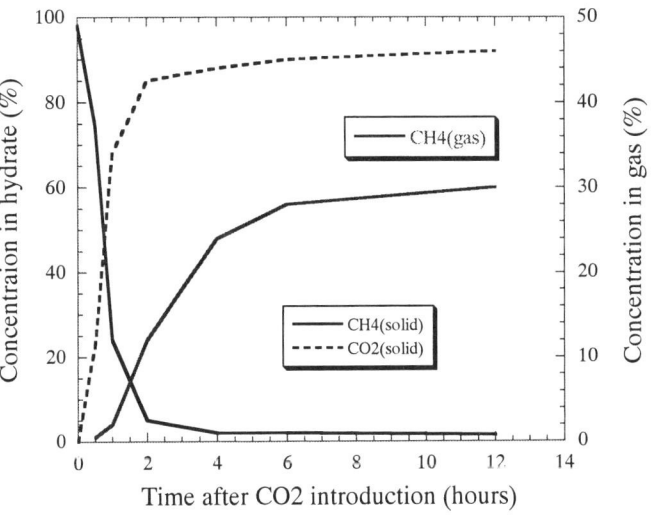

FIGURE 8. Trend curves for components in gas atmosphere and in solid gas hydrate sample.

gas should occur in porous media in the absence of a mixing mechanism, it may be possible to use the *in situ* replacement into CO_2 hydrate for the production of gas hydrates from the reservoirs. More systematic research is required to develop the production system of gas hydrates.

CONCLUSIONS

Experiments on the processes of gas hydrate reformation and the behavior of replacement were carried out under nonequilibrium conditions. The process was found to exhibit typical hysteresis behavior due to the supercooling effect. The environmental conditions for hydrate reformation were strongly affected by the states of the liquid phase, the number of runs and the restart temperature. The results of the bulk-scale experiment showed that the replacement of CO_2 gas hydrate could be achieved within a short period in pure samples of CH_4 gas hydrate that had been synthesized as solid crystals, if the pressure and temperature were precisely controlled in a pressure cell. The replacement of guests in CO_2 and CH_4 gas hydrates may be quite an innovative procedure from both viewpoints of the global environment and the stability of marine sediment, as well as from the promotion of production from the reservoirs.

ACKNOWLEDGMENTS

The authors thank Dr. M. Takahashi of NIRE for performing the Raman spectra analysis, and Mr. H. Hoshino and Mr. T. Kawamura, graduate students at Hokkaido University, for assistance with the experimental work on gas hydrates and with data analysis.

REFERENCES

1. HOLDER, G.D. 1988. Rev. Chem. Eng. **5:** 1–70.
2. KOMAI, T. & Y. YAMAMOTO. 1997. 213th ACS Meeting. Fuel Chem. **42:** 568.
3. SCHROETER, J.P. 1982. Gas Processors Association. 10.
4. SLOAN, E.D. & R.L. CHRISTIANSEN. 1994. First Conf. Natural Gas Hydrates. 715, 283.
5. YAMAMOTO, Y. & A. WAKISAKA. 1996. Second Int. Conf. Natural Gas Hydrate. 355.
6. SUM, A.K., R.C. BURRUS & D. SLOAN. 1996. Second Int. Conf. Natural Gas Hydrate. 51.
7. UCHIDA, T., A. TAKAGI, T. HIRANO, H. NARITA, J. KAWABATA, T. HONDO & S. MAE. 1996. 2nd Int. Conf. Natural Gas Hydrate. 335.

A New Class of Kinetic Hydrate Inhibitor

MALCOLM A. KELLAND,[a,b] THOR M. SVARTAAS,[c]
JORUNN ØVSTHUS,[c] AND TAKASHI NAMBA[d]

[b]*Glow, Stokkamyrveien 17, N-4300 Sandnes, Norway*

[c]*RF-Rogaland Research, Norway*

[d]*Nippon Shokubai Co., Ltd, Japan*

> ABSTRACT: Low dosage hydrate inhibitors (LDHI) offer a recently developed hydrate control technology that can be more cost-effective than traditional practices, such as the use of thermodynamic inhibitors (e.g., methanol and glycols). One class of LDHI, called kinetic inhibitors, is already being successfully used in the field. This paper describes efforts to develop a new class of kinetic inhibitor that shows various improvements over existing commercial technology. The polymer chemistry of the inhibitors and experiments carried out in high pressure cells and wheel/loops is described.

BACKGROUND

Gas hydrates are clathrates in which water molecules form a hydrogen bonded network enclosing roughly spherical cavities that are filled with gas molecules. Gas hydrate can form several structures but a typical natural gas mixture containing C_1–C_4 components will preferentially form structure II hydrates. FIGURE 1 shows an example of a pressure versus temperature (PT) phase diagram for a gas hydrate. The curve represents the equilibria between gas, water, and the gas hydrate. Under PT conditions to the left of the curve, gas hydrates may form. Plugs caused by gas hydrate formation are a menace to the oil and gas industry in production lines, during drilling (especially in deep water), and in workover operations.[1]

Multiphase production is a simple, and, therefore, often a very cost-effective solution for development of marginal or remote fields if it can be applied successfully. The potential for gas hydrate formation in the lines is one of the major problems that needs to be addressed, even more so for long tie backs, deep water, and shut-in situations in general since the fluids then have time to cool well into the hydrate region.

Methods to avoid hydrate plugs include raising the temperature/heating (e.g., insulation, bundles, electric, or hot water heating), lowering the pressure, removing the water, and shifting the equilibrium for hydrate formation by adding antifreeze chemicals. These techniques are often very expensive (such as heated pipelines or methanol regeneration facilities), or do not offer a complete solution (e.g., subsea water separation). Hence, there is a clear need for cheaper technologies. New low

---

[a]Telecommunication. Voice: 47-51639324; fax: 47-51639330.
mk@glow.rafoss.no

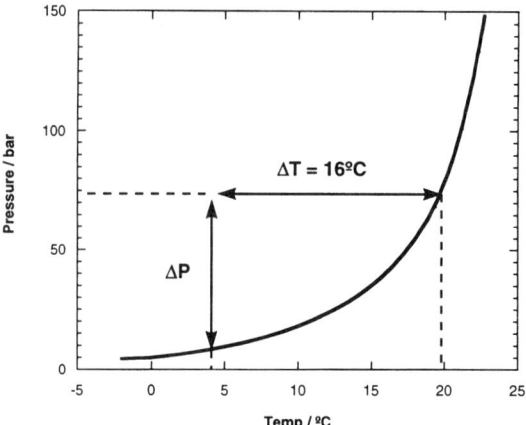

FIGURE 1. *PT* diagram for a gas hydrate illustrating the subcooling (ΔT) at 4°C (a typical subsea temperature) and 75 bar.

dosage chemical technologies that can be significantly cheaper have been developed in recent years. These are as follows.

Kinetic Inhibitors. These delay hydrate nucleation and crystal growth so that there is sufficient time to transport the fluids to the process facilities before hydrates build up in the line.

Anti-Agglomerants (AA). These prevent the agglomeration and deposition of hydrate crystals so that a transportable hydrate slurry is formed. (Surfactants designed to attack hydrate surfaces also disrupt the hydrate growth process.)

Both classes of additives are added at low concentrations, often about 0.3–0.5 wt.% active concentration. This can be contrasted with the 10–50 wt.% needed for "antifreezes" such as methanol, glycols, or salts. The two new types of additives have different field application ranges related to field conditions, fluid properties, and the properties of the additives, not least of which is their environmental impact.

Kinetic inhibitors are already being used in full-field applications today and have given operators very significant savings over alternative hydrate control methods.[1–3] The key ingredients in these products are polymers or copolymers containing primarily

FIGURE 2. Monomer units, vinyl pyrrolidone (**left**) and vinylcaprolactam (**right**).

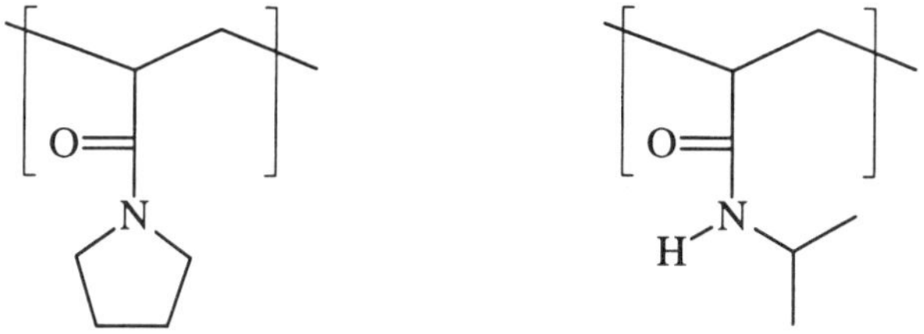

FIGURE 3. Monomer units, acryloyl pyrrolidine (**left**) and isopropylacrylamide (**right**).

vinyl lactam monomers, specifically the monomers vinyl pyrrolidone and vinyl caprolactam (see FIGURE 2). As far as we are aware, there is no commercially used kinetic inhibitor based on other types of polymers.

NEW KINETIC INHIBITORS

This paper discusses a new class of kinetic inhibitor developed through a joint industry program. The new kinetic inhibitor class is based on alkylacrylamide polymers and blends of alkylacrylamides with other synergists. We have concentrated on two polymers in particular, shown in FIGURE 3, based on the monomers acryloyl pyrrolidine (AP) and isopropylacrylamide (IP). We have found that these polymers show high performance as kinetic hydrate inhibitors.

These polymers inhibit hydrate formation in a similar way as do the vinyl lactam polymers, by interaction of the pendant alkylamides groups with the hydrate surface. The hydrate surface can be imagined as being made up of various open cavities in which small hydrocarbon molecules would normally be placed in filled cavities in gas hydrate. The pendant alkylamide groups interact in two ways with the hydrate surface. First, the small alkyl group penetrates an open cavity, and then the amide group hydrogen bonds to the hydrate surface via the carbonyl group "locking" the alkylamide to the surface. Hydrogen bonding between the nitrogen of the amide and surface water molecules is less important, based on our molecular modelling studies. Several of the alkylamide groups must interact with the hydrate surface for any significant kinetic inhibitor effect to be observed.

FIELD APPLICATION ISSUES

The main issues to be addressed for field applications of low dosage hydrate inhibitors are indicated in FIGURE 4. In evaluating the economics, the overall capital and operation costs of using low dosage inhibitors must be calculated against alternative hydrate control strategies. For example, if glycol injection is the alternative, the cost of the glycol regeneration unit must be taken into account, the space it uses on the platform, and its maintenance.

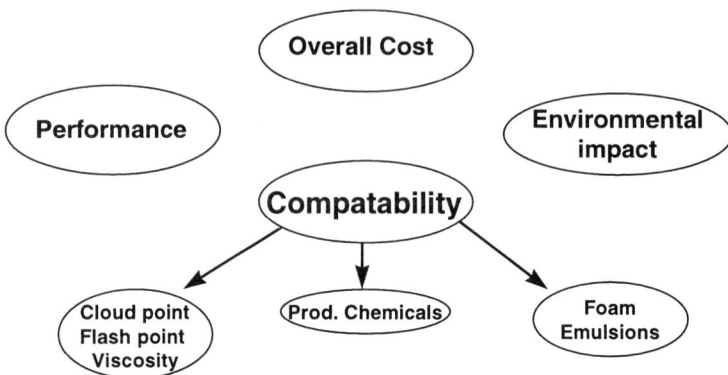

FIGURE 4. Application issues for kinetic hydrate inhibitors.

There are a variety of issues related to compatibility that are common to many production chemicals. For offshore use, the flash point should preferably exceed 56°C. Furthermore, the viscosity of the product must not be so high that it cannot be pumped to the injection point (an important issue for kinetic inhibitors, which are mostly polymeric materials). It is also important that the product does not cause foaming problems, nor produce emulsions that are difficult to break. The product must be compatible with other production chemicals added to the well stream. Finally, the cloud point is a critical issue for kinetic inhibitor polymers. This is discussed below in more detail.

Cloud Point and Deposition Point

One of the challenges in using polymers as kinetic hydrate inhibitors is that they may be incompatible with the well head temperature and produced water salinity. As the temperature of an aqueous solution of a polymer (and various surfactants) rises, phase separation of the polymer can occur causing the solution to exhibit turbidity or cloudiness. The temperature at which this occurs is often referred to as the cloud point, T_{cl}. The molecular reason for T_{cl} is related to the competitive strength of hydrogen bonding between the water phase and the dissolved molecule, and internal hydrogen bonding in the molecule as one varies the temperature. (An alternate description preferred by some is related to the hydrophobic interactions between the molecule and the water phase). However, the temperature at which the polymer precipitates from solution—the deposition temperature, T_{dp}—rather than T_{cl}, is the critical temperature for oilfield operations. T_{dp} decreases as the ionic strength of the water produced increases. T_{dp} often lies 5–15°C above T_{cl}, but in some cases, and using some manufacturing techniques, it can be significantly higher. Since T_{cl} is very easy to measure, it can be used to obtain a rough idea of the aqueous solubility characteristics of a polymer.

Kinetic inhibitors in use today are based mainly on vinyl caprolactam polymers. For polyvinyl caprolactam in distilled water, $T_{cl} \approx 30\text{–}35°C$ depending on the polymerization process used and the molecular weight. Hence, it is rarely suitable by itself for injection into hot or highly saline well streams since it is liable to deposit. Vinyl caprolactam copolymers with more hydrophilic comonomers can be used, but there is a balance between the performance of the copolymer and the T_{cl}, based on the percentage of vinyl caprolactam in the polymer. It is possible to improve the solubility characteristics of these kinetic inhibitor polymers by adding surfactants, but this often has a negative effect on the inhibitor performance. Therefore, it would be beneficial to have a more compatible class of polymer altogether, with higher T_{cl} and T_{dp}.

TABLE 1. Molecular weights (M_n) and cloud points (T_{cl}) of polymers

Sample A	M_n	T_{cl}/°C in DI H$_2$O	T_{cl}/°C in 3.6% SSW
PVCap	1,300	33	30
Gaffix VC713	4,500	33	30
VP:VCap	4,000	70	60
VIMA:VCap	2,000	74	62
AP-1000L	1,200	60	53
AP-1000 270K	43,000	43	38
AP-1000 ACM	6,100	>95	93
AP-1000 ACL	2,100	>95	>95
AP-1000ACB9	870	>95	>95
IP1000-11	1,100	42	32
IP1010ASN-26	2,600	76	46
IP-1020ASN-11	1,100	>95	>95
IP-1020DAQ-16	1,600	62	43
IP-1035DAQ-16	1,500	>92	78
IP-1010AA-10	1,000	39	36
IP-1020AAN-11	1,100	>95	66
AS-1000	8,400	>95	>95
GP-1000	4,300	>95	~80

NOTE: The sample codes refer to the following materials. *PVCap* is a polyvinylcaprolactam homopolymer. *Gaffix VC713* is a terpolymer of N-vinylcaprolactam, N-vinyl pyrrolidone, and dimethylaminoethylmethacrylate from International Specialty Products. *VIMA:VCap* is a 1:1 copolymer of N-vinylcaprolactam and N-methyl-N-vinyl acetamide. *VP:VCap* is a 1:1 molar ratio copolymer of N-vinylcaprolactam and N-vinyl pyrrolidone. All *AP-1000* prefixed polymers are polyacryloylpyrrolidine homopolymers. *IP-1000-11* is polyisopropylacrylamide homopolymer. *IP-1010ASN* and *1020ASN* are 9:1 and 4:1 copolymers of isopropylacrylamide and AMPS neutralized with NaOH. *IP-1010DAQ, 1020DAQ,* and *1035DAQ* are, respectively, 9:1, 4:1, and 6.5:3.5 copolymers of isopropylacrylamide and dimethylaminoethylacrylate reacted with HCl. *IP-1010AA* and *1020AAN-11* are 9:1 and 4:1 copolymers of isopropylacrylamide and acrylic acid. The sodium salt of the acrylate was used in *1020AAN-11* copolymer. *GP-1000* is polyglyoxyloylpyrrolidine. *AS-1000* is pyrrolidine modified poly(succinimide).

Acryloyl pyrrolidine (AP) homopolymers and copolymers have much higher T_{cl} values than vinyl caprolactam polymers if the correct manufacturing process is used. The same is true of copolymers of isopropylacrylamide (IP). In general, we found that for AP polymers, T_{cl} increases as the molecular weight decreases, and the effect was most significant in the low molecular weight range.

TABLE 1 shows the T_{cl} values for various alkylacrylamide polymers and two polymers containing the carbonylpyrrolidine as side chains attached to a heteroatom backbone. Nearly all M_n (number average molecular weight) values were measured by gel permeation or size exclusion chromatography (GPC or SEC). The T_{cl} values are for approximately 1% polymer solutions.

KINETIC HYDRATE INHIBITOR PERFORMANCE TESTING

The performance of these alkylacrylamide and related polymers as kinetic hydrate inhibitors in comparison with earlier inhibitors, was determined by tests in high pressure sapphire cell equipment at RF, and larger scale tests were performed on the wheel facilities at Petreco in Stjørdal.

Sapphire Cell Equipment

The sapphire cell high pressure test equipment is illustrated in FIGURE 5 and has been described previously.[4]

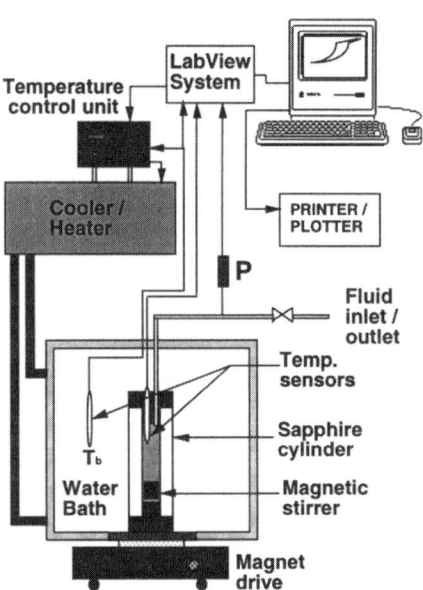

FIGURE 5. Sapphire cell high pressure test equipment.

All tests were performed on fresh synthetic sea water (SSW = 3.6%) and a recombined blend of a clear, colorless North Sea condensate and synthetic natural gas (SNG, containing C_1, C_2, C_3, i-C_4, CO_2 and N_2). The same procedure for preparing the experiment and filling the cell was followed in all experiments. Thus, the chemical to be tested was dissolved in synthetic sea water (SSW) to the desired concentration and added to the cell at room temperature. The data logging and video recording were started, and the cell was loaded with the recombined hydrocarbon fluid to the desired pressure while stirring at 700 rpm. When the temperature and pressure in the cell had stabilized, stirring was stopped. The closed cell was then cooled to the experimental temperature, resulting in a small decrease in pressure. When the temperature and pressure had again stabilized, stirring at 700 rpm was started. The induction time, t_i, for hydrate formation was measured from the time stirring started at the experimental temperature. The time from start of hydrate formation to the time when rapid growth of hydrate ensues is called the crystal growth delay time, St-1. The results of all experiments were recorded visually and by plotting temperature, pressure (or the total gas consumption, TGC), and torque as a function of time. Subcooling was used as an adequate measure for the system driving force in these experiments, although system chemical potential, $\Delta\mu$ is more accurate.[5]

Sapphire Cell Performance Tests on Single Polymers

TABLE 2 shows a series of cell results with single component polymer systems as kinetic inhibitors. All tests were all carried out between 89–91 bar pressure at various constant temperatures, T. The results indicate that some types of AP polymers and also IP polymers can perform as well as, or better than, the vinyl caprolactam polymers tested.

As expected, the lower the temperature the smaller the induction time (t_i) and growth delay (St-1) for hydrate formation. Compare for example tests 3 and 4, and 6 and 8. We generally observe more scatter in the induction times than in the slow growth stage. The induction time reproducibility is normally within 30–50%, and for the slow growth stage 10–20%. This scatter is normally good enough to determine significant differences in performance especially when the inhibitor is tested at a range of conditions; for example, by varying the subcooling, pressure, and inhibitor concentration.

In the field the performance criterion is preferably zero hydrate formation. However, it may be possible that small amounts of hydrates formed during the early stages of the slow growth period do not accumulate in the pipeline, as we often observe in the laboratory. Thus the induction time to hydrate formation provides a good conservative estimate of performance. In the laboratory, however, the induction time and growth time appear to be codependent, thus, it is worth recording and using both figures in performance analysis.

In our experience, increasing the concentration of the alkylacrylamide polymers gives better performance up to a level of about 6,000–10,000 ppm. For example, see tests 6 and 7.

The results show that there is an optimum molecular weight range, for AP or IP polymers, that gives the best performance, as has been shown previously for vinyl caprolactam polymers.[6,7] Below a certain weight there are too few interactions with

TABLE 2. Performance tests on single component systems

Test	Sample	Conc. ppm	M_n	T °C	t_i mins	St-1 mins	Total mins
1	none			7.6	<2	<2	<2
2	PVCap	5000	7,500	7.6	49	70	119
2	PVCap	5000	1,300	3.7	72	136	208
3	Gaffix VC713	5000	4,500	5.8	28	32	60
4	Gaffix VC713	5000	4,500	3.9	3,9	0.6	3,9
5	VIMA:VCap	5000	2,000	3.9	1,3	4,5	5,8
6	AP-1000L	5000	1,200	5.7	33	148	181
7	AP-1000L	6000	1,200	5.7	250	120	370
8	AP-1000L	5000	1,200	3.7	2	22	24
9	AP-1000 270K	6000	43,000	5.5	14	3	17
10	AP-1000 ACM	5000	6,100	5.5	620	180	800
11	AP-1000 ACM	5000	6,100	3.9	158	73	231
12	AP-1000 ACL	5000	2,100	3.9	402	30	432
13	AP-1000ACB9	5000	870	2.7	74	26	100
14	AP-1000ACB3	5000	300	2.7	0	7	7
15	AP:VP 7:3	5000	20,000	5.7	0	3	3
16	APAM-1050-09	5000	900	3.7	65	75	140
17	IP-1000-11	5000	1,100	3.9	66	64	130
18	IP-1000	5000	580	3.9	32	30	62
19	IP1010ASN-26	5000	2,600	3.6	108	62	170
20	IP-1010ASN-11	5000	1,100	2.7	325	38	363
21	IP-1010ASN-26	5000	2,600	2.7	50	62	112
22	IP-1020ASN-11	5000	1,100	2.7	46	10	56
23	IP-1030ASN-13	5000	1,300	2.7	29	8	37
24	IP-1010DA-12	5000	1,200	3.6	60	50	110
25	IP-1020DAQ-16	5000	1,600	3.6	45	88	133
26	IP-1020DAQ-16	5000	1,600	2.7	54	33	87
27	IP-1035DAQ-16	5000	1,500	2.7	35	12	47
28	IP-1010AA	5000	1,800	3.6	58	72	130
29	IP-1020AA-11	5000	1,100	3.6	45	70	115
30	IP-1020AAN-11	5000	1,100	3.6	58	32	90
31	IP-1010VP-11	5000	1,000	3.6	65	85	150
32	IP1050VI-12	5000	1,100	3.6	20	43	63
33	GP-1000	5000	4,300	5.4	909	28	937

NOTE: Polymer codes are as explained in TABLE 1, except: *AP:VP 7:3* is a 7:3 molar ratio acryloylpyrrolidine:vinyl pyrrolidone copolymer; *APAM 1050-09* is a 1:1 acryloylpyrrolidine:acrylamide copolymer.

the hydrate surface per polymer chain to cause any inhibition. For example, the polymers in tests 13 and 14 were made by a similar process but gave different molecular weights. Test 13 gave a good performance whereas test 14 was very poor, almost as bad as no additive. Second, as the molecular weight increases, the number of polymer strands in solution decreases and some of the alkylamide side chains may become less available for interaction with hydrate surfaces. Thus, high molecular weight polymers also have lower performance. See, for example, tests 7 and 9.

Another important observation was that increasing the cloud point of an AP polymer did not necessarily decrease the inhibitor performance. It can be argued that a polymer with a cloud point in the water phase at near the experimental temperature will be attracted more to hydrate surfaces than a polymer with higher cloud point. This does not seem to be the case with AP homopolymers, and certain IP copolymers.

Some comonomers drastically reduced the performance of the copolymer inhibitor compared with the homopolymer. Others did not but could actually enhance the performance. See for example, test 15 with a vinyl pyrrolidone:AP copolymer (albeit of rather high M_n) which gave a very poor result of zero induction time, compared with test 16 using an acrylamide:AP copolymer, which performed significantly better. This is surprising since polyvinyl pyrrolidone is itself known to work fairly well as a kinetic inhibitor whereas polyacrylamide has almost no inhibition effect at all.

For the IP polymers, we found that neutral, anionic and cationic copolymers could give similar or sometimes better performance than neutral IP homopolymers. As expected a high percentage of ionic comonomer (above 10–20%) worsened the performance, assuming the comonomer itself interacts poorly with hydrate surfaces (e.g., tests 19–27). This is related to the number of available alkylamide groups in the copolymer that can interact with hydrate surfaces.

Sapphire Cell Performance Tests with Blends of Polymers

Having identified the potential of alkylacrylamide polymers as kinetic hydrate inhibitors, we sought to improve their performance by blending them with other components—so-called synergists. Based on molecular modelling results, it occurred to us that since vinyl lactam polymers interact differently with hydrate surfaces than with alkylacrylamide polymers, there may be a synergistic effect from using blends of two or more polymers in these classes. The results in TABLE 3 indicate that this is indeed the case.

Although in practice, the performance of a kinetic inhibitor depends on many factors, the synergy results above indicate that significant improvements over vinyl lactam polymer based systems can be achieved by careful selection of the correct blends. Several other observations can be drawn from the results.

- Addition of small amounts of polymers containing the *N*-vinyl caprolactam monomer or *N*-vinyl pyrrolidone monomer to a larger amount of an alkylacrylamide polymer results in a significant increase in inhibition performance (particularly for the induction time), much greater than that using the same total amount of an alkylacrylamide polymer. This synergy is greatest when the alkylacrylamide polymer was of low molecular weight.

- Addition of small amounts of low or high molecular weight alkylacrylamide polymers to a larger amount of a N-vinyl lactam polymer does not give as significant an increase in gas hydrate inhibition performance.
- PVCap is a better synergist than PVP for alkylacrylamide polymers.
- There is an optimum molecular weight for both PVP and PVCap polymers as synergists. This molecular weight is higher than the molecular weight of the alkylacrylamide polymers.
- Certain triple blends of polyalkylacrylamide, PVP and PVCap polymers perform better than double blends. These results imply that the alkylacrylamide, vinyl caprolactam, and vinyl pyrrolidone groups inhibit hydrate in three separate processes.
- Copolymerizing AP and VP monomers produces a polymer with much poorer performance than a blend of AP and VP homopolymers.

The explanation of the results in TABLE 3 depends on several factors including the interaction energies of the various alkylamide groups in the polymers with hydrate surfaces, the polymer molecular weights (mobility), and their solubility characteristics.

Wheel Simulator Tests

To confirm our findings and gain a better feel for the performance of the new kinetic inhibitors, a series of wheel tests was carried out at Petreco in Stjørdal, Norway. The wheel simulator used consists of a 2″ stainless steel pipe with two windows. A camera is placed on one of the windows for visual inspection. The diameter of the wheel is 2 meters giving a total pipe length of about 6.3 m. The maximum pressure rating for the wheel with windows is 100 bars. The water cut was 25%, with a total liquid filling of 40%. The wheel can be rotated at varying speeds to simulate flow along a pipe. In the two examples given below, a rotation speed of 1 m/s was used.

In the first test, a 6,000 ppm active solution of a blend of polymers in 3.6% brine was added to the wheel. Exxsol D60 was added and the wheel pressurized with SNG (SNG is the mixture, $C_1:C_2:C_3:iC_4:CO_2 = 81:7:5:2:2$) and then cooled to 4°C without rotation to a pressure of 84 bar. This gave a subcooling ratio, ΔT, of about 13.6°C. Once stable, the wheel was then rotated. Foaming was encountered during rotation, which affected the torque measurements (the chemical system was not optimized to deal with foam problems). Hydrate crystal were visually observed to be present about 740 minutes after start-up and the wheel was plugged by "foamy" hydrates about 30 minutes later.

A repeat experiment was run at about 70 bar and 4.5°C ($\Delta T \approx 11.9°C$, experimentally determined in the wheel). Far less foam was created than for the experiment at 84 bar. The results are illustrated in FIGURE 6. No hydrate formation was observed in the system either visually or from the measured TGC (gas uptake) and torque responses during the period between the first start-up and the second shut-in (about 17 hrs). During the second shut-in (approximately 5.5 hrs), small amounts of hydrate particles could be seen in the region just above the water-hydrocarbon interface but the amount remained relatively constant during the second shut-in period. After the

TABLE 3. Synergy results

Test	Sample A	Concn ppm	Sample B (1000 ppm)	Sample C (1000 ppm)	T °C	t_i mins	St-1 mins	Total
33	AP-1000L	5000	AP-1000L		5.7	250	120	370
34	AP-1000L	5000	PVCap		5.7	>1347		>1347
35	AP-1000L	5000	PVCap		3.9	110	170	280
36	AP-1000L	5000	VP:VCap		5.7	>1161		>1161
37	AP-1000L	5000	VP:VCap		3.9	843	68	911
38	AP-1000L	5000	VIMA:VCap		3.9	93	77	170
39	AP-1000L	5000	Gaffix VC713		3.7	5	160	165
40	Gaffix VC 713	5000	AP-1000L		3.7	0.5	21	21
41	PVCap	5000	AP-1000L		3.7	6.5	0.8	7.3
42	AP-1000L	5000	PVIMA		5.7	60	159	219
43	AP-1000L	5000	PVP 15k		5.7	705	130	835
44	AP-1000L	4000	PVCap1	PVP 15k	3.9	765	133	898
45	AP-1000 ACL	4000	PVP 15k	PVCap 19k	2.7	130	90	220
46	AP-1000ACB21	4000	PVP 15k	PVCap 19k	2.7	137	73	210
47	AP-1000ACB21	4000	Antaron P904	PVCap 19k	2.7	310	160	470
48	IP-1000-11	4000	Antaron P904	PVCap 19k	2.7	250	115	365
49	IP-1010DA-12	5000	PVCap 19k		3.7	600	90	690
50	IP-1020DA-16	5000	PVP 15k		3.7	565	128	693
51	IP-1020DAQ-16	5000	VCap/VP=4/1		3.7	>1113		>1113
52	IP-1020DAQ-16	5000	VP:VCap 1:1		3.5	170	49	219
53	IP1010AS-26	5000	PVCap 19k		3.6	>1066		>1066
54	GP-1000	4000	PVP 15k	PVCap 19k	5.4	>2700		>2700
55	AS-1000	4000	PVP 15k	PVCap 19k	5.4	>5310		>5310

NOTE: Polymer codes are as in TABLES 1 and 2, except *Antaron P-904*, which is a butylated PVP polymer from International Specialty Products (ISP); and *PVP 15k*, a polyvinylpyrrolidone from BASF.

second start-up a slight elevation of the TGC level was observed and the TGC signal remained constant at this new level throughout the rest of the experiment. During the second start-up period some minor amounts of hydrate crystals could be seen but no rapid hydrate growth occurred and therefore no plugging. The second start-up period was run for about 20 hours before the experiment was stopped.

Although the fluids in the wheel experiments were not identical to the cell experiments, we observed a quite similar inhibition performance with respect to the subcooling and pressure.

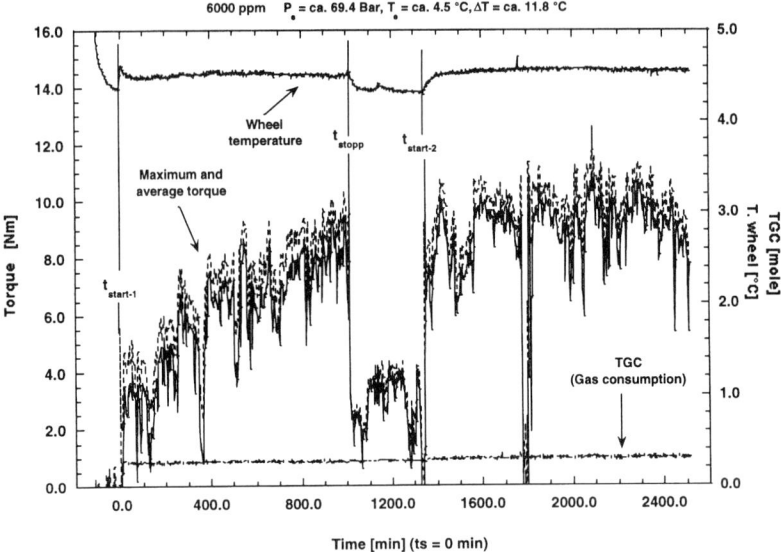

FIGURE 6. Wheel experiment with a kinetic inhibitor polymer blend.

Furthermore, the wheel experiments indicate that shut-ins, and their length, are critical for determining kinetic inhibitor performance, not just flow conditions as reported previously.[8] This may be due to the location of the inhibitor in relation to the water–hydrocarbon interface where hydrate growth is most pronounced; that is, "fresh" inhibitor diffuses slowly to the interface during shut-in, but can move easily during turbulent flow.

CONCLUSIONS

A new class of kinetic hydrate inhibitors has been developed based on alkylacrylamide polymers and related polymers with heteroatom backbones. The molecular weight was shown to be critical for optimal performance. High performance polymers were prepared that exhibit excellent compatibility with water produced at high temperatures. Blends of these polymers with smaller amounts of higher molecular weight vinyl lactam polymers gave significant performance enhancements. Wheel and cell experiments were shown to exhibit similar performance. Performance under stagnant conditions must be investigated if shut-ins are expected in the field application. Absolute performance of the polymers and their blends must be determined for each field application separately since there are a wide range of factors that must be taken into account. Even so, for practical applications the kinetic inhibitors appear to be able to handle a subcooling of 10–12°C, and even greater if the residence time is short.

ACKNOWLEDGMENTS

We thank Statoil, Conoco, BP-Amoco, Mobil, Elf, Norsk Hydro, Agip, TR Oil Services/Clariant and the Norwegian Research Council for their support of this work. We also thank Dr. Mark Rodger for carrying out molecular modelling of the alkylacrylamides, and Petreco for the wheel experiments.

REFERENCES

1. SLOAN, E.D. 1998. Clathrate Hydrates of Natural Gas. Marcel Dekker, New York.
2. ARGO, C.B., R.A. BLAIN, C.G. OSBORNE & I.D. PRIESTLY. 1997. Commercial deployment of low dosage hydrate inhibitors in a Southern North Sea 69 km wet-gas pipeline. Society of Petroleum Engineers International Symposium of Oilfield Chemistry. Paper 37255. Houston, 18–21 February, 1997.
3. PHILLIPS, N. 1998. Case study on ETAP. Controlling Hydrates, Waxes and Asphaltenes, IBC Conference. Oslo, 7–8 December 1998.
4. KELLAND, M.A., T.M. SVARTAAS & L. DYBVIK. 1994. Control of hydrate formation by surfactants and polymers. Society of Petroleum Engineers Annual Technical Conference, Paper 28506. New Orleans, 1994.
5. KELLAND, M.A., T.M. SVARTAAS & L. DYBVIK. 1995. A new generation of gas hydrate inhibitors. SPE 30695. Society of Petroleum Engineers Annual Technical Conference. Dallas 22–25 October 1995.
6. SLOAN, E.D., S. SUBRAMANIAN, P.N. MATTHEWS, J.P. LEDERHOS & A.A. KHOKHAR. 1998. Ind. Eng. Chem. Res. **37:** 3124–3132.
7. LARSEN, R. 1997. Clathrate Hydrate Single Crystals. Ph.D. Thesis, NTNU, Norway.
8. URDAHL, O. & A. LUND. 1998. Assessing the effectiveness of the recently developed kinetic gas hydrate inhibitors versus the hydrate growth inhibitors to mitigate hydrate formation. Hydrates, Waxes and Asphaltenes, IIR Conference. Aberdeen 17–18 June, 1998.

Hydrate Plug Properties

Formation and Removal of Plugs

TORSTEIN AUSTVIK,[a] XIAOYUN LI, AND LARS HENRIK GJERTSEN

Statoil Research Centre, Postuttak, 7005 Trondheim, Norway

ABSTRACT: The properties of hydrate plugs have been demonstrated to vary from case to case. Plugs formed in a gas/HC liquid/water system consist of hydrate particles and pores with gas and liquid constituting both water and HC at different portions. The composition of the gas and liquids trapped in the pores plays a major role in the probability of plugging as well as the behavior of plugs during melting.

INTRODUCTION

Hydrate plugs are formed in wells, pipelines as well as in processing equipment. Consequently, conditions for the formation of hydrates vary significantly. In the North Sea hydrate plugs are most frequently experienced in wells, during well interventions, and topside in processing equipment. However, plugs in pipelines are not rare.

A research program on the formation and removal of hydrate plugs is ongoing at the Statoil R&D Centre. The research is based on results from laboratory experiments, field trials, and field experience. The program combines fundamental and applied research. This paper presents an overview of some of the results and theories generated to date.

CONTENT OF HYDRATE PLUGS

Systematic studies have been carried out on fluid systems ranging from light gas condensates to heavy oils and with varying water-cuts. High pressure experiments are performed in a flow simulator shaped as a rotating wheel. The apparatus and experimental procedures have been described previously.[1]

A total of 44 cases of plugging in the flow simulator have been analyzed. The plugs developed from lumps in the bulk that eventually were sticking to the wall. The portion of water converted to hydrate during the hydrate plug formation was estimated using flash calculations based on temperature and pressure measurements. The water conversion at the time of plugging for the different experiments are compiled in TABLE 1. The water conversion is defined as the ratio of the amount of water in the form of hydrates to the total amount of water in the system. The cage occupancy

---

[a]Telecommunication. Voice: (47) 73584435; fax: (47) 73584628.
TAUS@Statoil.com

TABLE 1. Conversion (%) of water to hydrates at plugging in 44 experiments in the flow simulator

Water conversion at plugging	0–4%	5–8%	9–13%	46%	Total number of plugs (N_t)
Number of plugs (N_p)	34	5	4	1	44
N_p/N_t (%)	77	11	9	2	

of the hydrate lattice was 98% in the simulations. In cases where the hydrate cages have a different occupancy than the model predicts, the amount of water converted to hydrate is changed accordingly. However, the water conversion is still very low.

The majority of the plugs were formed when less than 4% of the water had been converted to hydrates. Had the plugs consisted of only hydrates, the amount of hydrates in the system would be far too small to block the wheel at such low conversions. Visual observations from video recordings, however, show that practically all the liquid water was trapped within the plug pores in all the experiments. In addition, parts of the hydrocarbon liquid phase and some gas were trapped within the pores. Hence, the plugs were porous, consisting of hydrate crystals, liquid water, hydrocarbon liquids, and gas. The length of one plug was measured and an estimate of the plug content was made. It was demonstrated that the plug contained 3 vol% hydrate, 54 vol% water, and 43 vol% oil and gas. The plug was in fact 1.7 times longer than the initial water column in the system. Experiments also demonstrate that the approximate portion of hydrocarbon liquid trapped in the plugs depend on the fluid in question. In some experiments practically all the HC liquid was located within the plug pores, leaving a system of solids and gas. In these systems the pores of the plugs were filled with more oil than water.

A hydrate reactor, depicted in FIGURE 1, is currently used for studies on plug properties. The reactor consists of a 900-mm long test section made of PVC, allowing for visual sighting. The inner pipe diameter is 153 mm and the maximum pressure allowed is 100 bar. Hydrates can be formed by injecting water droplets, gas, and hydrocarbon liquids through nozzles at the top of the test section. The size of the water droplets can be varied by using different types of injection nozzles. The injection rates for both water and HC liquid can be controlled, allowing control of the water to hydrate conversion. Some experiments were performed for which the goal was to obtain as high a water to hydrate conversion as possible. In more than 30 experiments, the water conversion varied between 6% and 71%, with the majority lying below 25%. It should be mentioned that only loose hydrate particles were formed in most of these experiments, and there was no visible free water to glue the hydrate particles together.

An initial attempt at forming hydrate plugs in the hydrate reactor was carried out with a gas mixture of C_1 to iC_4, and tap water in the absence of hydrocarbon liquids. Hydrates were initially formed by injecting water droplets into the gas atmosphere at a pressure of 60 bar. A total of 13.5 liters of water were injected during 1.5 hours at a subcooling of 13°C. Gas was periodically circulated through the plug from the bottom of the reactor and the plug was periodically allowed to mature for several days without disturbance. After a stagnant period four more liters of water were

FIGURE 1. Rig used for experiments on plug properties. The actual test section is indicated by the *arrow*.

injected at the top of the plug. Most of the water was sucked into the plug, working as a sponge, and remained in the plug for the rest of the experiment. Although the entire experiment lasted for 20 days, the plug remained porous. The plug length was 40 cm. When a pressure difference was introduced over the plug the video recordings demonstrated that gas bubbled easily through the internal channels in the plug. Hence, the nature of the channels determined the permeability of the plug. Gas bubbled through the channels until more hydrates were formed and blocked further migration. The gas then found new channels until these were also blocked and the plug gradually became less permeable. By measuring the gas released during melting and the water volume after complete dissociation, the plug content was calculated. The volume percentage of hydrates, free water, and free gas in the plug was 12%, 38%, and 50%, respectively. Hence, the hydrate content in the plug was low and the porosity of the plug was 88%.

Based on numerous experiments in different systems it can be concluded that hydrate plugs generally consist of small particles with pores in between. The only exception is probably hydrates formed in gas systems with little water in which the hydrates may grow directly on the pipe wall forming ice-like layers. Large ice-like crystals never formed in experiments with a HC liquid phase present. The results indicate that the water phase located in between the hydrate particles plays a crucial role, creating a glue-like effect. When lumps and plugs are formed almost all the free water seems to disappear into the pores of the hydrates. Water bridges formed between the hydrophilic particles and capillary forces created by the different gases and liquids are believed to be the dominating factors gluing the hydrates together to form plugs.

FORMATION OF PLUGS

The events taking place between hydrate formation and plugging differ from case to case. For example, plug formation in wells is significantly different from plugging in pipelines at flowing conditions. Nevertheless, the mechanisms of agglomeration of hydrate particles forming a porous hydrate mass seem to be comparable.[2]

All the experiments carried out with gas/water/oil in the flow simulator demonstrate a similar macroscopic development during hydrate formation. In the *initial stage* hydrate crystals are suspended in the water, usually as small particles or sometimes transported as hydrate covered droplets. Hydrate films on droplets, however, are usually broken into particles at an early stage, allowing water to come into contact with surrounding hydrocarbons. The agglomeration tendency at this stage is low. When more water is converted to hydrate the *first intermediate stage*, dominated by agglomeration of sticky particles is reached. At this stage the system is dominated by the stickiness between the particles forming lumps in the bulk, by the stickiness to the pipe wall causing hydrate layers to develop, or by a combination of these effects. In some systems deposits form sufficiently rapidly at the intermediate stage that this causes all the hydrate particles to deposit without forming transportable lumps. These hydrate layers normally trap most of the free water in the pores of the deposits, preventing further hydrate formation downstream. It is believed that the wetting condition at the pipe wall is an important parameter in determining the stickiness to the wall. The hydrate particles themselves are strongly hydrophilic. Hydrophobic conditions on the pipe wall prevent efficient water bridging to the wall and thus reduces the formation of deposits. Most of the plugs are formed either in the first intermediate stage or early in this *agglomerated stage*. When hydrate lumps are transported in the system hydrate formation occurs on the surface of the lumps. The gluing effect of the water phase is then lost. The reduced water wetting near the surface introduce a flux of water from the interior of the lump to the surface due to capillary forces. Internal pressure changes and loss of water bridging cause a gradual disintegration of the plugs in the *second intermediate stage*. A system in which most of the water has been converted into hydrates normally forms a fine powder that is easily transported in HC liquids, the *disintegrated stage*. The stickiness between the particles and of particles to the pipe wall is then lost.

Most systems tested do not form lumps when the system is completely at rest. A thin hydrate film is initially formed preventing further contact between the hydrocarbon phase and water. FIGURE 2 shows the formation of a lump at shut in conditions in a North Sea oil flow simulator. Time-lapse video recordings showed that hydrates were formed in a dispersion at the water-hydrocarbon interphase, gradually pushing the lump into the water phase. Water was drawn through the capillaries of the lump and into the HC. After a period of six hours a large portion of water phase was filled with hydrates. At restart the lump initially broke into smaller lumps, that then rapidly

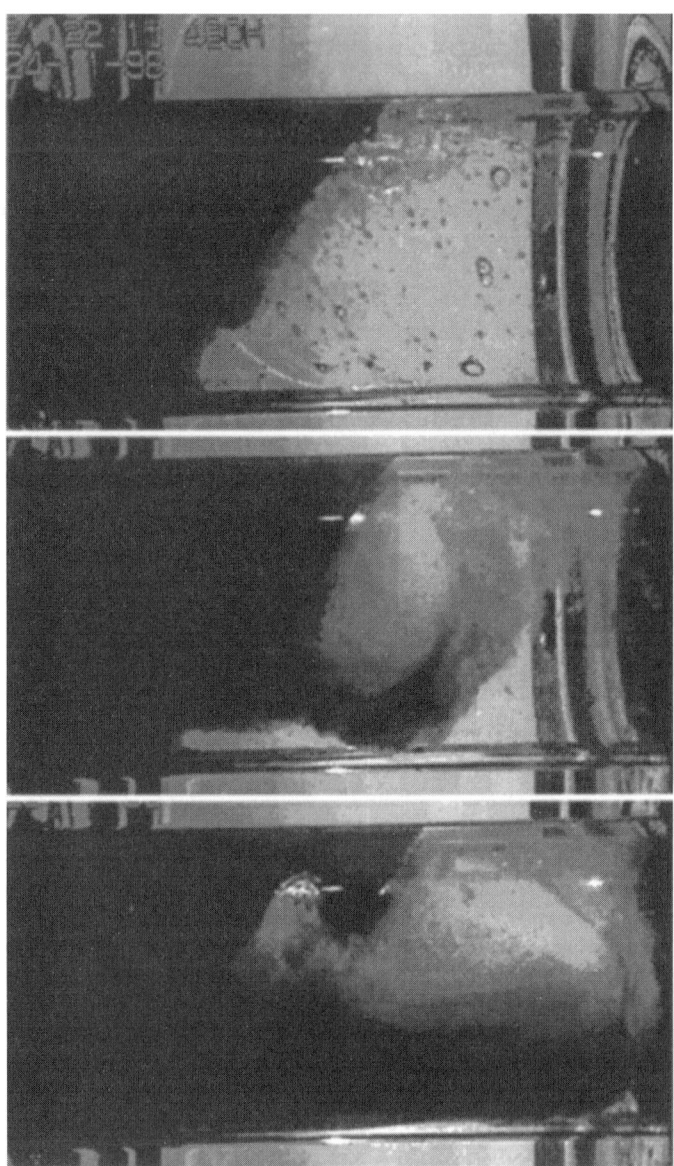

FIGURE 2. Formation of hydrate lumps under stagnant conditions. Before formation is initiated (*upper*), after 1 hour (*middle*), and after 4 hours 40 minutes (*lower*). Eventually most of the water in the system is trapped within a hydrate lump.

agglomerated under formation of a plug. At that time practically all the HC liquid as well as water was trapped within the pores of the plug. The experiment was repeated twice with a similar development, demonstrating that the special behavior was due to the composition of the fluids tested. The percentage of water converted to hydrate during the stagnant period was calculated to be 5.2% and 5.3% for the two experiments.

The hydrate behavior is, to a large extent, determined by the nature of the liquid phases filling the pores of the hydrate mass. The gas composition obviously affects the equilibrium conditions but does not significantly influence the hydrate behavior as such. In gas/condensate systems deposits as well as lumps in the bulk phase are typically formed during the first intermediate stage. The plugging tendency is normally high. Recall that the behavior varies among different condensates. Practically all the water is trapped between the hydrate particles at the agglomerated stage, however, a large portion of the condensate phase is not trapped.

The variation in behavior between different oil systems is more significant than for condensates. Some oil systems even form transportable hydrates without forming larger lumps in the bulk or hydrate layers on the pipe wall. Other systems rapidly form sticky hydrates in which practically all the water and liquid HC are trapped in the pores. Plugs formed in oil systems are often longer than they are in condensates as a result of the fact that the amount of liquid HC incorporated in the pores is frequently significantly larger. Due to the low percentage of water converted to hydrates when the first intermediate stage is reached, the plugging sometimes occurs rapidly after initial hydrate formation. The oil systems usually form less deposits than condensate systems but more sticky lumps in the bulk. The behavior depends on a number of parameters, such as the presence of surface active components, waxes, and asphaltenes. The surface active components in some systems seem to cause the formation of particles with hydrophobic surfaces that are transported through the system, in the HC liquid phase, without agglomeration to other particles or to the wall. In other systems with hydrophobic hydrate particles there seems to be a gluing effect caused by the formation of oil bridges between the particles. Some emulsified systems are blocked during hydrate formation under stagnant conditions, with all the emulsified droplets covered with hydrates.

Low contents of salts, methanol, or glycols in the water phase have been found to increase the stickiness of the hydrates formed and thus to increase the probability of plugging. This was observed for methanol in a field trial at Tommeliten and later verified in the laboratory.[3,4] Subsequent field experiences in the North Sea seem to support this conclusion.

PLUG REMOVAL

For plug removal the conditions within the plug are of great importance. Processes in the interior of the plug may have a substantial influence on the plug characteristics. Fluid slowly transported in the channels during pressure changes, or simply due to capillary forces, may cause more hydrates to form. The pressure changes may also cause gas to be released from oil located in the pores. During melting the temperature drops locally to the equilibrium temperature of the hydrates. An experiment carried out in the apparatus described in Reference 5 resulted in a temperature of −10°C, locally in the melting zone, in contact with a MEG/water mixture. This may, for example, cause wax precipitation within the pores. Experiments indicate that the low temperatures may result in viscosities of glycols that are sufficiently high as to practically prevent further melting.[6]

The porosity and permeability of plugs often vary considerably along the pipe length as well as over time. Pockets without hydrates are often observed between hydrate plugs in the field. This can be detected by sudden changes in the pressure during the melting procedure when parts of the plug have melted. Even more conclusive is the detection in wells in cases where tools are used to determine the position of the plug. A heating tool developed by Petroleum Engineering Services (PES), shown in FIGURE 3, was used to melt a plug in a production well in the North Sea. The tool was brought to the top of the plug by a wireline arrangement. The temperature at the lower section of the tool was in the range 90–95°C. A total of six plugs were melted, each with a length between 1 m and 2 m. There were pockets with lengths of the order of 1 m in between the plugs. The total melting time was 20 hours, demonstrating that even at such high temperatures removal of plugs is a tedious process. It should, however, be remembered that the length of the melting zone is important for the melting efficiency. Using a heating tool from the end of the plug results in a relatively small melting zone.

Field experience has demonstrated that the efficiency of thermodynamic inhibitors in plug melting varies significantly. On several occasions the time needed to

FIGURE 3. Heating tool (PES) used for removal of hydrate plugs in wells.

remove plugs in wells using methanol was three to four days, whereas methanol in other situations had no noticeable effect. In another well the melting rate was approximately 0.5 m/hour using MEG. If the inhibitor does not properly penetrate the plug, the melting zone becomes short, resulting in poor melting. When methanol is used in vertical columns, water released during melting may form a film, preventing further contact between the inhibitor and hydrates. Even though methanol and water are completely miscible, significant mechanical disturbance may be required to achieve good mixing.[6] In cases where the inhibitor efficiently penetrates the plug, the water content of the pores of the plug tend to dilute the inhibitor, again reducing melting efficiency. Efficient melting occurs when a good flow of inhibitor is obtained, constantly bringing concentrated inhibitor into the plug.

Methanol was initially used to melt a plug in a gas injection well in the North Sea. There was no liquid HC in the well. The melting efficiency was measured by monitoring the pressure development at the top of the well. No noticeable effect was detected from the pressure readings. The pressure readings also indicated a low permeable plug. Glycol (MEG) was then injected without significant effect. After thorough safety evaluations it was determined that melting by depressurization was applicable due to the special geometry of the well. The pressure was reduced to atmospheric pressure at the top of the well with a static pressure of about 15 bar at the top of the plug. The pressure beneath the plug was 195 bar. The pressure at the Xmas tree increased to 6.4 bar within a few minutes and stabilized. FIGURE 4 shows the pressure development. The procedure was repeated several times during the next 12 hours with a similar response from the plug. Gas was thus released during melting causing the pressure to increase. When the equilibrium pressure of the hydrates was reached the melting stopped, resulting in the development seen in FIGURE 4. After 12 hours following the described procedure the pressure gradually increased until the pressure over the plug equalized (see FIGURE 5). Fragments of the plug coming

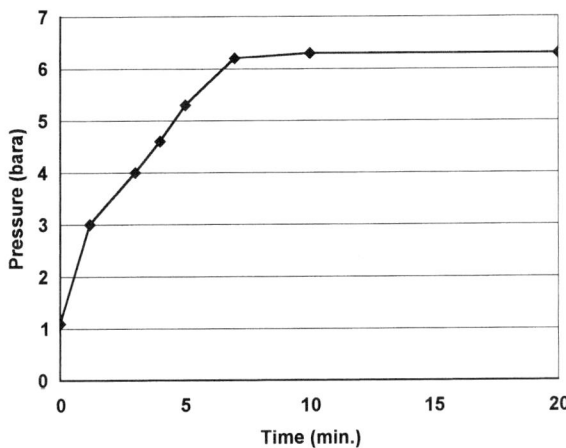

FIGURE 4. Pressure development at Xmas tree after depressurization to atmospheric pressure.

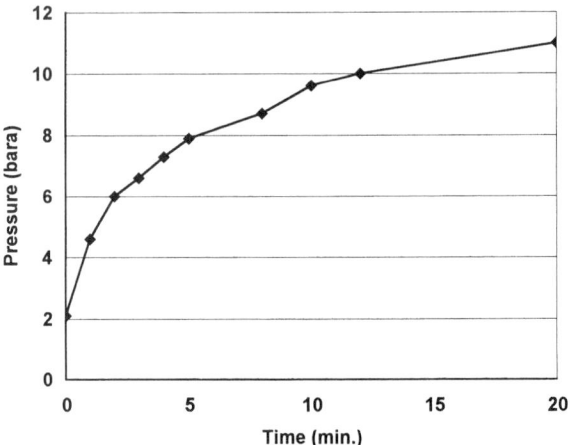

FIGURE 5. Pressure development at the Xmas tree when communication is obtained over the plug.

up through the well could be heard. It is interesting to note that the equilibrium pressure reached periodically during the melting procedure was as low as 6–7 bar. This pressure is, with the addition of the static pressure, approximately consistent with the equilibrium pressure of the hydrates in contact with pure water. The fact that the well above the plug is filled with inhibitor did not seem to have an impact on the behavior.

Depressurization is usually very efficient for plug melting. FIGURE 6 shows a picture of a plug taken with a thermocamera during depressurization from both ends. The plug was located topside at Edda in the Tommeliten system during a field trial.[3] The entire plug was colder (dark color) than the surroundings showing that melting occurred along the entire length of the plug. Gas flowing through plugs during one-sided depressurization, however, may cool the plug sufficiently, due to expansion, so

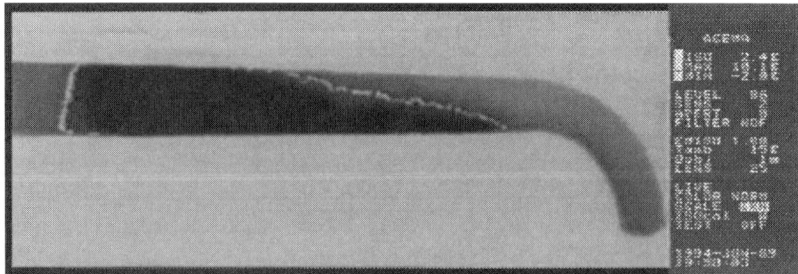

FIGURE 6. Hydrate plug visualized through a thermocamera topside at Edd (Tommeliten pipe). *Dark colors* represent colder areas.

as to minimize the length of the melting zone or even to completely hinder melting.[3] This is especially true for high GOR systems. Laboratory experiments with a pressure drop of 30 bar over a 0.5 m long plug resulted in a downstream temperature of $-34°C$.[5] However, field experience indicates that plugs in pure gas/water systems are often very compact with low through-flow. In systems with a high content of a HC liquid the cooling effect does not normally seem to predominate. The plug in a subsea pipeline at Gullfaks described in Reference 7 is an example of a system partly containing oil, in which one-sided depressurization was successfully carried out.

CONCLUSIONS

Hydrate plugs are frequently extremely complex, with a number of parameters influencing their behavior. The plugs are normally porous containing hydrate particles, liquid water, liquid HC, and gas. Most of the water in the system is trapped between the hydrate particles. To large extent the composition of the HC liquid and water phase, such as the salinity, determines the hydrate behavior. The majority of the plugs formed in the wheel shaped flow simulator had a hydrate content below 4%.

Gas/condensate systems formed deposits on the pipe wall as well as lumps in the bulk. The probability of plugging was high. Gas/oil systems may produce transportable powder-like hydrates in systems that form sticky plugs shortly after hydrate formation is initiated.

The complex nature of hydrate plugs causes a large variation in melting efficiency for different methods. Methods that are efficient for the removal of one plug may be without significant effect when used on other plugs. Each plug hence has to be individually treated, based on the responses learned during the procedures of removal.

REFERENCES

1. URDAHL, O., A. LUND, P. MØRK & T.N. NILSEN. 1995. Inhibition of gas hydrate formation by means of chemical additives. I: The development of an experimental set-up for characterization of hydrate inhibitor efficiency with respect to flow properties and deposition. Chem. Eng. Sci. **50**(5): 863–870.
2. AUSTVIK, T. 1992. Hydrate Formation and Behavior in Pipes. Dr. Ing. Thesis. NTH, Norway.
3. AUSTVIK, T., E. HUSTVEDT, L.H. GJERTSEN & O. URDAHL. 1997. Formation and removal of hydrate plugs: field trial at Tommeliten. Proceedings of GPA Annual Conference. San Antonio, USA.
4. GJERTSEN, L.H., T. AUSTVIK & O. URDAHL. 1996. Hydrate plugging in underinhibited systems. Proceedings of the Second Int. Conf. on Natural Gas Hydrates. Toulouse, France.
5. BORTHNE, G., L. BERGE, T. AUSTVIK & L.H. GJERTSEN. 1996. Gas flow cooling effect in hydrate plug experiments. Proceedings of the Second Int. Conf. on Natural Gas Hydrates. Toulouse, France.
6. LI, X., L.H. GJERTSEN & T. AUSTVIK. 1999. Thermodynamic inhibitors for hydrate plug melting. Ann. N.Y. Acad. Sci. **912**: this volume.
7. GJERTSEN, L.H., T. AUSTVIK, O. URDAHL & R. DUUS. 1997. Removal of a gas hydrate plug from a subsea multiphase pipeline in the North Sea. Multiphase '97. 515–526.

Hydrate Dissociation in Pipelines by Two-Sided Depressurization

Experiment and Model

DAVID PETERS,[a] M. SAMI SELIM, AND E. DENDY SLOAN, JR.[b]

Center for Hydrate Research, Colorado School of Mines, Golden, Colorado 80401, USA

ABSTRACT: Experimental data were obtained on the dissociation of short methane hydrate plugs in a simulated pipeline. The hydrate plugs were dissociated by the method of two-sided depressurization. Results indicated that plug dissociation occurred radially and not axially. This results in extreme safety concerns, listed herein. When the system was depressurized to atmospheric pressure, ice was formed from the dissociating hydrate plug, which aided in the dissociation process. A model describing hydrate dissociation assumes that heat is conducted radially into the plug from the surroundings. The model is in quantitative agreement with the data using no fitted parameters. A rapid pressure reduction to atmospheric pressure on both ends of the hydrate plug leads to the optimal dissociation rate.

INTRODUCTION

The purpose of this work was to investigate the phenomena of hydrate dissociation in pipelines. A model was developed that simulates the dissociation of hydrates in pipelines. An experiment was devised to verify and obtain data to correct and to refine the model. The goal of this work was to develop an optimal strategy for the removal of hydrate plugs in pipelines when depressurizing from both ends of the plug.

MATHEMATICAL MODEL

The mathematical model developed in this work was an extension of previous modeling, by Kelkar *et al.*,[1] to a finite media and using cylindrical coordinates. The resulting system of equations required a numerical solution.

In this model, the hydrate and subsequent ice plug were assumed to be porous. Recent laboratory work by Lysne[2] has shown that hydrate porosity varies between 33% and 84%. Later field experiments confirmed that hydrates plugs in pipelines were also porous.[3] Since hydrate plugs are porous, they are able to transmit pressure throughout the plug while acting as an impediment to normal pipeline flow.

---

[a]Present address: Shell E&P Technology Co., 3333 Hwy. 6, S. Houston, TX 77082, USA.
[b]Telecommunication. Voice: 303-273-3723; fax: 303-273-3730.
esloan@gashydrate.mines.edu

The temperature of the hydrate during dissociation was assumed to be constant throughout the plug and in equilibrium with the prevailing pressure. If the system pressure is decreased below about 2.7 MPa, the equilibrium temperature of hydrate is below the ice point and allows for the possibility of ice formation. Once the equilibrium temperature is below the ice point any water formed from dissociating hydrate is quickly converted to ice.

In this model, the pressure is reduced in a step change on either side of the hydrate plug. The resulting hydrate equilibrium temperature is below that of the surroundings, causing heat to flow radially inward to melt the hydrate. It was demonstrated experimentally (as is shown later) that the hydrate dissociated radially and that all axial dissociation could be neglected. FIGURE 1 shows a schematic of the hydrate dissociation model.

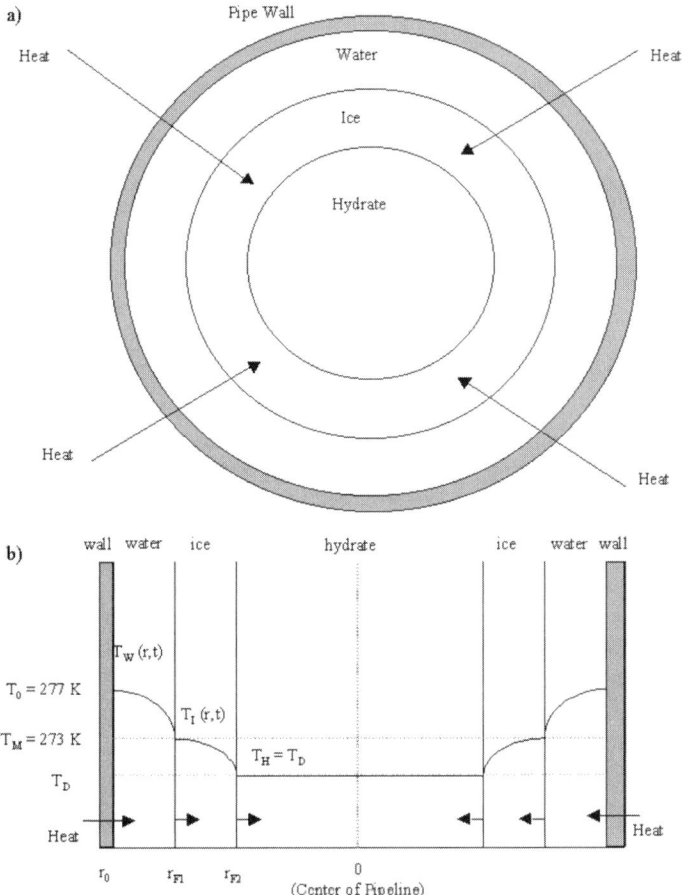

FIGURE 1. Hydrate dissociation schematic with water, ice, and hydrate present. (**a**) Cross-section of pipeline with all three phases present and radial heat flow into system. (**b**) Temperature profiles in each phase and moving boundary motion, constant wall temperature, and water/ice interface temperature.

FIGURE 1a shows an inner hydrate core surrounded by an ice layer, which is enclosed in a water layer adjacent to the pipe wall. Both the hydrate and ice layer shrink as a function of time until only water is present in the pipeline. FIGURE 1 represents the case when ice is formed during dissociation. The schematic for the case when no ice is present is similar, except the ice layer is removed and only water and hydrate are present.

FIGURE 1b shows a representation of the temperature profiles in the water, ice and hydrates layers. The temperature profiles are determined according to Fourier's law of heat conduction in cylindrical coordinates.

$$\frac{\partial T_W}{\partial t} = \alpha_W \left[\frac{1}{r}\frac{\partial T_W}{\partial r} + \frac{\partial^2 T_W}{\partial r^2} \right], \quad r_{F1} < r < r_0, \quad t > 0. \tag{1}$$

$$\frac{\partial T_I}{\partial t} = \alpha_I \left[\frac{1}{r}\frac{\partial T_I}{\partial r} + \frac{\partial^2 T_I}{\partial r^2} \right], \quad r_{F2} < r < r_{F1}, \quad t > 0. \tag{2}$$

The boundary conditions for this system are:

$$T_W = T_0, \quad r = r_0, \quad t > 0. \tag{3}$$

$$-k_W \frac{\partial T_W}{\partial r} = -k_I \frac{\partial T_I}{\partial r} + (1-\varepsilon)\rho_I \lambda_I \frac{dr_{F1}}{dt}, \quad r = r_{F1}, \quad t > 0. \tag{4}$$

$$T_W = T_I = T_M, \quad r = r_{F1}, \quad t > 0. \tag{5}$$

$$-k_I \frac{\partial T_I}{\partial r} = (1-\varepsilon)\rho_H \lambda_H \frac{dr_{F2}}{dt}, \quad r = r_{F2}, \quad t > 0. \tag{6}$$

$$T_I = T_D, \quad r = r_{F2}, \quad t > 0. \tag{7}$$

Boundary conditions (3), (5), and (7) are due to constant temperatures at the pipe wall, the water–ice interface, and the ice–hydrate interface, respectively. At the water–ice interface ($r = r_{F1}$), boundary condition (4) indicates that the heat conducted through the water layer is equal to the heat conducted into the ice, as well as heat to melt the ice. At the ice–hydrate interface ($r = r_{F2}$), boundary condition (6) equates the heat conducted through the ice layer to the heat to dissociate the hydrate.

For the case when the hydrate temperature is above the ice point and no ice is present, Equation (1) still determines the water temperature, but the boundary conditions change the following.

$$T_W = T_0, \quad r = r_0, \quad t > 0. \tag{8}$$

$$-k_W \frac{\partial T_W}{\partial r} = (1-\varepsilon)\rho_H \lambda_H \frac{dr_{F1}}{dt}, \quad r = r_{F1}, \quad t > 0. \tag{9}$$

$$T_W = T_D, \quad r = r_{F1}, \quad t > 0. \tag{10}$$

Boundary conditions (8) and (10) are due to constant temperatures at the pipe wall and the water–hydrate interface, respectively. Boundary condition (9) equates the heat conducted through the water layer to the heat to dissociate the hydrate.

There was no analytical solution to either system of equations and therefore, they had to be solved numerically, using a finite difference scheme. The main difficulty in solving these problems lies in the incorporation of the nonlinear boundary conditions **(4)**, **(6)**, and **(9)**.

EXPERIMENTAL PROCEDURE

FIGURE 2 is a schematic of the experimental apparatus. The experimental apparatus consisted of a 350 ml stainless steel reactor. The reactor was 0.2 m (8″) long and had an internal diameter of 0.048 m ($1^7/_8$″). The reactor was located in a temperature-controlled ethylene glycol/water bath. The bath temperature was monitored using a platinum resistance temperature detector. The temperature of the hydrate was monitored at five different axial positions within the reactor with type T thermocouples. The temperature of the gas exiting both ends of the reactor was also measured using type T thermocouples. The pressure in the system was monitored with a pressure transducer. All temperature and pressure readings were continuously monitored and stored on a personal computer.

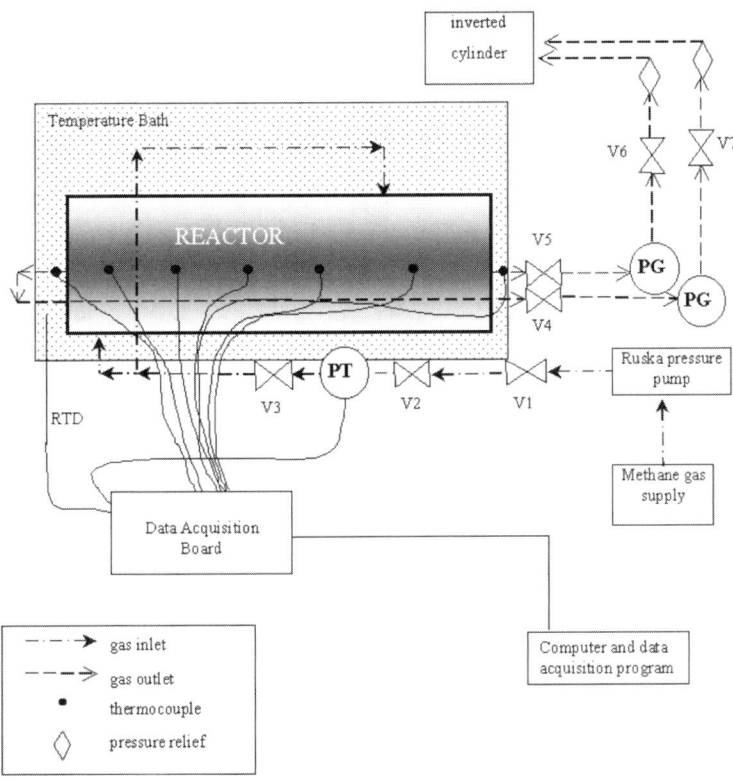

FIGURE 2. Schematic diagram of experimental apparatus.

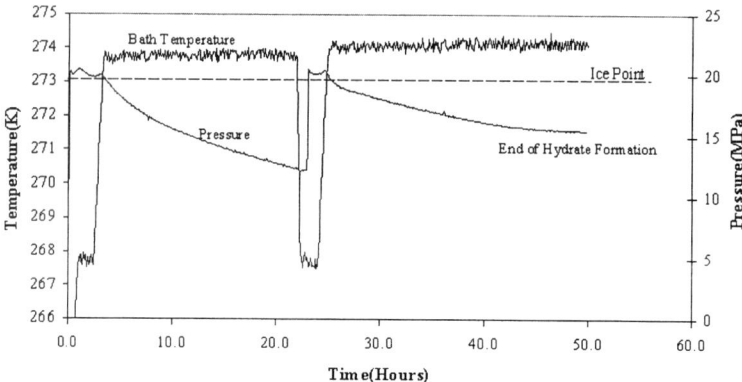

FIGURE 3. Formation of a methane hydrate plug. The pressure drop was used as an indication of hydrate formation. When the pressure stopped decreasing, that was considered the end of hydrate formation, and the dissociation experiments began just after that time.

The hydrate formation procedure was modified from the method developed by Stern et al.[4] In this procedure, 850-μm particles of ice were loaded into the stainless steel reactor. Once the reactor was loaded with ice, it was pressurized with methane gas to about 21 MPa. The temperature of the reactor was then increased above the ice point to initiate hydrate formation. The amount of hydrate conversion was monitored by measuring the pressure drop in the system (see FIGURE 3). Once the pressure stopped decreasing, the conversion of ice to hydrate was considered to have finished.

The dissociation of the hydrate plug was initiated by decreasing the pressure at both ends of the plug. The hydrate temperature was initially at the bath temperature (about 4°C). Once the pressure was reduced to a value below the equilibrium pressure at the bath temperature, the hydrate began to dissociate. As the hydrate was dissociating, several items were measured; the amount of methane gas released from the dissociating hydrate (using a water displacement device), and the temperature of the hydrate at the center of the plug at five different axial positions.

In the dissociation experiments, it was possible to maintain a constant dissociation pressure for the duration of the experiment at either atmospheric pressure or a pressure in the range of 2.4 to 5.2 MPa. The elevated pressure was maintained with a pressure relief valve.

RESULTS AND DISCUSSION

The formation experiment results indicate that it was possible to make a reproducible hydrate plug. In each experiment, approximately 90% of the ice was converted to hydrate. This was less than the conversion reported by Stern et al.,[4] because the size of ice used to form hydrates in this work was larger, due to limitations in the experimental setup.

The model indicates the optimal dissociation pressure is the lowest possible. As the pressure was decreased, the total dissociation time (time to remove ice and hydrate) also continued to decrease. This result indicates that the formation of ice helps to remove a hydrate plug—the more ice that formed, the better. There are two possible reasons for this: (1) the thermal diffusivity of ice is an order of magnitude higher than that of water, which means the transfer of heat through the ice layer is more efficient, relative to the water layer, and (2) with ice present, there is a higher thermal gradient for heat conduction into the system.

This suggests that when removing a hydrate plug from a pipeline, it is best to depressurize the pipeline as quickly as possible and to as low of a pressure as possible. When the pressure is decreased quickly it results in a Joule-Thomson cooling of the gas, which further cools the hydrate and leads to ice formation.[5] There is some question concerning the effect of ice formation in the dissociation process. It has been previously suggested that the formation of ice lead to a hydrate self-preservation effect.[6] This effect is the result of ice forming an impermeable barrier that prevents the release of gas. The conclusions for this paper are valid if the ice shell is porous and allows gas to escape from the dissociating hydrate. There is some discrepancy concerning the effect of ice formation during dissociation requires more experimentation.

An important result to come from the dissociation experiments was the verification of the assumption of radial dissociation. The benefit of a radial dissociation model is that the plug length is irrelevant, only the plug radius is important. FIGURE 4 shows three different dissociation experiments. Each hydrate plug was allowed to dissociate for predetermined time and then the reactor was dismantled to observe the remaining hydrates. These pictures clearly show the radial dissociation of hydrates.

The experiments also indicate that when the dissociation pressure results in an equilibrium temperature above the ice point, the hydrate remains at this equilibrium temperature for the duration of the experiment (see FIGURE 5). However, when the dissociation pressure results in an equilibrium temperature below the ice point, and ice forms in the system, the temperature of the hydrate is held at, or very near, the ice point (see FIGURE 6). Only one temperature profile is included in FIGURES 5 and 6, but all other thermocouples followed this same trend. The formation of ice from dissociating hydrate buffered the solid temperature to between 0°C and −1°C in most cases, as was first reported by Lysne.[2]

The results of the dissociation experiments agree with the model (see FIGURE 7). Both the experiment and the model predict the fastest dissociation time at lower dissociation temperatures. The trends in dissociation time are observed in both the model and the experiment.

An additional comparison can be made between the experiment and the model by comparing the amount of gas evolved from the dissociating hydrate (FIGURE 8). The predictions can be converted to the amount of gas evolved and compared to the experimentally determined gas evolution. Both the model and the experiment predict very similar dissociation curves. Note that the predicted dissociation time was within about 5% of the experimental dissociation time.

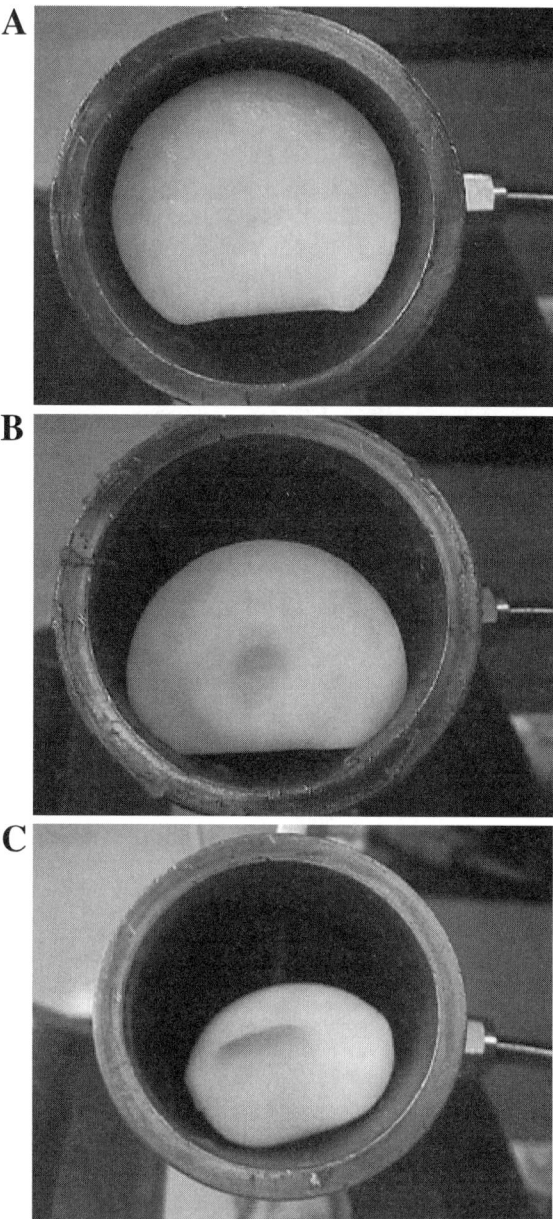

FIGURE 4. Time sequence of dissociating hydrate plugs. Each picture is of a different dissociation experiment. In each experiment, the hydrate plug was allowed to form normally, but during dissociation, each experiment was taken apart at some predetermined amount of time and photographed; (**A**) 1 hour, (**B**) 2 hours, and (**C**) 3 hours. It is important to note that in (**C**), approximately 95% of the hydrate had dissociated so the remaining solid was an ice plug. Also, note in (**C**) that there is only a limited degree of axial dissociation, which further verifies this assumption in the model.

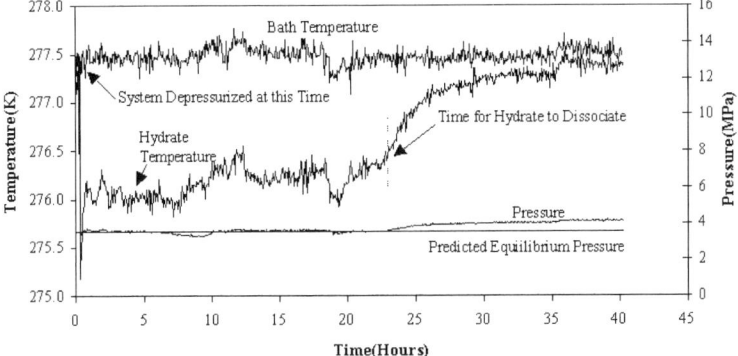

FIGURE 5. Hydrate dissociation experiment when no ice was formed. The pressure and temperatures both drop simultaneously. The time for hydrate dissociation is determined by the deflection in temperature and by monitoring the end of gas evolution. The predicted equilibrium pressure is based on the average hydrate temperature before hydrate completely dissociates.

This generalization of hydrate plug dissociation is only valid for a system with a solid hydrate plug. The dissociation characteristics of a plug with occluded water or liquid hydrocarbon may change the dissociation mechanism considerably. More work needs to be done in this area to determine how this affects hydrate dissociation.

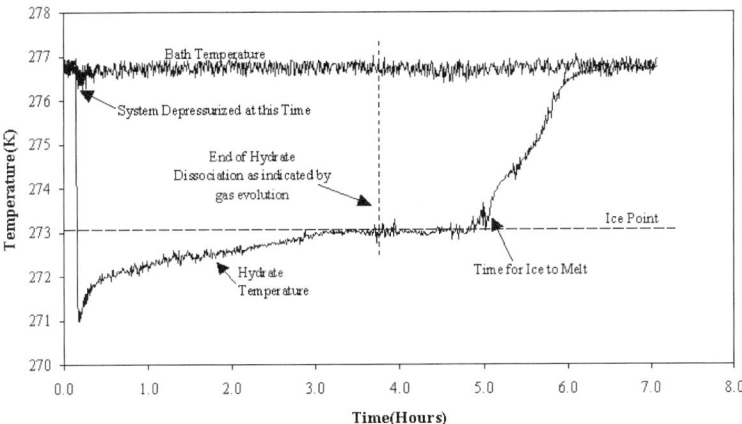

FIGURE 6. Hydrate dissociation experiment with atmospheric final pressure. Temperature decrease occurs at the same time as depressurization. Note the pressure curve is not shown, but is similar to the pressure curve in FIGURE 5 except that the final pressure is atmospheric. The hydrate temperature is not in equilibrium with the pressure (at atmospheric pressure, the equilibrium temperature is −81°C), as in FIGURE 5. The formation of ice buffers the hydrate temperature to remain near the ice point.[2]

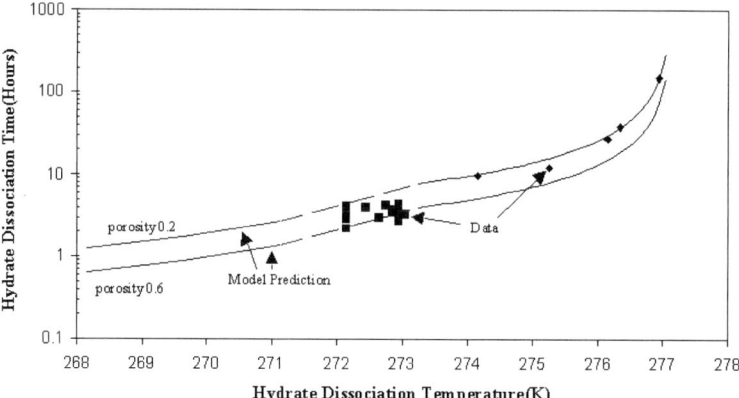

FIGURE 7. Comparison of hydrate dissociation model with data. The model used a range of porosities between 0.2 and 0.6. This range brackets all of the experimental data, which had porosities between 0.3 and 0.5.

SAFETY CONCERNS RELATED TO HYDRATE DISSOCIATION

A significant result of this work indicates that when hydrates dissociate in a pipeline, they do so radially. This implies that as the hydrate plug dissociates, it releases from the pipe wall first and then continues to shrink inward. A concern resulting from this dissociation mechanism is that any pressure gradient across the ends of the plug will result in the movement of the hydrate plug as soon as the plug

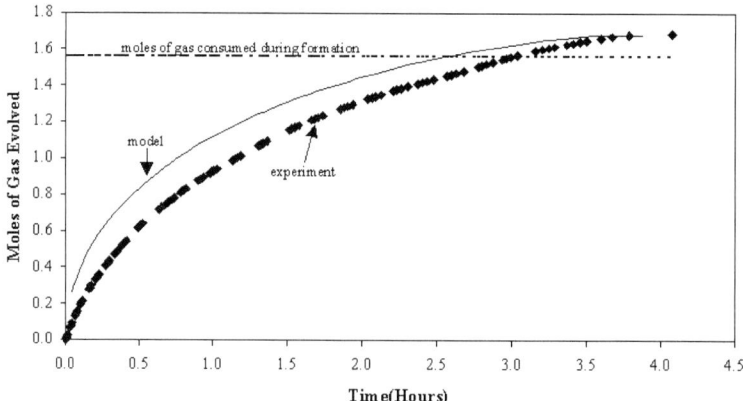

FIGURE 8. Comparison of gas evolved between experiment and model. Note that numbers of the moles of gas consumed during formation was within about 10% of the number of moles evolved during gas evolution (or hydrate dissociation).

is released from the pipe wall. Several cases have been reported, where a plug was depressurized from one end and the plug subsequently broke free and caused millions of dollars in damage and loss of life.[7] Field tests have shown that during hydrate dissociation, a hydrate plug may reach velocities up to 83 m/s when depressurized from one side.[8]

CONCLUSIONS

A model was developed with no fitted parameters that gives an *a priori* prediction of dissociation times of hydrate plugs in a simulated pipeline. The model verified experimental results. During dissociation, the formation of ice was beneficial to the dissociation of hydrate. There is some question concerning the effect of ice formation in a real system and this issue needs to be addressed in future research.

With the results from this work, it is possible to present a short, hydrate dissociation scenario. The model and experiments indicated that ice formation is beneficial and aids in the removal of hydrates. This implies that the pressure should be reduced as quickly as possible to atmospheric pressure on both ends of the plug. However, for safety reasons, such as the existence of multiple plugs, the pressure must be safely reduced to atmospheric pressure on both ends of the plug. Care should be taken to prevent any pressure gradient across the plug. The results of this work are for the case of two-sided depressurization. Current research is underway to extend the experiment and model to simulate one-sided depressurization.

ACKNOWLEDGMENTS

The authors wish to acknowledge the financial support provided by the DeepStar consortium of 22 energy companies.

REFERENCES

1. KELKAR, S.K., M.S. SELIM & E.D. SLOAN. 1998. Hydrate dissociation rates in pipelines. Fluid Phase Equilibria. **150–151:** 371.
2. LYSNE, D. 1995. An Experimental Study of Hydrate Plug Dissociation by Pressure Reduction. Ph.D. Thesis. Norwegian Institute of Technology, University of Trondheim.
3. AUSTVIK, T. *et al.* 1996. Tommeliten gamma field hydrate experiments. 7th International Conference on Multiphase Flow, Cannes. A. Wilson, Ed.: 539–552.
4. STERN, L.A., S.H. KIRBY & W.B. DURHAM. 1996. Peculiarities of methane clathrate hydrate formation and solid-state deformation, including possible superheating of water ice. Science **273:** 1843.
5. BORTHNE, G., L. BERGE, T. AUSTVIK & L. GJERTSEN. 1996. Gas flow cooling effect in hydrate plug experiments. *In* Proceedings of the 2nd International Conference on Natural Gas Hydrates, Toulouse, 2–6 June 1996. J.P. Monfort, Ed.: 381.
6. ERSHOVE, E.D. & V.S. YAKUSHEV. 1992. Experimental research on gas hydrate decomposition in frozen rocks. Cold Regions Science and Technology **20:** 147.
7. KENT, R.P. & M.E. COOLEN. 1992. Hydrate in natural gas lines. Mobil Internal Report.
8. HATTON, G.J., A.M. BARAJAS & C.A. KUHL. 1997. Hydrate Plug Decomposition Test Program, Final Report, SwRI Project 04-8217, October 1997, Prepared for DeepStar 3204 Subcommittee.

Comparison of Laboratory Results on Hydrate Induction Rates in a THF Rig, High-Pressure Rocking Cell, Miniloop, and Large Flowloop

LARRY D. TALLEY,[a] GARRICK F. MITCHELL, AND RUSSELL H. OELFKE

Exxon Production Research Company, P.O. Box 2189, Houston, Texas 77252-2189, USA

ABSTRACT: This paper compares experimental data obtained on three high-pressure devices and one atmospheric apparatus that measure hydrate formation onset and dissociation temperatures. The high-pressure devices are rocking sapphire tubes, a 0.5-inch diameter miniloop, and a 4-inch diameter flowloop. High-pressure miniloop results are compared to atmospheric pressure, tetrahydrofuran (THF) rig results in which chemically similar inhibitors ranked in different order. Although high-pressure, stirred tank apparatus is considered by many to be effective in obtaining data of this kind, this paper does not include any stirred-tank data. Many kinetics experiments are insensitive to the high-pressure apparatus used. However, results of kinetic experiments obtained in different types of screening apparatus may not agree if the methods of hydrate detection are different. An example of a gas/condensate/brine system that would be difficult to characterize in a rocking-cell or stirred-tank apparatus is discussed.

INTRODUCTION

At the first International Conference on Natural Gas Hydrates, important issues and questions were posed concerning the measurement of hydrate formation rates. Sloan[1] noted that the state of the art in hydrate research did not permit quantitative determination of hydrate formation rates inside the hydrate stability envelope. Meyer *et al.*[2] posed several critical questions including: "How should kinetic data be taken on the early stages of hydrate nucleation so that the results are clearly interpretable and can be translated to gas pipeline conditions?"

This paper discusses advances in the determination of hydrate induction rates in the presence of kinetic hydrate inhibitors. The results presented here suggest that data from a wide variety of laboratory tests can yield the same hydrate induction kinetics.

NOTATION

T_{eq}, hydrate equilibrium (dissociation) temperature (°F).
T_{onset}, hydrate onset temperature (°F).

[a]Telecommunication. Voice: 713-431-4406; fax: 713-431-6180.
ldtalley@mindspring.com

t_{ind}, induction time (hours).
T_{exp}, experimental temperature (°F).

EXPERIMENTAL ISSUES

Hydrate Detection

The first issue in designing a hydrate induction rate experiment is to establish how hydrates will be detected. The following four events were commonly used by presenters at the Second International Conference on Natural Gas Hydrates to establish the onset of hydrates. (For example, see Gaillard et al.,[3] Narita and Uchida,[4] Svartas et al.,[5] Lekvam and Ruoff,[6] Tohidi et al.,[7] and Herri et al.[8])

1. Detection of a gas volume decrease in excess of vapor-liquid equilibration in a closed, constant-pressure system.
2. Detection of a pressure drop in excess of vapor-liquid equilibration in a closed, constant-volume system.
3. Detection of an exotherm due to the hydrate heat of formation.
4. Detection of an increased differential pressure drop in a flowing system.

We define hydrate induction rates and onset temperatures based on hydrate detection by any of the above mentioned events. Each apparatus may have the capability of detecting more than one event. In those cases, our experience has been that no single event was always the first observed sign of the onset of hydrate formation.

Induction rate experiments *start* when the system drops below the hydrate equilibrium temperature and *end* when hydrates are first detected. We define T_{onset} as the temperature of a system where hydrates are first detected. For experiments in which the system temperature is ramped down below the hydrate equilibrium temperature and held at constant temperature for an induction period, t_{ind} is defined as the induction time required to observe hydrate formation onset at the hold temperature.

It is known that hydrate nucleation events actually take place before a system reaches T_{onset} or at a time shorter than the experimental t_{ind}. Thus, experimental hydrate detection rates set a lower bound to the actual induction rates. The art of the experiment is to reduce randomness of nucleation in order to achieve repeatability in measured hydrate detection rates. This is best done by using large, well mixed systems, in order to minimize statistical fluctuations in the test.

Measurement of Hydrate Dissociation Temperature

The second issue is how to accurately measure hydrate dissociation temperatures. Hydrate dissociation temperature is commonly regarded as the closest observable temperature to the hydrate equilibrium temperature. Measuring dissociation temperatures is generally simpler than measuring the induction rate. The following measurements are commonly used.

1. Gas volume changes at constant pressure.
2. Gas pressure changes at constant volume.
3. Differential pressure drop across an element in the flowing system.
4. Observance of an endotherm.

T_{eq} is defined as the temperature at which a system of fixed composition and pressure is at the hydrate equilibrium. Experimental methods for determining T_{eq} were presented at the Second International Conference on Natural Gas Hydrates, by Tohidi et al.,[9] Mei et al.,[10] Herri et al.,[11] and by Narita and Uchida.[4]

We measure T_{eq} experimentally as the temperature at which either a gas volume returns to its nonhydrated equilibrium value (in a constant pressure apparatus), a differential pressure returns to its nonhydrated value, or a pressure returns to its nonhydrated equilibrium value (in a constant volume apparatus). Endotherms are not temporally well defined and are not used in this work for detection of hydrate dissociation.

When measuring T_{eq}, caution must be exercised in multicomponent experiments to ensure that the observed hydrate decomposition occurs at constant water, hydrate, and gas phase compositions. If the salinity, gas composition, or hydrate type were to change after the onset of hydrate formation, the measured value of T_{eq} would be shifted relative to the actual value for the initial compositions. Experimentally, if the water and gas fractions in hydrates are small, then the compositions of the bulk gas and water phases will be constant during the measurement of T_{eq}. In the work presented here, excessive hydrate growth was not permitted for T_{eq} determination.

We prefer to measure T_{eq} before adding kinetic hydrate inhibitor or other chemicals to a system. The presence of kinetic inhibitor causes the system to equilibrate slowly and makes the measurement of T_{eq} difficult.

In this paper, we follow the convention that subcooling is defined as the difference between the hydrate equilibrium temperature and the experimental temperature, $|T_{eq} - T_{exp}|$. The maximum subcooling achievable in a system is $|T_{eq} - T_{onset}|$. Coupling reliable measurement of hydrate dissociation temperatures with reliable detection of the onset of hydrate formation allows the experimentalist to rank kinetic hydrate inhibitors by the maximum subcooling achievable in comparable systems. Kinetic hydrate inhibitors may also be ranked by comparison of t_{ind} as a function of subcooling and inhibitor concentration.

Choice of Experimental Composition

The third issue is choosing a composition to rank kinetic hydrate inhibitors. Generally, one wants to rank inhibitors for the compositions found in a particular field application. This has led to the establishment of screening/testing capabilities in many laboratories. Our experience leads us to conclude that the composition used to compare inhibitors is more important than the choice of apparatus, if used properly.

Part of the composition issue is verification that the kinetic hydrate inhibitor is chemically and biologically stable. Samples of inhibitors are monitored for aerobic and anaerobic bacterial degradation over time. Occasionally, some degradation is observed. Water or brine used in the experiments may need to be sterilized before storage or immediate use. This can be done with heat or by chemical treatment, such as the addition of bleach or commercial biocide.

Reproducible Driving Force

After choosing a system composition, the last issue in the design of kinetics experiments is the method of applying the driving force for hydrate formation. We have found it convenient to use a method that applies a variable driving force to the system via a reproducible path. The preferred method is described below.

For onset and induction time experiments, the compositions of all initial phases (gas, liquid hydrocarbon, and liquid water) should be maintained constant and as nearly equilibrated as possible. This condition makes the thermodynamic path experimentally reproducible from test to test, and from laboratory to laboratory. To satisfy this requirement, the following conditions should be met insofar as possible.
1. Temperature is maintained uniform and temperature changes are gradual before the onset of hydrate formation.
2. Pressure is constant.
3. Surface area is large and continually renewed by sparging, stirring, or pumping.
4. Water structure is initialized by heating the system above the temperature at which residual hydrate structure melts.
5. There is a large excess of water and/or each hydrate-forming component after initial hydrate formation, such that water and gas compositions are not altered before hydrate detection occurs.

RESULTS

The tests compared here include the tetrahydrofuran (THF) tumbling cells, a rocking-tube, high-pressure sapphire apparatus, a 0.5-inch diameter pressurized miniloop, and a 4-inch diameter pressurized flowloop. FIGURE 1 shows a comparison of miniloop and THF rig results for a series of alkyl-substituted polyacrylamides, including methyl-, ethyl-, and diethylacrylamide. It is obvious that subcooling measured in the miniloop does not correlate with ballstop time measured in the THF rig. We believe the main reason the THF test does not correlate with the high-pressure tests that use natural gas and hydrocarbon liquids under field conditions is because

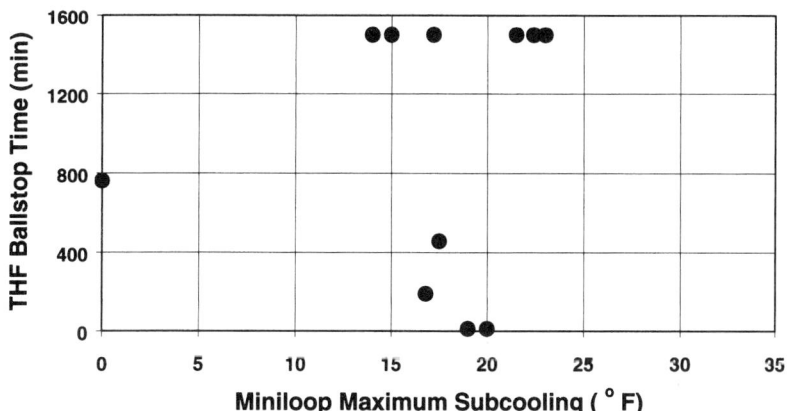

FIGURE 1. Comparison of miniloop maximum subcooling values versus THF ballstop times for a series of substituted and unsubstituted polyacrylamides.

the inhibition mechanism in THF is significantly different from that of natural gas systems.

FIGURE 2 shows a comparison of miniloop subcooling data to subcooling data obtained in the sapphire apparatus and in the 4-inch flowloop for various kinetic hydrate inhibitors. The correlation between miniloop and 4-inch flowloop data is excellent. There is greater scatter in the correlation between the miniloop and the sapphire apparatus. We believe the scatter is due to lower precision of hydrate onset and dissociation measurements made in the sapphire apparatus. Note that the intercept of the linear regression of the saphire rig data is 12.7°F, whereas the flowloop data intercept is 0.6°F. We conclude that flowloops give subcoolings that differ from rocking cell tests.

The important observation made with high-pressure apparatuses is that the relative ranking of inhibitors was independent of apparatus within experimental error limits. Flowloops have enabled Exxon to correlate structure with performance.[13] Relying on flowloop data instead of THF rig screening results led to our experience that molecular modeling was more productive for inhibitor performance optimization than the Edisonian approach.[14]

One of the advantages of flowloop experiments is the ability to measure the pressure drop across an obstruction. In many cases, differential pressure is the earliest detectable sign of the onset of hydrate formation. We have sometimes observed that differential pressure across the pump is the only sure sign of onset of hydrate formation in kinetically-inhibited systems. We have found that a gear pump with close

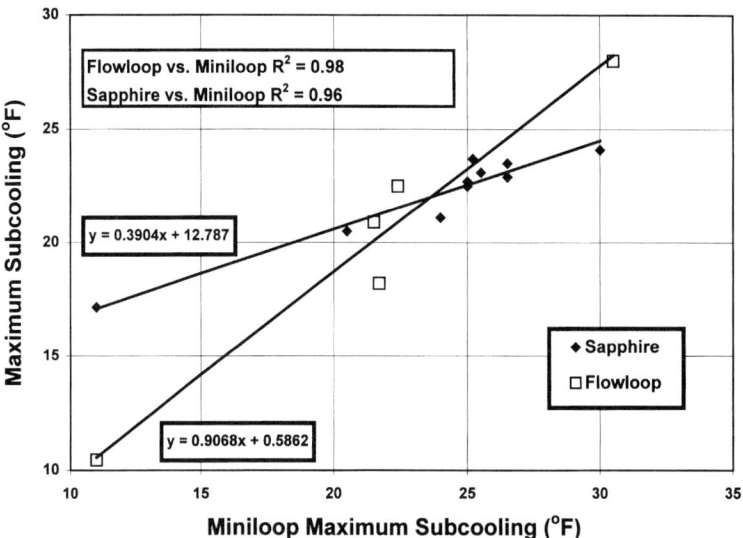

FIGURE 2. Comparison of miniloop maximum subcooling values versus sapphire apparatus and 4-inch diameter flowloop maximum subcooling values. All data are for 0.5% polymer in synthetic sea water, King Ranch gas condensate, and Green Canyon Gas or similar natural gas compositions.

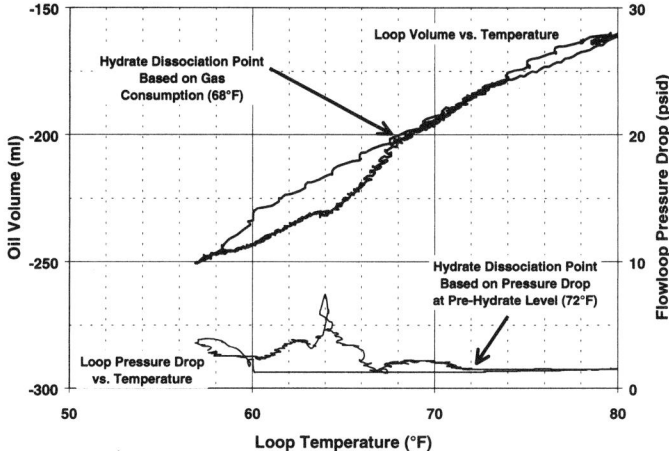

FIGURE 3. Miniloop data for a hydrate dissociation temperature experiment. Detection of last hydrate dissociation is by gas volume change and by pressure drop across the loop pump.

tolerances is an excellent device where one can measure differential pressure to reliably detect the onset of hydrate formation.

FIGURE 3 represents a system where differential pressure across the pump is more sensitive to hydrate dissociation than is gas consumption. Gas volume change data predicts a value for T_{eq} of 68°F. Differential pressure data indicate that there is a small amount of hydrate solids circulating through the miniloop at $T_{eq} = 72$°F. Because of the small amount of this hydrate, it is essentially undetected by gas

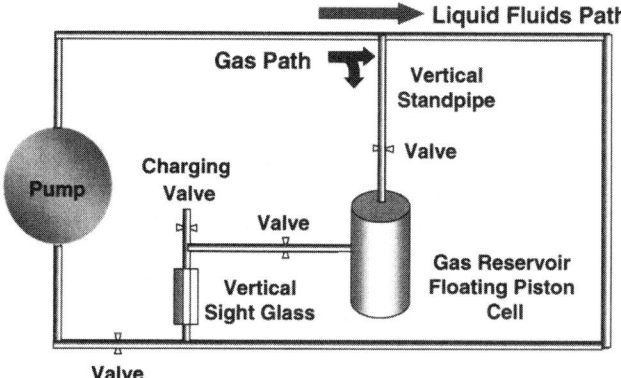

FIGURE 4. Flowloop diagram showing liquid fluid loop path and gas reservoir with connecting inlet and outlet to loop. The vertical sight glass is for verification of gas flow. The entire system inside the dashed line is temperature controlled.

consumption. However, in other gas and oil systems differential pressure and gas consumption often indicate the same value for T_{eq}.

Some experimental procedures reported in the literature for hydrate growth kinetics measurements maintain constant gas composition by adding the hydrate-forming components as the system consumes gas. An example of this technique was reported by Behar et al.,[12] in a study of antiagglomerants. The disadvantage of this technique is that the system cannot be cycled repeatedly at constant composition.

This paper focuses on the hydrate induction kinetics, where the relevant experimental data are obtained between the time the system enters the hydrate stability region and the onset of hydrate formation. In such experiments, a flowloop such as that shown in FIGURE 4 is used to circulate excess gas between a gas reservoir and the experimental fluids in the loop. This enables one to maintain compositions nearly constant as the water and liquid hydrocarbon temperatures drop. Maintaining constant compositions allows multiple cycles to be performed at constant composition. After the induction kinetics endpoint is measured at the onset of hydrate, the phase compositions do not need to remain constant.

Flowloop experiments are intrinsically more complicated than stirred-tank experiments. From our experience, flowloop characteristics that need to be verified include the following.

1. Temperature is uniform inside the loop (minimize pump mechanical heat).
2. Maximum subcooling in the uninhibited system is less than 7°F.
3. Maximum subcooling in the inhibited system is minimized with respect to stirring rate.
4. Ramping the system temperature down to a point above T_{onset} followed by slow warming shows no gas consumption hysteresis or differential pressure hysteresis characteristic of hydrate formation.
5. Verification that hydrate detection occurs while gas volume change is much less than the total gas volume.
6. Verification that hydrate detection occurs while water volume that forms hydrates is much less than total water volume.
7. Verification that T_{eq} and T_{onset} do not change with repeat cycles (due to unequilibrated gas in the reservoir).

CONCLUSIONS

THF tests are not reliable indicators of hydrate formation kinetics for natural gas systems. This is because the THF system has different mechanisms of hydrate formation and inhibition relative to natural gas systems.

Flowloops are recommended for ranking kinetic hydrate inhibitors according to subcooling and induction times as a function of concentration. This is based on the preferred conditions of reproducible driving force and reliable hydrate detection. The miniloop is of adequate scale to achieve these preferred conditions. Flowloops have enabled Exxon to correlate structure with performance.

We did not use a stirred-tank, high-pressure test apparatus in this work, although we believe these can be effective in reproducing the relative ranking of inhibitors determined in other types of pressurized test apparatus.

REFERENCES

1. SLOAN, E.D. 1994. International Conference on Natural Gas Hydrates. Ann. N.Y. Acad. Sci. **715:** 1.
2. MEYER, H.S., J.L. SAVIDGE & K.E. WOODCOCK. 1994. International Conference on Natural Gas Hydrates. Ann. N.Y. Acad. Sci. **715:** 24.
3. GAILLARD, C., J.P. MONFORT & J. PEYTAVY. 1996. Proceedings of the 2nd International Conference on Natural Gas Hydrates, June 2–6, Toulouse, France. 183.
4. NARITA, H. & T. UCHIDA. 1996. Proceedings of the 2nd International Conference on Natural Gas Hydrates, June 2–6, Toulouse, France. 191.
5. SVARTAS, T.M., L. DYBVIK, K. LEKVAM & M.A. KELLAND. 1996. Proceedings of the 2nd International Conference on Natural Gas Hydrates, June 2–6, Toulouse, France. 199.
6. LEKVAM, K. & P. RUOFF. 1996. Proceedings of the 2nd International Conference on Natural Gas Hydrates, June 2–6, Toulouse, France. 207.
7. TOHIDI, B., A. DANESH, K. OSTERGAARD & A. TODD. 1996. Proceedings of the 2nd International Conference on Natural Gas Hydrates, June 2–6, Toulouse, France. 229.
8. HERRI, J., F. GRUY & M. COURNIL. 1996. Proceedings of the 2nd International Conference on Natural Gas Hydrates, June 2–6, Toulouse, France. 243.
9. TOHIDI, B., A. DANESH, R.W. BURGASS & A.C. TODD. 1996. Proceedings of the 2nd International Conference on Natural Gas Hydrates, June 2–6, Toulouse, France. 109.
10. MEI, D.H., J.L. LIAO, J.T. YANG & T.M. GUO. 1996. Proceedings of the 2nd International Conference on Natural Gas Hydrates, June 2–6, Toulouse, France. 123.
11. HERRI, J., F. GRUY, M. COURNIL, D. DI BENEDETTO & P. BREUIL. 1996. Proceedings of the 2nd International Conference on Natural Gas Hydrates, June 2–6, Toulouse, France. 251.
12. BEHAR, E., A.-S. DELION, A. SUGIER & M. THOMAS. 1994. International Conference on Natural Gas Hydrates. Ann. N.Y. Acad. Sci. **715:** 94.
13. TALLEY, L.D. & R.H. OELFKE. 1999. Method for Predetermining a Polymer for Inhibiting Hydrate Formation. US Patent 5,900,516.
14. PANCHALINGAM, V. & E. DENDY SLOAN. 1996. Proceedings of the 2nd International Conference on Natural Gas Hydrates, June 2–6, Toulouse, France. 171.

Flow Properties of Hydrate-in-Water Slurries

VIBEKE ANDERSSON AND JÓN STEINAR GUDMUNDSSON

Department of Petroleum Engineering and Applied Geophysics,
Norwegian University of Science and Technology, 7491 Trondheim, Norway

ABSTRACT: Natural gas hydrates have been proposed as a means to capture associated gas produced on offshore oil platforms. Hydrates are produced by bringing the gas into contact with liquid water, resulting in a hydrate-in-water slurry. It is further suggested that the hydrates be mixed with the crude oil, resulting in an hydrate-in-oil slurry that might be transported to shore in shuttle tankers, or in long-distance pipelines. A hydrate laboratory has been built to obtain the data necessary to evaluate such processes. The laboratory contains a high-pressure tube viscometer in which the flow properties of water-based and oil-based hydrate slurries can be studied under laminar and turbulent pipe flow conditions. In this paper, the experimental equipment is described, and experiments on the flow properties of hydrate-in-water slurries are presented. It was found that the slurry viscosity increases with increasing hydrate concentration, and also that the hydrate slurries approach the same frictional pressure drop as the carrying water in the turbulent flow regime, regardless of hydrate concentration.

INTRODUCTION

The question of what to do with gas associated with oil produced offshore when there is no gas pipeline available has become prevalent in the oil industry. Flaring the gas is not an option because of environmental considerations. The common practice of reinjecting the gas is also increasingly being questioned; therefore, ways are being sought to bring the gas to market. An option is to convert the gas into a phase that can be more readily transported. This includes products such as liquefied natural gas (LNG), methanol, synthetic crude (syncrude), and *natural gas hydrates* (NGH). Hydrates have been proposed as a means to transport bulk gas from offshore gas fields and to capture associated gas produced on offshore oil-platforms.[1] One proposed scheme is to convert the gas into hydrates subsequently mixing them with the crude oil. The resulting slurry can then be frozen, depressurized, and transported in insulated, low-pressure shuttle tankers. FIGURE 1 shows a simplified flow diagram of the offshore hydrate slurry process. Alternatively, the slurry can be transported under pressure in a pipeline.[2]

In the NGH slurry process, natural gas is brought into contact with liquid water at typically 2–10°C and 60–90 bar, and is *trapped* inside the resulting hydrate structures. The basis for the technology is the potential high gas content of NGH (180 Sm^3 gas per m^3 hydrates), and that NGH stored in large, insulated tanks are relatively stable at atmospheric pressure when refrigerated to temperatures below 0°C.[3] Although this means transporting significant amounts of water along with the gas, the economics appear promising for such processes.[1,2]

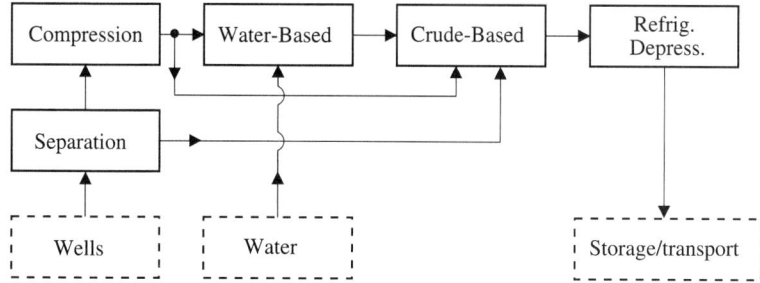

FIGURE 1. Simplified flow diagram of the hydrate slurry process.

Research has been ongoing at the Norwegian University of Science and Technology (NTNU), Department of Petroleum Engineering and Applied Geophysics, since 1990 to consider the use of NGH, focusing on large-scale transport and storage of natural gas. In 1997, a joint industry project, NGH at NTNU, was established to obtain the engineering data necessary to evaluate the feasibility of hydrate processes. NGH at NTNU was a three year program (1997 to 1999), with Aker Engineering as the main cooperative partner and with further support from several international oil companies.

During this program, a hydrate laboratory was built at NTNU. In the laboratory, natural gas hydrates are produced continuously in a liquid-based system, and may be filtered semicontinuously. The hydrate rig enables investigation of rates of formation and dissociation, thermophysical properties (a high-accuracy, low temperature, high pressure calorimeter is available for analyzing solid hydrate samples), and flow properties. In the following sections, the hydrate laboratory at NTNU is described, with a focus on the tube viscometer used to investigate the flow properties of hydrate slurries. The theory and experimental investigation of the flow properties of hydrate-in-water slurries are also described.

THE HYDRATE LABORATORY AT NTNU

The main units of the hydrate laboratory at NTNU are a 9-liter continuously stirred baffled tank reactor, an 18-liter tank separator, a shell-and-tube heat exchanger, a centrifugal pump (0–100 liter/minute), and a mass meter, forming a closed circulation loop. The mass meter is based on the Coriolis measuring principle and has mass flow rate and density as output. FIGURE 2 shows a schematic drawing of the hydrate laboratory. The laboratory is placed in a constant temperature room at 0–20°C, with further possibilities to reduce the process temperature to −25°C through the heat exchanger. The equipment is designed for operation at up to 120 bar, and is Ex-II classified. The signals from all the instruments are transferred to a PC-based data acquisition system for on-line and later analysis.

The hydrates are produced in the reactor, where gas of optional composition is injected and converted into hydrates when brought into contact with liquid water

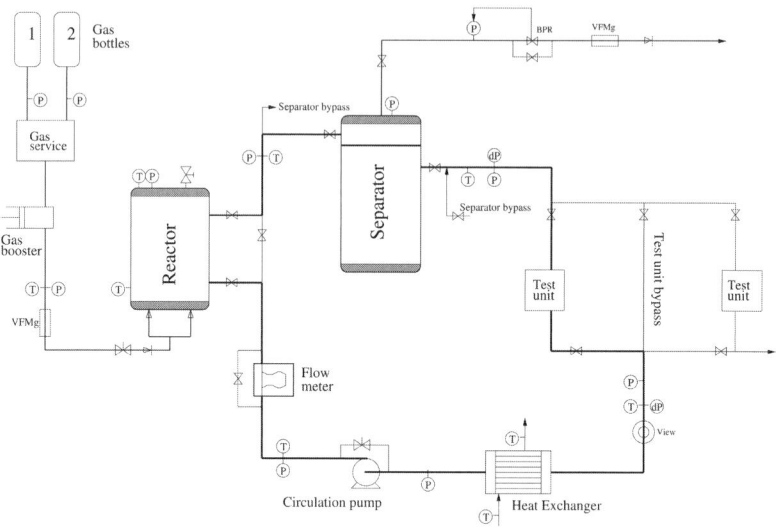

FIGURE 2. Flow diagram of the hydrate laboratory.

under hydrate forming conditions. From the reactor, the flow goes to the separator, where excess gas (gas that is not converted into hydrates nor dissolved in the liquid phase) is separated and vented. Thus, the flow leaving the separator is a gas-free hydrate slurry where the continuous phase can either be water or oil, depending on the experimental conditions.

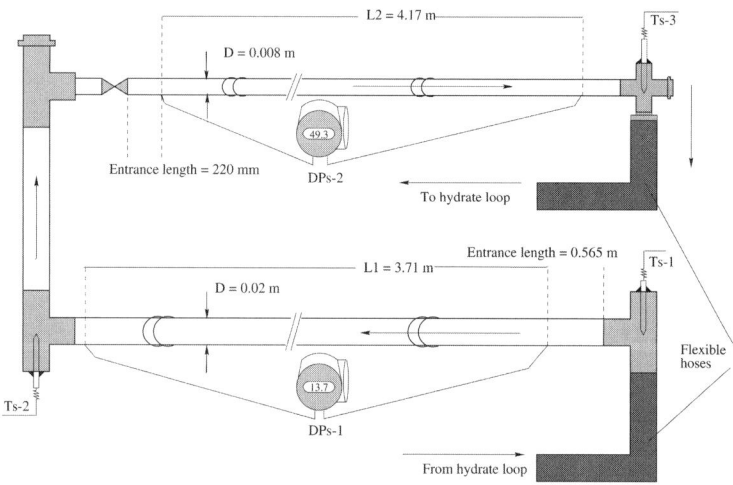

FIGURE 3. Flow diagram of the tube viscometer.

The hydrate laboratory was built so that various types of hydrate properties can be tested by connecting different equipment to two test ports. The hydrates can, for example, be separated in a filter arrangement, or high pressure samples of the hydrate slurry can be taken with a specially designed sampling device. In addition, the flow can be diverted to a horizontal flow section, where the flow properties of the hydrate slurries can be investigated under laminar and turbulent flow conditions by means of tube viscometer experiments.

A flow diagram of the high-pressure tube viscometer is shown in FIGURE 3, with the relevant sizes included. The tube viscometer consists of two pipe lengths of different diameters connected in series, and differential pressure transducers connected to each of the two pipes. The temperature is measured at the inlet and at the outlet of both pipes. One of the advantages of the tube viscometer is the direct connection with flexible hoses to the hydrate forming rig. This enables measurements on pressurized hydrate slurries on-line, revealing flow mechanisms not found from, for example, rotational viscometer experiments.

THE FLOW PROPERTIES OF SLURRIES

To evaluate the hydrate slurry processes, the flow properties of the slurries need to be determined. By *flow properties* we mean the flow behavior, and a quantitative expression for the frictional pressure drop in pipelines. Both water-continuous and oil-continuous slurries are of interest. As hydrates are known to block pipelines and process equipment, great care in *producing* the slurries was exercised, resulting in systems in which the hydrate particles flow along with the carrying fluid and no deposition took place.

The continuum equations governing single-phase fluids are assumed to be valid for water–hydrate slurries. This is based on the fact that the slurries are nonsettling due to small particle sizes and small difference in fluid-to-solid density. Visual observation of the hydrate-in-water slurries confirmed that they behave homogeneously. That is, the hydrate particles were dispersed throughout the water volume given a minimum of mixing energy. To obtain the apparent viscosity of such slurries, experiments can be performed in instruments of two basic types; rotational viscometers and tube viscometers.[5] The horizontal flow section described above works as a tube viscometer, enabling the viscosities of hydrate slurries to be determined over large ranges in shear rates, pressure, and temperature.

For viscosity-determination in a tube viscometer, the pressure drop ΔP over a pipe of length L and diameter D is recorded together with the corresponding volumetric flow rate, Q. The wall shear stress τ_w is calculated from

$$\tau_w = \frac{\Delta P}{4L}D, \tag{1}$$

and the wall shear rate $\dot{\gamma}_w$ is obtained from

$$\dot{\gamma}_w = \frac{dv}{dr} = \frac{32Q}{\pi D^3}, \tag{2}$$

Adjustment of Equation (2) for non-Newtonian behavior can be made using the Rabinowitch technique.[5]

Plotting τ_w versus $\dot{\gamma}_w$ results in a *rheogram*, in which the secant slope at every point is the apparent viscosity of the slurry. The determination of (non-Newtonian) viscosities from tube viscometer experiments is described in detail elsewhere.[5]

The underlying equations for viscosity determination of fluids are valid only for laminar flow. Extending the flow into the turbulent regime, the concept of a *hydraulic gradient* is often convenient in characterizing slurry flow:

$$i = \left|\frac{\Delta P}{L}\right| \cdot \frac{1}{\rho_w g}, \qquad (3)$$

giving the friction expressed in head of clear water per unit length of the pipe.

Several experiments and field tests on the flow behavior of hydrates in pipes have been performed during the last few years, mainly for investigating hydrate deposition in pipelines. Lately, the investigation of antiagglomerates has also been of interest in pipe flow behavior of hydrate slurries. The main difference from the present investigation is the gas-free medium, which enables the flow to be treated as a continuum and, thereby, use of the governing equations for single-phase fluids, and the concept of viscosity.

EXPERIMENTAL PROCEDURES AND RESULTS

When producing hydrate-in-water slurries, a certain amount of gas was injected into a water-filled system at hydrate subcooling temperatures of 2.5–6.5°C. The pressure was kept constant at either 60 or 90 bar. The gas phase used to produce the slurries was either pure methane (sI hydrates), or a mixture consisting of 92 mol% methane, 5 mol% ethane, and 3 mol% propane (sII hydrates). The water phase was deaerated tap water. Injection of gas resulted in a slurry with a certain concentration of hydrates. Several pressure drop measurements were then performed on the gas-free water–hydrate slurry in the tube viscometer, with corresponding recording of the mass flow rate and the slurry density. Simultaneously, a high-pressure sample of the slurry was taken with the sampling device. After the pressure drop experiments, injection of gas was continued to increase the concentration of hydrates, and the entire procedure was repeated.

The concentration of gas in the slurry was found by melting the slurry sample in a constrained volume and recording the pressure build-up. From this gas concentration, together with the measured slurry density (from the Coriolis meter), the volumetric hydrate concentration in the slurry was calculated.

The pressure drop data and the mass flow data from the laminar regions were then analyzed according to the procedure described above, resulting in more than 50 rheograms of which four are shown in FIGURE 4. The rheograms show shear stresses versus shear rates for different hydrate concentrations.

The relation between shear rate and shear stress in FIGURE 4 is linear, and the Bingham rheological equation,

$$\tau_w = \tau_o + \mu_{app}\dot{\gamma}_w, \qquad (4)$$

properly describes the data in the concentration range investigated. Here τ_o is the yield stress, and μ_{app} is the apparent viscosity. From the figure it can be seen that the slope of the lines, the apparent viscosity, increases with increasing concentration. In

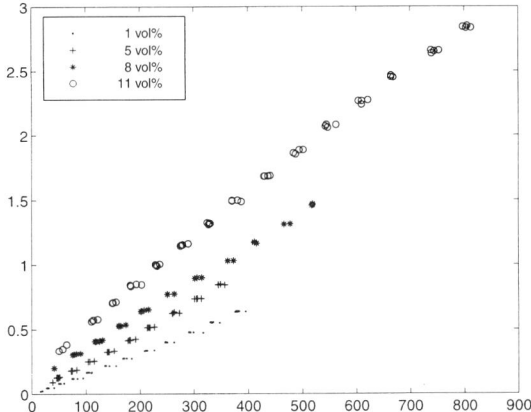

FIGURE 4. Illustrative rheograms showing wall shear stresses versus shear rates for different concentrations.

addition, the lines are slightly shifted upwards with increasing concentration, as a result of increasing yield stresses.

The behavior illustrated by FIGURE 4 was observed for most of the water–hydrate slurries investigated. In FIGURE 5, the estimated Bingham apparent viscosities are

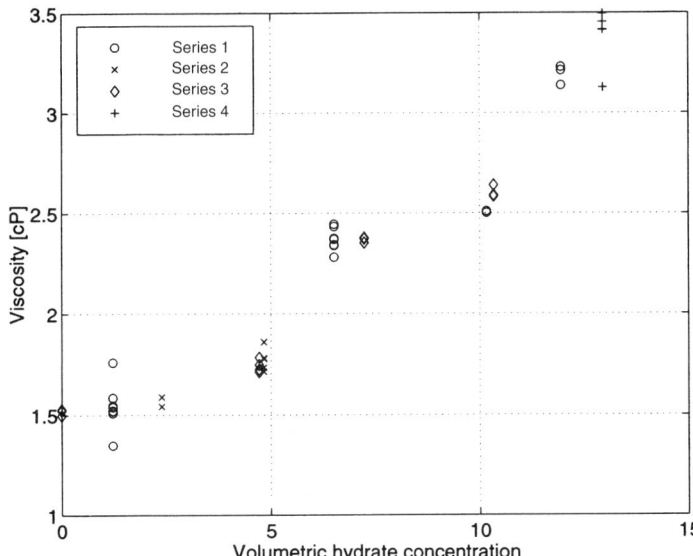

FIGURE 5. Slurry viscosity as a function of volumetric hydrate concentration. The viscosities are corrected to a temperature of 5.0°C.

plotted versus the volumetric hydrate concentrations for the investigated slurries. Series 2 and 4 are sII hydrates, whereas Series 1 and 3 are sI hydrates. There was no obvious difference between sI hydrate slurries and sII hydrate slurries in terms of viscosity.

For higher concentrations than those shown in FIGURE 5, it was evident that the stress/rate relationship no longer remained linear. The slurries were shear-thinning, and a power-law type of equation modeled the data better. However, the volumetric flow rate measurements with the Coriolis mass flow meter became unreliable with high hydrate concentrations, rendering it difficult to perform accurate and repeatable viscosity measurements on slurries with hydrate concentrations higher than 15 vol%.

The experiments showed that the resistance to flow increases with increasing hydrate concentrations, illustrated through the increasing viscosities in FIGURE 5. However, this was not the case in the turbulent region. FIGURE 6 shows the slurry hydraulic gradient, Equation (3), versus the water Reynolds number, $N_{Re,w} = D\rho_w v_m/\mu_w$, for the same slurries as those in FIGURE 4. The solid lines shown are the theoretical predicted water lines in the laminar and the turbulent regions. In the laminar region, it is seen that the frictional pressure drop expressed through the hydraulic gradient increases with increasing concentration, FIGURE 6, due to increased viscosity.

For all concentrations, however, it can be seen that the slurry hydraulic gradient curve approaches the theoretical pure water curve for higher velocities, starting at the onset of turbulence. Hence, in the turbulent regime, the effective viscosities for all the different concentrations of the hydrate slurries are equal, and resemble the

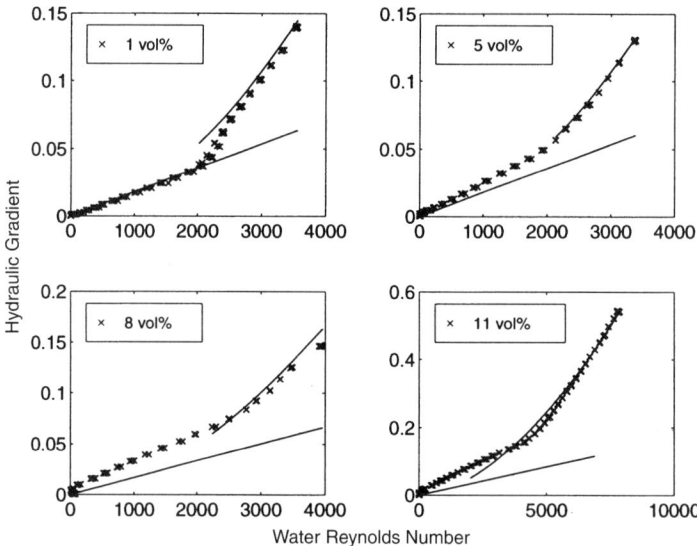

FIGURE 6. Slurry hydraulic gradient versus water Reynolds numbers. The hydrate concentration is increasing from *top left* to *bottom right* as indicated. The *solid lines* are the theoretical predicted laminar and turbulent water lines.

viscosity of water. This means that the frictional pressure drop of turbulent water–hydrate slurry pipe flow is identical to the frictional pressure drop of pure water. The reason for this may be the behavior of the hydrate particles near the walls; the velocity gradient causes particle rotation and migration away from the regions of high shear, such that a liquid film at the wall determines the frictional pressure drop.[6] This effect in the turbulent regime was observed for all the water–hydrate slurries investigated (0–21 vol%). Preliminary experiments show the same behavior for oil–hydrate slurries.

CONCLUSIONS

1. Hydrate-in-water slurries are readily produced in the NTNU hydrate laboratory, and the hydrates flow as homogeneous slurries.

2. The apparent viscosities of the hydrate-in-water slurries increase with increasing hydrate concentrations.

3. In the turbulent regime, the frictional pressure drops of the hydrate-in-water slurries are determined by the properties of the carrying water phase alone.

ACKNOWLEDGMENT

This paper was written as part of the joint industry project NGH at NTNU, supported by the following companies: Aker Engineering, Amerada Hess, ARCO, Neste Petroleum, Phillips, Shell, and Total. The NGH at NTNU project has the web-address: <http://www.ipt.ntnu.no/~ngh>. The Dr.Ing. work of Vibeke Andersson was further supported by the Research Council of Norway through UTBYGG contract 32706/211. Thanks to the staff at the mechanical and electrical workshops at the department. Vibeke Andersson would also like to thank Geir Ultveit Haugen for appreciated encouragement during my experimental work, and for helpful discussions afterwards.

REFERENCES

1. GUDMUNDSSON, J.S., V. ANDERSSON & O.I. LEVIK. 1997. Gas storage and transport using hydrates. *In* Offshore Mediterranean Conference, Ravenna, Italy. 1075–1083.
2. GUDMUNDSSON, J.S., V. ANDERSSON, O.I. LEVIK & M. PARLAKTUNA. 1998. Hydrate concept for capturing associated gas. *In* SPE European Petroleum Conference, The Hague, the Netherlands. SPE paper 50598.
3. GUDMUNDSSON, J.S., M. PARLAKTUNA & A.A. KHOKHAR. 1994. Storing natural gas as frozen hydrate. *In* SPE Production & Facilities. 69–73.
4. GUDMUNDSSON, J.S., F. HVEDING & A BØRREHAUG. 1995. Transport of natural gas as frozen hydrate. *In* 5th International Offshore and Polar Engineering Conf. The Hague, the Netherlands. 282–288.
5. WILSON, K.C., G.R. ADDIE, A. SELLGREN & R. CLIFT. 1997. Slurry Transport Using Centrifugal Pumps. Blackie Academic & Professional, 2nd edit.
6. CROWE, C., M. SOMMERFELD & Y. TSUJI. 1998. Multiphase Flows with Droplets and Particles. CRC Press.

Flow Loop Experiments Determine Hydrate Plugging Tendencies in the Field

PATRICK N. MATTHEWS,[a] PHIL K. NOTZ,
MARK W. WIDENER, AND GABRIEL PRUKOP

Texaco Upstream Technology, 3901 Briarpark, Houston, Texas 77042-5301, USA

ABSTRACT: Flow loop studies were conducted on live reservoir fluids from the Werner Bolley gas condensate well in Southern Wyoming to determine hydrate formation and plugging tendencies of the hydrocarbon fluids under field conditions. Field tests conducted by DeepStar in 1997 at Werner Bolley were summarized and the results were compared to data from flow loop studies. Subcooling measured in the flow loop was found to be comparable to that measured in the field near hydrate plug formations. Flow loop results indicated both the plugging tendency and the nature of plugging mechanism observed in the field. Used in conjunction with transient flow simulation, flow loop results indicated the most probable locations for plug formation.

INTRODUCTION

Hydrates have plagued the petroleum industry for decades, occurring commonly in inadequately protected transmission lines. As development focus pushes into more difficult operational environments, hydrate plug formation becomes a real menace to flow assurance. Economic pressures and environmental concerns motivate efforts to find more effective inhibitors and improved system solutions. One of the most pressing challenges to hydrate specialists in the petroleum industry today lies in the implementation of new technologies in the field. Critical to the successful implementation of new technologies will be tools that can accurately predict the field behavior of mixed hydrocarbon systems and potential hydrate inhibitors. The best tools currently available for performing this task are high pressure flow loops. A good flow loop will allow study of hydrate formation and plugging tendencies under field conditions in multiphase flow. Opportunities to prove field transferability of flow loop results have been few and far between because of the small amount of quality field data. Most field tests have focused on proving the capabilities of specific hydrate inhibitors, yielding data that is inappropriate for scaling studies. In January and February of 1997, DeepStar conducted field trials at Devon Energy's Werner Bolley well (in the Powell Pressure Maintenance Unit field) in Southern Wyoming for the express purpose of studying hydrate plug formation and dissociation. The results of the field trials provide a good opportunity to investigate transferability of results from flow loop to field.

[a]Telecommunication. Voice: 713-954-6026; fax: 713-954-6911.
matthpn@texaco.com

EXPERIMENTAL PROCEDURES

Flow Loop Specifications and Procedures

Texaco's Hydrates Flow Loop was completed in 1996 to allow the study of hydrate formation and plugging tendencies of hydrocarbon fluids in high pressure multiphase flow. The flow line was constructed from 316 ss schedule 80 pipe with an ID of 1.925 inches, and a pressure rating of 2000 psig. The loop layout, shown in FIGURE 1, is 6 feet wide by 46 feet long (four 90-degree elbows) for a total length of 104 feet and a total volume of approximately 2.29 ft^3. Multiphase flow is implemented with a Leistritz twin helical screw pump with variable rates up to 200 gpm (22 fps velocity). Temperature control of the flow loop is implemented using gravity fed water circulation through insulated 16-inch diameter PVC pipe that encompasses the loop. Heating and cooling are achieved by using a series of chillers and a flow-through water heater. The current temperature range is 38–150°F, with a maximum continuous heating/cooling rate of about 4°F/hr. Chilled water from a large secondary tank can be dumped into the water system to cool the flow loop at rates nearing 350°F/hr. Automated with a data acquisition and control system, the loop is monitored by four temperature sensors and six pressure sensors arrayed around the loop. Fluids inside the loop can be viewed and recorded by means of borescopes inserted

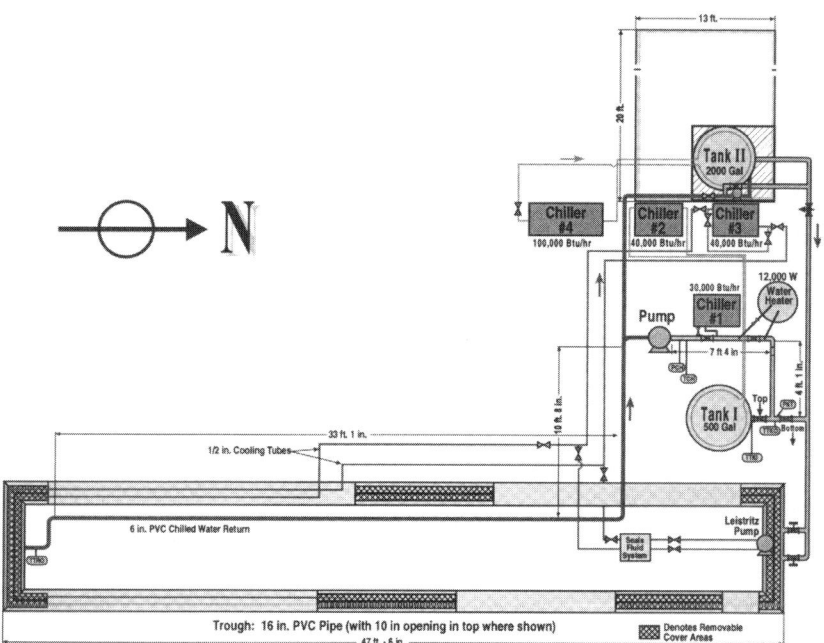

FIGURE 1. Texaco's Hydrate Flow Loop viewed from above. The loop is 1.925 inches ID, 104 feet in total length and has a maximum pressure rating of 2,000 psi. Circulation of multiphase fluids is achieved with a twin screw Leistritz pump.

into optical blocks positioned at the Leistritz pump outlet, and in the middle of the East and West sides of the loop.

Experiments are conducted with live reservoir fluids. Reservoir fluids are constructed in the loop by charging a combination of synthetic gas with stock tank oil. In addition, a volume of water is charged to achieve a target water cut. Based on reservoir fluids, an estimate of the composition of the hydrocarbons in the hydrate phase is made. Hydrate bound hydrocarbon (HBH) gas is made up under pressure to match the entrapped hydrocarbon composition and later injected for pressure maintenance as hydrate formation depletes the vapor phase. The amount of HBH injected is monitored and used as a measure of the amount of hydrate formation.

Equilibrium runs are conducted via temperature ramping with cooling at 3.0°F/hr to hydrate onset and subsequent heating at a rate of 1.8°F until the thermodynamic hydrate point is determined. Plugging runs involve a cooling cycle to a desired temperature and pressure maintenance via water and/or HBH gas injection at the hydrate onset pressure. Plugging experiments typically run for a few hours after hydrate formation has stopped and may also involve an extended shut-in period. Increases in pressure drop and/or visual observation determine plugging of the loop.

Field Configuration and Procedures

The Werner Bolley well produced primarily gas, some condensate, and water (4 mmscfd gas, 100 BPD condensate, and 10 BPD water). Near the wellhead, fluids from the producing well were heated, separated, metered, and then recombined for transportation. Recombined fluids were transported in a buried 4-inch diameter, 17,319 ft sch 40 flow line that terminated in a 6-inch transmission line. The flow line elevation profile is shown in FIGURE 2, with the numbered spots indicating sensor locations (pressure, temperature, and in some cases gamma ray densitometers). The

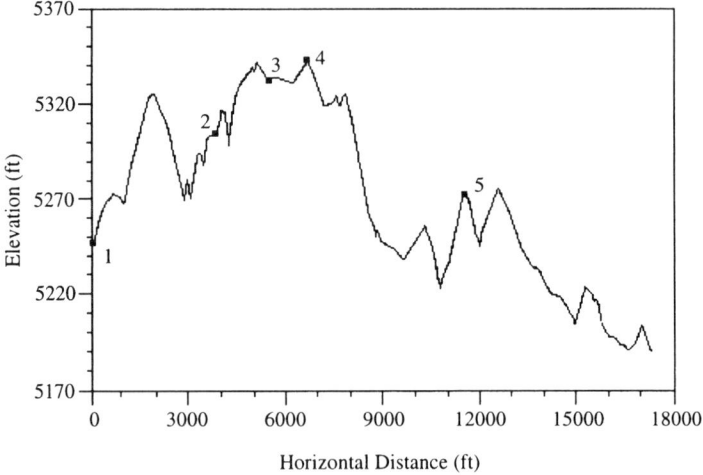

FIGURE 2. Werner Bolley flow line elevation profile had many low spots for liquids to gather.

TABLE 1. Composition of Werner Bolley fluids

Component	Composition (Mole%)	
	Werner Bolley Wellstream	HBH Gas[a]
N_2	0.43	0.33
CO_2	1.51	1.18
C_1	78.20	60.87
C_2	13.59	10.58
C_3	3.61	23.50
$i\text{-}C_4$	0.39	3.00
$n\text{-}C_4$	0.69	0.54
$i\text{-}C_5$	0.25	0.00
$n\text{-}C_5$	0.21	0.00
C_6	0.24	0.00
C_{7+}	0.88	0.00
Total	100	100

[a]HBH is the make up gas added to maintain pressure and reservoir fluid composition as hydrate forms. The HBH composition is the estimated composition of the hydrocarbon components trapped in the hydrate phase.

Werner Bolley fluid compositions are given in TABLE 1. The typical gas rate during the tests was about 4.0 mmscfd with typical liquid inventory of 4.5 to 5% of the total line volume. Estimated water cut of the liquid inventory varied from 0.5 to 1.5%. However, the pipeline profile had several low spots that allowed liquids to gather. Liquid hold up and water cut in these low spots were greater than the average pipeline values. For a more detailed account of the field facilities or results, refer to the summary report for this DeepStar project.[1]

In each field test, methanol injection was shut off, the line was pigged, and then the system was allowed to return to steady state without methanol. Hydrate plugs began to form as liquid inventories in the pipeline increased. Hydrate plug formation was indicated by increases in pipeline pressure drop and/or decreases in flow rate. Once plugs were formed, they were dissociated via depressurization. Plug length measurements were made as plugs pushed past a gamma ray densitometer.

RESULTS

Field Test Results

DeepStar conducted a total of five field tests at the Werner Bolley site. In four of the five tests, hydrate plugs were formed, studied, and dissociated with no hydrate inhibitors present. A fifth test was aborted before a plug hydrate formed since the project time for testing expired. Field conditions during testing, summarized in TABLE 2, created mild hydrate forming conditions. Inlet conditions were typically

TABLE 2. Field and flow loop data

Test	Field Tests					Flow Loop Tests		
	1	2	3	4	Equilibrium	Plugging #1	Plugging #2	Plugging #3
MP Flow (fps)	7–8	7–8	7–8	7–8	8	8	8	8
Cooling Rate (F/min)	NA	NA	NA	NA	0.06	0.9	1.2	6.5
Water Injection	NA	NA	NA	NA	No	No	Yes	Yes
Initial Water Fraction (Frac. of Loop Volume)	NA	NA	NA	NA	0.20	0.20	0.05	0.05
Water Fraction at Plugging (Frac. of Loop Volume)	NA	NA	NA	NA	NA	NA	0.24	0.18
Subcooling (F)	≥6	≥6	≥6	≥6	6.6	5.7	7.1	12.2
Plug (Y/N)	Y	Y	Y	Y	NA	N	Y	Y
Time to First plug (hrs)	67	63	40	135	NA	NA	3.9	2.2
Initial Plug Location[a] (sites)	2–3 and 4–5	1–2[d]	2–3	2–3 and 4–5	NA	NA	Northwest Corner	Northwest Corner
Plug ΔP[b] (psi)	112–173	NA	271	240–475	NA	NA	9.1	52.0
Plug ΔP ratio[c] (psi)	22–35	NA	54	48	NA	NA	6.5	37.0

[a]Partial hydrate plugs often began to form, then broke free and pushed down the flow line several times before creating a solid plug.
[b]ΔP developed across section of line containing hydrate plug before dissociation process initiated. Not necessarily an indication of the maximum ΔP plug might hold.
[c]ΔP ratio is defined as the ΔP after /ΔP before hydrate formation.
[d]Field test #2 was conducted on an unusually cold day and was aborted early when a hydrate plug formed in what was judged to be an unsafe location.

about 78°F and 1050 psig. Down line, typical temperatures and pressures, 53–65°F and 900–950 psig, respectively, created mild hydrate subcooling conditions of between 1–12°F (average in line length was about 6°F). Although hydrate plugs formed in each test, plugs required 2–4 days to form. Hydrates often created partial blockages between sites 2 and 3, broke loose, and pushed down the line. Mobile plugs often got stopped further down the line and created persistent blockages that had to be removed with depressurization.

Hydrate plugs formed at Werner Bolley displayed a wide range of characteristics. Some plugs were porous, whereas others were much less permeable. Plugs also appeared to have very different strengths, holding from 30 psi up to 475 psi. Estimated plug lengths ranged from 25 ft to 300 ft, but the length included all liquids pushed along in front of the plug, as well as the hydrate plug itself.

Flow Loop Results

A total of four flow loop runs were completed on Werner Bolley fluids under conditions simulating those observed in the field tests. Equilibrium conditions were measured in a traditional temperature ramping experiment and found to be 65.8°F at 989 psia, while predicted equilibrium using the *Infochem* hydrates model was 64.7°F at 989 psia.[2] In two of the three plugging experiments, water was used in addition to HBH for pressure maintenance and to more accurately simulate field conditions. Plugging was observed only in experiments with water injection, as indicated in TABLE 2, and did not occur until loop water fractions reached 0.18 to 0.24. The rate of the hydrate formation was extremely low and a surprisingly small amount of gas consumption was observed. Visual evidence from the East side of the loop suggests

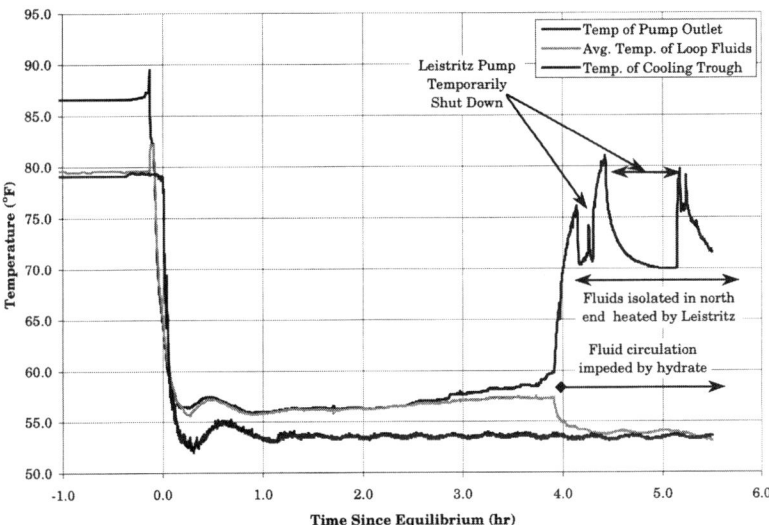

FIGURE 3. Evidence of hydrate plugging in the flow loop. Isolated fluids near pump were heated as indicated by a local temperature increase.

that the injected water was stratified along the bottom of the flow loop and slowly running to the South end of the loop. The South and West side view ports were mostly covered with hydrate, but in two instances hydrate/water slugs were observed passing the West side view port. Visual observations from the Leistritz pump outlet revealed no evidence that water or hydrate was getting back to the pump. Hydrate plugs solidified in the Northwest corner of the loop as a result of hydrate slugs getting caught in a 90° elbow or at the pump inlet. When solidified, hydrate plugs cut off flow of fluid to the pump inlet, starving the pump. Once circulation in the loop had stopped, the fluids nearest the Leistritz warmed due to heat generated in the pump, whereas fluids in the rest of the loop dropped to the temperature of the cooling trough, as illustrated by the temperature traces in FIGURE 3. As injected water fractions reached as high as 45% of the total loop volume, no liquid was observed circulating through the Leistritz pump.

Hydrate onsets occurred at an average subcooling of 6.5°F at cooling rates approximating, or less than, those observed in the field. This excludes the third plugging experiment in which cooling rates were much greater than anticipated. In that experiment, the subcooling was 12.2°F.

DISCUSSION

Results from flow loop experiments correlate well with hydrate plugging behavior observed at Werner Bolley. Flow loop studies indicated that hydrate plugs should develop at Werner Bolley and indicated that the availability of water and amount of

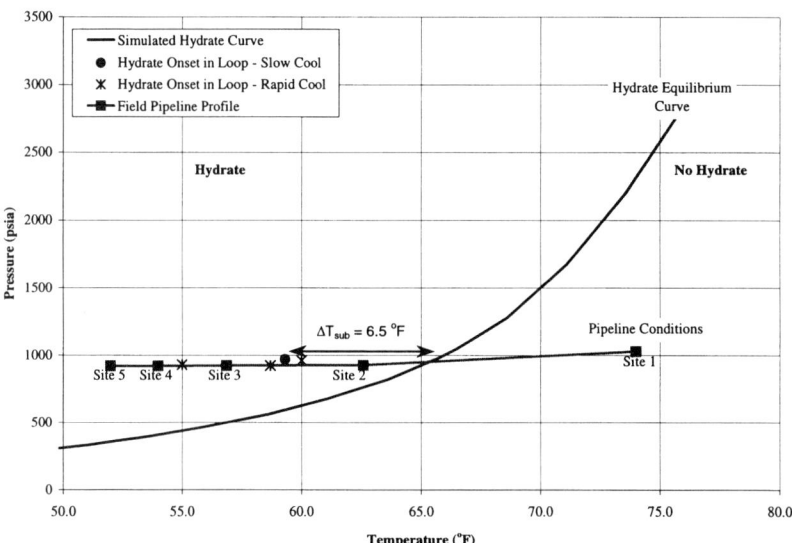

FIGURE 4. Werner Bolley pipeline profile and flow loop measured hydrate onset points. Plugs in field typically occurred between sites 2 and 3, or sites 4 and 5.

mixing between gas and liquid phases would limit hydrate formation. Field tests showed slow formation of hydrate plugs. Werner Bolley fluids required approximately six times longer to plug the flow loop than average.

As shown in FIGURE 4, hydrate plugs typically formed in the field between sites 2 and 3, or sites 4 and 5. The only exception being field test number two, when hydrates formed between sites 1 and 2. Formation in that case was induced in an above ground section of the pipeline on a colder than average day. Subcooling measured in the flow loop matches that typically encountered in the field between sites 2 and 3. Average measured subcooling for the flow loop was 6.5°F, whereas subcooling in the field between sites 2 and 3 was typically about 6°F. Although hydrate formation is certainly possible at higher subcooling (i.e. sites 4–5), the flow loop tended to predict the earliest point of hydrate formation.

Visual evidence from the flow loop indicates that liquids accumulated in low spots, liquid water partially converted to hydrate, then liquid/hydrate masses were pushed down the line until collecting enough mass to get caught in a tight spot. At Werner Bolley, hydrate blockages began to form and obstruct flow, then gave way and were caught down line to form more solid plugs.

Transient multiphase flow simulation was performed with OLGA to determine where water might have accumulated in the Werner Bolley line.[3] Simulations were conducted using the Werner Bolley pipeline profile, flow conditions, and fluid compositions with the assumption that the liquid water and condensate phases could slip. Simulation results, FIGURE 5, indicate that water could accumulate in low spots in the Werner Bolley profile. When plotted simultaneously with the flow loop results for subcooling at onset, the result is a prediction of probable hydrate formation locations.

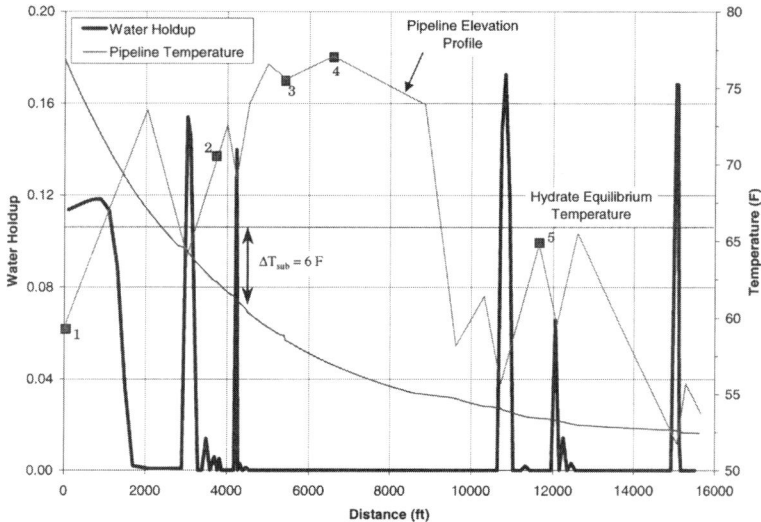

FIGURE 5. A combination of flow loop measured subcooling and flow simulation predicts the location of plugs formed at Werner Bolley field.

Results from the field varied widely depending on the location and the nature of the hydrate plugs. The observed pressure drop across hydrate plugs does not appear to be scalable between loop and field. This is not surprising given the fact that hydrate plugs at Werner Bolley had a very large variation in mass as indicated by the plug length. The mass of plugs in the field was usually many times that of the entire fluid mass in the flow loop.

SUMMARY

Experiments were conducted with Werner Bolley fluids in Texaco's high pressure hydrate flow loop under temperature, pressure, and flow conditions designed to duplicate tests conducted in the field. The results of four hydrate plugging trials at the Werner Bolley field were summarized and compared to flow loop data. It was found that plugging tendencies in the field can be predicted in a flow loop. Plugs were formed in two of three cases in the flow loop, in all completed experiments in the field. Flow loop data also indicated the manner in which hydrate plugs would form. Subcooling for which hydrate plugs formed in the field was comparable to that measured in the flow loop under similar cooling rates (field avg. 6.0°F, loop avg. 6.5°F). Flow loop results, when combined with flow simulation, correctly indicated where hydrate plugs formed in the field. Field–pressure measurements varied widely from test to test, rendering attempts at correlation pointless.

ACKNOWLEDGMENTS

The authors would like to recognize the hard work of the DeepStar engineers who carried out the Werner Bolley field tests, and thank both DeepStar and Texaco Inc. for permission to publish the data and results.

REFERENCES

1. HATTON, G.J. et al. 1997. Hydrate Plug Decomposition Test Program. Internal DeepStar 3204 Subcommittee Report.
2. INFOCHEM COMPUTER SERVICES, LTD. 1998. Multiflash Software v2.7.08. London, UK.
3. SCANDPOWER INTL. 1996. OLGA v1.1. Houston, USA.

Pilot Loop Tests of a Threshold Hydrate Inhibitor

T. PALERMO[a,b] AND S.P. GOODWIN[c,d]

[a]*Institut Français du Pétrole, 1 et 4, avenue de Bois-Préau, 92852 Rueil-Malmaison Cedex, France*

[c]*BP Amoco, Upstream Technology Group, Chertsey Road, Sunbury on Thames, Middlesex, TW16 7LN, United Kingdom*

ABSTRACT: The main objective of this work was to investigate the efficiency of a threshold hydrate inhibitor (THI 178) in terms of hydrate formation temperature under multiphase flow conditions. THI has been developed by BP and is commercialized by TROS (TROS trade name: Hytreat). THI is a mixture of polymers in a solvent carrier that inhibits the formation of hydrate nuclei in solution and at surfaces. A THI blend 178 was successfully tested on the Ravenspurn to Cleeton 16-inch gas line (Ravenspurn trial). Pilot loop tests were carried out to test the performance of THI to its limits and establish the range of its applications. The construction and operating characteristics are designed to make the loop represent a production system as closely as possible in a controlled, recirculating system. Its main outlines are a diameter of 2", a length of 140 meters and a maximum pressure of 10 Mpa (100 bara). It has been shown that THI has a marked effect on the formation conditions at concentrations of 500 ppm and above. A relationship between the subcooling temperature corresponding to hydrate appearance and THI concentration has been determined. The effect of the pressure on hydrate formation conditions was highlighted. This work has been carried out in the framework of JIP EUCHARIS.

INTRODUCTION

We present results obtained for the kinetic inhibitor THI 178 in a water and gas system on a multiphase pilot loop. THI 178 was the blend used in the Ravenspurn to Cleeton field trial conducted previously by BP.[1]

The main objective of this work was to determine the maximum achievable subcooling temperature as a function of THI concentration. The maximum subcooling temperature is defined as the difference between the temperature of dissociation and the actual temperature of formation. The work performed on THI and dry gas system is a part of a larger research programme supported by JIP Eucharis. In the framework of JIP Eucharis, supplementary tests have been carried out with other systems: gas condensate and crude oil systems with dispersant additives. These results can be found in previous papers.[2,3]

[b]Telecommunication. Voice: 33 1 47 52 67 89; fax: 33 1 47 52 70 58.
Thierry.palermo@ifp.fr
[d]goodwins@bp.com

TEST CONDITIONS

Multiphase Pilot Loop

A more detailed presentation as well as examples illustrating capabilities of the loop, are given in previous papers.[2–4] The main characteristics are reported in TABLE 1.

Fluids Composition

The system studied was composed of natural gas as the gas phase and water (non-deionized) containing additives as the liquid phase. THI is a mixture of polymers in a solvent carrier that inhibits the formation of hydrate nuclei in solution and at surfaces. This kinetic inhibitor was developed by BP and commercialized by TROS Ltd.

The natural gas is provided from the gas network. It is mainly composed of methane with small amounts of CO_2, N_2, and hydrocarbon molecules (C_2 to C_{6+}). Five different gases were used for this work. The molar concentration of propane was about 0.5% or higher leading to a preferential formation of structure II hydrate crystals. Composition of the gases are given in TABLE 2 with the corresponding

TABLE 1. Main characteristics of the pilot loop

Major specifications of the pilot loop:
    diameter: 2" (internal diameter: 49.3 mm)
    length: 140 m
    maximum working pressure: 10 MPa (100 bar)
    temperature range: 0 to 50°C

Flow specifications:
    compressor: 500 to 2000 Nm3/h
    gas velocity: 1 to 6 m/s
    volumetric pump (moineau type): 1.5 to 20 m3/h
    liquid velocity: 0.2 to 3 m/s

Fluid composition:
    crude or condensate
    non-corrosive gas
    pure or formation water

Specific equipment:
    vortex flow meter for the gas phase
    mass flow meter (Coriolis type) for the liquid phase
    gamma-ray densitometers
    transparent windows
    data acquisition and processing by a computer

TABLE 2. Composition of the five gases used for this work

	1	2	3	4	5
CO_2	0.4355	0.2766	0.3687	0.2781	0.2325
N_2	1.4733	1.6498	1.8068	1.6360	1.5676
C_1	92.1242	95.5936	94.8829	95.3613	96.0983
C_2	4.6509	1.8579	2.1969	2.0308	1.5485
C_3	0.9739	0.4483	0.5382	0.5017	0.4020
$i\text{-}C_4$	0.1432	0.0661	0.0750	0.0707	0.0582
$n\text{-}C_4$	0.1351	0.0730	0.0891	0.0820	0.0611
$i\text{-}C_5$	0.0263	0.0142	0.0168	0.0150	0.0138
$n\text{-}C_5$	0.0167	0.0105	0.0128	0.0120	0.0103
C_{6+}	0.0210	0.0121	0.0127	0.0123	0.0070
T dis (°C)					
Pa = 76 bar	15.5	13.87	14.14	14.22	13.65
Pa = 56 bar			11.97		

temperature of dissociation (calculated from the equiphase) obtained for an absolute pressure of $P_a = 76$ bara (and $P_a = 56$ bara for gas No. 3).

The kinetic inhibitor THI 178 was injected in the water phase at concentrations in the range 500–5000 wt ppm. Moreover, associated with the THI, a corrosion inhibitor C_{70} was also added at a constant concentration, 50 vol ppm. For blank tests, neither THI nor C_{70} have been injected. Both THI 178 and C_{70} were provided by TROS Ltd.

Experimental Procedure

The loop is progressively cooled at a constant pressure and under gas/liquid flow regime. During the cooling process, the temperature at which the onset of hydrate formation occurs is determined and corresponds to the so-called temperature of formation.

Pressure. Except for two tests, all the experiments were performed at a relative pressure of 75 barg. For the two others, the pressure was kept constant at 55 barg. It should be noted that, because of the flow, the pressure was not strictly constant throughout the loop. The pressure indicated above corresponds to that measured in the separator. Depending on the pressure drop in the line (DP), the pressure may be slightly higher (about 1 bar before hydrate formation) at the inlet of the line.

Temperature. The temperature in the loop was decreased at a rate between 6 and 10°C/hr from an initial temperature of about 20°C. After hydrate formation occurs, the cooling power was maintained until the loop blocks. In this case, because of the exothermic transformation of water into hydrate, the cooling of the loop generally become slower. In some cases, a heating corresponding to an exothermic peak of the temperature were observed.

The temperature was not strictly the same throughout the loop. The temperature reported in this paper was measured at the outlet and bottom of the line and corresponds to the lower temperature. It should be noted that the temperature inside the separator was not controlled. Only the separator was insulated.

Flow regime. To simulate more realistic flow conditions for dry gas systems, it was initially planned to carry out dry gas tests under a mist flow regime. Such a flow regime can be obtained at high gas velocity and low liquid velocity. To be able to generate low liquid velocity (lower than 0.1 m/s), a bypass was connected to the liquid pump. In this case, the liquid hold-up was very low and only a small amount of water flowed in the line. After several attempts, it was concluded that it was not possible to determine the temperature of formation with sufficiently accuracy. Under such conditions, too little hydrate is formed to allow correct identification of the beginning of hydrate formation.

Consequently, a stratified flow regime was chosen. This flow regime was obtained for liquid and gas superficial velocities of about 0.7 m/s and 3 m/s, respectively. The corresponding liquid hold-up in the line was about 30–35%.

DETECTION OF HYDRATES FORMATION

The onset of hydrate formation can be determined from the occurrence of several events: exothermic peak of the temperature, increase in pressure drop, beginning of the water conversion into hydrates (as determined from gas uptake), decrease of liquid density, and variation of the electrical load of the pump driver. According to results obtained for many tests, it seems that the more reliable variable is the beginning of water conversion determined from gas consumption. Thus, the onset of hydrate formation can be determined with a precision of ±3 minutes (±0.05 hr). For a cooling rate of 6°C/hr, this corresponds to a precision of ±0.3°C for the measurement of the temperature of formation.

The gas consumption (*cons*) is calculated from the pressure decrease in the gas tank (*DPbot*) corrected by thermodynamic factors. Under our conditions, we have:

$$cons/DPbot \approx mol/bar$$

The gas consumption is due both to the decrease in the temperature and to the formation of hydrate crystals. In order to evaluate the conversion of water into hydrates, at part due to the temperature effect is subtracted by extrapolating the relationship between the gas consumption and the temperature before hydrate formation occurs. The ratio of water molecules to hydrocarbon molecules to form a crystal is 5.75 (structure II with all the cages occupied). The conversion (*conv*) is expressed in terms of percentage with respect to the initial quantity of water in the loop and is plotted in the summary figures. For all the tests carried out in this work, the quantity of water in the loop corresponds to a volume of 350 liters.

EXPERIMENTAL RESULTS

Blank Tests

Two tests (DG-18m and DG-18a) were carried out without additive at $P = 75$ barg. The same gas used, gas 1. These two tests were performed one after the other with the same system. Therefore, in contrast to test DG-18m, test DG-18a was made with a system in which hydrates had been previously formed. The first test involved primary nucleation; the second was expected to involve secondary nucleation. These differences would not have a significant effect on the condition of formation, since it is expected that nucleation in the loop is heterogeneous (as in a real flow line).

The main results are shown in FIGURES 1 and 2. As can be seen, very similar results were obtained for the two tests. The temperature of formation was determined to be $T_f \approx 14°C$, corresponding to a subcooling of 1.5°C ($T_{dis} = 15.5°C$). These results indicate that the conditions of formation (without additive) can be obtained with a very good reproducibility and do not depend on previous formation of hydrates in the system.

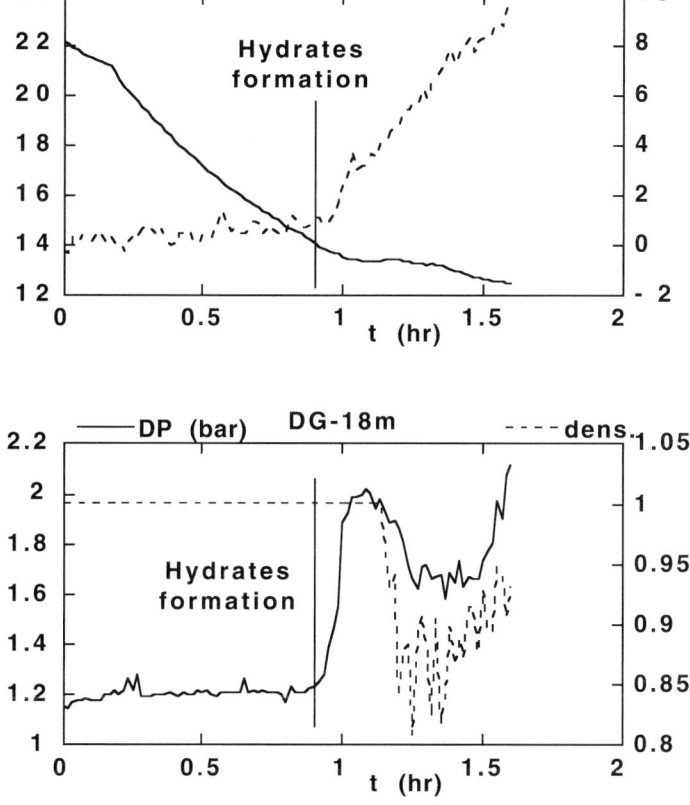

FIGURE 1. Main results for blank test DG-18m.

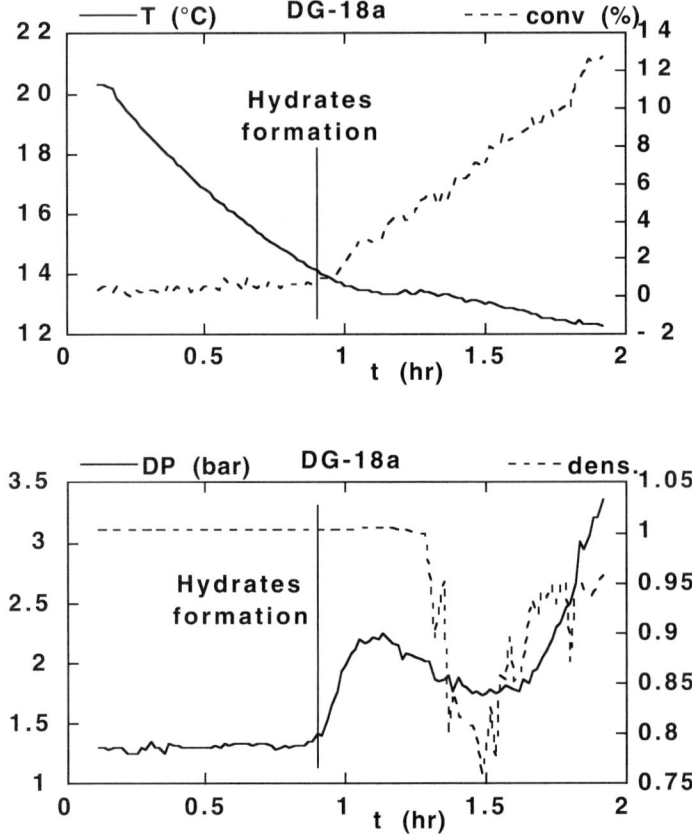

FIGURE 2. Main results for blank test DG-18a.

As soon as hydrate formation begins, the pressure drop (DP) increases abruptly. The rise of the pressure drop is due both to the formation of thick deposits on the wall and to the formation of hydrate plugs in the bulk fluid that were observed through the transparent windows installed in the line. After a short period, a decrease of the pressure drop is observed. This seems to coincide with a decrease of the liquid density (*dens*) which is measured between the outlet of the liquid pump and the inlet of the multiphase flow line by a corriolis mass flow meter. This abrupt decrease in the liquid density is interpreted as the result of a dispersion of the gas phase into the liquid phase. This phenomena was already discussed in a previous paper.[3]

500 ppm THI

Two tests (DG-20 and DG-21) were carried out with 500 ppm of THI (and 50 vol ppm of corrosion inhibitor C70) at $P = 75$ barg. The main results are shown in FIGURES 3 and 4.

The system used for these tests was the same as that used for blank tests (same liquid phase and same gas, gas 1). Therefore, hydrates were previously formed in the loop with this system. For test DG-20, technical problems met with the temperature control led to a slower cooling of the loop. This did not seem to have any effect on the conditions of formation.

As was the case with the blank tests, good reproducibility was observed. The temperatures of formation were 9.5°C (DG-20) and 9.7°C (DG-21) corresponding to subcooling temperatures of 6°C and 5.8°C, respectively. According to these results, THI additive starts to have a significant effect on the temperature of formation from 500 ppm.

A short while after hydrate formation occurs, a rapid, large increase in the pressure drop was observed, leading to rapid blockage of the loop. As previously noticed for the blank tests, thick deposits on the wall and hydrate in the bulk fluid were

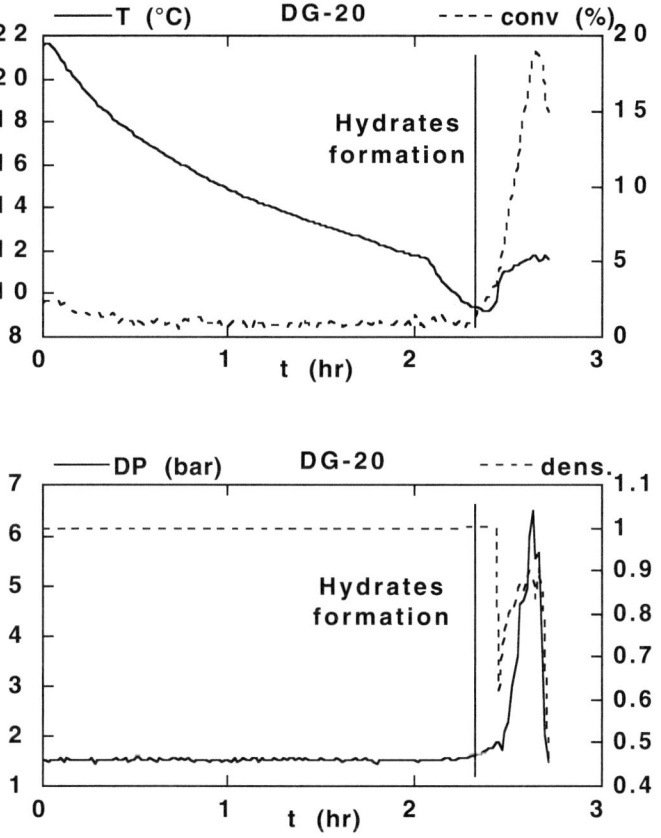

FIGURE 3. Main results for test DG-20 at 500 ppm THI.

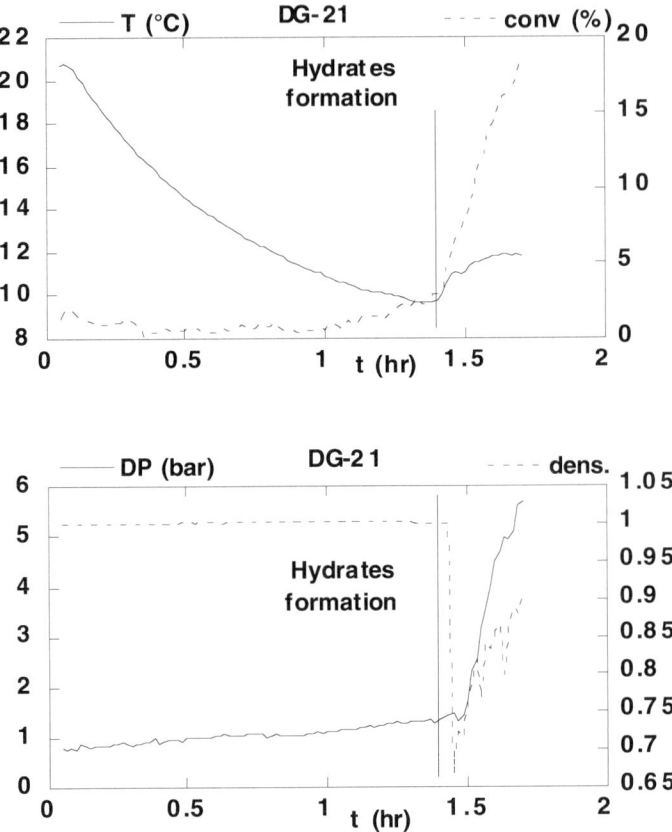

FIGURE 4. Main results for test DG-21 at 500 ppm THI.

formed. Moreover, abrupt decrease of the liquid density was also observed when hydrates are forming.

1000 to 5000 ppm THI

Tests have been carried out for THI concentrations in the range 1000 to 5000 ppm (and 50 vol ppm of C70), at $P = 75$ barg. Test conditions and the corresponding results are presented in TABLE 3. To make the comparison easier, conditions and results for previous tests are also reported. Comparison between test DG-7/3 and DG-17/3 indicates that neither previous formation of hydrates in the system, as for the blank tests, nor a change in the gas composition has a significant effect on the measured subcooling temperature.

The main result to report concerns the variation in subcooling temperature with THI concentration. As it was expected, subcooling continuously increases with the concentration and reaches a value of 11.6°C at 5,000 ppm. Although a high THI

TABLE 3. Composition and results for tests at different THI concentrations ($P = 75$ barg)

Test	Conc (ppm)	Gas	Hydrates previously formed	Temp. Formation T_f (°C)	Subcooling Temp. T_{sub} (°C)
DG-18m	0	1	No	14.0	1.5
DG-18a	0	1	Yes	14.0	1.5
DG-20	500	1	Yes	9.5	6.0
DG-21	500	1	Yes	9.7	5.8
DG-7/3	1000	2	No	6.1	7.8
DG-17/3	1000	4	Yes	6.0	8.2
DG-8/3	2000	2	Yes	4.7	9.2
DG-10/3	3000	2	Yes	3.7	10.2
DG-27/3	5000	5	Yes	2.0	11.6

concentration is necessary, the results show that the subcooling can exceed 10°C, which was previously believed to be an upper limit.

Pressure Effect

Two tests (DG-13/3 and DG-14/3) were carried out at $P = 55$ barg and for a THI concentration of 1000 ppm (and 50 vol ppm of C_{70}). The conditions and results are shown in TABLE 4. Reference results obtained at $P = 75$ barg are included for the same THI concentration. These tests confirm that it does not matter whether the fluids under test are fresh, or have previously formed hydrates that have remelted, the temperature of formation remains the same. Since the gas composition and the pressure are not identical from one test to another, it is preferable to discuss the results in terms of subcooling temperature rather than in terms of temperature of formation. For the lowest pressure, a higher subcooling temperature is reached (about 1°C more). That means that the THI efficiency may slightly depend on the mean pressure in the line. However, because of the small difference, this effect has to be confirmed.

TABLE 4. Conditions and results for tests at $P = 75$ barg and $P = 55$ barg (THI conc, 1000 ppm)

Test	P (barg)	Gas	Hydrates previously formed	Temp. Formation T_f (°C)	Subcooling Temp. T_{sub} (°C)
DG-7/3	75	2	No	6.1	7.8
DG-17/3	75	4	Yes	6.0	8.2
DG-13/3	55	3	No	2.9	9.1
DG-14/3	55	3	Yes	2.6	9.4

CONCLUSIONS

All of the results discussed above are gathered together in FIGURE 5.

THI has a marked effect on the formation conditions at concentrations at 500 ppm and above. At 75 barg, this concentration allows a subcooling of 6°C to be achieved, whereas subcooling of 9°C and more than 11°C can be reached at 2,000 ppm and 5,000 ppm, respectively.

A very good reproducibility in the measurements was obtained even when comparing results for fresh fluids with nonfresh fluids (no change of the fluids after previous tests). This confirms that heterogeneous nucleation is probably the most important mechanism involved in hydrate crystallization in the loop (and probably in a real flow line).

Tests carried out at a lower pressure ($P = 55$ barg) indicate that a higher subcooling (lower temperature of formation) may be reached.

As expected, a relationship between the subcooling temperature corresponding to hydrates appearance and THI concentration has been highlighted. By fitting experimental results obtained for different gases, under stratified water/gas flow conditions and at a relative pressure of 75 barg, we have:

$T_{sub} = (A + A'_{conc}) + B \exp(-C_{conc}); A = -8.08; A' = -7\ 10^{-4}; B = 6.46; C = 1.9\ 10^{-3}$

The curve described by this equation represents a design curve giving subcooling achievable (operating limit) as a function of THI concentration. However, this curve is probably not universal. It may depend on the presence of a hydrocarbon liquid phase and on the flow regime. As it has been shown, it also may depend on the pressure in the line. Moreover, hydrate formation is a time dependant process. Therefore, maximum subcooling temperatures achieved with respect to THI concentrations can depend on the cooling rate. In particular, for smaller cooling rates, lower subcooling temperatures are expected.

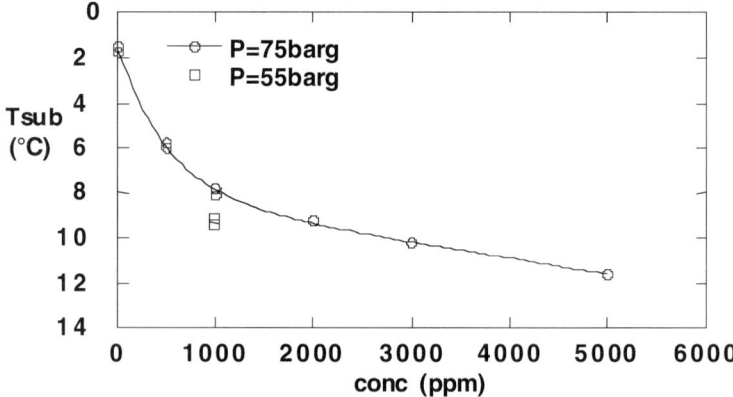

FIGURE 5. Overview of results. Subcooling temperature versus THI concentration.

ACKNOWLEDGMENTS

The authors would like to thank members of project EUCHARIS (BP Expl., British Gas, Conoco UK, Elf UK, Texaco Britain Ltd, Total Oil Marine, The UK Health & Safety Executive) for their financial support and for permission to publish these results. Special thanks to the flowloop team at the IFP Solaize facility: N. Bonnet, J.C. Declerc, and Y. Lagrange. Thanks also to Martin Grainger and Stephen Hamilton of TR Oil Services for formulation and supply of THI reagents.

REFERENCES

1. CORRIGAN, A., S.N. DUNCUM, A.R. EDWARDS & C.G. OSBORNE. SPE 30696.
2. PALERMO, T. et al. 1997. Proceedings of the Multiphase'97 Conference, Cannes. BHR Group.
3. PALERMO, T. & P. MAUREL. 1999. Proceedings of the Multiphase'99 Conference, Cannes. BHR Group.
4. PALERMO, T. & P. MAUREL. 1998. Proceedings of the IBC Conference on Controlling Hydrates, Waxes and Asphaltenes, Oslo, December, 1998.

Hydrate Plug Computer Tool

GUNNAR BORTHNE[a] AND BEN BLOYS[b]

[a]*SINTEF Petroleum Research, N-7465 Trondheim, Norway*

[b]*ARCO Exploration and Production Technology, Plano, Texas, USA*

> ABSTRACT: Petroleum transportation pipelines are occasionally plugged by gas hydrate. Such plugs can be very costly because of delayed production and complicated removal procedures. The objective of this work was to develop a prototype computer program, with a collection of calculation modules, that could be useful in connection with the removal of hydrate plugs, and that could be extended in future to a more comprehensive tool. The program presently consists of five calculation modules. Most of the focus of this paper will be on the first of these—a module that can be used to analyze pressure time series to monitor plug dissociation progress.

HYDRATE PLUG CHARACTERIZATION AND REMOVAL MONITORING

As indicated in FIGURE 1, the hydrate plug is characterized by the parameter k/L (plug permeability divided by plug length). If k increases or L decreases (i.e., k/L increases), this indicates that the amount of hydrate has decreased.

It is assumed that a permeable plug exists in a closed pipeline with a gas volume on both sides of the plug. Furthermore, a pressure change is applied on one side of the plug, and the pressures equalize by flow through the plug. The main principle in the derivation of an expression for k/L is that the gas flow through the plug (Darcy's law) is equal to the gas flow out of (or into) the closed pipe volumes adjacent to the plug (mass balance):

$$w = \dot{m}. \tag{1}$$

The flow and the mass balance sides of the equation are developed below.

Starting with the ideal gas law,

$$\rho = \frac{m}{V} = \frac{pM}{zRT}, \tag{2}$$

using $p = (p_u + p_d)/2$ as the mean pressure (to calculate a mean density), and combining this with Darcy's law (multiplied by density on both sides), gives

$$w = \rho q = -\frac{M}{zRT}\frac{Ak}{\mu L}\frac{(p_u + p_d)(p_d - p_u)}{2}, \tag{3}$$

where ρ is the density; m, the mass; V, the volume; p, the pressure; M, the molecular weight; z, the compressibility factor; R, the universal gas constant; T, the temperature; w, the mass flow rate; q, the volume flow rate; A, the cross-sectional area; μ, the viscosity; p_d, the (time series of) pressure in the volume downstream of the plug; and p_u is the (time series of) pressure in the volume upstream of the plug.

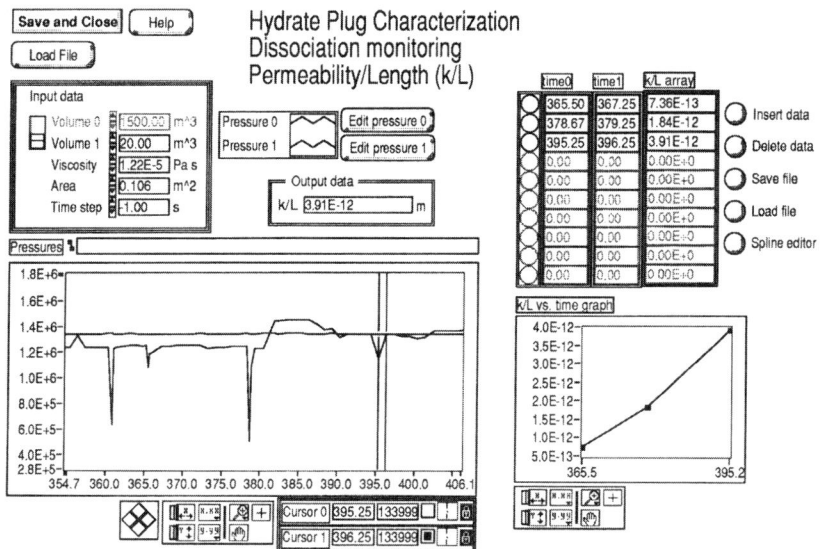

FIGURE 1. Hydrate plug characterization.

To develop the right hand side of (**1**), consider a mass balance for the closed pipe volume (V_u) at the upstream end of the plug:

$$\frac{\partial \rho_u}{\partial t} V_u + \dot{m} = 0, \qquad (4)$$

where $\dot{m}$ is the mass flow out through the volume boundaries, ρ_u is the density of the fluid in V_u, and t is time. The mass balance also holds when integrated from time 1 to time 2:

$$V_u \int_1^2 d\rho_u = V_u \Delta \rho_u = -\int_1^2 \dot{m} dt. \qquad (5)$$

Substituting the density expression from the gas law (**2**) into the mass balance, assuming constant z and T, gives

$$V_u \frac{M}{zRT} \Delta p_u = -\int_1^2 \dot{m} dt. \qquad (6)$$

The flow of expression is introduced using Equations (**1**) and (**3**):

$$V_u \Delta p_u = \int_1^2 \frac{A k}{\mu L} \frac{(p_u + p_d)(p_d - p_u)}{2} dt. \qquad (7)$$

It is assumed that k/L is constant during each test (integration period), which is short compared to the dissociation time during which k/L of course will change. Rearranging the previous equation results in an expression for k/L:

$$\frac{k}{L} = -\frac{(p_{u_2} - p_{u_1})V_u \frac{2\mu}{A}}{\int_1^2 (p_u^2 - p_d^2)dt}. \tag{8}$$

A similar expression can be derived based on the pipe volume downstream of the plug:

$$\frac{k}{L} = -\frac{(p_{d_2} - p_{d_1})V_d \frac{2\mu}{A}}{\int_1^2 (p_u^2 - p_d^2)dt}. \tag{9}$$

Consecutive pressure tests show the time development and indicate if there is progress in plug removal. The validity of the Darcy equation in this context might be questioned. However, even if there are uncertainties concerning the absolute values of the estimated parameter, the trend should be evident, and thus the method is valuable for monitoring purposes.

When running the program, the test intervals are selected interactively by the user, by dragging cursors in the time series graphs. Accepted results can be automatically stored and plotted as a function of time.

HYDRATE PLUG DISSOCIATION

As shown in FIGURE 2, the hydrate plug dissociation rate is taken to be proportional to the heat transfer rate, estimated from a steady-state formulation using an overall heat transfer coefficient. The calculation is applied repeatedly for time intervals, one for each pressure in an input pressure time series, to obtain a depiction of

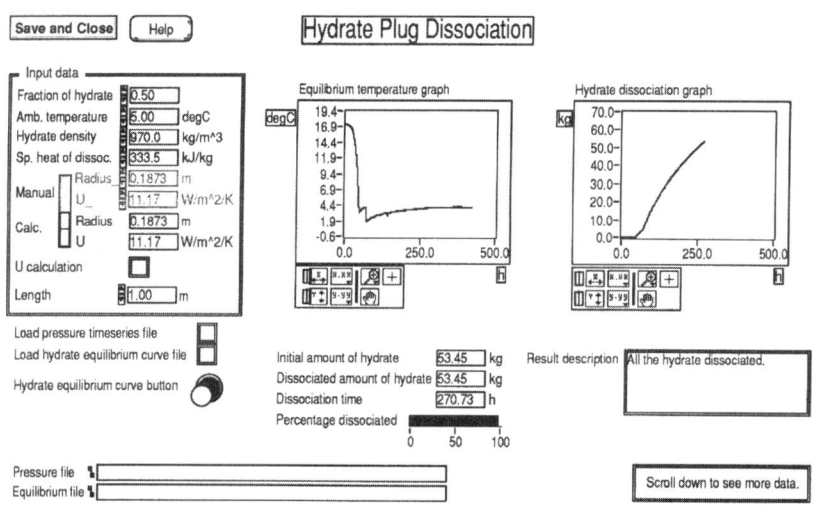

FIGURE 2. Hydrate plug dissociation.

the dynamic behavior. The dissociation temperature is interpolated for each step from hydrate equilibrium data.

It is assumed that the plug is to be depressurized from both sides (usually regarded as the preferred method), and that it is sufficiently porous and permeable so that the pressure becomes approximately the same throughout the plug within a short time compared to the dissociation time. The dissociation mode will then be radial and hydrate will dissociate simultaneously along the whole plug while heat is transferred to the plug from the surroundings.

PLUG CENTER LOCALIZATION

This module uses a simple method based on hydrate plug permeability. The pressure must be reduced or increased on one side of the plug. While the system is shut in, the applied pressure drop then will, over a period of time, and equalize through out the plug. Volumetric considerations from the recorded pressures give an estimate of the plug centre location.

PLUG END LOCALIZATION

As shown in FIGURE 3, the plug end is estimated by the "back-pressurization" method. Here the pressure must be reduced or increased on one side of the plug. Plug permeability is a disadvantage in this test, which should be performed fast enough

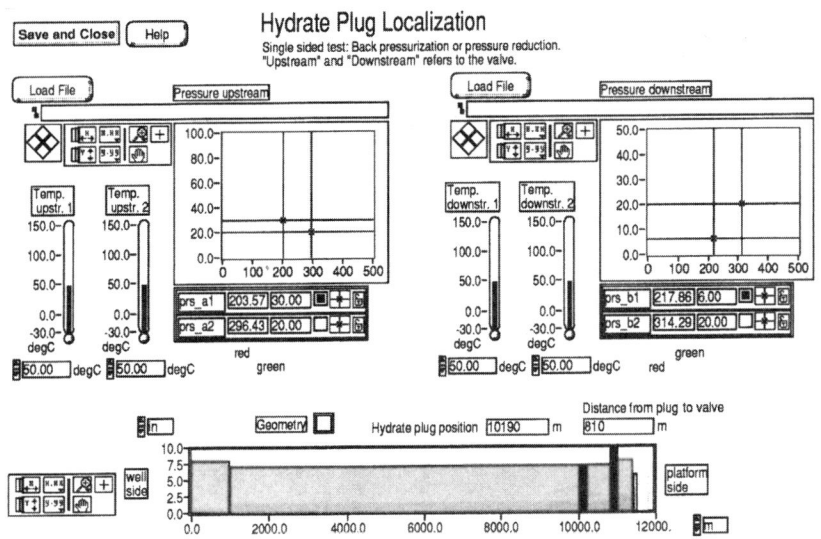

FIGURE 3. Hydrate plug localization.

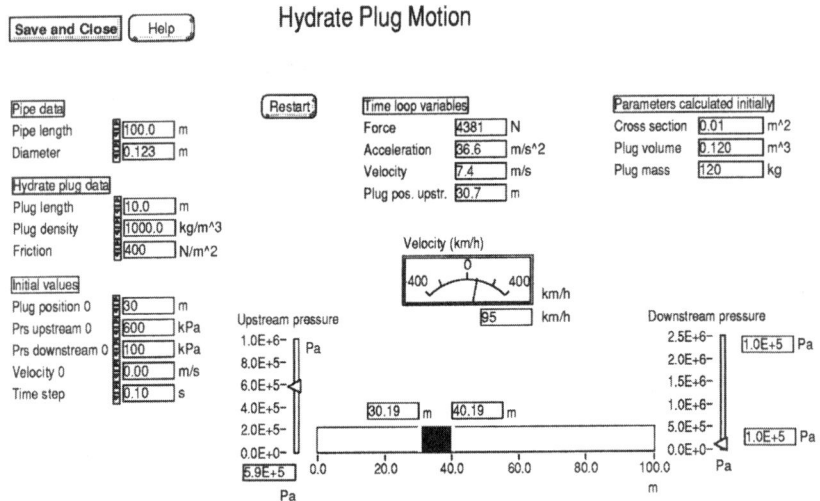

FIGURE 4. Hydrate plug motion.

to minimize loss through the plug. Volumetric considerations give an estimate of the plug end location.

For both localization modules the pressure input can be achieved interactively by dragging cursors in pressure time series graphs. The pipeline and equipment geometry is input as sections of given length and diameter, or of known volume. The liquid holdup in each section can be specified. In the output the plug position is shown graphically.

HYDRATE PLUG PROJECTILE VISUALIZATION

FIGURE 4 illustrates the module that performs estimation and visualization of a plug as it moves as a projectile in the line after detaching from the pipe wall. This can occur for plugs that are partially melted and subjected to a differential pressure due to single sided pressure reduction. A gas filled, closed pipeline is assumed. The maximum pressure is calculated.

Although the calculation modules presented are quite simple, such programs can often be useful, especially for first approximations. This work could be considered a basis for further development of new modules and functionality. There is a need for a comprehensive tool that can assist the user with the tasks of hydrate plug prevention and removal.

Flow Loop Tests on a Novel Hydrate Inhibitor to be Deployed in the North Sea ETAP Field

TH. PALERMO,[a,b] C.B. ARGO,[c,d] S.P. GOODWIN,[c,e] AND A. HENDERSON[f,g]

[a]Institut Français du Pétrole, 1 et 4, avenue de Bois-Préau, 92852 Rueil-Malmaison Cedex, France

[c]BP Amoco, Upstream Technology Group, Chertsey Road, Sunbury on Thames, Middlesex, TW16 7LN, United Kingdom

[f]BP Amoco, ETAP, Farburn Industrial Estate, Dyce, Aberdeen, AB21 7HN, United Kingdom

BACKGROUND

The Eastern Trough Area Project (ETAP) comprises an integrated development of seven oil and gas accumulations in the UK Central North Sea that lie between 130 and 145 miles East of Aberdeen in water depths of around 85 to 95 metres. Five of the fields (Egret, Heron, Machar, Monan, and Skua [future]) are subsea satellites operated by remote control from a central production platform. A sixth field (Mongo) is a not normally attended platform also operated from the central platform. Oil, gas, and produced water from the fields are transported via multiphase flowlines to the central production platform for processing.

During the concept engineering phase, three of the fields (Machar, Mungo, and Monan) were identified as potential candidates for the deployment of a novel low dosage hydrate inhibitor known as THI within BP Amoco. Extensive laboratory trials using hydrate cells were used to confirm this potential. Prior to deployment a final trial, using a flow loop, was commissioned at the Solaize test loop operated by IFP. The main purpose of this trial was to confirm the boundaries of operation for the selected inhibitor under winter conditions.

As the 16 NB, 35 km Machar flow line is one of the longest uninsulated black oil multiphase tiebacks in the North Sea, its operating conditions were selected as the basis for the flow loop trial. Construction work and production predrilling on the development commenced in early 1996, and first oil was achieved in July of 1998. All three candidate fields are currently operating using low dosage hydrate inhibitor.

[b]Telecommunication. Voice: 33 1 47 52 67 89; fax: 33 1 47 52 70 58.
thierry.palermo@ifp.fr
[d]argocb@bp.com
[e]goodwins@bp.com
[g]hendersaj@bp.com

INTRODUCTION

In this paper, we report on an investigation of the kinetic inhibitor THI 370 (TROS tradename: Hytreat 530) with the recombined Machar crude on the multiphase pilot loop installed at the IFP research center in Solaize. Prior to the flow loop trials extensive laboratory test work had identified potentially effective THI formulations. Laboratory autoclave cell test procedures for evaluation of hydrate inhibitors were reported previously.[1,2]

The main objectives of the flowloop tests were first, to demonstrate the effectiveness of the BP/TR Oil Services threshold hydrate inhibitor product Hytreat 530 in suppressing hydrate formation and second, to assess the possibilities for reduction of the THI dose rates. Pilot loop tests were carried out under conditions that simulate those expected in the Machar—Marnock black oil flow line during shut-in and restart.

Two worst cases (in terms of temperature) were considered:

- Winter conditions, shut-in and restart at the worst case sea temperature of 4.5°C.

- Summer conditions, shut-in and restart at the worst case sea temperature of 6.5°C.

The most severe Winter conditions expected in the shut-in flow line (40 barg pressure and 4.5°C temperature) were run with 2,000 wt ppm Hytreat 530. To test the possibilities for dose reduction, two options were run: Winter conditions with the dose reduced to 1,500 wt ppm and less severe Summer conditions (40 barg, 6.5°C) with 500 wt ppm and 1,000 wt ppm THI.

MULTIPHASE PILOT LOOP

A more detailed presentation, as well as examples illustrating the capabilities of the loop is given in previous papers.[3–5] The main characteristics of the multiphase pilot loop is described in TABLE 1.

TEST CONDITIONS

Fluids Composition

Hydrocarbon phase. The hydrocarbon phase was obtained by combining Machar tank crude oil with a gas phase, with the composition adjusted in order to simulate real conditions for Machar fluid composition. After inserting the liquid phases (tank crude oil and water), spiking gases, composed of CO_2, C_2, C_3, iC_4 and nC_4, were injected in the loop. The pressure was then increased up to the working pressure ($P = 40$ barg) by adding domestic gas provided from the gas network. The weight of spiking gases added in the loop, typical composition of the domestic gas, and Machar gas composition at 40 barg and 4.5°C are given in TABLE 2. Since the composition of the domestic gas could changed slightly from one test to another, the gas was sampled at the end of each test at $P = 40$ barg and $T = 4.5°C$ (or 6.5°C) in order to verify the agreement between actual and expected gas composition. Synthetic

TABLE 1. Main characteristics of the pilot loop

Major specifications of the pilot loop:
    diameter: 2" (internal diameter: 49.3 mm)
    length: 140 m
    maximum working pressure: 10 MPa (100 bar)
    temperature range: 0 to 50°C

Flow specifications:
    compressor: 500 to 2000 Nm3/h
    gas velocity: 1 to 6 m/s
    volumetric pump (moineau type): 1.5 to 20 m3/h
    liquid velocity: 0.2 to 3 m/s

Fluid composition:
    crude or condensate
    non-corrosive gas
    pure or formation water

Specific equipment:
    vortex flow meter for the gas phase
    mass flow meter (Coriolis type) for the liquid phase
    gamma-ray densitometers
    transparent windows
    data acquisition and processing by a computer

TABLE 2. Weight of spiking gases added in the loop, typical composition of domestic gas provided from the gas network, and Machar gas composition at 40 barg and 4.5°C

	Spiking gases kg	Domestic gas Mol%	Machar dry vapor Mol%
N_2		0.8844	0.4053
CO_2	5.1434	0.1219	2.1754
C_1		97.4894	88.9031
C_2	14.7819	1.1097	6.3354
C_3	11.5228	0.2830	1.5687
iC_4	2.9675	0.0413	0.1673
nC_4	6.85	0.0509	0.2884
iC_5		0.0113	0.0534
nC_5		0.0091	0.0545
C_6		0.0340(C_{6+})	0.0273
C_{7+}			0.0211

Machar gas is very close to actual field gas compositions. In some of the tests, the methane content was slightly lower, and ethane was slightly higher, than the target composition. These differences were not significant in terms of hydrate severity for the flow loop tests.

Sea-water containing THI 370. In order to simulate sea-water composition, salt (NaCl) was added to the water phase. The NaCl concentration was 35 g/l. To simulate injection of THI at the hot wellhead, the THI reagent was heated for about 5–10 minutes at 90°C in a sealed vessel (no air contact) prior to injection into the flow-loop.

Experimental Procedure

Prior to commencing the experiments, the fluids were circulated in the loop at $P = 40$ barg and $T = 37°C$ for several hours in order to ensure equilibrium conditions were obtained. The experimental procedure applied for the tests is illustrated schematically in FIGURE 1. The procedure was separated into three stages:

Stage 1. Corresponds to a decrease in temperature from 37°C to 4.5 or 6.5°C. The rate of the cooling was about 3°C/hr, leading to a duration of about 11 hours. Fluids circulated under slug flow regime.

Stage 2. Corresponds to the shut-in. The temperature was kept constant at the working temperature: $T = 4.5°C$ and $6.5°C$ for Winter and Summer conditions, respectively. The length of the shut-in was about 37 hours, corresponding to a total of 48 hours from the beginning of the test. Normally, the shut-in should have been simulated by a complete stoppage of the flow. Unfortunately, due to problems encountered with the control and/or the measurement of the temperature under such conditions, it was preferred to simulate the shut-in by a liquid flow at low flow rate.

Stage 3. Corresponds to the restart under slug flow regime for about six hours. The temperature was the same as that for the shut-in.

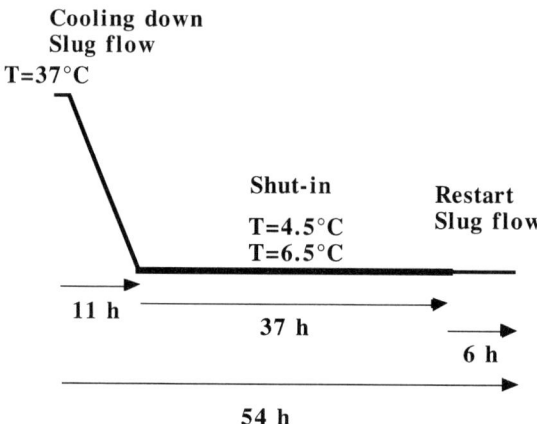

FIGURE 1. Test procedure.

The total duration of a test was 54 hours. At the end of the test, the loop was heated and depressurized. Then it was emptied. New fluids were inserted in the loop for each test.

Pressure. The pressure was kept constant at 40 barg throughout a test. It should be noted that, because of the flow, the pressure was not strictly constant throughout the loop. The pressure indicated above corresponds to that measured in the separator. Depending on the pressure drop in the line (DP), the pressure could be slightly higher (about 1 bar before hydrate formation) at the line inlet.

Temperature. The temperature decreased during stage 1 from 37°C to 4.5°C for Winter condition or 6.5°C for Summer conditions at a constant rate of 3°C/hr. The temperature was not strictly uniform throughout the loop. The temperature reported here is the minimum value. It was measured at the outlet and bottom of the line.

The separator was insulated, but the temperature inside the separator was not controlled. Moreover, to maintain the temperature at a certain level in the line, the temperature of the cooling fluid (glycol) circulating in the jacket around the flow line needed to be maintained at a lower temperature. The greatest temperature difference occurred during the shut-in stage and was about 3°C.

Flow regime. During cooling (stage 1) and restart (stage 3), the test was carried out under slug flow regime with a superficial liquid velocity $v_l \approx 0.6$ m/s and a superficial gas velocity $v_g \approx 2$ m/s. During the shut-in, for which a liquid flow was maintained, the liquid velocity was about 0.3 m/s.

Water cut. Tests were carried out at 20% or 30% water cut. Due to a separation process between water and oil which took place in the separator, the actual water cut in the flow line may be higher.

HYDRATE DETECTION

Results on hydrate formation for a blank test (without additive) are presented in FIGURE 2. The loop was progressively cooled from 37°C under slug flow regime and for a water cut of 20%. Because of the decrease in temperature, gas was delivered from the gas tank to maintain a constant pressure in the loop at 40 barg. The pressure in the gas tank continuously decreased. At $t \approx 10.5$ hr, corresponding to a temperature $T \approx 8$°C, a change in slope was observed. Simultaneously, the density of the liquid phase and the electrical load of the pump driver varied rapidly. This situation corresponds to the onset of hydrate formation. In this test, no exothermic peak temperature, or a variation in pressure drop were observed.

It should be noted that the temperature of formation ($T \approx 8$°C) measured during the blank test was significantly lower than the temperature of dissociation evaluated in a thermodynamic cell ($T = 12$°C at $P = 40$ barg). The gap between the two temperatures has already been observed for crude systems and was reported in a previous paper.[5]

Hydrate formation during a shut-in is more difficult to detect by analyzing only the pressure variation in the gas tank. FIGURE 3 presents results obtained for a test simulating Machar operating conditions for a non-water containing system. In this case, since no water was present, no hydrates were formed. During the shut-in, although the temperature was held roughly constant at $T = 4.5$°C, the pressure in the gas tank strongly decreased. No clear explanation can be given for this phenomenon. Since no gas leakage was detected and since this phenomenon was reproducible

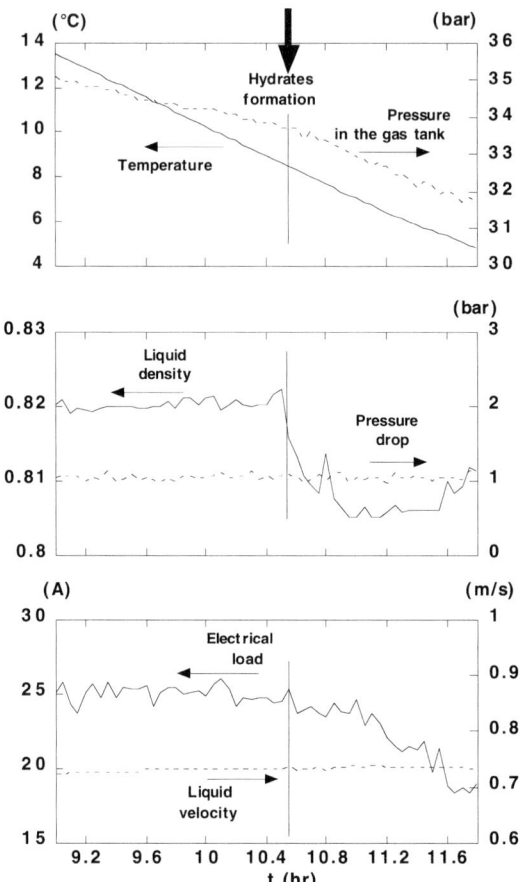

FIGURE 2. Hydrate formation for a blank test.

from one test to another, it is suspected that it may be the consequence of a low dissolution of the gas phase into the liquid phase. Moreover, at the beginning of the restart, an abrupt pressure drop in the gas tank occurred. This was the result of a rapid change in the temperature of the gas phase when it was reinjected in the flow line to generate slug flow conditions.

EXPERIMENTAL RESULTS WITH THI

2000 wt ppm THI, Winter Conditions

A test for a 2000 wt ppm THI containing system was examined at 20% water cut and under winter conditions (shut-in, $T = 4.5°C$). The results are shown in FIGURE 4.

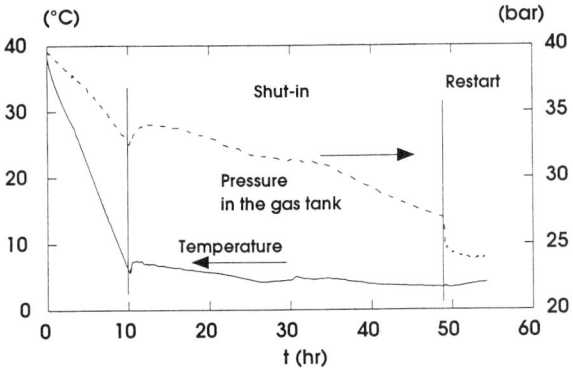

FIGURE 3. Test for a non-water containing system.

No evidence of hydrate formation was noticed. Indeed, no hydrate crystals were visible through the windows and no simultaneous events, associated with hydrate formation, were observed.

Under shut-in conditions (corresponding to a liquid flow at $v_l = 0.3$ m/s), slight fluctuations of the temperature in the line were observed. These varied over the range 2.6°C to 4.8°C, meaning that the temperature fell below the operating temperature ($T = 4.5$°C for winter conditions). These fluctuations were due to variations in the external temperature, mainly between night and day.

As previously discussed, the pressure in the gas tank decreased during the shut-in and the restart. It cannot be associated with hydrate formation, even when abrupt changes in the slope (probably related to temperature variations) were observed.

Variation in the liquid density was also observed. In particular, a rise in density occurred at the beginning of the shut-in. This was a result of a change in the water cut due to a lower liquid velocity that led to an increase in the residence time in the separator. During the restart, although identical flow conditions to those before the shut-in were restored, similar magnitude of the density was obtained. This confirms that no change in system properties was associated with the presence of hydrate particles. Moreover, no specific variation of the electrical load can be related to hydrate formation. This is well correlated with the liquid velocity. Consequently, according to the results reported above, the Winter test was successful, with 2,000 wt ppm of THI.

1,500 wt ppm THI, Winter Conditions

To test for the possibility of a dose reduction, a lower concentration (1,500 wt ppm) was investigated under Winter conditions. Two tests were performed, the first at 20% water cut and the second at 30% water cut. Results at 30% water cut are presented in FIGURE. 5.

In the two tests, hydrate formation could easily be observed. Several events occurred that can be interpreted as the result of hydrate formation: variation of the liquid density, variation of the pressure drop, variation of the electrical load and, to

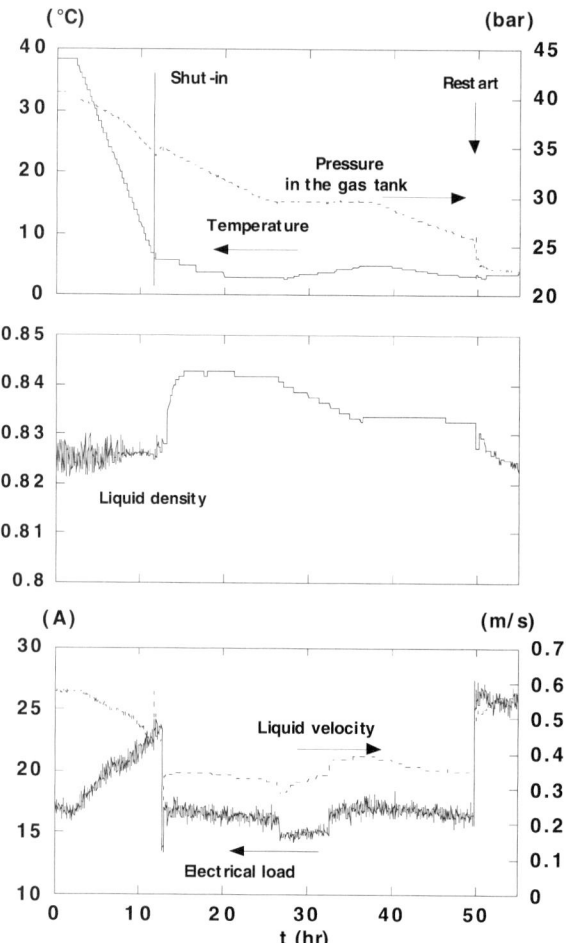

FIGURE 4. 2,000 wt ppm, Winter conditions. No hydrate formation.

a lesser extent the exothermic peak of the temperature and change in the slope of pressure in the gas tank. Moreover, the lower value of the liquid density during the restart indicates that a change in system properties, associated with hydrate formation, occurred. The presence of hydrates was also confirmed by observation through the windows of white-colored aggregates in the loop. At no point did the loop become blocked with hydrates, although a significant increase pressure drop across the loop was observed.

Because the temperature reached during the test at 20% water cut was too low (temperature excursion down to 2°C), hydrate formation occurred earlier 25 hr, whereas at 30% water cut, it occurred at 30 hr. This confirms that the lower the temperature, the shorter the duration of hydrate prevention.

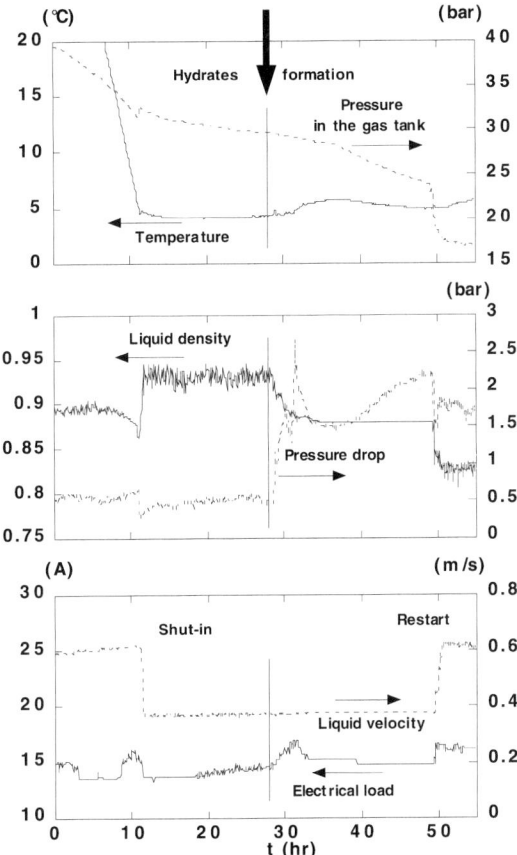

FIGURE 5. 1,500 wt ppm, Winter conditions. Hydrate formation.

Concerning these results, mainly those obtained at 30% water cut for which the temperature conditions are closer to Machar operating conditions, we can conclude that a concentration of 1500 wt ppm successfully suppresses hydrate formation for about 30 hours, but does not ensure a complete prevention for 54 hours.

500 wt ppm and 1,000 wt ppm THI, Summer Conditions

Under Summer conditions, corresponding to a higher temperature for the shut-in ($T = 6.5°C$), it was expected that a lower concentration of THI might effect a complete prevention of hydrate formation. Thus, tests were carried out at 500 wt ppm and 1,000 wt ppm. At 500 wt ppm, no evidence of hydrate formation was observed by analyzing recorded data. Variations in the main variables similar to those noticed for test at 2,000 wt ppm under Winter conditions were obtained. Nevertheless, aggregates were observed through the window installed at the outlet of the line. It is, therefore, suspected that a small quantity of hydrate crystals formed during the test. Consequently, 500 wt ppm cannot be considered effective under Summer conditions.

A test at 1,000 wt ppm confirms that a lower concentration of THI cannot prevent hydrate formation, even under Summer conditions. In this test, hydrates formation occurred at 30 hr. However, it should be noted that hydrate prevention is no longer assured after a short temperature excursion below 6°C.

CONCLUSIONS

The main conclusions from our tests are as follows:

- During a blank test, hydrates formed readily in the loop at a temperature of about 8°C, 10.5 hours into the test.

- 2,000 wt ppm Hytreat 530 successfully suppressed hydrate for 54 hours under Winter conditions (4.5°C).

- 1,500 wt ppm did not pass the Winter test criteria. However, it was observed that prior to failure, 1,500 wt ppm successfully suppressed hydrate formation for about 30 hours.

- 1,000 wt ppm successfully suppressed hydrate formation under Summer conditions for 30 hours until a low temperature excursion to 5.5°C caused hydrate formation to occur in the test loop.

- 500 wt ppm is not considered effective for suppression under summer conditions.

Although experiments were designed to be as realistic as possible, they are severe (conservative) tests of shut-in and restart. For example, no extra heating was applied during the restart phase, which would occur in the field as warm fluids start enter the flow line. The temperature of the cooling fluid circulating around the flow line is generally 3°C lower than the fluids temperature inside the line. This generated cold points on the internal wall of the line, promoting formation of hydrate crystals at the surface.

ACKNOWLEDGMENTS

The authors wish to thank the ETAP partner companies (BP Amoco, Shell, Total, Agip, Esso, MOEX, and Murphy Petroleum Ltd) for permission to publish these results. Special thanks to the flowloop team at the IFP Solaize facility, N. Bonnet, J.C. Declerc, and Y. Lagrange. Thanks also to Martin Grainger and Stephen Hamilton of TR Oil Services for formulation and supply of THI reagents.

REFERENCES

1. CORRIGAN, A. *et al.* 1995. Paper SPE 30696. 1995 SPE Annual Technical Conference, Dallas, October 22–25.

2. ARGO, C.B. *et al.* 1997. Paper SPE 37255. 1997 SPE Oilfield Chemistry Symposium, Houston, 18–21 February.
3. PALERMO, T. & P. MAUREL. 1998. Proceedings of the IBC Conference on Controlling Hydrates, Waxes and Asphaltenes, Oslo, December, 1998.
4. PALERMO, T. *et al.* 1997. Proceedings of the Multiphase'97 Conference, Cannes, BHR Group.
5. PALERMO, T. & P. MAUREL. 1999. Proceedings of the Multiphase'99 Conference, Cannes, BHR Group.

Hydrate Challenges in Deep Water Production and Operation

AJAY MEHTA,[a] JOHN WALSH,[a] AND SUSAN LORIMER[b]

[a]*Shell E&P Technology Company, Westhollow Technology Center, Houston, Texas 77082, USA*

[b]*Shell Deep Water Development Inc., One Shell Square, New Orleans, Louisiana 70160, USA*

ABSTRACT: Hydrate control is an integral part of any deep water development since hydrates exert a major impact on all aspects of systems design and operation. This paper identifies major challenges for hydrate research including high-pressure phase behavior, mixed electrolyte prediction, and kinetics. Operability issues related to deep water multiphase flowlines that ensure hydrate-free production and emerging technologies, such as low dosage inhibitors, are discussed.

INTRODUCTION

The past decade has witnessed dramatic changes in the oil and gas industry with the advent of deep water exploration and production. Until recently, each new deep water development has only shattered records of water depths considered inaccessible, from below which oil and gas are produced. Discovering oil in remote, hostile deep water environments does not however guarantee its successful and economical production. A major challenge in deep water field development is to ensure unimpeded flow of hydrocarbons to the host platform or processing facility. Managing solids—hydrates, waxes, asphaltenes, and scale—is key to the viability of developing a deep water prospect. Hydrate control in terms of its prevention, inhibition, and remediation, is central to the flow assurance strategy since it impacts all aspects of system design and operability.

What makes hydrate control in deep water unique? Deep water conditions differ from onshore and shallow water fields in two major hydrate sensitive areas, temperature and pressure. Sea bottom temperatures of 36–40°F and pressures of up to 10,000 psig are commonly encountered in deep water wells and flowlines, considerably expanding the hydrate stability zone. These flowlines lie deep within the hydrate region, with subcooling (hydrate dissociation temperature at a given pressure minus the ambient temperature) on the order of 35°F. Deep water fields are often in hilly terrain with soft seabeds leading to the installation of pipelines with low spots—ideal locations for water and hydrate accumulations.

In the Gulf of Mexico, Shell produces oil and gas from deep water wells (located on tension leg platforms TLPs) from depths greater than 4,000 feet of water. The TLP infrastructure in the Gulf of Mexico allows development of smaller subsea wells tied-back to the host platform, with offset distances ranging from 20 to 60 miles. Hydrate-free operation of these complex and lengthy flowlines is critical, since the revenue loss of a shut-in well for hydrate mitigation is very high. Subsea intervention through coiled tubing or other means is also a costly solution and adds to the hefty penalty of remediating a hydrated well or flowline. These factors lead to hydrate control philosophies that require flowlines to remain out of the hydrate region at all times during steady state operation. Achieving that design premise includes making decisions on flowline characteristics (insulation and layout), inhibition methods (typically methanol or glycol), inhibitor pumping capacities, and detailed operational procedures. Gas lines are typically operated with continuous methanol injection, but oil lines (even at low water cuts) cannot be economically operated with continuous inhibitor injection. Consequently complex inhibitor injection strategies are devised to handle transient operations such as start-up and shut-in of wells and flowlines. A clear understanding of hydrate thermodynamics and phase behavior, hydrate kinetics in terms of when, where, and how fast they form, and techniques for plug remediation, form the basis for a successful deep water development.

HYDRATE PHASE BEHAVIOR

Accurate prediction of hydrate phase behavior is essential since a majority of subsequent system design and inhibition methods are based upon it. However, the availability of a wide variety of thermodynamic hydrate prediction software packages often leads to the generation of hydrate curves to be taken for granted. This could lead to expensive mistakes, since regions predicted to lie outside the hydrate region could well still be within the hydrate zone if an incorrect hydrate structure formation is assumed. Consider the example shown in FIGURE 1 for a methane-rich gas condensate exhibiting two different types of hydrate structures—Type I and Type H. For temperatures below 50°F, Structure-H (sH) hydrates are stable, whereas at higher temperatures Structure-I (sI) is the stable hydrate structure. The ability to predict the correct hydrate structure is critical to the operation of a flowline. For the above case, if it was incorrectly assumed that only sI hydrates can form, depressurization of a shut-in flowline at 40°F down to a pressure of 600 psia would imply that the system is out of the hydrate region. In reality, however, the system would still be within the sH stability region and would require further depressurization to 450 psia to get out of the hydrate region. Similar hydrate structural transitions from sII to sI hydrates are also known to occur as a function of pressure for certain Gulf of Mexico gas condensates. The composition of the reservoir fluid also plays a critical role in the resultant stable hydrate structure as demonstrated by *a priori* predictions made at Shell by Hendriks *et al.*[1] Their work indicated that mixtures of methane and ethane, previously assumed to be only sI formers, could form sI and sII hydrates depending on their relative composition. Evidence for certain gas mixtures forming dual hydrate structures dates back to the early work of Stackelberg.[2] Therefore, the importance of accurate thermodynamic hydrate prediction cannot be overstated. A major challenge for the oil industry is to obtain high-pressure data across a wide range of gas, gas condensate, and oil systems. Extrapolation of data from existing low-pressure curves to higher pressures could cause expensive mistakes that can be avoided.

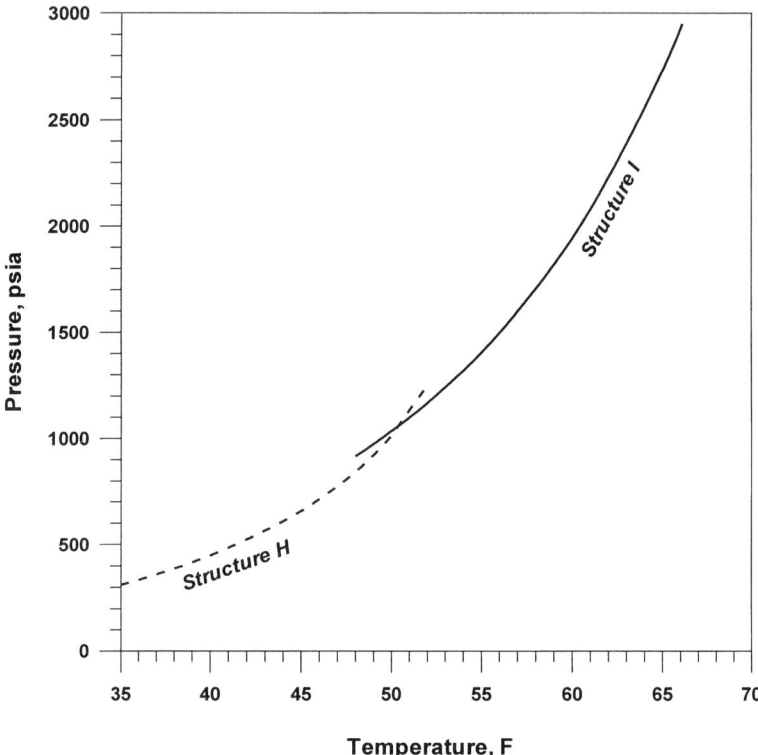

FIGURE 1. Hydrate curves for a methane-rich gas condensate.

A weakness of most prediction programs is their inability to predict hydrate inhibition in the presence of mixed electrolytes. Most deep water lines eventually co-produce formation water. The salinity of the brine produced ranges from a few hundred PPM to almost saturation. Whereas saline water provides additional hydrate protection in comparison to fresh water, most deep water flowlines still require the injection of methanol or glycol under extended shut-in and start-up conditions. The current state-of-the-art for mixed electrolyte hydrate inhibition is poor, with inaccuracies of up to ±12°F in hydrate prediction. The problem is compounded by a lack of consistent data across a range of concentrations. Joint-industry projects are currently underway to address this issue. It is also critical to have an inventory of all completion fluids and other oilfield chemicals that are used in deep water flowlines since their mutual compatibility with methanol or glycol must be tested prior to bringing a well online. Profitable subsea flowlines have been shut-in at high cost due to the inadvertent injection of methanol and glycol with completion brines, causing salt and glycol plugs rather than hydrates.

Another area of some confusion is the difference between the hydrate formation and the hydrate dissociation curve. In industry, the hydrate equilibrium curve as predicted by programs or measured in the laboratory is equivalent to the hydrate dissociation curve. Under flowing conditions it possible that hydrates may not start crystallizing soon after the reservoir fluid crosses into the hydrate stability region. Just as the freezing of liquid water into ice does not occur instantaneously at 32°F but often requires a degree of subcooling, so is the case with hydrates. Since hydrate nucleation is an inherently stochastic process it is, however, not possible to predict exactly when hydrates will form once they enter into the hydrate stability region. Subcooling to between 5 to 7°F is commonly quoted for the onset of hydrates, and there are even industry reports showing the so-called hydrate formation curve that factors in this expected subcooling. It is, however, risky to base operations on the formation curve since hydrate formation can be triggered by several factors (history of the fluid, precipitation of other solids acting as nucleation sites, etc.). Consequently, systems design should be based only on the hydrate dissociation curve, with the underlying assumption that once the fluid enters into the hydrate regime it may form hydrates immediately.

HYDRATE PLUG FORMATION AND DISSOCIATION

A major challenge for the hydrate community is to refocus its efforts on the neglected area of gas hydrate kinetics. Aside from the pioneering studies of Bishnoi et al.,[3] and subsequent bursts of sporadic interest elsewhere, hydrate kinetics remain poorly understood. The lack of systematic studies in this area has resulted in misguided conceptions in the industry about the rates of hydrate formation and dissociation.

Among the many myths surrounding hydrate formation, is an often-quoted belief in industry that oil systems do not form hydrates. Shell has had its share of hydrate plugs, including unequivocal evidence of hydrates in oil flowlines. There is sufficient gas dissolved in the oil available to form hydrates with free or emulsified water. It is speculative at this time to place a number on the low end of the percentage water cut below which hydrate plugs will not form. The formation of a hydrate plug results from a combination of several factors, including the liquid hold up in the line, the time-history of the fluid, and low spots in the line. Even fluids that show negligible or trace quantities of water in the oil have been reported to plug flowlines. In studies conducted in a 100′ long 1″ flow loop in our laboratory we have seen evidence similar to that reported by Lund et al.[4] showing the transition of hydrates from initial slurry-like to powder-like particles sweeping past the flowline. During the initial stages of hydrate formation, hydrate flakes especially are equally likely to stick anywhere along a flowline. If these hydrate flakes deposit in low spots in a line, or more susceptible cold spots in the line such as the riser base, they serve as excellent nucleation sites for further growth of hydrates and even possibly their conversion into solid plugs.

The early detection of hydrates is more difficult than it is for other solids, such as wax or scale. Scale builds up slowly and typically can be easily mitigated by chemical treatment. Wax deposition is largely driven by a thermal force between the pipe wall and the bulk fluid, with relatively slow build-up on the inner pipe wall that can be detected by a gradual increase in pressure drop across the flowline. Fairly sophisticated techniques for wax deposition have been developed at Shell to determine pigging frequencies of flowlines and to prevent a substantial build-up of wax. Moreover, wax deposition tends to occur only during the flow of reservoir fluid through a pipeline, and it is not likely during a shut-in. Hydrate formation on the other hand is much more complex, since it can form during continuous flow, shut-in, and restart of a flowline. Typically, there is not much advance warning prior to a hydrate plug blockage since the rate at which water can be converted into hydrates is extremely rapid once the initial hydrate deposits have formed. Much work remains to be done since even basic questions such as the rate of hydrate deposition, sticking properties of hydrates, porosity, and yield strength of plugs remain unanswered.

Although hydrate plug formation is not fully understood, hydrate plugs have to be remediated on a frequent basis. A thorough accounting of all the heat and mass transfer processes, as well as hydrate kinetics, is essential to building a comprehensive model for plug dissociation. It is not sufficient to have a continuous supply of heat to melt a hydrate plug, since plug dissociation, like plug formation, is not an instantaneous process. Hydrate dissociation is an endothermic process leading to a lowering of the core temperature. This, coupled with the sudden release and expansion of a substantial amount of gas trapped in the hydrate lattice, can cause Joule-Thomson cooling leading to a further cooldown and even recrystallization of hydrates. It is not unusual to expect a formation and remelting hydrate cycle during the dissociation process. Joule-Thomson cooling can also occur for gas flowing through a porous hydrate plug. We have observed this behavior in our laboratory, confirming results reported by Statoil[5] during their Tommeliten field tests. Temperatures substantially below the ice point were observed on the downstream side of the hydrate plug, which can easily freeze residual water present downstream of the plug. We are constructing a comprehensive plug dissociation model at Shell to mimic the physics of the melting process, and using field data to validate or modify the model as required.

OPERABILITY ISSUES

The basic design premise for any Shell operated deep water flowline is to keep it out of the hydrate region at all times during steady state operation. This entails continuous injection of methanol in gas and gas-condensate systems, and methanol injection during start-up and shut-in of oil systems. deep water developments, however, have their own set of unique operability issues with respect to hydrate mitigation, especially during system start-up and shut-in.

Since the continuous injection of methanol is not economically feasible for most deep water oil lines, these lines are typically provided with insulation to ensure that they stay out of the hydrate region during steady state operation. However, the transient conditions encountered during a cold well startup and extended shut-in require additional protection for the flowline in order to ensure that it stays out of the hydrate region. When a deep water system has to be shut-in, it is typically designed to provide at least three hours of no-reaction time for the well tubing, tree, jumpers, and manifold. Thus, if the system can be brought back on line within three hours, no additional measures need to be taken to ensure that hydrates do not form. The flowlines are designed to provide for up to 12 hours of no-reaction time. In addition to the cooldown time, systems are also designed to give a finite cooldown time, that is the time it takes for a system to reach the hydrate dissociation temperature after a shut-in. Flowlines are designed to give up to 24 hours of cooldown, whereas the tubing, tree, jumpers, and manifold have cooldown times of less than eight hours. The riser base is most susceptible to hydrate formation, and it typically has cooldown times of approximately 15 hours. The cooldown times, however, include the amount of time available to do something to the system by means of methanol displacement or depressurization.

If a system is to be shut-in for longer than the no reaction time, the wellbore, tree, and jumpers are displaced or "bullheaded" with methanol. This process pushes the reservoir fluids to below the subsurface safety valve (SCSSV) whose location below mudline is chosen such that it lies below the hydrate dissociation temperature at the shut-in tubing pressure. The long offset distances of the flowlines, however, make fluid displacement on shut-in very difficult. Consequently, for extended shutins, the flowlines are typically depressurized or blown down to a pressure below the hydrate dissociation pressure at ambient temperature (40°F). For typical deep water oil lines, this requires blowing down the system to pressures below approximately 350–400 psia. However, it may not be feasible to blow down some deep water flowlines with long offsets to these pressures due to the hydrostatic head of the liquid. Sometimes, initiating blow down soon after shut-in might be beneficial in removing a portion of the liquids that can be entrained in the gas. The rate of blow down is limited by the flare capacity to handle liquids. A careful evaluation of all the physical processes accompanying blow down must be considered to ensure successful depressurization of a flowline.

Similarly, several factors come into play during the start-up of a well. The rate at which the well warms up is a function of the degree of insulation in the tubing. If a well has been provided with a few thousand feet of vacuum-insulated tubing (VIT), it will warm up quickly on restart since heat losses are minimal. By the same token, however, the well will also cool down much quicker and require faster displacement with methanol or glycol upon shut-in. Another technique for bringing on a well is to outrun the hydrates. This procedure is based upon the assumption that, even though the well is initially brought on cold and it stays in the hydrate region for a finite period, there is not sufficient time available to form a hydrate plug. Consider the case shown in FIGURE 2 illustrating the start-up of a subsea oil well. The riser base lies in the hydrate stability region for less than two hours. Hydrate formation data are necessary in order to evaluate the risks of forming a hydrate plug during a cold-well start-up. Flowlines are often brought back on line by flushing out the fluids with dry or hot oil. Hot oiling has the advantage of allowing faster warm-up of the flowline which in turn reduces the volume of methanol that needs to be injected into the flowline on restart. Start-up and shut-in of wells and flowlines is an involved process and detailed logic charts are prepared for each sequence of events.

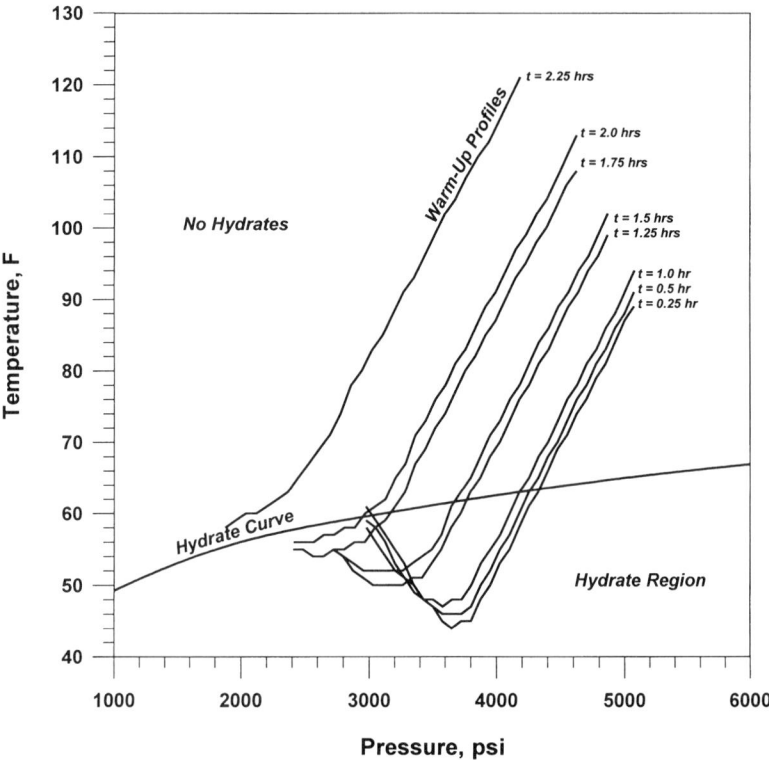

FIGURE 2. Cold well start-up of a subsea oil well.

EMERGING TECHNOLOGY

Two new hydrate control techniques that hold a lot of promise for future deep water developments are low dosage inhibitors and direct electrical heating. Low dosage inhibitors (LDIs) have been the subject of intense research over the past decade. Commercially viable LDIs are, however, just beginning to be implemented in the Gulf of Mexico and North Sea. The two major classes of LDIs are kinetic inhibitors and anti-agglomerants. Kinetic inhibitors work by suppressing the growth of hydrate crystals for a finite time designed to exceed the residence time of the fluid in the hydrate region. Kinetic inhibitors are currently in use in the North Sea for mild subcooling (below 15°F) and in the Gulf of Mexico for moderate subcooling (below 22°F) applications. Anti-agglomerants work by suspending hydrate crystals in the oil or condensate phase in a slurry, thus dispersing the hydrates and preventing their accumulation in larger solid plugs. Anti-agglomerants that are effective for subcooling of up to 30°F have been actively investigated at Shell with patents[6] issued for the most promising products. Based upon their ability to work at high subcooling and performance in extended shut-ins, anti-agglomerants are likely to find wide-scale applications in deep water field development. Another emerging area of hydrate control is in flowlines controlled by direct electrical heating (DEH). This entails switching on electrical current to keep the entire flowline above the hydrate temperature during a shut-in. Applications for plug mitigation are also envisaged. Initial field trials with DEH appear promising with field commissioning expected soon.

SUMMARY

Deep water production and operation has brought about new challenges for hydrate mitigation and control. Major efforts are required to fill several technology gaps in both hydrate thermodynamics and kinetics. The latter area in particular needs an infusion of fresh thinking and talent to address several critical plug formation and dissociation issues. New techniques, such as low dosage inhibitors and direct electrical heating, are likely to mature in the near future and possibly replace more conventional hydrate control methods.

ACKNOWLEDGMENTS

The authors thank Shell E&P Technology Company for permission to publish this work.

REFERENCES

1. HENDRIKS, E.M., B. EDMONDS, R.A.S. MOORWOOD & R. SZCEPANSKI. 1995. Hydrate structure stability in simple and mixed hydrates. *In* Proceedings of the Seventh International Conference on Fluid Properties & Phase Equilibria for Chemical Process Design.
2. STACKELBERG, M.V. 1949. Solid gas hydrates. Naturwiss. **37**: 327–359.
3. ENGLEZOS, P., N. KALOGERAKIS & P.R. BISHNOI. 1990. Formation and decomposition of gas hydrates of natural gas components. J. Incl. Phenomenon. **8**: 89–101.
4. LUND, A., D. LYSNE & T. AUSTVIK. 1991. Hydrate formation and behavior in flowing fluids. SINTEF Report STF21 F92034.
5. AUSTVIK, T., L.I. BERGE, D. LYSNE, E. HUSTVEDT & B. MELAND. 1995. Tommeliten Gamma field hydrate experiments. Presented at the Seventh International Conference MULTIPHASE 95, Cannes, France, June 1995.
6. KLOMP, U.C., V. KRUKA, R. REIJNHART & A.J. WEISENBORN. 1997. Method for inhibiting the plugging of conduits by gas hydrates. US Patent Number 5648575. July 15, 1997.

Can We Estimate the Amount of Gas Hydrates by Seismic Methods?

AKIO SAKAI

Japan Petroleum Exploration Co., Ltd. (Japex),
2-2-20 Higashi-shinagawa, Shinagawa, Tokyo 140-0002, Japan

ABSTRACT: Elastic wave velocity is the key parameter in determining the quantity of gas hydrate in sediment. Using data acquired in permafrost, elastic responses of gas hydrate bearing sediment are examined in detail by fitting observed elastic wave velocities of VSP and logging data with computed velocities based on two opposing gas hydrate bearing models. It is concluded that observed elastic wave velocity best fits a model of gas hydrate disseminated in pore space, without assuming cementation on grain boundaries in model simulations. This conclusion is important in the sense that shear wave velocity is crucial in determining gas hydrate bearing model and estimating gas hydrate saturation rate in pore space. The current case is the first example in model determination by reliable elastic wave velocity observed in the field.

INTRODUCTION

Seismic methods offer the most effective approach for indirectly estimating the spatial distribution of gas hydrates in locations where the existence of the gas hydrates is evidenced in the subsurface. By taking surveys in permafrost area, the geophysical methods were evaluated for data resolution and effectiveness in gas hydrate mapping. For gas hydrate estimation and prediction, the elastic grain composite models were examined using geophysical data.

From February to March 1998, the arctic well Mallik 2L-38 was drilled, with cores containing gas hydrate recovered below the permafrost zone in Mackenzie Delta, Canada. A multipolarized source vertical seismic profiling (VSP) with three-component receivers was conducted for well proximal seismic survey—the first field experiment related to the gas hydrate exploration in a permafrost area. Combined with surface seismic survey, these data can be spatially mapped with higher resolution. In the permafrost survey, the gas hydrate saturation rate was estimated from the velocity data combined with rock properties, such as porosity and mineralogy measured, and analyzed by using conventional wireline data and core samples. Two opposing elastic grain composite models for estimating gas hydrate saturation rates were examined. One method uses the gas hydrate disseminated in the sediment pore space, which is tentatively called the *compaction model* by analogy with the normally compacted sediment velocity trend. The increase of gas hydrate in the pore space has the weaker effect on the elastic wave velocity. The other method considers the sediment frame compliance stiffened by increasing gas hydrate as cementing agents

Telecommunication. Voice: +81-3-5461-7327; fax: +81-3-5461-7397.
akios@japex.co.jp

on the grain boundaries. This is tentatively called the *cementation model*. The increase of gas hydrates in the sediment has significant effects on the shear wave velocity as well as on compressional wave velocity. Computed elastic wave velocities were compared with the wireline logging elastic wave velocity, calibrated by the VSP survey.

Comparatively, marine surface seismic surveys have limitations in providing estimates of elastic wave velocity for quantitative gas hydrate saturation analysis, even if independent ground-truthing data are available. Elastic wave velocity can be estimated by converted shear wave travel time inversion and the image mapping method of multicomponent ocean-bottom seismographs (OBS) combined with high resolution reflection seismic surveys.[1,2]

TABLE 1. Specifications for the VSP survey of JAPEX/JNOC/GSC Mallik 2L-38 well

Period of the survey	March 24–25, 1998
Offset distances	
zero-offset VSP	33.9 m
offset VSP	400.7 m
Source of VSP	IVI MiniVibrator × 2
two for offset VSP	
one for zero-offset VSP of compressional and shear mode respectively	
Levels of zero-offset VSP	
depth interval	500–1145 m
receiver interval	5 m for compressional source
	15 m for shear source
Levels of offset VSP	
depth interval	240–1145 m
receiver interval	5 m for vertical source
Sweep	linear sweep
sweep time	12 sec
listen time	3 sec
sampling rate	1 msec
Sweep frequency bands	
zero offset compressional source	10–200 Hz
zero offset shear source	10–50 Hz
offset compressional source	10–100 Hz
Receiver cable	Schlumberger CSI × 2

DATA ACQUISITION AND ANALYSIS OF VSP DATA

Elastic wave velocity data are essential to the gas hydrate bearing sediment model. However, there have been very few reliable shear wave velocity data acquired in the field. In permafrost areas, shear wave velocity can be directly measured if polarized horizontal sources are used in the seismic survey. VSP on land is one method to directly measure shear wave velocity. One advantage of VSP is that VSP data acquisition is relatively free from adverse borehole conditions in comparison with conventional logging surveys that measure the velocity close to the perturbed well bore. Reliable *in situ* velocity data can be measured by VSP in the seismic frequency band.

The VSP survey at the Mallik 2L-38 well was conducted in March 1998. It used multipolarized sources and a three-component receiver tool for zero-offset and offset VSP (see TABLE 1). Shear and compressional wave velocities were measured by means of the travel time inversion of the first-breaks generated by a transversely polarized horizontal source and a vertical source for zero-offset VSPs. The horizontal source vibration frequency band was made as narrow as 10–50 Hz to enhance

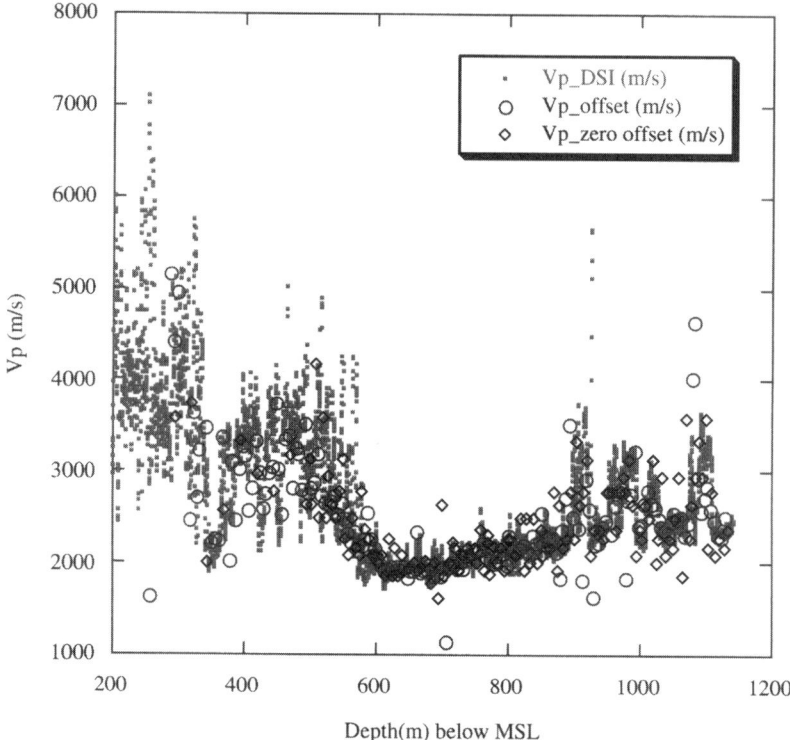

FIGURE 1. Compressional wave velocities estimated by several methods. Compressional wave velocities estimated by the wavefield inversion method for offset VSP are shown by *open circles,* and travel time inversion for zero-offset VSP are indicated by *open diamonds,* and monopole DSI sonic velocities are shown by *dots.*

source strength efficiency due to worse coupling condition of horizontal source base plate and ground surface. However, the quality of the acquired data was good enough to determine the shear wave velocity structure. Data quality of vertical sources was very good for zero-offset and offset VSP.[3]

In addition to the travel time inversion method, compressional and shear wave velocities were independently estimated by means of wavefield inversion method for offset-VSP waveforms. This was formulated in terms of the following governing equations for a linear and non-linear inversion scheme. The array of three-component seismic data over N neighboring receiver levels at a given depth z can be represented by the linear superposition of up- and down-oriented wavefields.[4,5]

$$\begin{bmatrix} \hat{u}_1(\omega) \\ \hat{u}_2(\omega) \\ \vdots \\ \hat{u}_N(\omega) \end{bmatrix} = \sum_{n=1}^{4} \begin{bmatrix} h_n \exp(i\omega s_n z_1) \\ h_n \exp(i\omega s_n z_2) \\ \vdots \\ h_n \exp(i\omega s_n z_N) \end{bmatrix} u_n(\omega),$$

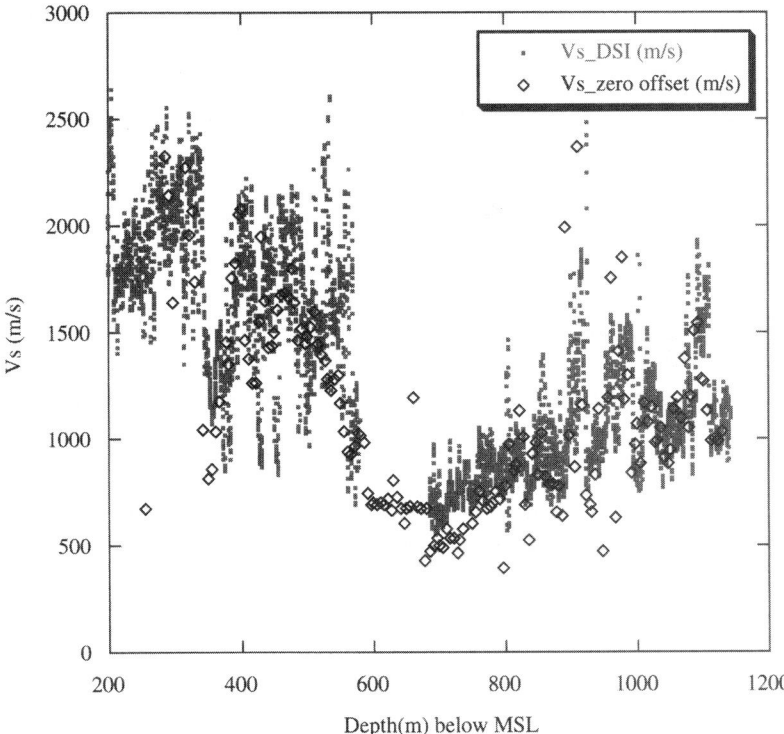

FIGURE 2. Shear wave velocities estimated by several methods. Shear wave velocities estimated by wavefield inversion method for offset VSP are shown by *open diamonds* and drift corrected DSI sonic velocities are shown by *dots*.

where $\hat{u}_k(\omega)$ is the frequency spectrum for three-component data, h_n is the unit polarization direction vector of the nth waveform, s_n is the apparent slowness of the nth waveform, and $u_n(\omega)$ is the frequency component of the nth waveform.

For our current data, starting from an initial velocity model, the separation of wavefields and the elastic velocity estimate were simultaneously executed over consecutive receiver levels. The velocity estimated by this method is consistent with the results of the travel time inversion for shear wave, as well as compressional wave velocities (see FIGURES 1 and 2).

DATA ANALYSIS OF LOGGING DATA

Log Calibration Using VSP Data

The conventional logging velocity values were calibrated by using VSP velocity values to reduce them to seismic frequency band. Logging velocity data were acquired by Schlumberger dipole seismic imager (DSI) tool in monopole P&S and upper dipole modes.[6] There were large velocity discrepancies between logging velocity and VSP velocity, especially at depths of approximately 677–889 m, and logging velocity values were higher than VSP velocity values over the entire measured depth interval.[3] This can be explained by either (1) DSI tool and analysis errors, (2) anisotropy effects, (3) velocity dispersion, or combinations of these factors. DSI data were filtered at 0.5–3 kHz and/or 1–3 kHz for upper dipole mode and reevaluated by slowness–time–coherence processing (STC) and a manual picking method. DSI analysis errors could still remain even after several reevaluated processing trials, since DSI is based on the analysis of dispersive flexural waves for a shear wave velocity estimate. Average drift values of DSI shear waves are about fifty times larger than the drift of DSI compressional wave velocity in the depth range 677–889 m. Deeper than 889m, drift values become smaller and are comparable for both compressional and shear waves.

Polarization analysis of shear wave mode VSP indicates that the mode of shear wave birefringence at depths shallower than 677 m is different from that at the deeper range. Shear waves split into a slower mode in the direction of shear source motion and a faster mode in the direction perpendicular to the source motion. This depth is close to the bottom of the permafrost layer, suggesting that the fracture orientation perpendicular to the vibrator motion prevails below the permafrost base.[5] In permafrost layers, fluid in fractures would be frozen and the azimuthal anisotropy mode is minimal. The faster mode of VSP shear wave in the transversely polarized source direction was used in the travel time analysis and it is the major energy component in wavefield inversion processing. Logging data was measured by using only upper dipole mode and, therefore, faster shear waves in both transverse and radial direction remain unseparated in the transmission path in DSI measurements.

Another possible explanation is velocity dispersion over seismic and logging source frequencies (approximately 1–30 kHz). Models of elastic medium responses with liquid coupling such as Biot, squirt, and others predict comparable rates of dispersion for both compression and shear waves. Unless there are unknown mechanisms for velocity dispersion, anisotropy effects provide the most likely explanation for the estimated velocity differences between VSP and conventional logging data below the permafrost zone.

Characteristics of Calibrated Velocity Data

A crossplot of the drift corrected compressional and shear wave velocity is illustrated in FIGURE 3. This crossplot is fitted by a linear regression curve over 685–1,140 m depth interval including the zone of gas hydrate occurrence.[5] Logging data scatters in a crossplot indicate that they can be fitted by a bilinear curve of slightly different gradients that is related to the gas hydrate saturation rate and lithology estimate.

$$V_s = -622.69 + 0.67032 V_p,$$

where V_p is the compressional wave velocity in m/sec and V_s is the shear wave velocity in m/sec.

Poisson's ratio computed by calibrated elastic wave velocities are fitted by a linear regression curve over the 685–1,140 m depth interval as illustrated in FIGURE 4. The velocity increment and Poisson's ratio decrement with depth from these quasi-

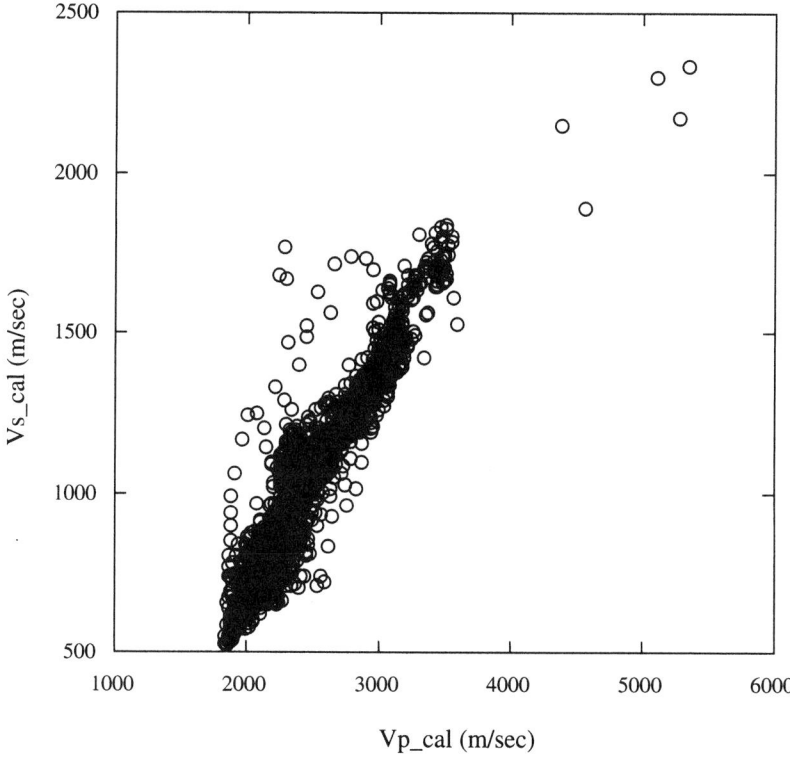

FIGURE 3. Crossplot of the drift corrected compressional and shear wave velocity of Mallik 2L-38 well.

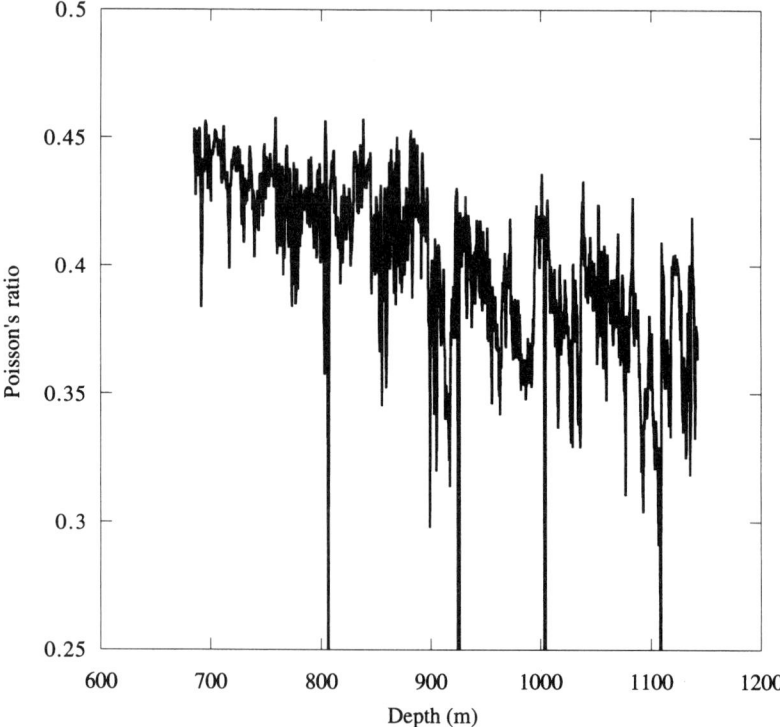

FIGURE 4. Poisson's ratio computed from the calibrated elastic wave velocities in the depth interval 685–1,140 m below KB.

linear trends indicate a hydrate saturation layer, which was evidenced by core sampling at depths of 886–952 m, where gas hydrate was recovered in sandy sediment with more than 70% pore space saturation rate. There was no indication of free gas from several *in situ* and logging measurement tools over the entire depth range.

Porosity, water saturation, and mineral composition were estimated from a neutron–density crossplot, Archie's relation, and others for conventional logging data. FIGURES 5 and 6 are relative histograms of estimated porosity and hydrate saturation rate over the 800–1,140 m depth interval. The hydrate saturated sediment is in clean to shaly sand facies with less than about 20% clay content in bulk and its porosity value averages 30–35%. Hydrate saturated layers exceeding 40% saturation rate in pore space are in almost clean sand facies. It is striking that the distribution of hydrate saturation rate in pore space is characterized by two isolated distribution peaks, one close to zero and the other as high as about 70% from conventional logging data estimation. In elastic wave velocity modeling, the hydrate saturated sediment can be approximated by clean sand of 30–35% porosity with more than 40% hydrate saturation rate in higher elastic wave velocity layer. Lower hydrate saturated sediment is contaminated by clay content in a lower or close to baseline velocity

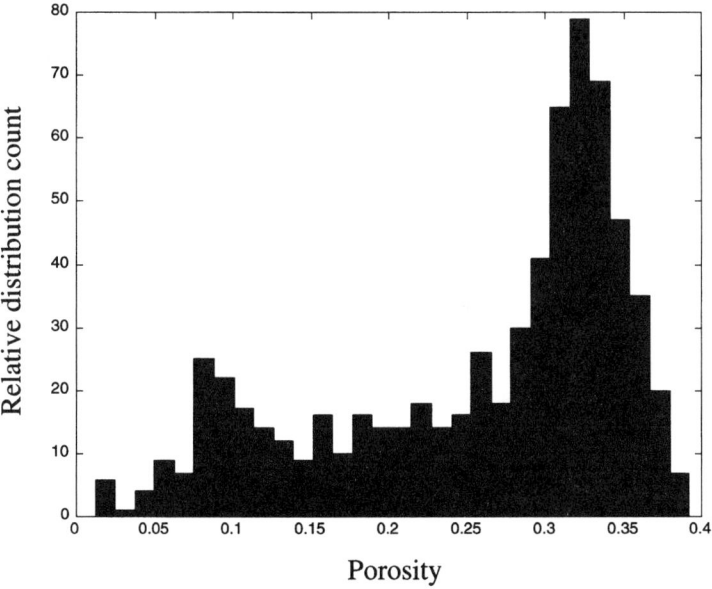

FIGURE 5. Histogram of the relative distribution of estimated porosity in the depth interval 800–1,140 m below KB.

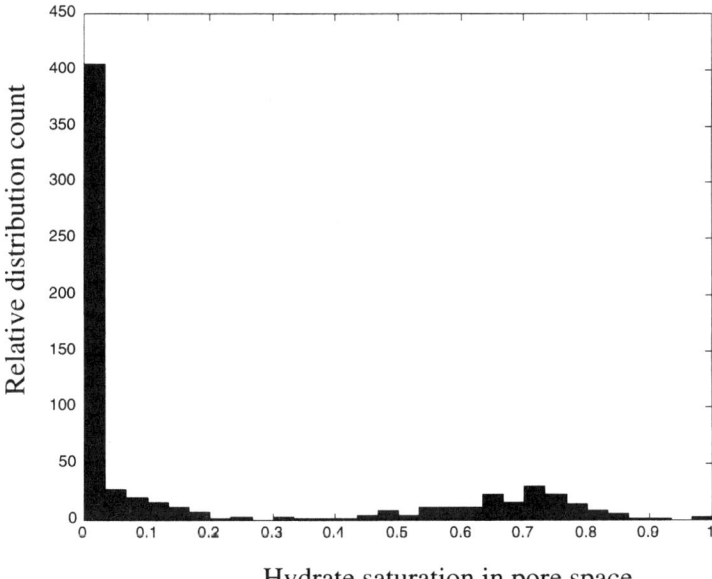

FIGURE 6. Histogram of the relative distribution of estimated gas hydrate saturation rate in pore space over the interval 800–1,140 m below KB.

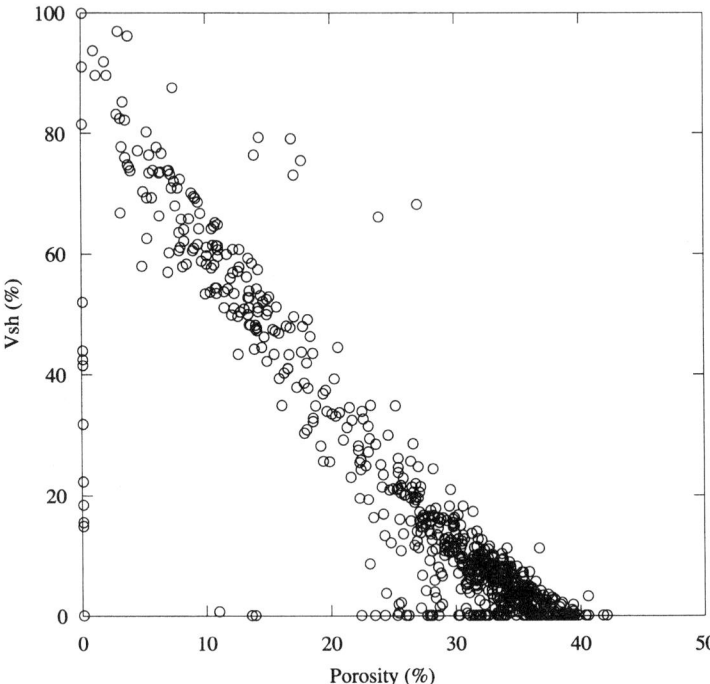

FIGURE 7. The relation between estimated clay content (V_{sh}) and porosity in the interval 800–1,140 m below KB.

trend. In FIGURE 7, clay content is illustrated as a function of porosity over the 800–1,140 m depth interval. Sediment with porosity less than about 30–35% is filled mainly by clay mineral. If the higher elastic wave velocity is presumed to indicate hydrate saturation, pure quartz composition of constant porosity is a good starting model for estimating hydrate saturation rates in the current dataset.

GAS HYDRATE BEARING MODELS

Empirical Approach versus Rock Physical Approach

The velocity baseline or reference trend is defined as the velocity of fully water saturated rock with no hydrate saturation. Seismic compressional wave velocity in a presumed gas hydrate free zone is used to provide a velocity baseline—a commonly known method for estimating gas hydrate quantity. However, there is uncertainty in determining the velocity baseline from seismic data by means of this empirical approach. A constructive approach based on the rock physics is effective in computing the elastic response with gas hydrate saturation. From other approaches, including time-average relations, it is hard to estimate gas hydrate saturation rates in a rational way without introducing empirical relations.

Compaction Model versus Cementation Model

Two opposing elastic grain composite models were examined in order to estimate the gas hydrate saturation rate. One model used the gas hydrate disseminated in the sediment pore space (compaction model). The other model used the gas hydrate as cementing agents on the grain boundaries (cementation model). The first model can be simulated by bounding models, such as the Hashin-Shtrikman-Hertz-Mindlin model.[7–9] This is a bounding model for elastic dry-frame moduli realized by connecting the point of zero porosity, where the elastic moduli are those of the solid phase, with the critical porosity, where unconsolidated rock is modeled by random packing of identical elastic spheres. At the critical porosity, the effective bulk and shear moduli of the rock are computed from the Hertz-Mindlin contact theory. The elastic moduli of the effective medium are approximated by a mixture of two hypothetical compositions with a lower Hashin-Shtrikman bound from porosity normalized at the critical porosity. Below the critical porosity, two hypothetical compositions used are the Hertz-Mindlin contact medium at the critical porosity and

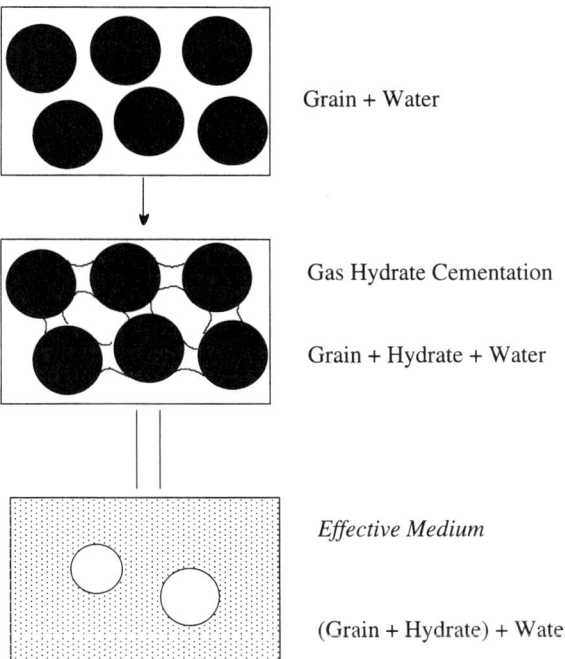

FIGURE 8. Schematic diagram of the cementation model constructed using the self-consistent approach, starting from small gas hydrate saturation rate to higher saturation rate in pore space.

the void medium with an upper Hashin-Shtrikman bound obtained by partitioning the concentration of the void and sphere packed critical phase.[10] Water saturation effects are represented by Gassmann equation with estimated dry-frame moduli.

For the cementation model, the model of Dvorkin and Nur[11] was applied in lower gas hydrate saturation conditions. For larger gas hydrate saturation conditions, hydrate saturated medium under lower saturation conditions was linked with the medium of hydrate filled in spherical voids in an effective medium realized by the self-consistent method.[12,13] The cementation model syntheses used are schematically illustrated in FIGURE 8.

ELASTIC WAVE VELOCITY ESTIMATES

Crossplot of Compressional and Shear Wave Velocities

The physical parameters used in our computations are shown in TABLE 2.

Compression and shear wave velocities are computed for compaction model and for the cementation model. FIGURES 9 and 10 illustrate crossplots of the compression and shear wave velocities in the cases of mineral composition of quartz, gas hydrate, and water at 10 MPa effective pressure with porosity ranging from 0.1 to 0.4 and gas hydrate saturation rate from 0 to 0.99 at 0.01 volume fraction interval for compaction models and cementation models, respectively. The crossplot of the cementation model is shifted to a larger shear wave velocity trend, even with a small amount of gas hydrate saturation in comparison with the calibrated crossplot. In compaction models, the compressional wave velocity increases at a larger rate than the shear wave velocity with the increment of gas hydrate saturation by changing the porosity of quartz grain models. The general trend of the crossplot for compaction models falls on the observed calibrated trend. Thus, models can initially be examined by using the crossplot trend of elastic velocities with an estimated gas hydrate saturation. If the critical porosity decreases, the crossplot follows a more compliant trend of the shear wave velocity. The effect of clay content in gas hydrate saturated sediment is examined by synthetic computations with clay and quartz volume fraction changing by 0.1 fraction intervals holding the porosities constant and hydrate saturation rate varying from 0–0.9 in 0.1 intervals. These are illustrated in FIGURE 11. Since in the sediment with higher hydrate saturation rate the clay content is small, the scatter in the synthetic V_p versus V_s in the higher velocity zone is an exaggerated estimate in comparison with the calibrated velocities. The scatter in lower velocities

TABLE 2. Physical parameters of sediment compositions

	Density (g/cm^3)	Bulk modulus (GPa)	Rigidity (GPa)
Quartz	2.65	36.0	45.0
Clay	2.58	20.9	6.85
Calcite	2.71	76.8	32.0
Gas hydrate	0.91	5.6	2.4
Water	1.02	2.29	0.0

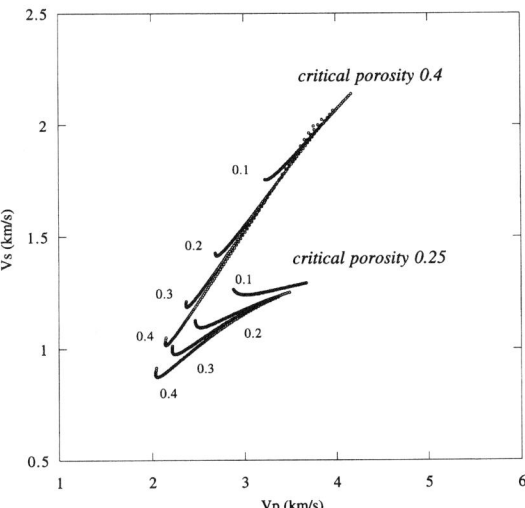

FIGURE 9. A crossplot of compression and shear wave velocity in a compaction model synthesized under the porosity condition 0.1, 0.2, 0.3, and 0.4 and gas hydrate saturation rate in the range of 0–0.99 at 0.01 fraction interval. Bulk density and solid bulk moduli are computed by assumed mineral composition of 100% quartz, gas hydrate, and water. Critical porosities and effective pressure are assumed to be 0.25, 0.4, and 10 MPa. Porosity values are shown at each branch of dot line segments.

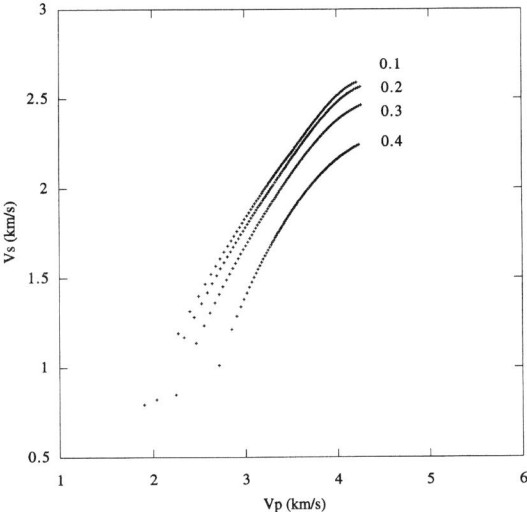

FIGURE 10. A crossplot of compressional and shear wave velocity in a cementation model synthesized under the porosity condition 0.1, 0.2, 0.3, and 0.4 and gas hydrate saturation rate in the range of 0–0.99 at 0.01 fraction interval. Bulk density and solid bulk moduli are computed by using an estimated mineral composition of 100% quartz, gas hydrate, and water. Effective pressure is assumed to be 10 MPa. Porosity values are shown at each branch of dot line segments.

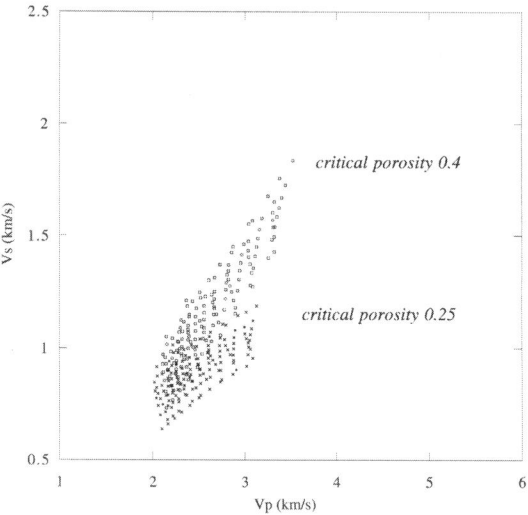

FIGURE 11. A crossplot of synthetic elastic wave velocities in which the partition of quartz and clay composition is varied under constant porosity conditions with gas hydrate saturation rate variation. The porosities used are 0.3 and 0.4. Gas hydrate saturation rate in pore space is varied from 0–0.9 in 0.1 volume fraction intervals.

is explained in terms of the mineral composition fluctuation. The variation of effective pressure estimate is not influential in the elastic velocity variations.

Comparison between Model and Calibrated Wave Velocities

Elastic wave velocities were fitted by a compaction model with varying critical porosity values and by a cementation model, synthesized for an estimated mineral composition, recomputed bulk density based on the estimated mineral composition, and computed effective pressure. The elastic moduli are computed as the Hill average of the elastic moduli of mineral composition.[14] The computed values are compared with the calibrated elastic wave velocity in the depth range of 880–925 m and 950–1,000 m, respectively, in FIGURES 12 and 13 for compaction models, and in the depth range 880–925 m for a cementation model in FIGURE 14. The overall fit with the calibrated velocities is poor for the cementation models. The synthetic elastic wave velocities of the cementation model are overestimated in the lower saturation interval, and in the higher saturation interval shear wave velocity is also overestimated. In the lower depth interval, gas hydrate core was recovered and hydrate saturation rate was estimated. If the critical porosity increases in compaction models, the fit of the velocities improves in the depth ranges 906–925 m and 950–1,000 m. In the shallower interval, the fit is improved by selecting a smaller critical porosity values.

Elastic velocities in the depth range 800–1,140 m are synthesized by compaction model as a function of gas hydrate saturation rate as shown in FIGURE 15, where computations using critical porosity values of 0.25 and 0.4 are overlaid with calibrated elastic wave velocities. If a critical porosity 0.4 is assumed, the data fit better over

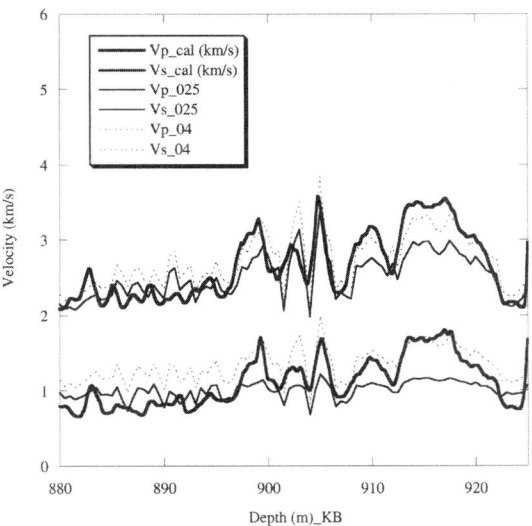

FIGURE 12. Calibrated elastic wave velocity of logging data and synthesized elastic wave velocity for compaction model in the depth interval 880–925 m below KB under conditions of estimated mineralogy. Bulk moduli, bulk density, and effective pressure are computed for the estimated mineralogical composition. Critical porosity is varied at 0.25 and 0.4. e.g., Vp_0.4 in figure legend means synthesized compressional wave velocity at critical porosity 0.4. Calibrated elastic wave velocities are shown as Vp_cal and Vs_cal.

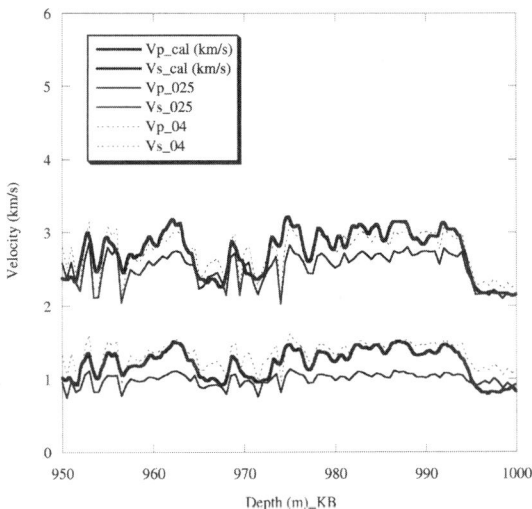

FIGURE 13. Calibrated elastic wave velocity of logging data and synthesized elastic wave velocity for compaction model in the depth interval 950–1,000 m below KB under the condition of estimated mineralogy. Bulk moduli, bulk density and effective pressure are computed by the estimated mineralogical composition. Legend has the same meaning as FIGURE 12.

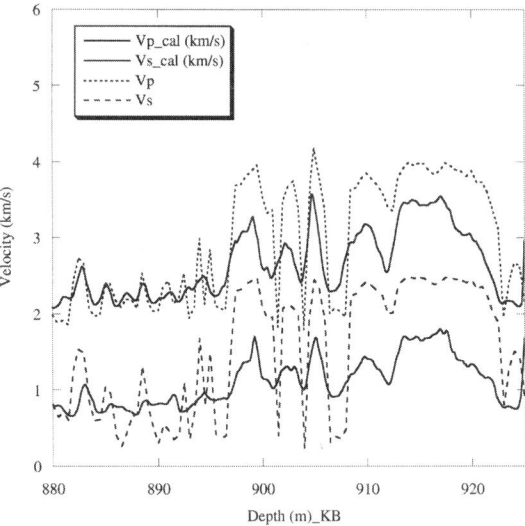

FIGURE 14. Calibrated elastic wave velocity of logging data and synthesized elastic wave velocity for cementation model in the depth interval 880–925 m below KB under the condition of estimated mineralogy. Bulk moduli, bulk density and effective pressure are computed by the estimated mineralogical composition. Calibrated elastic wave velocities are shown as Vp_cal and Vs_cal. Vp and Vs are the synthesized compressional and shear wave velocity, respectively.

a wide range of hydrate saturation rates. At lower hydrate saturation rates, calibrated shear wave velocities match with synthetic velocities of a compaction model with critical porosity 0.25 better than for a critical porosity of 0.4. This suggests that the clay content contributes to the velocity under lower hydrate saturation rate conditions.

Critical Porosity versus Mineral Composition

The critical porosity is defined as the transition porosity at the state of internal topology of grain aggregate between the state of supporting fluid and rarely contacting grains and the state of supporting solid frame. If an aggregate of identical elastic spheres is assumed, the critical porosity is determined as the porosity of random packing. Under the inhomogeneous distribution of aggregated grain radius, the critical porosity would be a random variable of the radius distribution. The reduction of critical porosity would then be illustrated by a simple binary model of sand and clay mixture as shown in schematic diagram of FIGURE 15.[5] This mechanism of critical porosity reduction partly agrees with larger clay content in shallower than 900 m. In the shallower range, gas hydrate saturation rate is smaller, and then a simple binary model of sand and clay mixture might approximate the critical porosity variation to fit elastic velocity assumed in simulation.

Another important point in the model identification is how to estimate mineral composition before conventional logging data is available. In the current dataset, the major component of the sediment is quartz and clay. Clay locates in the same place

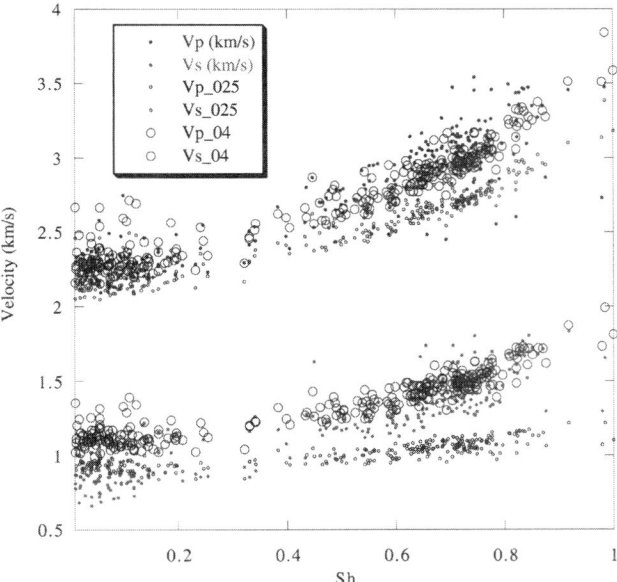

FIGURE 15. The relation between estimated gas hydrate saturation rate and computed elastic wave velocities synthesized by compaction model of critical porosity 0.25 and 0.4. e.g., Vp_0.4 in figure legend means synthesized compressional wave velocity at critical porosity 0.4. Calibrated elastic wave velocities are shown as Vp and Vs. Sh means the gas hydrate saturation rate in pore space.

as gas hydrate in the framework of quartz grain aggregate, in either the clay cementation mode or the clay compaction mode. The elastic moduli of clay are different from those of gas hydrate and this makes it possible to estimate clay and gas hydrate quantities by an inversion scheme when both quantities are relatively large in the solid phase. Such an approach was not taken since the sediment of higher saturation rate of clay mineral corresponds to the sediment of the lowest saturation rate of gas hydrate in the current dataset. In an other sedimentological setting, it might be necessary to analyze it in this way.

CONCLUSIONS

In order to determine unbiased elastic velocity distribution, a multipolarized source VSP survey was conducted with a three-component receiver cable incorporated with a conventional wireline survey. This was the first well-proximal seismic survey for hydrate research in permafrost.

Compression and shear wave velocities are estimated for offset VSP data by a wavefield inversion method. They prove to be consistent with the estimated results from a travel time inversion method applied to zero-offset VSP data.

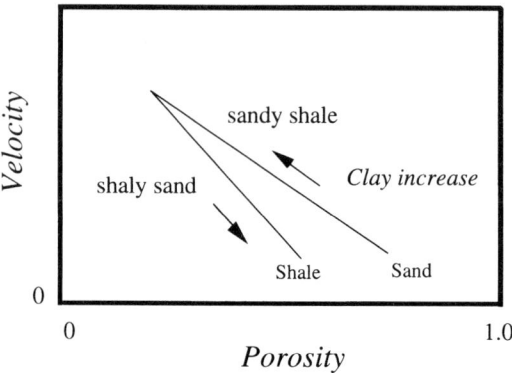

FIGURE 16. Critical porosity reduction of a binary system of sand and clay mixture.[5]

To estimate gas hydrate saturation rates, two extreme hydrate saturation models were examined. One is a compaction model, where hydrate is disseminated in pore space, and the other is a cementation model, where hydrate is cemented on grain boundaries and stiffens the sediment frame compliance. Combined with the analysis of wireline data, it was concluded that the model of hydrate disseminated in pore space simulates the estimated elastic velocity and hydrate saturation better than does the cementation models. In this analysis, it turns out that critical porosity values are crucial for estimating hydrate saturation rates and dependent on the clay content under a simple sand and clay binary mixture model.

Returning to the question of the effectiveness of the seismic or elastic wave velocity analysis method, it can be concluded that shear wave velocity data is decisive in model fixing and gas hydrate saturation estimation.

ACKNOWLEDGMENTS

I wish to express my sincere gratitude to Jack Dvorkin for his stimulating discussions on the rock physical approach for gas hydrate problems. Conventional wireline logging and coring operation was conducted under the Collaborative Research Program for drilling of the Japex/JNOC/GSC Mallik 2L-38 well. I thank all participants of this program. The operation of the VSP survey is financially supported by the Research Consortium of Methane Hydrate Exploration in Japan (JNOC, Japex and nine other organizations).

REFERENCES

1. SAKAI, A. 1997. Seismic studies on marine gas hydrates offshore Japan. Proceeding of the International Workshop on Gas Hydrate Studies at Tsukuba, Japan, March 1997.
2. SAKAI, A. 1997. Seismic studies related to gas hydrates offshore Japan. AGU Fall meeting abstract, December 1997.

3. SAKAI, A. 1998. Vertical seismic survey in the Mallik 2L-38-specifications, data acquisitions and data analysis. Proc. International Symposium on Methane Hydrates, Chiba, Japan, Oct. 20–22, 1998. 359–370.
4. LEANEY, W.S. 1990. Parametric wave-field decomposition and applications, SEG Expanded Abstract. 1097–1099.
5. SAKAI, A. 1999. Velocity analysis of vertical seismic profile (VSP) survey at JAPEX/JNOC/GSC Mallik 2L-38 gas hydrate research well, and related problems for estimating gas hydrate concentration. *In* Scientific Results from JAPEX/JNOC/GSC Mallik 2L-38 Gas Hydrate Research Well, Mackenzie Delta, Northwest Territories, Canada; Geological Survey of Canada, Bulletin **544:** 323–340.
6. KURKJIAN, A.L. & S.-K. CHANG. 1986. Acoustic multipole sources in fluid-filled boreholes. Geophys. **51:** 148–163.
7. MINDLIN, R.D. 1949. Compliance of elastic bodies in contact. J. Appl. Mech. **16:** 259–268.
8. HASHIN, Z. & S. SHTRIKMAN. 1963. A variational approach to the elastic behavior of multiphase materials. J. Mech. Phys. Solids **11:** 127–140.
9. DVORKIN, J. & A. NUR. 1996. Elasticity of high-porosity sandstones: Theory for two North Sea datasets. Geophys. **59:** 428–438.
10. DVORKIN, J. & A. SAKAI. 1997. Modification of the heuristic H-S model. Unpublished.
11. DVORKIN, J. & A. NUR. 1993. Rock physics for characterization of gas hydrates, the future of energy gases. USGS Professional paper 1570. 293–298.
12. WU, T.T. 1966. The effect of inclusion shape on the elastic moduli of a two-phase material. Int. J. Solids Struct. **2:** 1–8.
13. BERRYMAN, J.G. 1980. Long-wavelength propagation in composite elastic media. J. Acoust. Soc. Am. **68:** 1809–1831.
14. HILL, R. 1952. The elastic behavior of crystalline aggregate. Proc. Phys. Soc. (Lond.) **A65:** 349–354.

A Consistent Thermodynamic Model for Predicting Combined Wax-Hydrate in Petroleum Reservoir Fluids

A.R. TABATABAEI, A. DANESH, B. TOHIDI,[a] AND A.C. TODD

Department of Petroleum Engineering, Heriot-Watt University, Riccarton, Edinburgh EH14 4AS, United Kingdom

ABSTRACT: Low oil prices and the competitive nature of the oil industry worldwide demand more cost-effective techniques in the development and operation of offshore reservoirs. Extended-reach gathering networks and transportation of unprocessed well streams by subsea pipelines are two attractive options that could have significant impact in the development of marginal oil/gas fields. These subsea pipelines are prone to wax and/or hydrate formation that can potentially block the pipelines and lead to serious safety and operational problems. Wax and hydrate formation are both examples of solid deposition. They are modelled conventionally and are studied independently. In this work we describe a wax model that has been developed based on regular solution theory. The wax model was successfully coupled with an existing hydrate model. The integrated wax-hydrate model is capable of predicting five phase equilibria (i.e., water, liquid hydrocarbon, vapor, wax, and hydrate) as well as predicting the effect of wax formation on the hydrate phase boundary and vice versa. The results show that a reliable wax-hydrate model can minimize the risks involved in the transportation of fluids prone to wax and hydrate formation. The model can be used as a powerful tool in the design and operation of subsea and arctic pipelines and production facilities.

INTRODUCTION

Petroleum waxes are complex mixtures of high molecular weight saturated hydrocarbons, predominantly paraffins. Hydrates are crystalline compounds that form at elevated pressures and low temperatures when water is in contact with specific light gases. In order for the wax to be formed at typical operating conditions, there is a need for fairly high molecular weight hydrocarbons, usually above C_{16}. For the hydrate formation, the existence of light hydrocarbons or gases such as N_2 or CO_2 with the presence of water is necessary.

The majority of crude oil and gas condensate fluids contain a certain proportion of heavy hydrocarbon compounds that may precipitate as a waxy solid phase if the fluid is cooled below a certain temperature. Wax precipitation may take place at a temperature far above the freezing point of water, and is therefore a potential problem when petroleum mixtures are transported in pipelines (e.g., subsea), or treated in a process plant. Wax formation usually leads to increased fluid viscosity, whereas

---

[a]Telecommunication. Voice: +44 131 451 3672; fax: +44 131 451 3127.
bahman.tohidi@pet.hw.ac.uk

wax deposition generally results in an increase in pipeline roughness and reduction in effective cross sectional area. Therefore, wax formation/deposition results in an increase in pressure drop. When designing pipelines and separation plant it is, therefore, of importance to be able to determine the conditions under which wax precipitation takes place, and the amount of wax likely to form.

Gas hydrates are crystalline compounds formed by a combination of water with suitably sized gas molecules. The amount of gas absorbed depends on the pressure and composition of the gas, thus the composition of a gas hydrate is not constant. The hydrates resemble ice but unlike ice they can form at temperatures well above ice point. These compounds have been studied in detail by Sloan.[1]

To date many studies have been conducted on waxes and hydrates separately. These studies have addressed problems related to only one of the two solid phases. Although the characteristics of waxes and gas hydrates are very different, it is known that the formation of one will affect the thermodynamics and kinetics of the other. The objective of this paper is to develop a thermodynamic model that is capable of predicting wax and hydrate phase equilibria, as well as the impact of wax formation on gas hydrates and vice versa.

In this work, the regular solution theory model,[2,3] has been used for the wax phase modelling. A cubic equation of state (EoS) has been used to model water, liquid hydrocarbon, and vapor phases. Solid solution theory has been employed to model the hydrate phase.[4,5]

WAX MODELLING

For a system to be at equilibrium (vapor, liquid, solid) from a thermodynamic viewpoint, the chemical potential of each component throughout the system must be uniform. Hence, for an isothermal system the fugacity of each component throughout the system must be the same. Details of hydrocarbon, water, vapor, and hydrate modelling are discussed elsewhere.[4,5] We briefly discuss modelling of wax phases here.

The distribution of component i in the solid wax phase can be derived by using the uniformity concept. The fugacity of each component in the solid (wax) phase (f_i^S) is defined by:

$$f_i^S = \gamma_i^S s_i f_i^{oS} \exp\left[\int_0^P \frac{v_i^S dP}{RT}\right], \quad (1)$$

where γ is the activity coefficient, s is the mole fraction, v is the molar volume, R is the universal gas constant, and f is the fugacity. The superscript S refers to the solid phase, superscript o to the standard state and subscript i to the component index.

The standard state fugacity of solid is related to the standard state fugacity of liquid of the same component.[6]

$$\ln\left(\frac{f_i^{oL}}{f_i^{oS}}\right) = \frac{\Delta h_i^f}{RT}\left(1 - \frac{T}{T_i^f}\right) + \frac{\Delta C_{P,i}}{R}\left(1 - \frac{T}{T_i^f} + \ln\frac{T}{T_i^f}\right), \quad (2a)$$

where T^f is the fusion temperature, Δh^f is the heat of fusion and $\Delta C_P = C_P^L - C_P^S$ (C_P is the heat capacity and superscripts L and S refer to liquid and solid, respectively). In this work the standard state fugacity of liquid has been calculated by using a cubic EoS. The activity coefficient has been calculated from the regular solution theory as

$$RT\ln\gamma_i = v_i(\delta_i - \bar{\delta})^2, \tag{2b}$$

where $\bar{\delta} = \sum_i \phi_i \delta_i$, and $\phi_i = s_i v_i / \sum_j s_j v_j$ is the volume fraction of component i.

As shown in (2), the fusion temperature, T^f, and the heat of fusion, Δh^f, as well as the solubility parameters (δ) are required in order to calculate the fugacity of each component in the solid (wax) phase(s). In this work, based on the available experimental data,[7–9] new equations for T^f, Δh^f, and (δ) of each components were developed as

$$T^f = A + BC_n + CC_n^2 + \frac{D}{C_n} + E\ln C_n \tag{3}$$

$$\Delta h^f = A + BT^fM + C(T^fM)^2 + D(T^fM)^3 + E\ln(T^fM) \tag{4}$$

$$\delta^s = \frac{e^{(A + BT_r + CT_r^2 + DT_rP_c)}}{\sqrt{EP_c}} \tag{5}$$

where A, B, C, D, E, and F are constants, given in TABLES 1 to 3. C_n is the carbon number, M is the molecular weight and T_r is the reduced temperature, (T/T_c). Also, T^f is in K, Δh^f is in Kcal/mole and δ is in (cal/cc)$^{1/2}$. The comparison of the experimental data[7] with predicted T^f and Δh^f by different correlations, including those developed in this work, are presented in FIGURES 1 and 2, respectively. These correlations are mostly valid for carbon numbers above C_{10} to around C_{45}. As demonstrated in the figures, classifying hydrocarbon components by odd/even, the latent

TABLE 1. Coefficients of Equation (3)

Carbon numbers	A	B	C	D	E
α	−426.659	−16.560	0.1238	690.670	326.469
β	156.871	−2.086	0.01223	−775.598	76.219

TABLE 2. Coefficients of Equation (4)

Carbon numbers	A	B	C	D	E
α	−1.843	1.886×10^{-1}	-2.016×10^{-4}	—	7.284×10^{-1}
β	4.520	3.741×10^{-1}	-2.317×10^{-3}	6.903×10^{-6}	−2.901

TABLE 3. Coefficients of Equation (5)

Carbon numbers	A	B	C	D	E
α	5.670	−12.401	−1.342	0.1602	0.025
β	3.573	−3.749	−4.399	0.2151	0.053

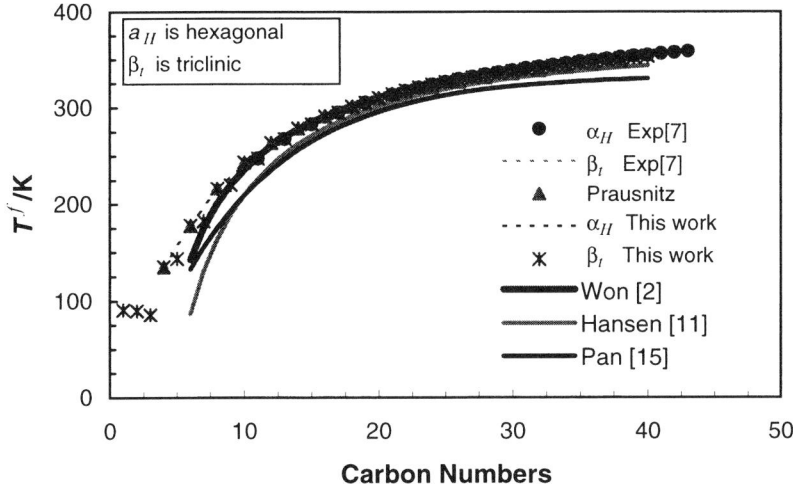

FIGURE 1. Comparison between experimental and correlated T^f from different authors and this work.

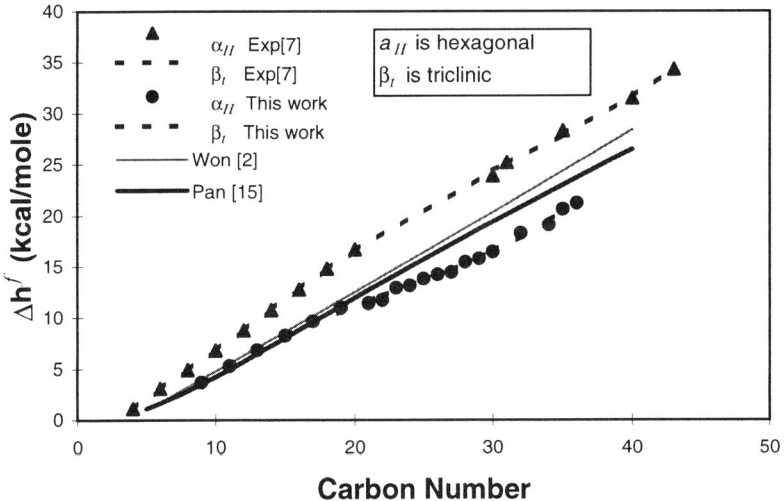

FIGURE 2. Comparison between experimental and correlated Δh^f from different authors and this work.

heat of fusion and the fusion temperature were best predicted by the correlations developed in this work.

RESULTS AND DISCUSSION

The wax model developed was used to predict the wax appearance temperature (WAT), the lowest temperature at which no solid (wax) can form, as well as the amount of wax formed in a number of multicomponent systems. FIGURE 3 shows experimental[10] and predicted WAT for a binary mixture of C_6–C_{18}. The predictions from our model are in good agreement with the experimental data.

FIGURE 4 presents the measured[11] and predicted amount of wax for a stock tank oil with a reported WAT of 289.15 K. As shown in the figure, the experimental and predicted WAT and the amount of wax are in good agreement, demonstrating the reliability of the developed model. The heavy parts of the fluid were characterized by using a semi-continuous approach.[12] A gamma distribution function was used to describe the continuous part.[12]

FIGURE 5 presents the experimental[3] and predicted amount of wax formed in a three-phase (wax–liquid hydrocarbon–vapor) system. This North Sea gas condensate mixture was reported to have a WAT of around 309 K at 13.61 atm.[3] The untuned model predicts a WAT of 313 K, slightly over-predicting the amount of wax formed. The model was tuned by adjusting the fusion temperature, T^f, of the heavy hydrocarbons, and the error in predicting the WAT was reduced to less than 0.5 K. As shown in the FIGURE 5, the tuned model gives reliable predictions on the amount of wax formed.

Generally, hydrocarbon systems are prone to both hydrates (when water is present) and wax formation (when heavier hydrocarbons are present). It is known

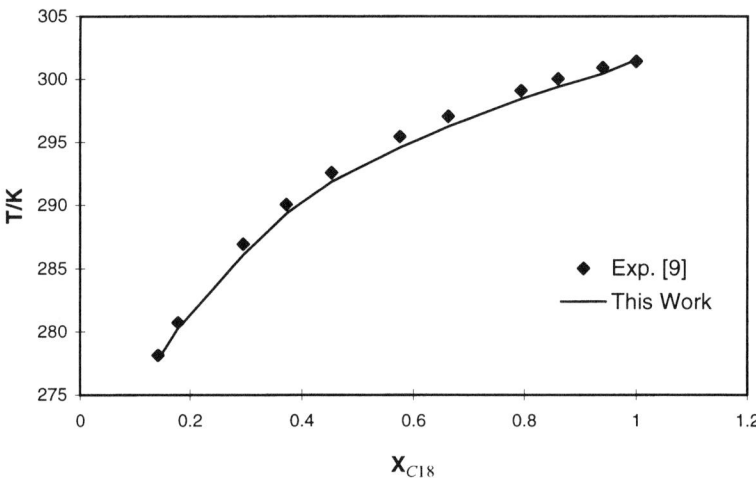

FIGURE 3. Experimental and predicted WAT for binary mixture of C_6H_{14} and $C_{18}H_{38}$.

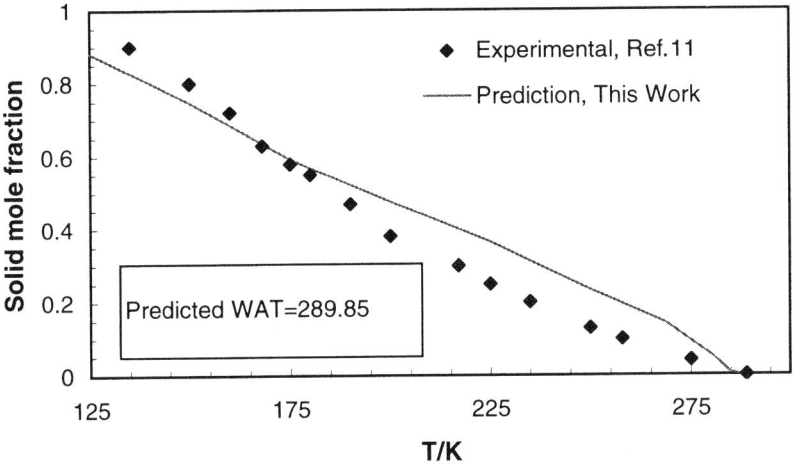

FIGURE 4. Comparison between experimental and predicted amount of wax for a stock tank oil at 1 atm WAT = 289.15 K.

that the formation of one phase will affect the kinetics and thermodynamic behavior of the other. The formation of gas hydrates removes lighter components of the fluid, which could reduce the solubility of waxy compounds in it, affecting the phase boundary of wax. On the other hand, wax formation will remove heavier compounds, increasing the concentration of lighter compounds in the remaining fluid, affecting

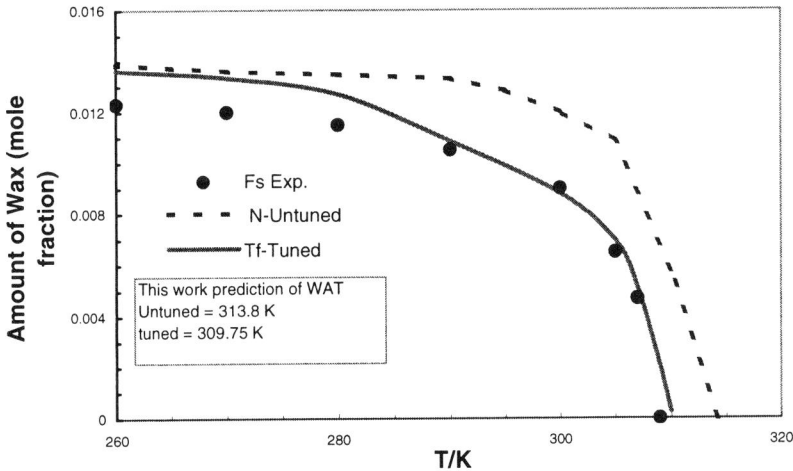

FIGURE 5. Comparison of predicted amount of wax before and after tuning T^f for a North Sea GC at 13.61 atm.

the hydrate phase boundary. More importantly, although beyond the scope of this work, it is likely that any wax or hydrate formation will provide a necessary nucleation site for other solid deposition, promoting the formation of hydrate or wax by reducing the required super-saturation. Traditionally, waxes and hydrates have been investigated separately. One of the objectives of this work was to develop an integrated wax–hydrate model, capable of predicting the impact of wax formation on the hydrate phase boundary (and vice versa) for different hydrocarbon systems.

The wax model developed, in this study is based on regular solution theory. The wax model uses the activity coefficient to solve the fugacities of solid (wax) components. The vapor and liquid phase calculations have been performed using a cubic EoS. The hydrate calculations are based on the Parrish and Prausnitz[13] approach of the extended van der Waals and Plateeuw thermodynamic model.[14] The wax model was successfully coupled with the existing hydrate model. The equality of the fugacities for each component in different phases were considered as the equilibrium conditions.[12] The resulting thermodynamic model was used to simulate different production scenarios, investigating a number of systems prone to wax and hydrate formation.

TABLE 4. The calculated composition and the amount of the five-phases for volatile oil at 280 K and 14.64 bar

Component	z	x_1	x_2	y	h	s
C_1	0.3863	0.0004	0.0362	0.8160	0.0676	0.0000
C_2	0.0626	0.0001	0.0358	0.1232	0.0141	0.0000
C_3	0.0312	0.0000	0.0453	0.0413	0.0298	0.0000
i-C_4	0.0055	0.0000	0.0151	0.0055	0.0053	0.0000
n-C_4	0.0076	0.0000	0.0337	0.0082	0.0014	0.0000
iC_5	0.0030	0.0000	0.0204	0.0018	0.0000	0.0000
nC_5	0.0049	0.0000	0.0354	0.0024	0.0000	0.0000
C_6-C_{10}	0.0356	0.0000	0.2493	0.0008	0.0000	0.1163
C_{11}-C_{15}	0.0228	0.0000	0.1300	0.0000	0.0000	0.1214
C_{16}-C_{19}	0.0124	0.0000	0.0684	0.0000	0.0000	0.0695
C_{20+}	0.0864	0.0000	0.3303	0.0000	0.0000	0.6928
water	0.3417	0.9995	0.0002	0.0007	0.8818	0.0000
		L_1	L_2	V	H	W
		0.1076	0.1069	0.4466	0.2651	0.0738

z, overall mole fraction;
x_1, mole fractions in liquid-water;
x_2, mole fractions in liquid-hydrocarbon;
y, mole fractions in vapor;
h, mole fractions in hydrate;
s, mole fractions in wax.
L_1, L_2, V, H, and W are mole fraction of water, liquid hydrocarbon, vapor, hydrate, and wax, respectively.

TABLE 5. Prediction of wax and hydrate mole fractions for different cases of a volatile oil at 280 K and 14.64 bar

Mole fractions	No wax	No hydrate	All included
L_2	0.2024	0.2041	0.1069
V	0.4308	0.4190	0.4466
L_1	0.1551	0.3421	0.1076
W	—	0.0348	0.0738
H	0.2117	—	0.2651

NOTE: L_1, L_2, V, H, and W are mole fraction of water, liquid hydrocarbon, vapor, hydrate, and wax, respectively.

A typical phase behavior as predicted by the developed wax-hydrate model for a volatile oil is presented in TABLE 4. The model is capable of predicting the amount and composition of different equilibrated phases, including wax and hydrate. The errors associated with predicting a multiphase equilibria with a wax only or hydrate only model, in comparison with an integrated wax-hydrate model, are presented in TABLE 5. In making our predictions the same volatile oil has been used. As shown in TABLE 5, the predictions from the hydrate only or wax only model could exhibit significant deviations in comparison with the combined wax-hydrate model, a fact that causes problems in transportation.

To further investigate the effect of wax formation on the hydrate phase equilibria and vice versa, two different mixtures were studied. Mixture-1 was rich in C_1 and C_2, with around 90% (in mole) light components, whereas Mixture-2 was rich in heavier compounds, with more than 50% (in mole) C_{7+}. Initially the hydrate and wax phase boundaries for the two oils were predicted, using the original composition. Next, at a chosen point (T, P), inside the wax or hydrate region, the wax or hydrate formed that was removed from the system and the hydrate and wax phase boundaries for the remaining system were recalculated. These two mixtures were chosen because of their characteristics, they represented two diverse cases. In one case the light components are dominant (i.e., the hydrate forming components dominates) and in the other case the wax forming components are dominant. Hence, these results cover the two extreme cases of the reservoir fluids, and result in understanding the impact of wax formation on hydrates and vice versa.

FIGURE 6 presents the effect of wax removal at three different temperatures and pressures for Mixture-1 (rich in lights), it can be seen that no major changes were observed in the hydrate phase boundary. When the hydrates were removed, a large reduction in the wax free zone of the remaining fluid was observed. The tests on the Mixture-2 (rich in heavies) show almost a reversal phenomenon. As shown in FIGURE 7, when hydrates were removed the reduction in wax free zone was very small. On the other hand, the removal of wax had much greater impact on the hydrate phase boundary of the remaining fluid, in comparison with Mixture-1. The impact was found to be a function of the temperature and pressure of wax removal conditions. The study showed a maximum shift of 1.5 K at 150 bar in the hydrate phase

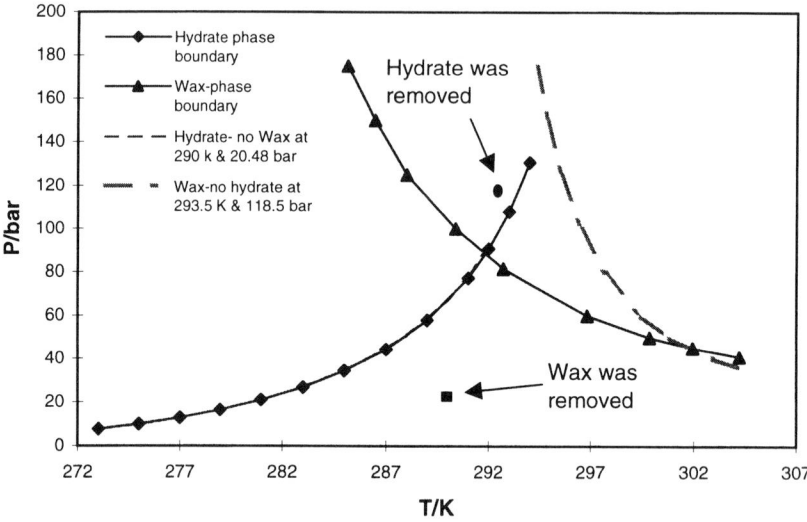

FIGURE 6. Comparison of the effect of the wax on hydrate or vice-versa for mixture-1.

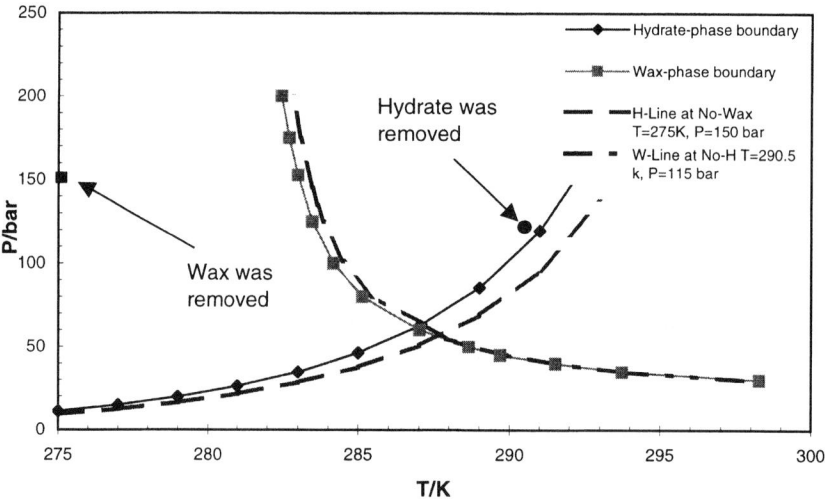

FIGURE 7. Comparison of the effect of the wax on hydrate or vice-versa for mixture-2 (heavier oil).

boundary of Mixture-2, whereas the maximum shift in the wax phase boundary of Mixture-1 was more than 10 K at 180 bar.

In a pipeline wax (or hydrate) formation will remove heavy (light) components from a reservoir fluids. According to the results of this study, this will affect the hydrate (or wax) phase boundary of the remaining fluid a fact that should be taken into account.

CONCLUSIONS

In this study a wax model based on regular solution theory was developed and tested on binary mixtures and multicomponent multiphase systems. In all cases we have been able to predict both the wax appearance temperature (WAT) and the amount of wax with good accuracy. It has been shown that the presence of both solid phases can significantly alter the amount of solidification of the phases and hence increases the transportation problems.

The wax model was successfully coupled with an existing hydrate model. It has been demonstrated that the model is capable of predicting the five phase (liquid hydrocarbon–liquid water–vapor–wax–hydrate) equilibria.

The resulting model was employed to investigate the impact of wax deposition on the hydrate phase equilibria and vice versa. Two oil systems (with different concentrations of light and heavy hydrocarbons) that are prone to hydrate and wax formation were investigated. For the systems investigated, the results showed that wax formation has little effect on the hydrate free zone of the remaining fluids. For light systems, hydrate formation could have significant impact on the wax phase boundary of the remaining fluid.

REFERENCES

1. SLOAN, E.D. 1990. Clathrate Hydrates of Natural Gases. Marcel Dekker Inc., New York.
2. WON, K.W. 1986. Thermodynamics for solid solution-liquid-vapour equilibria: wax phase formation from heavy hydrocarbon mixtures. Fluid Phase Equilibria **30:** 255–279.
3. WON, K.W. 1986. Continuous thermodynamics for solid-liquid equilibria: wax formation from heavy hydrocarbon mixtures. AIChE Spring National Meeting.
4. TOHIDI, B., A. DANESH, R.W. BURGASS & A.C. TODD. 1995. Measurement and prediction of the amount of gas hydrates. Proceedings of the BHR Group Conference, Multiphase 95, Cannes, France. 519–537.
5. TOHIDI, B., A. DANESH & A.C. TODD. 1995. Modelling single and mixed electrolyte solutions and its applications to gas hydrates. Trans. IChemE **73**(A): 464–472.
6. PRAUSNITZ, J.M., R.N. LICHTENHALER & E.G. DE AZEVEDO. 1986. Molecular Thermodynamics of Fluid-Phase Equilibria. 2nd edit., Prentice-Hall Inc. Englewood Cliffs, NJ.
7. BROADHURST, M.G. 1962. An analysis of the solid phase behaviour of the normal paraffins. J. Research of the National Bureau of Standard. **66**(3).
8. DOMANSKA, U., T. HOFMAN & J. ROLINSKA. 1987. Solubility and Vapour Pressures in Saturated Solutions of High-Molecular-Weight Hydrocarbons. Fluid Phase Equilibria **32:** 273–293.
9. MAZEE, W.M. 1949. On the properties of paraffin wax in the solid state. J. Inst. Petrol. **35:** 97–102.

10. DOMANSKA, U. & K. KNIAZ. 1990. Solid-liquid equilibria in some binary mixture. International Data Series **2:** 83–93.
11. HANSEN, J.H., A.A. FREDENSLUND, K.S. PEDERSEN & H.P. RONNINGSEN. 1988. A thermodynamic model for predicting wax formation in crude oils. AIChE **34**(12): 1937–1942.
12. TABATABAEI, S.A.R. 1999. Phase Behaviour Modelling of Petroleum Wax and Hydrates. Ph.D. Thesis, Heriot-Watt University, Scotland, UK.
13. PARRISH, W.R. & J.M. PRAUSNITZ. 1972. Dissociation pressures of gas hydrates formed by gas mixtures. Ind. Eng. Chem. Process. Des. Develop. **11**(1): 26–35.
14. VAN DER WAALS, J.H. & J.C. PLATEEUW. 1959. Clathrate solutions. Adv. Chem. Phys. **2**(1): 1–57.
15. PAN, H. & A. FIROOZABADI. 1996. Pressure and Composition Effect on Wax Precipitation: Experimental Data and Model Results. SPE 36740.

Hydrate Technology for Capturing Stranded Gas

J.S. GUDMUNDSSON,[a] V. ANDERSSON, O.I. LEVIK, AND M. MORK

Department of Petroleum Engineering and Applied Geophysics,
Norwegian University of Science and Technology, 7491 Trondheim, Norway

ABSTRACT: The use of natural gas hydrate (NGH) technology in stranded gas applications is presented in terms of a process description and a capital cost estimate. The cost estimated is based on a NGH hydrate slurry process developed in a joint industry project at Norwegian University of Science and Technology (NTNU). The slurry process is suitable for offshore production of oil and gas; for example, on a floating production storage and offloading (FPSO) unit.

INTRODUCTION

Stranded gas refers to gas located far away from an existing gas pipeline, and in situations where a gas pipeline cannot be built and operated economically. In some cases, the gas may be close to markets, but the reserves are too small to justify large investments for a pipeline or a liquefied natural gas (LNG) plant.

Associated gas produced offshore in fields at remote locations cannot be piped to gas markets. Flaring of such gas is no longer acceptable; instead, the gas is reinjected into reservoir structures. However, because of the commercial value of stranded gas, methods are being sought to bring this gas to market. Gas-to-liquid (GTL) technologies are being developed for this purpose.

For stranded gas in Norway, it was stated by Helgøy *et al.*[1] that offshore LNG was not considered a competitive solution. Instead, both methanol and syncrude (synthetic crude) are considered to offer a viable solution to the stranded gas problem. Syncrude options for remote locations have been presented by Singleton.[2] GTL technologies and projects have been presented by Knott,[3] Skrebowski,[4] and Thomas.[5]

Natural gas hydrate (NGH) technology provides an attractive option to solve the stranded gas problem in the oil industry.[6] Whereas the oil industry is familiar with hydrate deposits in pipes and equipment, the industry is less familiar with the business opportunity offered by NGH technology. Aspects of the NGH slurry technology under development at NTNU are presented in this paper.

[a]Telecommunication. Voice: 0047-7359-4952; fax: 0047-7394-4472.
jsg@ipt.ntnu.no

DRY HYDRATE AND SLURRY HYDRATE

NGH technology can be used in several types of applications, some of which are described in the APPENDIX. In Northern Norway and the Barents Sea, several promising fields have been discovered. The Snøhvit field is a major discovery in the region, with several other fields located nearby. The field predominantly contains gas; the nearby fields are reported to be similar. The distance from the Snøhvit field to gas markets in continental Europe is about 3,500 nautical miles (6,475 km)—too far for a gas pipeline. Therefore, either LNG technology or other technologies are needed to transport the oil and gas.

Using the Snøhvit field as an example, Gudmundsson and Børrehaug[7] reported that dry hydrate (see APPENDIX) technology was 24% lower in capital cost than LNG technology for 400 MMscf/d natural gas processing and 3,500 nautical miles transport by ship to continental Europe. The Snøhvit and other fields in the Barents Sea region can be developed using NGH technology. Dry hydrate and slurry hydrate technologies can be floater-based or land-based. FPSO-based technology is most appropriate for isolated fields at great distance from processing facilities.

If the distances are not too large, gas and oil from the various fields in the Snøhvit area can be piped to land in two-phase flow lines and processed to hydrate products, dry hydrate, and/or hydrate slurry. The NGH product can be transported by ship to continental Europe for processing, supplying crude oil and natural gas to established markets. NGH products can also be shipped to the East coast of the United States.

NGH technology has the potential to be much lower in capital cost than syncrude and methanol production from natural gas.[6] This result opens the possibility of producing hydrate slurry and mixing it with the crude oil for transport by pipeline. The temperature and pressure of the mixture need to be adjusted to make sure that the hydrates will be stable in the pipeline (at pipeline pressure the hydrates need not be refrigerated).

Hydrate particles produced as slurry in the NTNU hydrate laboratory have been shown to be small, with diameters in the tens of microns range. Because of their small size and the small difference in fluid-to-solid density, hydrate particles intentionally produced and mixed with crude oil, can be treated as a homogeneous mixture. The hydrate-crude mixture will flow in a pipeline with an effective viscosity similar to that of the fluid (crude oil) phase.

OFFSHORE SLURRY PROCESS

Work in the NTNU hydrate laboratory in Trondheim has lead to improvements in the design of an offshore slurry process. We give a brief description of a hydrate slurry process suitable for conditions world-wide; for example, FPSO-based oil production. For a complete NGH slurry chain, a shuttle tanker and a land-based melting process (regasification) are also required.

The main units in a hydrate slurry process are shown in FIGURE 1. The fluids enter the process from the production wells and are piped to a separator. For the sake of simplicity, it is assumed that no water is produced. It is possible to use the water produced in the hydrate process, but this option will not be discussed further. In the

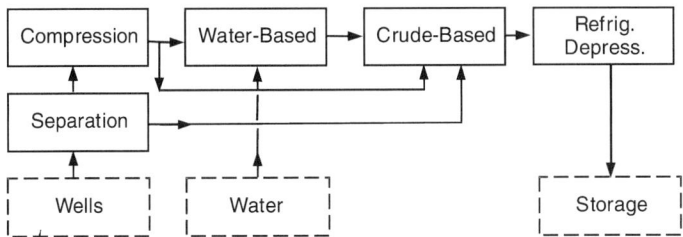

FIGURE 1. Block diagram of the hydrate slurry process.

separator, the gas phase and liquid phase are separated. The quality of this separation is not crucial.

The operating pressure and temperature of the separator depend on the field and well conditions. Both variables have an impact on the hydrate process. A separator pressure of 20 bara and temperature 30°C were assumed in the cost estimate presented below. The higher the pressure, the less the gas compression required; the higher the temperature, the more the cooling required. The gas from the main separator enters a compressor and the pressure is increased to 90 bara.

The crude oil from the main separator at 20 bara and 30°C enters a heat exchanger and is cooled as low as is practicable. The crude oil is under pressure and contains gas in solution. Thus, operational difficulties due to low viscosity and waxing are expected to be manageable through traditional means. The heat exchange system will be designed to minimize difficulties arising from handling cold crude oil. Assuming that the hydrate reactors and cooling units operate at 90 bara, the crude oil must be pumped to the same pressure.

In the hydrate slurry process, natural gas hydrate is formed/produced from associated gas and liquid water. In principle, the water used in the process can be fresh water, seawater, or produced water. Using fresh water has several advantages. Fresh water can be supplied from a shuttle tanker returning from a slurry receiving terminal, preferably at low temperature and saturated with natural gas. The water can be stored in tanks on the FPSO and further cooled before it enters the hydrate reactors and cooling units.

The streams entering the hydrate reactors and cooling units are separated gas, cooled crude oil and cooled fresh water. Several reactors and cooling unit designs are possible and are under development at Aker Engineering in Oslo and NTNU in Trondheim. The function of the reactors and cooling units is to bring gas and water into intense contact at 90 bara and low temperature, to form hydrate.

The storage tanks on the FPSO need to have some insulation, to maintain the crude/hydrate slurry at −10°C or lower. The pressure in the tanks should be close to atmospheric pressure. Typically, the crude/hydrate slurry will be transferred from the FPSO tanks to a shuttle tanker at regular intervals. Slurry pumping will be used for this purpose.

A shuttle tanker will transport the slurry product to a receiving terminal on land. The shuttle distance can be short or long, depending on the situation. At the receiving terminal the crude/hydrate slurry will be pumped to a recovery process. The slurry is pumped to storage (insulated tank or rock cavern) and from there to a heating (melting) process.

Several receiving terminal process designs are possible. In the heating process, the crude/hydrate slurry is heated to a temperature suitable for hydrate melting and subsequent three-phase separation. The recovery process delivers natural gas saturated in water vapor, crude oil saturated in gas and water, and liquid water saturated in natural gas.

FPSO SLURRY VERSUS GAS PIPELINE

When to use NGH technology depends on many factors, including cost comparison. In a 1995 hydrate slurry study,[8] the cost comparison suggested gas reinjection. In a 1996 study on dry hydrate,[9] the cost comparison led to LNG technology. In a 1998 associated gas study,[6] the cost comparison encouraged alternative technologies, including methanol and syncrude. In the following, FPSO-based slurry technology is compared to gas transport by a subsea pipeline.

For the offshore slurry process described above, the capital cost was estimated for a typical FPSO-based situation with an oil production rate of 100,000 bbl/day (16,000 Sm^3/day), a gas–oil ratio (GOR) of 100 Sm^3/Sm^3, giving a gas production rate of 1,600,000 Sm^3/day. Gas for fuel purposes was assumed to be 10% of the total gas rate. Therefore, the gas converted to hydrate was 1,440,000 Sm^3/day.

The gas volume 1,440,000 Sm^3 corresponds to 8,000 m^3 of hydrate, if the specific gas content is taken as 180 Sm^3/m^3. The volume of the slurry is the oil volume 16,000 m^3 and the hydrate volume 8,000 m^3, in a total 24,000 m^3. If the effective specific gas content of NGH is taken as 150 Sm^3/m^3, the hydrate volume is 9,600 m^3 and the total slurry volume 25,600 m^3. Experimental work at NTNU has shown that hydrate-crude slurries are pumpable.

The capital cost estimate was made on the assumption that a NGH slurry process was placed on a FPSO, built to produce oil and gas in a conventional manner. The capital cost estimated for the NGH slurry process was the marginal cost; that is, the additional cost associated with placing the slurry process on the deck of the FPSO. Previous work by Aker Engineering has shown that there is sufficient space for a hydrate slurry process on the deck of the types of FPSO built for service offshore United Kingdom and offshore Norway. The marginal capital cost of the NGH slurry process was estimated to be 160 M$ (exchange rate, 1$ = 7.5 NOK).

Oil tankers in class 800,000 bbl (127,200 m^3) cost about 110 M$ for North Sea service. Gas (LPG) carriers of similar size cost about 125 M$. It is expected that a NGH shuttle tanker will have a price in the range between the two types of ships. A carrier sailing 15 knots (nautical miles) per hour will cover a distance of 666 km/day.

A hydrate slurry tanker, 127,200 m^3 in size, will be able to transport five days' production, assuming 25,600 m^3 of slurry per day (crude oil and hydrate). Assuming it takes one day to fill the tanker (at the FPSO) and one day to empty the taker (at the receiving terminal), the tanker has three days sailing per trip. This corresponds to 1.5 days sailing each way, which means 999 km (speed 15 knots). Normally it takes 12–15 hours to fill and empty shuttle tankers; therefore, the above assumption is conservative.

The storage tanks on a typical FPSO will be about 145,000 m³ in volume. Therefore, spare storage volume on the FPSO will be less than one day's production if one shuttle tanker is used, and the distance is 1,000 km (999 km). For a distance of 2,000 km, at least two shuttle tankers will be required. The operators of offshore production units and shuttle tankers want spare capacity, but this has not been taken into consideration directly.

The capital cost of a receiving terminal depends on the infrastructure already there. It is estimated that the cost of new receiving terminal will be about 60 M$.[8] If the receiving terminal is to be integrated into an existing oil refinery and storage facility, the total capital cost will be about 30 M$. The total capital cost of the above NGH chain (FPSO, shuttle tanker, and new receiving terminal) was, therefore, 345 M$. If the tanker cost is the same as an oil tanker, and if the slurry is transported to an existing refinery facility, the total cost will be 300 M$.

The capital cost of a hydrate slurry chain can be compared to gas transport by pipeline, both FPSO-based. If the pipeline is to be connected to a FPSO, the export riser system and the pipeline landing cost will be about 15 M$. Gas compression and drying of 1,440,000–1,600,000 Sm³/day of natural gas will cost about 20 M$. The additional cost, therefore, will be about 35 M$ in total. The cost of a 16″ subsea gas pipeline is about 1 M$/km.

A subsea gas pipeline, 100 km in length, will cost 100 M$, and a longer pipeline proportionally more. These costs can be compared to the NGH slurry costs per transport distance. The capital cost (capex in M$) for a NGH slurry chain and a gas pipeline are shown in FIGURE 2. The gas pipeline costs start (0 km) at 35 M$ and increase with distance. The hydrate slurry costs start at 220 M$ (FPSO process and receiving terminal) and increase linearly with distance (based on 999 km and 125 M$ for shuttle tanker). It is assumed that other tanker sizes and numbers can be used for other distances.

The gas pipeline and hydrate slurry lines in FIGURE 2 cross at about 220 km. This means that, if the transport distance is less than 220 km, a gas pipeline costs less than a NGH chain. Similarly, for distances greater than 220 km, a NGH chain will have a

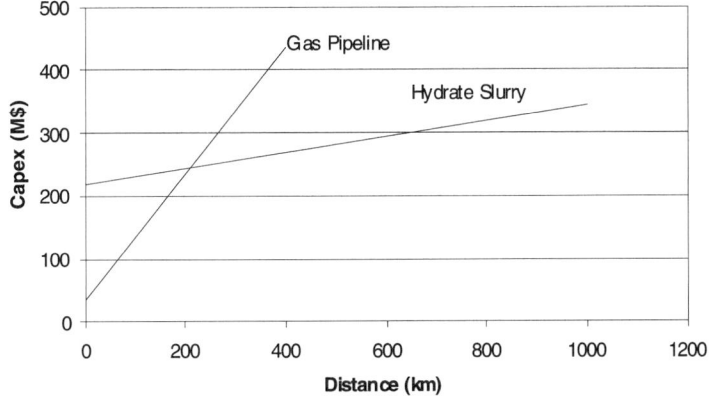

FIGURE 2. Capital costs for gas pipeline and NGH slurry chain.

lower capital cost than a gas pipeline. The annual operating costs of a gas pipeline will be about 2% of the capital cost; the annual operating costs of a NGH slurry chain will be higher, perhaps by 10%. The real cross-over distance, therefore, is greater than 220 km. It should be noted that the capital cost of the NGH slurry chain in FIGURE 2 is the marginal cost.

CONCLUDING REMARKS

Technology for the use of hydrates for the storage and transport of natural gas is being developed. The use of hydrates to capture associated gas on offshore platforms, especially when the gas is stranded, is a candidate for an early application of hydrate technology. The cost of hydrate technology will, in the end, determine whether it will be used instead of the competing technologies. Therefore, it is important to perform cost studies for comparison purposes.

For offshore applications, NGH technology is expected to be much lower in cost than the gas-to-liquid technologies reported in the literature. For land-based applications, it is less clear how hydrate technology stands in comparison with established and new technologies. It is expected that hydrate technology will compete with liquefied natural gas technology in certain markets, especially for small- and medium-sized applications (less than one typical LNG train). The disadvantage with hydrate technology is that it is not yet proven.

The use of hydrate technology on an FPSO is an attractive alternative for stranded gas applications. In situations where hydrate slurry can be transported to a refinery complex with access to an established gas market, the prospects for an early application is promising. Such situations are found in several locations in Europe and elsewhere. When the capital cost of hydrate technology is referenced to marginal costs for a slurry plant on a FPSO, its competitive advantage becomes important for distances of hundreds of kilometers.

ACKNOWLEDGMENTS

We thank Knut Arne Dahl of Fortum Petroleum for useful discussions. The writing of this paper was supported by the joint industry project NGH at NTNU and the Research Council of Norway through Dr. Ing. contracts 32706/211 and 125482/212. The NGH at NTNU project has the web address <http://www.ipt.ntnu.no/~ngh>.

REFERENCES

1. HELGØY, T., D. SCHANKE, A.H. DAHL & T. LÆRGREID. 1997. New options for dealing with gas at marginal fields. Floating Production Systems, IBC Conference, December 4–5, London.
2. SINGLETON, A.H. 1997. Advances make gas-to-liquids process competitive for remote locations. Oil Gas J. August 4, 68–72.
3. KNOTT, D. 1997. Gas-to-liquids projects gaining momentum as process list grows. Oil Gas J. June 23, 16–21.
4. SKREBOWSKI, C. 1998. Gas-to-liquids or LNG? Petroleum Rev. January, 38–39.

5. THOMAS, M. 1998. Water into wine: gas-to-liquids technology the key to unlocking future reserves. Euroil May, 17–21.
6. GUDMUNDSSON, J.S., V. ANDERSSON, O.I. LEVIK & M. PARLAKTUNA. 1998. Hydrate concept for capturing associated gas. SPE Paper 50598, EUROPEC, The Hague, The Netherlands, October 20–22.
7. GUDMUNDSSON, J.S. & A. BØRREHAUG. 1996. Frozen hydrate for transport of natural gas. Proc. 2nd International Conf. Natural Gas Hydrates, June 2–6, Toulouse, 415–422.
8. GUDMUNDSSON, J.S., K. KORSAN & A. BØRREHAUG. 1995. Crude oil/gas hydrate slurry—concept evaluation. Department of Petroleum Engineering and Applied Geophysics, Norwegian Institute of Technology (now Norwegian University of Science and Technology), Trondheim.
9. BØRREHAUG, A. & J.S. GUDMUNDSSON. 1996. Gas transportation in hydrate form. EUROGAS 96, June 3–5, Trondheim. 35–41.

APPENDIX: HYDRATE APPLICATIONS

Natural gas hydrates contain up to 180 Sm^3 of gas per m^3 of hydrate and can be used to store and transport natural gas. The pressure suitable for making hydrates (formation pressure) is in the range 60–90 bar, depending on temperature. The following are examples of hydrate applications:

Hydrate Slurry. In situations where associated gas is produced in locations without a pipeline: (1) associated gas is converted into frozen hydrate, mixed with refrigerated crude oil, and transported as slurry at atmospheric pressure in shuttle tankers; and (2) associated gas is converted into hydrate, mixed with crude oil, and transported as slurry under pressure in a pipeline.

Dry Hydrate. In situations where gas fields are located far away from gas markets, natural gas is converted to frozen dry hydrate. The frozen hydrate is transported to market at atmospheric pressure in large bulk carriers. The hydrate is melted and the natural gas recovered. Natural gas hydrate (NGH) technology competes in the same market as liquefied natural gas (LNG) technology.

Gas Storage. In situations where gas storage is required, natural gas is converted to hydrates and stored at atmospheric pressure and refrigerated. The storage operations can be small or large, and can be land-based or offshore. Land-based storage competes in the same market as conventional gas storage operations. Offshore-based storage competes with gas storage by reinjection into reservoir formations.

Natural Gas Processing. In situations where natural gas and associated gas contain a lot of nitrogen, carbon dioxide, and hydrogen sulfide, hydrate technology is used to separate these gases from the source gas. This because gas hydrates are thermodynamic equilibrium products. Mass transfer operations can be designed to carry out separation and cleaning processes.

Desalination and Water Treatment. In situations where saline and brackish water need to be cleaned, gas hydrates are produced and separated from the concentrated solution. This is because gas hydrates consume just water and gas, but not other constituents, such as dissolved salts and biological materials.

VOC Recovery. In situations where volatile organic compounds (VOC) need to be recovered, for example on oil tankers and receiving terminals, the hydrate forming gases are captured in the form of hydrate. The hydrate is stored and then melted when the VOC gases are ready to be used as fuel or blended with other hydrocarbons.

Carbon Dioxide Disposal. In situations where carbon dioxide disposal is needed, hydrate technology is used to capture the gas in the form of a hydrate. Carbon dioxide hydrate is transported by shuttle tankers and released at depth into the ocean. Because it is heavier than seawater, the carbon dioxide hydrate sinks to the bottom. Provided the depth is greater than 250 m, the carbon dioxide remains stable for practical purposes.

Gas Hydrates and Offshore Drilling

Predicting the Hydrate Free Zone

K.K. ØSTERGAARD, B. TOHIDI,[a] A. DANESH, AND A.C. TODD

Department of Petroleum Engineering, Heriot-Watt University, Riccarton, Edinburgh EH14 4AS, United Kingdom

ABSTRACT: A well-recognized hazard in offshore drilling is the formation of gas hydrates in the event of a hydrocarbon flow into the well bore from the reservoir (e.g., a kick). This could potentially block the BOP stack, kill lines and chokes, obstruct the movement of the drill string, and cause serious operational and safety concerns. Currently, salts are added to the drilling fluids to inhibit hydrate formation in offshore and arctic drilling. For deep water drilling, even saturated saline solutions may not provide the required protection, unless combined with chemical inhibitors. The reported experimental data on gas hydrate formation in drilling fluids are very limited and in some cases inconsistent. The available predictive methods are generally empirical correlations based on limited data and with limited application. In this presentation, a thermodynamic model capable of predicting the hydrate free zone in the presence of salts (NaCl, KCl, $CaCl_2$, NaBr, Na-formate, etc.) and chemical inhibitors (methanol, ethanol, ethylene glycol, glycerol, etc.) is presented. The model developed has been employed to predict the hydrate free zone in drilling fluids designed for offshore and deep water applications. The predictions are compared with experimental data and an empirical correlation, demonstrating the reliability of the thermodynamic model.

INTRODUCTION

In offshore drilling, the pressure of the drilling fluid and the relatively low seabed temperature could provide suitable thermodynamic conditions for the formation of hydrates in the event of a kick. This can cause serious well safety and control problems during the containment of the kick. Hydrate formation incidents during offshore drilling are rarely reported in the open literature, partly because they are not recognized. However, two cases have been reported,[1] where the losses in rig time were 70 and 50 days, respectively.

A recent study[2] showed that oil based drilling fluids inhibit hydrate formation by 2.7 to 5.5°C in a pressure range of 500 to 4,600 psig (lower inhibition at higher pressures), but they adversely affect the required degree of subcooling for hydrate formation, the rate of hydrate formation, and possibly the extent of hydrate formation. These adverse effects, together with the cost and the environmental concerns in using oil based drilling fluids, significantly affect the applicability of oil based drilling fluids in offshore operations.

[a]Telecommunication. Voice: +44 131 451 3672; fax: +44 131 451 3127. bahman.tohidi@pet.hw.ac.uk

The formation of gas hydrates in drilling fluids can cause problems in at least two ways:[3]

1. Gas hydrates can form in the drill string, blow-out preventer (BOP) stack, choke, and kill lines. This could result in potentially hazardous conditions—flow blockage, hindrance to drill string movement, loss of circulation, and even abandonment of the well.

2. Since gas hydrates consist of more than 85% water, their formation can remove significant amounts of water from the drilling fluid, hence, changing the properties of the fluid. This could result in salt precipitation, an increase in fluid weight, or the formation of a solid plug, particularly for water based drilling fluids.

The hydrate formation conditions of a kick depend on the composition of the kick fluid, as well as on the pressure and temperature of the system. For a typical natural gas composition, it can be equivalent to 120 m of water at a seabed temperature of 4°C. Salts are generally added to avoid hydrate formation in drilling fluids. As a rule of thumb, the inhibition effect of a saturated saline solution is not adequate for avoiding hydrate formation in water depths of greater than 1,000 m.[1] Therefore, a combination of salts and chemical inhibitors that can provide the required inhibition should be used.

The experimental data on gas hydrate formation in drilling fluids are very limited and in some cases inconsistent. The available predictive methods are generally empirical correlations based on limited data and with limited application. In production, transmission and processing of reservoir fluids, the available thermodynamic methods can predict hydrate formation temperature with an accuracy better than 1 K,[4] whereas in predicting hydrate formation conditions in drilling fluids, the available empirical methods have much lower accuracy (by almost one order of magnitude).

The error in predicting the hydrate free zone in drilling fluids can result in either formation of hydrates or over-inhibited drilling fluids. The former could have serious operational and safety consequences and the latter could increase the cost of operation, reduce the flexibility in preparing the drilling fluid, and increase the possibility of salt precipitation and correlation. Recent studies showed that the economics and safety of offshore drilling operations can be significantly enhanced by improving the accuracy of predictive methods in estimating hydrate formation conditions in water and oil based drilling fluids.

In this presentation, available experimental data on water freezing point depression of salts and chemical inhibitors, together with available vapor–liquid equilibrium (VLE) data, have been used to extend an in-house thermodynamic model to salts and chemical inhibitors commonly used in drilling fluids. The predictions of the resulting thermodynamic model are compared with experimental data and an empirical correlation to demonstrate the reliability and superiority of the thermodynamic model.

THERMODYNAMIC MODELING

The Heriot-Watt university hydrate model, HWHYD, uses the Valderrama modification of the Patel and Teja (VPT) equation of state (EoS) with a non-density dependent (NDD) mixing rule to model all fluid phases. Ice is modelled as a subcooled liquid. The hydrate phases are modelled by the solid solution theory of van der Waals-Platteeuw, as implemented by Parrish and Prausnitz. The Kihara model for spherical molecules is applied in order to calculate the potential function for compounds forming the hydrate phases. The general multiphase flash routine, as described by Cole and Goodwin, is adapted in the model.

Salts are modelled by combining the EoS with a modified Debye-Hückel electrostatic term with only one binary interaction parameter. Water vapor pressure depression data at 373.15 K and freezing point depression data of single electrolyte solutions have been used to optimize water–salt interaction parameters, which have been correlated with salt concentration and temperature. No binary interaction parameters between salts and chemical inhibitors have been employed. A detailed description of the numerical modelling approach is provided elsewhere.[5]

Binary interaction parameters (BIP) between each chemical inhibitor and water have been optimized, using freezing point depression and VLE data. Where available, VLE data have been used to optimize BIP between chemical inhibitor and components, which can be present in reservoir fluids at high concentrations (i.e., methane, ethane, propane, carbon dioxide, and nitrogen).

RESULTS AND DISCUSSION

TABLE 1 lists the salts and chemical inhibitors available in the HWHYD model. The list covers components common to sea and formation water, as well as inhibitors frequently used in drilling and production. The hydrate inhibition effects of the above compounds are compared in FIGURE 1. The figure presents the predicted hydrate inhibition effect of 10 wt% aqueous solutions of the compounds listed in

TABLE 1. Salts and chemical inhibitors in HWHYD

Salts	Chemical inhibitors
NaCl	methanol
KCl	ethanol
$CaCl_2$	monoethylene glycol (MEG)
Na_2SO_4	diethylene glycol (DEG)
NaF	triethylene glycol (TEG)
KBr	glycerol
$MgCl_2$	
$SrCl_2$	
$BaCl_2$	
NaBr	
Na-Formate	
K-Formate	

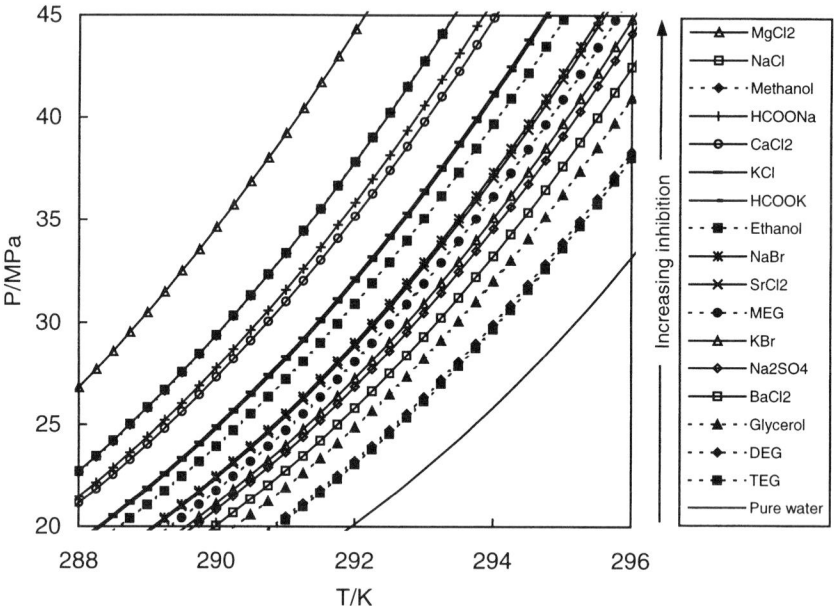

FIGURE 1. Effect of 10 wt% of salt or chemical inhibitor on the hydrate free zone of methane hydrates.

TABLE 1. As shown in the figure, the inhibition effect ranges from about 1 K for 10 wt% TEG to almost 6.5 K for 10 wt% $MgCl_2$.

Below, the predictions of the developed thermodynamic model (HWHYD) are compared with experimental data and available correlations. The objective is to evaluate the accuracy and reliability of the thermodynamic model in predicting the hydrate-free zone in the presence of salts and chemical inhibitors commonly used in offshore drilling.

Hydrate Prevention in Drilling Using Formate Salts

Formate brines have been extensively investigated for their use as drilling and completion fluids.[6] They were developed and proved to meet the design criteria required for deep slim-hole drilling; for example, cover large density range, very low crystallization temperatures, low corrosion potential, and are easily recyclable. Furthermore, formate brines also have the advantage of inhibiting hydrate formation.

FIGURE 2 presents the experimental methane hydrate free zone in the presence of pure water and 10, 20, and 30 wt% aqueous solution of potassium formate. Some scattering in the experimental data is observed, especially for 10 wt% potassium formate at high pressure. Two sets of predictions are also shown in FIGURE 2, that of the HWHYD model and values predicted by Fadnes *et al.*[8] A good agreement can be seen between the predictions from the HWHYD model and the experimental data, even at high salt concentrations.

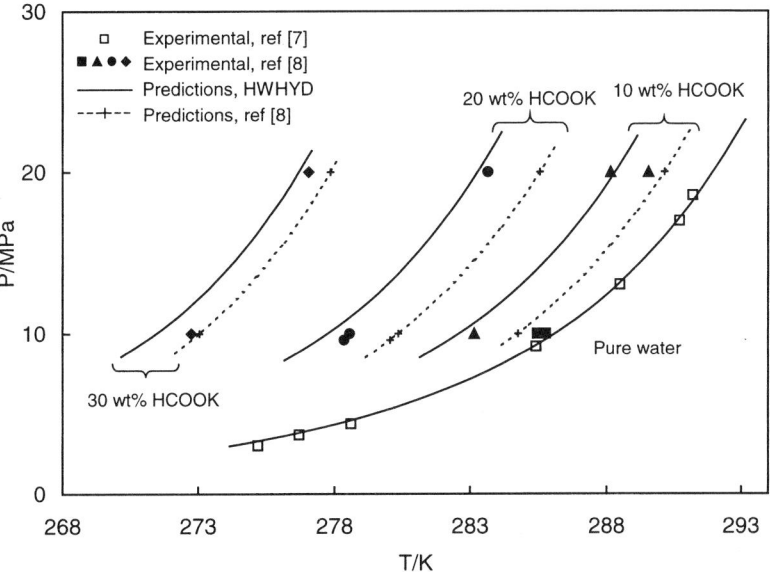

FIGURE 2. Experimental and predicted hydrate free zone of methane in the presence of potassium formate.

Hydrate Prevention in Drilling Using Glycerol

Glycerol is commonly used in water based drilling fluids to suppress the formation of gas hydrates. Drilling fluids based on salt–glycerol mixtures have proven to meet property requirements for deep water drilling as well as being cost-effective.[9]

FIGURE 3 presents the experimental and predicted hydrate free zones for methane hydrates in the presence of pure water, and glycerol/water solutions with 25 wt% and 50 wt% glycerol. A good agreement between the experimental data and the HWHYD predictions can be observed, demonstrating the success of the thermodynamic model.

Comparison of HWHYD and Empirical Correlation

Yousif and Young[11] proposed a correlation to predict the inhibition of gas hydrates in drilling fluids due to the presence of salts and glycerol. They correlated the hydrate temperature suppression with mole fraction and degree of ionization for chemical inhibitors and salts, respectively. The correlation is based on experimental hydrate equilibrium data of drilling fluids containing various amounts of NaCl, NaBr, KCl, $CaCl_2$, and glycerol measured in five different laboratories. The correlation proposed by Yousif and Young includes higher concentrations of inhibitors than the original Hammerschmidt equation.[12]

FIGURE 4 shows the experimental and predicted hydrate free zone for a natural gas in the presence of saline water. Both the thermodynamic model developed in this

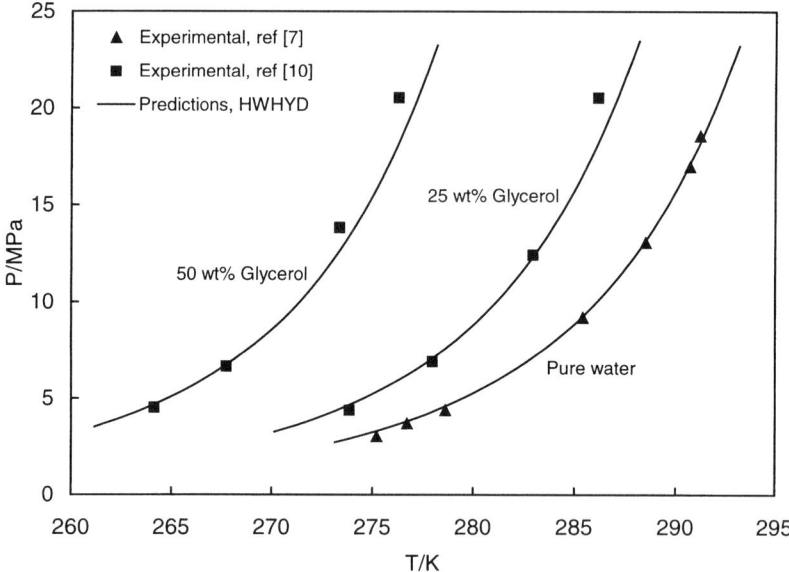

FIGURE 3. Experimental and predicted hydrate free zone of methane in the presence of glycerol.

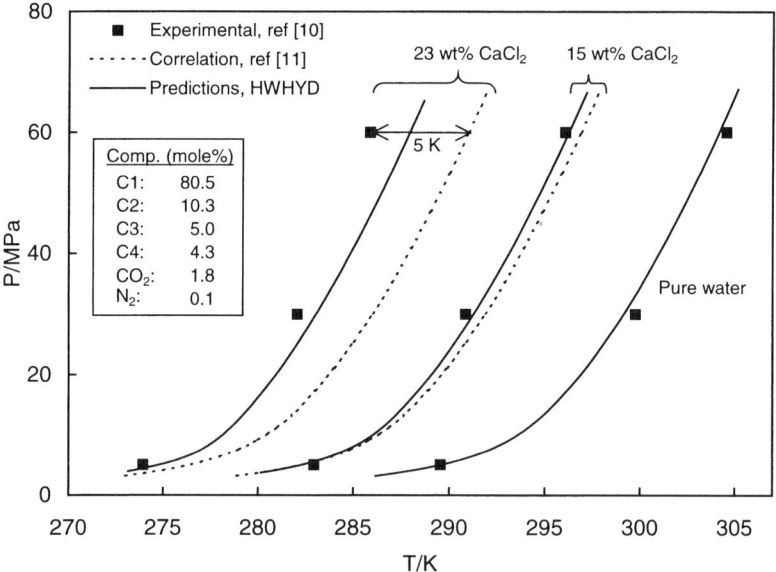

FIGURE 4. Experimental and predicted hydrate free zone of a natural gas in the presence of saline water.

work (HWHYD) and the correlation proposed by Yousif and Young have been used to predict the hydrate free zone. The predictions from the correlation are in good agreement with the experimental data for the 15 wt% $CaCl_2$ solution, whereas significant deviations are observed for the 23 wt% $CaCl_2$ solution (about 5 K). The HWHYD predictions are in good agreement with the experimental data for all concentrations, demonstrating the reliability of the thermodynamic model.

FIGURE 5 presents the experimental and predicted hydrate-free zone for a gas mixture in the presence of 2.5 wt% NaCl and 10, 35 and 60 wt% glycerol. The HWHYD thermodynamic model predicts the hydrate free zone consistently and with a maximum deviation of about 1 K. The predictions from the correlation are inconsistent and show a deviation of about 5.5 K at 2.5 wt% NaCl and 35 wt% glycerol, whereas they are in good agreement with the experimental data at 2.5 wt% NaCl and 10 and 60 wt% glycerol. The latter concentration of inhibitor, however, is well above the concentrations used to develop the correlation.

The error in prediction of the hydrate free zone in drilling fluids could result in either over-inhibited drilling fluids (FIG. 4) or formation of gas hydrates (FIG. 5). The latter could have serious operational and safety consequences, and the former could increase the cost of operation, reduce the flexibility in preparing the drilling fluid (e.g., low density mud), and increase the possibility of salt precipitation and corrosion. FIGURE 6 shows examples of estimated quantities of salt or chemical inhibitor required to compensate for the inaccurate predictions in the hydrate free zone, per 100 barrels of water in drilling fluid. For example, in the case of 5.5 K deviation in

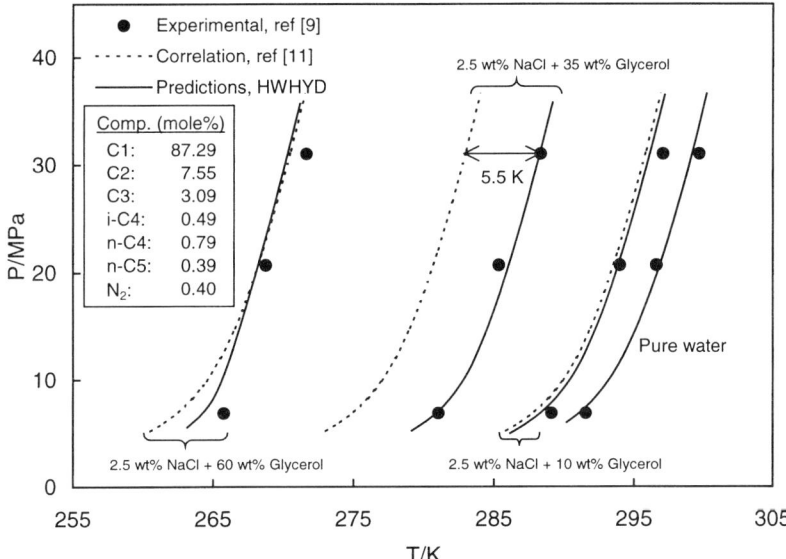

FIGURE 5. Experimental and predicted hydrate free zone of a gas mixture in the presence of NaCl/glycerol solutions.

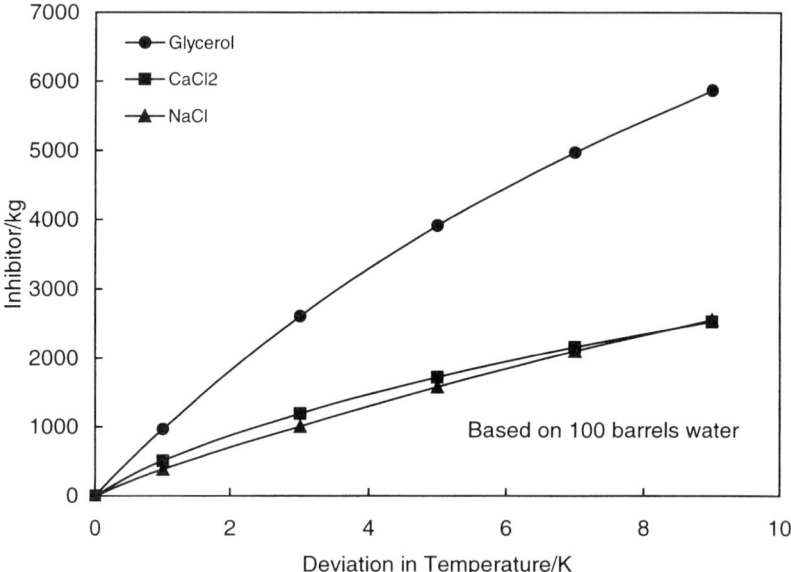

FIGURE 6. Amount of inhibitor required to compensate for deviation in predictive method.

FIGURE 5, an extra 4,000 kg of glycerol per 100 barrel of water is required to compensate for the error.

The correlation due to Yousif and Young[11] is based on thermodynamic principles. It is believed that some of the discrepancies observed between the correlation predictions and the experimental data at some salt/chemical inhibitor concentrations are partly due to inaccurate experimental data. Recently, a number of publications[13–15] have recommended experimental techniques that could improve the reliability of experimental data, particularly hydrate phase free zone measurements in the presence of salts and chemical inhibitors.

CONCLUSIONS

A comprehensive thermodynamic model, capable of predicting the hydrate-free zone in the presence of salts and chemical inhibitors commonly used in drilling fluids, has been described. Easily available water freezing point and/or vapor pressure depression data have been used to evaluate the parameters required for modelling salts and chemical inhibitors.

The predictions of the thermodynamic model have been compared with experimental hydrate dissociation data for gas mixtures in the presence of high concentration of salts and/or chemical inhibitors, simulating drilling fluids for deep water operations, demonstrating its reliability.

The predictions of the thermodynamic model have also been compared with a correlation designed to predict hydrate point suppression in drilling fluids. The comparison demonstrated the superiority of the comprehensive thermodynamic approach over the empirical correlation. However, it is likely that the correlation has been partly based on inconsistent experimental hydrate data.

Improved accuracy in predictive techniques could reduce the cost and improve the safety of drilling operation for deep water applications.

REFERENCES

1. BARKER, J.W. & R.K. GOMEZ. 1989. Formation of hydrates during deepwater drilling operation. J. Petrol. Technol. **41**(3): 297–301.
2. GRIGG, R.B. & G.L. LYNES. 1992. Oil-based drilling mud as a gas-hydrates inhibitor. SPE Drilling Engineering March, 32–38.
3. OUAR, H., S.B. CHA, T.R. WILDEMAN & E.D. SLOAN. 1992. The formation of natural gas hydrates in water-based drilling fluids. Trans. IChemE **70**(A): 48–54.
4. TOHIDI, B., A. DANESH & A.C. TODD. 1997. Predicting pipeline hydrate formation. The Chemical Engineer **642:** 32–37.
5. TOHIDI, B., A. DANESH & A.C. TODD. 1995. Modelling single and mixed electrolyte solutions and its applications to gas hydrates. Trans. IChemE **73**(A): 464–472.
6. HOWARD, S.K. 1995. Formate brines for drilling and completion: state of the art. SPE 30498. The SPE Annual Technical Conference & Exhibition, October 22–25, Dallas, Texas, USA.
7. VERMA, V.K. 1974. Gas Hydrates from Liquid Hydrocarbon-Water Systems. Ph.D. Thesis, University of Michigan, Ann Arbor. (See also Ref. 16.)
8. FADNES, F.H., T. JAKOBSEN, M. BYLOV, A. HOLST & J.D. DOWNS. 1998. Studies on the prevention of gas hydrates formation in pipelines using potassium formate as a thermodynamic inhibitor. SPE 50688. European Petroleum Conference, October 20–22, The Hague, the Netherlands.
9. HALE, A.H. & A.K.R. DEWAN. 1989. Inhibition of gas hydrates in deepwater drilling. SPE/IADC 18638. 1989 SPE/IADC Drilling Conference, February 8–March 3, New Orleans, Louisiana.
10. NG, H.-J. & D.B. ROBINSON. 1994. New developments in the measurement and prediction of hydrate formation for processing needs. Ann. N.Y. Acad. Sci. **715:** 450–462.
11. YOUSIF, M. & D.B. YOUNG. 1993. A simple correlation to predict the hydrate point suppression in drilling fluids. SPE/IADC 25705. SPE/IADC Drilling Conference, February 23–25, Amsterdam, the Netherlands.
12. HAMMESCHMIDT, E.G. 1934. Gas hydrate formation. Gas **15**(5): 30–34, 94.
13. TOHIDI, B., R.W. BURGASS, A. DANESH & A.C. TODD. 1994. Experimental study on the causes of disagreements in methane hydrate dissociation data. Ann. N.Y. Acad. Sci. **715:** 532–534.
14. TOHIDI, B., A. DANESH, A.C. TODD & R.W. BURGASS. 1997. Hydrate-free zone for synthetic and real reservoir fluids in the presence of saline water. Chem. Eng. Sci. **52**(19): 3257–3263.
15. TOHIDI, B., R.W. BURGASS, A. DANESH, K.K. ØSTERGAARD & A.C. TODD. 1999. Improving the accuracy of gas hydrate dissociation point measurements. 3rd International Conference on Gas Hydrates, 18–22 July, Park City, Utah.
16. SLOAN, E.D. 1998. Clathrate Hydrates of Natural Gases, 2nd edit. Marcel Dekker Inc., New York.

A Simple Model for Natural Gas Production from Hydrate Decomposition

GOODARZ AHMADI,[a] CHUANG JI,[a] AND DUANE H. SMITH[b]

[a]*Department of Mechanical and Aeronautical Engineering, Clarkson University, Potsdam, New York 13699-5725, USA*

[b]*Federal Energy Technology Center, Department of Energy, Morgantown, West Virginia 26507-0880, USA*

ABSTRACT: This paper describes a model for natural gas production from the decomposition of methane hydrate in a confined reservoir by a depressurizing well. The one-dimensional linearized model suggested by Makogon is used in the analysis. For different well pressures and reservoir temperatures, distributions of temperature and pressure in the porous layer of methane hydrate and in the gas region are evaluated. The distance of the decomposition front from the well to the natural gas output are evaluated as functions of time. Time evolutions of the resulting temperature and pressure profiles in the hydrate reservoir are also studied.

INTRODUCTION

Gas hydrates are solid molecular compounds of water with natural gas that are formed under certain thermodynamic conditions. World reserves of natural gas trapped in the hydrate state have been estimated to be several times the known reserves of conventional natural gas.[1] Therefore, developing methods for commercial production of natural gas from hydrates is attracting considerable attention.

Makogon[1] and Sloan[2] reported extensive reviews of hydrate formation and decomposition processes. Thermodynamic modeling of the hydrate decomposition process by depressurization has been studied by a number of authors. Assuming that the process of hydrate decomposition by a pressure decrease is analogous to the process of solid melting, Makogon[1,3] used the classical Stefan melting problem, to describe the process of hydrate decomposition. Verigin *et al.*[4] considered the effect of water flow and developed a more accurate model. Holder and Angert[5] considered the variation of temperature during the hydrate decomposition in their study. Burshears *et al.*[6] extended the model of Holder and Angert[5] and considered the influence of water transport in the layer, in addition to the natural gas flow. Kamath[7] studied the process of hydrate dissociation by heating. Recent studies on geological aspects of hydrates were reported by AGU.[8]

Makogon[1] summarized the study of Bondarev and Cherskiy in which the effects of heat transfer in the porous medium were included. The energy equation was used to describe the thermal condition of natural gas in the porous layer. Conductive and convective heat transfer, as well as the effects of the throttling process, were included. Makogon[1] reported analytical expressions for the one-dimensional temperature and pressure profiles that were obtained after linearization of the governing equations.

In the present work, we are concerned with the production of natural gas by depressurization through drilling a well into a hydrate reservoir. The parameters that control the natural gas production rate are also studied. We used the combined models of Verigin et al.,[4] and Bondarev and Cherskiy as reported by Makogon.[1] In this model, the fluid energy equation is used to describe the temperature and pressure distributions of the natural gas in the porous layer. The conductive and convective heat transfer, as well as the effects of the throttling process, were also included. Assuming that the hydrate layer also contained pressurized natural gas, Makogon[1] linearized the governing equations and obtained a set of self-similar solutions for temperature and pressure distributions in the reservoir. The results lead to a system of coupled algebraic equations for the location of the decomposition front, and the temperature and pressure at the front. In the present work, this system is solved by an iterative scheme. For several well pressures and reservoir temperatures, numerical results for time evolution of pressure and temperature profiles in the hydrate reservoir, as well the location of the front and the natural gas production rate are obtained. The results are presented in graphical form and discussed.

HYDRATE DECOMPOSITION MODEL

In this paper, the decomposition of methane hydrate in a reservoir due to depressurization is considered. Chemical reaction of methane with water to form hydrate is represented by

$$(CH_4)_{gas} + 6(H_2O)_{water} \overset{30\ atm,\ 273\ K}{\longleftrightarrow} (CH_4 \cdot 6H_2O)_{solid}$$

When the pressure decreases or the temperature rises, the reaction reverses, and the hydrate decomposes into CH_4 and water.

Suppose there is a large methane hydrate reservoir underground. It is assumed that solid hydrate and natural gas exist in the porous layer at the reservoir pressure P_e and reservoir temperature T_e. Hydrate is initially stable at this pressure and temperature. When a well is drilled into the reservoir, the pressure in the well drops to a certain value $P_G < P_D < P_e$, where P_D is the decomposition pressure of the hydrate at the dissociation temperature, T_D, and P_G is the well pressure. At this stage the hydrate near the well becomes unstable and begins to decompose into natural gas and water. The process of hydrate decomposition, then expands outward with time. It is assumed that the hydrate decomposition in a porous medium does not occur in the entire volume, but in a narrow region that can be treated as a surface, the so-called decomposition front. This moving front separates the volume of the reservoir into two zones with different phases. In the near-well zone only natural gas and water exist, whereas in the zone further from the well, only the solid hydrate and natural gas exist. Pressures and temperatures in these two zones gradually decrease, and the natural gas moves towards the well because of the pressure gradient, whereas the decomposition front moves in the opposite direction.

It should be emphasized that the model proposed by Makogon[1] and used in this study involves several important assumptions. One is that the pressure and temperature at any point on the decomposition front are the equilibrium pressure, P_D, and temperature, T_D, for dissociation of methane hydrate. Furthermore, the hydrate reservoir is assumed to be porous and to contain free natural gas. As the dissociation front moves outward, heat must be supplied to the front because of the endothermic nature of the hydrate decomposition process. Makogon[1] suggested that heat conduction is negligible when compared with convection and heat must be supplied from the reservoir for dissociation to continue. The other assumptions are that during hydrate decomposition the front is a source of mass that releases water and methane gas, the movement of water in the porous medium is negligible, and the permeability is constant.

MATHEMATICAL MODEL

The mathematical formation suggested by Makogon[1] is used in this study. A hydrate reservoir with a well as shown in FIGURE 1 is considered. For a one-dimensional model, the distribution of pressure in the layer is described using an analog of the classical Stefan problem for melting. In FIGURE 1, region 1 corresponds to $0 < x < l(t)$, and region 2 corresponds to the region $l(t) < x < \infty$, where $l(t)$ is the distance of dissociation front from the origin (located at the well). The boundary conditions are:

$$P_1(0, t) = P_G \tag{1}$$

$$P_2(x, 0) = P_2(\infty, t) = P_e$$

$$P_1(l(t), t) = P_2(l(t), t) = P_D(T_D) \tag{2}$$

$$T_2(x, 0) = T_2(\infty, t) = T_e \tag{3}$$

$$T_1(l(t), t) = T_2(l(t), t) \tag{4}$$

where P_n is pressure in zone 1 or 2, T_n is temperature in zone 1 or 2, P_D is pressure at the decomposition front, and T_D is temperature at the front. As noted, it is assumed

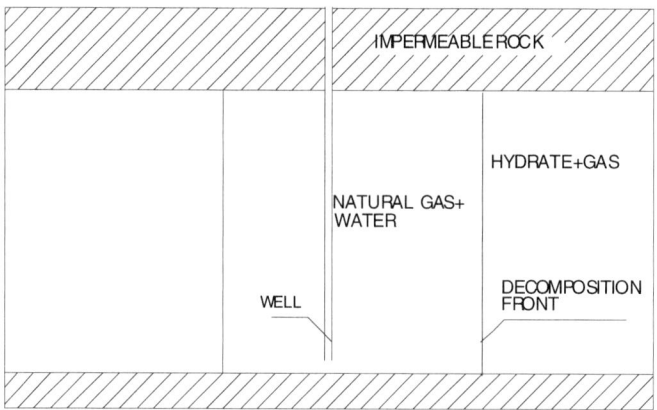

FIGURE 1. Diagram of a hydrate reservoir.

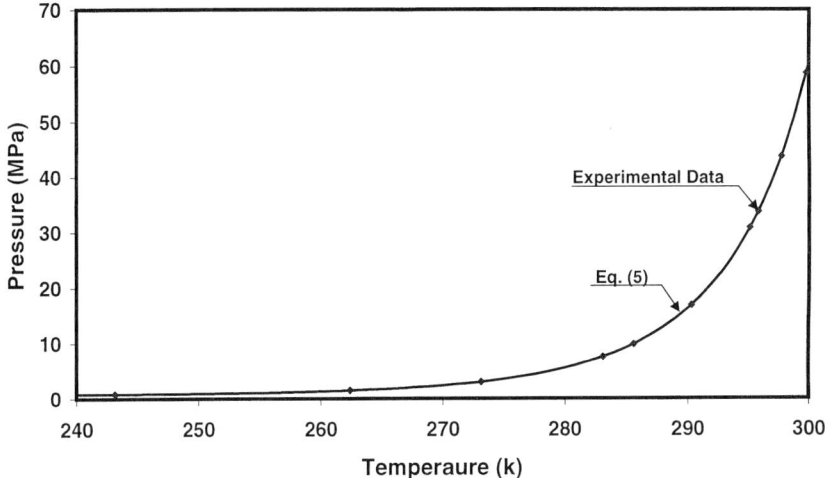

FIGURE 2. Comparison of Equation (5) with temperature–pressure equilibrium data of Makogon and Sloan,[1] and Marshal et al. (1964).

that P_D and T_D are the equilibrium pressure and temperature for dissociation of methane hydrate.

The relation between temperature T_D and pressure P_D on the decomposition front in terms of the phase equilibrium between natural gas and hydrate is

$$\log P_D = a(T_D - T_0) + b(T_D - T_0)^2 + c, \tag{5}$$

where T_0 is 273K and a, b, c are empirical constants that depend on the hydrate composition. Values for a, b, and c are obtained from the equilibrium pressure-temperature data of methane hydrate.[1] FIGURE 2 shows Equation (5) with the data of Makogon and Sloan,[3] and by Marshal and Kobayashi (1964). It is observed that the fit is in good agreement with the data in the range of variables considered. Self-similar solutions of the system of governing equations, as obtained by Magokon[1] were adopted for further analysis.

RESULTS

This section presents numerical results for time evolution of pressure and temperature profiles in a hydrate reservoir under various conditions. In addition, time variations of methane gas production and location of the dissociation front are evaluated. An initial reservoir pressure of 15 MPa is used in the simulation. For various values of well pressure and initial reservoir temperature, values of dissociating temperature and pressure at the front and the parameter γ (that controls the location of decomposition front) are evaluated with the results listed in TABLE 1.

For given reservoir pressure and temperature and well pressure, the linearized one-dimensional model leads to fixed values of dissociation-front pressure and

TABLE 1. Values of dissociating temperature and pressure and parameter γ for given reservoir and well conditions

P_e (MPa)	T_e (K)	P_G (MPa)	T_D (K)	P_D (MPa)	γ (m^2/sec)
15	287	2	277.14	4.27	0.0052
15	287	3	277.18	4.28	0.0034
15	287	4	277.38	4.36	0.00087
15	285	2	274.93	3.53	0.0029

temperature. TABLE 1 also shows that when the well pressure changes, the dissociation pressure and temperature change only slightly. The value of parameter γ controls the movement of the front and the gas production rate decreases sharply with increase in well pressure. The dissociation pressure and temperature, however, are sensitive functions of reservoir temperature. A decrease of 2 K in the reservoir temperature drops the dissociation pressure by about 0.8 MPa and reduces parameter γ by about 45%.

For a well pressure of 2 MPa, FIGURE 3 shows variations of reservoir pressure at different times. The pressure decreases from the reservoir pressure to the decomposition pressure over a distance of about 200 m near the front, and then decreases gradually toward the well, reaching its minimum value at the well. As noted previously, the hydrate reservoir is divided into two zones by the decomposition front and the pressure variations in the two zones are quite different. FIGURE 3 also shows that the pressure profiles for different times are self-similar in each zone and expand outward as the decomposition front moves away from the well.

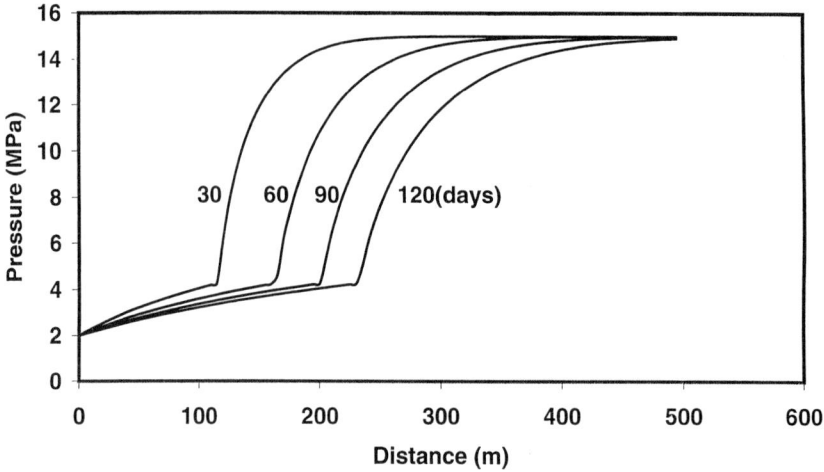

FIGURE 3. Time variation of pressure in the reservoir for a well pressure of 2 MPa and a reservoir temperature of 287 K.

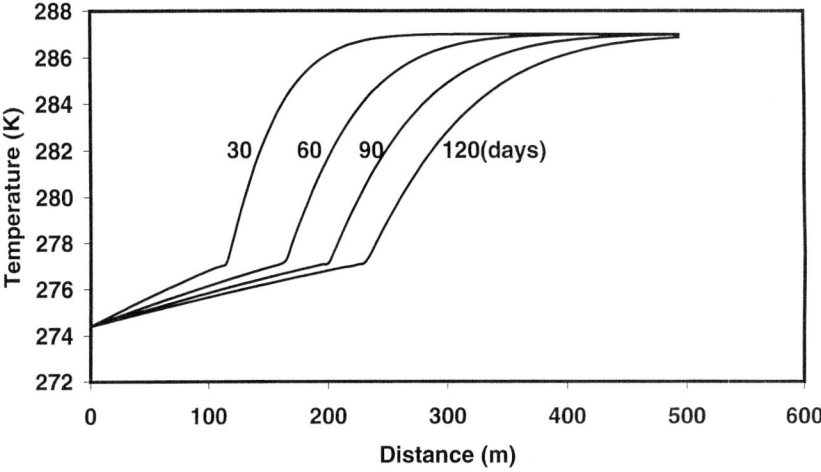

FIGURE 4. Time variation of temperature in the reservoir for a well pressure of 2 MPa and a reservoir temperature of 287K.

The corresponding temperature profiles for different times under the same conditions as in FIGURE 3 are presented in FIGURE 4. It is seen that the temperature decreases from the undisturbed reservoir value far from the front to the decomposition temperature at the front. The temperature gradient in the hydrate zone is largest near the front. In the gas zone, the temperature varies smoothly and decreases to its

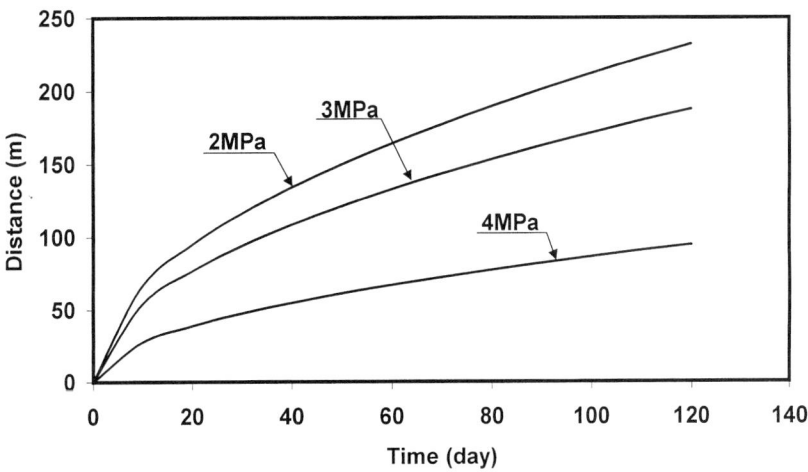

FIGURE 5. Variation in location of decomposition front for different well pressures.

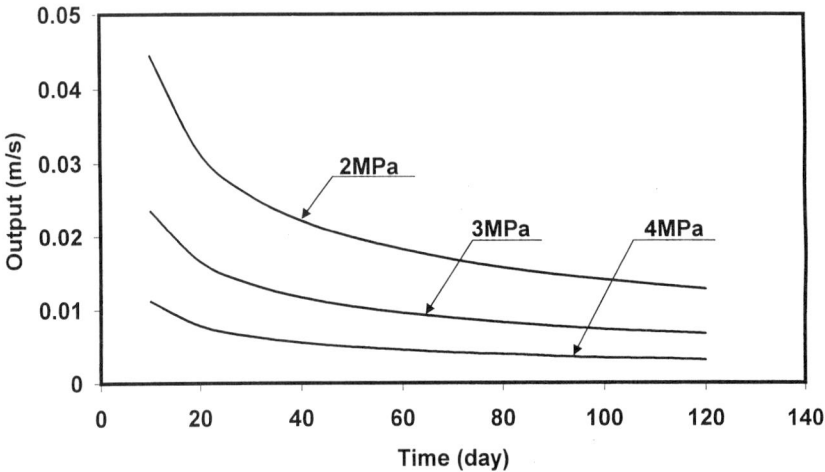

FIGURE 6. Variation in natural gas output for different well pressures.

minimum value at the well. The temperature profiles in the hydrate and the gas zones are also self-similar and evolve with time as the decomposition front moves outward.

FIGURE 5 shows the movement of the decomposition front for different well pressures. Here, the reservoir conditions are kept fixed at 15 MPa and 287 K. As expected, the distance of the front from the well increases proportionally to the square root of time. As the well pressure increases, the motion of the front decreases, especially as the well pressure approaches the decomposition pressure.

FIGURE 6 shows time variations of natural gas flow rate per unit height (and per unit width) of the well for different well pressures. As expected the flow rate decreases with inverse square root of time. The well output is a sensitive function of well pressure and the production rate decreases sharply as well pressure increases for a fixed reservoir condition.

CONCLUSIONS

Natural gas production from decomposition of methane hydrate in confined, pressurized reservoirs is studied. Evolution of pressure and temperature profiles in one-dimensional reservoirs are analyzed. On the basis of the results presented, the following conclusions may be drawn:

1. Under favorable conditions natural gas can be produced from hydrate reservoirs by depressurization.
2. The reservoir conditions and the well pressure control the natural gas production rate.
3. For an infinite homogenous hydrate reservoir containing natural gas, the dissociation pressure and temperature are fixed and depend only on the reservoir conditions and the well pressure.

4. For a fixed reservoir pressure and temperature the well output decreases and the motion of the decomposition front slows as the well pressure increases.

5. For fixed reservoir and well pressures the gas production rate decreases significantly as the reservoir temperature decreases.

6. The distance of decomposition front from the well increases in direct proportion to the square root of time and the production rate decrease in inverse proportion.

ACKNOWLEDGMENT

The support of the Office of Fossil Energy, U.S. Department of Energy is gratefully acknowledged.

REFERENCES

1. MAKOGON, Y.F. 1997. Hydrates of Hydrocarbons. Penn Well, Tulsa, OK.
2. SLOAN, E.D., JR. 1998. Clathrate Hydrates of Natural Gases. Marcel Dekker, New York.
3. MAKOGON, Y.F. & E.D. SLOAN, JR. 1994. Phase equilibrium for methane hydrate from 190 to 262 K. J. Chem. Eng. Data. **39:** 351.
4. VERIGIN, N.N., I.L. KHABIBULLIN & G.A. KHALIKOV. 1980. Linear problem of gas hydrate dissociation in a porous medium. Izv. Akad. Nauk. SSSR, Mekhanika Zhidkosti Gaza. **1:** 174–177.
5. HOLDER, G.D. & P.F. ANGERT. 1982. Simulation of gas production from a reservoir containing both gas hydrates and free natural gas. Proceedings of 57th Society of Petroleum Engineers Technology Conference, New Orleans, September 26–29, SPE11105.
6. BURSHEARS, M., T.J. O'BRIEN & R.D. MALONE. 1986. A multi-phase, multi-dimensional, variable composition simulation of gas production from a conventional gas reservoir in contact with hydrates. Society of Petroleum Engineers, Proceedings of Unconventional Gas Technology Symposium, Louisville, KY, May 18–21, SPE 15246.
7. KAMATH, V. 1983. Study of Heat Transfer Characteristics during Dissociation of Gas Hydrates in Porous Media. Ph.D. Thesis, University of Pittsburgh, Pittsburgh, PA.
8. AGU. 1999. American Geophysical Union, Mineralogical Society of America, and Geochemical Society, Spring Meeting, Boston, MA, June 1–4.
9. LYSNE, D. 1993. Hydrate plug dissociation by pressure reduction. *In* International Conference on Natural Gas Hydrates. E.D. Sloan, Jr., J. Happel & M.A. Hnatow, Eds. Acad. N.Y. Acad. Sci. **715:** 714–717.

Mathematical Models of Gas Hydrates Dissociation in Porous Media

GEORGE G. TSYPKIN[a]

Institute for Problems in Mechanics, Russian Academy of Sciences, Avenue Vernadskogo 101, 117526, Moscow, Russia

ABSTRACT: In strata hydrates may exist in different forms: as hydrates alone, hydrates in conjunction with free gas, or with gas and water in state of thermodynamic equilibrium, hydrates with excess ice or water, and other forms. A series of new mathematical models of gas production from gas hydrate reservoirs are proposed. The purpose of this work is to investigate dissociation problem for various initial states of gas hydrates fields under thermal stimulation and depressurization. The simplest case of a stratum in which the gas hydrate completely fills the pore space is proposed. It is shown that there are two different regimes of hydrate dissociation. In the strata with a positive initial temperature, the calculated dissociation temperature at the front may fall below the water crystallization temperature. This means that the hydrate does not satisfy the assumed decomposition into gas and water and the mathematical model has internal thermodynamic contradiction. In this case the hydrate dissociates with formation of an ice–gas mixture zone. The gas hydrate dissociation problem coexisting with gas is considered. It is shown that for high permeability the front solution does not apply, because the local temperature of the hydrate in the zone ahead of the dissociation front is higher than the local phase transition temperature calculated from the pressure distribution. In this case there are three zones: gas–water, gas–hydrate–water, and gas–hydrate. If the stratum initially contains gas, water, and hydrate in thermodynamic equilibrium the frontal model always gives a contradiction. It is shown that hydrate dissociation takes place in the extended gas–hydrate–water zone. For media with high permeability the amount of gas hydrate dissociated in this zone exceeds the amount of hydrate decomposed in the total dissociation zone by several orders.

INTRODUCTION

The problems associated with gas hydrate formation and dissociation in natural reservoirs have become of great interest to the hydrocarbon industry. Coupled processes of heat and mass transfer are major factors that affect the evaluation of gas hydrate accumulation in the earth.[1–4] We can describe gas hydrate dissociation in the Earth's sediments as a process in which gas and water are produced at an unknown moving boundary. The mathematical description is constructed as a generalization of the classical Stefan problem and takes into account the effect of the motion of the components constituting the gas hydrate.[4–10] In some cases the behavior

[a]Telecommunication. Voice: 095-434 3352; fax: 095-938 2048.
tsypkin@ipmnet.ru

of the physical system can be described within the framework of a frontal approach. In other cases there exist regimes of gas hydrate dissociation that are characterized by dissociation in an extended region or in the presence of two (or more) unknown moving boundaries, because the classical frontal approach leads to thermodynamic contradictions.[7-10]

We suggest a complete system of equations for correlating the phenomena of heat and mass transfer with gas hydrates dissociation. An essential part of the mathematical description is the correct specification of boundary conditions. The formulation of the problem with phase transitions contains the unknown moving interface. This moving boundary is the surface at which one or several parameters of the model exhibit a discontinuity or discontinuous derivatives. The conservation laws may be written in term of a *jump* at the phase transition front. We assume that all phases and components are locally at the thermodynamic equilibrium. These relations constitute the complete system of boundary conditions at the front.

It is vital to emphasize that in all zones it is possible to apply a linear approach and derive the analytical solutions. However, in general the problems are nonlinear. The treatment of simple physical situations for which solutions exist in a closed analytical form is very useful for illustrating essential features of gas hydrate dissociation in porous media.

FORMULATION OF THE PROBLEM AND BASIC EQUATIONS

We assume that the hydrate-containing reservoir is a multiphase and multicomponent system, whose basic components consist of the rock particles forming the skeleton of the porous medium, gas hydrate, natural gas, and water (or ice). For simplicity, we assume that the skeleton of the porous medium, hydrate, and ice are incompressible and motionless; water is incompressible; the natural gas is a perfect gas; and the thermophysical parameters of the component are constant. With these restrictions, the general system of equations, consisting of mass conservation laws for gas, water (ice), and hydrate, the generalized Darcy laws, the energy conservation law for the mixture, the Clapeyron equation for gas, the thermodynamic relations, and the equation for the hydrate–gas–water (ice) equilibrium curve may be written as follows.

$$m\frac{\partial}{\partial t}(1-\nu-S)\rho_g + \mathrm{div}\rho_g \vec{v}_g = M_g, \quad m\frac{\partial}{\partial t}S\rho_{w(i)} + \mathrm{div}\rho_{w(i)}\vec{v}_w = M_{w(i)},$$

$$m\frac{\partial}{\partial t}\nu\rho_{g0} = -M_g, \quad m\frac{\partial}{\partial t}\nu\rho_{wo} = -M_{w(i)}, \quad \vec{v}_j = -\frac{k}{\mu_j}f_j(S,\nu)\mathrm{grad}P, \quad j = w, g,$$

$$\frac{\partial}{\partial t}(\rho e)_m + \mathrm{div}(\rho_w h_w \vec{v}_w + \rho_g h_g \vec{v}_g) = \mathrm{div}(\lambda_m \mathrm{grad}T), \quad P = \rho_g RT,$$

$$e_g = h_g - \frac{P}{\rho_g}, \quad dh_g = C_P dT, \quad de_l = C_l dT, \quad l = s, w, i, h, \quad \ln P = A - \frac{B}{T}, \quad (1)$$

$$\lambda_m = m\nu\lambda_h + mS\lambda_{w(i)} + m(1-\nu-S)\lambda_g + (1-m)\lambda_S,$$

$$(\rho e)_m = m\nu\rho_h e_h + mS\rho_{w(i)}e_{w(i)} + m(1-\nu-S)\rho_g e_g + (1-m)\rho_S e_S.$$

Here, T is the temperature, P is the pressure, S is the water (ice) saturation, ν is the hydrate saturation, v is the filter velocity, M is the phase transition intensity, k is the permeability, μ is the viscosity, ρ is the density, e is the intrinsic energy density, h is the enthalpy density, λ is the thermal conductivity, m is the porosity, and C is the heat capacity. The subscripts denote: w, water; g, gas; h, hydrate; S, porous medium skeleton; and m, effective value.

In the pore space saturated with gas hydrate the effective densities of the water (ice) are calculated as the masses of the corresponding components divided by the volume occupied by hydrate. For methane hydrate with a degree of occupancy $n = 6$ we have $\rho_{wo(io)} = 783.87$ kg/m^3 and $\rho_{go} = 116.13$ kg/m^3 ($\rho_h = 900$ kg/m^3).

The system of basic equations for gas–water (ice) and gas–hydrate zones can be obtained from system **(1)** by formally setting $\nu = 0$ or $S = 0$, respectively. In these cases the equation for thermodynamic equilibrium of gas–water (ice)–hydrate is not considered.

The formulation of the problem allows the existence of hydrate dissociation and phase transition fronts. Boundary conditions at these interfaces can be obtained from the mass conservation laws for H$_2$O and gas and the energy conservation laws at discontinuities of the water and hydrate saturation functions

$$[\rho(V_n - u_n)]_-^+ = 0$$

$$[\rho h(V_n - u_n) + \lambda(\text{grad}\,T)_n]_-^+ = 0. \tag{2}$$

Here, V is the discontinuity velocity, u is the water or gas velocity, and the subscript n denotes the normal; plus and minus signs denote the quantities to the right and left of the front, respectively. The system of boundary conditions at the moving boundary must be supplemented with the thermodynamic relation between the pressure and the gas hydrate dissociation temperature.[4–11]

$$\ln P_* = A - \frac{B}{T_*} \tag{3}$$

$$T_+ = T_- = T_*, \quad P_+ = P_- = P_*.$$

An asterisk denotes the values of the quantities at the front.

After identical transformations, the system of basic Equations **(1)** reduces to a system of equations in the temperature, pressure, and saturations (of hydrate and water or ice). We consider the solutions of the problems in a linear approximation when the saturation, pressure, and temperature variations in each zone is small. We represent these functions in the form

$$S = \hat{S} + S', \quad \nu = \hat{\nu} + \nu', \quad P = \hat{P} + P', \quad T = \hat{T} + T',$$

where $\hat{f}$ is the undisturbed value, and f' is the perturbation of a function.

Consider the one-dimensional semi-infinite problem. Initially, let the stratum be characterized by the hydrate saturation ν_0, the water (ice) saturation S_0, the pressure P_0, and the temperature T_0. We assume that on the stationary wall $x = 0$ corresponding to the assumption that in the extracting well the pressure drops to a quite small value $P^0 < P_0$ (or the temperature in the well grows to a quite high value $T^0 > T_0$). Then the dissociation front $x = X(t)$ propagates to the right from the surface $x = 0$.

If initial and boundary conditions are constant, then the problems have self-similar solutions of the form

$$T = T(\zeta), \quad P = P(\zeta), \quad S = S(\zeta), \quad v = v(\zeta),$$

$$X_j(t) = 2\gamma_j\sqrt{at}, \quad j = 1, 2, \quad \zeta = \frac{x}{2\sqrt{at}}.$$

An analytical solutions in all regions can be expressed in terms of linear combinations of probability integrals. By substituting these solutions into the boundary conditions at the moving fronts we obtain the systems of transcendental equations for T_*, P_*, S_+, v, γ, and so forth, which are solved numerically to obtain the following characteristic values of parameters $\rho_w = 1,000$ kg/m^3, $\rho_h = 900$ kg/m^3, $\rho_i = 910$ kg/m^3, $\rho_S = 2,000$ kg/m^3, $R = 520$ J/(kg·K), $C_w = 4,390$ J/(kg·K), $C_h = 2,500$ J/(kg·K), $C_i = 2,090$ J/(kg·K), $C_S = 1,000$ J/(kg·K), $\lambda_w = 0.678$ W/(m·K), $\lambda_h = 2.11$ W/(m·K), $\lambda_i = 2.23$ W/(m·K), $\lambda_S = 2$ W/(m·K), $q_{hw} = 5 \times 10^5$ J/kg, $q_{hi} = 1.66 \times 10^5$ J/kg, $q_{iw} = 3.34 \times 10^5$ J/kg, $\mu_w = 1.8 \times 10^{-3}$ Pa·s, $m = 0.25$, $\mu_g = 1.8 \times 10^{-5}$ Pa·s, $A_w = 49.32$, $B_w = 9459$, $A_i = 24.38$, and $B_i = 26556$.

Below we consider a few simple examples that illustrate some typical properties of hydrate dissociation in hydrate saturated strata.

GAS HYDRATES DISSOCIATION IN SATURATED STRATUM

Consider the simplest one-dimensional model of a stratum, where the gas hydrate completely fills the pore space, and when the thermodynamic conditions at the stationary wall ($x = 0$) correspond to conditions under which the gas and water are in a free state. Then the dissociation front $x = X(t)$ propagates to the right from the surface $x = 0$.

Computations have shown that in strata with a positive initial temperature the calculated dissociation temperature at the front may fall below the water crystallization temperature. This means that the assumed hydrate decomposition into gas and water is not satisfied. In this case gas hydrate dissociation takes place with the formation of ice and gas. The heat influx from the surrounding rock, or hole, leads to the appearance of an ice melting boundary that moves in the same direction as the dissociation surface. The pressure is a monotonic function that falls from the value at the front to boundary value at the hole. FIGURE 1a shows a typical temperature distributions. Curve 1 corresponds to the noncontradictory solution. As may seen from FIGURE 1a, in another case, the water temperature (curve 2) lies below the water crystallization temperature. In this case, we assume that there exist two unknown moving phase transition boundaries that separate three regions (gas–hydrate, gas–ice, and gas–water), in which the mixture of gas and H$_2$0 remain in different states (FIG. 1b).

The dimension of the ice region depends principally on the permeability and the values of pressure and temperature in the hole. As the permeability increases or the pressure in the hole drops, the hydrate decomposition accelerates, which results in a greater velocity of the dissociation boundary. A decrease in permeability, or greater heating, leads to a reduction of the region saturated with ice. As the permeability reaches 10^{-16} m^2, the width of the ice containing region shrinks toward zero, which indicates that the boundaries and transition merge to the single front regime.

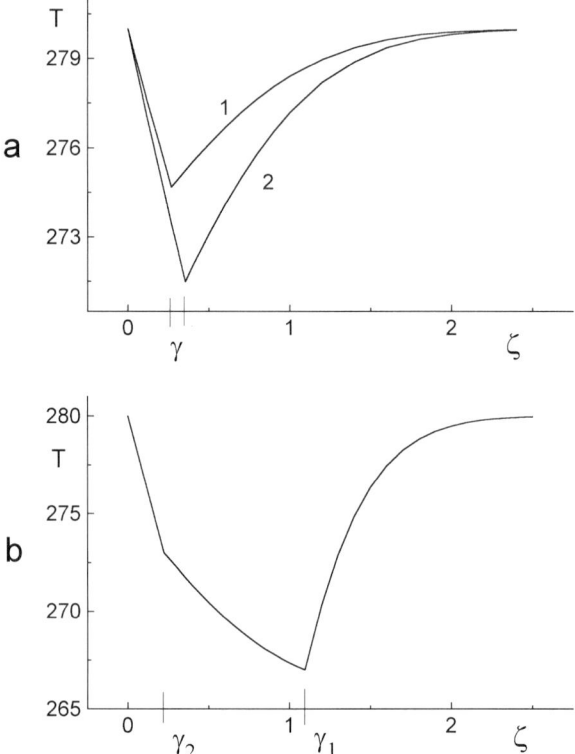

FIGURE 1. (a) Typical temperature distributions ahead of and behind the dissociation front. *Line 2* shows drop in the temperature below of water crystallization temperature. (b) The distribution of temperature of hydrate decomposition takes into account formation of two unknown phase transition boundaries and the ice region. Initial and boundary conditions correspond to the data of regime 2. $T_0 = T^0 = 280$ K, $k = 10^{-14}$ m², $P^0 = 3 \times 10^6$ Pa (*regime* 1), $P_0 = 2 \times 10^6$ Pa (*regime* 2).

DISSOCIATION OF GAS HYDRATES COEXISTING WITH GAS

Now consider the problem of gas hydrate dissociation in a stratum initially saturated with gas and hydrate mixture. The calculated results presented in FIGURE 2 reflect three fundamentally different situations. Thus, the solution shown in FIGURE 2a is thermodynamically consistent. Clearly, the dissociation temperature curve (line 2) calculated from Equation **(3)** lies above the temperature curve (line 1) in the hydrate zone. The results of numerical experiments show that such solutions can be constructed only in the case of low permeability $k \leq 10^{-16}$ m². For permeability above this limit the front solution has a thermodynamic contradiction (FIG. 2b). Here, the dissociation temperature curve (line 2) in the hydrate zone lies below the temperature curve (line 1). Physically, this corresponds to superheating of hydrate. Hence,

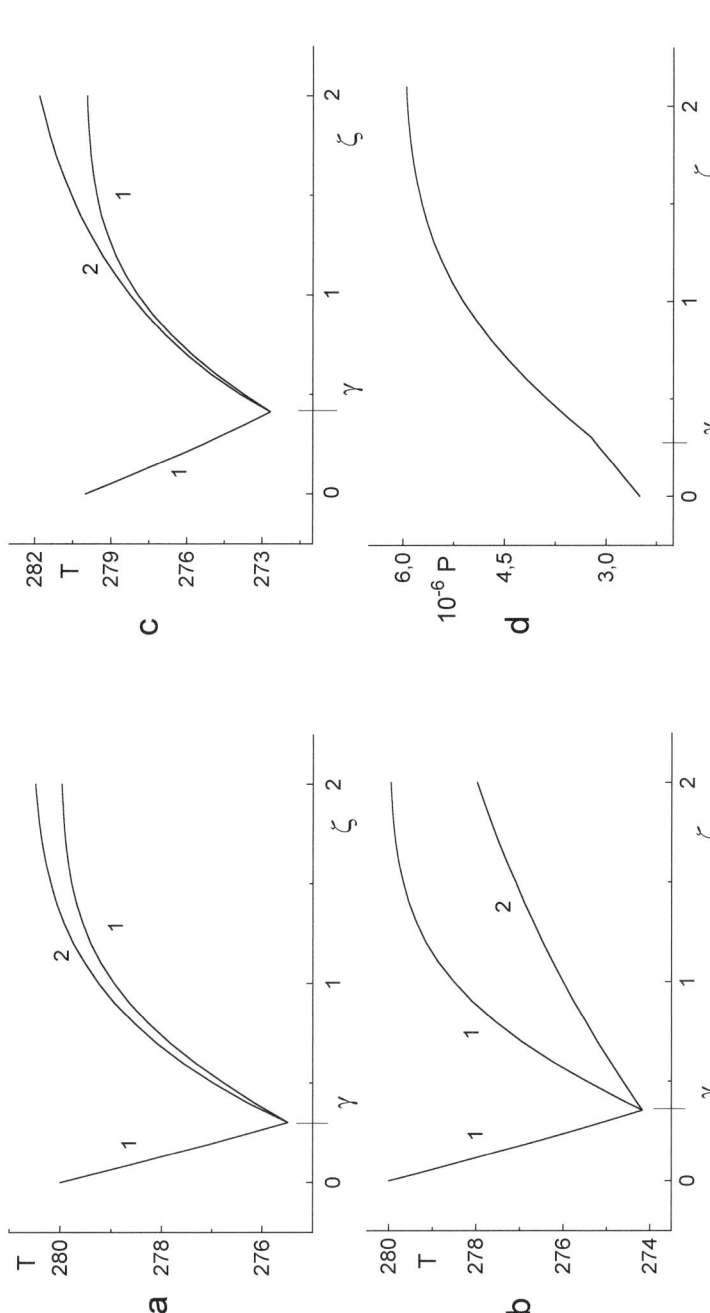

FIGURE 2. Three regimes of hydrate dissociation in strata: $v_0 = 0.5$, $T_0 = T^0 = 280$ K. (**a**) Noncontradictory regime $k = 10^{-16}$ m^2, $P_0 = 6 \times 10^6$ Pa, $p^0 = 2.5 \times 10^6$ Pa; (**b**) regime of hydrate decomposition described by a thermodynamically contradictory solution for which the equilibrium condition for existence of hydrate is not satisfied. Superheating of hydrate $(T > T_D)$ in the zone ahead the front (*line 1* and *line 2*, respectively) $k = 10^{-15}$ m^2, $P_0 = 6 \times 10^6$ Pa, $p^0 = 2.5 \times 10^6$ Pa; (**c**) regime for formation of the ice region $k = 2 \times 10^{-16}$ m^2, $P_0 = 8 \times 10^6$ Pa, $p^0 = 10^6$ Pa.

the thermodynamic condition for hydrate existence is not satisfied. Thus, the mathematical model has a contradiction and the dissociation process cannot be described within a framework of a frontal approach. Therefore, formation an extended hydrate dissociation domain should be taken into account. FIGURE 2c presents a situation that is similar to that in the hydrate saturated strata with a positive initial temperature. The calculated dissociation temperature at the front falls below the water crystallization temperature. This means that the assumed hydrate decomposition into gas and water is not satisfied. In this case gas hydrate dissociation takes place with the formation of ice. FIGURE 2d shows a typical pressure distribution function.

DISSOCIATION OF GAS HYDRATES COEXISTING WITH GAS AND WATER

We assume that the hydrate reservoir is a porous medium saturated with hydrate, water, and gas in a state of thermodynamic equilibrium. Acting on the reservoir for the purpose of dissociating the hydrate and subsequently extracting the gas leads to the formation of a region that is free of hydrate and saturated with a mixture of gas and water. If it is assumed that phase transitions take place only at the dissociation front, as shown in Reference 7, we necessarily arrive at a thermodynamic contradiction. Therefore we start by introducing an extended phase transition zone and a surface of partial dissociation that makes it possible to eliminate superheating of hydrate ahead of the dissociation front. Here we are extending the model to the case of arbitrary hydrate and water saturation, when the mobility of the liquid phase should be taken into account. Numeric experiments show that for low hydrate and water saturation the water immobility assumption is perfectly permissible, but may lead to considerable errors in the opposite case.

In the model proposed, the gas hydrate dissociation process is not localized in a narrow zone, but extends over the entire hydrate–gas–water region ahead of the dissociation front. This is expressed as a complication of the boundary conditions resulting from the introduction at the right of the front a quantity v_+, determined in the process of solving the problem. The computations show (see FIGURE 3) that the amount of hydrate decomposed in the mixture zone considerably exceeds the amount of hydrate dissociated in the zone to the left of the front.

CONCLUSIONS

We have proposed a number of new mathematical models to describe the dissociation of gas hydrate in strata. These models allow formation of an extended dissociation zone or formation of an ice–gas region located between hydrate dissociation and ice melting fronts.

It is shown that for media with high permeability, the amount of gas hydrate dissociated in the extended zone exceeds the amount of hydrate decomposed in the total dissociation zone by several orders.

The formation of ice during gas hydrate dissociation leads to a reduction in gas permeability that significantly affects the gas production volume. The presence of ice also changes the properties of the stratum, as in a multiphase system. This

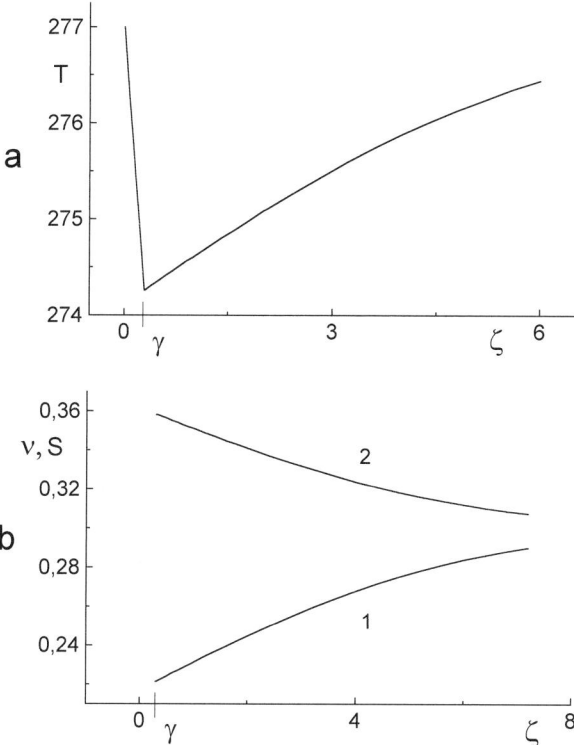

FIGURE 3. (a) Distribution of temperature. (b) Water saturation function (*line 2*) and hydrate saturation function (*line 1*) in the dissociation zone that lie ahead of the dissociation front. $S_0 = 0.3$, $\nu_0 = 0.3$, $k = 10^{-16}$ m^2, $T_0 = T^0 = 277$ K, $P_0 = P_0(T_0) = 3.88 \times 10^6$ Pa, $P^0 = 2.5 \times 10^6$ Pa.

circumstance should be taken into account when applying physical methods to the diagnostics of gas hydrate deposits.

In a further study both the influence of interfacial forces between rock minerals, hydrates, fluids, gases, and also the chemical action should be emphasized.

ACKNOWLEDGMENTS

This work was supported by the Russian Foundation for Fundamental Research (Project Nos. 96-01-00521 and 99-01-00272).

REFERENCES

1. HOLDER, G.D. *et al.* 1982. A thermodynamic evaluation of thermal recovery of gas from hydrates in the Earth. J. Petrol. Technol. **34:** 1127–1132.

2. ULLERICH, J.W. *et al.* 1987. Theory and measurement of hydrate dissociation. AIChE J. **33:** 747–752.
3. KAMATH, V.A. *et al.* 1987. Evaluation of hot-brine stimulation technique for gas production from natural gas hydrates. J. Petrol. Technol. **39:** 1379–1388.
4. SELIM, M.S. *et al.* 1989. Heat and mass transfer during the dissociation of hydrates in porous media. AIChE J. **35:** 1049–1052.
5. CHERSKY, N.V. *et al.* 1972. On thermal method of gas-hydrate production. Doklady AN SSSR. **203:** 550–552 (in Russian).
6. VERIGIN, N.N. *et al.* 1980. Linear problem of gas hydrate dissociation in a porous medium. Izvestiya. Akad. Nauk SSSR. Mekh. Zhidk. Gaza **1:** 174–177 (in Russian).
7. BONDAREV, E.A. *et al.* 1989. Mathematical modeling of the dissociation of gas hydrates. Dokl. Akad. Nauk SSSR. **308:** 575–578 (in Russian).
8. TSYPKIN, G.G. 1991. Effect of liquid phase mobility on gas hydrate dissociation in reservoirs. Izvestiya Akad. Nauk SSSR. Mekh. Zhidkosti i Gaza. **4:** 105–114 (in Russian).
9. TSYPKIN, G.G. 1992. Appearance of two moving phase transition boundaries in the dissociation of gas hydrates in strata. Dokl. Ross. Akad. Nauk. **323:** 52–57 (in Russian).
10. TSYPKIN, G.G. 1998. Gas hydrate decomposition in low-temperature strata. Izvestiya Ross. Akad. Nauk, Mekh. Zhidkosti i Gaza. **1:** 101–111 (in Russian).
11. BONDAREV, E.A. *et al.* 1976. Mechanics of hydrate generation in gas flow. Nauka. Novosibirsk (in Russian).

Department of Energy Methane Hydrate Research and Development Program

An Update

EDITH ALLISON[a]

United States Department of Energy, Office of Natural Gas and Petroleum Technology, FE-32, 1000 Independence Avenue, SW, Washington, DC 20585, USA

ABSTRACT: The objective of this paper is to summarize the program planning activities and current research being conducted by the U.S. Department of Energy Methane Hydrate Program.

INTRODUCTION

In 1998, the U.S. Department of Energy Office of Fossil Energy (DOE/FE) began working with other government agencies, the U.S. Geological Survey (USGS), the Department of Defense, Naval Research Laboratory (NRL), and the National Science Foundation (NSF) as well as the Ocean Drilling Program (ODP), academic, and industry researchers to develop a new federally funded research and development program. The purpose of the program is to maximize the potential contribution of the huge methane hydrate resource to reliable supplies of a cleaner fuel with reduced impacts on global climate, while mitigating potential hydrates risks for marine safety and seafloor stability. The program has four main objectives:

1. Determine the location, sedimentary relationships, and physical characteristics of methane hydrate resources so as to assess their potential as a domestic and global fuel resource.

2. Develop the knowledge and technology necessary for commercial production of methane from oceanic and permafrost hydrate systems.

3. Develop an understanding of the dynamics and distribution of oceanic and permafrost methane hydrate systems that is sufficient to quantify their role in the global carbon cycle and climate change.

4. Develop an understanding of hydrate systems in near-seafloor sediments and sedimentary processes, including sediment mass movement and methane release so that safe, standardized procedures for hydrocarbon production and ocean engineering can be assured.

[a]Telecommunication. Voice: 202-586-1023; fax: 202-586-6221.
edith.allison@hq.doe.gov

HISTORY OF DOE METHANE HYDRATE RESEARCH AND DEVELOPMENT

A 10-year program, established in 1982, at the DOE Morgantown Energy Technology Center (now the Federal Energy Technology Center, FETC) supported laboratory studies of physical and chemical properties and the mechanisms of formation and dissociation of hydrates, and studies of the geological, geophysical, and geochemical characteristics of marine and Arctic hydrate formations. DOE-supported studies were instrumental in developing a foundation for basic knowledge about the location and thermodynamic properties of gas hydrates. The DOE program:

- established the existence of hydrates in Kuparuk Field, Alaska;
- completed studies of 15 offshore hydrate basins;
- developed production models for depressurization and thermal production of gas from hydrates;
- developed preliminary estimates of gas in-place for gas hydrate deposits;
- built the Gas Hydrate and Sediment Test Lab Instrument (GHASTLI); and
- sponsored the first two International Conferences on Methane Hydrate Research and Development, in 1991 and 1994.

In 1992, the ten-year, eight million dollar program was terminated as government policy shifted from long-term, high-risk research and development to near-term exploration and production research and development. Although DOE funding ceased, work has continued at USGS, NRL, NSF, ODP, universities, other laboratories, and overseas.

PROGRAM PLANNING ACTIVITIES

DOE sponsored two workshops in 1998. The first workshop on *The Future of Methane Hydrates Research and Resource Development* was held in Denver, Colorado, on January 21–22, 1998. More than 100 senior scientists and managers from industry, academia, National Laboratories, government agencies, and international organizations attended the meeting. The objective of this meeting was to gather expert opinions on research needs and priorities as a preliminary step to developing a program plan.

The second workshop, conducted on May 12, 1998, in Washington, DC, reviewed the draft program plan and discussed organizational and budget issues. In August 1998, the plan, *A Strategy for Methane Hydrates Research and Development* was published. A more detailed science plan, *National Methane Hydrate Multi-Year Research and Development Program Plan*, was recently developed, reviewed by government, industry, and academic stakeholders, and published.

These two documents provide a framework for a ten-year research and development program. Detailed implementation plans and budgets that are consistent with these documents will be developed subsequently.

RESEARCH ACTIVITIES IN FISCAL YEARS 1997–1999

The DOE/FE redirected about $300,000 per year of natural gas research and development funds in fiscal years (FYs) 1997 and 1998 to begin work on the methane hydrate program. DOE has budgeted $500K in FY 1999 and requested $1.985 million in FY 2000 for methane hydrate Research and Development. Under this funding DOE has supported:

- NRL research on gas hydrates in deep-sea sediments;
- USGS preparation, testing, and sample analysis of the Malik well in Mackenzie Delta, Canada;
- USGS laboratory and Gulf of Mexico seismic studies;
- Idaho National Engineering and Environmental Laboratory microbial studies of samples collected at the Malik well; and
- Gas Hydrates Consortium at Colorado School of Mines.

The results of this work are reported elsewhere in this volume.

FUTURE RESEARCH AND DEVELOPMENT PLANS

The following multi-year efforts will be initiated in FY 2000 and FY 2001. The schedule and level of effort depends on the levels of Congressional appropriations.

Reservoir Characterization

- Laboratory studies to relate hydrate/sediment properties to seismic and well log signals.
- Geologic, geochemical, and thermodynamic studies of methane flux.
- Oceanographic sample collection and geologic/geochemical analysis.
- Database development documenting hydrate locations and research results.
- Collection and analysis of Arctic and marine hydrates.

Production

- Laboratory tests and model development of hydrate dissociation.
- Production tests in industry well(s) of opportunity.
- Preliminary assessment of alternative production methods.

Global Change

- Monitor subsea hydrate site, including sensor development.
- Ice core studies to define relation of atmospheric methane to global climate changes.

- Microbiological and chemical studies of the fate of methane in the ocean and atmosphere.
- Study application to greenhouse gas mitigation.

Safety and Seafloor Stability

- Document historic slump and collapse sites.
- Seismic and well logging for evaluation of subsea hydrate zone structure and strength.

PROGRAM COORDINATION

To effectively address the technological complexity of the proposed Research and Development, the methane hydrates program will marshal the resources of the petroleum industry, academia, National Laboratories, and a broad base of government programs with concurrent interests in methane hydrates research. These organizations will exchange research plans and results and coordinate activities through a Management Steering Committee.

Japan, India, Canada, United Kingdom, Germany, Brazil, Norway, and Russia currently have active methane hydrates Research and Development programs. DOE and the Management Steering Committee will explore cooperative work with these countries.

Growth and Inhibition of Ethylene Oxide Clathrate Hydrate

ROAR LARSEN,[a] CHARLES A. KNIGHT,[b]
KEVIN T. RIDER,[c] AND E. DENDY SLOAN, JR.[c,d]

[a]*SINTEF Petroleum Research, Multiphase Flow Laboratory, N-7465 Trondheim, Norway*

[b]*National Center for Atmospheric Research, P.O. Box 3000, Boulder, Colorado 80307, USA*

[c]*Center for Hydrate Research, Colorado School of Mines, Golden, Colorado 80401, USA*

ABSTRACT: Growth and inhibition of single crystals of ethylene oxide clathrate hydrate provide experimental accessibility to similar phenomena in methane clathrate hydrates. Ethylene oxide hydrate single crystals were grown in a simple experimental setup. The crystals grew from the subcooled melt as rhombic dodecahedra, macroscopically exhibiting {110} as the slowest growing faces. The addition of minute amounts of poly(N-vinylcaprolactam) (PVCap) or a random terpolymer of PVCap, poly(N-vinylpyrrolidone), and dimethylamino-ethylmethacrylate (VC-713) caused small-scale branching of crystals, but with a seemingly uniform crystal orientation. A few tenths of a weight percent of these additives caused complete crystal growth inhibition at several degrees of subcooling.

INTRODUCTION

Structure I (sI) gas hydrate is recognized as a potential energy source, with huge amounts identified in permafrost regions and in subsea sediments.[1] However, most prior research on hydrates has been motivated by efforts to provide flow assurance in production and transport systems that can be blocked by crystal growth.[2]

Interest has concentrated on avoiding hydrate formation. Kinetic inhibitors of hydrate crystallization are additives that suppress crystallization by sorption on the crystal.[3] Inhibitors include poly(N-vinylpyrrolidone) (PVP), poly(N-vinylcaprolactam) (PVCap), and a random terpolymer of the previous two with dimethylaminoethylmethacrylate (VC-713).

Our previous study of tetrahydrofuran (THF) hydrate[4] provided insights into growth and inhibition of hydrates of structure II (sII). From a practical viewpoint, the present study of EO hydrate crystal growth can be viewed as an analog of methane hydrate crystallization and inhibition, because both systems form sI.

Ethylene oxide has a boiling point of 283.85 K, and its hydrate sI is a body centered cubic crystal[5] like methane hydrate, but with a much higher melting point (284.2 K at 1 atm) and a molar ratio of EO:H_2O of 3:23.[6] Ethylene oxide is one of the few chemicals available, for which the hydrate composition can be matched to

---

[d]Address for correspondence: E. Dendy Sloan, Jr., Center for Hydrate Research, Colorado School of Mines, Golden, Colorado 80401, USA.

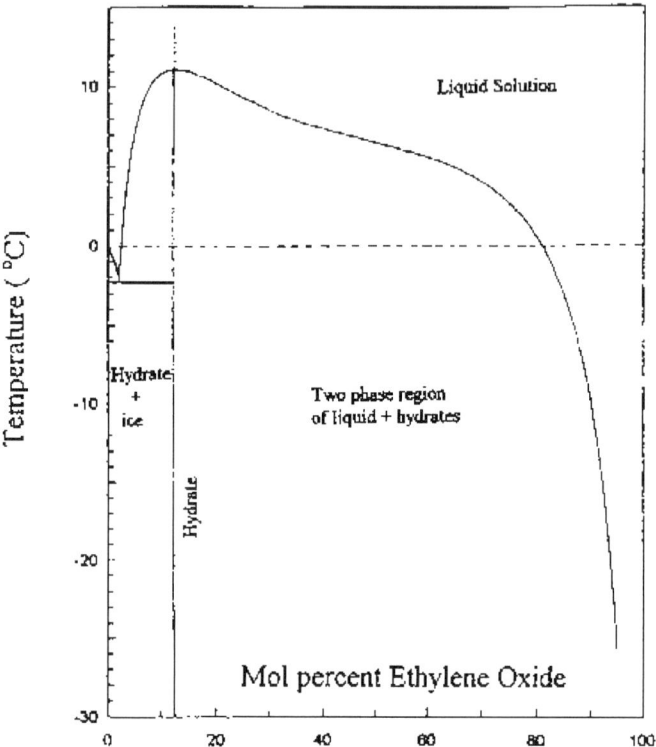

FIGURE 1. Partial phase diagram for ethylene oxide at atmospheric pressure. Adapted from Reference 6. Experiments were conducted at various temperatures along the hydrate composition line.

the liquid composition at temperatures above the ice point, simultaneously eliminating ice formation and mass transfer considerations. This enables a relatively simple experimental setup. FIGURE 1 is adapted from Reference 6 and shows a partial phase diagram for EO at atmospheric pressure.

EXPERIMENTAL PROCEDURE

FIGURE 2 is a diagram of the single crystal growth apparatus. The growth chamber was a test tube containing hydrate melt with a glass pipette inserted. This test tube and pipette were contained in a Plexiglas chamber with coolant circulated to maintain a constant temperature. Because EO hydrates are denser than their melt, the pipette tip was constricted from its original size of 2 mm outer diameter to about 20% of that size, so that the initial crystals in the pipette would generally be too large to sediment into the bulk solution. Growth is initiated high in the pipette, and the crystals grow enough to be too large for the opening by the time they reach the lower end.

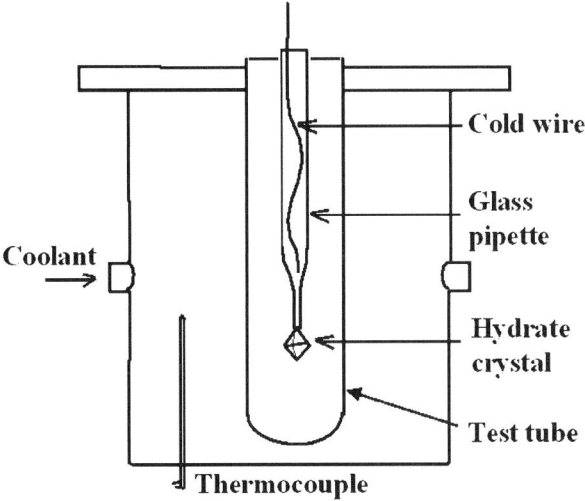

FIGURE 2. Main growth cell, original figure from Makogon.[9]

EO hydrates were readily nucleated with wires cooled in liquid nitrogen or dry ice and they grew slowly at temperatures between 283.7 and 284.2 K. Usually, a single crystal grew into the bulk solution when the crystal growth front reached the tip of the pipette. A second type of experiment was carried out in a flat capillary on a cooled bed under a microscope. All crystals grew freely in the melt and their shape development and growth rates were recorded on videotape or photographically with a 35-mm reflex camera.

RESULTS

Growth from the Melt

Single crystals of sI EO hydrates grown from the melt (i.e. about 13 mol% EO in water) exhibited the {110} crystallographic planes, forming a dodecahedron (12 sides) with rhombic faces. FIGURE 3 shows a crystal grown at a subcooling of 0.9 K. No other crystallographic planes were identified on the sI crystals, but it should be noted that crystals were chosen for exhibiting regular features; that is, non-single or twinned or irregular crystals often led to stopping the experiment and starting growth with a new crystal. Growth irregularities were often observed in the form of vertices becoming edges and edges of different lengths in the same crystal.

For subcooling as low as 1.0 K, pitting of faces was common. This growth form is known as skeletal growth, in which the edges grow faster than the face centers because of more rapid heat dissipation at the edges. FIGURE 4 shows a skeletal dodecahedron that developed at a slightly higher, but still moderate subcooling (2 K).

FIGURE 3. Hydrate structure I melt growth at subcooling of 0.9 K. Rhombic dodecahedra. *Scale:* upper end of pipette is approximately 1.6 mm across.

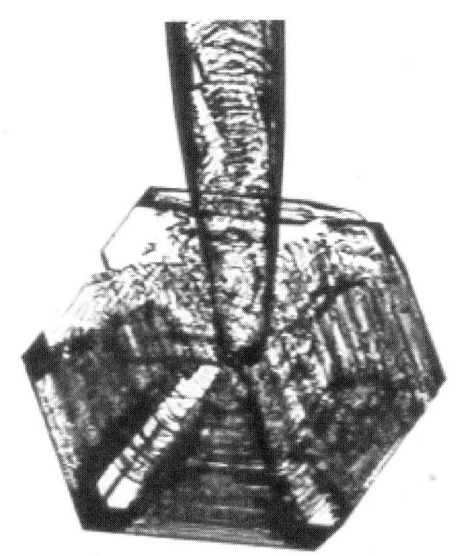

FIGURE 4. Melt growth at 2.0 K of subcooling. Skeletal growth of dodecahedra *Scale:* same as in FIGURE 3.

At high degrees of subcooling (3 K and higher), the growth of the crystals is best characterized as dendritic, where—to some extent—the edges and especially the vertices outgrew the crystal faces with a considerable difference in linear growth velocity. This phenomenon is generally associated with heat or mass transfer limitations.

Growth with Small Amounts of Inhibitor

Our previous work on THF hydrates[4] showed that any inhibitor influence on the solution freezing point was negligible. In the present work, the molecular weight (mw) of the inhibiting polymers was mainly in the range of 2,000–10,000 g/mol.

Most mws of PVP did not cause any habit changes in EO hydrate, but the growth rate of the crystals was reduced by more than 50% at 0.1 wt% of polymer. However, some early experiments with a "high molecular weight" (hmw) PVP (no further information on the mw was available) induced the same growth pattern as described for the pure melt at large subcooling—dendrite-like growth of the crystal vertices. This growth occurred already at 0.5 K subcooling with the hmw PVP, indicating that heat transfer limitations were enhanced by the presence of inhibitor.

When PVCap or VC-713 in small amounts were added to the stoichiometric EO/water solutions, the morphological changes were striking. Intermediate concentrations of the inhibitors (0.05–0.25 wt%) produced rapid small-scale branching of the crystals, resulting in spherical globules composed of flimsy branches. An example of this growth habit is seen in FIGURE 5. Lowering the concentration below about

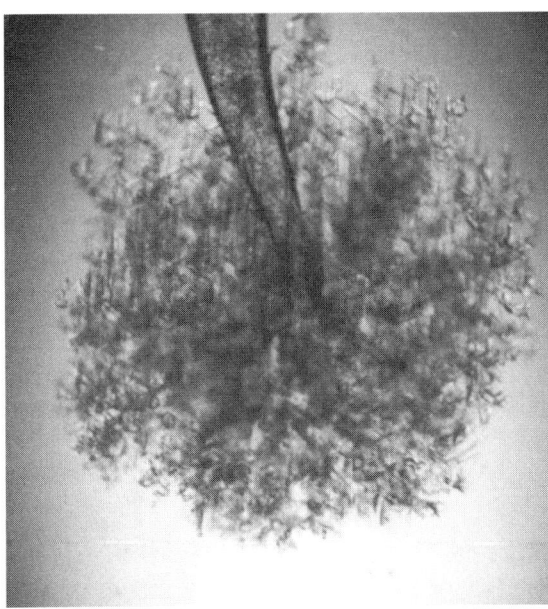

FIGURE 5. Branching structure I growth at subcooling of 0.9 K and 0.1 wt% PVCap. *Scale:* same as in FIGURE 3.

0.05 wt% at a subcooling of 0.5 K still resulted in branches, but with sturdier, larger limbs making a faceted, sponge-like structure. This dendritic growth was different from that observed in the pure melt at a subcooling higher than 2.0 K, or with the hmw PVP. That branching was an extension of skeletal growth whereas branching in the presence of the PVCap or VC-713 inhibitors is less regular and very prolific at a much smaller scale.

FIGURE 6 shows three different results from experiments in the flat capillary. FIGURE 6a shows large facets in the pure system. At 0.08 wt% PVCap (FIG. 6b), the crystal orientation was uniform throughout the highly branched growth. In FIGURE 6c (0.2 wt% PVCap), the branching became less regular and facets were not seen at the available magnification. In this last picture, based on the transition result at lower concentrations of inhibitor, we believe that even this extreme branching conserves the crystallographic orientation and is in effect still a single crystal. That

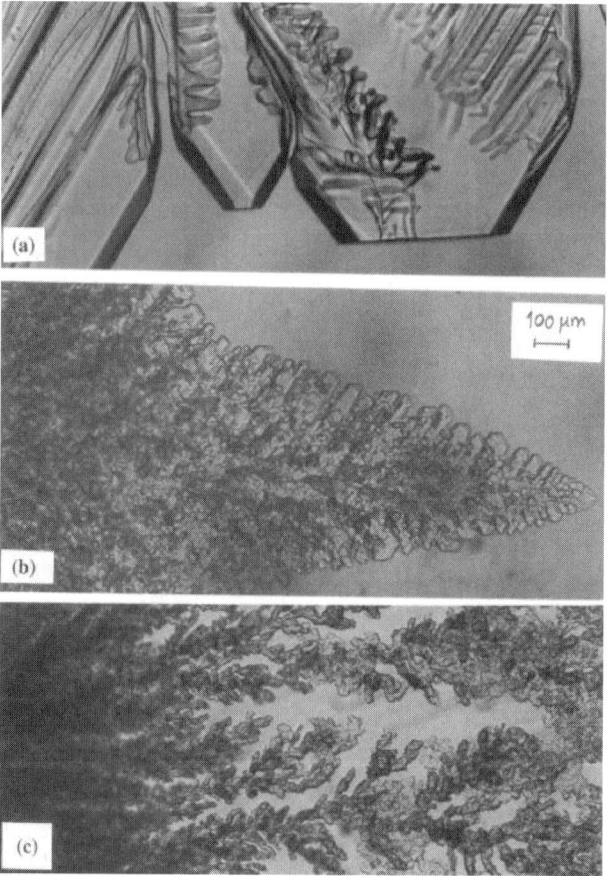

FIGURE 6. Structure I growth in thin capillaries at subcooling of 0.5 K. (**a**) Uninhibited, (**b**) 0.08 wt% PVCap, and (**c**) 0.2 wt% PVCap. Scale: constant in all photographs.

is, we assume that the branching does not change its nature at higher concentrations, only its "frequency" or density of occurrence.

Completely Inhibited Growth

The habit changes at low concentrations of inhibitor represented a transition between the dodecahedral habit and complete growth inhibition at higher concentrations. *No-growth* was defined as no progression of the crystal faces being detectable at 50× magnification over a period of no less than six hours, but more often 20–24 hours. This means that at most 0.02 mm of growth was possible with an approximate 1mm resolution on the video screen where each experiment was monitored. For six hours, this gives a very conservative estimate of an average maximum rate of 9.3×10^{-10} m/s, which is more than three orders of magnitude smaller than the smallest measured growth rates without inhibitors.

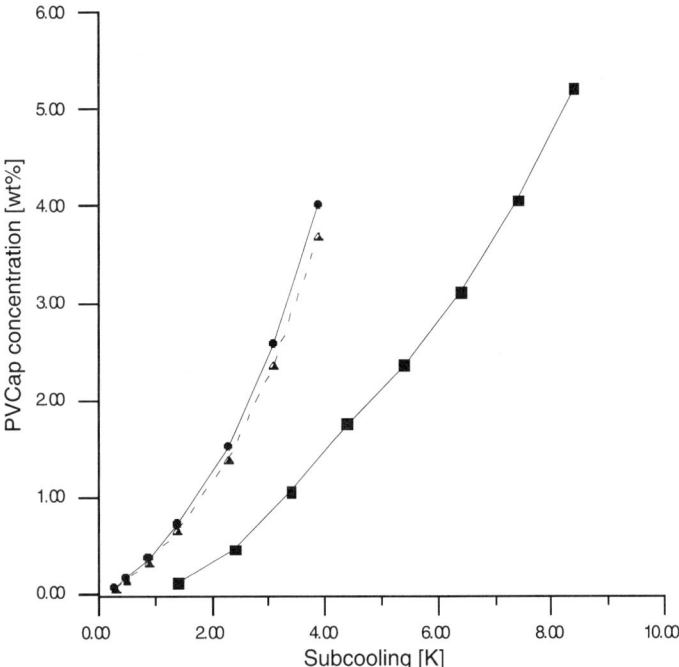

FIGURE 7. Amount of PVCap required to stop growth at a given subcooling, for both structure I (ethylene oxide) and structure II (tetrahydrofuran) hydrate. *Solid lines* (with *circles* for structure I ethylene oxide hydrates, and *squares* for tetrahydrofuran structure II hydrates) show concentrations based on the amount of water in the system. The *dotted line* shows the relative displacement of the structure I data compared to structure II when both concentrations are based on the total amount of liquid in the system. Data points for larger subcooling were not obtainable—it appears that the curves continue almost vertically. Key: ●, structure I data; ▲, adjusted structure I data; ■, structure II data.

Complete inhibition was achieved with both PVCap and VC-713 inhibitors. The concentration needed for complete inhibition was a strong function of the subcooling, the type of inhibitor, and the molecular weight of the inhibitor. FIGURE 7 shows the minimum concentration of PVCap needed to stop growth, as a function of subcooling. Total inhibition is only achievable up to a certain level of subcooling. Above that level, the growth rate of the crystals is too high to allow the efficient adsorption of the polymers, and exclusion from or inclusion in the crystal will most likely take place for the polymer molecules This figure also includes data for the THF hydrate sII system, as an extension of our previous work.[4] The data were collected using PVCap at a molecular weight of about 1,700. PVP was not able to inhibit crystal growth completely, even at a concentration of 5 wt%, which was the highest used in our experiments.

DISCUSSION AND CONCLUSIONS

Recent results from hydrate experiments in a high-pressure methane/water system[7] indicated that the EO hydrate system was a good model for sI methane hydrates, since the growth habits and morphology are the same for single crystals of methane sI hydrate and for sI EO hydrate. It should be noted, however, that specific surface characteristics have not been compared between the two systems.

The generally supposed mechanism for dendrite-like growth from the melt, is that it has a much higher surface area per unit volume, providing more efficient dissipation of the heat of crystallization. Mass transfer limitations should be eliminated in our system, because the melt and crystal compositions are identical, and crystal kinetic processes may in this case be more rapid than the heat dissipation.

It seems clear from the present study and our earlier sII experiments[4] that PVP acts only as a diffusion barrier without strong adsorption to the crystal structure. PVCap and VC-713 show very different effects from PVP in both sI and sII hydrate crystal systems, and we believe this to be caused by much stronger adsorption to the surface.

It is not known whether adsorption and inhibition were specific to the {110} faces, but it does appear that there was adsorption on {110}, probably interfering with step growth across these faces and leading to the development of the chaotic branching seen in FIGURES 4, 5, and 6. It can be argued that adsorption to any other plane would slow the growth of that plane and thereby—with sufficient adsorption—make it visible. No such planes were seen in the experiments. The branching was dissimilar to the planar habit for THF hydrate with PVCap and VC-713.[4]

A crystal growth morphology that is similar to the spherical, branched EO hydrate growths is spherulitic growth.[8] However, spherulites are not single crystals (i.e. each new branch differs slightly in crystallographic orientation). The EO hydrate growths had similar spherical shape symmetry but we believe them to be single crystals, and thus not true spherulites.

It is reasonable to assume that branching is caused by a growth retardation effect. Adsorption of the inhibitor polymer molecules on the surface of the crystals is one such possibility. The explanation of the different effects of the same growth inhibitors on the melt growth of THF and EO hydrates (structures II and I) might be sought in the different molecular structures of the two growth surfaces expressed, namely {111} for sII and {110} for sI. It has been proposed that the attachment of the polymers to the clathrate hydrate surface is by adsorption of pendant groups (caprolactam or pyrrolidone) into incomplete, large cages on the hydrate surfaces.[9,10] Previous work from our group[11] has suggested speculative explanations for the observed growth changes. One possible schematic representation of the sI {110} faces is shown in FIGURE 8. This is a simplified view, since modern understanding of a developing surface would describe the surface more as a zone (of considerable extension compared to the individual molecules or hydrate cavities), throughout which a more or less continuous change from crystal structure to liquid melt takes place.[12]

Similar representations for the sII crystal and its {111} planes given in Reference 11 show that for sII, any adsorbing molecule has a near two-dimensional freedom (sites are only 60° apart) in choosing adjacent adsorption sites for its pendant groups, and these sites are perpendicular to the surface. It can be argued that the situation depicted in FIGURE 8 for the sI hydrate only allows a one-dimensional freedom—the adsorbing molecule is blocked by the ridge between the troughs, or by the 45° angle of its adsorption plane, from choosing adjacent sites other than those lining up in the same trough. Thus, more reorientation (and longer time) is needed than for sII, which may be one important factor in explaining the higher concentrations needed to stop growth, as seen in FIGURE 7. Higher concentrations increase the probability of finding a molecule that requires less reorientation. This concentration variation is

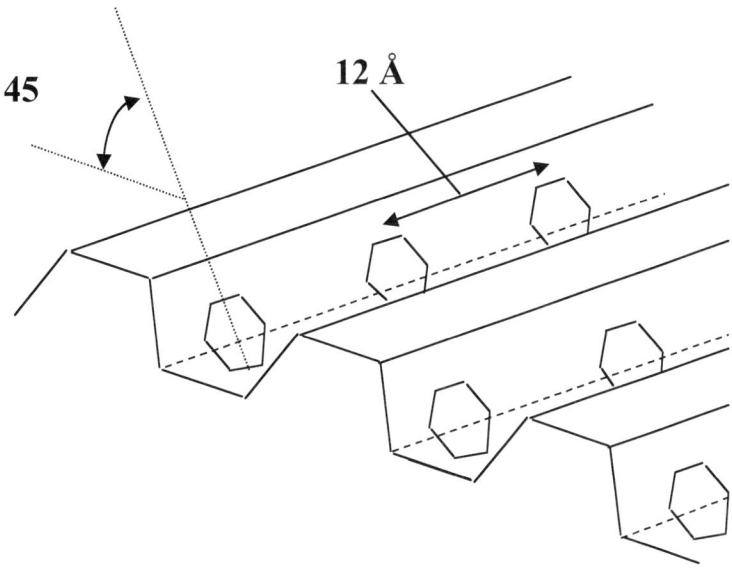

FIGURE 8. One possible schematic representation of the structure I {110} surface. Hexagonal rings from the large (inhibitor pendant group-accommodating?) cavities are present in the bottom of *troughs*, at 45° angles. Drawing not to scale.

also influenced by different growth rates and by different critical radii for growth on the surface.[11]

The above can be accompanied by relatively simple size and geometry considerations. The spacing between large cavities (assumed attachment points) is 12 Å in hydrate sI. Thus, every sixth pendant group (approximately 12.3 Å apart) from an inhibitor polymer fits well into adjacent cavities. The slight offset decreases the quality of the fit for the pendants as one moves along the polymer chain. The maximum number of well-fitted adsorbing pendants is assumed to be five, based on geometric considerations with molecular models. If this is the entire length of the polymer molecule, it corresponds to molecular weights of about 4,000 g/mol. This mechanism requires that the polymers remain straight in solution and not curl up on themselves. Measurements over a wide range of PVCap molecular weights, indicate that at 4,000 g/mol, the chains are stretched out in aqueous solution.[13]

In the simplistic view represented in FIGURE 8, the additional non-adsorbing pendants of the polymer chain come into conflict with the crystal structure between large cavities, whereas by viewing the surface as a region of continuous change, we can argue that these interspersed structures simply have not formed, or are unable to form because of the presence of the inhibitor molecule.

With preferred adsorption occurring in the parallel troughs, ridges and areas between adsorption sites are able to continue to grow and remain crystallographically unchanged. They may be envisioned as commencing branches between relatively widely spaced adsorbed growth impediments (the polymer chains). At high enough inhibitor concentrations, all growth is blocked, since the distance between neighboring (or next-neighbor) troughs is smaller than the required two-dimensional critical radius for growth of steps along the surface.

The authors are not completely comfortable with the hypotheses and speculations outlined, mainly because they have not been directly tested in the available laboratory facilities. The molecular level explanation of the effects of the inhibitors upon the growth of these materials—both sII and sI crystals—requires further examination.

ACKNOWLEDGMENTS

We would like to thank Dr. D. N. Glew for his valuable advice on EO hydrate crystal growth. This work was sponsored by Amoco, ARCO, Chevron, Conoco, Department of Energy (U.S.), Mobil, Oryx, Phillips, Shell, and Texaco. The National Center for Atmospheric Research is sponsored by the National Science Foundation.

REFERENCES

1. KVENVOLDEN, K.A. 1994. Proc. Intl. Conf. on Natural Gas Hydrates, New Paltz, New York, June 1993. Ann. N.Y. Acad. Sci. **75:** 232.
2. LONG, J.P. 1994. Ph.D. Dissertation, Colorado School of Mines (T-4265).
3. SLOAN, E.D., S. SUBRAMANIAN, P.N. MATTHEWS, J.P. LEDERHOS & A.A. KHOKHAR. 1998. Ind. Eng. Chem. Res. **37:** 3124.
4. MAKOGON, T., R. LARSEN, C.A. KNIGHT & E.D. SLOAN. 1997. J. Crystal Growth **179**(5): 258–262.

5. JEFFERY, G.A. 1984. *In* Inclusion Compounds, Vol. 1. J.L. Atwood, J.E.D. Davies & D.D. MacNicol, Eds.: 135. Academic Press, New York.
6. GLEW, D.N. & N.S. RATH. 1995. Can. J. Chem. **73:** 788.
7. SMELIK, E.A. & H.E. KING. 1997. Amer. Mineralogist **82:** 88.
8. PHILLIPS, P.J. 1994. Spherulitic crystallization in macromolecules. *In* Handbook of Crystal Growth, Vol. 2. D.T.J. Hurle, Ed.: 1168–1215. Elsevier.
9. MAKOGON, T. 1997. Ph.D. Dissertation, Colorado School of Mines (T-4969).
10. RODGER, P.M., G.B. DREW & T.J. CARVER. 1996. J. Chem. Soc. Faraday Trans. **92:** 5029.
11. LARSEN, R. 1997. Dr. Ing. Thesis, Norwegian University of Science and Technology, Trondheim, Norway.
12. KLUG, D.L. 1993. *In* Handbook of Industrial Crystallization. A.S. Myerson, Ed. Butterworth-Heinemann, Boston.
13. DORGAN, J. & V. PANCHALINGAM. 1994. Colorado School of Mines, CEPR Intradepartmental Memo. Unpublished.

Hydrogen and Oxygen Isotope Fractionation in Water during Gas Hydrate Formation

TATSUO MAEKAWA[a] AND NOBORU IMAI

Geological Survey of Japan, 1-1-3, Higashi, Tsukuba, Ibaraki, 305-8567, Japan

ABSTRACT: Isotope fractionation of hydrogen and oxygen in water caused by methane hydrate formation in saline water was investigated experimentally. The isotope fractionation factors were calculated by measuring the concentration of salts and hydrogen and oxygen isotopic ratios of water in both initial solution and residual solution after formation of methane hydrate. The amount of methane hydrates was estimated using two assumptions: complete desalination and gas consumption. As methane hydrate forms in saline water, the concentration of salt in solution increases and gas pressure decreases. We estimated the amount of methane hydrate not only from the increase of concentration of salts in the solution, but also from the decrease in gas pressure. The isotope fractionation factors of hydrogen and oxygen in the water between methane hydrate and saline water were obtained as 1.016 to 1.020 and 1.0028 to 1.0032, respectively. These are similar to the factors between ice and water.

INTRODUCTION

In deep-sea sediments, ^{18}O enrichment observed in the interstitial water is often associated with low salinity related to natural gas hydrates.[1] This is because the interstitial water was diluted with the water released from natural gas hydrates in the sampling process. It is suggested that natural gas hydrate crystals concentrate ^{18}O from water in their lattice and exclude dissolved salts in ambient water. Davidson *et al.*,[2] experimentally investigated the oxygen isotope fractionation in water between clathrate hydrate and liquid water by synthesizing THF hydrate, which is a structure II hydrate; however, THF hydrate differs in structure from natural gas hydrates, which generally have structure I. There are no experimental determinations of the fractionation factor of oxygen in water between structure I hydrate and water. Moreover, hydrogen isotope fractionation in water between clathrate hydrate and liquid water have not previously been reported.

The object of the present work is to estimate the isotope fractionation factors of hydrogen and oxygen in water between methane hydrate and liquid water. In the present experiments, gas hydrates were formed from methane gas and 3 wt% NaCl solution, which resembles to the concentration of chloride in natural pore water.

[a]Telecommunication. Voice: +81-298-54-3720; fax: +81-298-54-3533. maekawa@gsj.go.jp

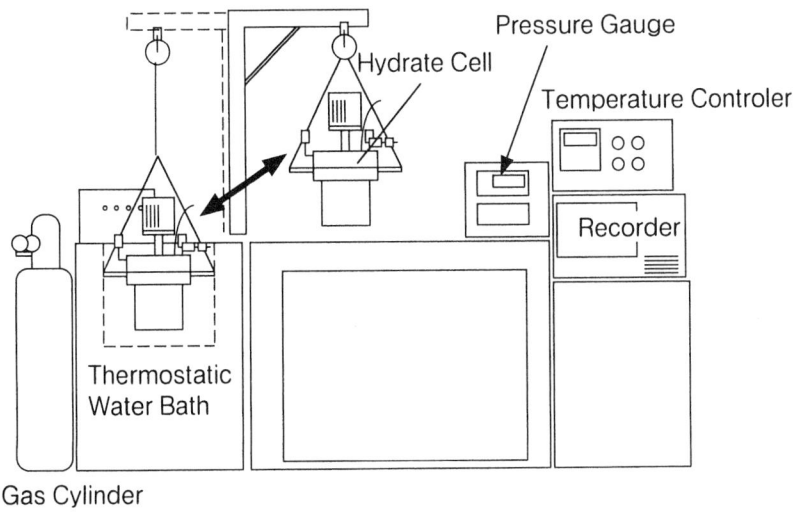

FIGURE 1. Schematic diagram of experimental apparatus.

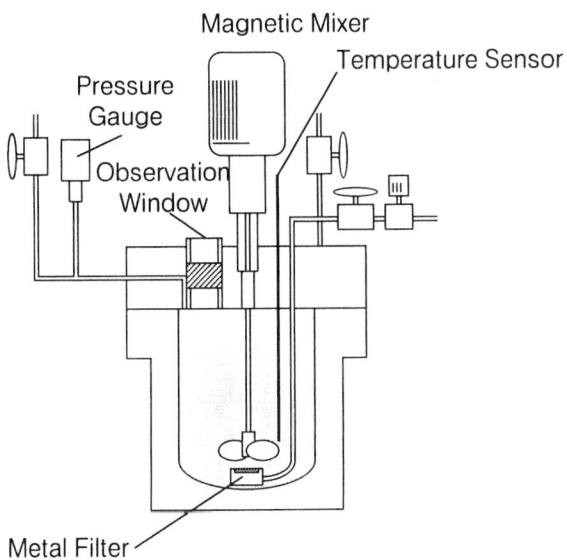

FIGURE 2. Schematic diagram of hydrate cell.

EXPERIMENTAL APPARATUS

The experimental apparatus used in the present work includes of a high-pressure hydrate cell, a glycol–water temperature controlled bath and a recorder (see FIGURE 1). The pressure cell with a volume of 705 ml (see FIGURE 2) is immersed in a glycol–water bath in which the temperature is controlled by a cooling unit. The temperature of the solution in the cell is measured with platinum resistance thermometers immersed in the liquid phase and the gas pressure is detected with a pressure gauge coupled with semiconductor transducer. The solution is agitated by a magnetic stirring mixer. The cell has two glass windows on the upper sides of the cell for visual observation.

EXPERIMENTAL AND ANALYTICAL METHOD

In this work, gas hydrate was formed from pure methane gas and 3 wt% NaCl solution. Initially, 250 to 450 cc of the solution was charged into the cell and the pressured methane gas was introduced at a specified pressure. After lowering the temperature of the solution to 273.5 K, the formation of methane hydrate was initiated by increasing the stirring rate. Methane hydrate was formed at constant temperature of 273.5 K. The stirring rate was 600 rpm in each experiment. We confirmed the formation of methane hydrate in the cell by measuring the pressure. After formation of methane hydrate, we intermittently sampled the liquid solution from the outlet at the bottom of the cell. The initial water and the water sampled after methane hydrate formation were subjected to measurement of the concentration of salt and isotopic compositions of hydrogen and oxygen.

Hydrogen and oxygen isotopic ratios of sampled water were determined by means of mass spectrometer (Finnigan Delta S) using the equilibrium method. In the case of hydrogen isotope measurements, hydrogen standard gas of known isotopic composition was equilibrated with a specified volume of the sampled water in a glass flask at constant temperature, then the hydrogen isotopic ratio of hydrogen gas was measured. For oxygen isotope measurements, carbon dioxide standard gas with a known isotopic oxygen composition was used. The isotopic compositions are expressed using the conventional delta notation in per mil relative to the SMOW standard. Chloride concentration of experimental water was measured by an ion chromatography PIA-1000 of Shimadzu.

ESTIMATION OF FRACTIONATION FACTOR

The salt concentration and isotopic compositions of water in the liquid phase was assumed to be homogeneous and the solution equilibrated with only the surface layer of solid methane hydrate. The fractionation factor α was calculated from

$$\ln \frac{R}{R_0} = (\alpha - 1)\ln f, \qquad (1)$$

where R is the ratio D/H or $^{18}O/^{16}O$ of water in solution and R_0 is that of water in initial solution. The factor f is defined as the ratio of the number of water molecules in the residual solution to the total number of water molecules in initial solution, and which is derived from the amount of gas hydrate formed.

$$f = \frac{N}{N_0}, \qquad (2)$$

where N is the number of water molecules in liquid solution and N_0 is that in the initial solution. Before methane hydrate formation, the factor f is unity. As methane hydrate formation proceeds, the f factor decreases. Using the conventional delta notation, Equation (1) can be written as

$$\Delta\delta = \delta - \delta_0 = 1000 \cdot (\alpha - 1)\ln f, \qquad (3)$$

where δ is the isotopic composition expressed in the delta notation in per mil of δD or $\delta^{18}O$ of water in the solution, and δ_0 is that in the initial solution.

We tried to estimate the factor f by using two different assumptions: complete desalination and gas consumption. Initially, we estimate the amount of methane hydrate by assuming that methane hydrate crystals completely expelled the dissolved salt into the residual liquid solution. As methane hydrate forms and expels the salt into the liquid, the concentration of salt in the residual solution increases. We obtained the concentration of salt in both initial solution and residual solution after methane hydrate formation, and calculated the factor f by using the following equation

$$f = \frac{C_0(100 - C')}{C'(100 - C_0)}, \qquad (4)$$

where C_0 wt% is the concentration of salt in the initial solution and C' is that in the residual solution after methane hydrate formation. However, the experiments suggested that the liquid solution was being entrapped in the aggregate of methane hydrate crystals. Consequently, the assumption that the methane hydrate completely expelled the salts to the residual solution leads to an overestimated amount of methane hydrate formed, and the value of the factor f is a minimum.

On the other hand, we calculated the amount of methane hydrate formed from the gas consumption. Methane hydrate formation consumes methane gas, which results in the pressure decrease. We can estimate the amount of methane hydrate from the pressure difference of gas during methane hydrate formation. We calculated the amount of methane gas in gas phase using Soave-Redrich-Kwong equation of state. Since there are more than six water molecules per methane molecule in methane hydrate,[3] we believe the result obtained from gas consumption to be the minimum amount of methane hydrate formed, the value of the factor f obtained is consequently a maximum.

RESULTS

Five methane hydrate formation experiments were carried out. The concentration of salt and the hydrogen and oxygen isotopic ratios of water in both initial solution and residual solution after formation of methane hydrate were measured in each experiment. We calculated the isotope fractionation factors of hydrogen and oxygen in the water using the concentration of salt (C', C_0) and the differences of isotopic

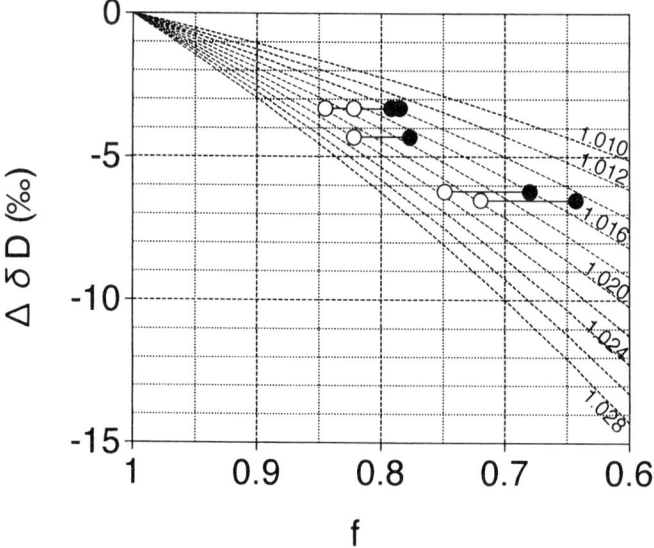

FIGURE 3. Hydrogen fractionation in water during gas hydrate formation. Factor f is the ratio of amount of water in the residual solution to initial amount of water. It is associated with the amount of gas hydrate formed. Factor f shown by *closed circles* is calculated from the assumption of complete desalination, and that shown by *open circles* is calculated from the gas consumption during gas hydrate formation.

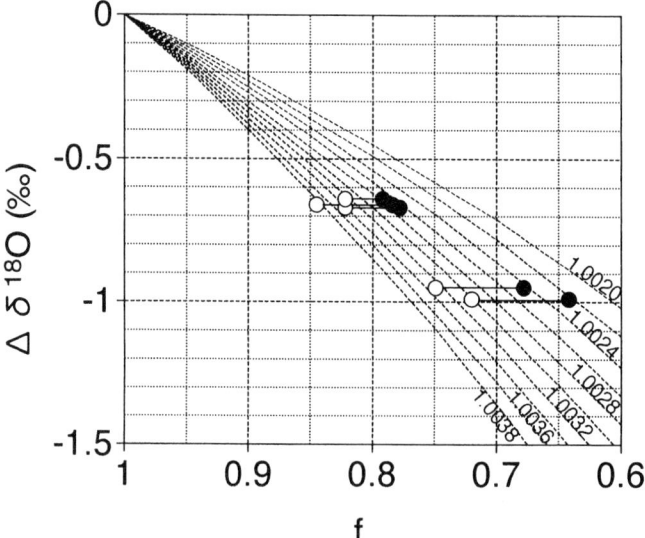

FIGURE 4. Oxygen fractionation in the water during gas hydrate formation.

compositions of hydrogen and oxygen [δ(D), Δδ (^{18}O)]. The range of the factor f was limited by the two assumptions—complete desalination and gas consumption. FIGURES 3 and 4 show the results of isotopic fractionation of hydrogen and oxygen in the water, respectively. Dotted lines represent variations in the difference of isotopic compositions at a specific fractionation factor calculated from Equation (**3**). From these figures, we estimated the approximate isotopic fractionation factors of oxygen in the water between methane hydrate and 3 wt% NaCl solution to be 1.0028 to 1.0032. In the case of hydrogen isotope, we need to consider the hydrogen isotopic fractionation between water and methane. However, in this work, we neglected this, assuming that methane dissolved during hydrate formation is trapped to methane hydrate crystals without equilibration with water. We roughly estimated the isotopic fractionation factors of hydrogen in the water between methane hydrate and 3 wt% NaCl solution to be 1.016 to 1.020.

DISCUSSION

The isotope fractionation factors of hydrogen and oxygen in the water between methane hydrate and saline water are determined to be 1.016–1.020 and 1.0028–1.0032, respectively. These values of the isotopic fractionation factor α are compared with the values reported previously (TABLE 1). For oxygen isotope fractionation, Davidson et al.[2] first presented the oxygen isotope fractionation factor in the

TABLE 1. Comparison of isotopic fractionation factors of hydrogen and oxygen in water between clathrate hydrate and water, and ice and water

	Oxygen Isotope Fractionation	
	α(^{18}O)	Reference
clathrate–water	1.0028–1.0032	This work: methane hydrate—3% NaCl solution
	1.00268	Davidson et al. (1983): THF hydrate–water
	1.0037	Handa (1984): THF hydrate–water
	1.0038–1.0040	Matsumoto (1998): natural gas hydrate–interstitial water
ice–water	1.0029, 1.0031	O'Neil (1968)
	1.0028	Suzuoki and Kimura (1973)
	1.0032	Handa (1984)
	1.0027	Craig and Hom (1969): ice–sea water
	Hydrogen Isotope Fractionation	
	α(D)	Reference
clathrate–water	1.016–1.020	This work: methane hydrate–3% NaCl solution
ice–water	1.0178, 1.0195	O'Neil (1968)
	1.0206	Suzuoki and Kimura (1973)
	1.0203	Craig and Hom (1969): ice–sea water

water between the clathrate hydrate and liquid water. They formed the THF hydrate from THF solution directly and determined the oxygen isotope fractionation factor as 1.00268. Handa[4] obtained the fractionation factor between THF hydrate and water as 1.0037 using isotope effects on heat capacities, enthalpies of fusion, and temperature of fusion. On the other hand, Matsumoto[5] estimated the oxygen fractionation factor using the geological samples collected from Blake Ridge by ODP Leg164. Matsumoto[5] reported the deference in oxygen isotope ratios between natural gas hydrate and the ambient interstitial water. Chloride concentration and oxygen isotopic composition of interstitial water are observed to vary in a remarkable zigzag pattern with spiky anomalies, reflecting methane hydrate dissociation during the sampling process. Matsumoto[5] estimated the oxygen isotopic fractionation factor between methane hydrate and ambient water as 1.0038, from the results of isotopic studies of the geological samples of natural gas hydrate and interstitial water. The fractionation factors obtained in this work are found to be in the range of literature values, and similar to the fractionation factors for water between ice and water.[6–8]

Hydrogen isotope fractionation factors in water between clathrate hydrate and water have not yet been reported. We found that the fractionation factor between methane hydrate and 3 wt% NaCl solution is in the range 1.016 to 1.020. Several researchers have determined the hydrogen isotopic fractionation between ice and water.[6–8] The fractionation factor between methane hydrate and saline water obtained in this work is similar to that between ice and water.

CONCLUSIONS

We experimentally estimated the isotope fractionation of hydrogen and oxygen in the water between gas hydrate and liquid water. We investigated the change in concentration of salt and isotopic composition of hydrogen and oxygen in the water during methane hydrate formation. In this work, the methane hydrate was formed from methane gas and 3 wt% NaCl solution at constant temperature 273.5 K. We sampled the solutions before and after methane hydrate formation, and measure the concentration of salt and the hydrogen and oxygen isotopic compositions of sampled water. We estimated the amount of methane hydrates from the increase of the concentration of salt in the solution and decrease in gas pressure.

We roughly estimated the isotope fractionation factors of hydrogen and oxygen in water between methane hydrate and saline water as 1.016 to 1.020 and 1.0028 to 1.0032, respectively. These values were found to be similar to the factors between ice and water. The isotopic fractionation factors estimated in the present work, however, have comparatively large errors. This is suggested to be caused by entrapment of liquid solution in methane hydrate crystals during methane hydrate formation. If methane hydrate is formed very slowly, with a lower reaction rate, methane hydrate completely expels dissolved salts into the solution and equilibrates with liquid water isotopically. It should, therefore, be possible to obtain the fractionation factors of hydrogen and oxygen in water more precisely in future.

REFERENCES

1. HESSE, R. & W.E. HARRISON. 1981. Gas hydrates (clathrates) causing pore-water freshening and oxygen isotope fractionation in deep-water sedimentary sections of terrigenous continental margins. Earth Planet Sci. Lett. **55:** 453–462.
2. DAVIDSON, D.W. *et al.* 1983. Oxygen-18 enrichment in the water of a clathrate hydrate. Geochim. Cosmochim. Acta **47:** 2293–2295.
3. HANDA, Y.P. 1986. Compositions, enthalpies of dissociation, and heat capacities in the range 85 to 270 K for clathrate hydrates of methane, ethane, and propane, and enthalpy of dissociation of isobutane hydrate, as determined by a heat-flow calorimeter. J. Chem. Thermodynamics **18:** 915–921.
4. HANDA, Y.P. 1984. Enthalpies of fusion and heat capacities for $H_2{}^{18}O$ ice and $H_2{}^{18}O$ tetrahydrofuran clathrate hydrate in the range 100–270 K. Can. J. Chem. **62:** 1659–1661.
5. MATSUMOTO, R. 1998. Methane hydrate estimates from the chloride and oxygen isotopic anomalies—Examples from the Blake Ridge and Nankai Trough sediments. *In* Proceedings of the International Symposium on Methane Hydrates—Resources in the Near Future? Chiba, Japan, October 20–22, 1998.
6. O'NEIL, J.R. 1968. Hydrogen and oxygen isotope fractionation between ice and water. J. Phys. Chem. **72:** 3683–3684.
7. SUZUOKI, T. & T. KIMURA. 1973. D/H and $^{18}O/^{16}O$ fractionation in ice-water system. Mass Spectroscopy **21:** 229–233.
8. CRAIG, H. & B. HOM. 1968. Relationships of deuterium, oxygen-18, and chlorinity in the formation of sea ice. Trans. Am. Geophys. Union **49:** 216–217.

Methane Hydrate Fuel Storage for All-Electric Ships

An Opportunity for Technological Innovation

M.D. MAX,[a] V.T. JOHN,[b] AND R.E. PELLENBARG[c]

[a]*Naval Research Laboratory, Code 7420, Washington, DC 20375, USA*

[b]*Department of Chemical Engineering, Tulane University, New Orleans, Louisiana 70118, USA*

[c]*Naval Research Laboratory, Code 6100, Washington, DC 20375, USA*

ABSTRACT: New industrial processes for forming methane hydrate, combined with the knowledge that small changes in gas hydrate-forming mixtures can greatly expand the pressure–temperature field of hydrate stability, suggest that virtually solid gas hydrates can be formed and stored in fuel tanks of any shape at near sea level ambient conditions. Development of this technology may allow a gas hydrate fuel transport medium to be devised and applied. Reduction-to-practice of this special fuel storage medium could have a significant impact upon the design and engineering of future Navy vessels and concurrently upon the future of naval operations.

INTRODUCTION

The U.S. Navy is committed to a revolution in the design of naval warships.[1] Within a 10–15 year period, surface warships are planned to have vastly increased firepower built around vertical launch missiles and *smart* projectiles, smaller crews, on the order of 1/2 to 1/4 of those on present day cruisers, augmented by more automatic ship systems, longer deployments supported by new engineering, and increased survivability through implementation of new stand-off antimissile weapons. Supporting this revolution, an integrated electric drive and associated clusters of technologies are identified as the principal future means of propulsion for the only new major surface combatant vessel in the twenty first century U.S. fleet,[2] other than nuclear-powered aircraft carriers. There will be a substantial impact of an all-electric warship on ship design, weapons, and survivability.

The precise implementation of all-electric ship systems is not yet fully known. Implementation will largely represent an evolution of technologies, such as main engine room turbo-electric generators, that have been used for some time but can be improved, through modern computer controlled weapons systems that will be direct developments of existing technology, to the possible use of new fuel cells and directed energy weapons. In addition to these largely evolutionary technologies, we suggest that a truly revolutionary implementation in the all-electric warship may

Telecommunication.
[a]max@qur.nrl.navy.mil [b]vj@mailhost.tcs.tulane.edu [c]pellenba@ccf.nrl.navy.mil

be made through consideration of methane fuel, stored abroad in novel physical media. Such fuels would have significant implications for engineering and combat applications.

Early this century, the nation's petroleum economy will likely begin transition to one based on methane, as diminishing oil reserves become increasingly expensive,[3,4] although some argue that oil will remain relatively abundant for the foreseeable future.[5] As oil becomes more expensive after the first decade of this century, the economic equation suggests that the use of methane will become more attractive. Even though it would make good sense to base future fuel for ships on a cheap and widely available fuel (e.g., methane) that will not likely to be subject to shortages, this utility alone does not justify an immediate transition to methane fuel from oil. Supplies of methane world-wide are much larger than oil reserves especially in, and adjacent to, North America. However, if the research and engineering objectives for developing a special industrial methane hydrate fuel storage and transport media[6] can be achieved, the all-electric warship that will probably become the backbone of the U.S. surface fleet well into the twenty first century, will benefit markedly from using methane as the primary fuel for generating electricity, regardless of future availability or price of petroleum.

In order that methane be considered as a naval fuel, it is necessary to develop a new gas fuel storage and transport medium. This technology would allow methane to be carried in the most highly condensed form possible, rendering it relatively inert and safe from explosion and fire hazards. In addition, achievement of this new fuel medium will allow a revolutionary reconsideration of warship design that may strongly affect overriding strategic and tactical issues. We outline the acknowledged changes in Navy vessels that are anticipated, presenting the case for development of an artificial industrial hydrate that has no natural analog, but which could provide the basis for new hydrate fuel storage technology. We also discuss potential engineering benefits.

ALL-ELECTRIC SHIPS

Electric-drive ships are not new to the U.S. Navy. The first of three 32,000-ton displacement, New Mexico class turbine powered battleships ordered in 1914 was given a main turbo-electric propulsion system comprising two 125,500 kW 2-phase bipolar alternators directly driven by steam turbines. Following W.W.I, Maryland and Tennessee Class battleships and the aircraft carriers *USS Lexington* and *USS Saratoga* had four steam turbine electric generators of 40,000 kW turbo-generators driving individual electric motor-shaft propellers.[2] The battleship *USS Colorado*, commissioned in 1923, had two main turbines of 11,000 kW utilizing steam at 250 lbs pressure. Each individual electric generator was rated at 13,400 kva to 4,325 v (series), 1,550 a at 82% power. Either or both generators drove four individual electric motors rated at 7,000 hp at 170 rpm. Such electric drive technology was also transitioned to merchant marine ships in the 1920s. An interesting non-Navy application of electric ships took place from 17 December, 1929 to 16 January, 1930, when the *USS Lexington* supplied power to Tacoma, Washington from her main propulsion generators because local hydroelectric generators were off-line.[7]

The early application of electric drive on naval vessels featured several large generators confined in near-traditional configuration engine rooms amidships. The generators were coupled to steam turbines that were placed in close proximity to oil-fired, steam generating boilers. The twenty first century embodiment of the electric ship, however, is not constrained by boiler technology. Present high-technology surface vessels, such as the R/V *Alliance*, owned and operated by NATO at the SACLANT Undersea Research Center in La Spezia, Italy, generate electricity from acoustically shrouded diesel engines that feed the power to electric motors at the stern of the ship, immediately inboard of the propellers. There is effectively no drive shaft, much engineering space is opened up, and much weight is saved. Although diesel-electric power is well understood, more efficient power generation may be achieved with high speed gas turbine- electric generating units. These can be powered from liquid or gas fuels, and special gas turbines optimized to operate on methane already exist. These turbines are small, often being less than 0.5 m diameter and 1–2 m long. They can be specially designed to meet the particular demands of a warship,[8] and are light-weight-power-per-volume, especially in contrast to diesel electric units. Turbines are particularly suited to a distributed generating capability. In addition, experimental fuel cells for naval application offer the promise of even more efficient electricity generation.

The U.S. Navy is studying the evolving concept of the all electric ship. Electric drive propulsion is regarded as a key to twenty-first century surface combatants (Walsh, 1989). However, these new ships will not have a central engine room and long drive shafts. Electricity will be supplied from gas turbines or fuel cells designed specially for the demands of naval vessels. The ships will depend on redundancy of distributed electric generating capacity for combat survival, rather than protection of a centralized engine room. The ability to change power allocation to meet demands should have an impact on ship design because a robust mechanical drive train, especially since propeller shaft and transmissions will not be necessary. It is envisioned that power and propulsion systems may be relocated and additional space created for fuel, weapons, and other new, high technology applications. The final form of these new dispositions is not yet fixed, but does not yet take into account the possibility of methane as a fuel, at least partly because there is no existing methane storage medium that can be regarded as a safe alternative to the traditional fuel, liquid petroleum.

METHANE HYDRATE, A POTENTIAL FUEL STORAGE MEDIUM

The suitability of some gaseous fuels for warships has been evaluated by the U.S. Navy. Hydrogen, stored as a liquid or metal hydride presents a number of difficulties that appear to rule it out as a fuel for warships.[9] Alternative liquid fuels such as alcohols and ammonia offer no advantage over fuel oil and would require additional traditional-configuration bunkerage because of their lower energy densities. Conventional storage of both gaseous hydrogen or methane have suffered from the same weakness. Large quantities of flammable gas present a considerable fire and explosion hazard, especially aboard a warship. Both compressed gas and liquefied gas, as transport media, also are potentially explosive. A potential third alternative for bulk methane transport, which is relatively energy dense, potentially stable, and nonflammable is as solid gas hydrate. Gas hydrates are crystalline substances in which a water-*ice* lattice is stabilized by hydrogen bonding of gas molecules. The compounds are found in nature in deep oceans and associated with permafrost because high pressures and/or low pressures thermodynamically stabilize the naturally occurring compound crystal structure.[10]

Clathrates, the generic term for hydrate, particularly methane and other hydrocarbon gas hydrates, have been known as laboratory curiosities since chlorine hydrate ($Cl \cdot 6H_2O$) was reported.[11] In the 1930s and 1940s the natural gas industry had problems with the formation of a crystalline, wax-like substance in natural gas transport pipes. This material clogged the lines and research was focused on understanding the origin and physical chemistry of the material so that its appearance in pipelines could be minimized. Methane hydrates are now recognized as being very widespread in marine sediments and in permafrost regions, and they may constitute the largest store of fixed carbon on earth.[12] Our present knowledge about methane hydrate physical chemistry, and the potential large volumes of recoverable methane from naturally occurring sources argues strongly that methane is likely to be the fuel of the future, especially if the technique of condensing methane within a hydrate crystal lattice can be used on an industrial scale.

Mixed-gas hydrates form spontaneously over a broad range of pressures and temperatures, some of which are sea level near-ambient. Indeed, the older literature contains many references to gas hydrates forming spontaneously in natural gas transmission pipelines and often blocking them; this potential situation requires the drying of gas prior to pipeline transportation.[10,13] Current experimental results show that gas hydrates can be fabricated with specific physical property ranges in solid and pseudofluid (slush) forms that are stable at near ambient pressures and temperatures. Naturally occurring hydrates of hydrocarbon constituents of natural gas can form at temperatures in excess of 40°C.[14] These hydrates have been produced for experimental study of physical properties and scientific and research applications. Because it is likely that the hydrates discussed here may not closely resemble those natural gas hydrates found in nature, they are referred to in this section as hydrates, fabricated hydrates, or special hydrates. It should be emphasized that this is a research area, rather than a description of existing materials.

Storage of methane in hydrate form, however, may significantly alter the safety aspects of methane storage to the extent that it would be safer than fuel oils if hydrate could be formed and stored at near-ambient temperatures and pressures. As well as having an attractively high energy density (TABLE 1), hydrate offers the attributes of being a solid rather than a gas or liquid. The compression factor is obtained because methane molecules are forced together in the crystalline hydrate lattice. Although the energy density of methane hydrate is low compared with common liquid fuels (TABLE 1), hydrate holds sufficient methane to be considered as a fuel storage medium. In fabricating an industrial hydrate it is reasonable to aim at exceeding the common naturally occurring 90% fill of available guest lattice sites (164 m^3 methane/m^3 hydrate). 100% methane guest site occupancy of the hydrate water lattice will yield 180 m^3/m^3 hydrate and substantially raise the energy density of the hydrate. If industrial hydrate can be fabricated to use the maximum available lattice space, which may itself positively expand the *P–T* stability field, it will be roughly 1.4 times

TABLE 1. Energy density of different forms of fuels[a]

Fuel and Form	Density (g/cc)	Energy Density (Btu/ft^3)
Hydrogen (liquid, cryogenic temps)	0.071	229,000
Methane		
gas (STP)	6.66×10^{-4}	1,012
liquid (LNG)	0.42	470,000
solid (natural hydrate)	0.91	161,920
solid (industrial hydrate[b])	0.92	181,800
Reference Liquid Fuels		
gasoline	0.74	876,000
JP-4 (jet fuel)	0.78	910,000
#2 Diesel fuel	0.78	995,000

[a]Based on data from Hoess and Stahman,[15] revised from Max and Pellenbarg.[6] Natural hydrate calculated with 160 m^3 per m^3 hydrate.

[b]Industrial hydrate is the theoretical target 100% of available guest sites filled completely with zero porosity in the mass.

more efficient for holding methane than LNG, comparing energy balances for producing LNG versus producing crystalized hydrate.

It may also be possible to raise the energy density of the industrial hydrate by flooding the structure with higher density hydrocarbon gases, which are known to positively expand the P–T stability field of methane hydrate,[10] or hydrogen, which would not be bound to the hydrate crystal lattice in the same way as the hydrocarbon molecules in the guest lattice sites. No experimental data exists at this time that bears on the possibility of increasing energy density by non-lattice site occupants, but increasing energy density of an industrial hydrate fuel storage medium is a worthwhile objective. Any means by which energy density can be increased in a synthetic hydrate without compromising safety would be beneficial.

SYNTHETIC METHANE HYDRATE FUEL STORAGE MEDIA

It must be emphasized that the precise nature of *s*ynthetic *m*ethane *h*ydrate *f*uel (SMHF) is not known because it has yet to be designed and fabricated.[6] Thus, the energy density, energy losses upon fabrication and subsequent gasification, and the equivalent energy of methane after conversion, in addition to the cost of the conversion and other engineering necessary for an SMHF system, need to be estimated before a commercial value can be placed on SMHF media. The potential energy content of naturally occurring methane hydrate is high enough to allow for some system or usage diminution and still remain an attractive new fuel storage and transport media. Thus, when an energy-efficient means for gasefying SMHF can be found, it may prove to be a more energy-efficient means of condensing methane than liquefaction achieved by energy-intensive compression and cooling.

Naturally occurring methane hydrates are not stable at sea level ambient temperatures and pressures. However, it is not intended to use pure methane hydrate as the basis for the new fuel transport and storage medium. Current experimental results show that hydrates can be fabricated both from natural gases more dense than methane[16] with variable physical property ranges that are stable well above the normal methane hydrate $P–T$ stability field.[10] Also, in the course of producing synthetic methane hydrate, metastabilities about the *liquidus* line exist,[17] a fact that may point toward controlling metastability ranges of methane hydrate rather than expanding the methane hydrate stability field.

It is known that small additions of higher density hydrocarbon gases alter the pressure–temperature stability field at which methane hydrates are stable to near normal sea level atmospheric pressures and temperatures, but the emphasis in hydrate research has been the gasification of unwanted hydrates and the recovery of methane from naturally occurring deposits. These low pressure hydrates have been regarded as undesirable in that they are mainly known from blocked gas pipelines and drill holes. Where bonding interaction between guest and host molecules might be enhanced somewhat, gas that normally does not hydrate, such as hydrogen, may participate in forming special hydrates. If host cavities were to be lined with groups having a high hydrogen bonding character, such as hydroxyl or amino groups, other factors, such as the solubility parameter of the host, could be of less importance. Increased hydrogen bonding affinity might also be induced by charging guest molecules prior to exposure to hydrate lattice, or through the use of magnetic field charging (e.g., moving the fuel in a field, pulsing a field, or moving a field with respect to the orientation of the hydrate). Increased energy densities of the special fuel may be achieved by co-storing hydrogen in such hydrates, by a process akin to hydrogenation. Research into this possibility will be fruitful.

Formation of Methane Hydrate Storage Media

Current experimental results show that methane hydrates can be fabricated with specific physical property ranges in solid and pseudo-fluid (slush) forms that are stable at near ambient pressures and temperatures. We propose that specially fabricated methane hydrates could have widespread industrial applications by allowing methane fuel to be safely compressed and transported as hydrate. The synthetic material will have applications both as a transport medium (e.g., for transporting methane from source to market where pipelines are inappropriate) and as a fuel on ships and possibly other large vehicles. We envisage a SMHF that does not closely approximate the pressure–temperature conditions of natural hydrate, such as has been proposed for replacing liquid natural gas (LNG) as a transport medium.[18,19] We also do not envision the hydrate gas storage media as being carried in a few large tanks arranged in a conventional manner. Indeed, it is very reasonable to consider storing the large-volume SMHF between the hulls of a twin-hull configuration for a large vessel. New approaches to manufacturing the industrial SMHF are therefore required. These methods should include hydrate processing technologies that involve very rapid hydrate formation and the designed buildup of hydrate layers. Additionally, the technology should also incorporate methods to rapidly dissociate hydrates so that the fuel can be produced on demand. A proposed technology is described in a later section, PRESENT STATE OF ARTIFICIAL HYDRATE MANUFACTURING AND FUTURE TRENDS.

Dissociation of Methane Hydrate Storage Media

Release of fuel gas could be induced through heating, lowering of pressure, or electronic stimulation that would produce effects similar to that of microwaving food, where the frequency of the applied microwave radiation is specific to the water molecule. The intention would be to dissociate only that hydrate required to maintain pressure in the gas fuel lines from the hydrate tanks to the turbine generating units. This approach would be primarily as a safety issue, because the methane in hydrate will not burn or explode as easily as would fuel in gaseous or liquid phase. If damage should occur to the hydrate-fueled naval vessel, therefore, the opportunity for fire is both minimized and localized. Gas delivery lines will have redundant courses with leak-stopping check valves and thus the fuel delivery system as a whole will have built-in increased survivability.

Naval apparatus requires some method for producing variable amounts of methane including gas, for high bursts of energy, for instance when all generators are producing power under combat conditions. Depressurization with thermal equilibration by pumping sea water or heating, or something more solid state like microwave stimulation are likely candidates for enhanced gas evolution.

Concerning solid hydrate versus small crystallites, we find that surfactant addition does make a significant difference. Without surfactant a fairly dense hydrate layer is built up; with surfactant, we get more fluffy hydrates that can easily flow as a slurry. Note that the nozzles that are used to initially deposit hydrates can also facilitate their dissociation to fuel gas, even for burst transitions. After hydrate deposition, these nozzles eventually become immersed in the hydrate layers. Now, if high temperature steam/water were pumped in through the same nozzles, the hydrate layers around the nozzles would dissociate very rapidly. In other words, one could get fuel bursts from rapid dissociation by means of steam injection through existing dual-use nozzles. Steam injection into hydrate reservoir data could be used as an analogy, to estimate transition rates. A similar analysis might hold for gas evolution during steam injection into the hydrate layers of a twin-hulled vessel.

PRESENT STATE OF ARTIFICIAL HYDRATE MANUFACTURING AND FUTURE TRENDS

How can hydrates be manufactured and stored in the void space between the hulls of a twin-hulled vessel, and how can the gas be released on demand? Such questions are vital to the realization of the proposed technology. Typically, hydrates are made simply by contacting gas and water at pressures above the hydrate equilibrium pressure at a specified temperature or by immersing crushed water ice in a methane atmosphere at suitable pressures and temperatures. Such manufacture of hydrates in bulk aqueous media has several drawbacks. Since hydrates are less dense than water, the hydrates soon form a layer on the water surface inhibiting gas access to liquid water. Hydrate formation then becomes highly mass-transfer limited. Although system agitation during the formation process should help minimize the formation of hydrate blocks, it is intuitively not economically viable to incorporate impellers for agitation that would then become frozen and risk damage.

The solution then, is to build up the hydrate layers from the bottom of the hull. Gudmundsson[20] has devised a method of spraying fine water droplets into a gas phase maintained at hydrate formation conditions. The high surface area of the water droplets leads to good contact with the gas and it is proposed that hydrates form rapidly and efficiently. The method is certainly viable, but may be prone to specific drawbacks. Since hydrates are formed from droplets injected into the gas phase, the hydrate crystals may adhere to hull surfaces, thus forming growth layers along the hull surface. Additionally, continuous spraying is necessary to keep the nozzles free from blockage by hydrate formation in the nozzles. The principal problem, however, is that the tank must first be fully charged with gas and kept charged as gas is condensed into the hydrate. The advantage of this method is that the hydrate will form a solid mass, which is one of the design requirements.

We propose another scheme based on our earlier work with hydrate formation in liquid hydrocarbon systems containing a microaqueous phase.[21–23] The vessel is designed to incorporate coaxial injection tubes vertically positioned at specific positions in the interhull tank spaces. The tubes are varying lengths to allow recovery of gas from hydrate layers. The hull voidage is first partially filled with a liquid hydrocarbon phase; ordinary diesel fuel may be adequate. The interhull tank space is then pressurized with methane, the hydrate forming gas, and brought to hydrate formation conditions. At this stage the system is free of hydrates, with the injection tubes being immersed in the liquid hydrocarbon and being wetted by the liquid hydrocarbon. Additionally, the hydrate forming gas saturates the liquid phase. Since gas and liquid phase are both hydrocarbons, the hydrate forming gas typically has a high solubility in the liquid hydrocarbon phase. For example, the mole fraction of methane in isooctane at 273 K and 2 MPa is 0.1, whereas that of ethylene is even greater (0.55).

Precooled water is then pumped through the inner tube of the coaxial injection tubes. Since the water is injected as droplets on entering the liquid hydrocarbon phase, hydrates form extremely rapidly. Experiments with ethylene hydrate formation conducted in a viewing cell indicate that the water droplets are completely converted to the hydrate form even before they settle to the bottom of the cell. The viscosity of the gas saturated liquid hydrocarbon retards the velocity of the water droplets that are being converted to hydrates. The visual observation is a gradual deposition of hydrates at the bottom of the view cell. We expect these observations to hold even for large scale hydrate production in the hull voidage. As pumping is continued, the level of the liquid hydrocarbon phase rises since it is displaced by the volume of the hydrates. Thus, almost all of the liquid hydrocarbon phase can be recovered. If the hull design involves compartmentalizing the tank space, the liquid hydrocarbon can be bled off into a subsequent compartment where it can be reused for the next batch of hydrate production. The normal bunkerage location in the hull floor can also be retained for the hydrocarbon liquid required for this process, and used for back-up fuel itself if necessary. We note that this method of hydrate formation does not lead to blockage of the injection tubes, since the tubes are continually wetted by the liquid hydrocarbon. The hydrates formed as the water is ejected from the nozzles do not collect on the sides of the vessel. Rather, the hydrate crystallites sink to the bottom of the vessel resulting in an evenly distributed buildup of the hydrate layers. Finally, the injection technique produces small crystals that can be fused to a solid mass by natural annealing recrystallization that will reduce crystal interface area. At the end of the hydrate formation process, the hull contains hydrate crystals. Gas saturated liquid hydrocarbon filling the voidage between the crystals can be reduced to a minimum by controlling the rate of crystal formation.

In another variation of the technique, the liquid hydrocarbon phase contains surfactants, typically those that form water-in-oil microemulsions. AOT, bis(2-ethylhexyl) sodium sulfosuccinate, is an example of such a surfactant. When water is injected into these surfactant containing hydrocarbon phases, the hydrates that are formed are very different than the hydrates formed from direct water injection into liquid hydrocarbon. The hydrate crystals that are formed in these microemulsion systems appear to have a flaky texture and they tend to form a loose aggregate. Additionally, the crystals do not adhere to the container walls. As a consequence, these hydrate crystals are easy to disperse perhaps even through natural vessel motion. In other words, this system leads to the formation of hydrate slurries that can be pumped in their entirety from one hull compartment to another. A method of solidifying them in the hull tanks is required so as to form a solid mass in the inter-hull tank.

Gas recovery can be accomplished by injecting steam again through the inner tube of the coaxial injection tubes. The system pressure then drives dissociated gas + water through the annulus between the two tubes. Alternatively heaters or microwave emanators can be located so as to dissociate hydrate at a variety of locations near the collector tubes. The gas will then naturally separate from water. It is in this context that initial injection tube placement and height becomes subject to design and optimization. However, the fact that both hydrate formation and gas recovery can be effected through the same tube arrangement is intrinsically appealing to the development of the technology.

IMPACT OF USING HYDRATE FUEL ON SHIP DESIGN

Basic Design

What could these new hydrate-methane fueled ships look like? A modern frigate carries about 8,000,000 gallons or 1,070,000 ft^3 of #2 fuel oil weighing about 52 lbs/ft^3 and containing about 936,000 btu ft^3. Because the energy density of the present Navy standard fuel is about 5.4 times that of naturally occurring methane hydrate and 4.8 times that of the target industrial hydrate fuel (TABLE 1), the bunkerage or fuel storage space requirement for hydrate would be this much larger (about five times). Hydrogenation or other energy density increases might bring the fuel storage requirements of the industrial hydrate fuel to possibly 4.5 times that of present fuel oils. Within the constraints of conventional bunkerage design, there appears to be no convenient way to carry the required hydrate fuel load in conventional storage systems. An unconventional fuel storage medium calls for unconventional engineering solutions.

If the concept of hydrate fuel can be realized and mated to standardized, dispersed, relatively small electric generators, the engineering character of a surface platform would be dramatically altered. Currently, surface warships have large internal engine rooms from which power is transferred to propellers through large reduction gears and long propeller shafts. Power derived from a number of distributed, and hence, redundant generating capacity can be specifically optimized for methane fuel to allow for efficiency and environmental concerns.[8] A number of experimental fuel cells could be adapted for naval use.[24] Even if methane must first be decomposed to hydrogen to be used in fuel cells, the greater efficiency of the fuel cell may more than compensate for the energy used to carry out the decomposition.

There are other benefits. Turbine electric generation and electric drive strongly reduces acoustic radiation, which is increasingly desirable in the new stealthy, warfighting environment. Ships are no longer armored, thus distributed power and weapons systems enhance survivability. Power generated from a number of physically separated generating units would have a number of favorable attributes, especially in the case of physical damage to the ship, a likely occurence in a warship. In effect, these hydrate-fueled ships would be turned inside out from present designs with the fuel and dispersed electric generators located outside of the inner hull housing personnel, weapons, and all other mechanical and electrical equipment.

Conventional ship designs are long, narrow monohulls that allow for the highest speed and maximum energy efficiency. This, in many ways, is a hold-over from when naval gunfire was the primary anti-ship weapon, and speed and maneuverability were a part of gunnery tactics. Today, and increasingly into the future, however, the defensive weapons of ships assume greater importance because anti-ship weapons are missiles that may be fired from hundreds of miles away and home to the target with a variety of sensors. Therefore, increasing the capability of a ship to hide, defend itself, or survive a successful attack, offers a good strategy for survival of surface naval forces.

The most obvious solution to increasing the offensive capability is to look to ship designs that carry more weapons. This means increasing the missile launch capability and probably also introducing airborne reconnaissance or attack capability. At the same time, the defensive capability must be increased, including, if possible, restoring to modern ships some of the defensive capabilities traditionally offered by armor. A non-radiating shorter hull, low in the water, offers the smallest target for any sensor. Taking into account offensive and defensive capabilities and target characteristics, a multihull design suggests itself. If the inner and outer-hull concept for hydrate tank disposition for this new fuel storage medium is achievable, then the bulkier hulls would not be detrimental to the new design and the required larger fuel storage volume can easily be achieved. In fact, in the situation where a navy must send its ships world-wide, a longer duration (higher fuel capacity) design using the hydrate fuel concept could be twin- or even tri-hull. Not only could the ships thus carry more firepower, and with larger deck space even carry aircraft and AAVs, but the exterior of each hull would be a hydrate fuel store. Holding the fuel in such a manner could have a very positive defensive effect.

Potential Reactive Armor Effect

In addition to a special hydrate's potential ability to store methane in a safe manner, the fuel store itself may provide an element of combat damage resistance not found on any active ship today. The methane hydrate fuel is envisaged as being stored in hull fuel tanks. In this position, especially with the smaller superstructures planned for future warships, the fuel is likely to receive hits from sea-skimming missiles, if the ship is hit at all. The hydrate-filled hull tanks themselves may offer benefits not normally associated with fuel storage, which is always a hazard to warfare explosives. The hydrate fuel is in the form of a sand-like moist slurry that will initially absorb shock rather than as a liquid that transfers shock hydraulically, and it will then react as an explosive, whose blast could be engineered to counteract an incoming missile explosive.

It is possible that the hydrate fuel storage tanks could be designed to display the same response to impact from missiles and projectiles as reactive armor. We suggest that with innovative design, the fuel tanks can be blow-out vented to allow a missile impact on a fuel storage cell to burn off in a particular manner. In addition, uncontrolled evolution of methane from an exploded fuel tank will evolve superheated steam, which should have a significant quenching action, allowing only partial ignition of any flammable.

We regard the hull hydrate storage as a potential naval analog for reactive armor, which was specially designed to protect armored vehicles from modern antitank weapons. Reactive armor is composed of a series of individual plates, fitted closely together and lain over a vehicle hull. Each of the plates is highly explosive with a very high burning rate and is ignited by the very high impact/heat of a cruise-type warhead or an antitank missile. Many antitank missiles operate through shaped charges that form narrow thermal lances, although missile arrival alone can be fatal to a ship. Deflecting or disrupting the blast of a shaped charge, or the inward progress of the missile (the forward part of which is largely converted to thermal energy at impact), is the function of reactive armor. The inward momentum of the explosion then splashes off the vehicle's normal armor. It may be possible to shape the hydrate storage tanks so that a sideways component can be added to the initial blast, effectively deflecting incoming energy so that it is diffuse when it reaches the inner hull. This concept is unproved for ships but is theoretically possible and may be a potentially productive area for new research that could yield very large benefits and restore a level of survivability to ships that has been lost in modern, unarmored ships.

Decreased Pollution

Methane hydrate fuel is essentially non-polluting in the marine environment. If contained, for instance, within a sinking ship, methane fuel will not spread over the surface, but will remain stable within the vessel because increased hydrostatic pressure increases stability of the stored solid hydrate fuel. Unintentional methane release would be small because it is envisaged that only a small amount of methane will be available in gaseous form at any one time.

Both the methane and the water evolved from hydrate dissociation are available for use. Water evolved during methane release in excess of that required for injection can be dumped or returned to the array of fuel tanks for use in the next hydration/fueling. This recycling approach would minimize water usage and scavenge dissolved methane. Maintaining ballast in some applications can be advantageous for stability. In ships retaining evolved water there would not be the same need for ballasting with sea water, and the positive environmental impact of not pumping fuel tanks should remove a large volume of oil pollution from the sea.

Enhanced Safety

The hydrate fuel medium would be a nonexplosive, stable compound crystalline solid from which no gas would leak. If ignited, for instance in the case of catastrophic damage, the hydrate fuel medium will release methane at a relatively slow rate because the process of dissociation itself consumes energy (57 joules/mole methane[10]). In addition, methane gas will not spread over sea surface as does liquid fuel. Spilled fuel oils can ignite and spread along the sea surface.

The only gas on the ship will be in gas generation units that will be part of the hydrate fuel tanks and in small-bore gas lines to the generators. These gas lines can be embedded in the outer shell of the ship so that there is no possibility of gas leakage into the body of the ship. The generators can be placed along the upper outer hull between a strong, main inner hull and outer shell hull in what amounts to a blow-out configuration.

Availability of Methane and Its Use in Fueling

Methane is a renewable resource in that it is a common by-product of microbiological decay of organic matter in addition to its wide availability as a conventional hydrocarbon. The ability to produce methane in remote locations may reduce the need to transport navy liquid fuels to those localities.

Methane hydrate fuel can be compounded any place that methane is produced either by standard hydrocarbon recovery techniques or through local industrial plants using microbiological action (sewage-slurry treatment plants, for instance). Methane, water, and suitable doping gas or fluid can be used to compound hydrate fuel wherever the appropriate platform (vessel) goes; hydrate is the safest way to store methane. It is envisaged that each large platform will contain apparatus to make hydrate fuel from raw components as well as to store and selectively gasify the fuel.

RESEARCH REQUIREMENTS

Virtually all research to date has involved the elimination of unwanted formation of hydrates. In direct contrast, the research called for here is twofold. First, a synthetic hydrate stable at near ambient conditions and that can be manufactured at an industrial scale must be perfected. Second, a robust system for forming hydrates, transporting the hydrate as a viscous fluid or a slurry to irregularly shaped tanks where the material is rendered solid and is available for dissociation (gasification) on demand, must be engineered.

ACKNOWLEDGMENTS

M.D.M. thanks the Federal Energy Technology Center (FETC), Morgantown, WV (U.S. Department of Energy), for support via funding element DE-AT26-97FT34344, and the Office of Naval Research, Program Element No. 0601153N. R.E.P. acknowledges the support of the Chemistry Division, Code 6100, NRL in this work. Funding from DARPA Grant MDA972-97-1-0003 is gratefully acknowledged by V.T.J.

REFERENCES

1. MURPHY, D. 1997. Sea-based firepower. U.S. Navy plans a measured revolution at sea for its surface warships. Armed Forces J. Internat. November: 34–37.
2. WOOD, G. 1995. Electric propulsion in warships—then and now. Naval Forces (Naval Technology) **3/95:** 20–24.
3. CAMPBELL, C.J. & J.H. LAHERRÈRE. 1998. The end of cheap oil. Scientific American Special Report, 77–83.
4. KERR, R.A. 1998. The next oil crisis looms large—and perhaps close. Science **281:** 1128–1131.
5. EMERSON, S.A. 1997. Resource plenty: why fears of an oil crisis are misinformed. Harvard Internat. Rev. **19**(3): 12–15.
6. MAX, M.D. & R.E. PELLENBARG. 1998. Hydrate-based fuel storage and transport media: potential impact. Proceedings Fuel Chemistry Division of the American Chemical Society, San Francisco, CA, April 13–15, 1997. **24:** 463–466.
7. JOLLIFF, J.V. & K.B. SCHUMACHER. 1989. The declining years. In R.W. King, P. Palmer & B.I. Meader, Eds.: 119–158. Naval Engineering and American Seapower. The Nautical & Aviation Publishing Company of America, Inc., Baltimore, MD.
8. PRESTON, C.E.M. 1992. The ICR Trent, a third generation gas turbine for warships. Naval Forces, **18:** 32–35.
9. CARHART, H.W., W.A. AFFENS, B.D. BOSS, R.N. HAZLETT & S. SCHULDINER. 1974. Hydrogen as a navy fuel. Naval Research Laboratory Report 7754.
10. SLOAN, D. 1998. Clathrate Hydrates of Natural Gases, 2nd Edit. Marcel Dekker, New York.
11. FARADAY, M. 1823. On fluid chlorine. Philos. Trans. Roy. Soc. (Lond.) **22A:** 160–189.
12. KVENVOLDEN, K.A. 1993. A primer on gas hydrate. In The Future of Energy Gases. D.G. Howell et al., Eds.: 279–291. USGS Professional Paper 1570.
13. DOE. 1987. Gas hydrates technology status report. Department of Energy DOE/METC 870246 (DE8700127).
14. SCOTT, M.I., P.L. RANDOLPH & J.B. PANGBORN. 1980. Assessment of methane hydrates. Final report for period December 1978 through June 1980. Department of Energy Report GRI-79/0070, Contract #5011-310-0097.
15. HOESS, J.A. & R.C. STAHMAN. 1969. Unconventional thermal, mechanical, and nuclear low-pollution-potential power sources for urban vehicles. Soc. Appl. Engin. Trans. Paper 690231.
16. DE BOER, R.B., J.J.H.C. HOUBOLT & J. LAGRAND. 1985. Formation of gas hydrates in a permeable medium. Geologie en Mijnbouw **64:** 245–249.
17. STERN, L.A., S.H. KIRBY & W.B. DURHAM. 1996. Peculiarities of methane hydrate formation and solid-state deformation, including possible superheating of water ice. Science **273:** 1843–1848.
18. GUDMUNDSSON, J.S., F. HVEDING & A. BORREHAUG. 1995. Transport of natural gas as a frozen hydrate. Proceedings 5th International Offshore and Polar Engineering Conference, The Hague, June 11–16. **I:** 282–288.
19. GUDMUNDSSON, J.S. & A. BORREHAUG. 1996. Frozen hydrate for transport of natural gas. Proceedings Second International Conference Natural Gas Hydrates, June 2–6, Toulouse. 415–422.
20. GUDMUNDSSON, J.S. 1994. Method for production of gas hydrates for transportation and storage, U.S. Patent 5536893.
21. NGUYEN, H., J.B. PHILLIPS & V.T. JOHN. 1989. Clathrate hydrate formation from reversed micelles. J. Phys. Chem. **93:** 8123.
22. NGUYEN, H.T., N. KOMMAREDDI & V.T. JOHN. 1993. A thermodynamic model to predict clathrate hydrate formation in water-in-oil microemulsion systems. J. Colloid Interface Sci. **155:** 482–487.
23. HOLDER, G.D. & V.T. JOHN. 1983. Thermodynamics of multicomponent hydrate forming mixtures. Fluid Phase Equilibria **14:** 353–361.

24. HIRSCHENHOFER, J.H., D.B. STAUFFER & R.R. ENGLEMAN. 1994. Fuel Cells. A Handbook (Revision 3). Gilbert/Commonwealth, Inc. Reading, PA 19603, DOE contract DE-AC01-88FE61684, DOE/METC-94/1006.

Methane Hydrate
Melting and Memory

P. MARK RODGER[a]

Department of Chemistry, University of Warwick, Coventry, CV4 7AL, United Kingdom

ABSTRACT: We present the results of a long timescale molecular dynamics simulation of a methane hydrate/methane gas interface formed along the [001] hydrate surface. The simulations were performed at 15–20°C above the stable hydrate temperature so that we were able to observe melting under conditions that were sufficiently gentle to allow any residual order associated with the memory effect for hydrate nucleation to be identified. The simulations have been analyzed using a set of novel order parameters designed specifically to quantify the microscopic molecular structure associated with the different phases of water. The simulations do show an enhanced level of ice- and clathrate-structure in the liquid water that forms when the hydrate decomposes, but there is no evidence of significant clusters of the ordered water.

INTRODUCTION

During recent years there has been considerable interest surrounding the question of nucleation mechanisms for natural gas hydrate formation. In part, this interest stems from the desire to exert better control over nucleation: if one understands the nucleation process then one is in a better position to design additives that will either inhibit or promote the process in a controlled manner. However, the question is also of scientific interest, since the reaction to form the hydrates, which is essentially one between two immiscible reactants, is intriguing.

There have been a number of mechanisms conjectured, usually involving clathrate-like solvation cages around aqueous methane, and the formation of labile clusters from these dissolved supermolecules. However, the experimental evidence against which to test these conjectures is relatively thin on the ground. In this context, one of the most significant phenomenon associated with clathrate hydrates is the memory effect, in which nucleation occurs much more slowly the first time through a freeze-thaw cycle in a closed system than it does in second and subsequent cycles. This is believed to be due to residual order in the melt, with metastable clathrate clusters surviving for up to 24 hours and thus making subsequent nucleation events more likely to occur once the thermodynamic state is again shifted into the hydrate-stable region. However, there is currently very little microscopic evidence to support the idea of residual order, and the nature of such residual order, if it exists, remains conjectural. We note also that recent simulations suggest that hydrophobic hydration, such as that encountered around methane in solution, does not lead to well defined structured water; so the mere

---

[a]Telecommunication. Voice: (44) (0)1203 523239; fax: (44) (0)1203 524112.
p.m.rodger@warwick.ac.uk

presence of dissolved methane will not automatically cause the formation of the first clathrate cages.[1,2]

In this paper we present the results of molecular dynamics simulations designed to look for and characterize any residual order in a methane hydrate melt. In view of the experimental evidence that the memory effect is limited to temperatures close to the hydrate/water/hydrocarbon three-phase stability line, the simulations must allow for relatively gentle melting processes. We have therefore chosen to simulate melting at a methane hydrate/methane gas interface at a superheating of no more than about 20°C. This has necessitated long simulations in order to see significant melting.

Simulation Details

Molecular dynamics simulations were performed on a system of up to 4,140 water molecules and 250 methane molecules. The simulation cell consisted of 3×3×10 unit cells of fully occupied type I methane hydrate under an atmosphere of methane with interface oriented along the (001) direction. The system was periodic in the xy plane, but aperiodic in the z direction. The stability of the system in the z direction was maintained by immobilizing the bottom layer of hydrate, and introducing a reflective boundary above the gas. The reflective boundary was located 150 Å above the hydrate surface, which was found to be sufficient to produce a homogeneous bulk gas phase above the hydrate. The frozen hydrate layer initially consisted of one layer of unit cells. From our early simulations, it became clear that the effect of the frozen boundary on neighboring stable hydrate regions was short ranged, and it was found that a layer of at least 20 Å of mobile stable hydrate was sufficient to exclude the possibility of the frozen boundary influencing the behavior of the interface. In subsequent simulations, the frozen boundary was maintained 25–40 Å below the hydrate side of the hydrate/gas interface.

Simulations were performed in the $[N,V,T]$ ensemble using the Hoover thermostat and the DL_POLY molecular dynamics code.[3] Initial hydrate coordinates were taken from crystal structure data; gas phase coordinates were taken from previous simulations. The system was equilibrated by first immobilizing all the hydrate and allowing the gas to evolve for 10 ps, then simulating the dynamics of the full system but with water masses set to 10 times their proper value (5 ps), and then simulating the full system for 5 ps. This was found to be sufficient to relieve any short timescale relaxation in the system energy, pressure, and density profiles. A further 600 ps trajectory was then accumulated to study the melting process.

As with earlier simulations from this laboratory, water was modelled with the SPC potential and methane represented by a single Lennard-Jones site. Details of the potential are given elsewhere.[4] All intermolecular interactions were truncated at a distance of 12 Å (i.e., one unit cell length), with the truncation being applied to the intermolecular separation. This neutral-group truncation for water has been found to give equivalent results to simulations employing the EWALD electrostatic summation in other systems.

Simulations were performed at 290 K and 1.5 MPa. From earlier simulations[5] and calculations this is about 15–20°C above the melting temperature. The pressure was established by comparing the density of the gas phase with the equation of state for a Lennard-Jones fluid.[7]

ANALYSIS

In order to characterize the microscopic structure across the interface, we have developed several novel methods. In this paper we present results from an analysis based on three order parameters that have been specifically developed to distinguish between water-like, clathrate-like, and ice-like structures. A full definition of the order parameters is given elsewhere[2,5] and only a brief outline presented here. One of the parameters, F_3 is a three-body order parameter that measures the average angle between three water oxygens from within a hydrogen bonded triplet; more specifically it measure how this angle deviates from the tetrahedral angle, so that low values characterize well defined hydrogen bonding networks as found in ice and hydrates. The other two parameters characterize the O...O...O...O torsion angles found in a hydrogen bonding network. $F_{4\varphi}$ is derived directly from the torsion angle of this chain, and F_{4t} is the average triple product for the O→O→O→O system.

Average order parameters for methane hydrate (270 K), xenon hydrate (270 K), ice (270 K), and liquid water (275 K) are presented in TABLE 1. It can be seen that the combination of the three order parameters does provide a clear distinction between the different environments. Indeed, the uniqueness of the set of order parameters has made it possible to characterize the local structure around each individual water molecule. To do this, order parameters were calculated for each molecule by averaging over all triplets (F_3) and all quartets (F_4) to which the molecule belonged. These values for each molecule were then compared with the average bulk-phase values. Where the instantaneous (F_3, $F_{4\varphi}$, F_{4t}) were sufficiently close to a set of bulk phase values, the water molecule was assigned to that phase. (To define "sufficiently close" the covariance matrix for the three parameters was calculated in each bulk phase. An eigenvalue decomposition was then used to define the distance of the instantaneous values from the average. The order parameters were deemed to be "sufficiently close" when all three distances were simultaneously within three standard deviations of their respective bulk average.) For example, instantaneous values of (0.011, 0.70, 0.45) would identify a molecule as being in a hydrate environment, whereas values of (0.010, −0.30, 0.25) would identify it as ice-like. The method was tested on bulk-phase simulations of type I and II hydrates, ice I_h and liquid water, and correctly identified in excess of 95% of the molecules in each case.

TABLE 1. Average order parameters determined from various bulk phase simulations (values in parentheses indicate standard deviations)

Bulk Phase	F_3	$F_{4\varphi}$	F_{4t}
methane hydrate (270 K)	0.0148 (0.0122)	0.690 (0.203)	0.474 (0.098)
krypton hydrate (270 K)	0.0136 (0.0105)	0.729 (0.186)	0.472 (0.101)
ice I_h (270 K)	0.0106 (0.0077)	−0.386 (0.141)	0.287 (0.120)
liquid water (275 K)	0.101 (0.036)	−0.011 (0.313)	0.257 (0.106)

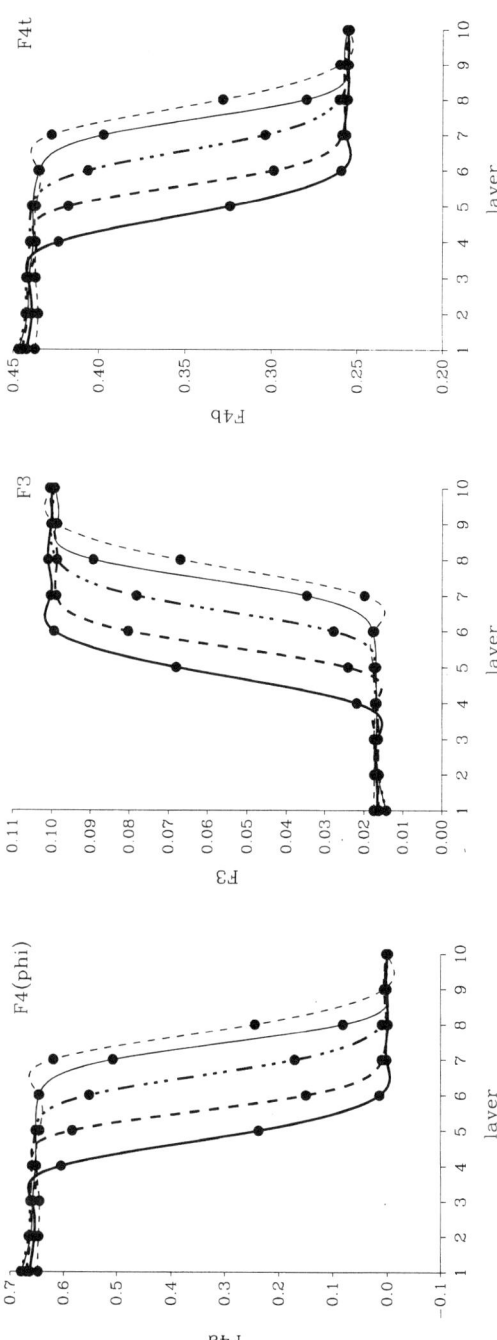

FIGURE 1. Average order parameters calculated from successive layers across the interface. Each layer is 6.0 Å thick, corresponding to half a crystallographic unit cell. The original interface lies at the junction of layers 10 and 11, with layers 1–10 being in the hydrate region. Key: ——— at 600 ps; — — — at 500 ps; —·—·— at 400 ps; ——— at 300 ps; and — — — at 200 ps.

RESULTS

Average order parameters were calculated across the interface and are presented in FIGURE 1. Note that the type I hydrate structure consists of a stack of equivalent layers, each 6.0 Å (half of a unit cell) thick. Order parameters were therefore averaged over all water molecules within 6.0 Å thick layers in order to remove the crystallographic variation in these profiles. The layers are numbered such that the

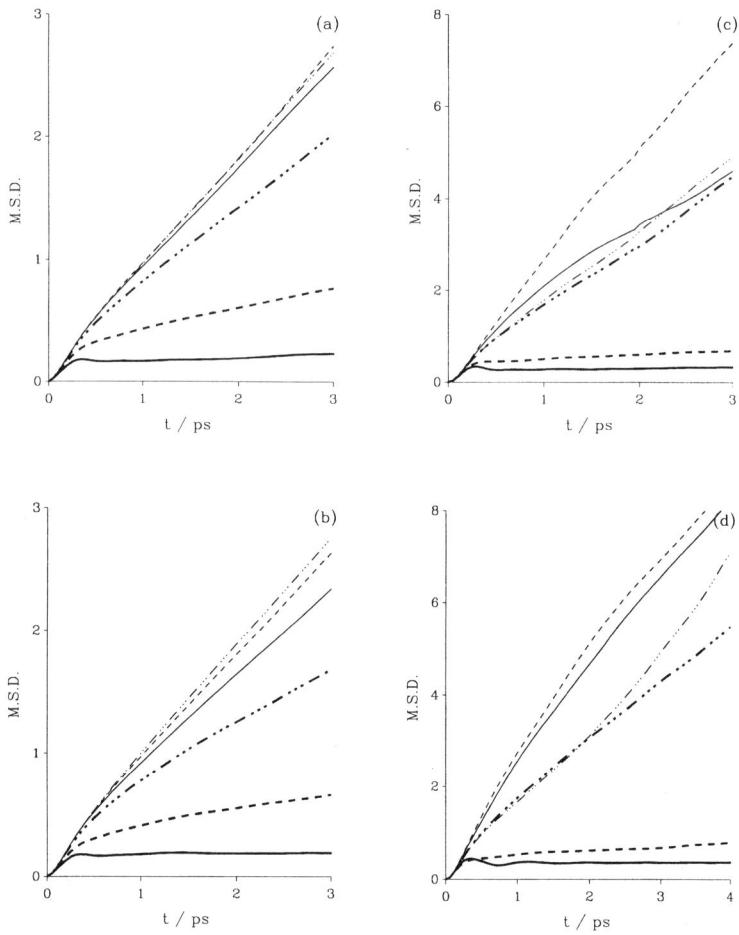

FIGURE 2. Mean square displacements of (**a**) water oxygens in the x direction, (**b**) water oxygens in the z direction, (**c**) methane in the x direction, and (**d**) methane in the z direction at 500 ps. The different curves correspond to molecules that commenced in different layers: ———— in layer 4; – – – – in layer 5; —-—-— in layer 6; ———— in layer 7; – — — – in layer 8; and — – – — in layer 9. At 600 ps the hydrate interface is located in layer 5.

original interface occurs at the boundary between layers 10 and 11, with layers 1–10 starting as hydrate. As can be seen, the order parameters present a consistent picture, with the interface being 10–15 Å thick in each case, and each order parameter locating the interface at the same place. FIGURE 1 also gives a clear representation of the progression of melting over time and shows that the profile of the interface does not vary substantially over the duration of the simulations. The location of the interface is seen to shift about 30 Å during the course of the simulation.

FIGURE 1 also reveals that a layer of water rapidly forms above the hydrate on melting. The order parameters in this region are completely consistent with liquid

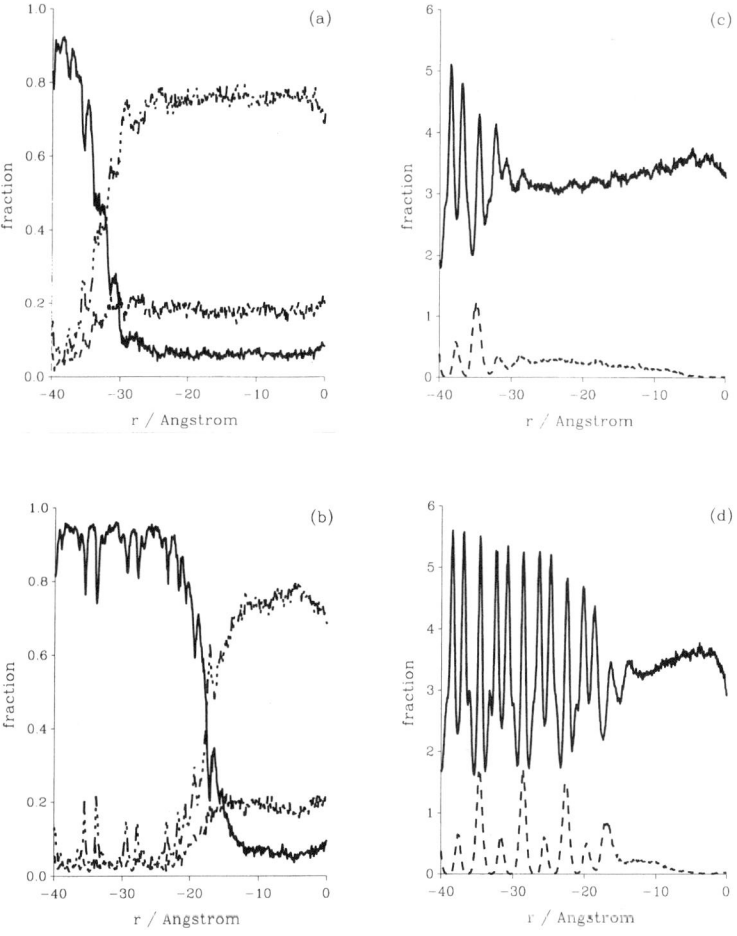

FIGURE 3. Density profile of water and methane molecules across the interface at 600 ps (**a** and **c**) and at 300 ps (**b** and **d**). **a** and **b** give a decomposition of the water into clathrate, ————— ; ice, – – – – –; and liquid water, — – – —; local structures. b and d give the total density profile for water, ————— ; and for methane, – – – – –.

water, despite the fact that there remains an abnormally high concentration of methane in this region.

The profiles presented in FIGURE 1 also make it possible to identify suitable regions from which to calculate other useful properties. In particular, the mean square displacement (MSD) of water molecules that originated in the various layers is presented in FIGURE 2. MSDs are given only at 600 ps, since curves calculated at other times agreed very well with these curves when FIGURE 1 was used to compare like with like (e.g., layer 8 at 200 ps should be compared with layer 5 at 600 ps). At 600 ps, the interface is located across layer 5. It is interesting to note that the dynamic transition from hydrate to water occurs over a longer range than does the geometric transition. It is also interesting to note that the dynamic behavior of water converges to liquid-like behavior within two layers of the interface, whereas for methane it takes up to three layers to converge; even then, the diffusion of methane in the layer below the surface is substantially slower than the diffusion of methane in other regions of the melt. We conclude that the presence of the interface is "felt" by methane molecules over a significantly longer distance than is the case for water molecules. In particular, the dynamics of methane escaping from the decomposing hydrate are significantly retarded, even under temperatures for which decomposition gives liquid water.

FIGURE 3 depicts the density profile of water across the interface, and decomposes this into fractions of waters in a hydrate-like, ice-like, or liquid-water-like local environment. The general shape of these curves is consistent with the above discussion, the change from predominantly hydrate to predominantly water structures coinciding with the interface as identified in FIGURES 1 and 2. It should be noted that for bulk water, only 2% of water molecules were classified as hydrate-like, and another 2% as ice-like. The values in FIGURE 3 are significantly above this level for

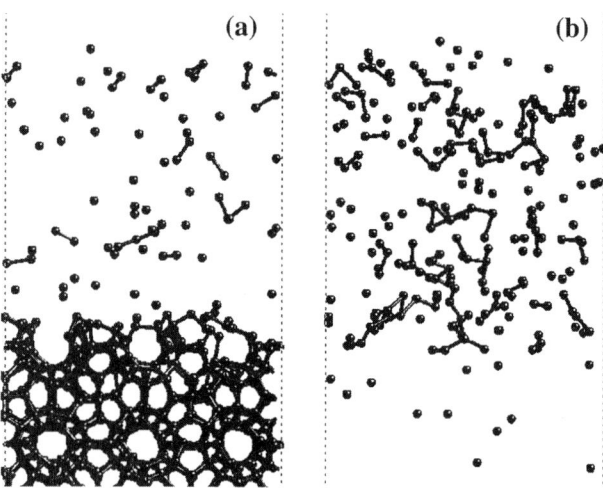

FIGURE 4. Snapshots of the residual clathrate (**a**) and ice (**b**) structures in the hydrate melt after 600 ps.

both the hydrate and the ice structures. Thus, we conclude that there is evidence of enhanced structure remaining in the hydrate melt. However, it is particularly interesting to note that the ice-like structure is the dominant form, with hydrate-like structures being only a secondary component. Furthermore, we note that there is very little clustering of the structured water molecules. Snapshots of these configurations taken from around 600 ps are presented in FIGURE 4 with bonds used to denote clusters of water with similar local structure—it is rare to observe more than three or four molecules in such clusters. Thus, although our simulations do show evidence of residual structure, there is no evidence of hydrate-like clusters persisting in the melt. We must conclude from this that the memory effect relates more to the persistence of a high concentration, and retarded diffusion, of methane in the melt than it does to the persistence of metastable hydrate precursors.

CONCLUSIONS

We have presented an analysis of molecular dynamics simulations of the methane hydrate methane gas interface under thermodynamic conditions that lead to a gentle melting of the hydrate. These simulations were analyzed using recently developed order parameters specifically designed to distinguish between liquid-, ice-, and clathrate-water structures. These order parameters have been shown to provide a clear and consistent characterization of the hydrate interface.

Throughout the melting process, the interface was found to show a constant behavior, with geometric parameters varying from bulk hydrate to bulk liquid water over a distance of 10–15 Å. The interface moved at a near constant rate of 5 Å/100 ps throughout the simulations. The dynamic behavior was found to vary over a significantly longer distance, with some retardation of methane diffusion occurring over the interval 20–25 Å.

The order parameters have also been used to characterize and quantify the extent of residual order in the hydrate melt. For these simulations, which were performed at 15–20°C above the melting temperature, there was a significant increase in both ice-like and water-like local structures in the melt compared with bulk water. However, there was no evidence of clusters of structured water molecules and so we must conclude that the memory effect in hydrate nucleation is not related to long-lived metastable hydrate precursors.

ACKNOWLEDGMENT

Computer time on the EPP CRAY-T3D was provided through the EPSRC materials MPP Consortium.

REFERENCES

1. MENG, E.C. & P.A. KOLLMAMAN. 1996. Molecular dynamics studies of the properties of water around simple organic solutes. J. Chem. Phys. **100:** 11460–11470.
2. FIDLER, J. & P.M. RODGER. 1999. Solvation structure around aqueous alcohols. J. Phys. Chem. In press.

3. SMITH, W. & T.R. FORRESTER. 1995. DL_POLY—a package of molecular simulations. Copyright C.C.L.R.C. Daresbury Laboratory, Nr. Warrington.
4. RODGER, P.M. 1990. Stability of gas hydrates. J. Phys. Chem. **94:** 6080–6089.
5. RODGER, P.M., T.R. FORRESTER & W. SMITH. 1996. Simulations of the methane hydrate/methane gas interface near hydrate forming conditions. Fluid Phase Equilibria **116:** 326–332.
6. WESTACOTT, R.E. & P.M. RODGER. 1996. Full-coordinate free-energy minimisation for complex molecular crystals: Type I hydrates. Chem. Phys. Lett. **262:** 47–51.
7. NICHOLAS, J.J., K.E. GUBBINS, W.B. STREETT & D.J. TILDESLEY. 1979. Equation of state for the Lennard-Jones fluid. Molec. Phys. **37:** 1419–1454.

Eutectic Freeze Crystallization Using CO_2 Clathrates

R.J.C. VAESSEN,[a] F. VAN DER HAM, AND G.J. WITKAMP

Laboratory for Process Equipment, Delft University of Technology, Leeghwaterstraat 44, 2628 CA Delft, the Netherlands

ABSTRACT: Highly soluble salts can be separated from aqueous solutions by eutectic freeze crystallization (EFC). This technique delivers significant energy savings in comparison with evaporative crystallization. The eutectic temperature can be raised by crystallizing CO_2 clathrates instead of normal ice. Application of this process is especially efficient for salt solutions with eutectic points at low temperatures, such as $CaCl_2$ ($-55°C$). In this work the shift of the eutectic point is predicted by use of a model for NaCl and $CaCl_2$ solutions. Experiments are executed under eutectic conditions, with these substances proving the feasibility of the eutectic clathrate freezing process. Energy calculations show that by using eutectic clathrate crystallization instead of regular EFC the electric energy consumption can be reduced by 32%.

INTRODUCTION

Solutions of highly soluble salts in water are widely present in numerous process and waste streams found in the chemical industry. Conventional separation techniques for processing salt solutions have disadvantages, such as a high energy requirement (evaporative crystallization), limited yield (cooling crystallization), and a limited operating range (reverse osmosis).

On the other hand, a maximum conversion into pure salt and pure ice with low energy consumption can be obtained with eutectic freeze crystallization (EFC). It can be applied to a wide range of salt solutions. The field of viable applications can even be widened to systems requiring low operating temperatures if process conditions are shifted to higher temperature levels by crystallizing clathrates instead of ice.

The aim of this work is to demonstrate the feasibility of eutectic freeze crystallization using CO_2 clathrates. The principle of EFC is discussed together with recent experimental results. A model is presented for calculation of the shift in eutectic conditions when clathrates are crystallized. A process flow sheet is proposed and energy requirements of regular EFC are compared to eutectic clathrate freezing. In an experimental setup, the theoretical principle of eutectic clathrate freezing is demonstrated. Reference experiments were performed on NaCl solutions for which data can be found in literature. Eutectic clathrate experiments were also conducted on $CaCl_2$ solutions.

[a]Telecommunication. Voice: +31-15-2786605; fax: +31-15-2786975.
R.J.C.Vaessen@wbmt.tudelft.nl

EUTECTIC FREEZE CRYSTALLIZATION THEORY

Separation Principle

Eutectic freeze crystallization processes are operated under the eutectic conditions of solutions. The main characteristic of this crystallization process is the simultaneous crystallization of two pure phases, salt and ice. The mother liquor surrounding the crystals contains a minimum of solute. This principle is shown in the wt%–T phase diagram of a salt solution, FIGURE 1. In the diagram the solubility line of the salt, the ice line, and the eutectic point are defined. If a solution of composition A (FIG. 1) is cooled (indicated by the *arrow*), ice is formed when the temperature reaches the ice line. With continued cooling, more ice crystallizes and the solution becomes more concentrated, eventually ending in the eutectic point. A solid/solid separation between the ice and the salt can be established based on the large difference in density. The ice moves upward forming an ice slurry layer at the top of the crystallizer, whereas the salt settles at the bottom.

Cooled Disk Column Crystallizer

An apparatus has been developed, called the cooled disk column crystallizer (CDCC), in which both crystallization and gravitational separation can be realized. The device consists of a column partitioned into several compartments, separated from each other by cooling elements. Cooling fluid is pumped through these elements, which are equipped with heat conducting cooling plates, in order to withdraw the crystallization heat from the solution. To prevent scaling of ice on the cooling plates, rotating wipers are installed. Holes in the cooling elements allow the transport of the ice upwards and the settling of the salt. In FIGURE 2, a photograph displays the CDCC prototype.

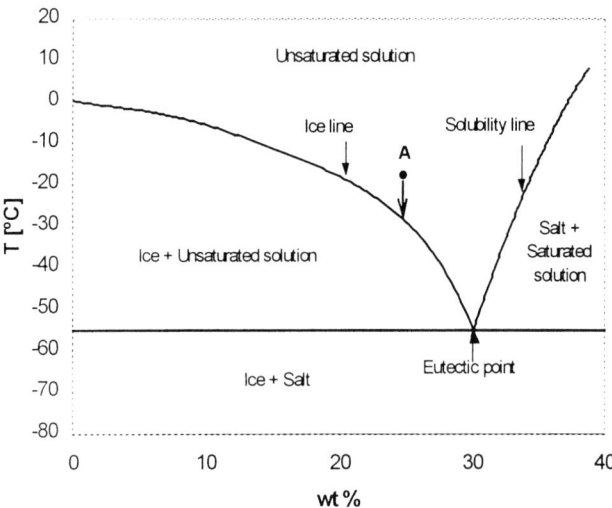

FIGURE 1. Phase diagram of a salt solution.

FIGURE 2. CDCC prototype.

Salt can be obtained in its pure form by filtering the salt slurry stream. The filtrate can subsequently be recycled to the crystallizer. The ice needs to be washed in a wash column before pure water can be extracted from the process. The enthalpy stored in the ice can be reintegrated in the process and used to precool the crystallizer feed. The energy can also be used in a two-stage cooling system for condensing a fraction of the evaporated cooling cycle medium. The remaining vapor needs to be processed in a condenser operated at ambient temperature.[1] Energy calculations performed for a variety of salt solution systems showed that energy reductions up to 70% can be achieved when EFC is compared to evaporative crystallization.

Experiments conducted on the CDCC prototype with $CuSO_4$ solutions delivered proof of the separation principle. The $CuSO_4$ content found in the ice after washing is less than 10 ppm.

EUTECTIC CLATHRATE CRYSTALLIZATION: THEORY AND PREDICTIONS

Implications of Clathrate Crystallization

The position of the eutectic point of a solution is fixed, depending on the solute. However, if the process is pressurized in the presence of an appropriate guest molecule, conditions can be created under which clathrates will form. The eutectic point is then determined by the point of intersection of the clathrate equilibrium line and the salt solubility line. This shifted eutectic point is typically situated at a higher temperature than under atmospheric conditions, thus minimizing the energy requirement for cooling equipment. Although equipment investments are higher due to the pressurized process operation, application of eutectic clathrate crystallization could be feasible for solutions with low eutectic points ($T < -15°C$ at 1 bar). Carbon dioxide is used as the clathrate former for its low corrosivity, low toxicity, and flammability levels. The gravitational separation of salt and clathrate can only be achieved if the density of the clathrate lies below the density of the solution at eutectic composition. Since CO_2 forms structure I hydrates, the salt solution density should be larger than 1.13 g·cm^{-3}, the maximum density of CO_2 clathrate.[2] Systems complying with this criterion include, for example, $CaCl_2$, NH_4NO_3, and $NaCl$.

Eutectic Shift Calculation Model

By calculating the clathrate equilibrium conditions for salt solutions at different compositions, the hydrate line can be constructed in the phase diagram and the shifted eutectic conditions can be determined. Suppression of the clathrate freezing point due to the presence of electrolytes is calculated using an equation provided by Pieroen, which was later revised and applied by Nasrifar et al.[3] In the model, the shift of the eutectic temperature is expressed as a function of the water activity, the enthalpy of clathrate formation, and the equilibrium temperature of clathrates in pure water:

$$\ln a_{water} = -\frac{\Delta H_{clathrate}}{nR}\left(\frac{1}{T_{clathrate, salt}} - \frac{1}{T_{clathrate, water}}\right) \quad (1)$$

The activity of water is calculated using the Pitzer model.[4]

$$\frac{\Delta H_{clathrate}}{nR} = \frac{\alpha_0 I^{\alpha_1}}{1 + \beta_0 P + \beta_1 \ln P} \quad (2)$$

$\Delta H_{cl}/nR$ can be calculated from (2) using parameters fitted on experimental data. The values of the parameters used in Equation (2) are listed in TABLE 1.

TABLE 1. Values of constants in Equation (2)

α_0	1000.0
α_1	0.01237
β_0	-1.205×10^{-2}
β_1	4.073×10^{-2}

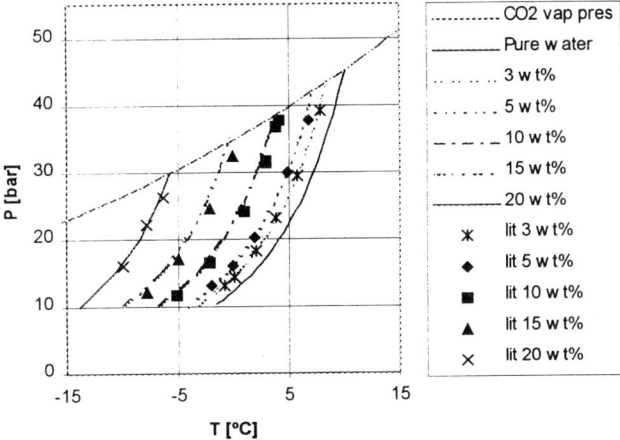

FIGURE 3. Clathrate P-T phase diagram, NaCl.

Experimental data for carbon dioxide clathrate equilibrium conditions for NaCl and $CaCl_2$ can found in Reference 5. Results of model calculations using (**1**) and (**2**) under similar conditions agree well with these experiments; see FIGURES 3 and 4. The discrete data points represent literature data, whereas the lines depict model calculations.

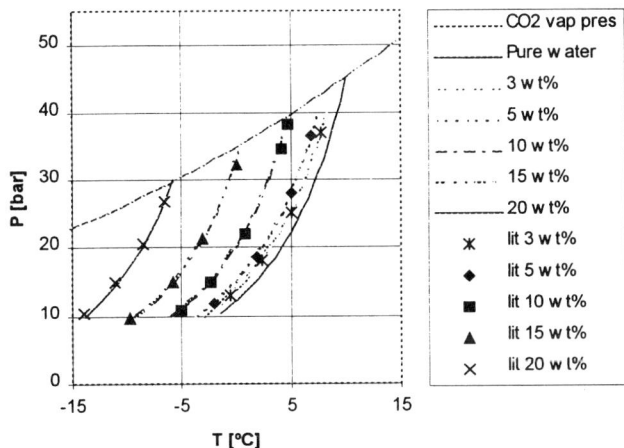

FIGURE 4. Clathrate P-T phase diagram, $CaCl_2$.

Prediction of Shifted Eutectic Conditions

The shift in temperature under eutectic conditions depends on the applied process pressure. The larger the pressure, the larger the shift in temperature. Since the objective is to operate the process with gaseous CO_2, the pressure is limited to the vapor pressure of liquid CO_2. Model calculations over a wide range of salt compositions, varying from zero to a composition that is approximately 5 wt% above the normal eutectic point, were performed, from which the maximum temperature and pressure at a certain composition were derived. By plotting these equilibrium data for each composition in a T–wt% phase diagram, the new calculated eutectic point is defined by the point of intersection of the clathrate line and the solubility line. The effect of CO_2 in the system on the solubility of the salt in water was not taken into account. Graphs obtained by using this procedure for both NaCl and $CaCl_2$ are depicted in FIGURES 5 and 6. Predictions for the shifted eutectic conditions are listed in TABLE 2 next to the regular eutectic points. It should be noted that eutectic conditions also occur at pressures lower than the CO_2 vapor pressure. The equilibrium lines at lower pressures are situated below the saturated pressure line. Therefore, these lines intersect the solubility line at a different temperature and a different composition. Eutectic clathrate data for NaCl[6] are different from the predicted eutectic conditions, indicating the eutectic point to be situated at $-9.6°C$ and 26.82 bar with a composition of 24.2 wt%. This difference of 2.9°C and 2.1 bar can be explained in terms of neglect of the effect of dissolved CO_2 on the NaCl solubility. The presence of CO_2 in the system lowers the solubility of NaCl. The solubility line shifts to the left in the phase diagram, causing an increase in eutectic temperature and a decrease of NaCl content. Although this specific model prediction on the eutectic shift is not exact, the model serves as a useful tool for predicting process conditions in eutectic clathrate crystallization.

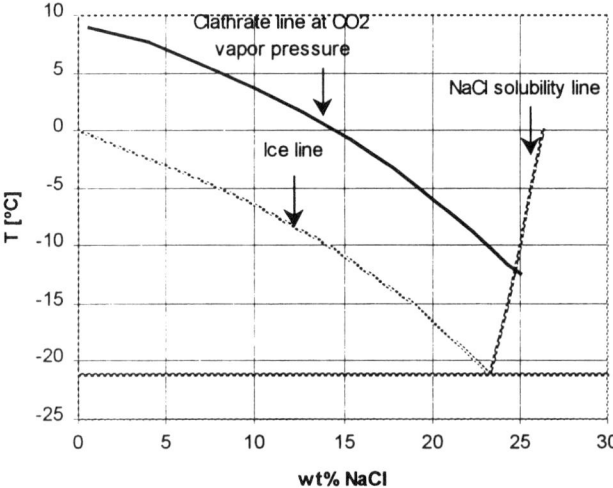

FIGURE 5. Eutectic shift phase diagram, for NaCl.

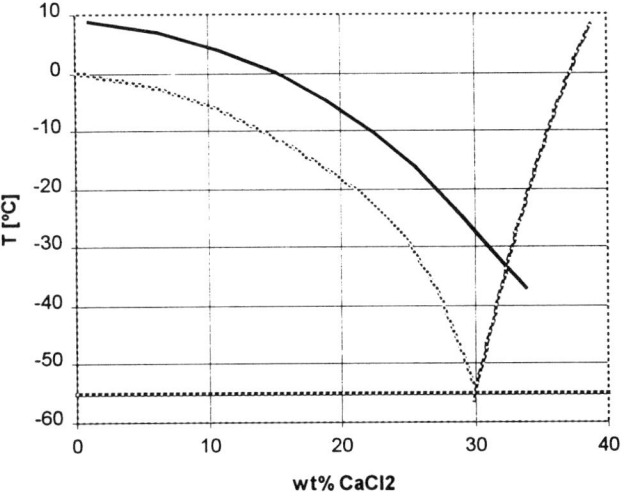

FIGURE 6. Eutectic shift phase diagram, for $CaCl_2$.

Energy Calculations

Energy calculations on NaCl and $CaCl_2$ systems provide the energy consumption reduction in comparison with normal EFC. The calculations for NaCl are performed at −10°C, using the experimental values from Reference 6. For $CaCl_2$ solutions, the operating temperature was taken at −33°C according to the model prediction. The calculations are based on flow sheets, in which the energy stored in the clathrate is recovered. Calculations showed that energy recovery through decomposition at constant pressure is more efficient than through lowering pressure. The CO_2 is recycled in the process after the clathrate is decomposed. Similar to the energy recovery principle in EFC, the feed can be precooled or the cooling system can be split into two stages. A flow sheet for a process using feed precooling is shown in FIGURE 7. This figure indicates process conditions that are valid for the eutectic clathrate freezing of $CaCl_2$ solutions. In the NaCl system, a recrystallization is necessary to form NaCl from $NaCl \cdot 2H_2O$. The energy involved with this process is also integrated in the calculations. Since the presence of CO_2 is required for clathrate formation in any case, direct cooling by evaporating liquid CO_2 is considered.

TABLE 2. Eutectic conditions compared to predicted eutectic clathrate conditions

	Atmospheric conditions		Clathrate eutectic prediction		
Substance	T [C]	wt%	P [bar]	T [C]	wt%
NaCl	−21.2	23.3	24.7	−12.5	25.0
$CaCl_2$	−55	29.9	12.7	−33.3	32.3

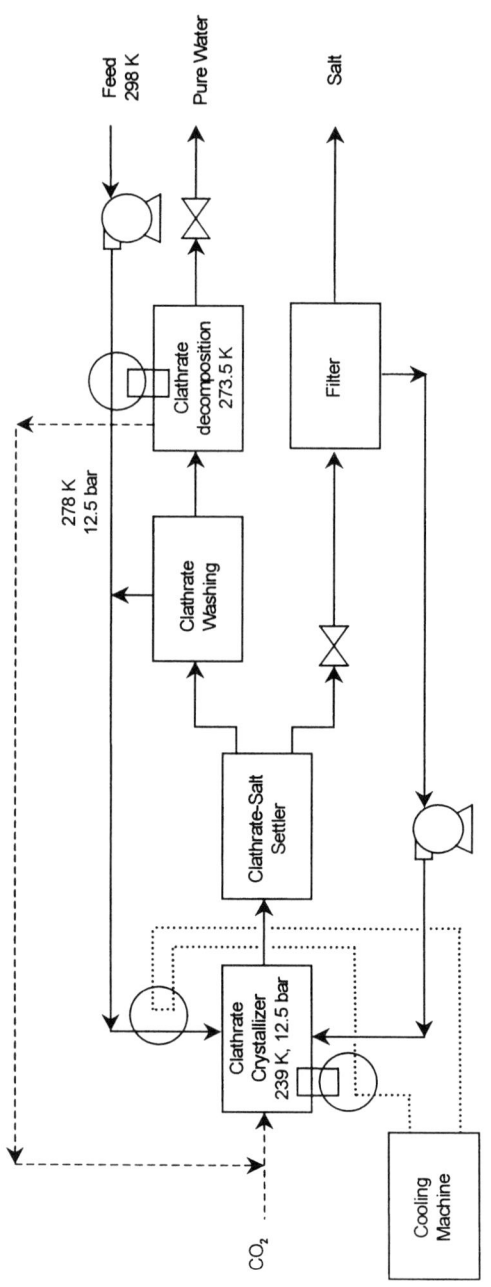

FIGURE 7. Flow sheet eutectic clathrate crystallization including process conditions for eutectic clathrate.

TABLE 3. Energy calculation results electric energy consumption

Product	Cooling principle (kJ/kg)				Reduction (%)
	Direct Cooling	Feed Precooling	Two-Stage Cooling	Regular EFC Feed Precooling	
NaCl	808	737	605	889	31.9
$CaCl_2 \cdot 6H_2O$	355	298	329	437	31.8

The results of the energy calculations are summarized in TABLE 3, where two different methods for using the clathrate energy and direct CO_2 cooling are compared with normal eutectic freeze crystallization. The energy requirement for EFC is calculated for a process in which the crystallizer feed is precooled by ice. In all cases, the process feed is assumed to have the eutectic composition. The relative energy savings are expressed by the reduction percentage. Required energies are calculated in the electric energy consumption per kg of solid salt product.

In the NaCl system, application of a two-stage cooling step is most efficient whereas precooling of the feed by clathrate decomposition is the best option for $CaCl_2$. The similarity between the reduction factors is purely coincidental. Although the operating temperature for $CaCl_2$ is shifted over a larger temperature range than for NaCl (22°C vs. 11°C), the fact that the process is operated at a lower temperature (−33°C vs. −10°C) reduces the energy saving due to less efficient operation conditions for the cooling machines.

EUTECTIC CLATHRATE CRYSTALLIZATION: EXPERIMENTS AND RESULTS

Experimental Setup

An experimental setup has been constructed in which clathrate crystallization experiments were performed in a batchwise operation. A schematic representation of the setup is shown in FIGURE 8. The main piece of equipment in the setup is the crystallizer—a 1 liter (effective volume) jacketed vessel. A technical drawing of the crystallizer is shown in FIGURE 9. The vessel is equipped with a stirrer and inside four removable baffles are installed. An in-line image probe, consisting of an Optem Zoom 70 lens attached to a CCD camera, is used to observe small particles inside the crystallizer. An optical fiber connected to a stroboscope can be inserted in the vessel for illumination.

The crystallizer is cooled by a Lauda RK8KP thermostat that pumps an ethylene glycol-water mixture through the vessel jacket. The temperature of the crystallizer is measured and controlled by thermostat using the external input of a PT-100 that is submerged in the crystallizer contents. The crystallizer is pressurized by adding gaseous CO_2 directly from a cylinder containing liquid CO_2. The pressure is controlled with a reduction valve on the cylinder. CO_2 is led into the vessel through a pipe that, during operation, is submerged in the salt solution and ends just below the stirrer. The gas, therefore, bubbles through the solution and the stirrer ensures good mixing of the gas phase in the liquid phase. The pressure inside the crystallizer is measured with a 0–40 bar abs PTX 610 pressure transducer.

Experimental Procedure

A solution is prepared using demineralized water and pure salts. The prepared solution contains a larger quantity of salt than for the expected eutectic composition. As a result, salt crystallizes before clathrate is formed. After thorough rinsing, the vessel is filled with the salt solution. The temperature desired in the crystallizer is set on the thermostat. After this temperature is reached, the crystallizer is pressurized to a level above the equilibrium pressure but below the CO_2 vapor pressure. Solid salt and clathrate are formed and when a considerable quantity of crystals is observed, the CO_2 feed is closed. The system is left to equilibrate. The pressure remaining constant for about one hour served as an indication that equilibrium pressure had been reached. The pressure is lowered slightly, causing decomposition of some clathrates. However, after some time equilibrium is reached and the constant pressure is taken to be the equilibrium pressure.

Results of Clathrate Experiments with NaCl Solutions

Preliminary tests using 5–20 wt% NaCl solutions showed that producing clathrates was possible in the experimental setup. In these experiments, when only clathrate was

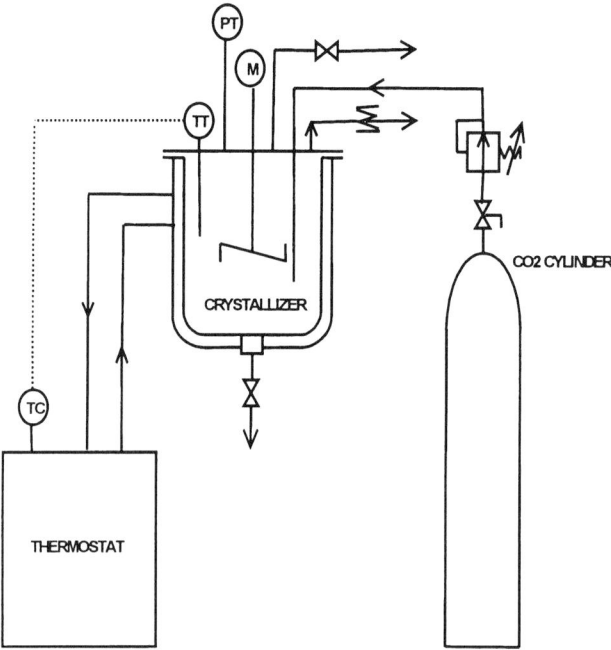

FIGURE 8. Schematic flow sheet experimental setup.

formed it was noticed that clathrate crystallization was induced more readily by pressurizing a solution at the desired temperature than by cooling a pressurized system. Clathrate formation occurred more rapidly at high stirring rates (750 rpm) causing more turbulence, which is in accord with published observation.[7] In a first set of experiments, CO_2 was led into the crystallizer above the liquid level. Even at high stirring rates clathrate formation was found to be slow. A clathrate film only covered the gas-liquid interface and slowed down mass transport phenomena. Adjustment of the equipment in order to bubble the gas through the liquid phase led to a considerable improvement in clathrate formation.

Eutectic experiments were performed with a 25.0 wt% NaCl solution in a temperature range from -11 to $-15°C$. Equilibrium data are provided in TABLE 4, showing good agreement with literature values.[6] The accuracy of the measurement at $-11°C$ is less than that at lower temperatures. This was concluded from the fact that the period over which the pressure remained constant was shorter than that in experiments at lower temperatures. This is possibly due to the fact that at high pressure, leakage of CO_2 from the vessel influenced the pressure conditions.

The $NaCl \cdot 2H_2O$ crystals formed had a faceted shape and a size of approximately 200 µm. The clathrate crystals have a flat plate like shape with a relatively large diameter of 3–5 mm. These crystals have the tendency to coagulate. When the pressure decreased rapidly CO_2 bubbles formed at the decomposing clathrate crystal

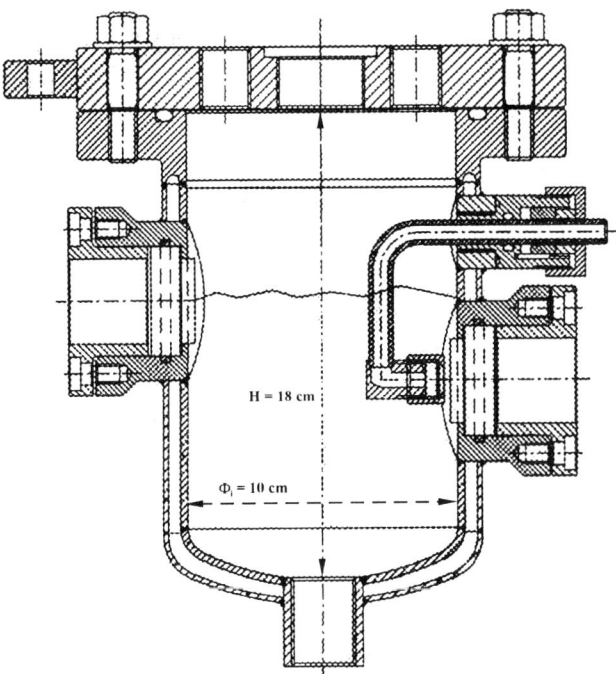

FIGURE 9. Technical drawing of the crystallizer.

TABLE 4. Experimental NaCl clathrate equilibrium conditions

T [°C]	P [bar]
−11	21.50
−12	18.40
−13	15.86
−14	13.73
−15	11.89

surfaces. The density of the eutectic NaCl solution is proven to be higher than the gas hydrate density, since the NaCl crystals fall whereas clathrates move upwards after stopping agitation. Due to the high agitation rates applied, settling of salt and clathrate was obstructed. To achieve gravitational separation in a continuous process, lower stirring rates should be applied or larger crystal sizes should be obtained. Alternatively, in contrast to normal eutectic freezing, settling can be performed in separate equipment.

Results of Clathrate Experiments with $CaCl_2$ Solutions

Clathrate formation in the presence of $CaCl_2 \cdot 6H_2O$ was observed at −29°C and 12.20 bar. This does differ substantially from the model prediction of −33°C and pressure of 12.7 bar.

The $CaCl_2 \cdot 6H_2O$ crystals are shaped like needles with a length of 500 to 1,200 μm and a width of 100 to 350 μm. The clathrates have a faceted blocked form, differing considerably from the structure produced in NaCl experiments. They have an average length of 220 μm and an average width of 85 μm. After stopping the stirrer $CaCl_2$ crystals were seen to settle to the bottom of the vessel and an upward movement of gas hydrate was observed. However, the settling velocity is lower than in the NaCl system due to the lower density of $CaCl_2 \cdot 6H_2O$ and the higher viscosity of the solution at these lower temperatures.

CONCLUSIONS

Eutectic freezing with clathrates is feasible. Reproduction of experimental results on the NaCl system showed the suitability of the experimental setup for clathrate experiments. The possibility of separating clathrate from salt crystals by gravity has been demonstrated. The model presented for the prediction of the clathrate equilibria is a useful tool in predicting the eutectic shift of salt solutions. An energy calculation and comparison showed that an energy consumption reduction of 32% is attainable for both NaCl and $CaCl_2$ if clathrates crystallization is applied instead of EFC.

ACKNOWLEDGMENT

This research was subsidized by the Ministry of Economic Affairs, the Ministry of Housing, Spatial Planning and Environment and the Ministry of Education and Science of the Netherlands.

REFERENCES

1. VAN DER HAM, F. *et al.* 1998. Chemical Engineering and Processing **37:** 207–213.
2. UCHIDA, T. 1997. Waste Management **17**(5/6): 343–352.
3. NASRIFAR, K.H. *et al.* 1998. Fluid Phase Equilibria **146:** 1–13.
4. PITZER, K.S. 1991. Activity Coefficients in Electrolyte Solutions, 2nd edit. CRC Press.
5. DHOLABHAI, P.D. *et al.* 1993. J. Chem. Eng. Data **38:** 650–654.
6. BOZZO, A.T. *et al.* 1973. Fourth International Symposium on Fresh Water from the Sea. **3:** 437–451.
7. PARENT, J.S. *et al.* 1996. Chem. Eng. Comm. **144:** 51–64.

A Unified Nucleation Theory for the Kinetics of Hydrate Formation

BJØRN KVAMME

Physics Department, University of Bergen, Allegt. 55, 5007 Bergen, Norway

ABSTRACT: In this work I present a nucleation theory for the kinetics of hydrate nucleation and growth. The first stage of the theory is the kinetic transport of hydrate formers towards the surface, calculations of average surface density, and composition of adsorbed molecules. The second step is molecular agglomeration and clustering, with a matrix approach for evaluation of cluster distribution, and a subsequent cluster stability analysis. Growth rates for metastable clusters are obtained from an extended diffuse interface theory. The final step is the calculation of growth rates for stable cores according to an explicit solution of the classical nucleation theory. Preliminary results from the theory are very promising. The deviations from the experiments are also in directions that are to be expected from the simplifications made at this stage in the development of the theory.

INTRODUCTION

This work represents an extension of previously published theory for the kinetics of hydrate nucleation and growth.[1,2] The scheme is based on the assumption that the initial nucleation takes part at the hydrate former/water interface, toward the gas side of the interface. This is in accordance with recent findings in NMR studies.[3,4] The revision of the theory is described in the next section, followed by some examples illustrating the theory and our conclusions so far.

KINETIC THEORY

In a system of hydrate former from a gas phase in contact with a liquid water phase we assume that the hydrate formation is taking place on the interface, toward the gas-phase. The transport of gas to the interface is represented by a theory for gas diffusion, in this version the simple Enskog theory is applied. In the present version of the theory I have replaced the adsorption description in the original theory[1,2] by a modified version the 2-dimensional adsorption theory due to Monson.[5] The original theory is modified from ideal gas phase behavior to general fluids using the SRK equation of state to represent the hydrate former phase. This means that the adsorption theory can be also used to estimate selectivity in surface adsorption when the hydrate formers are in the liquid phase. The adsorption theory is well described in the original paper by Monson[5] and is not be repeated here. Briefly it makes the assumption that two-dimensional interactions between the adsorbed molecules at the surface do not change the interactions between the molecules above the surface and the surface itself. In order to apply this concept here we approximate the water surface by a plane, with water molecules distributed in accord with the experimentally measured structure of bulk water.[6] These water molecules are smeared over spherical shells corresponding to the maxima of the correlation functions. The final calculation of the corresponding one-dimensional contribution to the chemical potential is very similar to the integral involved in the calculation of Langmuir contants. The two-dimensional contribution to the chemical potential of the adsorbed components is calculated by using the two-dimensional equation of state due to Henderson.[7] For CO_2 we apply the Lennard-Jones parameters employed by Kvamme and Tanaka,[8] and the same mixing rules as used by Monson[5] for cross-interactions. OPLS[9] parameters are used for methane and TIP4P[10] parameters are used for water. The iterative solution is quite similar to the scheme suggested by Monson[5] in which the iterative solution of conditions for equilibrium between gas and adsorbed layer gives the density of the adsorbed layer as well as the equilibrium distribution of the individual components on the surface. The corresponding individual densities, relative to the corresponding individual densities in the gas phase, gives the driving force for transport flux of each component towards the surface.

The next step in the hypothesis is the nucleation of water and hydrate formers into liquid like clusters. In the "sticky collision" approach the forward nucleation rates are governed by collisions between clusters and monomers. This approach is described in detail elsewhere[11] and is not be repeated here. In that work it was assumed that the clusters followed a Boltzmann distribution and this replaces the need for a theory describing the evaporation rates. I use a similar approach, but instead of assuming a Boltzmann distribution I use the internally consistent distributions according to Wilemski and Wyslouzil.[12] In the solution we neglect cluster-cluster collisions. This simplification is not necessary, but it reduces the memory need as well as the CPU time. Some calculations for CO_2/water, with these collisions included, indicate that the effect of this approximation may be less than two percent of the nucleation rate. Activity coefficients for the hydrate formers are recalculated from solubility data and fitted to the Wilson equation. Any activity coefficient model that is able to reproduce these solubilities can of course be used. The corresponding matrices that represent the different cluster compositions are solved for increasing matrix size until a further increase does not alter the results significantly. For CO_2/water at 30 bar and 273.15 K a matrix of 300 CO_2 and 600 water molecules is needed. The total calculated flux of carbon dioxide condensing with water as clusters is 2.3×10^{23} molecules/(m^3s) carbon dioxide under this condition. Clusters containing only one CO_2 molecule dominate for small number of water molecules in the clusters. In FIGURE 1 we have therefore plotted the flux of clusters with more than one CO_2, relative to the flux of corresponding clusters (i.e., same number of water molecules) with only one CO_2 molecule. The number of water molecules in the plot has been cut off at 120 curves and only for up to six CO_2 molecules are included in the figure. If, for instance, we consider clusters with two CO_2 molecules, then for roughly 60 water molecules the flux of CO_2 with this size of clusters is equal to the flux of clusters with only one CO_2. Since this approach follows classical nucleation theory in the sense that uniform properties within the nuclei are assumed, we recalculate the fluxes for all clusters using the DIT theory.[13] The limited space here does not permit a detailed discussion of the theory and the reader is directed to the original paper for

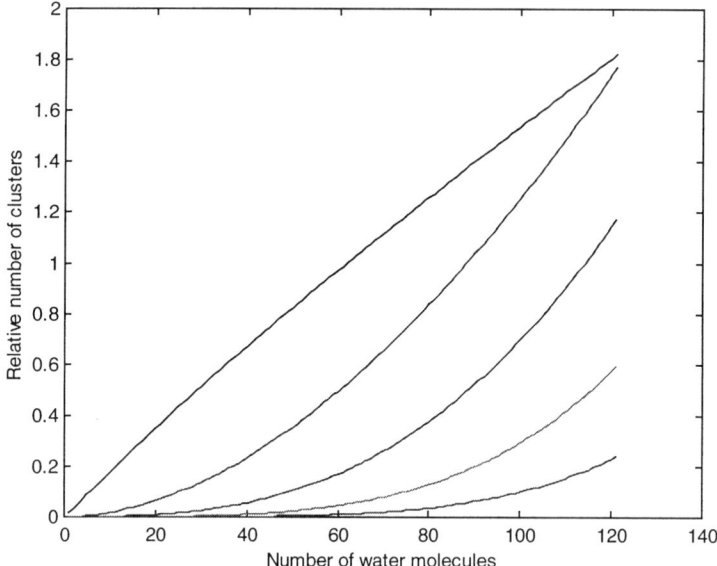

FIGURE 1. Relative number of water molecules around clusters containing two to six carbon dioxide molecules. The values are relative to the clustering rate around a single carbon dioxide molecule. The *upper curve* is for two carbon dioxide molecules in a cluster with a number a given by the horizontal axis, the *next curve* is for three carbon dioxide molecules, and so on to the *lower curve* included in the plot which is for six carbon dioxide molecules. The temperature is 273 K and pressure is 30 bar. The system was solved for 300 CO_2 and 600 water molecules.

more details. For binary nucleation, the appropriate expressions for the enthalpy and entropy terms involved in the work terms are given by:

$$\Delta h(\vec{r}) = \rho_k^l(\vec{r}) \left\{ x_i(\vec{r}) u_i^l(\vec{r}) + x_j(\vec{r}) u_j^l(\vec{r}) - k_B T^2 \left[x_i(\vec{r}) \frac{\partial \ln \gamma_i(\vec{r})}{\partial T}_{P,\vec{N}} + x_j(\vec{r}) \frac{\partial \ln \gamma_j(\vec{r})}{\partial T}_{P,\vec{N}} \right] \right\}$$

$$+ \rho_V \left\{ -y_i u_i^v - y_j u_j^v - k_B T^2 \left[y_i \frac{\partial \ln \phi_i(\vec{r})}{\partial T}_{P,\vec{N}} + y_j \frac{\partial \ln \phi_j(\vec{r})}{\partial T}_{P,\vec{N}} \right] \right\} + P_V \left[1 - \frac{\rho_k^l(\vec{r})}{\rho_V} \right] \quad (1)$$

$$\Delta s(\vec{r}) = \rho_k^l(\vec{r}) \{ x_i s_i^l(\vec{r}) + x_j s_j^l(\vec{r}) - k_B [x_i(\vec{r}) \ln x_i(\vec{r}) \gamma_i(\vec{r}) + x_j(\vec{r}) \ln x_j(\vec{r}) \gamma_j(\vec{r})] \}$$

$$- \rho_k^l(\vec{r}) k_B T \left[x_i(\vec{r}) \frac{\partial \ln \gamma_i(\vec{r})}{\partial T}_{P,\vec{N}} + x_j(\vec{r}) \frac{\partial \ln \gamma_j(\vec{r})}{\partial T}_{P,\vec{N}} \right]$$

$$+ \rho_V \{ -y_i s_i^v - y_j s_j^v + k_B T [y_i \ln y_i \phi_i + y_j \ln y_j \phi_j] \} \quad (2)$$

$$+ \rho_V k_B T \left[x_i \frac{\partial \ln \phi_i(\vec{r})}{\partial T}_{P,\vec{N}} + x_j \frac{\partial \ln \phi_j(\vec{r})}{\partial T}_{P,\vec{N}} \right]$$

where h is enthalpy, s is entropy, T is temperature, p is pressure, y is mole-fraction in the gas phase, and x is the mole-fraction in a cluster. γ is the activity coefficient, ϕ is the fugacity coefficient, and k_B is Boltzmann's constant. The r vector is the volume vector involved in the integral of the work for a specific growth (as given by the limits of the integral, see the original paper by Granasy).[13] The expansion to multicomponent mixtures is straightforward. The pure component gas phase molar energies and entropies in Equations (1) and (2) are calculated from the SRK equation. Similarly for the pure liquid component values for CO_2 and CH_4 at saturation temperature and the heat capacities are used to provide the actual values. Pure liquid water energies are estimated by using the calculated values in Kvamme and Tanaka.[8]

The fluxes from clusters to hydrate cores are calculated as follows. In Equations (1) and (2) the activity coefficients are replaced by the corresponding expression derived from the excess chemical potential for water in hydrate as function of size as derived by Kvamme.[1,2] The corresponding gas phase values are replaced by the corresponding liquid state cluster values (see (1) and (2)). The surface tension needed in the DIT theory for this purpose is approximately the value that can be calculated from the chemical potential for pure liquid water,[8] the chemical potential of the hydrate core[1,2] as function of size, and the actual surface area of the specific cluster. This also gives the fluxes for clusters beyond the sizes that were included in the matrix approach, all the way up to the critical sizes. This then gives the number of cores that enter the region of stable growth as function of time. The equations for this region have been described elsewhere and are not repeated here—see Kvamme[1]

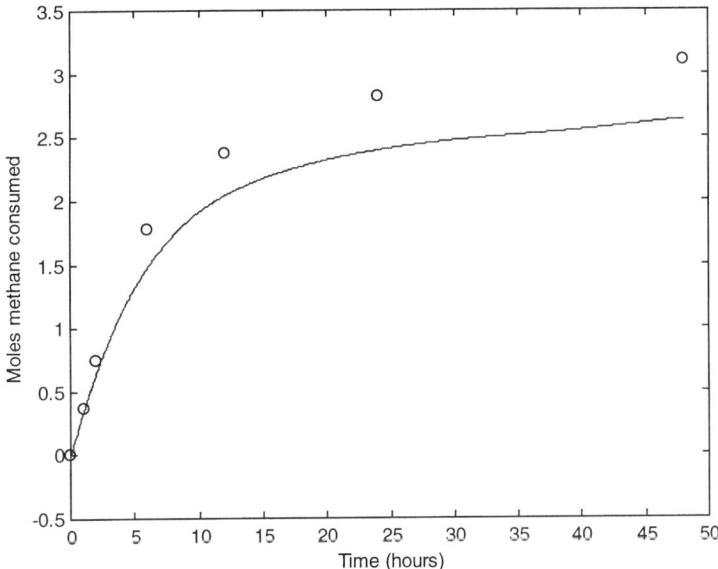

FIGURE 2. Consumption of methane as function of time in a stationary cell without stirring. The circles are experimental data are from Makogan[14] for temperature of 273.75 K and a pressure of 56.68 bar. *Solid line* indicates estimated results for the same conditions.

Equations 12–20 or Makogan[14] Equations 3.101–3.110. Transport limitations are calculated by using an approach due to Muelan[15] and Muelan and Friedman[16] as described by Kvamme.[2]

The same methodology was applied to a methane/water system at 56.68 bar and 273.75 K, experimentally studied by Makogan[14] in a cell with a cross sectional area of 90 cm^2. The necessary size for the cluster analysis of this system was 300 CH_4 and 900 water molecules. The consumption of methane as function of time is compared with experimental results in FIGURE 2. As expected the predicted results are lower than the experimental results. The actual—wave corrected interface—area is larger than the geometric cell area. The dynamics of the waves at the surface is also expected to facilitate nucleation and should theoretically increase the nucleation rate.

CONCLUSIONS

I have proposed a revised nucleation theory for hydrate initiation and growth. The theory consists of four stages: (1) kinetic transport of hydrate formers towards the surface using calculations of average surface density and composition of adsorbed molecules to determine individual density gradients; (2) agglomeration and clustering with a subsequent matrix approach to evaluate cluster distribution, cluster stability, and growth rates from an extended diffuse interface theory; (3) flux of clusters to hydrate cores, also calculated according to diffuse interface theory; and (4) growth of stable cores according to an explicit solution of classical nucleation theory. Preliminary results from the theory are very promising. The deviations from the experiments are also in directions that are to be expected from the simplifications made.

REFERENCES

1. KVAMME, B. 1996. A new theory for the kinetics of hydrate formation. Proceedings of the Second International Conference on Natural Gas Hydrates, June 2–6, Toulouse, France. 131–146.
2. KVAMME, B. 1998. Kinetics of hydrate formation. Proceedings of the 5th Asian Thermophysical Properties Conference, Seoul, Korea, August 30–September 2, 1998.
3. SONG, G. & R. KOBAYASHI. 1996. NMR studies of hydrate formation. Submitted for presentation at the Sixth International Offshore and Polar Engineering Conference, Honolulu, Hawaii, 1996.
4. KOBAYASHI, R. 1997. Rice University, Houston, U.S.A. Private communication.
5. MONSON, P.A. 1987. On the molecular basis of adsorbed solution behavior. Chem. Eng. Sci. **42**: 505–513.
6. SOPER, A. 1993. Unpublished data.
7. HENDERSON, D. 1977. Mol. Phys. **34**: 301.
8. KVAMME, B. & H. TANAKA. 1995. J. Phys. Chem. **99**: 7114.
9. JORGENSES, W.L., J.D. MADURA & C.J. SWENSON. 1983. J. Am. Chem. Soc. **106**: 6638.
10. JORGENSEN, W.L., J. CHANDRASEKHAR, J.D. MADURA, R.W. IMPEY & M.L. KLEIN. 1983. J. Chem. Phys. **79**: 926.
11. VEHKAMAKI, H., P. PAATERO, M. KULMALA & A. LAAKSONEN. 1994. J. Chem. Phys. **101**: 9997.
12. WILEMSKI, G. & B.E. WYSLOUZIL. 1995. Binary nucleation kinetics. I. Self-consistent size distribution. J. Chem. Phys. **103**: 1127.

13. GRANASY, L. 1996. J. Phys. Chem. **100:** 10768.
14. MAKOGAN, Y.F. 1997. Hydrates of hydrocarbons. PennWell Books.
15. MUELAN, Y. 1976. Water Resource Res. **12:** 513.
16. MUELAN, Y. & S.P. FRIEDMAN. 1991. Water Resource Res. **27:** 2771.

The Occurrence of Methane Hydrate in Ternary and Quaternary Systems of Methane, Water, Certain Organics, and Sodium Chloride

M.M. MOOIJER-VAN DEN HEUVEL, M. REUVERS,
R.M. DE DEUGD, C.J. PETERS,[a] AND J. DE SWAAN ARONS

Delft University of Technology, Faculty of Applied Sciences, Laboratory of Applied Thermodynamics and Phase Equilibria, Julianalaan 136, 2628 BL Delft, the Netherlands

ABSTRACT: From literature it is known that sodium chloride (NaCl) exerts a strong pressure enhancing effect on the three-phase equilibrium liquid water–hydrate–vapor (L_w–H–V) in the ternary system water + methane + sodium chloride (H_2O + CH_4 + NaCl). However, on the other hand, it recently became apparent that certain water-soluble organic components, present in low concentrations in the aqueous phase, exert the opposite effect on the equilibrium pressure of the three-phase equilibrium L_w–H–V. Compared to the hydrate equilibrium pressure of the organic free liquid phase, pressure reductions as high as 80% have been observed. This paper reports on experimental results of the competitive effect on the equilibrium pressure of the three-phase equilibrium L_w–H–V, of both NaCl and the water-soluble organic component 1,3-dioxolane. We also consider the effect on the hydrate equilibrium pressure of organic components that are very poorly soluble in water, for example, tetrahydropyran, cyclobutanone, methylcyclohexane, fluoroform (CHF_3), and tetrafluoromethane (CF_4). These poorly water-soluble components are referred to as water-insoluble.

INTRODUCTION

A few decades ago industry initiated an investigation of the possibility for storing natural gas as gas hydrates in salt caverns. Knowledge about the influence of sodium chloride (NaCl), present at saturation conditions, on the three-phase equilibrium liquid water–hydrate–vapor (L_w–H–V) was of crucial importance. The research executed at that time showed that NaCl exerts a strong hydrate inhibiting effect, resulting in a significant increase of the pressure of the three-phase equilibrium L_w–H–V. For instance, the hydrate equilibrium pressures may increase from 10 MPa in the absence of NaCl to values as high as 80 MPa in the presence of certain concentrations of NaCl, depending, of course, on temperature. A practical consequence of this finding was that storage of natural gas in salt caverns was not considered to be an attractive option.[1]

Recently it was discovered[2-5] that low concentrations of certain organic components (e.g., acetone, 1,4-dioxane, and tetrahydrofuran) show the opposite effect—

---

[a]Telecommunication. Voice: +31 (0)15 2782660; fax: +31 (0)15 2788047.
Cor.Peters@stm.tudelft.nl

these components show a hydrate promoting effect. The promoting effect frequently results in a significantly reduced hydrate equilibrium pressure when compared to the pressure of L_w–H–V for the system $H_2O + CH_4$. From the experimental results it was concluded that the maximum pressure reduction occurred at organic component concentrations of approximately 6 mole%, relative to water. However, at higher organic concentrations in the aqueous solution, a steady increase in hydrate equilibrium pressure was observed, that is, less reduction and eventually inhibition of the hydrate equilibrium occurs. In this respect the behavior of water-soluble additives differs from the influence of NaCl on the hydrate equilibrium pressure. The latter exhibits a continual pressure increase with increasing concentration.

Use of gas hydrates for storage and transportation of natural gas has not yet been put into practice, because hydrate equilibrium pressures were too high. The presence of specific organic components seems to be able to reduce these pressures and might enable gas hydrates to offer a practical option. Examples of practical cases in which salts and hydrate formers may be present are storage of natural gas in emptied salt caverns and decomposition of deposits of natural gas hydrates by hot brine injection. It can be advantageous to store natural gas in the form of hydrates to improve the load factor of natural gas supply systems. Benesh[7] already mentioned this idea in 1942. Under standard conditions hydrates occupy 150 to 170 times less volume than the corresponding highly compressed gas.[8] If an appropriate equilibrium pressure-reducing additive for natural gas hydrate can be found it would, in principle, be possible to store natural gas at ambient temperatures and low pressures, in contrast to the conditions required for liquefying natural gas.

All of those new developments in gas hydrate research require a good insight into the phase behavior of different kinds of hydrate systems, with or without NaCl or other salts. The main purpose is to find organic components that reduce the hydrate equilibrium pressure, acting as promoters, and to investigate the competing effects they exhibit with the hydrate inhibiting effect of NaCl. The organic components chosen for this paper were 1,3-dioxolane, which is soluble in water, and tetrahydropyran, cyclobutanone, and methylcyclohexane, which are very poorly soluble in water (referred to as insoluble). We also studied the influence of fluoroform (CHF_3) and tetrafluoromethane (CF_4), which are insoluble in water when liquefied, but are soluble in the gaseous state under the conditions used here. Only for the water-soluble organic 1,3-dioxolane has the competing effect with NaCl been studied at present and this is presented in this paper.

EXPERIMENTAL

The experiments were carried out in a so-called Cailletet apparatus. In this apparatus phase equilibria for mixtures of known overall composition can be determined visually to pressures as high as 15 MPa. For details of this experimental equipment, see References 9 and 10. Temperatures were determined to an accuracy better than 0.05 K and the pressures were measured to an accuracy better than 0.005 MPa. The different types of systems, with water-soluble and water-insoluble additives, require different measuring regimes and are described below:

1. Systems with $H_2O + CH_4$ + water-soluble additive. These systems show three phases when hydrates are present, which means that there are two degrees of freedom in a system with three components. In the pT-diagram the conditions are described in terms of a region that consists of different pT-curves for different compositions of the system. To describe the hydrate equilibrium the temperature of hydrate disappearance is determined at a preset pressure and for several concentrations of the additive. The line for hydrate disappearance is considered to be the hydrate equilibrium line and represents the following phase transition:

$$H + L_W + V \rightarrow L_W + V.$$

The pressure is preset, using a dead-weight pressure balance, and the sample is then cooled below the equilibrium temperature. When the hydrate phase precipitates, the temperature is increased gradually until the last hydrate crystal disappears. This temperature is considered to be the equilibrium temperature for the preset pressure value. The system of this type we considered is $H_2O + CH_4$ + 1,3-dioxolane.

2. Systems with $H_2O + CH_4$ + water-soluble additive + NaCl. When hydrates are present, this system has three degrees of freedom. These are only three phases, because both the additive and the NaCl are dissolved in the water phase. This means that the hydrate equilibrium pressure depends on temperature and on the concentrations of additive and NaCl. The same transition of phases is determined as above—that is, the hydrate disappearance line—and the same measuring regime is followed. The system of this type that was considered is $H_2O + CH_4$ + 1,3-dioxolane + NaCl.

3. Systems with $H_2O + CH_4$ + water-insoluble additive. When hydrates are present, four clearly distinct phases are present, which means that this system has only one degree of freedom. Therefore, when the temperature is set, the pressure will adjust itself and the univariant equilibrium conditions can be measured directly. The water-insoluble additives that were investigated in the presence of water and methane were: tetrahydropyran, cyclobutanone, methylcyclohexane, CHF_3, and CF_4.

When the conditions for hydrate disappearance are measured, it is assumed that at the temperature at which the hydrate phase is about to disappear, all methane is present only in the vapor phase. Furthermore, it is assumed that this phase consists of pure methane. Under these assumptions, the compositions of the fluid phases for

TABLE 1. **Purity of the chemicals used**

Chemical	Supplier	Purity (mole%)
water	—	Distilled
methane	Air Products	99.99
NaCl	Baker	99.6
1,3-dioxolane	MERCK-Schuchardt	> 99
tetrahydropyran	Acros	> 99
cyclobutanone	MERCK-Schuchardt	99.5
methylcyclohexane	MERCK-Schuchardt	> 99
fluoroform	Air Products	99.5
tetrafluoromethane	Air Products	99.5

both the ternary and quaternary systems are known and, as a good approximation, the three-phase L_w–H–V equilibrium behaves as a univariant system. The purity of the chemicals used to prepare the samples is summarized in TABLE 1.

EXPERIMENTAL RESULTS

Experimental results for the various types of systems with the additives used are discussed in this section. The pT-equilibrium lines for the systems with water-soluble additives seem quite complicated, showing a dependence on the concentration of additive. For these systems the pressure values are normalized or even double-normalized to elucidate the simultaneous influence of the additive and NaCl on the methane hydrate equilibrium and the influence of the additive alone, respectively. The relations that are used to normalize or double-normalize the pressure data are:

$$P_{normalized} = \frac{P_{experimental}}{P_{water + methane}}, \quad P_{double\text{-}normalized} = \frac{P_{experimental}}{P_{water + methane + NaCl}}.$$

Normalization and double-normalization is normally applied to the px-curves at various temperatures. Normalization of pressure follows from division of the experimental pressure of the ternary or quaternary system by the equilibrium pressure of the system $H_2O + CH_4$ at that particular temperature. The results of this operation are given in the FIGURES 2A–5A. Double-normalization of pressure follows from division of the experimental pressure of the quaternary system by the equilibrium pressure of the system $H_2O + CH_4 + NaCl$ at that particular temperature. Results of this operation are given in the FIGURES 2B–5B. The normalized curves can cross the value one, in particular in the systems with NaCl, which means that the hydrate equilibrium is inhibited rather than promoted.

The system $H_2O + CH_4 + 1,3$-dioxolane has been investigated by de Deugd.[11] The presence of additive shows a significant hydrate pressure reduction up to 1,3-dioxolane concentrations of about 5–6 mole% in the water phase. At higher concentrations the pressure reduction again decreases. The highest pressure reduction retrieved is approximately 80% compared to the hydrate equilibrium pressures of the organic-free system ($H_2O + CH_4$). The reduction in hydrate equilibrium pressure was promising enough to study the influence of 1,3-dioxolane in the presence of NaCl. Experimental data were collected by Reuvers for the system containing $H_2O + CH_4 + 1,3$-dioxolane + NaCl, with several concentrations of NaCl in the liquid water phase up to saturation—approximately 8 mole%. For each concentration of NaCl several concentrations of 1,3-dioxolane dissolved in the water phase were examined.

FIGURE 1 represents the primary experimental results of the quaternary system with 2 mole% NaCl and various concentrations of 1,3-dioxolane in a pT-diagram. In FIGURE 2 the normalized and double-normalized px-diagrams of the data are shown, in which the simultaneous influence of 1,3-dioxolane and NaCl and of 1,3-dioxolane alone can be observed. FIGURES 3 to 5 show the normalized and double-normalized hydrate equilibrium pressures for higher concentrations of NaCl. The figures show clearly that at concentrations lower than approximately 6 mole% the equilibrium pressure decreases with increasing mole fraction of 1,3-dioxolane present in the water phase. When the concentration of 1,3-dioxolane is further increased, the pressure-reducing effect decreases again. This value for the concentration of the organic at maximum hydrate equilibrium pressure reduction was also observed with other water-soluble additives. The same trend is seen in the presence of NaCl. That is, the minimum in the curves (indicating the highest reduction) is located between 5 and 6 mole% of 1,3-dioxolane, although the present concentration of NaCl varies. In the normalized diagrams of FIGURES 2A–5A it can be observed that the promoting effect of 1,3-dioxolane is able to compete with the inhibiting effect of NaCl up to a concentration of 6 mole% NaCl, which is just below saturation. At higher NaCl concentrations, the 1,3-dioxolane still reduces the hydrate equilibrium but not to such an extent that the hydrate equilibrium pressure is lower than the equilibrium pressure of the system $H_2O + CH_4$. The diagrams with the double-normalized pressure show that the influence of 1,3-dioxolane is the same for varying concentrations of NaCl. Thus, synergistic and antagonistic effects of the additive component and NaCl cannot be observed. In FIGURE 6 the influence of NaCl can be seen in a px-diagram, where the pressure dependency on the mole fraction of NaCl is given for several temperatures at a mole fraction of 1,3-dioxolane of 0.05.

For systems with an additive that is insoluble in water, the dependency on concentration of the additive is not important to the phase equilibrium. The phase equilibrium is represented by an univariant line and can be measured directly. Therefore, normalization or double-normalization of the pressures is not really required to obtain a clear insight into the influence of the additives. The hydrate equilibrium lines of the three additives that are in the liquid state, over the pressure and temperature range considered, are shown in FIGURE 7. This is a pT-diagram and the solid line gives the line of the methane hydrate equilibrium. It can be clearly observed that

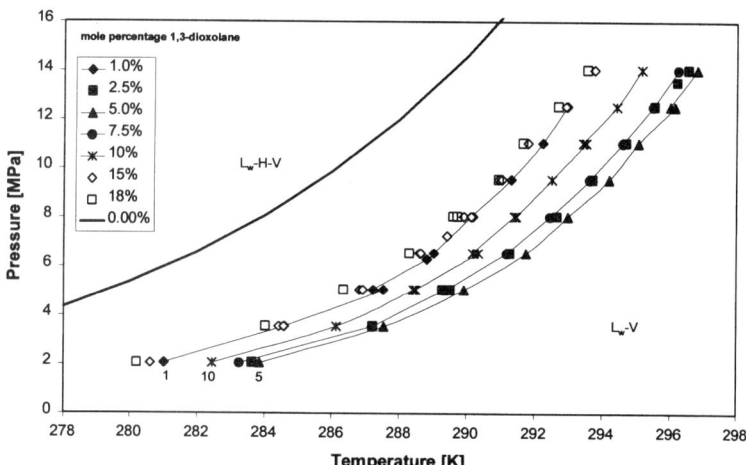

FIGURE 1. pT-diagram of the system with $H_2O + CH_4$ + 1,3-dioxolane + NaCl. The concentration of NaCl is 2 mole% and the pT-curves are given for various concentrations of 1,3-dioxolane; 1 mole% (◆), 2.5 mole% (■), 5 mole% (▲), 7.5 mole% (●), 10 mole% (✳), 15 mole% (◇), and 18 mole% (□). The *solid line* is the hydrate equilibrium curve of the system $H_2O + CH_4$.

the pressure reduction by tetrahydropyran and cyclobutanone is significant, namely greater than 70%. The promoting effect of methylcyclohexane is considerably smaller, approximately 25%. It should be noted that this last component is expected to form sH hydrate,[12] in contrast to the presence of the other two additives, which are believed to form sII. For gas storage and transportation purposes sH might be a more attractive structure because of the larger potential storage capacity[13] and a reduction of 25% in the hydrate equilibrium pressure is still considerable.

The other two additives that are insoluble in water were CHF_3 and CF_4. These fluoroalkanes were chosen because of their known formation of gas hydrates in systems with water. Both components are in the vapor phase over the pressure and

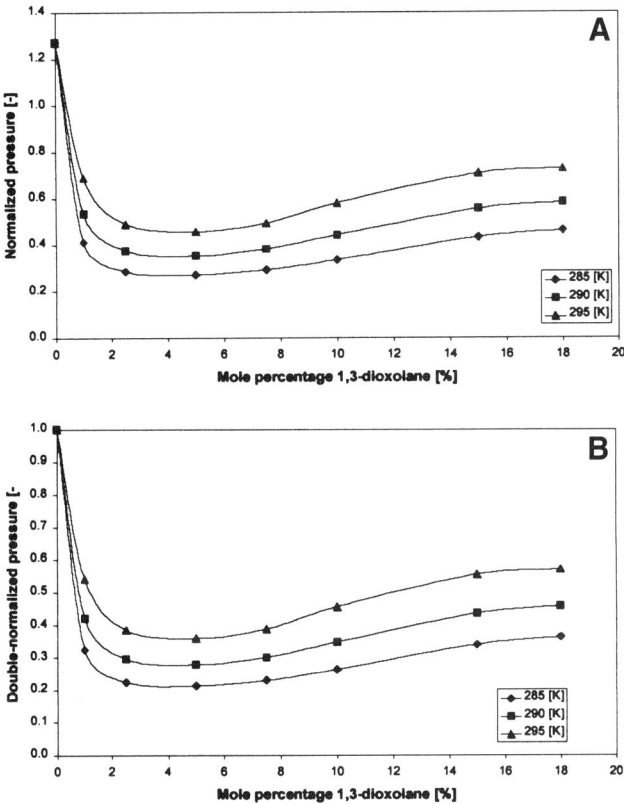

FIGURE 2. A. Normalized px-diagram of the system with $H_2O + CH_4 + 1,3$-dioxolane $+$ NaCl. The normalized pressure is given as a function of the mole fraction 1,3-dioxolane at various temperatures; 285 K (♦), 290 K (■), and 295 K (▲). The content of NaCl is 2 mole%. **B.** Double-normalized px-diagram of system with $H_2O + CH_4 + 1,3$-dioxolane $+$ NaCl. The double-normalized pressure is given as a function of the mole fraction 1,3-dioxolane at various temperatures; 285 K (♦), 290 K (■), and 295 K (▲). The content of NaCl is 2 mole%.

temperature range considered; in fact, CF_4 is in a supercritical state. The CHF_3 flashed in the temperature range studied; although it is insoluble in water no clear distinct fourth phase (liquid additive phase) could be observed. Therefore, hydrate disappearance conditions were measured at various compositions of the vapor phase, 0.25, 0.50 and 0.75 mole fraction of CHF_3. The hydrate equilibrium of the system $H_2O + CHF_3$ was also measured. This was also previously determined by Kubota et al.[14] Both data sets showed similar results with hydrate equilibrium pressures that were lower than those for the system $H_2O + CH_4$. The systems with the mixtures of CH_4 and CHF_3 showed hydrate equilibrium pressures between those for the systems $H_2O + CHF_3$ and $H_2O + CH_4$. All these lines are shown in FIGURE 8. The mixture

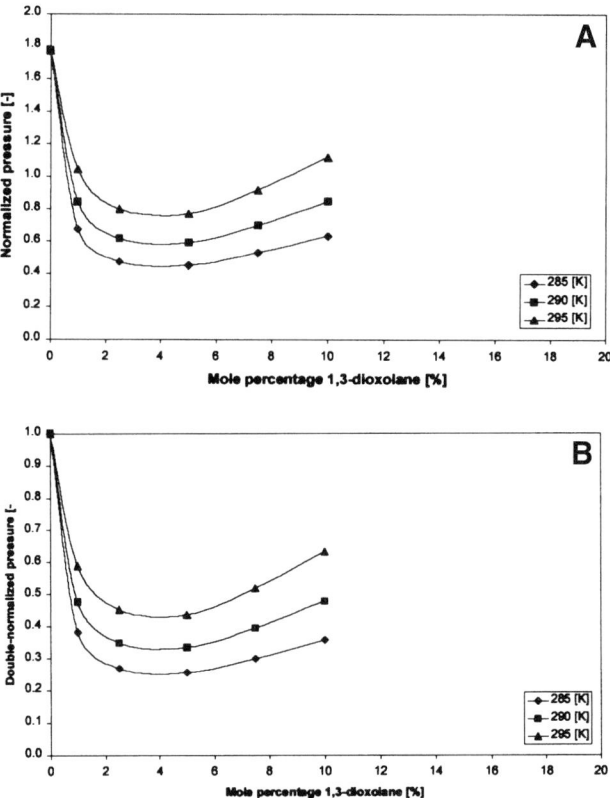

FIGURE 3. A. Normalized px-diagram of system with $H_2O + CH_4$ + 1,3-dioxolane + NaCl. The normalized pressure is given as a function of the mole fraction 1,3-dioxolane at various temperatures; 285 K (◆), 290 K (■), and 295 K (▲). The content of NaCl is 4 mole%. **B.** Double-normalized px-diagram of system with $H_2O + CH_4$ + 1,3-dioxolane + NaCl. The double-normalized pressure is given as a function of the mole fraction 1,3-dioxolane at various temperatures; 285 K (◆), 290 K (■), and 295 K (▲). The content of NaCl is 4 mole%.

with the relative small fraction of CHF_3 (circles in the figure)—an attractive feature for technological applications of gas hydrates with natural gas—shows reductions of nearly 60%.

The supercriticality of CF_4 makes it impossible to observe a fourth distinct phase and, therefore, the same experimental regime as for CHF_3 was followed. All the hydrate equilibrium pressures are compared to the equilibrium of the system $H_2O + CH_4$, and also to the measurements performed on the system $H_2O + CF_4$. The hydrate equilibrium pressures for the system $H_2O + CF_4$ show higher values than those for

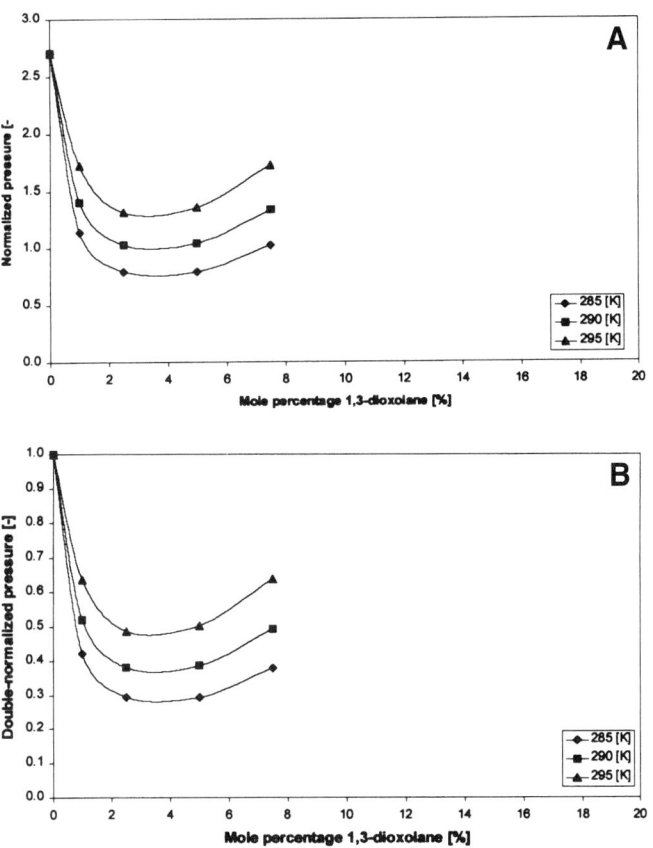

FIGURE 4. A. Normalized px diagram of system with $H_2O + CH_4 + 1,3$-dioxolane + NaCl. The normalized pressure is given as a function of the mole fraction 1,3-dioxolane at various temperatures; 285 K (♦), 290 K (■), and 295 K (▲). The content of NaCl is 6 mole%. **B.** Double-normalized px-diagram of system with $H_2O + CH_4 + 1,3$-dioxolane + NaCl. The double-normalized pressure is given as a function of the mole fraction 1,3-dioxolane at various temperatures; 285 K (♦), 290 K (■), and 295 K (▲). The content of NaCl is 6 mole%.

the system $H_2O + CH_4$. Therefore, CF_4 can assumed to be an inhibitor or very weak promoter. The systems with mixtures of CH_4 and CF_4 in the vapor phase that were considered had compositions of 0.25, 0.5, and 0.75 mole fraction of CF_4. The mixture with the smallest fraction of CF_4 shows reduction of the methane hydrate equilibrium pressure slightly larger than 20%. The equimolar mixture of CH_4 and CF_4 shows less reduction and the mixture with 0.75 mole fraction CF_4 shows elevation of the methane hydrate equilibrium pressure. This information is summarized in FIGURE 9.

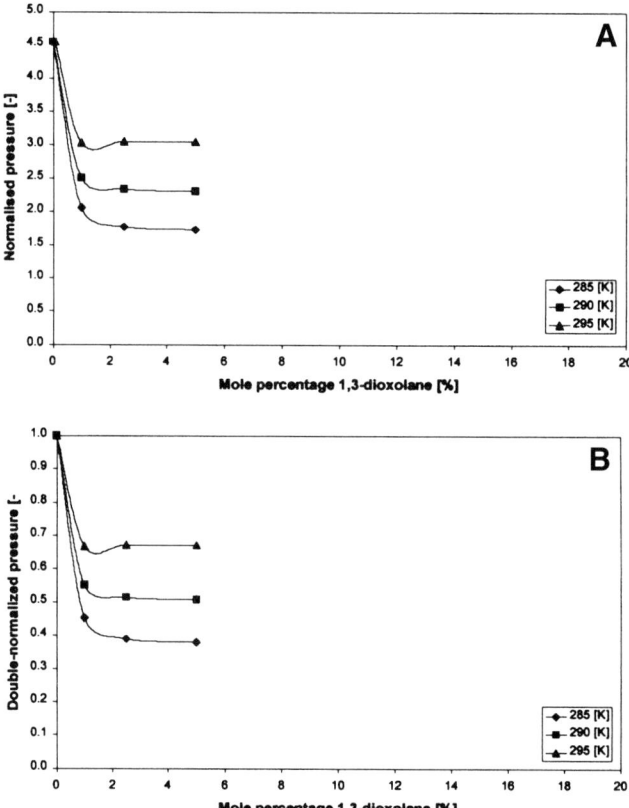

FIGURE 5. A. Normalized *px*-diagram of system with $H_2O + CH_4$ + 1,3-dioxolane + NaCl. The normalized pressure is given as a function of the mole fraction 1,3-dioxolane at various temperatures; 285 K (♦), 290 K (■), and 295 K (▲). The content of NaCl is 8 mole%. **B.** Double-normalized *px*-diagram of system with $H_2O + CH_4$ + 1,3-dioxolane + NaCl. The double-normalized pressure is given as a function of the mole fraction 1,3-dioxolane at various temperatures; 285 K (♦), 290 K (■), and 295 K (▲). The content of NaCl is 8 mole%.

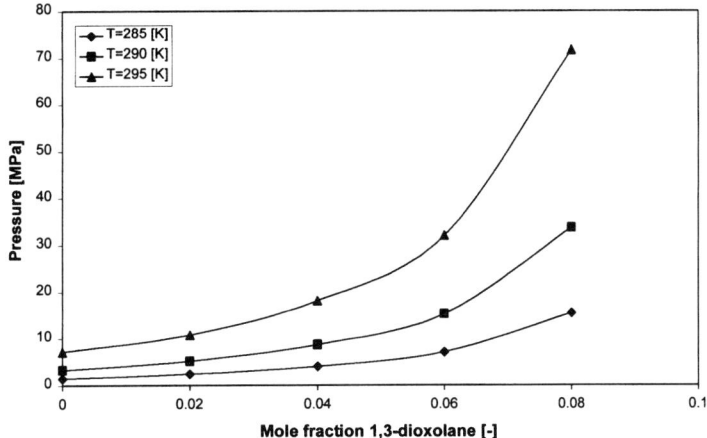

FIGURE 6. px-diagram of system with H_2O + CH_4 + 1,3-dioxolane + NaCl. The pressure is given as a function of the mole fraction NaCl at various temperatures; 285 K (♦), 290 K (■), and 295 K (▲), and a content for 1,3-dioxolane of 5 mole%.

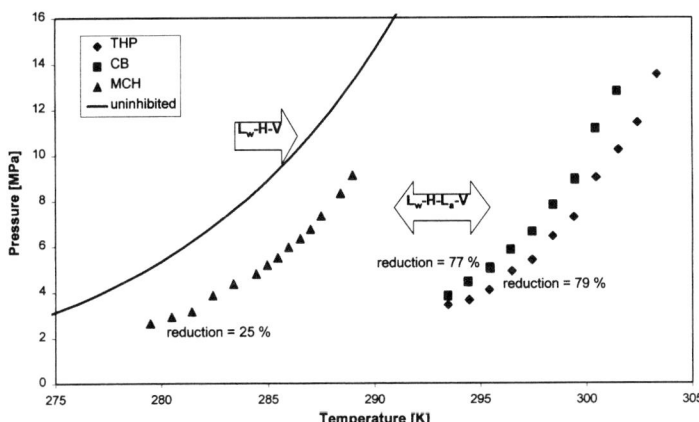

FIGURE 7. pT-diagram of systems with H_2O + CH_4 + additive with the four-phase hydrate equilibrium given for the additives: tetrahydropyran (♦), cyclobutanone (■), and methylcyclohexane (▲). The hydrate equilibria are compared to the hydrate equilibrium of the system H_2O + CH_4 (*solid line*).

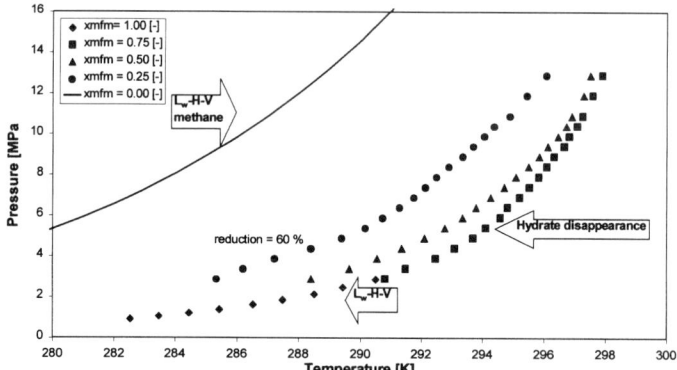

FIGURE 8. pT-diagram of systems with $H_2O + CH_4 + CHF_3$ with the hydrate equilibrium pressures for various mole fractions of CHF_3 in the vapor phase: 1.00 (◆), 0.75 (■), 0.50 (▲), and 0.25 (●). The hydrate equilibria are compared to the hydrate equilibrium of the system $H_2O + CH_4$ (*solid line*).

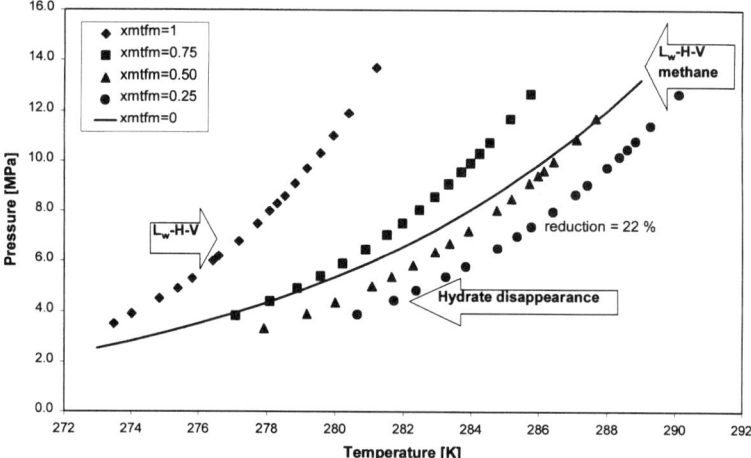

FIGURE 9. pT-diagram of systems with $H_2O + CH_4 + CF_4$ with the hydrate equilibrium pressures for various mole fractions of CF_4 in the vapor phase: 1.00 (◆), 0.75 (■), 0.50 (▲), and 0.25 (●). The hydrate equilibria are compared to the hydrate equilibrium of the system $H_2O + CH_4$ (*solid line*).

DISCUSSION AND CONCLUSIONS

L_w–H–V equilibrium lines for the systems $H_2O + CH_4$ + 1,3-dioxolane and $H_2O + CH_4$ + 1,3-dioxolane + NaCl have been measured. The measurements were carried out for sodium chloride concentrations of 2, 4, 6, and 8 mole% and at 1,3-dioxolane mole fractions ranging from 0.01 to 0.18. The operational pressure of the Cailletet equipment ranged from 2 to 14 MPa. Equilibrium temperatures at these pressures ranged from 270 K to 297 K. From the experimental results it was concluded that the organic compound 1,3-dioxolane, which is soluble in water, shows a hydrate promoting effect at low concentrations. In all cases, the lowest equilibrium pressure was measured at a 1,3-dioxolane concentration of approximately 6 mole%. Adding NaCl to the system increases the equilibrium pressure significantly, though the competing effect of 1,3-dioxolane is large enough to overrule this effect up to NaCl concentrations of 6 mole%. The minimum hydrate equilibrium pressure at 6 mole% of additive might be due to the formation of sII hydrate.[11] The value of approximately 5 mole% complies with the concentration of large cavities relative to the water molecules. Confirmation of this result should be obtained, for example, by using spectroscopic methods.

The comparable competing effect between NaCl and water-insoluble additives as obtained with the water-soluble additives remains to be investigated. However, thus far the promoting effects of the water-insoluble additives tetrahydropyran, cyclobutanone, methylcyclohexane, and CHF_3 on the methane hydrate equilibrium show promising results and justify further study for practical applications that require hydrate equilibrium pressure reduction. If the hydrate pressure reducing capacities of the water-insoluble additives are comparable to those of the water-soluble ones, then they have one significant advantage for practical application. Their insolubility in water could reduce the number of separation steps in processes that use application of gas hydrate technology. The behavior of CHF_3 and CF_4, which have a molecular structure similar to CH_4, seems comparable to that of CH_4 and thus study of the hydrate aspects of this type of component might be of importance for a fundamental understanding of gas hydrate formation and their stability.

REFERENCES

1. DE ROO, J.L., C.J. PETERS, R.N. LICHTENTHALER & G.A.M. DIEPEN. 1983. AIChE J. **29**(4): 651.
2. NG, H.-J. & D.B. ROBINSON. 1994. Ann. N.Y. Acad. Sci. **715**: 450.
3. SAITO, Y., T. KAWASAKI, T. OKUI, T. KONDO & R. HIRAOKA. 1996. Proceedings of the 2nd International Conference on Natural Gas Hydrates, Toulouse, France. 459.
4. MAINUSCH, S., C.J. PETERS, J. DE SWAAN ARONS, J. JAVANMARDI & M. MOSHFEGHIAN. 1996. Proceedings of the 2nd International Conference on Natural Gas Hydrates, Toulouse, France. 71.
5. JAGER, M.D., R.M. DE DEUGD, C.J. PETERS, J. DE SWAAN ARONS & E.D. SLOAN. 1999. Fluid Phase Equilibria **165**: 209.
6. SLOAN, E.D. 1998. Clathrate Hydrates of Natural Gases, 2nd edit. Marcel Dekker, New York.
7. BENESH, M.E. 1942. U.S. Patent No. 2,270,016.
8. GUDMUNDSSON, J.S., M. PARLAKTUNA & A.A. KHOKHAR. 1994. SPE Production and Facilities.

9. DE LOOS, TH.W., H.J. VAN DER KOOI, W. POOT & P.L. OTT. 1983. Delft Progress Report. **8:** 200.
10. PETERS, C.J., J.L. DE ROO & J. DE SWAAN ARONS. 1993. Fluid Phase Equilibria. **85:** 301.
11. DE DEUGD, R.M. & J. DE SWAAN ARONS. 1999. AIChE J. Submitted for publication.
12. MEHTA, A.P. & E.D. SLOAN. 1996. AIChE J. **42**(7): 2036.
13. KHOKHAR, A.A., J.S. GUDMUNDSSON & E.D. SLOAN. 1998. Fluid Phase Equilibria. **150–151:** 383.
14. KUBOTA, H., K. SHIMIZU, Y. TANAKA & T. MAKITA. 1984. J. Chem. Eng. Japan **17**(4): 423.

Control of Gas Hydrate Formation Using Surfactant Systems

Underlying Concepts and New Applications

GLEN IRVIN,[a] SICHU LI,[a] BLAKE SIMMONS,[a] VIJAY JOHN,[a,b]
GARY McPHERSON,[a] MICHAEL MAX,[c] AND ROBERT PELLENBARG[c]

[a]*Department of Chemical Engineering, Tulane University, New Orleans, Louisiana 70118, USA*

[c]*Naval Research Laboratory, Code 7420, Washington, D.C. 20375, USA*

ABSTRACT: We describe some new approaches to the rapid formation of gas hydrates using surfactant systems that form reverse micelles. The thermodynamics of hydrate formation in surfactant containing reverse micellar systems indicate the possibility of controlling hydrate deposition. The use of reverse micellar systems also allows the deposition of hydrates in the form of small crystallites that do not aggregate. Surfactants also appear to displace agglomerated hydrates from surfaces, thereby facilitating flow. Direct injection of water into gas saturated liquid hydrocarbons leads to rapid hydrate formation. On the other hand, water droplets injected into liquid carbon dioxide indicate a gradual conversion to the hydrate form with the initial formation of a hydrate skin. An interesting application of rapid hydrate formation is its use in the restriction of inorganic nanocluster growth.

INTRODUCTION

The crystalline inclusions of gas and water known as clathrate hydrates or gas hydrates are of significant interest for their relevance to natural gas recovery and processing.[1] The discovery of vast hydrate deposits indicates a significant natural gas resource.[2] Research over the last decade indicates the possibility that the gradual melting of geological hydrate deposits may have potential implications to climate change through methane release to the atmosphere.[3] An excellent reference on the range of gas hydrate science and technology is the recent monograph by Sloan.[1]

Recently, there has been significant interest in the use of surfactants in gas hydrate technology. Surfactants have been proposed as hydrate inhibitors, and flow loop experiments indicate that surfactants are effective in preventing blockage of flow lines by hydrate plugs.[4] Concurrently, there are also new and interesting technologies that seek to use hydrates as a method to store and transport natural gas.[5,6] Our research in gas hydrates attempts to address these newer technologies in gas storage and transportation using hydrates, and in the development of entirely new hydrate technologies.

[b]Address for correspondence. Voice: 504-865-5883.
vijay.john@tulane.edu

In this paper, we report our work on gas hydrate formation with the objective of finding methods to form hydrates extremely rapidly and with a degree of control. To begin, we review the well-studied experiment that demonstrates hydrate equilibrium thermodynamics. Here, a bulk aqueous phase is contacted with a gas phase under appropriate conditions for hydrate formation. In this three-phase system of gas (G), liquid water (L_1) and hydrate (H), the system follows univariant thermodynamics if pure water and a single gas species are the only two components. Hydrate formation occurs until the pressure reaches the equilibrium (dissociation) pressure at a specified temperature. Hydrate formation in such bulk aqueous systems is significantly mass transfer limited. This is due to the fact that hydrate layers partition to the gas–water interface and inhibit contact between the gas and the aqueous phase. Conversion of an entire bulk aqueous phase to hydrates is extremely difficult.

Our research was, therefore, motivated by the following questions. Can hydrates be induced to form extremely rapidly? Can hydrates be rapidly nucleated at pressures higher than the three-phase equilibrium pressure to increase the fractional occupancy of the hydrate cavities? Can hydrate formation be controlled for hydrate deposition? Can rapid hydrate formation be induced in an essentially quiescent system without the need for significant agitation to minimize mass transfer limitations? If these criteria for hydrate formation can be met, then it may be possible to develop new technologies for gas storage through hydrate formation. Additionally, if the hydrates can be deposited as discrete small crystallites, it may be possible to slurry them easily for applications to gas transport. We attempt to address these aspects in the following review of our work, which we describe rather phenomenologically to bring out the underlying concepts. We also describe new experiments on the formation of hydrates of carbon dioxide, and the development of potential new technology for nanomaterials synthesis using hydrate formation.

DESCRIPTION OF RESEARCH

Hydrate Formation from Reverse Micelles

To meet the criteria for rapid hydrate formation, we hypothesized that hydrate formation would be facilitated by maximizing the contact surface area of gas and water. The direction we have taken to maximize the interfacial area for gas water contact is to use water-in-oil microemulsions, or reverse micelles as they are commonly called. These are microdroplets of water suspended in a bulk organic solvent (e.g. isooctane) and stabilized by a surfactant, typically the double-tailed anionic bis(2-ethyl) sodium sulfosuccinate (AOT). In AOT based reverse micelles, the water to surfactant molar ratio, w_0, is an indication of micelle size. A rough correlation between micellar core radius R (in nm) and w_0 is $R = 0.15\ w_0$.[7] FIGURE 1a depicts hydrate formation in the water core of a reverse micelles, and is meant to convey the idea that hydrate unit cells (1.2 nm for Structure I and 1.73 nm for Structure II) can fit in the core of the micelle. The micelle can easily swell to w_0 values of 30 or so, simply by the addition of water, and remain stable. At these water contents, the micelle radius is about 4.5 nm and the microaqueous phase approaches the properties of bulk water. We can perform a quick calculation of the surface area in a micellar

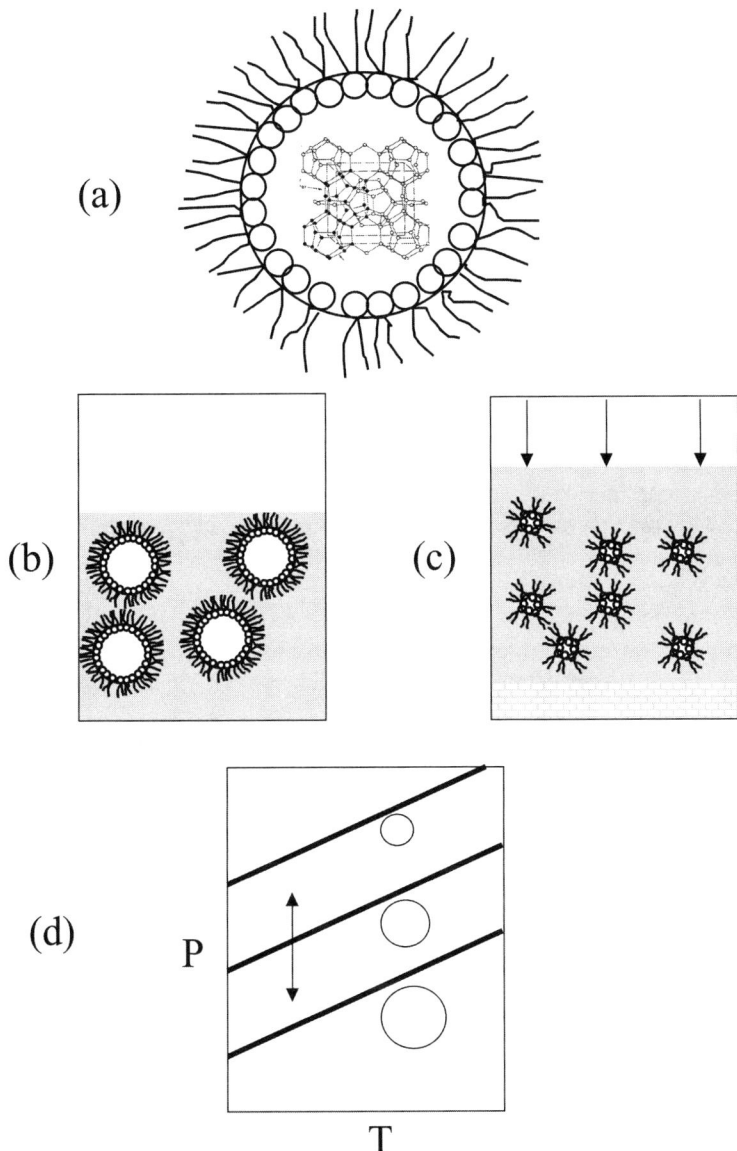

FIGURE 1. Hydrates in surfactant systems of reverse micelles. (**a**) Schematic of a reverse micelle. (**b**) Micellar system before hydrate formation. (**c**) Micellar system after hydrate formation. The micelles shrink in size with an increase in number density. Hydrates are deposited very uniformly. (**d**) Schematic of the thermodynamics of hydrate formation from reverse micelles.

solution containing 0.1 M AOT at $w_0 = 20$, for example, (2M H_2O), with micelles of radius 3 nm. The surface area is of the order 3.6×10^4 m^2/liter of solution.

In our previous papers dating to 1989, we have studied the thermodynamics of hydrate formation in reverse micelles[8,9] and have brought out some possible applications to biotechnology and advanced materials.[10] We very briefly review the concept of hydrate formation in reverse micelles through the illustrations of FIGURE 1 as relevant to the experiments conducted here. FIGURE 1b illustrates reverse micelles of large w_0 (e.g., 20). The micellar solution is contacted with a hydrate forming gas (e.g., methane or ethylene) and brought to hydrate forming conditions. The hydrate forming gas is soluble in isooctane (the bulk hydrocarbon liquid phase) and is therefore accessible to the water in the micelles. With hydrate formers such as ethylene, the liquid phase swells due to the appreciable solubility of the gas (e.g., ethylene has a solubility equivalent to a mole fraction of 0.5 in isooctane at 274 K and 2 MPa). The observation is that after an initial induction period, hydrates form very rapidly in these micellar solutions. Initially the solution is very clear; the macroscopic observation is that when hydrate formation occurs, the solution turns intensely turbid. The system exists in dynamic equilibrium with exchange of micellar constituents. As a consequence of such micellar exchange, the hydrate crystals grow and precipitate out of the micellar solution (hydrates being more dense than the bulk hydrocarbon solvent). The situation is shown schematically in FIGURE 1c.

Thermodynamic concepts of hydrate formation in reverse micelles are explained through FIGURE 1d.[8] We have found that hydrate formation in reverse micelles leads to the removal of intramicellar water and thus to a reduction in w_0 and micelle size. At each w_0, the system exhibits univariant equilibria as shown in FIGURE 1d. As w_0 decreases, a higher pressure is required to form hydrates at a specified temperature. This can be intuitively understood. At large w_0 (greater than 20, for example), the intramicellar water has thermodynamic activity approaching that of bulk water. At smaller w_0, the intramicellar water has a reduced thermodynamic activity due to interactions with the ionic environment of the micellar head groups and the hydrated counter ions. The thermodynamic driving force (chemical potential gradient) required to reorient water to the crystalline hydrate phase therefore increases with decreasing w_0, and this is manifested in a higher hydrate dissociation pressure. The consequences of these observations are interesting. As shown in FIGURE 1d, if the temperature is fixed and the pressure progressively increased, then hydrates progressively form and drop out of solution, leaving the micelles with a reduced water content (w_0). In other words, pressurization leads to reduction of micellar size and the formation of hydrates. If the pressure is subsequently decreased, the hydrates melt and the water is spontaneously reincorporated into the micelles. A specific example of methane hydrate formation thermodynamics is the formation of hydrates from micelles with $w_0 = 15$ at 274 K at a pressure of 2 MPa. A pressure of 5 MPa is required for methane hydrate formation from micelles with $w_0 = 5$. In other words, if micelles with $w_0 = 15$ are pressurized from 2 MPa to 5 MPa, 66% of the intramicellar water is removed in the form of hydrates.

We propose that this phenomenon can be exploited for technologies related to gas storage and transportation in the form of hydrates. The hydrates formed from reverse micelles are of the form of small crystallites since they are nucleated in the confined microenvironment of the micellar water cores and do not grow to a significant extent. Since the micellar environment is uniform (the droplet size is extremely monodisperse and each micelle has the same microenvironment) hydrate formation in these systems is very homogeneous. The small hydrate crystals pack easily without large occlusions, and the hydrates appear to have the consistency of fine snow. FIGURE 2a illustrates the fluffy nature of the hydrates. The hydrates do not form an agglomerated mass and can be easily slurried for gas transport applications. FIGURE 2b illustrates the hydrate phase after a significant amount of ethylene hydrate has precipitated from a concentrated water-swollen micellar solution. The view cell is kept inclined and one can see the hydrate phase submerged in the partially water-depleted micellar phase (the meniscus of the liquid phase can be seen as the horizontal line). The high-pressure view cell also contains an insertion tube for fluid injection or recovery. The important observation here is the rather uniform density of the hydrate phase. FIGURE 2c illustrates the flow diagram to develop technology for gas storage in the form of hydrates. After hydrates are deposited from the micelles, the water-depleted micelles are removed from the system. Water is then added to these micelles to swell them, and they are then recycled to the hydrate forming step. There are some inherent considerations to the process. The first is the loss of surfactant. Analysis of the micellar phase after hydrate formation indicates that although some surfactant does adsorb to the hydrate crystals and to the view cell surfaces, more than 90% of the surfactant is left in the supernatant water-depleted micellar phase.[11] Second, in the micelle recovery and water addition step there are liquid hydrocarbon losses due to partial depressurization, and these have to be accounted for in the process economics. It should be possible to simply use gas pressure to drive the entire recycling step without the need for depressurization and external pumping. The process does appear technically feasible and may have applications in novel technologies such as those proposed by Max and Pellenbarg[6] where hydrates may be stored as fuel sources in double-hulled naval or commercial vessels.

A direct extension of these studies is the concept of forming hydrates by directly injecting water into the liquid hydrocarbon phase containing water-free micelles and brought to hydrate formation conditions. If water uptake into the micelles is significantly faster than hydrate formation, it was our hypothesis was that the micelles would take up water and swell. Subsequently, hydrate nucleation from the water in the micelles would, in theory, lead to hydrate deposition. However, as shown in FIGURE 3, this hypothesis is incorrect. FIGURE 3a shows a schematic of the system, in which water is injected directly into "dry" AOT micelles in isooctane, pressurized with ethylene, and brought to pressures exceeding the dissociation pressure of ethylene hydrates. FIGURES 3b–e is a set of photographs taken of the process in a view cell. The experiment is one in which "dry" AOT micelles are solubilized in isooctane at a concentration of 0.5M and contacted with ethylene at a temperature of 275 K and pressure of about 2.4 MPa, significantly above the dissociation pressure of ethylene hydrates at 275 K. (0.7 MPa). Water is then injected through a nozzle inserted into the liquid hydrocarbon phase. We find that hydrate formation is extremely rapid, and a hydrate bridge develops from the lower surface of the view cell to the tube tip as shown in FIGURES 3b and 3c, with hydrates forming at the point of water ejection from the tube. Two additional observations are recorded. The

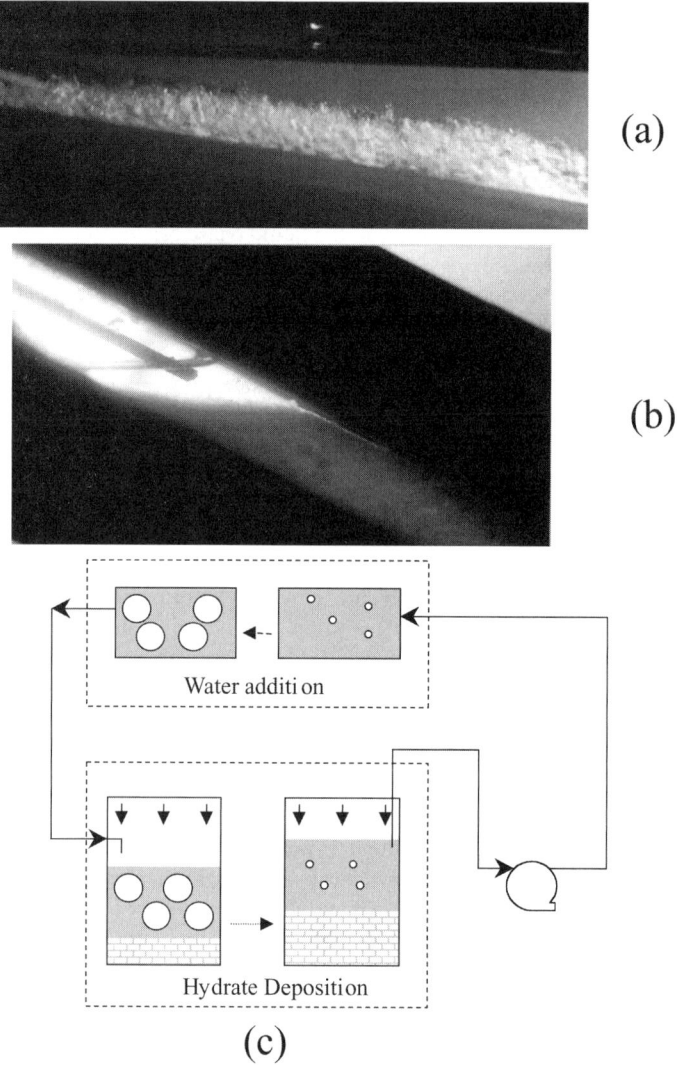

FIGURE 2. (a) Hydrates formed from reverse micelles with no system agitation. The rather fluffy nature of the hydrates can be observed. (b) Hydrates formed with high volume fractions from reverse micelles. The liquid meniscus of the micellar solution can be seen. The 1/8″ injection tube provides an idea of the scale. (c) Schematic of a process to continuously form hydrates from reverse micelles.

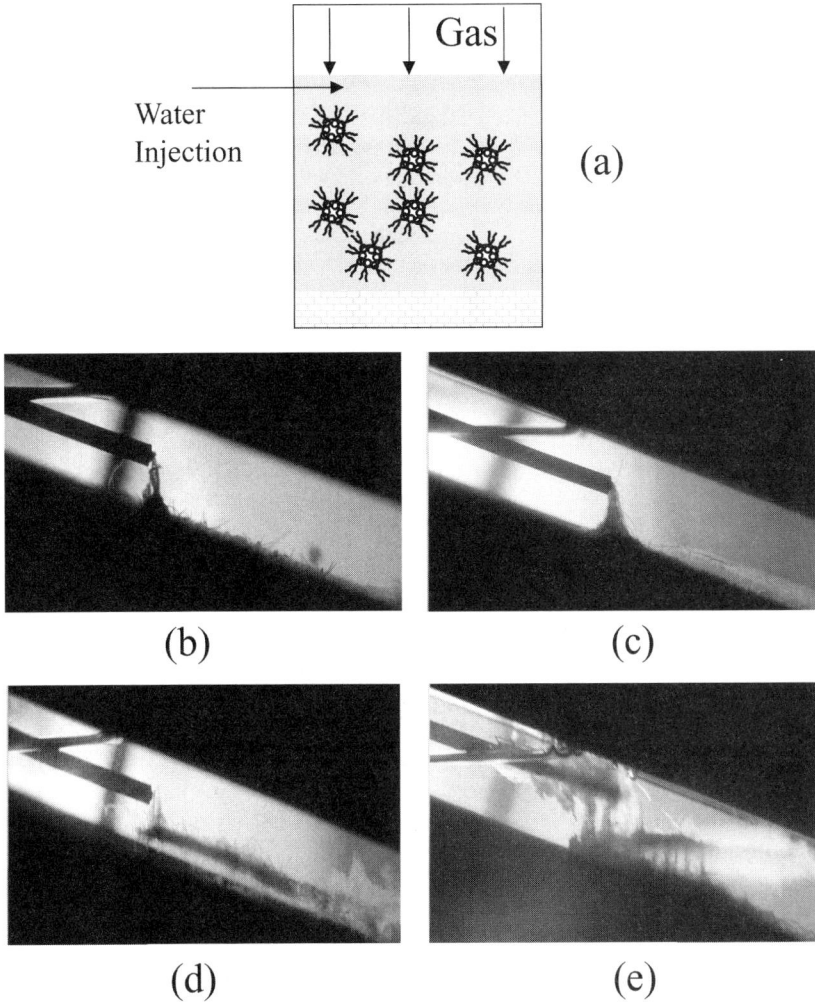

FIGURE 3. (a) Schematic of a process to form hydrates by water injection into a dry reverse micelles (b)–(e) The injection of water leads to very rapid hydrate formation and bridge between the view cell and the tube outlet. A gradual liftoff of hydrate layers is observed. The hydrates remain anchored to the bottom of the view cell.

first is the loss of some of the hydrates formed initially and indeed the hydrate bridge eventually disappears. This is explained intuitively. The very rapid hydrate formation implies that the pure water injected through the nozzles forms hydrates upon contact with dissolved ethylene. However, the hydrates exist in contact with an electrolyte containing system of the anionic surfactants. A small amount of hydrate dissociation therefore takes place with water becoming incorporated into the dry micelles until the thermodynamic equilibrium stated by

$$\mu_W^H(T, P) = \mu_W^{RM}(T, P) = \mu_W^0(T, P) + RT \ln a_W^{RM}(T, P)$$

is reached. Here, μ_W^H is the chemical potential of water in the hydrate phase, μ_W^{RM} is the chemical potential of water in the reverse micelle, μ_W^0 is the chemical potential of pure water, and a_W^{RM} is the water activity in the micelles. Hydrate formation is, therefore, more rapid than water incorporation into the micelles and does not follow the homogeneous nucleation process seen during formation from swollen water-containing micelles. Thus, our hypothesis that water would be incorporated into the micelles prior to hydrate formation is incorrect, but the outcome is the observation that hydrates can be induced to form extremely rapidly. The second observation is the gradual lifting of the hydrate layer from the bottom surface of the view cell. This is seen in FIGURES 3d and 3e. Over the course of a few hours, we observe that hydrates deposited on the surface gradually rise, although they remain anchored to the surface through tendril like crystalline needles. We speculate that surfactant adsorption to the surface renders the surface hydrophobic and inhibits hydrate adherence. These observations may have implications to the development of technology to break up hydrate plugs.

CO_2 HYDRATE FORMATION

The rapid hydrate formation observed by injecting water into a gas saturated liquid hydrocarbon led to a conjecture about the formation of CO_2 hydrates by the injection of water into liquid CO_2. The sequestering of CO_2 by injection of liquid CO_2 at ocean depths greater than 500 m has been proposed by several researchers.[12] At these depths, hydrates of CO_2 can form, as modeled by Holder and coworkers for the process of liquid CO_2 droplets injected into water.[13] These authors indicate that a thin hydrate layer forms on the droplet surface creating diffusion limitations to CO_2-water contact and inhibiting further hydrate formation. We have studied the phenomenon using the reverse technique of injecting water droplets into liquid CO_2 in light of our experiments with very rapid ethylene hydrate formation. The sequence of photographs in FIGURE 4 clearly shows the fascinating skin formation. As a water droplet is slowly injected into liquid CO_2 at 275 K, we observe the droplet initially growing upwards. This is due to the fact that the CO_2 hydrate skin has a greater density than water[13] and thus supports the liquid water that is slowly ejected through the nozzle. Eventually, the greater density of the droplet (compared to liquid CO_2) results in the settling of the droplet to the view cell bottom surface where it assumes an almost hemispherical shape with no spreading. The sequence of photographs shows a liquid water surface at the top of the droplet (FIGS. 4a through 4d), which becomes entirely converted to the hydrate skin (FIG. 4e). However, this surface hydrate layer eventually collapses (FIG. 4f). The transition from (FIG. 4e) to (FIG. 4f) takes up to an hour to occur. The observations appear to validate the previous work indicating diffusion-limited hydrate skin formation upon contact between liquid CO_2 and water at hydrate forming conditions.[13] At this point, it would be speculation to attempt to explain differences between CO_2 hydrate formation and the rapid ethylene hydrate formation observed in the previous experiments.

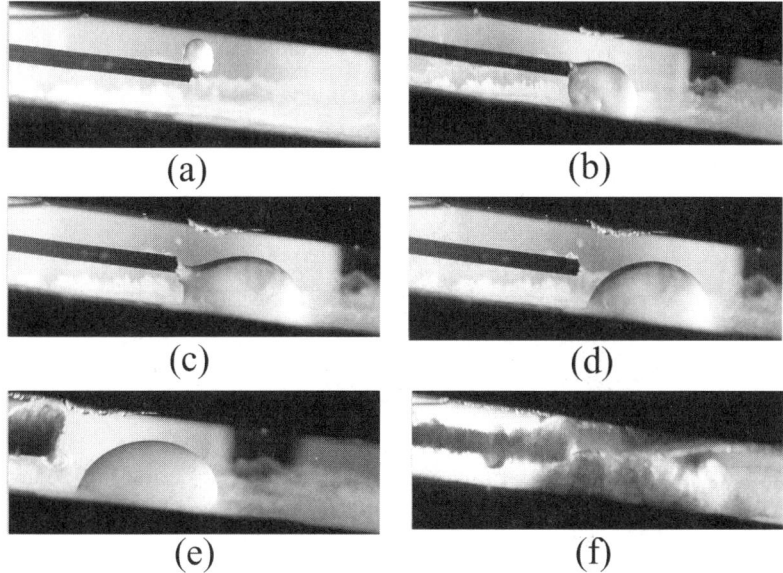

FIGURE 4. Hydrates of CO_2 formed by slowly injecting water into liquid CO_2. The growth of the skin is visualized in (**b**) through (**d**). The collapse of the water droplet with a hydrate skin is seen in (**e**).

A NOVEL APPLICATION TO MATERIALS SYNTHESIS

We have attempted to exploit the observations of rapid hydrate formation to the synthesis of nanomaterials. The concept is described through the schematic of FIGURE 5a. Two precursor salts are dissolved in water and injected through separate opposing nozzles. For example, if $FeSO_4$ and NH_4OH are contacted in solution, iron hydroxides form that subsequently oxidize to ferrites (iron oxides) and iron oxyhydroxides. The objective here is to contact these precursors and at the same time convert much of the aqueous phase to the solid hydrate form. It was hypothesized that removal of much of the aqueous phase to the solid hydrate form would quench growth of the inorganic particles thus restricting particle size to the nanoscale. As shown in FIGURE 5a, this is done by injecting the precursors into a liquid hydrocarbon phase (isooctane) saturated with a hydrate forming gas (ethylene) brought to pressures significantly above the VL_1H curve (to maximize the driving force for hydrate formation). The specific experiment was conducted at about 20 bar. Our observations indicate rapid hydrate formation and the formation of a dark green precipitate (iron hydroxides). The inorganic material is the recovered by using system pressure to force the dissociating hydrates and the particles out through an exit valve over a filter where the particles are collected. The recovery step is the bottleneck to the process and is not very efficient due to the Joule-Thompson effect associated with dissociating hydrates being forced out over a pressure drop. The particles collected gradually oxidize to a reddish-orange color indicating the formation of iron

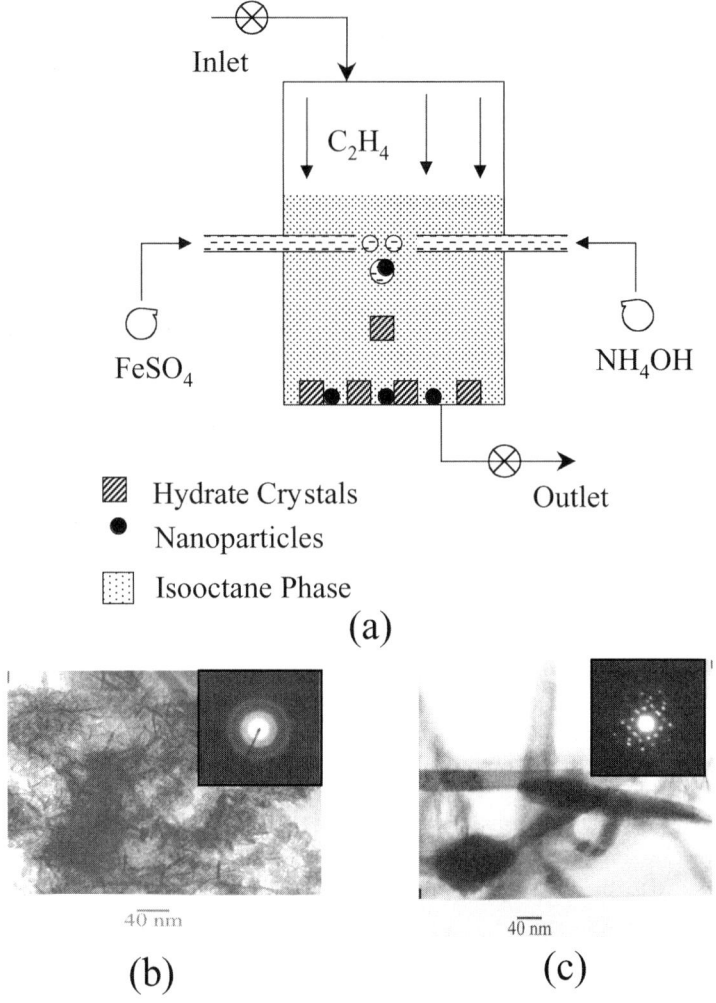

FIGURE 5. (a) The concept of nanoparticle synthesis by quenching growth through hydrate formation. (b) Iron oxides generated through the hydrate formation process. The *inset* shows the ring-type electron diffraction pattern characteristic of a lack of multiple diffraction planes. (c) The control experiment showing large acicular ferrites. The *inset* shows the spotted pattern, characteristic of large particles with multiple diffraction planes.

oxides and oxyhydroxides. FIGURE 5b shows the morphology of these particles obtained through transmission electron microscopy, as very small needles over a diffuse background. FIGURE 5c illustrates the morphology of the particles generated through the control experiment of simply mixing the two precursors, and illustrates that the acicular particles are significantly larger than those generated through the hydrate experiment. The electron diffraction patterns in the insets verify the observations. The concentric rings observed for the particles generated during hydrate formation indicate polycrystalline materials, whereas the discrete spotted pattern shown for the control samples indicate the presence of much larger particles. We are continuing these studies on inorganic nanoparticle synthesis during hydrate formation through extensive characterizations of materials properties.

CONCLUSIONS

Our experiments thus indicate that rapid hydrate formation can be induced by devising systems to increase the surface area for gas-water contact. This is done either by using swollen reverse micelles, or by injecting water droplets into gas saturated liquid hydrocarbon systems. The presence of specific double-tailed surfactants serve to form reverse micelles to encapsulate water, which can then be converted to hydrates leading to a very uniform deposition technique. Additionally, we observe some very interesting consequences of surfactant adsorption on surfaces resulting in the tendency of hydrate layers to lift off from such surfaces, but still remain anchored. These observations do have implications in our understanding of the flow of hydrates in pipelines. We also realize that CO_2 hydrates have some unique aspects related to hydrate skin formation on droplet surfaces.

The fact that rapid hydrate formation can be induced implies applicability to new technologies such as nanomaterials synthesis. The process for nanomaterials synthesis involves the simple procedure of injecting aqueous precursors into gas saturated liquid hydrocarbons. In principle, the process can be scaled up to large volume production and may have favorable economies of scale when compared to alternative laser ablation and vapor deposition processes. These need to be evaluated. The process of hydrate formation through water injection into gas saturated liquid hydrocarbons may also be a method of rapid screening of hydrate inhibitors by visual inspection. Combinatorial screening of hydrate inhibitors could have applications in the development of new compounds for hydrate inhibition. The control of hydrate formation has applications not only to gas recovery, storage and transportation, but also perhaps to the development of new technologies.

ACKNOWLEDGMENT

Funding from DARPA Grant MDA972-97-1-0003 is gratefully acknowledged by V.T.J. M.D.M. thanks the Federal Energy Technology Center (FETC), Morgantwon WV (U.S. Department of Energy), for support via funding element DE-AT26-97FT34344, and the Office of Naval Research, Program Element No. 0601153. R.E.P. acknowledges the support of the Chemistry Division, Code 6100, NRL in this work.

REFERENCES

1. SLOAN, E.D., JR. 1997. Clathrate Hydrates of Natural Gases. Marcel Dekker, New York.
2. KVENVOLDEN, K.A. 1988. Chem. Geol. **71:** 41.
3. MACDONALD, G.J. 1990. Climactic Change **16:** 247.
4. LEDERHOS, J.P., A.P. MEHTA, G.B. NYBERG, K.J. WARN & E.D. SLOAN. 1996. Chem. Eng. Sci. **51:** 1221.
   BEHAR, E., A.S. DELION, J.M. HERRI, A. SUGIER & M. THOMAS. 1994. Ann. N.Y. Acad. Sci. **715:** 94.
5. GUDMUNDSSON, J.S., V. ANDERSSON, O.I. LEVIK & M. PARLAKTUNA. 1998. Soc. Petrol. Eng. Conference. SPE 5058.
6. MAX, M.D. & R.E. PELLENBARG. 1998. Proc. Fuel Chemistry Div. Amer. Chem. Soc. **24:** 463.
7. PILENI, M.P., T. ZEMB & C. PETIT. 1985. Chem. Phys. Lett. **118:** 414.
8. NGUYEN, H., J.B. PHILIPS & V.T. JOHN. 1989. J. Phys. Chem. **93:** 8123.
   NGUYEN, H., W. REED & V.T. JOHN. 1991. J. Phys. Chem. **95:** 1467.
9. NGUYEN, H.T., N.S. KOMMAREDDI & V.T. JOHN. 1993. J. Colloid Interface Sci. **155:** 482.
10. JOHN, V.T., N. KOMMAREDDI, M. AYYAGARI, M. TATA, C. KARAYIGITOGLU & G. MCPHERSON. 1994. Ann. N.Y. Acad. Sci. **715:** 468.
    RAO, M., H. NGUYEN & V.T. JOHN. 1990. Biotech. Prog. **6:** 465.
11. JOHN, V.T., M. RAO, H. NGUYEN & J.B. PHILLIPS. 1991. J. Supercrit. Fluids. **4:** 238.
12. HERZOG, H., D. GOLOMB & S. ZEMBA. 1991. Environ. Prog. **10:** 64.
    LIRO, C.R., E.E. ADAMS & H.G. HERZOG. 1992. **33:** 667.
13. HOLDER, G.D., A.V. CUGINI & R.P. WARZINSKI. 1995. Environ. Sci. Technol. **29:** 276.

A Remote Station to Monitor Gas Hydrate Outcrops in the Gulf of Mexico

THOMAS M. McGEE[a] AND J. ROBERT WOOLSEY

Center for Marine Resources and Environmental Technology, University of Mississippi, 220 Old Chemistry Building, Oxford, Mississippi 38677, USA

ABSTRACT: Outcrops of gas hydrates will be monitored via a remote station installed on the continental slope of the Northern Gulf of Mexico. The project, driven by the need to initiate collection of a data base for assessing stability of the sea floor, will also address other factors associated with the formation and dissociation of gas hydrates. A group of experts has been organized to supervise the assembly and operation of the station, which will monitor physical and chemical parameters of sea water and sea floor sediments on a more or less continuous basis.

INTRODUCTION

The possibility of installing a multisensor station to monitor a gas hydrate mound on the sea floor of the Northern Gulf of Mexico has been discussed for some years. During March 22–26, 1999, a workshop entitled *New Concepts in Ocean, Atmosphere and Seafloor Sensor Technologies for Gas Hydrate Investigations and Research* was held in Biloxi, Mississippi. It was sponsored by the Center for Marine Resources and Environmental Technology (CMRET) of the University of Mississippi, Oxford, and the Institute for Marine Sciences (IMS) of the University of Southern Mississippi, Ocean Springs. Participation was international and included a delegation from the Russian Academy of Sciences.

One day of the workshop was devoted to discussion sessions that addressed practical aspects of assembling and deploying a remote station to monitor the formation/dissociation of gas hydrates and slope stability in the vicinity of a known hydrate mound. From those and subsequent discussions, a concept of a project to realize such a station was developed and is presented herein.

BACKGROUND INFORMATION

In the Gulf of Mexico, gas hydrate mounds form along the intersections of faults with the sea floor. They are edifices constructed largely of water from the sea and hydrocarbon gases that have migrated up the faults from buried reservoirs. In addition to gas hydrates, they also contain various minerals deposited by bacteria feeding on the hydrocarbons. The mounds are ephemeral, capable of changing greatly within

---

[a]Telecommunication. Voice: 662-232-7320; fax: 662-232-5625.
tmm@mmri.olemiss.edu

a matter of days. Many geoscientists familiar with recent geologic processes in the Gulf of Mexico think that events that produce changes in the hydrate mounds also trigger episodes of sea-floor instability.

Hydrates in direct contact with a relatively large volume of sea water are stable only marginally. Variations in the pressure, the distribution of temperature in the water column, the chemical composition of the gas and the rate of gas flow combine to determine whether hydrates within the mounds accumulate or dissociate. Major influences are the warm eddies of water that separate from the Loop Current and raise bottom temperatures in the Northern Gulf of Mexico by as much as 2–3°C. The result is a quasicyclicity of sea floor hydrate formation that is driven largely by these current-induced temperature variations of bottom waters. Changes in pressure, gas composition, and flow rate that can also contribute are not well understood but are probably due partially to tectonic activity associated with salt movement.

Hydrates contained in sediments are stable when the sediments are within the hydrate stability zone (HSZ) as defined by pressure, temperature, and chemical composition. If hydrocarbon gases migrating up faults encounter sediments of sufficient permeability lying within the HSZ, hydrates can form within the pore spaces and act to cement the sediment grains. This increases the sediment shear modulus and thereby its bearing capacity. The location of the lower boundary of the HSZ is determined by the geothermal gradient. As continuing sedimentation increases their depth of burial, hydrated sediments within the HSZ are subjected to successively higher temperatures and pressures until they eventually lie below the HSZ. Then the hydrates cementing them dissociate, their bearing capacity decreases, and a potential for sea floor instability is created. The same result can be produced by distortions of the geothermal gradient such as those caused by proximity to salt bodies or drilling activities. Common indicators of such instability are the speeds at which compressional (P) and shear (S) waves propagate below the sea floor and the efficiency of P-to-S conversion (PS) at reflecting horizons.

MONITORING SYSTEMS

Vertical Line Arrays

The primary sensory system of the station will be a net of vertical line arrays (VLAs) moored to the sea floor. Each VLA will consist of a number of hydrophones spaced at selected intervals. The signal from each hydrophone will be digitized and recorded individually. The set of signals will be processed by correlation and matched field processing (MFP) techniques that make use of time and amplitude information to provide estimates of:

- the distribution of temperature in the water column (by travel-time tomography),
- speeds of P-wave propagation in sea floor sediment/hydrate (by MFP), and
- three-dimensional images of geological structure beneath the sea floor (by MFP).

These estimates will be used to detect changes in the sea floor due to a triggering event and to provide an image of the subbottom geological structure after an event is detected.

Acoustic transducers will be placed at fixed locations on the sea floor so that the positions of individual hydrophones can be determined by triangulation. This will allow corrections to be made during data analysis for variations in the geometry of the VLA net.

After the station has been deployed, the site will be calibrated by determining an acoustic model of the station environment using shipboard sources fired at known locations and times. The noise of passing ships—that is, sources of opportunity—will then be tracked and employed to monitor changes to that model on a more-or-less continuous basis. The site will be recalibrated as necessary using controlled sources.

Horizontal Seismic Array

A horizontal seismic array (HSA) of four-component (4-C) sensors will be installed on the sea floor. Each 4-C sensor will consist of a hydrophone and a three-component seismometer or accelerometer. The hydrophone components of the HSA will augment the VLAs by improving the azimuthal resolution of the VLA tracking capability. The other three components will allow the identification of S waves and measurement of their amplitudes, thereby providing subbottom information not available from P waves alone. During site calibration, S-wave speeds immediately below the sea floor will be measured by recording signals from an S-wave generator towed on the sea floor and PS waves converted at deeper reflecting horizons will be generated by sources deployed on the sea surface. Between calibrations, the three-component instruments will monitor seismic activity and the noise of passing ships.

The complete set of HSA data will be useful to study the evolution in time of the gas hydrate stability zone, the underlying free gas zone associated with the hydrate mound, and the configuration of pathways through which gases and liquids migrate.

Current Measurements

An acoustic doppler current profiler will be installed to monitor water flow. Data from it will also be processed to estimate suspended particulate mass. In addition, a number of three-axis acoustic current meters will be installed near the VLA net to assist in determining the VLA receiver geometry.

Gas Bubble Observations

The sound of gas bubbles seeping from the sea floor will be recorded and analyzed. Since each bubble resonates at a characteristic frequency depending on its size and shape, it may be possible to infer the rate of seepage from their sounds. The sounds will be recorded by a broadband ambient noise measuring system placed near an area of known gas emissions. The size and distribution of gas bubbles will also be monitored using transmission and reflection of sound waves from an acoustic source.

Geoelectric Systems

The electrical resistivity of sediments in the vicinity of a gas hydrate mound is expected to be elevated due to the effects of free gas, hydrates, and particularly fresh pore waters and authigenic carbonates. During site calibration, a bottom-towed electromagnetic profiler will be used to determine the electrical resistivity in the upper

10–20 m of sediment. After calibration, the station will incorporate a number of remote probes to monitor the resistivity profile within the upper meter of sediment. These data will indicate changes in the sea floor resistivity with time and, when combined with acoustic and geochemical data, will be particularly useful for characterizing sediments near the sea floor.

Thermal Studies

Transects of heat flow measurements in the vicinity of the monitoring station will provide information about the background level of heat flow and its local variability. The interaction of temperature transients and hydrate formation/dissociation will be addressed by deploying an array of thermistors that spans a near-bottom interval from the water column into the subbottom sediments. Based on thermal conductivity/diffusivity determined by the heat probe measurements, the rate of propagation of oceanographic warming transients will provide valuable information about the dynamics of hydrate destabilization.

The subbottom thermal measurements will be accompanied by pore-fluid pressure measurements to provide information about the effects of hydrate dissociation on the physical/mechanical properties of the sediments. The effects will be described in terms of both absolute pressure (and how it changes due to a small amount of gas release) and load partitioning between sediment matrix and pore fluid during tidal cycles and meteorological events.

Optical Spectroscopy

Using a technique similar to that employed by the Mars Rover, an optical spectrometer will be used to identify and quantify hydrocarbon gases present in the sea water. Samples will be illuminated by laser light shining through an optic fiber and the back-scattered light collected by other optic fibers. Spectral analysis of the back-scattered light will provide information concerning the chemical composition of the gas in each sample.

Pore Water Chemistry

Pore water and sea water sampling will elucidate parameters affecting gas hydrate stability. Pore water analyses will address gases, broad based chemistry of major ions, and selective isotopic studies of pore water and solid phase substrates, including delta-13 carbon, radium, and radon. These will provide ground truth for acoustic data collection and the optical spectrometer observations. They will also validate and calibrate observations pertaining to sea floor stability and ecosystem health.

Underwater Vehicles

A variety of underwater vehicles will be used at various stages of the project: deep-tow and bottom-tow devices, tethered remotely operated vehicles (ROVs), and autonomous underwater vehicles (AUVs). They will carry sensors that are more effective when moved about, thus improving spacial resolution. The locations of underwater vehicle activities will be determined in relation to the long baseline navigation system.

Towed vehicles and ROVs will be deployed from surface ships during site selection and calibration. Towed vehicles will reconnoiter areas with survey profilers. ROVs will be used to investigate specific sites and to deploy and service instruments at fixed locations. Together, they will provide close-up images of the sea floor and measurements to identify hydrate outcrops, hydrocarbon seeps and chemosynthetic communities.

The AUV will be used primarily during periods when the station is not attended by surface ships. It will be guided by genetic algorithms and other software to search designated sectors for new targets. Given appropriate development, it will operate from a docking facility near the monitoring station where data can be downloaded, instructions received, and batteries recharged. The dock will be connected by optic fiber to a site, probably an oil platform, where the images and spectral data can be transmitted ashore and instructions received. Electric power for recharging batteries will be obtained from that site.

Data Recovery

The station will produce many channels of data on a more-or-less continuous basis. The total data volume will be large and its recovery is not a trivial problem. Present plans are to digitize each channel on site and transmit the digital signals via optic-fiber cable to a structure, such as an oil platform, where they can be telemetered to an onshore processing facility.

CONCLUSIONS

The concept of a monitoring station has emerged from discussions among a number of prominent scientists. A subset of that number constitute the board of scientific supervisors for the project. Members of the board (with areas of responsibility) are:

Bob Woolsey (general manager), CMRET, University of Mississippi, Oxford.

Harry Roberts (geologic framework), Coastal Studies Institute, Louisiana State University, Baton Rouge.

Ross Chapman (vertical line arrays), School of Earth and Ocean Sciences, University of Victoria, British Columbia.

Ingo Pecher (horizontal seismic array), Institute for Geophysics, University of Texas, Austin.

Rob Evans (geoelectric systems), Woods Hole Oceanographic Institution, Woods Hole.

Earl Davis (thermal studies), Geological Survey of Canada, Pacific Geoscience Centre, Sidney, British Columbia.

Ralph Goodman (gas bubble observations), Applied Research Laboratory, Pennsylvania State University, University Park.

Vernon Asper (current measurements), IMS, University of Southern Mississippi, Ocean Springs.

Jeff Chanton (pore water chemistry), Department of Oceanography, Florida State University, Tallahassee.

John Noakes (optical spectroscopy), Center for Applied Isotope Studies, University of Georgia, Athens.

Paul Higley (data recovery and technical advice), Specialty Devices Inc., Plano, Texas.

Tom McGee (geophysical site surveys), CMRET, University of Mississippi, Oxford.

Jean Whelan (geochemistry), Department of Marine Chemistry and Geochemistry, Woods Hole Oceanographic Institution, Woods Hole.

Roger Sassen (chemistry and geology), Geochemical and Environmental Research Group, Texas A & M University, College Station.

Growth Kinetics of Single Crystal sII Hydrates

Elimination of Mass and Heat Transfer Effects

P. BOLLAVARAM, S. DEVARAKONDA, M.S. SELIM, AND E.D. SLOAN, Jr.

Center for Hydrate Research, Colorado School of Mines, Department of Chemical Engineering and Petroleum Refining, Golden, Colorado 80401, USA

ABSTRACT: This work presents the results from experiments and modeling of tetrahydrofuran single crystal hydrate growth. The purpose was to study growth kinetics, independent of mass transfer and heat transfer. We used a single crystal apparatus, at stoichiometric concentrations of tetrahydrofuran and water, varying the fluid shear to decrease the boundary layer at the crystal surface. We found that with extreme precautions to totally eliminate mass transfer and to minimize heat transfer via high shear, it is very difficult to obtain reliable kinetic constants for the single hydrate crystal growth system. We eliminated mass transfer, but were only able to reduce the heat transfer resistance to a value of about 10% of the total resistance (i.e., 90% kinetic resistance) at the lowest value of subcooling. We found that growth rate increased with the driving force (i.e., subcooling) and established that the growth process occurred by a step mechanism. We only measured the fluid phases in order to obtain hydrate phase kinetics. The results of this work suggest that assessment of heat transfer, previously ignored in crystal growth kinetic studies is vital for accurate hydrate kinetics.

INTRODUCTION

Hydrate kinetics is a challenging and intriguing field. Questions relating to hydrate formation, dissociation, and kinetics of hydrate inhibition has yet to be addressed successfully. The kinetics of hydrate formation can be described as follows:

(a) formation of critically sized nuclei,

(b) growth of this critically sized nuclei to a crystal,

(c) diffusion of the components to the growing hydrate surface, and

(d) dissipation of the heat of crystal formation.

All the above steps are interrelated and a satisfactory quantitative model has yet to be found. The preliminary research on hydrate formation started in Russia.[1] In the past 10–15 years, the work by Bishnoi *et al.*, has set the path for modern kinetic studies. Englezos *et al.*,[2,3] obtained a model for single components of CH_4 and C_2H_6, and then extended the model to incorporate binary mixtures of $CH_4 + C_2H_6$. This model was the first of its kind to include both kinetic and mass transfer effects.

Skovborg and Rasmussen[4] extended the Englezos-Bishnoi model by restricting the hydrate formation process to be mass transfer limited. The Skovborg-Rasmussen model was simplified by assuming that all the hydrate particles were of the same size and have same growth rates. This assumption may not be true, as morphological

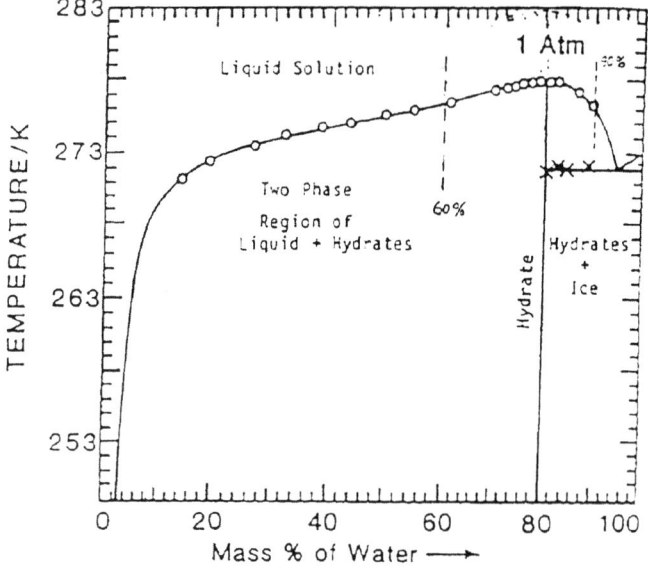

FIGURE 1. THF-water phase diagram. Dyadin et al.[8]

differences in crystals result in different growth rates. Neither the Englezos-Bishnoi nor Skovborg-Rasmussen models account for the heat transfer effects.

In this work, we have concentrated on obtaining insights of kinetics of hydrate crystal growth. FIGURE 1 shows a phase diagram of THF-water system at atmospheric pressure. We see that at 19 wt% THF equilibrium hydrate dissociation temperature is 4.4°C. Operation at this stoichiometric concentration results in the elimination of mass transfer effects. Heat transfer effects were reduced to a minimum by shearing the crystal face in order to obtain heat transfer limited growth rates.

EXPERIMENTAL PROCEDURE

A schematic of the experimental apparatus is shown in FIGURE 2. The water bath contained a round bottom flask with a U-tube connected to the side. A coolant, typically water in this case, was used to control the temperature of the water bath. The temperatures were measured by using thermocouples. The round bottom flask had a glass stirrer attached at the top that could be operated up to 5,000 rpm. The design of the stirrer was such that the diameter of the blades was almost equal to the diameter of the U-tube. The stirrer was used to circulate the fluid between the round bottom flask and the U-tube, and used to eliminate the heat transfer boundary layer across the crystal faces. A charge coupled device (CCD) was used to monitor the growth process. The CCD was attached to a video cassette recorder (VCR) to record the experiment, and used for experimental measurements.

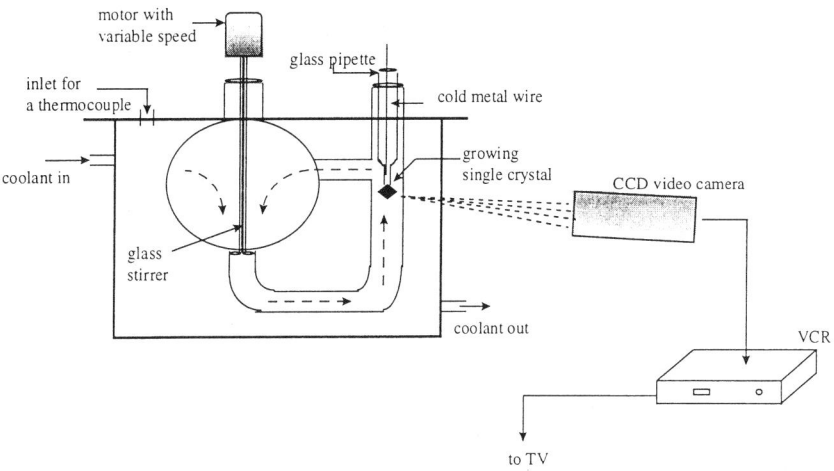

FIGURE 2. Schematic of the experimental apparatus.

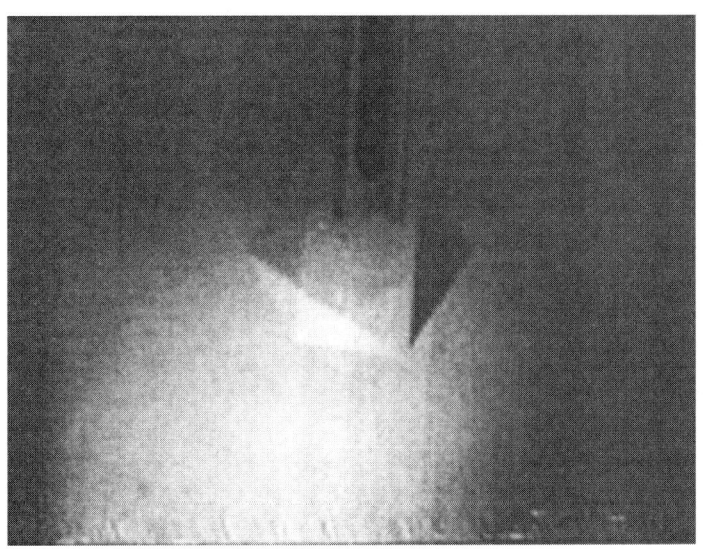

FIGURE 3. Single crystal grown at 1.5°C.

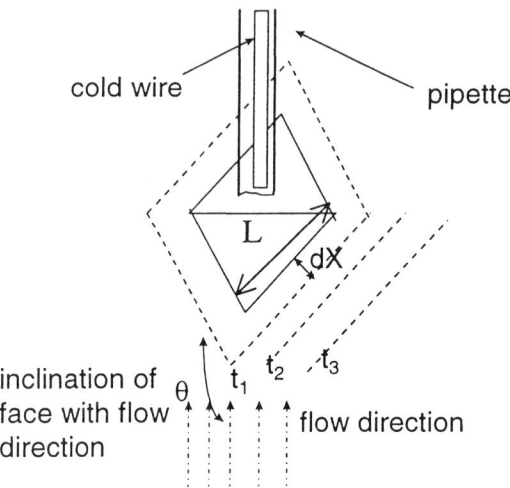

FIGURE 4. Schematic for calculating growth rates.

Single crystals were grown at the top of the U-tube with a pipette (2 mm O.D.). A cold wire was inserted in the pipette to act as a nucleation site. A single crystal grew out of the pipette into the melt. The experimental conditions were as follows:

1. atmospheric pressure,
2. fixed subcooling: defined as $T_{eq} - T_{bath}$,
3. 19 wt% stoichiometric THF–water solution,
4. equilibrium temperature $T_{eq} = 4.4°C$, and
5. fixed recirculation rate of either 300, 600, 850, or 1,200 rpm.

Single crystals thus grown in the melt were sII hydrates. The growth studies consisted of measuring the growth of a face with respect to time. FIGURE 3 shows a THF hydrate single crystal grown at a subcooling of 1.5°C. FIGURE 4 represents a schematic of the growth measurements. The figure shows movement of the face of a single crystal with time. The distance the face moves dX and length of the face L was noted with respect to time. The angle (θ) between the face made and the flow direction was measured. A plot of distance dX (μm) versus the time t (s) gives a straight line. The slope of this straight line gives the growth rate of the face.

The problems in growing single crystals from a melt were a result of the fact that it was nearly always necessary to control one or more of the following characteristics of the crystals that were to be grown: orientation, shape, and perfection. The challenges in conducting the experiments at high circulation and subcooling caused hindrance in having an insight of the growth rates and the process. We were not able to measure the growth rates at circulation rates higher than 1,200 rpm and subcooling greater than 1.5°C.

EXPERIMENTAL RESULTS AND DISCUSSION

Growth rates were obtained for fixed subcooling and circulation rates. Experiments were conducted at subcoolings of 0.4°C, 1.0°C, and 1.5°C. Growth rates at these subcooling values were conducted at different circulation rates of 300 rpm, 600 rpm, 850 rpm, and 1,200 rpm. The data at a subcooling of 0.4°C at a circulation rate of 600 rpm are shown in FIGURE 5, where distance (μm) on the ordinate and time (s) on the abscissa are plotted. Similar results were obtained for other experiments. The reader is advised to refer to the original thesis.[5] The reproducibility in the single crystal apparatus was found to be ±15%.

Growth rates of a crystal face depend on how fast the heat is removed from the surface and this depends on the rate of heat transfer at the interface. The rate of heat transfer is a function of the stirring in the system. We found that the inclination of the face to the flow direction plays a significant part by changing the flow patterns adjacent to the crystal faces. This change in flow patterns resulted in different heat transfer coefficients.

FIGURE 6 shows data for the case of subcooling at 0.4°C, with growth rates plotted on the ordinate and the angle of inclination to the flow direction on the abscissa. The data for circulation rates of 300 rpm, 600 rpm, 850 rpm, and 1,200 rpm are shown. The figure suggests that the growth rates are maximum when the angle of inclination to the face is equal to 90°, that is, when the flow is perpendicular to the crystal face—called impinging flows. The heat transfer coefficient is a function of shear stress on the crystal face and is different for each orientation of the crystal face. Shear stress is maximum for impinging flows, thus resulting in higher heat transfer coefficients, leading to higher growth rates of the crystal faces. Shear stress decreases as the inclination of the face increases; hence we should find a decrease in growth

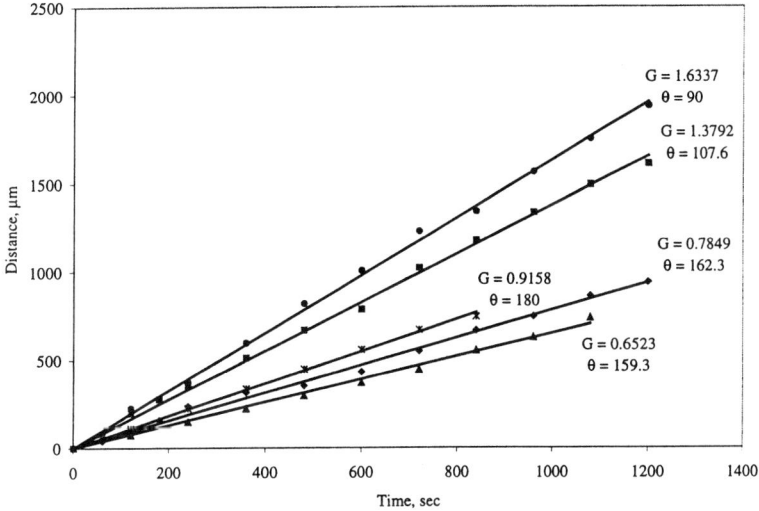

FIGURE 5. Growth rates at a subcooling of 0.4°C and circulation rate of 600 rpm.

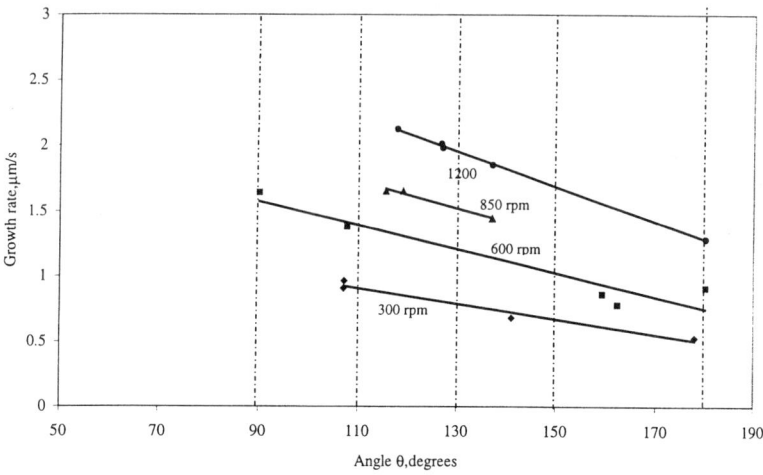

FIGURE 6. Growth rate versus angle of inclination of the face at a subcooling of 0.4°C.

rates. The same phenomenon was observed for the cases when the subcooling is equal to 1.0°C and 1.5°C for circulation rates of 600 rpm and 1,200 rpm at a subcooling of 1.0°C and 1.5°C, respectively.

Cross plotting the data of FIGURE 6 was done in order to obtain the effect of circulation rates on the growth rates at a fixed inclination of the faces to flow direction.

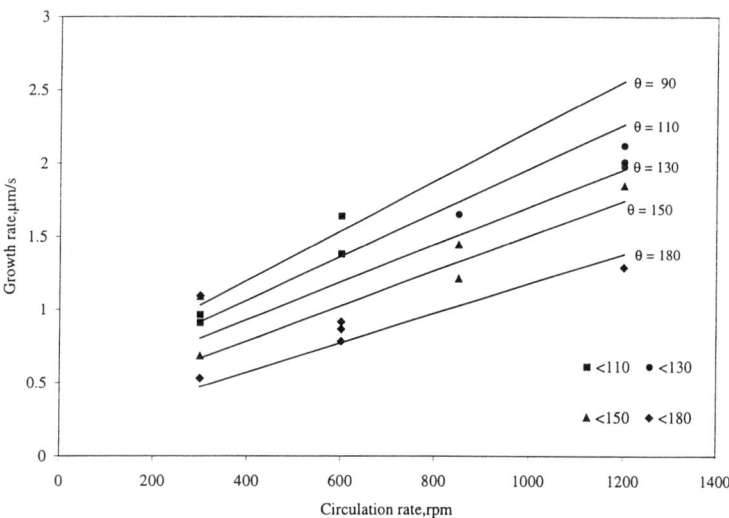

FIGURE 7. Growth rate versus circulation rates at a subcooling of 0.4°C.

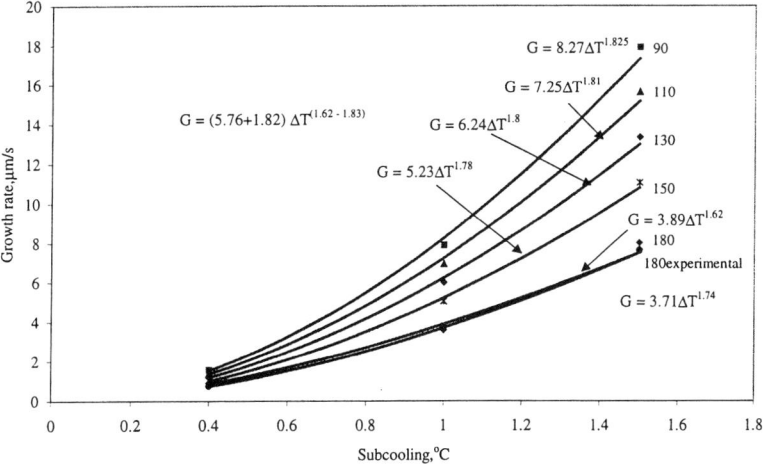

FIGURE 8. Growth rate versus subcooling at a circulation rate of 600 rpm.

FIGURE 7 shows the result of the cross plotting for a subcooling of 0.4°C. Growth rates are plotted on the ordinate and circulation rates on the abscissa. The points in FIGURE 7 represent data. It can be inferred from the figure that the growth rate increases with circulation rate for the same angle of inclination to the flow. The same trend is observed for the cases when subcooling is 1.0°C. Cross plotting at 1.5°C was obviated by a lack of data due to the experimental difficulties in growing single crystals.

We found that the growth rates increase with driving force (i.e., subcooling) in the single crystal experiments. FIGURE 8 shows a plot of growth rates as a function of subcooling at a constant stirrer speed of 600 rpm, varying the inclination of the face to flow direction. The figure shows that the growth rates increase with subcooling. As described,[6] a growth process is by step mechanism in which the growth rates are proportional to the square of the subcooling. In our case, we found that growth rates are proportional to a power of subcooling ranging from 1.62 to 1.83 (values close to 2.0). Hence, we propose that hydrate growth is controlled by step mechanism.

MODELING RESULTS AND DISCUSSION

As described, heat transfer effects play a vital role in hydrate crystallization. We have modeled a combined heat and kinetic growth process. We have modeled the single crystal growth as a moving boundary. The boundary here is defined as the hydrate interface, that is, the single crystal face. A schematic of the moving boundary is shown in FIGURE 9. The figure shows the temperature profile with distance across the crystal face. Clearly there are two phases, the solid hydrate phase and the bulk of the liquid phase. The surface temperature and bulk temperature in the liquid are represented by T_s and T_∞, respectively. $X(t)$ and $X(t + dt)$ represent the position of the moving front at times t and $t + dt$, respectively. The distance dX represents the movement of the front in time dt. The following assumptions are used in the model:

1. The surface temperature T_s is constant throughout the crystal face, that is, there are no temperature gradients on the crystal face.
2. The surface temperature T_s is assumed to be equal to the equilibrium temperature T_{eq} (= 4.4°C).
3. Temperature gradients in the bulk of the solution are neglected.

The mathematical model consisted of an energy balance at the moving boundary. During hydrate formation, the energy liberated due to the growth of the face is transferred to the bulk of the liquid. The energy balance is represented by

$$\rho_h A(dX)\lambda = KA\Delta T dt, \qquad (1)$$

where ρ_h is the hydrate density, A is the surface area of the face, λ is the latent heat of formation of hydrate, and K is the overall or the combined heat and kinetic coefficient given by

$$\frac{1}{K} = \frac{1}{k\Delta T^n} + \frac{1}{h_{avg}(t)}, \qquad (2)$$

where k represents the kinetic coefficient, h_{avg} is the average heat transfer coefficient across the crystal face and n is called as the power factor. Combining **(1)** and **(2)**, we obtain the following representation of the moving boundary model.

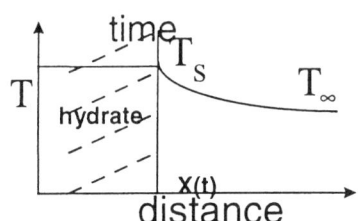

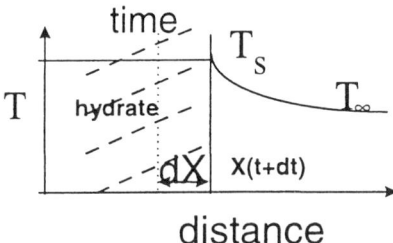

FIGURE 9. Moving boundary model for hydrate growth.

$$\rho_h \lambda \frac{dX}{dt} = \frac{\Delta T}{\frac{1}{k\Delta T^n} + \frac{1}{h_{avg}(t)}}, \tag{3}$$

where dX/dt is the growth rate of the crystal face.

Equation (3) was regressed to obtain the values of k and n. The average heat transfer coefficient was obtained by using wedge flow correlations.

The flow patterns in the system under consideration are typical for what are called *wedge-flows* and the boundary layer solutions corresponding to these potential flows are called the Falkner and Skan flows, as correlated to measurements.[7] The boundary layer solution is given by

$$\text{Nu Re}_x^{-0.5} = \text{constant}. \tag{4}$$

The constant in (4) was obtained from Reference 7. The constant is a function of Prandtl number and the angle of inclination to the flow direction. The heat transfer coefficient was obtained from (4). The values of k and n were obtained from (3) by a regression analysis. We found that the value of n varied from 0.62 to 1.0, that is, the growth rates is proportional to a power of 1.62–2.0 of the subcooling. This value agrees with experimentally obtained growth rate relationships. The effect of heat transfer and kinetics was calculated from:

$$\text{Effect of heat transfer} = \frac{1/h_{avg}}{1/K} \tag{5}$$

$$\text{Effect of kinetics} = \frac{1/k\Delta T^n}{1/K}. \tag{6}$$

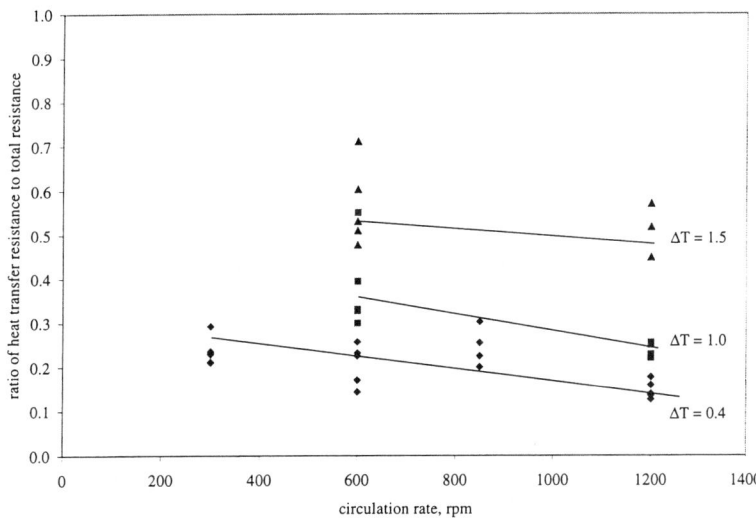

FIGURE 10. Effect of heat transfer with circulation rate.

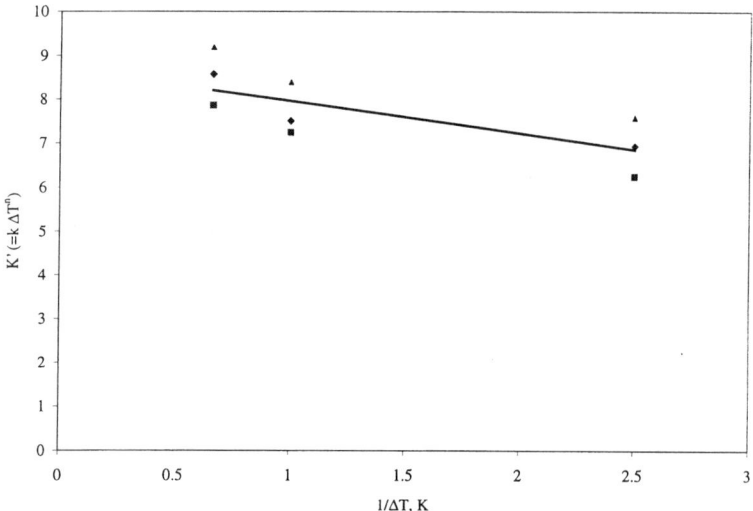

FIGURE 11. Arrhenius plot for sII hydrates.

The results from the regression are summarized in FIGURE 10. The figure represents the effect of heat transfer with changed circulation rate for the subcoolings investigated in the experiments. FIGURE 10 shows a straight line drawn roughly to indicate the relation between the heat transfer effect and circulation rate. The straight line has a negative slope, supporting the argument that increasing circulation rate reduces the heat transfer effects.

An important observation can be made with respect to subcooling. We could not reduce the heat transfer effects to the same values for all the subcooling values investigated. The reason can be attributed to the challenges of measuring crystal growth rates at higher circulation rates.

FIGURE 11 shows a Arrhenius plot for crystal growth. Kinetic constants (K') obtained from regression analysis (mean, minimum, and maximum values) are plotted against inverse of subcooling on a semilogarithmic plot. We obtained the following relationship:

$$K' = 5946.75 \exp(-0.733/\Delta T). \tag{7}$$

CONCLUSIONS

In the current work we show that, with extreme precautions to totally eliminate mass transfer and to minimize heat transfer via high shear, it is very difficult to obtain reliable kinetic constants for the single crystal hydrate growth system. The results of this work suggest that assessment of heat transfer, previously ignored in multiple crystal growth, is vital to accurate hydrate kinetics of single crystals. The situation should be exacerbated for multiple crystals.

The following summarizes our conclusions based on the experiments and modeling kinetics:

1. Mass transfer effects were eliminated by using stoichiometric solutions, although extreme measures required to reduce the heat transfer effects.

2. Growth rates strongly depend on shear across the crystal face, impinging flows result in fastest growth rates.

3. Growth rates of the crystal face increase with driving force (subcooling) in the system.

4. The growth process in sII hydrates may be controlled by a step mechanism (i.e., the hydrate growth rates are given by the expression, $G \alpha \Delta T^n$, where n varies from 1.62 to 2.0).

ACKNOWLEDGMENTS

For funding, we thank the Colorado School of Mines Hydrate Consortium: Amoco, ARCO, Chevron, Conoco, DoE, Mobil, Oryx, Phillips, Shell, and Texaco. We thank Dr. Allan Myerson from Brooklyn Polytechnic University for providing us with experimental apparatus.

REFERENCES

1. MAKOGON, Y.F. 1974. Hydrates of Natural Gas. Moscow. Nedra. Izadatelstro. 208.
2. ENGLEZOS, P. *et al.* 1988. AIChE J. **34:** 1718.
3. ENGLEZOS, P. *et al.* 1988. Fluid Phase Equil. **42:** 129.
4. SKOVBORG, P. *et al.* 1994. Chem. Eng. Sci. **49:** 1131.
5. BOLLAVARAM, P. 1999. Mesoscopic Growth Measurements of sII Hydrate Single Crystals. Master Thesis, Colorado School of Mines.
6. CHALMERS, B. 1967. Principles of Solidification. John Wiley and Sons. New York.
7. KAYS, W.M. & M.E. CRAWFORD. 1980. Convective Heat and Mass Transfer. McGraw-Hill, New York.
8. DYADIN, Y.A. *et al.* 1973. Dokl. Chem. **208:** 9.

Methane Hydrate Dissociation Rates at 0.1 MPa and Temperatures above 272 K

SUSAN CIRCONE,[a,b] LAURA A. STERN,[b] STEPHEN H. KIRBY,[b]
JOHN C. PINKSTON,[b] AND WILLIAM B. DURHAM[c]

[b]*U.S. Geological Survey, 345 Middlefield Road, MS 977,
Menlo Park, California 94025, USA*

[c]*Lawrence Livermore National Laboratory, Livermore, California 94550, USA*

ABSTRACT: We performed rapid depressurization experiments on methane hydrate under isothermal conditions above 272 K to determine the amount and rate of methane evolution. Sample temperatures rapidly drop below 273 K and stabilize near 272.5 K during dissociation. This thermal anomaly and the persistence of methane hydrate are consistent with the reported recovery of partially dissociated methane hydrate from ocean drilling cores.

INTRODUCTION

Naturally occurring sI methane hydrate has been recovered from numerous ocean drill cores in a partially dissociated state. Retrieved cores can contain several phases: hydrate, fine-grained sediment with altered textures, saline pore water that has been freshened by hydrate dissociation, methane gas in excess of *in situ* saturation levels, and occasionally ice. Drill core temperatures measured after recovery are depressed several degrees below the baseline core temperature in zones observed or inferred to contain gas hydrate.[1–3] Temperatures can be depressed as much as 2 K below the H_2O melting point at 273.15 K and are lower than any encountered during core recovery.[1,2] Such thermal anomalies have been attributed to the endothermic enthalpy of hydrate dissociation. An experimental study on synthetic methane hydrate dissociation at temperatures above 272 K may provide insights into these observations.

Moreover, temporary preservation of hydrate has been observed below 273 K and at 0.1 MPa, first noted in sI methane deuteriohydrate[4] and later in sI methane hydrate at high levels (more than 50% of the sample),[5,6] as well as in sII natural gas hydrates.[7] In experiments performed between 204 and 270 K,[6] we observe highly suppressed dissociation rates between 250 and 270 K, and 50–90% of the hydrate remains preserved after 24 hours, based on CH_4 evolution and X-ray diffraction analysis of samples. In contrast, below 250 K the time to 50% dissociation is several orders of magnitude smaller. We were interested in investigating methane hydrate dissociation kinetics at temperatures above 270 K following rapid depressurization to 0.1 MPa to better define the regime of anomalous preservation observed at lower temperatures.

[a]Telecommunication. Voice: 650-329-5674; fax: 650-329-5163.
scircone@isdmnl.wr.usgs.gov

The purpose of our study was fourfold: (1) to determine the dissociation rates of methane hydrate above 270 K, (2) to profile sample temperature during dissociation, (3) to identify factors that control dissociation rates in this temperature regime, and (4) to begin a preliminary investigation of the effects of natural impurities (sediment and sea water) on hydrate dissociation.

EXPERIMENTAL METHOD

Hydrate Synthesis

Polycrystalline methane hydrate, with expected stoichiometry $CH_4 \cdot 6.1\ H_2O$, was synthesized from granular water ice and pressurized CH_4 using the method of Stern et al.[8] Samples had at most trace amounts of ice remaining after synthesis and a porosity of about 30%. The 30 g hydrate samples were cylinders, 2.54 cm in diameter and approximately 9.3 cm in length. The sample axis was held vertical during synthesis and dissociation (see FIGURE 1).

Modifications were made to the hydrate synthesis procedure for several samples. A hydrate/sediment aggregate was made by mixing 150 ± 50 μm quartz sand with the ice (volume ratio 3:7) prior to synthesis. After synthesis, methane pressure (P_{CH_4}) was reduced to 4–5 MPa and temperature was held near 275 K. Sea water (SMOW) pressurized with CH_4 to above 10 MPa was introduced through a port above the sample and partially permeated the samples by gravity. The samples were annealed at 276 ± 1 K, 10 MPa for three days prior to dissociation. Additional hydrate growth did occur based on analysis of identically treated samples.

Dissociation Procedure

(1) Maintaining high P_{CH_4}, each sample was equilibrated at a temperature between 272 and 289 K by immersing the pressure vessel in a heated ethanol bath inside a freezer (FIG. 1). (2) P_{CH_4} was reduced over several minutes to at least twice the equilibrium pressure. (3) Sample vessels were rapidly depressurized to 0.1 MPa in about 15 s and then opened to the flow meter. This depressurization rate is an upper limit of the pressure–time pathway encountered during core retrieval. (4) Internal sample temperatures were monitored with one to four thermocouples, and CH_4 evolution was monitored with a custom-built flow meter (see FIG. 1). The external bath temperature (T_{ext}) was held constant during dissociation.

Temperature Measurement

Both T_{ext} and internal sample temperatures were monitored during dissociation (FIG. 1, left). T_{ext} was measured with an RTD located near the vessel. Sample temperatures were monitored using one of two setups. In setup #1, three thermocouples were located along the cylinder axis at the sample top (8.9 cm from sample bottom), middle (4.7 cm), and bottom (0.6 cm). The side thermocouple (4.8 cm) is a few mm in from the sample surface. In setup #2, two thermocouples were located along the cylinder axis, one about 0.5 cm into the sample top and the other 3.5 cm up from the

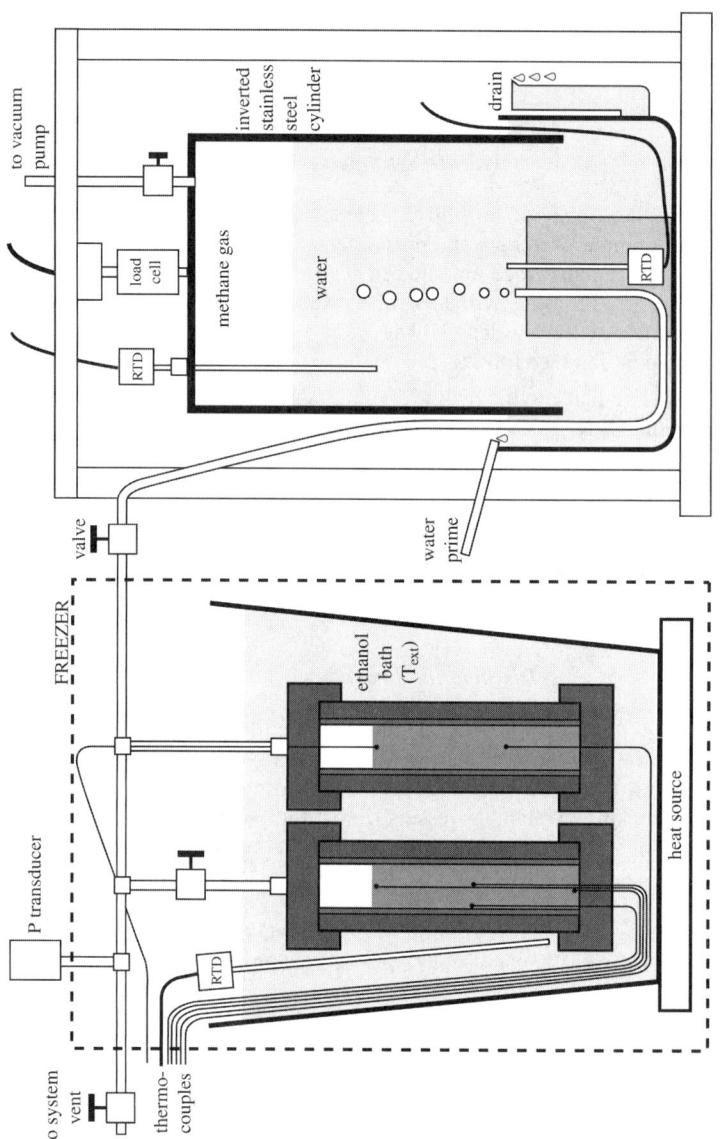

FIGURE 1. Schematic diagram (not to scale) of experimental apparatus showing sample thermocouple configurations used to monitor thermal evolution (*left*) and the custom-built flow meter that measures the amount of evolved CH_4 (*right*) during dissociation. Thermocouples, RTDs, load cell, and pressure transducer are interfaced with a data acquisition system.

bottom. The K-type thermocouples measured within ±0.1 K of 273.15 K in an ice/water calibration bath.

Custom-Built Flow Meter

The flow meter is based on the principles of the Torricelli tube and determines gas flow rate by monitoring the weight change of an inverted, water-filled cylinder as CH_4 displaces water (FIG. 1, right). The water column is drawn into the inverted cylinder under vacuum, then primed with flowing CH_4 to saturate the water and to minimize solution during measurement. As CH_4 displaces water in the cylinder, the external water level is maintained constant by a water priming and drain system. Water temperature and the partial pressure of water vapor in CH_4 are incorporated into the data analysis procedure. Because our laboratory site is near sea level, we assume that atmospheric pressure is constant at 0.1 MPa. The number of moles of CH_4 is calculated using the ideal gas law, which accurately predicts relative changes over the small pressure range of operation (based on comparison with a nonideal equation of state for CH_4). A 30 g hydrate sample yields about 6 L of gas. Flow meter volume capacity is about 8 L, and we measured flow rates ranging from 0.01 to 3,000 ml/min.

TABLE 1. Summary of rapid depressurization experiments above 272 K

Sample ID	Hydrate mass (g)[b]	T.C. Setup	T_{init} (K)	P_{init} (MPa)	ΔP rate (sec)	ΔT_{init}[a] (K)	Time, 100% reaction (min)	Yield (%)
C126	29.80	#1	272.8	5.8	14	−3.9	>605	101.7
C83	29.80	#2	273.6	9.3	16	−1.7	183	122.3[c]
C118	29.80	#1	273.8	6.1	12	−4.2	147	103.0
C77	30.31	#2[d]	276.4	9.3	12	−4.1	72	100.5
C114	29.80	#1	276.5	7.9	18	−5.5	85	96.0
C122	29.80	#1	277.1	8.5	12	−5.5	86	95.6
C129	29.80	#1	277.3	9.5	22	−8.9	35	125.3[e]
C130	26.93	#2[f]	277.3	7.6	35	−13.8	18	63.9[g]
C78	30.37	#2[f]	279.1	9.5	12	−7.0	48	101.3
C116	29.80	#1	281.4	10.1	16	−10.0	50	86.0
C81	29.80	#2	283.8	15.6	19	−12.1	46	96.8
C86	29.80	#2	288.8	26.9	17	−16.4	17	119.3[c]

[a]Initial sample temperature change, due to adiabatic cooling from CH_4 expansion and the onset of hydrate dissociation.
[b]Calculated from seed ice weight, assuming 100% conversion to $CH_4 \cdot 6.1 H_2O$.
[c]High yield probably due to excess pore gas remaining in sample after depressurization step.
[d]Only one thermocouple, in top position.
[e]Hydrate sample contained added sea water. High yield due to hydrate growth from sea water.
[f]Only one thermocouple, in middle position.
[g]Hydrate/sediment aggregate with sea water added. Low total is due to the longer depressurization step, in which the vented gas included that from hydrate dissociation.

RESULTS

Rates of Dissociation

A summary of the rapid depressurization experiments above 272 K is listed in TABLE 1. Methane gas yields are typically within ±5% of the expected values (based on sample mass and the assumed stoichiometry). This is acceptable given the possible presence of trace amounts of unreacted ice, the uncertainty in expected n (6.1 ± 0.1), and the possibility that pressurized gas remained trapped in pores (high yield) and/or that partial dissociation occurred during the depressurization step (low yield). Breakdown of methane hydrate commences immediately after depressurization, and CH_4 evolution is closely linked to the thermal history of the sample (see FIGURE 2). The rate of hydrate dissociation increases systematically with increasing T_{ext} (see FIGURE 3). At 272.8 K, the experiment reached 80% reaction after 10 hours. The sample required heating through 273 K to achieve 100% reaction in a reasonable time. Dissociation is complete after three hours at 273.6 to 273.8 K and only 18 minutes at 288.8 K. Above 273 K, the reaction time decreases exponentially with increasing T_{ext} (see FIGURE 4). The significantly longer reaction time at 272.8 K links the anomalous preservation region at 250–270 K[6] to the T_{ext} > 273 K region explored here.

Reconnaissance experiments (TABLE 1) show that introducing impurities significantly increases the methane hydrate dissociation rate at about 277 K (FIG. 4). Rates increase in the order: porous hydrate < hydrate + sea water < hydrate + quartz sand + sea water, such that the time to 80% dissociation is four and thirteen times shorter, respectively. This trend is seen in a similar suite of experiments at 268 K (unpublished data). Higher dissociation rates drive the system as low as 263 K in the sediment/sea water sample, thermal buffering at 272.5 K (see below) is not observed, and H_2O melting is depressed about 2 K due to the presence of sea water.

Thermal Buffering

Two-stage thermal buffering is observed in the experiments above 272 K, first at 272.5 K as methane hydrate dissociates and then at 273.15 K as ice reaction product melts (see FIGURE 5). After plummeting below 273 K on depressurization, temperatures partially rebound and are buffered at 272.5 ± 0.2 K within minutes, regardless of T_{ext}. When T_{ext} < 274 K, all thermocouples record thermal buffering, although the effect later disappears along the sample surface while continuing in the interior. The effect is observed only by top and middle thermocouples for T_{ext} > 274 K. Side and bottom thermocouples show that temperatures plunge below 273 K after depressurization, then climb steadily back through 273 K. The buffering temperature is maintained until dissociation nears completion, is reproducible and distinct from the buffering at the H_2O melting point, and is observed in other hydrate dissociation experiments (heating from low T through 273 K at 0.1 MPa, unpublished data). The buffering temperature is not constant, increasing very slowly over time from 272.3 to 272.7 K during dissociation. Peters et al.,[9] observe similarly depressed temperatures between 272 and 273 K in their experiments. Thermal buffering at the ice point may continue after dissociation is complete, indicating the presence of an H_2O ice/water mixture. After dissociation ceased, vessels were opened in

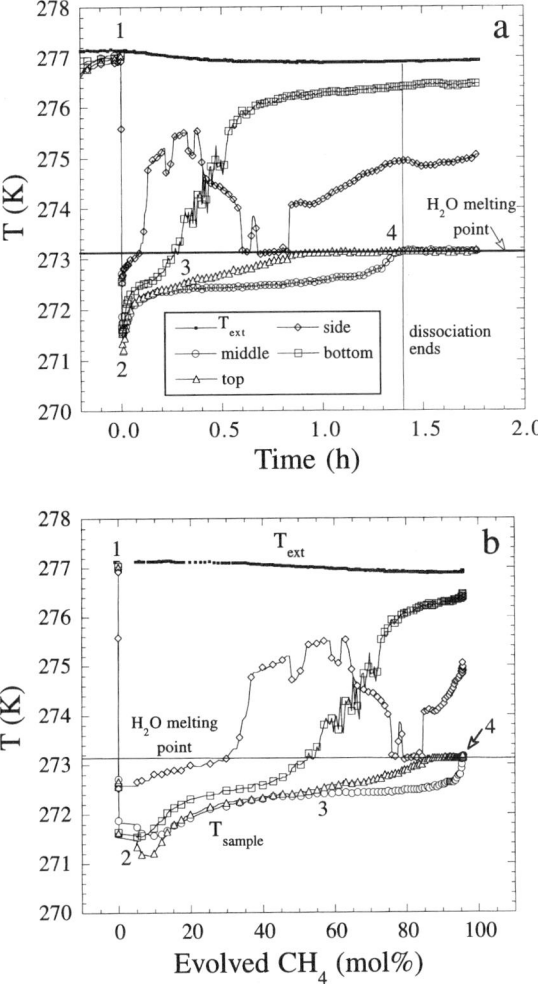

FIGURE 2. Thermal history (**a**) and CH_4 evolution (**b**) of a dissociating methane hydrate sample at 277 K (Run C122, see TABLE 1). T_{ext} remains approximately constant. *Point 1*: at time = 0, internal sample temperature and T_{ext} are nearly isothermal. Pressure is then dropped to 0.1 MPa (see text). *Point 2*: sample temperatures plunge within seconds to a minimum below 273 K. As dissociation proceeds, the sample temperature is buffered at about 272.5 K (top and middle thermocouples). Temperature evolves continually at the bottom and side thermocouple locations as CH_4 evolves and ice (reaction product) melts. *Point 3*: temperatures at the top and middle begin to diverge as dissociation nears completion at the sample top. *Point 4*: as dissociation ends, the sample interior warms above 272.5 K. In the four-thermocouple experiments, the top and middle thermocouples are buffered at the ice point of 273.15 K until the ice melts. Only every fourth point has been plotted for symbol clarity.

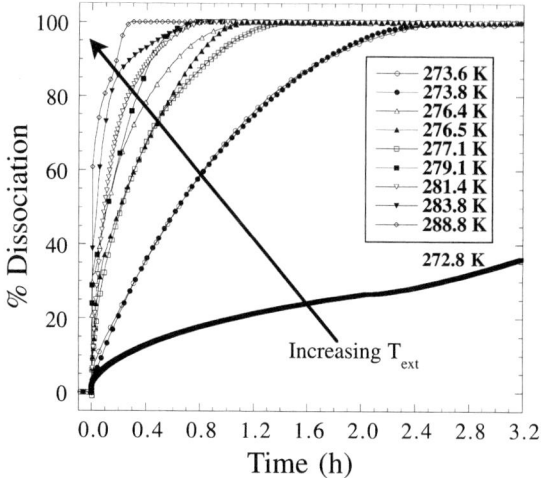

FIGURE 3. Methane evolution with time for all experiments, showing that dissociation rates systematically increase with increasing T_{ext}. Percent dissociation (the extent of reaction) is the measured amount of CH_4 normalized to the expected amount. Note the reproducibility between the two experiments conducted at 273.7 ± 0.1 K. Every fifth point has been plotted.

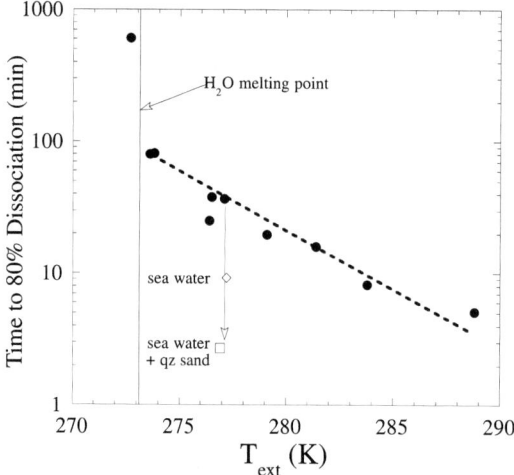

FIGURE 4. The exponential effect of T_{ext} on the time elapsed for 80% dissociation of the hydrate to occur. The scatter in the data is likely due to minor differences in the depressurization procedure. In the early experiments, depressurization was more rapid (under 10 sec) and involved a larger pressure drop (more than twice the equilibrium pressure). This difference appears to produce dissociation rates that are initially slightly higher than those in our later experiments. Curves for 90, 95, and 100% dissociation (not shown) are shifted up in time but have similar slopes. Preliminary results for hydrate + sea water and hydrate/sediment aggregate + sea water samples at about 277 K show the compounding effect of adding impurities (indicated by the *arrow*).

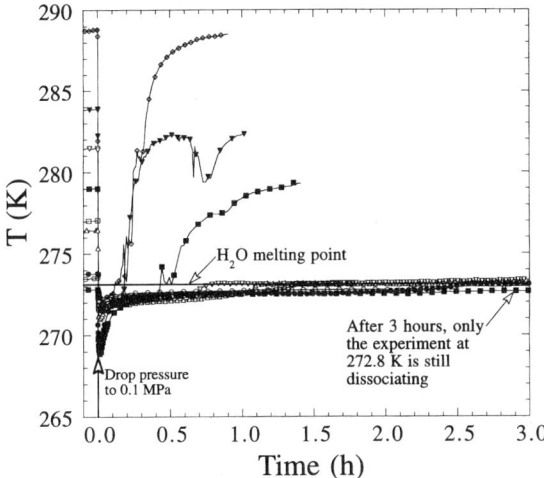

FIGURE 5. Thermal histories recorded by the middle thermocouples, showing the two-stage thermal buffering at 272.5 ± 0.2 K and 273.15 K (see text). Top thermocouples (not shown) record a similar history. The initial sample temperature is seen to the left of time 0, and is isothermal with T_{ext} at the start of rapid depressurization. The minimum T depends on the depressurization procedure and initial dissociation rates. After the material surrounding the thermocouple has dissociated and the ice product has melted, T rises toward T_{ext}. Every fifth point has been plotted, and symbols are equivalent to those in FIGURE 3.

experiments C114, C116, and C118, revealing a plug of wet, granular ice around the center thermocouples and surrounded by a gas gap. Liquid water had pooled at the vessel bottom.

DISCUSSION

Dissociation Rates and Thermal Buffering

Temperatures in some hydrate-bearing drill cores are not depressed below 273 K because the mass of dissociating hydrate is insufficient to depress temperatures to this extent in the surrounding sediments and pore fluids. In our experiments on methane hydrate, the temperature drops below 273 K when the pressure is reduced to 0.1 MPa, regardless of T_{ext}, and is self-maintained at not more than 272.7 K until the dissociation reaction nears completion (above 95%). This thermal buffering effect occurs in actively dissociating parts of the samples. When $T_{ext} > 274$ K, neither buffering at this temperature nor at the ice point occurs along the side or bottom surfaces. The sample dissociates and the ice product melts preferentially and quickly along these surfaces. Dissociation rates are increased significantly by impurities present in natural systems. The effect of sea water may be extreme in our experiments because of the high sample porosity and the high thermal conductivity of sea water relative to CH_4. A detailed study of the effects of impurities is in progress.

A key question arises: what causes the thermal buffering near 272.5 K? An apparent buffering effect could occur if the reaction rate and hence heat consumption rate were balanced by the heat flux rate from the surroundings, holding the sample temperatures within a narrow interval. However, the heat flux and buffering temperature should vary with T_{ext}, and we observe a narrow range of temperatures. It is more likely a thermal property of a reaction, like that for a phase change, but what that reaction might be is unclear. No methane hydrate phase boundary, including the metastable extension of hydrate $\rightarrow CH_4 + H_2O(l)$, corresponds to the buffering temperature at 0.1 MPa. The temperature is close to the melting point of H_2O, a correspondence that is probably not coincidental. Buffering is not observed in the isothermal portions of the rapid depressurization experiments below 272 K,[6] where experimental conditions are farther from the H_2O melting curve. However, we are not observing a simple case of H_2O melting or fusion: the buffering temperature is markedly lower than the ice point (FIG. 5), which is recorded after dissociation ceases at the thermocouples. Clearly, the source of the narrow buffering near 272.5 K is an interesting question, but one that cannot be resolved with the data from this study.

Although the values of the isothermal T_{ext} span 16 K, all of the reactions proceed at the same temperature after a few minutes of dissociation. However, the rate of dissociation increases monotonically with T_{ext}. We infer that the rate is a function of the heat flow supplied from the surroundings (bath) into the sample through the vessel. This influx supplies heat for the endothermic hydrate dissociation reaction. In the next section, we compare the cumulative heat budget for dissociation and the calculated heat flow into the system.

Heat Budget

Using the measured extent of reaction and the system temperature evolution over time, we calculated the cumulative heat budget for the dissociation reaction and the expected heat flow from the bath into the vessel interior. We considered the simplest problem of steady-state heat flow, $\Delta Q/\Delta t$, across a material (304 stainless steel) with thermal conductivity k (0.3 J/sec·K·cm[10]), cross section A (area of vessel surface), thickness Δx, and temperature difference ΔT (sample temperature − T_{ext}), where $\Delta Q/\Delta t = -kA(\Delta T/\Delta x)$. We assumed that the vessel wall is continuous and that the indium jacket lining the sample has no significant effect. We ignored vessel wall curvature, using the inner and outer wall surface areas as boundary conditions. We considered only radial heat flow through the vessel sides or axial heat flow through the vessel bottom (the top is insulated by a layer of CH_4). The heat budget for methane hydrate dissociation was calculated using the starting amount of hydrate, the enthalpy of dissociation (18.13 kJ/mol for $CH_4 \cdot 6H_2O$[11]), and the reaction progress with time (FIG. 3).

The calculated total radial heat flow (vessel sides) by the end of hydrate dissociation exceeds that needed to dissociate the hydrate by more than 50-fold (see FIGURE 6). There would be *ample* heat to also melt 100% of the ice product by the end of dissociation, but this is not observed experimentally. Axial heat flow (vessel bottom) is comparable to that needed to dissociate hydrate on the reaction time scale (FIG. 6). At 273.8 K, the calculated heat flow is insufficient for complete dissociation. In higher temperature experiments, heat flow lags behind heat consumption for

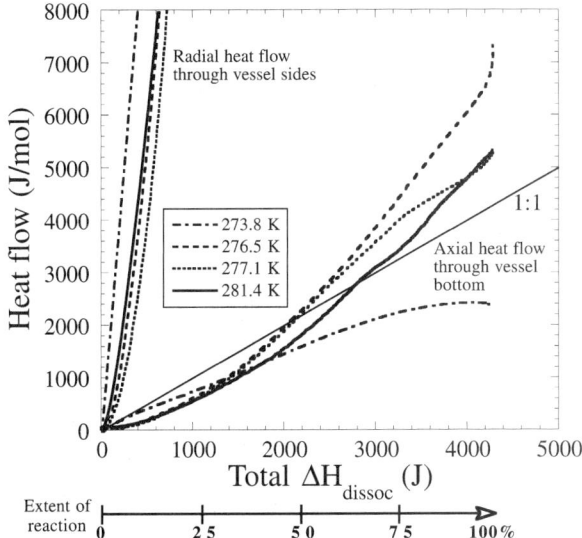

FIGURE 6. Comparison of the cumulative heat needed to dissociate hydrate near 272.5 K and the calculated heat flow through the vessel from the surroundings (ethanol bath) into the samples. Radial heat flow moves heat too quickly into the sample: dissociation should be finished in minutes and all ice should be melted at the end of the experiments, both predictions inconsistent with observations (see text). Axial heat flow is more consistent with observed dissociation rates (see text). Note that curves were placed at the lower bound (heat flow based on the inner stainless steel surface area) for the radial heat flow results; heat flow would be slightly higher for a true cylindrical conduction model. The samples are believed to be largely insulated from radial heat flow by a gas layer (see text and FIG. 7c).

about 50% of the reaction. At the end of dissociation, however, the available excess heat is sufficient to melt 10–30% of the ice product, consistent with observed vessel contents.

Of course, both radial and axial heat flow occurred in the experiments. The calculations can be reconciled by using the thermal evolution data (see FIGURE 7). In warmer tests, sample temperatures near the cylinder surface exceeded 273 K in minutes. Thus, this zone was hydrate- and ice-free, leaving only CH_4 and H_2O (l) (FIG. 7c). The water segregated by gravity, leaving CH_4 along the sides. Methane gas has a thermal conductivity 0.1% that of stainless steel.[10] As the gap between the steel and sample exceeds 13 µm, CH_4 becomes an insulating layer and greatly reduces the radial heat transfer. The pooling water is a better conductor (2% that of stainless steel[10]), although heat transfer will be affected when the water depth exceeds 0.5 mm. In the 273.8 K experiment, radial heat flow occurred over a longer time interval, since the side and bottom thermocouples indicated that hydrate and/or ice was present along the sample perimeter throughout the reaction (FIG. 7b). These observations are broadly consistent with the independent results of Peters et al.[9]

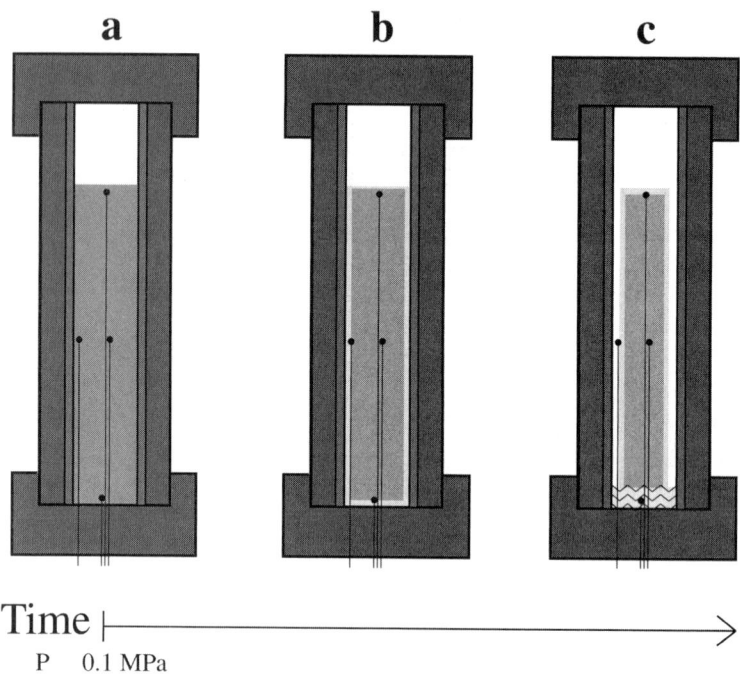

Time ⊢─────────────────────→
P   0.1 MPa

FIGURE 7. Schematic diagram of sample assembly, phases, and their distribution as dissociation proceeds: stainless steel synthesis/reaction vessels (*dark gray*), sample (□, CH_4; ■, hydrate; □, H_2O [s→l]; ▨, H_2O [l]), and thermocouples (positions shown). Diagrams are approximately to scale. (**a**) Prior to depressurization: $T_{hydrate} = T_{ext}$, porous hydrate + CH_4. (**b**) After depressurization: $T < 273$ K, hydrate + CH_4 + H_2O (s→l). Dissociation is greater along sample sides and bottom, and ice product may begin to melt. (**c**) Several minutes after depressurization ($T_{ext} > 274$ K): $T_{side, bottom} > 273$ K, $T_{top, middle} \approx 272.5$ K, hydrate + CH_4 + H_2O (s→l). An insulating CH_4 gap has formed and H_2O (l) pools at the vessel bottom (see text). A similar configuration is observed after dissociation ceases, when only melting ice remains around the center thermocouples (see text).

Their vessel axis was horizontal, and they observed preferential ice melting and water pooling along the bottom surface, clearly indicating that the direction of maximum heat flow was perpendicular to the vessel axis over a narrow, lower sector of the vessel circumference for the latter part of the experiment.

CONCLUSIONS

Temperature depression below 273 K has been observed in hydrate-bearing ocean drilling cores, suggesting that the process observed in our experiments is similar to that occurring in the field. A significant finding of our study is that at temperatures relative to natural settings, hydrate dissociation is a thermal process that is largely dependent on the heat flow into the system. Our results may help improve procedures for natural hydrate sample recovery. For instance, core retrieval systems could reduce heat flow into the drill core by using insulating gas gaps or by segregating the more thermally conductive fluid phases (pore water and/or the water dissociation product) from the hydrate. These issues will be pursued in a future paper.

ACKNOWLEDGMENTS

We gratefully acknowledge support by a grant from the LDRD Program, Lawrence Livermore National Laboratory, Livermore, CA. Work was performed under the auspices of the U.S. Department of Energy by the Lawrence Livermore National Laboratory under contract W-7405-ENG-48.

REFERENCES

1. WESTBROOK, G.K. *et al.* 1994. Proc. ODP, Init. Rep. **146**: 611.
2. KASTNER, M. *et al.* 1995. Relation between pore fluid chemistry and gas hydrates associated with bottom-simulating reflectors at the Cascadia Margin, sites 889 and 892. Proc. ODP, Sci. Res. **146**: 175–187.
3. PAULL, C.K. *et al.* 1996. Proc. ODP, Init. Rep. **164**: 623.
4. DAVIDSON, D.W. *et al.* 1986. Laboratory analysis of a naturally occurring gas hydrate from sediment of the Gulf of Mexico. Geochim. Cosmochim. Acta **50**: 619–623.
5. YAKUSHEV, V.S. & V.A. ISTOMIN. 1992. Gas-hydrates self-preservation effect. *In* Physics and Chemistry of Ice. N. Maeno & T. Hondoh, Eds.: 136–140. Hokkaido University Press, Sapporo.
6. STERN, L.A. *et al.* 1998. Thermal decomposition of methane hydrate at 0.1 MPa: short-term preservation by rapid depressurization. EOS Trans. AGU **79**: 462.
7. GUDMUNDSSON, J.S. *et al.* 1994. Storing natural gas as frozen hydrate. SPE Prod. and Fac. 69–73.
8. STERN, L.A. *et al.* 1996. Peculiarities of methane clathrate hydrate formation and solid-state deformation, including possible superheating of water ice. Science **273**: 1843–1848.
9. PETERS, D.J. *et al.* 2000. Hydrate dissociation in pipelines by two-sided depressurization: experiment and model. Ann. N.Y. Acad. Sci. **912**: this volume.
10. WEAST, R.C. 1985. CRC Handbook of Chemistry and Physics. CRC Press, Inc., Boca Raton.
11. HANDA, Y.P. 1986. Compositions, enthalpies of dissociation, and heat capacities in the range 85 to 270 K for clathrate hydrates of methane, ethane, and propane, and enthalpy of dissociation of isobutane hydrate, as determined by a heat-flow calorimeter. J. Chem. Thermodyn. **18**: 915–921.

Determination of the Intrinsic Rate of Gas Hydrate Decomposition Using Particle Size Analysis

MATTHEW CLARKE AND P. R. BISHNOI[a]

Department of Chemical & Petroleum Engineering, University of Calgary, 2500 University Drive NW, Calgary, Alberta, Canada T2N 1N4

ABSTRACT: An experimental technique to determine the intrinsic kinetics of gas hydrate decomposition is presented and a new mathematical model is developed. The experimental technique uses an on-line particle size analyzer. The new mathematical model accounts for the distribution of particle sizes in the hydrate phase when determining the intrinsic rate constant of decomposition. Results from an experiment for the decomposition of ethane gas hydrates are presented.

INTRODUCTION

The kinetics of gas hydrate formation have been studied for a few systems.[1-7] However, the decomposition of gas hydrates has only been examined by a handful of researchers. Ullerich *et al.*,[8] described the decomposition of a synthetic core of methane hydrate as a moving boundary heat transfer problem. Kim *et al.*,[9] developed a model for the intrinsic rate of decomposition and determined the rate constant from experimental data for methane. Their experimental work consisted of measuring the amount of methane collected over time by decomposing the hydrate slurry in water and maintaining a constant temperature and pressure in the reactor. In order to determine the intrinsic rate constant of their model, it was assumed that all particles had the same diameter before decomposition. Jamaluddin *et al.*,[10] modelled the decomposition of a core of methane gas hydrate by coupling the intrinsic kinetics with the rate of heat transfer. From their simulations, they found that by changing the system pressure, it was possible to move from a heat transfer controlled regime to a regime where both heat transfer and kinetics have a significant effect on the global rate of decomposition.

In the present work, the experimental setup of Kim *et al.*,[9] is modified to include a laser particle size analyzer in order to determine the initial particle size distribution. The population balance is incorporated in the model of Kim *et al.*,[9] to account for the distribution of particle sizes in the hydrate phase.

[a]Address for correspondence. bishnoi@ucalgary.ca

APPARATUS AND PROCEDURE

FIGURE 1 shows a simplified schematic of the high-pressure experimental apparatus. It is the same as that used by Kim et al.,[9] except that an in-line particle size analyzer has been added.

A Galai 2010 laser particle analyzer is used to measure the particle size distribution. The hydrate/water slurry is circulated through the high-pressure microflow cell, which was originally designed and used by Nielsen et al.,[11] to study Asphaltene precipitation. The entire particle size analyzer circuit is chilled by circulating glycol from the reactor bath through an annular tube cooling arrangement.

The procedure for a typical experiment is similar to that followed by Kim et al.[9] The major change in the procedure is the particle analyzer. The reactor is charged with 310 mL of water and the hydrates are formed in the reactor at a pressure above the three-phase equilibrium pressure. This corresponds to point A in FIGURE 2. Once the formation has proceeded for a suitable length of time, the gas flow supply to the reactor is terminated. The reactor is then depressurized to a pressure of about 0.5 bar above the three-phase equilibrium pressure, point B in FIGURE 2. This is accomplished by bleeding the gas to the collection reservoir R3, thus keeping the system closed during the experiment. At this point, the particle size distribution of the circulating slurry is measured.

The decomposition of hydrates is then initiated by further depressurizing the reactor to point C in FIGURE 2. The gas produced due to the decomposition is collected in reservoir R3. During the entire experiment, the mass is well stirred, at

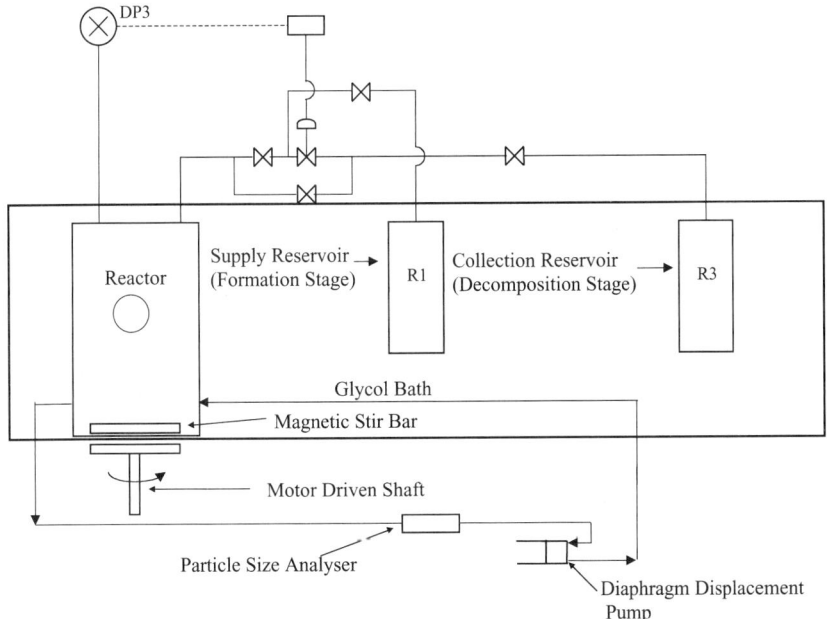

FIGURE 1. Schematic of experimental apparatus.

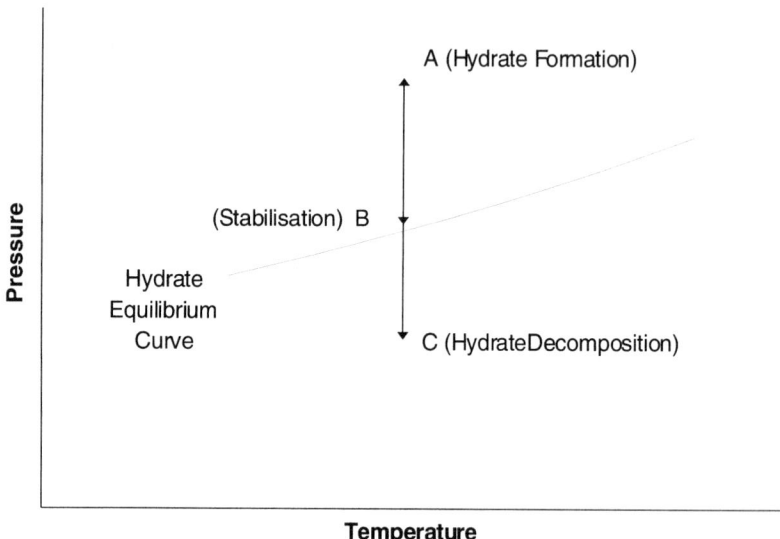

FIGURE 2. Experimental process path.

400 rpm. At any time during the experiment, the number of moles remaining in the hydrate phase is

$$n_H = n_0 - m(t) + m(0). \tag{1}$$

RESULTS AND DISCUSSION

FIGURE 3 shows the results for a typical experimental run. Prior to $t = 1,500$ s, ethane is being absorbed into the water. At $t = 1,500$ s, point 1 in FIGURE 3, there is an abrupt change in the slope of the curve. This corresponds to the onset of turbidity, resulting from the appearance of hydrate nuclei.

After hydrate formation has proceeded for a suitable period of time, the formation process is terminated by turning off the gas supply to the reactor. At time $t = 3000$ s, point 2 in FIGURE 3, the pressure in the reactor is reduced to a pressure that is approximately 0.5 bar above equilibrium and the solution is agitated for a sufficient time to allow the system to stabilize. This corresponds to point B in FIGURE 2. At this point, the particle size analyzer is used to determine the initial particle size distribution of the circulating hydrate slurry. FIGURE 4 shows the measured particle size histogram.

At $t = 3,500$ s, point 3 in FIGURE 3, the pressure in the reactor is further reduced to a pressure below the three-phase equilibrium pressure to initiate the decomposition of hydrates. The data from the decomposition section of the experiment is used to calculate the intrinsic rate constant, K_d—given below in Equation (**2**). Currently, additional experiments are in progress to collect extensive data on ethane to obtain a reliable value for the intrinsic rate constant, K_d.

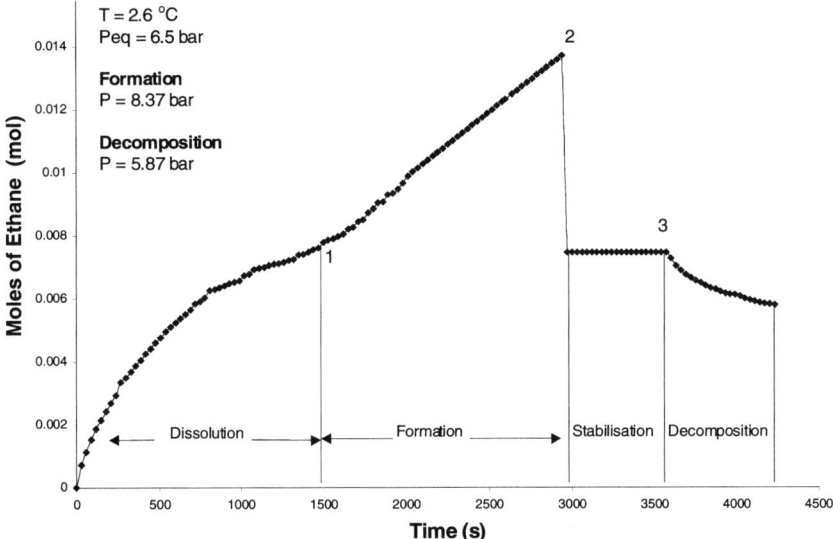

FIGURE 3. Moles of ethane consumed during formation and remaining in the hydrate phase during decomposition.

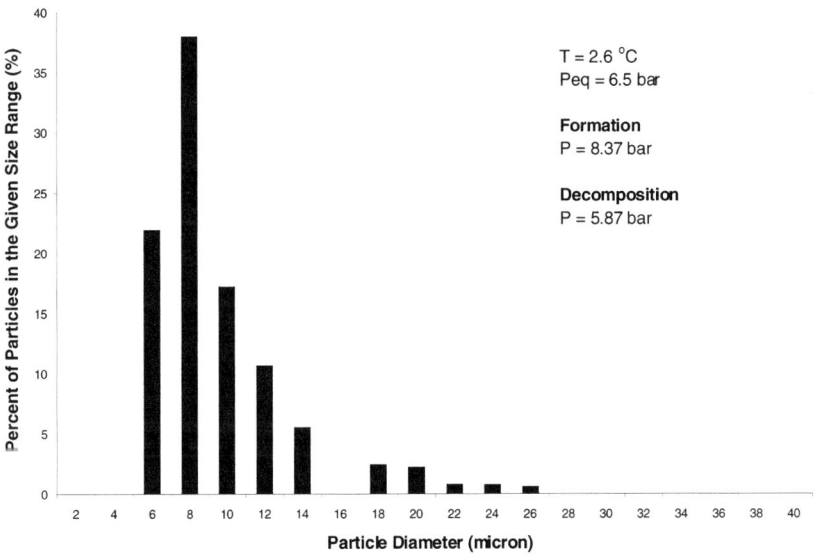

FIGURE 4. Particle size histogram for ethane hydrates.

Modelling the Data

FIGURE 5 shows the proposed mechanism of a decomposing hydrate particle. A cloud of gas surrounds the decomposing hydrate. Since the driving force for decomposition is small, it is assumed that the particle surface temperature is the same as the measured bulk temperature. Kim et al.,[9] wrote the rate of hydrate decomposition as

$$-\left(\frac{dn_H}{dt}\right)_p = K_d A_p (f_{eq} - f_g^V). \quad (2)$$

To complete the model, it is necessary to express the particle area as a function of the number of moles of hydrate. In the experimental work of Kim et al.,[9] all of the particles were assumed to have the same initial diameter prior to decomposition. The value for the diameter was obtained by measuring the settling time of the hydrate particles and applying Stokes law.

In the current work, the initial particle area is obtained from direct measurements using the particle size analyzer and at any time it is predicted from the population balance. A global rate of gas hydrate decomposition may be obtained by integrating (2) with respect to radius, over all particle sizes

$$R_y(t) = \frac{-\pi K_d (f_{eq} - f_g^V) \mu_2(t)}{\Psi}. \quad (3)$$

The second moment of the particle size distribution, μ_2, is obtained from the population balance for a semibatch system. Assuming that there is no breakage or agglomeration, the population balance becomes

$$\frac{\partial \varphi}{\partial t} + \frac{\partial}{\partial r}(G\varphi) = 0. \quad (4)$$

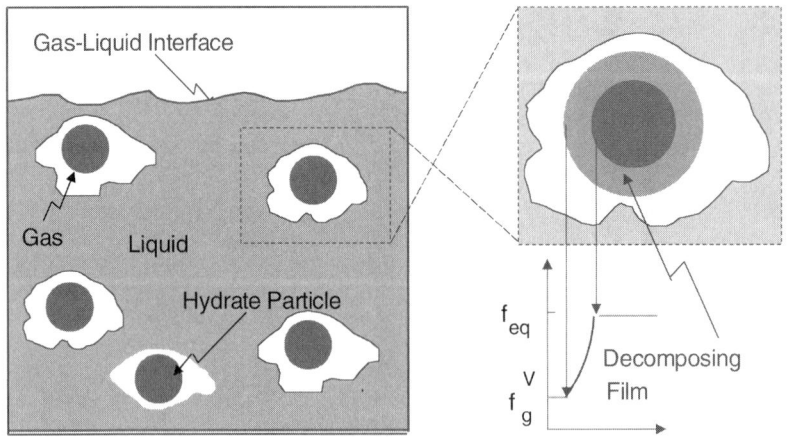

FIGURE 5. Schematic of the mechanism for hydrate decomposition.

Furthermore, if the linear growth rate, which is negative quantity for decomposition, is independent of size, a moment transformation gives the following set of ordinary differential equations

$$\frac{d\mu_0}{dt} = 0 \qquad \mu_0(0) = \mu_0^o$$

$$\frac{d\mu_1}{dt} = G\mu_0 \qquad \mu_1(0) = \mu_1^o \qquad (5)$$

$$\frac{d\mu_2}{dt} = 2G\mu_1 \qquad \mu_2(0) = \mu_2^o$$

The above set of differential equations can be solved (see Appendix) to give

$$\mu_2(t) = \mu_0^o G^2 t^2 + 2\mu_1^o G t + \mu_2^o. \qquad (6)$$

The expression for the second moment can be combined with the expression for the rate of decomposition to give

$$n(t) = n_0 - \frac{\pi}{\psi} K_d V (f_{eq} - f_g^V) \left(\frac{1}{3} \mu_0^o G^2 t^3 + \mu_1^o G t^2 + \mu_2^o t \right). \qquad (7)$$

The linear shrinkage rate is given by

$$G = -\frac{dL}{dt} = -\frac{K_d M (f_{eq} - f_g^V)}{3\rho} \frac{\pi}{\psi \phi_v} \left(\frac{6\phi_v}{\pi} \right)^{2/3}. \qquad (8)$$

The initial value of the moments and the number of moles remaining in the hydrate phase during decomposition are determined experimentally. Thus, the rate constant, K_d, can be determined by minimizing the sum of the square of the errors between the experimental data and the calculated data. Setting the sphericity, ψ, to unity and the volume shape factor, ϕ_v, equal to $\pi/6$ corresponds to the case of spherical particles. The definition of sphericity was given by Kim et al.[9]

CONCLUSION

The experimental procedure of Kim et al.[9] is modified to utilize an on-line particle size analyzer and a new model for the determination of the intrinsic kinetics of gas hydrate decomposition is developed. The mathematical model is based the work of Kim et al.[9] The population balance is incorporated to describe the particle size distribution in the hydrate phase, which should allow for a better estimate of the intrinsic rate constant, as compared to the previous work.[9] The results of one experiment with ethane are presented.

ACKNOWLEDGMENTS

Funding for this work was provided by Shell E & P Technology Company, Houston and by the Natural Sciences and Research Council of Canada (NSERC).

REFERENCES

1. VYSNIAUSKUS, A. & P.R. BISHNOI. 1983. A kinetic study of methane hydrate formation. Chem. Eng. Sci. **38**: 1061–1072.
2. ENGLEZOS, P., P. DHOLABHAI, N. KALOGERAKIS & P.R. BISHNOI. 1987. Kinetics of methane and ethane hydrate formation. Chem. Eng. Sci. **42**: 2647–2657.
3. DHOLABHAI, P., N. KALOGERAKIS & P.R. BISHNOI. 1993. Kinetics of methane hydrate formation in aqueous electrolyte solutions. Can. J. Chem. Eng. **71**: 68–74.
4. LEKVAM, K. & P. RUOFF. 1993. A reaction kinetics mechanism for methane hydrate formation in liquid water. J. Am. Chem. Soc. **115**: 8565–8570.
5. SHINDO, Y., P.C. LUND, Y. FUJIOKA & H. KOMIYAMA. 1993. Kinetics and mechanism of the formation of CO_2 hydrate. Int. J. Chem. Kin. **25**: 777–782.
6. CHUN, M.K. & H. LEE. 1996. Kinetics of formation of carbon dioxide clathrate hydrates. Kor. J. Chem. Eng. **13**(6): 620–626.
7. MALEGOANKAR, M.B., P.D. DHOLABHAI & P.R. BISHNOI. 1997. Kinetics of carbon dioxide and methane hydrate formation. Can. J. Chem. Eng. **75**: 1090–1099.
8. ULLERICH, J.W., M.S. SELIM & E.D. SLOAN. 1987. Theory and measurement of hydrate dissociation. AIChE J. **33**: 747–752.
9. KIM, H.C., P.R. BISHNOI, R.A. HEIDEMANN & S.S.H. RIZVI. 1987. Kinetics of methane hydrate decomposition. Chem. Eng. Sci. **42**: 1645–1653.
10. JAMALUDDIN, A.K.M., N. KALOGERAKIS & P.R. BISHNOI. 1989. Modelling of decomposition of a synthetic core of methane gas hydrate by coupling kinetics with heat transfer rates. Can. J. Chem. Eng. **67**: 948–945.
11. NIELSEN, B., W. SVRCEK & A.K. MEHROTRA. 1994. Effect of temperature and pressure on asphaltene particle size distributions in crude oils diluted with n-pentane. Ind. Eng. Chem. Res. **33**: 1324–1331.

APPENDIX: SOLUTION OF MOMENT EQUATIONS

The differential equations for the first three moments are:

$$\frac{d\mu_0}{dt} = 0 \qquad \mu_0(0) = \mu_0^o$$

$$\frac{d\mu_1}{dt} = G\mu_0 \qquad \mu_1(0) = \mu_1^o \qquad \text{(A1)}$$

$$\frac{d\mu_2}{dt} = 2G\mu_1 \qquad \mu_2(0) = \mu_2^o$$

These equations can be written in matrix form as

$$D\mathbf{x} = \mathbf{A}\mathbf{x}, \quad \mathbf{x} = \begin{bmatrix} \mu_0 \\ \mu_1 \\ \mu_2 \end{bmatrix}, \quad \mathbf{A} = \begin{bmatrix} 0 & 0 & 0 \\ G & 0 & 0 \\ 0 & 2G & 0 \end{bmatrix}. \qquad \text{(A2)}$$

The eigenvalues of the matrix $\mathbf{A}$ are

$$\det(\mathbf{A} - \mathbf{I}\lambda) = \begin{vmatrix} -\lambda & 0 & 0 \\ G & -\lambda & 0 \\ 0 & 2G & -\lambda \end{vmatrix} = (-\lambda)^3 = 0. \qquad \text{(A3)}$$

The eigenvalues of this matrix are 0, with multiplicity 3. To proceed, we must calculate $(\mathbf{A} - \mathbf{I}\lambda)^2$ and $(\mathbf{A} - \mathbf{I}\lambda)^3$.

$$(\mathbf{A} - \mathbf{I}\lambda)^2 = \begin{bmatrix} 0 & 0 & 0 \\ G & 0 & 0 \\ 0 & 2G & 0 \end{bmatrix} \begin{bmatrix} 0 & 0 & 0 \\ G & 0 & 0 \\ 0 & 2G & 0 \end{bmatrix} = \begin{bmatrix} 0 & 0 & 0 \\ 0 & 0 & 0 \\ 2G^2 & 0 & 0 \end{bmatrix} \quad (A4)$$

$$(\mathbf{A} - \mathbf{I}\lambda)^3 = \begin{bmatrix} 0 & 0 & 0 \\ G & 0 & 0 \\ 0 & 2G & 0 \end{bmatrix} \begin{bmatrix} 0 & 0 & 0 \\ 0 & 0 & 0 \\ 2G^2 & 0 & 0 \end{bmatrix} = \begin{bmatrix} 0 & 0 & 0 \\ 0 & 0 & 0 \\ 0 & 0 & 0 \end{bmatrix}. \quad (A5)$$

The eigenvectors associated with $(\mathbf{A} - \mathbf{I}\lambda)^3$ are

$$\mathbf{v} = \begin{bmatrix} 1 \\ 0 \\ 0 \end{bmatrix} \quad \mathbf{w} = \begin{bmatrix} 0 \\ 1 \\ 0 \end{bmatrix} \quad \mathbf{u} = \begin{bmatrix} 0 \\ 0 \\ 1 \end{bmatrix} \quad (A6)$$

The solution of the set of differential equations is

$$\mathbf{x}(t) = c_1 \mathbf{h}_1(t) + c_2 \mathbf{h}_2(t) + c_3 \mathbf{h}_3(t) \quad (A7)$$

$$\mathbf{h}_1(t) = \mathbf{v} + t\mathbf{A}\mathbf{v} + \frac{1}{2}t^2\mathbf{A}^2\mathbf{v} = \begin{bmatrix} 1 \\ tG \\ t^2G^2 \end{bmatrix} \quad (A8)$$

$$\mathbf{h}_2(t) = \mathbf{w} + t\mathbf{A}\mathbf{w} + \frac{1}{2}t^2\mathbf{A}^2\mathbf{w} = \begin{bmatrix} 0 \\ 1 \\ 2Gt \end{bmatrix} \quad (A9)$$

$$\mathbf{h}_2(t) = \mathbf{u} + t\mathbf{A}\mathbf{u} + \frac{1}{2}t^2\mathbf{A}^2\mathbf{u} = \begin{bmatrix} 0 \\ 0 \\ 1 \end{bmatrix}. \quad (A10)$$

Therefore, the general solution is

$$\mathbf{x}(t) = \begin{bmatrix} \mu_0 \\ \mu_1 \\ \mu_2 \end{bmatrix} = c_1 \begin{bmatrix} 1 \\ tG \\ t^2G^2 \end{bmatrix} + c_2 \begin{bmatrix} 0 \\ 1 \\ 2Gt \end{bmatrix} + c_3 \begin{bmatrix} 0 \\ 0 \\ 1 \end{bmatrix}. \quad (A11)$$

The constants are found from the initial conditions. Therefore,

$$\begin{bmatrix} \mu_0 \\ \mu_1 \\ \mu_2 \end{bmatrix} = \mu_0^o \begin{bmatrix} 1 \\ tG \\ t^2G^2 \end{bmatrix} + \mu_1^o \begin{bmatrix} 0 \\ 1 \\ 2Gt \end{bmatrix} + \mu_2^o \begin{bmatrix} 0 \\ 0 \\ 1 \end{bmatrix}. \quad (A12)$$

Mechanisms of Methane Hydrate Crystallization in a Semibatch Reactor

Influence of a Kinetic Inhibitor: Polyvinylpyrrolidone

J.-S. PIC, J.-M. HERRI,[a] AND M. COURNIL

*Ecole Nationale Supérieure des Mines de St-Etienne,
Centre SPIN 158 Cours Fauriel, 42023 St-Etienne Cedex, France*

ABSTRACT: In a previous paper,[1] we proposed a complete model for methane hydrate crystallization from pure water and methane gas in a semibatch reactor. This model takes into account the crucial importance of the rate-determining mass transfer at the gas/liquid interface, coupled with primary nucleation and growth. The validity of this model has been proved on a large number of experimental results. However, due to the complexity of the equations, only a numerical solution is possible, so that comparison between theory and experiment is not straightforward. In this paper, we consider a simple model of primary nucleation/growth and we propose an analytical solution for the time evolution of the total number of particles, mean diameter, and methane gas consumption rate during the first time of the crystallization. This model is used to analyze experimental results obtained in the presence of a kinetic additive: polyvinylpyrrolidone (PVP), in order to understand the origin of its effect.

INTRODUCTION

Most of the mechanisms for gas hydrate formation are relative to an simplified experimental system consisting of pure water and a gaseous hydrocarbon, generally methane, ethane, or propane, located in a batch or semibatch reactor. They sometimes differ in the rate determining step.[2–6] However, the gas/liquid interface has been identified as a critical zone due to the high level of methane concentration.

Skovborg and Rasmussen[6] proposed a simplified model in which the limiting step is the mass transport of gas in the film region. This conclusion was confirmed by Gaillard[7] and Herri.[8] This implies that the population density function (PDF) is not accessible from measurements of gas consumption, whereas it is clear that the kinetic models of gas hydrate formation should be based on crystallization models using the population balance equation.[2,9–11]

In a previous paper,[1] we developed a theoretical model, based on experiments carried out in an isothermal and isobaric stirred tank reactor. By monitoring the PDF via turbidimetry, this model explains the complex influence of the stirring rate on the evolution of the particles concentration by number N_p, their mean radius $\bar{R}$, and the gas consumption rate r, during methane hydrate crystallization from pure water. It

---

[a]Address for correspondence. Voice: 33 4 77 42 02 92; fax: 33 4 77 42 00 00.
herri@emse.fr

takes into account primary nucleation, growth, attrition, and agglomeration. We have shown that $\bar{R}$ and N_p depend on a complex coupling between the crystals nucleation/growth and the gas/liquid mass transfer. We have shown that attrition plays a role at high stirring rates and agglomeration at low stirring rates, but their influence is only observable after a long experimental time.

The model solution is obtained from a numerical approach based on the Runge-Kutta method[1] and, thus, the comparison between the theory and the experiment is not straightforward. The aim of this work is to retain a simplified model in order to propose an analytical solution that can explain the influence of the different crystallization steps (nucleation, growth and mass transfer) on the time evolution of N_p and $\bar{R}$. This paper is organized as follows: the first part is devoted to the proposal of the analytical resolution of a simplified crystallization model; the second part deals with the connection between this model and the experimental results.

MODEL

Methane Dissolution in Water

In previous papers,[1,8,12] we confirmed that the methane consumption rate r verifies the following relation

$$r = k_L a (C_{ext} - C_b), \quad (1)$$

where a is the mass transfer surface area per unit volume of liquid, k_L is the mass transfer coefficient, C_b is the methane concentration in the bulk of the liquid, and C_{ext} is the interfacial concentration imposed by the gas/liquid equilibrium.

Two-Zone Model

In the stirred reactor, two zones can be distinguished (see FIGURE 1), as mentioned by Vysniauskas[2] and Jones.[13]

The first zone is the interface layer. Its thickness is given by:

$$\delta = \frac{D}{k_L}, \quad (2)$$

where D is the diffusivity of the gas molecules in the liquid water and k_L the mass transfer coefficient, which strongly increases with the stirring rate.[1,12] In the system investigated, this thickness is a few tens of micrometers. Because of the high supersaturation level in this zone compared to the rest of the reactor, primary nucleation is particularly active and acts as a continuous source of nuclei for the bulk of the reactor. Moreover, we consider that the residence time of the nuclei is zero, so we need not write the population balance equation in the interfacial film.

The second zone is the bulk zone, in which the concentration is supposed to be uniform. This zone is assumed to be the main place for the crystallization process, which should include all the classical steps of primary nucleation, growth, agglomeration, and secondary nucleation.

The area of the gas/liquid interface is assumed to be independent of time. However, floating hydrate particles could reduce the interfacial area available for methane absorption. This has not been considered here because, in our experimental conditions, hydrates are formed in relatively small quantity and probably do not perturb absorption.

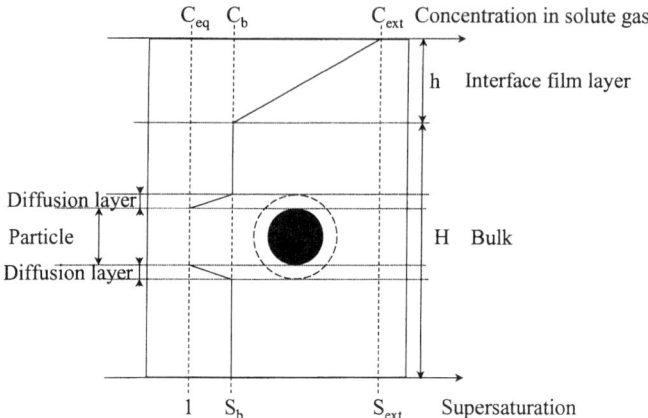

FIGURE 1. Schematic representation of the concentration profiles in the reactor.

Primary Nucleation

We retain the simplified model of nucleation[9]

$$B_I = k_1 \cdot G^n. \tag{3}$$

This quite simple expression gives the direct correlation between the growth rate G and the nucleation rate B_I.

Growth Rate

Growth rate G should take into account the consecutive steps: (1) gas transport from the bulk of the solution to the liquid/crystal interface, and (2) incorporation of gas molecules into the hydrate structure. Moreover, it should assess for the possible existence of a dead zone, which can appear in the presence of kinetic additives.

Thus, we propose to retain the expression:

$$G = k_g(C_b - C^*)^p, \tag{4}$$

where C^* is a critical concentration under which crystal growth is impossible. If the limiting step is the gas transport from the bulk of the solution to the liquid/crystal interface, then the p value is 1. If the limiting step is the incorporation of gas molecules into the hydrate structure, we assume that the exponent p should not be very different from 1 or 2 in order to be coherent with the classical BCF model.[9] For the moment we cannot presuppose which is the growth model to retain. Thus, we consider p as an unknown of the system.

Mass and Population Balances

The main variables of the system are the population density function in the bulk zone $f(R, t)$ and the concentration in dissolved methane in the bulk $C_b(t)$; t denotes the time and R the equivalent particle radius. Two equations can be written to represent the time evolution of the system. The first equation is obtained from the population balance equation of the hydrate crystals.[14] In the simplified model that we present in this paper, we consider only primary nucleation and growth:

$$\frac{\partial f}{\partial t} + G\frac{\partial f}{\partial R} = (B_{I,1} + B_{I,2})\delta(R). \tag{5}$$

In this expression, it is assumed that the growth rate $G = dR/dt$ is independent of the grain size. $\delta(R)$ is the Dirac function (it is assumed that new particles created by primary nucleation have initial size zero). $B_{I,1}$ is the production rate of nuclei coming from the interfacial zone, and $B_{I,2}$ is the primary nucleation rate in the bulk zone. The initial conditions are $f(R, 0) = 0$ and $C_b(0) = C_{eq}$.

The second equation is the mass balance equation in dissolved methane in the bulk zone:

$$\frac{dC_b}{dt} = k_L a \cdot (C_{ext} - C_b) - \frac{4\pi}{v_m} \cdot GM_2, \tag{6}$$

where v_m is the molar volume of the hydrate particles and M_2 is the second moment of the population density function. This expression can be transformed to:

$$\frac{d\Delta C}{dt} = k_L a(\Delta C_M - \Delta C) - k_v k_g \Delta C^p \cdot M_2 \tag{7}$$

with $\Delta C = C_b - C^*$, $\Delta C_M = C_{ext} - C^*$, $k_v = 4\pi/v_m$, and $p = 1$ or 2 depending on the adopted growth model.

The method of the moments has been applied to solve the problem. The jth order moment of the PSD is defined by

$$M_j = \int_0^\infty R^j f(R) dR. \tag{8}$$

The system of equations is transformed to the system of ordinary differential equations:

$$\frac{dM_0}{dt} = B_I = B_{I,1} + B_{I,2} \tag{9}$$

$$\frac{dM_1}{dt} = k_g \Delta C^p \cdot M_0 \tag{10}$$

$$\frac{dM_2}{dt} = 2k_g \Delta C^p \cdot M_1 \tag{11}$$

$$\frac{d\Delta C}{dt} = k_L a(\Delta C_M - \Delta C) - k_v k_g \Delta C^p \cdot M_2. \tag{12}$$

Solving the System of Equations

$B_{I,2}$ is the primary nucleation rate in the bulk zone. As the concentration level in the bulk is constant, it can be expressed simply as:

$$B_{I,2} = k_1 G^n = k_1 k_g^n \Delta C_b^{np}, \tag{13}$$

where $B_{I,1}$ is the nucleation rate due to the interfacial zone. The concentration level in this zone is not constant but varies from C_{ext} to C_b (i.e., ΔC_M to ΔC_b). Due to the low thickness (δ) of this zone, the concentration gradient is supposed to be constant. So, $B_{I,1}$ is expressed as:

$$\frac{1}{V}\Sigma\int_0^\delta k_1 k_g^n \Delta C^{np} dz \text{ with } \Delta C(z) = \Delta C_M + \frac{\Delta C_b - \Delta C_M}{\delta}z, \qquad (14)$$

where Σ is the surface of the gas/liquid interface and V is the volume of the liquid suspension. That is

$$B_{I,1} = \frac{1}{V}\Sigma\int_0^\delta k_1 k_g^n \Delta C^{np} dz$$

$$= k_1 k_g^n a\delta \int_0^1 (\Delta C_M + (\Delta C_b - \Delta C_M)x)^{np} dx \text{ with } a = \frac{\Sigma}{V}. \qquad (15)$$

This equation becomes

$$B_{I,1} = \frac{k_1 k_g^n a\delta(\Delta C_b^{np+1} - \Delta C_M^{np+1})}{(np+1)(\Delta C_b - \Delta C_M)} \text{ for } \Delta C_b \neq \Delta C_M, \text{ and} \qquad (16)$$

$$B_{I,1} = k_1 k_g^n a\delta \Delta C_M^{np} \text{ for } \Delta C_b = \Delta C_M. \qquad (17)$$

Near $\Delta C = 0$, the nucleation rates become

$$B_{I,1} \xrightarrow[\Delta C_b \to 0]{} B_{I,1\min} = \frac{k_1 k_g^n a\delta \Delta C_M^{np}}{np+1} \qquad (18)$$

$$B_{I,2} \xrightarrow[\Delta C_b \to 0]{} k_1 k_g^n \Delta C_b^{np} \to 0. \qquad (19)$$

Thus, near $\Delta C = 0$, the system of equations becomes

$$\frac{dM_0}{dt} = B_I = B_{I,1} = b_0 \qquad (20)$$

$$\frac{dM_1}{dt} = k_g \Delta C^p \cdot M_0 \qquad (21)$$

$$\frac{dM_2}{dt} = 2k_g \Delta C^p \cdot M_1 \qquad (22)$$

$$\frac{d\Delta C}{dt} = k_L a \Delta C_M - k_v k_g \Delta C^p \cdot M_2. \qquad (23)$$

Analytical Solution for $t \to \infty$

We can easily prove that $\Delta C \xrightarrow[t \to \infty]{} 0$. Indeed M_2 (which is proportional to the total surface of particles) is an increasing function of time. Approaching infinity, if ΔC does not reach zero, then the product $k_v k_g \Delta C^p M_2 \xrightarrow[t \to \infty]{} \infty$ and $d\Delta C/dt = k_L a(\Delta C_M - \Delta C) - k_v k_g \Delta C^p M_2 \xrightarrow[t \to \infty]{} \infty$, which is impossible because ΔC cannot be negative. Hence, if $\Delta C \xrightarrow[t \to \infty]{} 0$, we also have $d\Delta C/dt \xrightarrow[t \to \infty]{} 0$.

If we suppose that G is written in the form $G = Kt^a$ (series expansion), we obtain the following relation

$$M_2 = \frac{A}{G} = \frac{A}{Kt^\alpha} \text{ with } A = \frac{k_L}{k_V} a\Delta C_M. \qquad (24)$$

Using successively **(22)**, **(21)**, and **(20)**, we can write:

$$M_1 = \frac{-A\alpha}{2K^2 \cdot t^{-(2\alpha+1)}} \quad \text{and} \tag{25}$$

$$M_0 = \frac{A\alpha(2\alpha+1)}{2K^3 \cdot t^{-(3\alpha+2)}}. \tag{26}$$

Finally

$$b_0 = \frac{-A\alpha(2\alpha+1)(3\alpha+2)}{2K^3 \cdot t^{-(3\alpha+3)}}. \tag{27}$$

Since b_0 must be constant, we deduce $\alpha = -1$ and thus

$$K^3 = \frac{A}{2b_0} = \frac{k_L a \Delta C_M}{2k_v b_0}. \tag{28}$$

Finally, we can express: $\bar{R} = M_1/M_0 = K$ with $K = M_0 = b_0 t$ and obtain:

$$\bar{R} = \left(\frac{k_L a \Delta C_M}{2k_v b_0}\right)^{1/3}$$

and, using **(18)** and **(1)**, we obtain

$$\bar{R} = \left(\frac{np+1}{2k_v} \frac{k_L}{k_1 k_g^n \delta}\right)^{1/3} \Delta C_M^{\frac{1-np}{3}}. \tag{29}$$

EXPERIMENTAL RESULTS

FIGURE 2 shows the cumulative methane gas consumption versus time. We distinguish different steps:

- The first one is the saturation of water with gas molecules. Its kinetics are given by $r = k_L a (C_{ext} - C_b)$.
- After complete saturation of water, we can observe an induction period with no methane consumption.
- Then, the first crystals appear and grow. A turbidity signal is detected and the rate of methane consumption (r) increases.
- For pure water (FIG. 2, Curve a), the methane consumption rate reaches its maximum value (r_c) just after nucleation, keeps constant for several hours and then begins to decrease. This behavior is also observed in the presence of PVP, when it is added to the solution after the appearance of the first crystals.
- If PVP has been introduced in water prior to nucleation (FIG. 2, Curve b), the behavior is very different. In fact, the methane consumption rate increases continuously, reaches an asymptotic value (r_c) after several hours (as predicted by the model), and then begins to decrease. This decrease is not predicted by the model, but it can be attributed to the flotation of the particles, which reduces the interfacial area available for methane absorption. Hence, this final part of the curve cannot be theoretically exploited by using the simplified model.

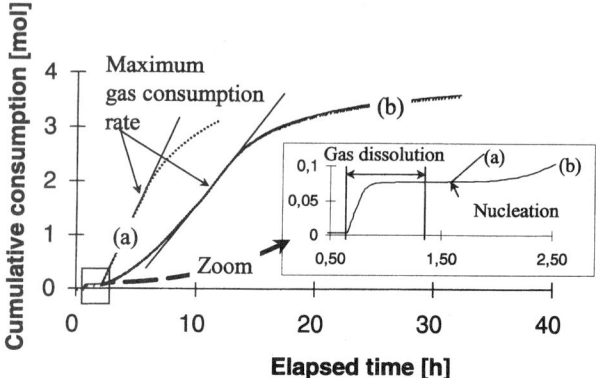

FIGURE 2. Evolution of the cumulative gas consumption versus time. *Curve a*: pure water or water with PVP injected just after nucleation. *Curve b*: water with PVP injected before nucleation. (45 bar, 1°C, 400 rpm, 1 wt% PVP.)

From the first part of the cumulative consumption curve, we can retrieve the value of the constant $k_L a$ as a function of pressure, stirring rate, and PVP concentration. It appears that $k_L a$ strongly increases with the stirring rate (see FIGURE 3), but is independent either on the PVP concentration and on pressure. If no bubbles are incorporated into the liquid phase (stirring rate is less than 500 rpm), the a value can be estimated from visual observation and then the k_L value can be calculated.

FIGURE 4 shows the influence of PVP on the evolution of the total number of particles versus time at a stirring rate of 300 rpm. We observe that the particles number decreases considerably as the PVP concentration increases from 0% to 0.8%. For a concentration of 0.8%, crystallization is practically stopped and the number of particles remains constant. Moreover, the maximum methane consumption rate (r_c) is

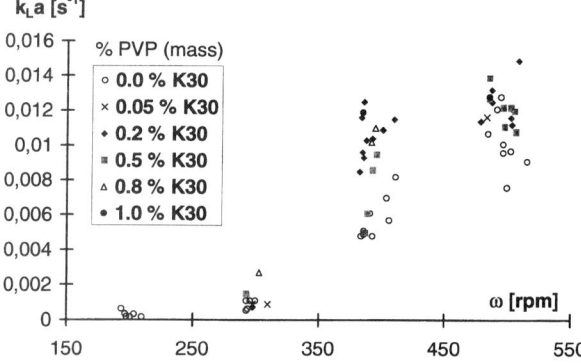

FIGURE 3. Influence of the stirring rate on $k_L a$ with different PVP concentrations.

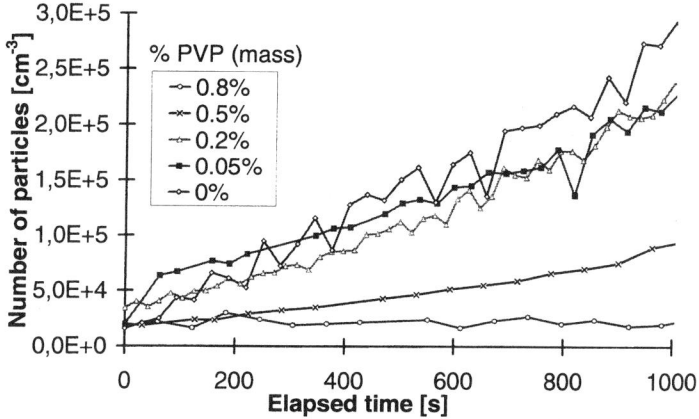

FIGURE 4. Effect of PVP on the total number of particles (45 bar, 1 °C, 300 rpm).

largely dependent on the PVP concentration and on the stirring rate (see FIGURE 5): the higher the stirring rate, the higher the methane consumption rate. The higher the PVP concentration, the lower the methane consumption rate.

FIGURE 6 shows the mean diameter of the particles at the moment of maximum consumption rate. It appears that the mean diameter is practically independent on the PVP concentration, but it is highly dependent on the stirring rate. Using a high pressure injector, PVP can be added during the crystallization process. FIGURE 7 shows the effect of successive injections on the cumulative methane gas consumption. We observe that the slope of the curve (which is the methane gas consumption rate) is not modified after the different injections. This is a very surprising result: if PVP is

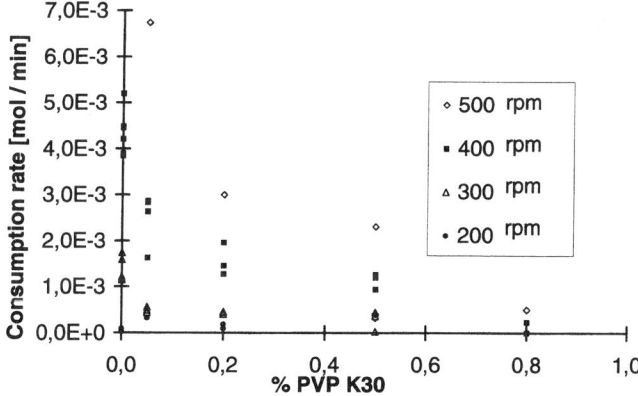

FIGURE 5. Effect of PVP on the methane consumption rate (45 bar, 1 °C).

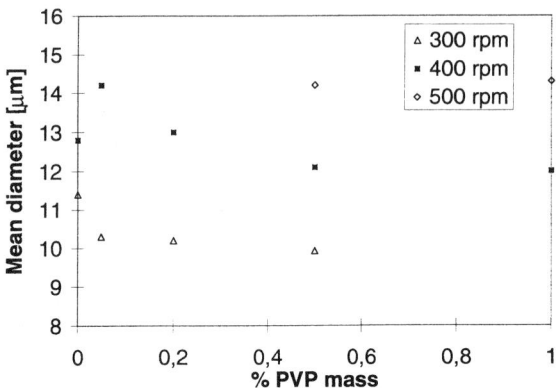

FIGURE 6. Effect of PVP on the particle mean diameter (45 bar, 1°C).

injected before nucleation, the methane gas consumption rate is considerably reduced as the PVP concentration increases (FIG. 5); however, if PVP is injected after the appearance of the first crystals, it has no effect (FIG. 7).

INTERPRETATION

From measurement of the methane dissolution rate $k_L a$ (FIG. 3) and of the gas consumption rate r_c (FIG. 5), we can determine the C^* value:

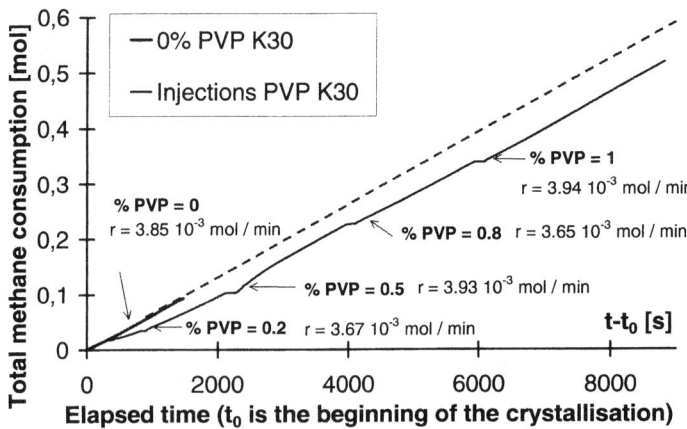

FIGURE 7. Influence of PVP after successive injections during the crystallization (45 bar, 1°C, 400 rpm).

TABLE 1. Experimental influence of PVP on the solute concentration C^*

$10^2 C^*$ (mol.l^{-1})	0% PVP	0.05% PVP	0.2% PVP	0.5% PVP	0.8% PVP
300 rpm	5.77	7.29	7.39	7.64	—
400 rpm	7.19	7.39	7.67	7.89	8.07
500 rpm	—	−7.16	7.68	7.96	8.07
mean value	6.48	7.33	7.58	7.83	8.07

$$C^* = C_{ext} - \frac{r_c}{k_L a} \quad (30)$$

This is the concentration that defines the dead zone $[C_{eq}, C^*]$ in which nucleation and growth are impossible. TABLE 1 shows the evolution of C^* as a function of the stirring rate and of the PVP concentration. We observe that the higher the PVP concentration, the higher the value of C^*. We also note that C^* is practically equal to C_{ext} for a 0.8% PVP concentration. In this condition, crystallization is practically stopped.

From measurement of the total number of particles (N_p), we can compare the experimental nucleation rate $B_{I,1\,min}$ to its theoretical value, given by:

$$\frac{dN_p}{dt} \xrightarrow[\sigma \to 0]{} B_{I,1\min} = \frac{k_1 k_g^n a \delta \Delta C_M^{np}}{np + 1} \quad (31)$$

FIGURE 8 plots the value of $\ln(dN_p/dt)$ as a function of $\ln(\Delta C_M)$. The straight line obtained confirms, with remarkable agreement, the validity of the theoretical relation **(31)**. From its slope, we can deduce the value of the ΔC_M exponent (i.e., $np \cong 1$). Moreover, we conclude that the value $k_1 k_g^n a \delta/(np + 1)$ is independent of the PVP concentration. This certainly signifies that the constant k_1 for nucleation and the constant k_g for growth are also independent of the PVP concentration.

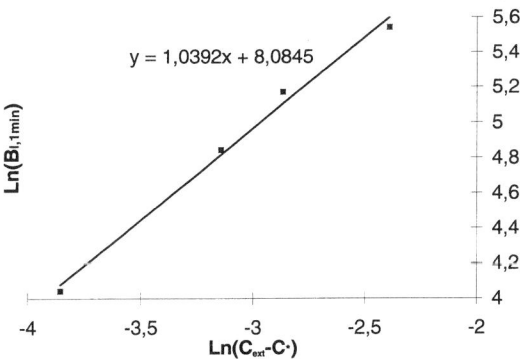

FIGURE 8. Experimental evolution of the minimum nucleation rate.

From the experimental measurement of the mean diameter, we verify this value of np. Indeed, we have seen

$$\bar{R} = \left(\frac{np+1}{2k_v} \frac{k_L}{k_1 k_g^n \delta}\right)^{1/3} \Delta C_M^{\frac{1-np}{3}}. \tag{29}$$

In FIGURE 6, we observe that the mean diameter is practically independent of the PVP concentration, that is to say it is practically independent of ΔC_M. This confirms the former result $np = 1$.

CONCLUSION

We conclude on the action mechanism of PVP on the methane hydrate crystallization in a semi-batch reactor, at a moderate stirring rate of 300 rpm. In this system we have shown that the experimental results can be interpreted in a simple model of nucleation at the gas/liquid interface and growth into the bulk liquid water. We have proposed an analytical model that allows us to explain completely the influence of the PVP concentration on the nucleation and growth of the methane hydrate particles. We have shown that the growth can be modeled using the relation $G = k_g \cdot \Delta C^p$ where ΔC is the growth driving force expressed by $\Delta C = C_b - C^*$. C_b is the dissolved methane concentration and C^* is a limit concentration. This limit concentration C^* is highly dependent on the PVP concentration. We have shown that the action mechanism of PVP is to initiate a temporary dead zone in which growth is totally inhibited, and thus nucleation, expressed by $B = k_1 \cdot G^n$. Moreover, the higher the PVP concentration, the wider the dead zone. It appears that the values of the constants k_1 (primary nucleation) and k_g (growth) are not affected by PVP.

In a very surprising way, we have observed that this inhibition is only effective if PVP is injected prior to the appearance of the first crystals. If PVP is injected after nucleation, it has no effect, even at a high concentration of 1%. These results are consistent with the adsorption mechanism proposed by Lederhos[15] and Larsen:[16] PVP is efficient as long as the number of adsorbing polymeric sites remains higher than that of crystalline cyclic surfaces. Afterwards, growth is no further affected by the inhibitor.

REFERENCES

1. HERRI, J.-M., F. GRUY, J.-S. PIC & M. COURNIL. 1999. Methane hydrate crystallisation mechanism from in situ particle sizing. AIChE J. **45**(3): 590–603.
2. VYSNIAUSKAS, A. & P.R. BISHNOI. 1983. A kinetic study of methane hydrate formation. Chem. Eng. Sci. **38**: 1061–1072.
3. VYSNIAUSKAS, A. & P.R. BISHNOI. 1985. A kinetic study of ethane hydrate formation. Chem. Eng. Sci. **40**: 299–303.
4. ENGLEZOS, P., N. KALOGERAKIS, P.D. DHOLABHAI & P.R. BISHNOI. 1987. Kinetics of formation of methane and ethane gas hydrates. Chem. Eng. Sci. **42**: 2647–2658.
5. ENGLEZOS, P., N. KALOGERAKIS, P.D. DHOLABHAI & P.R. BISHNOI. 1987. Kinetics of gas hydrate formation from mixture of methane and ethane. Chem. Eng. Sci. **42**: 2659–2666.
6. SKOVBORG, P. & P. RASMUSSEN. 1994. A mass transport limited model for the growth of methane and ethane gas hydrates. Chem. Eng. Sci. **49**: 1131–1141.

7. GAILLARD, C. 1996. Cinétique de Formation de l'Hydrate de Méthane dans une Boucle de Laboratoire. Ph.D. Thesis, Institut National Polytechnique de Toulouse, Toulouse, France.
8. HERRI, J.-M. 1996. Etude de la formation de l'hydrate de méthane par turbidimétrie in situ. Ph.D. Thesis, Université Paris VI, Paris, France.
9. RANDOLF, A.D. & M.A. LARSON. 1988. Theory of Particulate Processes, 2nd edit. Academic Press, New York.
10. MONTFORT, J.-P. & N. ZIHOU. 1993. A light scattering kinetics study of cyclopropane hydrate growth. J. Crystal Growth **128:** 1182–1186.
11. BYLOV, M. & P. RASMUSSEN. 1997. Experimental determination of refractive index of gas hydrates. Chem. Eng. Sci. **52:** 3295–330.
12. HERRI, J.-M., F. GRUY, J.-S. PIC, M. COURNIL, B. CINGOTTI & A. SINQUIN. 1999. Interest of *in situ* turbidimetry for the characterization of methane hydrate crystallization. Application to the study of kinetic inhibitors. Chem. Eng. Sci. **54**(12): 1849–1858.
13. JONES, A.G., J. HOSTOMSKY & Z. LI. 1992. On the effect of liquid mixing rate on primary crystal size during the gas-liquid precipitation of calcium carbonate. Chem. Eng. Sci. **47:** 3817–3824.
14. RANDOLF, A.D. 1964. A population balance for countable entities. Can. J. Chem. Engin. **42:** 280–281.
15. LEDERHOS, J.P., J.P. LONG, A. SUM, R.L. CHRISTIANSEN & E.D. SLOAN. 1996. Effective kinetic inhibitors for natural gas hydrates. Chem. Eng. Sci. **51:** 1221–1229.
16. LARSEN, R., C.A. KNIGHT & E.D. SLOAN. 1998. Clathrate hydrate growth and inhibition. Fluid Phase Equilibria **150–151:** 353–360.

Kinetics of Ethane Hydrate Growth on Latex Spheres Measured by a Light Scattering Technique

PHILLIP SERVIO,[a] PETER ENGLEZOS,[a,b] AND P. RAJ BISHNOI[c]

[a]*Department of Chemical and Biological Engineering, University of British Columbia, Vancouver, British Columbia, Canada V6T 1Z4*

[c]*Department of Chemical and Petroleum Engineering, University of Calgary, Calgary, Alberta, Canada T2N 1N4*

ABSTRACT: A high pressure, temperature-controlled sapphire equilibrium cell is used to observe the nucleation and growth behavior of gas hydrates in bulk water or on spheres suspended in water. Gas hydrate crystals are grown on positively and negatively charged latex spheres. The experiments have been conducted on an ethane-water system that is known to form structure I hydrate. Experiments were carried out at temperatures between 278.0 K and 278.6 K and pressures ranging from 1,300 kPa to 1,500 kPa. A light scattering apparatus was used to monitor the hydrate formation process. In particular, the photomultiplier voltage was recorded over time in order to observe the effects of latex spheres on the nucleation and growth of gas hydrates. More experiments need to be performed to better ascertain the hydrate nucleation and growth behavior in the presence of latex spheres.

INTRODUCTION

Hydrates are nonstoichiometric crystalline compounds. They occur when water molecules hydrogen bond and form cavities that can be occupied by a gas or volatile liquid. Without the presence of the gas or volatile liquid the host lattice is thermodynamically unstable. There is no chemical reaction or chemical bonding between the host lattice and guest molecule, just weak van der Waals forces. A comprehensive account of gas hydrates by Sloan is available.[1]

There are two fundamental issues that must be addressed with respect to the kinetics of gas hydrate formation. The first is the amount of time required in order for a hydrate to reach a critical size nucleus. The second is related to the rate at which a hydrate grows after the critical size nucleus has been achieved.[2] Hydrate nucleation refers to the process in which small hydrate particles called nuclei, grow until they attain a critical size for continued growth. Growing clusters of water and gas molecules is the first step of hydrate nucleation. If the size of the nuclei is less then the critical size, the nuclei are unstable and may continue to grow or break in the aqueous solution.[3] If the growing nuclei reach the critical size they then become stable, which leads to the formation of hydrate crystals. This period during which the

[b]Address for correspondence. Voice: 604-822-6184; fax: 604-822-6003.
englezos@interchange.ubc.ca

hydrate nuclei are forming and dissolving in a supersaturated solution to the time when the nuclei reach the critical size is called the induction time. The induction time is believed to be stochastic in nature.

Hydrate growth is the process in which the hydrate nuclei have achieved the critical size and continue to grow and form hydrate crystals. A basic problem posed by experimental work on the kinetics of gas hydrates is limited observability. Usually, only the rate of gas consumption is monitored. However, a detailed description of the system requires knowledge of the number of hydrate crystal particles, the particle size distribution, and the rate of growth of these particles. The knowledge of the hydrodynamics, heat, and mass transfer effects are also needed, if they are relevant.

A mechanistic model for the intrinsic kinetics of hydrate growth has been developed at the University of Calgary.[4] This model was initially developed for single component methane and ethane gas and was later modified to account for mixtures of methane and ethane. The model successfully describes the rate of hydrate growth of the above mixture without using any adjustable parameters.

Work to obtain more information than the gas consumption curve began in the early 1990s. Nerheim et al.,[5] Monfort et al.,[6] and Rasmusen et al.,[7] have all employed light scattering techniques to observe hydrate growth. Parent et al.,[8] studied the hydrate nucleation behavior using light scattering techniques. Measuring the rate of growth from hydrate nuclei using light scattering techniques is difficult due to the extremely small sizes of these nuclei.

An alternative approach is proposed in this work. The effects of hydrate growth on positively charged latex spheres having a diameter equal to 600 nm and negatively charged latex spheres having a diameter of 692 nm is monitored. Hydrate growth experiments are conducted at the University of Calgary on the ethane–water system which is known to form structure I hydrate. Hydrates are expected to grow preferentially on the particles since the energy barrier for such nucleation is expected to be smaller than that for nucleation in a gas–liquid system free of particles. It is noted that Raman spectroscopy has been employed to provide kinetic spectra describing the transition from dissolved methane to methane hydrate.[9] Such molecular level information coupled with light scattering growth data will aid in the refinement of mechanistic kinetic models for engineering applications.

EXPERIMENTAL APPARATUS AND PROCEDURE

Parent[8,10] assembled a laser light scattering apparatus at the University of Calgary to perform experiments. He studied the nucleation behavior of clathrate hydrates using light scattering techniques to characterize the early stages of crystallization. In the following paragraphs the apparatus that Parent used will be described in detail including the modifications that have been made to the apparatus.

Apparatus

The light scattering apparatus that Parent used consisted of a 10mW helium-neon laser (Melles Griot model 05LHP991). The detector is a photomultiplier tube (Oriel model 77348) that is biased by a variable voltage, 0 to $-1,000$ volt range, power supply (Oriel model 70705). The current produced by the photomultiplier, PMT, is amplified to a 0–10 volt signal via a current pre-amplifier (Oriel model 70710) before being processed by the data acquisition computer. The cell, constructed of sapphire and having the following dimensions 3/4" I.D. × 1-1/4" O.D. × 4-1/2" long, is aligned with the laser and detector with the aid of a custom made optical bench. There are two commercial optical rails (Melles Griot model 07 0RN 001 and 07 0RN 005) secured to a specially designed rotation stage where the sapphire cell rests. The detector rail is attached to a bearing mounted stage that allows the detector 180 degrees of movement about the optical axis. Agitation inside the sapphire cell is achieved with a stir bar that is coupled to a magnetic stirrer on the outside of the reactor. All the light scattering equipment is placed on a vibrational isolation table that absorbs the natural disturbances of the surrounding environment as well as the vibrations caused by the stirring of the contents in the sapphire cell.

The first modification was to replace the existing cooling system, which used CP grade carbon dioxide, with a system based on liquid nitrogen. Liquid nitrogen is much less costly and reduces the time required to reach the experimental temperature. A 35L dewar (Cole-Palmer model 03773-58) that dispenses liquid nitrogen at −190°C was purchased. This liquid nitrogen is then heated to approximately −150°C with the aid of heating tape. The flow rate is then controlled by a control valve that adjusts the flow rate according to the desired temperature in the sapphire reactor. Finally, stainless steel baffles have also been added to the sapphire cell to avoid the formation of vortices during agitation. FIGURE 1 shows a schematic of the modified light scattering apparatus.

The previous light scattering setup could not maintain a constant experimental pressure, therefore the apparatus was modified accordingly. This involved attaching the existing gas feed line to a supply reservoir, R1. The supply reservoir is then able to deliver gas to the sapphire equilibrium cell, in order to maintain a constant pressure. A data acquisition program for the light scattering apparatus was written and

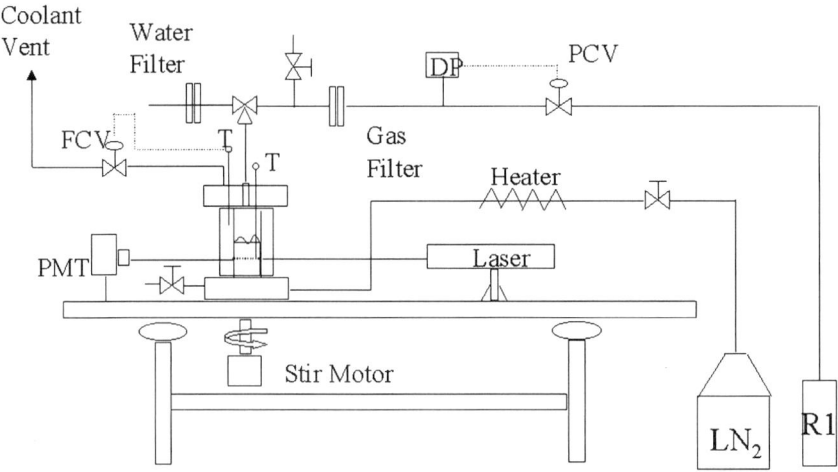

FIGURE 1. Light scattering apparatus.

installed in order to monitor the experimental temperature and pressure. The program also records the voltage from the photomultiplier that could later be used to calculate the hydrodynamic particle growth using Mie theory.[11]

Procedure

In order to measure the rate of hydrate growth on latex spheres the following procedure is used. Reservoir, R1, is filled to approximately four bars above the target experimental pressure. R1 is the supply reservoir that delivers gas to the sapphire cell reactor in order to maintain a constant pressure throughout the growth experiment and hence a constant driving force for nucleation and growth. The bias reservoir, R2, is then filled to approximately half a bar below the experimental pressure. This is to ensure the accurate measurement of the pressure in the supply reservoir, R1. The reactor is then charged with water and a known amount of monodispersed latex particles. The cell is then pressurized to the experimental pressure. The temperature in the sapphire cell reactor is kept constant by flowing nitrogen through the annular region around the sapphire cell reactor. The laser is then focused near the gas-liquid interface, because this is where nucleation will initially begin due to the higher concentration of dissolved gas at the interface. The photomultiplier is fixed in place perpendicular to the incident light from the helium neon laser. The laser is then turned on and the data acquisition program commences to log the photomultiplier voltage, pressure, and temperature during the experiment.

RESULTS AND DISCUSSION

Five experiments were performed on the ethane–water system; two with positively charged latex particles, two with negatively charged latex particles, and one without latex particles. The experimental conditions along with the nucleation times for each experiment are presented in TABLE 1. For each experiment the photomultiplier voltage was logged versus time. The results for the five experiments are all very similar, therefore only the results of experiments 1, 2, and 4 are given and can be seen in FIGURES 2, 3, and 4.

TABLE 1. Experimental conditions and induction times

Exp.	T (K)	P (kPa)	Particle Charge	Particle Size (nm)	Particle Number per milliliter	Induction Time (min)
1	278.3	1300	—	—	—	130
2	278.3	1300	Positive	600	1.92×10^8	154
3	278.1	1300	Positive	600	3.84×10^8	167
4	278.5	1300	Negative	692	2.69×10^8	94
5	278.2	1500	Negative	692	2.69×10^8	85

NOTE: At 278.8 K the ethane hydrate equilibrium pressure is 950 kPa.[12]

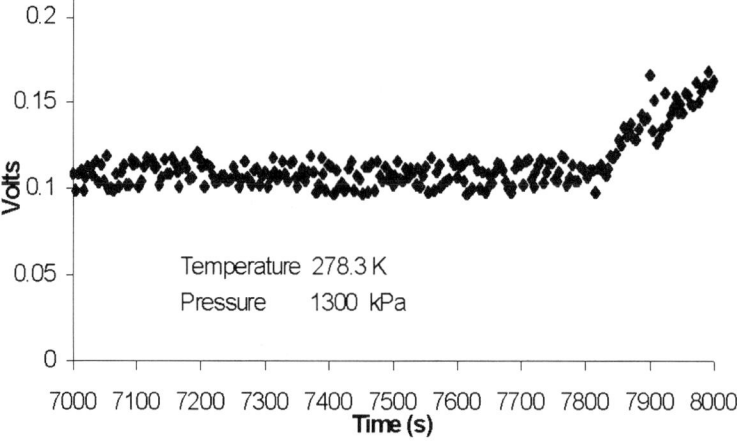

FIGURE 2. PMT response versus time (absence of particles).

In FIGURE 2 the photomultiplier voltage versus time for a hydrate formation experiment in the absence of latex particles is plotted. The figure shows a baseline photomultiplier voltage of 0.11 volts. The voltage remains constant until hydrate nucleation, approximately 7,800 seconds after the experiment was started, where the voltage begins to increase rapidly.

FIGURE 3 shows the photomultiplier voltage versus time for a hydrate formation experiment in the presence of positively charged latex particles. The baseline voltage is now slightly higher than in FIGURE 2 due to the presence of the latex particles, and

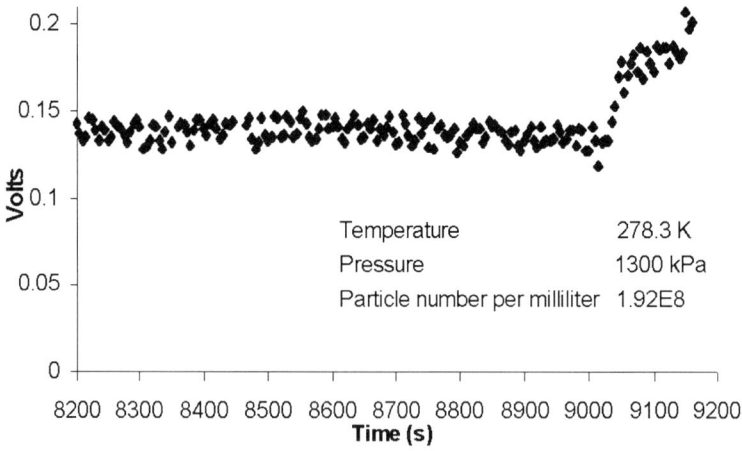

FIGURE 3. PMT response versus time (positively charged particles).

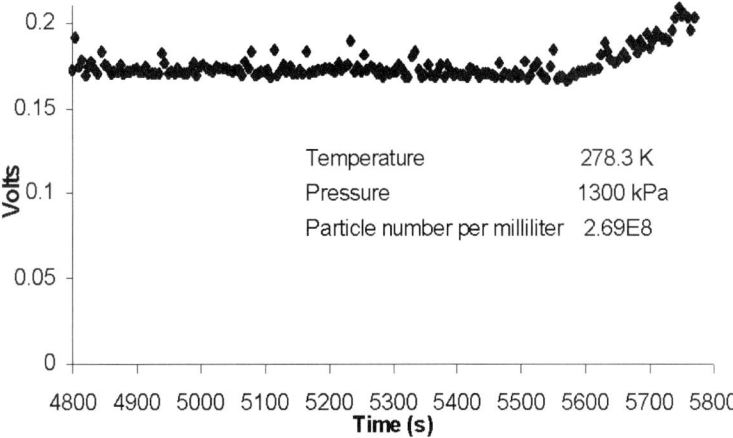

FIGURE 4. PMT response versus time (negatively charged particles).

is approximately equal to 0.14 volts. Once again the voltage remains constant until hydrate nucleation, approximately 9,000 seconds after the experiment began. The voltage then begins to increase in the same manner as in FIGURE 2.

The photomultiplier voltage versus time for a hydrate formation experiment in the presence of negatively charged latex particles is shown in FIGURE 4. The baseline voltage for this experiment is approximately 0.17 volts, which is higher then the baseline voltage in FIGURES 2 and 3 because of the higher concentration of latex particles. The voltage is constant until hydrates appear, 5,600 seconds into the experiment. The voltage then increases with approximately the same rate as that of FIGURES 2 and 3.

A comparison of FIGURES 2, 3, and 4 indicates a need to perform more experiments in order to distinguish between the hydrate nucleation and growth behavior in the absence and presence of latex particles.

The formation of a hydrate layer at the gas–liquid interface was observed in all the experiments performed. The hydrate layer occurred approximately 200 s after hydrate nucleation. The presence of a hydrate layer hinders the hydrate growth. Hence, it is necessary to obtain better agitation in the liquid phase in order to eliminate the formation of this hydrate layer at the gas–liquid interface.

CONCLUSIONS

Experiments have been performed on the ethane–water system in the presence of positively and negatively charged latex particles, each having a diameter of 600 nm and 692 nm, respectively. The experiments were carried out at temperatures between 278.0 K and 278.6 K with pressures ranging from 1,300 kPa to 1,500 kPa. Work is under way in order to differentiate between the hydrate nucleation and growth behavior with and without latex particles.

ACKNOWLEDGMENTS

Funding for this work was provided by the Natural Sciences and Research Council of Canada (NSERC). Special thanks to Matthew Clarke for his help and discussions.

REFERENCES

1. SLOAN, E.D. 1998. Clathrate Hydrates of Natural Gases. Marcel Dekker, New York.
2. ENGLEZOS, P. 1996. Nucleation and growth of gas hydrate crystals in relation to "kinetic inhibition". Rev. Inst. Fr. Pet. **51:** 789–795.
3. BISHNOI, P.R. & V. NATARAJAN. 1996. Formation and decomposition of gas hydrates. Fluid Phase Equilibria **117:** 168–177.
4. ENGLEZOS, P., N. KALOGERAKIS, P.D. DHOLABHAI & P.R. BISHNOI. 1987. Kinetics of formation of methane and ethane gas hydrates. Chem. Eng. Sci. **42:** 2647–2658.
5. NERHEIM, A.R., M.T. SVARTAAS & E.K. SAMUELSEN. 1992. Investigation of hydrate kinetics in the nucleation and early growth phase by laser light scattering. Proceedings of the Second International Offshore and Polar Engineering Conference, San Francisco.
6. MONFORT, J.P. & A. NZIHOU. 1993. Light scattering kinetics study of cyclopropane hydrate growth. J. Crystal Growth **128:** 1182–1186.
7. BYLOV, M. & P. RASMUSSEN. 1996. A new technique for measuring gas hydrate kinetics. Second International Conference on Natural Gas Hydrates. Toulouse, France, June 2–6, 1996.
8. PARENT, J.S. & P.R. BISHNOI. 1996. Investigations into the nucleation behavior of natural gas hydrates. Chemical Engineering Communications (CEC) **144:** 51–64.
9. SLOAN, E.D., S. SUBRAMANIAN, P.N. MATTHEWS, J.P. LEDERHOS & A.A. KHOKHAR. 1998. Quantifying hydrate formation and kinetic inhibition. Ind. Eng. Chem. Res. **37:** 3124–3132.
10. PARENT, J.S. 1993. Investigation into the Nucleation Behaviour of Clathrate Hydrates of Natural Gas Components. M.Sc. Thesis, University of Calgary, Alberta.
11. MIE, G. 1908. Ann. Physik. **25:** 377.
12. HOLDER, G.D. & J.H. HAND. 1982. Multiple-phase equilibria in hydrates from methane, ethane, propane, and water mixtures. AIChE J. **28:** 440–447.

Microscopic Measurements and Modeling of Hydrate Formation Kinetics

S. SUBRAMANIAN AND E.D. SLOAN, JR.[a]

Center for Hydrate Research, Colorado School of Mines, Golden, Colorado 80401, USA

ABSTRACT: The frequency of the Raman band for methane dissolved in water indicates clustering of water molecules around methane molecules. By monitoring trends in the position of this band with pressure, it was determined that the clusters resemble CH_4 trapped in the 20-coordinated small 5^{12} cavity, the basic building block of clathrate hydrates. These methane–water clusters act as precursors to hydrate cavity formation and their transformation to methane trapped in cavities of sI hydrate was monitored using time resolved Raman spectroscopy. The ratio of number of large to small cavities was obtained as a function of time. The formation of the large $5^{12}6^2$ cavity was determined to be slower than for the small 5^{12} cavity. Based on these results, a mechanistic model of methane hydrate formation from the aqueous phase is proposed.

INTRODUCTION

Clathrate hydrates of natural gases are crystalline ice-like compounds that consist of molecules like methane, ethane, propane (guests) trapped in hydrogen bonded cages of water molecules (host).[1] Methane is commonly the dominant component of clathrate gas hydrates formed either in nature or in oil and gas processing. Due to its small molecular size, methane can serve as a guest in all the three known gas hydrate structures sI, sII, and sH.

Few attempts have been made to model the kinetics of gas hydrate formation with limited success. The models by Englezos and Bishnoi[2,3] and Skovborg and Rasmussen[4] were based upon macroscopic measurements of the associated fluid phases to provide indirect measurements of the growing hydrate phase. Fundamental tools like NMR and Raman spectroscopy have been used to study hydrate formation mechanisms at a molecular level. Pietrass *et al.*,[5] used hyperpolarized ^{129}Xe NMR to monitor formation of Xenon sI hydrate on ice. Long[6] monitored formation of tetrahydrofuran (THF) sII hydrate from a miscible THF/water solution in real-time by observing the Raman spectrum for THF. The objective of this work is to study microscopic thermodynamic and kinetic phenomena in methane hydrate formation using Raman spectroscopy.

[a]Address for correspondence.

EXPERIMENTAL METHOD

Apparatus

The Raman Spectrometer is a Renishaw Inc. MK III multichannel fiber optic based system equipped with a 2,400 grooves/mm holographic grating and an internally mounted −70°C Peltier cooled CCD detector. The excitation source is an Ar-ion laser emitting a 514.53 nm green line providing about 10 mW at the sample. The laser is delivered to a high efficiency fiber probe through fiber optic bundles, and focused on the sample using a 20× microscope objective (focal depth, 6.09 μm; field of view, 1 mm). Scattered light collected in a 180° back scattering geometry was used to generate the Raman spectrum. Routine calibration of the monochromator was done by using Neon emission lines. Spectral data was analyzed using GRAMS/32® software.

The time-invariant Raman experiments were carried out in a custom designed high-pressure optical cell capable of operating at 70 MPa. The time-resolved Raman experiments were carried out in a similar cell capable of operating at 34 MPa. The cells consisted of two brass plates on either side of a central brass body that housed the sample chamber. Sapphire windows brazed onto the brass body provided optical capability in both cells. Coolant was circulated through grooves around the sample chamber and the temperature of the cell was monitored using a RTD placed in a hole drilled into the central brass body. A Heise Model 623 pressure transducer was used to record pressure.

Procedure

Two types of experiments were carried out with the methane (CH_4) sI hydrate forming system: (1) time-invariant measurements of methane dissolved in water, and (2) time-resolved Raman measurements of methane hydrate formation from dissolved methane.

For the time-invariant measurements, the sample chamber was first loaded with 1 cm^3 of double distilled water. The temperature of the cell was held constant at 30°C. The cell was charged with CH_4 to the desired pressure and then shaken for about 30 minutes to ensure equilibration. The cell was then mounted on a precision *XY* stage and the laser was focused on the cell contents at an arbitrary spot in the bulk aqueous phase. Raman spectra of CH_4 dissolved in water were collected by adding 20 spectra each integrated over 20 seconds. The pressure was then raised and a similar procedure was used to obtain Raman spectra.

For time-resolved Raman spectra, the cell was loaded with water and charged with CH_4 to the desired pressure (31.7 MPa). The cell was then cooled at constant pressure from 24°C (outside the hydrate stability region) to 2.5°C (within the stability region) at a rate of 0.1°C per minute. The laser was focused on the cell contents at an arbitrary spot at a distance of about 1 mm below the gas-liquid interface. Raman spectra were collected on a real-time basis during the temperature ramp with a time resolution of 25 seconds. The system was not stirred or shaken during this experiment.

EXPERIMENTAL RESULTS AND DISCUSSIONS

Time Invariant Raman Studies

The trend in frequency with pressure of the Raman band for dissolved methane was used to draw inferences about the local environment around dissolved methane molecules. FIGURE 1 compares the Raman spectrum of the ν_1 totally symmetric C-H stretching vibration of methane dissolved in water at 31.7 MPa with previously measured[7] Raman spectra of CH_4 in gas phase (at 2,917.3 cm^{-1} at 3.4 MPa), CH_4 in large $5^{12}6^2$ and small 5^{12} cages of sI hydrate at 2,904.8 cm^{-1} and 2,915 cm^{-1}, respectively.

The dissolved methane peak (2,911.3 cm^{-1}) is almost midway between the peaks for methane in the large and small sI cages (2,905 cm^{-1} and 2,915 cm^{-1}). The width (full width at half maximum, FWHM) of this dissolved methane band is much larger than the two bands for the hydrate cages. This indicates that methane dissolved in water may be in a local environment consisting of relatively disordered water molecules, thus causing the methane to explore a wide range of vibrational energy states. The position of the peak almost midway between the sI hydrate cages indicates that these clusters on the average may be comparable to 20 (5^{12}) or 24 ($5^{12}6^2$) coordinated hydrate cavities.

FIGURE 2 shows the variation in frequency of this ν_1 C-H stretching vibration of methane dissolved in water with pressure over the range from 1.7 to 63.3 MPa. The temperature was held constant at 30°C for all measurements. The equilibrium

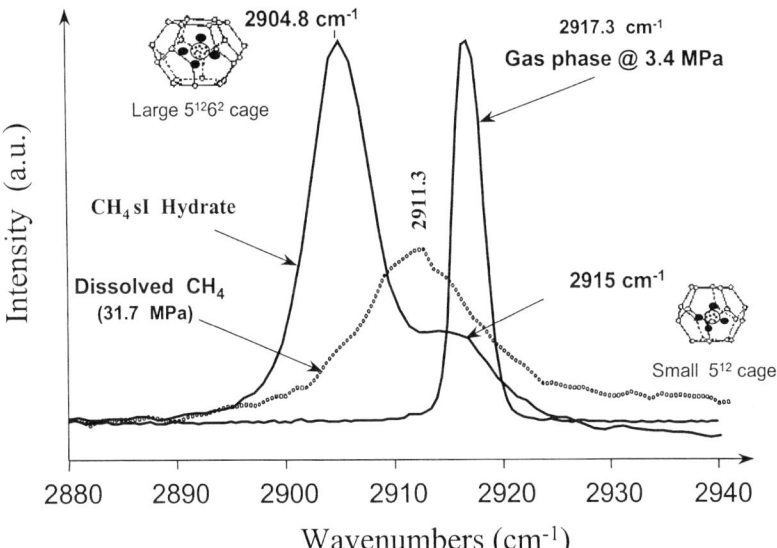

FIGURE 1. Comparison of Raman spectra of the ν_1 totally symmetric C-H stretching vibration of CH_4 in methane vapor at 3.4 MPa, CH_4 dissolved in water at 31.7 MPa, and CH_4 enclathrated in the cavities of pure sI CH_4 hydrate.

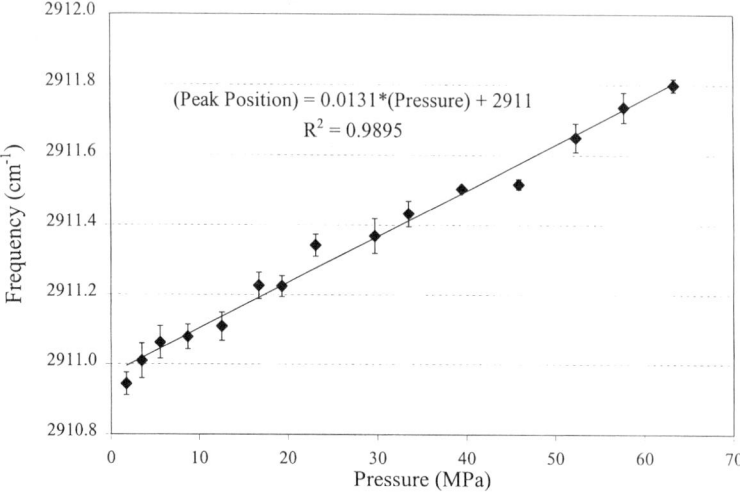

FIGURE 2. Variation in frequency of the ν_1 C-H stretching vibration of CH_4 dissolved in water with pressure at a constant temperature of 30°C. The increasing trend in frequency suggests that methane–water clusters resemble 5^{12} cavities in hydrates.

pressure for CH_4 hydrate formation at this temperature is predicted by CSMHYD to be around 65.8 MPa.[1] The trend indicates an increase in frequency from 2,910.9 cm^{-1} at 1.7 MPa to 2911.8 cm^{-1} at 63.3 MPa. This 0.9 cm^{-1} increase is fairly small, but it is reproducible. The error bars in FIGURE 2 indicate the standard deviation in frequencies (average $\sigma = 0.04$ cm^{-1}) of the C-H stretch based on four different spectra obtained at a given pressure. The direction of shift of the dissolved methane peak with pressure indicates an approach to the frequency (2,915 cm^{-1}) for methane trapped in the small 5^{12} cage in sI hydrate. This implies that with increasing pressure, the clustering of water around dissolved methane resembles the 20-coordinated small 5^{12} cage in sI hydrate. The existence of these 5^{12} cavity-like clusters implies that precursors resembling the basic building block of hydrates (5^{12} cavity) exist in the aqueous phase even before *P-T* conditions become favorable for forming hydrates.

Time-Resolved Raman Measurements of Kinetics of CH_4 Hydrate Formation

The time-invariant peaks for methane in the cavities of sI hydrate served as the end points for time-resolved hydrate formation experiments. The dissolved CH_4 peak served as the starting point.

FIGURE 3 shows the real-time Raman spectra for CH_4 sI hydrate formation from dissolved CH_4. In the figure, peak frequency (cm^{-1}) is plotted on the abscissa, against peak intensities in arbitrary units (a.u.) on the ordinate, with time (s) on the final, orthogonal axis. FIGURE 3 depicts a smooth transition from CH_4 dissolved in water (one Raman band) to CH_4 in the two cavities of sI hydrate (two Raman bands

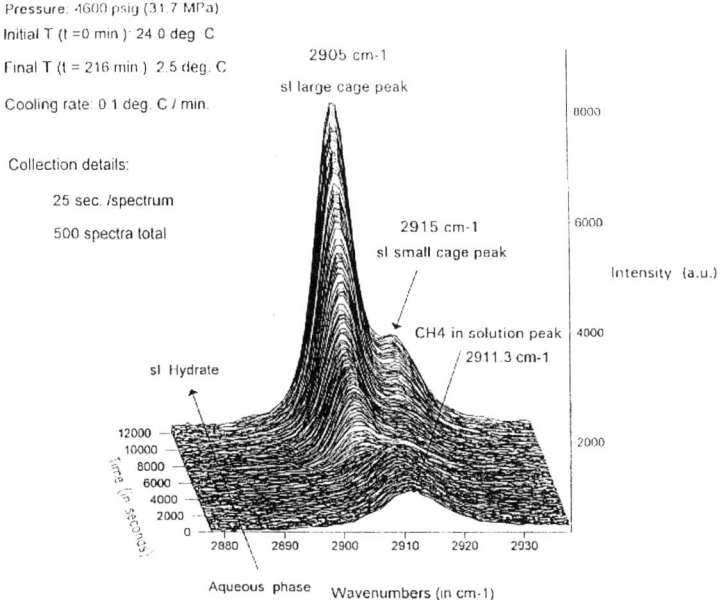

FIGURE 3. Real-time Raman spectra monitoring the transformation of dissolved CH_4 to CH_4 in the large $5^{12}6^2$ and small 5^{12} cages of sI CH_4 hydrate. Note that the temperature is decreasing with time and pressure is held constant.

at 2,905 cm^{-1} and 2,915 cm^{-1} for large and small cage, respectively) about 10°C into the temperature ramp.

Resolution and integration of hydrate cavity peaks yields areas that are proportional to the extent of occupation of a particular cavity by methane. Ratio of peak areas corresponding to the cages yields the relative amounts of methane occupying each type of cage. By repeating this procedure for every spectrum containing hydrate peaks shown in FIGURE 3, and obtaining ratios of areas of peaks for the large and small cavities, the relative number of large to small cavities was obtained as a function of time as sI CH_4 hydrate formed. This cavity ratio versus time profile is plotted in FIGURE 4, after the appearance of the first hydrate peaks in the spectral collection. The cavity ratio profile shows large to small cavity ratios as low as 0.5 and 1.0 in the initial stages, indicating that formation of the large $5^{12}6^2$ cavity is much slower than formation of the small 5^{12} cavity in sI CH_4 hydrate. The cavity ratio developed with time to a final value of about 3.2, which is close to the completely occupied expected equilibrium value of 3.0.

Similar trends in the cavity ratio versus time profile have been observed by researchers at the National Research Council, Canada using hyperpolarized ^{129}Xe NMR while monitoring the formation of xenon sI hydrate from powdered ice at low temperatures. FIGURE 5 shows a comparison between the cavity ratio versus time profile obtained from the time-resolved Raman studies on CH_4 hydrate formation

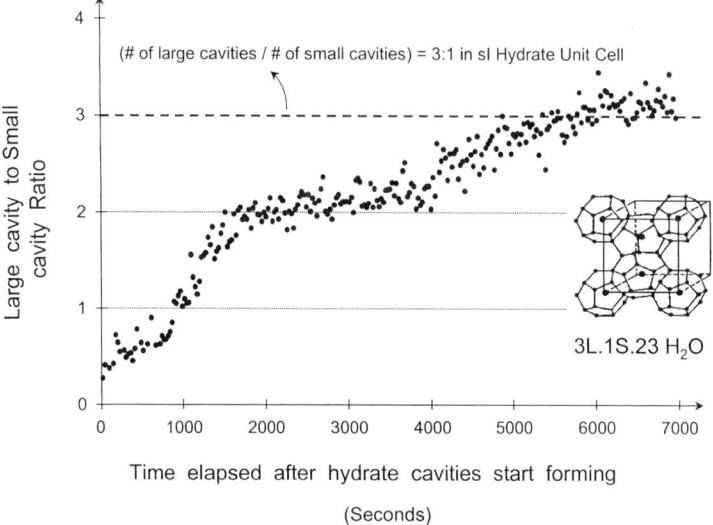

FIGURE 4. Ratios of numbers of large to small sI hydrate cavities forming as a function of time during sI CH_4 hydrate formation from CH_4 dissolved in water.

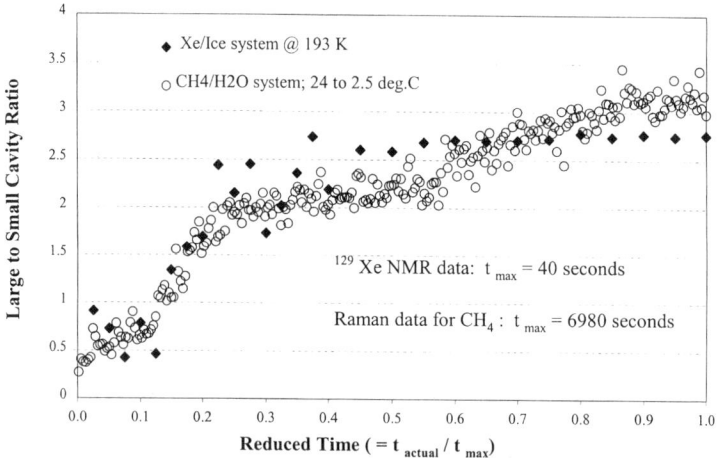

FIGURE 5. Comparison of large to small cavity ratio versus time profiles obtained from Raman studies of CH_4 sI hydrate formation from dissolved CH_4 (this work) and hyperpolarized ^{129}Xe NMR experiment monitoring xenon sI hydrate formation on powdered ice at 193 K.[8]

from dissolved CH_4 (this work) and the time-resolved ^{129}Xe NMR studies of sI xenon hydrate formation on the surface of powdered ice at 193 K.[8] Since the two experiments were carried out over significantly different time scales (40 seconds for the ^{129}Xe NMR experiment versus 6,980 seconds for the Raman experiment), a reduced time scale, as defined in FIGURE 5, had to be used to compare the cavity ratio versus time profiles. From FIGURE 5, it can be seen that both experiments appear to yield similar cavity ratio versus time profiles despite significant differences in P-T conditions and guest molecule type. Both profiles suggest the abundance of small 5^{12} cavities in the initial stages of hydrate formation indicating that the large $5^{12}6^2$ cavity formation is slower than the small 5^{12} cavity. Both profiles also seem to exhibit well-defined plateaus at cavity ratios of 0.6 and 2.0. These plateaus are not yet completely understood but are thought to result from competing rates of formation of the large and small cavities at the hydrate surface during the course of attaining a stable hydrate lattice. Deciphering the cause of the plateaus may provide insight into the mechanisms involved in combinations and rearrangements of cavities at the growing hydrate surface.

Macroscopic Observations and Inferences from Hydrate Formation

Visual observations made during CH_4 hydrate formation from dissolved methane in the high pressure optical cell show that hydrates first form as a film at the vapor–liquid interface. Following the rapid formation of this hydrate film, further hydrate growth occurs as fine needles extending into the bulk aqueous phase (FIGURE 6). For needle growth to progress into the bulk liquid, new methane guest molecules could

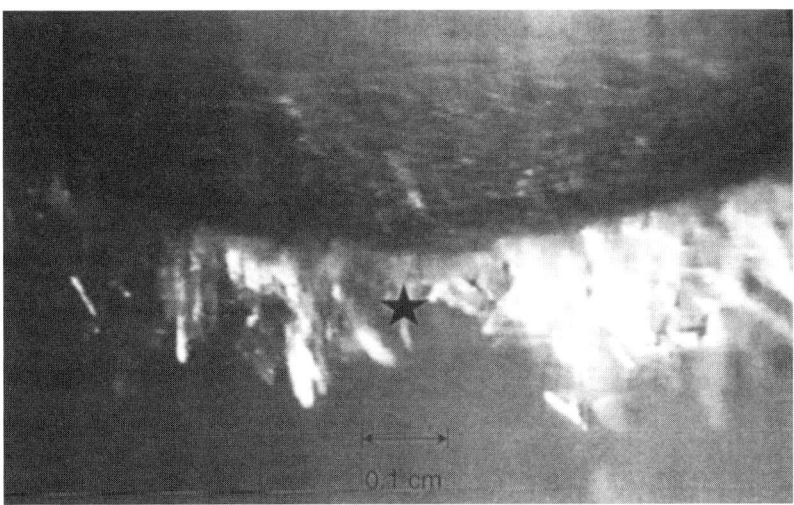

FIGURE 6. Snapshot of the cell contents at the end of a methane hydrate formation experiment from dissolved methane without any stirring in the cell. Note the hydrated V–L interface and the hydrate needles growing into the bulk aqueous phase. The point at which the laser is focused is indicated by a *star*.

be supplied either by the vapor phase or the aqueous phase. We hypothesize that the former may occur by slow diffusion of CH_4 through open microscopic cracks/pores in the hydrated V–L interface. The latter may occur by diffusion of methane from the bulk aqueous phase (dissolved methane) to the hydrated V–L interface. At this point, it is unclear as to where the CH_4 comes from. Nevertheless, the slow growth of hydrate needles past the laser spot used for Raman excitation yields time resolved information about molecular hydrate growth phenomena.

CONCEPTUAL MODEL FOR METHANE HYDRATE FORMATION

Based on the above results from microscopic Raman studies and macroscopic visual observations during formation of hydrates in the cell, a conceptual model tying in all the facts is proposed. FIGURE 7 depicts the proposed conceptual model for methane hydrate formation from dissolved methane. The emphasis is on the possible steps involved in hydrate growth once the V–L interface has been replaced by a hydrate film, and thus focuses on how the hydrate needles grow into the bulk liquid phase.

A methane molecule that ends up being enclathrated in the growing needles may either come from the vapor phase by diffusing through the cracks in the film to dissolve in the aqueous phase, or it may come from the aqueous phase where it is already dissolved. Nevertheless, in the aqueous phase, water molecules tend to

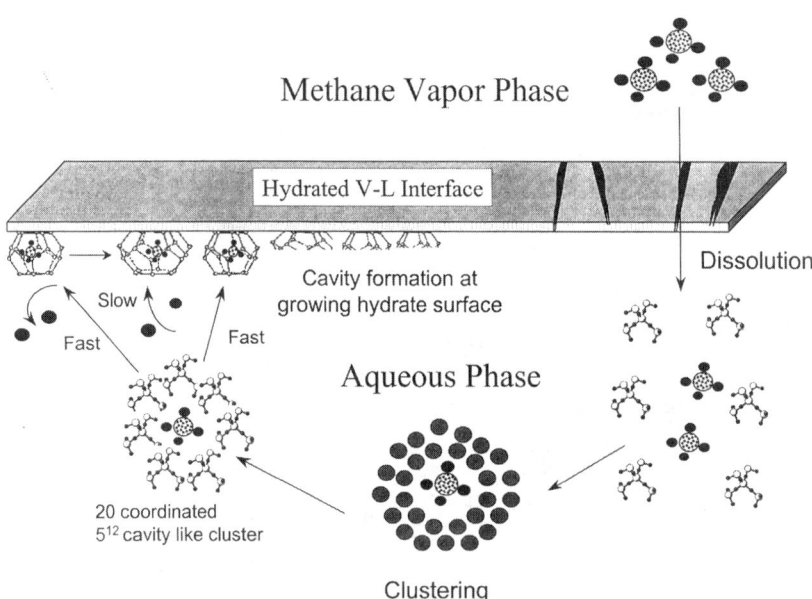

FIGURE 7. Conceptual mechanistic model for methane hydrate formation from dissolved methane.

cluster around the methane molecule in order to maximize the hydrogen bonding around the hydrophobic solute methane. These are relatively disordered clusters of water molecules that resemble the 20-coordinated small 5^{12} cavity in sI hydrate.

The next step is the incorporation of these clusters into the growing hydrate surface as well defined cavities with methane molecules trapped in them. This step occurs readily with expulsion of a few water molecules when forming the small 5^{12} cavity on the hydrate surface. This is because of the similar nature and configuration of the clusters and the cavity. The O-O-O angle is 108° in the case of pentagonal faces constituting the 5^{12} cavity. This angle is very similar to the 109.5° angle corresponding to the tetrahedral coordination in water. The large 24-coordinated $5^{12}6^2$ cavity, however, requires the formation of two hexagonal faces that force the water molecules forming those faces to be in a 120° angle configuration. The stretching of the O-O-O angle to 120°, well above the 109.5° angle corresponding to tetrahedral coordination of water, introduces some strain in formation of the large $5^{12}6^2$ cavity. Thus, formation of the large $5^{12}6^2$ cavities must be a slower step compared to the formation of the small 5^{12} cavities. This reasoning agrees well with the inferences from the time-resolved Raman data. It is interesting to note that despite the difficulty in formation of large $5^{12}6^2$ cavities, CH_4 forms sI hydrate which has an abundance of large cavities compared to the easier to form small 5^{12} cavities.

It is not clear at this point whether the large $5^{12}6^2$ cavities form directly from a methane–water cluster, or if they form by addition of water molecules to existing 5^{12} cavities on the growing hydrate surface. Further experiments using an internal intensity standard are needed to address the above question.

CONCLUSIONS

This work focuses on both time independent studies of methane dissolved in water and time dependent measurements of dissolved methane transforming to methane sI hydrate. There is evidence from Raman spectroscopy for the clustering of water molecules around dissolved CH_4, which suggests that the CH_4–H_2O clusters resemble the small 20-coordinated 5^{12} cavity in sI CH_4 hydrate. The cavity ratio versus time profile obtained for sI CH_4 hydrate formation indicates that formation of the large $5^{12}6^2$ cavity is much slower than formation of the small 5^{12} cavity in sI methane hydrate. The results from these studies have been used to generate a conceptual model of methane hydrate formation from methane dissolved in water.

ACKNOWLEDGMENTS

The National Science Foundation provided funding for this work under Grant CTS-9634899. We are very grateful to Dr. Robert C. Burruss for his support and advice. Dr. Vu Thieu designed and fabricated the high pressure optical cell.

REFERENCES

1. SLOAN, E.D. 1998. Clathrate Hydrates of Natural Gases, 2nd edit. Marcel Dekker, New York.
2. ENGLEZOS, P., N. KALOGERAKIS, P.D. DHOLABAI & P.R. BISHNOI. 1987. Kinetics of formation of methane and ethane gas hydrates. Chem. Eng. Sci. **42:** 2647–2658.
3. ENGLEZOS, P., N. KALOGERAKIS, P.D. DHOLABAI & P.R. BISHNOI. 1987. Kinetics of gas hydrate formation from mixtures of methane and ethane. Chem. Eng. Sci. **42:** 2659–2666.
4. SKOVBORG, P. & P. RASMUSSEN. 1994. A mass transport limited model for the growth of methane and ethane gas hydrates. Chem. Eng. Sci. **49:** 1131–1143.
5. PIETRASS, T., H.C. GAEDE, A. BIFONE, A. PINES & J.A. RIPMEESTER. 1995. Monitoring xenon clathrate hydrate formation on ice surfaces with optically enhanced ^{129}Xe NMR. J. Am. Chem. Soc. **117:** 7520–7525.
6. LONG, X.P. 1994. Raman Study of Tetrahydrofuran Clathrate Hydrate. Masters Thesis T-4458, Colorado School of Mines, Golden, Colorado, USA.
7. SUM, A.K., R.C. BURRUSS & E.D. SLOAN. 1997. Measurement of clathrate hydrates via Raman spectroscopy. J. Phys. Chem. B. **101:** 7371–7377.
8. MOUDRAKOWSKI, I. & J.A. RIPMEESTER. 1999. Personal Communication.

In Situ Observations of Methane Hydrate Formation Mechanisms by Raman Spectroscopy

T. UCHIDA,[a,b] R. OKABE,[c] S. MAE,[c] T. EBINUMA,[b] AND H. NARITA[b]

[b]Hokkaido National Industrial Research Institute, 2-17-2-1 Tsukisamu-higashi, Toyohira-ku, Sapporo 062-8517, Japan

[c]Faculty of Engineering, Hokkaido University, N13 W8 Kita-ku, Sapporo 060-8628, Japan

ABSTRACT: We compared time-resolved Raman spectroscopic measurements of CH_4-saturated aqueous solution with CH_4 hydrate crystal formed from the same solution. At 287.1 K and 6.6 MPa the C-H stretching mode for CH_4 molecules in the solution (ν_{sol} = 2911.7 cm^{-1}) was intermediate between that in the large cage (ν_L = 2905.1 cm^{-1}) and that in the small cage (ν_S = 2914.1 cm^{-1}), thus indicating an intermediate molecular environment. Comparison of these results with previous work on the CO_2–H_2O system indicated that water molecules in the type-I small-cage-like structure surrounded CH_4 molecules in the solution. Furthermore, the free energy barrier for CH_4 hydrate nucleation is likely to be formation of the large cage structure, which is probably larger than that for CO_2 hydrate formation.

INTRODUCTION

Methane hydrate ($CH_4 \cdot n\,H_2O$) is a crystalline molecular complex that includes a large quantity of methane molecules and is stable at high pressure and low temperatures. Its unit cell consists of 46 water molecules that construct two small cages (pentagonal dodecahedron, 5^{12}) and six large cages (tetrakaidecahedron, $5^{12}6^2$).[1] Since at most one CH_4 molecule can occupy each cage, eight or fewer CH_4 molecules are expected per unit cell. If the hydrate is fully occupied by CH_4 molecules, the number of moles of water reacted with one mole of CH_4, n, is 5.75.

CH_4 hydrates are expected to be an unconventional natural gas resource in the near future. They exist under deep-sea floor or underground in the permafrost zone. Furthermore, CH_4 hydrate has several unique characteristics for potential engineering use. Because it contains natural gas in high concentration, one such application is for natural gas storage and transport. To develop CH_4 hydrates as an energy resource or as an engineering utility material, it is very important to know its gas content and to understand its formation and dissociation processes.

Many studies have been done on the hydrate formation process. When CH_4 hydrate is formed from solutions, the steps are probably as follows:[1]

[a]Telecommunication. Voice: +11-857-8965; fax: +11-857-8989.
uchida@hniri.go.jp

1. molecular dissolution into water,
2. water molecules form the crystal-like construction around the gas molecules,
3. nucleation of CH_4 hydrate, and
4. crystal growth of the bulk CH_4 hydrate.

Many studies have reported an induction period between dissolution (Step 1) and hydrate nucleation (Step 3). To use CH_4 hydrate in engineering applications, control of the induction period is necessary. Understating the mechanisms during induction might lead to control of the induction period, and therefore, more control of hydrate formation.

Another problem is that there are relatively few quantitative studies on gas concentration in the hydrate. The high pressure and low temperature stability region of gas hydrate make it difficult to apply the usual measurement techniques. Recently, the spectroscopic techniques, NMR[2,3] and Raman spectroscopy,[4,5] were shown to be useful for quantitative measurement of gas concentration in hydrates. Furthermore, Raman spectroscopy has the advantage that it can be used *in situ*. This technique can detect molecular properties both in the hydrate and in the solution. In the present study, *in situ* observations of CH_4-hydrate formation were made by Raman spectroscopy to study the encaging process of CH_4 molecules (Step 2), and the change in gas concentration during hydrate during formation.

To understand hydrate formation from solution, we compared results obtained for the CH_4-water system with those obtained for the carbon dioxide (CO_2)-water system. Both have the type-I clathrate structure, but the gas molecules are encaged differently: CH_4 molecules occupy both the small and large cages, but CO_2 molecules mostly occupy the large cages. Comparison of these two systems allowed us to better understand the formation of gas hydrates and to estimate the free energy barriers for hydrate nucleation.

EXPERIMENTAL PROCEDURES

FIGURE 1 shows schematics of the reaction vessels and measurement configurations. The reaction vessel for macroscopic observation (FIG. 1a) was equipped with four glass windows (white-sapphire), a pressure transducer (Kyowa PHS-100KA; accuracy ±0.2 MPa) and a T-type thermocouple (accuracy ±0.4 K). A 100 mW Ar-ion laser light (wavelength 514.5 nm) was directed into either the top or side of the vessel, and 90-degree scattered light was focused to the high-sensitivity Raman spectrometer (SPEX RAMAN 750M) equipped with a 1800-groovs/mm grating. The Raman shift was calibrated by the light of the Ne lump. Its accuracy was ±1.5 cm^{-1} using the full width at half maximum of Ne light. This experimental set up allows observation of the Raman spectra of the CH_4 solution without any superimposed spectra. The spectrometer had a single monochromator and a CCD detector with a 1024 × 256 pixel array. We obtained spectra for intervals of several minutes throughout the experiment.

We also used the microscopic observations system shown in FIGURE 1(b). We observed the formation reaction and measured the spectral change through a microscope (Olympus BH2-UMA) and one sapphire window; hence, 180-degree scattered light was measured. Image magnification was 20× and the space resolution was

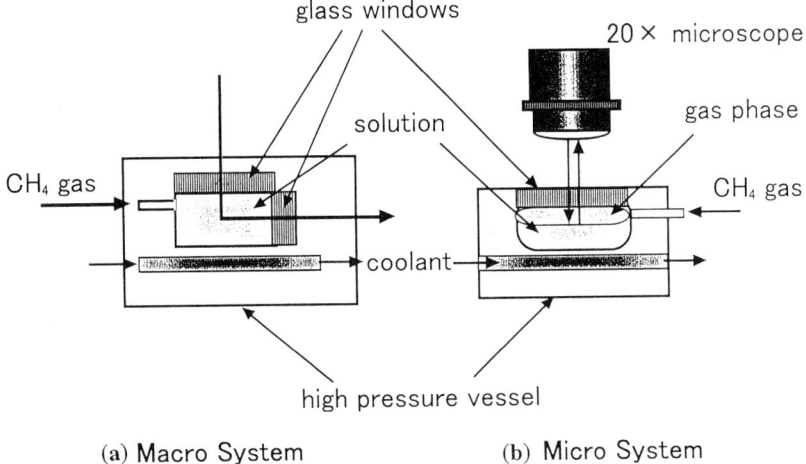

FIGURE 1. Schematics of sample chambers and observation systems: **(a)** macroscopic setup, **(b)** microscopic setup.

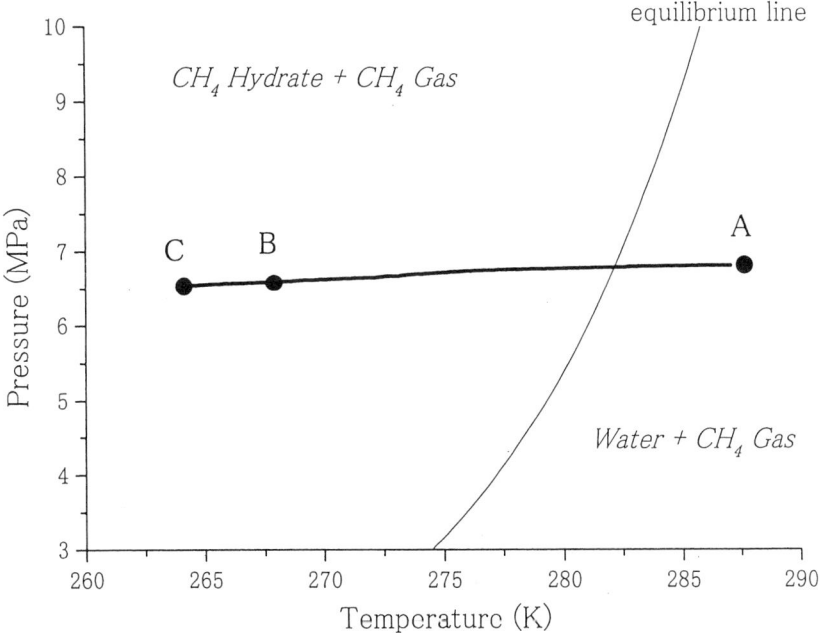

FIGURE 2. Experimental conditions for hydrate formation by cooling. The *thick line* indicates the *P-T* conditions of the system and the *thin line* denotes equilibrium between CH_4 hydrate and solution.[1] Points *A*, *B*, and *C* are typical conditions of the gas-water system, the supercooling solution, and the hydration point, respectively.

approximately 8 μm. Because the microscope had a short focus length, this setup could make Raman spectra measurements near the gas-water interface. However, the Raman spectra thus obtained included the spectrum from CH_4 gas. To study the variations of the spectra of aqueous CH_4 molecules, the peaks were separated using the commercial software (Microcal Origin ver.5.0).

To observe the change in Raman spectra due to hydrate formation, the reaction vessel was cooled from 287.1 K and pressure 6.6 MPa to 264.2 K and 6.3 MPa with the cooling rate of 0.1 K min^{-1}. The thick line in FIGURE 2 shows this process on a P-T diagram; the equilibrium conditions of CH_4 hydrate are shown as a thin line.[1] Several solid circles on this thick line (denoted A, B, and C) indicate typical measurement conditions for the solution and hydrate.

Water samples used for this study were distilled and filtered through three kinds of filters; the electric resistance was approximately 18.2 MΩ. The CH_4 gas, supplied from Daidoh-Hoxsan Co., Ltd., had a purity of 99.99%.

RESULTS AND DISCUSSIONS

Raman Spectra of Dissolved Methane Molecules

FIGURE 3 shows the Raman spectra of CH_4 saturated water (T = 273.4 K and P = 7.0 MPa). The peak position of CH_4 gas is also shown by the arrow in this figure. The C-H stretching (v_1) mode of CH_4 molecules dissolved in solution is the small peak near that of the gas, whereas the O-H stretching mode of water is the large plateau above 3,000 cm^{-1}. Compared to the gas peak, v_g = 2,917 cm^{-1} and full width at half maximum Δ = 4.2 cm^{-1}, the dissolved gas peak shifted to lower wave number, v_{sol} = 2,911.5 cm^{-1}, and broadened, Δ = 11.3 cm^{-1}. This result agrees with that obtained by Sloan et al.[6]

The peak wave number of the dissolved gas was close to that of the CH_4 molecule in the small hydrate cage; thus, interactions between dissolved CH_4 molecules and the surrounding water molecules were probably similar to those in the small cage (v_s = 2,913 cm^{-1}).[4,5] This interpretation is supported by a study of water clusters. Yamamoto et al.,[7] showed that a stable water cluster in the solution is the 20-bodies cluster and proposed that it had the 5^{12} arrangement (which also has 20 water molecules).

Since occupancy of the large cage is three times that for the small cage for CH_4 hydrate, the surroundings of the CH_4 molecule in the solution must change to that of the large cage to form the bulk hydrate crystals. However, Raman spectra of CH_4 hydrate indicates that the Raman peak offset for CH_4 in the large cage is greater than that in the small cage,[4,5] thus indicating that motion of the CH_4 molecule in the large cage is slightly more restricted than in the small cage. Therefore, the difficulty of reconstruction from water clusters to the large cages could be the free energy barrier for CH_4 hydrate nucleation.

Raman spectra of H_2O molecules in the solution, such as the O-H stretching mode (approximately from 3,000 to 3,800 cm^{-1}), did not change significantly compared with those of pure liquid water even if the P-T was in the hydrate stability region. The O-H stretching mode might change if the water molecules interacted with dissolved CH_4 molecules, but the change was too small to detect by Raman

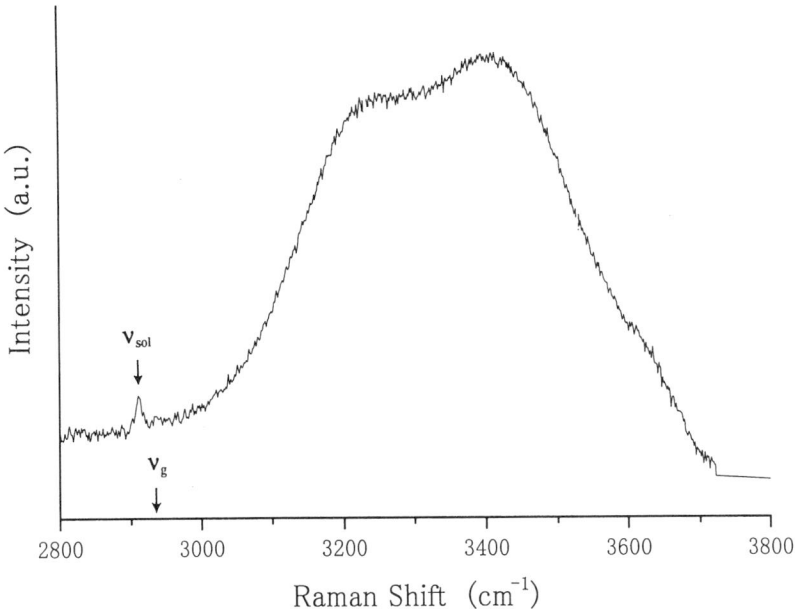

FIGURE 3. Raman spectra of CH_4 dissolved in water as recorded by the macroscopic setup ($T = 273.4$ K and $P = 7.0$ MPa). The *arrow* denotes the peak position for the C-H stretching (ν_1) mode of CH_4 molecule in the gas phase. The ν_1 mode of CH_4 in aqueous solution was observed slightly lower than the gas phase. The large, broad peak from the O-H stretching mode was observed above 3,000 cm^{-1}.

spectroscopy. This is mainly because of the small fraction of water molecules that can interact with dissolved CH_4 molecules and is consistent with the very small solubility of CH_4 in water.

Raman Spectra Change in C-H Stretching Mode during the Hydrate Formation

A change in the ν_1 mode of CH_4 molecules in solution was observed during cooling, under the conditions shown by large circles in FIGURE 2. The resulting Raman spectra are shown in FIGURE 4. Since these spectra came from the experimental setup of FIGURE 1(b), a large signal from the gas phase spectrum was also measured. The thick lines in FIGURE 4 are the ν_1 spectra after the peak separation calculation and were used to investigate changes in the Raman spectra of dissolved CH_4 molecules.

Under condition A ($T = 287.1$ K and $P = 6.6$ MPa), there was no hydrate phase. Water was saturated with CH_4 gas and the ν_1 mode of the dissolved gas is $\nu_{sol} = 2{,}911.7$ cm^{-1} and $\Delta = 9.9$ cm^{-1}. These values agree with those obtained from the macroscopic setup (FIG. 2).

The system started to cool at the rate of 0.1 K min^{-1} until the bulk hydrate formed. At condition B ($T = 268.2$ K, $P = 6.5$ MPa), the solution was supercooled although bulk hydrate was not observed. The stretching mode of the dissolved CH_4

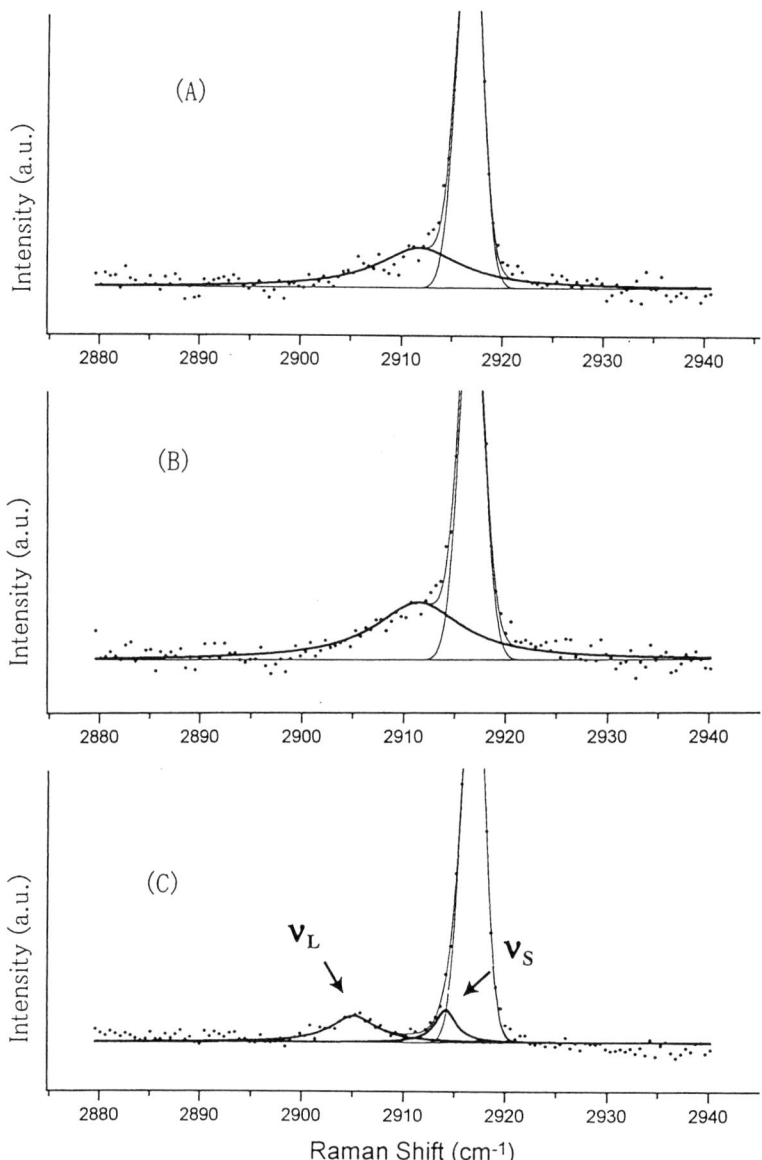

FIGURE 4. Change in Raman spectra of CH_4 molecules in solution and in the hydrate phase (*thick lines*). The large peak at 2,917 cm^{-1} (*thin line*) is the gas phase spectrum. Spectra in (**A**), (**B**), and (**C**) were measured under conditions labeled identically in FIGURE 2. In (**C**), ν_L and ν_S denote the peak wave numbers of CH_4 molecules in the large and small cage, respectively.

molecules, v_{sol}, peaked at 2,911.4 cm^{-1} with a half width, Δ, of 10.5 cm^{-1}. Compared to the Raman spectra at condition A, v_{sol} was almost the same and Δ increased slightly. The peak intensity also slightly increased. These peak changes might be due, not only to the decrease of temperature, but also to the increase of conditions of CH$_4$ molecules in the solution.

Finally, bulk hydrate formed in the vessel and the Raman spectra split into two peaks, $v_L = 2,905.1$ cm^{-1}, $\Delta_L = 5.3$ cm^{-1}, and $v_S = 2914.1$ cm^{-1}, $\Delta_S = 2.5$ cm^{-1}, respectively. The former peak corresponds to the CH$_4$ molecules in the large cage and the latter to the small cage. Within the accuracy of the measurements, these values coincide with those observed in the literature.[4,5] The change in spectra supports the assertion that the bulk hydrate formation requires a change in interaction between methane molecules and the surrounding water from that of the solution.

Since there are three times more large cages as small cages in a unit cell, the park intensity of v_L should be three times larger than that of v_S. This peak intensity ratio occurs in bulk CH$_4$ hydrate crystals,[4-6] but FIGURE 4(C) shows that these peak intensities were nearly the same. This might be because we were also measuring signals from solution. During the final stage of the crystallization, the peak intensities should become those of bulk hydrate.

Comparison of Raman Spectra with Those in the $CO_2 + H_2O$ System

The C-O stretching mode of CO$_2$ molecules in the hydrate phase and in the solution without hydrates is at approximately 1,309 cm^{-1} and 1,319 cm^{-1}, respectively.[8] Uchida et al.,[8] by including the interaction between CO$_2$ and surrounding water molecules, estimated the spherical cavity radius a in water that surrounds a CO$_2$ molecule in each phase. According to their calculation, $a = 2.5$ Å for aqueous CO$_2$ without hydrate phase whereas the van der Waals radius of a CO$_2$ molecule is 2.56 Å: water molecules closely surround CO$_2$ molecules dissolved in water. However, when CO$_2$ molecules occupy hydrate cages, $a = 3.1$ Å, the average free space of a large cage in structure I.[1] These results suggested that water molecules surrounding a CO$_2$ molecule in aqueous solution do not form cages, but almost contact the molecule. Raman shifts of these phases indicate that the CO$_2$ in aqueous solution has the largest interaction from surrounding water phase, and that the motion of CO$_2$ molecules becomes less constrained when the molecule is in the hydrate cage. These situations are different from those in CH$_4$ + H$_2$O system.

We considered the condition of gas molecules in aqueous solution and their formation of hydrates qualitatively by comparing results obtained from Raman spectra. For CH$_4$, liquid water molecules surround a CH$_4$ molecule with the small-cage-like structure. Due to the similar surroundings, the motion of the CH$_4$ molecule in aqueous solution is similar to that in the small cage. However, since the bulk CH$_4$ hydrate has three times as many large cages as small cages to form bulk hydrate, water molecules must reconstruct the large-cages from the small-cage-like structures. The motion of the CH$_4$ molecule in the large cage is, however, more restricted than that in the small cage. This reconstruction process is, therefore, thought to be a free energy barrier for the hydrate nucleation.

On the other hand, water molecules closely surround CO$_2$ molecules in the solution and their motion is strongly restricted. This could be due to mainly the straight configuration of the CO$_2$ molecule. After hydrate formation, CO$_2$ molecules primarily occupy the large cage of the structure I. This cage is somewhat flattened and better fits the CO_2 molecule shape. Hence, the motion of encaged CO_2 is freer than that in solution. Thus, the free energy barrier for formation of CO_2 hydrate from the solution should be smaller than that for CH_4 hydrate.

CONCLUSIONS

Raman spectra of CH_4–water system was measured to better understand the conditions of dissolved CH_4 molecules and the formation of CH_4 hydrates. The Raman shift of aqueous CH_4 molecules was similar to that of CH_4 molecules in the small cage of the hydrate. However, the full width at half maximum of v_1 for CH_4 in solution was more than twice that of the small hydrate cage.

Comparison of the change of Raman spectra upon crystallization between CH_4 and CO_2 molecules indicated that the dissolved conditions were different. The difference in the free energy barrier for hydrate nucleation from the solution probably results from the difficulty in constructing the hydrate flame from the cluster configurations, and this depends on interactions between guest and host molecules. The free energy barrier for CO_2 hydrate nucleation should be smaller than that for CH_4 because the CO_2 motion in the hydrate is freer than it is in solution. For CH_4 hydrates, the free energy barrier is probably formation of the large cage because it restricts the motion of the CH_4 molecule.

ACKNOWLEDGMENTS

This study was part of the project organized by New Energy and Industrial Technology Development Organization (NEDO): Preliminary Research and Development on Technology for Using Gas Hydrates as Resources. Authors acknowledge Dr. K. Haraguchi for supplying water sample. They also thank Dr. J.A. Ripmeester, Dr. C.A. Tulk, and Dr. S. Subramanian for their discussions, and Ms. Y. Hirata for help with preparing this manuscript.

REFERENCES

1. SLOAN, E.D., JR. 1998. Clathrate Hydrate of Natural Gases, 2nd edit. Marcel Dekker, Inc., New York.
2. RIPMEESTER, J.A. & C.I. RATCLIFFE. 1988. Low-temperature cross-polarization/magic angle spinning 13C NMR of solid methane hydrates: structure, cage occupancy, and hydration number. J. Phys. Chem. **92:** 337–339.
3. COLLINS, M.J., C.I. RATCLIFFE & J.A. RIPMEESTER. 1990. Nuclear magnetic resonance studies of guest species in clathrate hydrates: line-shape anisotropies, chemical shifts, and the determination of cage occupancy ratios and hydration numbers. J. Phys. Chem. **94:** 157–162.
4. UCHIDA, T., A. TAKAGI, T. HIRANO, H. NARITA, J. KAWABATA, T. HONDOH & S. MAE. 1996. Measurements on guest-host molecular density ratio of CO_2 and CH_4 hydrates by Raman spectroscopy. Proc. 2nd Int. Conf. on Natural Gas Hydrates. Toulouse, France. 335–339.
5. SUM, A.K., R.C. BURRUSS & E.D. SLOAN, JR. 1997. Measurements of clathrate hydrates via Raman spectroscopy. J. Phys. Chem. B **101:** 7371–7377.

6. SLOAN, E.D., S. SUBRAMANIAN, P.N. MATTHEWS, J.P. LEDERHOS & A.A. KHOKHER. 1998. Quantifying hydrate formation and kinetic inhibition. Ind. Eng. Chem. Res. **37:** 3124–3132.
7. YAMAMOTO, Y., A. WAKISAKA, T. SAITO & T. AKIYA. 1996. Measurements of the hydrate cluster by the liquid molecular beam mass spectrometry. Proc. 2nd Int. Conf. on Natural Gas Hydrates. Toulouse, France. 355–362.
8. UCHIDA, T., A. TAKAGI, S. MAE & J. KAWABATA. 1997. Dissolution mechanisms of CO_2 molecules in water containing CO_2 hydrates. Energy Convers. Mgmt. **38:** S307–S312.

Calorimetry to Study Metastability of Natural Gas Hydrate at Atmospheric Pressure and Temperatures below 0°C

O.I. LEVIK[a] AND J.S. GUDMUNDSSON

Department of Petroleum Engineering and Applied Geophysics,
Norwegian University of Science and Technology (NTNU), 7491 Trondheim, Norway

ABSTRACT: Natural gas hydrate is metastable when stored at atmospheric pressure and temperatures below 0°C. The hydrate is regarded as metastable since, under these pressure–temperature conditions, the rate of dissociation is very low. Isothermal calorimetry was used to study the rate of dissociation, which is a measure of the metastability. Low temperatures and large samples improve the stability. The isothermal method is being developed to quantify metastability. Scanning calorimetry was used to determine the hydrate number and amount of free water. In the scanning method, the natural gas hydrate sample was pressurized with methane. The calorimeter was operated below the methane hydrate equilibrium line and above the natural gas hydrate equilibrium line. Prior to analysis, the sample was conditioned in a separate heating-cooling cycle. This was necessary to eliminate undesirable thermal responses due to desorption as the ice melted. Desorption occurred because the sample was refrigerated (-20°C) under a high natural gas pressure, but analyzed under a relatively low methane pressure. The scanning method is being developed to analyze natural gas hydrate that contains large amounts of free water.

INTRODUCTION

When natural gas hydrate (NGH) is subject to temperatures below 0°C at 1 bara it remains metastable and can be stored for prolonged periods without large losses of gas.[1] However, there will always be situations where heat from the surroundings reaches the metastable hydrate, leading to temperature increase or dissociation. The aim of this work was to develop methods to quantify metastability properties of NGH for different rates of heat supply at temperatures below 0°C and at a pressure of 1 bara. This is to assess the effects that heat-flow to the sample, storage temperature, and particle size have on hydrate number and amount of free water. Differential scanning calorimetry (DSC) with three heating-cooling cycles was used. Storage of hydrate was simulated in a calorimeter by operating it isothermally. The DSC experiments were carried out using a Setaram BT2.15 heat-flow calorimeter capable of operating at elevated pressures.

[a]Telecommunication: Voice: (+47) 73 59 49 45; fax: (+47) 73 94 44 72.
levik@ipt.ntnu.no

HEAT FLOW CALORIMETRY

In heat-flow calorimeters two cells are placed inside a thermal block. One of the cells is the sample cell; the other is the reference cell, which is empty. Both cells are surrounded by heat flow meters. The heat signal is registered for both cells. The output heat flow is the difference between the two cells. This difference is, in principle, solely due to the sample. However, since it is impossible to make two identical cells, there is also a signal due to the difference between the cells. This signal may be ignored in many cases, or it can be corrected by subtracting the differential heat signal of a blank run (both cells empty). The logged sample temperature is in reality based on signals from thermocouples just *outside* the sample cell. This largely explains why isothermal melting processes may appear to take place over a temperature interval. The calorimeter is otherwise calibrated to yield the sample temperature with sufficient accuracy.

NGH metastability may be regarded as a kinetic phenomenon. Special precautions must be taken in calorimetric kinetic experiments on solids. The results are sensible to temperature, scanning rate, heat flow, and sample mass. In addition, parameters such as sample type (granulates or powder) and atmosphere are also important. The kinetic parameters obtained from such experiments are usually system dependent and are not general compound data.[2] Reactant concentration is a central parameter in kinetics. However, for solid crystalline systems, like NGH, concentration is a term that bears no definite meaning. In addition, solid reactions do not always proceed uniformly throughout the sample, but often in local zones of enhanced reactivity, such as the surface.[3]

Kinetic investigations by calorimetry may be carried out either by scanning procedures or isothermal procedures. An isothermal procedure was tested in this work. When a sample is subject to an isothermal procedure, it is the same as storing the sample in the calorimeter cell while recording the heat flow to the sample. In case of a NGH sample, there is an endothermic signal if it is undergoing dissociation. If the heat flow to the sample increases, then the rate of dissociation increases proportionally.

PREPARATION AND HANDLING OF SAMPLES

Hydrate samples were obtained by in-line filtration of a hydrate-in-water slurry in a flow-loop. The gas that was used was a synthetic natural gas (92 mol% methane, 5 mol% ethane, 3 mol% propane). The water was deaerated tap-water. The hydrate production took place in a nine-liter continuous stirred tank reactor. The flow-loop and the production process was described in detail by Gudmundsson *et al.*[4] The filters were cylindrical cup filters (0.5 dm^3) made from porous (80 μm) polymer placed inside a pressure vessel. After filtration, excess water inside the vessel was displaced by gas that had the same composition as the hydrate forming gas. Yet, the filter cake contained a considerable amount of free water. This water was saturated in the hydrate forming gas at the *PT* conditions that applied during hydrate production. While still pressurized, the filter vessel was left overnight in a freezer room at about $-20°C$. The pressure was then released, the vessel opened, and the filter cake retrieved. The cake with the hydrate looked a lot like an ice cylinder filling the filter cup. Samples were prepared by crushing the filter cake in a mortar. Samples of different representative sizes were sorted (±0.2 mm) by sieving. Samples could then be stored or subjected to calorimetric analysis immediately. The sample cell was cooled before a sample was introduced. Prior to the calorimetric analysis, samples were at no time exposed to temperatures above 0°C.

LABORATORY SETUP FOR NGH STORAGE

To investigate effects of representative sample size (D) and storage temperature (T_s), NGH samples were stored for prolonged times in desiccators. The humidity of the air inside the desiccators was maintained at about 92–98%RH by placing crushed ice on the bottom. Temperature was controlled by placing the desiccators in freezer cabinets with temperature regulation (±2°C). Air humidity and temperature were logged routinely. The hypothesis has been set forth, that for metastable NGH, hydrate number (n) and amount of free water (f) change little during storage. A calorimetric method was developed to measure these quantities in order to test the hypothesis.

THREE-CYCLE DSC

Hydrate number and amount of free water may be determined by DSC. Handa[5] used such a method. A key element of the method was to pressurize the sample by the hydrate forming gas and then heat the sample above 0°C. In this way, the frozen free water melted while the hydrate remained thermodynamically stable. The amount of free water was calculated from mass and energy balances. The method was successfully demonstrated on samples of xenon and krypton hydrates, and was the starting point in the development of the three-cycle method developed in this work. Handa's samples contained little free water. Hydrate formation during the analysis was negligible. However, the samples in this work contained a lot of free water. When the samples were pressurized with the hydrate forming natural gas, hydrate formation during the analysis was *not* negligible. This was concluded on grounds of exothermic signals. To get around this, the samples were instead pressurized with methane. The pressure had to be between the hydrate curves of methane (sI) and the hydrate forming gas (sII). For the temperatures that were applied during analysis, 17 bara was suitable. Under these conditions, no hydrate should form or dissociate. Methane hydrate would not form since the pressure was too low. NGH already present would not dissociate since the pressure was too high (see FIG. 4).

The thermal response due to mass transfer (desorption) between the free water and the pressurized gas (methane) above the sample was significant. This thermal response coincided with the thermal response of melting the frozen free water, which provided the basis for back-calculations to determine amount of free water and ultimately the hydrate number. It was not possible to separate the two thermal responses. Thus, it was necessary to condition or pretreat the sample, or actually the free water, so that desorption would not occur. The solution was to melt the frozen free water

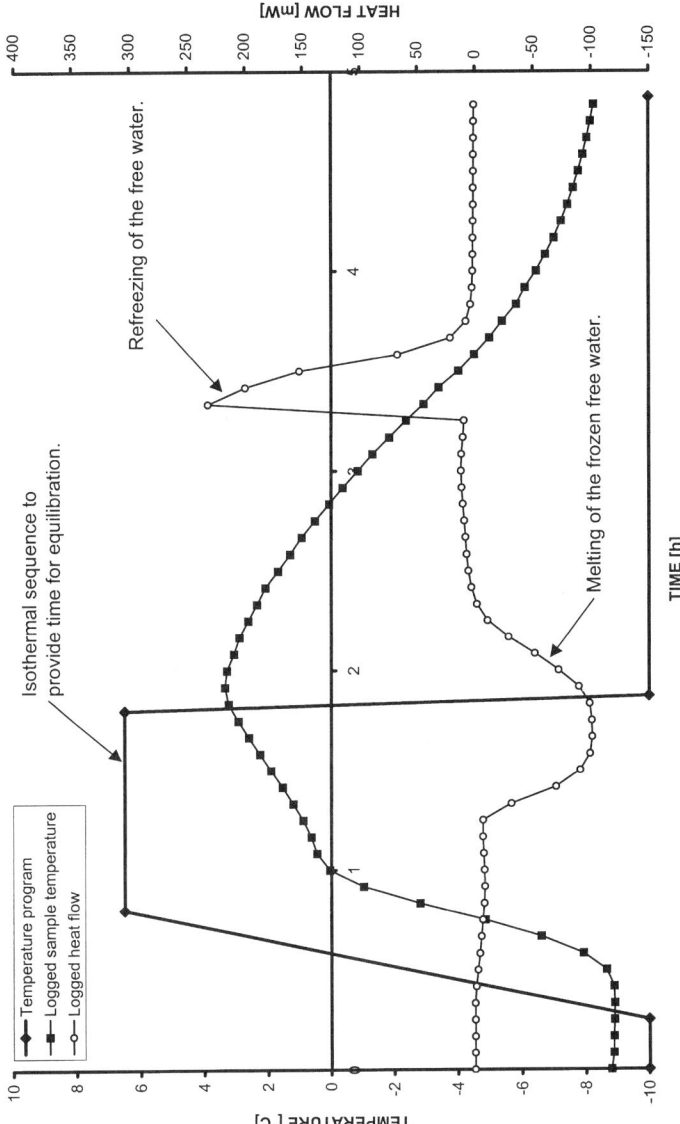

FIGURE 1. Ice conditioning. Melting and refreezing of the free water (ice) at 17 bara. The purpose is to pretreat or condition the free water (ice) by equilibrating it with the enveloping gas (methane, 99.95 mol%) and then refreeze it. Refer also to the corresponding operating line in FIGURE 4.

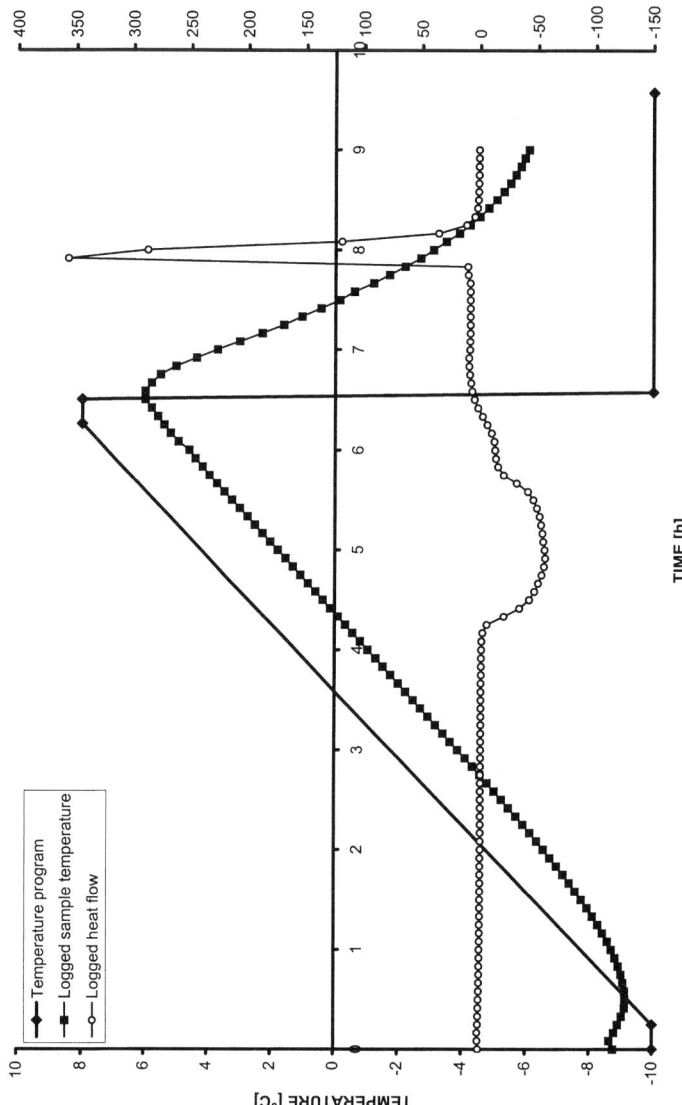

FIGURE 2. Ice melting. Melting and refreezing of the free water (ice) at a methane pressure of 17 bara. The purpose is to collect raw data to establish the mass and energy balance to calculate amount of free water and hydrate number. Refer also to the corresponding operating line in FIGURE 4.

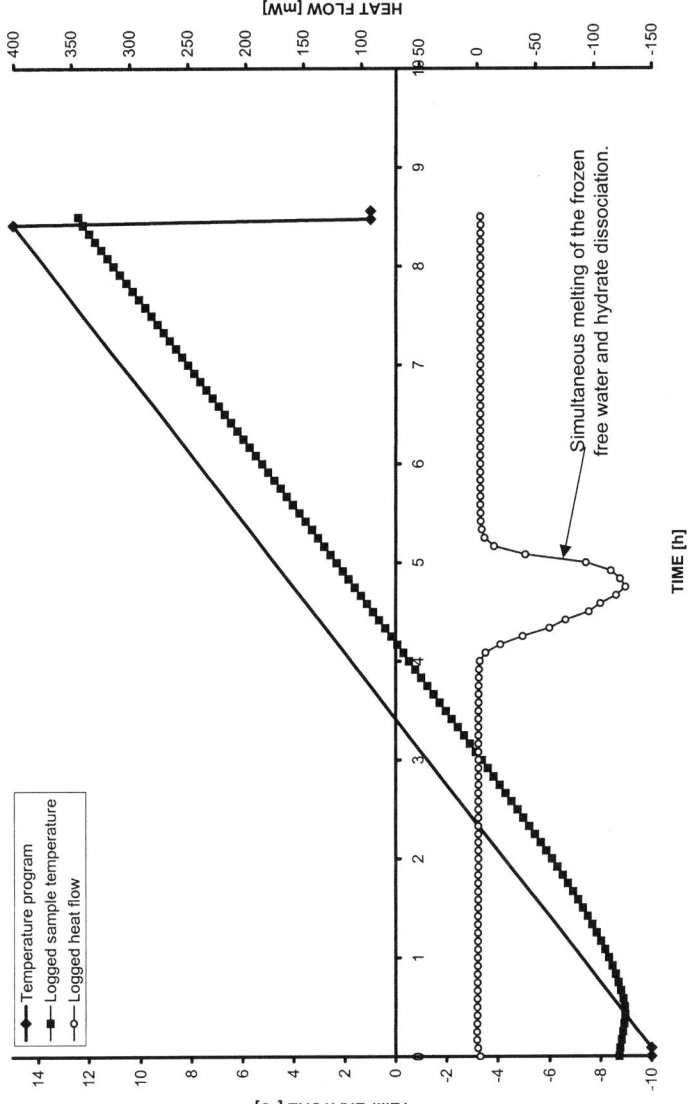

FIGURE 3. Hydrate dissociation. Combined free water (ice) melting and hydrate dissociation at 1 bara in an open calorimeter cell. The purpose is to collect raw data to establish the mass and energy balance to calculate specific enthalpy of hydrate dissociation. Refer also to the corresponding operating line in FIGURE 4.

and allowing it to attain phase equilibrium with the enveloping gas (methane, 17 bara) and then refreeze the free water by taking the sample back down to −10°C (see FIGURE 1).

When the frozen free water was melted again during the second heating-cooling cycle (see FIGURE 2), the water that formed at 0°C was already equilibrated with the enveloping gas. Thus, the thermal response was due to melting of frozen free water only. The amount of free water could then be calculated by taking the ratio of the integral thermal response to the specific enthalpy of the melting process. The melting process that took place was melting of ice under a methane pressure equal to the methane pressure at which the ice formed, 17 bara. The specific enthalpy of this melting process was determined in separate experiments to be 342.6 J/g.

Prior to the third heating-cooling cycle (see FIGURE 3), the sample was cooled to −10°C again, and the pressure taken down to 1 bara, in that order. The calorimeter cells were left open to the surrounding air. On heating, one single peak showed around 0°C. This peak was due to the melting of the frozen free water, as well as hydrate dissociation. The melting process that took place was the melting at 1 bara of ice that had formed under a methane pressure of 17 bara. The specific enthalpy of this melting process was determined in separate experiments to be 344.4 J/g. The thermal response due to the melting of frozen free water could then be calculated since amount of free water was already found in heating–cooling cycle number 2.

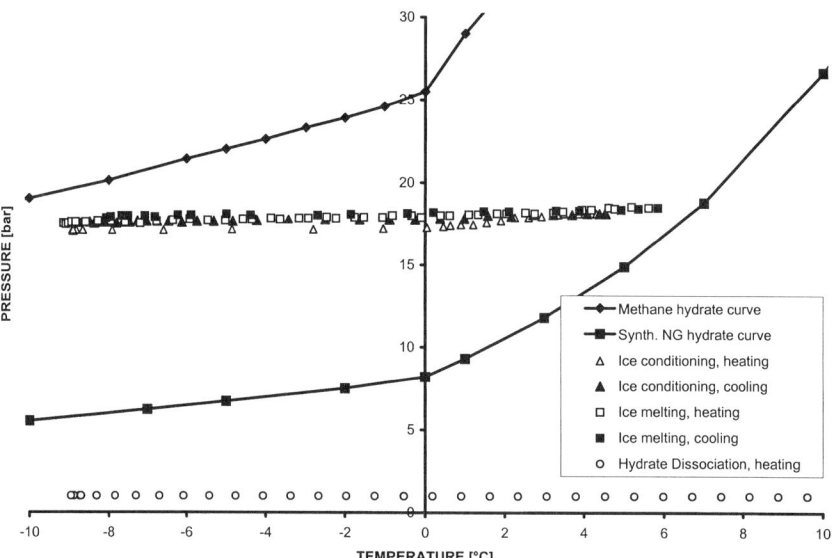

FIGURE 4. Equilibrium lines for methane and NG (92 mol% methane, 5 mol% ethane, 3 mol% propane) according to the CSMHYD code. Logged operating lines for each of the three heating-cooling cycles (ice conditioning in FIGURE 1, ice melting in FIGURE 2, hydrate dissociation in FIGURE 3) are plotted. It is seen that the two first cycles operate between the two hydrate equilibrium curves.

The residual thermal response was due to hydrate dissociation, thus the specific enthalpy of hydrate dissociation at 1 bara (Δh_{diss}) may be calculated. There is also some dissociation prior to the peak, as demonstrated in the isothermal experiments described later. This shifted the heat flow down to a more negative value. Thus, peak integration will yield a number that is too small. It is possible to get around this by manipulating the data and applying a model for the specific heat capacity of ice, but that is not the topic of this paper.

The three-cycle procedure is illustrated in FIGURES 1, 2, and 3. Each shows three curves: temperature according to the temperature program, logged sample temperature, and heat flow. FIGURE 4 shows the operating lines and the hydrate curves of methane and the synthetic natural gas used in this work. Note that the pressure is lowest in the heating sequence of the ice conditioning cycle, marked with open triangles. As the ice melts, desorption and equilibration leads to an increased pressure between 0 and 2°C.

ISOTHERMAL CALORIMETRY FOR SOLIDS DISSOCIATION

A given solid (hydrate) sample has its total enthalpy of dissociation (ΔH_{diss}). At some time during a dissociation process, only a portion of the sample will have dissociated. This portion is referred to as degree of reaction (α). By time t, an amount of heat equal to ΔH_t will have been supplied to the sample. These quantities relate as follows: $\alpha = \Delta H_t/\Delta H_{diss}$ (definition) and $\alpha \to 1$ as $\Delta H_t \to \Delta H_{diss}$ for $t \to \infty$. The higher the heat flux to the sample, the higher the rate of dissociation. It follows that metastability may be evaluated in terms of the (low) dissociation rate: $r = d\alpha/dt = [d\Delta H_t/dt]/\Delta H_{diss}$. It must be noted that solid-state kinetics is a complex issue. Two laboratories may arrive at very different results even if no mistake is made by either. System dependency is strong. The characteristics of isothermal kinetics are often highly dependent on the changing surface geometry of the sample as the degradation proceeds—edges on the sample tend to round, and this reduces the specific surface of the sample. If the system behaves according to a particular rate equation, this does not necessarily provide information about the processes, notably the chemistry, that the sample undergoes.[6]

RESULTS

Hydrate Number and Free Water Determined by Ramping DSC

n and f were determined by applying the three-cycle calorimetric method to a reference sample that had been stored for six months, at about −46°C to prevent dissociation (TABLE 1). According to theory, n for a sII sample with no empty cages is 5.67. The value obtained was 4.7, but taking into consideration the propagation of random errors according to Gauss,[7] the error interval (±1.8) covers the ideal value. The amount of free water in the sample was as high as 59 ± 1 mass% and illustrates how difficult it is to obtain pure hydrate by filtering hydrate-in-water slurries.

TABLE 1. Hydrate number and free water determined in a test of the three-cycle calorimetric method; specification of the sample history

Hydrate number [-]	4.7 ± 1.8
Free water [mass%]	59 ± 1
Storage temp [°C]	−46 ± 2
Storage pressure [bara]	1
Hydrate type	sII
Production pressure [bara]	90 ± 5
Sample mass [g]	0.92 ± 0.01

Rate of Dissociation at Constant Heat Flow Determined by Isothermal DSC

The rate of hydrate dissociation for different T_s and D may be determined by operating the calorimeter isothermally. During these experiments a constant specific heat flow (mW/g) was supplied to the samples. It was assumed that a constant heat flow corresponds to a constant rate of dissociation. For $T_s = -35°C$ and $T_s = -5°C$ the specific heat flows (q) are maintained constant except for periodic fluctuations according to the regulation characteristics of the calorimeter (see FIGURE 5). For $T_s = -1°C$ the fluctuations do not exhibit a periodic wave pattern. This may be so

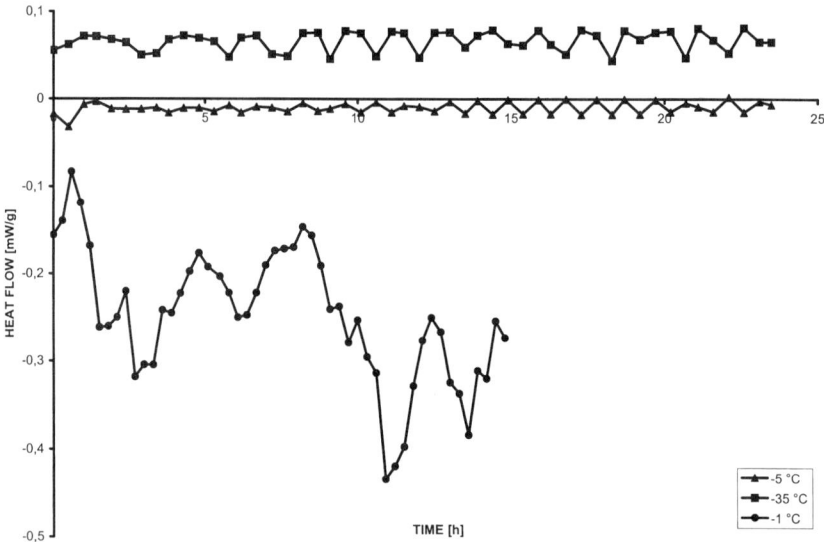

FIGURE 5. Effect of temperature on metastability properties. Heat flow to crushed hydrate samples of comparable representative diameter at different storage temperatures. The heat flow is proportional to the rate of dissociation. Thermal response due to free water (ice) sublimation was shown to be negligible in a separate test run.

because at $-1°C$ the rate of dissociation has become appreciable. The larger changes in the specific heat flow may be explained by nonuniformity of the sample, which becomes more pronounced at higher dissociation rates. According to conventions, the heat flow to the sample is large for a large negative value of q. It is then readily seen that rate of dissociation is lowered if temperature is lowered.

The effect of representative sample size on the rate of dissociation was studied in two test runs at $T_s = -5°C$ (see FIGURE 6). It is seen that a granulated sample ($D \sim 10$ mm) is more stable than a finely crushed sample ($D < 1$ mm). The difference between neighboring curves in FIGURE 5 is larger than in FIGURE 6, so it seems that for the T_s and D values used in these experiments, the effect of T_s dominates that of D.

In these tests, all q measurements were corrected by subtracting the heat signal from a blank run carried out at $-5°C$. To obtain better accuracy, the blank has to be run at the *same* temperature as the corresponding experiment. These curves (FIGS. 5 and 6) provide a basis for estimating the rate of dissociation, according to principles outlined above. Only the curve for $T_s = -35°C$ has to be disregarded, since it shows a positive value, $q \approx +0.05$ mW/g (FIG. 5). This could at first be taken as an indication of hydrate formation. However, this is not possible because of the absence of any hydrate forming gas. A likely explanation is that the blank signal recorded at $-5°C$ differs too much from what would have been seen at $-35°C$. A similar type of error, though oppositely directed and smaller, is probably encountered in the curve for $-1°C$. This error is assumed negligible.

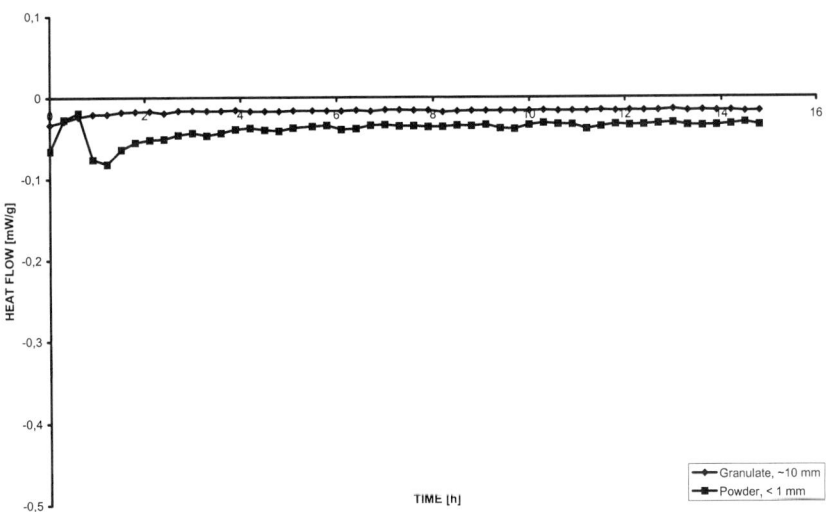

FIGURE 6. Effect of representative diameter on metastability properties. Heat flow to hydrate samples of different representative diameters (D) at $T_s = -5°C$. The heat flow depends on the sample diameter. Thermal response due to free water (ice) sublimation was shown to be negligible in a separate test run.

TABLE 2. Rates of NGH dissociation for different heat flows to the sample

Sample	D [mm]	T [°C]	$-q$ [mW/g]	r_{diss} [1/h][a]	α (t = 10 d)[b]
Powder	<1	−5	0.035	0.0029	0.27
Granulate	~1	−5	0.014	0.0011	0.11
Granulate	~10	−5	0.027	0.0022	0.21
Granulate	~1	−1	0.25	0.020	1

[a]$r_{diss} = (-3{,}600q)/(\Delta h_{diss} \cdot 0.4)$ since samples are assumed to contain 40% mass hydrate. q is the specific heat flow to the sample as shown in FIGURES 5 and 6.
[b]$\alpha = -qt/\Delta h_{diss}$.

The average heat signal for −5°C is about −0.02 mW/g and for −1°C it is about −0.3 mW/g. The rate of dissociation at −1°C is then about 15 times that of the rate at −5°C. This places a value on how important temperature control is when storing metastable NGH. Rough estimates about the absolute rate of dissociation may be made if three assumptions are accepted. (1) The amount of free water is 60 mass% (this assumption is known to be reasonable from other tests). (2) The specific enthalpy of hydrate dissociation is $\Delta h_{diss} = 215.59 \times 10^3 - 394.945T$ J/kg for 248 K < T < 273 K which yields 110 J/g at −5°C.[8] (3) The effect of the (small) gauge pressure that developed during the different tests is negligible (the gauge pressure did establish since the tests were carried out using closed cells). The results (TABLE 2) showed that the rate of dissociation at $T_s = -5$°C is in the range $r = 0.0011$ to 0.0029 gram hydrate per hour per gram initial hydrate, 1/h. Rate data for any other temperature may be obtained according to the same manner. It is also seen that the forecast degree of reaction (dissociation) after 10 days at $T_s = -5$°C is in the range $\alpha = 0.11-0.27$. After 10 days, about 11 to 27 mass% of the hydrate will have dissociated at this temperature, which is unrealistically high for large-scale NGH storage.

DISCUSSION

Methods to study kinetics in NGH-ice systems have been developed and tested in laboratory-scale. However, kinetics of solid-state reactions (transformations) are known to be highly system dependent. It is therefore necessary to extend the experiments to large-scale. The samples used in this work were small amounts of small granulates that were stored in container volumes that were much larger (1,000×) than the sample volume. It is expected that during large-scale experiments, where the sample volume compares to the container volume, the results will be different. The results from this work are regarded as conservative, since the overall rate for a huge sample (tons) will be lower than that of a small sample (grams) due to differences in specific bulk area.

The uncertainty in the hydrate number is large. This is the main limitation of the three-cycle method as it stands now. The reason is the large amount of free water that imposes constraints on the sample mass. If the sample mass exceeds about 0.9 g the melting peak during the second calorimeter cycle (*ice melting*) becomes so wide that the operating line crosses the NGH equilibrium line in FIGURE 4. This equilibrium line is based on an assumption of no fractionation during hydrate formation. Again, this is conservative practice, since the sample temperature during endothermic isothermal melting processes may be overestimated and since the equilibrium line in case of fractionation would be lower. Baseline correction was not regarded as necessary for the scanning experiments.

CONCLUSIONS

1. Hydrate number. A three-cycle calorimetric method was developed to determine the hydrate number for NGH samples that may contain large amounts of free water. The method can be applied to find out whether the hydrate number changes during metastable storage, but improvement with respect to accuracy is desirable.
2. Rate of dissociation. An isothermal calorimetric method was developed to produce heat flow data to calculate rate of dissociation and degree of dissociation during metastable storage.
3. Scaling effects. It is necessary to extend this work to large-scale storage experiments, since the metastability properties of hydrate are system dependent.

ACKNOWLEDGMENTS

This work was supported by the partners in the NGH at NTNU joint industry project: Aker Engineering AS, Amerada Hess Ltd., Atlantic Richfield Company, Fortum Petroleum AS, Phillips Petroleum Company, Shell International Exploration and Production B.V., and Total Norge A.S.

REFERENCES

1. GUDMUNDSSON, J.S., V. ANDERSSON & O.I. LEVIK. 1997. Gas storage and transport using hydrates. *In* Proceedings of the Offshore Mediterranean Conference, March 19–21. Ravenna, Italy. 1075–1083.
2. HÖHNE, G.W.H., W. HEMMINGER & H.-J. FLAMMERSHEIM. 1996. Differential Scanning Calorimetry—An Introduction for Practitioners. Springer, Berlin.
3. GALWEY, A.K. 1985. Solid state decompositions: the interpretation of kinetic and microscopic data and the formulation of a reaction mechanism. Thermochim. Acta **96:** 259–275.
4. GUDMUNDSSON, J.S., M. PARLAKTUNA, O.I. LEVIK & V. ANDERSSON. 2000. Laboratory for continuous production of natural gas hydrates. Ann. N.Y. Acad. Sci. **912:** this volume.
5. HANDA, Y.P. 1986. Calorimetric determination of the compositions, enthalpies of dissociation and heat capacities in the range 85 to 270 K for clathrate hydrates of xenon and krypton. J. Chem. Thermodynam. **18:** 891–902.
6. BAMFORD, C.H. & C.F.H. TIPPER, Eds. 1980. Comprehensive Chemical Kinetics. **22.** Reactions in the Solid State. Elsevier, Amsterdam.
7. MILLER, J.C. & J.N. MILLER. 1994. Statistics for Analytical Chemistry, 3rd edit. Prentice Hall, New York.
8. SELIM, M.S. & E.D. SLOAN. 1990. Hydrate Dissociation in Sediment. SPE Res. Eng. May. 245–251.

A Generalized Model for Calculating Equilibrium States of Gas Hydrates: Part II

S.-Y. LEE AND G.D. HOLDER

Chemical and Petroleum Engineering Department, University of Pittsburgh, 1249 Benedum Hall, Pittsburgh, Pennsylvania 15216, USA

INTRODUCTION

Gas hydrates are crystalline molecular complexes formed by the physical combination of water and low molecular weight gases. They have the general formula $M_n(H_2O)_p$, where n hydrate forming gas molecules, M, called "guests" associate with p "host" water molecules.[1–3] There is no chemical association between gas and water molecules and the interactions are physical in nature.

Hydrates were first discovered by Davy in 1810.[2,4] Between 1810 and 1900, most of research related to hydrates were focused on finding out the composition of gas and water in hydrate.[4] After Hammerschimmit discovered the pipeline-plugging potential of hydrates, the importance of gas hydrates as a subject of investigation increased.[2,4]

The first model for predicting hydrate formation conditions was developed by Wilcox, Carson, and Katz.[5] This method was based on distribution coefficients for hydrate formers, which came from the idea that hydrates were a solid solution that can be treated similarly to an ideal liquid solution.[1,4] In the early 1950s, Claussen, Pauling, and Marsh determined the two different types of hydrate structures, structure I and structure II, using the diffraction experiments of von Stackelberg and coworkers.[4] Using this information, van der Waals and Platteeuw (vdWP) derived the basic statistical thermodynamic equations for gas hydrates.[2] Parrish and Prausnitz first generalized this model for prediction of the conditions under which hydrates form by using the Kihara potential with modified Kihara parameters.[2,3] John and Holder determined the contribution of more distant water molecules by assigning "second" and "third" water shells[6] and later they developed a generalized model that could use the Kihara parameters obtained from virial coefficient data. In this model they used a corresponding states correlation to obtain *true* Langmuir constants from ideal values.[7] In 1993, Hwang, Zele, and Holder developed a distortion model for predicting hydrate equilibria based on the idea that the cavities are not rigid.[9] In 1987, structure H (sH), a third hydrate, was discovered by Ripmeester et al.[4]

In recent years, the interest in gas hydrates has increased not only because of their pipeline-plugging potential but also because of the discovery of naturally formed hydrates in the Earth.[1,4] Estimates of the amount of gas trapped beneath the ocean in the form of hydrate are enormous, twice as large as the combined fossil fuel reserves in coal, oil, and conventional gas combined. Therefore, the prediction of equilibrium conditions for hydrates has received greatly increased interest.

THERMODYNAMIC MODEL

Simple Hydrates

The assumptions for the new model modified from those of earlier workers are:
1. Each cavity can contain at most one gas molecule (vdWP model).
2. The cavity and guest molecules are spherical (vdWP model).
3. Pair potentials can describe the interaction between guest and host molecules (vdWP model).
4. The cavity is not rigid and can be distorted due to the size of the host molecule (Hwang et al., 1993, Zele, 1994, and Lee-Holder model).
5. There is nonlinear composition based distortion in hydrates formed from mixtures (Lee-Holder model).

In this new model, the Kihara potential is used to describe the pair potential between gas and water molecules, and Kihara parameters obtained from virial coefficient are used.[9] When the hydrate phase is in equilibrium with the water phase, the chemical potential for both phases is the same. If the chemical potential of hydrate at 0°C and zero pressure were subtracted from both terms, the equilibrium equation would be[1,3]

$$\mu^\beta - \mu_H = \mu^\beta - \mu_w \Rightarrow \Delta\mu_H = \Delta\mu_w, \quad (1)$$

where μ^β is the chemical potential of the water in the empty hydrate at 0°C, zero pressure, μ_H is the chemical potential of hydrate, and μ_w is the chemical potential of water. The chemical potential difference for hydrate can be expressed by van der Waals equation[2]

$$\Delta\mu_H = RT \sum_{i=1}^{2} v_i \ln\left(1 - \sum_j C_{ij} y_j \phi_j P\right). \quad (2)$$

In this equation v_i is the ratio of i type cavities to the number of water molecules in hydrate phase and C_{ij} is the Langmuir constant. C_{ij} is solved numerically using smooth-cell potentials

$$C = \frac{4\pi}{kT}\int_0^R e^{-(W_1(r) + W_2(r) + W_3(r))/(kT)} r^2 dr, \quad (3)$$

where W_1, W_2, and W_3 are the smoothed-cell potential contributions of the first, second, and third shells of water.[6]

The chemical potential difference for water phase is calculated from the equation developed by Holder et al.[1]

$$\frac{\Delta\mu_w}{RT} = \frac{\Delta\mu_w^o}{RT_o} - \int_{T_o}^{T} \frac{\Delta h_w}{RT^2} dT + \int_0^P \frac{\Delta V_w}{RT} dP - \ln\gamma_w X_w. \quad (4)$$

The first term on the right is the experimentally determined reference chemical potential. In this model (with lattice distortion), each simple gas hydrate has a different value of $\Delta\mu_w^o$ due to the change of the cavity size. The optimized $\Delta\mu_w^o$ values for each simple gas hydrate are given in TABLES 1a and 1b. The second term in (4) gives the temperature dependence at zero pressure and the third term are for the correction of chemical potential change due to pressure change. The last term is the chemical potential correction because of the impurities in the water phase (i.e. salt or methanol or dissolved gases).

From TABLES 1a and 1b, empirical equations for calculating the $\Delta\mu_w^o$ values that cannot be obtained from experiment, such as methane in hydrate structure II, can be obtained.

$$\text{For structure I: } \Delta\mu_w^o = 133.39 e^{0.0213a}, R^2 = 0.906. \quad (5)$$

$$\text{For structure II: } \Delta\mu_w^o = 171.91 e^{0.0101a}, R^2 = 0.881. \quad (6)$$

In (5) and (6), a is the Kihara parameter (pm).

In Equation (4), the temperature dependency of the enthalpy term, Δh_w, is expressed by[1]

$$\Delta h_w = \Delta h_w^o + \int_{T_o}^{T} \Delta C_{p_w} dT, \quad (7)$$

where Δh_w^o is the experimentally determined reference chemical potential difference between the theoretical empty hydrate lattice and pure water phase at the reference temperature (0°C) and pressure (zero). This value is the same for hydrates with the same structure regardless of the gas that is enclathrated, if an undistorted lattice model is used. However, in our model each simple gas hydrate has different Δh_w^o value because we assume that the difference of chemical potential between the undistorted theoretical value and the true distorted value is due primarily to the

TABLE 1a. $\Delta\mu°$ for hydrate structure I[12]

Guest Gas	Guest Diameter (Å)	Kihara parameter a (pm)	Cavity		$\Delta\mu°$ (cal/mol)
			Small	Large	
N_2	4.1	34.1	O	O	266
CH_4	4.36	26	O	O	259
Xe	4.58	25.2	O	O	194
CO_2	5.12	67.7	O	O	542
C_2H_6	5.5	57.4		O	417
Ethylene	6.1	53.3		O	419

TABLE 1b. $\Delta\mu°$ for hydrate structure II[12]

Guest Gas	Guest Diameter (Å)	Kihara parameter a (pm)	Cavity		$\Delta\mu°$ (cal/mol)
			Small	Large	
Ar	3.8	21.7	O	O	248.3
Kr	4.0	23.2	O	O	196.4
c-C_3H_6	5.8	65.3		O	300.0
C_3H_8	6.28	74.5		O	354.0
i-C_4H_{10}	6.5	85.9		O	451.3

enthalpy change associated with the distortion. This assumption is quite reasonable, because molecular simulation suggests that the hypothetical empty hydrate must be differently sized according to the guest molecule and this means that enthalpy of the corresponding empty lattice for each gas hydrate at 0°C, zero pressure should have different values.[8,10] Therefore, in our new model each hydrate has a different value for Δh_w^o.

The other term in (7) is the heat capacity,[1]

$$\Delta C_{p_w} = \Delta C_{p_w}^o + b(T - T_o), \tag{8}$$

where $\Delta C_{p_w}^o$ is the reference heat capacity difference and b is a constant fitted to experimental data. For constant b, the values calculated by Holder and Manganiello were used.[11] The value of b has only a small effect in most situations.

Equations (2) and (4) are substituted into (1) and this can be arranged to give X and Y values after performing the appropriate integration.[12]

$$Y = X\left(\frac{1}{T_F} - \frac{1}{T_o}\right), \tag{9}$$

where

$$X = -\Delta h_w^o + \Delta C_{p_w}^o T_o - \frac{b}{2}T_o^2$$

$$Y = \frac{\Delta \mu^o}{T_o} + (\Delta C_{p_w}^o - T_o)\ln\frac{T_F}{T_o} + \frac{b}{2}(T_F - T_o) + \int_0^P \frac{\Delta V_w}{T_F}dP \tag{10}$$

$$- \ln\gamma_w x_w + R \sum_{i=1}^{2} v_i \ln\left(1 - \sum_j C_{ij} y_j \phi_j P\right).$$

Using (10), Δh_w^o can be obtained from the slope of the graph X versus Y. Δh_w^o values for methane hydrate and propane hydrate are shown in FIGURE 1. The Δh_w^o for methane hydrate is 335.66 cal/mole and that for propane is 339.23 cal/mole.

The efficiency of the model for correlating dissociation pressures of simple hydrates is shown in FIGURE 2. The results are quite good and demonstrate the efficiency of the model. No Kihara parameter adjustment is necessary.

Binary Mixture Hydrates

The fifth assumption above is important for the calculation of the dissociation pressure of binary mixtures. For simple gas hydrates (one gas component + water), the cavity can be distorted due to the size of guest molecules and every simple gas hydrate has different value for $\Delta \mu_w^o$ and Δh_w^o. Distortion for mixture gas hydrates (multi gas component + water) is based upon a nonlinear combination of the $\Delta \mu_w^o$ and Δh_w^o for all species present.

For the new model, an ideal mixture is defined. The ideal mixture is one for which $\Delta \mu_{mix}^o$ is linear in composition. It is expressed as

$$\Delta \mu_{mix}^{o,ideal} = \sum_i \Delta \mu_i^o Z_i, \tag{11}$$

where Z_i is the water-free mole fraction of gas in hydrate phase. Real mixtures have a excess Gibbs potential term due to nonlinear or secondary distortion. $\Delta \mu^o$ for mixture hydrates can be expressed by

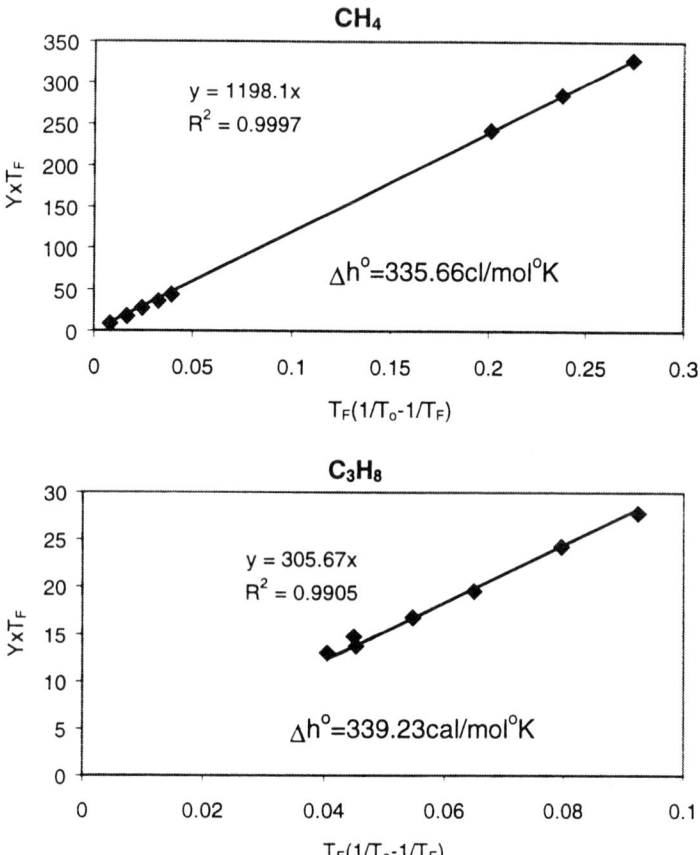

FIGURE 1. Δh_w^o values for methane hydrate and propane hydrate.

$$\Delta \mu_{mix}^o = \Delta \mu_{mix}^{o,\,ideal} + \Delta \mu_{mix}^{o,\,excess}. \qquad (12)$$

The effect of secondary distortion can be obtained from experimental data. The model for calculating the dissociation pressure for binary mixtures is similar to regular solution theory and it can be expressed by

$$\Delta \mu^{excess} = Z_1 Z_2 \{A + B(Z_1 - Z_2)\}, \qquad (13)$$

where A and B are experimentally determined constants and their dimensions are the same as $\Delta \mu$. Z_1 and Z_2 are the mole fraction of components 1 and 2 in hydrate phase, respectively.

FIGURE 3 shows how this model works for methane and isobutane mixture hydrates. The x coordinate is the mole fraction of methane in the hydrate phase and y coordinate is $\Delta \mu_{mix}^{o,\,real}$ value. The difference between $\Delta \mu_{mix}^{o,\,ideal}$ and $\Delta \mu_{mix}^{o,\,calculated}$ is the $\Delta \mu_{mix}^{o,\,excess}$ value.

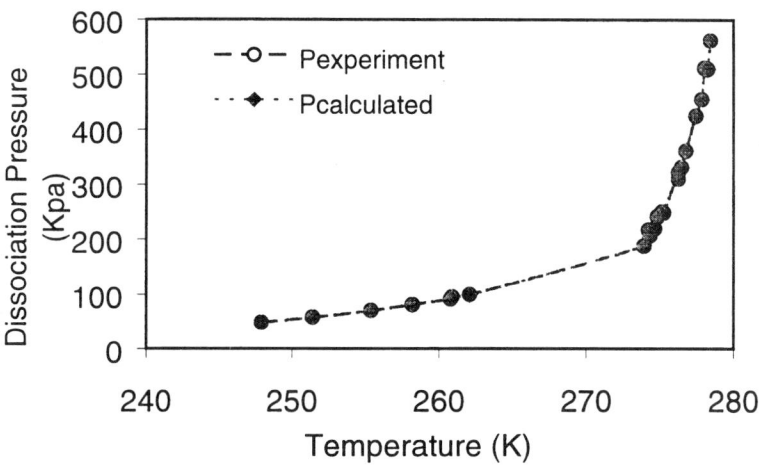

FIGURE 2. Dissociation pressure calculation for propane hydrate using the lattice distortion model.

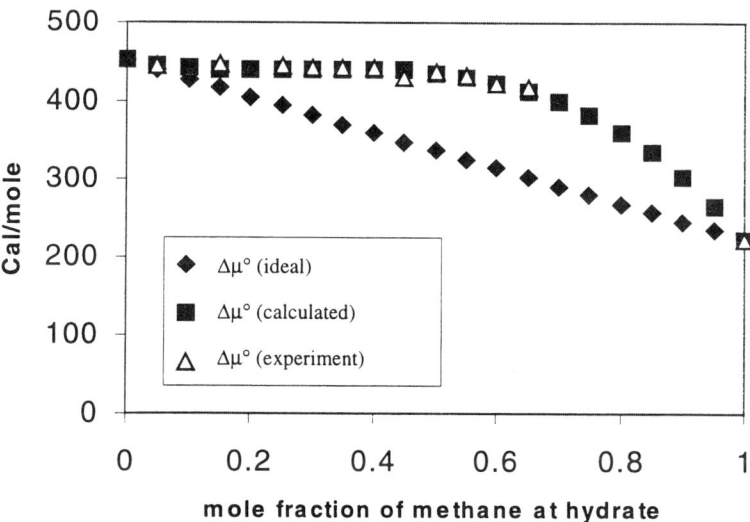

FIGURE 3. $\Delta\mu°$ for methane and isobutane mixture hydrates.

TABLE 2. A and B parameters for the excess Gibbs potential model

Binary Mixture	Structure	A (cal/mole)	B (cal/mole)	R^2
$CH_4 + iC_4H_{10}$	II	395.78	296.64	0.96
$CH_4 + C_3H_{18}$	II	194.74	81.27	0.85
$CH_4 + CO_2$	I	49.41	613.27	0.78
$C_2H_6 + CO_2$	I	277.97	156.06	0.88
$C_3H_8 + iC_4H_{10}$	II	158.34	744.22	0.85
$C_3H_8 + N_2$	II	289.82	−212.56	0.95
$C_3H_8 + Kr$	II	221.51	−84.73	0.90
$iC_4H_{10} + CO_2$	II	−2638.4	−21182	0.74

Constants A and B can be obtained from each binary experimental data by fitting experimental and calculated chemical potentials to the model. Values for A and B are listed in TABLE 2.

CONCLUSIONS

Using a lattice distortion model, optimal $\Delta\mu_w^o$ and Δh_w^o values for simple gas hydrates are accurately calculated with no Kihara parameter adjustment and no modification of the Langmuir constant for restricted rotation. For binary mixture

TABLE 3. Error calculation for simple gas hydrates using the distortion model

	Guest	Temperature Range (K)	Error (%)
Structure I	N_2	202–283	2.52
	CH_4	208–318	7.24
	Xe	216–272	2.49
	CO_2	220–283	14.3
	C_2H_6	260–288	5.46
	ethylene	273–302	10.7
Structure II	Ar	148–291	10.0
	Kr	274–281	7.42
	C_3H_8	262–278	4.49
	$i\text{-}C_4H_{10}$	241–275	4.06

$$\text{Error}(\%) = 100 \times \frac{1}{N} \sum_1^N \frac{\left| P_{dissociation}^{experiment} - P_{dissociation}^{calculated} \right|}{P_{dissociation}^{experiment}}.$$

TABLE 4. Comparison of error values between Zele's model (1994) and Lee-Holder model

Binary Mixture	Structure	$\Delta\mu_1°$	$\Delta\mu_2°$	Error^c (%)	Error^e (%)
$CH_4 + C_2H_6$	I	259	407	16.47	15.4
$CH_4 + C_3H_8$	II	222.5 (328^b)	354	12.3	7.4
$CH_4 + N_2$	I	259	266	14	14.0
$CH_4 + i\text{-}C_4H_{10}$	II	222.5 (328^b)	451.3	60.3^d	17.7
$CH_4 + CO_2$	I	259	542	160.76^d	9.5
$C_2H_6 + C_3H_8{}^a$	I	407	—	7.7	4.2
$C_2H_6 + C_3H_8$	II	305.5 (234^b)	354	13.9^d	8.6
$C_2H_6 + CO_2$	I	407	542	50.3	9.1
$C_3H_8 + Kr$	II	354	196	43.4	9.6

^{a}Propane cannot form hydrate structure II.

^{b}Optimized value. Other values of $\Delta\mu°$ in column 3 are calculated value from Equation (5).

^{c}Optimized value is used for calculation of column 5.

^{d}$CH_4 + CO_2$, 20 data points among 40 diverged; $C_2H_6 + C_3H_8$, 8 data points among 35 diverged; $CH_4 + i\text{-}C_4H_{10}$, 8 data points among 50 diverged.

^{e}Error (%) defined as in TABLE 3 footnote; after using the excess potential method. If chemical potential difference data are missing, calculated value is used instead of optimized value. Equations (5) and (6).

hydrates, an excess Gibbs potential model (Lee-Holder model) is used with good results. With the Lee-Holder model, the average error is about 10% which is less than the Zele model which used variable chemical potential without the excess function.

REFERENCES

1. HOLDER, G.D., S.P. ZETTS & N. PRADHAN. 1988. Phase behavior in systems containing clathrate hydrates. Rev. Chem. Eng. **5:** 1–69.
2. VAN DER WAALS, J.H. & J.C. PLATTEEUW. 1959. Clathrate solutions. Adv. Chem. Phys. **2:** 1–55.
3. PARRISH, W.R. & J.M. PRAUSNITZ. 1972. Dissociation pressure of gas hydrates formed by gas mixtures. Ind. Eng. Chem. Proc. Des. Dev. **11**(1): 26–34.
4. SLOAN, E.D., JR. 1998. Clathrate Hydrates of Natural Gases. 2nd edit. Marcel Dekker, Inc., New York.
5. KATZ, D.L. et al. 1959. Handbook of Natural Gas Engineering. McGraw-Hill Book Company, New York.
6. JOHN, V.T. & G.D. HOLDER. 1982. Contribution of second and subsequent water shells to the potential energy of guest-host interactions in clathrate hydrates. J. Phys. Chem. **88:** 455.
7. JOHN, V.T. et al. 1985. A generalized model for predicting equilibrium conditions for gas hydrates. AIChE J. **31**(2): 252–259.
8. HWANG, M.-J. et al. 1993. Lattice distortion by guest molecules in gas-hydrates. Fluid Phase Equilibria **83:** 437–444.
9. TEE, L.S. et al. 1966. Molecular parameters for normal fluids: the Kihara potential with spherical core. Ind. Eng. Chem. Fund. **5**(3): 363.
10. ZELE, S.R. 1994. Thermodynamic Modeling of Gas Hydrates. Ph.D. Thesis. University of Pittsburgh.

11. HOLDER, G.D. & D.J. MANGANIELLO. 1982. Hydrate dissociation pressure minima in multicomponent systems. Chem. Eng. Sci. **37**(1): 9.
12. LEE, S.-Y. & G.D. HOLDER. 1999. A generalized model for calculating the equilibrium states: Part I. To be submitted.

Mechanisms for Methane Gas Accumulation under Hydrate Deposits in Sediments

KEVIN L. GERING, ROBERT S. CHERRY,[a] AND DAVID M. WEINBERG

Idaho National Engineering and Environmental Laboratory,
PO Box 1625, Idaho Falls, Idaho 83415-2203, USA

ABSTRACT: Equations that describe partial molal volume effects were used to produce a methane–water–hydrate phase diagram which includes T, P, and aqueous phase methane concentrations. This diagram allows prediction of gas exsolution as a result of local biogenesis or as methane-containing water simply rises through sediments. Such exsolved gas can be trapped in homogeneous sediments by relative permeability effects. Hydrate stability conditions when the system is not at liquid–vapor saturation can also be determined.

INTRODUCTION

Bottom simulating reflectors seen in seismic images of hydrate-forming areas are thought to be caused by the strong seismic impedance contrast of a free gas zone trapped under a layer of hydrate. Our interest in this paper is the origin of that free gas, particularly in relatively homogeneous sediments with no apparent lithologic traps. Despite the appeal of viewing the overlying hydrate layer as a trapping mechanism, it is not known how impermeable such layers actually are. In fact, the great metastability of hydrate formation may require the prior presence of both free gas and substantial subcooling in order to form hydrate crystals.[1]

Two types of methane sources can be conceived: local sources in the final hydrate and free gas zones, or remote sources of methane, which is transported by advection through homogeneous sediment or through faults in the sediment. Since the temperature near hydrate stability zones and BSRs is too low for thermogenesis of methane, the local sources would likely be methanogenic microorganisms. Although the temperatures are low (5–15°C) and the pressures high compared to what most organisms will tolerate, there is evidence that such organisms do exist in ocean sediments.[2] If those organisms are sufficiently active, the methane they produce may accumulate until gas bubbles form. The question then is how this free gas accumulates and is held in the sediments. The alternative source of methane is located deep in the sediments, where temperatures are high enough for true thermogenesis (on the order of 100°C or more) or at somewhat shallower depths where the temperature is typical of where microorganisms are normally active, 20–60°C. Under these circumstances, the methane generated at those depths is transported upward in the sediment column either as free bubbles or as soluble methane in rising water.[3,4] Bubbles rise through faults until they are trapped or escape the sea floor. In this discussion, we assume no

[a]Address for correspondence. Voice: 208-526-4115; fax: 208-526-0828.
chy@inel.gov

traps, other than perhaps a plug of hydrate that might form in the fault. However, for soluble methane in homogeneous sediment it is not immediately apparent how free gas might be formed and trapped. Previous work on hydrate formation has not addressed this related issue of free gas formation.[3,4]

In this work, we show that there are conditions of pressure, temperature, and geothermal gradient that can lead to the formation from deep source methane of free gas at depths near the hydrate stability region. This formation is not dependent on the presence of hydrates. We also discuss how exsolved gas from any source can remain trapped in homogeneous sediment, potentially forming hydrates and an accompanying BSR. Although these ideas are not intended to be a detailed explanation of BSR genesis, they do show that either deep or shallow methanogenic activity can potentially lead to shallow free gas deposits even in the absence of barriers such as a hydrate layer.

PHASE EQUILIBRIA FOR THE SEAWATER-METHANE SYSTEM

Under typical conditions found ocean sediments (temperature from 0 to 25°C, pressure from 1 to 500 bar) there are three primary regions of interest: liquid–vapor (L-V) region, liquid–vapor–hydrate (L-V-H) boundary, and the liquid–hydrate (L-H) region. Rigorous laboratory studies[5] have established the L-V-H boundary for water–methane and seawater–methane systems, and have led to a deeper understanding of clathrate structures.

Prediction of L-V phase behavior in this study follows work in which aqueous methane solubilities were assessed by equating expressions of methane chemical potentials for liquid and gas phases at equilibrium.[6] The resulting new expression allows for consideration of the salt content in the seawater:

$$\ln m_2 = \ln x_2 \phi_2 P - \frac{\mu_2^o}{RT} - 2\lambda_{2,Na}(m_{Na} + m_K + 2m_{Ca} + 2m_{Mg}) \\ -0.06 m_{SO_4} + 0.00624 m_{Na} m_{Cl} \quad (1)$$

where subscript 2 denotes methane, m is solute molality, x is mole fraction, ϕ is the fugacity coefficient of methane in the vapor phase, μ is chemical potential, and λ is an interaction parameter. Our method uses the Peng-Robinson equation of state to obtain vapor phase methane properties. Equation (1) provides surprisingly good predictions of methane solubilities within the liquid phase (i.e., methane saturation at equilibrium), with most predicted values falling within 5% of experimental values for the P-T ranges given above.

L-H phase equilibrium predictions are based on work by van der Waals and Platteeuw[7] and Handa[8] with two modifications. First, we assume that the cage occupancy of methane in the hydrate clathrate (Type 1) is constant over the seawater depths under consideration. Second, we explicitly incorporate the concentration of salts in the seawater in the derivation of partial molar volumes of water. By equating expressions of water chemical potential in the liquid and solid hydrate phases, Handa[8] obtained the result

$$\left(\frac{\partial \ln x_2}{\partial P}\right)_T = \frac{1}{RT}\left[\frac{V_e - \overline{V}_1 - \overline{V}_2/N}{1/N - x_2/(1-x_2)}\right], \qquad (2)$$

where V_e is the molar volume of water comprising a hypothetical empty clathrate cage (22.5 cm^3/g mol), $\overline{V}_1$ and $\overline{V}_2$ are the partial molar volumes of water and methane, respectively, and N is $1/\Sigma v_i \theta_i$. For cage type i, the fractional occupancy by methane is denoted by θ_i, and the number of cages per water molecule is given as v_i. Under our theoretical modifications, however, the system is actually defined as water–methane–sea salt. Calculation of the partial molar volume of water uses the expression

$$\overline{V}_1 = \frac{M_1}{1000}\left[V - \sum_i^j \left(\frac{m_i}{\rho}\left(M_i - V\left(\frac{\partial \rho}{\partial m_i}\right)_{m_{k \neq i}}\right)\right)_{i \neq 1}\right], \qquad (3)$$

where M is molecular weight, V is the system volume (based on 1 kg water), and ρ is the liquid density. Equation (3) is based on the fundamental thermodynamic relation $V = \sum n_j \overline{V}_j$ for j components. The partial molar volume of methane was obtained from a parameterized expression[6] and the terms related to the sea salt were derived by considering seawater density as a function of salt molality.[9] Sea salt is assigned an average molecular weight based on the known concentrations of the various ionic species present in seawater.

In FIGURE 1, several isopleths of equilibrium liquid phase methane concentration, x_{CH_4}, are shown in the L-V and L-H regions. In the upper right L-V region, saturation concentrations increase as pressure increases or as temperature decreases. In these figures, pressure is expressed as depth, calculated using a seawater density of 1.0227 kg/L at 20°C with a temperature dependence of density paralleling that of fresh water. In the L-H region, saturation concentrations decrease rapidly as temperature decreases at constant pressure. In contrast, pressure has only a very small effect, a trend towards reduced methane solubility that is attributable to the increased ordering of water molecules at higher pressures.[8]

The concentrations in FIGURE 1 are saturation concentrations. By the Gibbs phase rule, the number of degrees of freedom (independently specifiable variables) in a system equals two plus the number of components minus the number of phases. In a two component methane–water system with two phases present (L-V or L-H), there are two degrees of freedom that can vary independently (T and P). The salt is neglected as a component because over the range of conditions considered here it stays in the liquid water phase at essentially constant composition, in contrast to previous work in which a substantial amount of gas (10 mol% of the entire system) was available to consume water by forming hydrate.[4] On the three phase L-V-H boundary in FIGURE 1, there remains only one degree of freedom, thus choosing the temperature or pressure fixes the value of the other variable. If the liquid phase is not saturated, there are not two phases present (or potentially so, since being at saturation is equivalent to the physical presence of both phases); hence, three variables are required to fully define the system. Here, these variables are conveniently temperature, pressure, and methane concentration in the liquid phase. Therefore, as long as the concentration of methane in the liquid is also specified, we can use this diagram to predict the system phase behavior at a given T and P.

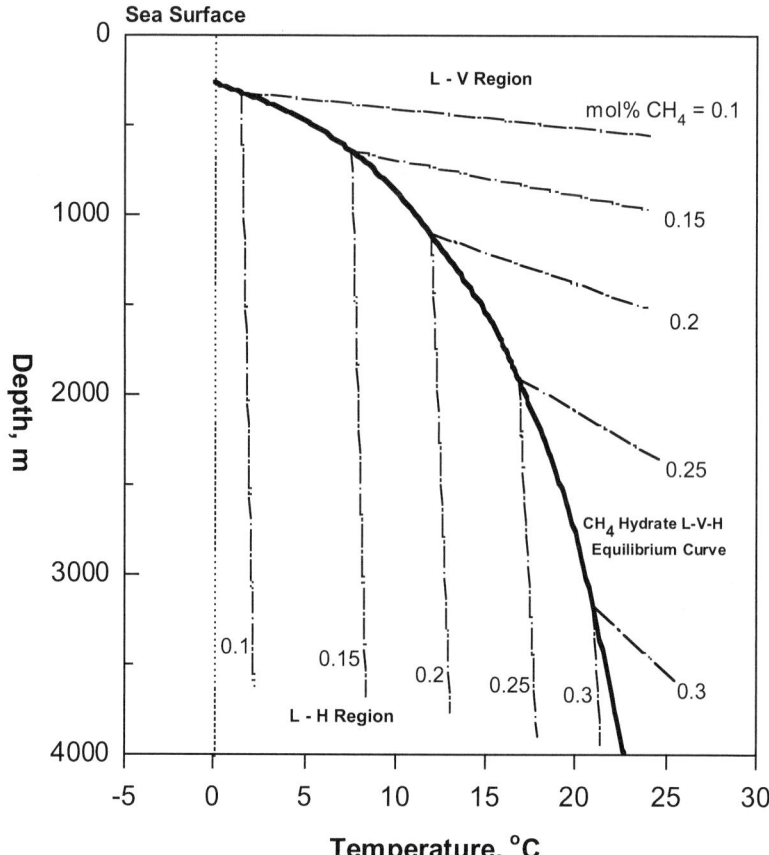

FIGURE 1. Phase equilibria for seawater–methane, showing the two phase regions and the hydrate boundary that separates them. L-V equilibria calculations are from Duan et al.[6] L-V calculations are from a modified approach of Handa.[8] Concentrations shown are liquid phase values.

We can further constrain the selection of temperature and pressure by noting that seafloor temperatures are typically 2–4°C and that the geothermal gradient in the sediment fixes the increase of sediment temperature with depth. Three such lines are shown in FIGURE 2, representing three different depths to the seafloor but the same geothermal gradient of 20°C/km in each case. Moving through the sediment corresponds to moving along one of these lines. This additional constraint removes a degree of freedom, allowing us to predict phase behavior as a function of depth if we define the liquid phase methane concentration. Many previous uses of graphs like FIGURE 2 have done this by implicitly assuming saturation in the liquid phase.

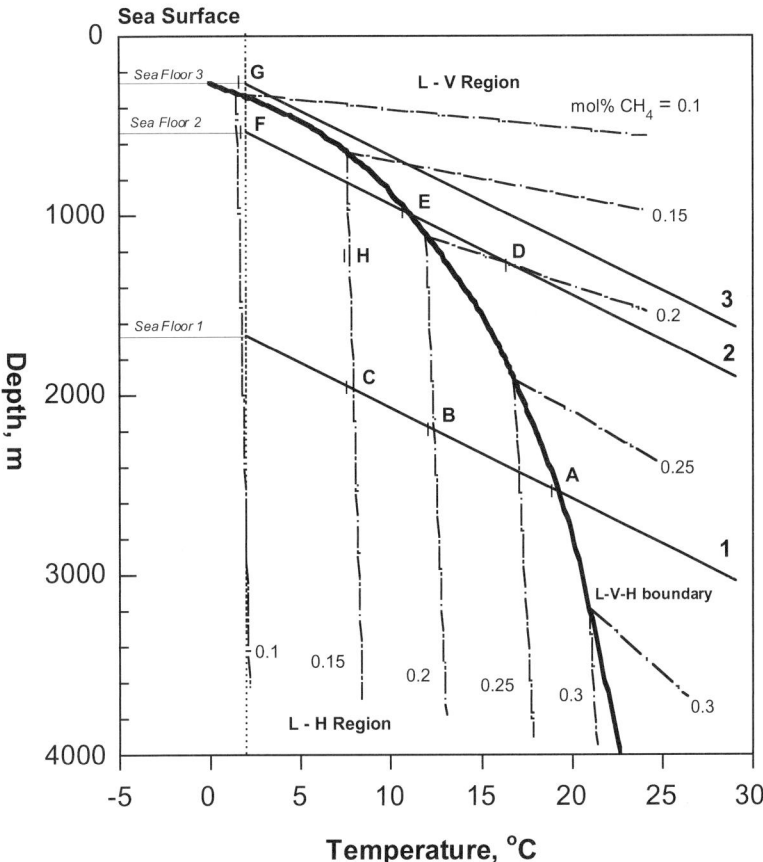

FIGURE 2. Results from superimposing three lines of equal geothermal gradient (20°C/km) onto phase diagrams of the seawater–methane system. Each path exhibits different thermodynamic behavior related to methane exsolution and hydrate formation. Seafloor $T = 2°C$.

PREDICTION OF GAS EXSOLUTION IN SEAWATER

Phase diagrams such as FIGURE 2 can be used to determine when methane will exceed its saturation concentration in the liquid phase and exsolve as bubbles. In the simplest case, the only source of methane is local production by methanogenic microorganisms. Microorganisms living at point D in FIGURE 2 release methane into the surrounding water, gradually increasing the local methane concentration as long as advection through the microbial zone is small compared to their rate of methane production. This methane remains in solution until the concentration reaches 0.2 mole percent, when the liquid becomes saturated. Any further methane production leads to formation of a growing bubble of free methane gas and does not increase the liquid phase concentration. In contrast, at point H the same process leads to formation of a hydrate phase. The liquid concentration reaches only 0.15 mole percent and the methane concentration is limited to that amount. Although the temperature is colder at point H than at D, the lower concentration of methane—a potentially inhibitory metabolic waste product for the microorganisms—might lead to better microbial growth at point H in the presence of hydrates than at point D.

This concept applies only if hydrate actually forms when the methane concentration reaches the saturation value. This requires either the growth of nearby existing hydrate crystals or the nucleation of new ones from solution. Hydrate nucleation can be notoriously slow and requires substantial supersaturation.[1,5] Until solid hydrate nuclei and crystals form, the methane concentration at point H will continue to increase. In principle, it could continue until it reaches 0.23 mole percent (based on the extension of the L-V saturation lines into the L-H region) at which time a gas bubble might form. However, the presence of free gas greatly enhances the nucleation of solid hydrates[1] and the system is then expected to rapidly form enough hydrate to drop the liquid phase concentration back to 0.15 mole percent. Once these first hydrate crystals form, they act as nuclei on which more hydrate from later rising water can grow quite easily.

Another interesting phase behavior results if we consider deep production of methane that is transported upward in solution. This corresponds to starting far to the right along the extension of the three geothermal paths in FIGURE 2. Starting with the lowest path (#1) and assuming a subsaturated methane concentration of 0.2 mole percent, rising water passes through regions where the saturation concentration is about 0.27 mole percent. The saturation concentration slowly increases as depth decreases, so the rising liquid remains subsaturated and no free gas forms. Continuing through the three phase boundary at point A, no hydrate can form because the liquid is still subsaturated. Finally, at point B, the liquid concentration of 0.2 mole percent equals the saturation concentration and hydrate can form (in the absence of nucleation problems). This depth at which hydrates can finally form is, in this example, about 350 meters above the level suggested by the three phase boundary and the geothermal gradient. Continuing from B to C on the 0.15 mole percent saturation line, an amount of hydrate corresponding to the concentration drop of 0.05 mole percent methane (or about 0.21 g hydrate/L solution) forms as water moves through about 200 meters between B and C.

The uppermost geothermal path in FIGURE 2 (#3) shows a very different behavior even though the only difference is the depth of the seafloor. Having started deep at an assumed concentration of 0.2 mole percent, the liquid reached saturation conditions at a depth of about 1,700 meters slightly outside the diagram. From there until it leaves the seafloor at G, gas comes out of solution at a monotonically increasing rate. Hydrate is never formed. However, although this is beyond simple phase behavior, in the real world at the seafloor (point G) the concentration of methane in the liquid becomes zero due to rapid microbial oxidation by aerobes and by sulfate-reducing bacteria. This reduction to zero over a short distance causes a diffusive flux of methane to the seafloor,[10] which in turn reduces the methane concentration for some distance into the seafloor. This reduction pulls the methane concentration below saturation and gas formation is prevented for some distance below point G.

The middle geothermal path in FIGURE 2 (#2) combines these behaviors. From the right edge to D, rising liquid containing 0.2 mol% methane is subsaturated. Between D and E, free gas is released into the sediment. Beginning at E, the liquid, being at saturation, can form hydrates over the entire distance to F (again, ignoring nucleation issues). Free gas evolution can occur in the L-V region when the geothermal gradient is less than the slope of the isopleths (in comparable units). For liquid phase methane concentrations of 0.1, 0.2, and 0.3 mol%, the maximum geothermal gradients at which gas will exsolve are 91, 30, and 11°C/km. In FIGURE 3 the effect of two different geothermal gradients bracketing the normal range is shown. At a seafloor depth of 880 m, the 20°C/km path generates a small amount of free gas (corresponding to perhaps 0.01 mole percent of methane exsolution) whereas the 40°C/km path does not. At shallower seafloor depths both gradients lead to gas exsolution, whereas deeper sea floors would prevent exsolution. The formation of

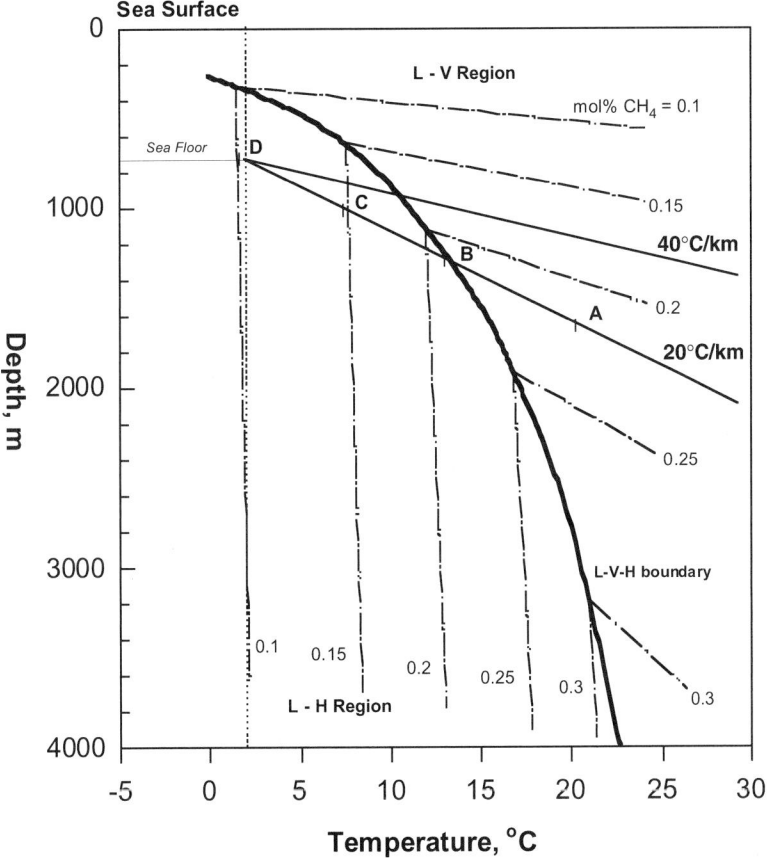

FIGURE 3. Results from superimposing two geothermal paths, having the same sea depth, onto phase diagrams of the seawater–methane system. Seafloor $T = 2°C$.

free gas from rising subsaturated water is thermodynamically limited to sea floors of less than about 1,000 m ocean depth with gas formation not deeper than about 500 mbsf for usual values of geothermal gradient. The finding of free gas or a BSR at other depths indicates that other mechanisms of free gas formation are active or, if the gas is shallower than expected from this analysis, that kinetic effects may be affecting thermodynamic predictions.

Finally, the effect of different presumed starting concentrations of methane is also examined in FIGURE 3 using the 20°C/km geothermal path. The value of methane concentration in the rising water can be measured directly from core water samples or it might be estimated from such information as sediment total organic carbon content, microbial activity estimates, and the flux of water through a presumed zone of methanogenic activity. However the number is obtained, if we use a value of 0.3 mole percent, free gas is exsolved from beyond the right edge of the diagram to point B, and hydrate forms from B to the seafloor D. If only 0.22 mole percent methane was in the water, gas would not form until point A was reached, and the behavior above that would include gas generation to B, then hydrate formation. With 0.15 mole percent methane in the liquid, no free gas is formed at all, and thermodynamically hydrate cannot form either until point C, and possibly not then either because of the kinetics of nucleation.

Although local production and deep production have been considered separately, they can occur together. The major additional comment in this case would be that the cumulative production of methane over a large range of sediment depths can lead to the concentrated production of free gas or hydrate in a rather narrow zone.

TRAPPING OF FREE GAS

To this point we have considered only how free gas might be formed, not how it might be retained in the sediments to form hydrate beds, free gas accumulations, and BSRs. The entrapment of oil and gas in porous media has long been the subject of study. The general phenomenon of capillary effects as stratigraphic traps has been explained with oil field examples[11] and more recently in unconsolidated sediments.[12] The introduction of a second immiscible fluid phase, such as methane gas, into sediment saturated with water creates what is known as the relative permeability effect when both fluids try to flow through the pore structure.[13] The relative permeability of the non-wetting (gas) phase is nearly zero until significant quantities of that phase are present (on the order of 10 vol% of the pore space). This can result in significant volumes of locally generated gas phase being trapped in the sediment, even though water can flow.

In simplest terms, for a gas bubble to move through the water-flooded sediment, it must squeeze through pore throats that are of much smaller diameter that the intraparticle pores themselves. These small throats require that the air–water interface of the bubble deform to make a highly curved, small diameter protrusion. Because of surface tension, the creation of a highly curved fluid surface requires an internal pressure in the bubble that is larger than the surrounding fluid pressure by $2\lambda/r$ where λ is surface tension and r is the (spherical) radius of curvature, which is roughly equal to the pore throat radius depending on wetting effects. This excess pressure is also known as the capillary entry pressure. Without a source of internal pressure generation, such as sediment compaction or continuing local methanogenesis or exsolution, the bubble cannot enter the small pores and is trapped in larger voids between sediment particles. Because of the inverse dependence of capillary pressure on r, capillary or relative permeability seals tend to occur in fine-grained sediments such as clays and muds or, as here, ocean sediments. This relative permeability trapping phenomenon has been used to explain the long term entrapment of deep deposits of geopressured gas.[14,15] This mechanism might also apply to the trapping of free gas formed in and near hydrate stability zones by microbial activity or by exsolution from rising water, even before a layer of solid hydrate forms to further plug the small channels between sediment particles.

ACKNOWLEDGMENT

This work was funded by Lockheed Martin Idaho Technologies Co., management contractor of the INEEL for the Department of Energy, under the Laboratory Directed R&D program.

REFERENCES

1. MAKOGON, Y.F. 1997. Hydrates of Hydrocarbons. Pennwell Publishing, Houston.
2. COLWELL, F., M.E. DELWICHE, D. BLACKWELDER, M. WILSON, R.M. LEHMAN & T. UCHIDA. 1999. Microbial communities from core intervals, JAPEX/JNOC/GSC Mallik 2L-38 gas hydrate research well. *In* Scientific Results from JAPEX/JNOC/GSC Mallik 2L-38 Gas Hydrate Research Well. S.R. Dallimore, T. Uchida & T.S. Collett, Eds. Geological Survey of Canada, Bulletin 544.
3. HYNDMAN, R.D. & E.E. DAVIS. 1992. A mechanism for the formation of methane hydrate and seafloor bottom-simulating reflectors by vertical fluid expulsion. J. Geophys. Res. **97:** 7025–7041.
4. ZATSEPINA, O.Y. & B.A. BUFFETT. 1998. Thermodynamic conditions for the stability of gas hydrate in the seafloor. J. Geophys. Res. **103:** 24,127–24,139.
5. SLOAN, E.D., JR. 1998. Clathrate Hydrates of Natural Gases, 2nd edit. Marcel Dekker, New York.
6. DUAN, Z., N. MOLLER, J. GREENBERG & J.H. WEARE. 1992. The prediction of methane solubility in natural waters to high ionic strength from 0 to 250°C and from 0 to 1600 bar. Geochim. Cosmochim. Acta **56:** 1451–1460.
7. VAN DER WAALS, J.H. & J.C. PLATTEEUW. 1959. Clathrate solutions. *In* Advances in Chemical Physics, 1–57. Interscience Publishers. New York.
8. HANDA, Y.P. 1990. Effect of hydrostatic pressure and salinity on the stability of gas hydrates. J. Phys. Chem. **94:** 2652–2657.
9. WEAST, R.C., Ed. 1978. CRC Handbook of Chemistry and Physics, 59th edit. CRC Press, Boca Raton.
10. BOROWSKI, W.A., C.K. PAULL & W. USSLER III. 1996. Marine pore-water sulfate profiles indicate *in situ* methane flux from underlying gas hydrate. Geology **24:** 655–658.
11. BERG, R.R. 1975. Capillary pressures in stratigraphic traps. Amer. Assoc. Petr. Geologists Bull. **59:** 939–956.
12. REVIL, A., L.M. CATHLES III, J.D. SHOSA, P.A. PEZARD & F.D. DE LAROUZIERE. 1998. Capillary sealing in sedimentary basins: a clear field example, Geoph. Res. Lett. **25:** 389–392.
13. AMYX, J.W., D.M. BASS, JR. & R.L. WHITING. 1960. Petroleum Reservoir Engineering—Physical Properties. McGraw-Hill, New York.

14. BENZING, W.M. 1994. The 'vapor-lock' pressure seal—a nonconformable seal creating potential buoyant forces within young, elastic basins. Collected Abstracts of the Amer. Assoc. Petr. Geologists Hedberg Conference, Abnormal Pressures in Hydrocarbon Environments, Denver, Colorado, June 8–10.
15. BENZING, W.M. & G.M. SHOOK. 1996. Study advances view of geopressure seals. Oil and Gas J. May 20, 62–66.

Modeling of Simultaneous Heat and Mass Transfer to/from and across a Hydrate Film

YASUHIKO H. MORI[a,b] AND TAKAAKI MOCHIZUKI[c]

[b]*Department of Mechanical Engineering, Keio University,
3-14-1 Hiyoshi, Kohoku-ku, Yokohama 223-8522, Japan*

[c]*Department of Technological Education, Tokyo Gakugei University,
Tokyo, 184-8501, Japan*

INTRODUCTION

When liquid water and a hydrate-forming fluid immiscible with liquid water are brought into mutual contact, an almost stationary interface separating the two bulk fluid phases sometimes forms. If this occurs under hydrate-forming pressure/temperature conditions, a hydrate phase preferentially forms at the interface and grows into a polycrystalline film that displaces the interface. The hydrate film thus formed appears to separate the two fluid phases from each other. However, experimental observations on drops (or bubbles) of CO_2 or a fluorocarbon in a liquid water phase have shown that the film or shell that encapsulates each drop does not prevent but only retards the dissolution of the *guest* substance (i.e., CO_2 or the fluorocarbon) into the water phase.[1,2] These observations indicate the existence of a mechanism by means of which the guest substance is transferred across a hydrate film to the liquid water phase. To date, various hydrate-film models have been developed to describe the above mechanism and to interpret experimental observations.[3]

Most of the existing hydrate-film models assume isothermal systems, neglecting temperature variation due to the formation and dissociation of hydrate crystals. In this paper, we attempt to extend our *water-permeable film* model[4] so that it can describe steady (or quasi-steady), one-dimensional, simultaneous heat and mass transfer, taking into account the exothermic hydrate formation and endothermic hydrate dissociation within the confines of a hydrate film.

ANALYTICAL MODEL AND ITS FORMULATION

General Description of the Model

FIGURE 1 schematically illustrates the binary, three-phase, isobaric system we consider hereafter. A hydrate film is held between the two fluid phases—a liquid water phase and a guest-fluid phase—both in steady (or quasi-steady) motion relative to the hydrate film. The bulk (or free-stream) temperatures in these fluid phases are held constant at the same level, T_∞, below the three-phase (hydrate/liquid-water/guest-fluid) equilibrium temperature, T_{tri}. The guest substance is not dissolved in the

---

[a]Telecommunication. Voice: +81-45-566-1522; fax: +81-45-566-1495.
yhmori@mech.keio.ac.jp

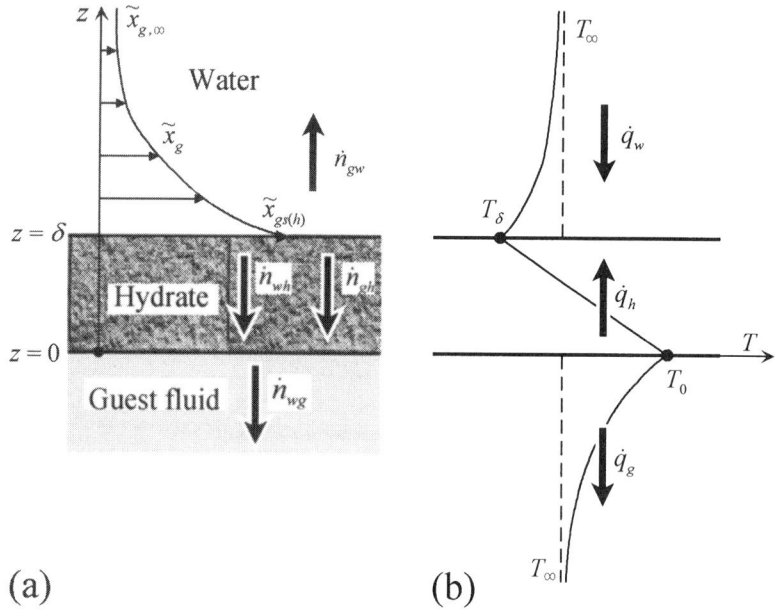

FIGURE 1. Conceptual illustrations of mass transfer (**a**) and heat transfer (**b**) to/from and across a hydrate film occupying the space between $z = 0$ and $z = \delta$, where z is the coordinate normal to the film with its origin fixed onto the guest-fluid-phase-side surface of the film. $\dot{n}_{wh}$ and $\dot{n}_{gh}$ are the molar fluxes of water and the guest substance, respectively, in the hydrate film due to the permeation of liquid water saturated with the guest substance into the film, and $\dot{n}_{gw}$ and $\dot{n}_{wg}$ are the molar fluxes of the guest substance in the liquid water phase and of water in the guest-fluid phase, respectively. $\dot{q}_h$, $\dot{q}_g$, and $\dot{q}_w$ are the heat fluxes in the hydrate film, the guest-fluid phase, and the liquid water phase, respectively.

liquid water phase to saturation; that is, the mole fraction of dissolved guest substance in the bulk of the liquid-water phase, $\tilde{x}_{g,\infty}$, is lower than the guest-in-water solubility $\tilde{x}_{gs(h)}$ at which the liquid water phase would be equilibrated with the hydrate. It is assumed that water is insoluble, or already dissolved to saturation, in the guest-fluid phase so that water dissolution into the guest–fluid phase no longer occurs ($\dot{n}_{wg} = 0$).

We assume a hydrate film to be a thin solid plate having numerous microperforations.[4] Each microperforation is simplified to be a tortuous cylindrical capillary into which liquid water is continuously sucked by a capillary force to flow toward the guest-fluid-phase-side surface of the hydrate film, resulting in the formation of hydrate crystals at this surface. Thus, the rate of hydrate formation per unit area of the guest-fluid-phase-side surface of the film, $\dot{n}_{hl}$, is related to the $-z$-directional molar flux of water permeating through the film, $\dot{n}_{wh}$, and to the hydration number, n, by

$$\dot{n}_{hl} = \frac{\dot{n}_{wh}}{n}, \tag{1}$$

where the hydrate is regarded as a compound with molar mass $M_h \equiv M_g + nM_w$. (M_g and M_w are the molar masses of the guest substance and water, respectively.) At the opposite surface, $z = \delta$, which is exposed to a steady flow of liquid water unsaturated with the guest substance, hydrate dissociation occurs. The rate of dissociation is assumed to be controlled by the convective transfer of the guest substance, which is released by hydrate dissociation and is dissolved in the liquid water, away from the surface. The rate of hydrate formation at $z = 0$ is in balance with the rate of dissociation at $z = \delta$, thereby maintaining the film thickness, δ, constant.

Hydrate-Film Surface Temperatures and Interphase Energy Balance

It is assumed that hydrate formation occurs at numerous discrete sites (around the openings of the capillaries) over the surface at $z = 0$, generating hydrate crystals at the three-phase equilibrium temperature, T_{tri}. This does not mean that the area-averaged temperature at the surface, T_0, is equal to T_{tri}; T_0 can be lower than T_{tri}, depending on the heat transfer from the surface into the guest-fluid phase and the hydrate film. The temperature at the other surface, T_δ, is also variable, depending on the heat transfer to the surface. The energy balance at $z = 0$ is written as

$$\dot{n}_{hl}[\tilde{h}_{hl}(T_{tri}) + (\tilde{c}_{p,h} - \tilde{c}_{p,g} - n\tilde{c}_{p,w})(T_{tri} - T_0)]$$
$$- (\dot{n}_{hl} - \dot{n}_{gh})\tilde{c}_{p,g}(T_0 - T_\infty) = \dot{q}_g\big|_{z=0} + \dot{q}_h\big|_{z=0}$$

or

$$\dot{n}_{hl}[\tilde{h}_{hl}(T_{tri}) + (\tilde{c}_{p,h} - \tilde{c}_{p,g} - n\tilde{c}_{p,w})(T_{tri} - T_0)]$$
$$- \dot{n}_{wh}\left(\frac{1}{n} - \frac{\tilde{x}_{gs(h)}}{1 - \tilde{x}_{gs(h)}}\right)\tilde{c}_{p,g}(T_0 - T_\infty) = \alpha_g(T_0 - T_\infty) + \dot{q}_h\big|_{z=0} \quad (2)$$

where $\tilde{h}_{hl}(T_{tri})$ is the heat of hydrate dissociation per mole of the guest substance at the three-phase equilibrium temperature; $\tilde{c}_{p,h}$, $\tilde{c}_{p,g}$, and $\tilde{c}_{p,w}$ are the molar specific heat capacities of the hydrate, the guest substance, and water, respectively; and α_g is the convective coefficient of the heat transfer from the surface to the guest-fluid phase. Note that the first term on the left side of (2) represents the rate of net enthalpy release due to the transformation of the guest substance and liquid water both at temperature T_0 into a hydrate at the same temperature; the second term represents the energy consumed in preheating the guest substance from T_∞ to T_0. Analogously the energy balance at $z = \delta$ is given by

$$\dot{n}_{hl}[\tilde{h}_{hl}(T_{tri}) - (\tilde{c}_{p,h} - \tilde{c}_{p,g} - n\tilde{c}_{p,w})(T_{tri} - T_\delta)]$$
$$+ \dot{n}_{wh}\left(\frac{1}{n} - \frac{\tilde{x}_{gs(h)}}{1 - \tilde{x}_{gs(h)}}\right)\tilde{c}_{p,g}(T_\infty - T_\delta) = \alpha_w(T_\infty - T_\delta) + \dot{q}_h\big|_{z=\delta} \quad (3)$$

where α_w is the convective coefficient of the heat transfer to the surface from the bulk of the liquid-water phase. The first term on the left side of (3) represents $\dot{n}_{hl}\tilde{h}_{hl}(T_\delta)$, the rate of net enthalpy absorption due to the dissociation of the hydrate at temperature T_δ, resulting in the release of liquid water and the guest substance at the same temperature. The second term on the same side represents the energy required to heat the guest substance that is released by hydrate dissociation and is dissolved into the liquid-water phase, from T_δ to T_∞. It should be noted that there is no *net* transfer of water at $z = \delta$ and hence we need not take into account the heating of water released by hydrate dissociation.

The overall energy conservation for the hydrate film requires that

$$\dot{q}_g\big|_{z=0} = \dot{q}_w\big|_{z=\delta}. \tag{4}$$

This requirement leads to the following constraint regarding the film surface temperatures:

$$\frac{T_\infty - T_\delta}{T_0 - T_\infty} = \frac{\alpha_g}{\alpha_w}. \tag{5}$$

Therefore, once T_∞, α_w, and α_g are specified, the surface temperatures, T_0 and T_δ, are no longer independent of each other.

Heat Transfer through the Hydrate Film

The heat transfer through the hydrate film is due not only to the conduction but also the countercurrent advections of hydrate crystals and liquid water saturated by the guest substance. The hydrate crystals once formed at the surface $z = 0$ are continuously displaced in the direction of increasing z with the molar flux given by $\dot{n}_{wh}/n$, whereas liquid water saturated (at temperature T_δ) by the guest substance is transferred in the opposite direction with the molar flux $\dot{n}_{w(g)h} \equiv \dot{n}_{wh} + \dot{n}_{gh} = \dot{n}_{wh}/(1 - \tilde{x}_{gs(h)})$, where the subscript $w(g)$ indicates the liquid water saturated by the guest substance, having effective molar mass $M_{w(g)} \equiv M_w(1 - \tilde{x}_{gs(h)}) + M_g \tilde{x}_{gs(h)}$. If we assume that the hydrate crystals and the liquid water are in thermal equilibrium at every z-axial location inside the hydrate film, the trans-film heat transfer is described by an energy conservation equation having the following form:

$$\lambda_h \frac{d^2 T}{dz^2} - \frac{\dot{n}_{wh}}{n}\tilde{c}_{p,h}\frac{dT}{dz} + \dot{n}_{w(g)h}\tilde{c}_{p,w(g)}\frac{dT}{dz} = 0, \tag{6}$$

where λ_h is the effective thermal conductivity of the hydrate film and $\tilde{c}_{p,w(g)}$ is the molar specific heat capacity of liquid water saturated by the guest substance. Once the hydrate film textural parameters—the radius r_c and tortuosity τ of the capillaries (trans-film microperforations) and the porosity ε—as well as the interfacial properties—the tension, σ, at the liquid-water/guest-fluid interface and its water side contact angle θ on the hydrate surface—are given, $\dot{n}_{w(g)h}$ is related to the hydrate-film thickness δ by

$$\dot{n}_{w(g)h} = \frac{\sigma \cos\theta}{4 M_{w(g)} v_{w(g)}} \frac{r_c \varepsilon}{\delta \tau^2}, \tag{7}$$

and δ is related to the mass transfer coefficient, $\alpha_{D,gw}$, as well as to the relevant guest-in-water concentration driving force as follows:

$$\frac{\delta \tau^2}{r_c \varepsilon} = \frac{\sigma \cos\theta}{4 \eta_{w(g)} n \alpha_{D,gw}} \frac{(1 - \tilde{x}_{gs(h)})^2 + n \tilde{x}_{gs(h)}^2}{\tilde{x}_{gs(h)} - \tilde{x}_{g,\infty}}, \tag{8}$$

where $v_{w(g)}$ and $\eta_{w(g)}$ are the kinematic viscosity and the dynamic viscosity, respectively, of liquid water saturated with the guest substance. The derivation of Equations (7) and (8) are given in Reference 4.

Equation (6) can be rewritten as follows:

$$\frac{d^2T}{dz^2} + E\frac{dT}{dz} = 0, \tag{9}$$

where E, a constant having the dimension of reciprocal length, is defined by

$$E \equiv \frac{1}{\lambda_h}\left(\dot{n}_{w(g)h}\tilde{c}_{p,w(g)} - \frac{\dot{n}_{wh}}{n}\tilde{c}_{p,h}\right) = \frac{\dot{n}_{wh}}{\lambda_h}\left(\frac{\tilde{c}_{p,w(g)}}{1-\tilde{x}_{gs(h)}} - \frac{\tilde{c}_{p,h}}{n}\right).$$

Integrating (9) under the boundary conditions

$$T = T_0, \text{ at } z = 0,$$
$$T = T_\delta, \text{ at } z = \delta, \tag{10}$$

we have the z-axial temperature distribution in the hydrate film

$$\frac{T_0 - T}{T_0 - T_\delta} = \frac{e^{-Ez} - 1}{e^{-E\delta} - 1}. \tag{11}$$

From (11) we can deduce the local temperature gradient, dT/dz, and then the z-directional heat fluxes at the surfaces of the hydrate film as follows:

$$\dot{q}_h\big|_{z=0} = -\lambda_h\frac{dT}{dz}\bigg|_{z=0} = \frac{E\lambda_h(T_0 - T_\delta)}{1 - e^{-E\delta}}, \tag{12a}$$

$$\dot{q}_h\big|_{z=\delta} = -\lambda_h\frac{dT}{dz}\bigg|_{z=\delta} = \frac{E\lambda_h(T_0 - T_\delta)}{e^{E\delta} - 1}. \tag{12b}$$

Note that the heat fluxes given by (12a) and (12b) should be in agreement with those given by Equations (2) and (3), respectively.

Solution Procedure

The quantities that we need to specify in advance of proceeding with the numerical solution of the problem formulated above are the system pressure p, the bulk temperature T_∞, the mass transfer coefficient $\alpha_{D,gw}$ and the heat transfer coefficient α_w on the liquid-water phase side, the guest-fluid phase side heat transfer coefficient α_g, the hydrate-film textural parameters (r_c, τ, and ε), and the liquid-water contact angle on the hydrate, θ. It should be noted, however, that $\alpha_{D,gw}$ and α_w are not independent; based on the analogy between the forced-convective heat and mass transfer, we can expect the following relation between these coefficients:

$$\frac{\alpha_w}{\alpha_{D,gw}} = \frac{\lambda_w}{D_{gw}}\left(\frac{\text{Pr}_w}{\text{Sc}_{gw}}\right)^{1/3} = \rho_w c_{p,w}\text{Le}_{gw}^{2/3} \tag{13}$$

where ρ_w, $c_{p,w}$, and Pr_w are the mass density, the specific heat capacity, and the Prandtl number, respectively, of liquid water; and D_{gw}, Sc_{gw}, and Le_{gw} ($\equiv \text{Sc}_{gw}/\text{Pr}_w$) are the binary diffusivity, the Schmidt number, and the Lewis number, respectively, for the guest-substance-dissolved water. Except for $\tilde{h}_{hl}$, $\tilde{x}_{gs(h)}$, and σ, all the physical properties are evaluated at temperature T_∞ under the system pressure p. $\tilde{h}_{hl}$ is evaluated at T_{tri}, which is predictable from p. $\tilde{x}_{gs(h)}$ and σ should be evaluated at T_δ and T_0, respectively, neither of which are known *a priori*. We use an empirical procedure to predict $\tilde{x}_{gs(h)}(p, T)$, which is given in Reference 5, and the Donahue–Bartell correlation[6] to predict σ from p- and T-dependent liquid–liquid mutual solubility data.

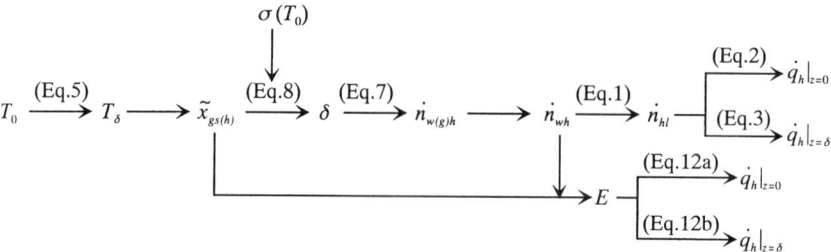

FIGURE 2. Flow chart of numerical solution procedure.

The iterative computation procedure we have contrived is outlined in FIGURE 2. The procedure starts by arbitrarily assuming T_0 ($< T_{tri}$) or alternatively T_δ ($< T_\infty$). We find two different ways to proceed from the above assumption to the evaluations of $\dot{q}_h|_{z=0}$ and $\dot{q}_h|_{z=\delta}$; one is to use (12a) and (12b), and the other is to rely on (5) or (6). These respective calculations should be repeated, changing the assumption of T_0 or T_δ each time, until the resultant values of $\dot{q}_h|_{z=0}$ or $\dot{q}_h|_{z=\delta}$ from the two different sets of calculations are in mutual agreement. The values of all of the quantities indicated in FIGURE 2 are determined in the final round of the above calculations.

ILLUSTRATIVE APPLICATION OF THE MODEL

The model described above has been applied to predict δ, $T(z)$ for $0 \le z \le \delta$, and $\dot{n}_{gw}|_{z=\delta}$ for hydrate films, possibly formed over the surface of a submarine liquid-CO_2 pond. The specific conditions assumed in the numerical calculations are summarized in TABLE 1. Most of these conditions follow those assumed, or estimated, by Hirai et al.,[7] in their simulation of CO_2 dissolution from a pond formed on a seabed at a depth of about 5,000 m. Using the property data given in TABLE 2, we

TABLE 1. Specification of geographic and thermodynamic conditions and of convective mass transfer at the surface of a submarine CO_2 pond

Location	lat. 9°16″ N., long. 146°42″ W.[a]
Depth from sea level	5182 m[a]
Pressure p	50.9 MPa[a]
Undisturbed temperature T_∞	1.4°C[a]
Mass transfer coefficient (water side) $\alpha_{D,gw}$	2.67×10^{-7} m/s,[a] 2.67×10^{-6} m/s[b]
Heat transfer coefficient (water side) α_w	31.0 W/(m²K),[c] 310 W/(m²K)[c]
Heat transfer coefficient (CO_2 side) α_g	(31.0×0.3 =) 9.29 W/(m²K)[b]

[a]Due to Hirai et al.[7]
[b]Arbitrary assumption.
[c]Deduced from $\alpha_{D,gw}$ by Equation (13).

TABLE 2. Physical properties (at $p = 50.9$ MPa and $T_\infty = 1.4°C$) used in calculations

$\tilde{c}_{ph}$	360 J/(mol K)	T_{tri}	13.6°C
D_{gw}	9.62×10^{-10} m²/s	λ_h	0.454 W/(m K)
$\tilde{h}_{hl}$	48.2 kJ/mol	θ	0 rad
n	7.67		

NOTE: Physical properties not specified in this table or mentioned in the text were evaluated by use of PROPATH, a commercially available program package for the thermophysical properties of fluids.

obtained the results shown in TABLE 3 and FIGURE 3. It is recognized that the effect of the advection of hydrate crystals and liquid water inside hydrate films practically arises, resulting in more or less curved temperature profiles inside the films, only when these films are highly porous and water-permeable.

TABLE 3. Thickness of, and molar guest-substance flux across, hydrate films intervening between liquid CO_2 and liquid water phases under the conditions indicated in TABLE 1

Case	r_c (μm)	ε	δ (mm)		$\dot{n}_{gw}\vert_{z=\delta}$ (mol/[m²s])	
			Isothermal model	Present model	Isothermal model	Present model
$\alpha_{D,gw} = 2.67 \times 10^{-7}$ m/s						
A-I	0.01	0.001	0.160	0.183	2.74×10^{-4}	2.74×10^{-4}
A-II	0.1	0.001	1.60	1.83	2.74×10^{-4}	2.74×10^{-4}
A-III	0.1	0.01	16.0	18.4	2.74×10^{-4}	2.73×10^{-4}
A-IV	1	0.01	160	187	2.74×10^{-4}	2.69×10^{-4}
A-V	1	0.1	1.60×10^3	1.88×10^3	2.74×10^{-4}	2.67×10^{-4}
$\alpha_{D,gw} = 2.67 \times 10^{-6}$ m/s						
B-I	0.01	0.001	0.0160	0.0183	2.74×10^{-3}	2.74×10^{-3}
B-II	0.1	0.001	0.160	0.183	2.74×10^{-3}	2.74×10^{-3}
B-III	0.1	0.01	1.60	1.83	2.74×10^{-3}	2.74×10^{-3}
B-IV	1	0.01	16.0	18.5	2.74×10^{-3}	2.72×10^{-3}
B-V	1	0.1	160	188	2.74×10^{-3}	2.69×10^{-3}

NOTE: The two film-textural parameters, the perforation radius r_c and the film porosity ε, are greatly varied to emphasize their effects on the thickness of, and the heat/mass transfer across, the films, whereas the perforation tortuosity τ is fixed at 2. The thickness δ and the molar guest-substance flux $\dot{n}_{gw}\vert_{z=\delta}$ are predicted both by Mori and Mochizuki's isothermal model[4] and the present model for comparison.

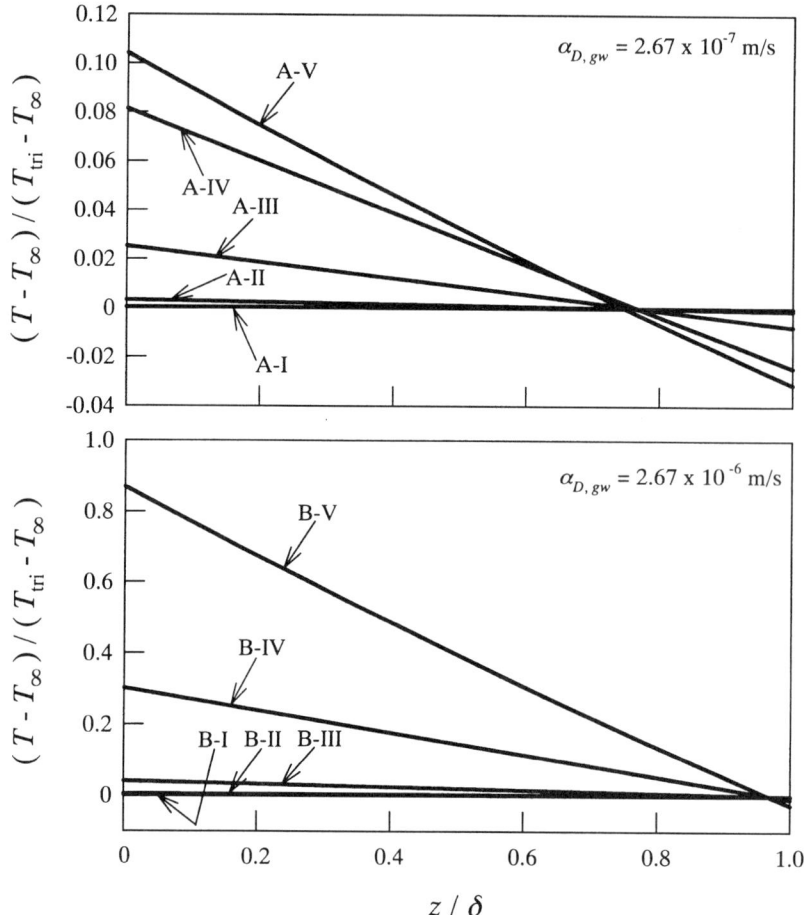

FIGURE 3. Predicted temperature variations across hydrate films intervening between the liquid CO_2 and liquid water phases. The labels **A-I** through **B-V** identifying temperature-profile curves denote the *cases* defined in TABLE 3.

ACKNOWLEDGMENT

This work was funded in part by a Grant-in-Aid for Scientific Research from the Ministry of Education, Science, Culture and Sports, Japan (Grant No. 10450088).

REFERENCES

1. AYA, I., K. YAMANE & N. YAMADA. 1992. Stability of clathrate-hydrate of carbon dioxide in highly pressurized water. ASME HTD **215:** 17–22.

2. SUGAYA, M. & Y.H. MORI. 1996. Behavior of clathrate hydrate formation at the boundary of liquid water and fluorocarbon in liquid or vapor state. Chem. Eng. Sci. **51:** 3505–3517.
3. MORI, Y.H. 1998. Clathrate hydrate formation at the interface between liquid CO_2 and water phases: a review of rival models characterizing "hydrate films". Energy Convers. Mgmt. **39:** 1537–1557.
4. MORI, Y.H. & T. MOCHIZUKI. 1997. Mass transport across clathrate hydrate films—a capillary permeation model. Chem. Eng. Sci. **52:** 3613–3616.
5. MORI, Y.H. & T. MOCHIZUKI. 1998. Dissolution of liquid CO_2 into water at high pressures: a search for the mechanism of dissolution being retarded through hydrate-film formation. Energy Convers. Mgmt. **39:** 567–578.
6. DONAHUE, D.J. & F.E. BARTELL. 1952. The boundary tension at water–organic liquid interfaces. J. Phys. Chem. **56:** 480–484.
7. HIRAI, S., K. OKAZAKI, Y. TABE, K. HIJIKATA & Y. MORI. 1997. Dissolution rate of liquid CO_2 in pressurized water flow and effect of clathrate film. Energy **22:** 285–293.

Numerical Simulation of Transient Heat and Mass Transfer Controlling the Growth of a Hydrate Film

TAKAAKI MOCHIZUKI[a,b] AND YASUHIKO H. MORI[c]

[b]*Department of Technology Education, Tokyo Gakugei University,
4-1-1 Nukuikitamachi, Koganei-shi, Tokyo 184-8501, Japan*

[c]*Department of Mechanical Engineering, Keio University, Yokohama 223-8522, Japan*

INTRODUCTION

When a hydrate-forming fluid, like carbon dioxide, is brought into contact with water under certain temperature/pressure conditions, a hydrate film may form to cover the interface between the two fluids in a relatively short period succeeding a hydrate nucleation at some location over the interface. The *initial* thickness of the hydrate film thus formed may or may not exceed the steady-state thickness that the film would have when the rates of hydrate crystal formation and dissociation inside it are just in balance. The film should grow or shrink asymptotically toward the steady-state thickness with time. Our attempt at modeling the mass transfer or the mass and heat transfer controlling the steady-state thickness is described elsewhere.[1,2] This paper reports our continuing effort directed to the above issue; it describes a simulation of the asymptotic process of growth, or thinning, of a hydrate film which is assumed to be formed instantly at the phase boundary between liquid water in steady convection and a quiescent guest fluid (a hydrate former) extending to a finite thickness. The three-phase (water/hydrate/guest-fluid) system is assumed to be uniform over the lateral extent of the hydrate film so that we can formulate the film growth (or the film thinning) process in the framework of one-dimensional heat and mass transfer with moving boundaries.

MATHEMATICAL FORMULATION AND SOLUTION PROCEDURE

Mathematical Formulation

Suppose that liquid water is flowing over a stagnant layer of a guest fluid that lies on a stationary, adiabatic, impermeable wall. The guest fluid is weakly soluble in water, whereas water is insoluble in the guest fluid. The system is isothermal (temperature T_∞) and isobaric (pressure p) in the hydrate-formable pressure–temperature region relevant to the guest fluid before the sudden formation, at time $t = 0$, of a hydrate film that separates the two fluid phases. The hydrate film is uniform in both

---

[a]Telecommunication. Voice: +81-423-29-7651; fax: +81-423-29-7604.
motizuki@u-gakugei.ac.jp

thickness, δ, and internal structure. The thickness of the guest fluid phase, $|\delta'|$, is reduced with time $t\ (> 0)$ due to the successive growth of the hydrate film thickness. We place the z coordinate normal to the hydrate film in such a way that its origin is fixed to the hydrate-film surface facing the guest-fluid phase (see FIGURE 1).

Assuming no convection inside the guest-fluid phase, we can write the energy conservation equation for the phase as follows:

$$\rho_g c_{p,g} \frac{\partial T}{\partial t} = \lambda_g \frac{\partial^2 T}{\partial z^2} \tag{1}$$

where ρ_g, λ_g, and $c_{p,g}$ are the mass density, the thermal conductivity, and the specific heat capacity, respectively, of the guest fluid. The energy conservation inside the hydrate film is written

$$\tilde{\rho}_h \tilde{c}_{p,h} \frac{\partial T}{\partial t} = \lambda_h \frac{\partial^2 T}{\partial z^2} - \dot{n}_{w(g)} \tilde{c}_{p,w(g)} \frac{\partial T}{\partial z} - \dot{n}_h \big|_{z=0} \tilde{c}_{p,h} \frac{\partial T}{\partial z} \tag{2}$$

where $\tilde{\rho}_h$, λ_h, and $\tilde{c}_{p,h}$ are the molar density, the thermal conductivity, and the molar specific heat capacity, respectively, of the hydrate film. $\dot{n}_h \big|_{z=0}$ is the rate of the hydrate formation per unit area of the film surface at $z = 0$, and $\dot{n}_{w(g)}$ is the molar flux of water saturated by the guest species, permeating the hydrate film in the direction of increasing z. (Note that $\dot{n}_{w(g)} < 0$; see caption to FIG. 1.) $\tilde{c}_{p,w(g)}$ is the molar specific heat capacity of water saturated with the guest species and is given by

$$\tilde{c}_{p,w(g)} = (1 - \tilde{x}_{g,z=\delta})c_{p,w}M_w + \tilde{x}_{g,z=\delta}c_{p,g}M_g, \tag{3}$$

where $c_{p,w}$, M_w, and M_g are the specific heat capacity of water, the molar mass of water, and the molar mass of the guest fluid, respectively. $\tilde{x}_{g,z=\delta}$ is the mole fraction of the guest species dissolved in liquid water at the water/hydrate interface, which is considered to be equal to the solubility of the guest species in liquid water. Considering the conservation of water inside the hydrate film, we can relate $\dot{n}_{w(g)}$ to $\dot{n}_h \big|_{z=0}$ by[1]

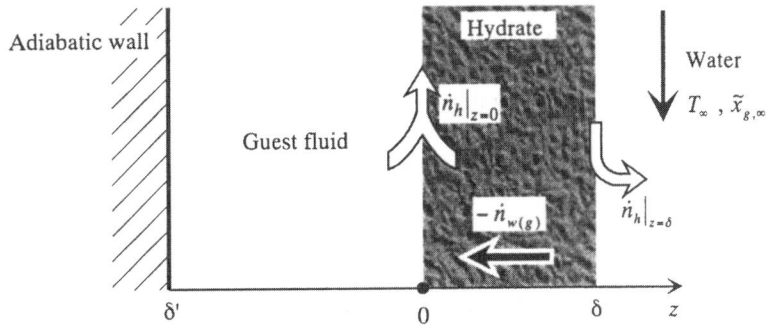

FIGURE 1. Schematic illustration of the model described in this paper. The origin of the z-coordinate normal to the hydrate film is laid on its surface contacting the guest-fluid phase. Note that all of the heat and mass fluxes used in the model formulation are defined to be positive in the direction of increasing z. The molar flux of liquid water (saturated with the guest species) permeating the hydrate film from its water-phase-side surface to the guest-fluid-phase-side surface is thus denoted by $-\dot{n}_{w(g)}$.

$$\dot{n}_{w(g)} = \frac{n\dot{n}_h\big|_{z=0}}{1-\tilde{x}_{g,\,z=\delta}} \qquad (4)$$

where n is the hydration number.

The boundary condition at the interface between the guest fluid phase and the hydrate film is written

$$-\lambda_g \frac{\partial T}{\partial z}\bigg|_{z=-0} + \lambda_h \frac{\partial T}{\partial z}\bigg|_{z=+0}$$
$$+ \dot{n}_h\big|_{z=0}[L + (\tilde{c}_{p,h} - c_{p,g}M_g - nc_{p,w}M_w)(T_{tri} - T\big|_{z=0})] = 0 \qquad (5)$$

where T_{tri} is the three-phase (guest-fluid/hydrate/liquid-water) equilibrium temperature and L is the heat of hydrate dissociation (per mole of the guest fluid) at T_{tri}.

The boundary conditions at the interface between the water phase and the hydrate film may be written as follows:

$$-\lambda_h \frac{\partial T}{\partial z}\bigg|_{z=\delta} + \alpha(T_\infty - T\big|_{z=\delta})$$
$$+ \dot{n}_h\big|_{z=\delta} + [L + (\tilde{c}_{p,h} - c_{p,g}M_g - nc_{p,w}M_w)(T_{tri} - T\big|_{z=\delta})] = 0 \qquad (6)$$

$$\dot{n}_h\big|_{z=\delta} = \alpha_D \frac{\rho_w}{M_w}(\tilde{x}_{g,\infty} - \tilde{x}_{g,\,z=\delta}), \qquad (7)$$

where α and α_D are the convective coefficients for heat transfer and mass transfer, respectively, between the hydrate/liquid-water interface and the bulk of the liquid-water phase wherein the temperature and the guest-species mole fraction are held constant at T_∞ and $\tilde{x}_{g,\infty}$, respectively.

We assume that $\dot{n}_h\big|_{z=0}$, the rate of hydrate formation at the guest-fluid/hydrate-film interface, is controlled by either the heat removal from the interface or the permeation of water through the hydrate film to the interface where water can react with the guest fluid. Either of the heat removal and the water permeation, which gives a lower rate of hydrate formation, is assumed to be the process actually controlling the rate of hydrate formation. The heat removal and the water permeation may alternately control the rate of hydrate formation during each process of hydrate-film growth. In the periods of heat-removal-controlled hydrate formation, $T\big|_{z=0}$ is assumed to be constant. The maximum rate of water permeation is given, as derived in Reference 1, by

$$\dot{n}_{w(g)} = \frac{\sigma\cos\theta}{4M_{w(g)}\nu_{w(g)}} \frac{r_c\varepsilon}{\delta\tau^2}, \qquad (8)$$

where σ, θ, $M_{w(g)}$, $\nu_{w(g)}$, r_c, τ, and ε are the interfacial tension between liquid-water and the guest fluid, the contact angle of water/guest-fluid interface on the hydrate measured through the water phase, the effective molar mass of water saturated with the guest species, the kinematic viscosity of water saturated with the guest species, the radius of microperforations in the hydrate film, the tortuosity of the microperforations, and the porosity of the hydrate film, respectively. The rate of hydrate formation limited by the water permeation is evaluated by combining Equations (4) and (8).

Solution Procedure

FIGURE 2 shows a schematic illustration of the algorithm to calculate the instantaneous hydrate-film thickness, $\delta(t)$. The following is an outline of the algorithm.

1. Specify p, T_∞, $\tilde{x}_{g,\infty}$, α, and the geometrical parameters of the hydrate film (r_c, τ, and ε). The physical properties of the fluids and the hydrate are evaluated at temperature T_∞ and pressure p. The analogy between forced-convective heat and mass transfer is used to evaluate α_D according to the following relation:

$$\frac{\alpha}{\alpha_D} = \rho_w c_{p,w} \left(\frac{\nu_w}{D_{gw}}\right)^{2/3}, \tag{9}$$

where ρ_w, $c_{p,w}$, ν_w, and D_{gw} are the mass density of water, the specific heat capacity of water (mass basis), the kinematic viscosity of water, and the diffusivity for the guest species in water, respectively.

2. Specify the initial thickness of the hydrate film and the guest-fluid phase.
3. Calculate the possible values of $\dot{n}_h|_{z=0}$ limited by the heat removal from, and the water permeation into, the guest-fluid/hydrate-film interface. Select the lower value for use in the later steps.
4. Evaluate $T|_{z=\delta}$ and $\dot{n}_h|_{z=\delta}$, using Equations (6) and (7).
5. Calculate $\partial\delta/\partial t$ and $\partial\delta'/\partial t$ as follows:

$$\frac{\partial \delta}{\partial t} = \frac{\dot{n}_h|_{z=0} + \dot{n}_h|_{z=\delta}}{\tilde{\rho}_h} \tag{10}$$

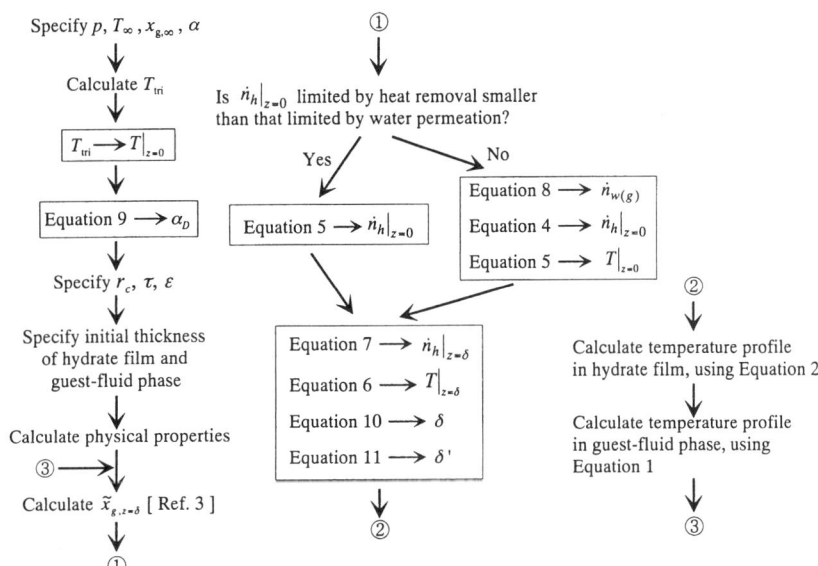

FIGURE 2. Algorithm for calculating the hydrate-film thickness varying with time.

$$\frac{\partial \delta'}{\partial t} = -\dot{n}_h\bigg|_{z=0} \frac{M_g}{\rho_g}. \tag{11}$$

6. Calculate the temperature profiles in the guest-fluid phase and the hydrate film, using Equations (1) and (2), respectively. Then go back to calculation step 3.

RESULTS AND DISCUSSION

The mathematical/numerical scheme described above has been applied to the hypothetical hydrate-forming situation specified in TABLE 1. The numerical data for the physical properties relevant to this situation are summarized in TABLE 2.

We first show, in FIGURES 3 and 4, the results of preliminary calculations based on the system specifications given in TABLES 1 and 2, but which neglect the water-permeation limit, Equation (8); that is, $T|_{z=0}$ is fixed at T_{tri} throughout the

TABLE 1. Specification of the hydrate-forming system assumed in calculations

Guest Fluid	Carbon Dioxide
Pressure p	30 MPa
Three-phase equilibrium temperature T_{tri}	12.1°C[a]
Undisturbed fluid-phase temperature T_∞	7.1°C
Initial temperature of guest-fluid phase	7.1°C
Mole fraction of guest species in water flow $\tilde{x}_{g,\infty}$	0
Hydration number n	7.67
Radius of perforations in hydrate film r_c	0.01 μm
Porosity of hydrate film ε	1.0×10^{-3}
Tortuosity of perforations in hydrate film τ	2.0
Water-side heat transfer coefficient α	258.6, 2965, 5931 W/(m^{2}K)
$|\delta'|$ at $t = 0$	1 mm
δ at $t = 0$	1 μm when $\alpha = 258.6, 2965$ W/(m^{2}K)
	0.1 μm when $\alpha = 5931$ W/(m^{2}K)

[a]Interpolated from data given in Ohgaki et al.[4]

TABLE 2. Physical properties used in calculations[a]

$\tilde{\rho}_h$	5.758 kmol/m^3	$\tilde{c}_{p,h}$	243.0 J/(mol K)
λ_h	0.5 W/(m K)	D_{gw}	1.15×10^{-9} m^2/s[b]
$\nu_{w(g)}$	1.37 mm^2/s[c]	θ	0 rad

[a]Physical properties not specified in this table were evaluated using PROPATH.[5]
[b]Estimated by the correlation due to Wilke and Chang.[6]
[c]$\nu_{w(g)}$ is assumed to be equal to ν_w.

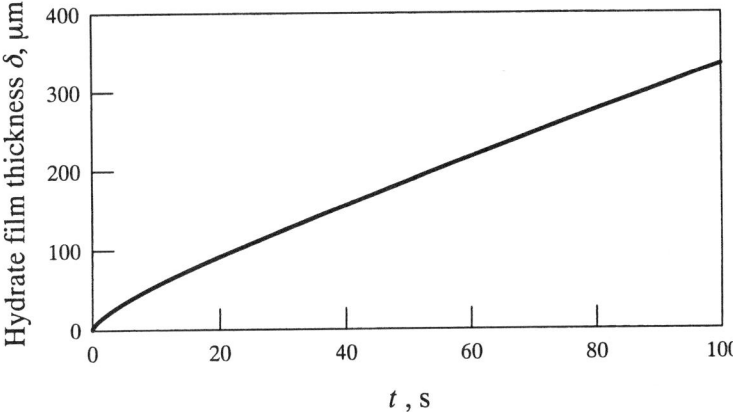

FIGURE 3. Growth of a hydrate film controlled exclusively by the heat removal from the film surface in contact with the guest-fluid phase. $\alpha = 258.6$ W/(m^{2}K).

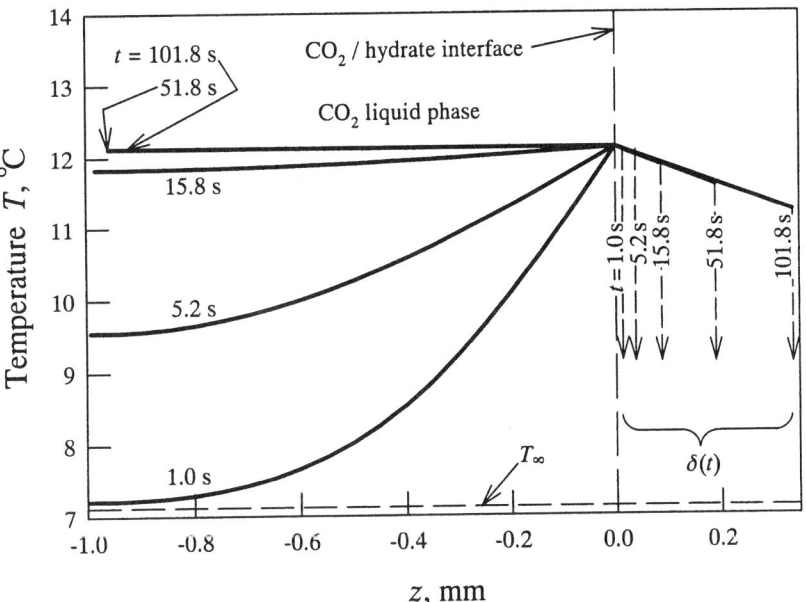

FIGURE 4. Instantaneous temperature profiles obtained by the calculation relevant to FIGURE 3. $\alpha = 258.6$ W/(m^{2}K).

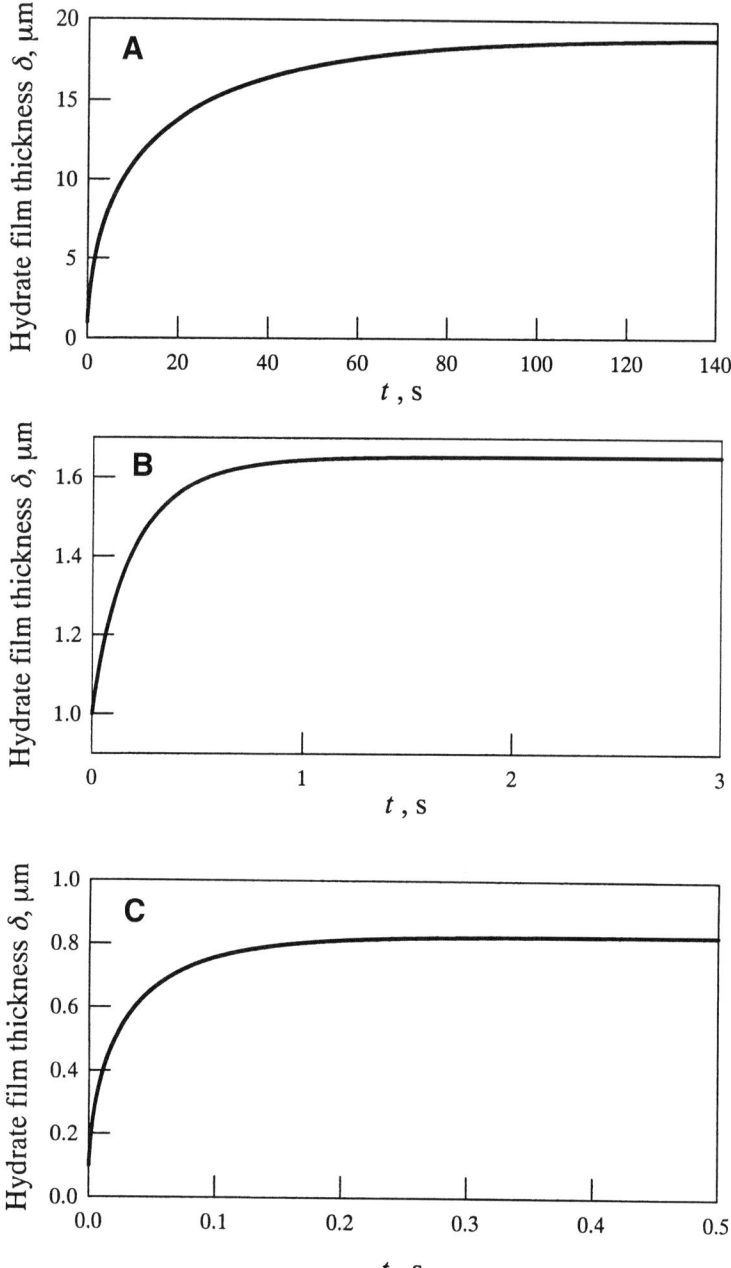

FIGURE 5. Growth of a hydrate film controlled by either the heat removal or the water permeation. (**A**) $\alpha = 258.6$ W/(m^{2}K). (**B**) $\alpha = 2965$ W/(m^{2}K). (**C**) $\alpha = 5931$ W/(m^{2}K).

calculations so that the rate of heat removal is maximized. FIGURE 3 shows a sharp, long-lasting film-thickening process leading to the thickness of about 0.4 mm within a few minutes. FIGURE 4 shows that the heat release into the guest-fluid phase almost vanishes within one minute and that the further film growth is exclusively sustained by the heat transfer into the film. These predictions are not in good agreement with existing experimental results indicating a hydrate-film thickness of the order of 10 μm.[7,8] This inconsistency between the predictions and the experimental results indicates that some mechanism, other than the heat removal, dominates the rate of hydrate formation.

FIGURES 5 and 6 show the results of calculations that exactly follow the formulation and the solution procedure described previously, incorporating the two alternative hydrate formation-rate-controlling mechanisms—heat removal and water permeation. In FIGURE 5, we observe that the time required for a hydrate film to grow to approach its final (steady-state) thickness decreases with an increase in the film-to-water heat and mass transfer coefficients. Throughout the calculations relevant to FIGURES 5 and 6, the dominant mechanism for controlling the rate of hydrate formation is water permeation. In short, we can safely neglect the heat removal restriction while we are dealing with hydrate films not thicker than about 0.3 mm.

In FIGURE 6, it is interesting to note that the direction of the heat transfer between the hydrate film and the guest-fluid phase is reversed in the course of the film growth. Except for the first several seconds, the heat generated by the hydrate formation at the guest-fluid-phase surface is entirely transferred to the opposite surface, where a substantial proportion of the heat is used to dissociate the hydrate, while the residual of the heat is convected away into the water phase.

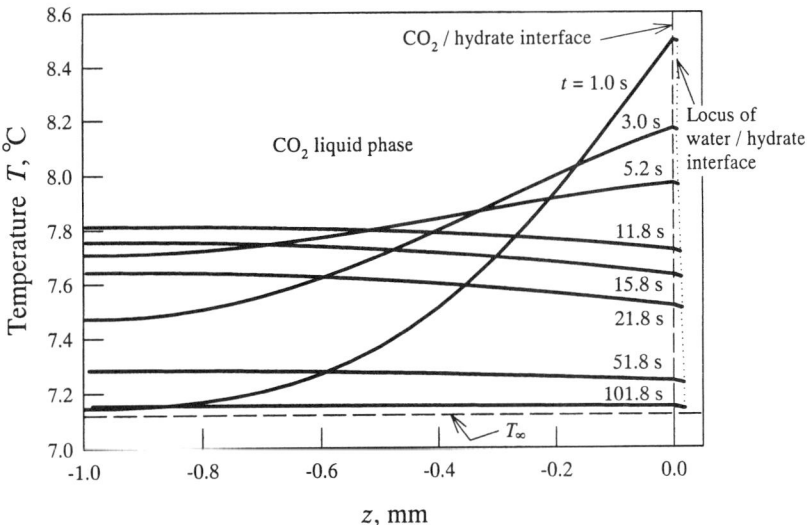

FIGURE 6. Instantaneous temperature profiles obtained by the calculation relevant to FIGURE 5(a). $\alpha = 258.6$ W/(m^{2}K).

REFERENCES

1. MORI, Y.H. & T. MOCHIZUKI. 1998. Mass transport across clathrate hydrate films—a capillary permeation model. Chem. Eng. Sci. **52:** 3613–3616.
2. MORI, Y.H. & T. MOCHIZUKI. 2000. Modeling of simultaneous heat and mass transfer to/from and across a hydrate film. Ann. N.Y. Acad. Sci. **912:** this volume.
3. MORI, Y.H. & T. MOCHIZUKI. 1998. Dissolution of liquid CO_2 into water at high pressures. Energy Conv. Mgmt. **39:** 567–578.
4. OHGAKI, K. & T. HAMANAKA. 1995. Phase-behavior of CO_2 hydrate–liquid CO_2–H_2O system at high pressure (in Japanese). Kagakukougaku Ronbunshu **21:** 800–803.
5. PROPATH GROUP. 1990. PROPATH—A Program Package for Thermophysical Properties of Fluids, Ver. 7.1. Corona Publishing, Tokyo.
6. WILKE, C.R. & P. CHANG. 1955. Correlation of diffusion coefficients in dilute solutions. AIChE J. **1:** 264–270.
7. SUGAYA, M. & Y.H. MORI. 1996. Behavior of clathrate hydrate formation at the boundary of liquid water and fluorocarbon in liquid or vapor state. Chem. Eng. Sci. **51:** 3505–3517.
8. UCHIDA, T. & J. KAWABATA. 1997. Measurements of mechanical properties of the liquid CO_2–water–CO_2 hydrate system. Energy **22:** 357–361.

An Engineering Approach to Kinetic Inhibitor Design Using Molecular Dynamics Simulations

E.M. FREER AND E.D. SLOAN, JR.

Center for Hydrate Research, Colorado School of Mines, Golden, Colorado 80401, USA

ABSTRACT: This paper examines the use of molecular dynamics simulations to predict the performance of kinetic inhibitor structures by simulating adsorption on the sII {1 1 1} hydrate growth plane. Inhibitor performance was observed to correlate to subcooling and van der Waals forces for the sample group of kinetic inhibitors used in the molecular dynamics simulations.

INTRODUCTION

Natural gas hydrates are crystals formed by water with natural gases and associated liquids, in a ratio of 85 mol% water to 15% hydrocarbons. The hydrocarbons are encaged in ice-like solids that do not flow, but rapidly grow and agglomerate to sizes that can block flow lines. Hydrates can form anywhere and anytime that hydrocarbons and water are present at low temperatures and high pressures.[1] The prevention of natural gas hydrate blockages in offshore pipelines has been predominately accomplished by means of thermodynamic inhibition. Thermodynamic inhibitors prevent hydrate formation by shifting equilibrium conditions (a lowering of the freezing point, as is similarly observed when salt is added to ice). The effectiveness of thermodynamic inhibitors for hydrate prevention is well established, but as reported by Sloan more than $500,000,000 is devoted annually to hydrate prevention by methanol injection.[2] This statistic as well as environmental concerns has provided the driving force for the development of kinetic inhibitors.

Kinetic inhibitors differ from thermodynamic inhibitors in that they do not shift the hydrate equilibrium temperature or pressure, but prevent hydrate growth for a finite period of time. These inhibitors are typically low molecular weight polymers that adsorb strongly to the hydrate surface. The early development of kinetic inhibitors for hydrate prevention followed an Edisonian approach, which led to the discovery of polymers that have favorable interactions with the hydrate surface. Recent improvements in molecular simulation have allowed studies of these inhibitors that help to aid our understanding of the inhibition mechanism and serve as a tool in the development of new kinetic inhibitors.

PROCEDURE

Adsorption to the sII {1 1 1} growth plane was simulated using Cerius^2 molecular simulation software on an IBM RS5000 workstation. The model consisted of the kinetic inhibitor and hydrate surface and was generated to ensure computationally inexpensive simulations.

The kinetic inhibitors used in these simulations are shown in FIGURE 1. This sample group was chosen to include inhibitors with differing structure and performance. The kinetic inhibitors were modeled using the Cerius^2 polymer builder package. This package constructs homopolymers using one monomeric unit, the specified tacticity and chain length were isotactic and eight monomeric units, respectively. Isotactic polymers were modeled for simplicity, and the chain length was chosen to be representative of polymers used in flow assurance experiments. This chain length corresponds to Makogon's conclusion that the length of the polymer chain sufficient to distinguish between good and poor inhibitors is eight monomeric units.[3] The initial conformation of the polymer was found by using appropriate minimization algorithms in sequence. These algorithms include steepest decent, conjugate gradient, and Newton-Raphson methods. This procedure does not guarantee that the conformation found is a global minimum, which may never be found for large molecules due to the complexity of the potential energy surface.[4]

The atomic charges of the polymers were calculated using the Gasteiger charge equilibration method.[5] This technique was chosen over quantum mechanical methods because the computation time increases only about linearly with the number of atoms, whereas with quantum mechanical computations the increase is with the third or fourth power of the number of atoms.

The sII hydrate surface was modeled using the Cerius^2 crystal builder package. A unit cell was generated using X-ray diffraction data obtained from McMullen and

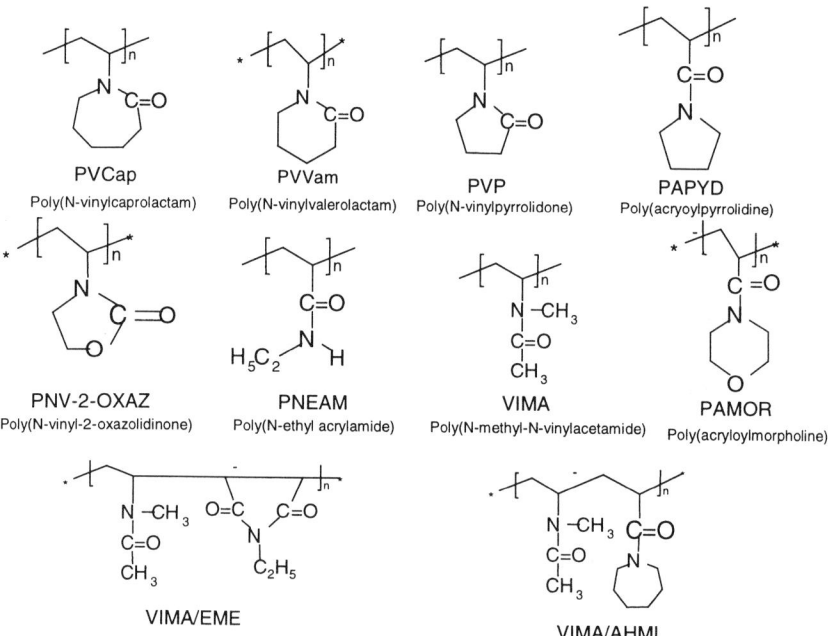

FIGURE 1. Sample group of kinetic inhibitors.

Jeffrey's detailed single crystal structural analysis.[6] The unit cell was then modified by cleaving a 7 Å slice from the {1 1 1} growth plane, which corresponded to Makogon's observation of a surface potential energy minimum. At this location a completed 5^{12} cavity and a partially built $5^{12}6^4$ cavity was observed.[3] The sII hydrate surface was composed of a 3×3 unit cell non-periodic superstructure, which was fixed to prevent movement of the lattice molecules (melting).

Adsorption to the sII {1 1 1} growth plane was simulated using NVT molecular dynamics. The theory behind molecular dynamics is discussed in Haile[7] and Frenkel and Smit.[8] The details of the molecular dynamics simulation are as follows: the simulation temperature was 277.15 K, 10 pico seconds was found to be a sufficient time for the simulation to equilibrate, the total simulation time was 30 pico seconds, and the time step was 1 femto second. The temperature was chosen to correspond to laboratory experiments and pipeline conditions. The simulation time, however, was chosen so that the properties did not change with time. The properties under investigation in this simulation were the energy interactions between the kinetic inhibitor and hydrate surface. These interactions stopped changing when the potential energy of the system reached a minimum and the perturbations were not strong enough to force the trajectory out of the potential energy well.

The molecular interactions were modeled using the Dreiding II force field,[9] which has been found to be accurate for predicting structures and dynamics of organic, biological, and main-group inorganic molecules. The accuracy of the Dreiding II force field was tested by comparing with 76 accurately determined crystal structures of organic compounds involving H, C, O, F, P, S, Cl, and Br, rotational barriers of a number of molecules, and relative conformational energies and barriers of a number of molecules.

The van der Waals forces were calculated using the Lennard-Jones potential, which is as follows:

$$E^{LJ} = D_o [\rho^{-12} - 2\rho^{-6}]$$

where $\rho = R/R_o$, R_o is the van der Waals bond length in angstroms, D_o is the van der Waals well depth in units of Kcal/mol. The hydrogen bonding potential was calculated using the following equation:

$$E_{hb} = D_{hb} [5(R_{hb}/R_{DA})^{12} - 6(R_{hb}/R_{DA})^{10}] \cos^4(\theta_{DHA})$$

where θ_{DHA} is the bond angle between the hydrogen donor (D), the hydrogen (H), and the hydrogen acceptor (A). R_{DA} is the distance between the donor and acceptor atoms in angstroms. Values of D_{hb} and R_{hb} depend on the convention used for assigning charges. Nonbond interactions were not calculated between atoms bonded to each other (1,2 interactions) or atoms involved in angle interactions (1,3 interactions) for the van der Waals forces.[9]

RESULTS AND DISCUSSION

Adsorption of kinetic inhibitors to the sII hydrate surface was studied extensively by Carver et al.,[10] and by Makogon[3] using Monte Carlo techniques. They found that kinetic inhibitors with lactam pendant groups, either PVP or PVCap, adsorb with the lactam ring sterically stabilized in the $5^{12}6^4$ cavity. The same results were observed

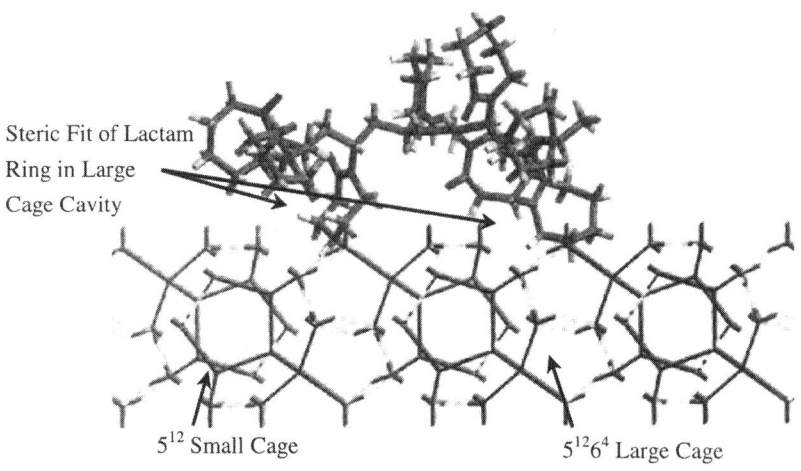

FIGURE 2. PVCap inhibitor adsorbed on sII hydrate surface.

in these simulations as shown for PVCap in FIGURE 2. Two of the lactam rings labeled in the figure adsorbed into the sII large cage cavity. The architecture of PVCap prevents every lactam ring from adsorbing into a sII large cage because the spacings between the lactam rings and each large cage are different. This observation introduces several questions about how polymers adsorb to surfaces and how this adsorption effects the molecular interactions.

The classical picture of polymer adsorption to solid surfaces includes polymers adsorbed in a tail (one end of the polymer adsorbed), train (lying flat with the backbone perpendicular to the surface), or loop (two sections of the polymer adsorbed) fashion. Makogon reported that the adsorption conformation of PVCap is preferably train with no simulations resulting in a loop conformation.[3] Carver *et al.*, however, found that PVP adsorbed in train and loop fashions with the carbonyl group of the lactam ring hydrogen bonded to water molecules forming the periphery of the partially built large cage cavity.[10] These observations indicate that both hydrogen bonding and long-range attractive molecular interactions play a key role in the adsorption of kinetic inhibitors.

The nonbond molecular interaction as described by the Dreiding II force field include both hydrogen bonding and long range attractive forces (van der Waals forces). Hydrogen bonding has been postulated to be a key functional for kinetic inhibitors. The magnitude of hydrogen bonding as seen in water clearly indicates that these forces are much stronger than typical nonbond interactions. The hydrogen bonding potential used by the Dreiding II force field, however, relies on the convention for assigning charges. Mayo *et al.* indicates that a better method for predicting accurate charges of large molecules is needed.[9] This limitation as well as the lack of a robust model for hydrogen bonding prevents a clear understanding of how hydrogen bonding affects the performance of kinetic inhibitors as interpreted from molecular simulation results.

Van der Waals forces or London dispersion forces between the hydrate surface and kinetic inhibitor are attributed to transient dipoles between the molecules and can vary in strength. These interactions seem to play a critical role in determining the effectiveness of a kinetic inhibitor as seen in FIGURE 3. This figure shows the van der Waals forces between the kinetic inhibitor and the hydrate surface versus experimentally determined subcooling. (Subcooling is defined as the difference between the temperature that hydrate formation is first observed ($T_{formation}$) and the equilibrium temperature (T_{eq}), which is essentially the driving force for hydrate formation; $T_{sub} = T_{formation} - T_{eq}$.) A regression line shows the trend of the data. Kinetic inhibitor performance was observed to increase with increasing van der Waals forces (negative energy indicates attractive interactions) for all but one kinetic inhibitor in the sample group. This inhibitor labeled in FIGURE 1 as PAMOR has an oxygen atom incorporated into the lactam ring, which as observed from experimental results reduces the inhibiting capability of the polymer. This indicates that van der Waals forces alone are not sufficient to describe the interactions between the polymer and hydrate surface, but that they do yield a reasonable approximation for the performance of this inhibitor.

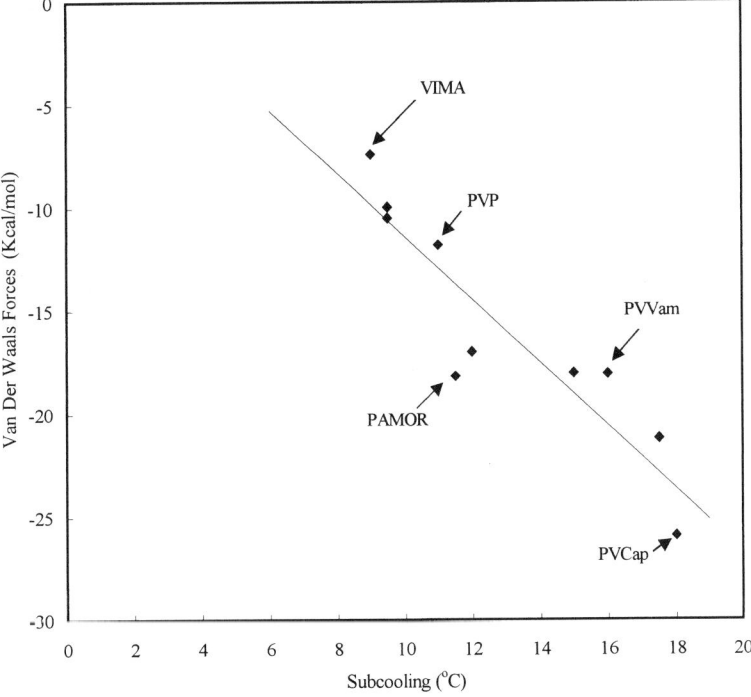

FIGURE 3. Hydrate surface and kinetic inhibitor potential energy difference versus experimentally determined subcooling.

Inhibitor performance is based on subcooling. The greater the subcooling the better the performer. The best inhibitor in the sample group, PVCap, differs from PVP and PVVam only in the size of the lactam ring. The van der Waals forces as calculated from these simulations increased with increasing ring size. This observation is consistent with the Lennard-Jones potential in that the potential energy increases due to attractive forces as the distance between the interacting molecules decrease. In this case, the interacting molecules are the water molecules representing the partially built sII large cage cavity and the carbon atoms representing the lactam ring. The increase in ring size leads to a reduced distance between the carbon atoms and water molecules, which results in more favorable interactions, up to an optimal size. If the distance between the interacting molecules is smaller than the van der Waals bond length, then the interactions become repulsive. The van der Waals bond length corresponds to the distance between the interacting molecules where the potential is at a minimum. This indicates that the favorable interactions between the lactam ring and the large cage cavity (steric fit) are highly dependent on the size of the lactam ring, a dependence that was observed in these simulations.

CONCLUSIONS

These simulations were performed in the hope of developing a tool that could be used in the design and screening of potential kinetic inhibitors. The results indicate that this technique could be used as an engineering tool for the development of kinetic inhibitors, which would save valuable time in the laboratory. The van der Waals forces were observed to play a key role in determining the performance of kinetic inhibitors, but the results also show that for certain molecules van der Waals forces are not sufficient in describing the molecular interaction between the hydrate surface and kinetic inhibitor. This leads to the conclusion that this method serves as a first approximation to determining the performance of kinetic inhibitors.

ACKNOWLEDGMENTS

We thank Amoco, ARCO, Chevron, Department of Energy (U.S.), Mobil, Oryx, Phillips, Shell, and Texaco for funding this project. We thank Dr. Surya Devarakonda for his guidance.

REFERENCES

1. SLOAN, E.D. 1999. Offshore Hydrate Engineering Handbook. ARCO.
2. SLOAN, E.D. 1999. Clathrate Hydrates of Natural Gases, 2nd Edit. Marcel Dekker, New York.
3. MAKOGON, T.Y. 1997. Ph.D. Thesis, Colorado School of Mines.
4. CERIUS2 USER GUIDE. 1997. Force field-based Simulations April 1997. Molecular Simulations Inc., San Diego.
5. GASTEIGER, J. & M. MARSILI. 1980. Iterative partial equalization of orbital electronegativity—a rapid access to atomic charges. Tetrahedron **36:** 3219.
6. MCMULLAN, R. & G. JEFFREY. 1965. Polyhedral clathrate hydrates. IX. Structure of ethylene oxide hydrate. J. Chem. Phys. **42:** 2725.

7. HAILE, J. 1992. Molecular Dynamics Simulation. John Wiley and Sons, Inc., New York.
8. FRENKEL, D. & B. SMIT. 1996. Understanding Molecular Simulation—From Algorithms to Applications. Academic Press, San Diego.
9. MAYO, S., B. OLAFSON & W. GODDARD III. 1990. A generic force field for molecular simulations. J. Chem. Phys. **94:** 8897.
10. CARVER, T., M. DREW & P. RODGER. 1995. Inhibition of crystal growth in methane hydrate. J. Chem. Soc. Faraday Trans. **95**(19): 3449.

Configuration-Biased Monte Carlo Simulations of Poly(vinylpyrrolidone) at a Gas Hydrate Crystal Surface

TIM J. CARVER,[a] MICHAEL G.B. DREW, AND P. MARK RODGER[b]

Department of Chemistry, University of Reading,
Whiteknights, Reading RG6 6AD, United Kingdom

ABSTRACT: In this paper we report the use of Monte Carlo simulation methods to study the properties of both isotactic and atactic PVP near a hydrate surface for polymers with molecular weights up to 12,000 Daltons. Information is presented about the conformation and particle size distribution found for PVP bound to a hydrate surface, and about the adsorption sites and energies involved. These results are found to correlate with the behavior already identified for the vinylpyrrolidone monomer. In particular, the same adsorption sites are evident. In the polymeric case the adsorption of any one unit is less optimal than was found for the monomer, but the trains of adsorbed polymer were found to involve partial occupancy of a succession of monomeric adsorption sites.

INTRODUCTION

As one of the first kinetic inhibitors identified, polyvinylpyrrolidone (PVP) has become the prototype for many investigations of hydrate inhibition. Yet despite this, much is still unknown about the way PVP interacts with hydrates and with water. There have been a number of studies of the vinylpyrrolidone monomer (and its analogues),[1-4] but most of the work on the polymeric PVP has been limited to macroscopic observable properties; thus our "knowledge" about the molecular detail of how PVP inhibits hydrate formation is still dominated by conjecture. Some information can be gleaned by analogy with PVP adsorbed on silica, as elucidated from Fourier transform infrared spectroscopy.[5] In that study the polymer was found to exhibit tails and loops that were not bound to the surface, as well as lengths of polymer (trains) in direct contact with the surface. One may then conjecture that the balance between trains and loops is important in determining how effective PVP is in inhibiting hydrate growth.

The purpose of this work is to study the behavior of PVP adsorbed onto a hydrate surface and thereby characterize how the polymeric system differs from the monomeric. In this sense the current study is complementary to, and a logical extension of, our earlier monomeric studies. As a first step we report here the results of grand

[a]Present address: Department of Material Science and Metallurgy, University of Cambridge, Pembroke Street, Cambridge CB2 3QZ, U.K.
[b]Address for correspondence: Department of Chemistry, University of Warwick, Coventry, CV4 7AL, U.K. Voice: (44) (0)1203 523239; fax: (44) (0)1203 524112.

canonical Monte Carlo (GCMC) simulations of PVP adsorbed onto a rigid hydrate surface under vacuum. Such a model is approximate, in that it ignores dynamic relaxation of the surface and allows for no solvent effects; however, it does provide a measure of the energetic factors that drive the adsorption of PVP onto hydrate surfaces. By understanding these influences we will be better placed to determine subsequently how PVP modifies the properties of the liquid water/hydrate interface.

COMPUTATIONAL DETAILS

General Methodology

Simulations have been performed using a GCMC method. In this method, particles are randomly added to or removed from the system, or moved about within the system, in a manner that is consistent with a specified chemical potential. For the present application, the *particles* are monomer units; *addition* corresponds to bonding another N-ethylenepyrrolidone unit to one end of the polymer; and particle movements include rotations of the pyrrolidone side group and twists of the polymer backbone. Note that the method is able to give a correct sampling of the equilibrium distribution of polymer size and conformation without having to mimic the real growth process. It is this equilibrium distribution, rather than the actual growth process itself, that is central to the present application. Conventional GCMC simulations are inefficient for sampling high density systems, as would occur in our system when the polymer is sufficiently entangled. We have, therefore, used a configurational biased Monte Carlo method (CBMC) that biases particle additions toward inserting the monomer at a place, and with an orientation, that is reasonably likely to lead to a successful addition. A good description of the GCMC and CBMC methods is given elsewhere.[6]

The System

For this study we have used the -0.01 Å cut of the (111) surface of a type II hydrate; this surface has been characterized previously[1] and identified as crucial in the growth process. A hydrate sample of dimensions $42.40 \times 42.96 \times 29.79$ Å ($x \times y \times z$, where z is the crystallographic [111] direction) was used, with periodic boundaries in the xy plane to give an infinite surface; this surface was shown in our earlier studies to be large enough to give a representative sample of the arrangement of water hydrogens and of the different surface microstructures.[7] No boundary conditions were used in the z direction. A representation of the surface (with adsorbed polymer) is depicted in FIGURE 1. The water positions within the hydrate were held fixed throughout each simulation. This is clearly an approximation, but one that has proved useful in previous studies.[1,2] Its main ramifications for the present study are that it is likely to reduce the quality of the match between the polymer and the surface: incorporation of pyrrolidone moieties into the surface potholes is likely to be limited by the entry process, and relaxation of the surface during entry counteracts some of the geometrical restrictions imposed on the active unit by the rest of the polymer to which it is attached.

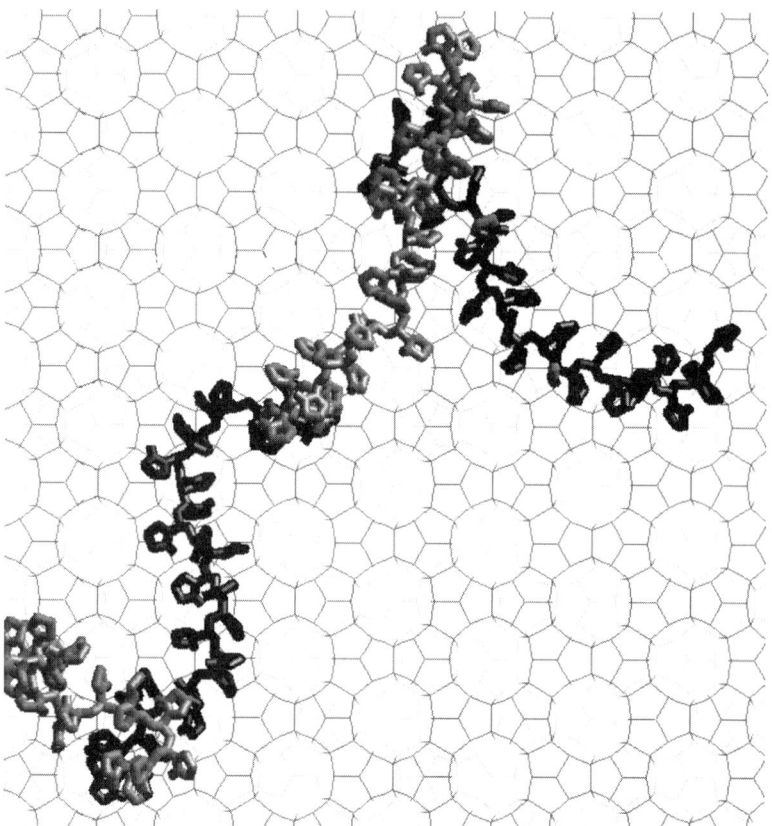

FIGURE 1. Top view of the adsorption of an atactic polymer on the −0.01 Å cut of the (111) surface of a type II hydrate. Adsorbed monomeric units are colored *black*, and others are *grey*. The *line drawing* of the hydrate surface includes only the uppermost portion of the substrate: the large, nearly hexagonal voids actually represent potholes on the surface that are about 7 Å deep. Note the way the adsorbed trains follow the succession of surface potholes.

Potentials

CHARMm parameters[8,9] were used to describe the polymer and SPC for water.[10] A minimum image truncation was used for polymer–water interactions, with imaging based on monomer and molecular positions for the polymer and water respectively. Polymer–polymer interactions were treated as aperiodic with no truncation at long range. This arrangement gives a single polymer on an infinite surface. Since the hydrate surface was kept fixed, no evaluation of hydrate–hydrate interactions was needed.

Protocol

Simulations were initiated from an N-ethylpyrrolidone molecule located at a favorable adsorption site on the hydrate surface. Each step of the simulation consisted of a trial conformational change (with probability 0.95), a monomer addition (0.025 probability) or monomer deletion (0.025 probability). The trial was then accepted or rejected in accordance with the CBMC method.[6] Conformational changes were made to either the backbone angles, or to the side group angle (defined by the C^a–N bond), with new angles being chosen from the energy-minima for that torsion angle, supplemented by a Gaussian random variable with standard deviation 0.3 radians. Monomer addition occurred at either end of the polymer with equal probability and amounted to removing one of the terminal hydrogens and replacing it with a bond to an N-ethylenepyrrolidone moiety; deletion was entirely analogous. The monomeric unit in PVP is chiral and so our simulations were designed to examine the effect of tacticity. This may be understood by considering the handedness of two adjacent monomeric units. If these have opposite handedness then the dyad is said to be *meso*, whereas if they have the same handedness the dyad is said to be *racemo*. The tacticity of the polymer then results from the way the dyads are joined together: an isotactic polymer may be described as *mmmmm...*, whereas an atactic polymer is a random arrangement of *m* and *r* dyads. We have performed simulations for the growth of both isotactic and atactic polymers. In the former case, trial additions were constrained to ensure that they only gave rise to *meso* dyads. For atactic simulations, the trial additions gave *meso* and *racemo* dyads in equal number, although subsequent energy considerations meant they were not necessarily accepted with equal probability.

Isotactic and atactic polymers were simulated at 290 K. In both cases separate simulations of 16,000 and 24,000 steps were performed; this generated about 800 (1,200) attempted additions or deletions. For each system, the polymer growth simulation was repeated 24 times to give a small ensemble from which to determine average properties. An effective chemical potential of 80 kcal mol^{-1} was used in all cases. The *effective chemical potential* is the simulation control parameter that determines the equilibrium polymer size distribution. It represents a composite of the monomer chemical potential and the energy needed to replace a hydrogen with a bond to the next monomer. Calculations were performed on SG R4000 workstations using a program written specifically for the project.[7] Each simulation typically took about six hours to perform.

RESULTS

Side views of the 24 polymers grown in the longer atactic simulations are given in FIGURE 2. It is apparent that a wide range of conformations are found. There are a few conformations for which essentially complete adsorption occurs, but more usually there are loops that bridge between adsorbed regions. In some cases the loops are quite large—about 60 monomer units. Some effect of surface topology is apparent, with adsorption occurring preferentially in the potholes and along the channels in the surface. A good example of this is seen in FIGURE 1. However, whereas the polymer tends to adsorb through the same sites as found in the monomer

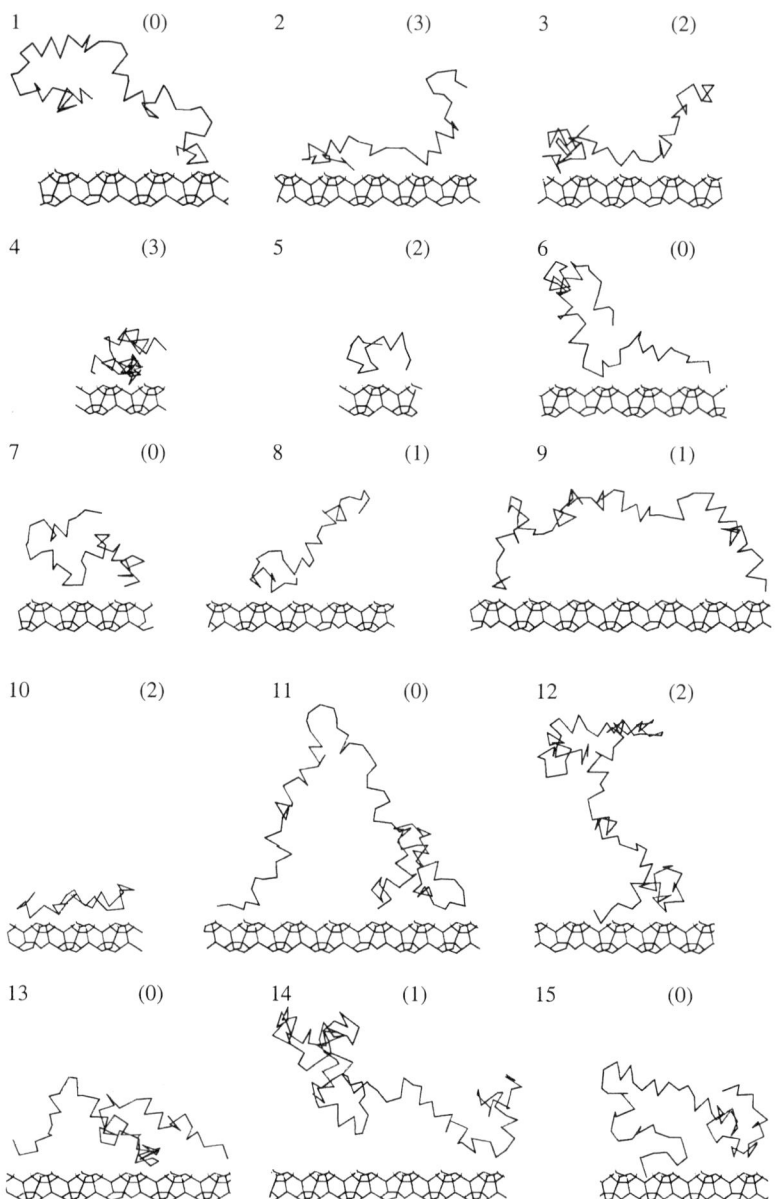

FIGURE 2. Continued on opposite page.

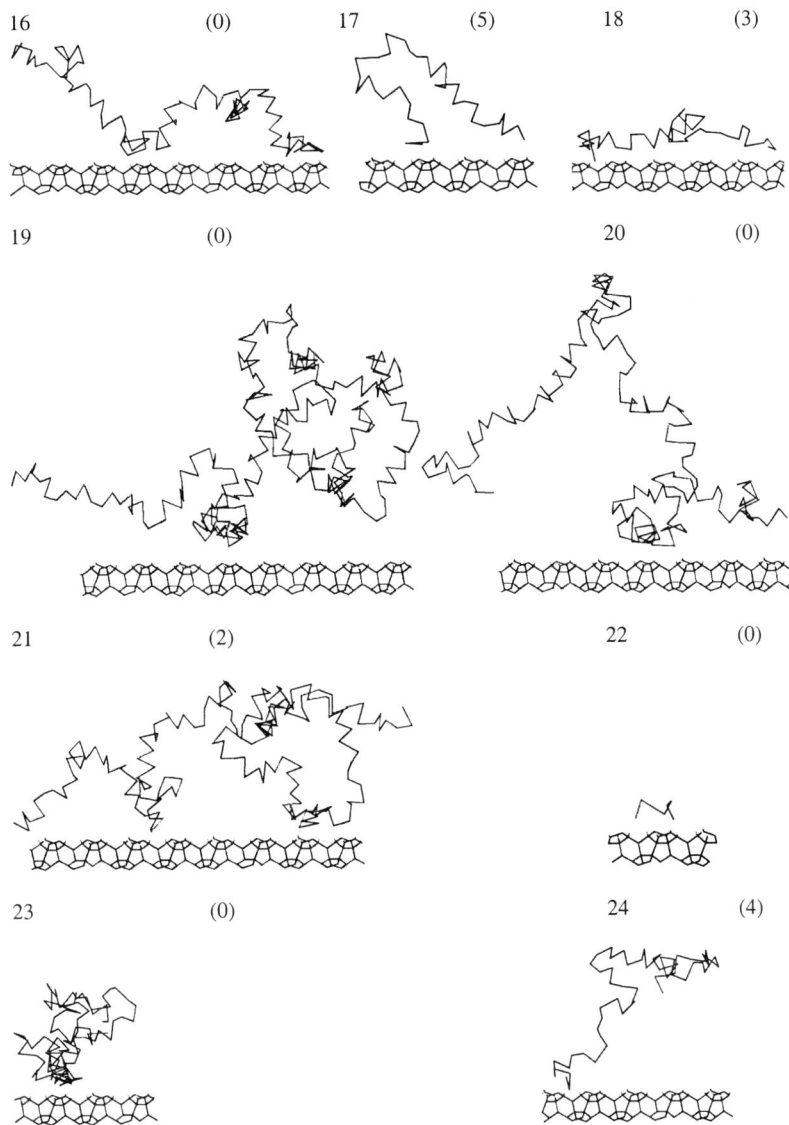

FIGURE 2. Side views of the adsorbed configurations for 24 atactic polymers grown on the −0.01 Å cut of the (111) surface of a type II hydrate. Each point in the chain is the center of mass of a monomeric unit. The numbers of PVP-hydrate hydrogen bonds (defined to occur when the O...H distance is less than 2.5 Å and the O-H...O angle exceeds 90°) are given in parentheses.

studies, the adsorption is not nearly so efficient and typically only a small fraction of the monomers completely occupy the surface potholes. This is also seen in the average number of hydrogen bonds between the polymer and the surface, which was only 1.5 for atactic PVP. This number is an underestimate for the real system, since both relaxation of the hydrate surface and the presence of liquid water are likely to lead to better hydrogen bonding with the PVP; however it does confirm that PVP oligomers adsorb competitively rather than synergistically,[2] and suggests that the conformational constraints between the active units in PVP may limit its effectiveness as a kinetic inhibitor.

We have analyzed various conformational properties for the simulated polymers. Two in particular are useful in determining the shape of the polymer. The first is the side chain angle, (O=)C–N–C_α–C_β, which defines the orientation of the pyrrolidone ring; corresponding distributions are presented in FIGURE 3. In all cases there is a broad distribution centred around 120° and a smaller peak at around 60°, but the distributions are sufficiently broad that the two peaks are not perfectly resolved. Angles of around 120° result in the pyrrolidone ring being perpendicular to the polymer backbone and so allow for parallel stacks of the pyrrolidone rings. From FIGURE 3 we see that angles near 120° are substantially more likely to occur in the isotactic polymers than in the atactic, which suggests that successive *meso* dyads are more favorable. The perpendicular arrangement of pyrrolidone rings also gives maximum access for this moiety to the surface potholes, and so is likely to lead to optimal surface adsorption for the polymer. The second angle is the backbone torsion angle, described by C_α–C_β–C_α–N. The distribution of these backbone angles showed no differences between the four simulations and, as expected, were strongly peaked around ±60° and 180° with the peaks being well resolved.

The loops and tails evident in FIGURE 2 are clearly a significant facet of the adsorbed structures, and so it is interesting to compare the polymer in the adsorbed and nonadsorbed regions. We have examined several different definitions for adsorbed regions of the polymer, and all of them gave consistent results. In this paper we report the analysis based on a simple distance criterion: monomer units with a centre of mass within 10 Å of the hydrate surface were deemed to be adsorbed. Summary results for the four ensembles are given in TABLE 1. Several points are worthy of note. First, the isotactic polymer appears to grow more readily, giving rise to longer polymers for the same length of simulation. An examination of the energies involved shows that this is largely due to steric considerations: although most of the energy contributions are the same in four simulations, the intramolecular van der Waals energy is 3.7 ± 0.2 kcal mol^{-1} per monomer in the isotactic simulations, compared with 4.3 ± 0.2 kcal mol^{-1} per monomer in the atactic simulations. Thus, there is a small driving force that favors the formation of the *mm* triads.

The second point to note is that the number of monomeric units adsorbed in each case is very similar, at about 25 units. This can be seen graphically in FIGURE 4 where the density of monomeric units across the interface is presented. All four systems show the same peak in the 0–10 Å range, whereas both the 24,000 step simulations show evidence of a plateau building up behind the adsorbed layer. Thus, even the smaller polymers grown in these simulations are large enough to saturate the hydrate surface, and these changes in polymer size affect mainly the length and number of loops and tails present to perturb the aqueous environments near to the hydrate

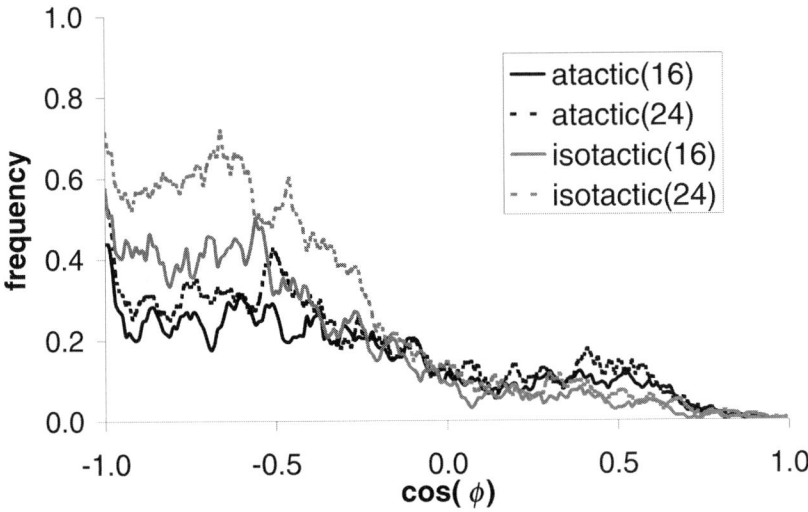

FIGURE 3. Distribution of side group torsion angles for the four groups of simulation.

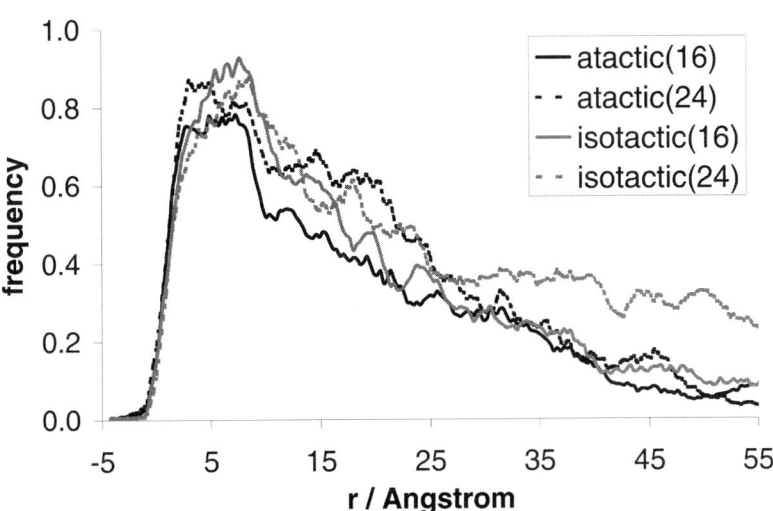

FIGURE 4. Distribution of monomer units across the hydrate interface for the four groups of simulation. Note that in each case the distribution is nonzero below the interface, indicating some adsorption of polymer into the surface potholes.

TABLE 1. Average properties for PVP grown on a (111) type II hydrate surface (values in parentheses give the standard deviation)

	Atactic							Isotactic					
	16,000 steps			24,000 steps				16,000 steps			24,000 steps		
	total	adsorbed	looped	total	adsorbed	looped		total	adsorbed	looped	total	adsorbed	looped
Average numbers													
No. monomers		64.9 (8.6)			77.8 (11.5)				78.1 (7.7)			112.1 (14.6)	
No. adsorbed monomer		22.4 (2.2)			24.8 (2.4)				24.4 (2.5)			23.2 (2.7)	
No. PVP-hydrate H-bonds		1.6 (0.4)			1.4 (0.3)				1.1 (0.3)			0.7 (0.2)	
Dyad probabilities (%)													
m	60.8 (1.8)	69.2 (3.9)	59.0 (4.0)	59.4 (2.0)	59.3 (3.2)	63.2 (4.5)		100	100	100	100	100	100
r	39.2 (1.1)	30.8 (1.7)	41.0 (2.7)	40.6 (1.2)	40.7 (2.7)	36.8 (2.6)		0	0	0	0	0	0
Triad probabilities (%)													
mm	41.2 (1.3)	53.1 (3.1)	38.1 (2.7)	38.8 (1.4)	38.5 (2.0)	43.7 (3.2)		100	100	100	100	100	100
rm	39.5 (1.2)	34.4 (2.1)	41.3 (2.8)	41.4 (1.3)	41.4 (2.8)	39.3 (2.7)		0	0	0	0	0	0
rr	19.3 (0.6)	12.5 (0.9)	20.6 (1.5)	19.8 (0.6)	20.1 (1.4)	17.0 (1.3)		0	0	0	0	0	0
Energies (kcal mol^{-1})													
$U_{intermolecular}$	−0.8 (0.1)	−2.2 (0.3)		−0.7 (0.1)	−2.0 (0.3)			−0.7 (0.1)	−2.6 (0.5)		−0.7 (0.1)	−1.9 (0.2)	
$U_{intramolecular}$ (VdW)	4.5 (0.5)			4.2 (0.3)				3.8 (0.6)			3.5 (0.4)		

surface. Combining this observation with the fact that experimental studies have found small oligomers (10–20 units) to be more active, we conclude that surface adsorption is more important than the loops and tails in determining activity.

The final point we note here is that adsorption and tacticity may be related. This is seen clearly in the shorter atactic simulation, where the adsorbed polymer has a much higher percentage of *meso* dyads than does the nonadsorbed polymer (TABLE 1); a *t*-test indicates that this is significant at the 95% level. The same effect is not seen in the longer simulation, and we are unable to tell whether this is simply hidden by statistical fluctuations, or a real difference relating to polymer size. These distinctions cannot be explained simply by considering the energy of adsorbing the separate monomeric units involved. The interaction energy between adsorbed monomeric units and the hydrate surface is given in TABLE 1 and shows no significant distinction between the four simulations. Thus, we conclude that this must be a cooperativeeffect between neighboring units in the polymer. This conclusion is supported by considering the relative frequency of the different triads, again listed in TABLE 1. In all cases, the triad frequencies are not consistent with a random arrangement of the constituent dyads. Instead they show a lower probability of observing the *rm* triad than expected, and a correspondingly higher probability for the *mm* and *rr* triads. The effect is small, but significant, and so we conclude that the adsorption of one unit is affected by the conformation of the polymer up to at least two units away. This is consistent with the results of our earliest studies on PVP.[2]

CONCLUSIONS

The calculations reported in this paper are the first of their kind for a kinetic inhibitor of gas hydrate crystal growth. The size of polymers simulated are comparable with those found to be most active in experimental studies. The results are consistent with those found in earlier monomer adsorption studies. In particular, the partially formed cages and channels on the hydrate surface are used by the polymer to adsorb tightly to the surface. Although the monomeric studies were found to overstate the structural match between hydrate and inhibitor when compared with any specific monomeric unit within the PVP, there is a clear correlation between the monomeric and polymeric behavior. A striking example is given in FIGURE 1, where the polymer is seen to lie along a surface channel that links successive surface potholes; although there is no strong adsorption within any one of these potholes, there is a clear tendency for the pyrrolidone groups to fill this chain of potholes. It is important to note that the present polymer growth calculations go well beyond the scope of the monomer calculations, for example giving the ability to study the effect of tacticity and polymer size on adsorption.

In the atactic polymers the probability of *meso* addition, P_m, was about 0.6. This is higher than found in synthetic PVP, but experiments indicate that the environment during synthesis does influence P_m (P_m = 0.44 or 0.55 for PVP prepared in water[11,12] or bulk monomer,[13] respectively). There is evidence that the isotactic polymer forms a better match to the hydrate surface than the atactic polymer. This can be understood in part as arising from steric considerations within the polymers, since the van der Waals energy per monomer is about 0.6 kcal mol^{-1} lower in isotactic PVP. However, the preference for the isotactic *mm* triad is significantly greater in the adsorbed region of the simulated polymers. This suggests that increasing the isotactic character of PVP may be lead to a more effective kinetic inhibitor.

ACKNOWLEDGMENTS

We wish to thank the EPSRC for providing computer resources under EPSRC Grant GR/H07207 and a studentship (TJC).

REFERENCES

1. CARVER, T.J., M.G.B. DREW & P.M. RODGER. 1996. Characterisation of the {111} growth planes of a type II gas hydrate and study of the mechanism of kinetic inhibition by poly(vinylpyrrolidone). J. Chem. Soc. Faraday Trans. **92:** 5029–5034.
2. CARVER, T.J., M.G.B. DREW & P.M. RODGER. 1995. Inhibition of crystal growth in methane hydrate. J. Chem. Soc. Faraday Trans. **91:** 3449–3460.
3. CARVER, T.J., M.G.B. DREW & P.M. RODGER. 1999. The structure of aqueous methylpyrrolidone solutions. Phys. Chem. Chem. Phys. **1:** 1807–1816.
4. KVAMME, B., G. HUSEBY & O.K. FORRISDAHL. 1997. Molecular dynamics simulations of PVP kinetic inhibitor in liquid water and hydrate/liquid water systems. Molec. Phys. **90:** 979–991.
5. DAY, J.C. & I.D. ROBB. 1980. Conformation of adsorbed poly(vinylpyrrolidone) studied by infrared spectroscopy. Polymer **21:** 408–412.
6. FRENKEL, D. & B. SMIT. 1996. Understanding Molecular Simulation. Academic Press, San Diego.
7. CARVER, T.J. 1997. A Study of the Kinetic Inhibition of Natural Gas Hydrates by Polyvinylpyrrolidone. Ph.D. Thesis, University of Reading.
8. BROOKS, B.R., R.E. BRUCCOLERI, B.D. OLAFSON, D.J. STATES, S. SWAMINATHAN & M. KARPLUS. 1983. CHARMm—a program for macromolecular energy, minimization, and dynamics calculations. J. Comp. Chem. **4:** 187–217.
9. MOMANY, F.A. & R. RONE. 1992. Validation of the general-purpose Quanta(r)3.2/CHARMm(r) force-field. J. Comp. Chem. **13:** 888–900.
10. BERENDSEN, H.J.C., J.P.M. POSTMA, W.F. VAN GUNSTEREN & J. HERMANS. 1981. *In* Intermolecular Forces. B. Pullman, Ed.: 331. Reidel, Dordrecht.
11. CHENG, H.N., T.E. SMITH & D.M. VITUS. 1981. Tacticity of poly(n-vinyl pyrrolidone). Polymer Lett. **19:** 29–31.
12. EBDON, J.R., T.N. HUCKERBY & E. SENOGLES. 1983. The influence of polymerization conditions on the tacticity of poly(n-vinyl-2-pyrrolidone). Polymer **24:** 339–343.

A Molecular Dynamics Study of the Mechanism of Kinetic Inhibition

MARK T. STORR[a] AND P. MARK RODGER[a,b]

Department of Chemistry, University of Reading, Reading RG6 6AD, United Kingdom

ABSTRACT: Over the past 10 years interest in the use of kinetic inhibitors for preventing gas hydrate blockages has gained momentum. This interest stemmed from the discovery of a protein in the fish Winter Flounder that depressed the freezing point of water as well as retarding the rate of ice nucleation and growth, so allowing the fish to survive in Arctic conditions. In this paper we present results of a series of molecular dynamics simulations aimed at investigating the action of a family of sulfonate compounds whose structures are related to a group of known quaternary amine kinetic inhibitors.

INTRODUCTION

Clathrate hydrates, or gas hydrates as they are more commonly called, are members of a family of compounds known as inclusion compounds, in which small guest molecules are encaged by a crystalline host lattice. Gas hydrates usually form in one of three main crystal structures: type I, which has a body centred cubic crystal structure; type II, with a diamond lattice structure; or structure H, which has a hexagonal structure. These structures are all favoured by relatively low temperatures and high pressures. Indeed conditions found during natural gas exploration and production can be ideal for the formation and growth of gas hydrates and so there is significant risk of pipeline blockages. Thus, there is considerable interest in developing better methods of inhibiting hydrate formation and growth.

Various options for controlling this problem have been tried. These include decreasing the pressure or increasing the temperature of the pipeline, removing water from the pipeline or adding inhibitors to the pipeline. Of these, the use of inhibitors is usually the only financially viable alternative. The inhibitors themselves can be separated into two categories: thermodynamic or kinetic. Thermodynamic inhibitors act by altering the phase diagram of gas hydrates; that is, altering the temperature or pressure at which the hydrate forms. Historically these have been the inhibitors of choice; however, they are required in large amounts—typically 10–50 wt%—so that the cost of using the most common inhibitor, methanol, is estimated at $500 million a year.[1]

Over the last decade, interest has turned to the kinetic inhibitors (KIs), which have also been called low dosage inhibitors (LDIs). KIs act to slow either the nucleation or the rate of growth of the hydrate crystals, and are believed to act via a mechanism that involves selective adsorption of the KI onto particular crystal growth

---

[a]Present address: Department of Chemistry, University of Warwick, Coventry CV4 7AL, UK.
[b]email: p.m.rodger@warwick.ac.uk

surfaces. This selective adsorption has been likened to the lock and key mechanism used to explain enzyme specificity towards substrates in biology. According to this model activity arises from a complementarity between the shapes of the inhibitor and surface adsorption sites, in much the same way as a key fits into a lock. The attraction of such inhibitors is that they are required in much smaller amounts than their thermodynamic counterparts and, therefore, afford the potential for significant financial savings.

In this paper we present results from work designed to investigate the mode of action of small zwitterionic kinetic inhibitors. Results are presented for a family of KIs based on quaternary ammonium sulfonates interacting with methane–propane double hydrate. This system adopts the structure that forms under most operating conditions—type II.

SIMULATION DETAILS

Molecular dynamics simulations (MD) have been performed on 408 water molecules and up to 137 methane molecules and 120 propane molecules contained within an orthorhombic supercell of dimensions $20.1 \times 24.0 \times 62.0$ Å; periodic boundary conditions were used to mimic an infinite system. System size effects were investigated in a previous paper[2] and indicate that a system of this size will display no substantive finite size effects. Water molecules were initially arranged to form an infinite film of type II hydrate, with the $\{1,1,1; -0.001\}$ surface exposed;[3] oxygen positions were taken from neutron diffraction data[4] and hydrogens assigned in a manner consistent with two-fold crystallographic disorder. Initially the small (large) cavities were filled with methane (propane), and a methane/propane gas (about 1:1 mixture) included to stabilize the hydrate film. Simulations were performed at 300 K and for a range of pressures. For pressures in excess of about 3.5 kbar, no decomposition of the hydrate was evident during the entire simulation. We therefore focussed on the 300 K, 3.5 kbar simulations in this paper.

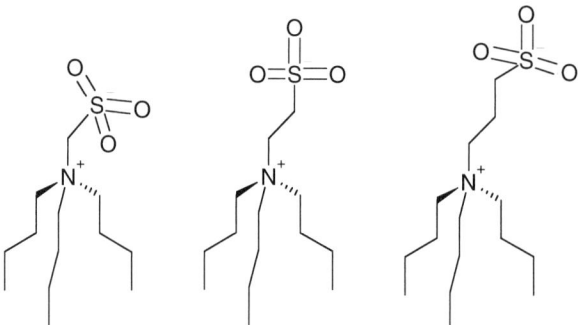

FIGURE 1. Structures of the sulfonate low dosage inhibitors. From *left to right*: tributyl amine ethyl sulfonate, TBAES; tributyl amine propyl sulfonate, TBAPS; and tributyl amine butyl sulfonate, TBABS.

Four sets of simulations were performed. The first (I) consists of just the natural gas hydrate and natural gas. In each of the other three sets (II–IV) a sulfonate inhibitor was also added. The structure of the three sulfonates can be seen in FIGURE 1.

As can be seen from FIGURE 1, three butyl chains are present in each of the sulfonates; the difference between the three molecules is the spacing between the nitrogen atom and the head group moiety. On moving from TBAES through TBAPS to TBABS the chain length increases by one CH_2 group at each step; this is in order to investigate the effect of charge-separation on efficiency.

Equilibration of the initial configurations was achieved in a series of stages. In the case of simulation I the gas hydrate lattice was fixed and the gas allowed to equilibrate for 5 ps; the water lattice was then allowed to move and the system equilibrated for a further 5 ps; finally a 200 ps production run was carried out. The same protocol was used for the other simulations except that the sulfonate was held fixed in position above the gas hydrate until the production run started.

Water was modelled using the SPC potential.[5] Methane and propane were modelled using the united atom model of Jorgensen et al.,[6] and parameters for the sulfonates were taken from the CHARMM force field;[7] Lorenz-Berthelot mixing rules were used for the cross-terms. Long range electrostatic interactions were evaluated with an EWALD summation. All other interactions were truncated for distances in excess of 9.5 Å. All of the simulations in this paper were carried out using DL_POLY[8] in the [N, V, T] Nose-Hoover ensemble, using a time step of 1 fs and a thermal parameter of 0.1.

RESULTS

Sulfonate Orientation

One of the simplest yet possibly most informative ways of analyzing the information produced by the MD runs is simply to consider the typical configurations encountered, and the process by which the molecule moves. For the long time-scale dynamics expected with KIs targeting the appropriate interface, this will be well illustrated by focussing on the initial and final configurations generated from our simulations.

As mentioned in the previous section, each of the sulfonates was initially located close to the hydrate surface. We report here our initial simulations, which were aimed at investigating whether the sulfonates have any effect upon the gas hydrate lattice structure. As such, no attempt has yet been made to correlate the initial orientation of the sulfonate with the resultant behavior of the hydrate. Instead we have sought to use sufficiently long simulations to ensure that the initial location of the sulfonate is not important; this will however be tested by simulations currently in progress.

At the start of the simulations TBAES was orientated perpendicular to the interface and with the head group just touching the hydrate surface. In contrast, both TBAPS and TBABS were orientated parallel to the interface, with two of the butyl chains running along the hydrate surface.

The final configuration of TBAES is shown in FIGURE 2a. From this it is apparent that the top of the gas hydrate lattice decomposed during the simulation and the sulfonate moved into the disordered water region produced by this melting. In the process strong hydrogen-bonding interactions between sulfonate and the water were produced, thus substantially lowering the energy of the sulfonate on immersion. The hydrogen bonding interactions were confirmed from the calculated water-oxygen–sulfonate-oxygen radial distribution functions (not reproduced here) which showed a sharp, distinct peak at 2.4 Å. It is not clear from these simulations whether the migration of the sulfonate into the water-rich region actually caused the surface-melting of the hydrate, or whether it merely exploited melting that was occurring anyway. Once the sulfonate group was immersed in the water, the TBAES reoriented so that one of the butyl chains partially submerged into the water-rich region, with the remaining butyl chains orientated in such a way as to lie along the surface of the water phase, albeit with the terminal methyl groups located on the gas-phase side of the interface. Animation of the MD trajectory indicated that the sulfonate group moved rapidly into the water phase and then proceeded to sample a number of different sulfonate/water relative orientations; in contrast, the migration of the butyl groups was much slower, and the partial immersion of the butyl group occurred gradually over the whole simulation.

Similar observations can be made from the TBAPS and TBABS simulations, FIGURE 2. Again there was surface-melting of the hydrate with consequent immersion and strong hydrogen-bonding (O–O distances of 2.4 Å) of the sulfonate in the water. For these two systems the immersion of one butyl chain in the water was more pronounced than with TBAES, the effect being greatest in TBABS; the remaining two butyl groups lay completely flat to the water surface. These trends are consistent with TBAES, and given the slow migration observed with the alkyl chains, the more pronounced effects observed with TBAPS and TBABS are likely to be consequences of the initial location of the KI being closer to the hydrate phase in these systems.

Z-Density Profile

The average number of the atoms found in each of a series of thin slabs oriented parallel to the interface was calculated from the simulations. The variation in average number across the stack of slabs is a measure of the density profile across the interface (i.e., the z-density profile). Z-density profiles for water-oxygen atoms and hydrocarbon-carbon atoms across the hydrate region are presented in FIGURES 3 and 4.

There are a number of points to note from the water-oxygen profiles presented in FIGURE 3. These plots all show a regular progression of peaks across the water-rich region; this is characteristic of crystalline structure, and the pattern indicates the presence of a stable gas hydrate lattice. In all cases there is some loss of structure on the edges, consistent with the surface-melting noted above.

The structure of the upper interface (10–20 Å, i.e., where the sulfonate was located) does reveal some interesting variations between the control and sulfonate systems. For the control, these plots show three peaks in this region, with a broad shoulder on the outer edge of the peak at 16 Å. In each of the sulfonate systems, however, this shoulder is resolved into a separate peak. Although these differences are small, they are substantially greater than the statistical uncertainties in the calculated density profiles. There are also differences between the sulfonates, with TBAES giving the best resolved shoulder peak, followed by TBAPS and then TBABS. Thus,

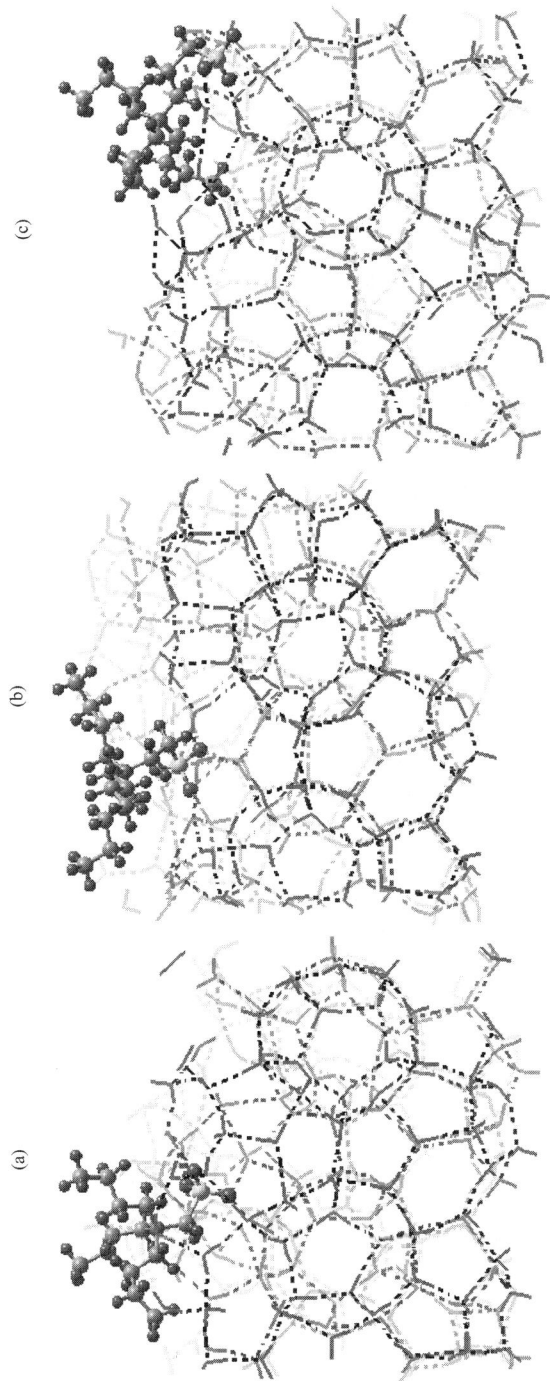

FIGURE 2. Final configuration of (**a**) TBAES, (**b**) TBAPS and (**c**) TBABS at the hydrate/gas interface. Hydrocarbon molecules have been omitted for clarity.

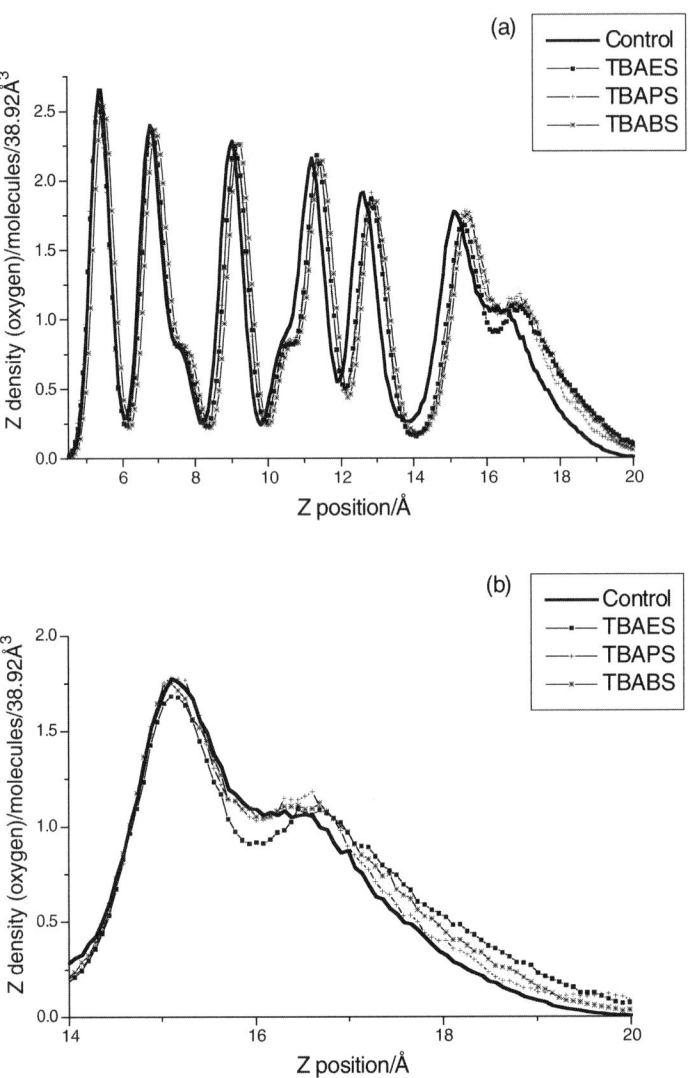

FIGURE 3. (**a**) Z-density profile for water–oxygen atoms from the center of the lattice to the surface. $Z = 0$ corresponds to the middle of the hydrate film, and the inhibitor was added to the $Z > 0$ surface. (**b**) is an expansion of (**a**) in the surface region. To aid comparison the inhibitor plots in (**b**) have been shifted horizontally to make the main peaks (at 15Å) coincide.

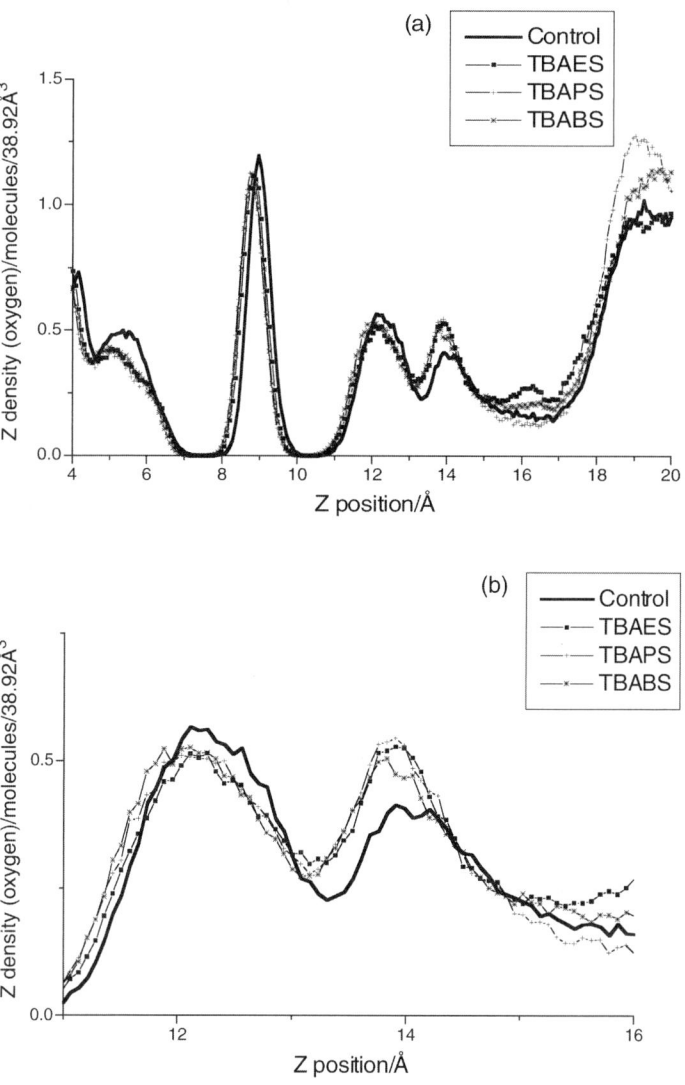

FIGURE 4. (a) Z-density profile for methane/propane carbon atoms in the gas hydrate region from the center of the lattice to the surface. $Z = 0$ corresponds to the middle of the hydrate film, and the inhibitor was added to the $Z > 0$ surface. (b) is an expansion of (a) in the surface region.

the presence of the sulfonates appears to enhance the crystalline nature of the hydrate interface. This is consistent with the lock-and-key mechanism by which LDIs are believed to work: adsorption of a molecule tailored to the hydrate topology should in fact stabilize the hydrate structure in the vicinity of the key. The reason that we see only a small effect in these simulations could be due to the fact that we only have part of the key, in essence a monomer. When one compares the butyl chains in the sulfonates to the carbon chains in the antifreeze protein they are negligible in length and so would be expected to produce a smaller effect than the antifreeze protein. One way to check this hypothesis would be to increase the length of the carbon chains in the sulfonates and calculate the z-density again and this work is currently being investigated; another alternative would be to consider higher KI concentrations.

The final observation is that the sulfonate systems show an apparent expansion in the hydrate region, with the interpeak distance that is significantly larger than in the control system. This effect is detailed in TABLE 1. Again the effect is small, but is statistically significant and reproducible. This observation can not be explained by a drop in the pressure of the system allowing the gas hydrate lattice to relax: analysis of the pressure profile during the MD runs indicates that the pressure is the same in the control and sulfonate simulations. The most likely explanation is that the increased order at the surface in the presence of the KI leads to more efficient packing of the water, and thus allows room for the hydrate to expand. If this is the case, then the disordered water surface would have to resemble low density amorphous ice rather than liquid water. A confirmatory analysis of the dynamic properties of this region is currently under way.

The z-density plots for guest atoms, (methane/propane carbon atoms), are given in FIGURE 4. Again the plots are regular in nature, and the stability of the hydrate film is evident from the complete confinement of the hydrocarbons in this region (i.e., the peaks are surrounded by regions where hydrocarbons are *never* seen). Perhaps the most interesting point to note is that the doublet of peaks at the top of the lattice (12–14 Å) alters when the sulfonates are added. When the sulfonate is present the outermost peak of the doublet increases in size compared with the control. We note

TABLE 1. Position of peaks in the oxygen Z-density for simulations I–IV

Peak	Z Displacement/Å			
	Control	TBAES	TBAPS	TBABS
1	5.36	5.36	5.36	5.51
2	6.71	6.93	6.94	6.94
3	8.96	9.19	9.19	9.19
4	11.36	11.36	11.36	11.51
5	12.56	12.94	12.94	12.94
6	15.11	15.26	15.49	15.49
7	16.31	16.69	16.84	16.84

Note: Positions were measured from the central trough, and were found to be completely symmetric to within the precision given in the table.

that carbon atoms from the sulfonates were not included in the calculation of these density profiles, and so the effect is a real reflection of the gas structure near the interface. It is likely that this effect is strongly correlated with the increased water structure noted above.

CONCLUSIONS

In this paper we have presented the initial results from a series of molecular dynamics simulations designed to elucidate the effect of quaternary ammonium sulfonate zwitterions on a hydrate/hydrocarbon interface; in all cases the zwitterion was based on a tributyl ammonium species. The results indicate that the zwitterions have a small but significant effect in increasing the structure of the hydrate surface, thus leading to a lower degree of surface melting for the hydrate. The zwitterions tend to adsorb at the interface with the sulfonate lying in the water-rich region and forming strong hydrogen bonds with the surrounding water. Two of the butyl groups tend to lie along the surface of the water-rich phase, with the third tending to be immersed within the water. The reorientations that lead to this arrangement of the butyl groups were found to be slow on the molecular dynamics time scale. The results of this study were consistent with a lock-and-key mechanism for the activity of low-dosage inhibitors, as suggested in more simplistic model calculations reported previously.[3]

ACKNOWLEDGMENTS

This work was supported through an EPSRC studentship.

REFERENCES

1. SLOAN, E.D., S. SUBRAMANIAN, P.N. MATTHEWS, J.P. LEDERHOS & A.A. KHOKAR. 1998. Quantifying hydrate formation and kinetic inhibition. Ind. Eng. Chem. Res. **37:** 3124–3132.
2. RODGER, P.M., T.R. FORESTER & W. SMITH. 1996. Simulations of methane hydrate/methane gas interface near hydrate forming conditions. Fluid Phase Equilibria **116:** 326–332.
3. CARVER, T.J., M.G.B. DREW & P.M. RODGER. 1996. Characterisation of the {1 1 1} growth plane of a type II gas hydrate and study of the mechanism of kinetic inhibition by poly(vinylpyrrolidone). J. Chem. Soc. Faraday Trans. **92:** 5029–5033.
4. MCMULLAN, R.K. & Å. KVICK. 1990. Neutron diffraction study of the structure II clathrate hydrate: $3.5Xe.8CCl_4.136D_2O$ at 13 and 100 K. Acta Cryst. **B46:** 390–399.
5. BERENSDEN, W.L., J.P.M. POSTMA, W.F. VAN GUNSTEREN & J. HERMANS. 1981. Intermolecular Forces. Reidel, Dordrecht.
6. JORGENSEN, J.L., J.D. MADURA & C.J. SWENSON. 1984. Optimized intermolecular potentials for liquid hydrocarbons. J. Am. Chem. Soc. **106:** 6638–6646.
7. BROOKS, B.R., R.E. BRUCCOLERI, B.D. OLAFSON, D.J. STATES & S. SWAMINATHAN. 1983. CHARMM: a program for macromolecular energy, minimisation and dynamics calculations. J. Comp. Chem. **4:** 187–217.
8. [SMITH, W. & T.R. FORESTER. 1996.] DL_POLY 2.0: a general-purpose parallel molecular dynamics simulation package. J. Mol. Graphics **14:** 136–141.

Molecular Dynamics Simulation of Dissociation Process for Methane Hydrate

KENJI YASUOKA[a] AND SUGURU MURAKOSHI

*Department of Mechanical Engineering, Keio University,
3-14-1 Hiyoshi, Kohoku-ku, Yokohama 223-8522, Japan*

INTRODUCTION

It is generally accepted that induction time (induction period) for the formation of a hydrate in a water/guest-substance system is related to the thermal history of water in the system.[1] The induction time required in a system that has experienced hydrate formation/dissociation in advance is generally shorter than that in a system having no prior experience of hydrate formation. This phenomenon, known as the *memory effect*, has been assumed by some research groups to be due to some hydrate cage-like structures of water molecules surviving after dissociation of the hydrate. However the cause of the memory effect has not been clarified yet at the molecular level. The nature and the mechanism of the memory effect in relation to the hydrate nucleation needs to be investigated in more detail.

Molecular dynamics (MD) computer simulation is suitable for studying the dynamics of formation and dissociation of hydrates at a molecular level. Attempts at applying MD simulation to such topics were reported recently. Báez and Clancy[2] studied the growth and dissolution of a single crystal of methane hydrate. In their simulation, a spherical crystal composed of 245 water and 42 methane molecules is implanted in a medium of methane/water melt to result in successive growth or dissolution. This simulation deviates greatly from the actual processes of hydrate formation or dissociation and, hence, their speculation that the dodecahedral cages preferentially survive in the melt during the process of hydrate dissociation and are responsible for the memory effect is highly questionable. Hirai *et al.*,[3] also adopted an MD technique to investigate the formation of CO_2 hydrate. However, this simulation is not realistic because the positions of carbon atoms in CO_2 are fixed during the calculation.

In this study we use a system consisting of 1,242 water and 216 methane molecules in performing an MD simulation to directly observe the dissociation process. The primary objective of this simulation is to reveal what kind (or kinds) of structures persist in the water/methane solution in the course of hydrate dissociation.

[a]Telecommunication. Voice: +81-45-566-1523; fax: +81-45-566-1495.
yasuoka@mech.keio.ac.jp

SIMULATION METHOD

Molecular dynamics simulation (MD) was performed on a fully occupied structure-I with a mixture of 1,242 water molecules and 216 methane molecules. Periodic boundary conditions were assumed for all three dimensions, and the leap-frog algorithm with quaternion is adopted for numerical integration of the classical equations of motion. The time step is 0.5 fs (0.5×10^{-15} s) in this simulation. The Ewald sum technique is used to take into account the Coulombic interactions.[4]

The water–water intermolecular interaction is described by Jorgensen's TIP4P 4-site potential.[5] In this model, a water molecule is represented by four interaction sites; two are on the hydrogen atoms (H-sites), one on the oxygen (O-site), and another on the hypothetical lone pair (M-site). Partial charges are assigned to the H- and M-sites, whereas the O-site has only a Lennard-Jones interaction,

$$u_{O-O} = 4\varepsilon_{O-O}\left[\left(\frac{\sigma_{O-O}}{r}\right)^{12} - \left(\frac{\sigma_{O-O}}{r}\right)^{6}\right], \quad (1)$$

where the particle diameter σ_{O-O} is 3.1536 Å and the potential depth ε_{O-O} is 0.64869 kJ/mol. Guest molecules are assumed to have spherical Lennard-Jones type potential with methane parameters ($\sigma_{CH_4-CH_4}$ = 3.730 Å, $\varepsilon_{CH_4-CH_4}$ = 1.231 kJ/mol); that is, Jorgensen's OPLS united atom potential model,[6] rotational and vibrational motions for methane are neglected. Jorgensen's geometric combining rule[6] is assumed between water and methane,

$$\sigma_{CH_4-O} = \sqrt{\sigma_{CH_4-CH_4}\sigma_{O-O}} \quad (2)$$

$$\varepsilon_{CH_4-O} = \sqrt{\varepsilon_{CH_4-CH_4}\varepsilon_{O-O}} \quad (3)$$

Thus a water molecule interacts with a methane molecule only with its oxygen site.

The unit cell size of the hydrate I was 12.03 Å × 12.03 Å × 12.03 Å. Six large ($5^{12}6^2$) and two small (5^{12}) cages are in this unit cell. The initial spacial arrangement of the oxygen atoms of the water molecules were specified on the basis of the data of X-ray scattering experiments.[7] Protons of the water were placed in an *ad hoc* manner according to the simple ice rule. The length of simulation box size was taken to be 36.09 Å. The system was initially equilibrated at 250 K. The run for the equilibrating process was about 1 ns in order to check the stability for crystal condition.

After equilibration, a high temperature (400 K) exceeding the critical dissociation temperature was imposed on the system using simple velocity scaling, resulting in the onset of hydrate dissociation. The system temperature was controlled by the Nosé-Hoover temperature constant method[4] during the process of dissociation. The total run after the equilibration process was 50 ps.

SIMULATION RESULTS

Snapshot

Examples of molecular configurations are shown in FIGURE 1, where the member molecules of a small, dodecahedral cage are pictured. In the initial configuration (just after the equilibration process), the methane molecule is enclosed in the

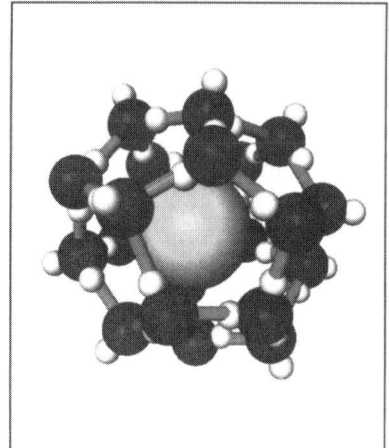
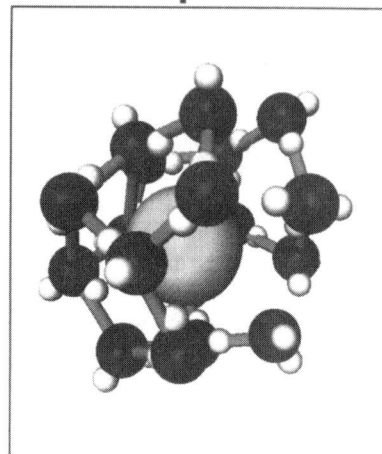
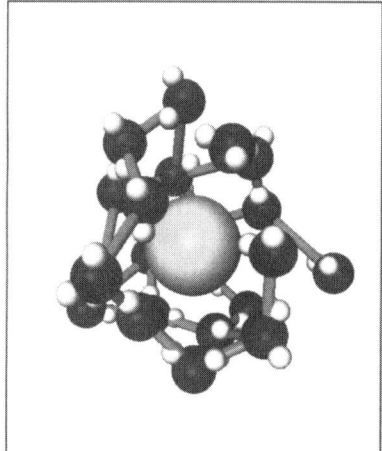
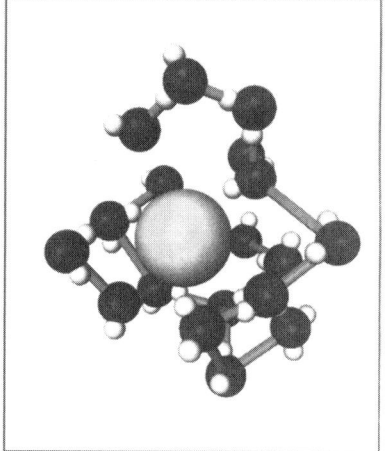

FIGURE 1. Four snapshots of the MD simulation. The member molecules of a small cage, dodecahedral, are shown.

dodecahedral cage, which is composed of hydrogen-bonded water molecules. We adopt an energy criterion for the hydrogen bond;[8] the molecules are defined to be connected by hydrogen bond if the potential energy between water molecules is less than −12 kJ/mol. Some cyclic water pentamers are shown in the second snapshot at 5 ps. The cagelike network structure formed by hydrogen bond does not survive after 15 ps.

Mean-Square Displacement

First, to clarify the diffusion of molecules in the dissociation process we calculate the mean-square displacement (MSD),

$$\langle [r(t) - r(0)]^2 \rangle = 6Dt \tag{4}$$

where $r(t)$ is molecule position and D is diffusion coefficient. FIGURE 2 shows the MSD results for the center of mass for host molecule (water) or guest molecule (methane) with time. The MSD value for water is similar to that for methane. The dissociation process may be classified into four stages. In the first stage (before time 5 ps), each water and methane molecule moves slightly from its initial position, which is a lattice point of the hydrate structure. This suggests that the conformation of hydrate is almost retained in this stage. From the time derivatives of MSD for water and methane, deduced from (4), we calculate the diffusion coefficients as $D_{H_2O}^{stage1} = 1.03 \times 10^{-10}$ m²/s and $D_{CH_4}^{stage1} = 0.49 \times 10^{-10}$ m²/s, respectively. In the third stage (10 ps < time < 15 ps), the molecules diffuse rapidly, which suggests considerable structure decomposition. The diffusion coefficients are $D_{H_2O}^{stage3} = 1.19 \times 10^{-9}$ m²/s and $D_{CH_4}^{stage3} = 1.10 \times 10^{-9}$ m²/s, respectively. The ratio of $D_{H_2O}^{stage3}$ to $D_{H_2O}^{stage1}$ is 11.6 for water, and $D_{CH_4}^{stage3} / D_{CH_4}^{stage1} = 22.4$ for methane. We conclude that the diffusion in the third stage is one order of magnitude faster than that in the first stage.

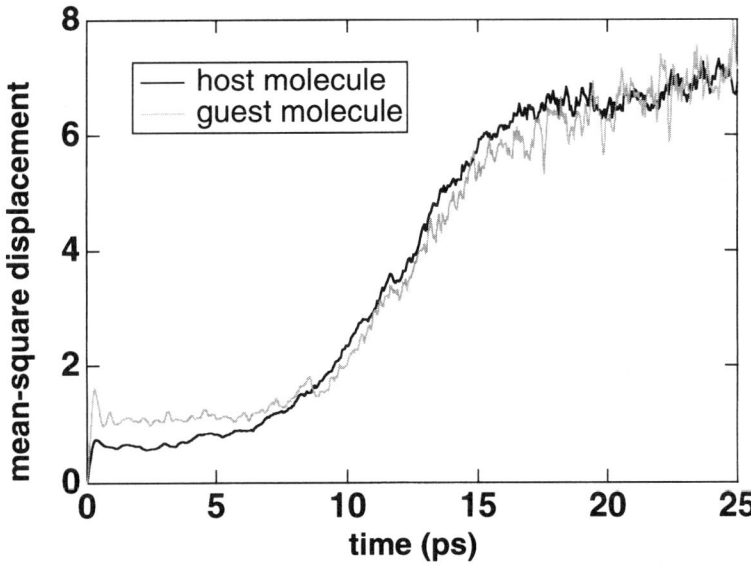

FIGURE 2. The results of mean-square displacement of the center of mass for host molecule (water) or guest molecule (methane) with time.

Fraction of Surviving Cyclic Pentamers and Hexamers for Water

The unit cell of structure I comprises two dodecahedrons and six tetrakaidecahedrons. Each dodecahedron is bound by twelve symmetry-related pentagonal faces (type A). The tetrakaidecahedron are bound by four pentagons of type A, which are shared with the dodecahedra, and by two hexagons (type C) and eight pentagons of type B, which are shared with other tetrakaidecahedron. In this simulation, there are 648 of type A, 648 of type B, and 162 of type C.

The decay of the fraction of surviving cyclic pentamers (type A and type B) and hexamers (type C) for water is shown in FIGURE 3. The fraction of type A is similar to that of type B and C. Thus, we see that the rate of dissociation of tetrakaidecahedral cages is almost the same as that of dodecahedral cages.

To estimate the decay time, the fraction of surviving type A faces is fitted to a single exponential function

$$f(t) = f(0)\exp\left(-\frac{t}{\tau}\right) \tag{5}$$

in stage 1 (before time 5 ps) and stage 3 (10 ps < time < 15 ps). Thus, in stages 1 and 3, this dissociation process conforms to a Markoff process. Least-squares fitting gives the decay time τ as

$$\tau^{(1)} = 25.1 \text{ ps in stage 1}, \tag{6}$$

$$\tau^{(3)} = 1.4 \text{ ps in stage 3}. \tag{7}$$

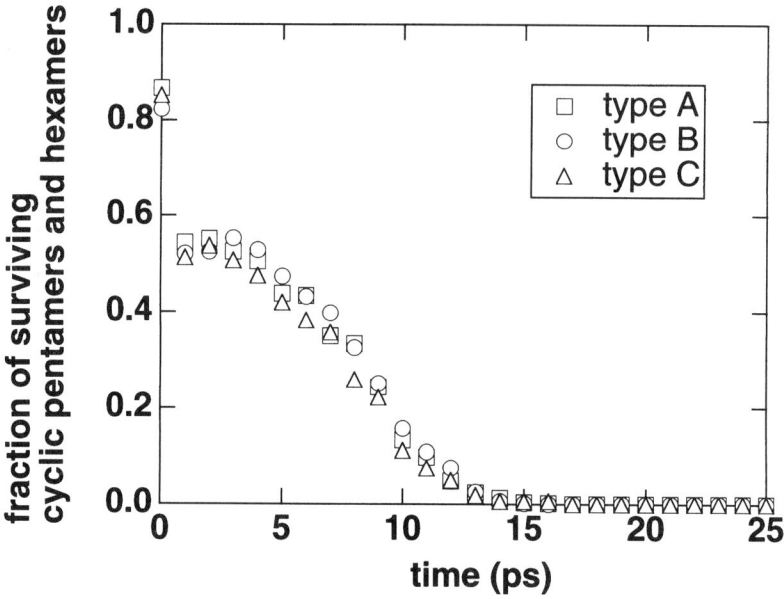

FIGURE 3. The decay of the fraction of surviving cyclic pentamers (type A and type B) and hexamers (type C) for water.

The ratio of $\tau^{(1)}$ to $\tau^{(3)}$ is 17.7. This result is similar to the mean-squares displacement result given in the previous section. In stage 4, all pentamers and hexamers completely vanish. Phase separation occurs after 15 ps.

Radial Distribution Function

In general, the structure of molecular aggregates is characterized by the distribution function for the atomic positions, the simplest of which is the radial distribution function $g(r)$. This function gives the probability of finding a pair of atoms a distance r apart relative to the probability expected for a completely random distribution at the same density. FIGURE 4 shows the methane–methane distribution function $g(r)$. At 5 ps (stage 1) the first peak point of $g(r)$ is 6 Å, the same as the case of initial hydrate structure, and a guest molecule is in each cage formed by hydrogen bonded water molecules. $g(r)$ at 10 ps has a small shoulder near 3.5 Å. The shoulder turns into the first peak at 15 ps. A few methane molecules move out the cage after 5 ps, and many other guest molecules move out and come close to each other after 10 ps.

CONCLUSION

Using a molecular dynamics computer simulation, we have studied the dissociation of methane hydrate. The system is a mixture of 1242 water and 216 methane

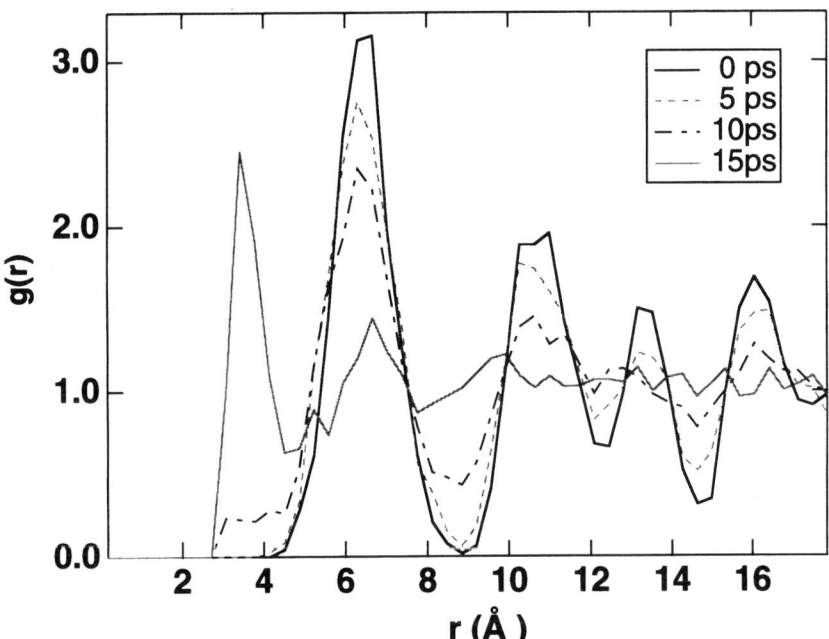

FIGURE 4. The methane-methane radial distribution function $g(r)$.

molecules, 27 unit cells of structure I, and the system temperature is controlled by the Nosé-Hoover method.

The dissociation process may be classified into four stages, based on the obtained variations in the mean square displacement for both host and guest molecules, the fraction of surviving cyclic pentamers and hexamers, and the radial distribution function for guest molecules. In the first stage, almost no structural change from the crystal hydrate occurs. In the third stage, the diffusion coefficients for water and methane molecules are maximized, and the hydrate structure undergoes substantial dissociation. The dissociation in the third stage is not a competitive chain reaction. The second stage represents a gradual transition from the first stage to the third. The last stage, stage four, is characterized by the separation between water and methane phases.

Our simulation shows that the 5^{12} cage and the $5^{12}6^2$ cages undergo dissociation processes similar to each other. This result contradicts Báez and Clancy's speculation[2] that 5^{12} cages preferentially survive. We presume that some structures (or molecular aggregates) remaining more or less intact in the dissociating system, before stage 4, are responsible for the *memory effect*. These structures are not necessarily 5^{12} cages or their fragments. More detailed analysis is required to obtain better understanding of such structures.

ACKNOWLEDGMENTS

We are grateful to Professor Y.H. Mori (Keio University) for stimulating discussion. Part of the calculation was carried out at the Institute of Physical and Chemical Research (RIKEN).

REFERENCES

1. SLOAN, E.D. 1998. Clathrate Hydrates of Natural Gases. Marcel Dekker. New York.
2. BÁEZ, L.A. & P. CLANCY. 1994. Computer simulation of the crystal growth and dissolution of natural gas hydrates. Ann. N.Y. Acad. Sci. **715:** 177–186.
3. HIRAI, S. *et al.* 1997. CO_2 clathrate-hydrate formation and its mechanism by molecular dynamics simulation. Energy Convers. Mgmt. **38:** S301–S306.
4. ALLEN, M.P. & D.J. TILDESLEY. 1987. Computer Simulation of Liquids. Oxford University Press, Oxford, UK.
5. JORGENSEN, W.L. *et al.* 1983. Comparison of simple potential functions for simulating liquid water. J. Chem. Phys. **79:** 926–935.
6. JORGENSEN, W.L., J.D. MADURA & C.J. SWENSON. 1984. Optimized intermolecular potential functions for liquid hydrocarbons. J. Am. Chem. Soc. **106:** 6638–6646.
7. MCMULLAN, R.K. & G.A. JEFFREY. 1965. Polyhedral clathrate hydrates. IX. Structure of ethylene oxide hydrate. J. Chem. Phys. **42:** 2725–2732.
8. RAHMAN, A. & F.H. STILINGER. 1971. Molecular dynamics study of liquid water. J. Chem. Phys. **55:** 3336–3359.

Molecular Orbital Calculations for Polyhedral Water Clusters Including Gas Molecules

AKIRA HORI[a] AND TAKEO HONDOH

*Institute of Low Temperature Science, Hokkaido University,
N19W8, Sapporo 060-0819, Japan*

ABSTRACT: Semiempirical molecular orbital calculations were performed for several types of polyhedral water clusters (12-, 14-, and 16-hedra) including a gas molecule. The stabilities of the polyhedral water clusters were evaluated from the differences in the cohesive energy between the clusters including a gas molecule and the corresponding vacant clusters. It was found that there is a correlation between the structure type of the hydrate crystal and the type of polyhedral water cluster that is energetically favored; a gas molecule for which the 14-hedral cluster is energetically more favorable than the 16-hedral one tends to form a structure I hydrate crystal. The vibrational frequencies of the guest molecules shifted due to the interaction between the gas molecule and the surrounding water molecules. Optical absorption spectra in the ultraviolet range were also calculated for polyhedral water clusters, including a gas molecule.

INTRODUCTION

Gas hydrates are known to be composed of several types of polyhedral water clusters (12-, 14-, and 16-hedra) including different gas molecules.[1–3] The structural types of clathrate hydrates—that is, structure I or II—depend on the kinds of included guest molecules. The only computational studies available are those that have been performed by the Monte Carlo (MC) or molecular dynamics (MD) methods. On the other hand, several studies on large water clusters have been recently reported.[4–8] Molecular orbital calculations revealed the stability of the polyhedral water clusters. However, the existence of a guest molecule included in the water cluster was not considered in those studies. Guest molecules play an important role in the formation of clathrate hydrate crystals. In this paper, we report on molecular orbital calculations performed for polyhedral water clusters, including a guest molecule, and the stability of these clusters and their optical properties.

METHOD

To estimate the cohesive energies of water clusters, molecular orbital calculations by the MNDO method were performed with the MOPAC 97[9] using the PM3[10] parameterization. For geometric optimization of dodecahedral water clusters, the

[a]Telecommunication. Voice and fax: +81-11-706-7351.
hori@hhp2.lowtem.hokudai.ac.jp

oxygen atoms of twenty water molecules were arranged with icosahedral symmetry. In the 14-hedral clusters, 24 oxygen atoms were arranged so that they had a hexagonal symmetry with respect to the axis intersecting the centers of two six-membered rings of water molecules. In the 16-hedral clusters, there are four six-membered rings of water molecules and four residual water molecules. These were tetrahedrally arranged. Water molecules were arranged so that they would meet the *ice rule* in all polyhedral clusters. Although reportedly[11] the total energy of a polyhedral water cluster depends on the proton arrangement, we chose a certain structure for each type of polyhedral water cluster. Thus, in this study, molecular orbital calculations were performed for polyhedral clusters with the same proton arrangement. The geometrical positions of all atoms of the gas molecules were fully optimized. The vibrational frequencies of the gas molecules were also calculated using MOPAC 97 for the same structures as in the total energy calculations. The absorption spectra of the polyhedral water clusters including a gas molecule were calculated for the optimized structure by the INDO/S[12] method using the program package MOS/F V4.[13]

RESULTS AND DISCUSSION

Cohesive Energy

Cohesive energies were defined as the difference between the heats of formation, ΔH of a water cluster and those of the constituents: $E_{coh} = \Delta H$ (cluster) $- n \times \Delta H$ (water) $- \Delta H$ (gas), where n is the number of water molecules in the polyhedral water clusters, and n = 20, 24, and 28 for the 12-, 14-, and 16-hedra, respectively. In TABLE 1, the cohesive energies of the clusters, including a gas molecule, per water molecule are shown. Although there are some exceptions (for example, SO_2, C_2H_4) by comparing the energies for the 14-hedra with those for the 16-hedra it was found that the former is lower than the latter in gas molecules forming a structure I crystal, whereas the former is higher than the latter in gas molecules forming structure II. Hence, the type of structure is closely related to the type of energetically favored cluster type. This suggests that energetically favored polyhedral water clusters formed during the initial stage of crystal growth of a clathrate hydrate and the type of polyhedra dominates the type of crystal structure. Additionally, for N_2 and O_2 molecules, the dissociation pressures of the clathrate hydrates are very high. The corresponding cohesive energies of the 12-hedra are high. Hence, the dissociation pressure is considered to reflect the stability of the polyhedral water clusters.

Molecular Vibration of the Guests

TABLE 2 shows the calculated frequencies of the guest molecule vibrations in the polyhedral water clusters. The difference between the vibrational frequencies of the free molecules and those of the enclosed molecules are also shown. In general, the calculated frequencies tend to be higher than the observed frequencies because anharmonic terms are ignored and the electron correlation is not fully considered. In CO_2 and CH_4, some vibrational mode splitting occurred due to the interaction between the guest molecules and the host clusters. The splitting causes a broadening of the vibrational spectra of the molecules in the hydrate crystals. In fact, the peak

TABLE 1. Cohesive energies of clusters including a gas molecule per water molecule

Molecules	Molecular Diameter (Å)	Structure Type	Dissociation Pressure at 273 K (10^5 Pa)	12-Hedron $E_{coh}/20$ (kJ/mol)	14-Hedron $E_{coh}/24$ (kJ/mol)	16-Hedron $E_{coh}/28$ (kJ/mol)
—				−22.649	−26.504	−25.675
N_2	4.10	II	141.00	−21.732	−26.165	−25.572
O_2	4.20	II	109.00	−22.437	−26.436	−25.648
CH_3F	4.30	I	2.12	−26.497	−27.38	−26.397
CH_4	4.36	I	26.26	−23.763	−27.107	−26.165
C_2H_5F	5.00	I	0.70	−25.561	−27.897	−26564
CH_3Cl	5.06	I	0.41	−26.985	−27.234	−26.331
CO_2	5.12	I	12.48	−25128	−26.313	−25.612
Cl_2	5.17	I	0.33	−22.802	−26.636	−25.785
CH_3Br	5.33	I	0.25	−27.078	−27.210	−26.366
C_2H_6	5.50	I	5.25	−26534	−27.276	−26.182
C_2H_2	5.50	I	5.76	−26.698	−27.044	−26.057
Br_2	5.68	I	0.06	−24.262	−27.400	−25.522
C_3H_8	5.28	II	1.74	−21.334	−26.490	−26.769
CCl_4	6.68	II		−16.211	−24.881	−25.559
H_2S	4.58	I	0.97	−22.661	−25.849	−25.899
SO_2	5.00	I	0.39	−11.907	−21.057	−25.792
C_2H_4	5.60	I	5.55	−23.678	−22.778	−26.290
C_2H_5Br	6.47	II		−23.025	−27.683	−26.596

Note: For the constituent gas molecules above the *thin line*, there is good correlation between the types of the crystal structure of the clathrate hydrates and those of the energetically favored polyhedral water clusters. Some of the exceptions are shown below the line. In particular, for N_2 and O_2, the cohesive energies of the 12-hedron are higher than that of the vacant 12-hedron. Therefore, they are considered to be unstable, which is the reason for the high dissociation pressure of N_2 and O_2 clathrate hydrates.

broadening of the vibrational spectra was observed in Raman spectroscopy measurements.[14] In CH_4, almost all of the frequencies were reduced. The differences are larger in the $2T_2$ and $1A_1$ modes than in the $1T_2$ and $1E$ modes. This is reasonable because the $2T_2$ and $1A_1$ modes are stretching modes whereas the $1T_2$ and $1E$ modes are bending modes. These bending modes are considered to be less affected by the water cluster enclosing a CH_4 molecule. In N_2, O_2 and CO_2, the differences in the stretching mode frequencies are larger for the small cages than for the large cages. This is also reasonable because the interaction between the guest molecule and the host cluster is considered to be stronger in the small cage than in the large cage. However, the calculated frequencies increase when enclosed in the cages, whereas the observed frequencies in a clathrate hydrate were reduced in N_2 and O_2.[15]

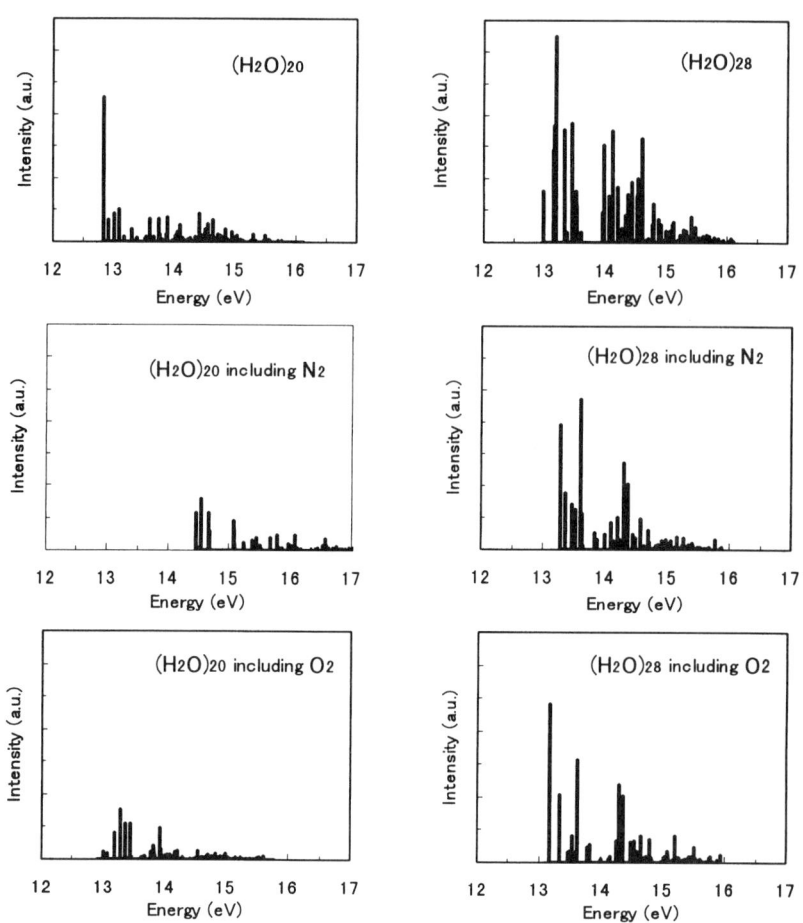

FIGURE 1(a). Optical absorption spectra of 12- and 16-hedral water clusters including N_2 and O_2 molecules. These molecules are known to form structure II clathrate hydrates. Those of the vacant 12- and 16-hedral clusters are also shown for comparison.

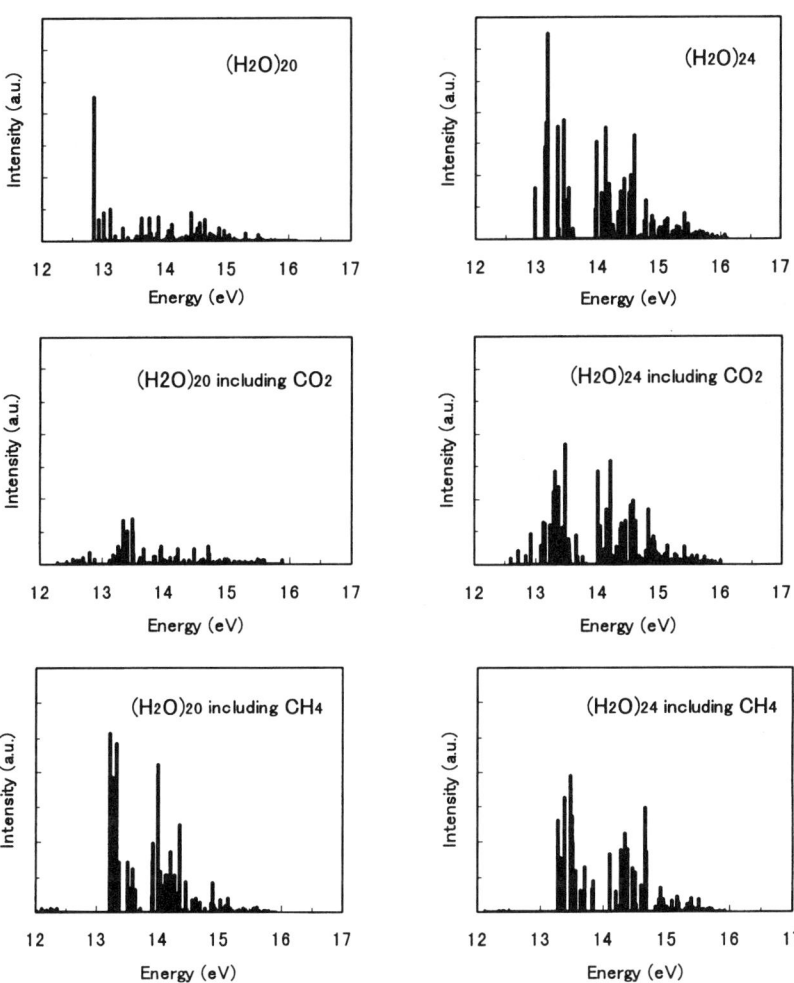

FIGURE 1(b). Optical absorption spectra of 12- and 14-hedral water clusters including CH_4 and CO_2 molecules. These molecules are known to form structure I clathrate hydrates. Those of the vacant 12- and 14-hedral clusters are also shown for comparison.

TABLE 2. Calculated vibrational frequencies of molecules included in polyhedral water clusters. Differences between the vibrational frequencies of the free molecules and those of molecules in the polyhedral water clusters are also shown

Molecule	Mode	Frequency (cm^{-1})	12–Hedron		14–Hedron		16–Hedron	
			Frequency	Difference	Frequency	Difference	Frequency	Difference
N_2	Σ_g	2639.70	2643.93	4.23			2641.31	1.61
O_2	Σ_g	2096.07	2099.16	3.09			2096.50	0.43
CH_4	$1T_2$	1362.16	1347.60	−14.56	1355.55	−6.61		
			1349.78	−12.38	1359.25	−2.91		
			1361.59	−0.57	1361.79	−0.37		
	1E	1451.09	1438.50	−12.59	1446.02	−5.07		
			1441.85	−9.24	1447.49	−3.60		
	$2T_2$	3207.39	3161.60	−45.79	3165.51	−41.88		
			3166.18	−41.21	3173.94	−33.45		
			3170.42	−36.97	3186.31	−21.08		
	$1A_1$	3310.94	3262.99	−47.95	3274.91	−36.03		
CO_2	π_u	520.83	512.43	−8.40	512.69	−8.14		
			513.75	−7.08	517.33	−3.50		
			517.67	−3.16	518.13	−2.70		
	Σ_g	1407.54	1412.24	4.70	1408.80	1.26		
	Σ_u	2385.50	2388.47	2.97	2385.05	−0.45		

Optical Absorption in the Electronic Transition Range

The optical absorption spectra of vacant polyhedral clusters are shown in FIGURES 1(a) and 1(b). In all these three clusters, the optical gap is about 13 eV. For a single water molecule, the calculated optical gap was estimated to be about 11 eV from the calculation. Hence, the formation of a polyhedral cluster caused a widening of the optical gap. In polyhedral water clusters including gas molecules, the absorption spectra are different from those of the vacant clusters. In particular, for N_2, O_2, and CO_2 molecules, the energetic levels attributed to these molecules are formed in

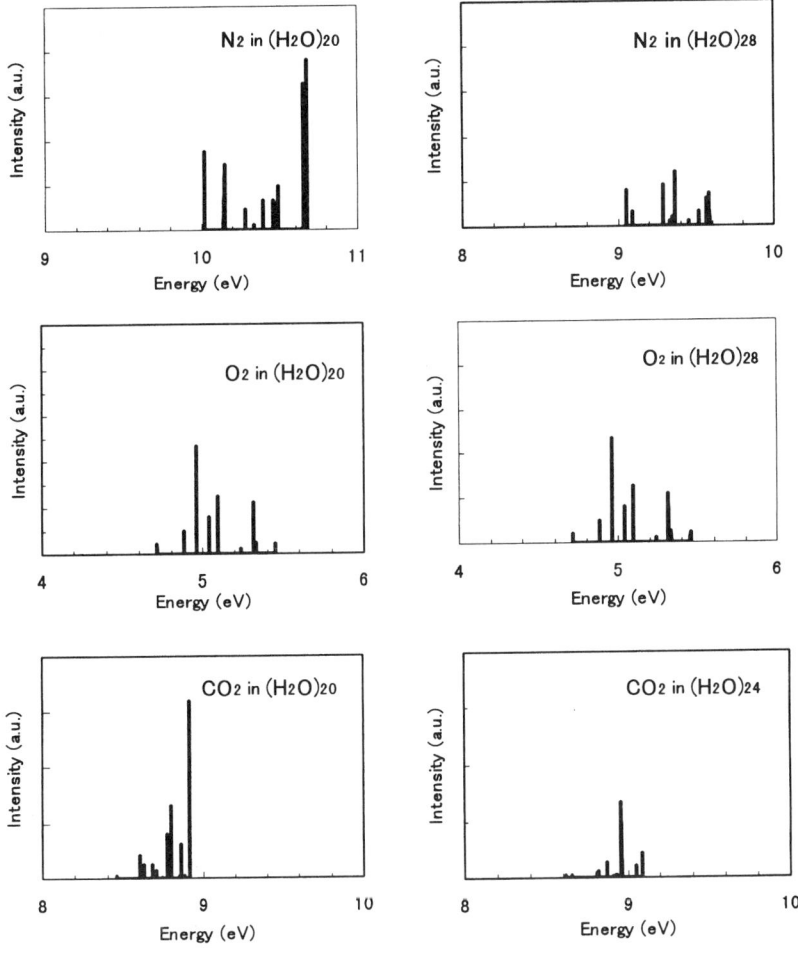

FIGURE 2. Absorption spectra of N_2, O_2, and CO_2 molecules in polyhedral water clusters. For the CH_4 molecule, the spectra overlapped around the absorption edges of the polyhedral water cluster as shown in FIGURE 1(b).

the HOMO-LUMO gap like impurity levels (see FIGURE 2). However, the spectra attributed to the polyhedral water clusters were similar to those of the vacant polyhedral water clusters (FIGS. 1(a) and 1(b)). For CH_4, the spectra overlapped around the absorption edge of the polyhedral water clusters. Since there are no experimental data reported at present, these results cannot be compared with experimental data.

CONCLUSION

Cohesive energy calculations using the semiempirical molecular orbital method revealed that there is a correlation between the structural types of clathrate hydrates and the type of energetically favored polyhedral clusters. The vibrational spectra of water clusters including a gas molecule were calculated and a splitting of the molecular vibration modes was found, which is considered to be the origin of the broadening of the observed spectra. The optical absorption spectra in the electronic transition range were also calculated. In N_2, O_2, and CO_2, several peaks attributed to the energy levels of the gas molecules were found in the HOMO-LUMO gap in the polyhedral water clusters.

REFERENCES

1. DAVIDSON, D.W. 1973. Water. A Comprehensive Treatise, 2nd edit. Plenum Press, New York.
2. SLOAN, E.D., JR. 1998. Clathrate Hydrates of Natural of Natural Gases, 2nd edit. Marcel Dekker, New York.
3. BERECZ, E. & M. BALLA-ACHS. 1983. Gas Hydrates. Elsevier, Amsterdam.
4. LAASONEN, K. & M.L. KLEIN. 1994. Structural study of $(H_2O)_{20}$ and $(H_2O)_{21}H^+$ using density functional methods. J. Phys. Chem. **98:** 10079–10083.
5. KHAN, A. 1994. Theoretical studies of the clathrate structures of $(H_2O)_{20}$, $H^+(H_2O)_{20}$ and $H^+(H_2O)_{21}$. Chem. Phys. Lett. **217:** 443–450.
6. KHAN, A. 1995. Examining the cubic, fused cubic and cage structures of $(H_2O)_n$, for $n = 8$, 9, 12, 16, 20, and 21: do fused cubic structures form? J. Phys. Chem. **99:** 12450–12455.
7. KHAN, A. 1996. Theoretical studies of tetrakaidecahedral structures of $(H_2O)_{24}$, $(H_2O)_{25}$, and $(H_2O)_{26}$ clusters. Chem. Phys. Lett. **253:** 299–304.
8. KHAN, A. 1997. Theoretical studies of large water clusters: $(H_2O)_{28}$, $(H_2O)_{29}$, $(H_2O)_{30}$, and $(H_2O)_{31}$ hexakaidecahedral structures. J. Chem. Phys. **106:** 5537–5540.
9. MOPAC97. 1997. J.J.P. Stewert & Fujitsu Limited, Tokyo.
10. STEWERT, J.J.P. 1989. Optimization of parameters for semiempirical methods I. Method J. Comp. Chem. **10:** 209–220.
11. LUDWIG, R. & F. WEINHOLD. 1999. Quantum cluster equilibrium theory of liquids: freezing of QCE/3-21G water to tetrakaidecahedral "Bucky-ice". J. Chem. Phys. **110:** 508–515.
12. RIDLEY, J. & M. ZERNER. 1973. An intermediate neglect of differential overlap technique for spectroscopy: pyrrole and azines. Theoret. Chim. Acta(Berl.) **32:** 111–134.
13. MOS/F V4. 1997. Fujitsu Limited, Tokyo.
14. SUM, A.K. *et al.* 1997. Measurement of clathrate hydrates via Raman spectroscopy. J. Phys. Chem. B **101:** 7371–7377.
15. FUKAZAWA, H. *et al.* 1996. Molecular fractionation of air constituent gases during crystal of clathrate hydrate in polar ice sheets. Proc. 2nd Int. Conf. Natural Gas Hydrates. 237–242.

Molecular Dynamics Simulations of Clathrate Hydrate

Intramolecular Vibrations of Methane

HIDENOSUKE ITOH[a] AND KATSUYUKI KAWAMURA

Department of Earth and Planetary Sciences, Faculty of Science, Tokyo Institute of Technology, 2-12-1 Ookayama, Meguro, Tokyo 152-8551, Japan

ABSTRACT: Intramolecular vibrational spectra of methane in clathrate hydrate structure I were obtained by molecular dynamics simulations using the Kumagai, Kawamura, and Yokokawa (KKY) potential model that allows unconstrained atomic motions. Peak frequency shifts between the power spectra of methane molecules in large and small cages were found in the GH stretching (v_1, v_3) modes, in contrast no frequency shifts were observed in the H-C-H bending (v_2, v_4) modes. We discuss an interaction between water and methane molecules from these frequency shifts and the GH distance depending on the cage size. An understanding of the interaction between guest and host molecules helps us to clarify the formation and dissociation processes of clathrate hydrates.

INTRODUCTION

Clathrate hydrates are nonstoichiometric compounds with a host framework formed by hydrogen bonded water molecules and enclathrated gas (guest) molecules. Clathrate hydrates are stabilized through the inclusion of a gas molecule at high pressure and low temperature. The crystal structure of clathrate hydrate is known as Stackelberg's structure I or II[1] depending on the guest molecule size. In the structure I there are two different types of cages, designated as a small cage (12-hedron) of diameter 5.0 Å and a large cage (14 hedron) of 5.9 Å. These values represent the free diameter of the cavities, meaning the van der Waals diameter of a water molecule subtracted from the average of the O-O distance across a cage.

It is known that a spherical molecule methane (CH_4) forms clathrate hydrate structure I because of its van der Waals diameter of 4.3 Å. The structure I hydrate is preferably formed when the maximum van der Waals diameter of the guest molecule is between 4.2 Å and 5.5 Å. For bigger and smaller molecules the structure II is required in order to enclathrate the molecule. The clathrate hydrate containing methane molecules, called CH_4 hydrate hereafter, has received considerable attention as an important material for energy resources in the future (see for example, Ref. 2). An understanding of the stability and the formation and dissociation processes of CH_4 hydrate is very important when CH_4 hydrate is extracted and put to practical use.

[a] Address for correspondence. Voice: +81-3-5734-2617; fax: +81-3-5734-2616.
hide@geo.titech.acjp

However, the microscopic dynamic behavior of CH_4 hydrate, which might help to clarify these topics, is not yet understood.

Spectroscopic studies on CH_4 hydrate were carried out using Raman spectroscopy by Sum et al.[3] It was found that the peak frequency shifts of the C-H stretching mode (υ_1) depending on the cage size were obtained from Raman spectra of $CO_2 + CH_4$ mixed hydrate. In the present study all intramolecular vibrational modes of CH_4 in the large and the small cages were directly calculated from their atom velocity data obtained by molecular dynamics (MD) simulations. We successfully revealed the different spectral characteristics of CH_4 depending on the cage size in our simulations.

COMPUTATIONAL DETAILS

Interatomic Potential Model: KKY Model

MD simulations of CH_4 hydrate were performed using various potential models for water and guest molecules.[4-6] In the previous MD simulations rigid molecular models such as SPC[7] and TIP4P[8] were used for an interaction between water molecules and Lennard-Jones potential between methane and water molecules. However, intramolecular vibrations of guest molecules were not obtained since intramolecular degrees of freedom are not taken into account in these models.

In the present MD simulations, the KKY potential model, developed by Kumagai, Kawamura, and Yokokawa,[9] was adopted in order to observe intramolecular vibrations of methane in two different types of cages. This potential model consists of two-body and three-body potential and takes into account the total degree of freedom of atomic motions. Two-body interactions include coulombic, short range repulsive, van der Waals interactions, and Morse interactions. A Morse interaction that represents O-H covalent bonding is used to express the characteristic structure of a water molecule. Two-body interactions are described as follows:

$$U_{ij}(r_{ij}) = \frac{Z_i Z_j e^2}{r_{ij}} + f_0(b_i + b_j)\exp\left(\frac{a_i + a_j - r_{ij}}{b_i + b_j}\right) - \frac{C_i C_j}{r_{ij}^6} \quad (1)$$
$$+ D_{ij}[\exp\{-2\beta_{ij}(r_{ij} - r_{ij}^*)\} - 2\exp\{-\beta_{ij}(r_{ij} - r_{ij}^*)\}],$$

where r_{ij} is the interatomic distance and f_0 is a constant for unit adaptations between these terms. The parameters, z, a, b, and c are of atom species, and D_{ij}, P_{ij}, and r_{ij}^* are of O-H pairs. A three-body potential, introduced to confine the H-O-H angle in a water molecule, is described as follows:

$$U_{HOH} = -f_K[\cos 2(\theta_{HOH} - \theta_0) - 1]\sqrt{K_1 \cdot K_2} \quad (2)$$

$$K_i = \frac{1}{\exp\{g_r(r_{OH(i)} - r_m)\} + 1} \quad (3)$$

where θ_{HOH} is the angle of H-O-H, and f_K and θ_0 are parameters. The Fermi-Dirac distribution function E defines the effective range of the three-body potential.

The potential parameters for H_2O and functions of the KKY potential model are refereed in our previous papers.[9] Our previous MD simulations using this model successfully revealed the structural and dynamic properties of ice[10,11] and clathrate

TABLE 1a. Two body potential parameters for methane determined in this study

	Weight	z	a (Å)	b (Å)	c (kJ$^{0.5}$ Å^{3}mol$^{0.5}$)
Carbon	12.011	−1.200	2.150	0.140	63.426
Hydrogen	1.01	0.300	0.035	0.100	0.000
	D (kJ/mol)	β (Å^{-1})	r_{ij}^* (Å)		
Morse term	226.0	2.500	1.05		

TABLE 1b. Three body potential parameters for methane determined in this study

	f_K (10^{-19} J)	θ_0 (degrees)	r_m (Å)	g_r (Å^{-1})
H-C-H group	1.19	104.5	1.35	7.0

hydrates.[12,13] In the present study, the potential parameters for CH_4 were determined empirically and are presented in TABLE 1. The reproducibility of density and vibrational spectra for methane is described in a later section.

Fundamental Cell

For parameter fitting, 27 CH_4 in a cell with a, b, and $c = 15$ Å were used to calculate power spectra of CH_4 corresponding to the normal modes of CH_4 (υ_1, υ_2, υ_3, and υ_4) shown in FIGURE 1. We used 216 CH_4 whose initial positions were randomly distributed to calculate the density of liquid CH_4.

The system used in the present MD simulations contains 368 H_2O for the structure I clathrate hydrate with dimensions a, b, and $c = 24.08$ Å (8 unit cells) shown in FIGURE 2. The initial positions of oxygen atoms were taken from the structure Pm3n obtained by X-ray diffraction experiments.[14] Hydrogen atoms were positioned in such a way in order to satisfy the ice rule[15] and to yield a proton-disordered structure. The 64 CH_4 were added at the center of each cage (i.e., the cages were fully occupied by guest molecules) and their orientations were randomly distributed.

Method of Calculation

We carried out MD simulations by using the computer program MXDORTO developed by Kawamura.[16] The present MD simulations were carried out using the Verlet algorithm for atomic motions and the Ewald method for electrostatic interactions. The time step used in the Verlet algorithm was 0.4 fs (femtoseconds). Temperature and pressure were kept constant at 100 K and 1 atm for liquid CH_4, and 100 K and 100 atm for CH_4 hydrate. Temperature and pressure were controlled by scaling atom velocities and cell length, respectively. All simulations were done on the fundamental cell with three-dimensional periodic boundary conditions.

The power spectra for each atom species are the Fourier transform of the velocity autocorrelation functions. The density of states (DOS) is

$$DOS = \int_0^\infty \exp(-i\omega t) \langle \mathbf{v}(t) \cdot \mathbf{v}(0) \rangle dt, \tag{4}$$

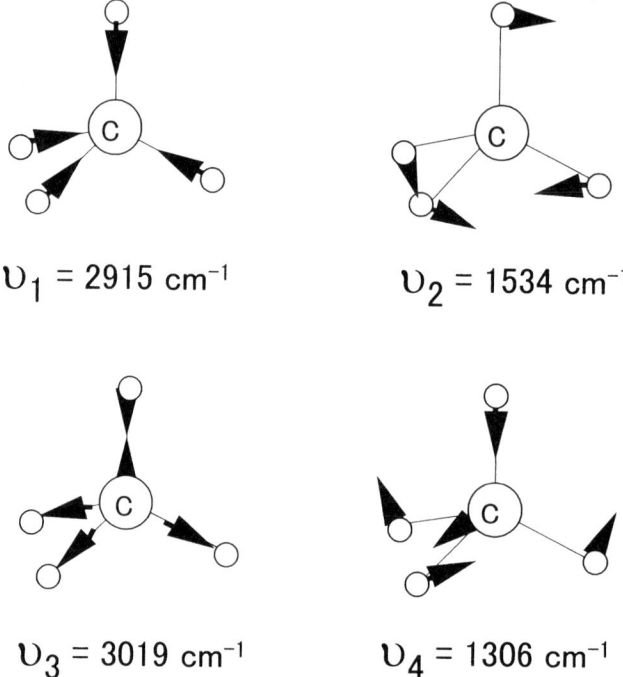

FIGURE 1. Normal modes for a CH$_4$ molecule. The four vibrational modes, the v_1, v_3 stretching modes and the v_1, v_3 bending modes, with the frequency values are illustrated.

$$\langle \mathbf{v}(t) \cdot \mathbf{v}(0) \rangle = \frac{1}{N} \sum_{i=1}^{N} \langle \mathbf{v}_i(t) \cdot \mathbf{v}_i(0) \rangle. \quad (5)$$

Integration of (4) was calculated with a time step of 0.4 fs for a duration of 4 ps (10,000 steps) after the initial equilibration period of 4 ps. The autocorrelation functions converge sufficiently for this time period. The power spectra of C and H atoms in small cages and those in large cages were independently obtained by classifying their atoms.

RESULTS AND DISCUSSION

Reproducibility of Liquid CH$_4$: Structure and Vibrational Spectra

The parameters for CH$_4$ were determined to reproduce the density of liquid and the normal modes. The density of liquid CH$_4$ at 100 K and 1 atm calculated in this MD simulation was 0.447 g/cm^3. This value coincides well with the experimental value of 0.438 g/cm^3. Four intramolecular vibrational modes (i.e., normal modes) of CH$_4$ shown in FIGURE 1 were reproduced with relative shifts of peak frequencies, as

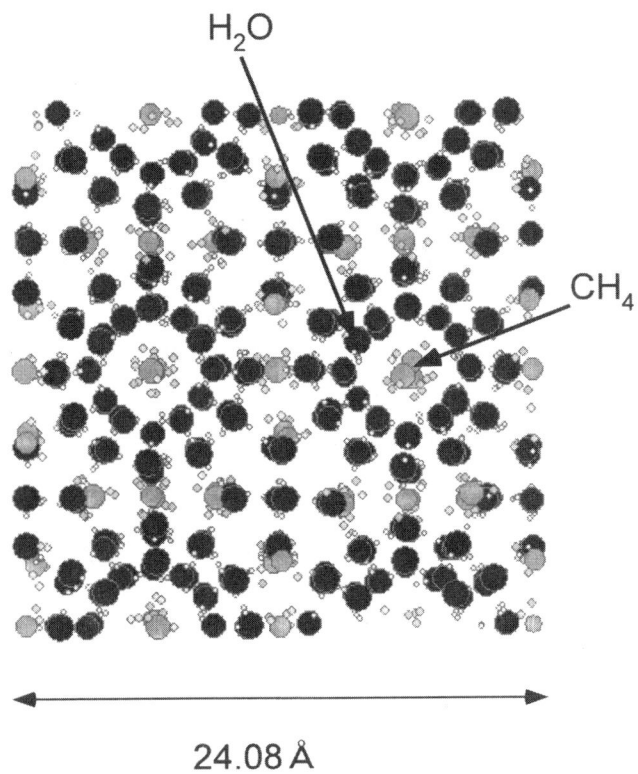

FIGURE 2. The simulated structure of CH_4 hydrate. The cages are fully occupied with the guest molecules of $CH_4 \cdot 368\ H_2O$ and 64 CH_4 molecules in a cell with dimensions a, b, and $c = 24.08$ Å (8 unit cells).

shown in FIGURE 3. This discrepancy was due to the result that the C-H distance in the simulated CH_4 molecule was slightly shorter than the experimental value.

Vibrational Spectra of CH_4 Hydrate

Spectroscopic studies on CH_4 hydrate by MD simulations were carried out by Tse et al.[5,17,18] They investigated the lattice vibrations of CH_4 hydrate and found the strong mode coupling between H_2O and CH_4.[18] However, no MD study of CH_4 hydrate has taken into account the intramolecular degrees of freedom. In the present study we have successfully obtained intramolecular vibrations of CH_4 hydrate by MD simulations using the KKY potential model.

Intramolecular vibrational spectra of CH_4 in the large and the small cages were calculated as shown in FIGURE 4. The peak at 1,255 cm^{-1} is attributed to the υ_4 bending mode because it is close to the υ_4 normal mode frequency (1,306 cm^{-1}), and both C and H atoms vibrate in phase. On the other hand, since only the H atom displacement gave rise to the peak at 1,395 cm^{-1}, as can be seen from FIGURE 4, it is assigned to the υ_2 bending mode. For the same reason as mentioned above, the peak at 2,990 cm^{-1} when the CH$_4$ is located in the large cage, is the υ_1 stretching mode, and the peak at 3,075 cm^{-1} is the υ_3 stretching mode.

The υ_1 stretching mode calculated in the present MD simulation is 85 cm^{-1} higher than that from the Raman spectrum by Sum *et al.*[3] This disagreement appears to be due to the shorter C-H distance in our simulations. The simulated C-H distance (1.08 Å) was about 0.01 Å shorter than that determined experimentally. This difference must give rise to the frequency shifts in liquid CH$_4$ and have a great influence on the shifts in CH$_4$ hydrate. The CH$_4$ in the cages strongly restrained by H$_2$O gave rise to the large frequency shifts.

Spectral Characteristics of CH$_4$ Depending on the Cage Size

To investigate the different spectral characteristics of CH$_4$ depending on the cage size, we calculated the power spectra of CH$_4$ in the large and the small cages

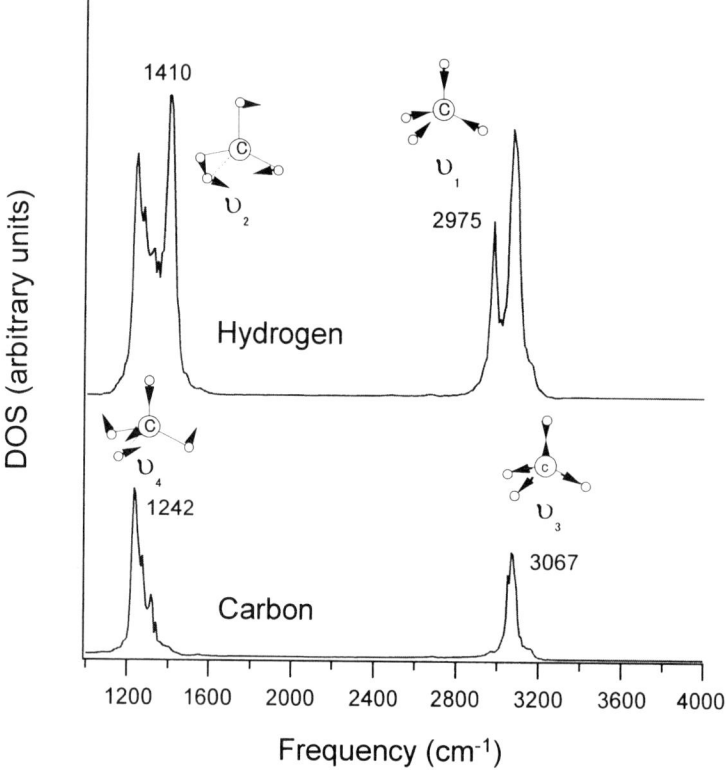

FIGURE 3. Power spectra of C and H atoms in liquid CH$_4$. The temperature was kept at 100 K and the pressure 1 atm. Vibrational modes corresponding to each peak are illustrated.

individually. The dashed line in FIGURE 4 shows the power spectra of C atoms in the large and the small cages, and the solid line the spectra of H atoms. As can be seen from FIGURE 4, the υ_1 and υ_3 stretching modes in the large cages are 10 cm^{-1} higher than those in the small cages. In contrast, the υ_2 and υ_4 bending modes locate at the same frequencies (1,255 cm^{-1} and 1,395 cm^{-1}) in both large and small cages.

The same frequency shift of the 1 stretching mode was found in Raman study by Sum et al.[3] In addition to this, the difference of another intramolecular vibration of CH$_4$ between the large and the small cages was found in the present MD simulations. We found that the υ_3 stretching mode in the large cages was also shifted to a lower frequency than that in the small cages. It is considered that these frequency shifts appeared because the force constant for the C-H stretching modes increases with strengthening restraint by surrounding H$_2$O molecules. Since this restraint, which is

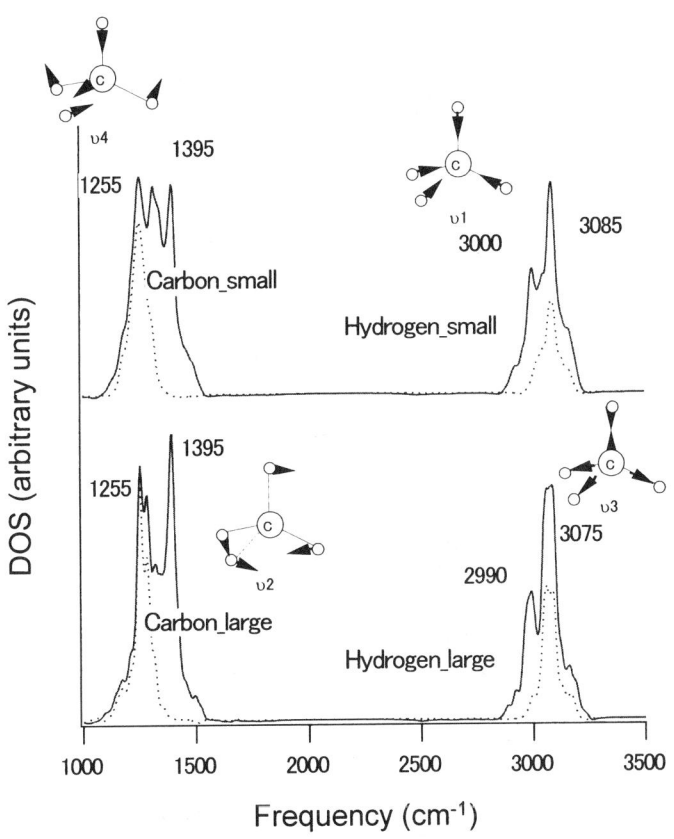

FIGURE 4. Power spectra of CH$_4$ in the clathrate hydrate structure I at 100 K and 100 atm. The dashed line represents the power spectra of C atoms in the large and the small cages and the solid line the spectra of H atoms. Vibrational modes corresponding to each peak are illustrated.

mainly dominated by the repulsive force, does not affect the bending modes significantly, there is no peak frequency shifts in the v_2 and v_4 bending modes.

CONCLUSIONS

MD simulations of CH_4 hydrate have been carried out to investigate the spectral characteristics of CH_4 depending on the cage size. We successfully obtained all intramolecular vibrational spectra of CH_4 in the large and the small cages by using the KKY potential model. It was found that (1) the v_1 and v_3 stretching modes of CH_4 in the large cages were located a lower frequency than those in the small cages, (2) the v_2 and v_4 bending modes of CH_4 in the large and the small cages were found in same frequencies. We consider that these peak frequency shifts are due to the significant restraint by surrounding H_2O molecules. This is mainly dominated the repulsive force between CH_4 and H_2O molecules.

ACKNOWLEDGMENTS

This work was supported by a Grant-in-Aid for Scientific Research and Grant-in-Aid for JSPS (Japan Society for the Promotion of Science) Fellows. The author H.I. acknowledges a conference fellowship from the United Engineering Foundation to attend the Third International Conference on Gas Hydrates.

REFERENCES

1. VON STACKELBERG, M. & H.R. MULLER. 1954. On the structure of gas hydrates. Z. Elektrochem. **58:** 25–39.
2. SLOAN, E.D. 1998. Clathrate Hydrate of Natural Gases, 2nd edit. Marcel-Dekker, New York.
3. SUM, A.K., R.C. BURRUSS & E.D. SLOAN, JR. 1997. Measurement of clathrate hydrates via Raman spectroscopy. J. Phys. Chem. B **101:** 7371–7377.
4. TSE, J.S., M.L. KLEIN & I.R. MCDONALD. 1983. Molecular dynamics studies of ice Ic and the structure I clathrate hydrate of methane. J. Phys. Chem. **87:** 4198–4203.
5. ROGER, P.M. 1990. Stability of gas hydrates. J. Phys. Chem. **94:** 6080–6089.
6. FORRISDAHL, O.K., B. KVAMME & A.D.J. HAYMET. 1996. Methane clathrate hydrates: melting, supercooling, and phase separation from molecular dynamics computer simulations. Mol. Phys. **89:** 819–834.
7. BERENDSEN, H.J.C., J.P.M. POSTMA, W.F. VAN GUNSTEREN & J. HERMANS. 1981. Interaction models for water in protein hydration. In Intermolecular Forces. B. Pullman, Ed.: 331–342. Reidel. Dordrecht.
8. JORGENSEN, W.L., J. CHANDRASEKHAR, J.D. MADURA, R.W. IMPEY & M.L. KLEIN. 1983. Comparison of simple potential functions for simulating liquid water. J. Chem. Phys. **79:** 926–935.
9. KUMAGAI, N., K. KAWAMURA & T. YOKOKAWA. 1994. An interatomic potential model for H_2O: applications to water and ice polymorphs. Mol. Sim. **12:** 177–186.
10. ITOH, H., K. KAWAMURA, T. HONDOH & S. MAE. 1996. Molecular dynamics studies of self-interstitials in ice Ih. J. Chem. Phys. **105:** 2408–2413.
11. ITOH, H., K. KAWAMURA, T. HONDOH & S. MAE. 1998. Polarized vibrational spectra of proton-ordered ice XI by molecular dynamics simulations. J. Chem. Phys. **109:** 4894–4899.

12. Itoh, H., S. Horikawa, K. Kawamura, T. Uchida, T. Hondoh & S. Mae. 1996. Molecular dynamics studies of structure I clathrate hydrate of carbon dioxide. 2nd Int. Conf. on Natural Gas Hydrates. 341–346.
13. Horikawa, S., H. Itoh, J. Tabata, K. Kawamura & T. Hondoh. 1997. Dynamic behavior of diatomic guest molecules in clathrate hydrate structure II. J. Phys. Chem. B **101:** 6290–6292.
14. McMullan, R.K. & G.A. Jeffrey. 1965. Polyhedral clathrate hydrates. IX. Structure of ethylene oxide hydrate. J. Chem. Phys. **42:** 2725–2732.
15. Bernal, J.D. & R.H. Fowler. 1933. A theory of water and ionic solution, with particular reference to hydrogen and hydroxyl ions. J. Chem. Phys. **1:** 515–548.
16. MXDORTO. Japan Chemistry Program Exchange, #029.
17. Tse, J.S., C.I. Ratcliffe, B.M. Powell, V.F. Sears & Y.P. Handa. 1997. Rotational and translational motions of trapped methane. Incoherent inelastic neutron scattering of methane hydrate. J. Phys. Chem. A **101:** 4491–4495.
18. Tse, J.S., V.P. Shpakov, V.V. Murashov & V.R. Belosludov. 1997. The low frequency vibrations in clathrate hydrates. J. Chem. Phys. **107:** 9271–9274.

Optimizing Thermodynamic Parameters to Match Methane and Ethane Structural Transition in Natural Gas Hydrate Equilibria

ADAM L. BALLARD[a] AND E.D. SLOAN, JR.

Colorado School of Mines, Center for Hydrate Research, Golden, Colorado 80401, USA

ABSTRACT: A simple natural gas mixture (methane and ethane) when mixed with water to form natural gas hydrates can exhibit complex phase equilibria. In this case, two structure I hydrate formers when combined form a structure II hydrate under certain conditions. Optimized Kihara parameters were derived from experimental data using the statistical thermodynamics model based on the van der Waals and Platteeuw theory and using the Kihara spherical core potential for the guest-water interactions. The parameters were based upon both pure component data and the structural transition point, with the intent of better predictions over the remainder of the phase diagram.

INTRODUCTION

Gas hydrates are ice-like crystalline structures that form in the presence of light gases such as methane or ethane; these guests are trapped in several different cages of water molecules that form via hydrogen bonding. Depending on the hydrocarbon guest molecules in the hydrate lattice, different structures can form (sI, sII, or sH). Several models have been developed to predict the formation and dissociation of gas hydrates as well as the equilibrium structure (Parrish and Prausnitz,[1] Chen and Guo[2]).

Pure methane and pure ethane gases both form structure I hydrates due to stability considerations involving sizes of the guests relative to cage sizes. In modeling it has been assumed that the combination of these two gases would also form a structure I hydrate. The lack of change in slope in the three phase (L_w-H-V) pressure and temperature data for this system does not suggest any phase except structure I hydrate. Holder and Hand,[3] however, noted that 90.4 mole percent methane and 9.6 mole percent ethane data could not be fitted well to sI predictions. Hendriks *et al.*[4] predicted a structural transition (sI to sII) when all parameters were optimized to fit hydrate equilibria data. Earlier, van der Waals and Platteeuw[5] had predicted structure II formation for the ternary mixture of hydrogen sulfide + difluoro-ethane, noting that hydrogen sulfide and difluoro-ethane each formed structure I hydrates.

Recently it was experimentally determined that the methane + ethane gas mixture can form structure II hydrates at certain compositions (Subramanian *et al.*[6]). These data were obtained using Raman and NMR spectroscopy, which specifically measure

[a]Telecommunication. Voice: 303-273-3561.
aballard@mines.edu

the hydrate structure. In this work, thermodynamic parameters were optimized to fit the available methane + ethane hydrate data using the recent spectroscopic data. The aim of the work was to fit the lower transition point and to observe the predictions of the model at the upper and lower transition points as a function of temperature and pressure.

MODEL

The hydrate prediction program used in this work, CSMHYD, developed at the Colorado School of Mines, is based on the van der Waals and Platteeuw theory.[5] The governing equation predicts the chemical potential of water in the hydrate lattice, μ_w^H, as

$$\mu_w^H = \mu_w^\beta + RT\sum_m \upsilon_m \ln\left(1 - \sum_J \theta_{Jm}\right), \tag{1}$$

where μ_w^β is the chemical potential of water in the *hypothetical* empty hydrate lattice, υ_m is the number of cavities of type m per water molecule in the lattice, and θ_{Jm} is the fractional occupancy of the Jth guest molecule in the mth hydrate cavity. This fractional occupancy can be expressed in a Langmuir type manner as

$$\theta_{Jm} = \frac{C_{Jm}f_J}{1 + \sum_m C_{Jm}f_J}, \tag{2}$$

where f_J is the fugacity of component J and C_{Jm} is the Langmuir constant of component J in cavity m. The Langmuir constants are temperature dependent functions that describe the potential interaction between the encaged guest molecule and the water molecules surrounding it. They are evaluated by assuming a spherically symmetrical potential, which is described as

$$C_{Jm} = \frac{4\pi}{kT}\int_0^R \exp\left(-\frac{\omega(r)}{kT}\right)r^2 dr, \tag{3}$$

where $\omega(r)$ is the potential function of guest J in cage m, and R is the radius of the cage. In this work, the Soave-Redlich-Kwong equation of state is used for the fugacity of each component, and the Kihara spherical core potential is used to calculate the Langmuir constants.

A sensitivity analysis showed that a structural transition in the methane + ethane + water system is most sensitive to the Kihara parameters. In particular, σ, the distance between core surfaces corresponding to zero potential energy, affected the transition point dramatically. As little as 0.2% change in σ changed the lower transition point (on a molar basis) by as much as 12.0%. The larger the σ for either molecule, the lower the transition point due to increased stabilty of the guest in the structure II hydrate cages. The depth of the intermolecular potential well, ε, is extremely sensitive to σ so these two parameters together were optimized for each guest component.

The Kihara parameters for each guest molecule were found by fitting the expression for the chemical potential of water in the hydrate lattice given by Equation (1) to the expression for the chemical potential of water in the aqueous liquid derived from classical thermodynamics by Marshall et al.,[7] and simplified by Holder et al.,[8] and Menten et al.,[9] as

$$\frac{\mu_w^L}{RT} = \frac{\mu_w^\beta}{RT} = \frac{\Delta \mu_w^o}{RT_o} + \int_{T_o}^{T} \frac{\Delta h_w}{RT^2} dT - \int_0^P \frac{\Delta v_w}{RT} dP + \ln(\gamma_w x_w), \qquad (4)$$

where $\Delta \mu_w^o$ is the difference in the chemical potential of water between the hypothetical empty hydrate lattice and ice at a reference temperature T_o (273.15 K) and zero absolute pressure, Δh_w and Δv_w are the enthalpy and volume differences between the empty hydrate and pure ice/liquid water phases, and γ_w and x_w are, respectively, the activity coefficient and mole fraction of water in the ice/liquid water phase. Note that the chemical potential of water is equal in all phases at equilibrium, so that equating (1) and (4) is valid for hydrate equilibrium calculations. The algorithm to predict hydrate phase equilibria used in this work is based on that described by Parrish and Prausnitz.[1]

OPTIMIZATION OF KIHARA PARAMETERS

In this work, we derived the optimum Kihara parameters, σ and ε, for both methane and ethane. These parameters were found using PT data for pure methane, pure ethane, and methane + ethane mixtures as well as recent Raman data indicating a structural transition in the methane + ethane system (72 mole percent methane in gas at 274.15 K). The objective of this work was to fit the lower structural transition and then use the program to observe the predicted upper transition as a function of temperature. The thermodynamic parameters used in the optimization are listed in TABLE 1 and given in Sloan.[10]

The approach to the optimization was to first predict pure methane and pure ethane data. This was accomplished via regression of pure component data to generate an infinite set of the Kihara parameters σ and ε for each component that best fit the pure component data. FIGURE 1 shows the infinite sets for both components. It should be noted that for any data point, an infinite set of Kihara parameters could be obtained that will precisely predict that dissociation pressure. This is analogous to

TABLE 1. Parameters used for hydrate modeling

Parameter	Units	Structure I		Structure II	
$\Delta \mu_w^o$	J/mole	1263.60		883.82	
ΔH_w^o (ice)	J/mole	1389.08		1025.08	
ΔH_w^o (water)	J/mole	−4623.27		−4987.27	
ΔV_w^o (ice)	Cm^3/mol	3.0		3.4	
ΔV_w^o (water)	Cm^3/mol	4.6		5.0	
ΔC_{pw}^o	J/mole K	−38.12			
Cavity Type	—	5^{12}	$5^{12}6^2$	5^{12}	$5^{12}6^4$
	nm	0.395	0.43	0.391	0.473

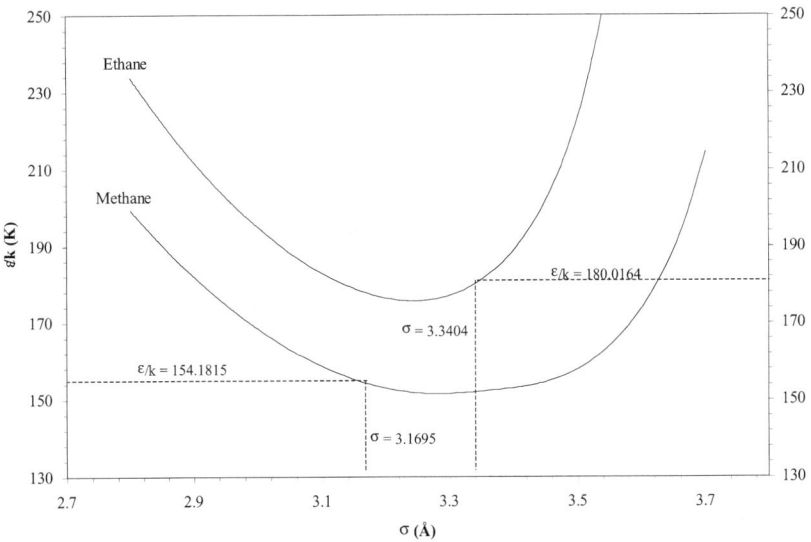

FIGURE 1. Infinite set of Kihara parameters for methane and ethane.

the solution of one equation with two unknowns. Generating N infinite sets for N data points will then allow for a best line fit all of the sets, thus producing an infinite set that will best fit the data.

With the pure methane and pure ethane Kihara parameter sets, a regression on the methane + ethane gas mixture data was performed. This involved simultaneously minimizing the error to the data using both pure component sets. It should be noted that the only assumption of hydrate structure is for the pure components. The methane + ethane data was fitted without assuming them to be either structure I or structure II. This technique resulted in unique values for the Kihara parameters of each component, which provide an *a priori* prediction of a structural transition at 52 mole percent methane in the gas mixture at 274.15 K. The experimentally determined structural transition is approximately 72 mole percent methane (Subramanian *et al.*[6]) so another optimization was performed to find the best set of Kihara parameters to achieve the structural transition at this gas composition. The Kihara parameters that satisfy the requirements are given in TABLE 2.

TABLE 2. Optimized Kihara parameters

Component	a (Å)	σ (Å)	ε/k (K)
Methane	0.3834	3.1695	154.1815
Ethane	0.5651	3.3404	180.0164

CHEMICAL POTENTIAL OF WATER IN THE HYDRATE

With parameters predicting a structural transition in the methane + ethane system, it is interesting to consider the chemical potential of water in the hydrate lattice as a way to understand the thermodynamics of the system. FIGURES 2 and 3 present the chemical potential of water in the hydrate at the lower and upper structural transition points, respectively. The chemical potential of water in the hydrate lattice was found by using a Gibbs energy minimization for a flash on the methane + ethane system to determine the chemical potential of water in the aqueous liquid phase at several nshydrate equilibrium pressures. Using **(4)**, the chemical potential of water in the hypothetical empty hydrate lattice was determined and then substituted into Equation **(1)** to determine the chemical potential of water in the hydrate lattice. This was done for both hydrate structures.

Thermodynamics indicates that the chemical potential of water in the stable hydrate lattice will be equal to that in the aqueous liquid water phase. The figures confirm this condition for equilibrium. The chemical potential of water in the unstable hydrate structure can also be followed to observe its movement with composition and pressure. The chemical potentials cross at the point of structural transition, as well as exhibiting a slight change of slope in the equilibrium pressure, as van der Waals and Platteeuw suggest.[5]

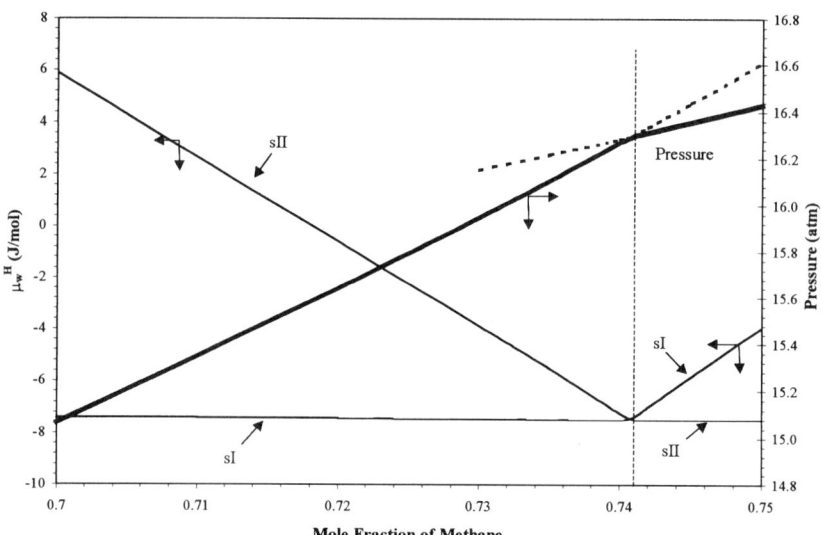

FIGURE 2. Trace of chemical potential of water in the hydrate phase at the lower transition point at 277.6 K.

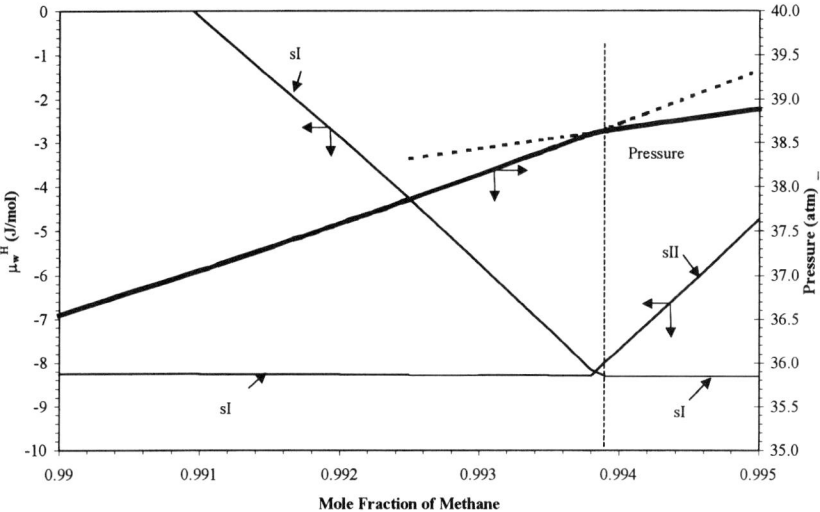

FIGURE 3. Trace of chemical potential of water in hydrate phase at the upper transition point at 277.6 K.

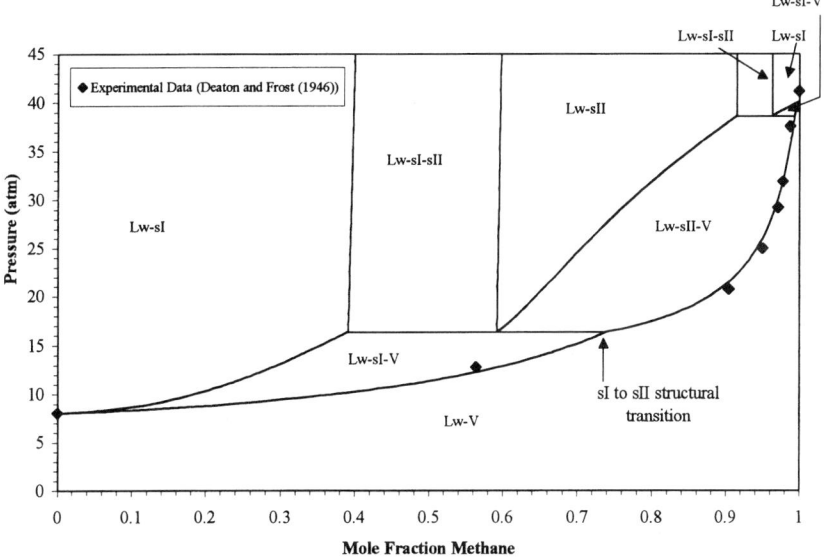

FIGURE 4. Predicted phase diagram for the methane/ethane/water system at 277.6 K.

RESULTS AND DISCUSSION

At 274.15 K it was found that the equilibrium hydrate structure for methane and ethane changed from structure I to structure II at approximately 72 mole percent methane in the water free gas phase. FIGURE 4 gives the pressure versus composition plot for this system at 277.6 K. This plot was constructed using a Gibbs energy minimization flash routine described by Bishnoi et al.[11] Note that it is necessary for the equilibrium structure to pass through a transition again (sII to sI) in order for the pure methane gas to form a structure I hydrate. We can also see regions above the formation conditions in which structure I and structure II coexist. Note that in order for these regions to exist, according to Gibbs phase rule, one of the other phases must diminish, namely the vapor phase, if there is excess water present in the system.

The optimized parameters predict all experimental data for pure methane, pure ethane, and methane + ethane mixtures with an average error in temperature of 0.4 K. FIGURE 5 shows hydrate formation data and predictions in pressure versus gas composition plots at several different temperatures for the methane + ethane system. FIGURE 6 shows data compared to the predictions in a pressure versus temperature plot. In both figures, notice the region mapped out in the thick, dashed line, which represents the envelope in which structure II will be the stable hydrate structure at formation.

With the data adequately represented relative to the structural transition, we considered the point of structural transition with increasing and decreasing temperature. FIGURE 7 shows in a temperature versus composition plot the range of gas

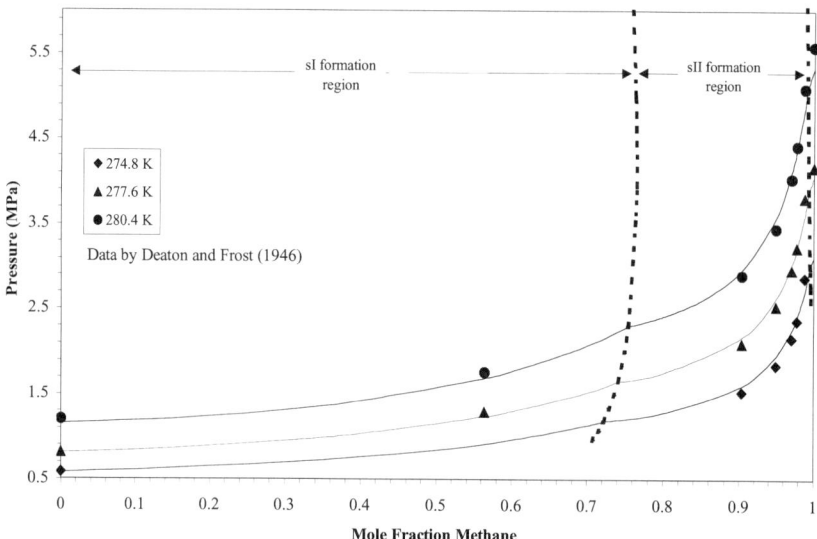

FIGURE 5. Prediction of sI and sII hydrate equilibrium including the transition region.

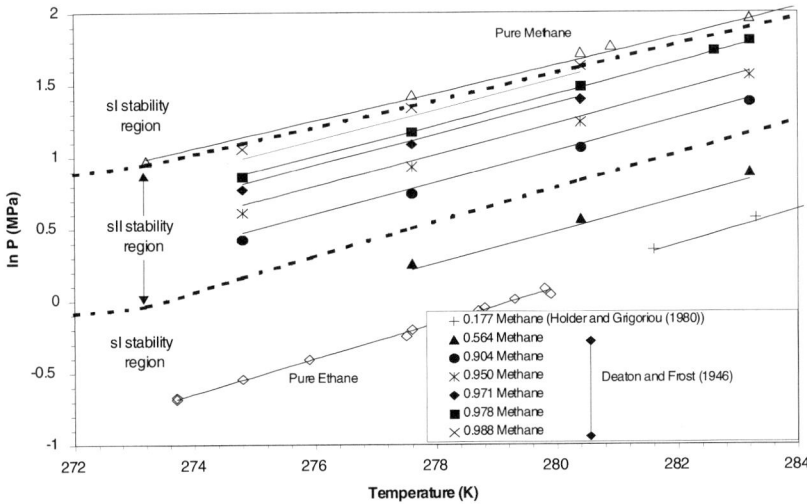

FIGURE 6. Prediction of sI and sII hydrate equilibrium for several different compositions including the transition region.

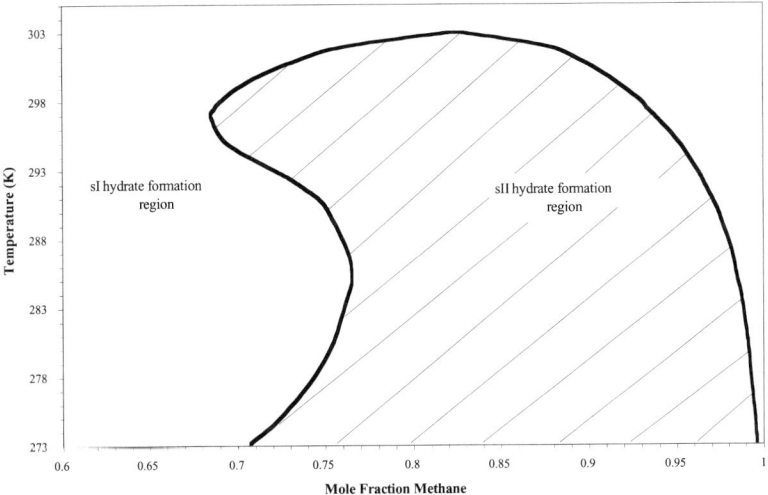

FIGURE 7. Predicted composition region for the structural transition to occur.

composition in which structure II will be the stable phase at hydrate formation conditions at several different temperatures. It is interesting to see that, for temperatures above 303 K, the stable structure is structure I over the entire composition range. This is most likely due to lack of stability because of a decrease in occupancy of ethane in the large cage of structure II. At 303 K the fractional occupancy of ethane in the large cage of structure II is approximately 74 percent and that of methane is 25 percent. Since methane is not a good structure II large cage stabilizer, those cages are not as stable as the structure I large cages.

Another interesting point on the plot is the change of concavity for the lower transition point from the ice point to 297 K. Multiple phase transitions are predicted to occur at methane compositions between 0.7 and 0.75 mole fraction methane by changing the temperature.

IMPLICATIONS TO INDUSTRIAL APPLICATIONS

At lower temperatures of the system, a small amount of ethane is needed to stabilize structure II hydrates. This can have many implications in deep sea and permafrost hydrates that are thought to be structure I because they are almost pure methane. From the predictions, at 273.15 K methane gas need be only contaminated with 0.3 mole percent ethane for the stable structure to be structure II.

Deep-sea hydrate deposits are usually found to contain about 99 mole percent methane. To determine the composition of methane in the hydrates, mass balance calculations must be performed assuming a hydrate structure. FIGURE 8 shows the composition of methane in the hydrate versus the composition of methane in the gas,

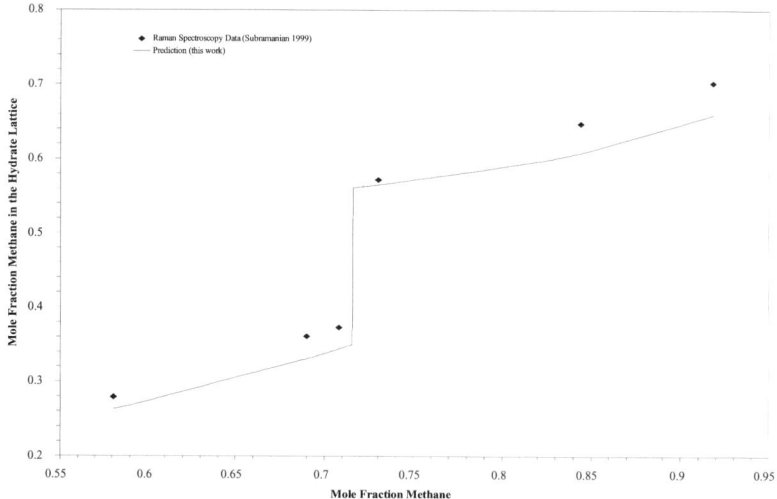

FIGURE 8. Predicted methane composition in the hydrate (water free basis). $T = 274.15$ K.

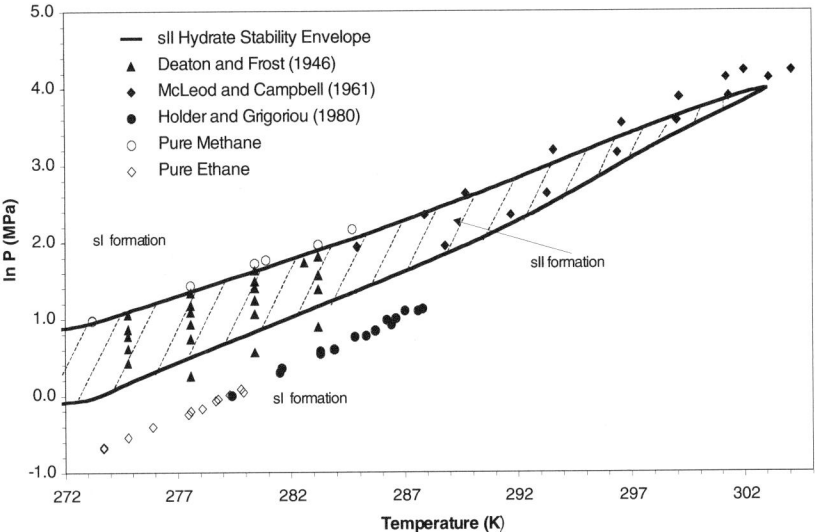

FIGURE 9. All available experimental data plotted with the structural transition envelope.

all on a water free basis at 274.15 K. The predictions are plotted against Raman spectroscopic data. There is a marked decrease in the amount of methane encapsulated in the hydrate relative to ethane as the hydrate goes from structure I to structure II. This jump could have serious implications in mass balance calculations used to determine the amount of methane gas trapped in the hydrate. There would also be effects in energy balance calculations due to the different heats of dissociation for the two hydrate structures.

FIGURE 9 shows all of the available hydrate experimental data for methane + ethane gas mixtures relative to the structure II stability envelope. Due to the practical implications of this structural transition, it may be necessary to reevaluate some of these data using such techniques as NMR, XRD, or Raman spectroscopy.

CONCLUSIONS

Optimized Kihara parameters for methane and ethane predicting a structural transition have been presented. Using these parameters in conjunction with the statistical thermodynamics model, pure methane, pure ethane, and methane + ethane mixture predictions fit the hydrate data with an average error in temperature of 0.4 K. The upper and lower transition points as a function of temperature were explored. There could be serious implications to fieldwork where gas could be produced from hydrates. Experimental data evaluating the solid hydrate phase needs to be obtained to evaluate the upper transition point, either to confirm the predictions or to increase the accuracy.

ACKNOWLEDGMENT

Financial support by Amoco, ARCO, Chevron, Department of Energy (U.S.), Mobil, Oryx, Phillips, Shell, and Texaco is greatly appreciated.

REFERENCES

1. PARRISH, W.R. & J.M. PRAUSNITZ. 1972. Dissociation pressures of gas hydrates formed by gas mixtures. Ind. Eng. Chem. Process Des. Develop. **11**(1): 26.
2. CHEN, G. & T. GUO. 1996. Thermodynamic modeling of hydrate formation based on new concepts. Fluid Phase Equilibria **122**: 43.
3. HOLDER, G.D. & J.H. HAND. 1982. Multiple-phase equilibria in hydrates from methane, ethane, propane and water mixtures. AIChE J. **28**(3): 440.
4. HENDRIKS, E.M., B. EDMONDS, R.A.S. MOORWOOD & R. SZCZEPANSKI. 1996. Hydrate structure stability in simple and mixed hydrates. Fluid Phase Equilibria **117**: 193.
5. VAN DER WAALS, J.H. & J.C. PLATTEEUW. 1959. Clathrate solutions. Adv. Phys. Chem. **2**(1): 1.
6. SUBRAMANIAN, S., R. KINI, S. DEC & E.D. SLOAN. 1999. Methane + ethane mixtures can form structure II hydrates. Ann. N.Y. Acad. Sci. **912**: this volume.
7. MARSHALL, D.R., S. SAITO & R. KOBAYASHI. 1964. Hydrates at high pressure: part I, methane-water, argon-water and nitrogen-water systems. AIChE J. **10**(2): 202.
8. HOLDER, G.D., G. CORBIN & K.D. PAPADOPOULOS. 1980. Thermo and molecular properties of gas hydrates from mixtures containing CH_4, argon and krypton. Ind. Eng. Chem. Fund. **19**(3): 282.
9. MENTEN, P.D., W.R. PARRISH & E.D. SLOAN. 1981. Effect of inhibitors on hydrate formation. Ind. Eng. Chem. Process Des. Develop. **20**: 399.
10. SLOAN, E.D. 1998. Clathrate Hydrates of Natural Gases, 2nd edit. Marcel Dekker, New York.
11. BISHNOI, P.R., A.K. GUPTA, P. ENGLEZOS & N. KALOGERAKIS. 1989. Multiphase equilibrium flash calculations for systems containing gas hydrates. Fluid Phase Equilibria **53**: 97–104.

Application and Extension of Aasberg-Petersen Model for Prediction of Gas Hydrate Formation Conditions in Mixtures of Aqueous Electrolyte Solutions and Alcohol

JAFAR JAVANMARDI,[a] MAHMOOD MOSHFEGHIAN,[a,b] AND ROBERT N. MADDOX[c]

[a]*Department of Chemical Engineering, Shiraz University, Shiraz, Iran*

[c]*Oklahoma State University, Stillwater, Oklahoma 74074, USA*

ABSTRACT: A new thermodynamic model for calculating the hydrate formation temperature of different hydrate formers in aqueous solutions of both electrolytes and alcohols is described. This method uses a generalization of the Aasberg-Petersen model for water activity. To calculate the activity of water in the presence of electrolytes, the effect of alcohols was taken into account without using any new fitting parameters. The results are in good agreement with published experimental data. Calculated values of the hydrate forming temperature in the presence of alcohols and electrolytes are compared with values calculated by using other models.

INTRODUCTION

Hammerschmidt[1] developed an empirical method for predicting the incipient hydrate formation temperature for natural gas constituents in the presence of alcohol. Subsequently several other models were suggested, for example Anderson and Prausnitz[2] and Moshfeghian and Maddox.[3]

In the presence of electrolytes, the method of Englezos and Bishnoi[4] is applicable. Their procedure uses the activity model developed by Pitzer and Mayorga[5] to predict water activity coefficients in the presence of electrolytes. Their model is not recommended for hydrate formers with high solubility in the water phase. Javanmardi and Moshfeghian[6] and later Javanmardi et al.,[7] took a different approach and, using the pure water hydrate formation temperature, predicted conditions for incipient hydrate formation in the presence of electrolytes. Englezos,[8] by using flash calculation algorithms and the fugacity model of Aasberg-Peterson et al.,[9] predicted the temperature for incipient hydrate formation in single aqueous electrolyte solutions. Javanmardi and Moshfeghian[10] used the same activity model to predict the hydrate formation temperature in mixed aqueous electrolyte solutions.

Telecommunication.
[b]Voice: +98-71-303071; fax: +98-71-52725. moshfeg@succ.shirazu.ac.ir
[c]Voice: 405-372-0402; fax: 405-372-3089. rmaddox@gibbs.cheng.okstate.edu

Yousif and Young,[11] based on modification of the Hammerschmidt model, developed a simple model to predict the hydrate temperature suppression of aqueous mixtures of salts and alcohols. The results of their correlation are restricted to prediction of the hydrate condition for a specified gas mixture. In that work, however, prediction of hydrate formation based on this correlation, is given for other systems. Based on the methods of Moshfeghian and Maddox[3] and Javanmardi et al.,[7] Nasrifar et al.[12] developed a combining rule for the prediction of hydrate temperature in the presence of mixed aqueous electrolyte solutions and alcohols.

In the present work, a new method is suggested for predicting hydrate temperature in the presence of both aqueous electrolyte solutions and alcohols. The model is based on the extension of the water activity model to the above systems. The suggested method is also applied to highly soluble hydrate formers, such as CO_2. In addition, the suggested model can handle the extreme conditions; that is, the hydrate temperature of the systems containing each of the above inhibitors individually.

THERMODYNAMIC MODEL

At equilibrium, the chemical potential of water in the hydrate phase is equal to that in each of the other coexisting phases. Parrish and Prausnitz[13] developed a thermodynamic model to describe this phenomenon. Later, the model was improved by Holder et al.[14] According to the improved model, when the hydrate phase is in equilibrium with pure water or ice, one can show that:

$$\sum_{m=1}^{2} \upsilon_m \ln\left(1 + \sum_{j=1}^{nc} C_{mj} f_j\right) = \frac{\Delta \mu_w^o}{RT^o} - \int_{T^o}^{T} \frac{\Delta h_w^l}{RT^2} dT + \int_0^P \frac{\Delta v_w^l}{RT} dP - \ln a_w, \quad (1)$$

$$\sum_{m=1}^{2} \upsilon_m \ln\left(1 + \sum_{j=1}^{nc} C_{mj} f_j\right) = \frac{\Delta \mu_w^{oi}}{RT^o} - \int_{T^o}^{T} \frac{\Delta h_w^i}{RT^2} dT + \int_0^P \frac{\Delta v_w^i}{RT} dP. \quad (2)$$

In the above equations, i and l refer to ice and liquid state, respectively; superscript o refers to temperature T^o. Definitions of the terms in (1) and (2) were given by Holder et al.[14]

Evaluation of the activity of water, a_w, depends on the system under consideration. For low solubility hydrate former and pure water, a_w can be replaced with x_W, the mole fraction of water. In the presence of a third component, for example electrolytes or alcohols, the activity coefficient of water deviates from unity. This illustrates the need for a suitable activity model of water. For aqueous electrolyte solutions, from the model developed by Aasberg-Petersen et al.,[9] it can be shown that:

$$\hat{f}_w^L = x_w \phi_w^{Asb} P \quad (3)$$

$$\ln \phi_w^{Asb} = \ln \phi_w^{EOS} + \ln \gamma_w^{el} \quad (4)$$

$$\ln \gamma_w^{el} = \frac{2 A h_{ws} M_m F(BI^{0.5})}{B^3} \quad (5)$$

TABLE 1. Parameters in Equation (10)

	a_1	a_2	a_3	a_4
NaCl	-4.7450×10^{-3}	-1.5564×10^{-5}	3	7.23×10^{-5}
KCl	-5.4850×10^{-3}	3.0462×10^{-4}	1	0.0
CaCl$_2$	-2.4776×10^{-3}	-4.0972×10^{-4}	2	4.162×10^{-5}

$$A = 1.327757 \times 10^5 \frac{d_m^{0.5}}{(\varepsilon_m T)^{1.5}} \qquad (6)$$

$$B = 6.359696 \times \frac{d_m^{0.5}}{(\varepsilon_m T)^{0.5}} \qquad (7)$$

$$F(BI^{0.5}) = 1 + BI^{0.5} - \frac{1}{1 + BI^{0.5}} - 2\ln(1 + BI^{0.5}) \qquad (8)$$

$$\varepsilon_m = x_N \varepsilon_N. \qquad (9)$$

Definitions of terms in (3) through (9) are given by Aasberg-Petersen et al.[9] In (4), to evaluate the fugacity coefficient, the salt free mole fraction must be used. In this work the following form for the interaction coefficient between the dissolved salt, NaCl, KCl, and CaCl$_2$ and water, suggested by Javanmardi and Moshfeghian,[10] has been used.

$$h_{ws} = a_1 + a_2 m_s^{a_3} + a_4 \exp\frac{p}{10}, \qquad (10)$$

where P is in MPa, and a_1 through a_4 were determined and are reported in TABLE 1.[10]

PROPOSED MODEL

In the present work, to include the simultaneous effect of aqueous electrolyte solutions and alcohols, the following form of water activity is suggested.

$$\ln \phi_w^{el-al} = \ln \phi_w^{EOS} + \ln \gamma_w^{el} + \ln \gamma_w^{al} \qquad (11)$$

In this equation, the third term accounts for the effect of alcohols. To evaluate $\ln \gamma_w^{el}$ and $\ln \gamma_w^{al}$, alcohols and salt-free mole fraction of water must be used, respectively. To evaluate $\ln \gamma_w^{al}$, any suitable activity model can be used; in this work the Margules model is employed. Note that Equation (11) needs no additional terms or fitted parameters. From the classical definition of activity we have:

$$a_w = \frac{\hat{f}_w^l}{f_w^{l\,0}} \qquad (12)$$

where $f_w^{l\,0}$ is the pure water fugacity at the temperature and pressure of the system. This term is equal to $\phi_w^{EOS°} P$. Thus, (12) can be written in the following form:

$$a_w = \frac{x_w \phi_w^{el-al}}{\phi_w^{EOS°}} \qquad (13)$$

Therefore, from (**1**), (**11**), and (**13**), one can predict the conditions for incipient hydrate formation for aqueous electrolyte solutions and alcoholic systems. It should be noted that in contrast to previous models, under extreme conditions—single inhibitor or pure water—Equation (**11**) is reduced to proper form. For example, in the absence of electrolyte and for low soluble hydrate former, the second term in (**11**) is equal to zero and therefore, $\ln\phi_w^{el-al}$ reduces to $\ln\gamma_w^{al}$. In this work, PR EOS (Peng and Robinson[15]) is used to evaluate fugacities.

RESULTS

The available experimental data concerning hydrate formation conditions in aqueous solutions of electrolytes (NaCl, KCl, and $CaCl_2$) and alcohols (methanol, ethylene glycol, and glycerol), were used to test the accuracy of the model. For this purpose, the results of two existing models that can handle the system of aqueous electrolytes and alcohols were included.

Model prediction for the CO_2 system are shown in FIGURE 1. The average absolute deviation for this model and those obtained by Yousif and Young[11] and Nasrifar et al.[12] are given in TABLE 2. The inhibitors of this system are mixtures of methanol and aqueous electrolytes. TABLE 3 shows the same results for CO_2–methanol system, as an example of a single inhibitor system. Since the Nasrifar et al. model is for mixtures of inhibitors, for this system, results from the Nasrifar et al.[12] model are not included.

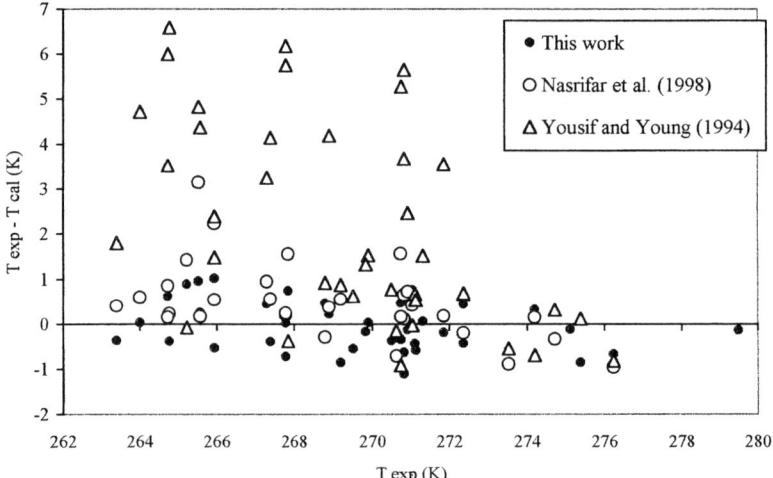

FIGURE 1. Absolute error of hydrate formation temperature for CO_2 in a system of aqueous electrolyte solution and alcohols. Experimental data: Dholabhai et al.[16] Aqueous solution contains: methanol, methanol + NaCl, methanol + KCl, and methanol + $CaCl_2$. Number of data points, 42: ionic strength, 0.0–7.77; methanol concentration, 0.0–15 wt%; pressure range, 0.910–3.0388 MPa.

TABLE 2. Average absolute error of hydrate formation temperature for CO_2 in the presence of electrolytes and methanol

	Yousif and Young[11]	Nasrifar et al.[12]	This work	Exp. Data
AAE (K)	2.78	0.54	0.48	16

TABLE 3. Average absolute error of hydrate formation temperature for CO_2 in the presence of methanol

	Yousif and Young[11]	This work	Exp. Data
AAE (K)	0.87	0.34	16

The ability of this model to predict the hydrate condition for the system of Yousif and Young[11] is illustrated in FIGURE 2 and TABLE 4. The composition of the gas mixture used to obtain these data is presented in TABLE 5. Mei et al.'s[17] experimental data were also used to test the accuracy of hydrate formation prediction for systems containing a single electrolyte, and mixed electrolyte in the presence of methanol. A summary of the results is shown in TABLE 6. Bishnoi et al.'s[18] experimental data for H_2S were also used to test the accuracy of hydrate formation prediction for systems containing single electrolyte, and mixed electrolytes in the presence of methanol. A summary of these results is shown in TABLE 7. As can be seen, both this work and that of Nasrifar et al.[12] essentially give results of comparable accuracy.

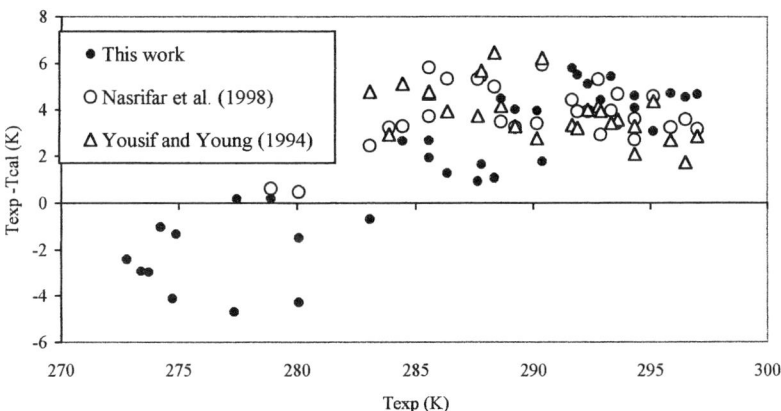

FIGURE 2. Absolute error of hydrate formation temperature for the gas mixture presented in TABLE 5 in mixtures of aqueous electrolyte solution and alcohols. Experimental data: Yousif and Young.[11] Aqueous solution contains: NaCl + methanol, NaCl + ethylene glycerol, NaCl + glycerol, KCl +glycol, $CaCl_2$ + glycerol, NaCl +KCl + glycerol. Number of data points, 37; ionic strength, 0.71–5.22; methanol concentration, 0.0–10.0; ethylene glycol concentration, 0.0–10.0; glycol concentration, 0.0–30.0 wt%; pressure range, 11.336–39.824 MPa.

TABLE 4. Average absolute error of hydrate formation temperature for the system shown in FIGURE 2

	Yousif and Young[11]	Nasrifar et al.[12]	This work	Exp. Data
AAE (K)	4.00	1.86	3.22	11

TABLE 5. Composition of the gas mixture used in FIGURE 2

Component	N_2	CH_4	C_2H_6	C_3H_8	i-C_4	n-C_4	i-C_5	n-C_5
Mole%	0.39	87.26	7.57	3.10	0.49	0.79	0.20	0.20

TABLE 6. Average absolute error of hydrate formation temperature for CO_2 in the presence of pure water, single, or mixed inhibitor

Aqueous Phase	NPTS	P Range (MPa)	Yousif & Young[11]	Nasrifar et al.[12]	This Work	Exp. Data
Pure Water	8	0.92–2.67	—	0.71	2.15	17
Na10	8	0.81–3.43	1.91	0.78	1.47	17
K10	6	0.78–2.81	3.20	1.35	2.63	17
Ca10	7	0.63–3.34	2.34	0.92	1.96	17
Na2K0.5Ca0.5	6	0.60–2.75	3.47	1.46	2.16	17
Me10	7	0.61–2.71	7.52	0.62	3.17	17
Me20	7	0.74–4.50	11.43	2.40	1.37	17
Me30	6	1.16–6.03	17.06	5.32	0.57	17
K10Me10	7	1.05–4.90	6.11	0.66	0.32	17
Na10Me10	7	1.60–3.20	6.02	0.58	0.78	17
Na10Me20	6	1.70–4.96	11.09	3.43	3.66	17
Ca10Me10	7	1.10–5.95	5.31	0.90	1.19	17
K10Me20	6	1.80–11.18	11.80	3.11	2.24	17
Ca10Me20	6	1.47–5.89	12.27	0.68	1.15	17
Na2K0.5Ca0.5Me20	7	1.47–7.88	10.44	1.82	1.78	17
Na2K0.5Ca0.5Me10	6	0.78–4.69	5.97	0.76	0.72	17
Overall	107		7.57	1.53	1.67	

Note: The gas composition is 97.25%C_1 + 1.42%C_2 + 1.08%C_3 + 0.25%i-C_4.
Na10 represents 10 weight percent NaCl, and Ca10Me20 represents 10 weight percent $CaCl_2$ + 20 weight percent methanol, and so forth.

TABLE 7. Average absolute error of hydrate formation temperature for H_2S in the presence of pure water, single, or mixed inhibitors

Aqueous Phase	NPTS	P Range MPa	Nasrifar et al.[12]	This Work	Exp. Data
Pure water[a]	3	0.33–0.80	0.34	0.98	18
Na20[a]	3	0.23–0.55	3.22	0.45	18
Na10Ca10[a]	3	0.32–0.89	2.94	1.27	18
Na10K10[a]	3	0.27–0.85	2.70	0.54	18
Na10Ca5K5[a]	3	0.34–0.98	2.83	1.01	18
Me10[a]	3	0.30–0.93	0.64	0.55	18
Me20[a]	3	0.32–1.07	2.80	1.60	18
Me10Na10[a]	3	0.40–1.07	0.60	2.43	18
Me5Na15[a]	3	0.44–0.95	0.27	2.31	18
Me15Na5[a]	3	0.41–0.89	0.88	2.46	18
Me10Ca10[a]	3	0.41–0.88	0.61	2.10	18
Me10Na10[b]	3	3.72–7.89	1.22	1.66	18
Me5Na15[b]	3	4.56–8.86	0.16	0.46	18
Me15Na5[b]	3	4.40–8.10	1.21	1.33	18
Me15Na5[b]	3	4.20–8.07	0.81	0.91	18
Me10Ca10[b]	3	4.15–7.44	0.57	0.63	18
Me10Na10[c]	3	2.43–5.88	0.82	1.85	18
Me5Na15[c]	3	2.56–6.69	0.33	1.59	18
Me15Na5[c]	3	2.35–5.94	0.55	1.34	18
Me10Ca10[c]	3	2.49–6.32	0.44	0.88	18
Me10Na10[d]	3	3.79–6.87	1.78	1.26	18
Me5Na15[d]	3	3.13–6.89	0.94	0.53	18
Me15Na5[d]	3	2.86–4.78	1.54	0.69	18
Me10Na10[d]	3	3.58–5.42	0.84	0.74	18
Overall	72		1.21	1.23	

[a]Gas phase was composed of: H_2S.
[b]Gas phase was composed of: 95% CH_4 + 5% H_2S.
[c]Gas phase was composed of: 85% CH_4 + 15% H_2S.
[d]Gas phase was composed of: 90% CH_4 + 5% H_2S + 5% C_3H_8
NOTE: Na10 represents 10 weight percent NaCl, and Ca10Me10 represents 10 weight percent $CaCl_2$ + 10 weight percent methanol, and so forth.

The following points should be noted from an overview of the above results:

- The correlation of Yousif and Young[11] was based on the gas mixture shown in TABLE 5. Therefore, the AAE of this model for the CO_2 system is very high.
- According to the correlation of Yousif and Young,[11] hydrate temperature suppression depends only on the concentration of inhibitors. For example, this correlation predicts that a solution of 10 wt% NaCl and 10 wt% methanol, either for hydrate formers such as CO_2 or the gas composition given in TABLE 5, have the same temperature suppression. Additionally, pressure has no effect on this depression. Obviously, this is not true in all cases.
- The model of Nasrifar et al.,[12] is applicable for mixed inhibitors; that is, in the presence of electrolytes and alcohols. Therefore, the results of this model for prediction of hydrate formation temperature in the presence of alcohol are not included in TABLE 3. However, this limitation can be avoided in model programming by considering these cases as a single inhibitor system.
- The experimental data of FIGURE 2 show some discrepancy.

CONCLUSIONS

To predict hydrate forming conditions in the presence of mixtures of electrolytes and alcohol, a simple mixing rule for calculation of water activity is presented. The new model is based on modification of the Aasberg-Petersen et al.,[9] model of water activity, which was applicable to single electrolyte system. The new model predicts water activity in the presence of both aqueous electrolyte solutions and alcohols. The results show the high accuracy of this model. However, in some cases it is not as good as the model of Nasrifar et al.[12] In addition, our model needs no adjustable parameters.

ACKNOWLEDGMENTS

The authors wish to express their appreciation to the School of Engineering for use of computer facilities and the financial support of the Vice-Chancellor for Research of Shiraz University. The authors also would like to thank Mr. Nasrifar for use of his software during this work.

REFERENCES

1. HAMMERSCHMIDT, E.G. 1934. Formation of gas hydrate in natural gas transmission lines. Ind. Eng. Chem. **26:** 851.
2. ANDERSON, F.E. & J.M. PRAUSNITZ. 1986. Inhibition of gas hydrates by methanol. AIChE J. **32**(8): 1321.
3. MOSHFEGHIAN, M. & R.N. MADDOX. 1993. Method predicts hydrates for high-pressure gas streams. Oil & Gas J. **30:** 78.
4. ENGLEZOS, P. & P.R. BISHNOI. 1988. Prediction of gas hydrate formation conditions in aqueous electrolyte solutions. AIChE J. **34**(10): 1718.

5. PITZER, K.S. & G. MAYORGA. 1973. Thermodynamics of electrolytes with one or both ions universal. J. Phys. Chem. **77**(19): 2300.
6. JAVANMARDI, J. & M. MOSHFEGHIAN. 1996. A thermodynamic model for gas hydrates in the presence of single or mixed electrolyte solutions. Proceeding of 2nd International Conference on Natural Gas Hydrate, Toulouse, France.
7. JAVANMARDI, J., M. MOSHFEGHIAN & R.N. MADDOX. 1998. Simple method for predicting gas-hydrate-forming conditions in aqueous mixed-electrolyte solutions. J. Energy Fuels **12**(2): 219–222.
8. ENGLEZOS, P. 1992. Computation of the incipient equilibrium carbon dioxide hydrate formation conditions in aqueous electrolyte solutions. Ind. Eng. Chem. Res. **31**(9): 2232.
9. AASBERG-PETERSEN, K.E. STENBY & A. FREDENSLUND. 1991. Prediction of high-pressure gas solubility in aqueous mixture of electrolytes. Ind. Eng. Chem. Res. **30**: 2180.
10. JAVANMARDI, J. & M. MOSHFEGHIAN. 1998. A new approach for prediction of gas hydrate formation conditions in mixture of aqueous electrolyte solutions. Submitted for publication.
11. YOUSIF, M.H. & D.B. YOUNG. 1994. A simple thermodynamic model to predict the hydrate suppression in aqueous solutions of salts and alcohols. Proceedings of the 73th GPA Annual Convention. 94–99.
12. NASRIFAR, KH., M. MOSHFEGHIAN & R.N. MADDOX. 1998. Prediction of equilibrium conditions for gas hydrate formation in the mixture of both electrolytes and alcohols. J. Fluid Phase Equilibria **146**: 1–13.
13. PARRISH, W.R. & J.M. PRAUSNITZ. 1972. Dissociation pressures of gas hydrates formed by gas mixtures. Ind. Eng. Chem. Proc. Dev. **11**: 26.
14. HOLDER, G.D., G. GORBIN & K.D. PAPADOPOULOS. 1980. Thermodynamic and molecular properties of gas hydrates from mixtures containing methane, argon, and krypton. Ind. Eng. Chem. Fund. **19**(3): 282.
15. PENG, D.Y. & D.B. ROBINSON. 1976. A new two constant equation of state. Ind. Eng. Fund. 59.
16. DHOLABHAI, P.D., J.S. PARENT & P.R. BISHNOI. 1996. Carbon dioxide hydrate equilibrium conditions in aqueous solutions containing electrolytes and methanol using a new apparatus. Ind. Eng. Chem. Res. **35**: 819.
17. MEI, D.-H., J. LIAO, J.-T. YANG & T.-M. GUO. 1998. Hydrate formation of synthetic natural gas mixture in aqueous solutions containing electrolyte, methanol, and (electrolyte + methanol). J. Chem. Eng. Data **43**: 178–182.
18. BISHNOI, P.R., D. DHOLABHAI & K.N. MAHADEV. 1996. Hydrate phase equilibria in inhibited and brine systems. Research Report RR-156, Gas Processors Association, Tulsa, Oklahoma.

Measurements and Predictions of Hydrate Equilibrium Conditions

LARS HENRIK GJERTSEN[a,b] AND FINN HALLSTEIN FADNES[c]

[b]*Statoil Research Centre, Postuttak, 7005 Trondheim, Norway*

[c]*Norsk Hydro, P.O. Box 7190, 5020 Bergen, Norway*

ABSTRACT: The main objective of this study was to demonstrate the importance of reliable hydrate equilibrium data for thorough mitigation and remediation of hydrate problems. Hydrate prevention strategies generally rely on extensive use of different hydrate simulators that, in turn, are based on experimental data. Erroneous hydrate predictions may severely jeopardize the established hydrate strategy and can have severe economic implications. In this study the hydrate equilibrium conditions for eight different hydrocarbon systems were investigated by experimental measurements and by hydrate simulator predictions. The results clearly verify the importance of having due control over the experimental parameters, especially the heating gradient. The error in the determination of the hydrate equilibrium temperature seems to increase with the complexity of the system; for example, when inhibitors are added to the aqueous phase.

INTRODUCTION

Determination of hydrate equilibrium data constitutes the basis for hydrate control in development and operation of oil and gas fields. The hydrate equilibrium conditions determine whether preventive measures are needed on a continuous basis or only during shut-ins. They are further important for the selection of remediation techniques, the amount of thermodynamic inhibitor needed, whether insulation can be used on a stand alone basis or whether a low concentration additive may be employed, and the pressure window available for depressurization.

Hydrate equilibrium data are usually calculated by use of a hydrate computer program simulator. Several simulators are available on the market and there is a scatter in the predictions from these tools. All of these simulators are predictive, but they contain parameters tuned to fit experimental data. Their accuracy is evaluated by comparison with available experimental data. If the experimental data used for tuning and verification are inaccurate, the simulator predictions may lead to incorrect conclusions; hydrate plugs are formed due to inaccurate measures or the method used for hydrate plug removal is less effective than it could have been. Recently, several hydrate equilibrium data sets have been made available for which the deviations between calculated and experimental data are unexpectedly high—in excess of 5–8°C.

[a]Telecommunication. Voice: (47) 73584687; fax: (47) 73584628.
LHG@Statoil.com

In this work the target has been to determine the impact of experimental procedures on the accuracy of experimental data and to evaluate the accuracy of a commercial hydrate simulator, PVTsim, by Calsep.

HYDRATE EQUILIBRIUM?

The problem of accurately determining the hydrate equilibrium curve can be illustrated by considering the simplest system possible, methane (C1) and pure water. In FIGURE 1 the difference between the calculated (PVTsim) and measured hydrate equilibrium temperature is illustrated. In addition to our own data, data from Sloan[1] and Nixdorf et al.,[2] are compared. The important factor is the range of the deviations, 0.6 to −0.05°C. The observed variation, which must be considered significant having in mind the simplicity of the system, could be due to different causes:

- purity of the experimental fluids,
- calibration of the pressure and temperature probes,
- experimental procedures and parameters, and
- interpretation of the measured data.

The influence of the experimental parameters can be demonstrated by observing the effect of varying the heating gradient in a constant pressure experiment. In this experiment hydrate formation and decomposition are revealed by deflections in a volume versus temperature plot, as shown in FIGURE 2, where the hydrate

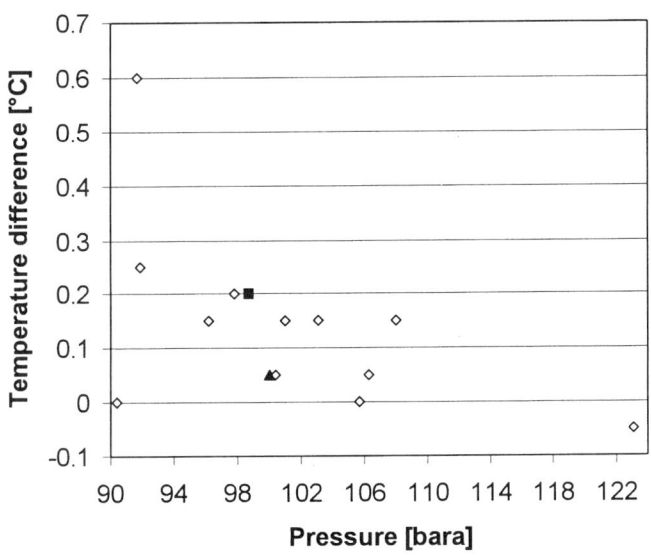

FIGURE 1. Difference between calculated and experimental hydrate temperature for methane.

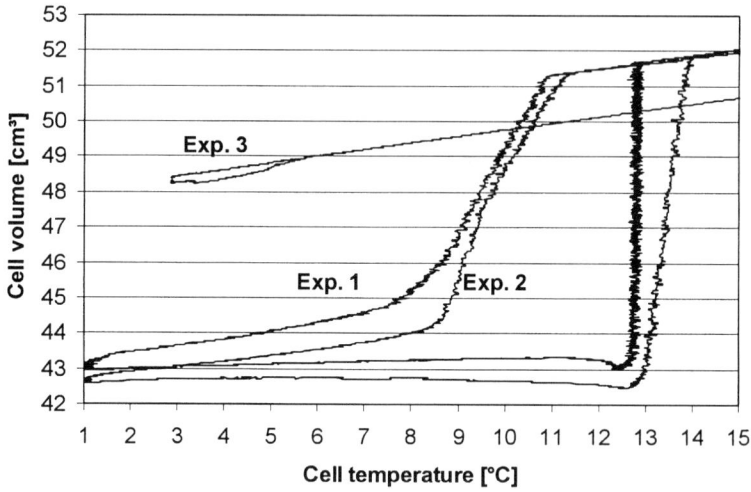

FIGURE 2. Examples of hydrate formation curves (*Exp. 1* and *Exp. 2*, different heating rates; *Exp. 3*, with inhibitor).

equilibrium temperature at 100 bara was determined. It is essential to keep the heating rate during decomposition as low as possible in order to assure thermal equilibrium within the system—or at least to be as close to equilibrium as possible. There has, however, to be a net heat flux into the cell to supply the heat needed for hydrate melting. The effect of varying the heating rate is illustrated in FIGURE 3. The

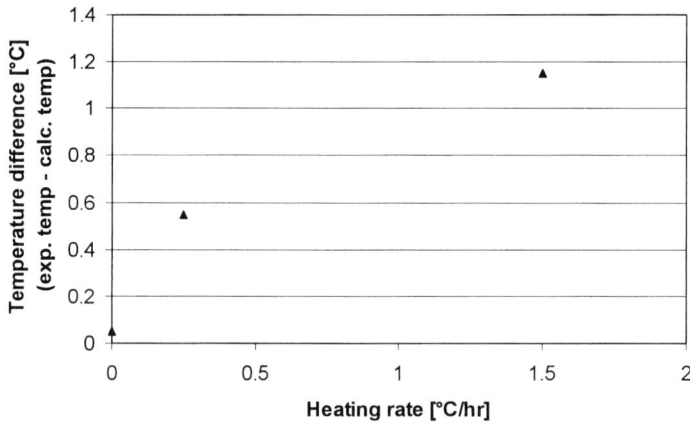

FIGURE 3. Difference between calculated and experimental hydrate temperature for methane at 100 bara, as a function of heating gradient.

deviation between the calculated and the measured hydrate equilibrium temperature increases with increasing heating rate.

For a system with a single hydrate former in pure water, hydrate melting should occur at one single temperature, that is, start of hydrate melting and complete dissociation should occur at the same temperature. The slope of the volume versus temperature curve should be vertical during melting, as in Exp. 1 in FIGURE 2. Any deviation from a vertical slope, as shown by Exp. 2 in FIGURE 2, indicates thermal nonequilibrium and consequently an erroneous determination of the hydrate equilibrium temperature. If the slope is not vertical, the start of melting could be used to define the equilibrium temperature, for such a simple system. This error in the determination of the hydrate equilibrium temperature is likely to be smaller.

When the experimental system consists of multiple hydrate forming components, or when the aqueous phase contains dissolved salts or inhibitors, the hydrate will not decompose at one single temperature, as shown Exp. 3 in FIGURE 2. In these systems it is more difficult to confirm whether the system was in thermal equilibrium during dissociation and the accuracy of the equilibrium temperature determination is also more difficult to assess.

EXPERIMENTAL

Apparatus

The experimental studies presented in this work were performed in two sets of basically identical, hydrate equilibrium apparatus, schematically illustrated in FIGURE 4. In both sets of apparatus a high pressure sapphire cell is the vital element. The cell is placed in an air bath in which the temperature can be controlled and varied between -40 and $+206°C$. The temperature stability is $0.1°C$. The accuracy of the pressure measurement is estimated to be within 0.2 bar. The cell volume is controlled and varied using a piston directly coupled to a computerized brushless motor. The estimated accuracy in volume is $0.005\ cm^3$. The maximum cell volume is $100\ cm^3$. Stirring is provided by a magnetically coupled stirrer, driven by a computer controlled, variable speed motor. Rheology changes for the experimental fluids are continuously monitored by measuring the effort required to keep the motor running at constant speed. All experiments are documented through collected data and video recordings.

Procedures and Methods

The sapphire cell was cleaned and evacuated prior to filling with the experimental fluids. All fluids were added volumetrically in the following sequence: water (degassed and purified by reverse osmosis)—eventually added inhibitor or salt, hydrocarbon liquid phase (when used) and finally the hydrocarbon vapor phase (C1 or synthetic gas mixture). The volumes of water and liquid hydrocarbons were adjusted to obtain the desired water cut. In some experiments the hydrocarbon composition was recombined to field composition prior to injection into the cell. All synthetic fluids were of analytical grade and all possible precautions were taken in order to reduce the degree of contamination of the real hydrocarbon fluid samples. Hydrocarbon and water compositions are given in TABLES 1 and 2, respectively.

TABLE 1. Experimental hydrocarbon compositions in mole percent

Fluid	N_2	CO_2	C_1	C_2	C_3	iC_4	nC_4	iC_5	nC_5	C_6	C_7	C_8	C_9	C_{10+}	MW C_{10+}	Dens. C_{10+}
A			74.21	16.93	8.86											
B	0.02	0.03	18.18	1.71	0.64	0.73	0.2	0.49	0.05	1.4	0.95	0.31	0.15	0.23	162.3	0.83
C			55.72	12.8	6.76	0.17	0.73	0.49	0.86	1.47	2.6	2.79	1.86	13.75	289.4	0.88
D			57.23	13.07	6.64	0.01	0.22	0.15	0.26	0.5	2.3	2.73	0.21	16.68	301.8	0.88
E	0.16	56.61	6.74	0.6	0.15	0.29	0.08	0.18	0.05	0.3	0.74	1.07	0.8	32.24	357.1	0.95
F	1.77	0.28	92.68	3.77	0.62	0.36	0.09	0.07	0.03	0.09	0.14	0.08	0.01			
G	1.51	0.35	44.48	9.45	4.84	6.39	1.74	4.27	0.47	12.2	8.3	2.66	1.33	2.02	133	0.8
H	0.2	1.73	43.34	7.39	4.54	1.03	2.59	1.16	1.58	2.45	3.65	4.42	3.7	22.22	257.9	0.86

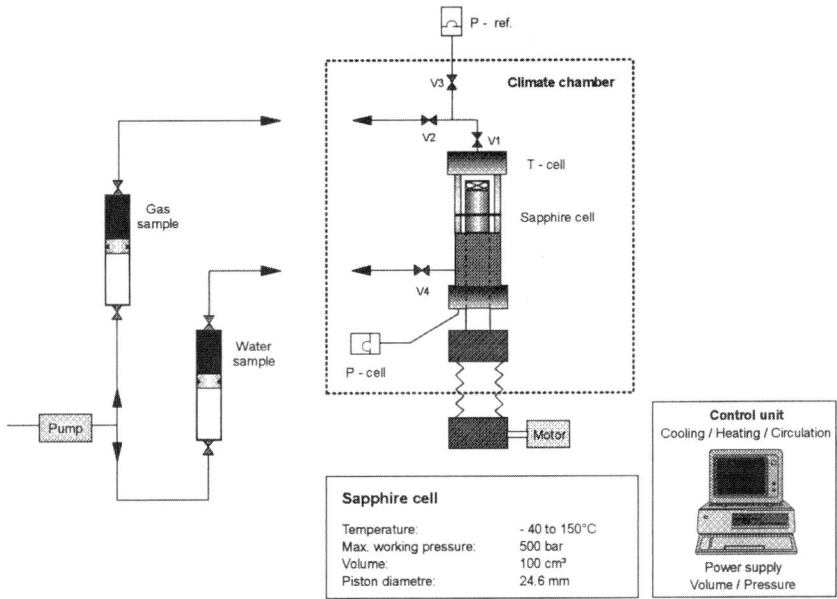

FIGURE 4. Schematic of the sapphire cell.

The most frequently used methods for determining the hydrate equilibrium conditions are the constant volume (isochoric) method and the visual method. In this study we also included a constant pressure (isobaric) method and a isothermal method in which hydrates are formed and melted by increasing and reducing the pressure. The isobaric, the isochoric, and the isothermal methods are essentially similar. In the isobaric method, which was the preferred method in this work, hydrate formation and decomposition was detected by a change in volume (as shown in FIGURE 2), whereas a corresponding change in pressure indicates hydrate formation and decomposition in the isochoric method.

We believe that the two latter methods are superior to the visual method because they allow automatic execution of the experiments and a more objective determination of the hydrate equilibrium point. The equilibrium point is interpreted as being the point where initial conditions are resumed after a hydrate formation/decomposition cycle. There are, however, several options for defining this point, as is discussed below.

TABLE 2. Experimental water composition in g/l

Saline Water	NaCl	CaCl$_2$	KCl	MgCl	SrCl$_2$
W1	27.67	1.19	0.88	5.36	0.01
W2	25.12				
W3	28.54	6.59			

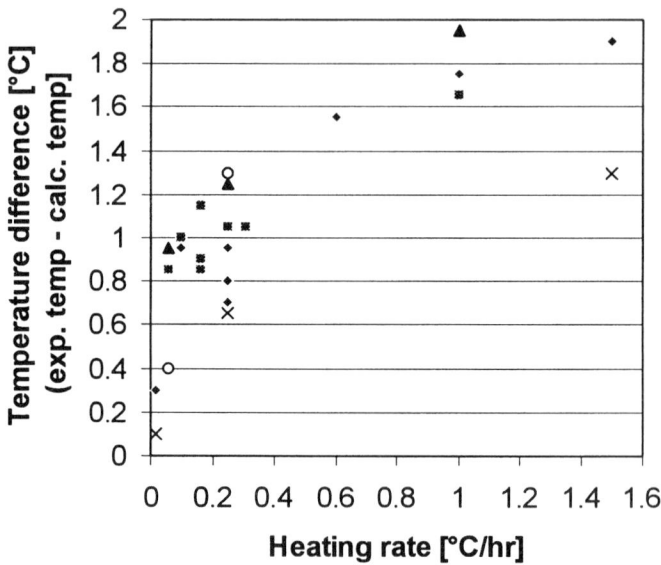

FIGURE 5. Effect of heating rate on the experimentally determined hydrate equilibrium temperature. ◆, Comp.A/dist, water, 70 bar; ■, Comp.A/dist, water, 100 bar; ▲, Comp.A/water W1; ×, C1/dist.water; ○, Comp.B/65 wt% MEG.

EXPERIMENTAL RESULTS

The hydrate equilibrium conditions have been determined for 14 different experimental systems. The hydrocarbon phases range from a simple C1 system to complex bottom-hole samples. The span of the aqueous phase composition is correspondingly broad, ranging from pure, distilled water to high concentration of inhibitors and inhibitor/brine mixtures. The experimental results shown in FIGURES 5–10 demonstrate the effect of different experimental parameters.

DISCUSSION

Experimental Procedure

To provide experimental data with satisfactory accuracy an experimental procedure has to be selected that includes an experimental method (isochoric, isobaric, or isothermal) and the gradient of the controlled parameter.

In FIGURES 3 and 5, the effect of changing the temperature gradient during heating is illustrated for experiments made under isobaric conditions. These results demonstrate that there is a significant impact from the heating rate employed. An increase from 0.25°C/hr to 1°C/hr results in an increase in the measured hydrate equilibrium temperature of 0.6–0.8°C for these rather simple systems. These results accord with expectations, since an increased temperature gradient will reduce the

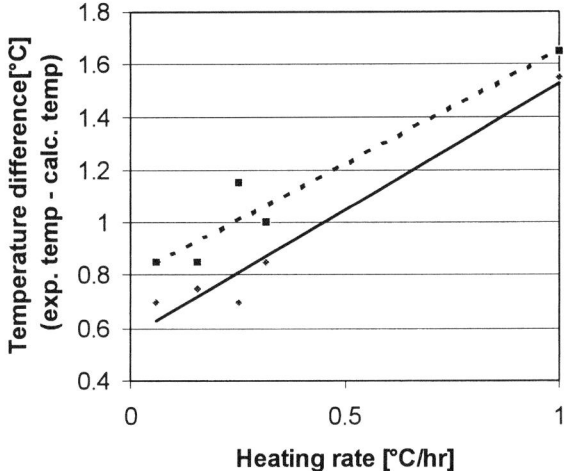

FIGURE 6a. Effect of the choice of experimental method on the experimentally determined hydrate equilibrium temperature—constant pressure (isobaric) versus constant volume (isochoric) experiments: ◆, constant volume; ■, constant pressure; *solid line*, linear (constant volume); *dashed line*, linear (constant pressure).

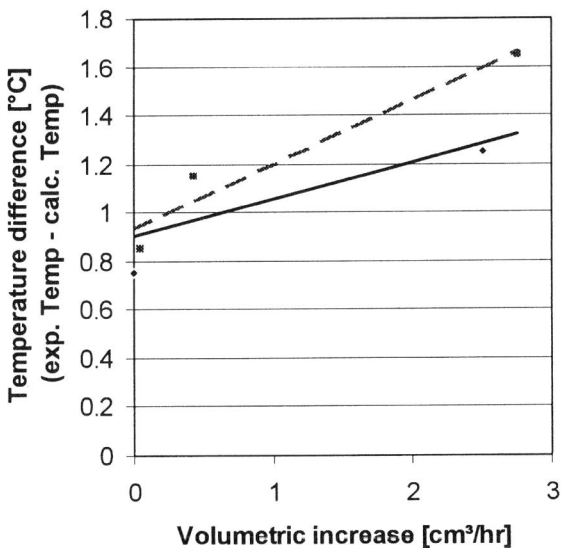

FIGURE 6b. Effect of the choice of experimental method on the experimentally determined hydrate equilibrium temperature—constant pressure (isobaric) versus constant temperature (isothermal) experiments: ◆, constant temperature; ■, constant pressure; *solid line*, linear (constant temperature); *dashed line*, linear (constant pressure).

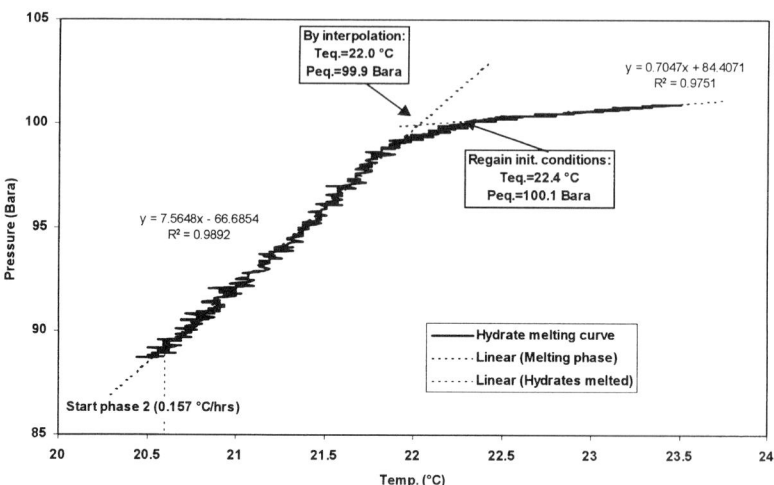

FIGURE 7. Effect of different interpretation of measured data.

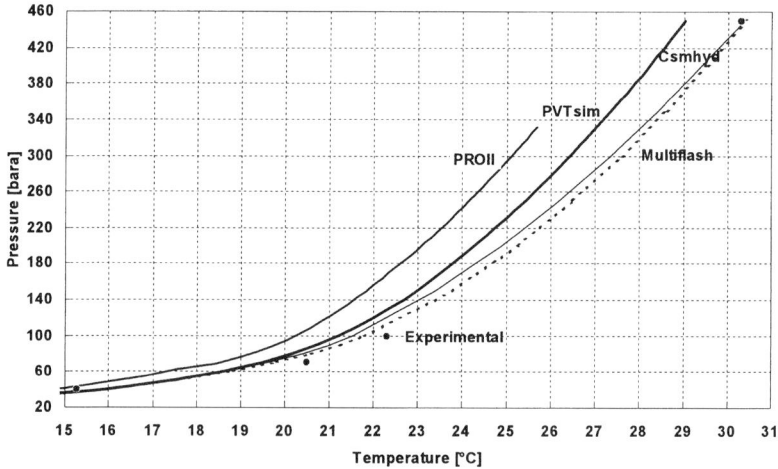

FIGURE 8. The differences between the experimental data and calculations by selected hydrate simulators (simple model system, HC-comp. A).

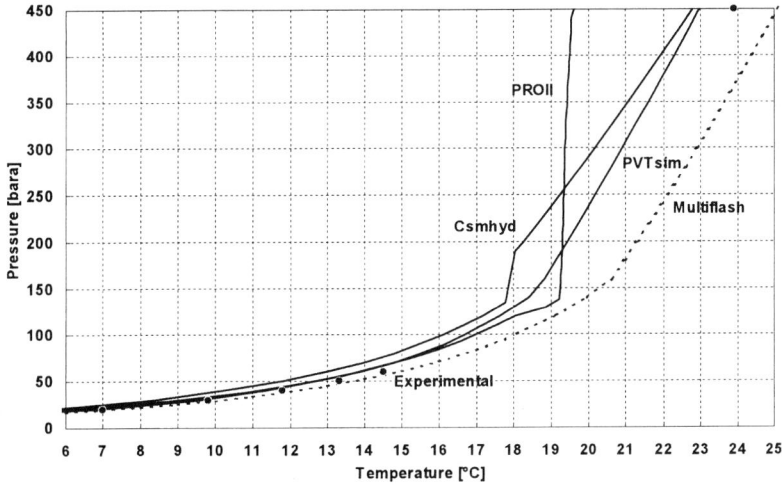

FIGURE 9. The differences between the experimental data and calculations by selected hydrate simulators (black oil system, HC-comp. H, water-comp. W1).

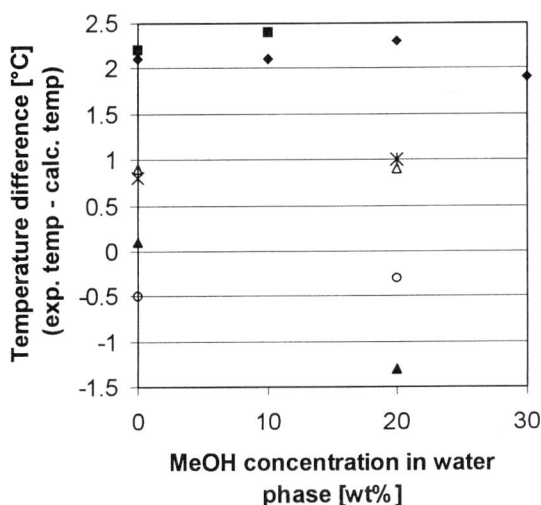

FIGURE 10. Effect of methanol combined with saline water on the hydrate equilibrium temperature: ◆, Comp. C/W2; ■, Comp. D/W2; ▲, Comp. E/dist. 101 bar; ○, Comp. E/dist. 51 bar; ∧, Comp. E/W3 101 bar; ✳, Comp. E/W3 51 bar.

possibility to obtain equilibrium conditions in the system. The results also indicate that a further reduction in heating rate below 0.25°C/hr will lower the measured hydrate equilibrium temperature even more. A further reduction to 0.02°C/hr reduces the measured equilibrium temperature by up to 0.9°C. The most significant reduction in the measured temperature was obtained for systems with an inhibitor present. This implies that the heating rate should be as low as possible in order to obtain equilibrium conditions. Too rapid an increase results in a measured equilibrium temperature significantly higher than the real one.

The effect of using different experimental methods, isobaric versus isochoric and isobaric versus isothermal, is demonstrated in FIGURES 6a and 6b, respectively. There is some scatter in the data, which to some extent reflects different interpretation of the measured data (to be discussed later). A trend observed is that the hydrate temperatures determined by the isochoric method are lower than the corresponding results obtained with the isobaric method. However, the differences are small, approximately 0.1–0.2°C. Furthermore, a general trend is that the deviation between experimental and predicted data decreases, with decreasing heat gradient. The difference between the results obtained with the isothermal and isobaric methods is even less at low gradients. All three methods may be employed, since the differences are barely significant. The interpretation of the experimental data from isothermal experiments has proved to be more difficult than for the other methods because the equilibrium conditions for melting seem to be less uniquely defined.

There will always be room for different interpretations of experimental data when the hydrate equilibrium point is to be determined. The thermodynamic equilibrium point is defined as the temperature, at a given pressure, where the last hydrate crystal melts. Using visual detection it is obvious that a certain degree of subjective evaluation will lead to some inaccuracy. In the non-visual methods used in this study, the equilibrium conditions were established in two different ways, as demonstrated in FIGURE 7. The temperature may be determined at the point where the pressure versus temperature plot regains the initial slope or one can fit straight lines to the steep part of the melting curve and to the curve after complete melting and determine the equilibrium point as the intersection of the extrapolation of these two straight lines. Using the first of these two methods, it is essential to have a very low heating rate, whereas the extrapolation method to some extent compensates for lack of thermal equilibrium in the system. The differences in the equilibrium temperatures determined by the two methods may be significant, as is demonstrated FIGURE 7.

Absolute Pressure

Experimental hydrate equilibrium temperatures for a simple fluid system at pressures varying between 40 and 450 bar have been compiled. The differences between the experimental data and the predictions made by selected hydrate computer simulators (PVTsim by Calsep, Csmhyd by CSM, Multiflash by Infochem, and PROII by Simsci) are illustrated in FIGURE 8. This plot demonstrates that the simulators give relatively uniform estimates of the equilibrium temperature at low pressures (40 and 70 bar), but at higher pressures the differences between the calculations are significant. At 450 bar the deviations are as high as 1.25°C.

At pressures in the range 70 bar and above the simulators seem to calculate equilibrium temperatures that are lower than the experimental values. At 40 bar the simulators calculate equilibrium temperatures that are higher than the experimentally determined values. Since the simulators, for such a simple system, seem to be conservative at high pressures and non-conservative at low pressures, the range of uncertainty of the simulators is higher, so that it is hard to tell whether the actual hydrate equilibrium temperatures are higher or lower than the predicted values. Again, reliable experimental data are vital. The experiments show that the accuracy of the calculations is pressure dependent, an evaluation of computer simulators should include a wide pressure range.

Experimental hydrate equilibrium temperatures for a black oil system at pressures varying between 20 and 450 bar have been compiled. In FIGURE 9, a comparison of several simulators has been made for this black oil system. For this system, at low pressures, the deviation between the models is higher than for the simple model system. At higher pressures the deviations are even greater. These results clearly demonstrate that when exact hydrate equilibrium data are crucial, experimental data are needed.

The Aqueous Composition

In FIGURE 10 the effect of using methanol in the range of 10–30 wt% in the aqueous phase is evaluated. For compositions C and D the differences between the experimental and the calculated equilibrium temperatures are in the range 1.9–2.4°C for all the experiments, with and without inhibitor. The magnitude of this difference can, to some extent, be attributed to the high heating rate applied. The results indicate that the effect of methanol is fairly well described in the computer tool and that the effect of methanol can be added to the effect of saline water. The same can be concluded for composition E, with saline water. For composition E, with distilled water the results are more scattered. By considering composition E with and without saline water, the simulator seems to overpredict the effect of the salt.

In TABLE 3 the effect of high concentrations of MEG in the water phase is illustrated. MEG concentrations of 48 wt% and 65 wt% have been evaluated. The experimental hydrate equilibrium temperatures were 0.3–0.4°C higher than the calculated values. This is in the expected range, taking into account the heating rate applied. At concentrations above 65 wt% MEG did freeze out as a solid[3] and the hydrate equilibrium temperature could not be determined. The simulator does not calculate precipitation of solid glycols, nor does it give warnings that this may occur at high concentrations.

TABLE 3. Effect of high concentrations of monoethylene glycol on the hydrate equilibrium temperature

Exp. No.	Pressure (bara)	Heating rate (°C/hr)	MEG conc. (wt%)	Hydrate equilibrium temperature (°C)	Dev. from predicted temperature (°C)	Comments
Kolmeg49	100	0.06	48.7	−2.9	0.4	Comp. F with dist. water/MEG
Kolmeg29	67	0.06	65	−21.8	0.3	Comp. G with dist. water/MEG

CONCLUSIONS

The main objective of the present investigation was to verify the basic assumption that reliable experimental hydrate equilibrium data are required for thorough mitigation and remediation of hydrate problems. This hypothesis is verified by the main conclusions:

- The importance of controlling the experimental parameters is crucial. A too high heating rate may lead to erroneous hydrate equilibrium predictions. The choice of different experimental methods for determination of hydrate equilibrium seems to be of less importance.

- The error in the determination of the hydrate equilibrium temperature depends on the actual experimental systems. The importance of a low heating gradient is amplified when inhibitors or salts are present in the aqueous phase.

- Hydrate simulators may show negative or positive deviations from experimental measurements, depending on the pressure. Significant deviations in the predictions performed by different simulators have been observed.

- The effect of thermodynamic inhibitors seems to be fairly well handled by the evaluated hydrate computer simulator.

- The effect of salts may not be precisely handled by the evaluated hydrate computer simulator.

- The overall accuracy of the hydrate simulators seems to be in the order of $\pm 2.0°C$.

REFERENCES

1. SLOAN, E.D. 1998. Clathrate Hydrates of Natural Gases. Marcel Dekker Inc.
2. NIXDORF, J. & L.R. OELLRICH. 1999. Fluid Phase Equilibria **39:** 325–333.
3. LI, X., L.H. GJERTSEN & T. AUSTVIK. 1999. Thermodynamic inhibitors for hydrate plug melting. To be published in NGH 99 Proceedings.

Effect of Surfactants on Hydrate Formation Rate

UĞUR KARAASLAN[a] AND MAHMUT PARLAKTUNA

Department of Petroleum and Natural Gas Engineering,
Middle East Technical University, 06531, Ankara, Turkey

INTRODUCTION

Natural gas hydrates exhibit two opposing properties. First, they are considered as a nuisance in oil industry since they plug pipelines requiring costly remedies. Second, there are several possible applications of hydrates such as storage and transportation of natural gas, desalinization of water, and recovery of rare gases. In addition, *in situ* hydrates are considered to be the commercial energy supply for the future.

These two aspects of gas hydrates stimulated hydrate studies from two opposite directions. The studies related to the problematic side of hydrates are trying to find methods for hydrate inhibition. Other studies are exploring for ways to promote hydrate formation.

The traditional hydrate inhibitors such as methanol and glycols have been in use for several decades, but as production platforms are becoming more remote within deeper water, use of these chemicals becomes more and more costly. Replacing these traditional thermodynamic inhibitors with a new generation of hydrate inhibitors can lead to very substantial cost savings. Kelland *et al.*[1] discuss the types and properties of these new generation gas hydrate inhibitors.

Surfactants and polymers are classes of chemicals that affect the hydrate formation process and are considered new hydrate inhibitors (Kelland *et al.*[2]). These chemicals affect hydrate formation through one of two processes, kinetic inhibition or anti-agglomeration. Although polymers are mainly considered kinetic inhibitors, surfactants exhibit diverse agglomeration characteristics.

Kalogerakis *et al.*[3] investigated experimentally the effect of surfactants on the kinetics of methane hydrate. They reported that surfactants do not influence the thermodynamics; however, they have a strong influence on the kinetics of gas dissolution in the water phase as well as increasing the overall rate of hydrate formation. They also observed that the hydrate particles formed in the presence of the various surfactants exhibit diverse agglomeration characteristics.

Urdahl *et al.*[4] developed an experimental set-up for characterizing of gas hydrate inhibitor efficiency with respect to flow properties and deposition. They reported that the chemicals used in their study are surfactants, polymers, and various patented chemicals. The surfactants among them are known to stabilize water in oil emulsions. These chemicals showed some positive effects, but at concentrations above 1 wt% of the aqueous phase. They did not prevent deposition of hydrates at the water

---

[a]Telecommunication. Voice: 90 312 2104896; fax: 90 312 2101271.
ugurk@metu.edu.tr

pipe wall. The isolation of the water phase by formation of a finely dispersed water/oil emulsion is suggested by the authors to provide a mechanism for suppression of gas hydrate formation.

The main focus of this paper is to investigate experimentally the effect of surfactants on the formation kinetics of gas hydrate, in particular to observe their ability as hydrate promoter (i.e., production of hydrates for storage and transport).

EXPERIMENTAL SET-UP AND PROCEDURE

In this study, a high-pressure system was used to produce gas hydrate. A schematic diagram of the experimental set-up is given in FIGURE 1. A cylindrical high-pressure reactor with inner diameter, 3.4 cm; length, 15 cm; and total available volume, 142 cm^3 (including connections) is used to produce gas hydrate. It is made of brass and tested up to 1000 psig pressure. The cell consists of glass windows that are used for visual observation of hydrate formation on two faces. The high-pressure cell is placed into a constant temperature bath made from plexiglass that permits easy observation of the system. A temperature controller and a refrigerated chiller are connected to the water bath to provide temperature control. A thermocouple and a pressure transducer are connected to the high-pressure cell to measure temperature and pressure inside the high-pressure cell. They are connected to a data-logger and a personal computer to record the temperature and pressure in the hydrate formation cell as functions of time. A magnetic stirrer, rotating a stirring bar within the cell, accomplishes the stirring of the system. The magnetic stirrer has a speed range 0–1,200 RPM and it was kept at 500 RPM throughout this study. Tap water, natural gas and surfactants were used as reagents in experiments. The names of the

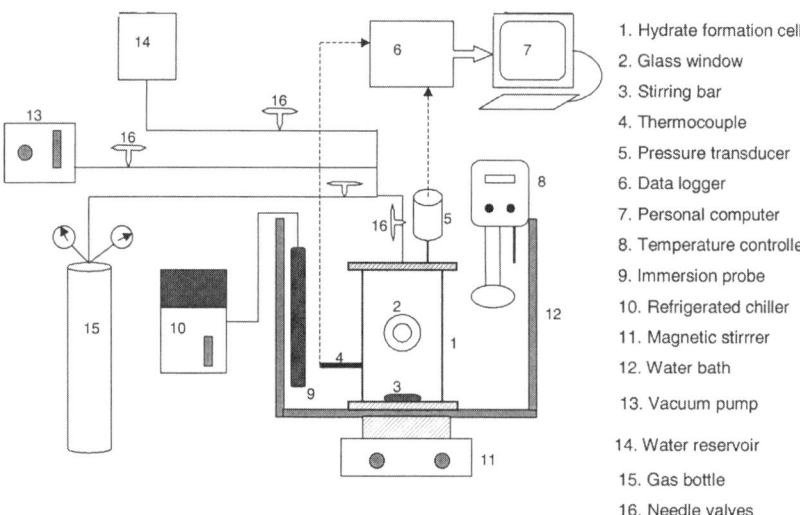

FIGURE 1. Experimental set-up for high pressure test.

TABLE 1. Names and types of surfactants used in this study

Commercial Name	Structure	Type
LABSA	Linear alkyl benzene sulphonic acid	Anionic
Dehyguard Dam (DAM)	Quaternary ammonium salt	Cationic
ETHOXALATE	Nonylphenol ethoxalate	Nonionic

TABLE 2. Chemical composition of gas used in experiments

Component	CH_4	C_2H_6	C_3H_8	$i\text{-}C_4H_{10}$	$n\text{-}C_4H_{10}$	$i\text{-}C_5H_{12}$	$n\text{-}C_5H_{12}$	N_2	CO_2
Mole%	89.47	5.39	1.89	0.40	0.62	0.28	0.19	1.58	0.18

surfactants and their types are given in TABLE 1. A natural gas from Değirmenköy Gas Field of Turkish Petroleum Corporation was used to produce hydrate during this study. The composition of natural gas is given in TABLE 2.

Three different surfactants, namely ETHOXALATE (nonionic), DAM (cationic) and LABSA (anionic) were used throughout the study to observe the effect of type and concentration of surfactants on the formation of natural gas hydrates. Thirteen experiments (four experiments for each type of surfactant and one experiment without surfactant) were carried out. The experimental conditions of these tests are presented in TABLE 3. The initial temperature and pressure of each test were adjusted to $18 \pm 1°C$ and 576 ± 2 psig, respectively.

TABLE 3. Experimental conditions of the tests

Experiment Number	Surfactant	Concentration (weight%)	Initial Temperature (°C)	Initial Pressure (psig)
1	—	—	18.0	576
2	ETHOXALATE	0.005	18.5	576
3	ETHOXALATE	0.01	18.3	576
4	ETHOXALATE	0.1	17.9	574
5	ETHOXALATE	1.0	18.2	576
6	LABSA	0.005	17.6	577
7	LABSA	0.01	18.4	576
8	LABSA	0.1	19.0	575
9	LABSA	1	19.0	575
10	DAM	0.005	18.5	576
11	DAM	0.01	18.6	577
12	DAM	0.1	18.8	576
13	DAM	1	18.5	578

After preparing system and reaching temperature and pressure equilibrium, the system was cooled and the temperature and pressure inside the high-pressure cell were recorded as a function of time at every 10 seconds. After forming hydrates and getting the cell temperature below 4°C, the experiment was ceased. This period is called hydrate formation, since the aim is to produce hydrates from the surfactant solution and natural gas. Details of experimental set-up and procedure can be found from Karaaslan.[5]

RESULTS AND DISCUSSION

Hydrate Formation Thermodynamics

A sample plot of the change in pressure and temperature of experiments during hydrate formation is shown in FIGURE 2 from Test 11, where the type of surfactant was DAM (cationic) with a concentration of 0.01%. As can be observed from this figure there are three regions of pressure drop with time during hydrate formation after the start of cooling. At the initial stage of experiment, the drop in pressure is due to cooling. The second region starts with a change (an increase) in the rate of pressure drop. This is an indication of the start of gas consumption for hydrate formation. The third region shows itself with a reduction in the rate of pressure drop. This is due to loosing the gas–water interaction because of a solid hydrate layer formed between them. Hydrate particles grow to a certain thickness and make a solid boundary on the water surface. Since hydrate is a less permeable substance it limits gas transportation into water phase.

Pressure–temperature plots for each test during hydrate formation were used to observe the effect of surfactants on hydrate formation thermodynamics (FIGURE 3). The temperatures at which the system entered Region 2, as given in FIGURE 2 (start

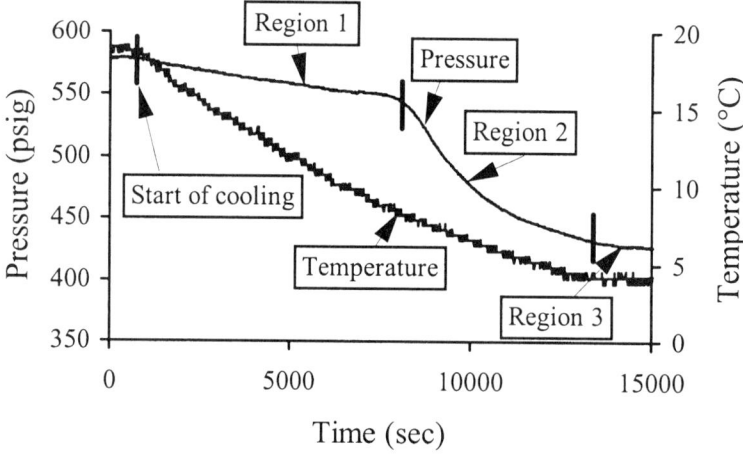

FIGURE 2. Change in pressure and temperature during hydrate formation (Test 11).

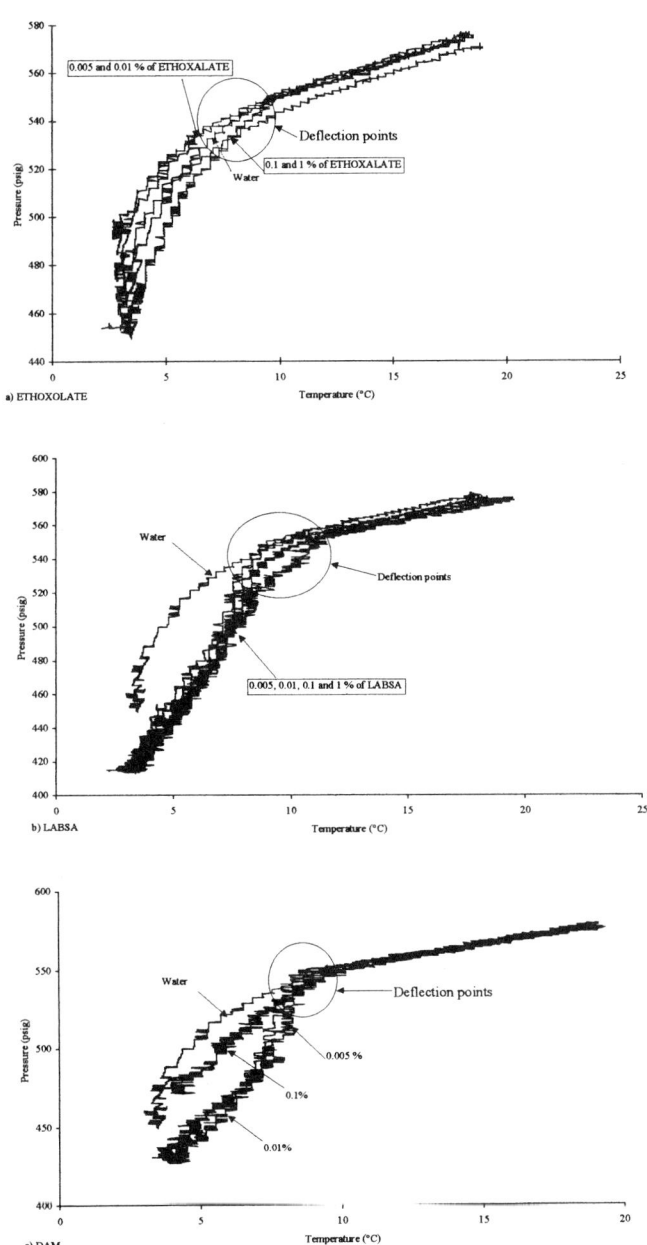

FIGURE 3. Pressure–temperature plots.

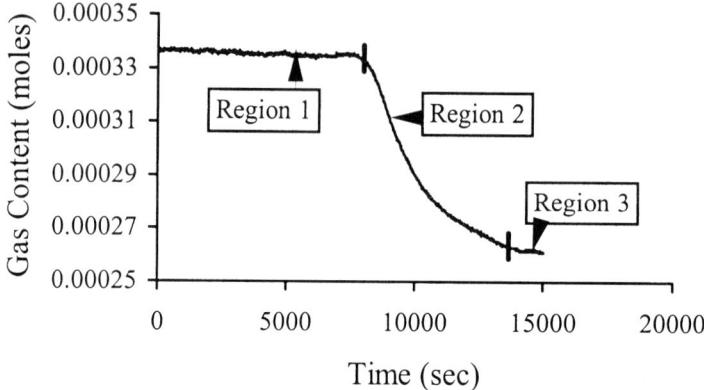

FIGURE 4. Change in free gas content during Test 11 (0.01 wt% DAM).

of gas consumption for hydrate formation), are encircled in these plots. It can be seen from these plots that the starting temperatures of hydrate formation do not differ significantly with the change in surfactant type and concentration. This observation agrees with the observation made by Kalogerakis et al.,[3] that the surfactants do not influence the hydrate formation thermodynamics.

Hydrate Formation Kinetics

The consumption rate of natural gas during the hydrate formation process was determined from the pressure–temperature data. The real gas law was utilized to compute the number of moles of free gas present in the cell at each time step. The compressibility factor was calculated by using the Lee and Kesler[6] compressibility

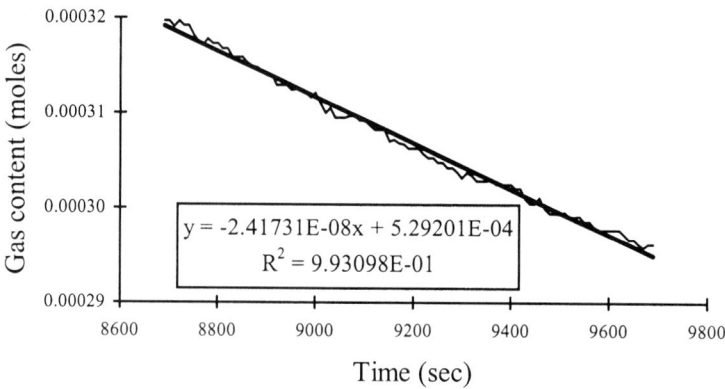

FIGURE 5. A typical change in gas consumption during hydrate formation (Test 11).

factor expression, and the free gas volume was assumed constant throughout the reaction (i.e., volume changes due to phase transitions were neglected).

A typical change of the gas content in the cell is shown in FIGURE 4 for Test 11. As observed there is a slight decrease in the gas content at the initial stage of experiment (Region 1), which can be considered as the dissolution of gas in hydrate forming solution. After a certain point, the slope changes drastically, which can be taken as the initiation point of hydrate formation. The change in gas content with time shows a linear behavior after the commencement of hydrate formation in Region 2. However, since the gas-water interaction reduces because of the hydrate layer, this linearity disappears gradually. It was assumed that the initial linear part of the data of Region 2 gives more realistic and deterministic results concerning surfactant

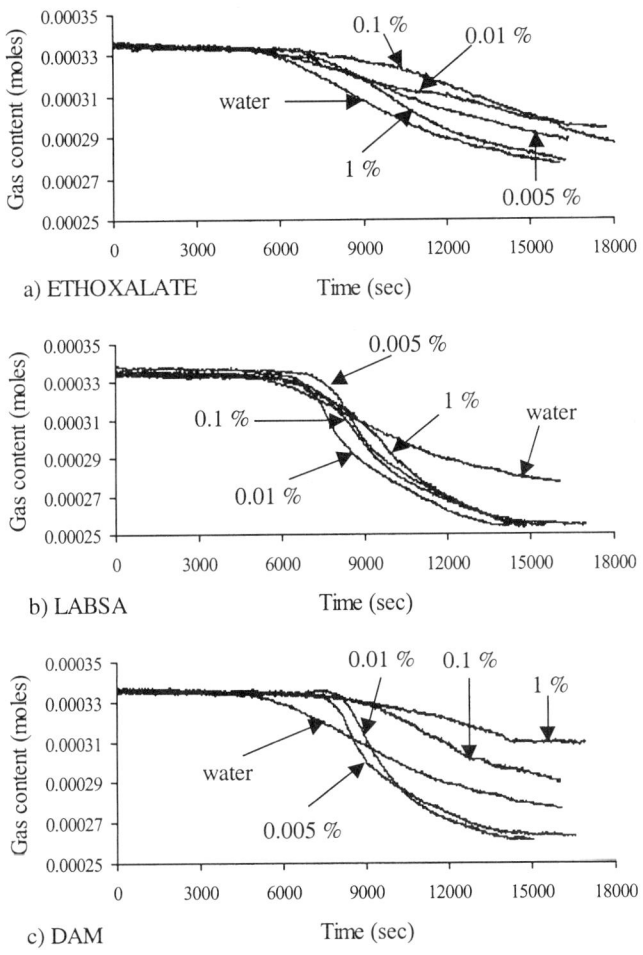

FIGURE 6. Change in gas content during hydrate formation.

TABLE 4. Gas consumption rates during hydrate formation (moles/sec)

Concentration (weight%)	ETHOXALATE (nonionic)	LABSA (anionic)	DAM (cationic)
0.0	7.143×10^{-9}	7.143×10^{-9}	7.143×10^{-9}
0.005	3.799×10^{-9}	2.452×10^{-8}	2.931×10^{-8}
0.01	4.393×10^{-9}	2.553×10^{-8}	2.417×10^{-8}
0.1	5.197×10^{-9}	1.773×10^{-8}	6.687×10^{-9}
1	8.832×10^{-9}	1.343×10^{-8}	2.285×10^{-8}

effect on hydrate formation. A plot of this data is shown in FIGURE 5 with a linear fit and the slope of this straight line is the rate of gas consumption.

The change in gas content for various surfactant concentrations of (a) ETHOXALATE, (b) LABSA, and (c) DAM are shown in FIGURE 6 as functions of time. The estimated initial gas consumption rates for all tests taken from Region 2 are presented in TABLE 4. The change in gas consumption rate as a function of surfactant concentration is given in FIGURE 7. Analysis of FIGURE 6 by eye inspection shows that all tests with LABSA at 0.005% and 0.01% concentration tests with DAM, have higher slopes compared to tests with water alone. These observations are quantitatively verified in FIGURE 7.

As can be observed from FIGURES 6 and 7, ETHOXALATE (nonionic) always has lower gas consumption rates with respect to water alone (0.00001% was taken as 0% for logarithmic plot purposes). There is a slight decrease in the global hydrate formation rate. Kalogerakis et al.[3] made the same observation for the nonionic surfactants that they studied. They had qualitative observations for the appearance of hydrate particles. The following is quoted from Kalogerakis et al.,[3] for the nonionic

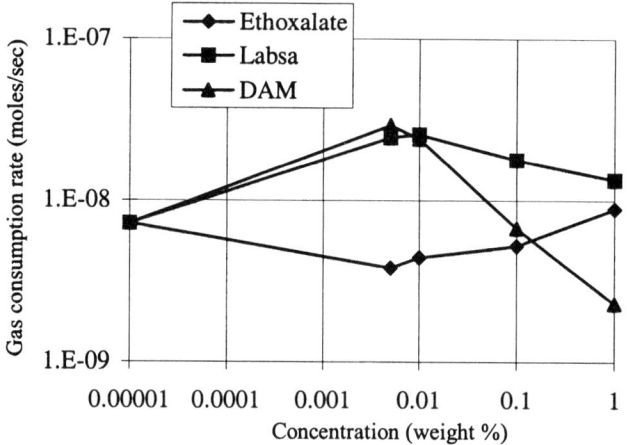

FIGURE 7. Gas consumption rates of ETHOXALATE, LABSA, and DAM.

surfactants: "the formed hydrate particles appeared to remain separated from each other at all times in the bulk"—an indication of the anti-agglomeration effect.

The presence of LABSA (anionic) has a pronounced effect on the global rates of hydrate formation. It always shows higher gas consumption rates (higher hydrate formation rates), compared to water alone, at all concentrations studied. This observation also agrees with an observation of Kalogerakis et al.,[3] where they observed a noticeable increase in the hydrate formation rate with anionic surfactant. It is therefore possible to consider this surfactant to provide a good candidate for a hydrate promoter.

DAM showed similar behavior with LABSA for lower concentrations (0.005 and 0.01 weight%), but this behavior is reversed for higher concentrations of DAM.

CONCLUSIONS

A series of experiments were conducted to investigate the influence of surfactants on the rate of hydrate formation. There is no appreciable effect of surfactants on the thermodynamics of hydrate formation, for surfactant concentrations studied. The overall hydrate formation rate is increases with the use of anionic surfactant for all concentrations tested. It is, therefore, possible to use this surfactant as a hydrate promoter. The effect of a nonionic surfactant is less pronounced compared to an anionic one. Cationic surfactants showed opposite behaviors at low and high concentrations. They are effective as promoters at low concentrations.

ACKNOWLEDGMENT

The authors wish to acknowledge the financial support by Middle East Technical University through the Research Fund Project AFP-96-03-06-01.

REFERENCES

1. KELLAND, M.A., T.M. SVARTAAS & L. DYBVIK. 1995. A new generation of gas hydrate inhibitors. SPE 30695. Proceedings SPE Annual and Technical Conference, Dallas, October 22–25. 529–537.
2. KELLAND, M.A., T.M. SVARTAAS & L.A. DYBVIK. 1994. Control of hydrate formation by surfactants and polymers. SPE 28506. Proceedings of the SPE 69th Annual Technical Conference and Exhibition, New Orleans, September 25–28. 431–438.
3. KALOGERAKIS, N., A.K.M. JAMALUDDIN, P.D. DHOLABHAI & P.R. BISHNOI. 1993. Effect of surfactants on hydrate formation kinetics. SPE 22188. Proceedings of the SPE International Symposium on Oilfield Chemistry, New Orleans, March 2–5. 375–383.
4. URDAHL, O., A. LUND, P. MORK & T.N. NILSEN. 1995. Inhibition of gas hydrate formation by means of chemical additives—I. Development of an experimental set-up for characterization of gas hydrate inhibitor efficiency with respect to flow properties and deposition. Chem. Eng. Sci. **50**(5): 863–870.
5. KARAASLAN, U. 1998. Effect of Surfactants on Hydrate Formation. M.Sc. Thesis. Middle East Technical University, Ankara.
6. LEE, B.I. & M.G. KESLER. 1975. A generalized thermodynamic correlation based on three-parameter corresponding states. AIChE J. **21**: 510–527.

Experiments Related to the Performance of Gas Hydrate Kinetic Inhibitors

T.M. SVARTAAS,[a] M.A. KELLAND,[b] AND L. DYBVIK[b]

[a]*Stavanger College, Institute for Petroleum Technology, PO Box 2557, Ullandhaug, N-4091 Stavanger, Norway*

[b]*RF-Rogaland Research, Norway*

ABSTRACT: (1) The hydrate formation process and growth is described through three different stages: (a) an induction period (nucleation), (b) a slow growth period prior to (c) a final stage described by a catastrophic, fast growth rate. In this paper the effect of kinetic hydrate inhibitors (KI) is characterized by the total delay of the catastrophic growth process at given subcooling ratios at given pressures. (2) The effects of a 20,000 M_w PVCap on a structure I (sI) ethane hydrate and a sII synthetic natural gas (SNG) hydrate have been examined in sapphire cells. A baseline was first established for each of two hydrate forming systems to describe the respective induction periods, slow growth periods, and initial growth rates. Without inhibitor, at similar subcoolings, and at similar pressures a longer induction period and a slower growth rate were observed prior to the catastrophic stage for the sII hydrate system as compared to the sI hydrate system. By the addition of 5,000 ppm PVCap to the aqueous phase the total delay of the catastrophic growth stage increased by a factor 12 for the sII SNG-hydrate system and by a factor 5 for the sI ethane-hydrate system. (3) The effect of different kinetic hydrate inhibitors (KI) at various pressures ranging from about 65 bar and up to about 200 bar has been examined in high pressure sapphire cells using a sII hydrate forming condensate–SNG–synthetic sea water system (SSW, 3.6% salt). For all the experiments reported the degree of subcooling (ΔT) with respect to the hydrate equilibrium properties of the fluid system used was kept at a similar magnitude. This was done to examine KI effects at similar degrees of subcooling within the different pressure regions. The experiments at constant ΔT showed that the KI effect (i.e., total delay of the catastrophic growth stage) decreased with increasing pressure and that a dramatic decrease was observed for increasing pressures above 90 bar. The *true* driving force of a hydrate forming system is described through a chemical potential, $\Delta \mu$, which is a function of pressure (i.e., fugacity) and absolute temperature (K). At a given ΔT the chemical potential is greater in the high-pressure region than in the low-pressure region.

INTRODUCTION

Effect of Hydrate Structure on KI Performance

The active groups of kinetic hydrate inhibitors (KI) are assumed to interact with specific hydrate surfaces through hydrogen bonds and/or penetration into open

---

[a]Telecommunication. Voice: 47-51832285; fax: 47-51831750.
Thor.M.Svartas@tn.his.no

cavities. Thus, for a given active group the interaction could be optimal for a defined hydrate surface or structure. To examine the effect of KI on different hydrate surfaces, experiments were conducted in a sapphire cell on sI ethane hydrate and a sII synthetic natural gas (SNG) hydrate using synthetic sea water (SSW, 3.6% salt) to simulate the water phase. The hydrate equilibrium properties of the pure ethane–SSW and the SNG–SSW systems were very similar in the pressure region about 20 to 25 bar. All experiments were therefore carried out in the pressure region 22.5 to 23.5 bar and at similar temperatures. Thus, the experiments were carried out at comparable driving forces with respect to the subcooling ratio. For the two systems temperatures producing subcooling ratios (ΔT) in the range 8 to 11°C were applied.

A PVCap having a molecular weight (M_w) of about 20,000 was selected for the experiments. The active concentration of PVCap was kept constant at 5,000 ppm (i.e., 0.5 wt%) in the SSW solution.

Effect of Absolute Pressure

The effect of kinetic inhibitors (KI) was examined in sapphire cells at various pressures and temperatures. Increasing the pressure above about 90 bar it was observed that the performance of the KI examined could be considerably reduced. Based on this observation, selected KI chemicals were examined more thoroughly at increasing pressures and at comparable degrees of subcooling to examine the effect of the absolute pressure.

EQUIPMENT, PROCEDURES, AND METHODS

Equipment and Test Fluids

The high-pressure sapphire cell test equipment used during the experiments is illustrated in FIGURE 1 and has been described previously.[1] All experiments were conducted using fresh synthetic seawater (SSW) with composition according to ASTM D1141-52. The hydrate forming hydrocarbon phase was either (1) pure ethane, (2) the synthetic natural gas mixture (SNG) as specified in TABLE 1, or (3) a condensate–SNG mixture as specified in TABLE 2. The ethane gas had a purity of 99.95% and was supplied by Norsk Hydro Special Gas Division, Rjukan, Norway. The SNG was recombined and analyzed by the same supplier (cf. TABLE 1). The condensate (dead fraction) that was recombined with our SNG came from a North Sea condensate field.

Stages of Hydrate Formation and Growth

The hydrate formation process may be separated into three main stages: (1) induction period, (2) slow growth period, and (3) a catastrophic fast growth period, as illustrated in FIGURE 2.

The induction time, $t_i = t_o - t_s$, was measured as the time elapsed from start of stirring until the first visual sign of hydrate crystal formation or the first sign of gas consumption by hydrate formation. These latter observations defined the onset of hydrate formation (t_o) in the system. Start of stirring defined zero time for the experiments (t_s). The slow growth period was measured as the time elapsed from onset of

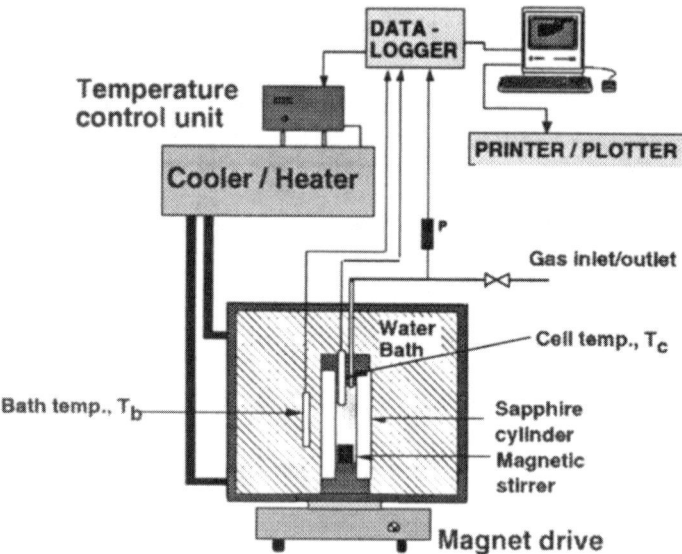

FIGURE 1. Schematic outline of the basic components of the sapphire cell system.

hydrate formation (t_o) until onset of the catastrophic growth process (t_a). The total delay, t_{tot}, of the catastrophic growth process was then defined by

$$t_{tot} = t_a - t_s \text{ (=induction + slow growth).} \quad (1)$$

Induction times and total delays were recorded visually and by plotting pressure or the total gas consumption (TGC) as a function of time (see FIG. 2). The discrepancy between the visually determined onset point of hydrate formation and the onset point determined by gas consumption was normally less than about two minutes for the experiments presented in this paper.

TABLE 1. Composition of the synthetic natural gas used in the experiments

Component	Concentration (mole%)
N_2	0.1
CO_2	1.8
C_1	80.5
C_2	10.3
C_3	5.0
i-C_4	1.6
n-C_4	0.7

TABLE 2. Composition of the North Sea condensate–SNG mixture used in the experiments

Component	Concentration (mole%)
Condensate	21.5
SNG	78.5

Experimental Performance

All experiments were conducted on fresh fluids. The cell was loaded with the actual hydrocarbon mixture to the desired pressure at a temperature just outside the hydrate region while stirring at 700 rpm. When the pressure and temperature had stabilized the stirrer was stopped and the system was cooled to a predetermined experimental temperature, at maximum system cooling gradient (about 25°C/h). The temperature had normally stabilized at the set point about 40 minutes after start of the cooling cycle. The stirrer was started at 700 rpm 60 minutes after start of the cooling cycle. This start procedure was used for all experiments. Start of stirring was defined as start of the experiment.

Determination of the Degree of Subcooling, ΔT

The hydrate equilibrium curves for all the test fluid systems (i.e., dissociation curves) were determined experimentally by a minimum of five different equilibrium points covering the actual pressure region examined. The experimental points were

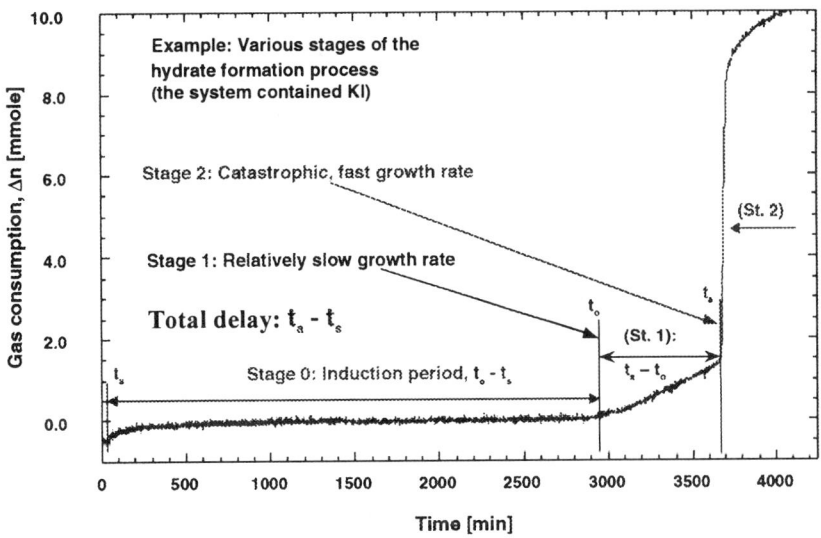

FIGURE 2. The three main stages of the hydrate formation process.

fitted to curves (polynomial- or exponential-curve fits) and the degree of subcooling was calculated according to:

$$\Delta T = T_{eq} - T_{expr}, \qquad (2)$$

where T_{eq} is hydrate equilibrium dissociation temperature at the given pressure and T_{expr} is the experimental temperature at the experimental pressure.

Different curve fits were valid for the pressure regions below and above about 90 bar. For temperatures above 0°C, plotting $\ln P_{eq}$ (P_{eq} is the equilibrium pressure at T_{eq}) as a function of the hydrate dissociation temperature, T_{eq}, two approximately straight lines with a point of deflection at about 90 bar were obtained. When 90 bar is exceeded, the slope of the $\ln P$ versus T curve increases. For fluid systems covering experiment in the pressure regions above and below 90 bar at least three different experimental hydrate dissociation points were measured for each of these pressure regions.

Driving Force: $\Delta\mu$ versus ΔT

A quantitative description of the real driving force for the hydrate formation process is needed to interpret and model kinetic experiments. Some researchers have chosen to use the extent of subcooling below the equilibrium temperature (ΔT) as a measure of the driving force. Bishnoi and his group chose to use ($f_i^{exp}/f_i^{eq} - 1$) as a measure of the driving force.[2,3] Christiansen et al.,[4] proposed to describe the driving force as the change in Gibbs free energy when hydrates form. Others have described the driving force as the change in the chemical potential, $\Delta\mu$, of the reacting species when hydrates form.[1,5,6]

In this paper we study the effect of absolute pressure on the effect of KIs at constant ΔT. The degree of subcooling, ΔT, is given by (2).

The driving force during the hydrate formation process is probably best described by the chemical potential difference, $\Delta\mu$, between the hydrate forming species (i.e., water and gas components). Skovborg[6] expressed a chemical potential of the form:

$$\Delta\mu = RT\left\{-\ln(\gamma_w x_w) + \sum_i v_i \ln\left(1 - \sum_K Y_{Ki}\right)\right\}$$

$$+ RT\left\{\frac{\Delta\mu_o}{RT_o} - \int_{T_o}^T \frac{\Delta h_o + \Delta C_p(T - T_o)}{RT^2}dT + P\frac{\Delta v}{R\overline{T}}\right\}. \qquad (3)$$

Lekvam[5] and Svartaas[1] described the driving force in terms of the chemical potential difference of the gaseous components in the aqueous liquid and the solid hydrate phase at onset of hydrate formation (i.e., at equilibrium):

$$\Delta\mu = RT \cdot \ln\frac{f_{exp}}{f_{eq}}, \qquad (4)$$

where subscript exp refers to experimental conditions and subscript eq to equilibrium conditions, respectively.

Within limited pressure and temperature regions below 90 bar there is an approximately linear relationship between ΔT and $\Delta\mu$. Above 90 bar this linear relationship is not valid. For the high pressure region the hydrate equilibrium temperatures are higher than in the low pressure region. The contribution from the absolute temperature on the chemical potential (4) is thus greater in the high-pressure region (i.e., $P > 90$ bar). Thus kinetic experimental results obtained in the low-pressure region (i.e., $P < 80–90$ bar) can not be directly extrapolated to the high-pressure region on the basis of ΔT values.

RESULTS

Effect of Hydrate Structure on PVCap Performance

The experimental results are shown in FIGURE 3. This figure shows that the basic kinetics of the sI and the sII hydrate forming systems were different. At similar pressures and subcooling ratios, the induction time was shorter for the sI ethane hydrate system than for the sII SNG hydrate system. Similarly the average growth rate during the slow growth period was much higher in the sI ethane hydrate system than in the examined sII hydrate system. This resulted in a shorter slow growth period for the sI system than the sII hydrate system. Over the range of subcooling ratios used to examine the total delay of the catastrophic growth process in the uninhibited sI and sII baseline systems the total delay was about a factor five higher for the sII hydrate system as compared to the total delay measured for the sI hydrate system. The main results are summarized below.

- PVCap gave increased delay of the catastrophic growth process for both the sI and sII hydrate systems.

- At 22.5 bar and a subcooling ratio of about 8.5°C the total delay $(t_a - t_s)$ of the

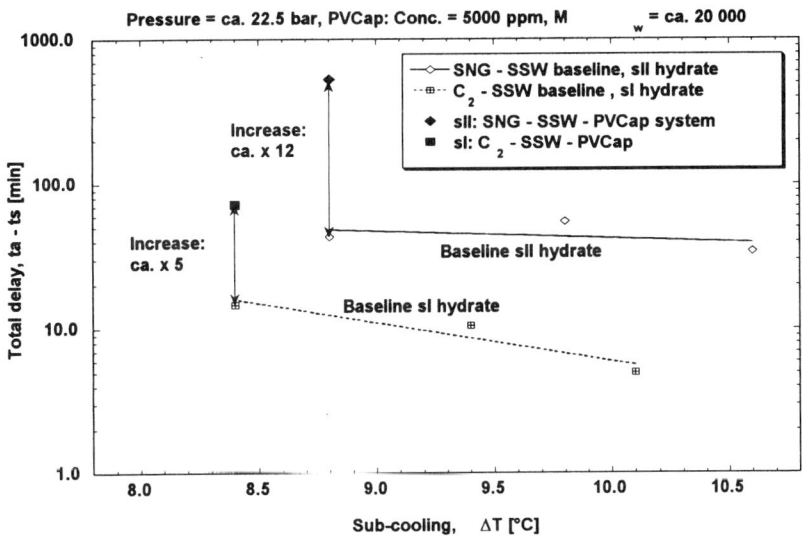

FIGURE 3. The total delay, $t_a - t_s$, of the catastrophic growth period for 5000 ppm PVCap at about 22.5 bar in sI and sII hydrate systems at given subcooling ratios.

catastrophic growth process increased by a factor five, adding 5,000 ppm PVCap to the sI hydrate forming system.

- At similar conditions and concentration in the sII hydrate forming system the total delay of the catastrophic growth process increased by a factor 12.
- PVCap showed a greater KI effect on the sII SNG hydrate system than for the sI ethane hydrate system.

Effect of Absolute Pressure on KI Performance

Variable Pressures, Constant Subcooling Ratio

Experiments were conducted at constant subcooling ratio (ΔT) covering the pressure range between 90 and 200 bar. Graphs showing the main results from these experiments are shown in FIGURE 4. In this figure two different KI systems (KI-1 and KI-2, 6,000 ppm active concentration) are compared. All the experiments were conducted at subcooling ratios in the range 14.4 to 14.7°C. The degree of subcooling was calculated on the basis of an experimentally determined hydrate equilibrium curve for the test fluid system (smoothed experimental curve). FIGURE 4 shows that both KI-1 and KI-2 gave reduced total delays (induction, slow growth, and total) as a function of increasing absolute pressures at constant subcooling ratio.

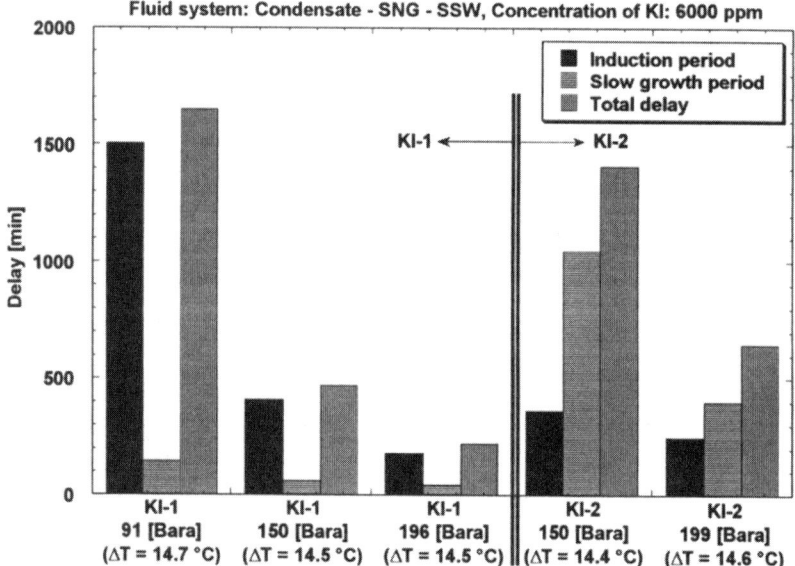

FIGURE 4. Measured induction times (*black bars*), duration of the slow growth period (*horizontal hatched bars*) and the total delay of the catastrophic growth period (*vertical hatched bars*) for two different KI systems (KI-1 and KI-2) at increasing pressures but constant subcooling ratios (ΔT).

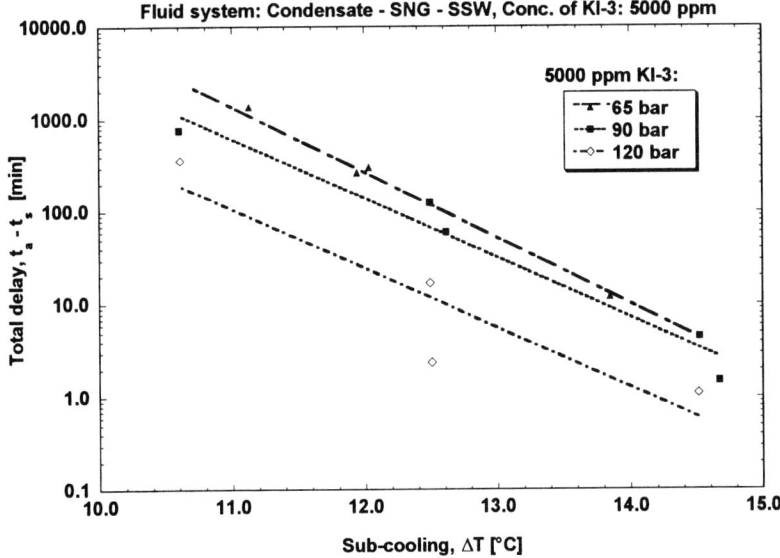

FIGURE 5. Measured total delay of the catastrophic growth period for hydrate inhibitor, KI-3, at various pressures and increasing subcooling ratios (ΔT).

Variable Pressures, Variable Subcooling Ratios

Experiments were conducted at variable subcooling ratios and at three different pressure regions covering the range from 65 bar to 120 bar for the condensate–SNG–SSW system containing 5,000 ppm of the kinetic hydrate inhibitor KI-3. The results from these experiments are presented in FIGURE 5. In this figure measured values of the total delay of the catastrophic growth process are plotted as functions of the subcooling ratio. This figure showed that the total delay decreased with increasing subcooling ratio and that the delay was reduced for increasing pressures. FIGURE 5 illustrates that the effect of increasing the pressure from 65 to 90 bar may be less dramatic on the kinetic performance of a KI-system than an increase from 90 bar to 120 bar.

CONCLUSIONS

Effect of Hydrate Structure on KI Performance

- The hydrate crystal growth rate of a sI hydrate forming system is greater than the growth rate of a sII hydrate forming system. This is probably due to a greater symmetry in the sI hydrate lattice.

- The effect of a KI on hydrate crystal growth is dependent on the interaction between the active groups on the KI molecule and defined growth sites on the hydrate surface.

- PVCap has a greater affinity to the sII hydrate surface than the sI hydrate surface.

Effect of Pressure and ΔT on KI Performance

- At constant initial pressures, P, the performances of the examined KIs decreased for increasing degrees of subcooling (ΔT).
- At constant ΔT the performances of the examined KIs decreased for increasing pressures above 90 bar.
- A driving force described by ΔT does not account for the increased solubility of the hydrate forming gas components in the aqueous phase at increased pressures.
- The chemical potential, $\Delta \mu$, gives a better description of the real driving force during hydrate formation and growth than ΔT.
- A *driving force* described by ΔT is easier to identify during the evaluation of KI effects on real fluids and at real conditions as compared to $\Delta \mu$.

ACKNOWLEDGMENTS

Esso Norge A/S is gratefully acknowledged for their financial support for parts of the work presented (effect of KI on sI and sII hydrates).

Statoil, Norsk Hydro, Norsk Agip, Elf, Conoco, Mobil, BP-Amoco, TR Oil Services/Clariant and NFR (The Royal Norwegian Council for Scientific and Industrial Research) are gratefully acknowledged for their financial support for parts of the presented work (effect of absolute pressure on KI performance).

REFERENCES

1. SVARTÅS, T.M. *et al.* 1996. Kinetics of formation of a multicomponent synthetic natural gas hydrate in an electrolyte solution—effect of kinetic hydrate inhibitors. 2nd International Conference on Natural Gas Hydrates, Toulouse, France.
2. ENGLEZOS, P. *et al.* 1987. Kinetics of formation of methane and ethane gas hydrates. Chem. Eng. Sci. **42**(11): 2647–2658.
3. DHOLABHAI, P.D., N. KALOGERAKIS & P.R. BISHNOI. 1993. Kinetics of methane hydrate formation in aqueous electrolyte solutions. Canad. J. Chem. Eng. **71**: 68–73.
4. CHRISTIANSEN, R.L., V. BANSAL & E.D. SLOAN. 1994. Avoiding hydrates in the petroleum industry: kinetics of formation. University of Tulsa Centennial Petroleum Engineering Symposium, August 29-31, Tulsa. 383–393.
5. LEKVAM, K. 1995. Kinetics of Natural Gas Hydrates. Ph.D. Dissertation, Stavanger College, Department for Technical and Physical Sciences. ISBN label: ISBN.82-7644-018-5.
6. SKOVBORG, P. 1993. Gas Hydrate Kinetics. Ph.D. Dissertation, Department of Chemical Engineering, Technical University of Denmark, Copenhagen, Denmark.

Kinetics of Gas Hydrates Formation and Tests of Efficiency of Kinetic Inhibitors

Experimental and Theoretical Approaches

J.P. MONFORT,[a] L. JUSSAUME, T. EL HAFAIA, AND J.P. CANSELIER

ENSIGC-INPT, Laboratoire de Genie Chimique,
UMR 5503, 31078 Toulouse Cedex, France

INTRODUCTION

Prevention of deposits of gas hydrates in transport lines and facilities in the oil and gas industries has lead to numerous studies of gas hydrate phase equilibrium properties as well as nonequilibrium properties. One of the major targets of research and development in this field has been to find alternative inhibitors to replace the traditional antifreeze agents such as methanol and glycols. To attain this goal, prior knowledge of the intrinsic mechanisms of hydrate formation in water-light hydrocarbon systems is necessary.

When dealing with hydrate kinetics, it is common to use the similarity with the crystallization process, by splitting hydrate formation into various steps. Two main steps are involved: nucleation in the early stage of the formation and crystal growth. Later, other crystallization steps may simultaneously occur such as agglomeration, flotation, and eventually attrition of hydrate particles. These last phenomena depend strongly on the hydrodynamic characteristics of the flow of liquid mixture undergoing hydrate formation in a transport line[1] or agitation of liquid in a crystallizer. For both preliminary steps, understanding the general phenomena remains largely incomplete because of experimental difficulties, which cause a paucity of reliable experimental information on the characteristics of specific gas hydrate–liquid mixtures. Although the kinetics of nucleation and growth have been studied during the past decade, agglomeration of hydrate particles is still unknown.

In this work we report on results of a research program targeted to the kinetics of hydrate formation in pure gas hydrate former–water systems and in chemically inhibited systems. The objective is first to get information on the growth step in order to model hydrate formation kinetics and second to better understand the behavior of inhibiting chemical additives.

In this paper, a short description of the experimental equipment and procedures is given. This is followed by the presentation of a complete experimental study of ethane and propane hydrate kinetics in a stirred reactor. The evaluation of the efficiency of some patented kinetic additives to prevent either nucleation or crystal growth and agglomeration of hydrate particles is finally proposed.

[a]Corresponding author.

EXPERIMENTAL METHODS

Gas Hydrate Stirred Reactor: Pure and Inhibited Systems

Equipment used in the hydrate formation kinetics measurements for pressure up to 30 bar is described in detail elsewhere.[2] The main characteristics are as follows:

- A stirred reactor with variable gas-liquid interfacial exchange area (gas bubbling) and recirculating flows (both gas and aqueous liquid). Gas consumption rates and particles size distribution during hydrate formation are measured in order to quantify the effects of the main driving forces (subcooling and pressure). The particle sizer is equipped with a high pressure cell (100 bar) and is positioned on the recirculating liquid phase, outside the reactor.

- The reactor is operated isothermally or a temperature ramp can be set, pressure is kept constant or can vary, and all parameters and data are controlled and collected by a computer system.

Multiple Additive Screening Equipment

The equipment is based on an original set up proposed by R.L. Christiansen and E.D. Sloan[3] for the multiple screening of chemical inhibitors. The test system is the tetrahydrofuran (THF)–water mixture which at 4°C and 20 wt% THF forms the type II THF-hydrate. The main characteristics are as follows:

A thermostated water bath where test tubes, each one with a ball inside, are filled with the THF-water solution and are agitated with a reciprocating gear motor. Temperature in each test tube is recorded along with the peak exothermicity, which is detected when hydrates form. The video recording of the tubes during stirring is used to visualize the two phenomena:

- appearance of crystals (turbidity), which indicates the induction time; and

- end of hydrate formation indicated by the stopping of the balls in the test tubes and corresponding to the complete blockage time.

The experimental procedure was established by studying the influence of the different parameters: subcooling (ΔT), material of the balls acting as a stirrer and nucleation site inside the tubes, number of hydrate formations, and rate of cooling. For each compound the induction time and the delay of growth/agglomeration of crystals were recorded. A compound is an efficient kinetic inhibitor if the induction time and/or the complete blockage time are longer than these values for the pure THF–water system.

EXPERIMENTAL RESULTS AND DISCUSSIONS

Ethane– and Propane–Water Systems

As a first step work was dedicated to the measurements of gas consumption rates and particle sizing distributions on pure water–ethane or propane systems.[2]

Gas Consumption Experiments

In order to develop a model equation for the hydrate formation rates a parametric study of the gas consumption rates has been carried out within the hydrate stability zone. It shows that the consumption rates are directly proportional to the hydrodynamic parameters, such as stirring velocity ω and gas–liquid interfacial exchange area a_S, and that the determining parameters are the subcooling $\Delta T = T_{eq} - T$ and the driving force, expressed in terms of fugacity difference $\Delta f = f_{eq} - f$, where: T_{eq} is the hydrate formation equilibrium temperature at the experimental pressure; T is the experimental temperature; f_{eq} is the three-phase equilibrium fugacity at the experimental pressure; and f is the gas fugacity at the experimental pressure. Gas fugacities are calculated with the Peng-Robinson equation of state.[4]

The influence of these parameters along constant paths with the range values of the operating conditions appear in TABLE 1.

We used as modelling equations for the rate of formation r, defined as the moles of gas consumed per unit time, the original expression of Vysniaukas and Bishnoi[5]

$$r = A_1 a_s \exp\left(-\frac{\Delta E_a}{RT}\right) \exp\left(-\frac{a}{\Delta T^b}\right) P^\gamma. \tag{1}$$

TABLE 1. Experimental conditions and influence of parameters on hydrate formation rates along constant paths

Parameter	Ethane	Propane
ω (rpm)	250–410	283–410
a_s (cm^2)	50–350	50–415
T (K)	274.15–281.96	273.4–276.71
P (bar)	11–17.35	3–4.75
ΔT (K)	2–8.17	0.6–3.77
Δf (bar)	2–8.15	0.5–1.90
r_{min}, r_{max} (mol/min)	0.0005–0.0038	0.0057–0.032
No. experimental points	52	55

	ΔT (K)	T (K)	P (bar)	Δf (bar)
Variable parameter				
ΔT (K)		⇓	⇑	⇑
T (K)	⇓		⇓	⇓
P (bar)	⇓	⇓		⇓
Δf (bar)	⇓	⇓	⇑	
r variations	⇓	⇓	⇑	⇑

Note: ⇓ decreasing effect, ⇑ increasing effect, r is always increasing with ω and a_s.

On the other part, from the previous analysis of rate consumption data, we propose a semiempirical expression in which two determining factors ΔT and Δf are used, as follows:

$$r = K(\Delta T^2 + \Delta f^2)^{b_1} \omega^{b_2} a_s^{b_3}. \qquad (2)$$

In both equations the distinct constants are estimated by a least squares multiple regression analysis of the data and are shown in TABLE 2. The comparison between calculated and experimental formation rates, is shown in FIGURES 1 and 2, where it appears that this new equation model can represent the formation rate of ethane and propane hydrate in a semi-batch reactor with good agreement to within ±10% of absolute deviation.

In the following section we propose a more detailed description of the growth crystal step during hydrate formation that could be useful for a comprehensive treatment of the gas hydrate formation rate.

Particle Sizing Measurements

The raw information obtained from the on-line particle sizer is the particle size distribution for various adjacent classes with the corresponding level of confidence of the measuring equipment. The detailed description of the measuring procedure as well the algorithm proposed to calculate, during the first stage of the particle size measurements, a mean crystal growth rate is reported elsewhere.[2]

From the cumulative number and distributions of particles the mean diameter of $(\bar{x}_{nl})_i$ particles for each size distribution i is given by:

$$(\bar{x}_{nl})_i = \frac{\sum x_i dn_i}{\sum dn_i},$$

TABLE 2. Results of the fitting of formation rate data with Equation (1)

Parameter	Ethane	Propane
ΔA_1	1.413×10^{-13}	5.29×10^{-12}
ΔE_a (Kj/mol)	-35	-39.99
a	4.65	3
b	1	0.35
γ	1.178	0.715
sd[a]	0.17	0.075
K	5.989×10^{-7}	1×10^{-5}
b_1	0.735	0.54
b_2	0.817	0.973
b_3	0.054	0.154
sd[a]	0.075	0.11

[a]sd represents the standard deviation of $|(x_{exp,i} - x_{calc,i})|/x_{exp,i}$ for each experimental point.

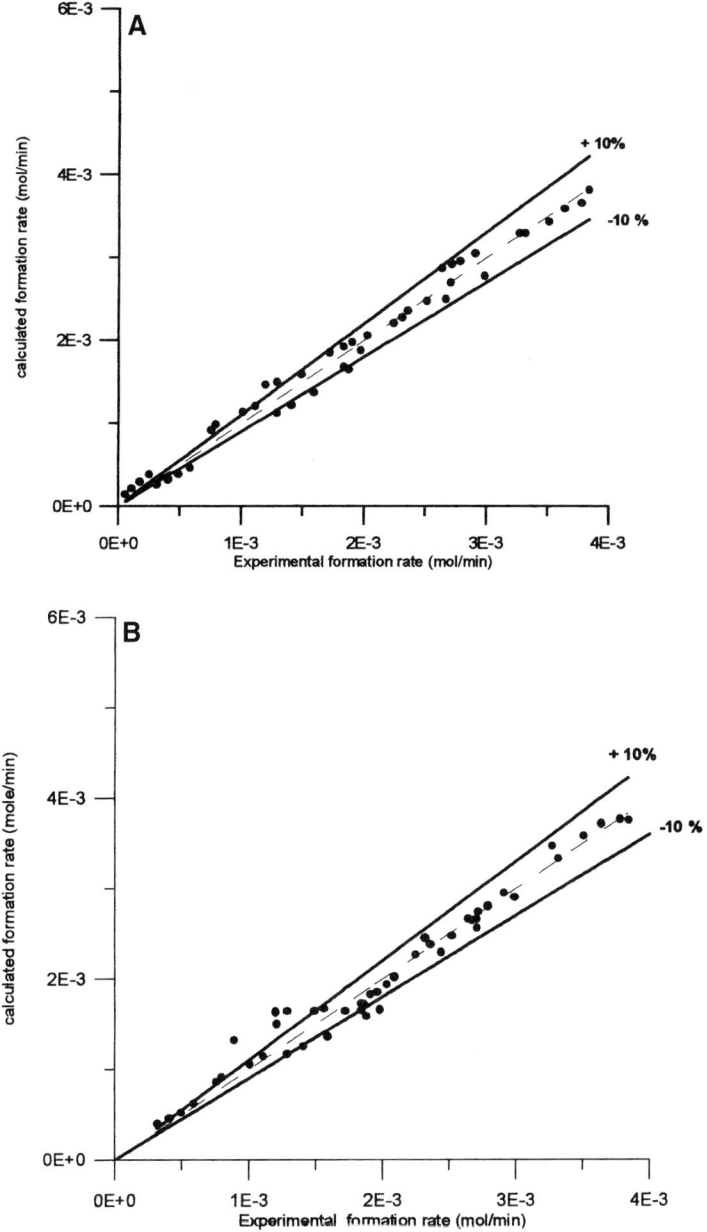

FIGURE 1. Experimental and calculated ethane hydrate formation rate, modelling equations:
(**A**) $r = A_1 a_s \exp\left(\dfrac{\Delta E_a}{RT}\right)\exp\left(-\dfrac{a}{\Delta T^b}\right)P^\gamma$ and (**B**) $r = K(\Delta T^2 + \Delta f^2)^{b_1}\omega^{b_2}a_a^{b_3}$.

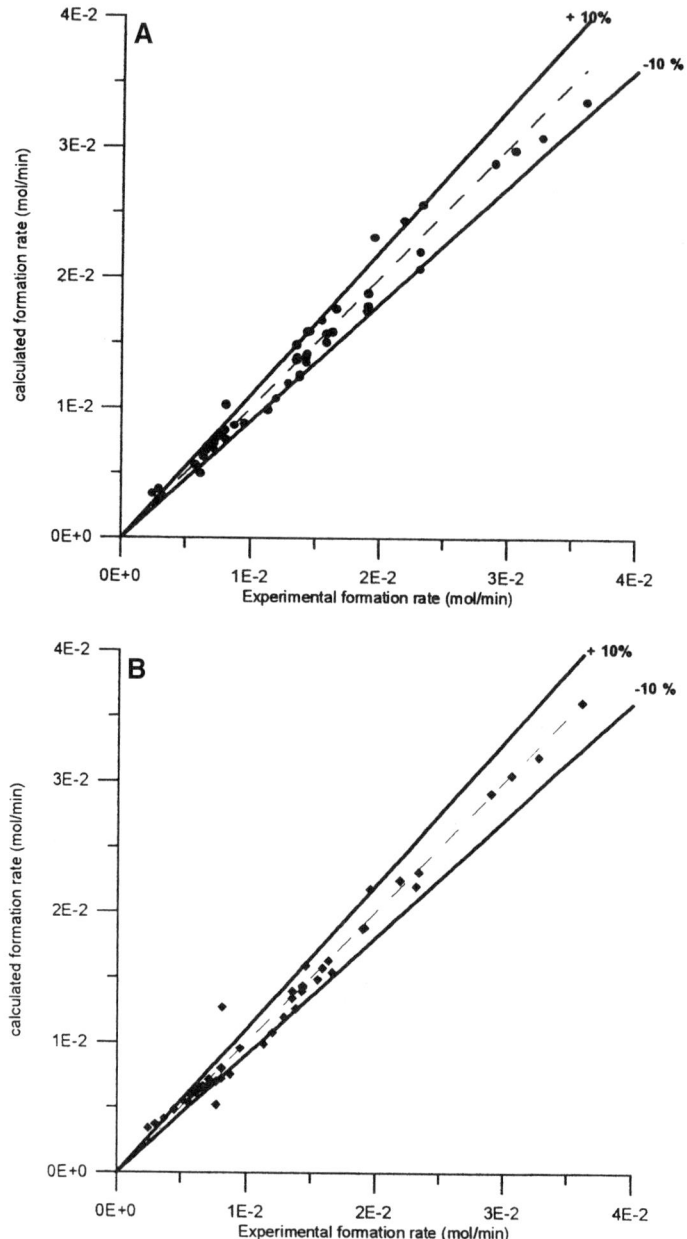

FIGURE 2. Experimental and calculated propane hydrate formation rate, modelling equations: **(A)** $r = A_1 a_s \exp\left(-\dfrac{\Delta E_a}{RT}\right)\exp\left(-\dfrac{a}{\Delta T^b}\right)P^\gamma$ and **(B)** $r = K(\Delta T^2 + \Delta f^2)^{b_1} \omega^{b_2} a_a^{b_3}$.

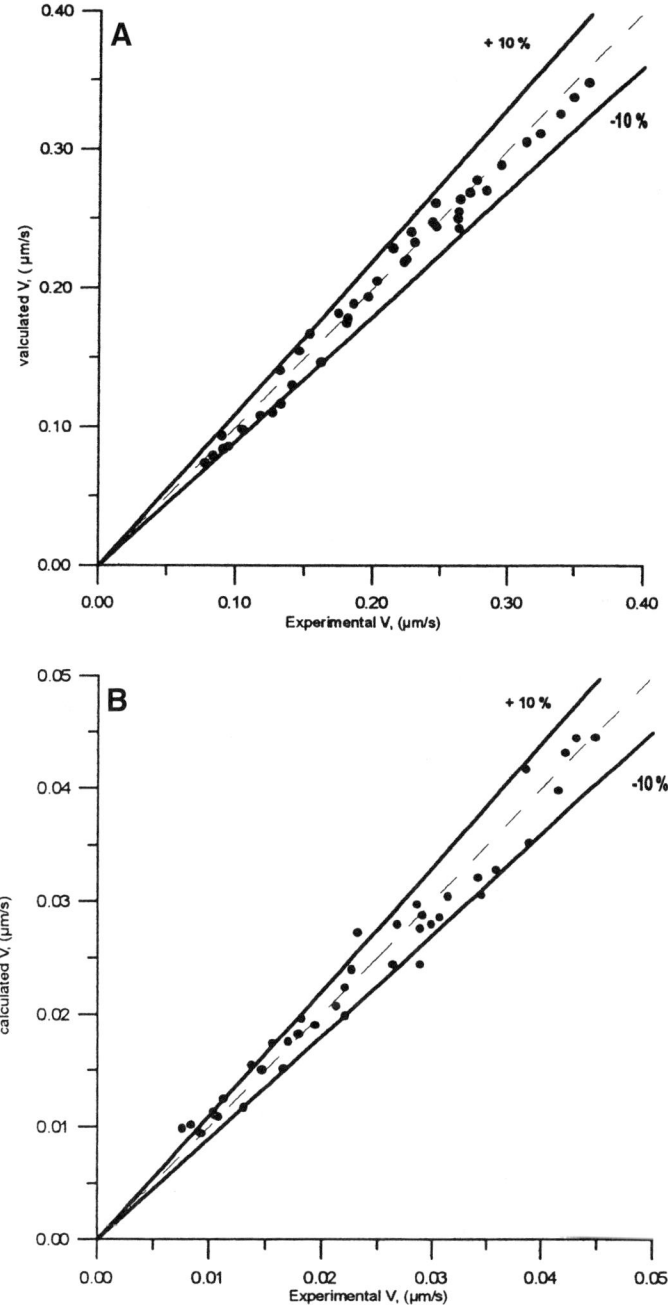

FIGURE 3. Experimental and calculated growth rates V for (**A**) ethane and (**B**) propane.

where x_i is the particles diameter whose number is n_i. The mean diameters $(\bar{x}_{nl})_i$ were calculated for a set of distinct operating conditions (T, P, and stirring velocity ω). It has been observed that the mean diameters increase linearly with time and that the size distribution of a fixed population of crystals moves towards classes of larger sizes with time.[2] It is thus hypothesized that the weak increase of the mean diameter of a population with time correspond to a crystal growth step that is minor compared to the agglomeration step.

The mean growth rate of particles $\bar{V}$ is expressed by the following expression:

$$\bar{V} = \frac{d(\bar{x}_{nl})_i}{dt}.$$

The mean growth rate of particles is therefore constant and is independent of the particle size during the reaction step, where no agglomeration is observed. The experimental results for the calculation of $\bar{V}$ corresponding to a large set of experimental conditions are reported in TABLE 3. We propose to apply the equation of Palwe et al.[6] to represent the linear growth rate $\bar{V}$. In this present case no crystal size contribution appears in the equation:

$$\bar{V} = K \exp\left(\frac{-\Delta E_a}{RT}\right) w^p \Delta f^g. \qquad (3)$$

The comparison between calculated and experimental crystal growth rates is shown in FIGURE 4, and values of experimental conditions and parameters appear in TABLES 3 and 4, respectively. For both hydrates rates the proposed equation gives a fairly good representation; it is used below for a general simulation treatment of the crystallization of ethane and propane hydrates.

Water–THF–Chemical Additives Tests

In order to test the efficiency of various kinetic additives in the reactor a preliminary selection of potential additives is carried out with the screening equipment described in the experimental section. More than forty compounds were tested, both hydrosoluble polymers and surface active agents. In parallel a molecular simulation study of several chemical structures showed specific trends,[7] that helped to select the molecules we had to test. The main results appear in FIGURE 4, they correspond to a 0.25 wt% additive concentration. In the case of PVCap, no hydrate formation occurs before six hours, which corresponds to the upper limit of the experiments.

TABLE 3. Experimental conditions at constant a_s (50.3 cm^2) corresponding to crystal growth measurements

Parameter	Ethane	Propane
ω (rpm)	202–341	202–341
T (K)	274.36–279.5	273.65–276.80
Δf (bar)	2–8.15	0.7–2.25
$\bar{V}_{min}$, $\bar{V}_{max}$ (µm/s)	0.078–0.349	0.0076–0.045
No. experimental points	46	40

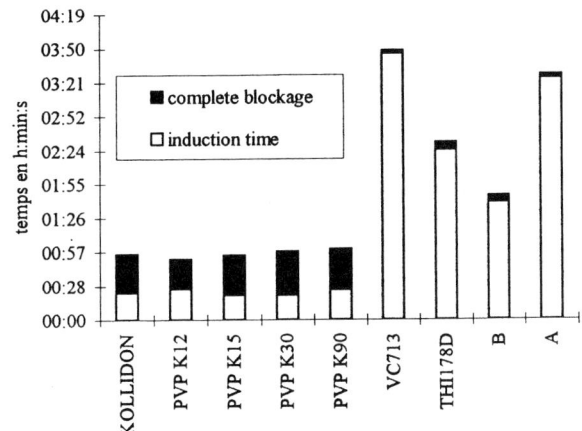

1) **PVPs**: polyvinylpyrrolidone

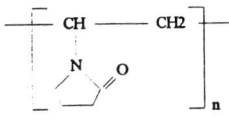

Kollidon: molecular weight 10000, manufactured by BASF
PVP K12: molecular weight 8000, manufactured by ACROS
PVP K15: molecular weight 10000, manufactured by Fluka
PVP K30: molecular weight 40000, manufactured by Fluka
PVP K90: molecular weight 360000, manufactured by Fluka

2) **VC 713** : terpolymer vinylpyrrolidone, vinylcaprolactam and dimethylaminoethhyl methacrylate

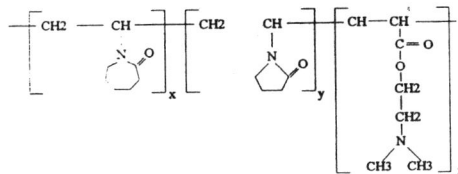

3) **THI178D** : mixing of different compounds including VC713.
4) **PVCap** : polyvinylcaprolactam

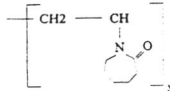

5) **A, B** . water soluble polymers

FIGURE 4. Efficiency of polymers in THF–water hydrate system.

TABLE 4. Results of the fitting experimental crystal growth rate data in µm/s with Equation 3

Parameter	Ethane	Propane
K	6.38×10^{-9}	4.81×10^{-11}
ΔE_a (Kj/mol)	−34.77	−39.99
p	0.392	1.31
g	0.851	1.14
sd[a]	0.032	0.06

[a]sd represents the standard deviation of $|(x_{exp,i} - x_{calc,i})|/x_{exp,i}$ for each experimental point.

The qualitative observations from the recording of experiments showed that morphology of crystals can vary if additives are present and in this case hydrate crystals are whiter and larger than in the pure water–HF system. From these results two types of additives have been chosen, PVP for comparison, its kinetic inhibitive properties have been established[8–10] and homopolymers A and B for industrial interest. Preliminary results are reported in the next section for additive A.

Water–Ethane–Chemical Additives Tests

Another point that has been verified is the existence of any correlation between the two types of efficiency results obtained either from the THF screening experiment or from the gas hydrate reactor experiment. Ethane as a gas former has been chosen for the reactor tests. In view of comparing additive hydrate inhibition efficiencies it is necessary to use a standardized reactor test procedure that takes into account the repeatability of experiments. This method is based on a controlled dissociation process and has been reported elsewhere;[8] general trends between these two tests are observed but no precise correlation can be drawn. The main conclusion is that the THF test is not very interesting to quantify the effect of crystal growth or agglomeration inhibitors; it is more appropriate for the study of nucleation inhibitors and it is probably useful to rank the efficiency of a polymer on the THF scale but it is not a precise method and a gas hydrate test must be done.

A typical gas consumption curve is shown in FIGURE 5, where efficiencies of three polymers are directly evaluated with respect to the pure ethane–water system. The next results are related to experimental measurements of crystal size distribution and concentration from the particle sizer.

Inhibitor A has been tested at 18.5 bar and 281.5 K and the influence of the driving force on the efficiency at 0.001 wt% of A evaluated. For this case study a more realistic definition of the driving force was used. It is expressed in term of the chemical potential difference $\Delta\mu$, as follows:

$$\Delta\mu = RT \ln\left(\frac{f}{f_{eq}}\right).$$

The results are presented in TABLE 5. It can be seen that an increase of the driving force with or without additive induces a higher particles concentration, as well as more consumed gas and hydrate formation is more violent. However, the inhibitor

TABLE 5. Influence of the driving force on pure and A inhibited system $P = 12$ bar at 0% wt A and 18.5 bar at 0.001% wt A

Driving Force (J/mol)	Creation of Particles (part/ml/min)		Growth of Particles (µm/min)		Gas Consumption (10^{-3} mol/h)	
	0%	0.001%	0%	0.001%	0%	0.001%
377	52		0.499		1.2	
615	2690		1.494		5.8	
680		2262		1.306		2.5
717	4200		1.929		8.9	
753		8556		0.702		7.6
838		14726		0.744		13.8

effect on particle growth, is opposite: growth rate is reduced when the driving force is increased. The general trend is that, with identical driving force, this inhibitor reduces the creation and the growth of particles.

Results of the inhibitor concentration effect on the inhibition, at constant value of $\Delta\mu$, appear in TABLE 6. For concentrations higher than 0.1% weight the induction time is extremely high and gas consumption variations are below the limit detection of the equipment. The additive concentration is observed to have a strong effect both on the nucleation phenomenon and on the growth/agglomeration steps. The activity of this kinetic inhibitor is demonstrated: the creation of particles is reduced by half at the minimum 0.001% weight concentration. Furthermore, the growth rate indicates that the existence of a critical concentration, corresponding to a delay of the growth of particles. Indeed, at the value of 0.001 wt% the growth rate is 1.941 µm/min even when it is 0.933 µm/min for pure system. The efficiency is still high at 0.1% weight concentration. This may correspond to a minimal covering of the crystal surface blocking sufficient growth sites. A special experimental test is necessary to verify this hypothesis.

Finally the effect of inhibitor concentration on the subcooling corresponding to the hydrate formation has been investigated. Indeed, pure water–gas systems were studied at different pressures and it appears that a fixed subcooling or driving force exists for each pressure corresponding to a limit hydrate formation. In other words, hydrate formation cannot occur at a lower subcooling, hydrates form at this fixed

TABLE 6. Influence of the concentration of A on pure and inhibited systems at $P = 18.5$ bar and $\Delta\mu = 680$ J/mol

Concentration of A (wt%)	Creation of particles (part/ml/min)	Growth of particles (µm/min)	Gas consumption (10^{-3} mol/h)
0	5400	0.933	23.9
0.001	2262	1.941	4.9
0.1	724	0.636	0.4

TABLE 7. Influence of the concentration on the limit hydrate formation

Concentration (wt%)	Driving Force (J/mol)
0	690
0.001	838
0.003	1061
0.01	1093
0.05	1316

temperature for a fixed pressure. It is observed that when more inhibitor is added the greater the driving force corresponding to limit hydrate formation. The results are presented in TABLE 7. This kind of experiment can be used to measure the efficiency of an hydrate inhibitor more quickly than the measurement of induction time.

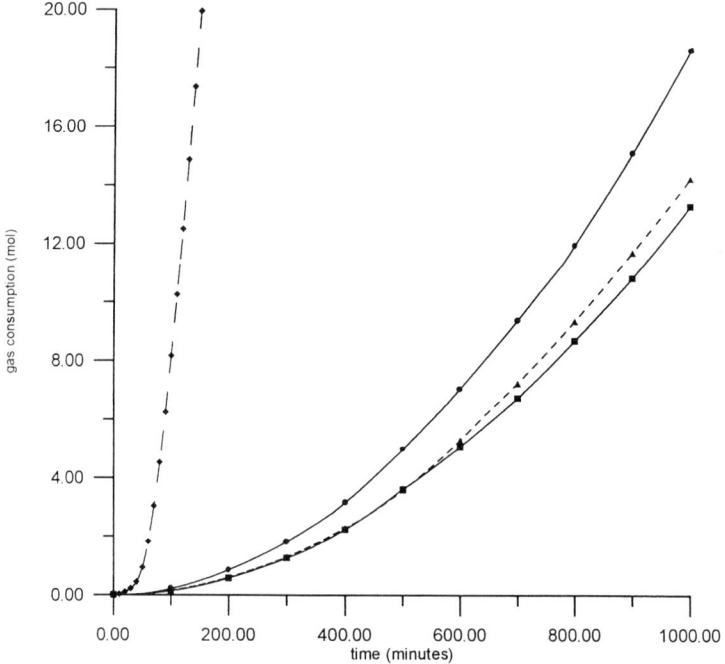

FIGURE 5. Gas consumption versus time for distinct 0.05 wt% polymers, $P = 12$ bar and $T = 277$ K. Key: ■, A; ●, VC7 13; ▲, PVP K 15; ◆, no inhibitor.

CONCLUSIONS

We have developed new correlations for an agitated semi-batch gas hydrate crystallizer that enables us first, to evaluate the crystal growth rate thanks to particle sizing analyzer and second, to estimate the formation rate of ethane and propane hydrates. This mathematical tool could be useful for testing the efficiency of kinetic inhibitors. Work on this is currently in progress.

On the other hand, an experimental procedure for quantifying the crystal growth—originally proposed for pure gas–water systems—has been successfully applied to chemically inhibited systems.

In conclusion, the measurement of hydrate particles sizes and crystal population is a complementary technique to an easier way of evaluating additives efficiencies by means of gas consumption measurements.

ACKNOWLEDGMENTS

The authors are thankful to Elf ep, CNRS programme ECODEV and Midi-Pyrénées Regional Authorities for the financial support for this research.

REFERENCES

1. GAILLARD, C., J.P. MONFORT & J.L. PEYTAVY. 1999. Investigation of methane hydrate formation in a recirculating flow loop: modelling of the kinetics and tests of efficiency of chemical additives on hydrate inhibition. Oil & Gas Science and Technology-Rev. IFP **54**(3): 365–374.
2. EL HAFAIA, T. & J.P. MONFORT. 1999. On-line gas hydrate particle sizing: measurement of hydrate growth in a crystallizer. 1999 AIChE Spring National Meeting, Houston, March 14–18.
3. CHRISTIANSEN, C. & E.D. SLOAN. 1994. Mechanisms and kinetics of hydrate formation. Ann. N.Y. Acad. Sci. **715**: 283–305.
4. PENG, D.Y. & D.B. ROBINSON. 1976. A new two-constant equation of state. Ind. Eng. Chem. Fundam. **15**(1): 59–64.
5. VYSNIAUKAS, A. & P.R. BISHNOI. 1985. Kinetics of ethane formation. Chem. Eng. Sci. **40**: 220–303.
6. PALWE, B.J., M.R. CHIVATE & N.M. TAVARE. 1985. Growth kinetics of ammonium nitrate crystals in a draft tube agitated batch crystallizer. Ind. Eng. Process Des. Dev. **24**: 914–921.
7. JUSSAUME, L., J.P. CANSELIER, J.P. MONFORT & M. RODGER. 1998. Molecular modelling of the interaction of additives on type II gas hydrate surfaces: development of an efficiency scale. *In* Applying Molecular Modeling And Computational Chemistry. AIChE Topical Conference, November 15–20, Miami Beach. 346–352.
8. JUSSAUME, L., J.P. CANSELIER & J.P. MONFORT. 1999. Efficiency of hydrate kinetic inhibitors: experimental procedure and testing method in a crystallizer. AIChE Spring National Meeting, Houston, March 14–18.
9. LEDERHOS, J.P., J.P. LONG, A. SUM, A.R.L. CHRISTIANSEN & E.D. SLOAN. 1996. Effective kinetic inhibition for natural gas hydrates. Chem. Eng. Sci. **51**: 1221–1229.
10. OKUI, T., Y. MAEDA & S. KONO. 1998. Kinetic control of methane hydrates in drilling fluids, in methane hydrates resources in the near future? Proceedings, JNOC International Symposium, October 20–22. 293–299.

Study of Methane Hydrate Inhibition Mechanisms Using Copolymers

B. CINGOTTI,[a] A. SINQUIN,[b] J.P. DURAND, AND T. PALERMO

Institut Français du Pétrole, Division Chimie et Physicochimie Appliquées, 1-4 Avenue Bois-Préau, 92 852 Rueil-Malmaison, Cedex, France

INTRODUCTION

In multiphase production and transport of hydrocarbon resources, gas, water, and eventually a liquid hydrocarbon phase (condensate or crude) are in contact under relatively high pressure. In cold areas and subsea pipelines, these systems can reach rather low temperatures leading to hydrate crystals that plug production facilities.

To prevent gas hydrate formation, different techniques are already used. For example, hydrate formation is avoided by physical actions on the system like insulation and/or heating the pipeline, depressurization, water removal, and others. These solutions are often not sufficient, especially in the case of a shut down in subsea production. A long residence time in a subsea pipe leads to a prolonged exposure of the system at very low temperatures. Another widespread solution is the injection of thermodynamic additives such as methanol in large amount (up to 30% with respect to produced water). These products present the drawbacks of being expensive in terms of the large quantities of additives used and also the resulting pollution.

To overcome this crucial industrial problem, a new category of additives has been developed, kinetic inhibitors. These additives are designed to act on the different steps of hydrate formation,[1] by delaying nucleation, slowing down their growth, and avoiding agglomeration. Kinetic inhibitors are generally water soluble compounds that are effective at low concentrations (less than 1% by weight with respect to water).

Using a semibatch reactor equipped with a turbidimetric sensor to determine the evolution of the particles size distribution (PSD),[2] we apply a new experimental procedure that has a reproducible induction time (t_i) even when no solid nucleation agents are present. This experimental procedure is based on the control of hydrate nuclei using the turbidimetric sensor. With this new experimental procedure, we have studied the effect of various hydrosoluble polymers on induction time and on the PSD.

Telecommunication.
[a]Voice: 01 47 52 71 55; fax: 01 47 52 70 58. beatrice.cingotti@ifp.fr
[b]Voice: 01 47 52 71 57; fax: 01 47 52 70 58. anne.sinquin@ifp.fr

EXPERIMENTAL SETUP

Our experimental setup is shown in FIGURE 1 and has been described in detail elsewhere.[3,4] The liquid phase (pure water or polymer solution) is placed in a stainless steel autoclave and pressurized with methane. The reactor is equipped with two sapphire windows to observe hydrate particles. The temperature of the system is controlled by circulating a water/glycol mixture in a cooling jacket surrounding the autoclave. Pressure control is achieved by the measurement of the differential pressure ΔP between the reactor and a ballast pressure reference P_{ref}. A flowmeter placed on the input gas line is directly connected to an integrator and gives the cumulated gas consumption during the test. The turbidimetric sensor is a UV-visible analyzer

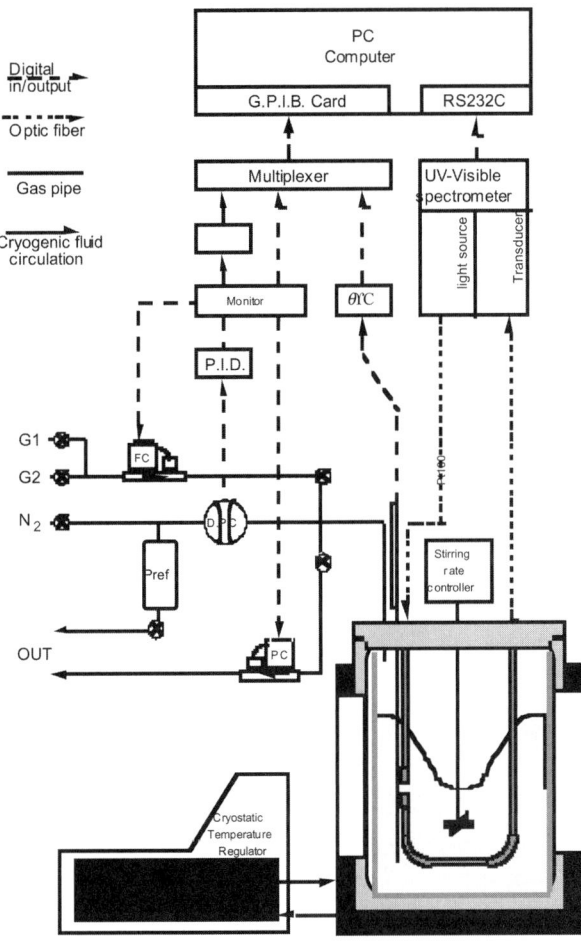

FIGURE 1. Experimental setup.

that measures, *in situ*, the attenuation of a polychromatic beam in the wavelength range 230–750 nm. Using this experimental setup, we can perform experiments with a maximum working pressure of 120 bar and at ambient temperature down to 0°C. The data recorded are temperature and pressure in the cell, cumulated gas consumption, and turbidity of the media at different wavelengths.

The turbidity measurements can be used as a particle sizing method[2] and an *in situ* PSD of the hydrate suspension can be obtained within the particle diameter range 10–100 μm. Turbidity measurements are possible only if the medium is not too concentrated in particles. Consequently, PSD evolutions are determined only at the early stage of the reaction when the number of solid particles is not too high and at moderate supersaturation.

All the experiments described in this paper were made with ultra-pure water (conductivity, 18.2 Ω·m) or ultra-pure water + kinetic inhibitors and methane with a purity of 99.9% provided by Air Liquide.

EXPERIMENTAL PROCEDURE

The aim of our procedure is mainly to obtain reproducible induction time (t_i) without adding any impurity, in order to estimate the efficiency of kinetic additives. Our new test procedure is based on the control of the number of nuclei in the solution. The idea is to substitute impurities by hydrate nuclei. The turbidimetric sensor is the essential tool to follow the formation and above all the dissociation of hydrates. Our procedure comprises three phases:[4] phase 1, first formation; phase 2, dissociation and *over dissociation*; phase 3, second formation.

FIGURE 2 illustrates the experimental procedure and FIGURE 3 shows the corresponding evolution of pressure and turbidity at the wavelength 500 nm versus time.

Phase 1: First Formation

The reactor is filled with ultra pure water at ambient temperature, purged three times with methane, the third time the pressure is maintained at 20 bar (point I in FIG. 2). The temperature is then decreased to 1.0°C under isobaric conditions, to reach the LP point (LP, low pressure in FIG. 2). The temperature is kept constant during the rest of the procedure. Then the pressure is increased to the very high pressure point (VHP point, FIG. 2) 70 bar and stirring is started at a very high rate. The driving force for hydrate formation is very high and hydrates are formed within a few minutes. Consequently, the turbidity of the medium increases sharply (FIG. 3).

Phase 2: Hydrate Dissociation and Over Dissociation

The dissociation step is achieved by a decompression from the VHP to LP (FIG. 2). Since the system is no longer in the hydrate formation zone, hydrates melt slowly and the turbidity of the medium decreases to zero (point V in FIG. 3). When no more particles are visible in the reactor, a time called the *over dissociation* time (t_{ov}) is measured. During this period, the hydrate melting process is still going on and crystals become nucleï. At the end of t_{ov}, the pressure is increased again to reach

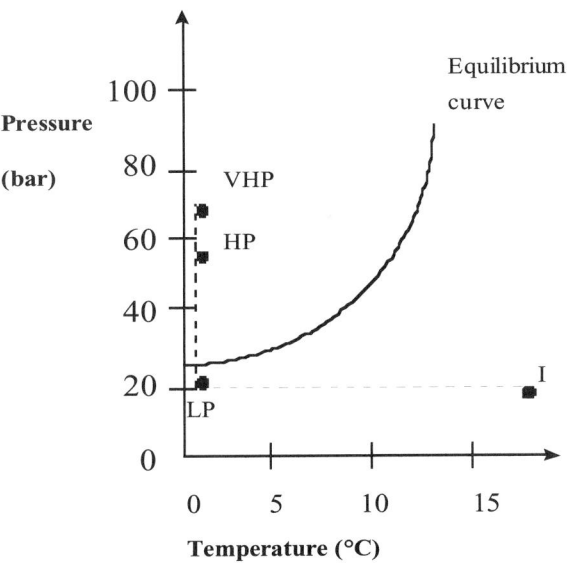

FIGURE 2. Schematic representation of the experimental procedure.

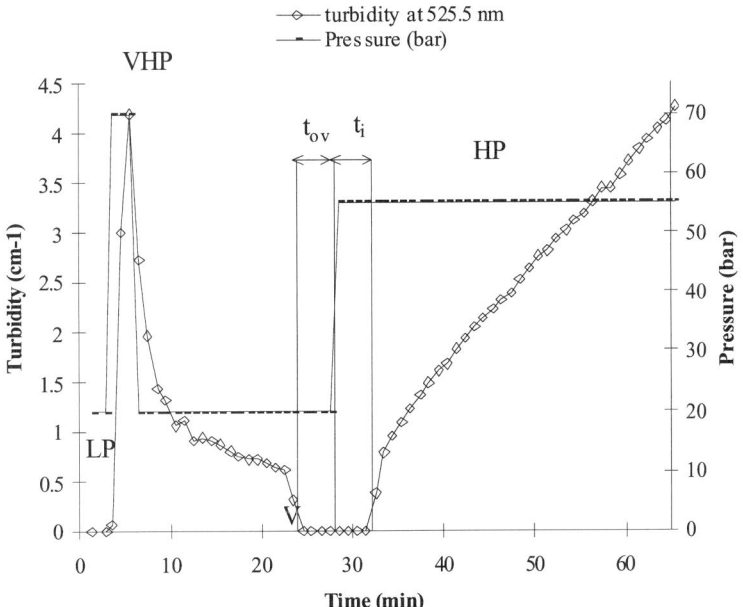

FIGURE 3. Evolution of pressure and turbidity.

the point HP in the hydrate formation zone. This last step represents the end of the dissociation phase.

Phase 3: Second Formation

After a period that corresponds to the induction time t_i, at the HP pressure, hydrates appear again. The variations of turbidity and gas consumption are then recorded. The end of phase 3 corresponds to a total extinction of the turbidimetric signal. New cycles of phase 2 and 3 can be repeated throughout the day.

EXPERIMENTAL RESULTS

Various copolymers were studied using this experimental setup and this procedure. These copolymers, provided by SNF Floerger (St. Etienne, France), include two different monomers, acrylamide (AA), and sodium acrylamidomethylpropanesulfonate (AMPS). The influence of AMPS molar ratio and the influence of copolymer concentration are described below. For each study, the influence of copolymers on t_i and on the PSD evolution is discussed, in comparison with a test without additive. All the experiments were performed using the experimental procedure described above with a t_{ov} of 5 min, at 400 rpm, 1°C and 45 bar.

Influence of the AMPS Molar Ratio

Copolymers of similar molecular mass with AMPS molar ratios varying from 25% to 80% were tested in the reactor at a concentration of 0.15% wt.

Induction Time

The results concerning t_i are given in TABLE 1. All the additives increase t_i but this increase is not linear with the AMPS molar ratio. An enhanced efficiency is observed for the copolymer containing 40% of AMPS, and in this case, t_i is multiplied by four regarding the blank test.

Granulometric Measurements

When hydrates start to form, the PSD evolution versus time were studied for all the copolymers and for the blank test. Typical results of mean diameter ($\bar{D}$) and

TABLE 1. Influence of AMPS molar ratio on induction time (45 bar, 1°C, 400 rpm)

Products	t_i (min)[a]	Standard Deviation
Water	4.5	1.1
25% AMPS	12.3	0.577
40% AMPS	20.5	2.12
50% AMPS	15.0	4.47
60% AMPS	13.0	1.73

[a]Each value is the mean of at least three experiments.

particle number (N_p) evolution are given FIGURE 4. These graphs show clearly that all the copolymers have an effect on the PSD. The first observation is that if an additive has an efficiency in delaying hydrate crystallization, it will also act on the size and the number of particles formed. In other words, AA/AMPS copolymers delay hydrate formation and interact strongly with crystals when they start to form. In FIGURE 4, we can see two types of behavior. The additives containing 25 and 60% of AMPS have constant $\bar{D}$ and N_p rapidly becomes higher than for the blank test. Due to the faster evolution of N_p, the upper detection limit is reached earlier and,

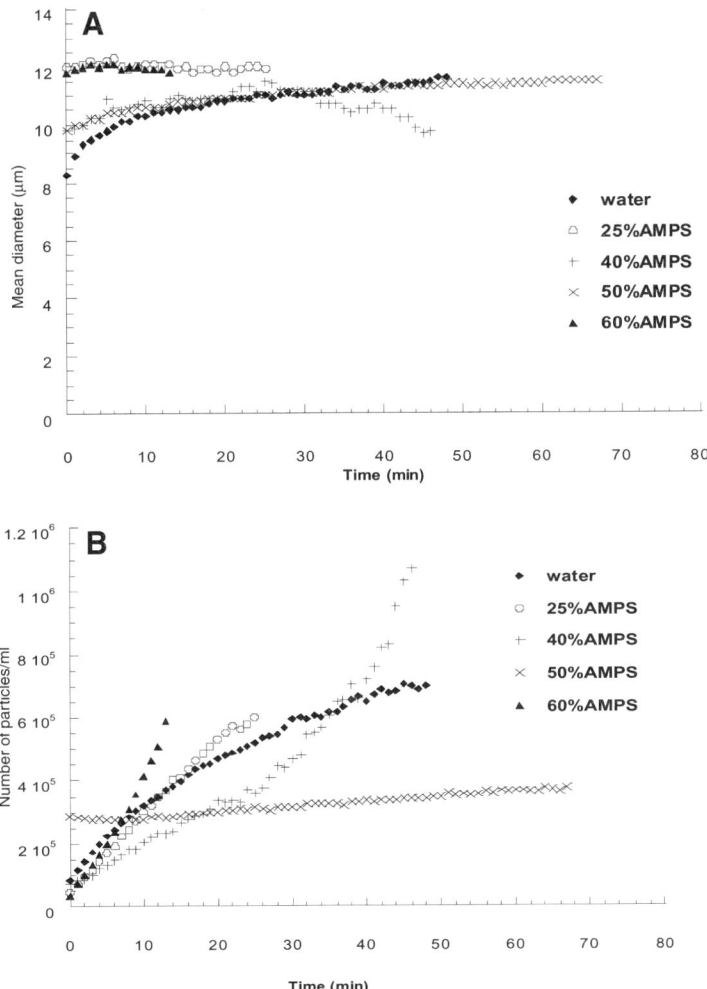

FIGURE 4. Influence of AMPS molar ratio on (**A**) the mean diameter ($\bar{D}$) and (**B**) the number of particles (N_p) (45 bar, 1°C, 400 rpm).

consequently, the PSD analysis is performed in a shorter time than in the blank test (less than 20 min for these copolymers and about 50 min for water). On the other hand, with the 40 and 50% AMPS copolymers, $\overline{D}$ is similar to that measured in the blank test at the beginning of the crystallization. After 30 min, with the 40%AMPS copolymer, $\overline{D}$ decreases strongly from about 12 to 8 μm. The corresponding N_p, which is notably inferior than without additive between 0 and 30 min, increases sharply and becomes greater than for the blank test. We conclude that this copolymer is no longer efficient on hydrate crystal growth after 30 min.

The evolution of $\overline{D}$ versus time for the 50%AMPS copolymer and blank test are identical but this copolymer exhibits a very surprising evolution of N_p. The number of particles seen by the sensor at the onset of the formation is very high (about 3×10^5) and remains almost constant throughout the growth stage. Conventionally, hydrate formation in the reactor begins in the thin layer of water at the gas/liquid interface. In this layer, gas concentration is maximum and primary nucleation starts. The few crystals formed are swept along in the bulk, by the stirring, to seed the bulk. Logically, the number of particles seen by the sensor is almost zero at the beginning and then increases more or less rapidly with the driving force, the stirring rate, and the additive growth inhibition. In the case of the 50%AMPS copolymer, the fact that N_p is very high at the beginning suggests that nucleation does not take place at the interface but directly in the bulk. Consequently, we emphasize that this additive blocks efficient primary nucleation at the interface and nucleation occurs in the bulk when its efficiency decreases or when the driving force increases. When crystals are formed, the fact that N_p remains constant during about 70 min indicates a very strong growth inhibiting effect of this copolymer, which avoids an explosive formation of new hydrate crystals.

In conclusion, all the AA/AMPS copolymers tested delay hydrate formation and the 40%AMPS copolymer is the most efficient additive on t_i. The copolymers with 25 and 60% molar ratio do not have any growth inhibiting effect. On the contrary, when hydrates start to form, the number of particles seen by the sensor is higher than in the blank test. The 40%AMPS has a growth inhibiting effect during 30 min. Finally, the 50%AMPS has a limited effect on t_i but blocks primary nucleation and inhibits growth very efficiently.

Influence of Concentration

The influence of average molecular weight was studied but is not presented in this paper. The influence of concentration has been studied with a 40%AMPS copolymer of lower molecular mass than that presented above, because of the higher concentrations (0.15–1% by weight). The increase of polymer concentration induces a significant increase of viscosity of the solution, but we have verified that there is no significant decrease in the mass transfer constant when viscosity increases from 1.73 mPa.s (water) to 37.8 mPa.s (1% by weight of copolymer). We emphasize that the changes in t_i and in the PSD are essentially due to the polymer/crystal interactions.

Induction Time

A first series of t_i measurements was made at a HP of 45 bar. The results are given in TABLE 2. At 0.3%, t_i is similar to the value obtained in the blank test. Beyond a

TABLE 2. Influence of concentration on induction time (45 bar, 1°C, 400 rpm)

Products	Viscosity of the Additive Solution at 1°C (mPa.s)	t_i (min)^a	Standard Deviation
water	1.73	4.5	1.1
0.15%	10.9	3.0	2.0
0.3%	16.5	4.0	1.3
0.5%	23.2	35	1.41
0.8%	30.6	39	1.41
1%	37.8	23.5	0.707

^{a}Each value is the mean of at least three experiments.

threshold concentration of 0.5% by weight, the copolymer dramatically increases induction time (35 min at 0.5%). The best t_i is obtained at 0.8%, t_i is almost ten times greater than without polymer. For 1%, t_i decreases again. Another series of experiments was performed at HP = 55 bar. Under these conditions, the t_i is detected at a threshold concentration of 1% by weight. From a comparison between induction times measured at 45 and 55 bar (see FIGURE 5), we conclude first that t_i depends strongly on the driving force of the system; and second, that viscosity has no determinant effect on induction time, since it is possible to find a threshold concentration beyond which the polymer is efficient when the pressure increases from 45 to 55 bar.

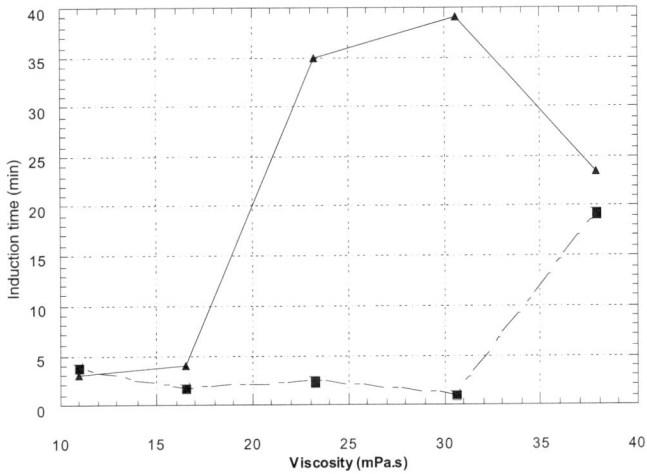

FIGURE 5. Induction time versus viscosity of the additive solutions at ▲ 45 and ■ 55 bar (1°C, 400 rpm).

Granulometric Measurements

FIGURE 6 shows the evolution of $\bar{D}$ and N_p versus time for the experiments performed at 45 bar. The PSD evolution at 0.15 and 0.3%, which corresponds to a t_i that is inferior or equivalent to the blank test, exhibits a decrease of $\bar{D}$ and a sharp increase in N_p, with N_p evolution similar to that determined with water. As with the 25 and 60% AMPS copolymers in the previous section, the acquisition times are much shorter than for the blank test. At higher concentrations (0.5 and 0.8%), the evolution of $\bar{D}$ is again similar to that for the blank test. The experiments can be

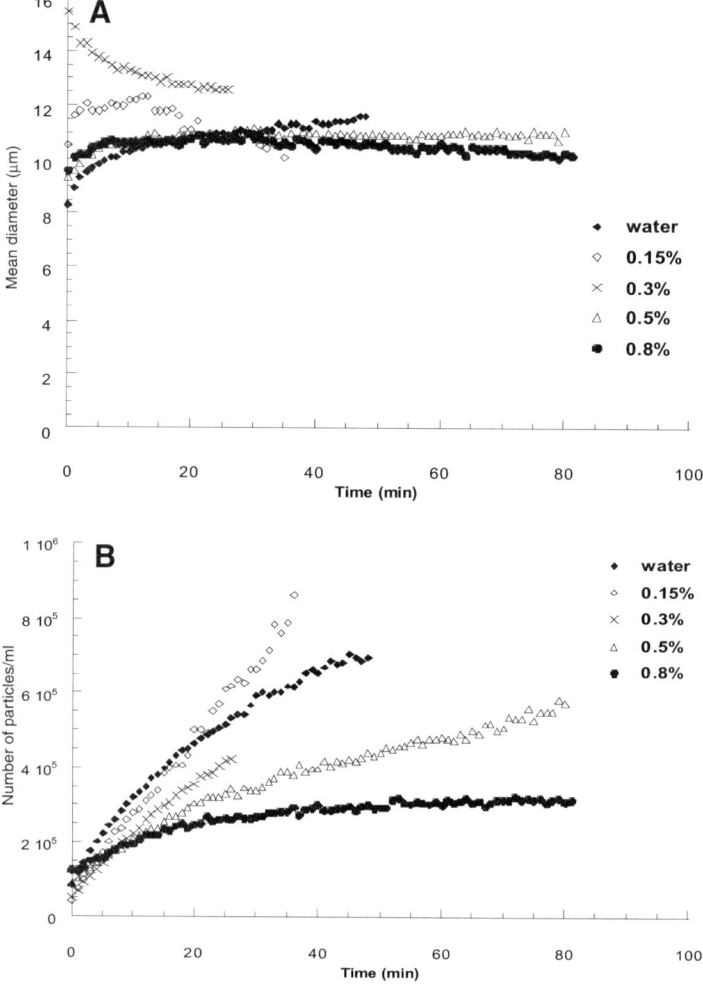

FIGURE 6. Influence of concentration on (**A**) the mean diameter ($\bar{D}$) and (**B**) the number of particles (N_p) (45 bar, 1°C, 400 rpm).

TABLE 3. Influence of concentration on induction time (55 bar, 1°C, 400 rpm)

Products	Viscosity of the Additive Solution at 1°C (mPa.s)	t_i (min)[a]	Standard Deviation
water	1.73	4.3	0.577
0.15%	10.9	3.7	1.53
0.3%	16.5	1.8	1.33
0.5%	23.2	2.3	0.577
0.8%	30.6	1	0
1%	37.8	19.3	8.74

[a]Each value is the mean of at least three experiments.

prolonged longer than with water, because of the much lower rate of increase in N_p. Logically, the increase in N_p is slower at 0.8% than at 0.5%.

In conclusion, we have evidence for a threshold concentration below which the copolymer does not delay hydrate formation. This threshold concentration increases with the driving force applied to the system. At 45 bar, the PSD study shows that below the threshold concentration, the hydrates formed have a greater $\bar{D}$ and N_p increases exponentially. At concentrations beyond the threshold, the evolution $\bar{D}$ of is similar to that for the blank test and the increasing rate of N_p versus time is much lower than with water. The higher the polymer concentration, the slower the increasing rate.

CONCLUSION

A new experimental procedure based on formation/dissociation cycles has been developed. This procedure, based on the control of the number of hydrates nuclei, allows us to attain reproducible induction times (t_i) without adding impurities to the medium.

Various AA/AMPS copolymers were studied using this procedure. Influence of the AMPS molar ratio and polymer concentration have been explored; first, on t_i, and second, on the PSD evolution versus time. At a given molecular mass, all the AA/AMPS copolymers delay hydrate formation and there seems to be an efficiency peak around 40–50% AMPS. The best inhibition performances, with respect to the mean diameter and the quantity of hydrates formed, are obtained with a copolymer containing 50% molar AMPS, which blocks primary nucleation. For a given copolymer composition, beyond a threshold concentration, the copolymer increases induction time, independently of viscosity. The higher the concentration, the higher the inhibiting effect on the PSD, as far as increasing concentration prevents the mean diameter from diminishing and the number of particles from increasing very rapidly.

ACKNOWLEDGMENTS

The authors express their thanks to Total, Elf, and SNF Floerger.

REFERENCES

1. LONG, J.P. *et al.* 1994. Kinetic inhibitors of natural gas hydrates. Proceedings of the Seventy-third GPA Annual Convention. New Orleans, LA.
2. HERRI, J.M. 1996. Etude de la Formation de l'Hydrate de Méthane par Turbidimétrie In Situ. Ph.D. Thesis, University of Paris VI, France.
3. HERRI, J.M. *et al.* 2000. Interest of *in situ* turbidimetry for the characterization of methane hydrate formation: application to the study of kinetic inhibitors. Chem. Eng. Sci. Accepted for publication.
4. CINGOTTI, B. *et al.* 1999. A new method to investigate kinetic hydrate inhibitors. Paper SPE 50757 presented at the International Symposium on Oilfield Chemistry. Houston, Texas, February 16–19.

Kinetics and Mechanisms of Gas Hydrate Formation and Dissociation with Inhibitors

Y.F. MAKOGON,[a,b] T.Y. MAKOGON,[c] AND S.A. HOLDITCH[b]

[b]Texas A&M University, College Station, Texas 77845, USA

[c]Mobil Technology Company, Dallas, Texas 75244, USA

ABSTRACT: A common chemical used in petroleum industry for preventing hydrates is methanol. Other alcohols and glycols (thermodynamic inhibitors) can also be used to shift hydrate formation to lower temperatures and higher pressures. A new family of chemicals called kinetic inhibitors delays the formation of hydrates, but does not change the equilibrium formation conditions. We have constructed several new types of apparatus and present results on the kinetics of hydrate formation and dissociation in static and dynamic conditions with fresh water and different solutions of water, seawater, and with thermodynamic and kinetic inhibitors. We also present new morphological forms of hydrate crystal growth in different static and dynamic conditions.

INTRODUCTION

Hydrates of natural gases are solid inclusion compounds, or clathrates. At certain pressures and temperatures hydrates form in gas production, transportation, processing, and storage facilities. Hydrates are widespread in nature, in near-polar continental regions and in deep-water oceanic locations.

The most important directions of the work on hydrates today are: (1) prevention of hydrate formation and removal of large hydrate plugs from production and transportation systems; (2) search, surveying, and mastering of gas deposits in a hydrate state; and (3) development of effective hydrate-based technologies. Neither an effective method for preventing hydrates nor a technology of using hydrates can be developed without revealing the laws of hydrate formation and decomposition kinetics in a free volume (well, pipeline) and in pore space (hydrate-saturated rock) and without the knowledge of properties of hydrates. This work presents several results from investigating the kinetics of hydrate formation and decomposition with fresh water, seawater, and several thermodynamic and kinetic inhibitors. This study was performed at the Laboratory of Phase Transitions at the Texas A&M University.

Hydrate accumulations can completely block a pipeline and stop the production of transported gas and oil for an extended period. Great expenses are incurred in order to prevent the formation of hydrate plugs. The cost of methanol injected into pipelines by petroleum companies is over one million dollars every day. Water cut in the transported fluid determines the amount of the injected methanol. Water is usually separated and removed from the flow prior to methanol injection. However,

---

[a]Telecommunication. Voice: 409-693-5814; fax: 409-693-5814.
Makogon@spindletop.tamu.edu

in many cases it is too costly to separate water from gas before adding methanol or another inhibitor; for example, when hydrate formation conditions exist between a well head and a platform in deep-water. In a number of situations the use of traditional (thermodynamic) inhibitors becomes overly expensive or environmentally unacceptable. Many laboratories are looking for other effective inhibitors of hydrate formation, such as kinetic inhibitors and antiagglomerants to substitute methanol and other thermodynamic inhibitors.

EXPERIMENTAL SETUP

Experiments were performed in three types of apparatus with volumes ranging from 120 ml to 1,200 ml and full visual control of the processes at pressures up to 210 bar. Pressures were measured with 0.07 bar accuracy by Omega pressure transducers. Temperatures were measured by thermocouples with 0.05°C accuracy. Temperatures in the cells were maintained within 0.1°C of a set point by a THERMOTRON programmable ramp air chiller with a 0.6 m^3 volume. Visual control was performed in different cells through 50 mm and 86 mm diameter windows from two sides in the 120 ml cell and from five or six sides in the large cells. The two cells with 890 ml volumes worked separately or in parallel, in static and dynamic regimes, with a full control of all the process parameters, including the mass of accumulated hydrate and the change of gas volume dissolved in water prior to and after the hydrate formation. 99.99% pure methane and a typical composition natural gas (TABLE 1) were used with pure water and standard composition seawater (TABLE 2), and with solutions of thermodynamic and kinetic inhibitors.

All experiments were run for 6 to 36 days. The initial pressure was 70–170 bar, and the initial temperature was 20–25°C. The cooling rate was maintained at 1°C/hr. Initial magnetic stirrer speed was 380–780 RPM. The stirring speed decreased with time depending on the viscosity, presence of microcrystals in water, composition and structure of fluid. Nucleation and growth of crystals at the interface it was recorded manually by a Nikon-70 camera and automatically by a high resolution RTI CCD 1300U-NS camera. This allowed to monitor crystal growth in dynamic and static conditions with low light.

TABLE 1. Typical natural gas composition

Component	CH_4	C_2H_6	C_3H_8	i-C_4H_{10}	n-C_4H_{10}	N_2	CO_2	C_5H_{12}
Mole%	87.2	7.6	3.1	0.5	0.8	0.4	0.1	0.3

TABLE 2. Standard sea-salt composition (41.953 gr/L)

Component	NaCl	$MgCl_2 \cdot 6H_2O$	Na_2SO_4	$CaCl_2$	KCl	$NaHCO_3$	KBr	H_3BO_3	$SrCl_2 \cdot 6H_2O$	NaF
Mass%	58.490	24.460	9.750	2.765	1.645	0.477	0.238	0.071	0.095	0.007

Note: Density of sea water =1.025 at 15°C, pH = 8.2.

RESULTS AND DISCUSSION

We emphasize that the kinetic inhibitors do not prevent the formation of hydrates, but cocrystallize with small hydrate particles and delay their further growth. Antiagglomerants also do not prevent hydrate formation, but act as a surfactant by dispersing water droplets in liquid hydrocarbon phase and prevent hydrate-coated water droplets from sticking together in the flow. These inhibitors shift the time and location of solid hydrate plug formation. The typical cost of mass-production kinetic inhibitors and antiagglomerants chemicals is comparable to the cost of thermodynamic inhibitors, but their consumption is significantly smaller, thus expenses for treating the hydrate problem decrease.

Knowledge of the details of kinetics of hydrate formation, accumulation, and decomposition processes is the key factor during selection of new inhibitors and creation of new technologies using hydrates. Unfortunately, studies on the kinetics itself of hydrate formation and decomposition processes are very rare. The studies of natural gas hydrate formation kinetics performed during the 1960s through the 1980s[1-7] have revealed a number of previously unknown factors that affect the process of hydrate formation. One such factor is the structural memory of water,[2] when clusters of water molecules arranged as parts of the hydrate lattice remain stable after hydrate decomposition. The presence of such clusters significantly simplifies and accelerates the formation of crystal nuclei and the growth of primary crystals. The stability of these clusters increases with pressure and decreases with superheating of water above the hydrate equilibrium after hydrate decomposition.

Several radically new technologies[8] can be developed based on unique properties of gas hydrates, such as storage and transportation of gas in hydrate state, desalinization of sea water, and compressorless pressure boosting. The effectiveness of such technologies is determined by the kinetic parameters of hydrate formation and decomposition processes.

There are three types of hydrate crystals: massive (see FIGURES 1–3), whiskery (see FIGURE 4), and gelly crystals (see FIGURE 5). Under certain conditions all three types of hydrate crystals can form and coexist. The massive crystals grow because of the adsorption of water and gas molecules on the newly forming crystal surface. The massive crystals can grow in the gas phase and in the liquid phase, but grow more easily in the gas phase. Porosity of the massive crystals can be up to 80–90%, depending on growth conditions. The whiskery crystals grow by adsorption of water and gas molecules moving by diffusion or convection to the base of the crystal through a oscillating fracture (tunnel) between the vessel wall and the base surface of the growing crystal. The whiskery crystals usually start at the points of maximum exerted (developed) capillary pressure, and form both in the liquid phase and in the gas phase. Whiskery crystals are most strong, have the highest density, and dissociate after all other crystals dissociate by increasing temperature. Massive or gel crystals do not grow on the surface of a whiskery crystal. The gelly crystals usually form in the liquid water phase during a decreasing pressure or temperature and a small excess amount of gas dissolved in the water. Gel crystals are very soft and their porosity is near 95–98%.

Nucleation of the crystals takes place at the free gas-water interface (water can be liquid or solid phase). After the coverage of the free gas–water interface the interfacial

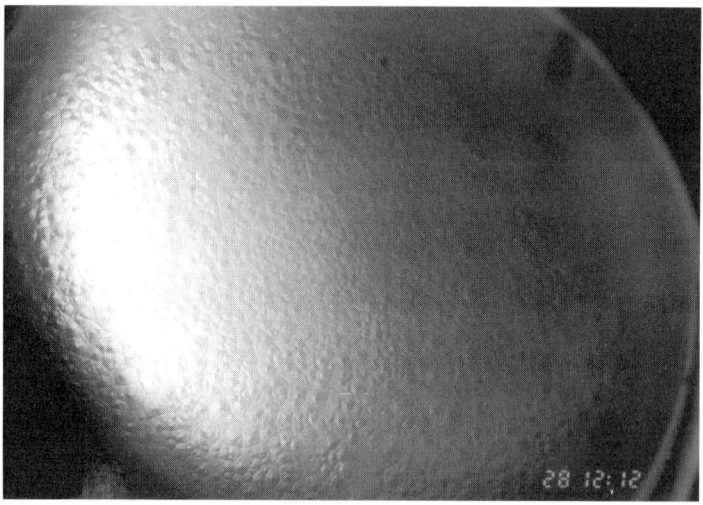

FIGURE 1. Primary methane hydrate film formed in static on a free gas-water surface. $P = 56$ bar, $T = 279.4$ K.

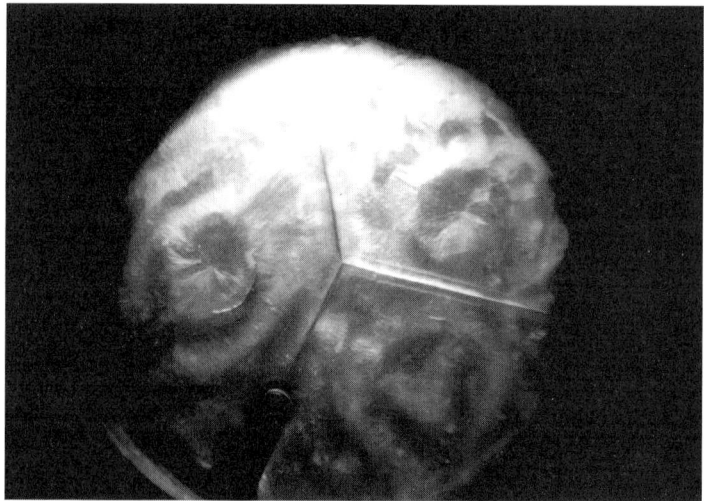

FIGURE 2. Methane hydrate massive crystals formed with 2% methanol in water solution on water surface. $P = 114.8$ bar, $T = 277.7$ K.

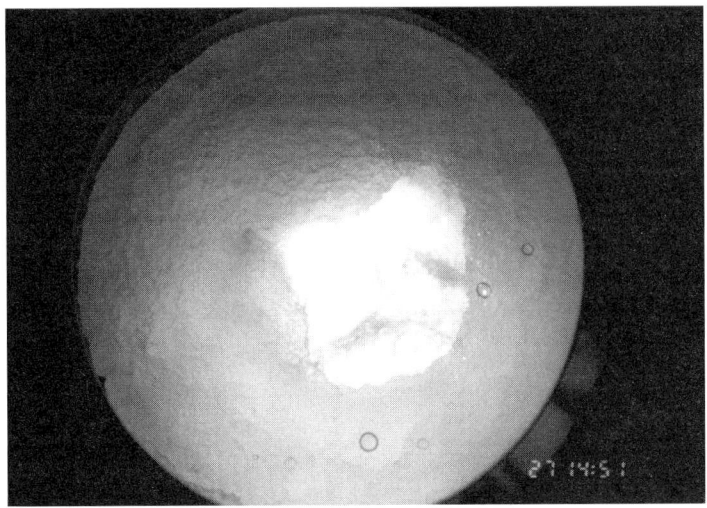

FIGURE 3. Methane hydrate massive crystals growth on cell walls.

FIGURE 4. Methane hydrate whiskery crystals growth with seawater. $P = 81$ bar, $T = 274$ K.

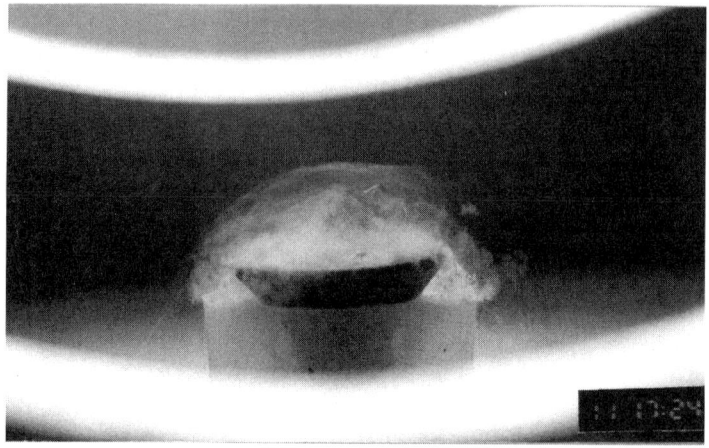

FIGURE 5. Soft gel massive methane hydrate crystals formed in water phase. $P = 93$ bar, $T = 280.6$ K.

process of hydrate formation becomes controlled by diffusion when the hydrate forming molecules diffuse through the porous hydrate film at the gas-water surface. During this process thickness of the primary hydrate film on the interface can remain constant or change, depending on the initial conditions of its formation. Direct measurements showed that the thickness of the primary hydrate film formed at the gas–liquid water interface is equal to the diameter of hydrate critical nuclei at any specified pressure and temperature conditions (from 10 up to 70 micron).

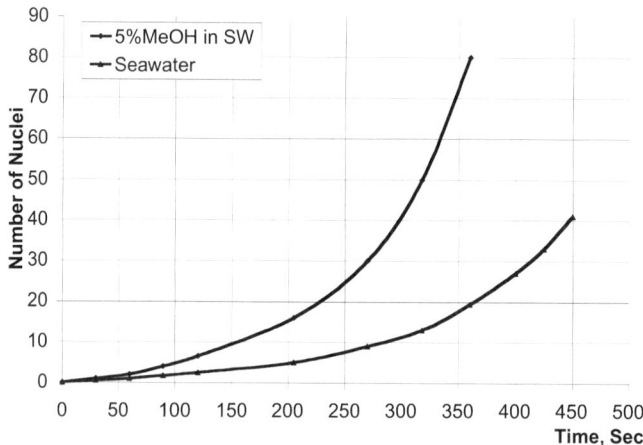

FIGURE 6. Methane hydrate nucleation with seawater and 5% methanol in seawater. S5-S6.II.99.

The process of hydrate formation depends on a number of variables. The major variables are the compositions and phase states of gas phase and water-rich phase, temperature, pressure, amount of supercooling, cooling rate, flow regime (turbulence), and renewal of free gas–water interface, presence and completeness of hydrogen-bonded clusters in water.

We emphasize that the commonly accepted terms of hydrate equilibrium correspond to hydrate melting conditions, and not to the conditions of hydrate formation. Usually a significant amount of supercooling is required for initiation and growth of hydrate crystals. The degree of supercooling depends on numerous parameters, such as pressure, temperature, fluid composition, and thermal history of water forming hydrate.

After making a large number of experiments, we have obtained the functional dependence of the onset of methane hydrate formation and natural gas hydrate formation with pure water, with and without thermodynamic and kinetic inhibitors, on pressure and amount of supercooling. The minimal degree of supercooling at which methane hydrate formed with water was about 1.2°C at $P = 57$ bar. During this experiment, the delay before the start of hydrate formation (the induction time) was 507 hours. Morphology of hydrate film for these crystals is shown in FIGURE 1. Multiple crystal nuclei formed at the free gas-water interface around which the hydrate film grew, with the radial rate of about 6.6×10^{-2} mm/min.

The rate of nucleation depends on gas and water composition, pressure, temperature, and cooling rate. FIGURE 6 shows the dependence of methane hydrate nucleation for seawater and seawater +5% methanol. FIGURE 7 shows the dependence of the radial growth rate of the massive type hydrate film at a free interface for methane with seawater and seawater +5% methanol.

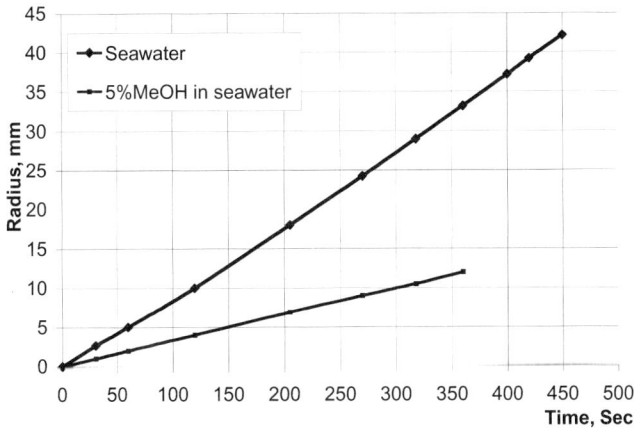

FIGURE 7. Rate of methane crystal growth on free surface of seawater and 5% methanol in seawater.

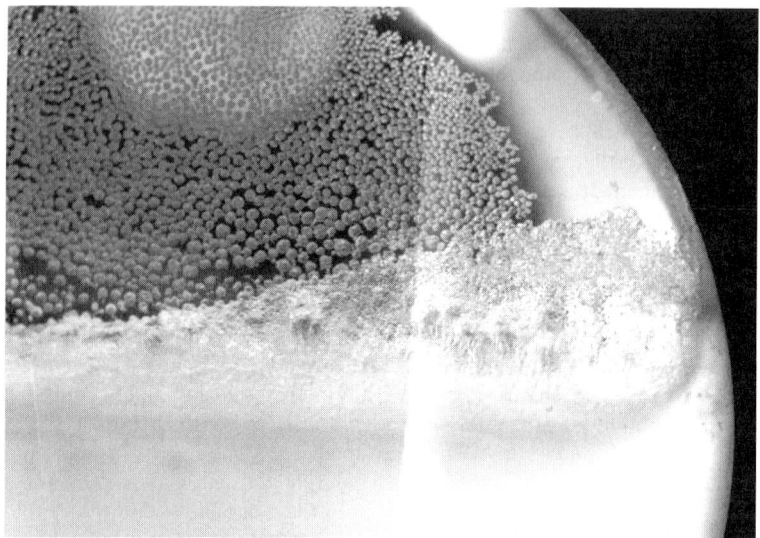

FIGURE 8. Natural gas hydrate formation with 0.5% DGT-75 in seawater in dynamic conditions in gas phase. $P = 86$ bar, $T = 277$ K.

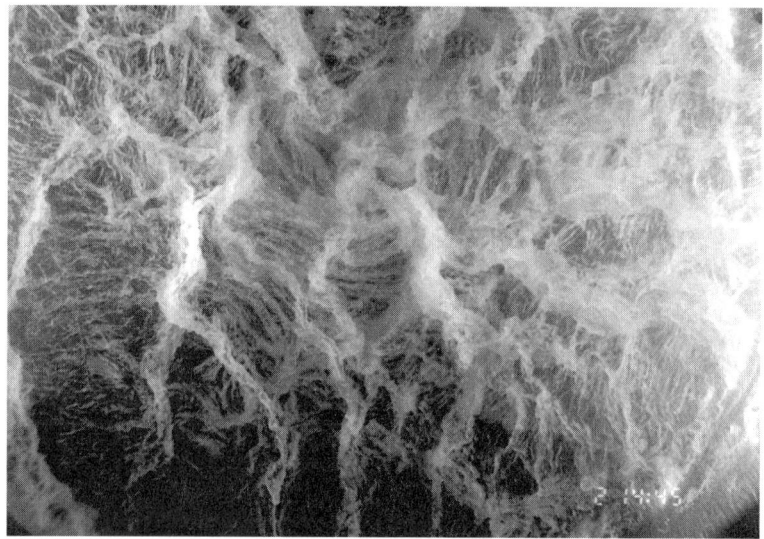

FIGURE 9. Natural gas hydrate formation with 0.5% DGT-75 in seawater in dynamics on gas-liquid interface. $P = 86$ bar, $T = 277$ K.

HYDRATE FORMATION WITHOUT INHIBITORS

Phase equilibria of gas hydrates are important for industry and academia. A pressure–temperature diagram usually describes thermodynamic conditions at which hydrates form from pure water. A univariant curve describes the equilibrium for hydrate formation from binary water–hydrate former mixture. The univariant curve is prescribed by the Gibbs' phase rule:

$$2 + C = F + P, \qquad (1)$$

where C is the number of components (e.g., CH_4, H_2O), F is the number of degrees of freedom, P is the number of phases (e.g., gas, liquid). The three phase line (V, Lw, H) for Structure I hydrate of methane will serve as an example:

$$2 + 2(\text{methane} + \text{water}) = 1 + 3(\text{vapor} + \text{hydrate} + \text{liquid}). \qquad (2)$$

This gives only one degree of freedom for hydrate formation equilibrium, which defines pressure for a specified temperature. Usually, as temperature of the system decreases, a lower pressure is required to form gas hydrate. This can be shown from the Clausius-Clapeyron equation:

$$\frac{d\ln P}{d(1/T)} = \frac{-\Delta_d H}{ZR}, \qquad (3)$$

where $\Delta_d H$ is the enthalpy of hydrate dissociation. Hydrate consumes heat on dissociation and $\Delta_d H$ is greater than zero.

This also can be explained from an entropic viewpoint. At lower temperature water molecules vibrate less and become more ordered. The entropy of the system decreases.

$$G = U + PV - TS. \qquad (4)$$

According to the Equation (4) if the volume stays constant and temperature decreases, then the pressure must also decrease in the closed system (U, constant) at equilibrium (G, constant).

HYDRATE FORMATION WITH INHIBITORS

There are four classical methods for preventing hydrate formation in a system containing a hydrocarbon gas: increasing temperature of the system, decreasing system pressure, removing water from the system, or adding inhibitors of hydrate formation. *Thermodynamic* hydrate inhibitors such as alcohol, glycol, or salt are commonly used to avoid hydrate formation in gas and oil industry for systems where dehydration or heating are impossible or not economic. Such inhibitors shift the thermodynamic stability boundary of hydrates to higher pressure or lower temperature by aggregating with water molecules and preventing their arrangement into a hydrate lattice.[6,9] This method is not always the best environmental or economic solution for preventing hydrate plugs. Makogon[10,11] reported that, "With an increase in concentration of alcohols in water, a breakdown is observed in the structural organization of water and in the clathrate-forming aggregates. As a result, the probability of hydrate formation is reduced". This observation suggests that the thermodynamic inhibitors change the structure of water away from that favoring hydrate formation as a part of their effect. A neutron diffraction study[12] of a 1:9 molar ratio methanol–water mixture showed the experimental evidence that water molecules form a disordered hydrogen bonded cage around the methanol molecule.

Salt readily ionizes in water and aggregates water molecules in solvation shells around ions. The presence of solvated ions near a hydrate crystal causes a hindrance for the water and guest molecules adsorbing on a hydrate surface.[9]

In 1991 the CSM Center for Hydrate Research proposed an alternative method of inhibiting gas hydrates. It was found that certain polymers, when added to water, would delay the conversion of gas and water into hydrate. The induction time for the onset of hydrate formation is generally unpredictable, or stochastic. However, the hydrate induction time is a function of supersaturation or pressure.[9,13] Kinetic inhibitors do not inhibit hydrate formation completely, as thermodynamic inhibitors do. Kinetic inhibitors can delay hydrate formation for a time longer than the residence time of the water phase in a pipeline.

Kinetic inhibition is currently applied in the industry together with thermodynamic inhibitors in wells and in pipelines. Generally, kinetic inhibitors are polymer molecules having an amide (–[C=O]–N=) linkage in their side groups giving them the ability to hydrogen-bond with hydrate. Several kinetic inhibitors have rings of carbon atoms as their side groups. It was determined that the ring structure is not imperative to an effective inhibitor. Hydrogen bonding ability is a necessary but not a sufficient property of a kinetic inhibitor. Poly(vinyl alcohol) (PVA) has a hydrogen bonding capability with the hydroxyl (OH) group; however, it is not a hydrate inhibitor. It was shown experimentally and via molecular modeling by Makogon[10] that kinetic inhibitors can adsorb on {111} surfaces of hydrate structure II and {100} surface of hydrate structure I and inhibit further crystal growth in that direction. This is a fundamental issue for understanding hydrate growth inhibition on a molecular scale. Preferential adsorption of kinetic inhibitor on hydrate {111} faces was recently proven experimentally by Larsen.[14]

KINETICS OF HYDRATE FORMATION

Studying the kinetics of hydrate formation allows one to determine two attributes of hydrate formation. One is how soon hydrate will start forming (induction time) since the system was placed in appropriate thermodynamic conditions. The other attribute is the growth rate at which liquid water or ice will be converted into a solid hydrate. Interest in this area of hydrate research has previously been purely academic. Today industry is seeking the new chemicals that allow operation of a gas-water system at conditions where hydrate would normally form, so that it stays in a metastable state without aggregating to a large hydrate mass—hydrate plugs. Experimental data for kinetics of hydrate formation is available in the literature for temperatures above and below the ice point. Falabella[3] studied the formation of hydrates of different gases and mixtures of gases at low subzero temperatures. The work on kinetics of hydrate formation has been reviewed by Sloan.[9]

Kinetics of gas hydrate formation may be affected by many different factors. Among these are:

1. Supercooling, or lowering the system temperature below the equilibrium value for a given pressure. Different shapes of hydrate crystals were obtained by Makogon[2] depending on the amount of supercooling.
2. Overpressurization or pressure above the system equilibrium value for a set temperature. This is another measure of supercooling, or an equivalent of supersaturation.
3. Rate of cooling, or gradient of temperature decrease of the system in time.
4. Stirring rate. Effect of turbulence in the hydrate system on kinetics of crystallization, the crystal size distribution and the duration of the induction period was described by Englezos.[15]
5. Previous temperature history of water available for hydrate formation. This effect was studied by Makogon.[2]
6. Presence of the sites for hydrate nucleation, such as steel walls of the reactor or pipeline, or particles of silica.
7. Presaturation of water with hydrate forming gas. Dissolution of gas in water is a diffusion process, if no mixing were applied to the system. Rate of gas dissolution may be monitored and subtracted from total gas consumption during the experiment.
8. Additives. Kinetics of hydrate formation is strongly affected by the so-called kinetic inhibitors.

It is quite important to determine and understand the effect of thermodynamic and kinetic inhibitors on the hydrate formation and decomposition processes. To accomplish this goal we have performed a large number of experimental studies of the formation and decomposition conditions of methane and natural gas hydrates. These

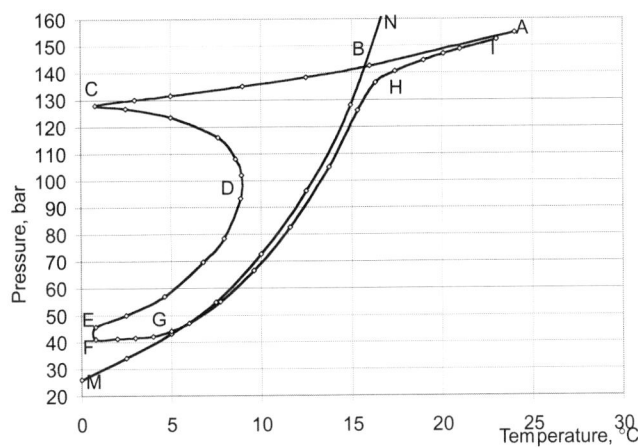

FIGURE 10. Methane hydrate formation and dissociation with fresh water in static. Supercooling (BC) is 15°C. At such supercooling hydrate forms quickly, the heat of crystallization is not removed by heat transfer and temperature increases (CD). New hydrate film on liquid surface is very permeable. Gradually the surface of diffusion channels gets covered by new hydrate layers, diffusion decreases and the processes of heat release by crystallization and heat removal equalize (DE), and the rate of hydrate accumulation decreases.

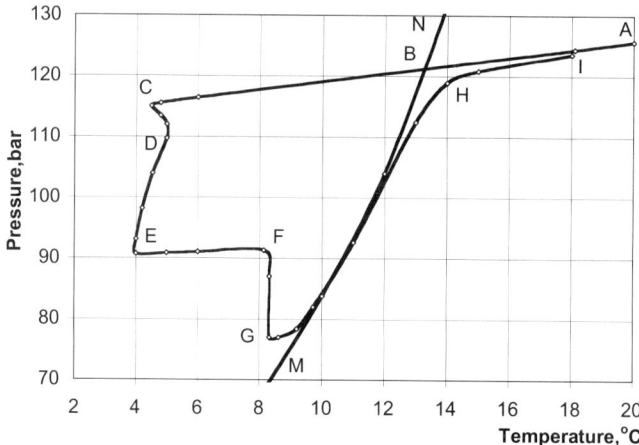

FIGURE 11. Methane hydrate formation and decomposition with 2% methanol in water in static. The induction time (time before onset of crystallization when temperature is in hydrate region) was about 22 hours. Supercooling required to start hydrate growth was 9°C, which is 6°C less than with fresh water. Hydrate forms actively (*FG*) after a temperature increase of the formed hydrate (*EF*), which we explain by increase in permeability of the hydrate film separating gas from water due to increased molecular motion. Further temperature increase (*GH*) causes hydrate to melt.

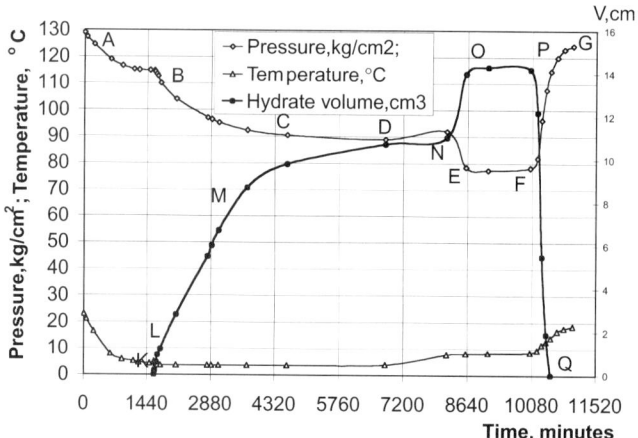

FIGURE 12. Kinetics of methane hydrate formation and dissociation with 2% methanol in water in static shown in FIGURE 11.

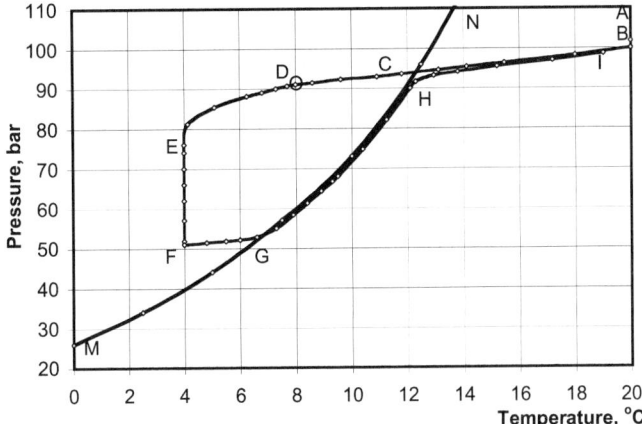

FIGURE 13. Methane hydrate formation and dissociation in fresh water dynamic conditions (S8.KI.99). Supercooling was 4.4°C, much less than in static. Initial speed of magnetic stirrer was 780 rpm. With hydrate accumulation water viscosity increased and stirrer slowed to a stop (*E*). Dynamic process became static with hydrate accumulation via water and gas diffusion through hydrate layer on water surface. The rate of hydrate accumulation in gas phase was over two orders of magnitude greater than in liquid phase, as defined by diffusive permeability of water and gas through hydrate.

tests with pure water, seawater, and experiments with solutions of thermodynamic inhibitors methanol and ethylene glycol, as well as four most advertised antiagglomerants and inhibitors of Shell, Champion, and Exxon were performed in parallel

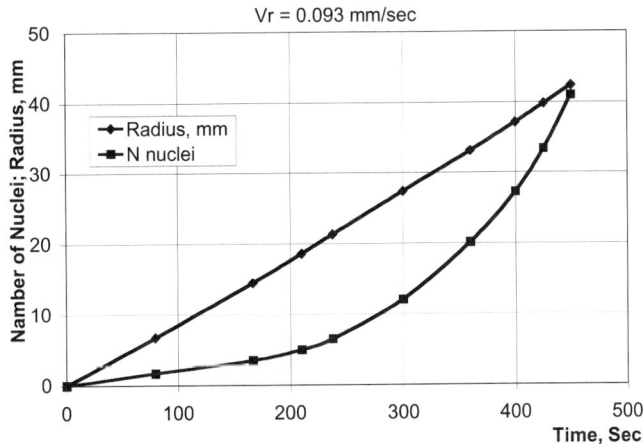

FIGURE 14. Methane hydrate nucleation and radial growth on seawater surface. 9.II.99.

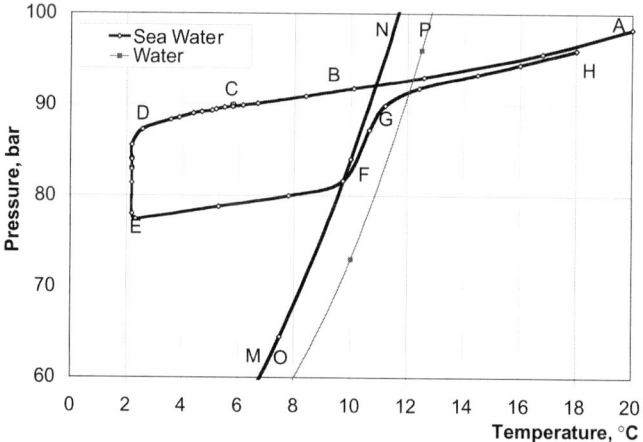

FIGURE 15. Methane hydrate formation and dissociation with seawater in static. Supercooling (*BC*) was 5.1°C.

in static and dynamic (stirred) conditions. The names of these chemicals are not published. FIGURES 10 through 22 show the characteristic pressure and temperature traces and the rate of hydrate formation and decomposition in time during these experiments. Based on these results the authors can recommend the selection and use

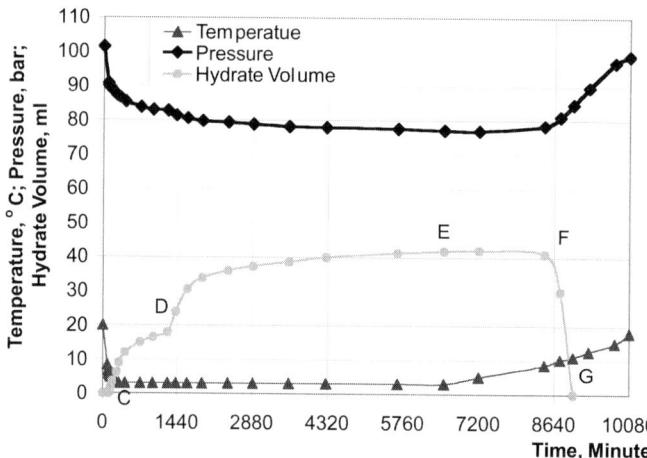

FIGURE 16. Kinetics of methane hydrate formation and dissociation with seawater in static shown in FIGURE 15. Two degrees of hydrate accumulation activity were observed (*CD* and *DE*), which correspond to curves (*CD* and *DE*) in FIGURE 15.

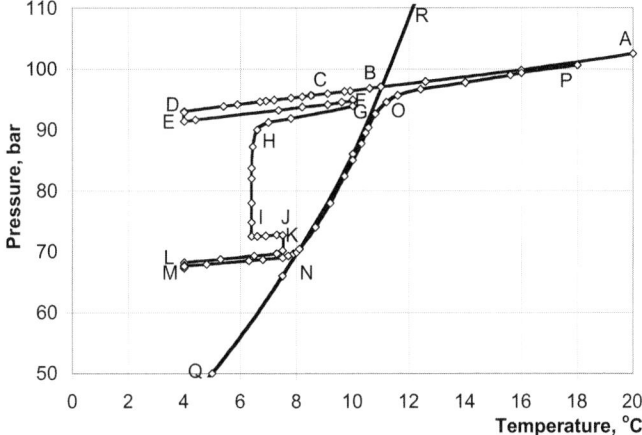

FIGURE 17. Methane hydrate formation and dissociation with 5% MEG in fresh water (S9.K2.99). Supercooling of 3.6°C was required to start hydrate formation. Two instances of secondary hydrate accumulation activity were observed (*FI* and *JM*), when temperature climbed to 1°C below equilibrium.

of the most effective and safe inhibitors for preventing hydrate plugs in wells and pipelines. Brief details of hydrate formation are given in captions of FIGURES 10–22.

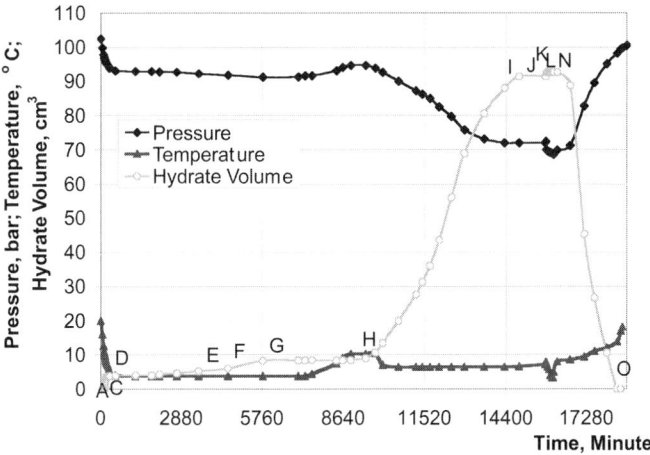

FIGURE 18. Kinetics of methane hydrate formation and dissociation with 5% MEG in fresh water shown in FIGURE 17.

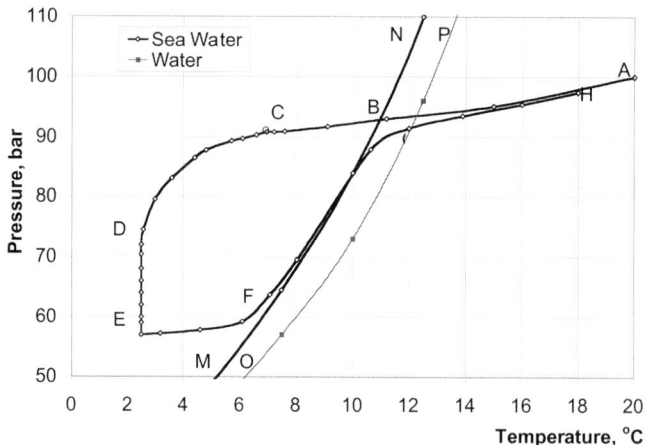

FIGURE 19. Methane hydrate formation and dissociation with seawater in dynamics. Supercooling of 4°C was required to start hydrate formation. Amount and rate of hydrate accumulation were significantly higher than in the process shown in FIGURE 16.

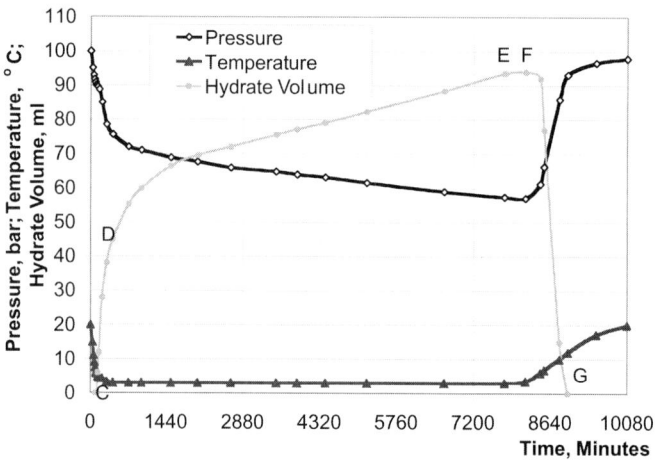

FIGURE 20. Kinetics of methane hydrate formation and dissociation with seawater in dynamics shown in FIGURE 19.

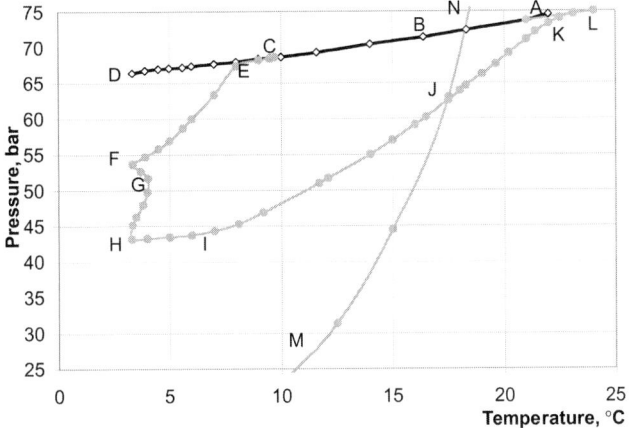

FIGURE 21. Natural gas hydrate formation and dissociation with 0.75% kinetic inhibitor A in water in dynamics. (K1dyn.9-26.VI.99).

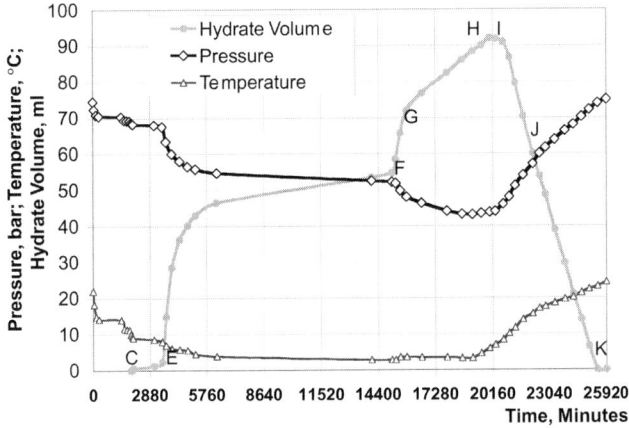

FIGURE 22. Kinetics of natural gas hydrate formation and dissociation with 0.75% kinetic inhibitor A in water in dynamics as shown in FIGURE 21. Natural gas composition is shown in TABLE 2. Supercooling of 8.8°C was required to start hydrate formation. Viscosity of solution increased, free surface of gas-water interface decreased, and rate of hydrate accumulation decreased with hydrate formation (*CEF*). The rate and amount of hydrate accumulation rose with increased stirring as the free gas-water interface appeared (*FGH*). As temperature of a multicomponent gas hydrate increased, the hydrate amount, properties, strength and composition partially changed. Presence of kinetic inhibitors on hydrate surface strengthened hydrate and increased its kinetic decomposition temperature. In this experiment 5°C superheating was required (*JK*) to fully melt hydrate at 0.2°C/hour heating ramp.

FIGURE 23. Pipeline steel corrosion by massive methane hydrate with seawater. Pipe on the right remained without hydrates.

FIGURE 24. Pipeline steel corrosion by massive methane hydrate with fresh water. Pipe on the right remained without hydrates.

CONCLUSION

Based on an analysis of the results obtained we draw the following conclusions:

1. Supercooling required to start hydrate formation increases with pressure.

2. Thermodynamic inhibitors present in the system lower the onset temperature of hydrate formation, decrease the required supercooling, and increase the rate of hydrate formation (accumulation).

3. Kinetic inhibitors or antiagglomerants present in the system do not lower the temperature of formation and they increase the rate of hydrate formation, but decrease mechanical strength of hydrate.

4. Higher supercooling is required for initial hydrate formation in static conditions than in dynamic (stirred) conditions.

5. Rate of hydrate accumulation is much higher in dynamics than in static due to constant renewal of free gas–liquid interface and active removal of the heat of hydrate formation.

6. The rate of hydrate accumulation in static is controlled by gas and water adsorption on the growing crystals and by the diffusive permeability of the hydrate film at gas–water interface.

7. Massive hydrate crystals formed in vapor phase have high porosity up to 80%. Massive hydrate crystals are good absorbents and actively absorb water, which converts into hydrate extremely slowly.

8. The rate of absorbed water transformation into hydrate state is controlled by diffusive permeability of massive hydrate to gas.

9. Kinetics of hydrate dissociation depends on composition of hydrates, time of existence of hydrates, type of crystals.

10. At the same conditions gel crystals dissociated first, then massive and the last to melt were the whiskery crystals.

11. Most important reasons for rate of dissociation are heat transfer, desorption, and diffusion.

12. Dissociation temperature of massive crystal hydrate formed with kinetic inhibitors is much higher then for pure water.

13. Hydrate formation and decomposition processes are highly corrosive to metal (see FIGURES 23 and 24).

This work will be continued to study the kinetics and morphology of hydrates and the corrosion of metals in order to find most effective means of preventing hydrates, corrosion, and to develop economical methods of producing gas from natural hydrate deposits, especially in subsea conditions. More details about rate of decomposition will be presented in the next paper.

ACKNOWLEDGMENT

This work was sponsored by ARCO, Chevron, Halliburton, and ONGC-India companies through the Texas A&M University hydrate consortium.

REFERENCES

1. MAKOGON, Y.F. 1966. Specialties of exploration of gas hydrate fields in permafrost conditions. TNTI, Mingasprom. Moscow.
2. MAKOGON, Y.F. 1974. Hydrate of Natural Gases. Nedra, Moscow. (1981, PennWell, Tulsa, USA.)
3. FALABELLA, B.J. 1975. A Study of Natural Gas Hydrates. Ph.D. Dissertation, University of Massachusetts.
4. BISHNOI, P.R. & A. VISNIAUSKAUS. 1980. Kinetics of Gas Hydrate Formation. GRI, Chicago.
5. HOLDER, G.D. & S.P. GODBOLE. 1982. AIChE J. **28:** 930.
6. VISNIAUSKAUS, A. & P.R. BISHNOI. 1983. Chem. Eng. Sci. **38:** 1961.
7. KOBAYASHI, R., K.Y. SONG & E.D. SLOAN. 1987. Petrol. Eng. Handbook, SPE, Richardson.
8. MAKOGON, Y.F. 1997. Hydrates of Hydrocarbons. PennWell, Tulsa, USA.
9. SLOAN, E.D. 1990; 1998. Clathrates of Natural Gases, Marcel Dekker, New York.
10. MAKOGON, T.Y. 1995. Center for Hydrate Research Annual Report, CSM, Golden, Colorado.
11. MAKOGON, T.Y. 1997. Ph.D. Dissertation, Colorado School of Mines, Golden, Colorado.
12. SOPER, A.K. & J.L. FINNEY. 1993. Hydration of methanol in aqueous solutions. Phys. Rev. Lett. **71**.
13. HERRY, J. *et al.* 1996. Kinetics of methane hydrate formation. Presented at 2nd International Natural Gas Hydrate Conference.
14. LARSEN, R., T.Y. MAKOGON, C.A. KNIGHT & E.D. SLOAN. 1996, The influence of kinetics on the morphology of clathrate hydrates. Presented at the 2nd International Natural Gas Hydrate Conference, Toulouse, France 15.
15. ENGLEZOS, P.N. *et al.* 1987. Kinetics of formation of methane and ethane hydrates. Chem. Eng. Science **42:** 2647.

Effect of Inhibitor Methanol on the Microscopic Structure of Aqueous Solution

YOSHITAKA YAMAMOTO,[a] KAZUSHIGE NAGASHIMA, TAKESHI KORNAI, AND AKIHIRO WAKISAKA

National Institute for Resources and Environment,
16-3 Onogawa, Tsukuba-shi, Ibaraki 305-8569, Japan

INTRODUCTION

Several experimental investigations have been carried out to study the stability of methane hydrates under various conditions.[1–7] Methanol has been used as an inhibitor of gas hydrate formation for a long time. It is well known that the equilibrium temperature of gas hydrates is reduced by the addition of methanol.[1] The addition of some cyclic ethers, such as THF, also changes the equilibrium condition of gas hydrates[1] and they are added mainly to promote gas hydrate formation. Recently, studies were carried out to clarify the effect of methanol on hydrate formation on a molecular level.[6,7] From a microscopic viewpoint, these compounds are expected to change the clustering structure of liquid phase from that of pure water, leading to the change of the crystallization properties of gas hydrates. Therefore, it is important to determine the correlation between the crystallization properties and the microscopic liquid structure of an aqueous solution in order to gain further understanding of the gas hydrate formation mechanism. We previously reported that the stable clusters that were formed from methanol–water and THF–water mixtures have rather different structures.[8] However, this measurement was carried out only for comparatively low concentration solutions.

In this study, we attempted to compare the correlation between crystallization properties of methanol aqueous solution and clustering structures of these mixtures in a more systematic manner. For this purpose, solid–liquid equilibrium properties under atmospheric and higher pressures of water–methanol mixtures (1–600 atm) of various mixing ratios were measured and compared with the clustering structures of these mixtures, measured using a liquid cluster beam mass spectrometer.[8–11]

To investigate the effect of the alkyl chain length of alcohol, the equilibrium properties and clustering structures of 1-propanol–water mixtures were also measured.

---

[a]Address for correspondence: Yoshitaka Yamamoto, National Institute for Resources and Environment, 16-3 Onogawa, Tsukuba-shi, Ibaraki 305-8569, Japan.

EXPERIMENTAL

Measurement of Equilibrium Properties

FIGURE 1 shows a schematic illustration of the experimental setup to measure solid–liquid phase transition properties under high pressure. The pressure in the cell was changed by moving the piston with a servomotor. Equilibrium conditions at various mixing ratios were determined by measuring the pressure and temperature under zero crystal growth. Morphologies of forming crystals were also observed with this optical cell. Differential scanning calorimetry (DSC) analyses were performed using a Seiko DSC220CU to obtain a understanding of the detailed equilibrium property. Crystallographic structures were measured using X-ray powder diffraction (XRD) analysis with a Rigaku RINT11500 (target, Co; tube voltage, 40 kV; tube current, 200 mA).

Measurement of Stable Clusters

Mass spectra of the clusters of water–methanol and water–1-propanol solutions were measured using a liquid cluster beam mass spectrometer.[8–11] Sample solutions were injected into a four-stage differentially pumped vacuum system through a heated nozzle. Mist particles (very small liquid droplets) that formed in the first chamber (0.2 torr), were disintegrated into clusters by adiabatic expansion in the second (10^{-3} torr) and third (10^{-6} torr) chambers. The resulting clusters were ionized by electron

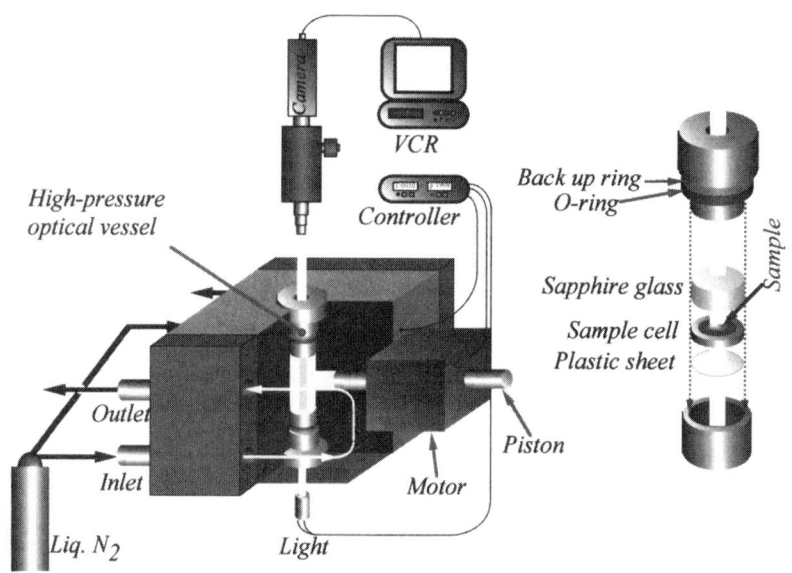

FIGURE 1. Experimental setup to measure solid–liquid phase transition properties under high pressure.

impact (30 eV) in the third chamber, and analyzed using a quadrupole mass spectrometer in the fourth chamber (10^{-7} torr). A total of 15 measurement points were obtained for 0–100 mol% solutions of methanol–water and 1-propanol–water.

RESULTS AND DISCUSSION

Equilibrium Properties

Equilibrium Properties of Water–Methanol Mixed Solutions

Some equilibrium measurements in methanol–water mixtures under atmospheric pressure are available.[12,13] However, there are few measurements under high pressure. Therefore, we measured solid–liquid equilibrium properties under atmospheric and higher pressures for this system and examined the effect of pressure.

Our optical cell observation results are summarized in FIGURE 2. Both chemical species formed crystals, and the pressure dependence of the equilibrium temperature (dP/dT) changed with methanol concentration.

In the water rich region (0 < [MeOH] < 50 mol%), hexagonal ice (I_h) crystals were formed. Crystals in this region melt with an increase in pressure. In the intermediate concentration region (50 < [MeOH] < 70 mol%), water–methanol intermolecular compound crystals with a fourfold axis formed, and pillar-shaped methanol crystals were formed in the methanol rich region (70 mol% < [MeOH]). Crystals in these regions grew with an increase in pressure.

As shown in FIGURE 3, the concentration of methanol at a eutectic point between ice (I_h) and intermolecular compound reduced with an increase in pressure. It became 49 mol% at 500 atm from 54 mol% at atmospheric pressure. This indicates that an intermolecular compound crystal is preferably formed, rather than an ice (I_h) crystal, under high pressure. This fact suggests that the water–methanol intermolecular compound crystal has a denser structure than the ice (I_h) crystal and is very different. Furthermore, dP/dT for the solutions in the intermolecular compound region was very close to dP/dT for pure methanol. Thus, it appears that the intermolecular compound crystal structure resembles that of methanol rather than the ice crystal.

X-ray Powder Diffraction Patterns

FIGURE 4 shows the XRD results for 47.4 mol% methanol solution. The sample mixture was prepared at room temperature. Then, it was cooled to −190°C, using a liquid-nitrogen bath, and crushed into pieces. Holding the sample room temperature at about −150°C, which was sufficiently lower than the melting temperature for this concentration mixture shown in FIGURE 2, the X-ray powder diffraction pattern of this sample was measured. As shown in FIGURE 4(a), very broad peaks were observed for this sample. This suggests that amorphous solids were formed.

It was reported that the crystallization of the methanol–water mixture progresses by annealing.[6] Therefore, we annealed the sample at a temperature close to the melting temperature of this mixture for four hours so as to accelerate crystallization. As shown in FIGURE 4(b)–(d), crystallization proceeded and the peaks became sharper with the passage of time.

FIGURE 5 shows the XRD patterns of methanol crystal, ice (I_h) crystal, and the crystal of the 47.4 mol% sample mixture (after annealing). In this figure, peaks that are marked *I* and *M* correspond to characteristic peaks of ice (I_h) and methanol, respectively. Other peaks (*W–M*) should correspond to methanol monohydrate crystal. These peaks agreed well with the characteristic peaks of methanol monohydrate reported by Nakayama et al.[6]

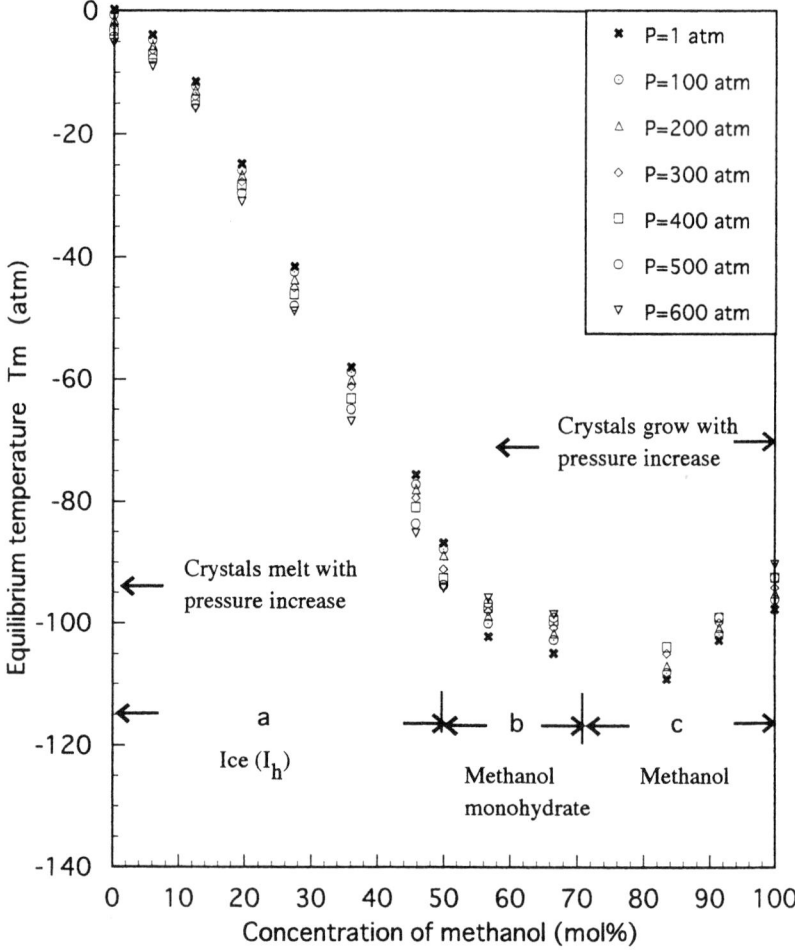

FIGURE 2. Equilibrium properties of water–methanol solutions under atmospheric and higher pressures.

Equilibrium Properties of Water-1-Propanol Mixed Solution

FIGURE 6 shows the crystallization properties of 1-propanol–water mixtures under atmospheric and higher pressures. In regions a (0 < 1-propanol < 7 mol%) and a' (7 mol% < 1-propanol < 50 mol%), crystals melted with an increase in pressure. This implies these that these crystals have a bulky and highly similar structure. They have a hexagonal shape. At present, we cannot confirm whether these crystals are all identical or not.

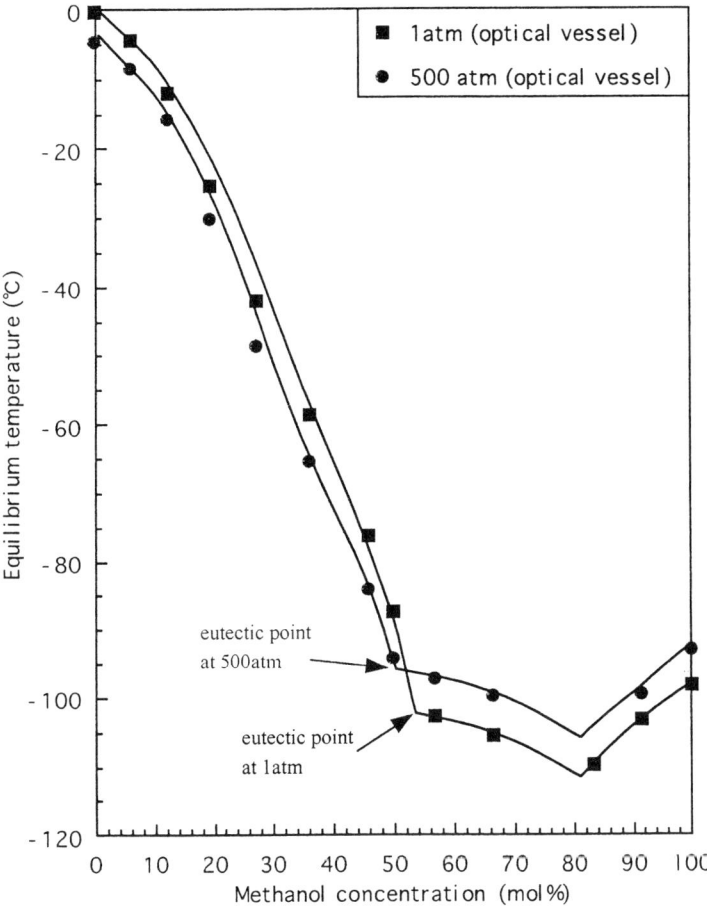

FIGURE 3. Comparison of equilibrium properties of methanol–water mixtures under atmospheric and at 500 atm. At 500 atm the eutectic point was shifted to a lower methanol concentration than that at atmospheric pressure.

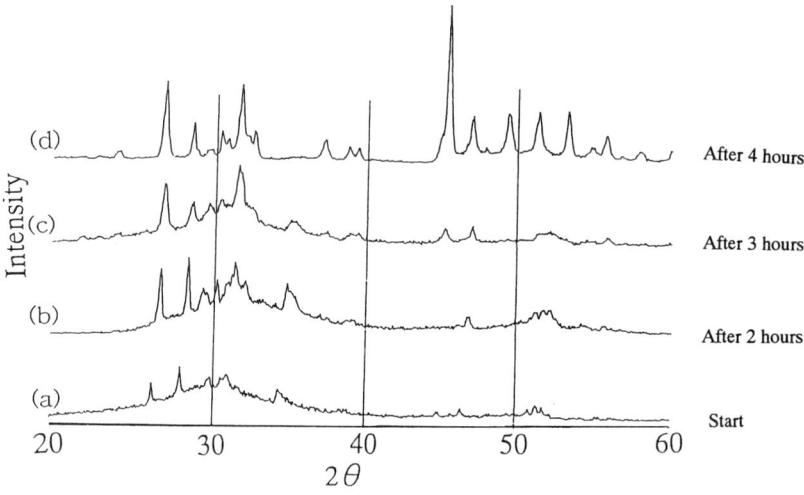

FIGURE 4. Effect of annealing on X-ray diffraction patterns for 47.4 mol% methanol solution.

On the other hand, crystals in region *b*, which have a fourfold axis, grew with an increase in pressure, the same as the 1-propanol crystal. This indicates that it might be a water–1-propanol intermolecular compound. The pressure dependence reveals this crystal of region *b* to more resemble 1-propanol than water.

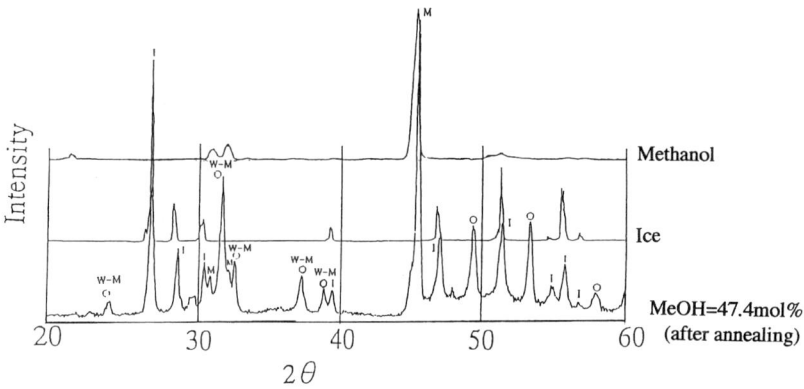

FIGURE 5. X-ray powder diffraction patterns for methanol, ice, and 47.4 mol% methanol aqueous solution.

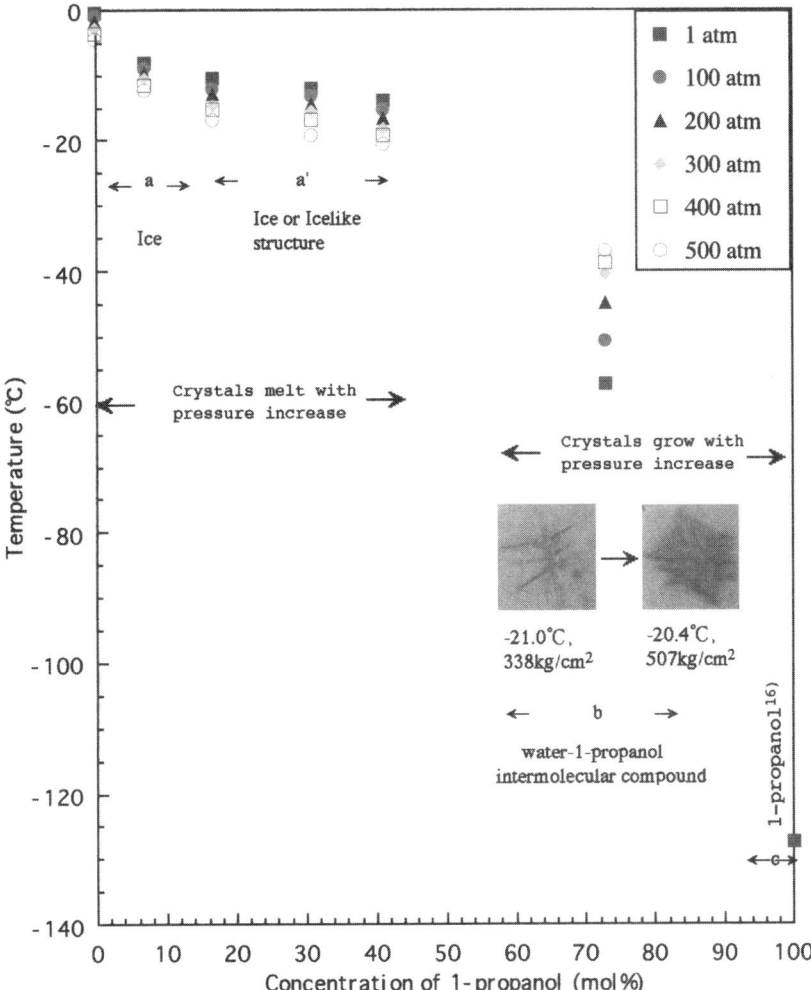

FIGURE 6. Equilibrium properties of 1-propanol–water mixtures under atmospheric and higher pressures.

Distribution of Stable Clusters

Clustering Structures of Water–Methanol Mixed Solutions

In order to obtain information on the molecular clustering structure, the mass spectrometric analysis of clusters isolated from liquid droplets was carried out using the specially designed mass spectrometer. Since the clusters are generated through the fragmentation of liquid droplets via vacuum adiabatic expansion, the strongly interacting molecules in the liquid droplets are thought to form clusters. We

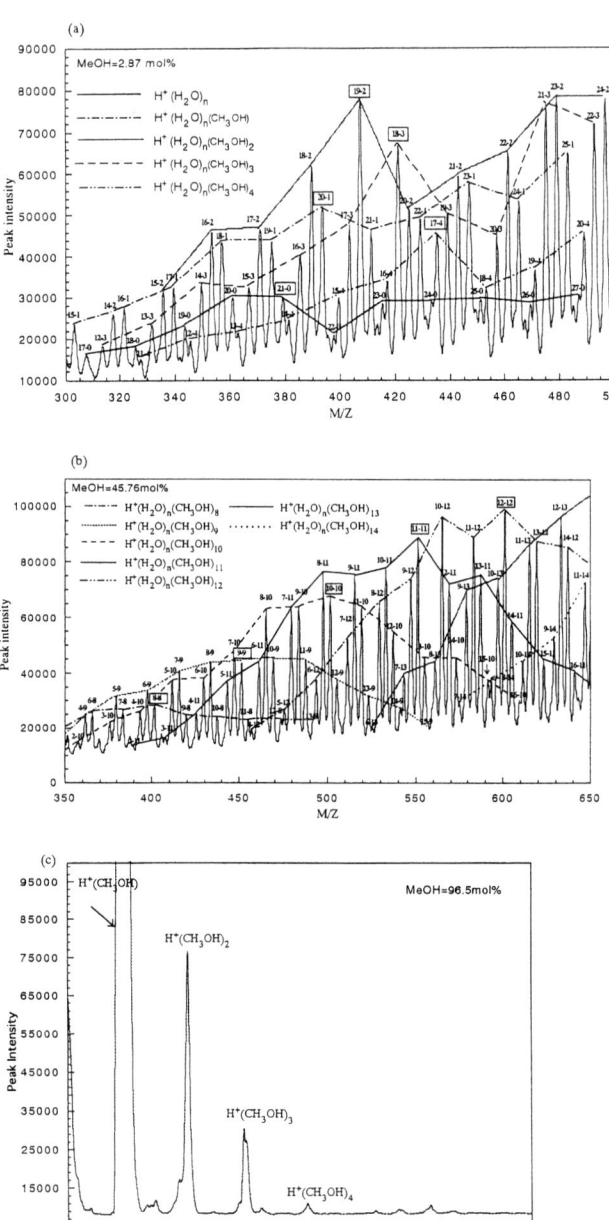

FIGURE 7. Mass spectra of clusters generated from adiabatic expansion of water–methanol solutions with methanol concentrations: **(a)** 2.87 mole%; **(b)** 45.76 mole%; **(c)** 96.5 mole%.

previously reported that the clustering structures observed by this mass spectrometry reflected the microscopic structures in solutions,[9] molecular self-assembling,[10] formation of intermolecular compound,[11] and so forth. Accordingly, it is noted that the cluster structures observed here are also related to the intermolecular interaction in solution.

FIGURE 7(a) shows the mass spectrum of clusters generated from the low concentration mixture (2.87 mol% MeOH). Stable clusters were mainly composed of $(H_2O)_n$ with few CH_3OH_n molecules. Similar to the $(H_2O)_{21}$ H^+ stable cluster of pure water,[14,15] $(H_2O)_{21n}CH_3OH_n$ clusters ($n = 0, 1, 2, 3, ...$) are extremely stable, indicating that methanol molecules, substitutionally interact with water clusters as follows:

$$H^+(H_2O)_{21} + nCH_3OH \rightleftharpoons H^+(H_2O)_{21-n}(CH_3ON)_n + nH_2O \ (n = 1,2,3,...)$$

However, the overall tendency of cluster distribution is very similar to that of pure water. It indicates that the hydrogen bonding networks of water molecules remain in this region.

FIGURE 7(b) shows the mass spectrum of clusters generated from a solution of the intermediate concentration region (45.76 mol% MeOH). The cluster distribution is significantly changed from that of pure water. Characteristic magic-numbered peaks of water such as $(H_2O)_{21}H^+$ cannot be observed in this spectrum. Water and methanol mixed clusters such as $(H_2O)_nH^+(CH_3OH)_m$ ($n \neq m$) were predominantly observed.

FIGURE 7(c) shows the mass spectrum of clusters generated from the high concentration region (96.5 mol% MeOH). In this region, $(CH_3OH)_n$, $n = 1,2,3...$ clusters were predominant. Hydrogen bonding networks of water molecules were completely eliminated and only methanol molecules could aggregate easily. The cluster size in this region was considerably smaller than that in the water rich region or the intermediate concentration region.

Clustering Structures of Water-1–Propanol Mixed Solutions

Similar to the methanol solution, the stable cluster distribution changed with 1-propanol concentration as follows:

1. $(H_2O)_n$ with few 1-propanol clusters.
2. Water/1-propanol mixed clusters.
3. $(CH_3CH_2CH_2OH)_n$, $n = 1, 2, 3$) clusters.

Compared with the methanol solution, each region tends to shift to the lower 1-propanol concentration, and the 1-propanol cluster region becomes wide.

CONCLUSIONS

It became clear that not only chemical species forming crystals, but also the stable cluster distribution of the water–alcohol mixed solution changes with alcohol concentration discontinuously. This clearly indicates that the microscopic structure of these aqueous solutions can be classified according to concentration. This change of clustering structure in the liquid phase should strongly influence the inhibition and the promotion of hydrate formation.

With respect to the nucleation process for gas hydrates, the basic structure, which is a composed of hydrogen bonded network of water molecules, will grow to yield the primary nuclei. The formation of a water–methanol molecular pair in aqueous solution, which inhibits the growth of the hydrogen bonded network of water molecules, seems to be important for inhibiting gas hydrate formation. It would be helpful to investigate such a microscopic structure of aqueous solutions in order to understand the inhibiting/promoting mechanism of gas hydrate formation.

REFERENCES

1. SLOAN, E.D. 1990. Clathrate Hydrates of Natural Gases. Marcel Dekker, Inc., New York.
2. HOLDER, G.D. 1980. Hydrate dissociation pressure of (methane + ethane + water)—existence of a locus of minimum pressures. J. Chem. Thermodyn. A **150**: 1093–1104.
3. HANDA, Y.P. 1990. Effect of hydrostatic pressure and salinity on the stability of gas hydrates. J. Phys. Chem. **94**: 2652–2657.
4. DAVIDSON, D.W. et al. 1981. The effect of methanol on the stability of clathrate hydrate. Can. J. Chem. **59**: 2587–2590.
5. WALLQVIST, A. 1992. On the stability of type I gas hydrates in the presence of methanol. J. Chem. Phys. **96**: 5377–5382.
6. NAKAYAMA, H. et al. 1997. Methanol–clathrate hydrate former or inhibitor. ACS Division of Fuel Chemistry. **42**(1): 516–520.
7. BLAKE, D. et al. 1991. Clathrate hydrate formation in amorphous cometary ice analogs in vacuo. Science **254**(25): 548–551.
8. YAMAMOTO, Y. et al. 1996. Measurement of the hydrate cluster by the liquid molecular beam mass spectrometry. Proc. 2nd Int. Conf. on Natural Gas Hydrate, Toulouse, France, June 2. 355–362.
9. WAKISAKA, A., T. SATORU & N. NISHI. 1995. Preferential solvation controlled by clustering conditions of acetonitrile–water mixtures. J. Chem. Soc. Faraday Trans. **91**: 4063–4069.
10. WAKISAKA, A. et al. 1996. Molecular self-assembly controlled by acid-base noncovalent interactions—a mass spectrometric study of some organic acids and bases. J. Chem. Soc. Faraday Trans. **92**: 3539–3544.
11. YAMAMOTO, Y. & A. WAKISAKA. 1997. Clustering of a hydrogen-bonding complex between indole and isoquinoline. J. Chem. Soc. Faraday Trans. **93**: 1405–1408.
12. CONRAD, F.H. et al. 1940. Freezing points of the system ethylene glycol–methanol–water. Ind. Eng. Chem. April. 542–543.
13. PUSHIN, N.A. et al. 1922. The equilibrium in systems composed of water and alcohols—methyl alcohol, pinacone, glycerol, and erythritol. J. Chem. Soc. (Lond.) **121**: 2813–2822.
14. NAGASHIMA, U., H. NISHI & H. TANAKA. 1986. Enhanced stability of ion-clathrate structure for magic number water clusters. J. Chem. Phys. **84**: 209–214.
15. WEI, S., Z. SHI & A.W. CASTLEMAN, JR. 1991. Mixed cluster ions as a structure probe—experimental evidence for clathrate structure of $(H_2O)_{20}H^+$ and $(H_2O)_{21}H^+$. J. Chem. Phys. **94**: 3268–3270.
16. CHEMICAL SOCIETY OF JAPAN, Ed. 1993. Handbook of Chemistry, 4th edit. Maruzen: I-386.

Calculation of Gas Hydrate Equilibrium in Presence of Aqueous Salt Solutions Using a New Predictive Activity Model

ASLE JØSSANG[a] AND ELLEN STANGE

Norsk Hydro ASA, 0246 Oslo, Norway

ABSTRACT: The objective of this work is to develop a general and predictive hydrate model for hydrates in equilibrium with brines. The model developed only needs the ionic radius, the charge on the ions, and the concentration of each salt to predict the inhibition effect of the salts in the brines. The model has proven to be consistent with experimental data.

INTRODUCTION

Formation water is present in most hydrocarbon reservoirs and the salts in this water depress gas hydrate formation. Often methanol or ethylene glycol is added to inhibit hydrate formation. The use of these inhibitors can be reduced or avoided if the depression of gas hydrate formation caused by salts is accounted for.

The equilibrium of gas hydrates when inhibitors are present in the water phase can be calculated from thermodynamic theory when the activity to water in the water phase is known. In this work a predictive water activity model for brine is developed from data on evaluation of water activities. The model is used to calculate water activities to brine, and the water activities are used in the calculation of hydrate equilibrium. The model is suitable for engineering purposes and shows which chemical properties of the salt affect the formation of gas hydrates.

DEVELOPMENT OF THE MODEL

Hydrate Model

The hydrate model due to van der Waals and Platteeuw[1] modified by McKoy and Sinanoglu,[2] is used. Today two models for the water phase are in common use, the model by Parrish and Prausnitz[3] and that due to Holder.[4] These models use water activity to predict the equilibrium when inhibitors are present in the water phase. In this work the model of Parrish and Prausnitz[3] is used.

[a]Telecommunication. Voice: +47 22 73 81 00; fax: +47 22 73 81 76.
asle.jossang@hydro.com

Activity Model

Different activity models can be used to predict the freezing point depression of gas hydrates caused by different salts.[5–7] These models give good predictions, but are only valid for a few salts. The activity models used need a parameter regression for each salt. In this work a general and predictive activity model is developed. This is to be able to predict hydrate equilibrium in presence of brines of any salt. The following assumptions have been used:

- Water is the continuous phase and the salt is dissolved to ions.
- Salt solution is homogeneous with no concentration gradients.
- The reference state is distilled water, without ions, at a pH of 7.0. The activity coefficient is, by definition, 1.00 in this state at all pressures and temperatures.
- Water activity is independent of pressure and temperature.
- Ionic radius of the salt crystal lattice is used as ionic radius in the water activity model.[8,9]
- Salts do not form solid solutions with the gas hydrates.

Conventions

As a convention the different salts will be referred to by the charge on its cation (M^+) and anion (X^-). The salts is then referred to as 1:1 salts, 2:1 salts, and so forth. As an example, NaCl (Na^+, Cl^-) is a 1:1 salt, and Na_2SO_4 ($2Na^+$, SO_4^{2-}) is a 1:2 salt.

Evaluation of Chemical and Physical Properties that Affect Water Activity

When salts are added to water they dissolves to ions. This requires a negative change in Gibbs energy, which can be caused by a negative change in enthalpy or positive change in entropy. The change in interaction forces between ions and water cause the change in enthalpy. The change in entropy is caused by increased disorder in the water phase and is accounted for by the change in water activity. The activity is defined as the activity coefficient multiplied by the mole fraction of the actual component.

To evaluate which chemical and physical properties of the salts affect the activity of water, the activities of water with LiCl, NaCl, and KCl[8,10–13] have been plotted at different concentrations, as is shown in FIGURE 1. The figure shows that the water activity increases with increased ionic radius. This change in water activity is general for all cations.

The reason for this change in activity with ionic radius is assumed to be due to different disorder in the water phase caused by different ions. This is a result of the difference in interaction forces between water molecules and ion. Small ions have a greater surface charge than larger ones. This results in stronger interactions between the ion and water molecules and more disorder in the water phase.

FIGURE 2 shows a decrease in water activity with an increase in ionic radius. This is opposite to the case for cations. For Cl, Br, and I the effect of increased surface area is larger than the effect of decreased surface charge. This results in a decrease in water activity with an increase in ionic radius. The ionic radii for Cl, Br, and I are much larger than the ionic radii of most cations. The change in surface charge

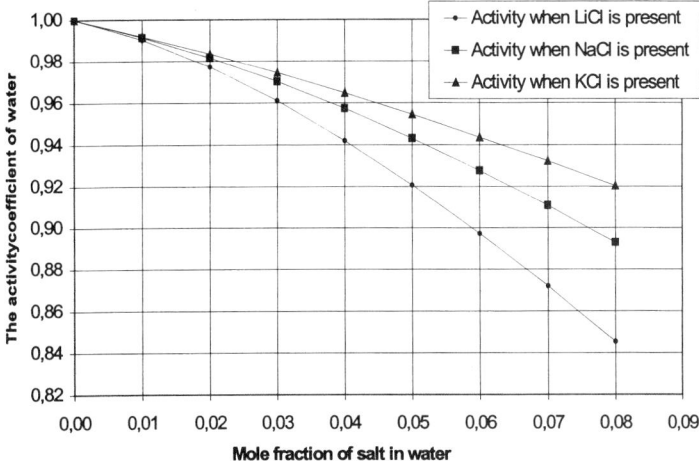

FIGURE 1. Activity coefficient of water in brines of LiCl, NaCl, and KCl plotted at different concentrations.

between anion is much less than between cations. The change in surface charge does not affect the water activity to the same extent.

The degree of disorder in the water phase is a function of how strong each water molecule is influenced and number of water molecules involved. How strongly each water molecule is affected is a function of the interaction forces. The number of molecules affected is a function of the ionic surface area and the distance from the

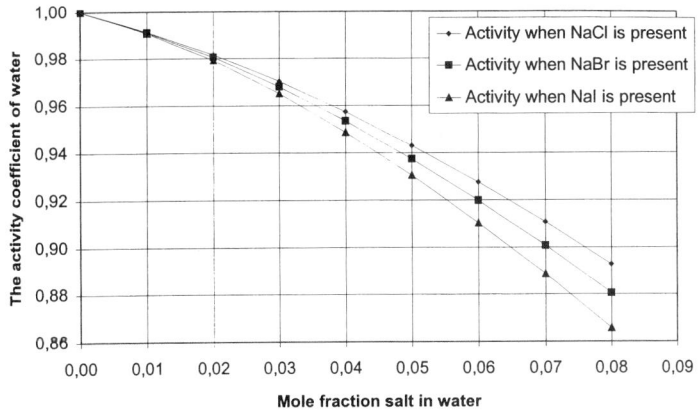

FIGURE 2. Activity coefficient of water in brines of NaCl, NaBr, and NaI[8,10–13] plotted at different concentrations.

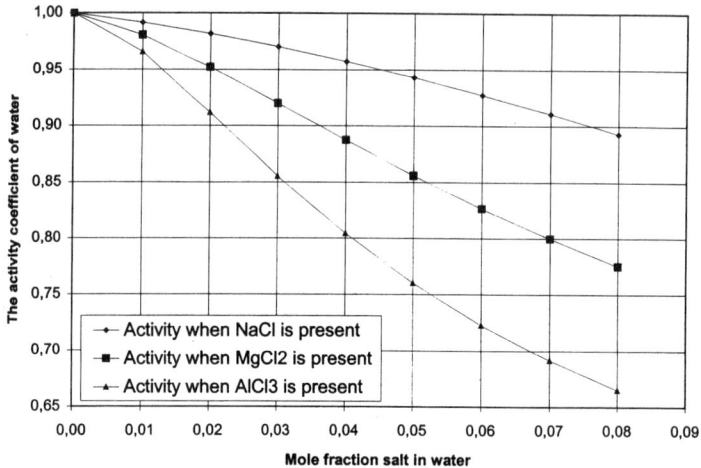

FIGURE 3. Activity coefficient of water in brines of NaCl, MgCl$_2$, and AlCl$_3$ plotted at different concentrations.[8,10–13]

surface of the ion that the water molecules are influenced. Here only 1:1 salts are described, but the same correlation is valid for other salt groups.

FIGURE 3 shows that the water activity decreases with increased charge on the ion. The effect of charge and ionic radius is expressed through the surface charge, which is the charge of the ion divided by the surface area of the ion. The ions are considered hard spheres. The water activity coefficient decrease with increased salt concentration. This is accounted for by the mole fraction of the ions. From FIGURES 1–3 it is possible to see that the activity change with concentration is slightly curved. An exponent on the mole fraction of the ion accounts for this curvature. This exponent is empirical.

Mixing Rule

Formation and seawater contains a mixture of different salts the model has to account for this. The mixing rule developed by Patwardhan and Kumar[14] is used. The activity model should be able to calculate the water activity of an aqueous mixture of salts, methanol, or ethylene glycol. The water activity of the mixture is calculated by multiplying the water activity caused by each inhibitor.

The New Activity Model

The activity model developed for the water phase in brine can be expressed as follows:

$$\ln a_w = \sum_{i=1}^{m} \left(\frac{m_k}{m_k^o}\right) \ln \left[\left(1 - \sum_{Anion}^{Kation} f(r_{M^+,i} + r_{X^-,i}, z_{M^+,i}, z_{X^-,i}) \cdot \frac{z_{ion,i}}{4\pi r_{ion,i}^2} \cdot x_{ion,tot}^{z_{M^+,i} : z_{X^-,i}}\right) \cdot x_{w,i}\right] \quad (1)$$

where the increase in solvatization of the ions with increased ionic radius is expressed by

$$f(r_{M^+,i} + r_{X^-,i}, z_{M^+,i} : z_{X^-,i}) = k_1(z_{M^+,i} : z_{X^-,i}) - \frac{k_2(z_{M^+,i} : z_{X^-,i})}{r_{M^+,i} + r_{X^-,i}} \quad (2)$$

$$E(z_{M^+,i} : z_{X^-,i}) = k_3(z_{M^+,i} : z_{X^-,i}) \quad (3)$$

The constants in these equations are given in TABLE A1.1 in APPENDIX 1.

If the water phase contains salts, methanol and ethylene glycol the following mixing rule was used:

$$a_{w,total} = a_{w,salt} \cdot a_{w,MeOH} \cdot a_{w,EG} \quad (4)$$

COMPARISON WITH EXPERIMENTS

The the model was evaluated in two steps. The first step compares the new activity model with experimental data for the activity of water. The second step compares the hydrate model and new activity model with experimental data for hydrate in equilibrium with brine.

The New Activity Model

The calculated activities of water in brines are compared with experimental data and results from established activity models. The water activity is converted from experimental data for salt activity by using Gibbs-Duhems equation. The data are

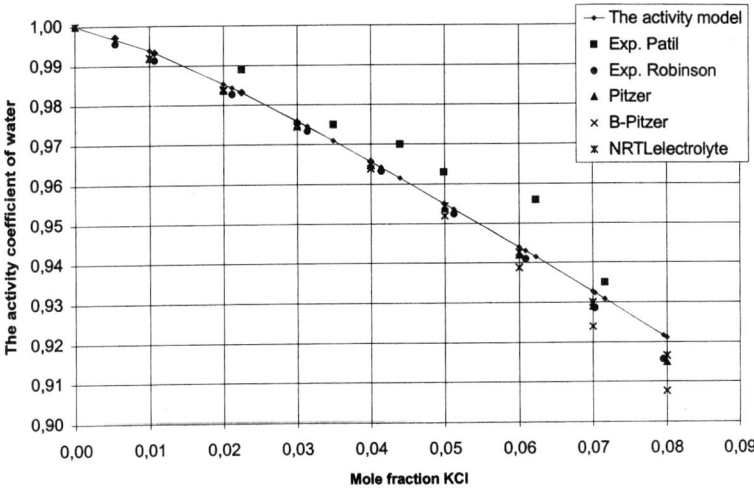

FIGURE 4. Calculated water activities compared to experimental data and calculated values from other activity models given by ASPEN PLUS.[17]

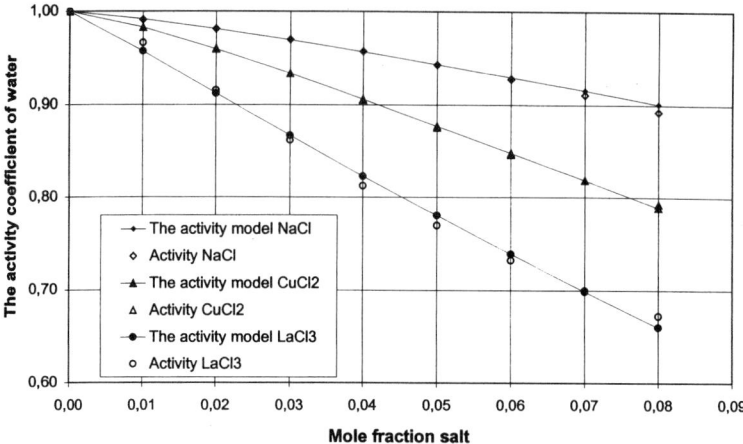

FIGURE 5. Calculated water activities compared to experimental data.

given by Patil[15] and Robinson and Stokes.[16] The Pitzer, Bromley-Pitzer, and NRTL electrolyte models are used with the parameter given by ASPEN PLUS.[17]

In FIGURE 4 are calculated water activities compared to experimental data for KCl and results from other activity models.

In APPENDIX 2, FIGURE A2.1 are shown calculated water activities in comparison with experimental data[8,13] for $BaBr_2$ and with results from other activity models.

FIGURE 5 shows calculated water activities in comparison with experimental data[8,10–13] for brines of NaCl, $CuCl_2$, and $LaCl_3$.

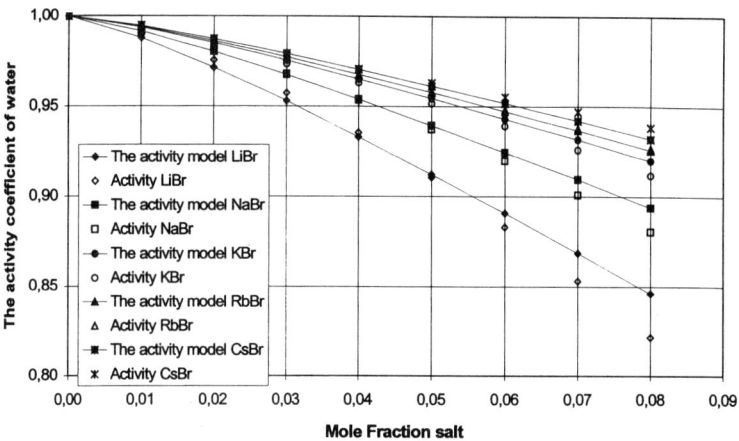

FIGURE 6. Calculated water activities compared to experimental data.

APPENDIX 2, FIGURE A2.2 compares calculated water activities with experimental data[8,11,13] for brines of HNO_3 and $NaNO_3$. FIGURE A2.3 shows calculated water activities in comparison with experimental data[8,13] for brines of NH4Cl and Na_2CO_3.

In FIGURE 6 we show calculated water activities in comparison with experimental data[8,13] for brines of LiBr, NaBr, KBr, RbBr, and CsBr.

This paper gives five graphical comparisons of the water activity model with experimental data. In the work with the model, experimental data for 46 different salt solutions has been compared with the calculated water activities showing good results.[18]

The range of validity retrieved from this comparison is shown in TABLE A1.2 in APPENDIX 1. The maximum deviation from experimental data, where the model is valid, is set to 0.01. This corresponds to a deviation in the hydrate temperature of 1 to 1.5°C.

The Hydrate Model

In FIGURES 7 to 9, and A2.4 to A2.6 the hydrate model, using the new water activity model, is compared with experimental hydrate equilibrium data.[19,20]

FIGURE 7 shows calculated hydrate temperatures for natural gas hydrates (composition: 0.10 N_2, 1.76 CO_2, 80.53 CH_4, 10.30 C_2H_6, 4.99 C_3H_8, 0.72 i-C_4H_{10}, and 1.60 n-C_4H_{10}; all in mole%) with brines of NaCl, compared with experimental data.

In APPENDIX 2, FIGURES A2.4 and A2.5 show calculated hydrate temperatures for natural gas hydrates with brines of KCl or $CaCl_2$, in comparison with experimental data. FIGURE A2.6 calculated hydrate temperatures for H_2S hydrates with experimental data.

FIGURES 8 and 9 show calculated hydrate temperatures for hydrates of a gas containing about 80 mole% CH_4 and 20% CO_2 in comparison with experimental data.

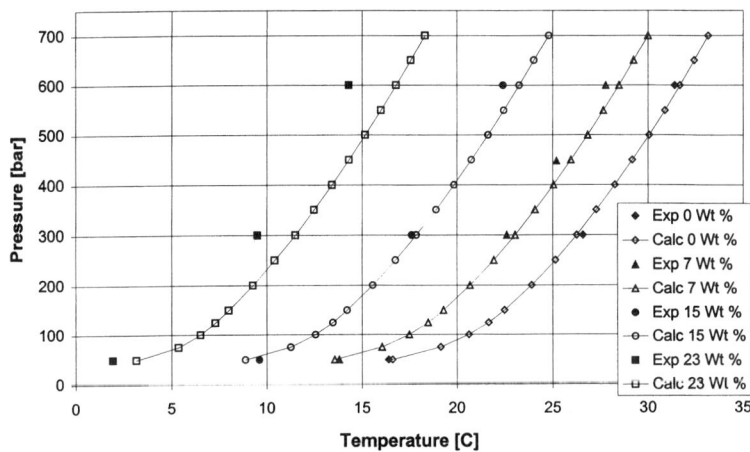

FIGURE 7. Calculated hydrate temperature compared with experimental measurements when NaCl is in the water phase.

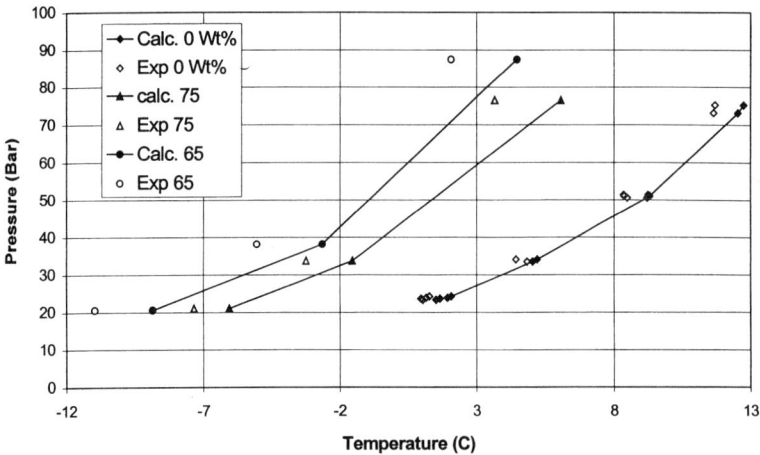

FIGURE 8. Calculated hydrate temperature compared to experimental measurement for 9.98 wt% NaCl, 9.99 wt% methanol (65); and 20 wt% ethylene glycol, 5.02 wt% $CaCl_2$ (75).

This paper gives six graphical comparisons of the hydrate model compared to experimental data. The model has also been compared to other experimental data with good results.[18]

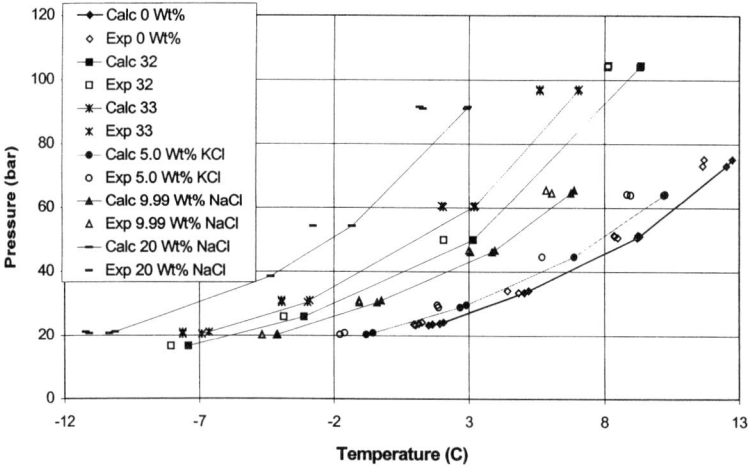

FIGURE 9. The inhibitors in the water phase are NaCl, KCl, or a mixture of 5.00 wt% NaCl and 10.00 wt% KCl (32), or 10.17 wt% NaCl and 5.08 wt% CaCl2 (33).

DISCUSSION

Activity Model

The water activity model shows why the water activity changes with different salts in the water phase at the same concentration. The development of the model is based on surface charge of the ion because the ion–water distances are equal for all ions when the ion have a defined radius and are treated as hard spheres. The strength of the interaction forces is then represented by the surface charge since the charge on the water molecules are the same. Calculation of the water–ion interactions is not necessary when using this model.

Few data on water activity is available in the literature. Most of these data are for salt activity in water converted to water activities with Gibbs-Duhems equation. A consistency test between the converted data and available experimental data for water activity[15,16] has been done.

The radius of ions in water solution has not been found in the literature. Therefore, the ion radius of an ion in a salt crystal lattice was used. It is assumed that this does not affect the results. For ions of more than one atom, a radius is predicted from the covalent and ionic radius of the different atoms. This method has proved to give results that accord with experimental data.

Hydrate Model

The model is predictive. The hydrate temperature at a given pressure can be estimated, without experimental data, for either water activity or hydrate equilibrium, in presence of a given salt. The model is suitable for evaluating the hydrate inhibition effect of various salts. The model only needs the ionic radius, the charge on the ions, and the concentration of each salt in order to predict the effect of the inhibitor.

When fitting the constants in the activity model, it has been found necessary not to under predict the water activity.

The hydrate model has proved to be consistent with all experimental data used in this work and the effect of the inhibitor is independent of the gas phase. This is in accord with theory, which states that the change equilibrium is a function only of the change in the water phase. The deviation from hydrate equilibrium is in accordance with the deviation of the water activity calculated.

FIGURE A2.5 shows a high deviation between experimental and calculated data at high salt concentration. This is in accord with TABLE A1.2 in APPENDIX 1, where it is stated that the model is only valid up to 18 wt% $CaCl_2$. The wt% limitation is given by a deviation in hydrate temperature of 1.5°C

The gas solubility in water changes with addition of salt to water. This effect is not considered. However, Tohidi[6] has shown that the solubility of the gas does not influence the hydrate equilibrium data.

Limitations of the Model

When going from pure water to pure salt there are different zones in which the mixture has different chemical and physical states. To simplify this, the salt solution is considered to be in three different states. One is salt in water continuous phase, the second is the intermediate state where there are both water–water interactions and ion–ion interactions, and the third state is water in salt continuous phase. Many activity models try to describe all three states. Because of the different chemical properties in these three states, this model considers only the first state. This gives a limitation in concentration as described in APPENDIX 1, TABLE A1.2.

The calculation of water activity when salt, methanol, and glycol are present does not account for the interaction effect between the different inhibitors. Because of this, the estimated equilibrium temperature is higher than the experimental at a given pressure and composition. This model will always be conservative when mixtures of inhibitors are present in the water phase.

CONCLUSIONS

Three general parameters have to be fitted to experimental data for each salt group (1:1, 1:2, etc.). After regression it is possible to calculate the activity of water containing other salts in the same salt group. This is possible without further regression for the particular salt. The model estimates water activities with an accuracy of 0.01.

The activity model is used together with the hydrate model developed by Parrish[3] to predict the hydrate formation temperature. The hydrate model has been compared with experimental data and it shows good agreement with the data available. TABLE A1.2 gives the concentration range of salt in brine where the model is valid. The model estimates hydrate temperature when brines are present with an accuracy of 1.5°C.

ACKNOWLEDGMENT

This work was supported by Hydro Technology and Projects, Norsk Hydro ASA.

REFERENCES

1. VAN DER WAALS, J.H. & J.C. PLATTEEUW. 1959. Adv. Chem. Phys. **2**: 1.
2. MCKOY, V. & O. SINANOGLU. 1963. J. Chem. Phys. **38**: 2946.
3. PARRISH, W.R. & J.M. PRAUSNITZ. 1972. Ind. Eng. Chem. Proc. Des. Dev. **11**(1): 26–35.
4. HOLDER, G.D., G. CORBIN & K.D. PAPADOPOULOS. 1980. Ind. Eng. Chem. Fund. **19**: 282–286.
5. MUNCH, J., S. SKJOLD-JØRGENSEN & P. RASMUSSEN. 1988. Chem. Eng. Sci. **43**(10): 2661–2672.
6. TOHIDI, B., A. DANESH & A.C. TODD. 1995. Chem. Eng. Res. Des. **73**(A4): 464–472.
7. BISHNOI, P.R., P.D. DHOLABHAI & N. KALOGERAKIS. 1995. Proc. Annu. Conv.—Gas Process. Assoc. **74**: 28–34.
8. LIDE, D.R. 1990–1991. CRC Handbook of Chemistry and Physics, 71st edit. CRC Press, Inc., Boston.
9. AYLWARD, G.H & T.J.V. FINDLAY. 1974. SI Chemical Data, 2nd edit. John Wiley & Sons, Hong Kong.
10. CONWAY, B.E. 1969. Electrochemical Data, 2nd edit. Greenwood Press, London.
11. KAGE, G.W.C. & T.H. LABY. 1995. Tables of Physical and Chemical Constants, 16th edit. Longman Group Limited, Harlow.

12. BORGE, G., R. CASTAÑO, M.P. CARRIL, M.S. CORBILLÓN & J.M. MADARIAGA. 1996. Fluid Phase Eq. **121:** 85–98.
13. DOBOS, D. 1975. Electrochemical Data. Elsevier Scientific Publishing Company, Amsterdam.
14. PATWARDHAN, V.S. & A. KUMAR. 1986. AIChE J. **32**(9): 1419–1428.
15. PATIL, K.R., A.D. TRIPATHI, G. PATHAK & S.S. KATTI. 1991. J. Chem. Eng. Data **36:** 225–230.
16. ROBINSON, R.A. & R.H. STOKES. 1970. Electrolyte Solutions. Butterworths, London.
17. ASPEN TECHNOLOGY INC. 1994. ASPEN PLUS, User Guide, Release 9.
18. JØSSANG, A. 1996. Calculation of the activity of aqueous salt solution for application in gas hydrate formation modelling. NTNU.
19. DB ROBINSON RESEARCH LTD. 1991. Hydrate formation of a typical North Sea gas mixture in the presence of selected aqueous solution. Alberta, Canada.
20. BISHNOI, P.R., P.D. DHOLABHAI & K.N. MAHADEV. 1996. Hydrate phase equilibria in inhibited and brine system. GPA, RR-156, Canada.

APPENDIX 1

Empirical constants developed for the water activity model, using different salt groups and concentration ranges, for which the activity model has proven to be valid.

TABLE A1.1. Empirical constants in the water activity model

Ion pair $(z_{M^+,i} : z_{X^-,i})$	$k_1 (z_{M^+,i} : z_{X^-,i})$ [m²/C]	$k_2 (z_{M^+,i} : z_{X^-,i})$ [m²/C]	$k_3 (z_{M^+,i} : z_{X^-,i})$ [m²/C]
1:1	4.9204	9.353×10^{-10}	1.27
2:1	12.747	9.9313×10^{-9}	1.30
3:1	5.6604	1.2042×10^{-9}	1.10
1:2	0.40	0.00	1.00
1:NO_3^-	0.40	0.00	1.00

TABLE A1.2. The highest concentrations for which the activity model has proven to be consistent with experimental data

Salt	Highest mole fraction	Highest wt%	Salt	Highest mole fraction	Highest wt%
KF	0.06	17.1	$AgNO_3$	0.07	41.5
NaF	0.08+	16.9	NH_4NO_3	0.08+	27.9
HCl	0.05	9.6	NH_4Cl	0.08+	20.5
LiCl	0.065	14.1	Na_2SO_4	0.06	33.5
NaCl	0.08	22.0	K_2SO_4	0.05	33.8
KCl	0.08+	26.5	Na_2CO_3	0.08+	33.9
RbCl	0.08	36.9	$MgCl_2$	0.02	9.7
CsCl	0.08+	44.9	$CaCl_2$	0.035	18.3
HBr	0.05	19.1	$SrCl_2$	0.04	26.8
LiBr	0.055	21.9	$BaCl_2$	0.03	26.4
NaBr	0.067	29.1	$FeCl_2$	0.02	12.6
KBr	0.08+	36.5	$CuCl_2$	0.08+	39.4
RbBr	0.08+	44.4	$ZnCl_2$	0.02	13.4
CsBr	0.08	50.7	$MgBr_2$	0.05	35.0
HI	0.02	12.7	$CaBr_2$	0.035	28.7
LiI	0.07	35.9	$SrBr_2$	0.035	33.3
NaI	0.06	34.7	$BaBr_2$	0.05	46.5
KI	0.08+	44.5	MgI_2	0.01	13.5
RbI	0.08	50.6	CaI_2	0.05	46.2
CsI	0.07	52.1	SrI_2	0.035	40.8
HNO_3	0.08+	23.3	BaI_2	0.035	44.1
$NaNO_3$	0.08+	29.1	$AlCl_3$	0.06	32.1
KNO_3	0.07	29.7	$LaCl_2$	0.08	54.2

APPENDIX 2

Calculated water activities and hydrate temperatures compared with experimental data.

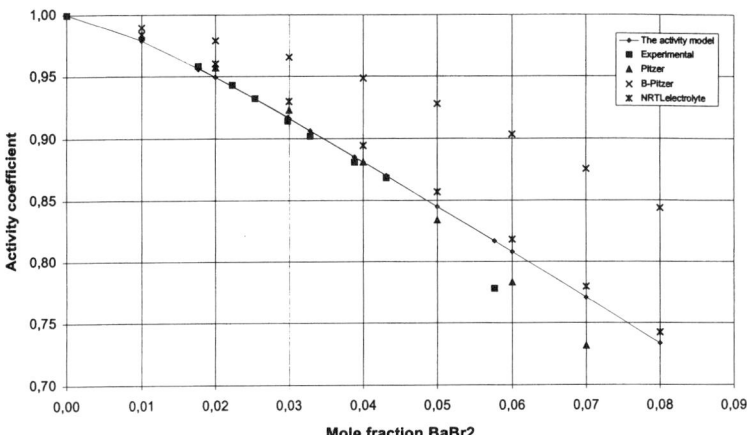

FIGURE A2.1. Calculated water activities compared with experimental data and calculated values from activity models given by ASPEN PLUS.[11]

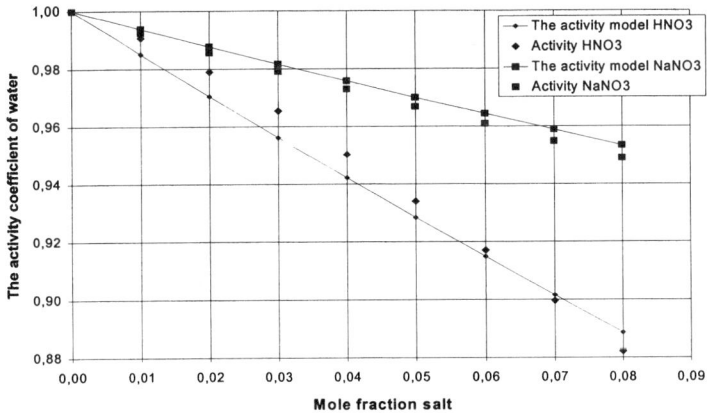

FIGURE A2.2. Calculated water activities compared with experimental data.

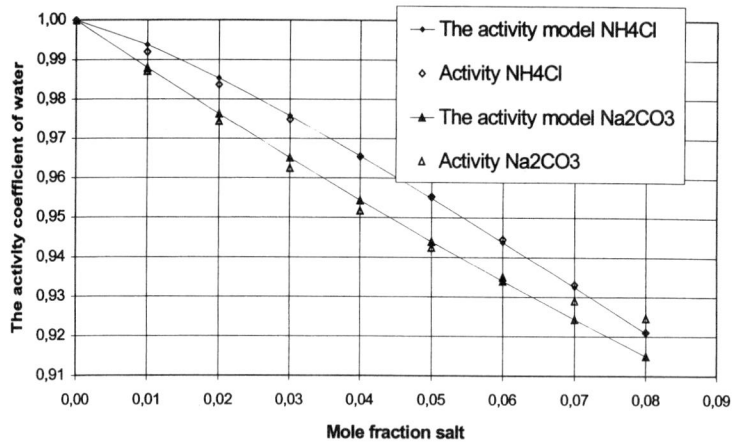

FIGURE A2.3. Calculated water activities compared with experimental data.

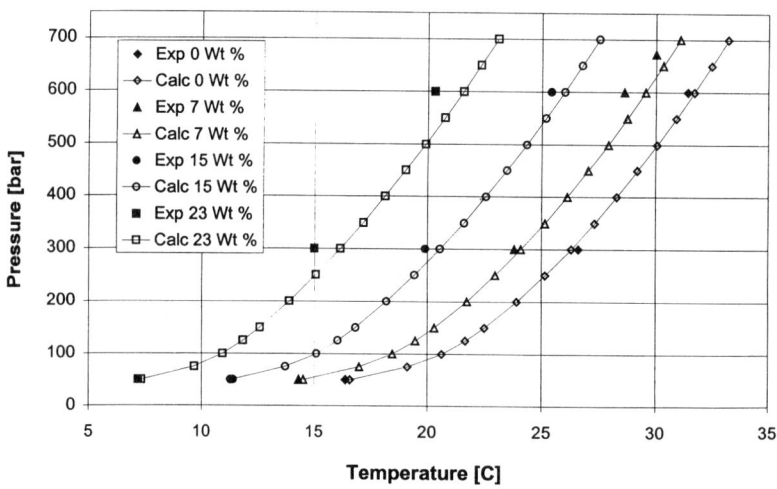

FIGURE A2.4. Calculated hydrate temperature compared with experimental data when KCl is in the water phase.

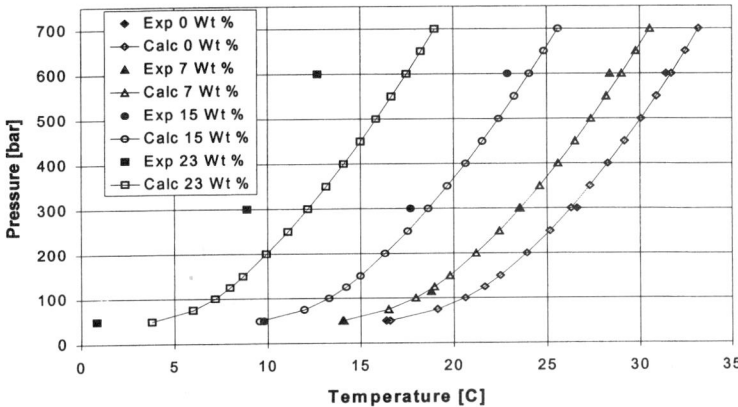

FIGURE A2.5. Calculated hydrate temperature compared with experimental data when $CaCl_2$ is in the water phase.

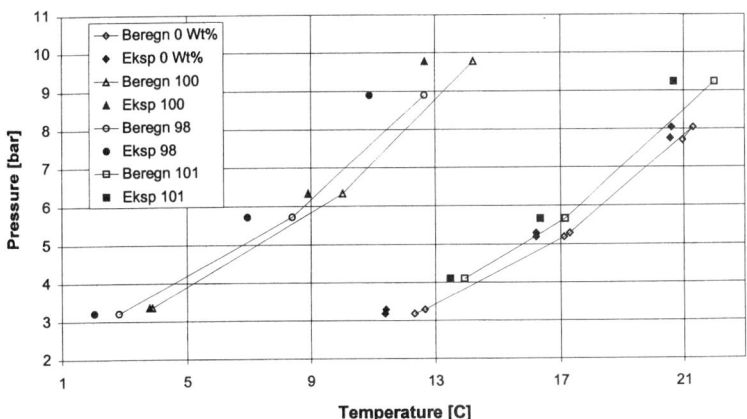

FIGURE A2.6. Calculated hydrate temperature compared with experimental data for synthetic saltwater consisting of 2.6430 wt% NaCl; 0.0794 wt% KCl; 0.3282 wt% Na_2SO_4; 0.2562 wt% $MgCl_2$; 0.0717 wt% $CaCl_2$; and 0.0020 wt% $SrCl_2$ (101). 10 wt% NaCl and 10 wt% $CaCl_2$ (98); and 10 wt% NaCl, 5.06 wt% KCl, and 5 wt% $CaCl_2$ (100).

Thermodynamic Inhibitors for Hydrate Plug Melting

XIAOYUN LI,[a] LARS HENRIK GJERTSEN, AND TORSTEIN AUSTVIK

Statoil Research Centre, Postuttak, 7005 Trondheim, Norway

ABSTRACT: The hydrate melting efficiency of thermodynamic inhibitors has been shown to depend on hydrate plug properties, inhibitor properties, and the turbulence of the liquid system in question. Inefficient mixing of inhibitor and water, lack of contact between inhibitor and hydrate, and a solid or gel precipitation in the melting region have been demonstrated to give low melting efficiency. MeOH seems to be the most efficient inhibitor for melting porous plugs, but may not melt a plug with low porosity. MEG has the ability to penetrate into a compact plug and cause melting. However, when used at high concentrations, MEG may freeze out as a solid or gel, which can lead to reduced hydrate melting efficiency. At high pressures and in the presence of a hydrocarbon liquid, MEG solutions were shown to freeze out at temperatures more than 30°C higher than values given in the literature. TEG seems to easily freeze and become inefficient for melting hydrate plugs.

INTRODUCTION

In the North Sea, hydrate plugs are frequently formed in wells, flow lines and topside. To remove hydrate plugs, thermodynamic inhibitors, such as methanol and glycols (MEG and TEG), are often employed. However, an inhibitor that is efficient for one case may be useless for the removal of the next hydrate plug. When inhibitors are successfully applied, the time between injection and response is usually short, normally a few hours. When failing, a plug may not be removed by the inhibitor for weeks or months. For example, it has been experienced several times that TEG did not melt a hydrate plug for more than a month. There are also cases where MeOH or MEG did not melt hydrate plugs effectively.

Low melting efficiency has been experienced in different systems such as oil producers, gas condensate lines, and in WAG injectors. Lab experiments as well as field experience have shown that hydrate plugs formed in different systems, and at different conditions may be very different concerning porosity and oil content.[1] The characteristics of plugs and inhibitor properties, in turn, influence the inhibitor efficiency for plug melting. The inhibitor distribution, effective contact between inhibitor and hydrate, and solid or gel precipitation during plug melting are believed to be crucial factors for hydrate plug melting efficiency when using inhibitors.

In this work the mixing between inhibitors and water, with inhibitor properties like density and viscosity as important parameters, has been studied. The conditions under which glycols may freeze out have also been studied and the melting efficiency of

[a]Telecommunication. Voice: (47) 73584453; fax: (47) 73584628.
XLI@Statoil.com

TABLE 1. Important selected properties of the studied inhibitors at 25°C and 1 bar

Fluid	Water	MeOH	MEG	TEG
Density (kg/L)	0.995	0.787	1.11	1.12
Viscosity (centipoise)	0.923	0.539	17.647	37.783
Freezing point (°C)	0	−98	−13	−5

three inhibitors on various plugs has been evaluated. The inhibitors studied were methanol (MeOH), monoethyleneglycol (MEG), and triethyleneglycol (TEG). Important, selected properties of the inhibitors,[2] at 25°C and 1 bar, are given in TABLE 1.

INHIBITOR DISTRIBUTION IN WATER

The thermodynamic hydrate inhibitors are miscible with water, but the rate at which inhibitors are completely mixed with water is not well defined. When an inhibitor fails to melt a hydrate plug, it is frequently asked whether the inhibitor is at all in contact with the hydrate plug. The inhibitor distribution in water was studied in a phase distribution rig to provide visual information. The rig was two meters high with a diameter of 0.23 m, a 550 ml glass cylinder, and small beakers. In the distribution rig, inhibitors were injected either from the top or bottom, the rig was already filled with water and condensate. Air could be bubbled in through the bottom of the

TABLE 2. MeOH concentrations along the aqueous phase at different sampling times when MeOH was added from the top to the condensate and water phases

Sampling point (number of liters from bottom)	MeOH concentration (wt%) in the aqueous phase at different times			
	at time 1[a]	at time 2[b]	at time 3[c]	at time 4[d]
47				60 wt%
41				41 wt%
35				33 wt%
30	<0.5 wt%			
28		4 wt%	19 wt%	30 wt%
18		4 wt%	18 wt%	28 wt%
14		4 wt%	18 wt%	27 wt%
4				27 wt%

[a] Just after the MeOH injection when MeOH and water were separated.
[b] After moderate bubbling of air from the bottom.
[c] After even longer and stronger bubbling of air from the bottom.
[d] After intense bubbling of air from the bottom.

rig to allow mixing of the fluids. Along the sides of the rig and the cylinder flask, sampling points were mounted. The inhibitor concentrations at different levels of the aqueous phase were sampled and analyzed by gas chromatography (GC). Ink was added to the inhibitors to provide visual information.

Experiments with Methanol

MeOH was injected from the top of the rig filled with condensate (heavier than MeOH) and water; the condensate separated the MeOH and water phases and prevented contact. Natural diffusion/convection and moderate air bubbling gave too low a driving force to obtain good mixing between MeOH and water. Even after intense bubbling, a higher concentration of MeOH at the top of the water phase was detected. In TABLE 2, the MeOH concentrations as a function of the height of the aqueous phase are listed. FIGURE 1 shows a picture of the phase distribution rig when air was bubbled through the water, condensate, and MeOH phases from the bottom. In the

FIGURE 1. Phase distribution experiment with MeOH on the top of water. The MeOH was dyed to a dark blue color during the experiment.

picture, the MeOH and the condensate phases are mixed mechanically, whereas the water phase remains almost undisturbed.

In another experiment, MeOH was injected from the top of a flask filled with condensate, lighter than MeOH, and water. The MeOH flowed easily through the condensate and accumulated on the top of the water phase. After three days, there was no noticeable mixing between the MeOH and water. The MeOH concentration at the upper and lower part of the aqueous phase was 90 and 0 wt%, respectively. Only powerful mechanical disturbance at the MeOH/water interface caused mixing.

To simulate the mixing of MeOH with melting water at the top of a compact plug, water was injected at the bottom of a flask filled with MeOH. Some gas was also bubbled through the bottom, simulating gas released during hydrate melting. Although some turbulence was created during the water and gas injection, very little MeOH was mixed with the water. The MeOH concentration at the bottom of the water phase was only 5 wt%, although 44 wt% should result from complete mixing. MeOH was also injected from the bottom of a water phase in one experiment. Gravity forced an almost complete and immediate mixing of the two phases.

Experiments with Glycols

In an experiment in which MEG was injected from the bottom of the distribution rig, MEG displaced the water at the lower part without mixing with the water. In the experiments where MEG was added from the top, the condensate did not hinder the downward motion of MEG. However, the degree of mixing between water and MEG depended very much on how MEG was injected and the amount of water employed. When MEG was carefully added to the top of a high water column, the MEG concentration in the upper part of the water phase was low. When TEG was added from the top into a water phase, most of the TEG flowed directly through the water phase and accumulated at the bottom. Powerful forced convection was necessary to obtain an even distribution of the inhibitors in most of the glycol experiments.

By using a simple numerical method to solve the Fick's second law of diffusion,[3] and using the diffusion coefficient in dilute water solutions,[2] the time needed to get an even inhibitor concentration in water has been calculated for some of the inhibitor distribution experiments. The results show that the diffusion time is in the range of years, confirming that inhibitor diffusion into water is a very slow process.

FREEZING OF GLYCOLS

The freezing point of pure MEG and TEG is $-13°C$ and $-5°C$, respectively. Furthermore, aqueous solutions of MEG and TEG have lower freezing points than do the pure substances.[1] However, field experience shows that at high pressures MEG solutions may freeze out at temperatures higher than those given in the literature. The freezing experiments with MEG solutions were performed in a high pressure sapphire PVT cell.[5] The freezing temperature of MEG solutions was observed to depend on both pressure and the composition of the HC liquid present. For an aqueous 75 wt% MEG solution in contact with a HC liquid at elevated pressure, the freezing point was more than 30°C higher than that given by Campbell.[4] The freezing points of MEG solutions from the present study and data from Campbell are plotted

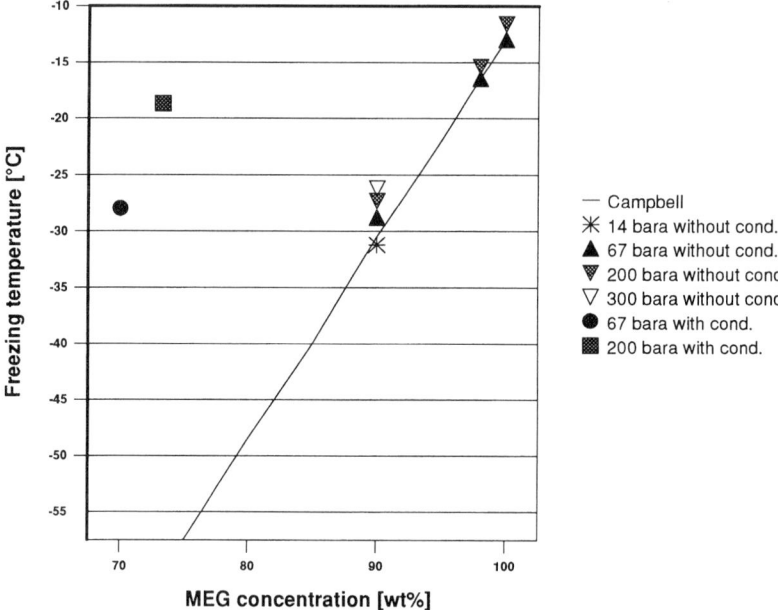

FIGURE 2. Freezing temperature of MEG solutions under various conditions.

in FIGURE 2. At an onshore facility, a 70 wt% MEG solution in contact with a condensate has been observed to freeze in the range of −26°C to −28°C at 67 bar. This result is in accord with the measured freezing point of −28°C in the present work. According to Campbell, the freezing temperature of a 70 wt% MEG solution is below −60°C. The difference in freezing temperatures of MEG solutions between our high pressure data with a HC liquid present and the data given by Campbell seems to increase with decreasing MEG concentrations.

Some qualitative experiments were performed where MEG and TEG were used to melt propane hydrate at −45°C and 1 bar. The equilibrium temperature for propane hydrate at 1 bar is −11°C. It was observed that the TEG added at 80, 90, and 100 wt% froze to a gel like phase in all the experiments. The frozen TEG gel was extremely viscous and very little hydrate was melted by TEG in these experiments. MEG did not freeze to a gel like phase, but a kind of half frozen white solid was observed in the MEG experiments. Furthermore, some hydrate was melted on contact with MEG.

MELTING OF HYDRATES AND ICE

PVTsim, by Calsep, was used for hydrate equilibrium calculations with MeOH, MEG, or TEG as the inhibitor. The depression of the hydrate equilibrium temperature for a model fluid, the composition of which is given in TABLE 3, at 200 bar has

TABLE 3. Composition of the model oil used for hydrate equilibrium temperature predictions

Component	Mol%
C1	55.155
C2	13.07
C3	6.535
C8	0.002
C9	0.036
C10	1.867
C11	5.766
C12	7.064
C13	7.261
C14	0.244

been calculated as a function of inhibitor concentration. The corresponding hydrate equilibrium temperature without inhibitor is 23°C. In FIGURE 3, a comparison of the inhibitor efficiency for three inhibitors is shown, with the inhibitor concentration in wt%, mol%, and vol%. TEG is most efficient on a mole basis. However, due to the high molecular weight, TEG is the least efficient inhibitor on a weight or volume basis except at very high concentrations. Up to 55 vol%, MeOH and MEG are almost equally good, but MEG is more efficient at higher concentrations according to the equilibrium predictions.

Four ethane hydrate plug melting experiments were performed in a specially designed steel cell at SINTEF,[6] two with MeOH and two with MEG. Liquids could be injected from the top of the cell, and a condensate lighter than MeOH was injected prior to inhibitor injection in all four experiments (compare with TABLE 4).

Experiments with MeOH. Even with gas circulation melting ceased only a short time after MeOH injection in the first experiment, where some water was injected prior to inhibitor injection. In the second experiment, where no water was injected and gas was not circulated either, the hydrate melted gradually.

TABLE 4. Information for the hydrate plug melting experiments in hydrate pipe

Inhibitor and Number	Extra water before inhibitor injection	Ethane circulated from bottom	Condensate
MeOH I	Yes	Yes	Yes
MeOH II	No	No	Yes
MEG I	Yes	Yes	Yes
MEG II	No	No	Yes

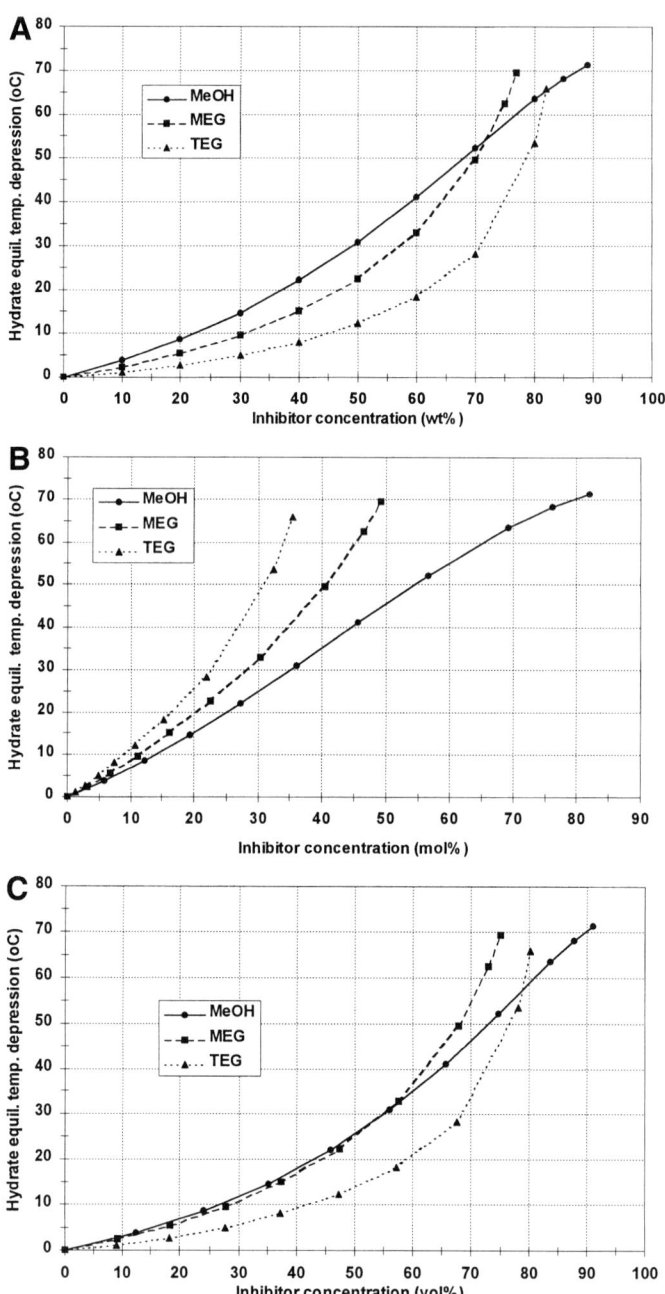

FIGURE 3. Hydrate equilibrium temperature depression as a function of inhibitor concentration in (**A**) wt%, (**B**) mol%, and (**C**) vol%.

Experiments with MEG. In the first experiment, with gas circulation, the hydrate melted gradually. In the second experiment, with no gas circulation, melting ceased some time after MEG injection.

Since all the inhibitors studied can melt ice in the same manner as melting hydrates, qualitative studies with pure MeOH, MEG, or TEG as the inhibitor were conveniently performed on porous snow and solid ice. The experiments were conducted in small beakers at atmospheric pressure and ink was added to the inhibitors.

Porous Snow Plugs. MeOH was clearly the most efficient inhibitor followed by MEG, although TEG was the least efficient inhibitor to melt porous snow on a volume basis. The experiments also revealed that a significant temperature gradient can exist in a plug. In snow melted by MeOH, the temperature in the upper part was $-30°C$ and only $-10°C$ in the lower part, with large temperature gradients lasted for a long period. For a snow plug melted by MEG, the temperature in the upper part was $-10°C$ and in the lower part $-24°C$. The melting temperature gradients in the snow plugs indicate inhibitor concentration gradients, with high MeOH concentration on the top in the MeOH experiments and high MEG concentration at the bottom in the MEG experiments.

Solid Ice Plugs. MeOH remained on the top of the ice plug after having melted some ice at the interface. MeOH was unable to penetrate into the ice plug and melting stopped after a short period. In the experiments with MEG and TEG, the inhibitors melted the outer layer of the ice plugs first, then they penetrated into the ice plug. MEG was the most efficient inhibitor for melting solid ice plugs.

DISCUSSION

Mixing of Inhibitor and Water

In common practice, an immediate mixing of water and inhibitor is expected when risers and wells are inhibited. However, our results demonstrated that two factors were very important in determining if the inhibitor and water may mix immediately.

Density Difference. When a lighter phase was carefully injected on the top of a heavier phase, no detectable mixing occurred unless a strong forced convection was applied. Some turbulence during the injection caused limited mixing, but the turbulence was normally not strong enough to overcome the density difference and provide good mixing. A third immiscible phase, like oil on the top of a water phase, may reduce the degree of turbulence during injection. Furthermore, a heavy oil phase may hinder the physical contact between inhibitor and water.

Viscosity. Viscous forces are frictional forces that attempt to make all parts of a fluid move at the same velocity. When the viscosity of a fluid is high, its ability to mix into another medium is low. The low viscous MeOH and water mixed easily when water was added from the top to an MeOH phase. Due to the high viscosity of glycols, it was more difficult to get a good mixing of glycols and water although the glycol was added from the top into the water phase. Furthermore, it was more difficult to mix the more viscous TEG with water than to mix MEG and water. When an inhibitor and water did not mix naturally, strong forced convection was necessary to obtain good mixing.

Inhibitor and Plug Properties versus Melting Efficiency

Theoretically, MeOH is an efficient hydrate inhibitor over the entire concentration range (vol or wt%). Even used at extremely high concentrations, MeOH will not freeze out due to its low freezing point. However, experiments indicated that special conditions may make MeOH an inefficient inhibitor for plug melting. An oil phase heavier than MeOH could hinder the physical contact between MeOH and the hydrate plug. Even when MeOH was in direct contact with a hydrate plug, a low melting rate could still be obtained. During the hydrate plug melting experiment with MeOH, where melting stopped early, the free water layer added prior to MeOH injection was believed to prevent contact between MeOH and the hydrate plug. For a compact plug, a water film released from melting could prevent MeOH from melting the plug beneath it, as was demonstrated in one ice plug melting experiment. MeOH is believed to be most efficient in melting rather porous plugs.

When TEG is used to melt a hydrate plug, its high viscosity makes it move very slowly towards and in the plug. As soon as some TEG comes into contact with the plug, melting starts. At high concentrations, the temperature in the melting region can be very low, which may cause the freezing out of TEG. The freezing out and the high viscosity of TEG at low temperatures hinders more TEG from coming into contact with the hydrate plug and the melting may stop. Furthermore, for a plug with a high water content, or when some hydrate has melted, the melting efficiency will be low since TEG is an especially poor inhibitor at low concentrations (vol or wt%). Experiments with TEG show that TEG froze out at low temperatures and was inefficient for melting both hydrate and ice/snow plugs.

MEG is believed to have a similar behavior to TEG with respect to hydrate plug melting but with a higher efficiency. At low concentrations (vol or wt%) the melting efficiency of MEG is considerably higher than that of TEG. Furthermore, experiments show that it was easier to get an even distribution of MEG in water than for TEG. The high density of MEG allowed it to penetrate to the bottom, indicating that a low permeable plug could be melted by MEG. However, the plug melting experiments with MEG revealed that high concentrations of MEG could accumulate at the bottom without melting the hydrate above it effectively. The viscosity and freezing point of MEG solutions are lower, which also makes MEG better than TEG for plug melting. However, the freezing out of MEG solutions at low temperatures may still be a problem, especially at high pressures and in the presence of a HC liquid phase.

There may be many explanations for the lack of ability to melt hydrate plugs efficiently with thermodynamic hydrate inhibitors. The freezing and slow motion of MEG and TEG at low temperatures is one. Although not studied in this work, local freezing of wax or gelling of oil at low temperatures can be another reason for low melting efficiency. The distribution of inhibitors and effective contact between the inhibitor and the hydrate plug are also important issues. When MeOH or MEG becomes inefficient due to lack of contact between the inhibitor and the hydrate plug, only very strong turbulence in the system may enhance the mixing and, hence, increase the melting rate. However, this is generally difficult to achieve in actual plug removal situations.

CONCLUSIONS

The hydrate melting efficiency of thermodynamic inhibitors depends on the plug properties, inhibitor properties, and the turbulence of the system. Inefficient mixing of the inhibitor and water, lack of contact between the inhibitor and the hydrate plug, and solid or gel precipitation in the melting region could cause low melting efficiency. MeOH is expected to be efficient for permeable/porous hydrate plugs, or when a high degree of turbulence can be obtained in the liquid system. However, if applied on a compact plug, MeOH may stay stagnant on the top of the hydrate plug without melting the plug beneath it. On a volume basis, MEG is at least as good as MeOH in melting hydrate according to equilibrium predictions. However, MEG is rather viscous, which may cause a low penetration rate into a plug. Applied at high concentrations, the low melting temperature may cause the freezing of MEG solutions and reduce the melting rate. At high pressures and in the presence of a HC liquid, MEG solutions froze out at temperatures more than 30°C higher than those reported in the literature. TEG may be inefficient for melting hydrate plugs. Due to the high viscosity and density of TEG, no other inhibitors can be squeezed into the plug after a failed attempt with TEG. TEG should not be used for hydrate plug melting.

REFERENCES

1. AUSTVIK, T., X. LI & L.H. GJERTSEN. 1999. Hydrate plug properties—formation and removal of plugs. To be published in NGH'99.
2. YAWS, C.L. 1995. Handbook of Transport Property Data. Gulf Publishing Company.
3. NOGGLE, J.H. 1996. Physical Chemistry. Harper Collins College Publishers.
4. CAMPBELL, J.M. 1994. Gas Conditioning and Processing, 7th edit. Campbell Petroleum Series.
5. GJERTSEN, L.H. & F.H. FADNES. 1999. Measurements and predictions of hydrate equilibrium conditions. To be published in NGH'99.
6. BORTHNE, G., L. BERGE, T. AUSTVIK & L.H. GJERTSEN. 1996. Gas flow cooling effect in hydrate plug experiments. Proceedings of the 2nd Int. Conf. on Natural Gas Hydrates. Toulouse, France.

A Novel Approach for Oil and Gas Separation by Using Gas Hydrate Technology

K.K. ØSTERGAARD,[a] B. TOHIDI,[a,b] A. DANESH,[a]
R.W. BURGASS,[a] A.C. TODD,[a] AND T. BAXTER[c]

[a]*Department of Petroleum Engineering, Heriot-Watt University, Riccarton, Edinburgh EH14 4AS, United Kingdom*

[c]*Uragaig Ltd., 9 Deemount Avenue, Aberdeen AB11 7UF, United Kingdom*

ABSTRACT: It is known that gas hydrates remove the light ends from reservoir fluids. Therefore, controlled hydrate formation in reservoir fluids could be an attractive option for separating oil and gas; that is, to replace conventional production facilities. In this communication we present the results of an integrated experimental and modelling study on the feasibility of the process, and the impact of the various parameters on the rate of hydrate formation. The study investigated the impact of parameters, such as mixing, water history, temperature, pressure, volume of reactor, heat removal requirements, and the quality of separated liquid. The work identified the major parameters and some of the technological requirements. Based on the experimental data, a simplified mass transfer model was constructed to simulate the kinetics of the separation process and to calculate the reactor volume and heat requirements at a specified degree of conversion. The results showed that it is possible to remove most of the lights from the liquid hydrocarbon phase by hydrate formation. The resulting liquid phase could be suitable for pipeline export or tanker loading after some treatment. Associated gas could be recovered locally from the hydrate phase. Alternatively, in cases where there is no infrastructure for transporting this gas, it might be exported as a hydrate slurry, as proposed by Gudmundsson and coworkers.

INTRODUCTION

The formation of gas hydrates is regarded as a source of problems in production, transportation, and processing of reservoir fluids, due to potential blockage of pipelines and process facilities. This work looks into the possibility of using hydrate technology as a means of separating oil and gas by controlled formation of gas hydrates in reservoir fluids.

Reservoir fluids are conditioned in production facilities to deliver a liquid and a gas that meet certain vapor pressure and pipeline specifications, respectively. Generally, light components are removed from reservoir fluids by a reduction of pressure and progressive flashing in several separators, operating at different pressures.

The objective of this integrated experimental and modelling study is to investigate the possibility of employing hydrate technology in separating oil and gas in

[b]Telecommunication. Voice: +44 131 451 3672; fax: +44 131 451 3127.
bahman.tohidi@pet.hw.ac.uk

reservoir fluids; that is, to eliminate or reduce the requirement for conventional oil/gas separators. The study investigates the effect of different parameters and their significance in the process. The use of hydrate technology in oil and gas separation could improve the economics of remote and marginal oil fields. It can also be regarded as a compliment to the novel natural gas transportation techniques suggested by Gudmundsson and coworkers.[1,2]

EXPERIMENTAL

A high pressure (69 MPa) hydrate rig was used to conduct the experiments (see FIGURE 1). A detailed description of the experimental set up is given elsewhere.[3] The experiments were performed using a North Sea black oil (BO) with a bubble point pressure of 9.411 MPa at 373.15 K. Distilled water was used in all the tests.

Once the cell had been loaded, the desired pressure was achieved by injecting mercury. With the cell static, the temperature was then reduced to achieve the required degree of subcooling. Then, the mixing was started simultaneously with the logging of pressure and temperature as functions of time. Generally, after each test the cell was heated to about 303.15 K for a number of hours to dissociate hydrates and to remove the water history. Five tests were performed under conditions outlined in TABLE 1.

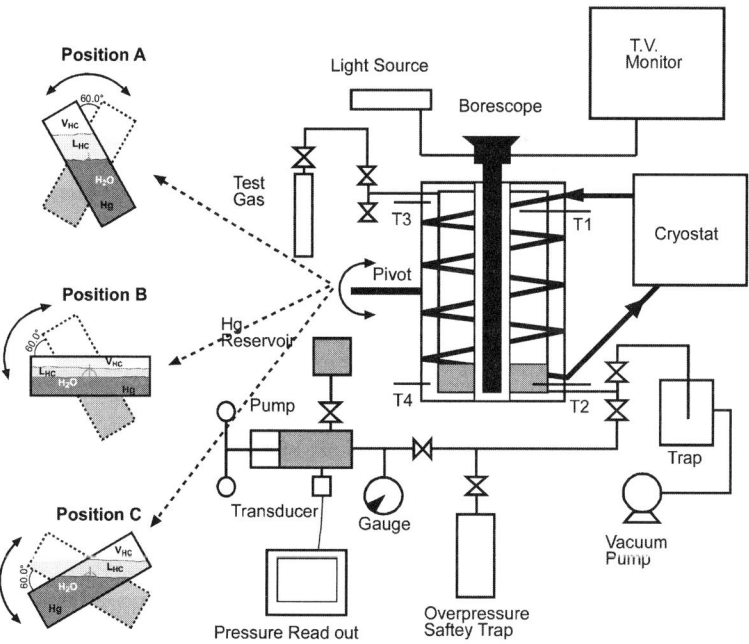

FIGURE 1. Schematic of experimental equipment.

TABLE 1. Main features of the hydrate formation tests

Test No.	Position[a]	T/K	P_i^b/MPa	ΔP^c/MPa	ΔT^d/K	Tuned K^e/cc·s^{-1}
1	A	281.15	4.151	2.301	5.61	0.15
2	B	281.15	4.151	2.301	5.61	2.5
3	C	281.15	4.137	2.287	5.59	11.0
4	C	275.75	2.730	1.880	8.10	9.0
5[f]	C	275.75	2.758	1.908	8.17	23.0

[a]See FIGURE 1.
[b]P_i is initial pressure.
[c]$\Delta P = P - P_{equilibrium}$.
[d]$\Delta T = T_{equilibrium} - T$.
[e]Values for methane.
[f]This test was performed from thawed ice.

THERMODYNAMIC AND KINETIC MODELLING

A detailed description of the thermodynamic modelling of hydrate phase equilibria is given elsewhere.[4] The mass transport limited model for the growth of gas hydrates by Skovborg and Rasmussen[5] was used in modelling hydrate kinetics tests. The tuning parameters of the model are the products of the mass transfer coefficients ($k_{i,L}$) for different components (i) and the hydrocarbon–water interfacial area ($A_{(HC-water)}$):

$$K_i = k_{i,L} A_{(HC-water)} \tag{1}$$

For a single component system, K can be tuned by matching experimental and calculated pressure drops versus time, due to hydrate formation. This procedure, however, is not possible for a multicomponent system unless the mass transfer coefficients for the different components are related to each other. Skovborg et al. suggest using similar K-values for all components in the fluid.[6] However, we decided to relate the mass transfer coefficients for the various components to that of methane, based on the diffusion coefficients in water according to a single film theory. The empirical methods for predicting diffusion coefficients in liquids proposed by Wilke and Chang, and Hayduk and Minhas have been used.[7]

RESULTS AND DISCUSSIONS

The overall mass transfer coefficients, K-values, for the five experiments in TABLE 1 were determined. Three of the tests (1, 2, and 3) were performed at 281.15 K and at an initial pressure, P_i, of about 4.14 MPa with different rocking angles, to investigate the effect of mixing efficiency. The last two tests (4 and 5) were conducted at 275.75 K and $P_i \approx 2.74$ MPa, with the objective of studying the impact of water history, that is, Test 4 with no water history and Test 5 in the presence of thawed ice. The tuned K-values of methane in the above experiments are presented in TABLE 1.

FIGURE 2 presents experimental temperature profiles and experimental and calculated pressure profiles for Test 2 and Test 5. The *K*-values have been tuned to match the pressure drop at the start of the hydrate growth, where the driving force is relatively high, simulating the condition in a continuous hydrate reactor for gas–oil separation.

TABLE 1 shows that mixing efficiency plays an important role in the rate of hydrate formation. The mass transfer coefficient has increased by a factor of about 70 between Test 1 and Test 3 (mixing positions A and C in FIG. 1, respectively) due to efficient mixing and surface renewal. The mass transfer coefficient could be further improved by better design of the reactor. This could significantly improve the rate of hydrate formation, reducing the reactor volume.

The water history is also seen to have a significant impact on the rate of formation. Between Test 4 (no water history) and Test 5 (thawed ice) the mass transfer has increased by a factor of about 2.5. It is unlikely that water history could cause such a change in the mass transfer coefficients, and thereby the formation rate. In other words, the kinetic model is too simplistic to take into account the effect of water history. Therefore, the tuned *K*-values are more likely to be fitting parameters, as also pointed out by Sloan.[8]

In Test 5, where the formation rate was relatively high, it was noted that the cell temperature increased, particularly at the beginning of the hydrate growth where the formation rate is the highest (FIG. 2). This meant that the heat released due to hydrate formation was not being completely removed by the cooling system. This is an important consideration in reactor design, as inadequate heat removal may limit rate

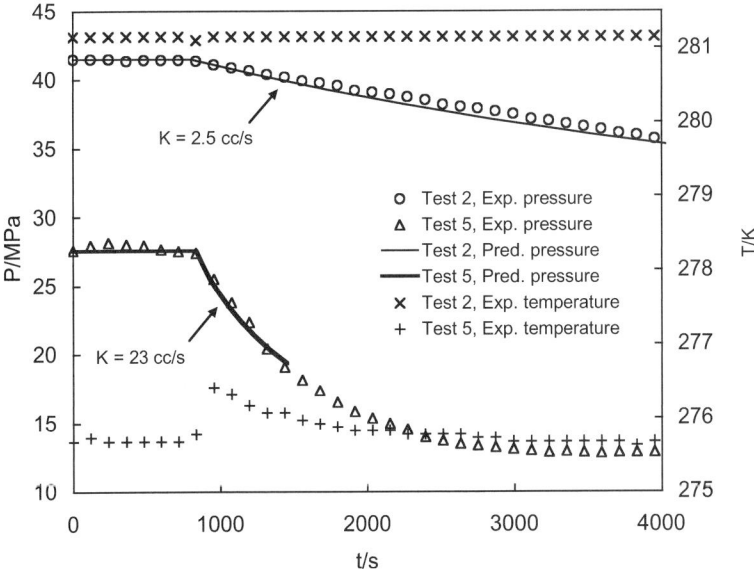

FIGURE 2. Pressure and temperature profiles for Test 2 and Test 5.

of hydrate formation. Also, the kinetic model, which is mass transfer limited, may not be applicable if the heat removal is not efficient. The use of an ice/water slurry, rather than heat exchangers, can be an effective way to remove the heat of hydrate formation.

CONTINUOUS HYDRATE REACTOR SIMULATION

The results (mass transfer coefficients) of the laboratory tests were used in the design of a simplified continuous stirred tank reactor (CSTR) model. The important difference between a conventional chemical reaction and hydrate formation is that in the former the reaction occurs with constant stoichiometric ratios whereas in the latter the stoichiometric of gas hydrates depends on temperature, pressure, and composition of the hydrate forming system. This adds to the complexity of hydrate reactor modelling.

CASE STUDY

The application of hydrate technology in oil–gas separation on a typical North Sea offshore production platform was investigated. The selected field was producing oil at 55,000 B/D with a calculated GOR of 950 SCF/B. Since the reservoir fluid was not available for determine the overall mass transfer coefficients, the values calculated for the black oil were used for simulation purposes.

The reactor model was employed to investigate the effect of various parameters on the volume of a hydrate reactor processing the above fluid. The main objectives were to investigate the effect of mixing efficiency, pressure, temperature and degree of conversion on the reactor size. The results are summarized below:

Effect of Mixing (K-Values). As shown in Equation (1) the K-values represent the rate of mass transfer in units of volume per time. This means that if K is doubled the reactor volume is halved. The maximum value of K obtained from the laboratory tests was 23 cc/s. It is believed, however, that the efficiency of mixing could be improved by better design.

Effect of Pressure. The driving force and the rate of hydrate formation increases with an increase in the reactor pressure. This results in a reduction in the reactor volume. An example of the effect of pressure on the reactor volume is shown in FIGURE 3 for an arbitrary set of input variables. The reactor volume is a stronger function of the reactor pressure closer to the hydrate phase boundary. At pressures above the bubble point of the hydrocarbon phase, changes in the operating pressure have only minor effects on the overall reactor volume. However, it should be mentioned that the increase in the reactor pressure results in higher GOR for the separated liquid hydrocarbon phase.

Effect of Temperature. The effect of temperature on the reactor volume is shown in FIGURE 4 for two different operating pressures, 3.792 MPa and 4.826 MPa. The figure shows that increasing the operating temperature by 10 K (e.g., from 273.15 K to 283.15 K) increases the reactor volume by an order of magnitude. This indicates the importance of cooling method on the efficiency of the process.

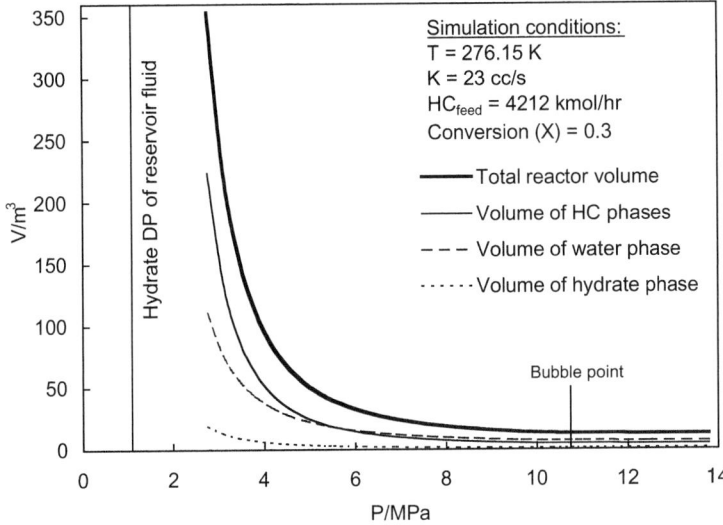

FIGURE 3. Example of reactor volume as function of pressure.

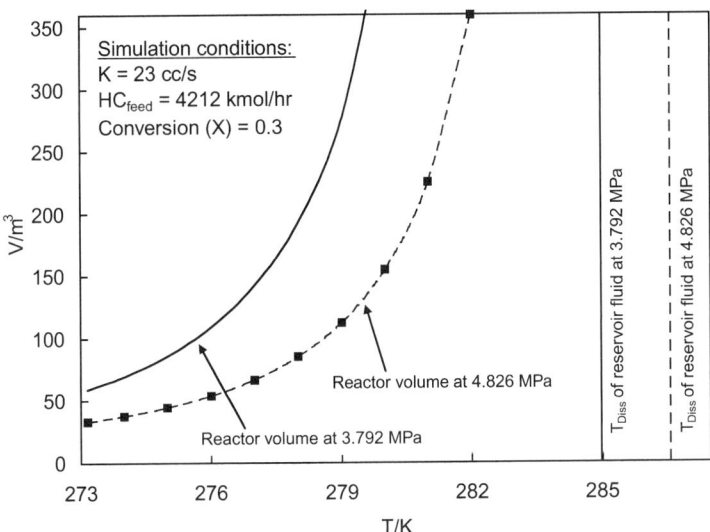

FIGURE 4. Example of reactor volume as function of temperature.

Effect of Conversion Factor (X). FIGURE 5 presents the calculated reactor volume as a function of the molar conversion of the hydrate forming components at 276.15 K and 273.65 K. As expected, the reactor volume increases with an increase in conversion factor. Interestingly, as shown in the figure, the calculated reactor volume is almost a linear function of the conversion factor, particularly at 273.65 K. As a result, the rate of formation, that is, the driving force, is almost independent of the conversion. This is illustrated in FIGURE 6 where the bubble point pressure and the hydrate dissociation pressure for the above systems are shown. The bubble point pressures decrease with increasing conversion, indicating an increase in the removal of the light ends, whereas the hydrate dissociation pressures are almost unchanged.

Effect of Water/HC Ratio. From Equation (1) it is obvious that the mass transfer coefficient, and thereby the rate of formation, is a function of the interfacial area. The interfacial area is a function of the amount of both water and hydrocarbon phases. Different ratios of water/HC have not yet been experimentally investigated but are believed to play an important role and should be optimized to provide maximum reaction rate. Currently, a number of experiments are being conducted at different water/HC ratios to investigate its effect on the rate of reaction and reactor size. The water/HC ratio chosen for the CSTR modelling reflects the value used in the experimental tests.

Hydrate Reactor Design for Oil and Gas Separation

A number of operational parameters were chosen based on the results of the hydrate reactor model and the fuel requirement on the platform (i.e., 15×10^6

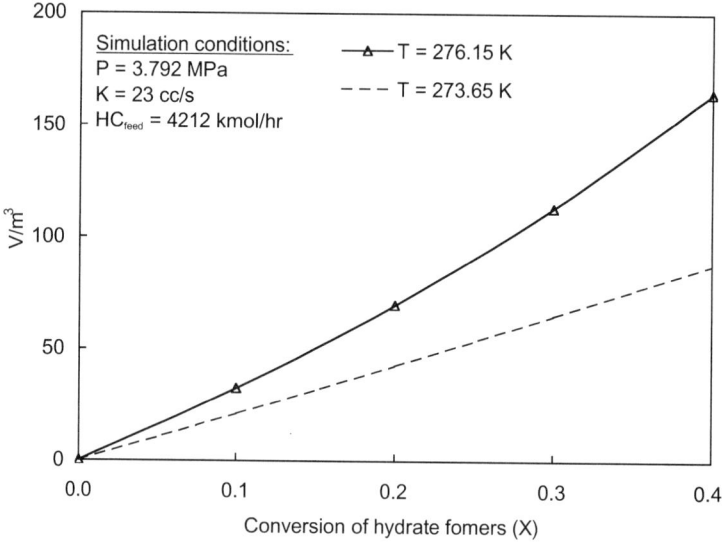

FIGURE 5. Example of reactor volume as function of conversion.

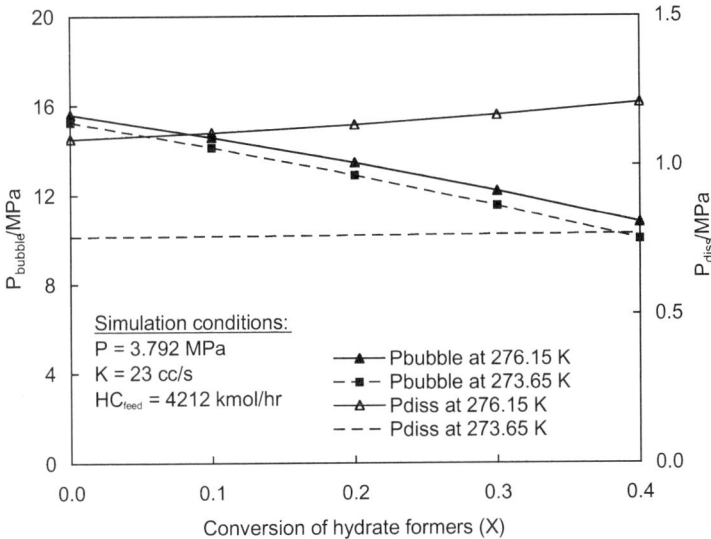

FIGURE 6. Hydrate DP and bubble point pressure as function of conversion.

SCF/day). The main objectives were to calculate the volume of the hydrate separator, the necessary heat load, quality, and composition of the separated oil and gas.

TABLE 2 summarizes the important design parameters for this study. An overall molar conversion of 0.68 (based on hydrate forming components) was chosen in order to match the fuel consumption on the platform. This conversion was achieved by using two reactors in series. FIGURE 7 shows the flow diagram of the reactors with molar flow rates as presented in TABLE 3.

TABLE 2. Design parameters for the case study

Design Parameter	Value
Feed rate of reservoir fluid, HC_{feed}	4212 kmol/hr (55MBD)
Temperature, T	273.65 K
Pressure, P	4.137 MPa
Water to oil feed ratio	10
Mass transfer coefficient, K	23 cc/s (methane normalized)
Overall conversion of hydrate formers, X	0.68
Conversion in first CSTR	0.45
Conversion in second CSTR	0.42

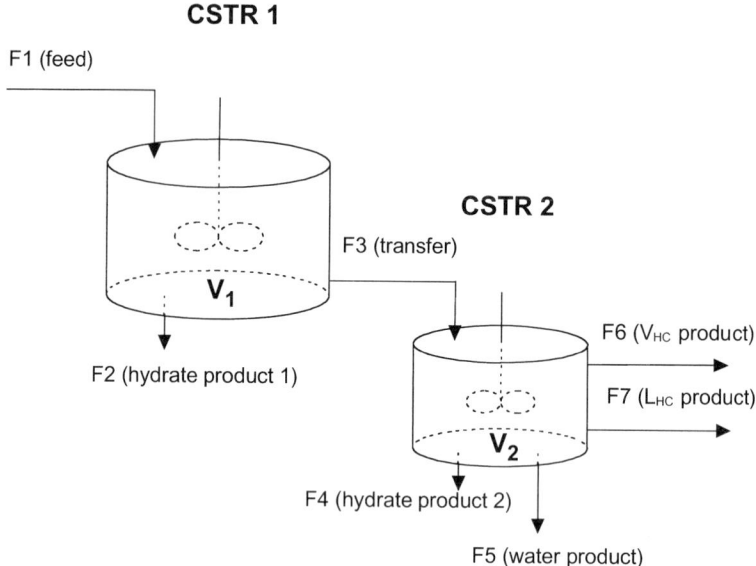

FIGURE 7. Flowchart of the case study (two CSTRs in series).

TABLE 3. Calculated molar flow rates of components based on the design parameters in TABLE 2

Component	F1 (kmol/hr)	F2 (kmol/hr)	F3 (kmol/hr)	F4 (kmol/hr)	F5 (kmol/hr)	F6 (kmol/hr)	F7 (kmol/hr)
CO_2	63.2	18.4	44.8	9.6	17.2	0.7	17.3
N_2	19.4	5.7	13.7	6.3	0.9	2.0	4.5
C1	1971.2	971.3	999.9	506.0	36.8	43.7	413.4
C2	251.1	123.7	127.4	51.4	1.3	1.3	73.4
C3	147.9	29.1	118.8	15.8	0.4	0.5	102.0
i-C4	33.3	1.6	31.7	0.9	0.0	0.1	30.7
n-C4	75.0	2.6	72.4	1.6	0.1	0.1	70.6
i-C5	36.6	—	36.6	—	0.0	0.0	36.6
n-C5	48.0	—	48.0	—	0.0	0.0	48.0
C6+	1566.3	—	1566.3	—	0.0	0.0	1566.3
H_2O	42120.0	7414.9	34705.1	3797.3	30907.6	0.0	0.3
Total	46332.0	8567.3	37764.7	4388.9	30964.3	48.4	2363.1

$F_{\text{out,gas in hydrates}} = F2_{\text{tot}} + F4_{\text{tot}} - F2_{\text{water}} - F4_{\text{water}} = 8567.3 + 4388.9 - 7414.9 - 3797.3 = 1744$ kmol/hr.

Reactor Volume. TABLE 4 lists the calculated total volume of the reactors as well as the calculated volumes of the different phases present in the reactors. The volume estimates should be regarded as tentative, mainly due to uncertainties in the scale up of the mass transfer coefficients from the experiments. It is likely that the reactor volumes can be reduced significantly in purpose designed reactors with more efficient mixing. The volumes can also be reduced further by using more reactors in series.

Cooling Load. The latent heat of hydrate dissociation/formation is estimated at $\Delta H_d \approx 65$ kJ/mol (gas in hydrates). By neglecting sensible heat, heat-losses to the surroundings, and so forth, the cooling load for the hydrate reactor can be calculated, using the data listed in TABLE 3:

$$Q = \Delta H_d \cdot R_{out,\,gas\,in\,hydrate} \Leftrightarrow$$

$$Q = 65 \left(\frac{kJ}{mol\,gas\,in\,hydrate} \right) \left(1744 \frac{kmol}{hr} \right) \left(\frac{1}{3600s} \right) \approx 31500\ kW.$$

CONCLUSIONS

An integrated experimental and theoretical study was conducted to investigate the possibility of employing hydrate technology to oil–gas separation, as an alternative to conventional separators. A series of hydrate kinetics experiments were conducted on a black oil to investigate the impact of different operational parameters on the rate of the reaction. A continuous hydrate reactor was simulated and employed to investigate the impact of different parameters (i.e., pressure, temperature, mixing efficiency, conversion factor, etc.) on the volume of the reactor.

In a case study, the applicability of hydrate reactor for oil and gas separation on a North Sea platform, processing 55 MBD oil with a GOR of 950 SCF/B, was investigated. The study showed that although it may not be economical to achieve a stock tank stabilized oil through hydrate formation, significant portion of the light ends could be removed (i.e., changing the GOR from 950 to 270 SCF/B), reducing the load on conventional separators significantly. The calculated reactor volumes were

TABLE 4. Calculated required reactor volumes for the two CSTRs in series based on the design parameters in TABLE 2

Parameter	Volume, CSTR 1 (m^3)	Volume, CSTR 2 (m^3)	Total reactor volume (m^3)
Reactor volume	77.9[a]	36.6[b]	114.5
Volume of water phase	32.9	19.1	52.0
Volume of HC phases	36.1	14.5	50.6
vapor	15.0	0.8	15.8
liquid HC	21.1	13.7	34.8
Volume of hydrates	8.9	3.0	11.9

[a]This corresponds to a cylindrical vessel with the dimensions, length = 11.7 m, diameter = 2.9 m.
[b]This corresponds to a cylindrical vessel with the dimensions, length = 9.1 m, diameter = 2.3 m.

comparable with conventional separators. The cooling requirements play an important role on the economics of the process. It is believed that the process could improve the economics of development in remote and marginal fields, where the use of hydrate technology for transportation of associated gas is being considered.

The economics of the process in comparison to conventional offshore separator are currently under investigation.

ACKNOWLEDGMENT

This work has been initiated and supported by BP Amoco Exploration Operation Company Ltd., which is gratefully acknowledged. Also, the authors wish to thank BP Amoco for the permission to publish this work.

REFERENCES

1. GUDMUNDSSON, J., F. HVEDING & A. BØRREHAUG. 1995. Transport of natural gas as frozen hydrate. Proceedings of the 5th International Offshore and Polar Engineering Conference, June 11–16: 282–283, The Hague, the Netherlands.
2. GUDMUNDSSON, J.S., V. ANDERSON, O.I. LEVIK & M. PARLAKTUNA. 1998. Hydrate concept for capturing associated gas. SPE 50598. Proceedings of the 1998 SPE European Petroleum Conference, October 20–22, The Hague, the Netherlands.
3. TOHIDI, B., A. DANESH, R.W. BURGASS & A.C. TODD. 1994. Hydrate formed in unprocessed wellstreams. SPE 28478. Proceedings of the SPE 69th Annual Conference and Exhibition, September 25–28, New Orleans, USA. Π157–167.
4. TOHIDI, B., A. DANESH & A.C. TODD. 1995. Modelling single and mixed electrolyte solutions and its applications to gas hydrates. Chem. Eng. Res. Des. **73**(A): 464–472.
5. SKOVBORG, P. & P. RASMUSSEN. 1994. A mass transport limited model for the growth of methane and ethane gas hydrates. Chem. Eng. Sci. **49**(8): 1131–1143.
6. SKOVBORG, P., H.-J. NG, P. RASMUSSEN & U. MOHN. 1994. Gas hydrate kinetics in model systems and real petroleum fluid systems. Report SEP 9406. Department of Chemical Engineering. The Technical University of Denmark.
7. REID, R.C., J.M. PRAUSNITZ & B.E. POLING. 1987. The Properties of Gases and Liquids, 4th edit. McGraw-Hill.
8. SLOAN, E.D. 1998. Clathrate Hydrates of Natural Gases, 2nd edit. Marcel Dekker, New York.

Feasibility of Storing Natural Gas in Hydrates Commercially

R.E. ROGERS[a] AND YU ZHONG

Department of Chemical Engineering, Mississippi State University, P.O. Box 9595, Mississippi State, Mississippi 39762, USA

ABSTRACT: A study to determine the feasibility of using gas hydrates to store natural gas for peak-load use in electric power plants and other commercial applications gives new impetus for practical utilization of unique gas-hydrate storage properties. Laboratory apparatus was constructed to evaluate storage capacity attainable with hydrates and to achieve formation/decomposition rates of the hydrates at a level allowing commercial use. A fiber optics-camera system was installed to view the action inside the hydrate test cell; photographs are presented that give insight into hydrate formation at the gas/liquid interface, as well as the formation mechanism and packing density. From a quiescent water solution in a cell that contained 286 ppm sodium dodecyl sulfate, storage capacity of 155 vol/vol of natural gas in hydrates was achieved at a constant pressure of 3.89 MPa (550 psig) and a temperature of 276 K (37°F); 86% of the theoretical maximum storage capacity was achieved within 2.5 hours of hydrate initiation; hydrate particles were adsorbed on the vessel walls; interstitial water of the particles formed hydrates to give high-density packing. Formation rates and process simplification were such that a complete formation/decomposition cycle of commercial scale could be achieved within a 24-hour period. The work suggests the possibility of storing gas in hydrates industrially by means of a simple, quiescent system that rapidly forms the hydrates and packs the mass for storage.

INTRODUCTION

The subject DOE grant is the first serious attempt to use the unique properties of hydrates to store natural gas for applications such as peak loads of electric power plants. The objective was to determine the technical feasibility of commercial storage and, if found feasible, develop a conceptual design and determine an economic viability of a large-scale process.

At the beginning of the work, hydrate state-of-the-art recognized positive factors concerning the feasibility of gas storage in hydrates: (1) more than 180 vol-gas/vol-hydrate could be stored; (2) safety of storing natural gas in hydrates was unmatched, for gas would be essentially encased in ice; (3) gas could be stored in the hydrates at relatively low pressures; (4) the slow release of gas from hydrates in the event of storage-tank rupture enhanced safety; (5) vast quantities of gas stored in naturally occurring hydrates were known.[1]

---

[a]Telecommunication. Voice: 601-325-5106; fax: 601-325-2482.
rogers@che.msstate.edu

At the beginning of the project, the following negative factors diminished feasibility prospects: (1) the formation of hydrates in a quiescent pure water–hydrocarbon gas was extremely slow at hydrate-forming temperatures and pressures; (2) a dynamic process was complex for scale-up, especially if a mechanically stirred reactor became necessary to achieve acceptable formation rates; (3) the slurry separation and packing of hydrate particles formed in the reactor seemed prohibitive; (4) excessive free water typically trapped between hydrate particles would increase the storage tank size and cost; (5) a commercial storage process for hydrates had never been demonstrated. These negatives had to be overcome in order to develop a feasible process.

EXPERIMENTAL

A stainless steel cell of 3,900 cm^3 was constructed and equipped with two temperature (RTD) probes, pressure transducer, and relief valve. Cooling coils were placed inside the cell and around the exterior walls; a gas mass flowmeter, a constant pressure regulator, and a computerized Omega data-collection system were installed. Video capture of hydrate formation was made with still photographs taken through a 2″ thick, 3″ diameter pure quartz disk and with an invasive fiber optics/camera system viewed through sapphire-windowed ports in the cell; activities inside the cell could be viewed on a monitor screen and/or filmed with VHF video cassette recording.

DISCUSSION OF RESULTS

Hydrate Storage Capacity and Formation Rate

For hydrates to be a feasible means of storing natural gas for industrial use, an adequate storage capacity must be achieved, and the rate of formation must be high enough to make the process practical. First, ultimate storage capacity of gas in the hydrates is limited to about 181 vol-gas/vol-hydrate when all cavities in the hydrate are filled with natural gas; as pressure of the system is increased, the fraction of occupied cavities approaches its maximum. Since costs of processing and storage increase with pressure, there must be a tradeoff between added costs of higher pressures and the cost benefits of a larger fraction of cavities filled. Second, hindering bulk storage capacity is a large amount of free water usually trapped between hydrate particles[2] and isolated from gas. Much of the space in a storage tank would be occupied by this benign interstitial water. Thus, the volume of gas per volume of bulk hydrate mass could be low even though the cavities in the hydrate crystals of that mass were mostly filled with occluded gas. Third, even if maximum capacity is attained, commercialization would require a high rate of attainment.

A surface active agent solves the problems of achieving high storage capacities in a short period of time. Consider the results presented in FIGURE 1 of three experimental runs made to determine attainable capacity. The temperature was held constant at 275.4 K (36°F), and a sodium dodecyl sulfate (SDS) surfactant concentration was maintained at 286 ppm; the procedure and other conditions except pressure were repeated. Hydrates were formed in the cell until all of the water became tied up in the hydrate structure, that is, interstitial water also formed hydrates. Individual runs were made at pressures of 3.89, 3.47, 3.11, and 2.76 MPa (550.2, 503.6, 451.2, and 400.5 psig), where the chosen pressure was held constant throughout each run with a constant pressure regulator, and the rate of gas occluded was measured with a gas mass-flowmeter. Such a constant pressure procedure in a semicontinuous process would be preferred in an industrial operation.

The results presented in FIGURE 1 show that 155 (vol gas)/(vol bulk hydrate) storage capacity is achieved at a processing pressure of 3.89 MPa (550 psig). Furthermore, as a result of surfactant in the water, this much gas is incorporated into the hydrates in less than three hours of processing time in a quiescent system, a rate of hydrate formation about 700 times faster than in a pure water system. The 155 vol/vol represents 86% of the theoretical storage capacity if all cavities were filled, including hydrates formed from interstitial water.

At a common time after hydrate initiation, 291 minutes for example, the storage capacity is seen to increase with pressure as shown in FIGURE 2. A processing pressure of approximately 3.89 MPa (550 psig) is suggested for a large-scale process, since processing costs increase rapidly at higher pressures.

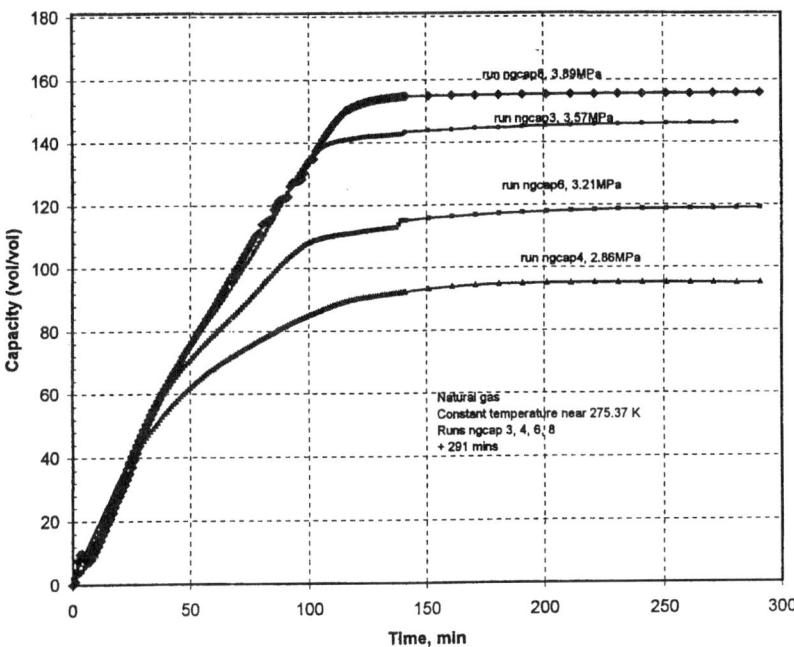

FIGURE 1. Capacity and rate of natural gas hydrate formation, 2.86 ppm SDS.

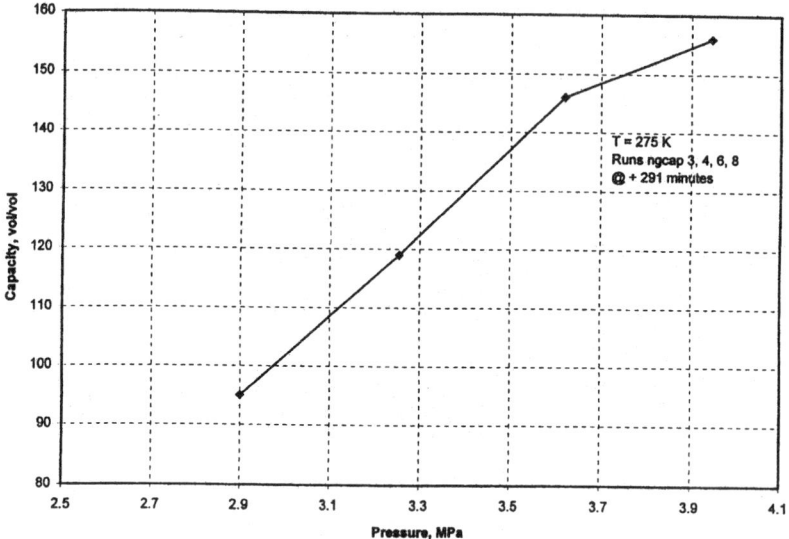

FIGURE 2. Pressure improves attainable natural gas hydrate capacity.

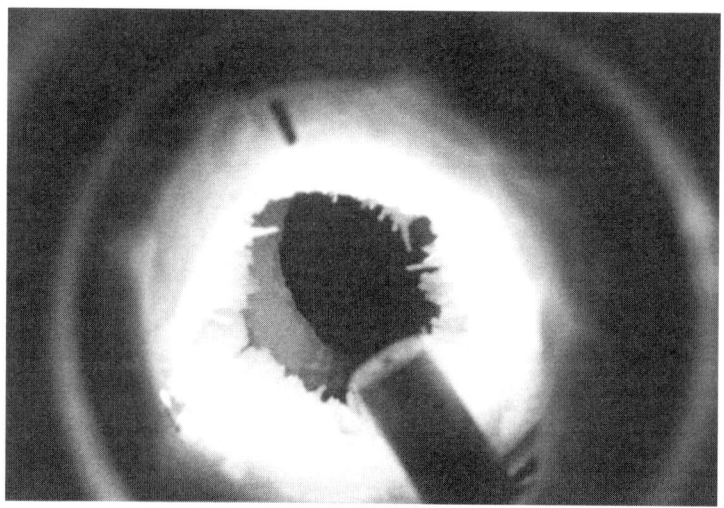

FIGURE 3. Hydrate particles pack concentrically with test cell.

Packing of Hydrate Particles

The inherent problem in scaling the process to industrial size was addressed. It was found that SDS surfactant facilitates the packing of hydrate particles as formed. By situating a camera at the water–gas interface, VHS film captured the formation of the hydrate particles subsurface and their movement (specific gravity is less than 1) from the water slurry to the surface of the water. Near the water surface, the particles moved to the stainless steel walls to be adsorbed, building inwardly from the walls in a concentric cylinder (see FIGURE 3). Practically, this means that an expensive processing step of separating particles from a water slurry and packing them in a storage container is preempted. Furthermore, storage space is maximized when the surfactant-laden particles build inwardly from the container walls.

The packing arrangement near ultimate capacity of natural-gas hydrates in our test cell may be viewed in FIGURE 4. The photograph was taken after the maximum capacity (155 vol/vol) of natural gas had been occluded in all of the water-SDS solution. This photograph of near theoretical capacity was taken 3 hours, 57 minutes after hydrate initiation. Note that the natural gas hydrates have symmetrically packed available space in the cell, filling the cell by growing inwardly to leave open only a small cylindrical void, slightly off-center, through which the bottom of the cell is visible.

Conversion of Interstitial Water

The conversion to hydrates of interstitial water was achieved in a semicontinuous process with a system containing natural gas, water, and 286 ppm of SDS. Gas was continuously added to the cell to replenish gas being occluded and to maintain a constant pressure during hydrate formation. At the higher constant pressure, a greater

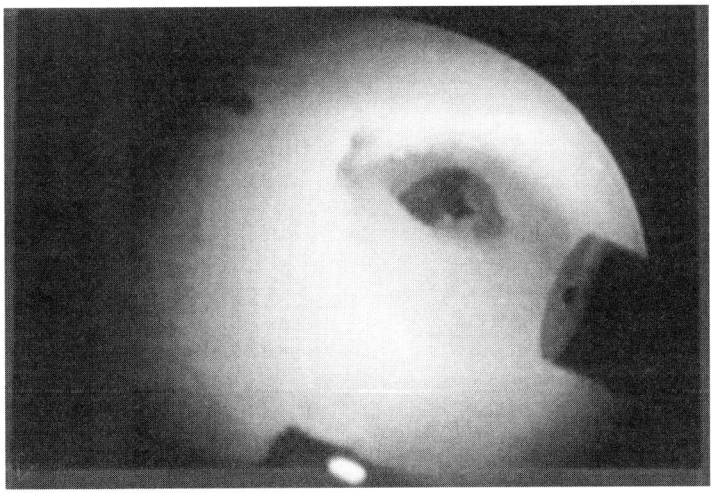

FIGURE 4. Symmetrical packing of hydrates at ultimate storage capacity.

fraction of the cavities fill than if gas was added batchwise and pressure allowed to drop. Natural gas in our test cell was maintained at 3.89 MPa (550 psig). Hydrates were allowed to form until all of the free water initially in the bottom of the cell was dissipated into the hydrate structure or as free water in the interstices; then the free water remaining in the interstices was converted to hydrates. Results showed that within three hours, 86% of theoretical capacity was achieved.

Process Simplification

The feasibility of a practical, economical hydrate storage process depends on simplicity. For the sake of simplicity, it would be preferable to have no moving parts in the hydrate formation storage tank in order to reduce maintenance, labor, operating difficulties, and capital investment.

Surfactant in the water solution simplifies the process four ways: (1) hydrates form rapidly, (2) hydrates form in a quiescent system, (3) hydrate particles migrate to the cell walls where they are adsorbed and packed, and (4) interstitial water forms hydrates. In the first simplification, hydrates form in the presence of surfactant about 700 times faster than in a quiescent, pure-water/gas system. Therefore, with proper design of the formation storage vessel, a formation decomposition cycle including turnaround time could be achieved within a 24-hour period. In the second simplification, surfactant causes hydrates to be formed in a quiescent system. Surfactant eliminates the need to impose water flow or movement and eliminates the need of mechanically moving parts in the equipment. As a consequence, complexity of the formation/storage tanks would be reduced; in fact, hydrate formation, storage, and decomposition could be accomplished in the same vessel. Surfactant allows reuse of the process water: after hydrate decomposition, the water and surfactant would remain in the storage tank, and the next formation cycle would proceed by repressurizing with gas. The simplifications greatly enhance the prospects for an economical large-scale process. In the third and fourth simplifications, difficult intermediate process steps of collecting and packing hydrate particles in a storage vessel are eliminated. With surfactant the particles form in the cell in an ordered and packed manner. After packing, the interstitial water is converted to hydrates at a high rate.

Critical Micellar Concentration for Hydrate Formation

Critical micellar concentration (CMC) refers to a threshold level of surfactant concentration in water necessary for micelles to form, and above which some physical properties of a solution abruptly change. To determine concentration levels needed to influence hydrate formation and to give insight into the mechanism of hydrate formation, surface tension was measured for numerous concentrations of SDS/water solutions; the CMC at 298 K (77°F) and 1 atm for our solution occurred near 2,700 ppm of SDS (see FIGURE 5).

However, when using the same surfactant in the chilled and pressurized test cell, hydrates rapidly formed at lower SDS concentrations than 2,700 ppm, indicating a lower CMC. In explanation, CMC is a function of temperature and the amount of gas dissolved in the water.[3] In order to get a better estimate of CMC at test conditions, multiple runs were made with a pressurized and chilled water/ethane system in the test cell while varying SDS concentration. The results are shown in FIGURE 6.

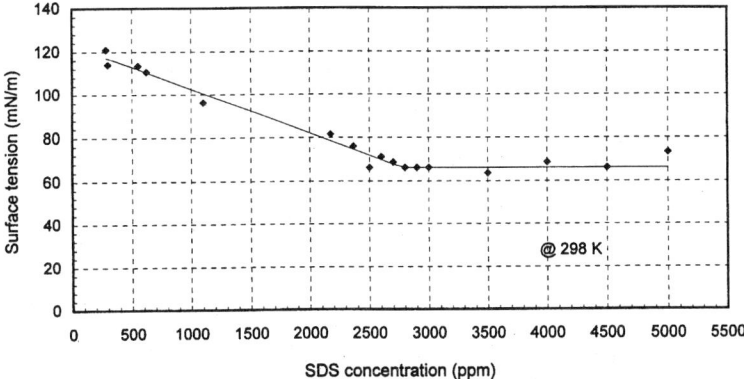

FIGURE 5. CMC of surfactant as measured at ambient conditions.

In FIGURE 6 the very sharp break in the curve at 242 ppm represents the CMC at ethane hydrate-forming conditions for SDS. Hydrate induction time gave the most sensitive indication of the breakpoint, although amount of gas occluded at a specific time would also have sufficed. The conclusion from FIGURE 6 is that a SDS surfactant solution concentration above the CMC greatly enhances the hydrate gas storage process.

CONCLUSIONS

A breakthrough occurred early in the work that provided answers to multiple problems regarding commercial storage of gas in hydrates. By using the surfactant sodium dodecyl sulfate in concentrations as low as 286 ppm in water, process benefits

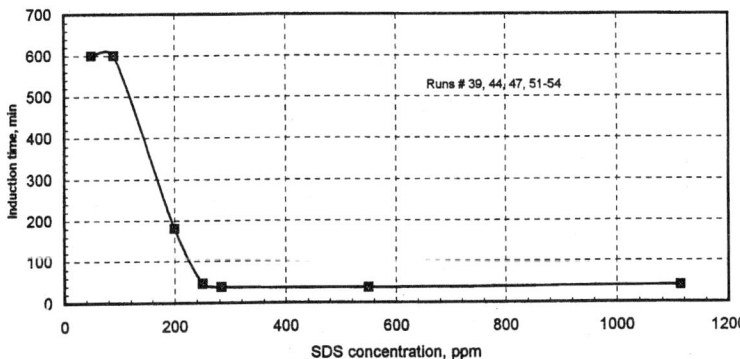

FIGURE 6. CMC of surfactant as measured at hydrate-forming conditions.

were achieved: (1) hydrates form rapidly—about 700 times faster than without surfactant, (2) free water trapped between particles is fully utilized to give a high bulk density, (3) particles move from the water solution as they form and deposit (pack) on the chamber walls, (4) hydrates form in a simple quiescent system, (5) 86% of the theoretical storage capacity can be achieved within three hours of hydrate initiation, and (6) 155 vol-gas/vol-hydrate can be stored at 3.89 MPa (550 psig) and 275.4 K (36°F). A practical design for large-scale natural gas storage in hydrates now seems feasible.

ACKNOWLEDGMENTS

The authors would like to acknowledge with appreciation the grant DE-AC26-97FT33203 from the Department of Energy that supported this work.

REFERENCES

1. COLLETT, T. 1996. Geologic assessment of the natural gas hydrate resources in the onshore and offshore regions of the United States. 2nd International Symposium on Gas Hydrates, Toulouse, France.
2. ENGLEZOS, P. 1966. Nucleation and growth of gas hydrate crystals in relation to "kinetic inhibition". 2nd International Symposium on Gas Hydrates, Toulouse, France.
3. ROSEN, M.J. 1978. Surfactants and Interfacial Phenomena. John Wiley & Sons, New York.

Laboratory for Continuous Production of Natural Gas Hydrates

JON STEINAR GUDMUNDSSON,[a] MAHMUT PARLAKTUNA,[b]
ODD IVAR LEVIK, AND VIBEKE ANDERSSON

Department of Petroleum Engineering and Applied Geophysics,
Norwegian University of Science and Technology, 7491 Trondheim, Norway

ABSTRACT: Natural gas hydrate technology is an attractive alternative for storing and transporting natural gas. A high-pressure laboratory has been built to provide experimental data for use in the design and development of hydrate-based processes for the oil and gas industry. In the laboratory, hydrate is produced from liquid water—and water-in-oil emulsions—and injected natural gas mixtures. The hydrate reactor and circulation loop can operate at pressures up to 120 bar and constant temperatures in the range 0–20°C. The hydrate slurry produced can be circulated at up to 100 liter/minute through 4-m long pipes equipped with differential pressure transducers and flow meters to determine their rheological characteristics under laminar and turbulent flow conditions. The laboratory can also be used for various flow assurance studies.

INTRODUCTION

Gas hydrates can be stored at atmospheric pressure, provided the temperature is a few degrees below the freezing point of water. This has opened the possibility for cost-effective storage and transport of associated gas in the form of hydrate.[1] A high-pressure laboratory for the study of continuous hydrate production has been built at the Norwegian University of Science and Technology (NTNU) in Trondheim. The main units are a nine-liter continuously stirred tank reactor (CSTR), an 18-liter separator, a centrifugal circulation pump and a shell-and-tube heat exchanger in a high pressure circulation loop with optional in-line filters.

NTNU's research on natural gas hydrate (NGH), for storage and transport of gas, dates back to the early 1990s. The basis for these activities was that NGH is metastable at conditions of 1 bara and temperatures below 0°C.[2] Metastability refers to the hydrate being thermodynamically unstable, yet it dissociates at such low rates that it appears stable for practical purposes. Furthermore, the highest theoretical gas content of NGH is as much as about 180 Sm^3/m^3 of hydrate, rendering hydrates a suitable medium for storage and transportation of natural gas.

Results of studies on long-distance transportation of natural gas as dry hydrate were reported by Gudmundsson *et al.*,[3,4] and Børrehaug & Gudmundsson.[5] The dry hydrate is produced from natural gas and refrigerated to typically −15°C and then

---

[a]Electronic mail: jsg@ipt.ntnu.no
[b]Present Address: Department of Petroleum and Natural Gas Engineering, Middle East Technical University, 06531 Ankara-Turkey.

conveyed to bulk carriers. The bulk carriers take the frozen hydrate to distant gas markets where the hydrate is melted and the gas recovered. The capital cost for a NGH chain to transport 400 MMscf/d of natural gas over 3,500 nautical miles was found to be 24 percent lower than the capital cost of LNG chain, assuming European conditions.[5] Another attractive application of NGH technology is to capture associated gas on FPSOs to solve the stranded gas and marginal gas problem in the oil industry.[6,7] A three-year joint industry project, called NGH at NTNU, was established at NTNU to provide the data necessary for evaluating the feasibility of hydrate processes, including the construction of a pilot plant, and eventually commercial plants for FPSOs and land-based plants.

The early work was carried out at NTNU using a 600 ml batch reactor, but some of the data needed to design a large-scale process could not be obtained with this setup. To approach a large-scale process, a hydrate laboratory featuring a nine-liter CSTR was designed. Issues that had to be studied in more detail included rate of formation and dissociation, flow properties of hydrate slurries, separation of hydrate particles, hydrate sample preparation, and thermophysical properties of frozen hydrate samples. Hence, the equipment was designed to perform experiments to provide quantitative and qualitative information on these issues. This paper describes the hydrate laboratory, its operation, and some of the experiences. Selected results are also reported.

NTNU HYDRATE LABORATORY

The NTNU hydrate laboratory is designed to operate at up to 120 bar pressure and at constant temperature in the range of 0–20°C. FIGURE 1 shows a schematic diagram of the NTNU hydrate rig. The main units of the rig are a reactor, a separator, a centrifugal pump, and a heat exchanger. The design capacity of the heat exchanger corresponds to 1 kg/hour of natural gas hydrate. The nine-liter reactor is a CSTR type reactor where mixing is achieved with a variable speed (0–2,500 RPM) magnetically driven straight blade impeller. Four baffles are mounted in the reactor to increase the mixing efficiency of the impeller. Hydrate forming gas is injected through two sparger units located at the bottom of the reactor. The sparger units are equipped with non-return valves to prevent back-flow into the gas lines. The main function of the 18-liter separator is to separate gas from the slurry. Circulation of the hydrate slurry in the loop is achieved with a variable speed centrifugal pump having a maximum flow rate of 100 l/min of water. A shell-and-tube heat exchanger is used to remove the heat evolved during hydrate formation and heat added from several components, such as the circulation pump.

There are two ports available in the flow loop to connect different type of test units to the main loop. The main test unit items are filters, samplers, a flow cell, and two 4-m long horizontal flow sections for flow property measurements. All of these test units are connected to the main loop with quick connection couplings. Filters are mainly used to obtain the samples of hydrate slurry at high pressures. Each filter unit has a volume capacity of 0.5 liters. These filters make it possible to obtain samples of hydrate slurry under pressure. Samples of dry hydrate are obtained from this hydrate slurry by displacing the liquid with high-pressure gas. The pressure is then

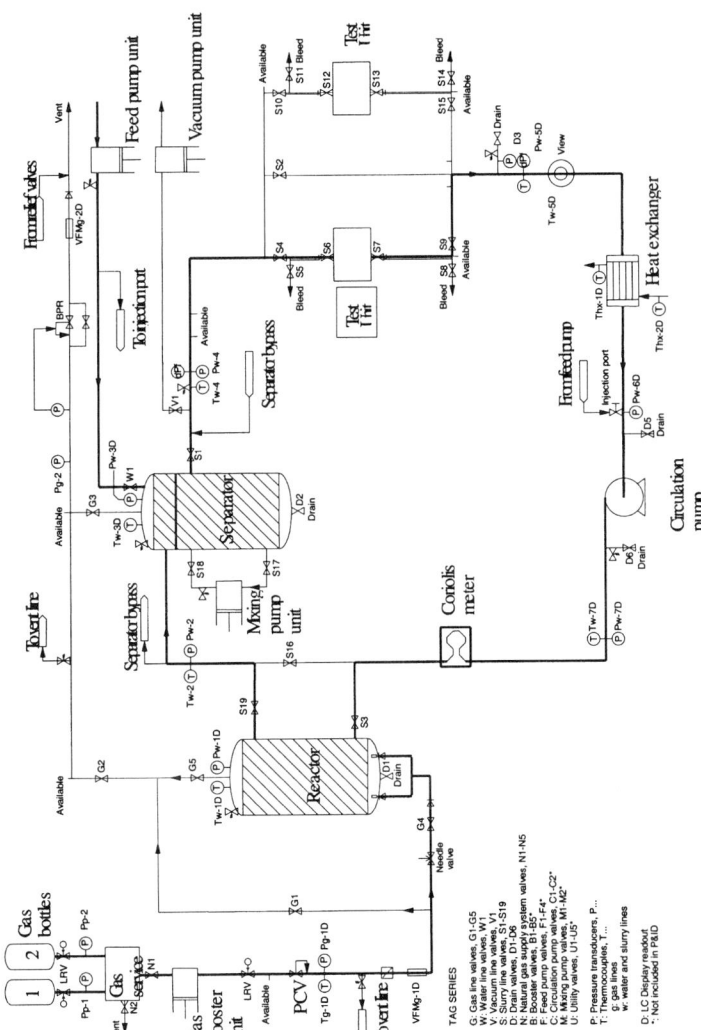

FIGURE 1. Schematic diagram of the NTNU hydrate rig.

released after the filter content is frozen under pressure in a freezer room which is operated at about $-25°C$. The other sampling unit is a pipe sampler, which is a 22 mm inner diameter pipe with ball valves attached to both ends of the pipe. The pipe sampler is attached to an evacuated container after sampling, and the hydrate sample is dissociated by simply opening the valves between pipe sampler and container.

Flow property studies of hydrate slurries are carried out using the horizontal flow sections. The sections are two pipes each of four meters long but having different diameters, with differential pressure transducers connected to each pipe. The temperature can be measured at the inlet and outlet of both pipes. One of the advantages of the sections is the direct connection, using flexible hoses, to the hydrate-forming rig, enabling measurements on pressurized hydrate slurries on-line.

The hydrate rig is also equipped with several auxiliary systems such as gas supply and gas booster; water supply and water feed pump; and a reciprocal mixing pump attached to the separator in order to mix the content of separator before sampling and flow property measurements. This is needed because the produced hydrate tends to be separated from the carrying fluid in the separator where the residence time can be significant. Vacuum pump, view ports, and bypass lines are other auxiliary units.

Safety is an important aspect of the NTNU Hydrate Laboratory. The laboratory is an explosion secure zone and all the instrumentation and equipment are, accordingly, Ex-II classified. In addition, the flow loop and high-pressure vessels are equipped with line rupture and pressure-relief valves. There are fans for ventilation of the room. A portable gas detector is also available to sense any gas leakage from the loop.

Several instruments are used to measure and regulate the process parameters. These instruments include pressure transducers and thermocouples to measure pressure and temperature, a Coriolis meter to measure the mass flow rate and density of the slurry, and gas flow meters to measure the gas injection and vent rates. A differential pressure (DP) transducer is attached to the inlet and outlet ports of test units. The main application of this DP cell is to observe the pressure build-up across the filter units while filtering. Regulating devices are used to control the process parameters while running experiments. One of these regulating devices is a back-pressure regulator (BPR in FIG. 1) located along the gas vent line. It is used to operate the system at constant pressure while producing hydrate. It consists of a solenoid valve activated by a signal from a high-accuracy pressure transducer. If the pressure inside the rig exceeds the set pressure by 0.1 bar, the BPR opens and to let the gas leave the system through a gas flow meter. Gas injection rate, on the other hand, is regulated by the combination of a constant pressure regulator, a needle valve, and a gas flow meter. The flow of hydrate slurry in the loop is regulated by the control of frequency of the circulation pump.

The output signals of pressure, temperature, and mass flow meter devices are recorded at an optional frequency by a PC-based data acquisition system. There are video cameras and a TV set to record and to observe events while hydrate is produced. There are two view ports within the loop to record events. One of these ports is a specially constructed thin layer flow cell with facing windows with variable flow opening. Data recording PC and TV set are located in a separate control room.

A freezer room and a calorimeter room are other facilities related to the hydrate laboratory of NTNU. The purpose of freezer room is to freeze the filtered hydrate samples. As mentioned previously, the hydrates are metastable at atmospheric pressure and temperatures below 0°C. Thermophysical properties of frozen hydrate samples are measured with a Setaram BT2.15 heat-flow calorimeter that has been adopted to operate at pressures up to 100 bar. The calorimeter is designed for temperatures between -196 and 200°C and samples up to 8 to 10 cm^3 depending on cell type.

OPERATIONAL EXPERIENCE

Considerable operational experience has been obtained in the NTNU hydrate laboratory since it opened in May 1997. The work has been carried out as a part of the NGH at NTNU joint industry project. To illustrate the laboratory work, some of the operational experience is reported below.

Positive Displacement and Centrifugal Pumps

There were initially two positive displacement pumps, identical to each other, as circulating and mixing pumps in the system. The circulating pump is the main pumping unit used to achieve circulation in the loop, whereas the mixing pump is used to mix the contents of the separator. The pulsing nature of positive displacement pumps created pressure fluctuations in the system, making it impossible to perform pressure drop measurements in the horizontal flow section for flow property measurements or to keep constant pressure during hydrate production. The pulsation also affected filtration by causing back flushing of filtrate. This despite the installation of pulse accumulators. As a result, for smooth operation, the positive displacement pump for slurry circulation was replaced with a variable speed high-pressure centrifugal pump.

Continuous Gas Feed and Vent

One of the operational procedures while producing hydrate is to keep the system pressure constant. This can be achieved by replenishing the rig with the amount of gas that was used for hydrate formation. Injecting the gas continuously with a higher rate than its consumption rate and venting the excess gas through a gas flow meter was the operation mode during constant pressure experiments. The main item along this process is the back-pressure regulator (BPR), which keeps the upstream pressure constant. A mechanical BPR was used at the initial stage but gave large pressure hysteresis. This problem was overcome with the use of a solenoid valve in connection with a high-accuracy pressure sensor. The solenoid valve is activated with signals from the pressure sensor and allows the gas to vent through the gas flow meter to the atmosphere. The difference between gas injection rate and gas vent rate is used as the gas consumption rate during hydrate formation. The pressure history of one of the hydrate formation tests is shows in FIGURE 2 as an example. This test was a methane hydrate formation test at 70 bara system pressure and 8°C room temperature. The gas injection pressure, measured at a point prior to needle valve and spargers,

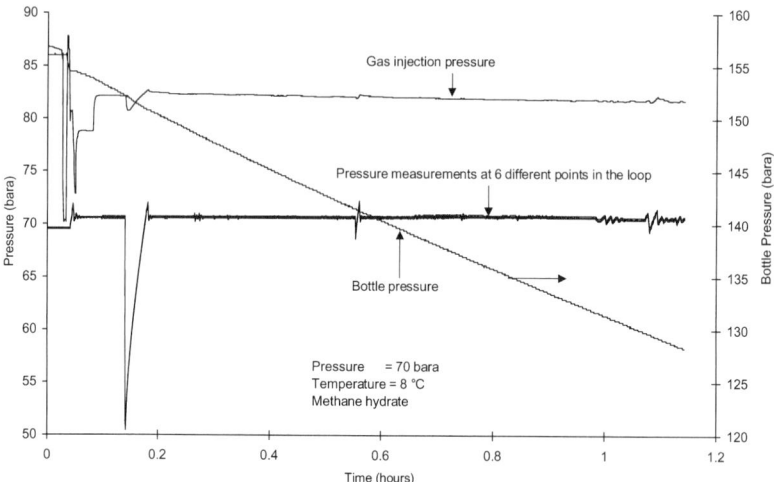

FIGURE 2. Change in pressure with time during hydrate production.

was set to 82 bara. The line with continuous decline is the pressure of gas bottle from which the gas was injected. The decline rate of gas bottle pressure is used to calculate the gas injection rate. It can be seen from the figure that the pressure in the loop was kept constant during hydrate formation with continuous gas injection.

Hydrate Formation Tests with Water

Hydrates are readily formed when the pressure and temperature are within the hydrate region. No plugging was observed within the main loop, consisting of one-inch diameter pipes. There were some cases that the plugging occurred along with the gas vent lines, between valve G3 and BPR (FIG. 1). This line has a smaller diameter (1/4 inch) and any water carried out this line by gas could cause hydrate formation. This problem was approached by heat tracing this line with band heaters and keeping the water level within the separator away from G3 by leaving a gas pocket in the separator.

SELECTED RESULTS

The experimental results are used by Dr. Ing. students, and other researchers in the NGH at NTNU joint industry project. The three-year project aims to provide data for the design of natural gas hydrate processes for use in the oil and gas industry. To illustrate the kind of work carried out in the NTNU hydrate laboratory, experiments on methane hydrate formation rate are described below.

Experiments were carried out to study the effect of gas injection rate on methane hydrate formation rate. Pure methane with deaerated tap water was used as the hydrate forming media. Continuous gas injection and vent mode was selected to

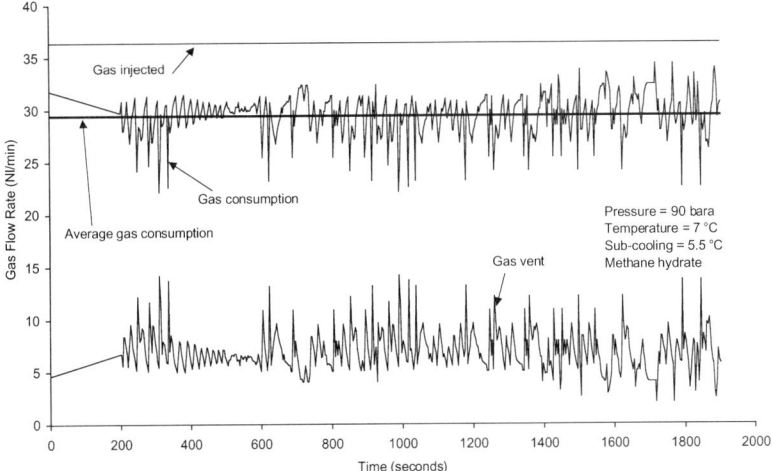

FIGURE 3. Change in gas flow rates with time during hydrate production.

perform the experiments and the liquid circulation rate, system temperature, and formation pressure were kept constant. During the course of an experiment, the laboratory was cooled to the test temperature before the rig was evacuated and filled with deaerated water. Then the gas was injected into the rig to a pressure approximately five bar lower than the hydrate formation pressure at room temperature. The rig was left overnight for gas dissolution in water and the content of reactor was stirred at

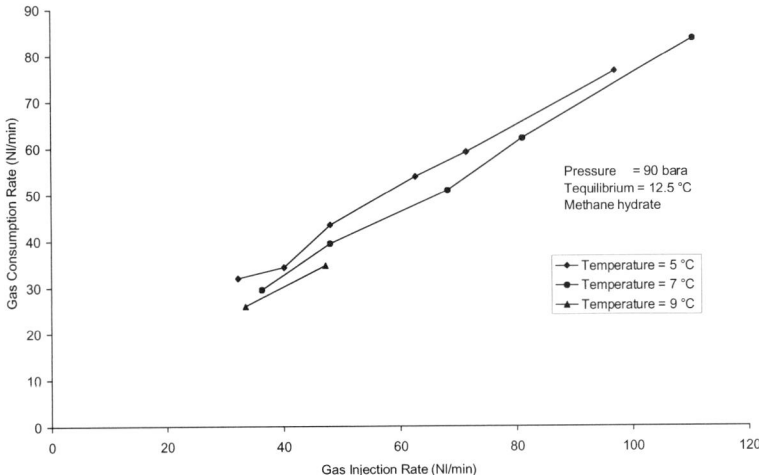

FIGURE 4. Gas consumption rate with gas injection rate and test temperature.

400 RPM. The following day, the stirring was stopped and the BPR set to test pressure followed by gas injection to test pressure. When the temperature and pressure equilibrium was achieved throughout the rig, liquid circulation, stirring of reactor content at 400 RPM, and gas injection were started at the same time. Temperature and pressure at different points of the rig, as well as bottle pressure and gas vent rate, were recorded.

The change in gas flow rates during one of the hydrate formation tests is shown in FIGURE 3. The uppermost line is the constant gas injection rate line; the curve at the bottom shows the gas vent rate from the system. The gas injection rate was obtained from the change in bottle pressure and the gas vent rate was measured with a mass flow meter. The difference between these two gas flow rates is the amount of gas used for hydrate formation. The fluctuation in the gas vent rate was due to opening and closing the BPR within ±0.1 bar pressure. The arithmetic average of gas consumption rate data of each test was taken as the overall result of this test. FIGURE 4 gives the result of a group of experiments with different gas injection rates and test temperatures. As indicated, the higher the gas injection rate, the higher the gas consumption rate for hydrate formation at a given temperature. On the other hand, decrease in temperature (higher subcooling) results with higher gas consumption for a given gas injection rate.

ACKNOWLEDGMENTS

The work reported was carried out within the NGH at NTNU joint industry project, supported by Aker Engineering, Amerada Hess, Neste Petroleum, Phillips, Shell and Total. We thank the workshop staff of the Department of Petroleum Engineering and Applied Geophysics at NTNU for their tireless efforts in the hydrate laboratory.

REFERENCES

1. GUDMUNDSSON, J.S. 1993. Method for Production of Gas Hydrate for Transportation and Storage. U.S. Patent No. 5,536,893.
2. GUDMUNDSSON, J.S., M. PARLAKTUNA & A.A. KHOKHAR. 1994. Storing natural gas as frozen hydrate. SPE Production and Facilities, February. 69–73.
3. GUDMUNDSSON, J.S., F. HVEDING & A. BØRREHAUG. 1995. Transport of natural gas as frozen hydrate. Proceedings 5th International Offshore and Polar Engineering Conference, The Hague, June 11–16. **1:** 282–288.
4. GUDMUNDSSON, J.S. & A. BØRREHAUG. 1996. Frozen hydrate for transport of natural gas. Proceedings 2nd International Conference on Natural Gas Hydrates, June 2–6, Toulouse. 415–422.
5. BØRREHAUG, A. & J.S. GUDMUNDSSON. 1996. Gas transportation in hydrate form. EUROGAS 96, June 3–5, Trondheim. 35–41.
6. GUDMUNDSSON, J.S., V. ANDERSSON & O.I. LEVIK. 1997. Gas storage and transport using hydrates. Offshore Mediterranean Conference, Ravenna, March 19–21.
7. GUDMUNDSSON, J.S., V. ANDERSSON, O.I. LEVIK & M. PARLAKTUNA. 1998. Hydrate concept for capturing associated gas. Proceedings SPE European Petroleum Conference, October 20–22, The Hague. 247–258.

The Application of Raman Spectroscopy to the Study of Gas Hydrates

C.A. TULK,[a] J.A. RIPMEESTER, AND D.D. KLUG

Steacie Institute for Molecular Sciences, National Research Council of Canada, Ottawa, Ontario, Canada K1A 0R6

ABSTRACT: Raman spectroscopy is reviewed with particular emphasis placed on its application to gas hydrates. Experimental examples discussed include studies of the totally symmetric C-H stretching vibration ν_1 (A_1) of methane in both synthetic and natural hydrate samples (from the JAPEX/JNOC/GSC Mallik 2L-38 research well); a comparison of the coupled O-H vibrations of water in the host lattice of CH_4 hydrate and ice I_h at low temperature; and local structural details of the relaxation of the hydrogen-bonded water on crystallization to structure II hydrate of amorphous tetrahydrofuran (THF) aqueous solutions. This paper is intended to be an introduction to Raman spectroscopy with specific examples from research at the National Research Council of Canada, and is aimed at those who wish to apply the technique as a tool to investigate gas hydrates.

INTRODUCTION

Raman spectroscopy is being exploited increasingly as a tool to study clathrate hydrates. Several laboratories have explored the technique[1-3] and prospects are such that many more, including industrial laboratories, should be alerted to the potential of Raman spectroscopy for studying structure, formation and decomposition processes, and molecular dynamics in gas hydrates. This paper will give a general overview of Raman spectroscopy with particular emphasis on its application to gas hydrates. Recent examples of work carried out at the National Research Council of Canada include studies of methane, both in synthetic and recovered natural gas hydrate samples, studies of the coupled and uncoupled internal water modes (O-H and O-D stretching vibrations), and, finally, the Femri resonance of CO_2 in structure I, II, and H clathrate hydrates. This is followed by a discussion of relevant experimental and theoretical challenges, and of the structural information that can be inferred. The most important, *take-home*, messages are given in separate paragraphs and in italics.

Perhaps the most accessible derivation of the light scattering process comes from the classical treatment of the interaction of the electrical field of an incident laser beam with the polarizability tensor of the scattering material. Although the classical treatment cannot account accurately for the intensity of the scattered light, it can provide an intuitive understanding of the phenomenon.

[a]Corresponding author.

THE CLASSICAL TREATMENT OF RAMAN SCATTERING

Many books have been dedicated to the detailed explanation of Raman scattering. The following brief introduction is not in any way intended to be a complete description and is presented only as background to the following sections. A much more complete description using the same notation can be found in the books by Herzberg or Long.[12]

Both elastic and inelastic scattering of light can be thought of as a coupling between the electromagnetic fields of the incident light with the polarizability tensor of the scattering material. Essentially the incident laser light induces oscillating dipoles that are modulated by molecular or lattice vibrations within the material. In the case of elastically scattered light the process is known as Rayleigh scattering. In the case of inelastically scattered light the process is known as either Raman scattering, when lattice or internal molecular vibrations are involved, or Brillouin scattering, when acoustic vibrations are involved.

When a material is placed in an electromagnetic field a dipole moment is induced that can be described by

$$\vec{P} = \alpha \vec{E}, \qquad (1)$$

where $\vec{E}$ is the electric field vector and $\vec{P}$ is the induced dipole moment vector. Since both $\mathbf{E}$ and $\mathbf{P}$ are vectors and, in the general case, the induced dipole moment is not in the same direction as the incident electric field, then the polarizability tensor α must be second rank with nine components, α_{ij} (it can be shown that the tensor is symmetric, that is, $\alpha_{ij} = \alpha_{ji}$). This leads to three equations of the form,

$$P_i = \sum_j \alpha_{ij} E_j. \qquad (2)$$

Diagonalizing the polarizability tensor by defining its elements in the principal axis coordinate system allows the polarizability tensor to be visualized in terms of the polarizability ellipsoid with principal axes α_a, α_b, and α_c (in such a system all of the off-diagonal elements of the polarizability tensor are zero). Along the principal axes of the polarizability ellipsoid the direction of the induced dipole moment $\vec{P}$ is the same as the direction of $\vec{E}$. It should be noted here that the CH_4 molecule forms a regular tetrahedron (has four threefold rotation axes, one along each C-H bond) and therefore has T_d symmetry. The totally symmetric vibration $\nu_1(A_1)$, discussed in the next section, maintains the T_d symmetry of the molecule (i.e., the polarizability ellipsoid is spherical during the vibration) and the light scattered at 90° is completely polarized.[5]

In the case of internal molecular vibrations the polarizability ellipsoid changes when the nuclei of the molecule are displaced from their equilibrium positions, as is the case during a molecular vibration. If the displacements are small compared to the internuclear distance then the change in each component of the polarizability tensor may be described by a series expansion in the normal coordinate system as

$$\alpha_{ij} = \alpha_{ij}^o + \sum_k \left(\frac{\partial \alpha_{ij}}{\partial \xi_k}\right)_o \xi_k + \dots . \qquad (3)$$

Here ij are the coordinates x, y, and z, the summation is over all the nuclear displacements and the derivatives are taken with respect to the equilibrium positions. The derivatives with respect to the normal coordinates in the above equation are known as the elements of the derived polarizability tensor. The linear term in ξ_k is responsible for first order scattering, whereas the quadratic and higher order terms. account for second and higher order effects. If harmonic vibrations of the molecule are considered, then the time dependence of the normal coordinate ξ_k, with frequency ν_k, can be described by

$$\xi_k = \xi_{ko} \cos(2\pi\nu_k t). \tag{4}$$

Next we consider the applied electric field to be an incident laser beam of frequency ν. The oscillating electric field can then be described by

$$\mathbf{E} = \mathbf{E}^o \cos(2\pi\nu t). \tag{5}$$

Substituting these into the above equation for the induced dipole moment gives

$$P_i = \sum_j \alpha_{ij}^o E_j^o \cos(2\pi\nu t) \\ + \frac{1}{2}\sum_{kj}\left[\left(\frac{\partial \alpha_{ij}}{\partial \xi_k}\right)_o E_j^o\right]\xi_k\{\cos[2\pi(\nu+\nu_k)t]+\cos[2\pi(\nu-\nu_k)t]\}. \tag{6}$$

From the above expression it can be seen that the induced dipole oscillates at the frequency of the incident radiation, leading to Rayleigh scattering, and with additional frequencies $\nu \pm \nu_k$, arising from the Raman effect.

The equation describing the induced dipole in the scattering process contains two terms: the first term describes the Rayleigh scattering process and the scattered light is at the same frequency as the incident radiation: the second term describes the Raman scattering process at the frequency of the incident laser light plus and minus the frequency of the Raman active vibrational modes of the material.

It is the derived polarizability tensor that gives rise to the Raman scattered light and for a molecular vibration to be Raman active at least one of the six components of the derived polarizability tensor must change during the molecular oscillation. The above derivation describes the vibrational frequencies well; however, the full derivation of Raman intensities requires a quantum-mechanical treatment. Some important points regarding Raman intensifies related to gas hydrates of methane can be illustrated based on this simple explanation.

Experiments described in the following sections were performed on samples held at very low temperature (generally between 10 and 180 K). Samples were mounted in a copper sample holder under cold nitrogen gas and thermally connected to the cryotip of an Air Products Displex closed cycle refrigerator. A gold–iron thermocouple was mounted on the cryotip just beneath the copper sample holder. The temperature did not rise above 80 K during the mounting procedure. The loading technique insured that no gas hydrate decomposed before testing. For a detailed discussion of the sample handling process and experimental set-up see Tulk et al.[6,22]

EXAMPLES FROM METHANE HYDRATE STUDIES

Totally Symmetric Stretching Mode of Methane—Synthetic Methane Hydrate

In the study of methane hydrate, Raman spectroscopy is most easily applied to (1) the totally symmetric ν_1 (A_1) vibration of the guest methane molecule. This

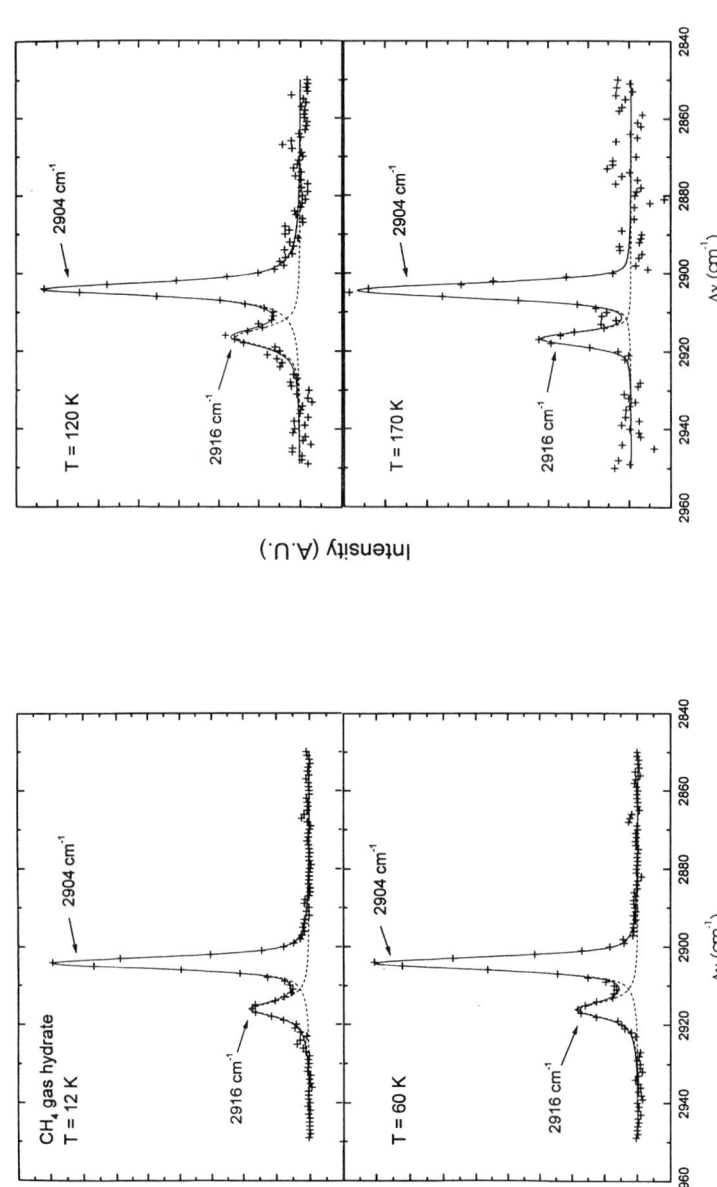

FIGURE 1. Raman spectra of the totally symmetric stretching vibration v_1 (A_1) of enclathrated methane, Str I hydrate, as a function of temperature. The sample was exposed to a dynamic vacuum during the data collection and began to decompose between 120 and 170 K, as evidenced by the decreased signal to noise ratio at higher temperature. *Smooth curves* represent the fitted peak functions and + indicates experimental data.

occurs at 2,914.2 cm^{-1} in the gas phase,[4] and between 2,904 and 2,916 cm^{-1} in the hydrate (see FIGURE 1); and (2) the internal vibrational modes of the water molecules making up the host lattice, which occur between 3,050 and 3,600 cm^{-1} (see FIGURE 3). (There is another very weak Raman active CH$_4$ mode, ν_3 (F_2), which occurs at 3,026 cm^{-1} in pure methane at 90 K.) Previously, it has been shown that the cage structure of clathrate hydrates causes the totally symmetric mode to split into two resolvable lines with distinct vibrational frequencies.[7] An attempt has been made to determine the cage occupancies based on an analysis of the relative Raman band intensities. Based on experiments with gas mixtures the band assignment (i.e., the Raman band appearing at 2,905 cm^{-1} resulting from methane in the large cage and the Raman band appearing at 2,915 cm^{-1} resulting from methane in the small cage) appears to be correct (Sum et al.[2]). In addition, it is quite reasonable to expect the Raman line intensities to be related to the cage occupancies. It is, however, necessary to investigate the possibility of a more complex, and as yet not well understood, relationship between the intensity ratio and the cage occupancy.

The vibrational frequencies of the symmetric stretching C-H vibration of methane in the different cages types (5^{12}, $5^{12}6^2$, and $5^{12}6^4$) of Str. I and Str. II are slightly perturbed by the local molecular environment and consequently have slightly different vibrational frequencies. This is evident as distinct lines in the Raman spectrum.

The electric fields produced by the disordered water molecules making up symmetrically distinct 5^{12} and $5^{12}6^2$ cages of Str. I hydrate are responsible for changing the electronic structure of the methane molecule in such a way that the force constants (k_1, k_2, and k' following the notation of Herzberg[4]) of methane in each of the two types of cage are slightly different; thus the frequency of the totally symmetric mode A_1 of methane in the 5^{12} cage (2,916 cm^{-1}) is different then that in the $5^{12}6^2$ cage (2,904 cm^{-1}) and both are different from the gas phase value (2,914.2 cm^{-1}). The interaction between water and methane and perhaps between methane in adjacent cages is significant, a fact that is reflected in the large coefficient of thermal expansion[9] that is thought to be related to the glass-like thermal transport properties of clathrates in general.[10,11] Whereas the spectral positions of Raman bands can be described adequately using classical arguments, the intensities of the Raman bands require a full quantum mechanical treatment. It is not within the scope of this paper to present the full quantum mechanical argument and the reader is referred to the book by Long.[12] However, in general terms, the intensity of an oscillating dipole is written as

$$I = \frac{\pi^2 c \nu_P^4 \mathbf{P}_o^2 \sin^2\theta}{2\varepsilon_o}, \quad (7)$$

where c is the speed of light, ν_P is the frequency of the oscillating dipole, $\mathbf{P}_o$ is the amplitude vector of the oscillating dipole, θ is the angle between the dipole axis and the direction of observation, and ε_o is the permittivity of free space. In the case of Raman scattered light the second term on the right side of (**6**) is used with (**7**), and $\nu_P = \nu \pm \nu_k$ representing anti-Stokes and Stokes scattering, and

$$\mathbf{P}_o = \frac{1}{2}\left(\frac{\partial \alpha}{\partial \xi_k}\right)_0 \mathbf{E}_o \xi_k, \quad (8)$$

where α is the polarizability tensor, ξ_k represents a particular normal coordinate. These equations are modified considerably when experimental details, such as

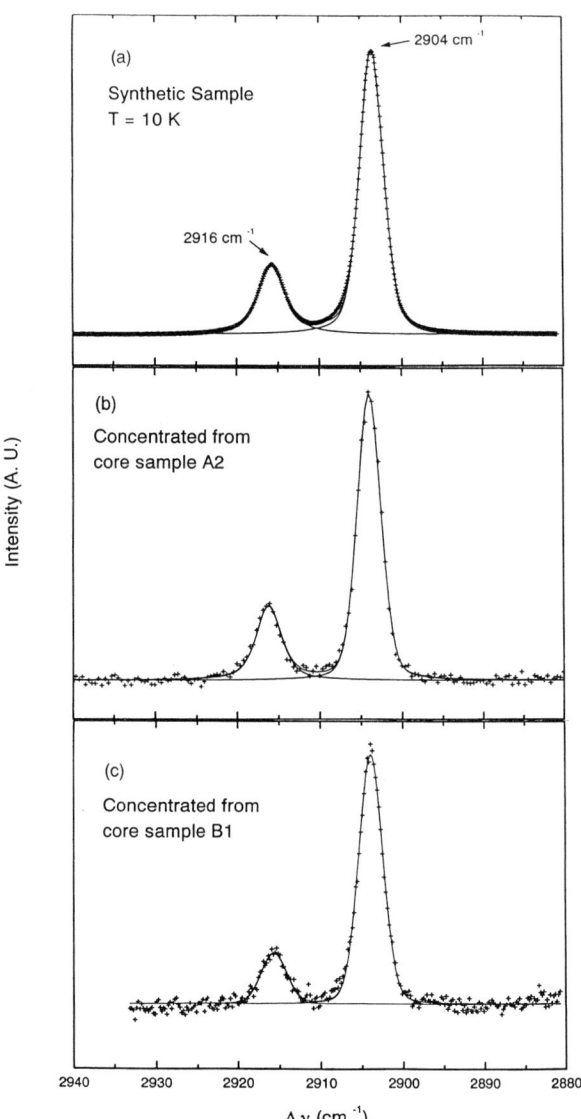

FIGURE 2. Raman spectra of the ν_1 vibration of methane in a synthetic sample (**a**) and two naturally occurring samples collected from the JAPEX/fNOC/GSC Mallik 21-38 research well (**b**) and (**c**). The gas hydrate concentrated from core sample A2 (recovered from 898.5 m) contains exclusively methane gas, whereas the gas hydrate concentrated from core sample BI (recovered from between 913.60 and 913.66 m) contains a mixture of methane with a small component of carbon dioxide and propane.

incident and scattered polarization states and the geometry of collection of the scattered radiation, are considered. Long has provided tables giving the functional form for the intensity of Raman scattered light for several scattering geometries. The main reason for introducing the above simple functional description of the intensity of

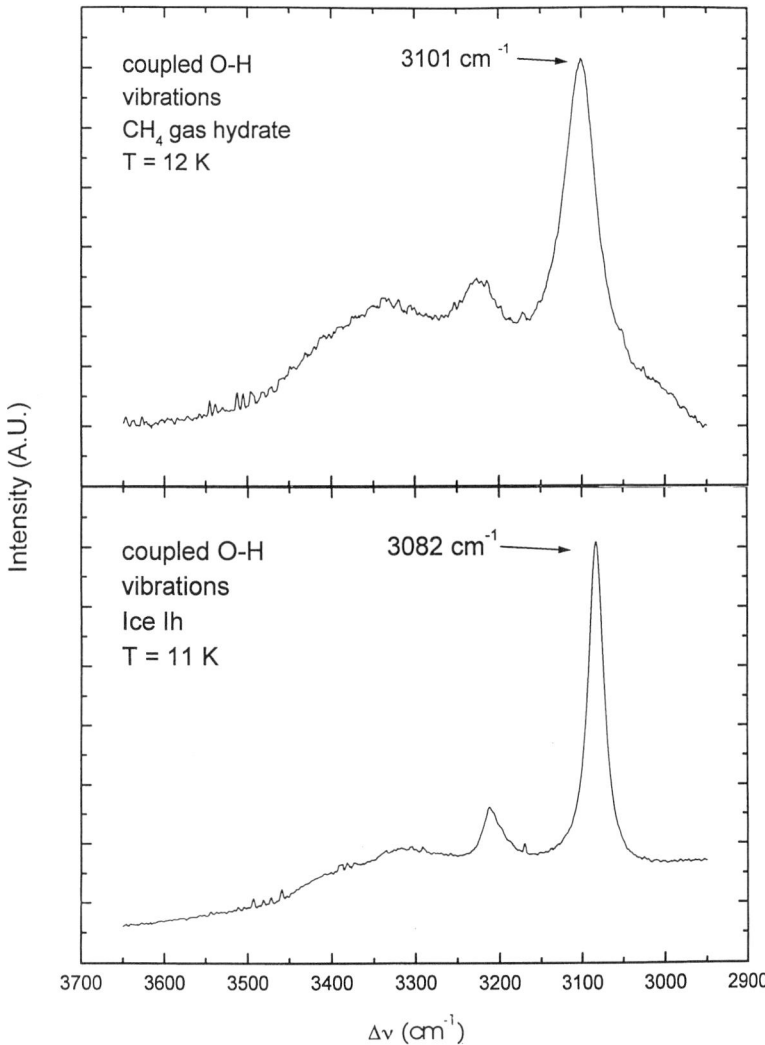

FIGURE 3. Spectra of the coupled O-H stretching vibration of the water molecules making up the host framework (**upper frame**, $T = 12$ K) and for comparison the same vibrational mode in crystalline ice Ih (**lower frame**, $T = 11$ K). The spectrum of the crystalline ice was collected after the gas hydrate sample had been annealed at 180 K for several hours and the sample had completely decomposed.

Raman scattered light is to show that the intensity is proportional to (1) v_P^4 and (2) the derivatives of the polarizability tensor.

The intensity of Raman scattered light from methane in each of the two structure I cage types is dependent on (1) the frequency of the scattered light raised to the power four (the shorter the wavelength, the more intense the scattered light) and (2) the values of the elements of the derived polarizability tensor.

Even though the derivatives of the polarizability tensor with respect to the normal coordinates are required for the above equations, it is often easier to measure and calculate Raman intensities using derivatives with respect to the symmetry coordinates of the molecule; a number of such studies have been conducted.[13–15] Symmetry coordinates have the symmetry properties of the point group (T_d in the case of methane). Symmetry coordinates can be chosen based on the molecular symmetry alone and greatly simplify any description of the vibrational problem. For example the symmetry coordinate describing the totally symmetric v_1 (A_1) vibration of methane is given as S_1 (A_1) = ($\Delta r_1 + \Delta r_2 + \Delta r_3 + \Delta r_4$)/2, where Δr_i is the change in the C-H distance along each C-H bond during the symmetric stretching vibration. Other symmetry coordinates describing Raman active modes are described in Reference 15.

In a recent paper by Ioannou and Amos,[5] density functional theory has been applied to calculate the derivatives of the frequency dependant polarizability tensor (the polarizability tensor is considered to be perturbed be the application of an oscillating electric field, as required by Placzek polarization theory). The derivative $\partial \alpha_{xx}/\partial S_1$, which is responsible for the Raman spectrum of the A_1 mode in methane, was found to be 2.74×10^{-30} $C^2 \cdot mJ^{-1}$ at a wavelength of 514.5 nm, a commonly used argon-ion laser line. As is also noted in the paper by Ioannou and Amos,[15] the values of the polarizability derivatives depend upon the C-H bond length. The two distinct Raman lines arising from the totally symmetric A_1 mode of enclathrated methane in methane hydrate, see FIGURE 1, indicate that the force constants of a methane molecule in the 5^{12} cage are different from those in the $5^{12}6^2$ cage; it is therefore reasonable to expect that the C-H bond lengths are different in the two cages and therefore the scattering cross sections are not expected to be identical.

Direct comparison of the intensities of Raman bands resulting from the totally symmetric vibration of methane in the small and large cage may be difficult; at the present time the proper method of comparison is not well understood.

A second point concerns the resolution of Raman contributions from cages in different hydrate structures. Published work[2] suggests that the v_1 Raman bands from methane in small cage and large cages, respectively, in structure I and II hydrates are very near to each other, thus limiting the use of Raman spectroscopy to the study of pure phases.

Some valuable qualitative information can, however, be obtained from the analysis of Raman spectra of clathrate hydrates. Representative spectra of the totally symmetric stretching mode of enclathrated methane collected from synthetic hydrate samples are shown in FIGURE 1. Two distinct Raman bands can be identified quite readily at 2,904 cm^{-1} and 2,916 cm^{-1}, and due to the fact that these line are well resolved (full width at half maximum of the Raman line at 2,904 cm^{-1} is 3.5 cm^{-1} and that of the Raman line at 2,916 cm^{-1} is 3.7 cm^{-1}) accurate values of the integrated intensity can be obtained by simple curve fitting routines. The resolution in this set of experiments was 1.5 cm^{-1} at the chosen slit width; therefore the Raman lines in FIGURE 1 are not instrumentally broadened and can be considered the natural line widths. When the instrumental resolution is approximately equal to that of the measured line width, a correction must be made before accurate line widths can be obtained.[16] The smooth curves shown in FIGURE 1 represent pseudo-Voight functions that were fitted to each of the lines in the spectrum and the + signs represent experimental data. From the fitted curves, the ratio of the integrated intensities (I_{2904}/I_{2905}) is found to be between 3.01 and 3.08. (Using previous band assignments [Sum et al.^{2}], assuming that the intensities can be compared directly and taking into consideration that there are three times as many large cages as small cages in the structure one unit cell, the occupancy ratio would be $\theta_s/\theta_l = 0.95$. This number is in fair agreement with such ratios determined by other techniques. For example, NMR data[17] have given $\theta_s/\theta_l = 0.92$.) The spectra were obtained up to 170 K without significant deviation in the intensity ratio. The ratio increases slightly at $T = 170$ K to 3.12, however, considering the signal to noise ratio, this is within the experimental error (the signal to noise ratio in the figure also indicates that the sample had started to decompose).

Totally Symmetric Stretching Mode of Methane Hydrate from the Mallik 2L-38 Research Well

The spectral intensities of the Raman bands resulting from the totally symmetric stretching mode of enclathrated methane in gas hydrate samples collected from the JAPEX/JNOC/GSC Mallik 2L-38 research well were measured and compared with those of the synthetic methane hydrate discussed above. The core samples recovered from the Mallik site can be characterized as being composed of approximately 85% consolidated sand of variable grain size (from fine grained sand, about 100 μm, to pebbly sand, about 1–3 mm) that was held together by a mixture of ice and gas hydrate, and comprised approximately 15% of the total mass. This low concentration of mostly invisible gas hydrate presented several challenges during the sample analysis. However, a technique has been developed to separate the ice/gas hydrate mixture from the consolidated sand without decomposition of the gas hydrate.[18]

It was impossible to obtain high-quality Raman spectra from the unprocessed core material. It was also noted that at relatively high laser power settings the samples decomposed rapidly, even with the bulk sample temperature at 11 K. This required the use of low laser power with subsequent loss of signal quality.

Raman spectra were collected from gas hydrate from two core samples, (1) composed of exclusively Str. I methane hydrate and collected at a depth of 898.50 m (core sample A2, see Ref. 18.) and (2) composed of methane gas hydrate with 2% CO_2 and propane as other guest components and collected between 913.60 m and 913.66 m (core sample B1, see Ref. 18). These samples were observed to have different decomposition thermodynamic profiles.

The Raman spectra of the totally symmetric mode of enclathrated methane from core samples A2 and B1 are shown in FIGURE 2 (b) and (c), respectively, and similar spectra collected from the synthetic sample is reproduced for comparison (a). The two distinct Raman bands easily can be identified at 2,904 cm^{-1} and 2,916 cm^{-1} and the integrated intensities of these lines were obtained by fitting two pseudo-Voight functions. Using the fitted curves, the ratio of the integrated intensities for the A2 core sample hydrate was found to be $I_{2904}/I_{2916} = 3.19$ (considering the difference in the numbers of cages in the unit cell and assuming a straightforward comparison of the intensities is valid, the occupancy ratio is $\theta_s/\theta_l = 0.94$). Considering possible variations in the hydrate formation conditions between the synthetic sample and the natural sample the A2 data agrees very well with the data from the synthetic sample. The ratio of the integrated intensities for gas hydrate from the B1 core, predominantly methane hydrate with low quantities of propane and CO_2, is somewhat different, $I_{2904}/I_{2916} = 3.89$ (or using the approximation above, $\theta_s/\theta_l = 0.77$).

This is considerably lower than that for pure methane hydrate and provides strong evidence that, in the case of core sample B1, either structure I gas hydrate has formed in the presence of the other guests, thus possibly displacing some of the methane from the clathrate cages, or some structure II gas hydrate is present with propane in the large cages, thus changing the overall experimental occupancy ratios observed by Raman spectroscopy. Another possibility is that the interaction between guests has an undetermined effect on the Raman scattering cross sections.

Coupled O-H Vibrations of Water in the Host Framework

The Raman spectrum arising from the coupled intramolecular O-H vibrations of the water molecules making up the host framework of CH_4 gas hydrate are presented in FIGURE 3, the analogous spectrum collected from crystalline ice I_h was collected after the sample decomposed and is given for comparison. Studies have been carried out on crystalline and vapor deposited amorphous ice and an excellent historical review has been given by Sceats and Rice.[19] The spectrum of the coupled molecular water vibrations, above 3,050 cm^{-1} for H_2O and above 2,300 cm^{-1} for D_2O in crystalline ice, is broad and relatively featureless and it has been studied in some detail by producing dilute solutions of HDO (5%) in either H_2O or D_2O. A number of empirical correlations between the hydrogen bond strength (length) and the isolated O-H (or O-D) vibrational frequency have been made[20,21] and, for example, a 1% change in the hydrogen bond length in ice will result in a 23 cm^{-1} shift in the peak position. Water molecules in the various phases of ice and in clathrate hydrates are tetrahedrally coordinated with nearly linear O-H–O bonds; therefore the application of the correlations is equally valid for either. A recent study has been conducted at the National Research Council of Canada illustrates the use of Raman spectroscopy of isolated (or *uncoupled*) O-D oscillations of HDO water molecules to study changes in the local bonding structure on crystallization of hyperquenched amorphous clathrate forming solutions (tetrahydroftiran + H_2O).[22] The spectra and frequency shift data as a function of annealing temperature have been reproduced here in FIGURES 4 and 5. This work may have implications for unfrozen (glassy) water that may be present in a gas hydrate + water + porous system.

Since the most intense peak in the Raman spectrum of water (pure H_2O or pure D_2O) in ice and clathrate hydrates, FIGURE 3, results predominantly from the coupled symmetric stretching vibration O-H, only qualitative information about the local hydrogen bond structure can be inferred. The spectra collected at 11.5 K contains almost entirely structure I gas hydrate and thus the spectral lines above

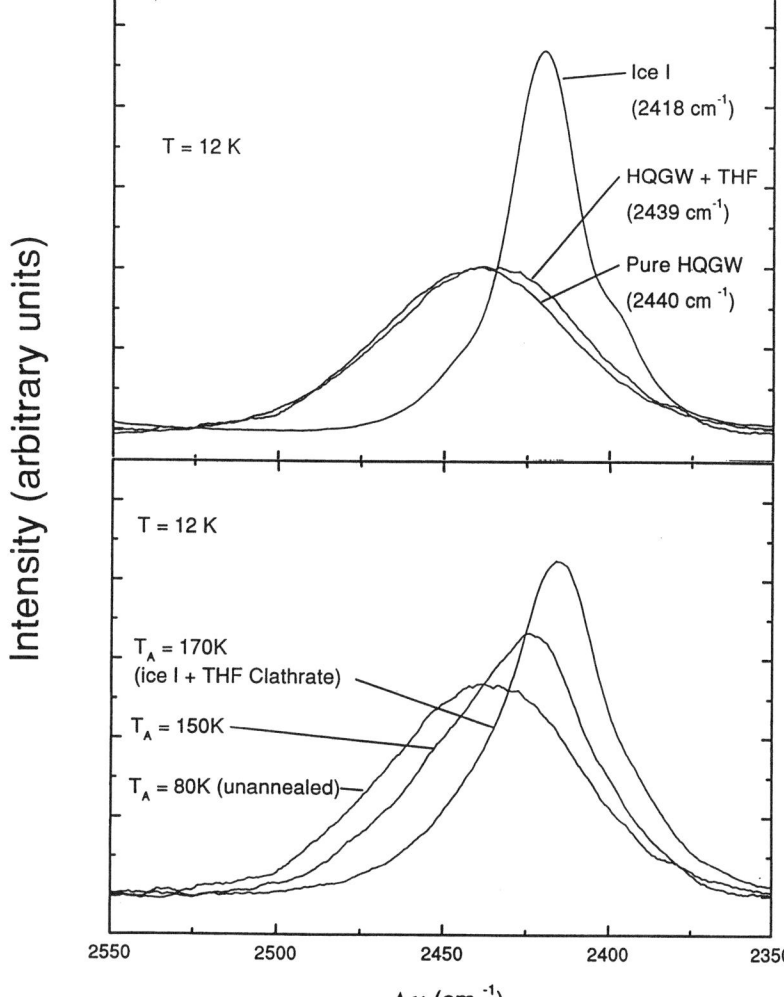

FIGURE 4. Raman spectra resulting from uncoupled O-D vibrations in hyperquenched clathrate gas hydrate forming aqueous solutions of water and tetrahydrofuran (THF). The hyperquenching process involves cooling solutions droplets at 10^6 K/sec onto a cold target, thus forming an amorphous material that is structurally related to the supercooled liquid above the homogeneous nucleation temperature. The broad Raman lines (resulting from the isolated O-D vibration) give good evidence that the material is not microcrystalline. The **upper frame** compares the hyperquenched solution with pure hyperquenched water, indicating similar local bonding structure of the H_2O molecules, and crystalline ice. The **lower frame** illustrates the relaxation of the hydrogen bonded water structure toward that of crystalline ice + hydrate as the amorphous material is annealed at higher temperature (the samples were annealed at T_A for 15 minutes before being crash-cooled back to 12 K for data collection).

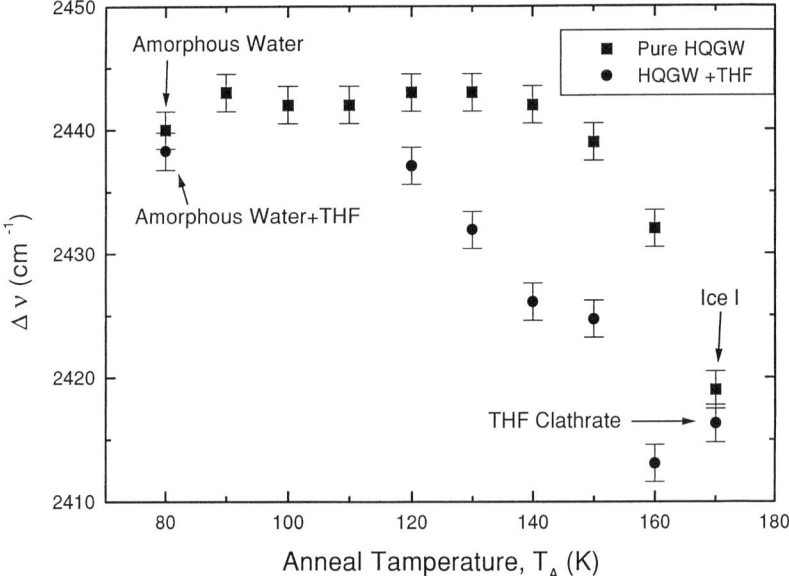

FIGURE 5. A comparison of the Raman peak position resulting from the isolated O-D vibration in hyperquenched amorphous solutions of water + THF with pure hyperquenched amorphous water as the samples are annealed at T_A. The amorphous water + THF solutions begin to relax toward the crystalline state at a much lower temperature ($T_A = 120$ K) then pure amorphous water. (Generally, dipolar guest molecules, or guest molecules that contain oxygen atoms, increase the dynamic reorientation rate of the host water molecules.)

3,050 cm^{-1} arise from water molecules in the clathrate host framework. The spectrum collected at $T_a = 180$ K resulted after the sample was annealed at 180 K for three hours to allow the gas hydrate to decompose completely. The sample was then cooled to 11.5 K and the coupled O-H vibrations of the ice I$_h$ recorded.

It is clear from FIGURE 3 that the Raman spectrum of water in the clathrate framework is distinctly different from that of ice I$_h$.

In the clathrate structure this line is centred on 3,101 cm^{-1} with a full width at half maximum FWHM equal to 54 cm^{-1}. It is clearly shifted to lower frequency by 19 cm^{-1} on dissociation of the gas hydrate into ice I$_h$ + gas with the leading O-H vibration Raman peak centred on 3,082 cm^{-1} with FWHM = 22 cm^{-1}.

This indicates that the average hydrogen bond strength in the Str. I clathrate is weaker than that in ice I$_h$. The half-width of the line is also considerably broader in the clathrate then in ice and is a good indication that the distribution of hydrogen bond strength in the clathrate is significantly different from that in ice.

It is well known from X-ray diffraction analysis that the hydrogen bond length in ice I$_h$I$_c$ is 2.76 Å with a very narrow distribution whereas the average hydrogen bond length in structure I gas hydrate is 2.79 Å with a much larger distribution of H-bond lengths.

CONCLUSIONS

A general review of the Raman technique has been presented along with specific discussion concerning its application to gas hydrates, as well as some new data. It has been shown that qualitative information regarding the cage occupancies can be obtained with the current state of knowledge (although the precise nature of the interaction between the frame work and the guest remains poorly understood); the local structure of the hydrogen bonds in water is readily accessible, as are the effects of subtle perturbations of the guest molecules due to the water frame-work. However, considerable caution must be applied when using the technique and the results should not be over interpreted. These include the following:

- The Raman scattering process is a very weak phenomenon that requires very intense monochromatic incident light and very carefully aligned collection optics, which can have significant affects on the shape and intensity of the observed spectra.

- A variety of materials characterization techniques, such as simple powder X-ray diffraction or NMR for phase determination, should be applied to ensure sample quality control.

- Experimental concerns, such as corrections of instrumentally broadened lines, should be addressed before quantitative analysis of data.

- The derivatives of the polarizability tensor ($\delta\alpha_{xx}/\delta S_1$) for the totally symmetric stretching vibration of methane should be calculated before any quantitative calculations involving intensity comparisons are made; experiments are under way to cross calibrate Raman and NMR data collected from the same samples.

- The low concentration of gas hydrate material mixed with the sand and other mineral matter in gas hydrate recovered from the JAPEX/JNOC/GSC Mallik 2L-38 research well makes collection of good quality Raman data difficult; the sample preparation and experimental details are described in Reference 18.

- In natural samples the absorption of laser light and the subsequent heating and decomposition of sample may occur due to other components; this was a problem in the present work and was overcome only by running the laser slightly detuned and correcting the spectra during analysis.

REFERENCES

1. UCHIDA, T., A. TAKAGI, J. KAWABATA, S. MAE & T. HONDOH. 1995. Energy Convers. Mgnt. **36:** 547.
2. SUM, A.K., R.C. BURRUS & E.D. SLOAN. 1997. J. Phys. Chem. **101:** 7371.
3. PAUER, F., J. KIPFSTUHL & W.F. KUHS. 1997. J. Geophys. Res. **102:** 26519.
4. HERZBERG, G. 1945. Infrared and Raman Spectra. Van Nostrand Reinhold, New York.
5. Spectra of powdered samples, as in the present case, are often completely depolarized probably because the microcrystals making up the sample destroy the polarization of the incident radiation.

6. TULK, C.A., D.D. KLUG, R. BRANDERHORST, P. SHARPE & J.A. RIPMEESTER. 1998. J. Chem. Phys. **109:** 8478.
7. SEITZ, J.C., J.D. PASTERIS & B. WOPENKA. 1987. Geochim. Cosmochim. Acta **51:** 1651.
8. DAVIDSON, D.W. 1971. Can. J. Chem. **49:** 1224.
9. TSE, J.S., W.R. MCKINNON & M. MARCHI. 1987. J. Phys. Chem. **91:** 4188.
10. TSE, J.S., B.M. POWELL, V.J. SEARS & Y.P. HANDA. 1993. Chem. Phys. Lett. **215:** 383.
11. TSE, J.S. 1994. Ann. N.Y. Acad. Sci. **715:** 187.
12. LONG, D.A. 1977. Raman Spectroscopy. McGraw-Hill, New York.
13. MONTERO, S. & D. BERMEJO. 1976. Molecular Physics **32:** 1229.
14. SVENDSEN, N. & T. STROYER-HENSEN. 1985. Molecular Physics **56:** 1025.
15. IOANNOU, A.G. & R.D. AMOS. 1997. Chem. Phys. Lett. **279:** 17.
16. GREGORYANZ, E., M.J. CLOUTER, N.H. RICH & R. GOULDING. 1998. Phys. Rev. B. **58:** 2497.
17. RIPMEESTER, J.A. & C.I. RATCLIFFE. 1988. J. Phys. Chem. **92:** 337.
18. TULK, C.A., C.I. RATCLIFFE & J.A. RIPMEESTER. 1999. Geological Survey of Canada Bulletin. **544:** 251.
19. SCEATS, M.G. & S.A. RICE. 1982. *In* Water, a Comprehensive Treatise. F. Franks, Ed. Plenum Press, New York.
20. WHALLEY, E. 1976. *In* The Hydrogen Bond. P. Schuster, G. Zundel & C. Sandorfy, Eds. North-Holland, Amsterdam.
21. NOVAK, A. 1974. Struct. Bond. **18:** 177.
22. TULK, C.A., Y. BA, D.D. KLUG, G. MCLAURIN & J.A. RIPMEESTER. 1999. J. Chem. Phys. **110:** 6475.

Structural Transition Studies in Methane + Ethane Hydrates Using Raman and NMR

S. SUBRAMANIAN, R.A. KINI, S.F. DEC, AND E.D. SLOAN, JR.

Center for Hydrate Research, Colorado School of Mines, Golden, Colorado 80401, USA

ABSTRACT: Methane and ethane are each known to form structure I (sI) hydrates. However, Raman and NMR measurements made on hydrates formed from gas mixtures of methane (CH_4) and ethane (C_2H_6) indicate that structure II (sII) forms in this system for certain gas compositions. Raman band frequencies for ethane obtained from different $CH_4 + C_2H_6$ hydrates at 274.2 K indicate a change in hydrate structure from sI to sII between 72.2 and 75 mole% of CH_4 in the vapor. A significant increase in the amount of methane incorporated into the hydrate is observed with change in hydrate structure from sI to sII. Solid-state ^{13}C CPMAS NMR measurements at 253 K confirm that $CH_4 + C_2H_6$ gas mixtures can form sII hydrates at certain feed compositions.

INTRODUCTION

There are three known hydrocarbon gas hydrate structures—sI, sII, and sH. TABLE 1 lists the cavities found in these hydrates.[1] Small molecules like CH_4 and C_2H_6 (4.36 and 5.5 Å in diameter, respectively[1]) form sI hydrates. The formation of sII hydrate from mixtures of sI formers has either been experimentally observed or predicted by researchers in the past. Von Stackelberg and Jahns observed that mixtures of sI hydrate formers like difluoroethane (CH_3CHF_2) and hydrogen sulfide (H_2S) formed sII hydrates.[2]

Mixtures of CH_4 and C_2H_6 are usually predicated to form sI hydrates. However, the model by Hendriks *et al.*, predicted sII for this system over a certain composition range.[3] Holder and Hand pointed out that their model, which assumed sI hydrate, failed to predict experimental data for a 90.6% CH_4 and 9.4% C_2H_6 feed.[4] In this work, Raman and NMR spectroscopic methods, two powerful tools for hydrates,[5–11] were used to obtain structural and compositional information for $CH_4 + C_2H_6$ hydrates.

APPARATUS

A Renishaw MK III fiber optic based Raman spectrometer equipped with a 2,400 grooves/mm grating and a CCD detector was used in this work. The spectral resolution was 4.5 cm^{-1}. The excitation source was a 514.53 nm Ar-ion laser delivered to a fiber probe fitted with a 20× microscope objective to focus the laser on the sample. Spectra were obtained from different spots within each sample by adding 20 spectra each integrated over 25 seconds.

TABLE 1. Types and sizes of cavities found in hydrocarbon gas hydrates

Property	Hydrate						
	Structure I (sI)		Structure II (sII)		Structure H (sH)		
Cavity type	5^{12}	$5^{12}6^2$	5^{12}	$5^{12}6^4$	5^{12}	$4^35^66^3$	$5^{12}6^8$
Radius Å	3.95	4.33	3.91	4.73	3.91	4.06	5.71
Cages/Unit cell	2	6	16	8	3	2	1
H_2O molecules/cell	46		136		34		
Crystal type	cubic		cubic		hexagonal		
Lattice constants (Å)	$a = 12$		$a = 17.3$		$a = 12.26; c = 10.17$		

Source: Reference 1, page 33.

A Chemagnetics CMX Infinity 400 spectrometer was used to record NMR spectra. All ^{13}C NMR spectra were recorded using a Chemagnetics variable temperature (VT) probe at 100.6 MHz with proton decoupling, and ^1H–^{13}C cross-polarization, magic-angle spinning (CP MAS).[12] All chemical shifts for ^{13}C NMR spectra are relative to tetramethylsilane.

PROCEDURE

High purity methane, ethane, propane, and deuterated propane (C_3D_8, 98% D atom) were used in this work. A synthetic gas mixture with a composition identical to the Gulf of Mexico Green Canyon natural gas was also used.[9] $CH_4 + C_2H_6$ mixtures were prepared gravimetrically and thermally agitated to ensure uniform composition.

Sample Preparation for Raman

Hydrate samples were prepared from $CH_4 + C_2H_6$ mixtures in a custom-designed thick walled Pyrex capillary cell described elsewhere.[10] The L_w-H-V equilibrium pressure at 274.2 K for each mixture was determined to within 0.014 MPa by pressure cycling and visual observation of hydrate dissociation. Raman spectra were obtained at 274.2 K and at pressures slightly above the determined L_w-H-V equilibrium point.

Sample Preparation for NMR

A Pyrex glass tube with a small bulb at one end was used to prepare hydrate samples. The bulb was partially filled with powdered ice, formed from deionized water, and pressurized with a gas or gas mixture to about 7 MPa. The sample cell was conditioned at 273 K for several days and subsequently at temperatures above the ice point. The glass bulb was sealed off from the tube using a flame torch and used as the NMR sample cell.

EXPERIMENTAL RESULTS

Raman Studies

Raman spectra for CH_4 in different hydrates and different phases have been extensively studied.[9,10] Pure CH_4 forms sI hydrate by occupying both the 5^{12} (small) and $5^{12}6^2$ (large) cavities. In the presence of a large molecule like THF-$d8$, sII hydrate is formed with CH_4 occupying both the 5^{12} (small) and $5^{12}6^4$ (large) cavities and THF-$d8$ occupying the $5^{12}6^4$ cavity. FIGURE 1 compares the Raman spectra of the ν_1 totally symmetric C-H stretching vibration for CH_4 in free gas, and in the large and small cavities of CH_4 sI hydrate, and CH_4 + THF-$d8$ sII hydrate.

There is no information in the literature about the frequencies of Raman bands for C_2H_6 trapped in the cavities of different hydrate structures. FIGURES 2 and 3 show the C-H and C-C regions of the Raman spectra for ethane trapped in sI and sII hydrates, obtained for the first time here, and compared to the spectra for free C_2H_6 gas. The sI spectrum in each figure corresponds to pure ethane hydrate with C_2H_6 occupying only the large $5^{12}6^2$ cavities in the structure. The sII spectra correspond to a C_2H_6 + C_3D_8 hydrate prepared from a gas mixture containing 65 mole% C_2H_6 and 35 mole% C_3D_8.[4] Ethane in this case occupies only the large $5^{12}6^4$ cavities.

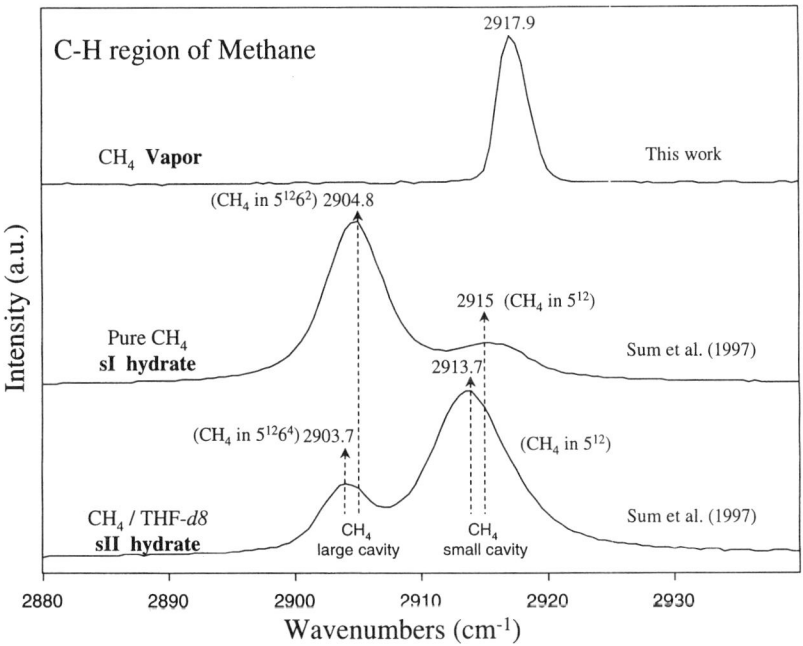

FIGURE 1. Raman spectra for CH_4 (ν_1 C-H symmetric stretch) in vapor (2.07 MPa, 298 K), pure CH_4 sI hydrate (2.97 MPa, 274.2 K), and CH_4 + THF-$d8$ sII hydrate (2.07 MPa, 273.2 K).

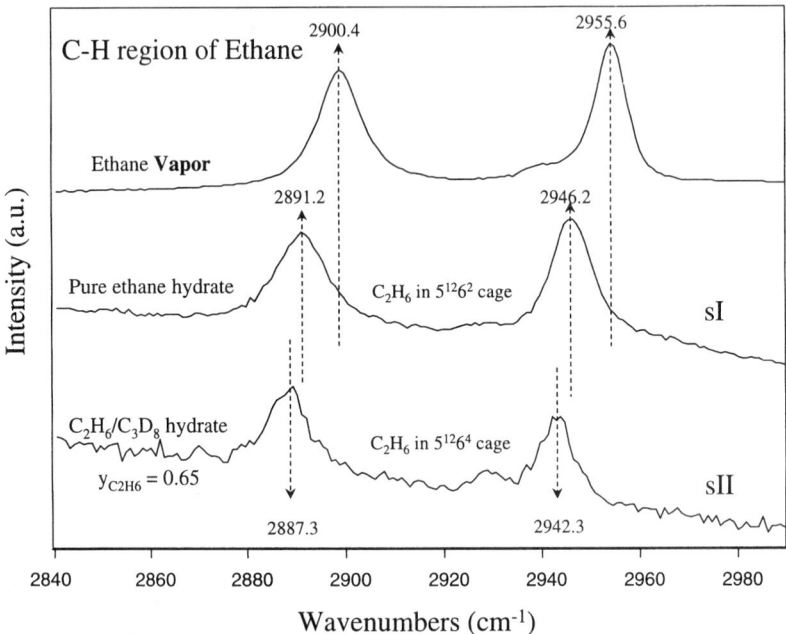

FIGURE 2. Raman spectra of the C-H region for C_2H_6 showing the strong resonance doublets and a weak shoulder due to the $(\nu_8 + \nu_{11})$ CH_3 or ν_{10} CH vibrations. C_2H_6 vapor (2.07 MPa, 298 K), pure C_2H_6 sI hydrate (1.03 MPa, 274.2 K), and $C_2H_6 + C_3D_8$ sII hydrate (0.8 MPa, 274.2 K) spectra are compared.

TABLE 2. Summary of $CH_4 + C_2H_6$ hydrate systems investigated in Raman studies

Feed Gas Composition $y_{CH_4}{}^a$	L_w-H-V Equilibrium Pressure P_{eq} (MPa)	Pressure at which Sample was Conditioned P_{actual} (MPa)	Sample Temperature (deg. K)	Equilibrium Vapor Composition $y_{CH_4}{}^b$	Number of Raman Spectra Collected
0.581	0.883	0.938	274.2	0.628 ± 0.014	4
0.690	0.958	0.9997	274.2	0.676 ± 0.009	5
0.708	0.972	1.027	274.2	0.722 ± 0.008	6
0.730	0.986	1.055	274.2	0.750 ± 0.007	5
0.844	1.165	1.255	274.2	0.851 ± 0.995	5
0.918	1.448	1.572	274.2	0.921 ± 0.002	4

[a]Obtained using gravimetric gas mixture preparation techniques.
[b]Obtained using peak areas in the Raman spectra of final vapors above the hydrates.

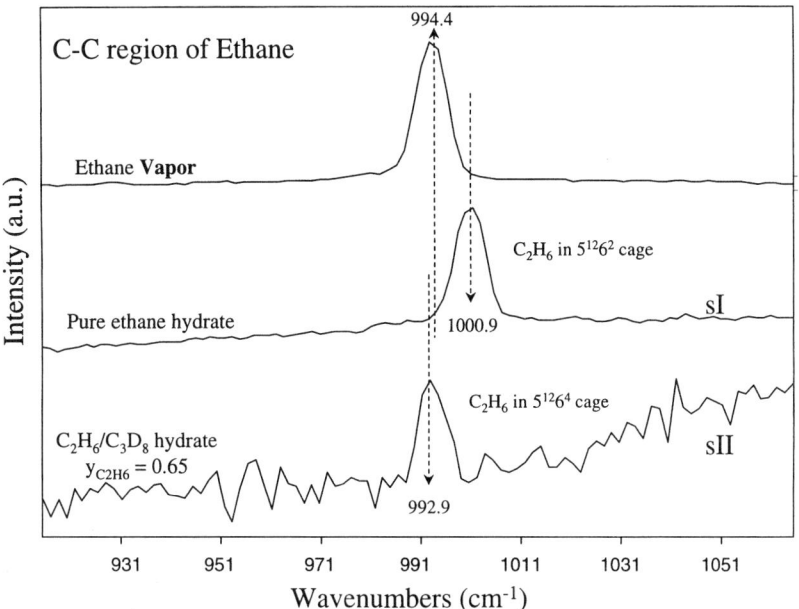

FIGURE 3. Raman spectra of the ν_3 C-C stretch for C_2H_6 vapor (2.07 MPa, 298 K), pure C_2H_6 sI hydrate (1.03 MPa, 274.2 K), and $C_2H_6 + C_3D_8$ sII hydrate (0.8 MPa, 274.2 K) are compared.

TABLE 2 lists the measured three phase L_w-H-V equilibrium pressures for six different $CH_4 + C_2H_6$ mixtures at 274.2 K, and the conditions at which Raman spectra were obtained. The compositions of the final vapor in equilibrium with the hydrates, as determined from vapor phase Raman spectra, are also listed in TABLE 2. FIGURE 4 shows the C-H region and FIGURE 5 shows the C-C region Raman spectra obtained from these $CH_4 + C_2H_6$ equilibrium hydrate samples. The C-H region spectra contain bands from both CH_4 and C_2H_6 in the hydrate cavities, whereas the C-C region spectra contain bands only from ethane. The compositions of the final vapor in equilibrium with the hydrates are indicated in the figures.

NMR Studies

FIGURE 6 compares ^{13}C CP MAS NMR spectra obtained at 253 K from hydrates of pure CH_4 (sI) and pure C_2H_6 (sI), $CH_4 + C_3H_8$ (sII), and a hydrate (sII) formed from a synthetic gas mixture[14] (Green Canyon gas, Gulf of Mexico). The chemical shift values of CH_4 and C_2H_6 in different cavities of sI and sII were obtained from these spectra.

Three different $CH_4 + C_2H_6$ gas mixtures were used to form hydrate samples for NMR measurements using the procedure described earlier. FIGURE 7 shows CP MAS ^{13}C NMR spectra for these hydrates and lists the compositions of gas mixtures used to prepare the hydrates.

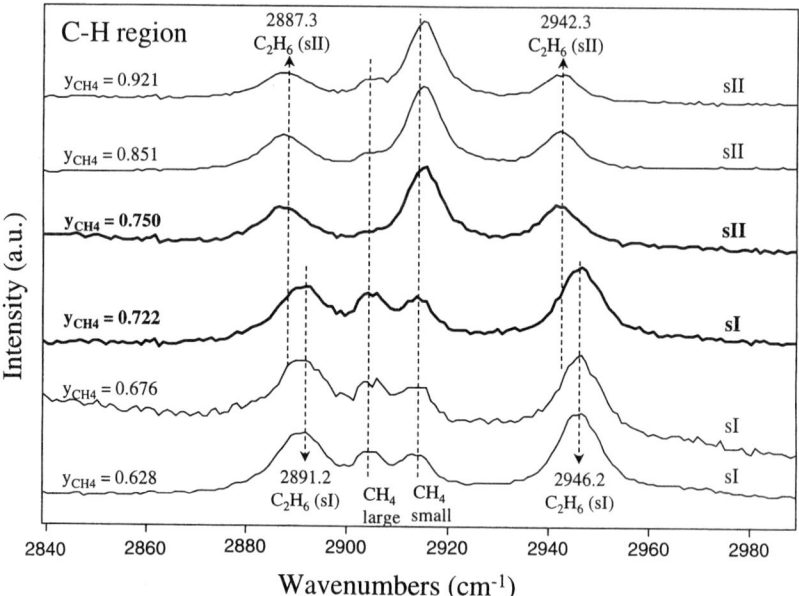

FIGURE 4. C-H region Raman spectra for six different $CH_4 + C_2H_6$ hydrates obtained at 274.2 K and L_w-H-V equilibrium conditions. Note the abrupt change in hydrate spectra as equilibrium vapor composition increases from 0.722 to 0.750 methane mole fraction (y_{CH_4}).

DISCUSSION

Raman Results

Raman spectra in FIGURE 1 indicate that the v_1 C-H stretching bands for CH_4 in sI and sII hydrates are distinct from that for the vapor phase. From the peak assignments of Sum et al.,[10] there is a large 10 cm^{-1} difference between large and small cavity frequencies in each hydrate structure (sI and sII) but only a 1 cm^{-1} difference between sI and sII frequencies for each cavity type (large and small). Thus, CH_4 Raman peak positions are excellent discriminators between hydrate cavities, but not between hydrate structures. From FIGURE 2, it is clear that peaks in the C-H region of the Raman spectra for C_2H_6 in sI and sII hydrates are shifted compared to peaks corresponding to C_2H_6 vapor. There seem to be three peaks in the C-H region of the Raman spectrum for ethane in each of the hydrate and vapor spectra—two of them are strong and the third is very weak. The strong peaks in the C_2H_6 vapor spectrum at 2,900.4 cm^{-1} and 2,955.6 cm^{-1} are assigned to the resonance doublet comprised of the v_1 C-H stretch and an overtone mode of a CH_3 deformation vibration ($2v_{11}$) and the weak band at 2,944.3 cm^{-1} may correspond to either ($v_8 + v_{11}$) CH_3 or v_{10} CH vibration.[13] The peaks corresponding to the vibrational modes in the resonance doublet for ethane trapped in sI hydrate $5^{12}6^2$ cage and sII hydrate $5^{12}6^4$ cage

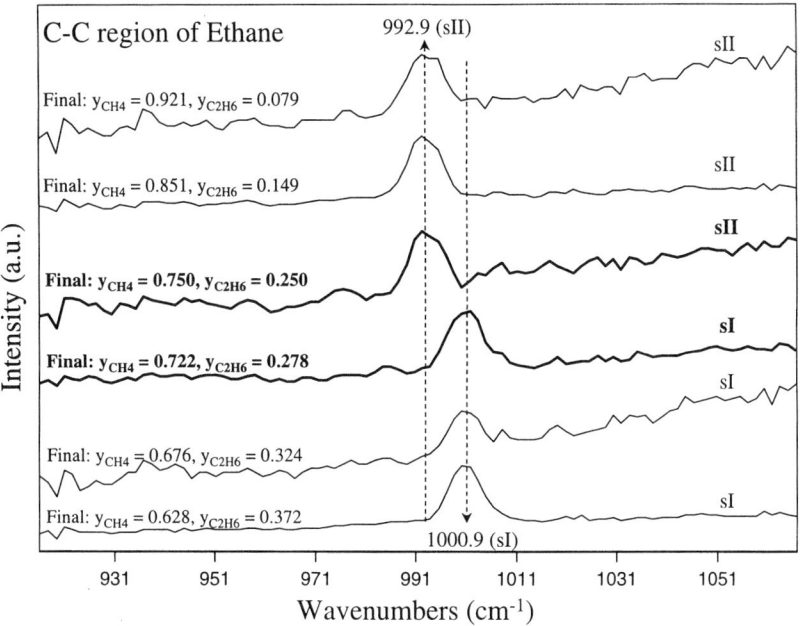

FIGURE 5. Raman spectra of the v_3 C-C stretch of C_2H_6 in the six different CH_4 + C_2H_6 hydrates. Note the abrupt shift in band frequency over the y_{CH_4} range of 0.722 to 0.750.

(FIG. 2) are given in TABLE 3. It should be noted that there is a 4 cm^{-1} difference in band frequencies between sI and sII. This is more pronounced than the 1 cm^{-1} difference in frequencies observed for the C-H bands of CH_4 between sI and sII hydrates.

In FIGURE 3, Raman spectra arising from the C-C stretching vibration of C_2H_6 in sI and sII hydrates are compared with C_2H_6 vapor spectrum. A single strong peak is observed at 994.4 cm^{-1} for ethane vapor that can be assigned to the v_3 C-C symmetric

TABLE 3. Frequencies of C_2H_6 bands in sI and sII hydrates

	Hydrate Structure: Type (Cavity)	
Vibrational modes	sI ($5^{12}6^2$)	sII ($5^{12}6^4$)
v_1 symmetric C-H stretch component of the resonance doublet Peak position (cm^{-1})	2891.2	2887.3
Overtone mode of one of the CH_3 deformation vibrations ($2v_{11}$) contributing to the resonance doublet Peak position (cm^{-1})	2946.2	2942.3
v_3 symmetric C-C stretch Peak position (cm^{-1})	1000.9	992.9

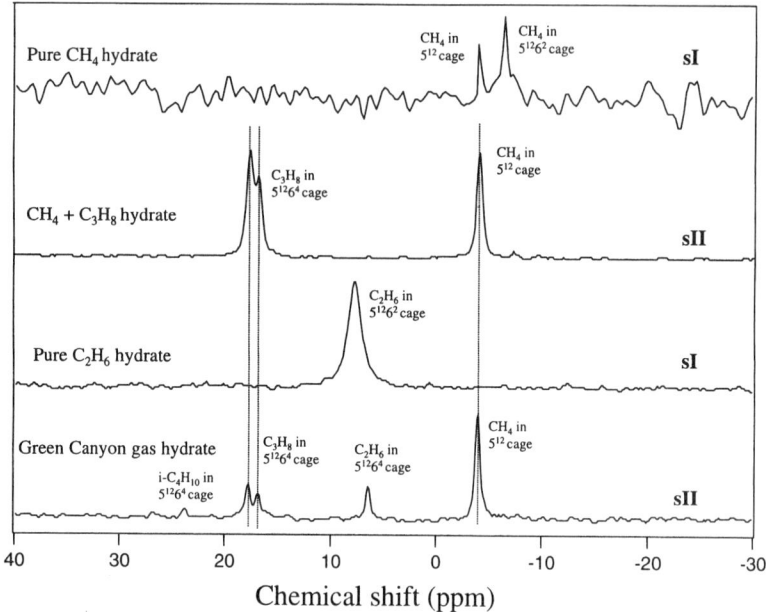

FIGURE 6. ^{13}C CP MAS NMR spectra obtained at 253 K for pure CH_4 sI hydrate, $CH_4 + C_3H_8$ sII hydrate, pure C_2H_6 sI hydrate, and Green Canyon gas sII hydrate.

stretch.[13] In the spectrum for pure C_2H_6 hydrate, this C-C stretch for C_2H_6 in sI occurs at 1,000.9 cm^{-1}. Similarly, the spectrum from $C_2H_6 + C_3D_8$ hydrate shows that the C-C stretch for C_2H_6 in sII occurs at 992.9 cm^{-1}. Thus, the hydrate peaks are shifted relative to the vapor peak and a larger cavity ($5^{12}6^4$) for ethane results in a lower C-C stretch frequency. It should be noted that there is an 8 cm^{-1} difference in C-C stretch frequencies of C_2H_6 between sI and sII hydrates. Differences of 4 cm^{-1} and 8 cm^{-1} in band frequencies for C_2H_6, in the C-H and C-C regions, respectively, between sI and sII hydrates, are much larger than the 1 cm^{-1} differences in the C-H region bands of CH_4 between sI and sII. Hence, for $CH_4 + C_2H_6$ hydrates, the hydrate structure type (sI or sII) can be unambiguously determined from ethane C-H and C-C Raman signatures.

FIGURES 4 and 5 show the C-H and C-C region Raman spectra, respectively, obtained from six different $CH_4 + C_2H_6$ hydrates at 274.2 K and near L_w-H-V equilibrium conditions. In FIGURE 4, each spectrum was deconvoluted into a total of five mixed Gaussian-Lorentzian bands to accurately obtain the frequencies corresponding to the resonance doublet for C_2H_6 occupying only the large cavity in the hydrate and frequencies corresponding to CH_4 occupying the small and large cavities in the hydrate. Two sets of vertical lines, one at 2,891.2 and 2,946.2 cm^{-1} and another at 2,887.3 and 2,942.3 cm^{-1}, corresponding to frequencies of the resonance doublet bands for C_2H_6 in sI and sII respectively, are shown in FIGURE 4 as guides to the eye.

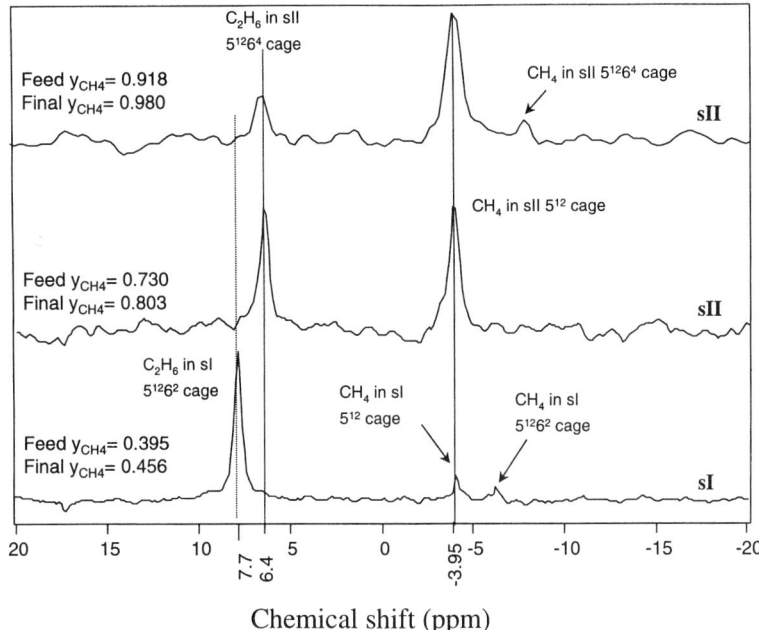

Chemical shift (ppm)

FIGURE 7. ^{13}C CP MAS NMR spectra obtained from three different $CH_4 + C_2H_6$ hydrates at 253 K. Note shifts in the ethane peaks as final vapor y_{CH_4} increases from 0.456 to 0.980.

C-H region Raman spectra in FIGURE 4 indicate that $CH_4 + C_2H_6$ hydrates in equilibrium with vapors containing 0.628, 0.676, and 0.722 CH_4 mole fractions, y_{CH_4}, have frequencies of the C_2H_6 resonance doublet Raman bands corresponding to C_2H_6 in sI hydrate. Also from FIGURE 4, hydrates in equilibrium with vapors containing a y_{CH_4} of 0.750, 0.851, and 0.921, have frequencies of the C_2H_6 resonance doublet Raman bands corresponding to C_2H_6 in sII hydrate.

FIGURE 5 shows the Raman spectra arising from the C-C stretching vibration of ethane for the $CH_4 + C_2H_6$ hydrate samples. These spectra indicate that hydrates in equilibrium with vapors having a y_{CH_4} of 0.628, 0.676, and 0.722 have a C_2H_6 C-C stretching vibrational frequency of 1,000.9 cm^{-1} corresponding to C_2H_6 in sI hydrate. However, hydrates in equilibrium with vapors having a y_{CH_4} of 0.750, 0.851, and 0.921 have a C_2H_6 C-C stretching frequency of 992.9 cm^{-1}, which is 8 cm^{-1} lower than the sI value and corresponds to C_2H_6 in sII hydrate.

Based on band frequencies in the C-H and C-C regions of C_2H_6 in FIGURES 4 and 5 for different $CH_4 + C_2H_6$ hydrates at 274.2 K, we conclude that sII hydrate can form in this system. The change from sI to sII occurs over the composition (y_{CH_4}) range of 0.722 to 0.750. This is particularly interesting because CH_4 and C_2H_6 both form sI hydrate by themselves and combinations of CH_4 and C_2H_6 were generally thought to form sI hydrate.

NMR Results

FIGURE 6 shows four ^{13}C CPMAS NMR spectra that were used to obtain the chemical shifts of CH_4 and C_2H_6 in sI and sII hydrates. The spectra of pure ethane and pure methane hydrate were used for the assignments of sI hydrates. The spectrum of $CH_4 + C_3H_8$ sII hydrate, formed from a gas mixture with $y_{CH_4} = 0.880$ was used to obtain chemical shift values of CH_4 in both small and large cavities of sII hydrate. The Green Canyon gas hydrate spectrum in FIGURE 6 was used as a reference for the chemical shift of C_2H_6 in sII hydrate. Due to the presence of C_3H_8 and i-C_4H_{10}, the hydrates formed from this gas mixture have been shown to be of type sII.[14]

TABLE 4 summarizes the ^{13}C chemical shifts, obtained from spectra shown in FIGURE 6, for various hydrocarbon guests in different types of hydrate cavities. The differences in chemical shift values between literature and the present work may be due to temperature effects and/or external chemical shift referencing differences.

^{13}C CP MAS NMR spectra of $CH_4 + C_2H_6$ hydrates formed from three different gas mixtures are shown in FIGURE 7. These spectra, from the very nature of the CP experiment, have intense contributions from only those CH_4 and C_2H_6 molecules that are trapped as guests in the hydrate cages. Due to large intensities and significant differences between ethane sI and sII signatures, C_2H_6 is a better indicator of hydrate structure. Therefore, chemical shift values of C_2H_6 in $CH_4 + C_2H_6$ hydrates are used for definitive identification of hydrate structure type.

As can be seen in FIGURE 7, the ^{13}C CP MAS NMR spectrum of the $CH_4 + C_2H_6$ hydrate in contact with a vapor having a CH_4 mole fraction of 0.456, shows that the chemical shift of C_2H_6 trapped in the large cavity of the hydrate is 7.7 ppm. This chemical shift value corresponds to C_2H_6 trapped in the $5^{12}6^2$ cavity of sI hydrate suggesting that this $CH_4 + C_2H_6$ hydrate is of type sI.

Similarly in FIGURE 7, the ^{13}C CPMAS NMR spectra of $CH_4 + C_2H_6$ hydrates in contact with vapors having a y_{CH_4} of 0.803 and 0.980 show a resonance line for C_2H_6

TABLE 4. ^{13}C NMR chemical shifts (ppm) of hydrocarbon guests in different phases

Component	sI Hydrate		sII Hydrate	
	small cage (5^{12})	large cage ($5^{12}6^2$)	small cage (5^{12})	large cage ($5^{12}6^4$)
Methane	−4 ppm (−2.84 ppm[a])	−6.1 ppm (−5.21 ppm[a])	−3.95 ppm (−2.73 ppm[a])	−7.7 ppm (−6.27 ppm[a])
Ethane	—	7.7 ppm	—	6.4 ppm (7.8 ppm[b])
Propane	—	—	—	17.7 ppm, 16.8 ppm (18.9 ppm[b])
Iso-butane	—	—	—	26.6 ppm, 23.7 ppm (28.1 ppm[b])

[a]Ripmeester and Ratcliffe.[6]
[b]Davidson et al.[14]

at 6.4 ppm, which corresponds to C_2H_6 trapped in the $5^{12}6^4$ cavity of sII hydrate. This confirms the formation of sII $CH_4 + C_2H_6$ hydrates at these compositions.

CHANGES IN RELATIVE GUEST DISTRIBUTION WITH STRUCTURE

The deconvolution procedure mentioned earlier yielded both the frequencies and areas of CH_4 and C_2H_6 bands in the Raman spectra of hydrates. The band areas were used to show that the distribution of guests within the hydrate changes significantly upon change in structure. The area $A_{C_2H_6}$ corresponding to amount of C_2H_6 in the hydrate was obtained by summing areas of the three ethane peaks in the C-H region. For sI hydrates, areas of bands at 2,891.2 cm^{-1}, 2,946.2 cm^{-1}, and 2926.5 cm^{-1} were used in the summation and for sII hydrates, areas of bands at 2,887.3 cm^{-1}, 2,942.3 cm^{-1}, and 2,928.8 cm^{-1} were summed. The area A_{CH_4} corresponding to the amount of CH_4 in the hydrate was obtained from summation of the areas under peaks corresponding to methane in the small and large cavities in the hydrate.

Ratio of areas ($A_{CH_4}/A_{C_2H_6}$) obtained from each $CH_4 + C_2H_6$ hydrate spectrum in FIGURE 4 can be considered to be proportional to the relative amounts of CH_4 and C_2H_6 in the hydrate. In FIGURE 8, $A_{CH_4} / A_{C_2H_6}$ values from the C-H region Raman spectra of $CH_4 + C_2H_6$ hydrates are plotted as a function of the composition of the vapor in equilibrium with the hydrates. FIGURE 8 shows that there is a sharp increase

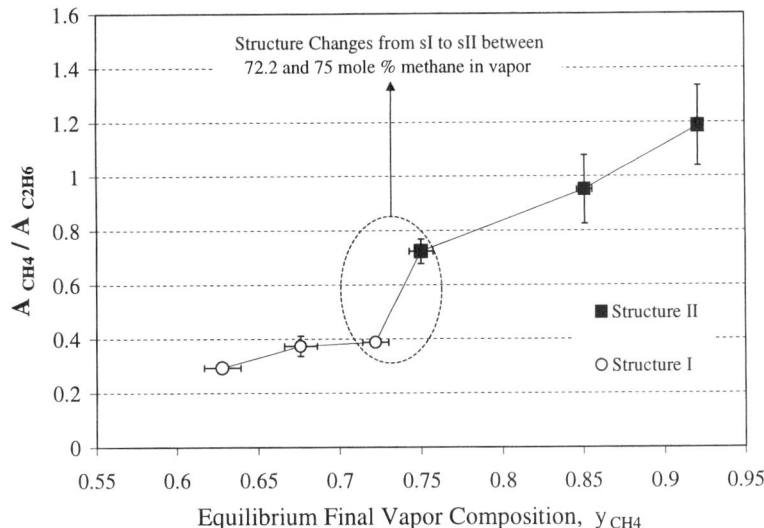

FIGURE 8. Ratios of methane and ethane band areas ($A_{CH_4}/A_{C_2H_6}$) from Raman spectra in FIGURE 4, plotted as a function of vapor composition (y_{CH_4}). Note the abrupt change in $A_{CH_4}/A_{C_2H_6}$ with change in structure.

in the $A_{CH_4}/A_{C_2H_6}$ value over the y_{CH_4} range of 0.722 to 0.750. This indicates a significant increase in the amount of CH_4 incorporated in the hydrate over this composition range due to change in structure from sI to sII.

IMPLICATIONS OF RESULTS

The formation of sII in the $CH_4+C_2H_6$ hydrate system was qualitatively predicted by Hendriks et al.[3] FIGURE 9 shows the predictions using their method at 274.2 K for this system with excess water. As can be seen from the L_w-H-V predictions, a change from sI to sII is predicted at a y_{CH_4} of 0.62. However, our data show that this transition actually occurs in the y_{CH_4} range of 0.722 to 0.750.

It is very important to change existing hydrate equilibrium prediction models so that they can correctly predict the sI to sII transition point in the $CH_4 + C_2H_6$ hydrate system. The method of Hendriks et al.,[3] also predicts a transition from sII back to sI at $y_{CH_4} = 0.993$ at 274.2 K. This prediction implies that hydrates formed in nature in either oceanic or permafrost environments, from gases containing CH_4 and as little as 0.7% C_2H_6 could potentially be sII hydrates instead of sI hydrates.

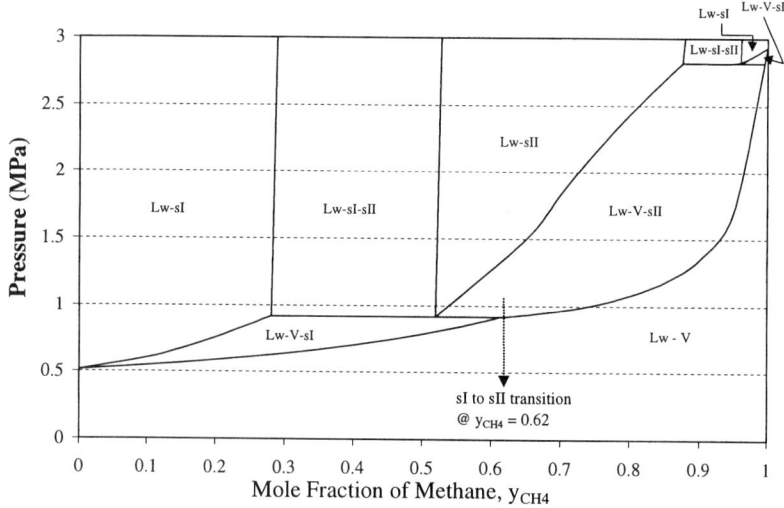

FIGURE 9. Predicted P-x diagram for the $CH_4 + C_2H_6$ hydrate system (w/excess water) using the method of Hendriks et al.[3] Note that structure changes from sI to sII along the L_w-H-V equilibrium curve at a y_{CH_4} of 0.62. This value differs from the transition point of $0.722 < y_{CH_4} < 0.750$ obtained in this work using Raman spectra.

CONCLUSIONS

Raman and NMR spectroscopic techniques were used in this work to demonstrate that sII hydrate can form from mixtures of CH_4 and C_2H_6, which are known sI hydrate formers.

Raman band frequencies for C_2H_6 were used to identify a change in structure of $CH_4 + C_2H_6$ hydrates from sI to sII over the vapor composition range (y_{CH_4}) of 0.722 to 0.750, under L_w-H-V conditions at 274.2 K. Band areas corresponding to CH_4 and C_2H_6, obtained from the Raman spectra, indicate a significant increase in the amount of CH_4 incorporated into the hydrate as structure changes from sI to sII. ^{13}C NMR chemical shifts of ethane obtained from $CH_4 + C_2H_6$ hydrates confirmed the formation of sII in this system.

The sI to sII transition point obtained from Raman spectra should be used to correct existing hydrate prediction models. It is also important to obtain the gas composition at which the hydrate structure reverts back to sI from sII as this may have important implications for structures of naturally occurring hydrates.

ACKNOWLEDGMENTS

The National Science Foundation provided funding for this work through the Research Grant CTS-9634899. The NMR spectrometer used in this work is funded by NSF grant CTS-9512228. We also thank Dr. Robert C. Burruss, U. S. Geological Survey, for helpful discussions regarding the Raman work.

REFERENCES

1. SLOAN, E.D. 1998. Clathrate Hydrates of Natural Gases, 2nd edit. Marcel Dekker Inc., New York.
2. VON STACKELBERG, M. & W. JAHNS. 1954. Feste gashydrate VI: die gitteraufweitungsarbeit. Z. Elektrochem. **58:** 162.
3. HENDRIKS, E.M., B. EDMONDS, R.A.S. MOORWOOD & R. SZCZEPANSKI. 1996. Hydrate structure stability in simple and mixed hydrates. Fluid Phase Equilib. **117:** 193–200.
4. HOLDER, G.D. & J.H. HAND. 1982. Multiple-phase equilibria in hydrates from methane, ethane, propane, and water mixtures. AIChE J. **28:** 440–447.
5. RIPMEESTER, J.A., C.I. RATCLIFFE & J.S. TSE. 1988. The nuclear magnetic resonance of ^{129}Xe trapped in clathrates and some other solids. J. Chem. Soc. Faraday Trans. 1. **84:** 3731–3745.
6. RIPMEESTER, J.A. & C.I. RATCLIFFE. 1988. Low-temperature cross-polarization/magic angle spinning ^{13}C NMR of solid methane hydrates: structure, cage occupancy, and hydration number. J. Phys. Chem. **92:** 337–339.
7. IKEDA, T., S. MAE & T. UCHIDA. 1998. Effect of guest-host interaction on Raman spectrum of a CO_2 clathrate hydrate single crystal. J. Chem. Phys. **108:** 1352–1359.
8. LONG, X. 1994. Masters Thesis. Colorado School of Mines, Golden.
9. SLOAN, E.D., S. SUBRAMANIAN, P.N. MATTHEWS, J.P. LEDERHOS & A.A. KHOKAR. 1998. Quantifying hydrate formation and kinetic inhibition. Ind. Eng. Chem. Res. **37:** 3124–3132.
10. SUM, A.K., R.C. BURRUSS & E.D. SLOAN. 1997. Measurement of clathrate hydrates via Raman spectroscopy. J. Phys. Chem. B **101:** 7371–7377.

11. UCHIDA, T., A. TAKAGI, T. HIRANO, H. NARITA, J. KAWABATA, T. HONDOH & S. MAE. 1996. Measurements on guest-host molecular density ratio of CO_2 and CH_4 hydrates by Raman spectroscopy. Proceedings of the Second International Conference on Natural Gas Hydrates, Toulouse, France. 335–339.
12. MEHRING, M. 1983. High Resolution NMR Spectroscopy in Solids, 2nd edit. Springer Verlag, Berlin.
13. HERZBERG, Z. 1951. Infrared and Raman Spectra. D. von Nostrand Company, New York.
14. DAVIDSON, D.W., S.K. GARG, S.R. GOUGH, Y.P. HANDA, C.I. RATCLIFFE, J.A. RIPMEESTER, J.S. TSE & W.F. LAWSON. 1986. Laboratory analysis of a naturally occurring gas hydrate from sediment of the Gulf of Mexico. Geochem. Cosmochim. Acta **50:** 619–623.

Formation of Natural Gas Hydrates in Marine Sediments

Gas Hydrate Growth and Stability Conditioned by Host Sediment Properties

M. BEN CLENNELL,[a,b] PIERRE HENRY,[c] MARTIN HOVLAND,[d]
JAMES S. BOOTH,[e] WILLIAM J. WINTERS,[e] AND MICHEL THOMAS[f]

[b]*Department of Earth Sciences, University of Leeds, Leeds LS2 9JT, United Kingdom*

[c]*ENS-Laboratoire de Géologie, CNRS URA 1316, 24 rue Lhomond, 75231 Paris, France*

[d]*Statoil, P.O. Box 300, N-4001, Stavanger, Norway*

[e]*United States Geological Survey, Woods Hole, Massachusetts 02543-1598, USA*

[f]*IFP, Physico-chimie Appliquée, 1-4 Av. de Bois Préau, 92852 Rueil-Malmaison Cedex, France*

ABSTRACT: The stability conditions of submarine gas hydrates (methane clathrates) are largely dictated by pressure, temperature, gas composition, and pore water salinity. However, the physical properties and surface chemistry of the host sediments also affect the thermodynamic state, growth kinetics, spatial distributions, and growth forms of clathrates. Our model presumes that gas hydrate behaves in a way analogous to ice in the pores of a freezing soil, where capillary forces influence the energy balance. Hydrate growth is inhibited within fine-grained sediments because of the excess internal phase pressure of small crystals with high surface curvature that coexist with liquid water in small pores. Therefore, the base of gas hydrate stability in a sequence of fine sediments is predicted by our model to occur at a lower temperature, and so nearer to the seabed than would be calculated from bulk thermodynamic equilibrium. The growth forms commonly observed in hydrate samples recovered from marine sediments (nodules, sheets, and lenses in muds; cements in sand and ash layers) can be explained by a requirement to minimize the excess of mechanical and surface energy in the system.

INTRODUCTION

The vast deposits of natural gas hydrates found in marine and permafrost regions are locked up in a porous sediment matrix. Clathrates have been encountered in coarse sediments, such as sand and sandstone; fine-grained sediments, such as clays; deep sea ooze and mudstones; and also occur filling in fractures within indurated sediments.[1–3] Although there is an apparent tendency for the hydrates to occur

---

[a]Present address for correspondence: Centro de Pesquisa em Geofísica e Geologia, -IGEO, Universidade Federal da Bahia-UFBa, Rua Caetano Moura, 123, Salvador, Bahia, 40210-340, Brasil. Voice: +55 (0)71 3329433; fax: +55 (0)71 2473004.
clennell@cpgg.ufba.br

preferentially in sediments that have higher porosity, larger absolute pore sizes, and high permeability, there is not yet enough data to assert a clear-cut relationship.[4] This paper examines how pore size and capillary behavior, permeability, and mechanical properties may affect gas hydrate stability, and the way in which hydrate is distributed on the microscale to mesoscale within sediments.

PHYSICAL PROPERTIES OF HYDRATE HOST SEDIMENTS

The permeability of calcareous clay-oozes (porosity 45–55%) typical of the Blake Outer Ridge, a classic hydrate accumulation,[5] were found to be in the range 10^{-16} to 10^{-17} m^2 (FIGURE 1A). Pore diameters in these sediments measured with mercury porosimetry range from tens of nannometres to a few microns across, with the mode at about 200 microns (FIG. 1B).

The Blake Ridge sediments are typical of deep marine muds found at many sites on the ocean margins where hydrates are encountered. In locations such as Central America Trench and the Cascadia Accretionary Margin, where the sediments are deformed by tectonic processes, the sediments hosting gas hydrates are typically more compacted and may be pervasively fractured.[6] The low permeability of sediments hosting hydrates means that gas and liquid transport rates may be very slow, limiting the extent of hydrate accumulations. The small size of pore throats in the sediments leads to high capillary entry pressures for free-phase gas transport. At a saturation lower than 15–20%, bubbles of free gas occupy isolated pore bodies, rather than connecting through the smaller throats to form a continuous stringer capable of moving through the pore network[7] (FIG. 1C). In many areas we envisage that the free gas layer beneath the hydrate stability zone forms such a mist, which is practically immobile.[7] Hydrate occurrences in rocks of good reservoir quality such as sands and sandstones (permeability 10^{-12} to 10^{-15} m^2, pore sizes from tens to hundreds of microns) are very much less common on a world scale, although these are understandably the main targets for commercial exploitation.

ANALOGY BETWEEN GAS HYDRATES AND PERMAFROST ICE

Gas clathrate hydrates share a number of structural and physical similarities with water ice.[8] Although there are no definitive measurements, it also appears that the interfacial energy between gas hydrate and liquid water is similar to that between ice and water.[9] These facts, and the close resemblance of certain sediment-hosted hydrate occurrences to permafrost ice-lenses led us (among others; see Refs. 9, 10, and references therein) to speculate that gas hydrates may interact with the host sediments according to the thermodynamic principles that underlie freezing of water in the pores of soils.[11]

In porous media, the temperature at which melting or freezing occurs depends pore size. The specific surface energy of ice crystallizing in a small pore, and its internal *capillary* pressure can be very large, meaning that the liquid phase is thermodynamically favored down to lower temperatures than would be the case under bulk conditions.[11] This result can be quantified using the Gibbs-Thomson equation,

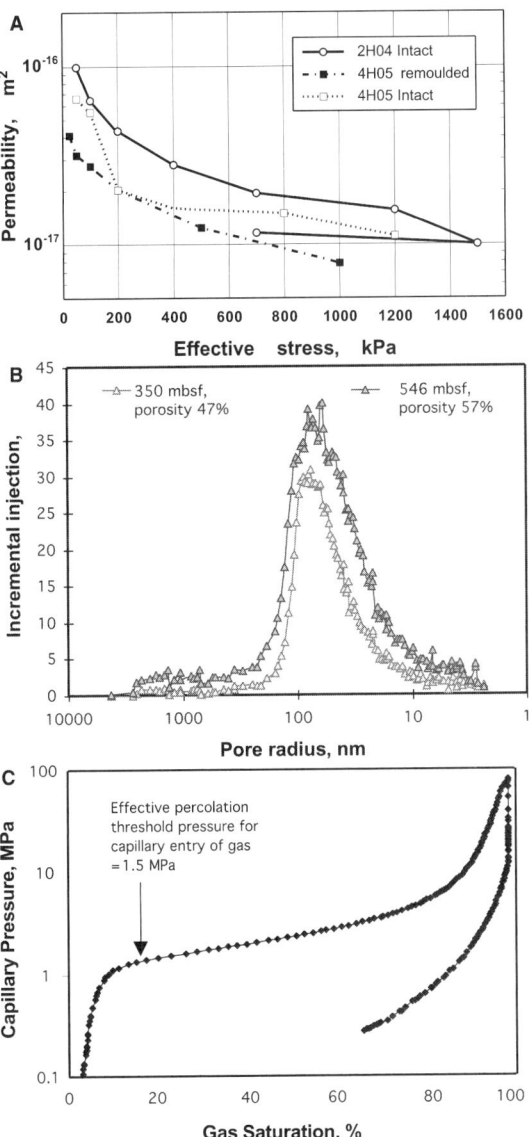

FIGURE 1. (**A**) Permeability of calcareous clay-ooze sediment sample from Blake Outer Ridge as a function of effective stress. (**B**) Mercury injection porosimetry curve for clay samples from the Blake Ridge, ODP Hole 995A, 350 and 546 mbsf. Pore size distribution (volume intruded through progressively smaller pore throats) determined using a cylindrical pore model and the differential of the cumulative volume intrusion curve; mercury surface tension 0.484 Nm^{-1}, entry contact angle 140°. Samples were freeze dried to prevent shrinkage. (**C**) Gas water capillary pressure curve calculated from the mercury injection curve, assuming a contact angle of 0° (water wet) and surface tension of 0.072 Nm^{-1}.

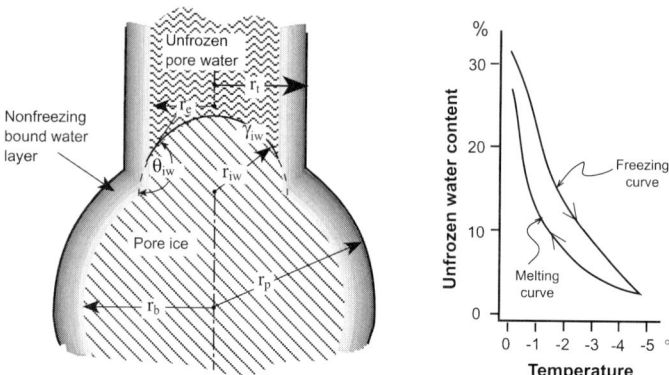

FIGURE 2. Capillary model of pore freezing. Ice exists within a large pore body (idealized as a spherical cavity), and with further depression in temperature adopts sufficient surface curvature to penetrate an adjacent pore throat (idealized as a cylinder); that is, $r_{iw} = r_e$. Inset shows the proportion of unfrozen water during freezing and melting; this curve exhibits considerable hysteresis because ice masses in pore bodies (effective radius r_b) are larger than the effective throat sizes (r_e) that must be penetrated for progressive freezing to occur. In our analysis the contact angle for ice θ_{iw} is assumed to be 180° (water wet pores; ice nonwetting).

which relates the curvature of solid surface to the thermodynamic energy required to form the solid-liquid interface. A suitable form of this equation is[12]

$$\Delta T_{i,pore} = \frac{2\gamma_{iw} T_{bulk} \cos \theta_{iw}}{\rho_i \Delta H_f r_e}, \qquad (1)$$

where ΔT is the depression of freezing melting temperature below the bulk melting temperature T_{bulk}, ρ_i is the density of ice, ΔH_f is the enthalpy change of melting (latent heat) per unit mass. Thus, in a fine-grained soil or sediment that has a range of pore sizes, water freezes progressively over a range of temperatures, penetrating through smaller pore throats, as the temperature is lowered (see FIGURE 2). When the temperature is raised again, the liquid water content profile follows a different path (hysteresis) because the effective pore size r_e now relates to the distribution of larger pore body sizes rather than throat apertures.

PORE-SIZE EFFECT ON GAS HYDRATE STABILITY

Henry *et al.*,[7] calculated the effect of pore size on the stability of the empty clathrate lattice, and used this to deduce the approximate temperature depression, for a given pressure of the methane clathrate equilibrium when the hydrate crystallizes from methane-saturated aqueous solution. They assumed that the surface energy of methane hydrate with liquid water is similar to that with water ice, viz. 26–30 mJm^{-2}. They also assumed that the enthalpy of formation of clathrate is similar to, or slightly lower than, that of bulk hydrate, although others[9,13] have found experimentally that the apparent value of ΔH_f in very small pores (tens of nanometers) can be significantly less than that for bulk hydrate.

In microporous media, the capillary pressure of free gas bubbles in equilibrium with water may be on the order of several MPa. Thus, the fugacity of methane is increased for any particular value of the pore water pressure, raising the equilibrium methane concentration in the water. This mechanism, termed capillary supersaturation,[6] promotes hydrate stability. Henry et al.[7] have shown that if both gas bubbles and hydrate inclusions exist in pores of a given size, then the net effect is to increase the stability of hydrate relative to bulk conditions.

WATER CHEMICAL POTENTIAL IN WATER-BUFFERED OR GAS-EXCESS SYSTEMS

In the marine subsurface, so long as open system conditions prevail, the pressure of the liquid water phase is fixed by the height of the water column. Water is a continuous phase and in the pore system there is an extensive reservoir available to react at a fixed chemical potential, dictated by the pressure, temperature, and salinity. We call this a water-buffered system.[6]

Under different conditions the availability of water can limit the hydrate-forming reaction. A typical case is in an experiment in which gas pressure is held constant, and there is a fixed amount of water in the pores the sample; we call this a gas-excess system (e.g., the experiments cited above[9,13,14]). Another case is where the system is partially closed to free transport of the water phase (low permeability sediments) and there a strong gas supply through a sediment pile. Now the gas pressure and methane fugacity are fixed, and the liquid phase pressure and water chemical potential become dependent variables as the system seeks capillary and thermodynamic equilibrium.

FIGURE 3 shows the situation in a fine grained sediment when water is extracted progressively (in order to form hydrate). At first, water in the main body of the pore space is removed. The chemical potential, or activity of this water is essentially the same as in an extensive reservoir of bulk water with the same temperature, pressure and salinity.[15,16] Removal of more water requires that the sediment will shrink and compact, until the shrinkage limit is reached. At this point, gas enters the pore space, and since the gas pressure is fixed, the capillary pressure of the water in the pores will drop below the fixed gas pressure. This capillary water has a lower chemical potential than the bulk water, so that further progress of the hydrate-forming reaction is inhibited (a lower equilibrium temperature is required). Further depletion of water requires an increasing thermodynamic drive, until only the bound-water layer remains. Hydrate formation is progressively inhibited by a combination of the depressed water activity and the surface energy excess of crystallizing solid hydrate in small pores.

Clarke et al.[10] conducted a similar analysis to Henry et al.,[7] but with a superior thermodynamic model for bulk hydrate stability and a more generalized treatment of surface curvature. These studies show that the main effect of free gas/water interfaces in a gas excess system is to destabilize hydrate by lowering the activity of water. For this situation, both models fit the experimental data of Handa and Stupin[9]

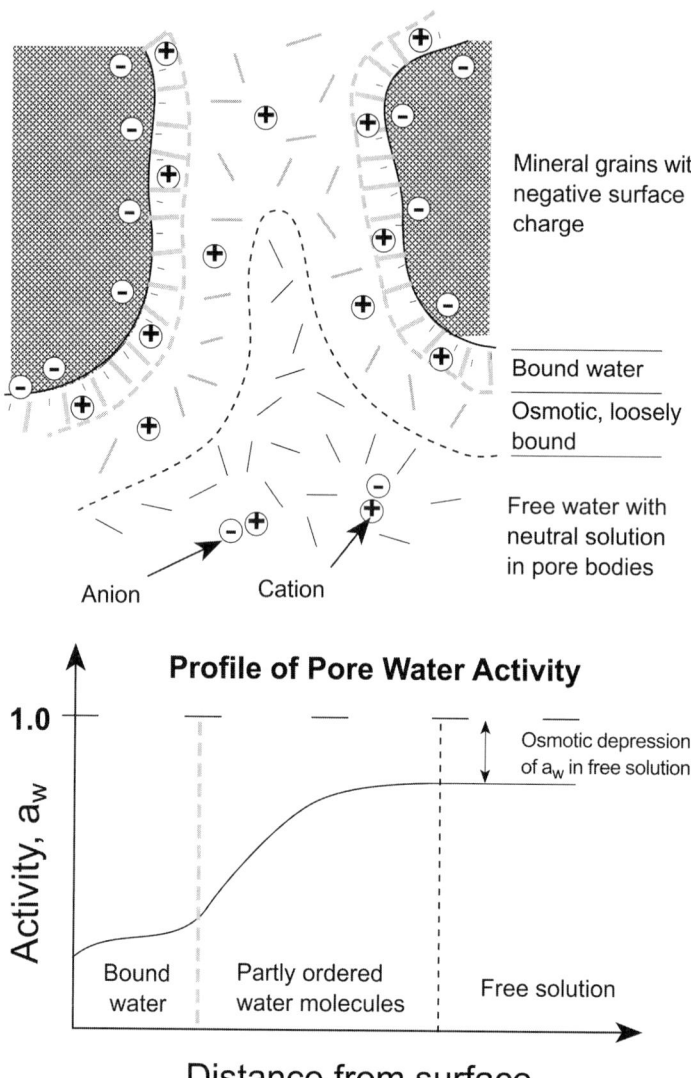

FIGURE 3. Distribution of capillary, bound and free water in sediment layers and nature of the diffuse bound water layer around clay particles. Adapted from Mitchell[14].

reasonably well, suggesting that capillary effects are of first-order importance. No experimental data are yet available to test the capillary models under conditions of two phase equilibrium (without free gas being present), where depressed methane fugacity, caused by hydrate-water interface effects, is predicted.

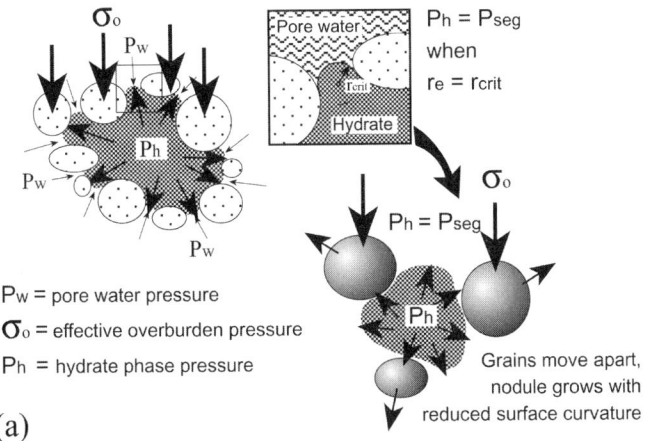
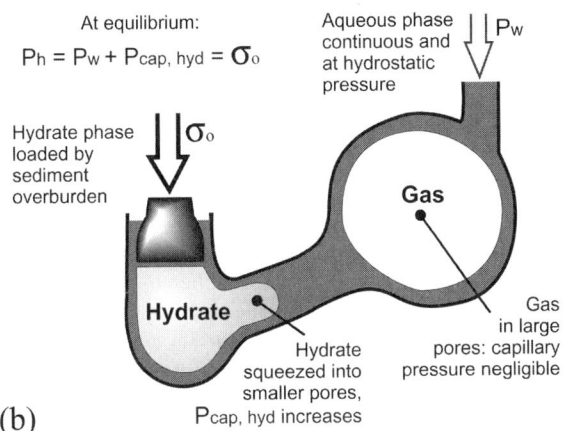

FIGURE 4. Explanation of gas hydrate segregation. (**a**) Hydrate grows as an interstitial phase when it can adopt a stable surface curvature that allows it to penetrate pore throats. When this curvature is greater than a certain value (i.e., all pore throats greater than a critical size have been penetrated), then the hydrate pushes aside the sediment grains and grows displacively. This happens when the surface curvature enforces a phase pressure inside the solid that exceeds the yield strength that can be mobilized by the sediment matrix. (**b**) Relationship between lithostatic load, pore size and hydrate phase pressure. The lithostatic pressure of overlying sediments is shown schematically as a weight pushing down on the segregated hydrate. At equilibrium the crystal generates a higher internal pressure to balance this force, enabling it to adopt greater surface curvature and extrude into smaller pores. Circumstance illustrated where gas bubbles, having higher surface energy, occupy larger pores than hydrate crystals, though the above process is not dependent on free gas being present.

MECHANICAL CONSTRAINT AND SEGREGATED GROWTH

The mechanical properties of the sediments and the solid growing within it are important, since an additional constraint is placed on the stability of the solid phase when the soil or sediment is placed under stress (e.g., the gravitational load due to subsurface burial). If we assume that the hydrate mass can, over a long time period, flow like ice to dissipate unequal stresses,[6,7,11,17–19] then the eventual configuration of the hydrate in the sediment will represent a balance between capillary and mechanical forces. The curvature of the clathrate–water interfaces will be just sufficient to penetrate pores of a certain size, commensurate with an internal capillary pressure that balances the effective mechanical load of the sediments above. This critical pore size defines a criterion for segregated growth of hydrate in sediments (see FIGURE 4). In sands and coarse sediments, it is generally more favorable for hydrate to grow wholly within pores, and cement the sediment, whereas in muds we expect the capillary forces to push the sediment grains aside, so that the hydrate forms a segregated mass such as a nodule or sheet.[6]

EVIDENCE FOR AN INFLUENCE OF HOST SEDIMENT PROPERTIES IN NATURAL GAS HYDRATE SYSTEMS

There is considerable evidence that in fine muddy sediments gas hydrates tend to form as discrete layers or lenses.[1,2,4,6] In sands and gravels hydrate is observed as a cement, and as massive layers.[1,2,4,19] However, at present so few observations on the microscale have been made that it is still equivocal as to which physical and chemical factors control the growth habits of natural gas hydrates in marine sediments and in permafrost environments.

If hydrates are inhibited by host medium effects, then the position of the seismic bottom simulating reflector (BSR) that marks the base of hydrate stability in sediments will be shifted to a position closer to the surface (cooler temperature) than would be predicted from bulk equilibrium stability. There is strong evidence that in parts of the world such a situation occurs,[2,5] but as pointed out previously,[6] the evidence is not yet conclusive as to what mechanism causes this inhibition.

IMPLICATIONS AND CONCLUSIONS

It is important to understand the mechanisms controlling hydrate growth on the pore scale, in order to appraise the natural gas hydrate reservoir on regional and global scales. Estimates of the amount of hydrates present in the subsurface depend upon geophysical data, such as seismic wave velocity and attenuation, electrical resistivity and mechanical compliance, and these physical properties depend not only on how much hydrate is present, but also how it is distributed in the pore space (e.g., cementing grains together, inside isolated pores, or as discrete nodules, pellets or layers). Seabed stability, important for planning subsea installations and drilling, and also for understanding the relation between climate change, continental margin slope failure, and catastrophic hydrate dissociation[1] also depends upon how the presence of hydrate changes the sediment physical properties. Finally, scenarios for exploitation of the gas hydrate reservoir must take into account the thermodynamic and physical influences that the host sediment plays in controlling hydrate stability, growth form, small scale distribution of hydrates within the reservoir, and permeability of the formation. More experimental and field studies are required in order to address these issues.

ACKNOWLEDGMENTS

We thank the Engineering Foundation for financial support for M.B.C. to attend the NGH3 meeting. The U.S. Geological Survey and U.K. Natural Environment Research Council, under Fellowship No. GT5/97-IS-ODP, supported scientific work and exchanges. We dedicate this paper to the memory of our friend and colleague Dr. Gabriel D. Ginsburg.

REFERENCES

1. KVENVOLDEN, K.A. 1993. Gas hydrates—geological perspective and global change. Rev. Geophys. **31:** 173–187.
2. GINSBURG, G.D. & V. SOLOVIEV. 1998. Submarine Gas Hydrates. VNIIOkeangeologia/Norma, St. Petersburg.
3. MAKOGON, Y.F. 1981. Hydrates of Natural Gas. Penwell, Tulsa.
4. BOOTH, J.S., W.J. WINTERS, W.P. DILLON, M.B. CLENNELL & M.M. ROWE. 1998. Major occurrences and reservoir concepts of marine clathrate hydrates: implications of field evidence. In Gas Hydrates. J.-P. Henriet & J. Mienert, Eds. Geol. Soc. Spec. Publ. **137:** 113–128.
5. PAULL, C.K., R. MATSUMOTO, P.WALLACE & LEG 164 SCIENTIFIC PARTY. 1997. Proceedings of the Ocean Drilling Program, Initial Reports, **164**. Ocean Drill. Program, College Station, Texas.
6. CLENNELL, M.B., M. HOVLAND, J.S. BOOTH, P. HENRY & W.J. WINTERS. 1999. Formation of natural gas hydrates in marine sediments. Part 1: conceptual model of gas hydrate growth conditioned by host sediment properties. J. Geophys. Res. In press.
7. HENRY, P., M. THOMAS & M.B. CLENNELL. 1999. Formation of natural gas hydrates in marine sediments. Part 2: thermodynamic calculations of stability conditions in porous sediments. J. Geophys. Res. In press.
8. SLOAN, E.D. 1998. Clathrate Hydrates of Natural Gases, 2nd. edit. Marcel Dekker, New York.
9. HANDA, Y.P. & D. STUPIN. 1992. Thermodynamic properties and dissociation characteristics of methane and propane hydrates in 70-Å-radius silica-gel pores. J. Phys. Chem. **96:** 8599–8603.
10. CLARKE, M.A., M. POOLADI-DARVISH & P.R. BISHNOI. 1999. A method to predict equilibrium conditions of gas hydrate formation in porous media. Ind. Eng. Chem. Res. **38:** 2485–2490.
11. EVERETT, D.H. 1961. The thermodynamics of frost damage to porous solids. Trans. Faraday Soc. **57:** 1541–1551.
12. JALLUT, C., J. LENOIR, C. BARDOT & C. EYRAUD. 1992. Thermoporometry: modelling and simulation of a mesoporous solid. J. Membrane Sci. **68:** 271–282.
13. UCHIDA, T., T. EBINUBA & T. ISHIZAKI. 1999. Dissociation condition measurements of methane hydrate in confined small pores of porous glass. J. Phys. Chem. B **103:** 3595–3662.

14. LU, H. & R. MATSUMOTO. 2000. Stability condition of gas hydrate in marine sediments: constrained from the experimental results. Ann. N.Y. Acad. Sci. **912:** this issue.
15. MITCHELL, J.K. 1993. Fundamentals of Soil Behavior, 2nd edit. John Wiley, New York.
16. NITAO, J.J. & J. BEAR. 1996. Potentials and their role in transport in porous media. Water Resour. Res. **32:** 225–250.
17. YAMADA, K., K. NAKAMURA, M. HYODO, Y. NAKATA & M. FUKUNAGA. 2000. Compressive strength of methane hydrate; effect of temperature and pressure. Ann. N.Y. Acad. Sci. **912:** this issue.
18. ZHANG, W., W.B. DURHAM, L.A. STERN & S.H. KIRBY. 2000. Experimental deformation of gas hydrates. Ann. N.Y. Acad. Sci. **912:** this issue.
19. WINTERS, W.J., J. KATSUBE, S.R. DALLIMORE, J.F. WRIGHT, F.M. NIXON, T.S. COLLETT, R.E. CRANSTON & T. UCHIDA. 2000. Relation between gas hydrate physical properties at the Mallik 2L-38 research well, N.W.T. Ann. N.Y. Acad. Sci. **912:** this issue.

NMR Imaging Study of Hydrates in Sediments

MARIT MORK,[a] GRETHE SCHEI,[b,c] AND ROAR LARSEN[b,d]

[a]Norwegian University of Science and Technology, N-7491 Trondheim, Norway

[b]Sintef Petroleum Research, N-7465 Trondheim, Norway

> ABSTRACT: In this study, hydrates were generated in synthetic sediments in a laboratory cell. After hydrate formation took place and the sediment solidified, the samples were investigated both visually and by the use of nuclear magnetic resonance (NMR) imaging. The hydrates in this initial study were formed from model systems at low pressure. The results show hydrates distribution effects. Scans in the NMR apparatus were also made of the unfrozen samples to serve as a basis for comparison. NMR here maps the mobility of hydrogen atoms and their distribution in a sample. The relevant factor is the *density of mobile H-atoms*, and this is shown to be about five times smaller for a volume of (for example) tetrahydrofuran (THF, C_4H_8O) hydrate than for the fluids of water and/or THF. This correlates very well with an observed signal decrease by a factor of six in an NMR-studied sample after hydrate formation had taken place.

INTRODUCTION

Sediments containing natural gas hydrates are found in permafrost areas onshore and in subsea sediments—where pressure and temperature conditions are inside the hydrate stability region and where natural gas is available. Hydrate-bearing sediments offshore have generally been found in waters deeper than 300 m and their zone of existence is from the seafloor downward—a few hundred meters, depending upon the local geothermal gradient. Kvenvolden[1] and others have tried to estimate the possible amounts of gas to be found in hydrated sediments on- and offshore. Some of these estimates claim that as much as 10^{16} Sm^3 (standard cubic meters) of methane gas may exist in this form. These gas hydrates are therefore considered to be a possible energy resource of a significant magnitude for the future. There has also been some concern in the oil industry about problems that may be caused by hydrate-bearing sediments, such as gas blowouts due to dissociation while drilling, foundation liquefaction for production installations or transport pipelines, and the possible triggering of landslides if a delicate balance is upset by man-made processes.

Much of the motivation for the current work stems from a wish to better understand how hydrates behave under undisturbed conditions. A first step towards gaining this knowledge has been made by demonstrating the usefulness of nuclear magnetic resonance (NMR) imaging techniques for hydrates under appropriate conditions.

Telecommunication.
[a]Voice: +47 73 59 49 51; fax: +47 94 44 72. mmork@ipt.ntnu.no
[c]Voice: +47 73 59 13 01; fax: +47 73 59 77 40. Grethe.Schei@iku.sintef.no
[d]Voice: +47 73 59 16 33; fax: +47 73 59 10 50. Roar.Larsen@iku.sintef.no

EXPERIMENTAL SET-UP AND PROCEDURE

The experimental work performed in this study consists of manufacturing of hydrated sediment samples, analysis by NMR and Computer Tomography (CT), and mechanical strength and deformation measurements on the samples. The hydrated sediment samples were manufactured by using quartz sand with a particle size distribution of 210–295 μm, distilled water, and tetrahydrofuran (THF, C_4H_8O) or freon-11 (R-11, CCl_3F). Both THF and R-11 form hydrate structure II. In total 12 samples were manufactured, seven samples using THF and five samples using R-11. The dimensions of the samples produced were 10 cm diameter and approximately 15 cm length. A schematic of the equipment used for the manufacturing and a photograph of the sample cell are shown in FIGURE 1. The main part of the apparatus is a sample cell made of Lexan (transparent), 10 cm diameter and 250 cm length. The equipment used for NMR imaging is an electromagnet with a field strength of 2.4 T and a proton frequency of 100 MHz. The inner diameter of the magnet is 20 cm. The mechanical rock characterization was performed by using a MTS load frame with maximum capacity of 10 kN.

Experimental Procedure

All chemicals were cooled to 275–277 K prior to sample manufacture. Additionally, the water was distilled, cooled below the freezing point and then melted before use. Pure THF and R-11 hydrate samples were made by adding a stoichiometric mixture of THF and water (volume ratio water/THF, 3.77) or R-11 and water (volume ratio water/R-11, 3.41) to the empty cell with the top lid removed. The R-11 and water were mixed vigorously, leaving a white, glossy, highly viscous suspension. After a few minutes, the samples had partially solidified. The top lid of the cell was

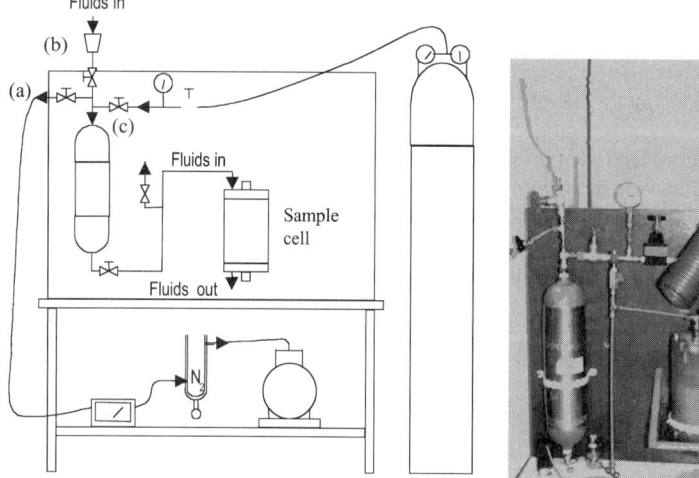

FIGURE 1. Equipment for manufacturing of hydrated sediment samples.

then put in place and the samples were left in a refrigerator (275–277 K) overnight or longer, to solidify.

The hydrated sediment samples were made by first adding dry quartz sand to the cell. The cell was then sealed and connected to a vacuum pump, which reduced the pressure in the cell to approximately 200 Pa. To manufacture sand/R-11 hydrate samples, the sand was first saturated with a known amount of degassed water and then R-11 was squeezed through the sample from the top of the cell, by a pressure differential on the order of 1–2 bar. The amount of added R-11 was equivalent to the stoichiometric ratio between R-11 and the remaining water. To manufacture sand/THF hydrate samples, a stoichiometric mixture of THF and water was added to the liquid container and then squeezed through the sand. The pressure on the topside was then increased to 0.3 MPa in order to compress the sand. NMR and CT investigations were performed before and after the samples were frozen. Mechanical characterization of the hydrated sediment samples was carried out on 0.038 m (1.5″) diameter cores, drilled out of the larger samples and cut to 78 mm length. The samples were preloaded axially up to 0.02 MPa with a load rate of 1 mm/min during strength testing. The axial load was then increased with a load rate of 0.005 mm/min until failure was observed.

For the tests where deformation during melting was observed, the samples were placed in a container with a diameter 2 mm larger than the core samples, and preloaded to a value below the failure point (0.1 MPa for THF hydrated sediment samples). The axial load rate was 0.005 MPa/sec. The reason for using a container around the core samples during testing was to prevent large radial deformations. Both the container and the pistons were cooled to 277 K before mounting.

RESULTS AND DISCUSSION

General

Pure samples (i.e., no sand) in the THF system grew to solid, very hard masses that filled the container, whereas the R-11 hydrates had the consistency of packed snow, that is, more polycrystalline in appearance. With sand present, induced nucleation was no longer necessary, and the THF system again produced very hard samples in the sediment cell. The R-11 system with sand generally took longer to solidify fully, partly because the correct ratio between water and hydrate former was difficult to maintain throughout the slow solidification process. One sample with R-11 hydrated sand was left to solidify for four weeks, with no major observable differences except a somewhat harder feel to it, qualitatively.

Samples with Tetrahydrofuran (C_4H_8O) Hydrate

Both frozen and unfrozen samples with sand, water, and THF were submitted to computer tomography (X-ray). However, the important density differences were so small (0.971 g/cm^3 for THF hydrate with 100% cage occupancy, and 0.978 g/cm^3 for the stoichiometric THF–water mixture), and the sand was generally so evenly distributed, that no useful information was gained from these investigations.

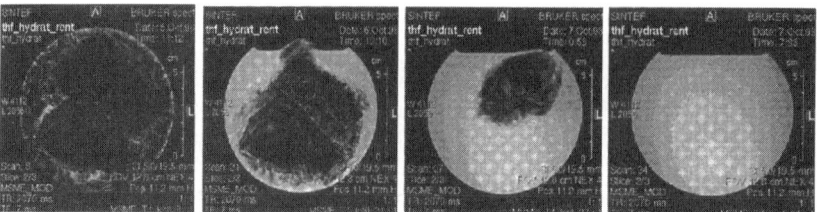

FIGURE 2. Melting process for pure THF hydrate.

FIGURE 2 shows an NMR image of a melting pure THF hydrate sample, the same position being imaged over a period of 18 hours. In short, NMR measures the mobility of hydrogen atoms in a substance. In these pictures, a strong signal (high hydrogen mobility) is seen as a bright region in the images. Water and liquid THF are highly mobile and also indistinguishable, since they mix perfectly. When water is tied up in the hydrate crystal structure, it becomes invisible at the magnetic field strengths and resolution used in these experiments. Despite the relative freedom of the THF molecule, in the hydrate cavities, its hydrogen atoms also become invisible when hydrated, under these experimental conditions. Shorter signal echo times are needed to presumably achieve enough signal strength from the THF guest molecules. The way to interpret the images is to regard darker areas as more solid (immobile) than lighter areas. Areas with intensities between the brightest and darkest arise due to signal averaging over a certain volume. This volume (one pixel in the images) was 0.4 mm square and 5 mm in depth, that is, 0.8 mm^3.

FIGURE 3a shows an NMR image of a sample consisting of THF, water, and sand before any hydrate had formed. It shows a fairly homogeneous sample, with

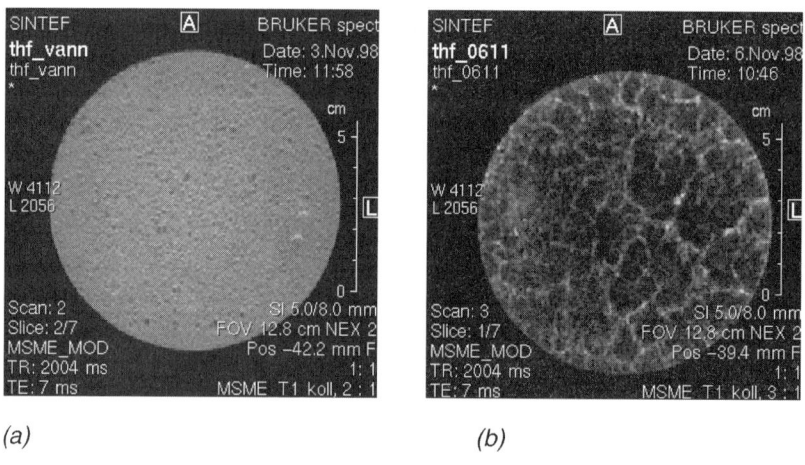

FIGURE 3. NMR imaging of **(a)** sand/water/THF and **(b)** sand/THF hydrate.

presumed air bubbles (or gas-filled volumes) as the only irregularities. FIGURE 3b shows the same sample, at almost the same position, after hydrate formation had taken place. These images clearly show a structural change in the interior of the sample after hydrate formation occurred. We see a three-dimensional network of bright channels, in which highly mobile phases exist; that is, liquid that has not formed hydrate. There are *lumpy* volumes of solidified sand and hydrate between these channels. One possible explanation for this pattern is that hydrate formation starts more or less simultaneously in many interior points in the sample, and that any deviation from stoichiometry (THF evaporation, etc.) leads to excess fluid between these lumps when neighboring lumps converge. There is a difference in lump size between different areas of the samples, and between samples, but no systematic effects were observed, either coupled to solidification times or other factors. This is probably just an indication of the fundamentally stochastic nature of the nucleation process. The relative distribution of sand and hydrate in the lumps could not be discerned from the NMR images, but samples that were opened were qualitatively seen to have the sand homogeneously spread through continuous hydrate masses. It could not be shown conclusively that the hydrate is able to push sand particles apart upon formation, but we believe this to be likely to occur, since sand particles seemed to be *suspended* in some of the solid hydrate particles.

Attempts were made to quantify the size of the dark areas in the images, or lumps. The resulting size distribution for three samples is shown in FIGURE 4. The size of each lump was decided by manually measuring its *representative* diameter directly from the images, from a number of slices for each sample. A 10-mm wide zone along the wall of the cell was excluded from these considerations, to minimize any wall effects on the results. This data set is far too small to draw any firm conclusions, but qualitatively, similar hydrate lump sizes can be seen in images from real hydrate sediment samples.[2]

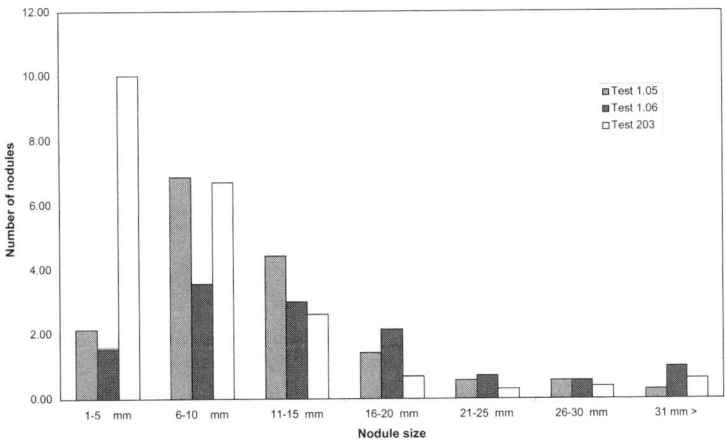

FIGURE 4. Size distribution of solidified sand/THF hydrate nodules.

Samples with R-11 (CCl_3F) Hydrate

For comparison, the same tests as described for THF hydrates and sand were also conducted for R-11 hydrates and sand. It could be argued that this system is somewhat closer to *real* conditions, since water and R-11 are immiscible, and the hydrate formation in this system is susceptible to some of the same interfacial effects as water and a hydrate-forming gas. Like THF, however, R-11 also forms structure II hydrate, not structure I, which is the predominant form in subsea sediments.[3] The melting test of pure hydrate showed a more homogeneous starting mass, probably due to a much more fine-grained polycrystalline structure. R-11 (CCl_3F), containing no hydrogen atoms, does not show up on the NMR images in either liquid or enclathrated form.

FIGURE 5a shows an R-11, water, and sand sample before hydrate formation. FIGURES 5b and 5c show different slices of the same sample after freezing has taken place. The unfrozen sample is indistinguishable from the corresponding THF case. The frozen sample, however, is different. The lack of a network of mobile phases between solid lumps, and the apparent inhomogeneity from one part of the sample to another, must be due to the differences between the THF and the R-11 system. The lower part of the cell shows higher mobility, indicating the presence of free water. This may also indicate that there are no continuous pore paths from this part of the cell to the top part, since water would have migrated upwards compared to the heavier R-11. The effect may also be due to the presence of considerably more water than the stoichiometric ratio would demand for this system. The fact that R-11 is invisible to NMR makes it difficult to determine whether dark areas are predominantly hydrate, sand, or liquid R-11. The immiscibility of the phases all but rules out homogeneous hydrate formation throughout the sample with the procedures used in these experiments. Thus, the absence of the structuring effect seen for THF may be explained by lack of widespread nucleation.

Rock Mechanical Characterization

Two test types were performed on THF and R-11 hydrated sediment samples, strength tests without confining pressure and deformation tests during melting. The stress–strain curve for the THF hydrated sediment sample (see FIGURE 6) shows a

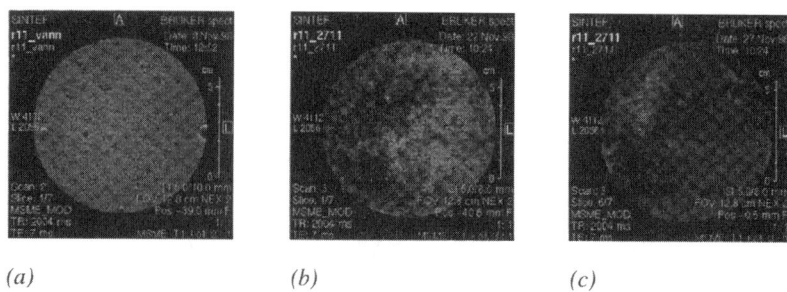

(a)   (b)   (c)

FIGURE 5. NMR imaging of (a) sand/water/R-11, (b) sand/R-11 hydrate near the bottom of the core, and (c) sand/R-11 hydrate near the top of the core.

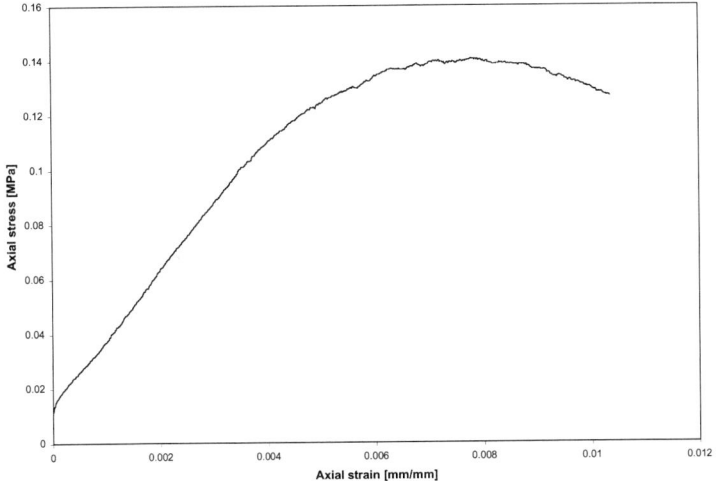

FIGURE 6. Axial stress versus axial strain for a THF hydrated sediment sample.

linear deformation until 0.1 MPa axial stress and core failure at 0.14 MPa axial stress. The observed stress–strain curve for the R11 hydrated sediment sample (see FIGURE 7) shows an s-shaped curve and core failure occurred at 0.56 MPa for this material. Hence, in these experiments, the R11 hydrated sediment sample is stronger than the THF hydrated sediment sample.

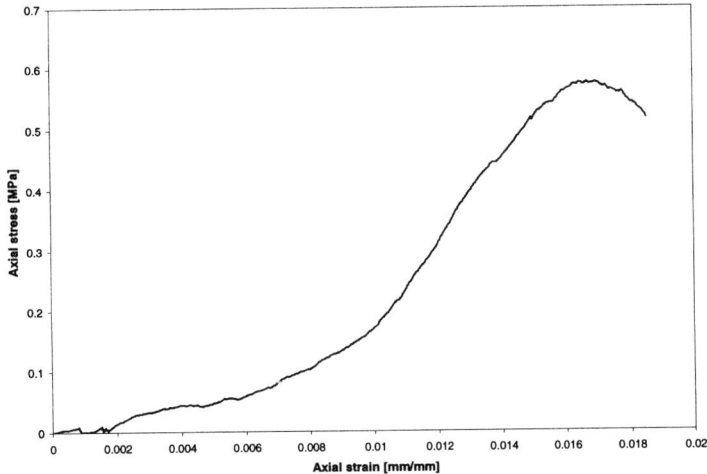

FIGURE 7. Axial stress versus axial strain for a R11 hydrated sediment sample.

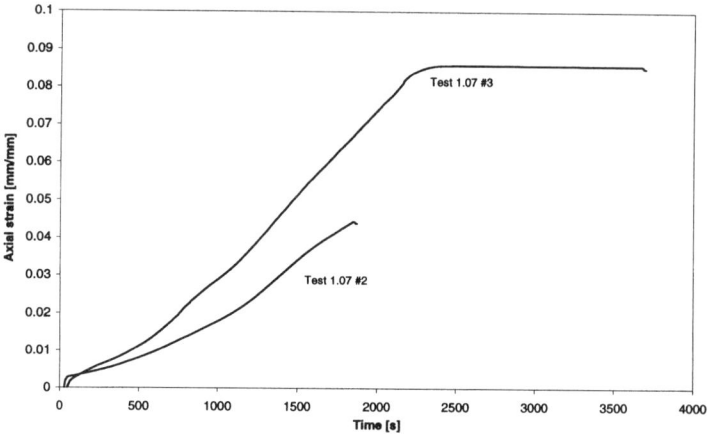

FIGURE 8. Axial strain versus time for two THF hydrated sediment samples.

The difference in deformation progress for the two hydrated sediment samples may be due to melting of the core ends during mounting. Since all tests were performed without temperature control (room temperature, about 293 K), it is possible that the samples started melting during mounting without any visual signs. Hence, variations in the time used for mounting may give variations in the deformation progress. This may be the reason for the low stiffness in the beginning of the test with the R-11 hydrated sediment sample.

Chuvilin and Yakushev[4] have performed uniaxial compression strength (UCS) tests at hydrated/frozen, fine-grained sediment samples. Results from these tests gave a UCS of 0.49 MPa for the hydrated sample and 0.05 for the frozen sample. Hence, the core failure values are in the same range as in this study.

FIGURE 8 shows axial strain versus time for two THF hydrated sediment samples, one sample was completely melted during the test and one test was interrupted before the melting period was finished. The difference in deformation progress for the samples may also be due to melting caused by variations in time used for mounting.

CONCLUSION

The experiments performed in this study were simplified in order to make preliminary investigations of methodology and equipment suitability. Hence, model systems at low pressure were used and the results are not expected to be directly comparable to *in situ* conditions. For instance, no pressure effects could be studied in the current setup. However, the results show clear solidification effects from hydrate formation in sand, and some structuring effects in systems where nucleation is widespread. The R-11 system used here is probably closest to *in situ* conditions due to the immiscibility of the phases, but use of a structure I hydrate former would have been preferable, and will be the goal for future high pressure studies. NMR spectral measurements of relaxation times will probably be needed to answer in any detail the remaining questions about how hydrate is distributed and how the dark areas in the samples correspond to hydrates, sand, and unreacted R-11.

The deformation during melting seems to be large (8–10%) and may give serious problems if melting of such sediments occurs in the sea floor. Additional tests, under near real conditions will, however, be necessary in order to determine a more exact relationship.

ACKNOWLEDGMENT

The authors would like to thank Trond Singstad, Tina Bugge Pedersen, and Tore Skjetne at SINTEF Unimed who have contributed to this study with their expertise in NMR- and CT-technology. We would also like to thank Norsk Hydro for sponsoring parts of the work.

REFERENCES

1. KVENVOLDEN, K.A. 1988. Methane hydrate—a major reservoir of carbon in the shallow geosphere. Chem. Geol. **71:** 41–51.
2. UCHIDA, T., S. DALLIMORE, M. NIXON & J. MIKAMI. 1998. Occurrence of gas hydrates obtained from the JAPEX/JNOC/GSC Mallik 2L-38 research well: video presentation. Proceedings of the International Symposium on Methane Hydrates, Chiba City, Japan, October 20–22.
3. BOOTH, J.S, M.M. ROWE & K.M. FISCHER. 1996. Offshore gas hydrate sample database. USGS report 96–272, Woods Hole.
4. CHUVILIN, E.M. & V.S. YAKUSHEV. 1998. Structure and some properties of frozen hydrate-containing soils. Proceedings of the International Symposium on Methane Hydrates, JNOC-TRC, Chiba, Japan, October 20–22.

Rheological Characterization of Hydrate Suspensions in Oil Dominated Systems

R. CAMARGO,[a,b] T. PALERMO,[b] A. SINQUIN,[b] AND P. GLENAT[c]

[b]Institut Français du Pétrole, Division Chimie et Physicochimie Appliquées, 1 et 4 Avenue Bois-Préau, 92852 Rueil Malmaison Cedex, France

[c]Tour TOTAL, Cedex 47, 92069 Paris–La Défense, France

ABSTRACT: In the work described in this paper, the rheological behavior of hydrate suspensions in an asphaltic crude oil was studied. A *P-T* cell adapted to operate as a double cylinder configuration rheometer was used. The results obtained show that the hydrate suspensions studied exhibit shear thinning behavior and thixotropy. This behavior is more visible in suspensions with higher particle content and seems to be related to colloidal interactions between the crystals, which causes aggregation between crystals at low shear rates.

INTRODUCTION

The development of offshore oil exploitation at increasing water depths has highlighted the need for better knowledge about hydrate formation and transportation in oil dominated systems. Indeed, in the deep subsea flowlines that link oil production wells to their platforms the conditions needed for hydrate formation are found, such as high pressures, low temperatures, and the presence of natural gas and water.

In oil dominated systems the injection of a thermodynamic inhibitor to avoid hydrate formation is not feasible, due to the substantial water volume produced in advanced phases of field production. Solutions may then involve the injection of low dosage hydrate inhibitors (LDHI) such as kinetic inhibitors and dispersant additives, which are presently in research and development phases. Dispersant additives do not avoid hydrate formation, but prevent crystal agglomeration and, consequently, allow crystals to be transported as a hydrate suspension.

It is well known that crude oils have natural surfactants, such as resins and asphaltenes that stabilize water-in-oil emulsions.[1,2] If these heavy polar components also exert a dispersing effect, hydrate crystals formed in such a water-in-oil emulsion would be transportable as a suspension, without the need for further chemical injection. This is the context in which studying the rheology of hydrate suspensions is seen as one of the possible methods that may allow the prediction of hydrate transportation capabilities of crude oils under realistic production conditions.

Once the transportation of a given hydrate suspension is considered feasible, a better knowledge of its rheological behavior is essential to improve the design of

[a]Telecommunication. Voice: 00 33 1 47 52 71 55; fax: 00 33 1 47 52 70 58. ricardo.camargo@ifp.fr

multiphase pipelines, since viscosity is a very important variable in pressure drop calculations.

The main purpose of the work is to analyze the rheological behavior of water-in-oil emulsions and of the hydrate suspensions formed from them, under the same pressure and temperature conditions. A crude oil, rich in heavy polar components, has been used.

EXPERIMENTAL SETUP

A pioneer device and an experimental procedure have been specially developed for this rheological investigation. The rheological cell is a pressure and temperature cell that has been adapted to allow the evaluation of rheological characteristics of hydrate suspensions.

The rheological cell allows one to work at a pressure up to 9 MPa and with temperatures ranging from 258.15 to 338.15 K. It has a double cylinder configuration and works with imposed shear rate and measured couple. The rotation speed of the mobile cylinder ranges from 0 to 600 rpm. The couple captors are able to perform measurements between 0 and 100 mN·m, but magnetic coupling is only capable of up to 35 mN·m.

FIGURE 1 shows a sketch of the rheological cell. The electric motor, at a constant angular velocity, drives the mobile external cylinder by means of a magnetic coupler, the couple needed to maintain this velocity being measured by two parallel couple captors. The 100 cm^3 transparent cell is located inside a thermoregulated compartment

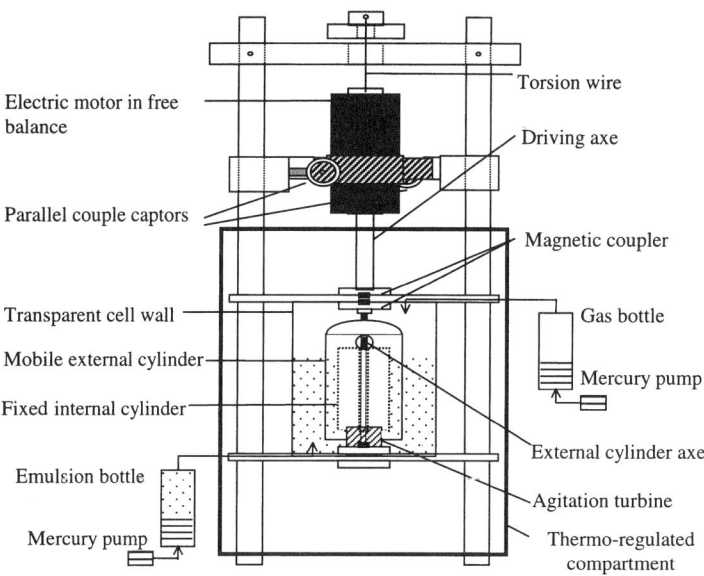

FIGURE 1. Rheological cell sketch.

that provides temperature control. The emulsion bottle is used to inject the previously prepared water-in-oil emulsion into the cell. The pressure in the cell is kept constant by injecting gas from the gas bottle to compensate for the pressure drop due to temperature reduction and hydrate formation.

One important characteristic of this rheometer is that there are in reality two gaps that contribute to the measured couple. There is one gap between the internal fixed cylinder and the external mobile cylinder (internal gap), and another gap between this external cylinder and the transparent cell wall (external gap). Thus, the viscosity η of a Newtonian fluid in mPa·s, is given by

$$\eta = \frac{C}{4\pi\omega_0(G_i - G_e)}, \quad (1)$$

where C is the couple in mN·m and ω_0 is the angular velocity in rad/s. G_i and G_e are geometrical factors related to the two gaps that are given by

$$G_i = \frac{h_i}{\left(\frac{1}{R_1^2} - \frac{1}{R_2^2}\right)} \text{ and } G_e = \frac{h_e}{\left(\frac{1}{R_3^2} - \frac{1}{R_4^2}\right)} \quad (2)$$

where the subscript i stands for the internal gap and e for the external gap. The height of liquid is represented by h, R_1 is the internal cylinder radius, R_2 and R_3 are, respectively, the internal and the external radius of the mobile cylinder, and R_4 is the transparent cell radius. All these values are given in meters.

Calibration

Since the rheological cell is an adaptation of a P-T cell and not a true rheometer, it was necessary to perform calibration experiments in order to estimate its accuracy

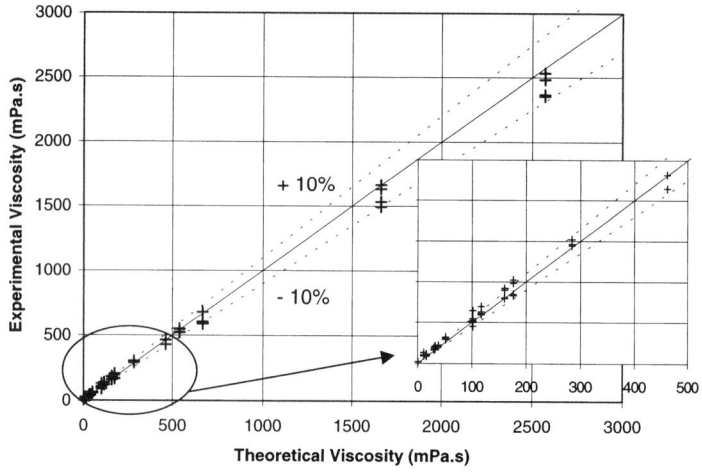

FIGURE 2. Results from rheological cell calibration with Newtonian fluids.

and to be acquainted with its limitations. Two sets of experiments were performed. The first was performed with calibrated Newtonian oils, the second with shear-thinning non-Newtonian liquids.

The results of the experiments with the Newtonian fluids are shown in FIGURE 2 where the experimental values were calculated using Equation (**1**). In this figure, each point is the result of a complete rheological test and represents the arithmetical mean of the values obtained for the viscosity at each angular velocity. These calibration experiments were carried out before starting the tests with hydrate suspensions and also between the hydrate tests in order to check that the device was still calibrated. Even though the rheological cell accuracy is far from what is expected for a true rheometer, it is more than satisfactory for the purposes of this research.

Three aqueous solutions of a high molecular weight polyacrylamide were used for calibration in a set of experiments with non-Newtonian liquids. In order to calculate the apparent viscosity μ as a function of the shear rate $\dot{\gamma}$ the power law model expressed by

$$\mu = k\dot{\gamma}^{n-1} \tag{3}$$

was assumed. In (**3**), k and n are constants that characterize the rheological behavior of a given fluid and were calculated from measured data. FIGURE 3 shows a comparison between the results obtained with the rheological cell and those obtained by a Carri-Med rheometer, confirming that the rheological cell is also able to determine the shear-thinning behavior of a given fluid.

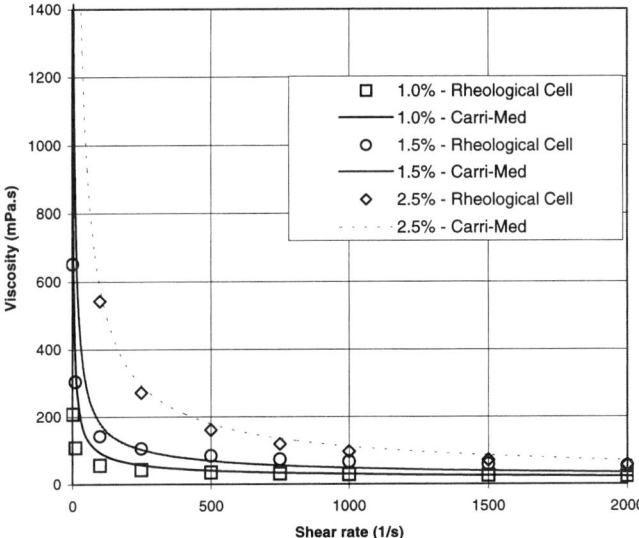

FIGURE 3. Comparison between the results obtained by the Carri-Med rheometer and by the rheological cell for non-Newtonian fluids (aqueous solutions of polyacrylamide with concentrations of 1.0, 1.5, and 2.5%).

TABLE 1. Crude oil main properties

Density at 288.15 K	g/cm^3	0.911
Dead oil viscosity at 293.15 K	mPa·s	135
Saturates + aromatics content	% weight	77.3
Resins content	% weight	17
Asphaltenes content (in heptane)	% weight	5.7

EXPERIMENTAL PROCEDURE

A complete experimental run includes emulsion preparation, its saturation in gas at 298.15 K, cooling the fluids to the desired temperature, one rheological test for the emulsion at that temperature (before hydrates formation begins), and the complete process of hydrate formation, during which several rheological tests are performed. A rheological test consists in recording the pair of values (C, ω_0) while the angular velocity slowly increases from 0 to 70 rad/s and, afterwards, decreases back to zero. Between the rheological tests, the mobile cylinder is maintained at a constant angular velocity, usually 35 rad/s. The values of couple, pressure, temperature, angular velocity, and gas consumption are recorded throughout the experiment as a function of time. The pressure and temperature conditions used in this work are 8 MPa and 280.65 K.

The systems studied have three phases: liquid hydrocarbon, water, and gas. The liquid hydrocarbon is an asphaltenic heavy crude oil from a Brazilian deep water field. Its main properties are shown in TABLE 1. The water phase is salted water

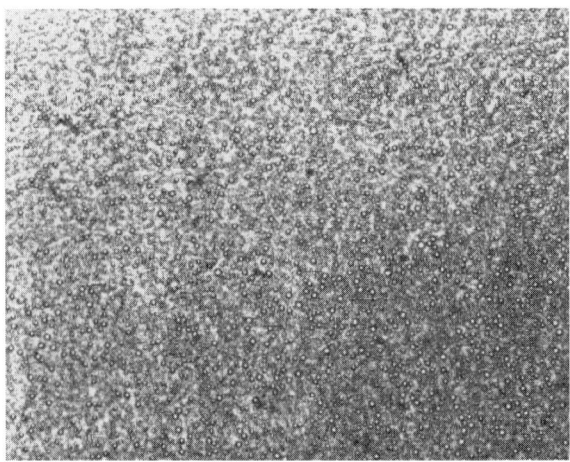

FIGURE 4. Photomicrograph of a water-in-oil emulsion. Water content, 30% weight. Amplification: 200×.

(33 g/l NaCl). A mixture of 90% molar of methane and 10% ethane was chosen as the gas phase.

The method used to prepare the emulsion was to agitate the crude oil with a high shear mixer Ultra-Turrax Model T25 running at 8,000 rpm for 180 seconds, while slowly pouring the water on it. This procedure results in a very tight emulsion, with a characteristic water-drop diameter of 2.5 µm (see FIGURE 4). Emulsions with 15, 30, and 50% of water content (by weight) were used.

It is very important to have a homogeneous fluid between the gaps. Thus, the emulsions prepared must be very stable and also present a very low droplet settling velocity. In order to check whether those requirements were satisfied, separation and sedimentation tests were performed with the prepared emulsions, letting them rest in graded test tubes maintained at 293.15 K. After a period of one month, neither separated water nor emulsion segregation were observed.

RESULTS AND DISCUSSION

While working with crude oils, hydrate crystallization is not a visible phenomenon, thus we must be sure that it really did occur. FIGURE 5 shows how we realize that hydrates were formed. After a period of time at the desired pressure and temperature conditions, we observe a sharp increase in the measured couple (and therefore in the viscosity) caused by crystal formation, concomitant with a pronounced decrease in cell pressure, due to the related gas consumption. The cell pressure decreases because its connection with the gas bottle, which permits automatic pressure control, is kept closed during hydrate formation.

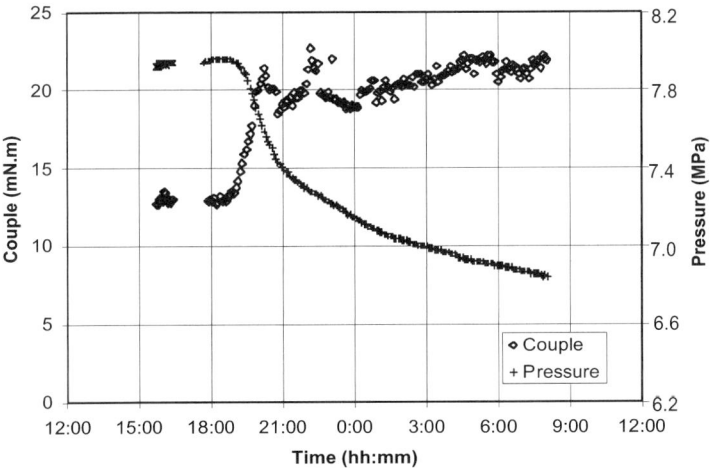

FIGURE 5. Hydrate formation evidence. Water content, 50%; T, 280.65 K; angular velocity, 35 rad/s.

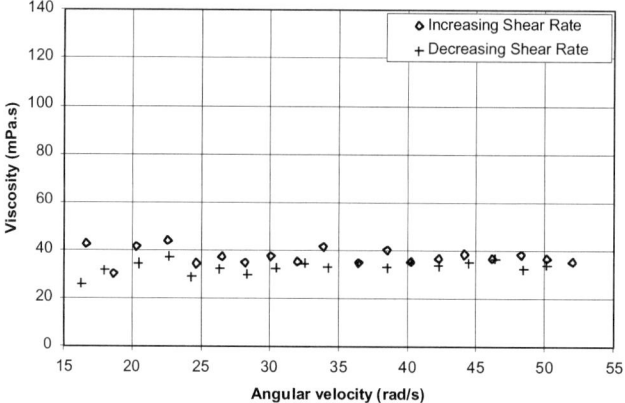

FIGURE 6. Rheological test for the pure oil. T, 280.65 K; P, 8 MPa.

As can be seen in FIGURE 6, the pure oil presents a Newtonian behavior, with a viscosity of roughly 35 mPa·s.

The results for the rheological tests with 15% of water content (see FIGURE 7) suggest a Newtonian behavior for both the emulsion and the hydrate suspension. Nevertheless, it is known that emulsions are in general non-Newtonian fluids.[1] Hence, the conclusion is that a possible non-Newtonian behavior for both emulsion and hydrate suspension is not strong enough to be detected by the rheological cell. Test A was performed at the firsts stages of hydrate formation. Tests B and C were performed after the end of the crystallization process (end of gas consumption). An increase in hydrate suspension viscosity was observed as more hydrates were

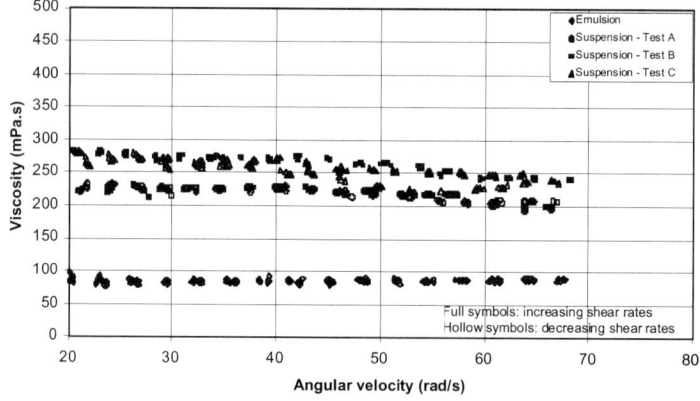

FIGURE 7. Rheological tests for the water-in-oil emulsion prepared with 15% water content. T, 280.65 K; P, 8 MPa.

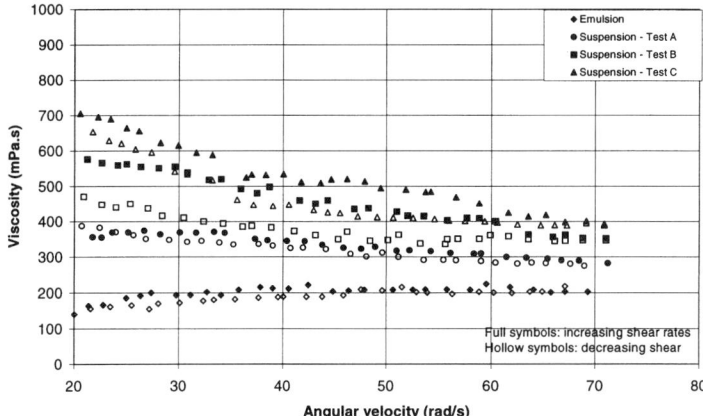

FIGURE 8. Rheological tests for the water-in-oil emulsion prepared with 30% water content; T, 280.65 K; P, 8 MPa.

formed. The emulsion viscosity is 85 mPa·s, whereas the hydrate suspension viscosity is 260 mPa·s.

The rheological behavior of the 30% water content emulsion also seems to be Newtonian (see FIGURE 8). However, a shear-thinning behavior for the hydrate suspension can be observed. This shear-thinning behavior, as well as the apparent viscosity itself, both increase with the quantity of hydrates formed (from tests A to C). An increasing thixotropic behavior—expressed by higher viscosity while increasing the angular velocity than while decreasing the angular velocity—was also observed.

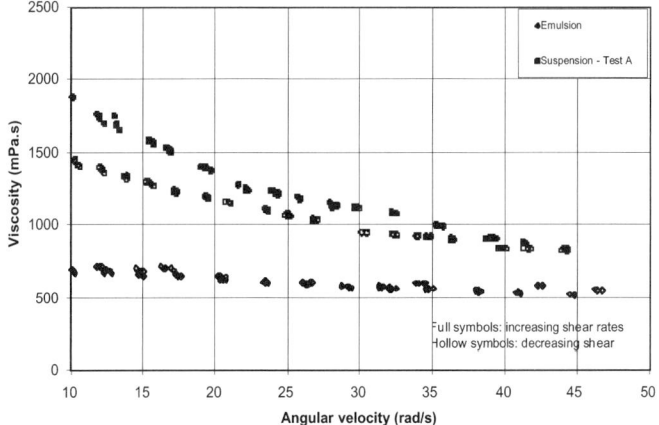

FIGURE 9. Rheological tests for the water-in-oil emulsion prepared with 50% water content. T, 280.65 K; P, 8 MPa.

This suggests an aggregation tendency between hydrate crystals at low angular velocities, forming a structure that is broken at higher shear stresses. Accordingly, at high angular velocities the apparent viscosity does not decrease any further and a Newtonian region is reached. The emulsion viscosity is 200 mPa·s. After the end of the hydrate formation process, the hydrate suspension had an apparent viscosity in the Newtonian region of 400 mPa·s (test C).

Concerning the emulsion prepared with 50% water content, there has not been any problem in performing rheological characterization of the emulsion. We can assume that this fluid has a Newtonian behavior and its viscosity is 600 mPa·s (see FIGURE 9). Nevertheless, at this water content, some instability problems were observed after hydrate formation (see FIGURE 10). It seems that the high solid concentration suspension formed from this emulsion is a very heterogeneous system, where direct contacts between particles play an important role in governing flow characteristics. It can also be observed that, after a period of time, the viscosity drops, reaching an even lower value than the emulsion viscosity. The most likely reason for this behavior is the sedimentation of the solid particles, since the contact between them enlarges their typical diameters and, consequently, increases their settling velocities. Of course, a rheological test during the instability or sedimentation phases would be meaningless. However, in FIGURE 9, test A was performed between the first and the second hydrate formation process, hence, before the problems referred to above began. The general tendencies of increasing the viscosity and more pronounced shear-thinning behavior and thixotropy at higher water contents have been confirmed in this rheological test.

In FIGURE 11, the results of apparent viscosity for each water content have been gathered. For the cases in which a shear-thinning behavior was found, the values at the Newtonian region have been used (Tests C). This value was obtained by extrapolation in the case 50% of water content and represents only a rough estimate.

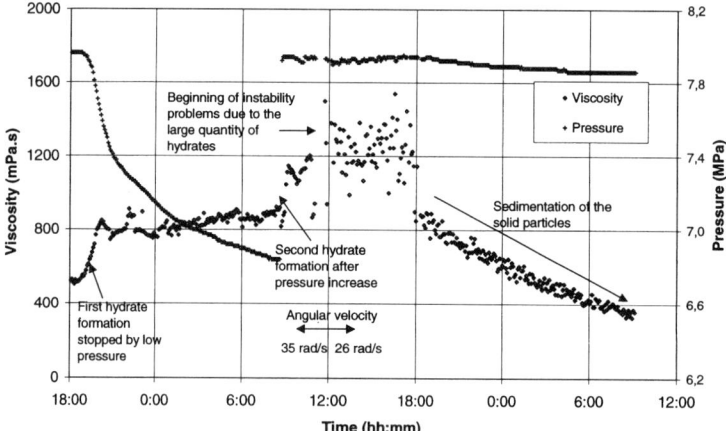

FIGURE 10. Apparent viscosity evolution during one experimental run with the water-in-oil emulsion prepared with 50% water content; T, 280.65 K.

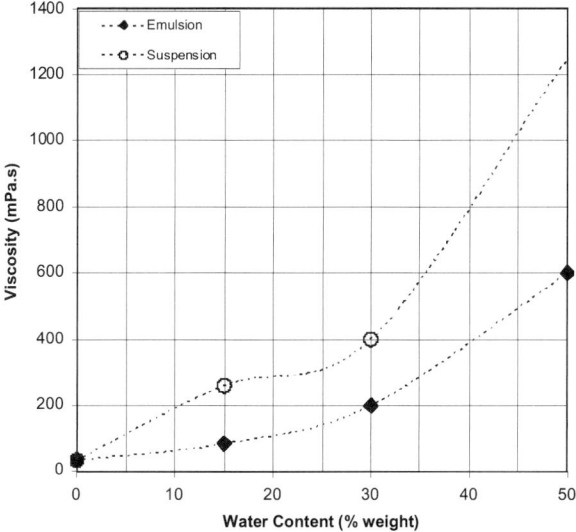

FIGURE 11. Comparison between emulsion and hydrate suspension apparent viscosity at different water contents; T, 280.65 K; P, 8 MPa.

The exponential growth in the hydrate suspension apparent viscosity with the water content suggests that its transportability is limited by high viscosity before being limited by the capacity of maintaining the crystal dispersion.

CONCLUSION

Despite the difficulties usually found in rheological characterization of emulsions and suspensions at high pressures, this work allows some important preliminary conclusions. Among them, the main conclusions are:

- Besides being a very good emulsifier, natural surfactants contained in the oil studied act also like a highly efficient dispersant additive for the hydrate suspension. Consequently, crystal transportation is expected under realistic production conditions, at least up to a given water content level.
- The water-in-oil emulsion viscosity increases as water is converted into hydrates.
- As expected, the larger the water content, the higher the viscosity of both the water-in-oil emulsion and the hydrate suspension. This suggests a limitation on transportability of hydrate suspensions in pipes due to high viscosity.
- Hydrate suspensions show a shear-thinning behavior that is more pronounced in more concentrated suspensions.

- The shear-thinning behavior seems to be related to colloidal interactions between the solid particles, which causes aggregation between them at low shear rates. This possibility is reinforced by a thixotropic behavior found in the hydrate suspensions analyzed.
- Shear-thinning behavior in the hydrate suspension means difficulties for restart operation after production shut down (very high viscosity at low shear rate).

ACKNOWLEDGMENTS

This work was performed within a collaboration program involving IFP, PETROBRAS, and TOTAL. The authors would like to thank these companies for their financial support and for permission to publish these results.

REFERENCES

1. MANNING, F.S. & R.E. THOMPSON. 1983. Oilfield Processing Volume Two: Crude Oil. PennWell Books, Tulsa.
2. MOURAILLE, O. *et al.* 1998. Stability of water-in-crude oil emulsions: role played by the state of solvatation of asphaltenes and by waxes. J. Dispersion Sci. Technol. **19** (2/3): 339–367.

A Model for Systems with Soluble Hydrate Formers

M.D. JAGER,[a,b] R.M. DE DEUGD,[c] C.J. PETERS,[c]
J. DE SWAAN ARONS,[c] AND E.D. SLOAN[b]

[b]*Colorado School of Mines, Department of Chemical Engineering and Petroleum Refining, Center for Hydrate Research, Golden, Colorado 80401-1887, USA*

[c]*Delft University of Technology, Faculty of Applied Sciences, Laboratory of Applied Thermodynamics and Phase Equilibria, Julianalaan 136, 2628 BL, Delft, the Netherlands.*

ABSTRACT: The objective of this study was to measure and model hydrate phase equilibria in a system containing a water-soluble hydrate former. The system investigated is methane + water + 1,4-dioxane in the pressure range between 2 and 14 MPa. Experimental results show that the stability of hydrates is a strong function of 1,4-dioxane concentration in the water phase. The hydrate phase equilibria data are modeled using the van der Waals and Platteeuw theory assuming that 1,4-dioxane is a soluble sII former. Activity coefficients of the liquid phase can be calculated from water + 1,4-dioxane vapor liquid equilibria. Under these assumptions, the predicted equilibrium pressures are within 5% of the experimental data up to a concentration of 20 mol% 1,4-dioxane relative to water.

INTRODUCTION

Clathrate hydrates are crystalline inclusion compounds formed at elevated pressures in mixtures of water and hydrocarbon molecules. The thermodynamic stability of a gas hydrate is determined by temperature and pressure, the structure of the hydrate and the nature of the compounds included. Gas hydrates are stabilized by a variety of small (below 9 Å diameter) molecules. Methane and 1,4-dioxane are both known to form hydrates. The main incentives for research in gas hydrates are plugging of pipelines by hydrates, natural gas hydrate resources and storage of natural gas as hydrates.

Interest in the possibility of storage of natural gas led to the research by de Roo et al.,[1] who studied the inhibiting effect of sodium chloride on methane hydrate formation. Ng and Robinson[2] observed that acetone, a hydrate former that is soluble in water, can promote or suppress formation of gas hydrates in systems of water and methane, depending on the concentration of acetone. Mainusch et al.[3] confirmed, using the Cailletet apparatus, the experimental results by Ng and Robinson.

Interest in water-soluble hydrate formers was renewed, when it was suggested that they could be used to stabilize gas hydrates for the purpose of storing natural

[a]Telecommunication. Voice: 303-273-3561; fax: 303-273-3730.
mjager@mines.edu

gas. Previous research by Saito et al.[4] on stabilization of gas hydrates showed qualitatively that a large group of soluble hydrocarbons stabilize gas hydrates when added in low concentrations to a water phase. One of the hydrocarbons Saito et al. investigated was 1,4-dioxane ($C_4H_8O_2$). The current work presents experimental data on gas hydrate formation in systems of water + methane + 1,4-dioxane, for seven different concentrations of 1,4-dioxane ranging from 1 to 30 mole% with respect to water. All mole fractions are with respect to water, unless otherwise indicated.

Thermodynamic calculations have been carried out to model the measured equilibrium data, using a model based upon the van der Waals and Platteeuw theory.[5] We use the model including extensions by Saito et al.,[6] and Parrish and Prausnitz.[7] The interaction between a guest molecule and the surrounding water molecules is described with a Kihara spherical core potential.[8] A computer program similar to the program we used is available with a monograph by Sloan.[9]

The model here presented is based on the prediction program due to Sloan. The major difference between the new model and previous models is that the new model can be used to calculate hydrate phase equilibria involving hydrate formers that are mainly present in the liquid water phase.

EXPERIMENTAL SECTION

The hydrate phase equilibrium temperatures were determined at nine different pressures between 2 and 14 MPa in the system water + methane + 1,4-dioxane. The measurements were carried out with a Cailletet apparatus, described previously by Peters et al.[10]

The experimental method used to determine hydrate equilibrium conditions was as follows. A glass tube with a sample of known composition was present in a glass cylinder, which holds a thermostat liquid. The temperature was measured with a platinum resistance temperature. The pressure inside the sample tube was controlled and measured by a dead weight gauge. The accuracy of the measurements was within 0.05 K and 0.005 MPa. The sample was stirred by magnetically moving a steel ball up and down. All chemicals were used without further purification.

The Cailletet tube was filled as follows: an amount of water + 1,4-dioxane mixture of known composition was weighted into the Cailletet tube. The liquid was degassed under vacuum. A known volume of methane was added by displacing mercury. The mole fraction error was less than 0.002.

The transition points were determined by visually observing the disappearance of the hydrate phase. The equilibrium conditions were reached when the transition from a system with a hydrate, liquid water, and a vapor phase to a system with only liquid water and vapor occurred. The temperature at which the last hydrate crystal disappeared was taken as the equilibrium temperature for hydrate formation.

EXPERIMENTAL RESULTS AND DISCUSSION

The experimentally determined hydrate phase equilibrium points are shown in FIGURE 1 as a pressure versus temperature plot. The hydrate equilibrium pressures

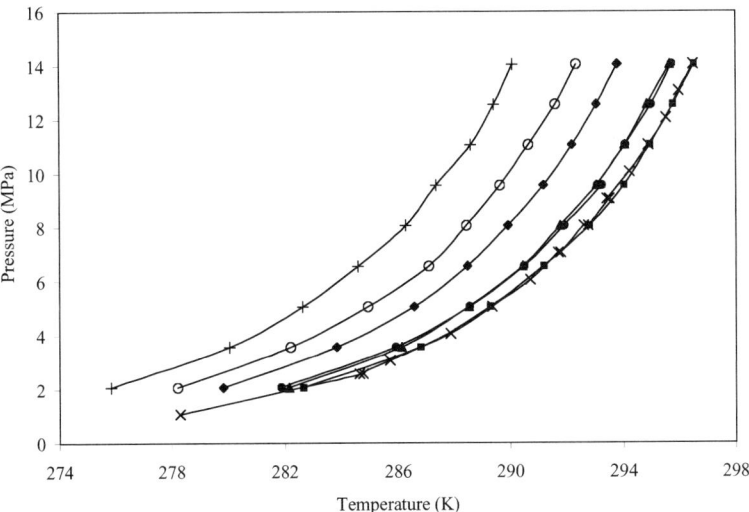

FIGURE 1. Pressure versus temperature plot of the hydrate equilibrium data measured for the ternary system of water + methane + 1,4-dioxane. Data: ♦, $x = 0.01$; ▲, $x = 0.02$; ×, $x = 0.05$; ■, $x = 0.07$; ●, $x = 0.1$; ○, $x = 0.2$; +, $x = 0.3$. x = mole fraction 1,4-dioxane.

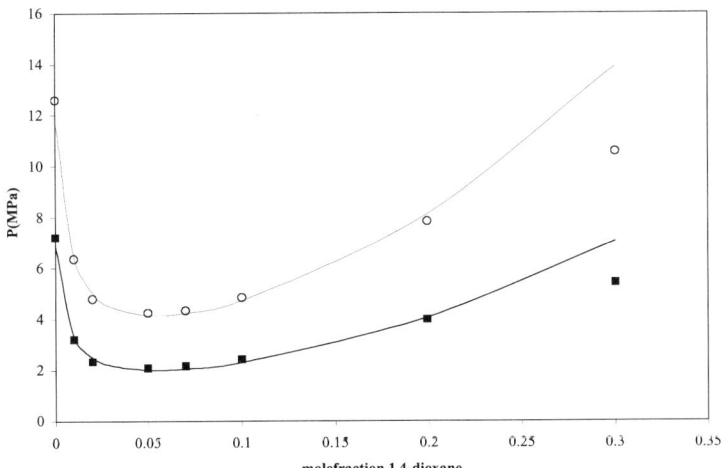

FIGURE 2. Predicted equilibrium pressures of hydrate formation at 283.15 and 288.15 K for the system water + methane + 1,4-dioxane. Data: ■, $T = 283.15$ K; ○, $T = 288.15$ K. *Solid lines* predictions.

of water + methane were taken from literature.[11,12] The data clearly shows that the mixed hydrates of 1,4-dioxane and methane are more stable than structure I methane hydrates. The equilibrium pressure of hydrate formation reaches a minimum when between 5 and 7 mole% of 1,4-dioxane is present in the liquid phase. At higher concentrations of 1,4-dioxane, the equilibrium pressure starts to rise again. One would expect this behavior from the ratio of large cages to water molecules in a structure II hydrate—1/17 (= 0.0588). From the isotherms shown in FIGURE 2, it is clear that the optimum concentration (the concentration at which the hydrate equilibrium pressure reaches a minimum) moves slightly with temperature. This means that the exact locus of the optimum is a function of the activity of the water phase.

MODELING

The results obtained for the system water + methane + 1,4-dioxane are similar to previous obtained data by Mainusch on water + methane + acetone, and data by de Deugd[13] for water + methane + 1,3-dioxolane and water + methane + tetrahydropyran. Since no rigorous calculations have yet been made for hydrate equilibrium conditions in systems containing a soluble hydrate former, our data were used to extend existing hydrate theory. Basically, it was assumed that the soluble hydrate former has only two effects on the hydrate equilibrium: it acts as a guest to the hydrate lattice and it changes the activity coefficient of water. An article by Mehta and Sloan was most helpful in developing the model for soluble hydrate formers.[14]

Equilibrium is defined by the chemical potential of water being equal in all phases. For the purpose of hydrate modeling, chemical potential differences (the difference between the chemical potential of a theoretical empty hydrate lattice and the respective chemical potentials) are compared:

$$\mu_w^\beta - \mu_w^H \equiv \Delta\mu_w^H = \Delta\mu_w^{L:} \equiv \mu_w^\beta - \mu_w^L. \tag{1}$$

The chemical potential of a hydrate depends on the fractional occupancies of the constituent hydrate cages and the fractional occupancy of a type i cage with a type k molecule, θ_{ki}, which can be expressed as a function of fugacity and the Langmuir constants C_{ki}. These can be evaluated by integrating the interaction potential between a hydrate guest and the lattice water molecules. We use the Kihara potential to describe these interactions[15] and the Soave-Redlich-Kwong[16] equation of state to calculate the fugacity of methane. The chemical potential difference of the liquid water phase can be calculated from the classical equation given by Holder et al.[17]

The fugacity of a highly water-soluble hydrate former, in our case 1,4-dioxane, was calculated from the water rich liquid phase, because the vapor composition of the compound is minimal. The equation for the fugacity of a compound in a liquid phase is given by

$$f_k = x_k \cdot \gamma_k \cdot \Phi_k^{sat} \cdot P_k^{sat} \cdot e^{\frac{V_k^L \cdot (P - P^{sat})}{R_e \cdot T}} \tag{2}$$

The vapor pressure of 1,4-dioxane can be calculated from an Antoine equation, see for example Reid et al.[18]

For the radius of the hydrate cage, the coordination number, the number of cages per water molecule, and the Kihara parameters for methane we use reported values,

TABLE 1. Structure parameters for sII hydrate and Kihara parameters for methane

Hydrate Structure	sII		Kihara parameters
Cavity type	5^{12}	$5^{12}6^4$	a = 0.3834 Å
Radius (nm)	0.39	0.473	σ = 3.165 Å
Coordination No.	20	28	ε = 154.54 K
Cavities per unit cell	16	8	
Cavities/water molecule	2/17	1/17	
Water molecules/unit cell	136		
Crystal type	Cubic		

given in TABLE 1. The values for the reference parameters in the Holder equation are given in TABLE 2.[10] The presence of a soluble hydrate former in the system water + methane + 1,4-dioxane can be modeled under the following assumptions:

1. A methane + 1,4-dioxane mixture forms sII hydrates; 1,4-dioxane is only present in the large cavity.
2. Due to low solubility of methane, the liquid phase can be considered a binary system of water and 1,4-dioxane. Activity coefficients are calculated from the VLE data reported by Ghmeling et al.[19]
3. The concentration of water in the liquid phase can be calculated by subtracting from 1 the mole fraction of 1,4-dioxane and the solubility of methane in pure water.

The critical and new assumption is that VLE data can be used to calculate activity coefficients for a hydrate forming system. The VLE data from literature was fitted to a van Laar equation. To introduce temperature dependence, it was assumed that the logarithm of an activity coefficient is proportional to inverse temperature as given by

$$\ln(\gamma) = a + \frac{b}{T(K)}. \qquad (3)$$

This equation relates the activity coefficient at a constant composition to temperature. The Langmuir constant for 1,4-dioxane in the large sII cage was also calculated

TABLE 2. Reference parameters for sII hydrate

Reference Parameter	Unit	sII
$\Delta\mu_w^0$	J/mole	883.82
ΔH_w^0 (ice)	J/mole	1025.08
ΔH_w^0 (water)	J/mole	−4987.27
ΔV_w^0 (ice)	Cm^3/mol	3.4
ΔV_w^0 (water)	Cm^3/mol	5.0
ΔC_{pw}^0	J/mole K	−38.12
B	J/mole K^2	0.141

TABLE 3. Calculated Kihara parameters for 1,4-dioxane

Concentration 1,4-dioxane	ε/k (K)	
0.01001	264.937	a = 1.094 Å
0.02011	265.275	σ = 3.3 Å
0.05007	263.966	ε/k = 264.7 K
0.07003	264.415	
0.10014	264.326	
0.19939	265.253	
0.29997	269.185	

from a Kihara potential. The Kihara core distance a for 1,4-dioxane was estimated with a method devised by Tee et al.[20] The position of zero potential energy and the depth of the energy well, σ and ε, were regressed from the measured equilibrium data. The results of these calculations are given in TABLE 3.

Based on this model, equilibrium lines for all measured concentrations of 1,4-dioxane were calculated. These prediction lines were used to construct a set of P, x-sections at 283.15 K and 288.15 K. The predicted equilibrium pressures show good agreement, when compared with the data, as is shown in FIGURE 2, except at the highest concentration of 1,4-dioxane.

Excluding the data points at 30 mol% of 1,4-dioxane, the model yields equilibrium pressures that are within 5% of the data. Probably, the assumption that the liquid phase is a binary mixture breaks down at concentrations above 20 mol% due to an increase of the solubility of methane. This means that at higher concentration the liquid phase should no longer be treated as a binary system.

CONCLUSIONS

Equilibrium pressures of hydrate formation were measured in the 2 to 14 MPa pressure window for the system water + methane + 1,4-dioxane. This was done for seven different mole fractions of 1,4-dioxane relative to water. A minimum occurs in a pressure–composition section at the stoichiometric ratio of large sII cages to water in a sII hydrate (1/17). These data were modeled with classical hydrate theory.

The hydrate transition in this system were modeled as a binary liquid phase of water and 1,4-dioxane in equilibrium with a sII mixed hydrate of 1,4-dioxane as large guest and methane as small guest molecule. Activity coefficients in the liquid phase were calculated from binary VLE data. The equilibrium pressures calculated are equal to measured data up to a concentration of 20 mol% 1,4-dioxane in water. This shows that hydrate phase equilibria of soluble hydrate formers can be modeled by combining VLE data with van der Waals and Platteeuw theory.

ACKNOWLEDGMENT

The authors would like to acknowledge the financial support by College van Bestuur Universiteit Delft, Universiteitsfonds Delft and Laboratory of Applied Thermodynamics and Phase Equilibria, University of Delft. We are thankful to Mr. W. Poot and Mr. L. Florusse for assistance in the measurements.

REFERENCES

1. DE ROO, J.L., C.J. PETERS, R.N. LICHTENTHALER & G.A.M. DIEPEN. 1983. AIChE J. **29**(4): 651.
2. NG, H.-J. & D.B. ROBINSON. 1994. Ann. N.Y. Acad. Sci. **715**: 450.
3. MAINUSCH, S., C.J. PETERS, J. DE SWAAN ARONS, J. JAVANMARDI & M. MOSHFEGHIAN. 1996. 2nd Int. Conf. Nat. Gas Hydrates. 71.
4. SAITO, Y., T. KAWASAKI, T. OKUI, T. KONDO & R. HIRAOKA. 1996. 2nd Int. Conf. Nat. Gas Hydrates.
5. VAN DER WAALS, J.H. & J.C. PLATTEEUW. 1959. Adv. Chem. Phys. **2**: 1.
6. SAITO. S., D.R. MARSHALL & R. KOBAYASHI. 1964. AIChE J. **10**: 734.
7. PARRISH, W.R. & J.M. PRAUSNITZ. 1972. Ind. Eng. Chem. Proc. Des. Dev. **11**: 26.
8. KIHARA, T. J. 1951. Phys. Soc. Japan **6**: 289.
9. SLOAN, E.D. 1998. Clathrate Hydrates of Natural Gases, 2nd edit. Marcel Dekker, New York.
10. PETERS, C.J., J.L. DE ROO & J. DE SWAAN ARONS. 1993. Fluid Phase Eq. **85**: 301.
11. DEATON, W.M. & E.M. FROST. 1946. U.S. Bureau of Mines Monograph. 8.
12. MCLEOD, H.D. & J.M. CAMPBELL. 1961. J. Petrol. Tech. **13**: 590.
13. DE DEUGD, R.M., C.J. PETERS & J. DE SWAAN ARONS. 1999. To be published.
14. MEHTA, A.P. & E.D. SLOAN. 1994. AIChE J. **40**(2): 312.
15. MCKOY, V. & O. SINANGOLU. 1963. J. Chem. Phys. **38**(12): 2946.
16. SOAVE, G. 1972. Chem. Eng. Sci. **27**: 1197.
17. HOLDER, G.D., G. CORBIN & K.D. PAPADOPOULOS. Ind. Eng. Chem. Fund. **19**(3): 282.
18. REID, R.C., J.M. PRAUSNITZ & B.E. PAULING. 1987. The Properties of Gases and Liquids, 4th edit. McGraw-Hill.
19. GMEHLING, J., U. ONKEN & W. ARLT. 1981. Dechema Data Series, I Part 1a. 288. Dortmund.
20. TEE, L.S., S. GOTOH & W.E. STEWART. 1966. I. & E.C. Fund. **5**(3): 363.

Improving the Accuracy of Gas Hydrate Dissociation Point Measurements

B. TOHIDI,[a] R.W. BURGASS, A. DANESH, K.K. ØSTERGAARD, AND A.C. TODD

Department of Petroleum Engineering, Heriot-Watt University, Riccarton, Edinburgh EH14 4AS, United Kingdom

ABSTRACT: Following our previous communications on the impact of the amount of water phase on the time required at each temperature-step[1] and the limitations of visual techniques in determining hydrate dissociation points,[2] a series of tests were conducted in this laboratory to investigate the impact of measuring techniques, mixing efficiency, heating method and heating rate on the accuracy of hydrate dissociation point measurements. The results showed that nonvisual techniques combined with stepwise heating and satisfactory mixing could save a significant amount of time while providing accurate hydrate dissociation data.

INTRODUCTION

Three procedures are in use for hydrate point measurements when a liquid water-rich phase is present.[3] They are isothermal, isobaric, and isochoric. In the first method the cell pressure is varied whereas in the last two methods it is necessary to decrease/increase the cell temperature for the formation/dissociation of gas hydrates. The hydrate point can then be determined by visual observation through a sight glass and/or nonvisual means such as monitoring pressure versus temperature plot in isochoric procedure.

It is known that, although hydrate formation conditions depend on many factors, such as rate of cooling, degree of subcooling, water history, presence of foreign material, and may not be repeatable, the hydrate dissociation point is a thermodynamic equilibrium point and is repeatable. In a typical test the hydrates are formed by increasing the cell pressure (isothermal) or decreasing the cell temperature (isobaric and isochoric), beyond the hydrate equilibrium point. The hydrates are then dissociated by reducing the cell pressure (isothermal) or increasing the cell temperature (isobaric and isochoric). Two methods, continuous or stepwise heating, are generally used in increasing the cell temperature and dissociating gas hydrates so as to measure the hydrate dissociation point.

In this communication, after a brief description of the experimental set-up, the two methods of heating are compared with each other. A natural gas was used as the test fluid, and the hydrate dissociation points were determined by using the two methods. This resulted in important conclusions and guidelines in performing hydrate experiments. The results lead to a series of recommendations that could

[a]Telecommunication. Voice: +44 131 451 3672; fax: +44 131 451 3127.
bahman.tohidi@pet.hw.ac.uk

reduce the time and the cost requirements for each hydrate test. The recommendations, which are applicable to different methods of hydrate point measurements, have been successfully applied to the isochoric technique in which a pressure versus temperature plot was used to determine hydrate dissociation points. Furthermore, the impact of efficient mixing on measuring hydrate dissociation point has been revisited. The results offer interesting guidelines in designing hydrate equilibrium cells and mixing mechanisms.

EXPERIMENTAL

The experiments were performed on a high pressure hydrate rig whose layout is shown in FIGURE 1. The hydrate cell consists of a metal body inside which an optically clear quartz glass tube is housed. The maximum effective volume of the cell is 540 cc and the maximum working pressure is 69 MPa. The effective volume is adjusted by injecting or withdrawing mercury. The temperature of the cell is controlled by circulating coolant round a jacket surrounding the cell. The cell is mounted on a pivot and mixing is obtained by rocking the cell. The system is monitored and controlled using a Macintosh computer through various temperature and

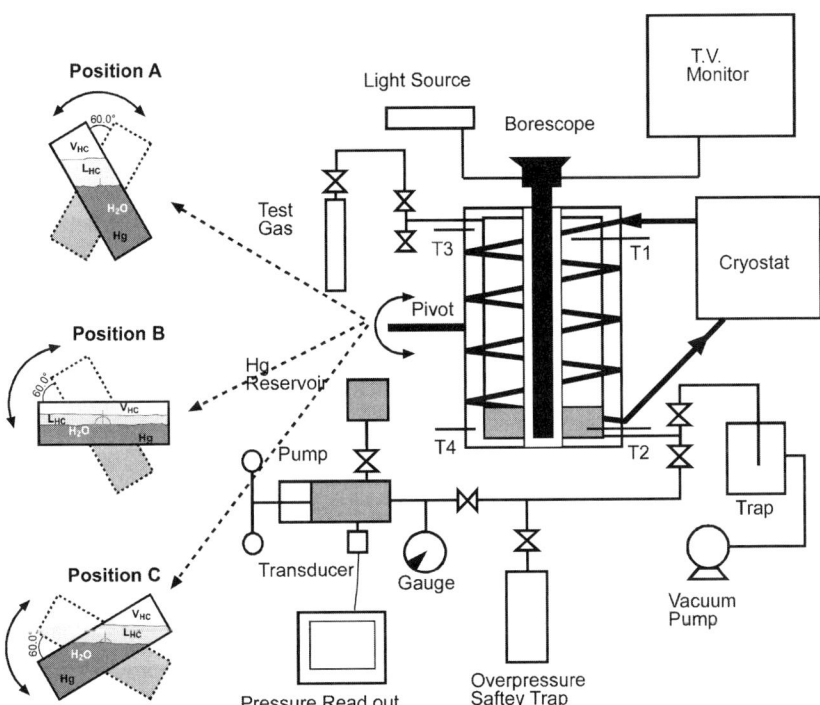

FIGURE 1. Schematic of experimental equipment.

TABLE 1. Measured composition of natural gas

Component	Mole%
CO_2	1.12
N_2	4.99
C_1	86.36
C_2	5.43
C_3	1.49
i-C_4	0.18
n-C_4	0.31
i-C_5	0.06
n-C_5	0.06
C_{6+}	<0.01

pressure probes positioned inside and around the cell. A borescope can be passed into the quartz glass tube in order to view the cell contents. A detailed description of the experimental set-up is given elsewhere.[4] The experiments were performed using a natural gas with the composition given in TABLE 1. Distilled water was used in all the tests.

The test fluids were loaded into the clean dry evacuated cell. Once the cell had been loaded the contents were mixed by rocking the cell and the effective volume was reduced by injecting mercury until the desired pressure was attained. The temperature was then reduced so as to be in the hydrate forming region. After forming hydrates, the temperature was increased by two different methods to simulate continuous or stepwise heating.

RESULTS AND DISCUSSIONS

To investigate the effect of insufficient equilibrium time on determining hydrate dissociation points, two series of tests were conducted on the natural gas. In both tests the hydrates were formed and then the cell temperature was increased stepwise in 0.1 K steps. In one test the cell was left for six hours at each step and in the second test for 10 minutes. The results of the two tests are presented in FIGURE 2. As shown in the figure, the two methods resulted in two distinctly different dissociation points. The measured dissociation point with six hour equilibrium time at each temperature step was 286.4 K and 4.51 MPa, whereas the results of 10-minute equilibrium time showed a dissociation point of 288.0 K and 4.54 MPa, a difference of 1.6 K. The data at high heating rate seem to be more scattered compared to the data at low heating rate, causing difficulties in accurate determination of the hydrate dissociation point, using graphical techniques.

The results clearly indicate that nonequilibrium conditions, as occurred in the above short equilibrium time test (expected to be more severe in continuous heating),

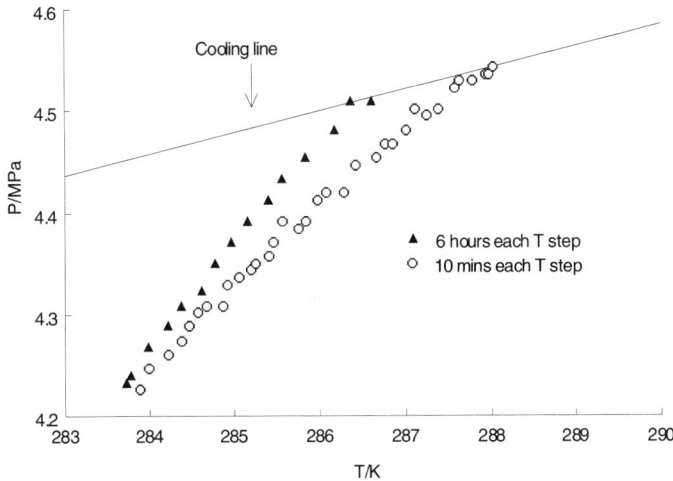

FIGURE 2. Comparison of dissociation points measured on natural gas hydrates with different heating rates.

generate unreliable hydrate dissociation data. In a number of recent publications,[5–7] measurements were made with a heating rate of 0.83 K (1.5°F) per hour, which is higher than the heating rate investigated in this work. Based on the results of this work, it is suspected that the data generated at such high heating rates are in error, as acknowledged by some investigators.[7]

One may argue that generating experimental data at a rate of six hours per temperature step requires a considerable amount of time and will be very time consuming and expensive. It is the objective of this work to demonstrate that it is possible to minimize the time and the cost associated with generating experimental data, while ensuring high accuracy and reliability of the generated data. In the following paragraphs the results of the above study have been applied to the isochoric method using non-visual technique (i.e., pressure versus temperature plot) for hydrate point determination.

This laboratory has been conducting hydrate dissociation point measurement tests for the last 12 years. Our results show that for non-visual techniques, where the hydrate dissociation point is determined by the intersection of cooling and heating curve, it is possible to reduce the time requirement significantly by increasing the size of the temperature steps. The results show that stepwise heating with large temperature steps, combined with adequate equilibrium time at each step and optimized amount of water,[1] could reduce the time requirement for each test significantly. Hence, improving the economics of the tests and the reliability of the results. It is possible to minimize the time required for dissociation pressure determination by employing the following steps:

1. Unless the point at which hydrates form is of interest, obtaining the cooling curve should take only a short amount of time. In fact only few points should be sufficient and in a well mixed cell it will take about one hour to measure three

points. The three points could be measured by step cooling, in large temperature steps (i.e., 3–5 K). Our experience indicates that, for most systems, the equilibrium (where there is no change in pressure with time) is achieved in less than 30 minutes.

2. Only three or four equilibrium points are required on the heating curve. For these points to be accurate, sufficient time should be given at each temperature step to ensure equilibrium. It is also essential that the cell contents are well mixed and different phases are in equilibrium, for example, avoiding corners where a significant mass of hydrates could be stuck and thus not be in contact with other phases. The time taken to reach equilibrium varies with the proximity of the temperature to the dissociation temperature. This point is obvious from looking at FIGURE 2 where the lower temperature points for the two heating methods are similar but differ significantly close to the dissociation point. This entails that, at the lower temperature steps, less time is required and thus the test can be speeded up. An example of dissociation point determination from six equilibrium temperature/pressure points is shown in FIGURE 3 and this agrees well with the point as determined from a large number of temperature steps as shown in FIGURE 2.

The results of the above study could be partly applied to isothermal and/or isobaric methods. The important message is that, adequate time is required to achieve equilibrium in the presence of gas hydrates, which could not be achieved by methods based on continuous temperature and/or pressure changes. Larger step changes in temperature and/or pressure combined with adequate equilibrium time at each step will reduce the time requirement and improve the reliability of the results. Obviously, when using visual techniques the temperature or pressure steps should be adjusted (i.e., reduced nearer to the dissociation point) to ensure the accuracy of the results.

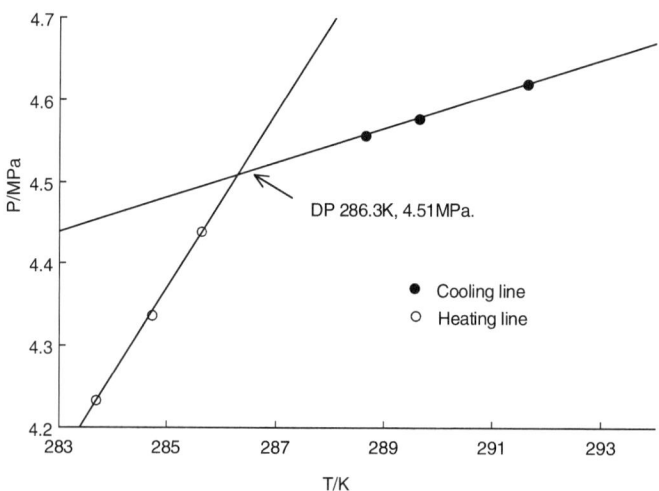

FIGURE 3. Dissociation point determination for the natural gas hydrates with minimum amount of data.

Another important factor in generating reliable hydrate data is effective and efficient mixing. If the cell contents are not well mixed, as might be the case where one stirring bar or propeller is situated in the base of an equilibrium cell then, as mentioned above, clumps of hydrates may stick to the wall, in a corner for example. This will reduce the mass transfer rate and significantly increase the time required to achieve equilibrium. If the hydrates become isolated from the other phases (i.e., inefficient mixing or by a mercury layer) then they will be more stable.

A simulation with our in-house thermodynamic model, HWHYD,[8] showed that hydrates formed under certain conditions require a higher temperature for dissociation when isolated from the hydrate forming system. For example, hydrates formed from the above natural gas at 285 K and 4.01 MPa are predicted to fully dissociate at 291.1 K at the same pressure, an increase of 6.1 K. This fact can be used to advantage in storage and/or transportation of natural gas in form of gas hydrates, as has been suggested,[9] where the separated hydrates will be more stable and can be stored or transported at significantly higher temperatures/lower pressures than conditions under which they were formed.

To demonstrate the stability of separated hydrates in the experimental rig used in this work the following test was carried out. Hydrates were formed with normal mixing then the mixing was changed so that in the lower part of the cell the hydrates adhering to the cell walls would only be in contact with the mercury and would not at any stage come into contact with the vapor or water phases (positions A or B in FIG. 1). The results of this test are given in FIGURE 4 and, as can be seen, the heating does not rejoin the cooling line until about 293 K.

The above results are in acceptable agreement with the predictions of the thermodynamic model, demonstrating the importance of efficient mixing, which might be difficult to achieve with some mixing techniques and/or equilibrium cell designs.

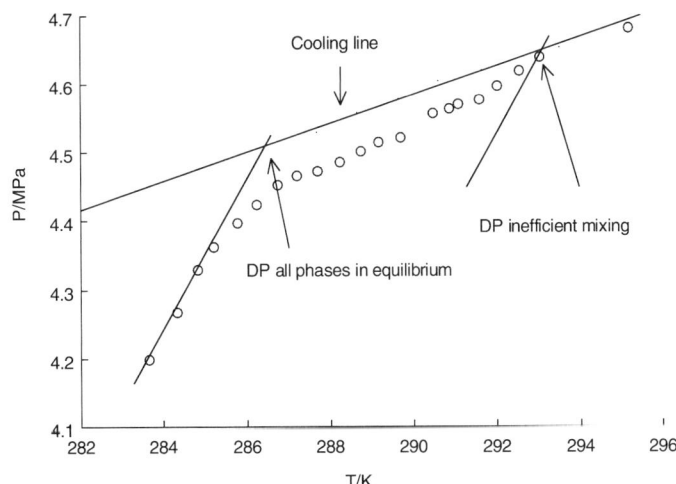

FIGURE 4. Effect of isolating some hydrate from other phases on measured hydrate dissociation point.

Therefore, to generate reliable hydrate dissociation data, it is important to avoid isolated pockets of gas hydrates, particularly when using nonreactive fluids like mercury.

CONCLUSIONS

As demonstrated in the above experiments and previous communications,[1,2] the time required to achieve equilibrium in the presence of gas hydrates could be few hours or few days, depending on the system under investigation, amount of water, and the efficiency of mixing.

The results show that step heating combined with adequate time at each temperature step results in generating reliable experimental data, as compared with continuous heating. The error associated with continuous heating is a function of the heating rate. The error could be reduced by reducing the rate of heating, but this will increase the time required for each tests significantly.

The findings were applied to isochoric methods using non-visual techniques. A procedure was suggested that could determine the hydrate dissociation point by measuring only few points on cooling and heating curves. The results showed that the time required to perform a test can be reduced significantly, by using larger temperature steps, while ensuring equilibrium at each temperature step.

The results can be partly applied to other hydrate dissociation point determination methods—isothermal and isobaric. However, when using visual techniques, the step change in pressure or temperature should be reduced nearer to the hydrate dissociation point, to ensure accurate measurement.

The importance of ensuring efficient mixing in the design and operation of hydrate rigs was highlighted. Experimental measurement and numerical simulation show that isolated hydrates, due to inadequate mixing or bad design, are more stable than those in contact with equilibrated phases, which could lead to unreliable data.

REFERENCES

1. TOHIDI, B., R.W. BURGASS, A. DANESH & A.C. TODD. 1994. Experimental study on the causes of disagreements in methane hydrate dissociation data. Ann. N.Y. Acad. Sci. **715:** 532–534.
2. TOHIDI, B., A. DANESH, A.C. TODD & R.W. BURGASS. 1997. Hydrate free zone for synthetic and real reservoir fluids in the presence of saline water. Chem. Eng. Sci. **59**(19): 3257–3263.
3. SLOAN, E.D. 1998. Clathrate Hydrates of Natural Gases, 2nd edit. Marcel Dekker Inc., New York.
4. DANESH, A., B. TOHIDI, R.W. BURGASS & A.C. TODD. 1994. Hydrate equilibrium data of methyl cyclopentane with methane or nitrogen. Trans. IChemE. **72**(A): 197–200.
5. KOTKOSKIE, T.S., B. AL-UBAIDI, T.R. WILDEMAN & E.D. SLOAN. 1992. Inhibition of gas hydrates in water-based drilling muds. SPE Drilling Engineering, June: 130–136.
6. EBELTOFT, H. & M. YOUSIF. 1997. Hydrate control during deep water drilling: overview and new drilling fluids formulations. SPE 38567. Proceedings of the 1997 SPE Annual Conference and Exhibition, October 5–8: Δ35–41. San Antonio, Texas.
7. GRIGG, R.B. & G.L. LYNES. 1992. Oil-based drilling mud as a gas-hydrates inhibitor. SPE Drilling Engineering. March: 32–38.

8. TOHIDI, B., A. DANESH & A.C. TODD. 1997. Predicting pipeline hydrate formation. The Chemical Engineer **642:** 32–37.
9. GUDMUNDSSON, J., F. HVEDING & A. BØRREHAUG. 1995. Transport of natural gas as frozen hydrate. Proceedings of the 5th International Offshore and Polar Engineering Conference, June 11–16, The Hague, the Netherlands. 282–283.

Equilibrium Conditions of Methane and Ethane Hydrates in Aqueous Electrolyte Solutions

TATSUO MAEKAWA[a] AND NOBORU IMAI

Geological Survey of Japan, 1-1-3, Higashi, Tsukuba, Ibaraki, 305-8567, Japan

ABSTRACT: In this work, we determined the equilibrium conditions for gas hydrates of methane and ethane in aqueous solutions containing various electrolytes, such as NaCl, NaBr, NaI, and Na_2SO_4. We have developed experimental apparatus in which a light beam through solutions was used to detect the formation and dissociation of gas hydrate. To model the experimental data of this work, a statistical thermodynamic model based on the classical adsorption model proposed by van der Waals and Platteeuw[1] is used. The water activity of the aqueous solutions containing electrolytes, which affects the gas hydrate stability, was evaluated by the model of Englezos and Bishnoi.[2]

INTRODUCTION

It is well known that addition of electrolytes to water shifts the equilibrium conditions of gas hydrate. Many experiments to determine the equilibrium conditions of gas hydrate in aqueous electrolyte solution have been reported.[3,4] Also we have performed many experiments on methane hydrate synthesis in saline water to determine the formation and dissociation conditions of methane hydrate, and to evaluate natural gas hydrate stability in deep-sea sediments.[5,6] Most of experiments using various chloride solution that were previously performed were used to investigate the effect of electrolyte on gas hydrate stability, but a few experiments considered the inhibiting effect of bromide, iodide, and sulfate.

The object of the present work is to determine, experimentally, the conditions for methane or ethane hydrate formation in aqueous electrolyte solutions, and to compare the data with those predicted by the statistical thermodynamic model.

EXPERIMENTAL APPARATUS

We improved the experimental apparatus used in our previous works.[5,6] The experimental apparatus used in the present work consists of a high-pressure hydrate cell of stainless steel in which gas hydrate is synthesized (see FIGURE 1). The hydrate cell has a volume of about 500 ml (see FIGURE 2). Glycol-water coolant is continuously circulated around the cell, whose temperature is controlled by a heater and a

---

[a]Telecommunication. Voice: +81-298-54-3720; fax: +81-298-54-3533.
maekawa@gsj.go.jp

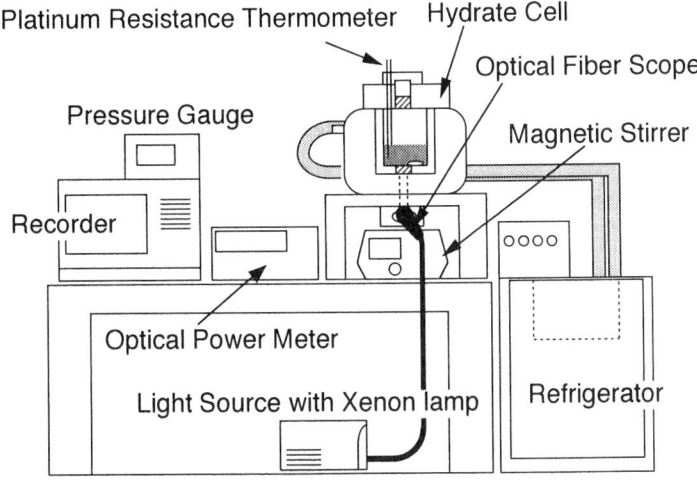

FIGURE 1. Schematic diagram of the experimental apparatus.

refrigerator. The temperature of the solution in the cell is measured with platinum resistance thermometer. The gas pressure is measured with a pressure gauge using semiconductor transducer. The solution is agitated by a magnetic stirrer. The cell has

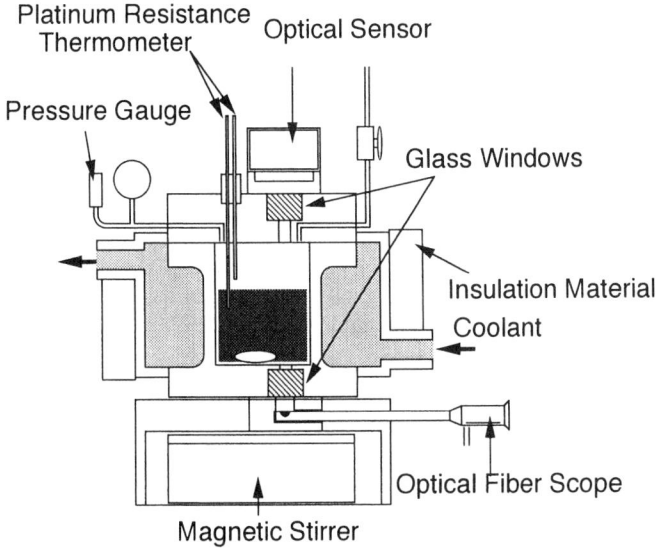

FIGURE 2. Schematic diagram of the hydrate cell.

two glass windows on the upper and bottom sides of the cell for visual observation and optical detection of gas hydrate formation and dissociation.

The formation and dissociation detector for gas hydrate consists of a xenon lamp, an optical sensor and an optical power meter. The light beam from the lamp is introduced into the solution through the window from the bottom side of the hydrate cell. An optical sensor on the upper side measures the intensity of the light penetrating through the solution. When gas hydrate forms on the surface of the solution exposed to the gas phase, the light is scattered by the gas hydrate, and the light intensity detected by the optical sensor sharply decreases. On the other hand, when the gas hydrate dissociates, light intensity also sharply increases for the same reason. Therefore, the formation and dissociation of gas hydrate can be detected by measuring the change of light intensity. The estimated accuracy of temperature and pressure measurements in this experiments is ± 0.1 K and ± 0.1 MPa, respectively, which is mainly caused by reading the records manually.

One advantage of the experimental apparatus is that optical detection unit using a light beam penetrating solution in the cell can detect gas hydrate formation and dissociation more clearly than visual observation. The light intensity changes very sharply at the formation and dissociation of gas hydrate; thus, we can precisely determine pressure and temperature conditions for the formation and dissociation of gas hydrate. The other advantage is that large amount of gas hydrate formation is not needed and there are no changes of the composition of the solution during gas hydrate formation. Conventional isochoric techniques using the hysteresis diagram of pressure and temperature is often used to determine the equilibrium conditions for gas hydrate. This technique needs to form large amount of gas hydrate in order to obtain a large pressure difference. With the formation of large amount of gas hydrate it is possible to change the concentrations of electrolytes in solution and composition of gas mixture. Compared to the conventional isochoric technique, only a small amount of gas hydrate is formed on the surface of solution (like a film) in our experiments, resulting in no change in concentration of electrolytes in the solution, which improves determination of gas hydrate conditions.

EXPERIMENTAL PROCEDURES

In this experiment, gas hydrates form from pressured gas and aqueous electrolyte solution in the hydrate cell. The pure gas samples used in this work were research grade methane and ethane. Appropriate quantities of each electrolyte were weighed and added to a weighed quantity of distilled water. The mixture was stirred to dissolve the electrolyte at room temperature. Initially, 200 ml of the aqueous electrolyte solution was added to the hydrate cell. Then, pressured methane or ethane gas was supplied to the cell to reach the starting pressure.

The formation and dissociation of methane hydrate was regulated by changing the temperature of the solution after a constant volume of gas was sealed in the hydrate cell. FIGURE 3 shows the typical changes in temperature, pressure, and light intensity in determination of the formation and dissociation of gas hydrate in electrolyte solutions. The temperature in the hydrate cell was lowered to form gas

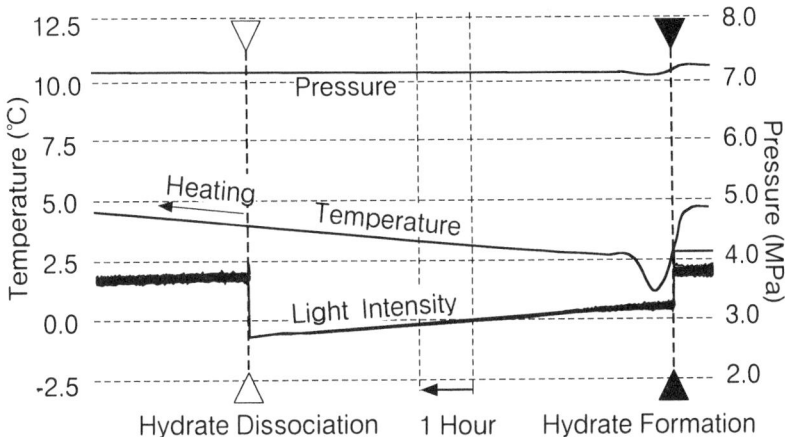

FIGURE 3. Typical changes of the temperature, pressure and light intensity in determination of the formation and dissociation of gas hydrate in electrolyte solutions.

hydrate, with a slight pressure decrease by cooling. When the gas hydrate formed, the light intensity detected by optical sensor sharply decreased. The formation of gas hydrate was also confirmed by visual observation using an optical fiber scope. The temperature was continuously raised by 0.25 K/ hr after the formation of gas hydrate. During the heating, the light intensity changed again and recovered to the original light intensity when gas hydrate dissociated. The condition for gas hydrate dissociation was determined by measuring the pressure and temperature at which the change of light intensity occurred during heating.

THERMODYNAMIC MODEL

The basic statistical thermodynamic equations for gas hydrate was presented by van der Waals and Platteeuw[1] by use of a clathrate model, which is similar to that of Langmuir for gas adsorption. Many thermodynamic studies on gas hydrate based on the van der Waals and Platteeuw model have been performed.[2,7–9] We also used this model to investigate of equilibrium conditions for gas hydrates in electrolytes solutions.

Under equilibrium conditions, the chemical potential of water in hydrate phase is equal to that in liquid phase. To calculate the chemical potential of water in the hydrate phase, the parameters of Kihara potentials for gas–water interactions presented in Avlonitis[10] were used. Gas phase fugacity was calculated using the Valderrama-Patel-Teja equation of state.[11,12] To calculate the chemical potential of water in liquid phase, the molar enthalpy difference and the heat capacity difference of water between empty hydrate lattice and liquid presented by Holder *et al.*[8] were used. The molar volume difference for water was calculated from the equation of Avlonitis.[10]

The presence of electrolytes in the liquid phase influences the activity of the water. Englezos and Bishnoi[2] calculated the water activity a_w by using Pitzer's activity coefficient model.

$$\ln a_w = \frac{-18\nu m}{1000}(1 + |z_-z_+|\theta_1 + m\theta_2 + m^2\beta_2)$$

$$\theta_1 = -\frac{A_\phi I^{0.5}}{1 + 1.2 I^{0.5}}$$

$$\theta_2 = \beta_0 + \beta_1 \exp(-2I^{0.5}),$$

where m is molality of the solution, ν is the stoichiometric number of moles of ions in one mole of electrolyte, I is ionic strength, A_ϕ is the Debye-Huckel coefficient, β_0, β_1, and β_2 are the parameters of Pitzer and Mayorga.[13]

RESULTS

In order to check the experimental apparatus and the experimental procedure, the dissociation conditions of methane or ethane gas hydrate in pure water were measured and the results were compared with previous data (FIG. 4). The experimental data of this work for methane and ethane hydrate in pure water are in good agreement with the previous data.[14–16]

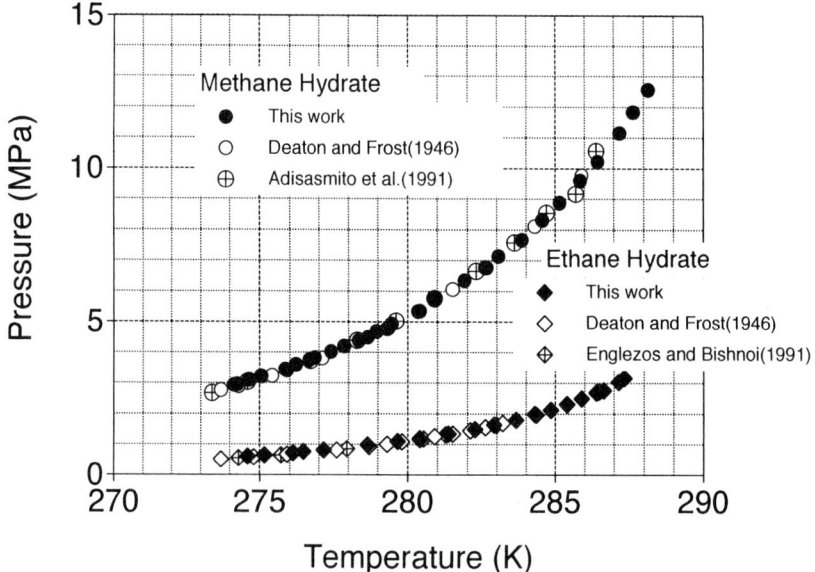

FIGURE 4. Comparison of experimental data obtained in the present work and data reported previously for methane and ethane hydrate in distilled water.

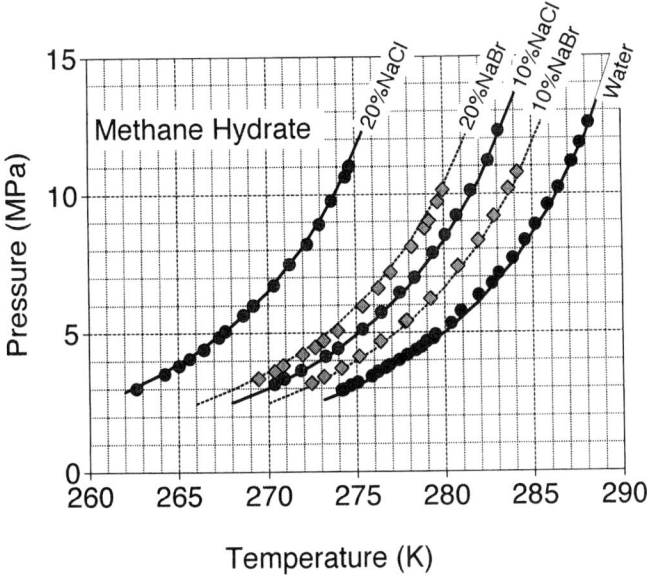

FIGURE 5. Experimental and predicted data for methane hydrate formation in the aqueous solutions containing NaCl and NaBr. Lines are predicted by using the thermodynamic model.

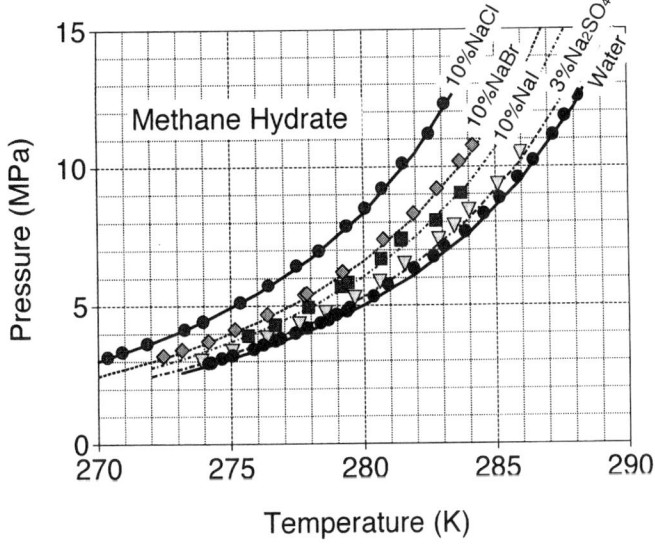

FIGURE 6. Experimental and predicted data for methane hydrate formation in the aqueous solutions containing NaI and Na_2SO_4.

In the present work, we determined the equilibrium conditions for methane hydrate in the aqueous solutions containing NaCl, NaBr, NaI, and Na_2SO_4 (see FIGURES 5 and 6), and those of ethane hydrate in the solutions containing NaCl, NaBr, and Na_2SO_4 (see FIGURE 7). Our results show that the equilibrium pressure of gas hydrate increases and the equilibrium temperature decreases with increasing concentration of each electrolyte in the solution.

For methane hydrate in the solution containing NaCl, de Roo et al.[3] obtained an empirical equation to explain their experimental results on methane hydrate dissociation in the NaCl solutions. The experimental data obtained in the present work agrees well with the data predicted by their empirical equation. We obtained the new experimental data on the equilibrium conditions of methane hydrates in the aqueous solutions containing 10 wt% and 20 wt% NaBr, 10 wt% NaI, and 3 wt% Na_2SO_4.

We performed the experiments of ethane hydrate formation in the solutions of 3.1 wt%, 10 wt%, and 20 wt% NaCl, and 10 wt% and 20 wt% NaBr. The experimental data obtained in the present work are in very good agreement with the data of Englezos and Bishnoi[16] for 20 wt% NaCl solution, and the data of Tohidi et al.,[17] for 10 wt% and 20 wt% NaCl solutions.

The predicted equilibrium calculated by the thermodynamic model of the present work agrees well with the experimental data of methane and ethane hydrate in the aqueous electrolyte solutions. In the prediction of gas hydrate equilibrium conditions, the calculations of water activities using the Pitzer's activity coefficient model were adapted for the solutions containing the electrolytes such as NaBr, NaI, and Na_2SO_4.

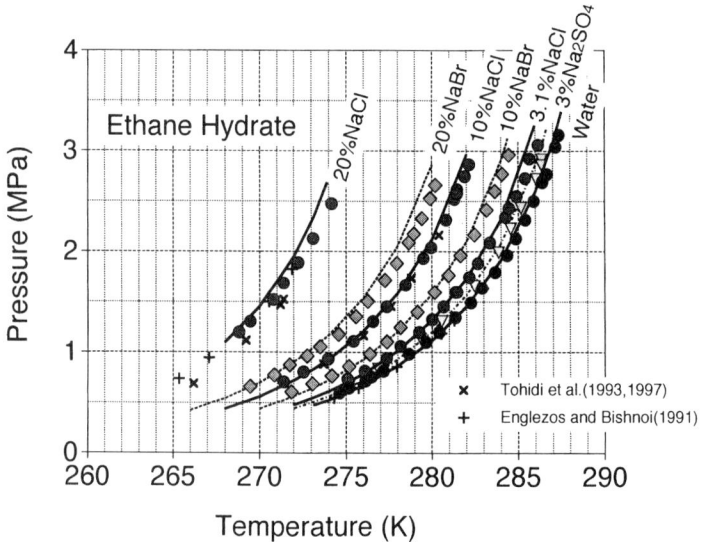

FIGURE 7. Experimental and predicted data for ethane hydrate formation in the aqueous solutions containing NaCl, NaBr, and Na_2SO_4.

CONCLUSIONS

We experimentally determined the dissociation conditions of methane hydrate or ethane hydrate in aqueous solutions containing various electrolytes as NaCl, NaBr, NaI, or Na_2SO_4. The experimental data were compared with the predictions by using the clathrate model proposed by van der Waals and Platteeuw and Pitzer's activity coefficient model. It was found that the predictions agree well with the experimental data and the inhibiting effect of various anions on gas hydrate formation were also calculated by using the method of Englezos and Bishnoi.[2]

REFERENCES

1. VAN DER WAALS, J.H. & J.C. PLATTEEUW. 1959. Clathrate solutions. Adv. Chem. Phys. **2:** 1–57.
2. ENGLEZOS, P. & P.R. BISHNOI. 1988. Prediction of gas hydrate formation conditions in aqueous electrolyte solutions. AIChE J. **34:** 1718–1721.
3. DE ROO, J.L. et al. 1983. Occurrence of methane hydrate in saturated and unsaturated solutions of sodium chloride and water in dependence of temperature and pressure. AIChE J. **29:** 651–657.
4. DHOLABHAI, P.D. et al. 1991. Equilibrium conditions for methane hydrate formation in aqueous mixed electrolyte solutions. Can. J. Chem. Eng. **69:** 800–805.
5. MAEKAWA, T. et al. 1995. Pressure and temperature conditions for methane hydrate dissociation in sodium chloride solutions. Geochem. J. **29:** 325–329.
6. MAEKAWA, T. & N. IMAI. 1996. Stability conditions of methane hydrate in natural seawater. J. Geol. Soc. Japan **102:** 945–950.
7. PARRISH, W.R. & J.M. PRAUSNITZ. 1972. Dissociation pressures of gas hydrates formed by gas mixtures. Ind. Eng. Chem. Proc. Des. Dev. **11:** 26–34.
8. HOLDER, G.D. et al. 1980. Thermodynamic and molecular properties of gas hydrates from mixtures containing methane, argon, and krypton. Ind. Eng. Chem. Fundam. **19:** 282–286.
9. ENGLEZOS, P. 1992. Computation of the incipient equilibrium carbon dioxide hydrate formation conditions in aqueous electrolyte solutions. Ind. Eng. Chem. Res. **31:** 2232–2237.
10. AVLONITIS, D. 1994. The determination of Kihara potential parameters from gas hydrate data. Chem. Eng. Sci. **49:** 1161–1173.
11. VALDERRAMA, J.O. 1990. A generalized Patel-Teja equation of state for polar and nonpolar fluids and their mixtures. J. Chem. Eng. Japan **23:** 87–91.
12. AVLONITIS, D. et al. 1994. Prediction of VL and VLL equilibria of mixtures containing petroleum reservoir fluids and methanol with a cubic EoS. Fluid Phase Equilibria **94:** 181–216.
13. PITZER, K.S. & G. MAYORGA. 1973. Thermodynamics of electrolytes II. Activity and osmotic coefficients for strong electrolytes with one or both ions univalent. J. Phys. Chem. **77:** 2300–2308.
14. DEATON, W.M. & E.M. FROST. 1946. Gas hydrates and their relation to the operation of natural-gas pipe lines. U.S. Bureau of Mines Monograph 8.
15. ADISASMITO, S. et al. 1991. Hydrates of carbon dioxide and methane mixtures. J. Chem. Eng. Data **36:** 68–71.
16. ENGLEZOS, P. & P.R. BISHNOI. 1991. Experimental study on the equilibrium ethane hydrate formation conditions in aqueous electrolyte solutions. Ind. Eng. Chem. Res. **30:** 1655–1659.
17. TOHIDI, B. et al. 1997. Hydrate-free zone for synthetic and real reservoir fluids in the presence of saline water. Chem. Eng. Sci. **52:** 3257–3263.

Crystal Growth, Structure Characterization, and Schemes for Economic Transport

An Integrated Approach to the Study of Natural Gas Hydrates

D. MAHAJAN,[a,b] T.F. KOETZLE,[c] W.T. KLOOSTER,[c]
L. BRAMMER,[d] R.K. McMULLAN,[c] AND A.N. GOLAND[b]

[b]*Department of Applied Science, Brookhaven National Laboratory,
Upton, New York 11973, USA*

[c]*Department of Chemistry, Brookhaven National Laboratory,
Upton, New York 11973, USA*

[d]*Department of Chemistry, University of Missouri-St. Louis,
St. Louis, Missouri 63121-4499, USA*

ABSTRACT: In this paper, two themes are specifically targeted for developing a cost-effective option to transport methane hydrates from distant locations. Under the first theme, data are presented on crystal growth techniques, sample preparation and neutron diffraction studies of $3.5Xe \cdot 8CCl_4 \cdot 136D_2O$, $xCH_4 \cdot 8CCl_4 \cdot 136D_2O$, $xH_2S \cdot 8CS_2 \cdot 136\ H_2O$, and $20Br_2 \cdot 172D_2O$. Under the second theme, the GTL option is selected wherein methanol is the product of choice for transport. For GTL, the processing of aqueous CH_4 by steam reforming is the preferred route to synthesis gas. Subsequent conversion of synthesis gas into methanol will require the formulation of advanced catalysts.

INTRODUCTION

Clathrate hydrates owe their existence to the ability of H_2O molecules to undergo hydrogen bonding and form framework structures with cavities that can serve as hosts to small gas (guest) molecules. Thus, it is not a coincidence that gas hydrates are primarily located in ocean floor sediments and in the Arctic permafrost, and may contain more CH_4 than all other known reserves combined.[1] At present, there is global interest in mining this unconventional CH_4 source to secure supplies of natural gas well into the next millennium, which includes recent experimental well drilling activities[2] at the Nankai trough in Japan (1997) and the Mackenzie Delta in Canada (1998). A recent report[3] released by the U.S. Department of Energy (DOE) emphasizes knowledge gaps at the fundamental level that must be addressed to develop an

[a]Address for correspondence: Dr. Devinder Mahajan, Building 815, Department of Applied Science, Brookhaven National Laboratory, Upton, New York 11973-5000, USA. Voice: 631-344-4985; fax: 631-344-7905.
dmahajan@bnl.gov

environmentally safe production method for the United States to develop gas hydrates and reduce reliance on imported oil.

Remote natural gas represents a significant fraction of conventional gas reserves, but economic transport remains a challenge.[4] CH_4 produced from gas hydrate deposits is considered "stranded", and economic transport of this gas must be a part of any gas hydrate development scheme. This paper outlines a comprehensive strategy to develop an economic transport method by addressing two research themes that are based on the work at Brookhaven National Laboratory (BNL). The first theme relates to neutron diffraction studies that provide the ability to accurately characterize the hydrogen-bonded clathrate water-cage structures, including the hydrogen (deuterium) atoms and their attendant disorder. This is crucial to understanding the factors governing clathrate thermodynamic stability. We have, therefore, carried out extensive studies on the growth of large single crystals by recrystallizing powders at pressures near the high end of the clathrate stability range (about 30 atm) while accurately controlling the temperature. Our aim is to obtain results for the natural gas hydrates, including methane and ethane hydrates, that will be of quality comparable to those obtained previously in our neutron diffraction study of $3.5Xe \cdot 8CCl_4 \cdot 136D_2O$.[5] Here, we briefly summarize that earlier work, to illustrate the diffraction methodology employed, and describe our crystal growth experiments and diffraction results for the systems $xCH_4 \cdot 8CCl_4 \cdot 136D_2O$, $xH_2S \cdot 8CS_2 \cdot 136H_2O$, and $20Br_2 \cdot 172D_2O$. The structural data are crucial to understanding stability and are a prerequisite for addressing the second research theme that relates to challenges in developing economic methods to transport mined gas hydrates.

EXPERIMENTAL

Crystal Growth Techniques

We have explored various methods and procedures for obtaining gas hydrate crystals of sizes large enough for single crystal neutron diffraction. In addition, techniques have been developed to transfer the crystalline samples to a neutron diffractometer at temperatures below 225 K. The equipment used in crystal-growth studies consists of pressure cells, a temperature controlled bath, and a control unit designed to impose fluctuations on the temperature. A vacuum system was used to facilitate gas transfers and volume measurements. Crystallization of materials of interest appears to involve rapid interaction between water and hydrocarbon vapor to give powder samples under laboratory conditions. Recrystallizing the powders to obtain large single crystals also occurs through the vapor phase, but is a slower process. Since decomposition pressures of clathrates such as the methane and ethane hydrates are high (e.g., approximately 30 atm for methane hydrate at 273 K), our preparations were carried out in thick-walled cells designed to permit visual monitoring of the crystal growth process. Temperatures of the cells were kept in the higher ranges of clathrate stability, since high water vapor pressures enhance crystal growth. Under static conditions of temperature and pressure, crystal growth was still unacceptably slow. To accelerate growth, the temperature was periodically varied between limits bracketing preset thermostatted values. This technique was clearly effective for ethylene hydrates and other nonhydrocarbon hydrates.

Preparation of Neutron Diffraction Samples

The clathrate compound $3.5Xe \cdot 8CCl_4 \cdot 136D_2O$ was prepared[5] from a degassed mixture of CCl_4 and D_2O in the mole ratio 1:17 under Xe gas at 1 atm. The crystalline powder formed on cooling slowly sublimed and recrystallized at 278 K yielding large plate-like crystals. Crystals of $6C_2H_4 \cdot 46D_2O$ were deposited at 277 K from gaseous C_2H_4/D_2O under 5 atm of ethylene. Similarly, crystals of $xCH_4 \cdot 8CCl_4 \cdot 136D_2O$ were produced at about 277 K from the vapors above CH_4/D_2O under CH_4. For deuterated samples, deuteriomethane can be prepared from the reaction of Al_4C_3 with D_2O:
$Al_4C_3 + 12D_2O \rightarrow 3CD_4 + 4Al(OD)_3$.

Neutron and X-ray Scattering Studies

For $3.5Xe \cdot 8CCl_4 \cdot 136D_2O$, a sample measuring $2.0 \times 2.4 \times 1.7$ mm was affixed to an aluminum pin at 213 K and sealed under helium gas inside an aluminum can.

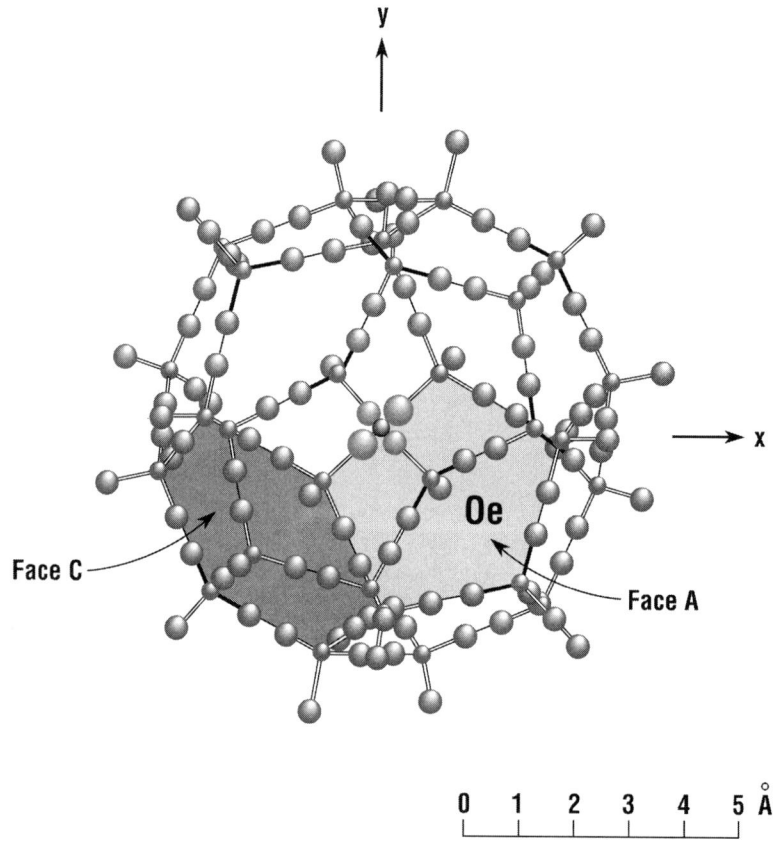

FIGURE 1. The hexakaidecahedron of $(D_2O)_{28}$ in $3.5Xe \cdot 8CCl_4 \cdot 136D_2O$ with D atoms shown in disordered positions.

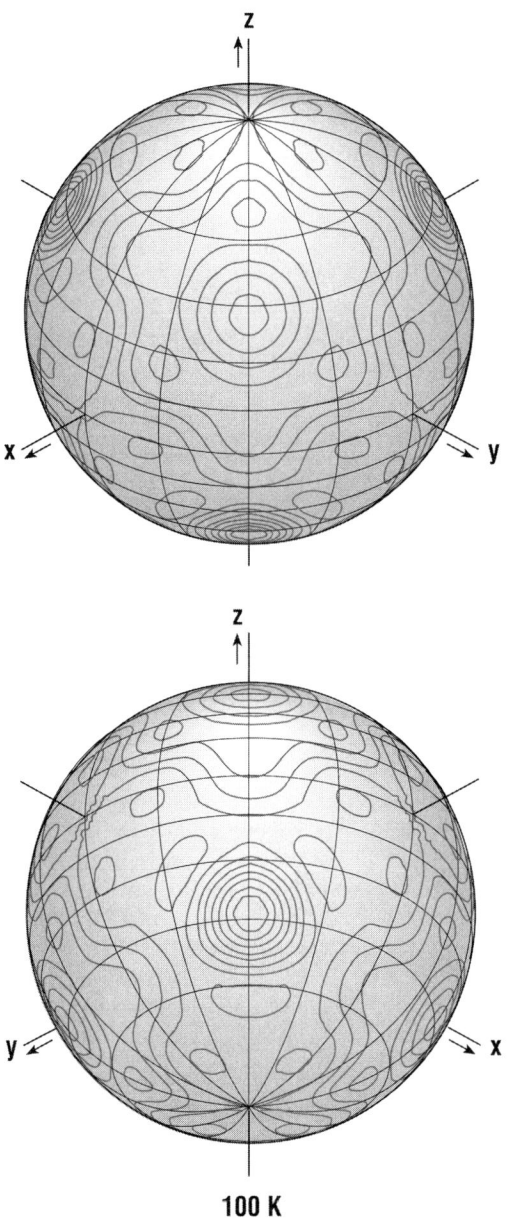

FIGURE 2. Nuclear scattering density at 100 K of the four Cl atoms in $3.5\text{Xe} \cdot 8\text{CCl}_4 \cdot 136\text{D}_2\text{O}$ on the surface of a sphere with radius equal to the C–Cl distance of 1.762 Å. Views are shown in both directions along the cubic body diagonal [111].

Neutron diffraction data were measured on a four-circle diffractometer at the Brookhaven high flux beam reactor (HFBR), using a neutron beam of wavelength 1.047 Å to a resolution limit of $\sin\theta/\lambda = 0.78$ Å^{-1} at two temperatures (13 K and 100 K). The temperature of the sample crystal was maintained at ±0.5 K using a closed-cycle DISPLEX® helium refrigerator. This resulted in 524 and 522 independent observations at 13 K and 100 K, respectively (space group $Fd3m$, $a = 17.192(1)$ Å (13K), $a = 17.240(1)$ Å (100 K)). The type II cubic clathrate structure (FIGURES 1 and 2) was refined against the neutron diffraction data by least squares methods, with expansion coefficients of spherical harmonics as parameters describing rotational disorder of the CCl_4 molecules based on the formalism of Press and Hüller.[6] The $wR(F^2)$ indices are 0.040, 0.039 (13, 100 K) for 64, 61 refined parameters.

Single crystal X-ray and neutron diffraction data have been obtained between 13 K and 100 K on two other cubic II hydrates, $xCH_4 \cdot 8CCl_4 \cdot 136D_2O$ and $xH_2S \cdot 8CS_2 \cdot 136H_2O$. Neutron diffraction data also have been obtained for the unusual tetragonal bromine hydrate $20Br_2 \cdot 172D_2O$. The inelastic neutron scattering (INS) data have been measured[7] near 4 K on a polycrystalline sample of $6C_2H_4 \cdot 46D_2O$, which crystallizes in the cubic I form. The compositions given are for the unit cell contents and have been determined from analyses of the diffraction data. For the cubic I ethylene hydrate, the composition has been assumed from the most probable number of sites available to the guest ethylene molecule in the water framework structure.

DISCUSSION

Structural Studies

The majority of clathrate hydrates crystallize into one of four structures. Two distinct cubic structures are formed[8] when either linkage of the cavities through their vertices create hydrate structure I or the dodecahedra are linked through their faces to create hydrate structure II. The third structure known as H hydrate[9,10] is hexagonal as indicated by NMR and powder diffraction evidence.[9] More recently single crystal diffraction data have also been obtained.[10] A fourth structure type that is tetragonal is observed for bromine hydrate, $20Br_2 \cdot 172D_2O$. This was first proposed by Allen and Jeffrey.[11] Subsequently, we were able unambiguously to confirm this supposition using single crystal low temperature neutron diffraction.[12] More recently, single crystal X-ray diffraction studies by Ripmeester et al. have established that this unusual structure type persists over a range of bromine hydrate compositions.[13] The type II structure of $3.5Xe \cdot 8CCl_4 \cdot 136D_2O$ (FIG. 1) shows that the D_2O molecules are disordered at both temperatures in six hydrogen bonded orientations of equal statistical weights as in ice Ih. There are six non-equivalent hydrogen-bonded interactions with O···D distances (at 13 K) between 1.738(3) and 1.802(1) Å and O··D–O angles between 174.8(1) and 180°. The covalent O–D lengths (uncorrected for thermal motion) are in the range 0.986(1)–1.001(2) Å and are decreased significantly from 13 to 100 K. The Xe atoms occupy statistically 22% of the dodecahedra and vibrate about the cage centers with rms displacements of 0.137 Å at 13 K and 0.179 Å at 100 K. The CCl_4 molecules exhibit large-amplitude librating motion about the C atom located at the center of the hexakaidecahedron. The nuclear scattering density mapping (FIG. 2) shows that there are seven preferred molecular orientations with the C–Cl bonds directed toward the O vertices of the $(D_2O)_{28}$ polyhedron. The C–Cl bond length from the radius parameter of the spherical distribution is 1.765(2) Å at 13 K and 1.762(3) Å at 100 K, compared with the gas-phase value of 1.766(3) Å.[14] Taken together, the details available from this neutron diffraction study provide the most comprehensive structural description of the type II clathrates that is presently available.

The INS study of $6C_2H_4 \cdot 46D_2O$[7] revealed a broad local mode at 7–8 meV, probably corresponding to a combined rotation–vibration of the encapsulated ethylene in this cubic I phase. In analysis of the cubic II $xCH_4 \cdot 8CCl_4 \cdot 136D_2O$ diffraction data, precise information on the CCl_4 guest distribution and ordering within the water host structure was obtained by expansion of the nuclear (or electron) scattering density in the symmetry restricted spherical harmonics in a manner similar to that done previously for $3.5Xe \cdot 8CCl_4 \cdot 136D_2O$ (vide supra). The secondary CH_4 guests were shown to occupy statistically dodecahedral sites within the water framework available to them at levels of 20% or less. There occurs no mixing of the primary and secondary guest molecules among their respective sites. Although translational motions of the guests are relatively small, the CCl_4 molecule exhibits large amplitude oscillations within the nearly spherical environments of water molecules (compare with FIG. 2). In the water framework, each proton (deuteron) occupies statistically two sites with equal weights near lines connecting pairs of oxygen atoms, as in common hexagonal ice (compare with FIG. 1). In the tetragonal bromine hydrate, $20Br_2 \cdot 172D_2O$, the bromine molecules are distributed among sites having three distinct water environments, two tetrakaidecahedra (T_A and T_B) and one pentakaidecahedron (P); dodecahedral sites are too small for occupation by Br_2. Based on the X-ray diffraction studies of Ripmeester et al., at 173 K, the bromine molecules show substantial disorder in all sites, occupying 14, 15, and 12 orientations in the T_A, T_B, and P cavities, respectively. Different occupancies then account for the range of compositions shown to exhibit this structure type ($Br_2 \cdot 8.62H_2O$ to $Br_2 \cdot 10.68H_2O$). By contrast, our neutron diffraction data obtained at 100 K for composition $20Br_2 \cdot 172D_2O$ (i.e. $Br_2 \cdot 8.62H_2O$ with fully occupied T_A, T_B, and P cavities) suggest much greater ordering of the bromine molecules, including only a single orientation in the T_A cavity. Furthermore, the neutron diffraction data indicate that in this tetragonal hydrogen bonded framework, the deuterons are distributed in a complex pattern of disorder with unique statistical weights at positions between pairs of oxygen atoms, in contrast to the patterns found in the cubic II and hexagonal ice structures.

Schemes for Economic Transport

To develop a scheme for the transport of methane hydrates, three available options for transportion of conventional remote natural gas are considered. These are: (1) liquefied natural gas (LNG), (2) gas to liquids (GTL), and (3) natural gas hydrates. Option 1 is preferentially practiced commercially with delivered energy efficiency of about 80% and is considered baseline. Option 2 involves initial conversion of natural gas to synthesis gas via partial oxidation (POX) or steam reforming (SR)

$$\text{POX: } CH_4 + \frac{1}{2}O_2 \to CO + 2H_2 \qquad (1)$$

$$SR: CH_4 + H_2O \rightarrow CO + 3H_2 \qquad (2)$$

In a subsequent step, synthesis gas is catalytically converted to Fischer-Tropsch liquids. Safety considerations make the room temperature liquid transport option attractive, but at present this two-step route has a delivered energy efficiency of less than 60%.[15] Option 3 involves gas hydrate formation for transport at 1 atm and $-15°C$ to $-5°C$. An economic study[16] of Option 3 that assumed large-scale (4×10^9 m^3 per year) and long-distance (5,500 km) transport from the SNØHVIT, Norway offshore gas field showed this route to cost about 25% less than the LNG option. It appears that CH_4 transport at a higher temperature of $-15°C$ more than offsets the additional weight of H_2O in gas hydrates.

We have focused our attention on developing an energy efficient, and thus cost effective, pathway to transport gas hydrates via the GTL scheme (Option 2). The conceptual scheme, shown in FIGURE 3, is based on our ongoing work in developing advanced catalysts for conversion of natural gas–derived synthesis gas to methanol and other oxygenates with targeted efficiency rivaling LNG.[17] In the scheme, four steps are coupled to transport hydrates as liquid fuels. The steps are: (1) Decomposition of CH_4 hydrates to yield "aqueous CH_4", (2) synthesis gas production via SR **(2)**, (3) conversion of synthesis gas to methanol via an atom-economic low temperature catalysis pathway, and (4) transport of methanol product. Step 1 requires precise structural and thermodynamic stability data. This aspect is one of the two themes discussed in this paper. In Step 2, one of the two commercially practiced processes need to be selected. Since CH_4 from hydrates is naturally wet, the SR route is preferred over POX that is less energy intensive but will require an expensive drying step. Step 3 involves methanol synthesis though the synthesis gas composition **(2)** allows production of CH_3OH and H_2 in a 1/1 mol ratio **(3)**. The extra H_2 could be used as a clean fuel to provide energy for the operation of the overall GTL plant.

$$CO + 3H_2 \rightarrow CH_3OH + H_2 \qquad (3)$$

At BNL, work is under way to optimize a highly active and selective methanol synthesis catalyst that operates in liquid phase and achieves more than 90% per pass conversion of synthesis gas into methanol.[17] The absence of gas recycle due to a high per pass conversion will allow a simpler unit design. Given that gas hydrates are dis-

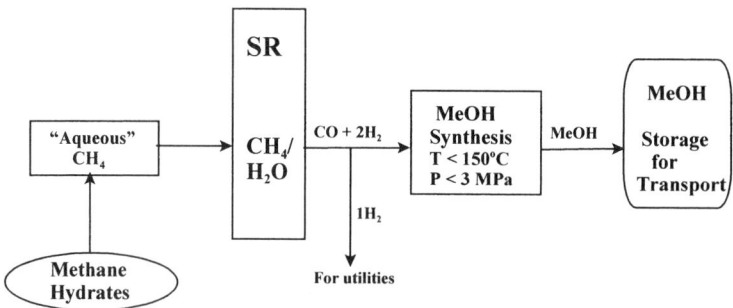

FIGURE 3. A conceptual baseline scheme for economic and environmentally safe transport of methane hydrates as methanol.

persed, a small off-shore floating methanol synthesis unit based on such a process would provide the basis for an economically attractive viable transport option. Once at the destination, methanol would be available as a building block for a variety of fuels and chemicals. FIGURE 3 describes a conceptual scheme for the economic transport of methane hydrates as methanol.

CONCLUDING REMARKS AND FUTURE DIRECTIONS

The future effort at BNL will be specific to natural gas hydrates. A bench-scale unit at BNL, normally used for catalytic studies, has been modified to conduct static and dynamic kinetic studies. The unit consists of two interchangeable pressure vessels (Strahman liquid level gauges) that have transparent windows for monitoring crystal growth/decomposition phenomena (specifications: Vessel 1, 1.9 cm I.D. × 35 cm length; volume, 175 mL; rated at 12 MPa @ 140°C. Vessel 2, 6.3 cm I.D. × 35 cm length; volume about 1.5L; rated at 7MPa @ 35°C). The pressure vessel is immersed in a bath that can be thermostatted from $-20°C$ to $+150°C$ (within $\pm 1°C$) with a Poly-Science Model 9100 refrigerated constant temperature circulator (working liquid: ethylene glycol/H_2O). An Omega mass flow meter (MFM) allows precise addition (0 to 5L min^{-1}) of CH_4 and other gases to the pressure vessel. An Omega OM205 data logger constantly monitors the pressure and temperature changes at various points in the unit. Further modifications to the unit will include: (1) a video camera/TV monitor to capture real-time kinetic events, and (2) an interface with a PC for data collection and analysis. The kinetic studies will determine experimental conditions for preparation of crystals for neutron diffraction measurements.

Future efforts in the area of X-ray and neutron diffraction studies will be concentrated on the hydrates of methane, ethane, and ethylene. These investigations will include high-pressure neutron diffraction studies on the methane hydrate phase of composition $xCD_4 \cdot 46D_2O$, which in its protio form represents the hydrate likely to comprise the largest part of the arctic region's "frozen natural gas" reserve. In our work, crystal-growth procedures for methane hydrates will be optimized. Neutron diffraction studies will be carried out as part of the research program of the Brookhaven Neutron Crystallography Collaboration (BNCC). Arrangements have been made to utilize the spallation neutron facilities at the Argonne National Laboratory (ANL) Intense Pulsed Neutron Source (IPNS) and the reactor neutron facilities at the Oak Ridge National Laboratory high flux isotope reactor (HFIR). The time-of-flight Laue technique employed at pulsed neutron sources such as IPNS is ideal for work at high pressures, since the sample is stationary while diffraction measurements are recorded, thus facilitating the mounting of high-pressure chambers in the neutron beam. Work undertaken by our BNCC team will complement research on gas hydrates by ANL scientists, including powder neutron diffraction studies of carbon dioxide hydrates.[18]

Lessons learned from the Messoyaka field[19] are useful in the application of the *depressurization* technique to recover CH_4 from gas hydrates, though controversy still exists on the free gas/gas hydrate equilibrium model.[1] However, the diversity of hydrate sources may require *site specific* production techniques. The structural and thermodynamic stability data will be coupled with the optimized synthesis gas to methanol catalyst system. These data will be incorporated in our GTL scheme to formulate a hydrate specific transport technology that is essential to deliver the product CH_4 to its intended destination, that is, the customer.

ACKNOWLEDGMENTS

Work at Brookhaven National Laboratory was carried out under contract DE-AC02-98CH10886 with the U.S. Department of Energy. We would like to thank the late Joseph Henriques and Conrad F. Koehler III for assistance in our neutron diffraction studies.

REFERENCES

1. COLLETT, T.S. & V.A. KUUSKRAA. 1998. Hydrates contain vast store of world gas resources. Oil & Gas J. May: 90–95.
   SLOAN, E.D., JR. 1998. Clathrate Hydrates of Natural Gases, 2nd edit. Marcel Dekker, New York.
   ENGLEZOS, P. 1993. Ind. Eng. Chem. Res. **32:** 1251–1274.
2. DALLIMORE, S.R., T. UCHIDA & T.S. COLLETT. 1998. JAPEX/JNOC/GSC Mallik 2L-38 gas hydrate research well: overview of science program. 311–318.
   TSUJI, Y., A. FURUTANI, S. MATSUURA & K. KANAMORI. 1998. Exploratory surveys for evaluation of methane hydrates in the Nankai trough area, offshore central Japan. Proceedings of the International Symposium on Methane Hydrates, Resources in the Near Future, Chiba City, Japan, October 20–22. 15–26.
3. OFFICE OF FOSSIL ENERGY/U.S. DEPARTMENT OF ENERGY. 1998. A strategy for methane hydrates research and development. Report # DOE / FE–0378.
4. Monetizing Stranded Gas Reserves 1998. San Francisco, CA, December 14–16, 1998. Sponsored by Zeus Development Corporation, Houston, TX.
5. MCMULLAN, R.K. & Å. KVICK. 1990. Neutron diffraction study of the structure II clathrate hydrate: $3.5Xe \cdot 8CCl_4 \cdot 136D_2O$ at 13 and 100 K. Acta Cryst. B **46:** 390–399.
6. PRESS, W. & A. HÜLLER. 1973. Analysis of orientationally disordered structures. I. Method. Acta Cryst. A **29:** 252–256.
7. MCMULLAN, R.K., J. ECKERT & S.M. SHAPIRO. Unpublished results.
8. JEFFREY, G.A. & R.K. MCMULLAN. 1967. The clathrate hydrates. *In* Progress in Inorganic Chemistry. F.A. Cotton, Ed. Interscience, New York, London, Sydney. **8:** 43–108.
9. RIPMEESTER, J.A., J.S. TSE, C.I. RATCLIFFE & B.M. POWELL. 1987. A new clathrate hydrate structure. Nature **325:** 135-136.
10. UDACHIN, K.A., C.I. RATCLIFFE, G.D. ENRIGHT & J.A. RIPMEESTER. 1997. Structure H hydrate: a single crystal diffraction study of 2,2-dimethylpentane. $5(Xe, H_2S) \cdot 34 H_2O$. Supramol. Chem. **8:** 173–176.
11. ALLEN, K.W. & G.A. JEFFREY. 1963. On the structure of bromine hydrate. J. Chem. Phys. **38:** 2304–2305.
12. BRAMMER, L. & R.K. MCMULLAN. 1993. Tetragonal bromine hydrate: structure determination by single crystal neutron diffraction at 100 K. Abstr. Am. Crystallogr. Assoc. Meeting, Albuquerque. Abstr. II01: 73.
13. UDACHIN, K.A., G.D. ENRIGHT, C.I. RATCLIFFE & J.A. RIPMEESTER. 1997. Structure, stoichiometry, and morphology of bromine hydrate. J. Am. Chem. Soc. **119:** 11481–11486.
14. BARTELL, L.S., L.O. BROCKWAY & R.H. SCHWENDEMAN. 1955. Refined procedure for analysis of electron diffraction data and its application to CCl_4. J. Chem. Phys. **23:** 1854–1859.

15. STODOLSKY, F. & D.J. SANTINI. 1993. CHEMTECH: 54–59. Remote Gas Strategies. April 1998. 4–5, 13–14.
16. GUDMUNDSSON, J.S., M. PARLAKTUNA & A.A. KHOKHAR. 1994. Storing natural gas as frozen hydrate. SPE Production Facilities: 69–73.
17. WEGRZYN, J.E., D. MAHAJAN & M. GUREVICH. 1999. Catalytic routes to transportation fuels utilizing natural gas hydrates. Catal. Today **50:** 97–108.
18. SCHULTZ, A.J., V. THIEU & Y. HALPERN. 2000. Neutron diffraction studies of CO_2 clathrate hydrate: formation from deuterated ice. J. Am. Chem. Soc. In press.
19. MAKOGON, Y.F. 1994. Russia's contribution to the study of gas hydrates. Ann. N.Y. Acad. Sci. **715:** 119–145.

Natural Gas Storage Properties of Structure H Hydrate

A.A. KHOKHAR,[a] E.D. SLOAN,[b] AND J.S. GUDMUNDSSON[c]

[a]*C-FER Technologies Inc., Edmonton, Alberta T6N 1H2, Canada*

[b]*Center for Hydrate Research, Colorado School of Mines, Golden, Colorado 80401, USA*

[c]*Norwegian University of Science and Technology, Department of Petroleum Engineering and Applied Geophysics, 7491 Trondheim, Norway*

> ABSTRACT: Methane gas storage properties of structure H (sH) hydrate were investigated in a Jerguson rocking cell reactor. It was assumed that the methane gas will occupy five small cages (three 5^{12} and two $4^3 5^6 6^3$) whereas the largest cage (one $5^{12} 6^8$) will be occupied by a large (e.g., neohexane) molecule in a unit structure. A theoretical gas storage comparison with this methodology of storing methane in small cages showed that one unit volume of sH hydrate will store 201 volumes of methane gas. Three types of experiments were conducted to measure the methane gas storage in sH hydrate. It was discovered that sH hydrate promotion occurs with 0.1 weight percent solutions of lecithin and polyvinylpyrrolidone (PVP). These promoters also showed hydrates stabilization as hydrates melted at higher temperature compared to pure water hydrate. The heats of dissociation (ΔH^{diss}) below 0°C were estimated using Clausius-Clapeyron equation and the experimentally measured equilibrium data for three hydrate structures (sI, sII, sH). Comparison of the estimated and calorimeter measured values (ΔH^{diss}) showed good agreement.

INTRODUCTION

The storage of natural gas in hydrates has been investigated since their discovery[1] because they store large quantities (e.g., up to 180 Sm^3 per m^3 of hydrate) of gas.[2–5] Different storage methods have been suggested: either keeping hydrate under low temperature[6,7] or under pressure.[8–10] Gudmundsson et al.[11] showed that sII hydrate can be stored at −15°C under atmospheric pressure for 14 days retaining almost all the gas. Later, they also published[12] a feasibility study showing a substantial cost saving (24%) for the transport of natural gas in hydrated form compared to liquefied natural gas (LNG) from the Northern Norway to Central Europe. A paper by Saito et al.,[13] addressed the storage of methane gas in sII hydrate using miscible tetrahydrofuran that occupy the large cages. The authors concluded[13] that the storage of methane gas was much lower than expected.

The discovery of structure H (sH) hydrate in 1987 by Ripmester et al.[14] and subsequent discovery of it in nature by Sassen et al.,[15] opened a new series of petroleum components that can form sH hydrate in a subsea pipeline.[16–18] Experimental data.[19,20] suggests that molecules as large as 9.293Å (1,3-dimethylcyclohexane) can form sH hydrate. Unit cells and types of cages for each hydrate structure are shown in FIGURE 1. This paper reports the experimental storage of methane gas in sH with lecithin and polyvinylpyrrolidone (PVP) as hydrate promoter. The sH hydrate work

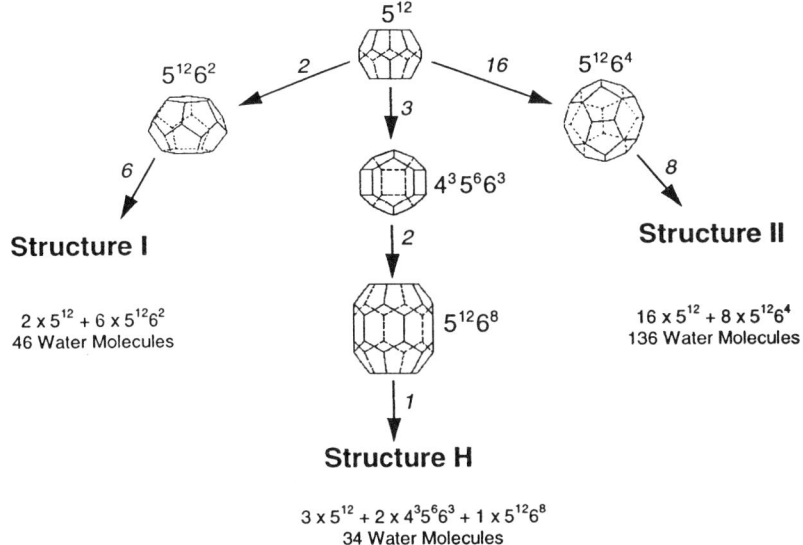

FIGURE 1. Hydrate unit structures and types of cages.

also shows that these two hydrate promoter are also hydrate stabilizer as they delay the dissociation of hydrate compared to pure water hydrate.

One of the important parameters in the study of gas storage in hydrate is the enthalpy of dissociation below the ice point. This is the minimum heat required to melt hydrate for recovery of gas stored.[19,21] Although heats of dissociation (ΔH^{diss}) have been predicted for sI, sII, and sH above 0°C,[22,16] using the hydrate equilibrium data and the Clausius-Clapeyron equation, there are neither experimental nor estimated ΔH^{diss} values below 0°C for sH hydrate. This work reports the first ΔH^{diss} for sH hydrate below 0°C. The necessary subzero equilibrium data were measured[21] and ΔH^{diss} for three hydrate structures were predicted using Clausius-Clapeyron equation. A comparison of ΔH^{diss} below 0°C between the predicted values from this work[21] and the reported four experimental values[23,24] showed a good agreement.

METHANE STORAGE IN HYDRATES

The best option for methane gas storage is sI hydrate since methane occupies both small and large cages, but at high formation pressure. To reduce high formation pressure in our theoretical study of methane storage, we adopted the approach of filling the large cage in sI and sII with a large soluble molecule. Similarly, for sH hydrate the largest cage ($5^{12}6^8$) was stabilized with, a yet to be found, a miscible molecule.[21] The results of this theoretical study are shown in TABLE 1. The maximum methane gas storage volumes were compared in three hydrate structures by assuming large

TABLE 1. The maximum methane storage potential in the three hydrate structures by stabilizing the large cavity with a large molecule

Small Cages Methane Occupy in Hydrate	Cages Ratio Small : Large	Methane Gas Storage Potential (Sm^3)
sI (5^{12})	2 : 6	56
sII (5^{12})	16 : 8	154
sH (5^{12} & $4^3 5^6 6^3$)	5 : 1	201
LNG at $-162°C$	—	600

cages occupied with large molecule and 100% occupancy of small cages by the methane gas. The third column of TABLE 1 shows a comparison of gas volume stored in hydrates at standard conditions. It is noted that methane gas storage in sH is 1/3 of the liquefied natural gas (LNG) capacity.[25] The standard volumes of methane gas per unit volume of LNG, which is kept at $-162°C$ during its transportation, are also given (TABLE 1).

Based on these results, an experimental study was conducted to investigate the storage of methane gas in sH hydrate. It was decided to stabilize the largest cage in sH with neohexane also known as 2,2-dimethylbutane. It should be noted here that sH hydrate can only be stabilized with two components, the large molecule of neohexane in $5^{12} 6^8$ cage and a help-gas like methane in 5^{12} and $4^3 5^6 6^3$ cages.

EXPERIMENTAL SETUP AND PROCEDURES

The apparatus for measuring methane gas storage in sH hydrate is shown in FIGURE 2. It consisted of a Jerguson sight glass cell that rocked about its axis. The cell was immersed in a refrigerated, constant temperature bath. Temperature of the cell was monitored with a platinum resistance probe accurate to ±0.01 K, and the pressure was measured using a Heise gauge, which was accurate to ±3.0 psia. The cell was charged with 50 cc of distilled deionized water and 8 cc (10% excess of stoichiometric ratio) of neohexane. There were 25 stainless steel balls in the cell (0.79 cm diameter) to agitate fluids during rocking.

After cell evacuation, it was pressurized with methane gas below the equilibrium of sI hydrate. This gas was supplied from a known volume (1055.1 cc) and pressure cylinder. On temperature stabilization, cell-rocking motion was started. The sH hydrate appeared on the sight glass of test cell in few minutes. Absence of a cell pressure drop was an indication of no hydrate formation. The pressure and temperature of two cylinders were monitored for hydrate formation. The methane gas storage experiments with and without hydrate promoters were also conducted in the rocking cell. The hydrate promotion was achieved by using small wt% of lecithin and PVP in pure water. The three experimental procedures used for monitoring were:

1. Constant-volume experiment, where the pressure was decreased as hydrate formed.

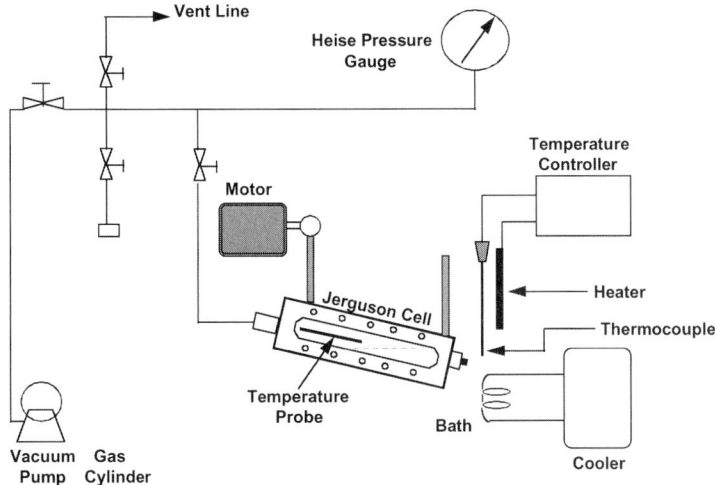

FIGURE 2. Jerguson rocking cell setup used for gas storage experiments.

2. Pressure-replenishment experiment, where gas to test cell was supplied when the cell temperature equals the water bath temperature after hydrate formation.

3. Constant-pressure experiment, where test cell pressure was maintained within ±138 kPa without any consideration to temperature increase during hydrate formation.

METHANE GAS STORAGE IN SH HYDRATE

The constant-volume hydrate formation results are shown with and without hydrate promoters or stabilizers in FIGURE 3. The test fluids were cooled in the cell from 20.0°C to 2.5°C at a rate of 2.5°C/hr. When a constant temperature and pressure were established in the cell, rocking of the cell started. On completion of hydrate formation, the hydrate cell was warmed to 25.0°C at a rate of 7.5°C/hr. The hydrate dissociation cycle was completed when the dissociation P-T line joined the cooling line.

TABLE 2. Water amount converted to hydrate based on each experiment pressure drop

Experiment Procedure	Pure Water % Hydrate	0.1 wt% PVP % Hydrate	0.1 wt% Lecithin % Hydrate
Constant-volume	21.14	23.66	28.07
Pressure-replenishment	30.08	40.18	42.62
Constant-pressure	17.48	16.38	not available

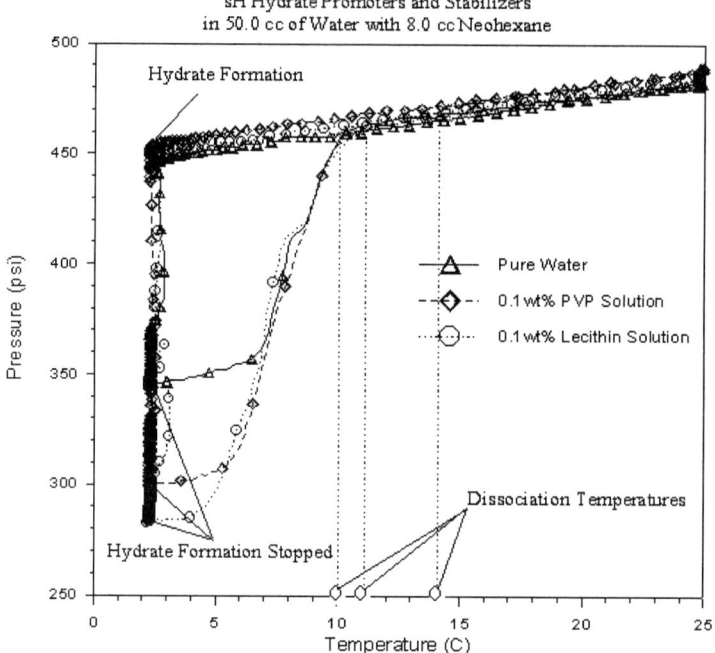

FIGURE 3. Constant-volume, sH hydrate formation with and without promoter/stabilizer.

The intersection of cooling and warming line is usually taken as the hydrate dissociation (P-T) conditions. The P-T conditions were also measured with 0.1 wt% solutions of PVP or lecithin. The maximum pressure drop in FIGURE 3 was observed with 0.1 wt% lecithin solution compared to 0.1 wt% PVP and pure water solutions. This means solutions of lecithin or PVP were able to convert more water to hydrate (TABLE 2) by changing the growth mechanism of hydrate particles from 3D to 2D and this provided a larger interface area during the hydrate formation. The possible hypotheses for such hydrate promotion are given elsewhere.[20,21] The FIGURE 3 also shows delay in hydrate dissociation with the solutions of lecithin or PVP compared to hydrate formed with pure water. TABLE 2 also shows the water percentage converted to hydrate (assuming 100% cage occupancy) with three experimental procedures.

HEAT OF DISSOCIATION BELOW 0°C

The purpose of the study was to estimate the minimum heat that is required to dissociate a laboratory produced hydrate. Because ΔH^{diss} estimated above 0°C by using Clausius-Clapeyron equation is within an acceptable accuracy, and since measurements via calorimeter are painstaking,[16,22,26] we decided to use the Clausius-

Clapeyron equation. This equation requires the hydrate equilibrium data below the ice point, which was measured experimentally and then plotted as ln P (pressure) versus inverse of T (temperature). The slope of this line is equal to the left-hand-side of the equation (Clausius-Clapeyron) shown below.

$$\frac{d \ln P}{d(1/T)} = -\frac{\Delta H^{diss}}{zR},$$

where P and T are the absolute equilibrium pressure and temperature respectively, z is the gas compressibility, and R is the universal gas constant. The above equation can be used to calculate ΔH^{diss} if hydrate equilibrium (P-T) data is known for a univariant system (e.g., three components and four phases, or two components and three phases).

The description of experimental apparatus used during this study, experimental procedures, and the hydrate equilibrium measured below 0°C are reported in the literature.[19,21] TABLE 3 shows the slopes of lines and the estimated values of ΔH^{diss} calculated through the Clausius-Clapeyron equation. The ΔH^{diss} for sI hydrated varied ±18.3% from the average value (TABLE 3), mainly due to a large deviation caused by methane gas. We do not have an explanation for this difference, but it is worthwhile to note that the methane gas occupies both small (5^{12}) and the large ($5^{12} 6^2$) cages when methane alone forms sI hydrate. The estimated deviation in the ΔH^{diss} for sII hydrates varied only by ±2.8% from the average value for the four systems considered. The variation (ΔH^{diss}) for sH hydrate binaries of methane + neohexane and xenon + neohexane was ±18%. However, it reduced to ±1.2% with methane binaries of methylcyclohexane and neohexane (TABLE 3). This difference may be small because the unit structure have five small cages (for xenon or methane) versus one large cage for the large guest. Therefore, the change in ΔH^{diss} is significant with the change of help-gas in sH hydrate.

TABLE 3. ΔH^{diss} for hydrate structures using Clausius-Clapeyron equation below 0°C

Structure Type	Guest Composition	Slope (1/K)	ΔH^{diss} (kJ/gas-mole)
sI	methane	−2437.44	−19.13
	ethane	−3043.83	−24.37
	methane (69%) + ethane (31%)[a]	−3193.20	−25.67
	carbon dioxide	−3011.72	−23.89
	hydrogen sulfide	−2916.28	−24.03
sII	propane	−3583.62	−28.96
	i-butane	−3544.45	−28.64
	methane (88%) + propane (12%)[a]	−3361.48	−27.53
	methane + n-butane	−3533.02	−28.17
sH	methane + methylcyclohexane[a]	−2847.49	−23.06
	methane + neohexane	−2872.07	−23.33
	xenon + neohexane	−3450.44	−28.43

[a]Measured for this work.[21]

TABLE 4. Comparison of predicted and calorimeter measured ΔH^{diss} values

Structure Type	Gas Constituents	$^a(\Delta H^{diss})_{predicted}$ (kJ/gas-mole)	$^b(\Delta H^{diss})_{calorimeter}$ (kJ/gas-mole)
sI	methane	−19.13	−18.13 ± 0.27
	ethane	−24.37	−25.70 ± 0.37
sII	propane	−28.96	−27.00 ± 37
	natural gas	−27.85^c	−27.80

^{a}Predicted values using Clausius-Clapeyron Equation (21).
^{b}Calorimeter experimental values from Handa.[23,24]
^{c}Average value from two natural gas mixture [TABLE 3].

TABLE 4 is a comparison between the estimated ΔH^{diss} values (third column) with the reported experimental values (fourth column) determined by calorimeter.[23,24] The good agreement between the estimated values and measured values showed that the estimated values of ΔH^{diss} are acceptable for the hydrate melting process when experimental values are not available. Excluding the ΔH^{diss} of methane hydrate, the similarity in the ΔH^{diss} among the three hydrate structures may be due to the fact that all hydrates are 85 mole% water and 15 mole% gas.

CONCLUSIONS

1. Methane gas storage appeared promising in sH relative to storage in sI or sII hydrates when methane occupies only the small cages. Methane storage capability in sH hydrate was verified with experiments.

2. Formation kinetics of sH hydrate were enhanced with the addition of lecithin or PVP in the test cell, as more methane gas was consumed during hydrate formation than with pure water.

3. The new hydrate promoters are also hydrate stabilizers as the melting temperatures with promoters were higher than the melting temperatures of pure water hydrate.

4. Hydrate heats of dissociation below 0°C predicted through equilibrium data and using the Clausius-Clapeyron equation are comparable with calorimeter values.

ACKNOWLEDGMENTS

Assistance for A.A. Khokhar from the NGH project at NTNU and materials support from the hydrate consortium at the Colorado School of Mines is greatly appreciated.

REFERENCES

1. HAMMERSCHMIDT, E.G. 1939. Oil Gas. J. 66.

2. SLOAN, E.D. 1998. Clathrate Hydrate of Natural Gases, 2nd edit. Marcel Dekker, New York.
3. SLOAN, E.D. 1990. Clathrate Hydrate of Natural Gases. Marcel Dekker, New York.
4. MAKOGON, Y.F. 1997. Hydrates of Natural Gas. PennWell Books, Tulsa.
5. MAKOGON, Y.F. 1981. Hydrates of Natural Gas. PennWell Books, Tulsa.
6. MILLER, B. & E.R. STRONG. 1945. A.G.A. Proceedings **27:** 50.
7. BENESH, M.E. 1942. U.S. Patent No. 2,270,016.
8. BERNER, D. 1992. Proc. of 2nd Internl. Offshore and Polar Eng. Conf., San Francisco, 14–19 June. 636–643.
9. CHERSKY, N.V. 1975. U.S. Patent No. 3,888,434.
10. ROGERS, R., G. YEVI & M. SWALM. 1996. Proc. 2nd Intnl. Conf. on Natural Gas Hydrates. Toulouse, June 2–6. 423–429.
11. GUDMUNDSSON, J.S., M. PARLAKTUNA & A.A. KHOKHAR. 1994. SPE Production and Facility, February.
12. BØRREHAUG, A. & J.S. GUDMUNDSSON. 1996. Proceedings Eurogas, June 3–5.
13. SAITO, Y. et al. 1996. Proc. 2nd Intnl. Conf. on Natural Gas Hydrates. Toulouse, June 2–6. 459–465.
14. RIPMEESTER, J.A. & C.I. RATCLIFFE. 1990. J. Phys. Chem. **94:** 8773–8776.
15. SASSEN, R. & I.R. MACDONALD. Org. Geochem. **22**(6): 1029–1032.
16. MEHTA, A.P. 1996. A Thermodynamic Investigation of Structure H Clathrate Hydrate. Ph.D. Thesis. Colorado School of Mines, Golden, Colorado.
17. THOMAS, M. & E. BEHAR. 1994. Structure H hydrate equilibria of methane and intermediate hydrocarbon molecules. 73rd GPA convention, New Orleans, March 7–9.
18. RIPMEESTER, J.A., C.I. RATCLIFFE & G.E. MCLAURIN. 1991. The role of heavier hydrocarbons in hydrate formation. AIChE meeting, Houston.
19. KHOKHAR, A.A., J.S. GUDMUNDSSON & E.D. SLOAN. 1997. Thirteenth International Symposium on Thermophysical Properties, Boulder, CO., June 22–27.
20. KHOKHAR, A.A., J.S. GUDMUNDSSON & E.D. SLOAN. 1998. Fluid Phase Equilibria **150-151:** 383–392.
21. KHOKHAR, A.A. 1998. Storage Properties of Natural Gas Hydrates. Ph.D. Thesis. Norwegian University of Science & Technology.
22. SLOAN, E.D. & F. FLEYFEL. 1992. Fluid Phase Equilibria **76:** 123–140.
23. HANDA, Y.P. 1986. J. Chem. Thermodynam. **18:** 915–921.
24. HANDA, Y.P. 1988. Ind. Eng. Chem. Res. **27:** 872–874.
25. KIDNAY, A.J. 1972. Liquefied natural gas. Min. Indust. Bull. **15**(2).
26. HANDA, Y.P. 1985. Chem. Thermodynam. **18:** 891.

Mechanical Properties of Water/Hydrate-Former Phase Boundaries and Phase-Separating Hydrate Films

R. OHMURA,[a] T. SHIGETOMI, AND Y.H. MORI

Department of Mechanical Engineering, Keio University,
3-14-1 Hiyoshi, Kohoku-ku, Yokohama, 223-8522, Japan

INTRODUCTION

The objective of this study is to measure two kinds of mechanical properties relevant to the formation and dissociation of clathrate hydrate films at the phase boundary between water and a hydrate former both in the liquid state. The properties of interest are the liquid–liquid interfacial tension before the formation and after the dissociation of each hydrate film, and the membrane force working in the hydrate film during its complete covering of the phase boundary. Based on continual observations of single sessile drop of a hydrate former immersed in a pool of liquid water, we have evaluated these properties during the course of prescribed temperature variations about the three-phase (liquid/liquid/hydrate) equilibrium temperature under constant pressure. The motivations for measuring these properties are, however, independent of each other and are described below.

A few recent attempts were made to obtain experimental evidence of the water-clustering hypothesis conceived to explain the so-called memory effect for hydrate formation. Sloan et al.[1] measured the time for a stainless-steel ball to travel one inch in an oscillating sapphire tube filled with liquid water and methane or natural gas. The ball-travel time was found to be longer after hydrate dissociation than before hydrate formation, indicating an increase in the effective viscosity of the liquid-water phase due to its experience of hydrate formation. Kato et al.[2] performed simultaneous measurements of the surface tension and kinematic viscosity of water exposed to xenon gas, using a surface laser-light scattering method. They observed an unusual elevation of these two properties just after hydrate dissociation on the water surface, followed by a relaxation process apparently lasting for several hours. In contrast to the above two studies that suggest a change in the nature of liquid water due to hydrate formation/dissociation, Bylov and Rasmussen[3] reported no detectable difference in the refractive index of liquid water between the time before hydrate formation and that just after hydrate dissociation at a water surface in contact with methane or natural gas. Stimulated by these pioneering studies to detect traces of hydrate formation/dissociation experience in liquid-water properties, we devised an experiment to successively measure the interfacial tension at the surface of each sessile drop of a dense hydrate-forming liquid immersed in a liquid-water pool by the drop-profile-

[a]Telecommunication. Voice: +81-45-563-1141; fax: +81-45-566-1495.
rohmura@mech.keio.ac.jp

fitting method, to examine if the interfacial tension as determined by such a completely static method shows traces of prior hydrate formation/dissociation.

Once a polycrystalline hydrate film covers the surface of a sessile drop taking the form of a thin-walled shell, what sustains the drop against gravity is no longer the liquid–liquid interfacial tension, but the membrane force, a tensile force acting tangentially in the hydrate film. How hydrate films grow themselves, allowing the membrane force (i.e., the product of the internal stress and the film thickness) to emerge spontaneously, may be an issue of crystallographic interest. Another significance of the membrane force evaluated lies in the fact that it indicates a safe measure of the tensile strength of hydrate films. Thus, we attempted to derive the membrane force from the profiles of hydrate-covered drops by a procedure completely analogous to that used in deriving the liquid–liquid interfacial tension from the profiles of hydrate-free drops.

THEORY AND NUMERICAL PROCEDURE FOR CALCULATING INTERFACIAL TENSION AND MEMBRANE FORCE

The meridian profile of an axisymmetric sessile drop schematically illustrated in FIGURE 1 is described by the Young–Laplace equation, which is given in a dimensionless form as follows:[4]

$$\frac{d\theta}{dS} + \frac{\sin\theta}{X} = \frac{2}{B} + Z, \qquad (1)$$

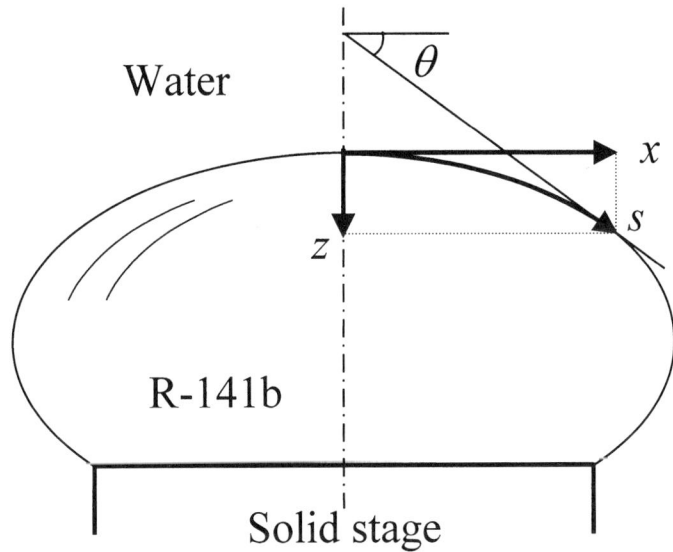

FIGURE 1. Axisymmetric, sessile drop and coordinates.

where $S = s/c$, $X = x/c$, $B = b/c$, $Z = z/c$, $c = \{\sigma/(\rho_F - \rho_w)g\}^{1/2}$, b is the radius of curvature at the apex of the drop, σ is the interfacial tension between the two fluids, ρ_F and ρ_w are the densities of the drop-forming fluid and the surrounding fluid, respectively, g is the acceleration due to gravity, and x, z, s, and θ are the coordinates indicated in FIGURE 1. Note that the dimensionless coordinates, X, Z, S, and θ, are interrelated so that

$$\frac{dX}{dS} = \cos\theta, \text{ and} \tag{2}$$

$$\frac{dZ}{dS} = \sin\theta. \tag{3}$$

Using Equations (2) and (3) to eliminate θ and S, we can rewrite (1) in the following form:

$$\frac{d^2Z/dX^2}{[1+(dZ/(dX))^2]^{3/2}} + \frac{dZ/dX}{X[1+(dZ/(dX))^2]^{1/2}} = \frac{2}{B} + Z. \tag{4}$$

Equation (4) may be numerically integrated under the boundary condition specifying the drop apex:

$$\frac{d\theta}{dS} = \frac{\sin\theta}{X} = \frac{1}{B} \text{ at } X = Z = S = \theta = 0. \tag{5}$$

In the present calculation the Runge–Cutta–Gil method was used to solve (4) for $Z(x)$, the dimensionless profile of each drop, by arbitrarily assuming values for σ and b. Each increment in S, in advancing the calculation, was set to 10^{-3}, whereas those in X and Z were automatically determined by (2) and (3), respectively. Considering that the possible error per increment in using the Runge–Cutta–Gil method was reported to be proportional to the fifth power of the size of the increment[5] and that the total number of increments in S was of the order 10^3 in the present calculations, we estimate the resultant error in $Z(X)$ to be of the order of 10^{-12}, which corresponds 10^{-8} µm. This error is negligible compared to the resolution limit in the experimental drop-image analyses (about 20–30 µm). The dimensionless drop profile thus obtained was then converted into a dimensional $z(x)$ profile. This solution procedure was repeated until the predicted $z(x)$ profile showed the best fit to the corresponding $z(x)$ data, obtained by the experimental procedure described in the next section. The σ value leading to the best fit was retained as the final result of one of the successive interfacial tension measurements.

If we assume the hydrate film encapsulating a sessile drop to be an elastic solid configured to an axisymmetric, thin-walled shell, we can apply the classical shell theory[6] to describe its profile. Here, we further assume that the shell is subjected to a uniform, isotropic membrane force, F. These assumptions reduce the problem of describing the shell profile to be mathematically identical with that of describing the profile of a film-free sessile drop sustained using the uniform interfacial tension σ.[6] Thus, we can use the formulation and the numerical solution procedure described above to determine F as well as σ, having just replaced σ with F in the formulation.

MATERIALS, APPARATUS, AND PROCEDURE USED IN THE EXPERIMENTS

Hydrochlorofluorocarbon R-141b (CH_3CCl_2F) was selected as the hydrate former. A sample of 99.9 wt% certified purity was used as received from the supplier (Daikin Kogyo Co., Osaka, Japan). Water, used together with the R-141b sample, was deionized and distilled. All of the experiments were performed at atmospheric pressure, and the temperature of the above liquids was controlled at a prescribed level to within 275–285 K. Under such conditions, R-141b may be in a liquid state that is denser than liquid water or in the state of hydrate.

A schematic of the experimental set up used in video recording and analyzing the drop profiles is shown in FIGURE 2. The set up was almost the same as that we previously used in our observational study[7] of the formation, growth, and dissociation of hydrate crystals at the water–R-141b interface. A rectangular Pyrex cell, 30×30 mm wide and 40 mm high, with a PMMA cover plate was used as the test cell, in which a water pool and an R-141b drop were held. The test cell was placed in a rectangular bath integrated with a jacket through which temperature-controlled water was circulated via a rectangular channel capping the bath and the external cooler/pump assembly. The bath, the jacket, and the channel were all made of transparent PMMA plates.

Each experimental run commenced by dripping a certain amount of R-141b onto the circular stage—a 10-mm dia. plano-concave lens, placed at the bottom of the cell. Because its density was higher than that of liquid water, the R-141b liquid dripped in this way turns into a single sessile drop lying on the stage. To ensure a highly axisymmetric drop shape consistent with the formulation described in the

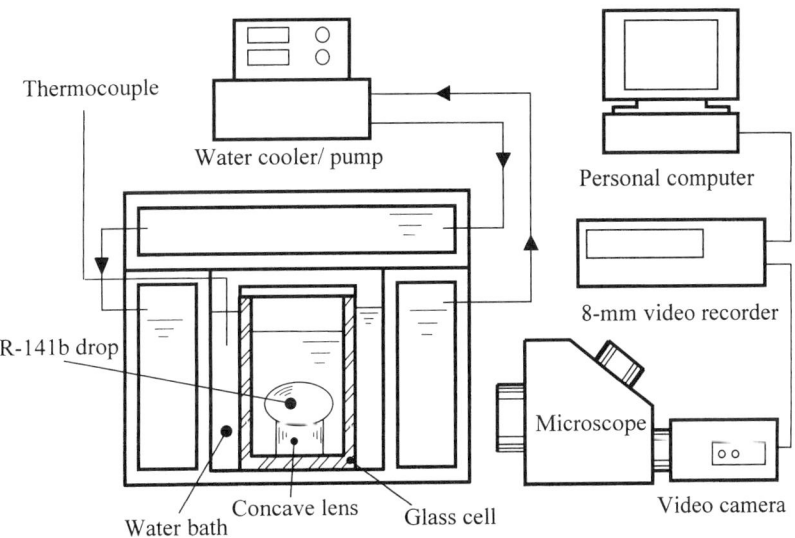

FIGURE 2. Schematic of experimental setup.

previous section, the drop volume was controlled to about 0.5 cm^3 so that the three-phase (R-141b/water/glass) contact line stably lies on the sharp-edged rim of the circular stage. The meridian profile of the drop was video-recorded using a CCD camera connected to a micrographic zoom lens, both of which were held horizontally.

In measuring the liquid–liquid interfacial tension, each run followed the stepwise temperature changes outlined below. The temperature in the test cell was adjusted at 0.7 K above the water/hydrate/R-141b three-phase equilibrium temperature, $T_{tri} = 281.6$ K. A video recording of the drop was made occasionally throughout this stage, which extended to several hours. The test cell was then cooled to 275.6 K (±0.1 K), which is 6 K cooler than T_{tri}. The formation of a hydrate film on the drop surface was triggered by sprinkling a trace amount of hydrate particles prepared in advance in an external crystallizer. After several hours of intermittent video-recordings of the hydrate-covered drop, the temperature was raised to its initial level, 0.7 K above T_{tri}, resulting in the dissociation of the hydrate. The video recording of the drop was performed successively for several hours after the dissociation of the hydrate film. In some experimental runs, the cooling–heating operation was repeated once again. This additional operation was done to examine the survival of the "memory" of the prior hydrate formation.

For the membrane-force measurements, a simpler temperature variation was prescribed. The test cell was cooled to 275.0 K (±0.1 K), then maintained at this level thereafter.

The videographs of the drops were stored in a digital memory in the form of 640 × 480 pixel images, which were then converted into binary (black and white) images. These binary images were processed to yield digitized data of the meridian drop profiles to which the numerical $z(x)$ solutions were to be fitted.

RESULTS AND DISCUSSION

Interfacial Tension

FIGURE 3 shows the results of the measurements of the liquid–liquid interfacial tension in three runs, in each of which warmer hydrate free periods and colder hydration periods were alternated. It should be noted here that the second hydration period began by spontaneous hydrate nucleation during the second quasi-stepwise cooling process in every run, whereas the first hydration period could be commenced only by artificially sprinkling hydrate particles onto the interface after the first cooling process had completed. (Some preliminary experiments revealed that in the absence of any artificial nucleation aid, the induction time before the first hydration period extends to several tens of hours or longer.) The above fact shows that "memory" of the first hydration period survived in the R-141b drop-in-water system throughout the warm hydrate-free period that intervened between the two hydration periods.

Despite the survival of the memory of the previous hydrate formation/dissociation thus confirmed, the interfacial tension measured in the second and the third hydrate-free periods showed no deviation (exceeding the measurement uncertainty) from that in the first hydrate-free period, in which the experimental system had no experience of hydrate formation. It follows, as a logical consequence, that the

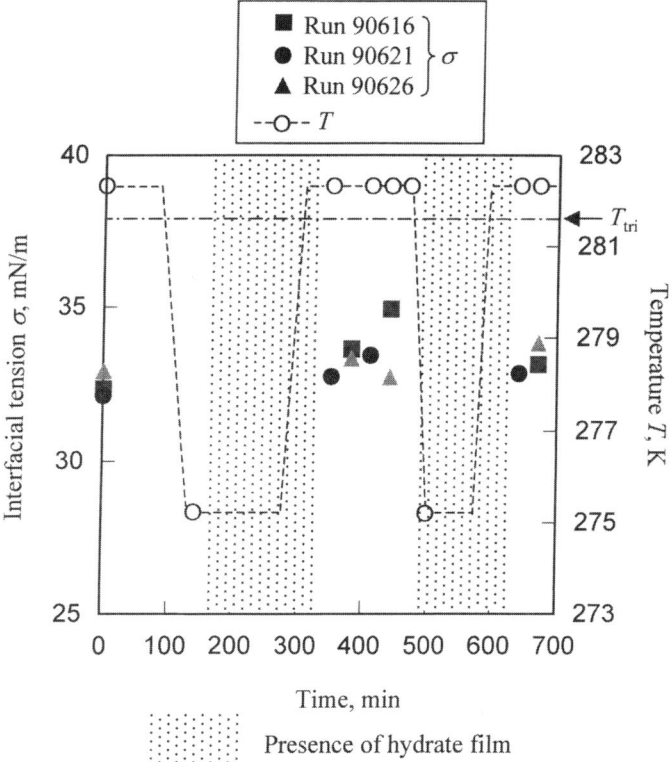

FIGURE 3. Chronological plot of R-141b–water interfacial tension data obtained over a time period in which the formation and dissociation of the hydrate films were repeated.

memory effect in the system we used did not originate from molecular structuring in the liquid water phase exerting an appreciable effect on the interfacial tension.

Membrane Force

FIGURE 4 shows videographs of a hydrate covered R-141b drop in a medium of water in which R-141b was not initially dissolved. Here we find little change in the appearance between the two pictures, one taken 20 hours after the other, except for the surface morphology of the hydrate film. FIGURE 5 illustrates how we determined the membrane force F, showing the digitized data of the meridian drop profile obtained from each picture given in FIGURE 4 and the numerical $z(x)$ solution of Equation (1) (in which we should read F for σ) best fit to the data. The agreement between the experimental data and the best-fit numerical solution of (1) is generally satisfactory. From this fact we conclude that the assumptions of isotropy and uniformity of the membrane force, which underlies Equation (1), are valid.

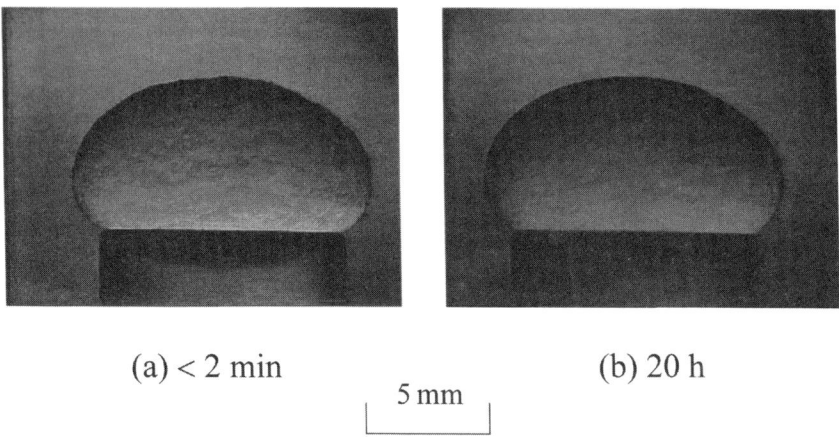

(a) < 2 min   5 mm   (b) 20 h

FIGURE 4. Videographs of a hydrate-covered R-141b drop. Indicated below each picture is the time after the formation of a hydrate film covering the drop surface.

FIGURE 6 shows the chronological variations in the membrane force working on hydrate films covering four different drops. The membrane force values thus determined mostly fall in the range 40–60 mN/m, showing greater scatter than the interfacial tension values demonstrated before. Note that the results shown in FIGURE 6 were obtained with R-141b drops immersed in liquid water that was not saturated with R-141b in advance. Hence, the hydrate films covering the drops must suffer *metabolism*, or continuous renewal, of film-forming hydrate crystals due to the hydrate crystal dissociation at the water-side surface of each film and the formation of new

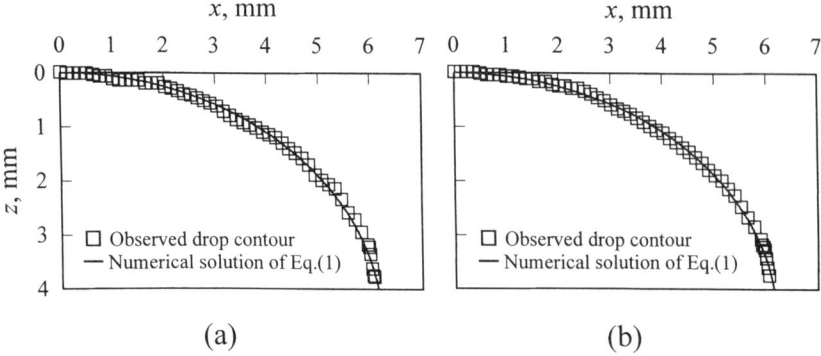

FIGURE 5. Comparson between digitized data of the observed meridian profiles of a hydrate-covered drop and the numerical solutions of Equation (**1**) best fitted to the observed profiles. Diagrams (**a**) and (**b**) in this figure are relevant to those in FIGURE 4.

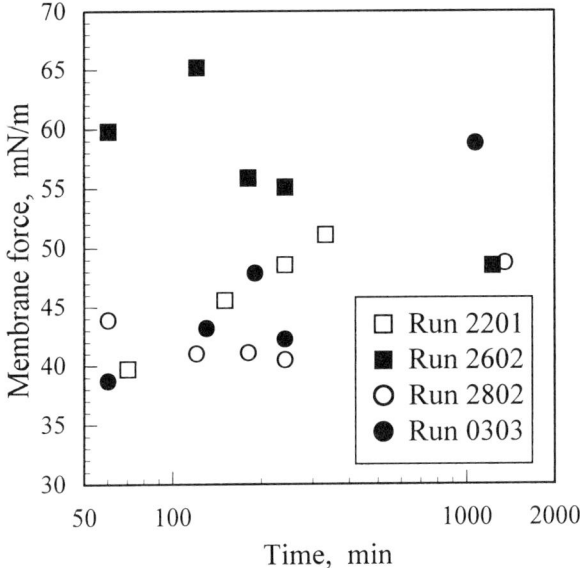

FIGURE 6. Chronological variations in membrane force working on hydrate films. All force values shown here were obtained with hydrate-covered drops immersed in water in which R-141b was not initially dissolved. The temperature was controlled at 275.0 ± 0.1 K.

crystals within the film, thus compensating for dissociation.[8] (Such *metabolism* of hydrate films is suggested by their morphological change exemplified in FIGURE 4.) Nevertheless, hydrate films exhibit no particular pattern of chronological change in the membrane force. A mechanistic or thermodynamic interpretation of this invariability of the membrane force working on hydrate films with internal crystal renewal is left for future study.

ACKNOWLEDGMENTS

This work was subsidized by the Keio Gijuku Fukuzawa Memorial Fund and by the Grant-in-Aid for Scientific Research from the Ministry of Education, Science and Culture of Japan (Grant No. 10450088). The authors are indebted to Mr. M. Isojima and to Dr. K. Ishida for providing us with a program package for numerically solving the Young–Laplace equation and image-analyzing computer software, respectively, that they devcloped.

REFERENCES

1. SLOAN, E.D., JR., S. SUBRAMANIAN, P.N. MATTHEWS, J.P. LEDERHOS & A.A. KHOKHAR. 1998. Quantifying hydrate formation and kinetic inhibition. Ind. Eng. Chem. Res. **37:** 3124–3132.

2. KATO, T., Y. YOKOTA & Y. NAGASAKA. 1998. Observation of the dynamic variations of interfacial tension and kinematic viscosity of Xe–water by the surface light-scattering method. Proc. 5th Asian Thermophysical Properties Conf. M.S. Kim & S.T. Ro, Eds. **2:** 379–382.
3. BYLOV, M. & P. RASMUSSEN. 1997. Experimental determination of refractive index of gas hydrates. Chem. Eng. Sci. **52:** 3295–3301.
4. HARTLAND, S. & R.W. HARTLEY. 1976. Axisymmetric Fluid–Liquid Interface. Elsevier, Amsterdam, Netherlands.
5. BOUCHER, E.A., M.J.B. EVANS & T.G.J. JONES. 1987. The computation of interface shapes for capillary systems in a gravitational field. Adv. in Colloid Interface Sci. **27:** 43–79.
6. TIMOSHENKO, S.P. & S. WOINOWSKY-KREIGER. 1959. Theory of Plates and Shells, 2nd edit. McGraw-Hill, Tokyo.
7. OHMURA, R., T. SHIGETOMI & Y.H. MORI. 1999. Formation, growth and dissociation of clathrate hydrate crystals in liquid water in contact with a hydrophobic hydrate-forming liquid. J. Crystal Growth **196:** 164–173.
8. SUGAYA, M. & Y.H. MORI. 1996. Behavior of clathrate hydrate formation at the boundary of liquid water and a fluorocarbon in liquid or vapor state. Chem. Eng. Sci. **51:** 3505–3517.

Double Gas Hydrates at High Pressures

The Highest Decomposition Temperatures

E.G. LARIONOV, A.YU. MANAKOV, YU.A. DYADIN,[a] AND F.V. ZHURKO

Institute of Inorganic Chemistry, Siberian Branch of the Russian Academy of Sciences, 630090, Novosibirsk, Russian Federation

ABSTRACT: CS-II double clathrate hydrates of tetrahydrofuran, $CHCl_3$, CCl_4, SF_6, and xenon as a "help gas" keep their structures in the pressure interval 1–15,000 bar. Decomposition temperatures of these hydrates rise monotonically in the pressure interval under consideration. The decomposition temperature of the double clathrate hydrate of SF_6 and xenon reaches 129.4°C at 14.8 kbar.

INTRODUCTION

$CHCl_3$, CCl_4, SF_6, and tetrahydrofuran (THF) form the clathrate hydrates $G \cdot 17H_2O$ of cubic structure II (CS-II).[1] Guest molecules fill only the large H-cavities in these hydrates. The free diameter of this type of cavity is about 6.6 Å, and the fraction of these cavities in the water framework of this structure is only one third. The remaining two-thirds of cavities are dodecahedrons (D-cavities) with a free diameter of about 5.2 Å. They are vacant and the hydrate structure is loose. Therefore, despite the good spatial correspondence of the H-cavities to the guest molecules under consideration, as a whole these hydrates have a low packing coefficient k (about 0.5)[2] and are comparatively unstable. Increase in pressure decreases decomposition temperature of CS-II clathrate hydrates with the $CHCl_3$, CCl_4, and THF guest molecules (hydrates with more dense packing are stable in these systems under high pressure[2]). Data on the SF_6–water system at high pressures are presented in this work for the first time (see FIGURE 3 below).

The size of a xenon molecule (we chose this gas as an auxiliary component) is very close to the size of cavities in the CS-I water clathrate framework. When a hydrate is formed, xenon almost completely fills large cavities and significant part (about 72%)[3] of small cavities ($k = 0.564$).[4] This is reflected in the stability of the xenon hydrate. First, it retains its structure when changing the pressure from 1 bar to 15 kbar[1] (unlike the hydrates of $CHCl_3$, CCl_4, THF, SF_6, and hydrates of argon and krypton, which form at least three hydrates in this pressures interval.[2,5,6] Second, it reaches extremely high (for simple clathrate hydrates) decomposition temperatures (79.2°C at about 9 kbar[4]).

[a]Address for correspondence. Voice: (7-3832)391346; fax: (7-3832)344489. CLAT@che.nsk.su

The aim of this work was to study the stability of the double hydrates (in which each type of cavity is occupied by guest molecules with the most appropriate size) under high pressures.

EXPERIMENTAL

The high-pressure apparatus and the cell for differential thermal analysis (DTA) at high pressures of a gas (fluid) were described previously.[4,5] The hydrate decomposition temperature was measured with a Chromel-Alumel thermocouple, calibrated against standard reference compounds (the temperature measurement was accurate to $\pm 0.3°C$). The pressure was measured by a Burdon manometer (up to 2.5 kbar) and a manganin manometer (up to 15 kbar). The manganin manometer was calibrated against the melting of mercury under pressure. The pressure measurement was accurate to within 1%.

The starting aqueous THF solution, with the 1:17 (THF:H_2O) molar ratio of components, was prepared using distilled water and tetrahydrofuran (purified from peroxides, dried, and then distilled) by weighing. Aqueous emulsions of $CHCl_3$ and CCl_4 were prepared as indicated in Reference 7. A known volume of these solutions (about 0.03 ml) was transferred by a micropipette into a plastic ampoule for DTA (the ampoule was filled with SiC powder with grain size about 0.01 mm^5). The gas reservoir was washed with xenon up to the moment of attaching on ampoule with solution (or emulsion) to avoid evaporation of organic substance and thus changing the composition of these solutions. When filling the reservoir with xenon, its pressure was chosen to ensure twofold excess of xenon with respect to stoichiometric composition of hydrate $G \cdot 2Xe \cdot 17H_2O$ (about 2 atm). In the case of SF_6 hydrate, about 0.03 ml of water was transferred into a DTA ampoule (the ampoule was also filled with SiC powder) and the high pressure cell was filled with the stoichiometric (SF_6:2Xe) mixture of gases.

RESULTS AND DISCUSSION

The experimental data on the decomposition of double hydrates are shown in FIGURES 1–3. The data on the double SF_6–H_2O system are presented for the first time and will be the subject of a special publication. In the double systems under consideration (water–$CHCl_3$, CCl_4, THF, and SF_6) three hydrates are formed in the pressure interval 1 bar–15 kbar (see References 2 and 7, also FIG. 3). The low-pressure species are CS-II clathrate hydrates.[1,2] The next hydrate belongs to CS-I in the case of THF.[2,8] The same situation, most probably, takes place in the water–$CHCl_3$, CCl_4, and SF_6 systems.[2,7] The nature of the high-pressure hydrates is unknown at present. Our data show, that CS-II double clathrate hydrates are the only stable clathrate phases in the 1 bar–15 kbar pressure interval (and the decomposition temperature of these hydrates significantly increase) if D-cavities in the CS-II clathrate frameworks of $CHCl_3$, CCl_4, THF, and SF_6 hydrates are filled by the xenon molecules. The maximal decomposition temperatures of the double hydrates discussed depend on the

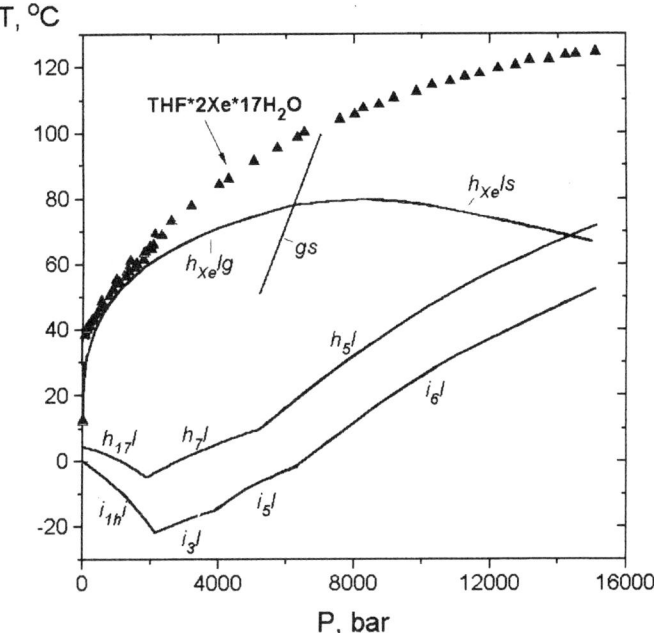

FIGURE 1. Decomposition curves of the hydrates formed in the THF–Xe–H$_2$O system. THF·2Xe·17H$_2$O, decomposition curve of the double hydrate; $h_{17}l$, melting curve of the hydrate THF·17H$_2$O (CS-II); h_7l, melting curve of the hydrate THF·7H$_2$O (CS-I); h_5l, melting curve of the hydrate THF·5H$_2$O (unknown structure); h_{Xe}/g, melting curve of xenon hydrate; gs, melting curve of the solid xenon; and $i_n l$, melting curves of ice.

spatial correspondence between the *H*-cavity (CS-II host framework) and each type of guest molecules (CHCl$_3$, CCl$_4$, THF, and SF$_6$).[2]

Consider the THF–Xe–H$_2$O system in more detail. As would be expected,[9] in the presence of xenon, the decomposition temperature of the double hydrate increases significantly from 4.3°C for the simple hydrate THF·17H$_2$O to 13.5°C for THF·xXe·17H$_2$O at a pressure of 1 atm. Under these conditions, the decomposition temperature of the CS-I xenon hydrate is −10.4°C.[1] The difference between decomposition temperature of the xenon and double hydrates increases with pressure. As already mentioned, the CS-II hydrate THF·17H$_2$O is destabilized by the pressure, and the difference in the melting (decomposition) temperatures of this hydrate and double hydrate at a pressure of 2 kbar reaches 74°C. This difference also increases in comparison with the THF·17H$_2$O (CS-I) hydrate, reaching 83°C at 5.2 kbar. In the case of high-pressure THF hydrate (unknown structure) the difference between decomposition temperatures decreases, but under 15 kbar it remains significant (52°C). Within the entire range of pressures studied, the double hydrate melts at temperatures higher than the xenon hydrate, and with the rise of pressure this difference noticeably increases. The relative sizes of guest molecules and cavities in which they

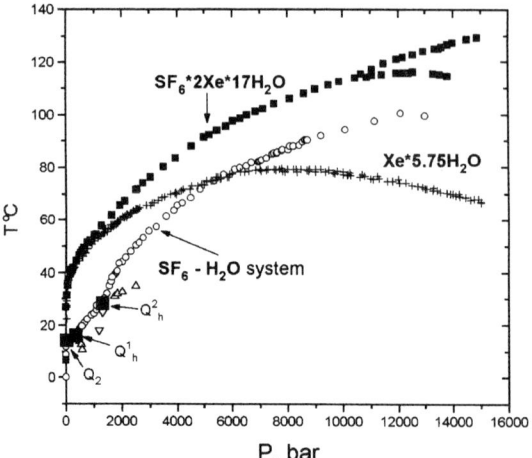

FIGURE 2. Decomposition curves of the hydrates formed in the SF_6–Xe–H_2O system. $SF_6 \cdot 2Xe \cdot 17H_2O$, decomposition curve of the double hydrate; and $Xe \cdot 5.7H_2O$, decomposition curve of xenon hydrate. SF_6–$17H_2O$ system, decomposition curves of SF_6 hydrates in the system: Q_2–Q_h^1 pressure interval, CS-II hydrate; Q_h^1–Q_h^2 pressure interval, h_1 hydrate (probably CS-I); and Q_h^2, 15,000 bar pressure interval for h_2 hydrate (probably CS-IV).

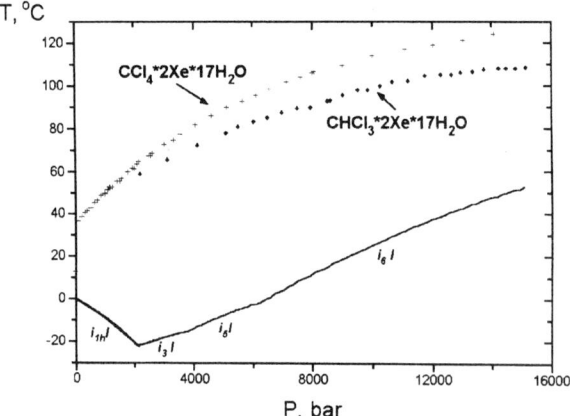

FIGURE 3. Decomposition curves of the double hydrates $CCl_4 \cdot 2Xe \cdot 17H_2O$ and $CHCl_3 \cdot 2Xe \cdot 17H_2O$. i_nl, melting curves of ice High pressure part of the decomposition curve of the $CCl_4 \cdot 2Xe \cdot 17H_2O$ hydrate coincides with the decomposition curve of the $THF \cdot 2Xe \cdot 17H_2O$ hydrate within experimental error.

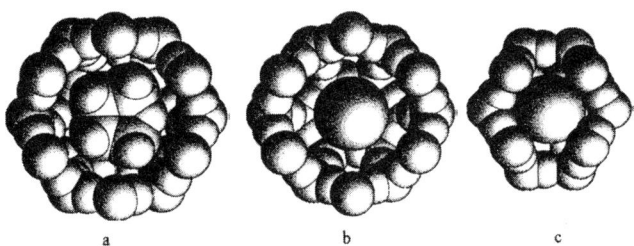

FIGURE 4. Allocation of guest molecules in cavities: **a**, tetrahydrofuran molecule in H-cavity; **b**, xenon in T-cavity; and **c**, xenon in D-cavity.

are located are shown in FIGURE 4. As can be seen, the size of the THF molecule corresponds better to the large H-cavity in CS-II structure than the size of the xenon molecule corresponds to the large T-cavity in CS-I clathrate framework. The xenon molecule effectively occupies small D-cavity in both structures. Evidently, the packing coefficient of the double hydrate should be higher than that of any clathrate hydrate in which different types of cavities are filled with the same type of guest molecules. If we assume that parameters of the unit cell of the double hydrate THF·2Xe·17H$_2$O are similar to those of the simple hydrate THF·17H$_2$O (17.29 Å at 0°C[10]), its packing coefficient $k = 0.619$[1] is really higher than the packing coefficient of the xenon hydrate and is highest for known hydrates of classical (polyhedral) structures.[2] To calculate the packing coefficient we used the following reference values: van der Waals radii of oxygen $R_o = 1.29$ Å, hydrogen $R_H = 1.16$ Å, and xenon $R_{Xe} = 2.18$ Å, the length of the H-bond $l_H = 2.80$ Å. Based on these initial data, the volume occupied by a water molecule in the hydrate $V_{H_2O} = 14.12$ Å^3,[2] and the volume of molecules $V_{Xe} = 43.40$ Å^3 and $V_{THF} = 73.24$ Å^3. This fact explains the efficient stabilization of the double hydrates by pressure and the highest temperature of decomposition of these hydrates among known gas hydrates.

ACKNOWLEDGMENTS

This work was supported by the RFBR Grant 97-03-33521a, and grant of Presidium SD RAS 97-18.

REFERENCES

1. DAVIDSON, D.W. 1973. Clathrate hydrates. *In* Water. A Comprehensive Treatise, Volume 2. F. Franks, Ed.: 115. Plenum Publishing Corporation, New York-London.
2. DYADIN, A.YU. *et al.* 1991. Clathrate hydrates at high pressures. *In* Inclusion Compounds, Volume 5. J.L. Atwood, J.E.D. Davies & D.D. MacNicol, Eds.: 213–275. Oxford University Press, Oxford.
3. DAVIDSON, D.W. *et al.* 1986. Xenon-129 NMR and the thermodynamic parameters of xenon hydrate. J. Phys. Chem. **90:** 6549–6552.
4. DYADIN, A.YU. *et al.* 1996. Clathrate hydrate of xenone at high pressure. Mendeleev Commun. 44–45.

5. DYADIN, A.YU. et al. 1997. Clathrate formation in the Ar–H_2O system under pressures up to 15,000 bar. Mendeleev Commun. 32–34.
6. DYADIN, A.YU. et al. 1997. Clathrate formation in the Kr–H_2O and Xe–H_2O systems under pressures up to 15 kbar. Mendeleev Commun. 74–76.
7. DYADIN, A.YU. et al. 1990. Clathrate formation in binary aqueous systems with CH_2Cl_2, $CHCl_3$, and CCl_4 at high pressures. J. Inclusion Phenom. **9:** 37–49.
8. ZAKRZEWSKI, M. et al. 1994. On the pressure-induced phase transformation in the structure II clathrate hydrate of tetrahydrofuran. J. Inclusion Phenom. **17:** 237.
9. GLEW, D.N. et al. 1968. Aqueous non-electrolyte solutions: Part VII. Water shell stabilization by interstitial nonelectrolytes. In Hydrogen-Bonded Solvent Systems. A.K. Covington & P. Jones, Eds.: 195–210. Taylor and Francis, London.
10. DAVIDSON, D.W. et al. 1986. Crystallographic studies of clathrate hydrates. Part I. Mol. Cryst. Liq. Cryst. **141:** 141–149.

In Situ Observation of CO_2 Hydrate by X-ray Diffraction

SATOSHI TAKEYA,[a,b] TAKEO HONDOH,[b] AND TSUTOMU UCHIDA[c]

[b]*Institute of Low Temperature Science, Hokkaido University, N19 W8, Sapporo 060-0819, Japan*

[c]*Hokkaido National Industrial Research Institute, 2-17 Tukisamu Higashi Toyohira-ku, Sapporo 062-8517, Japan*

ABSTRACT: *In situ* observations of CO_2 hydrate growth using high-energy X-rays were done at high-pressures and the growth rates of CO_2 hydrate from ice particles were measured. Assuming a model of diffusion through CO_2 hydrate layers, interdiffusion coefficients of CO_2 and H_2O molecules were calculated between 233 K and 272.5 K. The diffusion coefficient at 263 K was 7.4×10^{-16} m^2/s, and the activation energy of diffusion was 0.40 eV.

INTRODUCTION

Under quiescent conditions, hydrates nucleate and grow at the water–guest boundary. The boundary is soon covered with hydrate; hence the growth rate is determined by the diffusion of H_2O and guest molecules through the hydrate layer. However, little is known about the diffusion mechanism.

Lund *et al.*,[1] constructed mathematical model of a liquid CO_2 droplet covered with a CO_2 hydrate film in unsaturated water, and based on measurements, reported that the diffusion coefficient of water molecule was in the 10^{-11}–10^{-9} m^2/s range at about 30 MPa and 275 K. On the other hand, Salamatin *et al.*,[2] constructed a mathematical model for air hydrate growth in their laboratory experiments. They estimated the mass transfer coefficient for diffusion of air and H_2O molecules in air hydrate to be 1.9×10^{-14}–4.1×10^{-14} m^2/s (0.6–1.3 mm^2/yr) at 263 K with an activation energy less than 0.31–0.52 eV.

To understand the growth mechanism, we observed the growth process of CO_2 hydrate crystals around hexagonal ice (I_h) *in situ*, using time-resolved X-ray diffraction.

EXPERIMENTAL PROCEDURES

To observe the changes of diffraction profiles during hydrate formation, a high-energy X-ray generator (200 kV, 90 mA, W-target) was used. FIGURE 1 is a schematic diagram of the experimental apparatus. The cylindrical, high-pressure cell was made of duralumin (Al, Cu-Mn-Mg-Si alloy), which was specifically designed for the *in situ* observation using X-ray diffraction. This was installed on a four-circle

[a]Address for correspondence. Fax: +81-11-706-7351.
takeya@hhp2.lowtem.hokudai.ac.jp

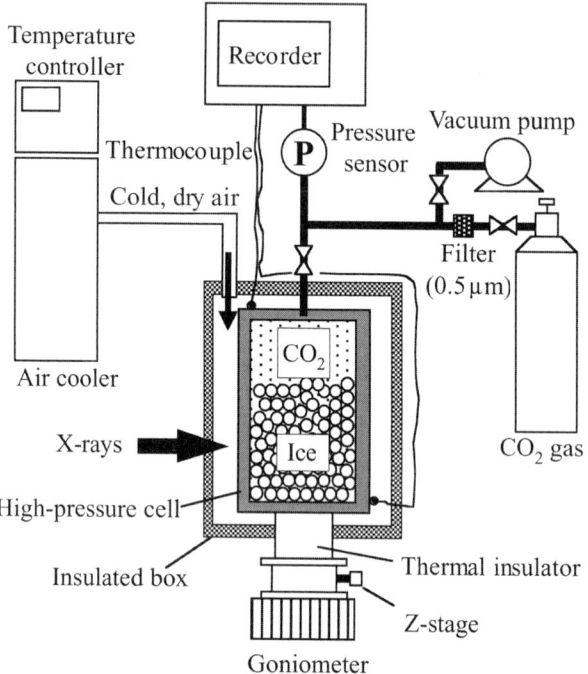

FIGURE 1. Schematic of the experimental equipment and the high-pressure cell for X-ray diffraction measurement. The vertical axis of the cell is the ϕ-axis.

goniometer. The inner diameter and the thickness of the cell were 11.0 mm and 2.0 mm, respectively. The X-ray transmissivity of the cell was 30% at 20 keV and 76% at 60 keV. The temperature of the cell was controlled to ±0.1 K by blowing cold, dry air around it. The energy dispersive spectra were measured by a Cd-Zn-Te solid state detector. The spectra accumulated continuously, and the data were retrieved every 3,600 s. The incident X-ray beam was collimated by a long-sleeve slit 0.40(W) × 2.00(H) × 110(L) mm. The diffracted X-ray was also collimated by two 0.40 mm width line slits to eliminate the diffraction by the cell. The 2θ angle was fixed at 5.00° for all experiments. The cell was rotated 90° about the ϕ-axis during each measurement to include many crystals.

Ice particles with average diameter about 150 μm were put in the high-pressure cell. The total ice particle volume was about 1.0×10^3 mm^3. These ice particles made by spraying high-purity water (18.3 MΩ·cm) on liquid nitrogen and then sieved out in a −20°C cold room. The initial bulk density of this porous mass of ice was about 0.50 g/cm^3. To avoid ice particle sintering, the experiments began within one hour of putting the ice particles into the cell. Just before putting compressed CO_2 gas into the cell, the samples were in vacuum for 30 s to remove air from the cell and pipe. *In situ* observations were made at 0.98 MPa with temperatures of 238, 244, 255, 262.5, and 269 K; and at 1.48 MPa and 272.5 K.

RESULTS

FIGURE 2 show the change in X-ray diffraction profiles from I_h to CO_2 hydrate. Each initial profile is that of I_h before putting CO_2 gas into the cell. After one or two hours, several diffraction peaks from the CO_2 hydrate appeared and continued to increase with time, whereas the ice diffraction peak heights decreased with time. Since the integrated intensity of X-ray diffraction is in proportion to the crystal volume, their rates of change are measures of the growth and transformation rates. Moreover, based on the experimental results, the growth rates were independent

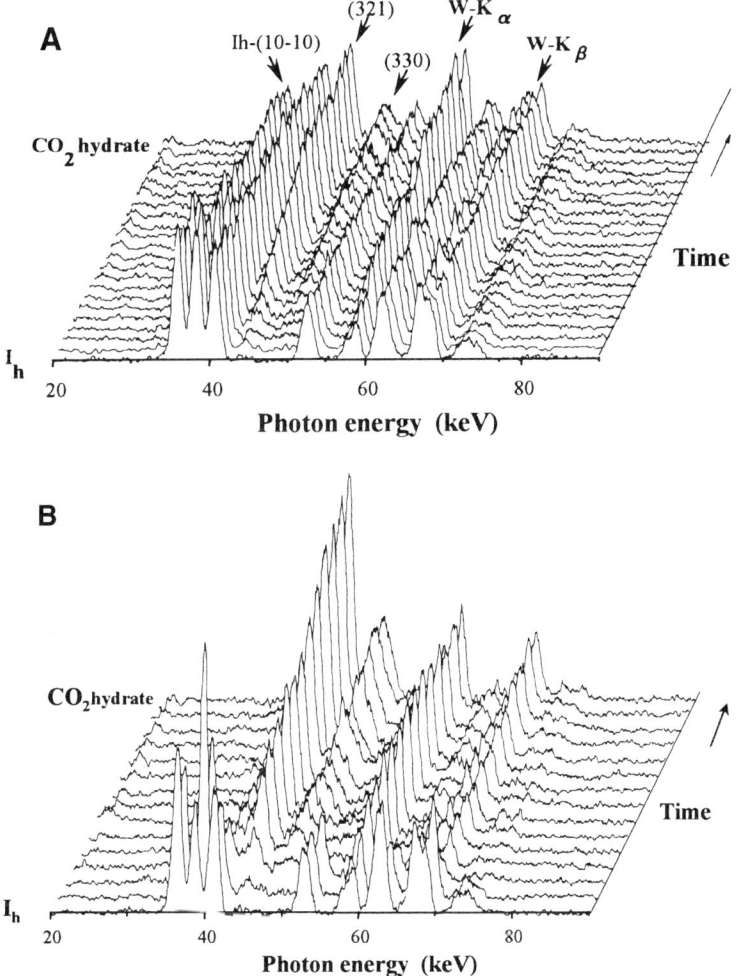

FIGURE 2. *In situ* time-resolved energy dispersive X-ray diffraction profiles during I_h-to-CO_2 hydrate growth. (**A**) 0.98 MPa and 238 K. (**B**) 1.48 MPa and 272.5 K.

of position in the cell. Therefore, it can be assumed that every ice particle transformed to CO_2 hydrate crystal monotonously of the diffusion of CO_2 gas in the high-pressure cell.

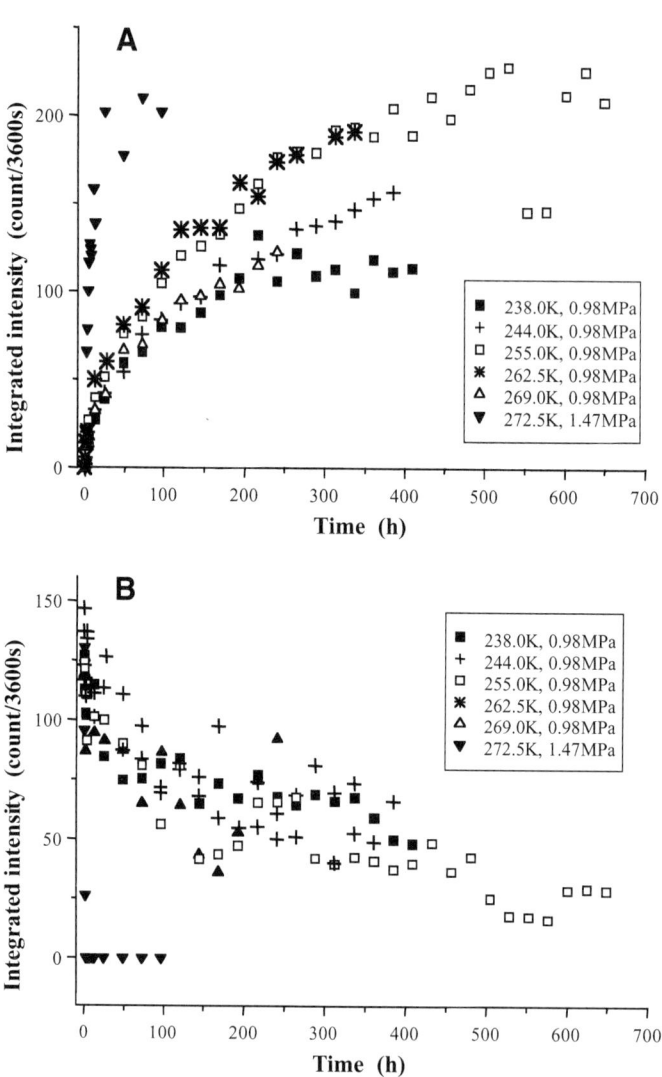

FIGURE 3. (**A**) Comparison of growth rates of CO_2 hydrate represented by the integrated intensity change of the (321) plane. (**B**) Comparison of transformation rates at different conditions. The integrated intensity change of the $(10\bar{1}0)$ plane of I_h. The integrated intensity at 272.5 K and 1.47 MPa (▼) went to zero within two hours of introducing of CO_2 gas.

The integrated intensities of the (321) plane of CO_2 hydrate and the ($10\bar{1}0$) plane of I_h versus time are plotted in FIGURE 3. This figure shows the growth rates of each CO_2 hydrate and transformation rate of I_h under various conditions. The figure indicates that each growth and transformation rate is similar except at 272.5 K. The quantitative analysis is discussed in the following section.

Next, the mass flux of H_2O molecule was estimated using the decreasing ice ratio and increasing CO_2 hydrate ratios determined by the integrated intensities (see FIGURE 4). At 262.5 K, the total mass of H_2O was nearly constant through out the experiment. The other experiments exhibited the same trend. Therefore, nearly every water molecule removed from ice must have solidified into hydrate. However, the total mass of H_2O at 272.5 K decreased and then increased during the first several tens of hours. Because the integrated intensity measures the total crystalline volume, some ice crystal must have melted and then solidified into CO_2 hydrate crystal. However, since CO_2 hydrate covered the ice particles completely at first, and the meltwater was contained inside the shell. Therefore, it would take 20–40 h to diffuse all of the water or CO_2 molecules through the hydrate shell to transform into CO_2 hydrate.

MODEL CONSTRUCTION

Based on the experimental results, we formulated the following two-stage model.
1. A CO_2 hydrate nucleates on an ice–CO_2 gas boundary; the ice particles are fully covered by CO_2 hydrate within several hours.

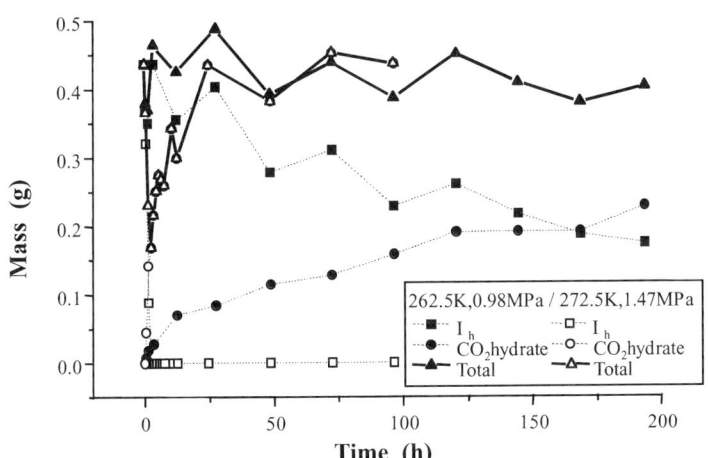

FIGURE 4. Mass flux of H_2O molecules during growth. Here the mass of I_h and CO_2 hydrate express the estimated mass of I_h and CO_2 hydrate in the high-pressure cell as determined by their integrated intensities respectively. The total mass is the summation of both.

2. The CO_2 hydrate layer increases its thickness by the diffusion of interstitial CO_2 and H_2O molecules through the layer.

The growth rate in Stage (1), υ_I can be expressed using the excess pressure, $p_a - p_d(T)$[3],

$$\upsilon_I = K_I(p_a - p_d(T)), \qquad (1)$$

where K_I is the reaction rate constant and $p_d(T)$ is dissociation pressure of CO_2 hydrate at temperature T K. The later is approximately[4]

$$p_d(T) = \frac{e^{18.594 - 3161.41/T}}{1000} \text{ [MPa]}. \qquad (2)$$

The volume change of ice in Stage (2) can be written according to a diffusion equation[2]

$$\frac{dV_i}{dt} = -4\pi D \frac{r_i r_h}{r_h - r_i} \left(\frac{p_a - p_d(T)}{P_a}\right) \frac{\rho_{hw}}{\rho_i}. \qquad (3)$$

Here t is reaction time, V_i is the volume of ice; and r_i and r_h are radius of ice particle and hydrate, respectively. ρ_i is the ice density (0.918 cm^3/g) and ρ_{hw} is the water density of the hydrate (0.796 cm^3/g). D is an inter-diffusion coefficient expressed by

$$D = (1 - y_d)\frac{D_{hc}}{y_d} + \nu c_d D_{hw}, \qquad (4)$$

where D_{hc} and D_{hw} are the diffusion coefficients of CO_2 and H_2O molecules in the hydrate crystal near equilibrium thermodynamic condition, respectively. ν is the number of cages per one H_2O molecule in a hydrate cell ($\nu = 4/23$ for structure I), and c_d is the local concentration of interstitial water molecules in the hydrate.

DISCUSSION AND CONCLUSIONS

Based on the model and experiments, we interpreted the growth mechanism of CO_2 hydrate during the surface coverage stage, Stage (1), and diffusion-controlled stage, Stage (2), as follows.

As FIGURE 3(A) shows, the integrated intensities increased linearly for the initial several hours and then the growth rate gradually decreased. Therefore, the surface probably became covered within the first few hours. We assumed that it took two hours to completely cover the ice particles with CO_2 hydrate. The initial reaction rate constant, K_I, was calculated by substituting the integrated intensity change rate of (321) and (330) planes for υ_1 in (1), respectively; plotted in FIGURE 5. This figure shows that the constants below 3.72×10^{-3} K^{-1}, that is, above 269 K, are extremely large. Hwang et al., reported that measurable formation rates of CH_4 hydrate were observed only when melting ice was involved.[5] Furukawa et al., reported that the critical temperature at which the quasi-liquid layer (QLL) was detected on the hexagonal ice surface is 271 K for the {0001} face, and 269–271 K for the {10$\bar{1}$0} face; moreover, the thickness of the layer rapidly increases as the temperature approaches the melting point under atmospheric conditions.[6] The existence of a liquid-like layer on the ice surface might cause such an increase of the initial reaction rate constant.

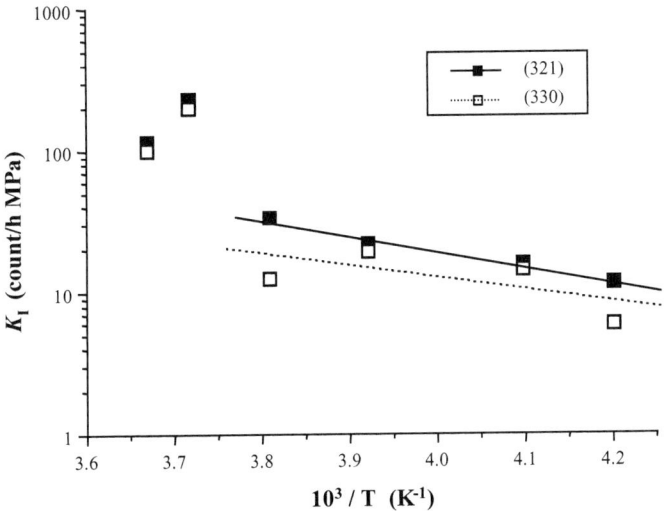

FIGURE 5. Comparison of initial growth coefficients.

The straight lines in FIGURE 5 were obtained by the four constants above 3.81×10^{-3} K^{-1} using the least-squares method. As a result, the activation energy for the reaction at the ice surface of surface coverage stage was about 0.2 eV. This value does not equal to the activation energy of diffusion of H$_2$O molecule in I$_h$, 0.6 eV, but is close to that in the QLL, 0.24 eV.[7] Moreover, Mizuno et al., measured mobile molecules at the surface down to 173 K.[7] Therefore, the nucleation and growth of hydrate on the ice surface might be determined by migration of H$_2$O molecules on the ice surface.

Next, we discuss the diffusion-controlled stage. Comparing the rate of I$_h$ decreased with the model prediction using Equation (3), the inter-diffusion coefficients were calculated. They are plotted in FIGURE 6. The finding is that the diffusion coefficients of CO$_2$ hydrate are one to two orders smaller than that of air hydrate. Between 238 K and 263 K, the slope of the curve in FIGURE 6 gave a diffusion activation energy of 0.40 eV. This value is consistent with the value for air hydrate reported by Salamatin et al.[2] This result might imply that the diffusion coefficient is determined by the diffusion rate of H$_2$O molecules. Because both CO$_2$ hydrate (Structure I) and air hydrate (Structure II) are composed of the same small cages connected by the same pentagonal and hexagonal faces, the diffusion of N$_2$, O$_2$, and CO$_2$ might give different activation energies because of their different molecular size.

On the other hand, FIGURE 7(a) shows the integrated intensity ratio change of CO$_2$ hydrate, $I_m(330)/I_m(321)$. This figure shows that the ratios decreased to an equilibrium value within about a hundred hours. In fact, the diffraction peak of the (330) from FIGURE 2(a) is the summation of the (400), (410), and (330) peaks; $I_m(330) = I_c(330) + I_c(400) + I_c(410)$, whereas $I_m(321) = I_c(321)$. Here $I_m(hkl)$ is the

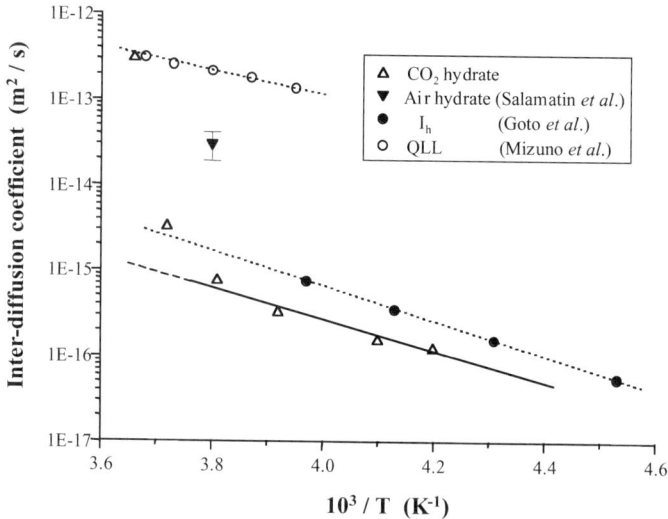

FIGURE 6. Comparison of diffusion coefficients.

measured integrated intensity of the (hkl) plane and $I_c(hkl)$ is the calculated value using atomic parameters determined by McMullan et al.[9] The intensity ratio was expressed as a function of the site occupancy of the small and large cages, and is plotted in FIGURE 7(b). This figure shows that the intensity ratios decrease as the site occupancy increases. Consequently, FIGURE 7(a) shows that site occupancy of the large and small cages are small at the beginning of growth and it takes several tens of hours to reach equilibrium. This result implies that the site occupancy of CO_2 hydrates come to equilibrium by the diffusion of guest molecules through its vacant cages after the formation of their crystallographic structure.

Based on these results, we estimated the respective diffusion coefficients, D_{hW} and D_{hC}, under the following assumption. The self-diffusion coefficient in I_h is larger than that in CO_2 hydrate; hence, the local concentration of interstitial water molecules of CO_2 hydrate can equilibrate with that in I_h at the CO_2 hydrate–ice boundary. Above 263 K, the crystalline structure of ice at the CO_2 hydrate–ice boundary might not be perfect. The local concentration of interstitial water molecules, c_d, at 273 K does not correspond to the value of I_h, $c_d = 2.8 \times 10^{-6}$,[8] but may be nearly equal to that of liquid water, $c_d \sim 1$. The increase of interstitial water molecules could have caused the increase of the self-diffusion coefficient. The extrapolated diffusion coefficient at 272.5 K from experimental results below 263 K is 1.2×10^{-15} m^2/s. The local concentration of interstitial water molecules correspond to the value of I_h at 273 K. On the other hand, the total site occupancy of CO_2 hydrate is greater than 0.9 based on the Langmuir isotherm.[10] However, as FIGURE 7 shows it might take a hundred hours to be equilibrium. Therefore, the total site occupancy, y_d, is assumed to be 0.8. Substituting these values in **(4)**, the diffusion coefficients of H_2O and CO_2 molecules in hydrate at 272.5 K are $D_{hw} \sim 10^{-12}$ m^2/s

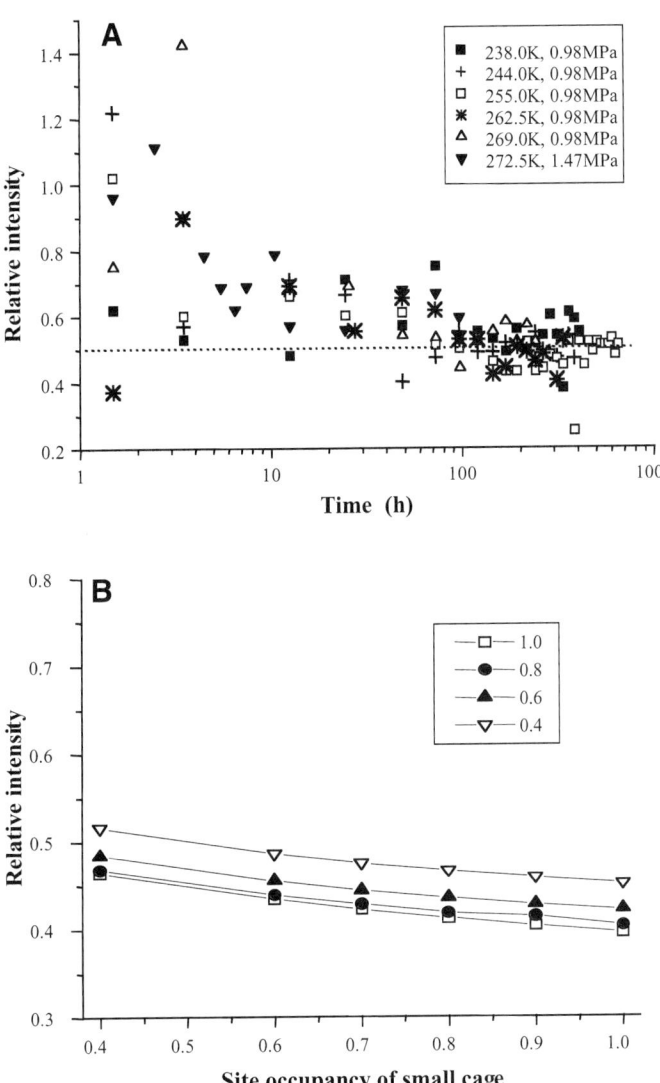

FIGURE 7. (**A**) Integrated intensity ratios of CO_2 hydrate, $I_m(330)/I_m(321)$. The dotted line is drawn as a guide. (**B**) Calculated intensity ratios of CO_2 hydrate, $(I_c(330) + I_c(400) + I_c(410))/I_c(321)$, as a function of small and large cage site occupancy. Each value of the signs expresses the site occupancy of the large cage. To calculate ideal diffraction intensity for each site occupancy, the CO_2 molecules were assumed to be in the center of each cage.

and $D_{hc} \sim 10^{-14}$ m²/s, respectively. Consequently, our experimental results suggest that the growth rate of CO_2 hydrate is determined by the diffusion rate of interstitial H_2O molecules in the hydrate.

ACKNOWLEDGMENTS

We thank Professor A.N. Salamatin for useful discussions on the model of diffusion of this research.

REFERENCES

1. LUND, P.C., Y. SHINDO, Y. FUJIOKA & H. KOMIYAMA. 1994. Study of the pseudo-steady-state kinetics of CO_2 hydrate formation and stability. Int. J. Chem. Kinet. **26**: 289–297.
2. SALAMATIN, A.N., T. HONDOH, T. UCHIDA & V.YA. LIPENKOV. 1998. Post-nucleation conversion of an air bubble to clathrate air-hydrate crystal in ice. J. Cryst. Growth **193**: 197–218.
3. UCHIDA, T., T. HONDOH, S. MAE, P. DUVAL & V.YA. LIPENKOV. 1994. Effects of temperature and pressure on transformation rate from air bubbles to air-hydrate crystals in ice sheets. Ann. Glaciol. **20**: 143–147.
4. SLOAN, E.D. 1998. Clathrate Hydrate of Natural Gases, 2nd edit. Marcel Dekker, Inc., New York.
5. HWANG, M.J., D.A. WRIGHT, A. KAPUR & G.D. HOLDER. 1990. An experimental study of crystallization and crystal growth of methane hydrates from melting ice. J. Incl. Phenomena **8**: 103–116.
6. FURUKAWA, Y. & H. NADA. 1997. Anisotropic surface melting of an ice crystal and its relationship to growth forms. J. Phys. Chem. B **101**: 6167–6170.
7. MIZUNO, Y. & N. HANAFUSA. 1987. Studies of surface properties of ice using nuclear magnetic resonance. J. Phys. Colloque Cl. Supplement. **48**: Cl-511–Cl-517.
8. GOTO, K., T. HONDOH & A. HIGASHI. 1986. Determination of diffusion coefficients of self-interstitials in ice with a new method of observing climb of dislocations by X-ray topography. Jpn. J. Appl. Phys. **25**: 351–357.
9. MCMULLAN, R.K. & G.A. JEFFREY. 1965. Polyhedral clathrate hydrates. IX. Structure of ethylene oxide hydrate. J. Chem. Phys. **42**: 2725–2732.
10. PARRISH, W.R. & J.M. PRAUSNITZ. 1972. Dissociation pressure of gas hydrates formed by gas mixtures. Ind. Eng. Chem. Fundam. **11**: 26–35.

High-Pressure Optical Cell for Hydrate Measurements Using Raman Spectroscopy

V. THIEU,[a] S. SUBRAMANIAN,[b] S.O. COLGATE,[c] AND E.D. SLOAN, JR.[b,d]

[a]Argonne National Laboratory, Argonne, Illinois 60439, USA

[b]Center for Hydrate Research, Colorado School of Mines, Golden, Colorado 80401, USA

[c]Department of Chemistry, University of Florida, Gainesville, Florida 32612, USA

ABSTRACT: A high-pressure optical cell was designed and fabricated for the measurement of hydrate phase equilibria. These measurements were carried out using Raman spectroscopy, which provided molecular microscopic evidence of hydrate dissociation, along with macroscopic temperature and pressure evidence. Initial measurements were made of hydrates formed from methane (CH_4) dissolved in deionized water. Distinct signatures of CH_4 in different phases (aqueous and sI hydrate) were the basis for this study. Time resolved Raman spectra monitored the dissociation of hydrate as cell temperature was increased. The spectra depicted the transition of CH_4 in the two cavities of sI hydrate to CH_4 dissolved in water. A pressure trace of the experiment indicated a slight jump in pressure of about 30 psi (0.2 MPa) at the hydrate dissociation point, which agreed with the Raman spectroscopic data. Experimental data obtained using this method agreed well with literature data for the simple methane system. The technique demonstrated that the optical cell is suitable for Raman measurements of hydrate phase equilibria at high pressures. Ultimately, this procedure will be used to measure thermodynamic conditions for more complex systems.

INTRODUCTION

Existing hydrate equilibrium databases do not extend beyond about 4,000 psia (27.4 MPa), because that is the highest pressure presently required for the gas processing industry. Exploration in deep water generates an increasing need for higher pressure equilibrium data, particularly in conjunction with inhibitors, such as methanol, glycol, and salts. A hydrate optical cell was designed and fabricated for the measurement of high pressure hydrate phase equilibria. These measurements were carried out using Raman spectroscopy, to provide molecular microscopic evidence of hydrate dissociation, together with macroscopic temperature and pressure evidence. The use of Raman as a means to study hydrate has recently been employed by several groups of researchers.[1–5] A simple methane hydrate system was studied initially and is reported here to confirm our technique. Ultimately, other systems of industrial interest (i.e., with inhibitors) will be examined.

[d]Address for correspondence: E.D. Sloan, Jr., Center for Hydrate Research, Colorado School of Mines, Golden, CO 80401, USA.
  esloan@gashydrate.mines.edu

EXPERIMENTAL METHODS

Optical Pressure Cell

A brass cell was designed to operate safely at pressures up to 10,000 psi (70 MPa) and this is unique in several ways. It is compact, about 7.60 cm by 3.80 cm in size. The hydrate-forming compartment is approximately 0.5 ml in volume. The cell contains sapphire windows to permit the use of laser light to probe the hydrate sample. Finally, the high pressure cell has built-in channels to allow for coolant circulation and to provide temperature control.

The cell was constructed from three major plates, machined to a diameter of 7.60 cm. The top and bottom plates were machined to a thickness of 1.27 cm and the middle plate to a thickness of 1.90 cm. The three plates are held together by Allen cap screws in a six-hole pattern. For viewing purposes, the top and bottom plates are bored to provide a slightly recessed opening. A cross section of the cell design is shown in FIGURE 1.

The middle plate comprises the heart of the apparatus. Two sapphire windows are brazed onto the brass body via a Kovar sleeve, creating a sample compartment with an approximate volume of 0.5 ml. A 0.32-cm well, drilled in the body of the middle plate, allows for a temperature-sensing device (RTD) to be placed within a few millimeters of the sample compartment wall. Two small ports allow for gas flow into the sample chamber. Two larger ports allow for coolant circulation throughout the cell body. FIGURE 2 shows a more detailed views of the middle plate. This unique and compact design permits very rapid heating and cooling of the entire visual cell.

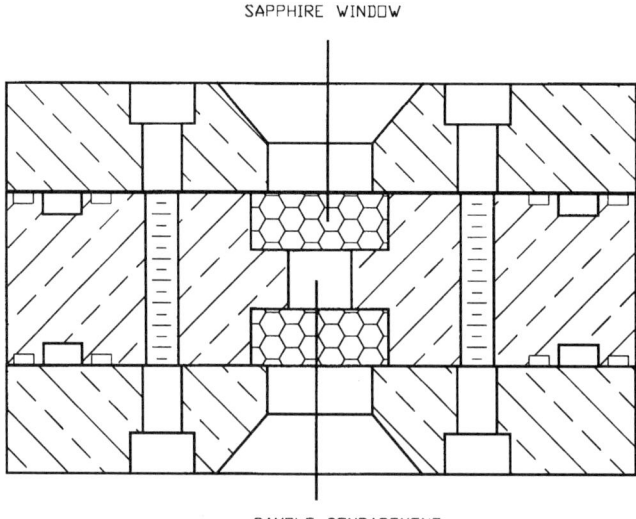

FIGURE 1. Cross section of the high-pressure optical cell.

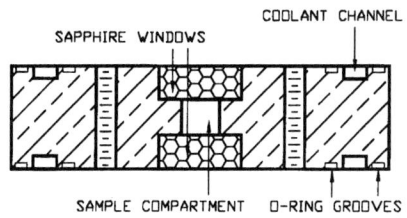

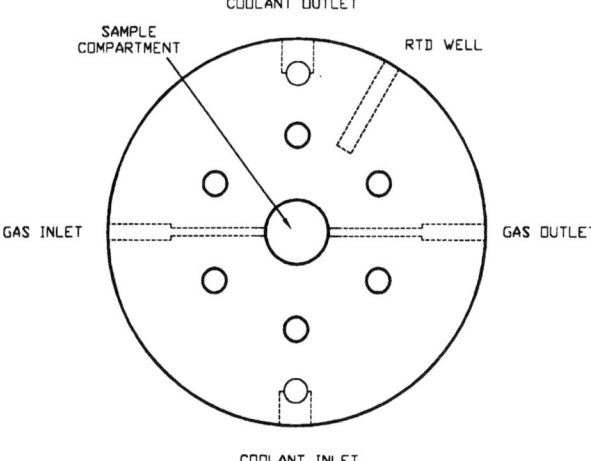

FIGURE 2. Detailed views of the middle plate of the optical cell.

Apparatus

The Raman Spectrometer is a Renishaw Inc. MK III multichannel fiber optic based system equipped with a 2,400 grooves/mm grating and a CCD detector. Excitation is provided by a 514.35 nm, 30 mW Ar-ion green laser, delivered to a fiber probe through fiber optic bundles. A 20× microscope objective on the probe focuses the laser onto the sample. The final excitation at the sample is 10 mW. The spectrometer interfaces with a computer to provide instrument and experimental control. The Raman spectra collected are analyzed using GRAMS32C spectral analysis software package. Routine calibration of the monochromator is performed by using 2,702.5 cm^{-1}, 2,852.6 cm^{-1}, and 2,973.3 cm^{-1} neon emission lines relative to the 514.532 nm plasma line. Details of this Raman Spectrometer were reported by Subramanian.[2]

Procedure

Initially, the cell was loaded with about 0.4 g (cell 80% full) of deionized H_2O. The cell was then charged with methane (Scott Specialty Gases, 99.99% purity) to the working pressure. The temperature was measured using a platinum RTD (Omega Engineering) with ±0.1 K accuracy. Pressure measurements were made with a Heise pressure transducer, having a range of 0–30,000 psi and accurate to 0.1% of full scale. Coolant circulation was commenced so as to cool the cell to about 0°C. Hydrates formed very quickly using this cooling procedure, coupled with manually shaking the cell. A schematic diagram of the experimental set up is shown in FIGURE 3. Each hydrate sample was formed and conditioned by cycling the temperature several times across its equilibrium temperature. Such conditioning was necessary to form a uniform hydrate sample, one with a relative cavity occupancy ratio of about 3 to 1 (large cavity to small cavity ratio for sI hydrate).

In these experiments, the cell was aligned vertically with the laser beam focused along an axis parallel to the liquid–gas interface. Moreover, the cell was affixed to an X–Y translational stage to allow for precise focusing of the laser beam on a specific hydrate mass in the sample and, hence, to optimize the scattering. The laser beam was normally focused at a spot within 1 mm below the liquid–gas interface. Using CSMHYD,[6] an equilibrium temperature was predicted for the hydrate sample at the working cell pressure. To start an experiment, the cell temperature was set at 5°C below the predicted hydrate equilibrium. The cell temperature was then raised slowly at a rate of 0.1°C/minute, without stirring, to a final temperature 5°C above the predicted equilibrium temperature. Temperature, pressure, and spectroscopic Raman data were collected throughout the experiment. For a typical experiment, temperature and pressure data were collected every 10 seconds, whereas a Raman

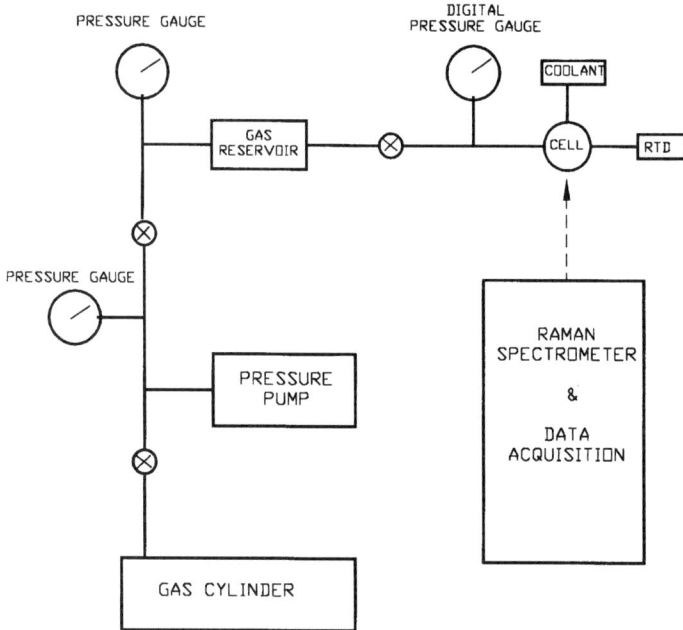

FIGURE 3. Schematic of the experimental setup.

spectrum was collected every 20 seconds. At the end of an experiment the cell pressure was normally increased using the same water loading for the next trial. Our experimental procedure involved the formation of polycrystalline hydrate samples, whereas Nakano et al.[3,4] formed single crystals of CO_2 and CH_4 hydrates in their pressure cell.

EXPERIMENTAL RESULTS AND DISCUSSION

Distinct signatures[1] of methane (CH_4) in different phases (gas, aqueous, and sI hydrate) were the basis for these studies, as is shown in FIGURE 4. The plot shows intensity in arbitrary units (a.u.) as a function of Raman shift (wavenumbers, cm^{-1}) characteristic of methane molecules in various environments. The frequency of the v_1 C-H symmetric stretch for dissolved CH_4 at 31.3 MPa and 298 K occurred at 2,911.3 cm^{-1}. The peaks corresponding to methane in the large $5^{12}6^2$ and small 5^{12} cages appear at 2,904.8 cm^{-1} and 2,915 cm^{-1}, respectively. The peak at 2,917.3 cm^{-1} corresponds to CH_4 in the gas phase at 3.4 MPa. FIGURE 5 shows the peak position of methane in the gas phase as a function of pressure. Over the pressure range 0–10,000 psig, the methane gas peak is shifted to lower wavenumbers with increasing pressure (see TABLE 1.) This calibration plot was important in two respects: (1) knowing the CH_4 gas peak position helps in the correct identification of hydrate peaks (for CH_4 in large $5^{12}6^2$ and small 5^{12} cages of sI hydrate the peaks are at 2,904.8 cm^{-1} and 2,915 cm^{-1}, respectively); and (2) methane gas peak position serves as a quick indicator of cell pressure.

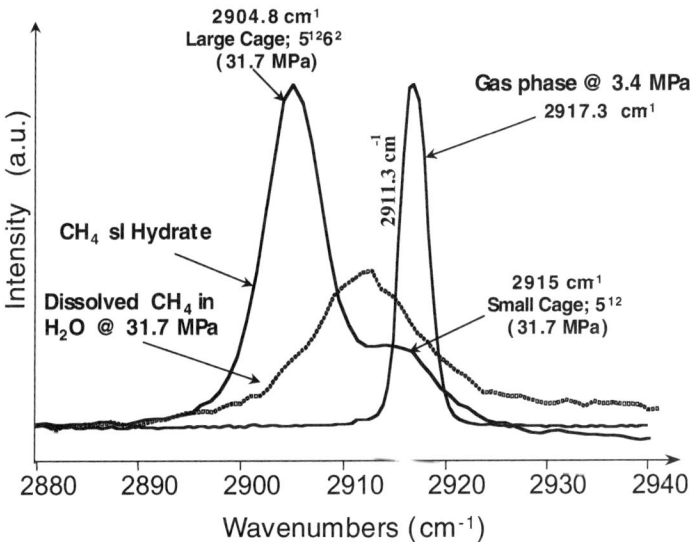

FIGURE 4. Raman spectra of methane in different environments (see text for details).

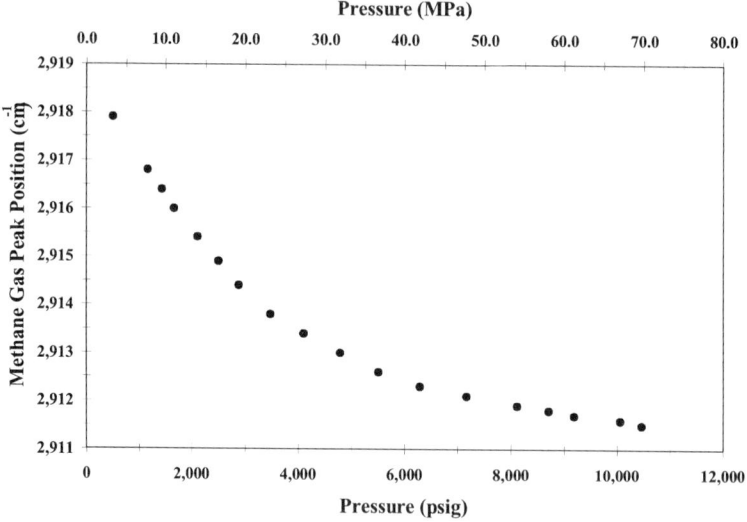

FIGURE 5. Methane peak position (gas phase) as a function of pressure at 298 K. Peak position shifts to lower wavenumbers with increasing pressure.

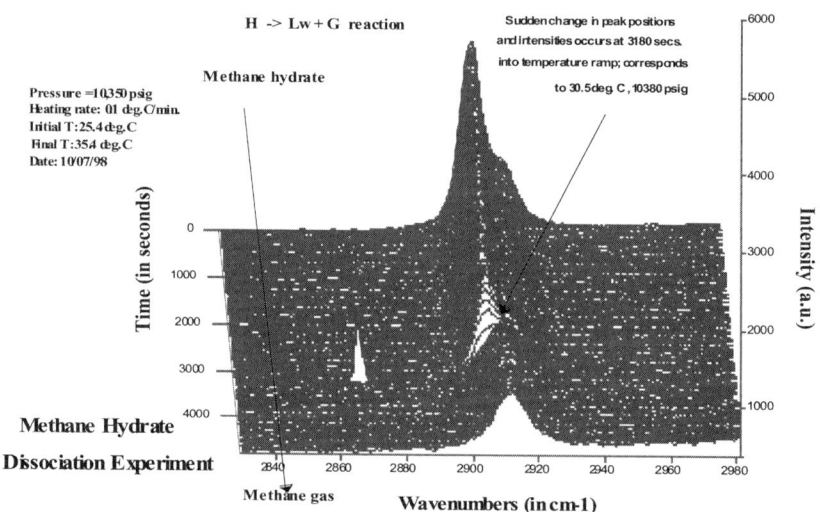

FIGURE 6. Raman spectra of methane hydrate undergoing dissociation (see text for details).

TABLE 1. Methane peak position (gas phase) as a function of pressure at 298 K

Pressure (MPa)	Peak Position (cm^{-1})
3.4	2917.3
8.0	2916.8
9.9	2916.4
11.5	2916.0
14.5	2915.4
17.2	2914.9
19.8	2914.4
24.0	2913.8
28.3	2913.4
33.0	2913.0
38.0	2912.6
43.4	2912.3
49.4	2912.1
55.9	2911.9
60.0	2911.8
63.3	2911.7
69.3	2911.6
72.1	2911.5

FIGURE 6 shows time-resolved Raman spectra monitoring the dissociation of hydrate as cell temperature was increased. The spectra depict the transition of CH_4 in the two cavities of sI hydrate to CH_4 dissolved in water. At the beginning of the experiment (time zero), the Raman signal indicates the presence of sI methane hydrate with a large peak at 2,904.8 cm^{-1} and a small shoulder at 2,915 cm^{-1}. As the cell was slowly heated, the intensities of these two peaks slowly decreased. A sudden change in peak positions and intensities occurred at about 3,200 seconds into the experiment—designating the hydrate dissociation point. The hydrate peaks abruptly disappeared, and a signal corresponding to CH_4 dissolved in water (2,911.8 cm^{-1}) emerged. In FIGURE 6, the hydrate dissociation point occurs at a temperature of 30.5°C and 10,380 psig (303.7 K and 71.6 MPa).

FIGURE 7 shows a typical pressure profile of a hydrate dissociation experiment. The pressure trace indicates a dramatic jump in pressure of about 30 psi (0.2 MPa) near to 299.5 K and 3,300 seconds. This is characteristic of hydrate dissociation. The point where these changes occur agree very well with the Raman spectroscopic data. The pressure increase is striking due to the small cell volume. However, this pressure jump was reproducible in all experiments. Experimental equilibrium data points for the methane hydrate system are provided in TABLE 2. A comparison of our work with literature data[7] for the methane system studied is shown in FIGURE 8. Data obtained

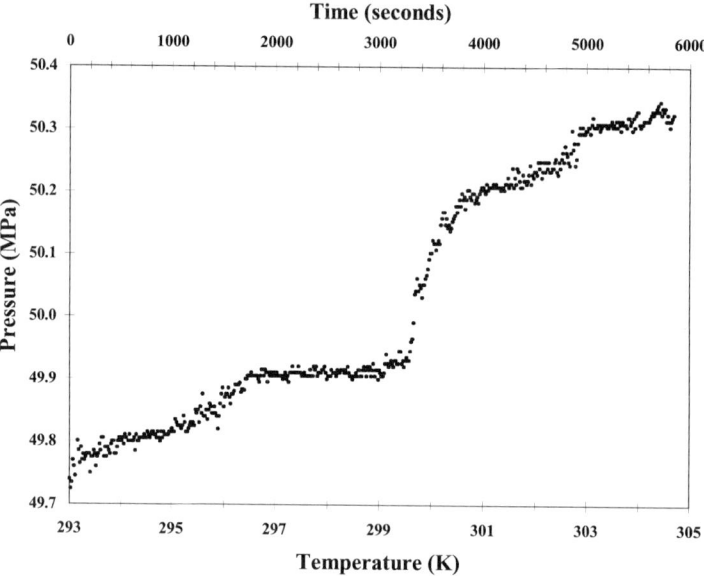

FIGURE 7. Typical pressure profile of a hydrate dissociation experiment. Hydrate dissociation was indicated by a jump in pressure at around 299.5 K and 3,300 seconds. This event coincided with changes in the Raman spectroscopic data.

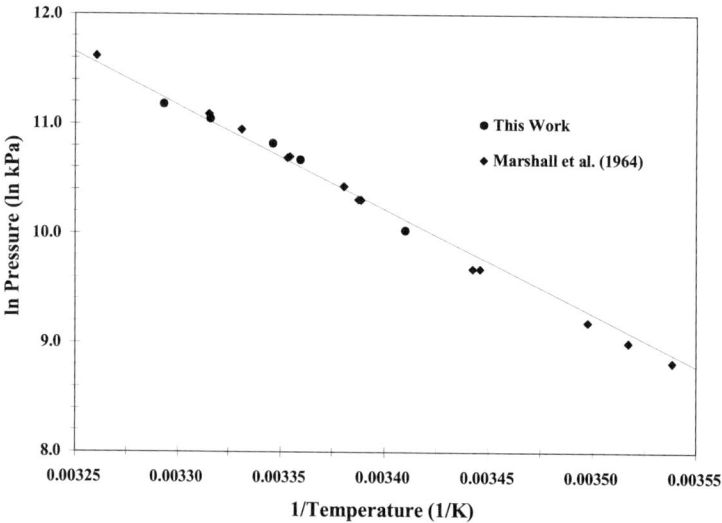

FIGURE 8. Comparison of our work with literature data for the methane hydrate system: ● this work; ◆ Marshall *et al*. (1964).

TABLE 2. Experimental phase equilibrium data (L_w-H-V) for methane hydrate

Temperature (K)	Pressure (MPa)
293.3	22.7
297.7	43.0
298.9	50.0
301.6	62.8
303.7	71.6

using this method are indicated by circles, literature data are shown by the diamonds. Our data agree fairly well with literature data over the temperature and pressure range covered.

CONCLUSIONS

The high sensitivity of cell pressure, along with definitive Raman evidence of changes at the molecular level when hydrates dissociate, demonstrated that the cell and this technique were suitable for Raman measurements of hydrate phase equilibria at high pressures. Ultimately, this procedure will be used to measure thermodynamic conditions for systems of industrial interest. Our current optical cell has a limiting working pressure of 10,000 psi (70 MPa). However, efforts are under way to extend the experimental procedure to 20,000 psi (140 MPa) by means of a new cell.

ACKNOWLEDGMENTS

The following companies are gratefully acknowledged for their financial support: Amoco, ARCO, Chevron, US Department of Energy, Mobil, Oryx, Phillips, Shell, and Texaco.

REFERENCES

1. SUM, A.K., R.C. BURRUS & E.D. SLOAN. 1997. Measurement of clathrate hydrates via Raman spectroscopy. J. Phys. Chem. B **101:** 7371–7377.
2. SUBRAMANIAN, S. & E.D. SLOAN. 1999. Molecular measurements of methane hydrate formation. Fluid Phase Equilibria **158-160:** 813–820.
3. NAKANO, S., M. MORITOKI & K. OHGAKI. 1998. High-pressure phase equilibrium and Raman microprobe spectroscopic studies on the CO_2 hydrate system. J. Chem. Eng. Data **43:** 807–810.
4. NAKANO, S., M. MORITOKI & K. OHGAKI. 1999. High-pressure phase equilibrium and Raman microprobe spectroscopic studies on the methane hydrate system. J. Chem. Eng. Data **44:** 254–257.
5. TULK, C.A., D.D. KLUG & J.A. RIPMEESTER. 1998. Raman spectroscopic studies of THF clathrate hydrate. J. Phys. Chem. A **102:** 8734–8739.

6. SLOAN, E.D. 1998. Clathrate Hydrates of Natural Gases, 2nd edit. Marcel Dekker, New York.
7. MARSHALL, D.R., S. SAITO & R. KOBAYASHI. 1964. Hydrates at high pressures. Methane-water and argon-water systems. AIChE J. **10:** 202–205.

Mechanical Stability of Gas Hydrates under Pressure

V.R. BELOSLUDOV,[a,b] V.P. SHPAKOV,[b] J.S. TSE,[c]
R.V. BELOSLUDOV,[d] AND Y. KAWAZOE[d]

[b]*Institute of Inorganic Chemistry, SB RAS, Novosibirsk, Russia, 630090*

[c]*Steacie Institute for Molecular Sciences, National Research Council of Canada, Ottawa, Ontario, Canada K1A 0R6*

[d]*Institute for Materials Research, Tohoku University, Sendai 980-8577, Japan*

INTRODUCTION

When clathrate hydrates are compressed at high pressure, the crystalline structures collapse into high-density phases. The nature of the phase transition was postulated[1] to be related to the onset of mechanical instability by analogy with the amorphization of ice under pressure. However, unlike ice, the dense structures revert back to the original crystalline hydrate structures when allowed to recover at ambient pressure. This *memory effect* was observed experimentally at $T = 77$ K in Structure II THF and SF_6 clathrate hydrates, and was confirmed by molecular dynamics calculation for the Structure I ethylene oxide (EO) and Structure II Kr hydrates.[1] Furthermore, the volume–pressure behavior near the phase transition was found differ for hydrate with different enclathrated guests. For example, the experiments show a very sharp transition at $P = 13$ kbar for THF hydrate, a much smoother volume decrease at $P = 15.8$ kbar for SF_6 hydrate, and the absence of sudden volume change for Xe hydrate when compressed to 18 kbar.[1] In order to investigate the mechanism for this novel transformations, the quasiharmonic lattice dynamics method[2,3] and a new geometry optimization scheme for the accurate description of structural details, are used to compute elastic constants for the examination of the mechanical stability of clathrate hydrate with different arrangements of proton positions under compression. For comparison, methane and xenon hydrates and a hypothetical empty Structure I hydrate were investigated.

THEORETICAL DETAILS

Lattice dynamics (LD) is a well established technique to calculate the vibrational spectra of molecular crystals.[4,5] The quasiharmonic approximation employed here gives the equation of state correct to first order in the anharmonicity of the potential. The anharmonicity affect the vibrational frequencies that depend on the structural parameters and directly on the temperature of the system. At a finite temperature T

[a]Telecommunication. Voice: 7-3832-343057; fax: 7-3832-344489.
bel@casper.che.nsk.su

under an applied stress $\sigma_{\alpha\beta}$ the equilibrium shape of crystal unit cell (external coordinates) is determined by the state equation:

$$\frac{1}{V_0}\left(\frac{\partial F_{qh}}{\partial \eta_{\alpha\beta}}\right)_0 = \sigma_{\alpha\beta}, \tag{1}$$

where $\eta_{\alpha\beta}$ is the strain tensor, and the free energy taken in quasiharmonic approximation is $F_{qh} = U + F_{vib}$. Here U is the potential energy of crystal and F_{vib} is the vibrational part of the free energy. The derivatives in **(1)** with respect to $\eta_{\alpha\beta}$ are calculated for the reference (equilibrium) configuration and V_0 is the corresponding volume. The variation of the equilibrium strain tensor under small change of the stress tensor or temperature can be obtained[4] as

$$\Delta\eta_{\alpha\beta}^{eq} = -\frac{1}{V}\sum_{\sigma\tau} C_{\alpha\beta\sigma\tau}^{(is)-1} F_{T,\sigma\tau}^{vib} \Delta T + \frac{1}{V}\sum_{\sigma\tau} C_{\alpha\beta\sigma\tau}^{(is)-1} \Delta\sigma_{\sigma\tau}, \tag{2}$$

where $F_{T,\sigma\tau}^{vib} = (\partial^2 F_{vib}/\partial T \partial \eta_{\sigma\tau})_0$ and $C_{\alpha\beta\sigma\tau}^{(is)-1}$ is the inverse of isothermal elastic matrix $C_{\alpha\beta\sigma\tau}^{(is)}$. The expression for $C_{\alpha\beta\sigma\tau}^{(is)}$ [6] can be taken in quasiharmonic approximation:

$$C_{\alpha\beta\sigma\tau}^{(is)} = \left(\frac{\partial^2 F_{ph}}{\partial \eta_{\alpha\beta}\partial \eta_{\sigma\tau}}\right)_0 - P(\delta_{\alpha\sigma}\delta_{\beta\tau} + \delta_{\alpha\tau}\delta_{\beta\sigma} - \delta_{\alpha\beta}\delta_{\sigma\tau}). \tag{3}$$

Equation **(2)** is the key expression used in the optimization of the unit cell. Under pressure (P), the new cell parameters can be calculated from $\Delta\eta_{\alpha\beta}^{eq}$ ($\Delta\sigma_{\sigma\tau} = -\Delta P\delta_{\sigma\tau}$). Similarly, the temperature variation of the unit cell, starting from the known point with isotropic stress tensor, can be computed. In the former case the correction to the strain tensor $\eta_{\alpha\beta}^{cor}$ is calculated from **(2)** as follows: $\Delta\sigma_{\alpha\beta}^{cor} = -P_0\delta_{\alpha\beta} - \sigma_{\alpha\beta}$, where the mean stress $P_0 = -1/3(\sigma_{11} + \sigma_{22} + \sigma_{33})$. In this way, after the correction, a new cell shape corresponding to the isotropic stress tensor (which is equal to $-P_0\delta_{\alpha\beta}$) can be obtained. In the above procedure, knowledge of isothermal elastic constants is required. The method calculation for elastic constants $C_{\alpha\beta\sigma\tau}^{(is)}$ of molecular crystals is described in detail elsewhere[2,3] and will not be presented here.

To optimize the internal coordinates, the Newton-Raphson[7] method is used. In this case, for each molecule, the internal coordinates are the positions of center mass and three rotation variables. Taking into account the large number of water molecules in the unit cell, the zero static internal stress approximation (ZSISA)[8] is employed. In this approximation, the contribution of the vibrational part of the free energy is ignored for internal coordinate optimization and, therefore, the optimization problem reduces to minimization of potential energy with respect to internal coordinates.

The general Born criteria are used in our calculation to monitor mechanical stability under pressure. These criteria consist of the requirement of positive definiteness of all principal minors D_k ($k = 1,...,6$) for elastic constant matrix C_{ij} ($i, j = 1,...,6$) written in Voight notation.[9] The principal minor D_k is the determinant of the matrix of principal minor C_{ij} ($i, j = 1,...,k$).

COMPUTATIONAL DETAILS

The reference configuration for quasiharmonic lattice dynamics calculations is a single unit cell of Structure I clathrate hydrate, which contains 46 water molecules and eight methane or xenon guests, with a lattice parameter $a = 11.82$ Å at $T = 10$ K. The positions of the oxygen atoms and the guests were taken from the X-ray analysis of ethylene oxide hydrate by McMullan and coworkers.[10] The protons were placed according to the Bernal-Fowler[11] rules, and the water molecules were oriented such that there is no net dipole moment in the unit cell. This procedure for the assignment of proton positions does not produce a unique structure. The change in the calculated properties with the choice of proton positions were examined for two structural models. The interactions between the water molecules were described with a modified TIP4P potential. The TIP4P can be improved by scaling the interaction parameters by a constant $K = 1.0066$. The scaling factor for the effective charges is K^{-2}. The parameters describing short-range interaction between the oxygen atoms, σ, are scaled by K^{-1} and the energy parameter, ε, by K^{-3}. Finally, all distances between the interacting centers on the water molecule is scaled by K^{-1}. By this scaling, the modified TIP4P potential significantly improves the agreement between the calculated cell parameters for ice Ih with the experimental values without deteriorating the other calculated quantities. The guests (xenon and methane) are considered spherically symmetric Lennard-Jones particles. The potential parameters for the methane–methane and xenon–xenon interactions are taken from other investigations.[12,13] The interactions between different types of atoms are evaluated from the Lorentz-Berthelot mixing rules. Full occupancy by the guests of eight empty cages is assumed in the calculations.

Starting from the reference state, the calculations proceeded as follows. Initially, the reference cell was obtained by optimization of the host water and guest positions in the initial unit cell. After that, the stress tensor and the elastic constants were computed. An optimization of cell shape was carried out if the calculated stress tensor contains nonisotropic components. In the ensuing step, a new shape of the unit cell under the variation of T or P was determined according to (**2**). In this way, evolution of hydrate structure as the function of temperature or pressure can be modeled. At all calculated (T, P) points the principal minors D_k ($k = 2,..., 6$) of the elastic constants matrix were calculated to test mechanical stability according to the Born criteria. The radial distributions of equilibrium oxygen positions around the centers of the large and small cavities were studied as functions of pressure.

RESULTS AND DISCUSSION

At Ambient Pressure

We first examined the adequacy of the lattice dynamics method and the potential parameters used in calculations. The mechanical instability point is independent of the proton preparation procedure both for methane (190–200 K) and for xenon (330–340 K) hydrates. To illustrate the results, the temperature variation of the principal minors of the elastic constants matrix for methane hydrate is shown in FIGURE 1. It is interesting to note that the calculated mechanical instability point for

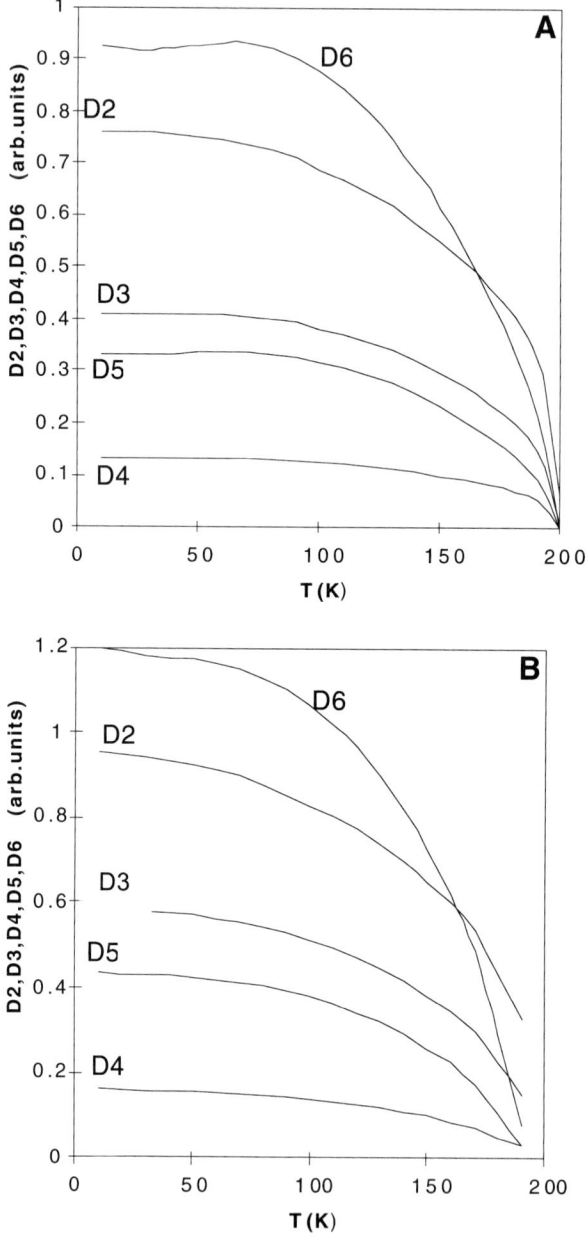

FIGURE 1. A comparison of the principal minors variation (**A**) for model A, (**B**) for model B, of methane hydrate versus T at ambient pressure.

methane hydrate is in a good agreement with the experimentally observed melting point ($T = 193$ K) and the calculated thermal expansion agrees well with experiment measurements.[3]

At High Pressure

The mechanical stability of methane and xenon hydrates at 10 K for two structural models (A and B) with zero dipole moment and different proton configurations were studied. In FIGURE 2 the pressure variation of principal minors is presented (1) for model A, and (2) model B of xenon hydrate. For model A the instability occurs at $P = 16.5$ kbar. In contrast, for model B, the transition pressure occurs at $P = 24.5$ kbar. A similar finding is obtained for the methane hydrate. In this case the calculated transition pressures are 13 kbar and 16.5 kbar, respectively. It is reasonable to speculate that the discrepancy in the calculated transition pressures is related to details in the differences between the two structures. To investigate this possibility, the radial distribution of the oxygen positions around the centers of the cavities for xenon hydrate were examined. The radial distributions, averaged on all large or small cavities in the unit cell, are shown in FIGURE 3. The radial distributions for the large cavities are the same at ambient pressure (FIG. 3a) but differ at higher pressure (FIG. 3b). In fact, the higher the pressure, the larger the difference in the profile of the radial distribution for the large cavities from the ambient pressure structure. For structural model B of xenon hydrate, the hydrate structure is stable up to $P = 24.5$ kbar (FIG. 3c). At 24.5 kbar, the profile of the radial distribution for large cavities differs significantly from the initial shape. In contrast, the oxygen atom around the centers of small cavities does not change significantly with increasing pressure (FIG. 3d). This observation leads to the conclusion that the water molecules forming the small cavities are more rigid and, therefore, the structural changes with pressure are likely related to remaining water molecules in the unit cell forming the hydrate framework.

Calculations on two empty lattices with different proton orders were performed. Contrary to the full guest occupancy, both of empty structures become mechanically unstable at the same pressure ($P = 11$ kbar). It is noteworthy that the calculated transition pressure is in good agreement with the result of a previous molecular dynamics calculation on the empty lattice using the TIP4P potential.[1] The calculated pressures at the onset of mechanical instability for methane and xenon hydrates are in the same range as observed experimentally for other hydrates.

Several discussions can be made from the results presented above. The calculated pressures at the onset of mechanical instability for methane and xenon hydrates depend on the arrangement of hydrogen atoms in the unit cell. The transition pressure seems to increase with guest size. The calculated results are rationalized as follows. Cubic Structure I can be considered to be a structure composed of dedecahedra forming the small cavities (see FIGURE 4). The centers of the dedecahedra form two primitive cubic sublattices that are displaced relative to each other along the vector $\{0.5a, 0.5a, 0.5a\}$ where a is the cell parameter. The distance between the centers of neighboring dedecahedra from the same sublattice is equal to a. The distance between the centers of neighboring dodecahedra from different sublattices is equal to $a\sqrt{3}/2$. The dodecahedra are held together by hydrogen bonding with the remaining water molecules (six in the unit cell of Structure I hydrate). The main

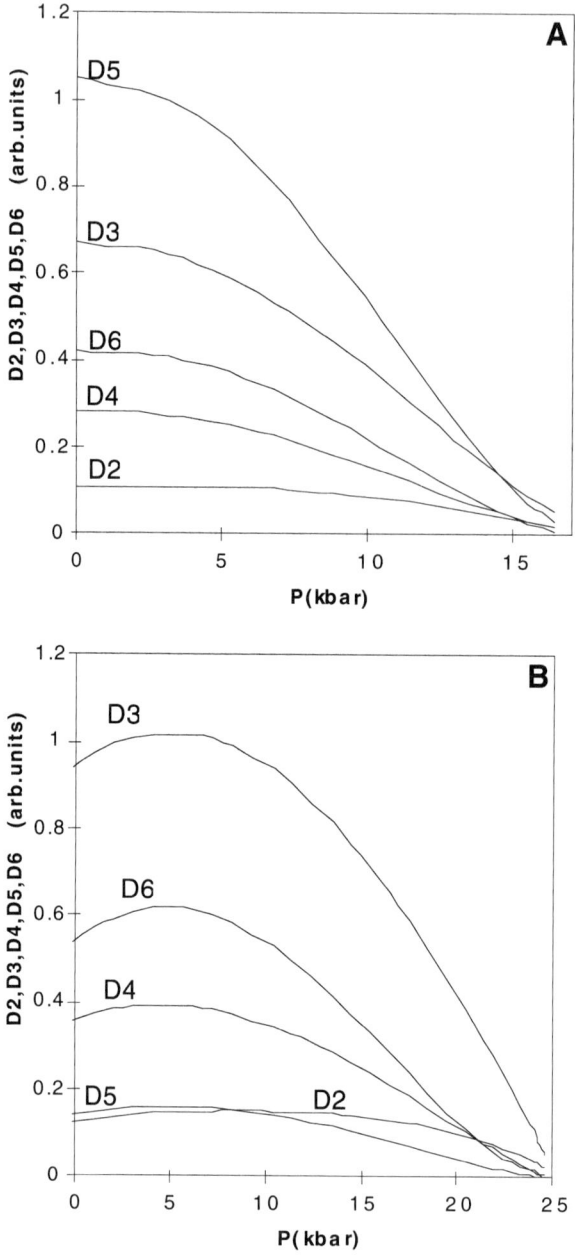

FIGURE 2. A comparison of the principal minors variation (**A**) for model A, (**B**) for model B, of xenon hydrate versus P at $T = 10$ K.

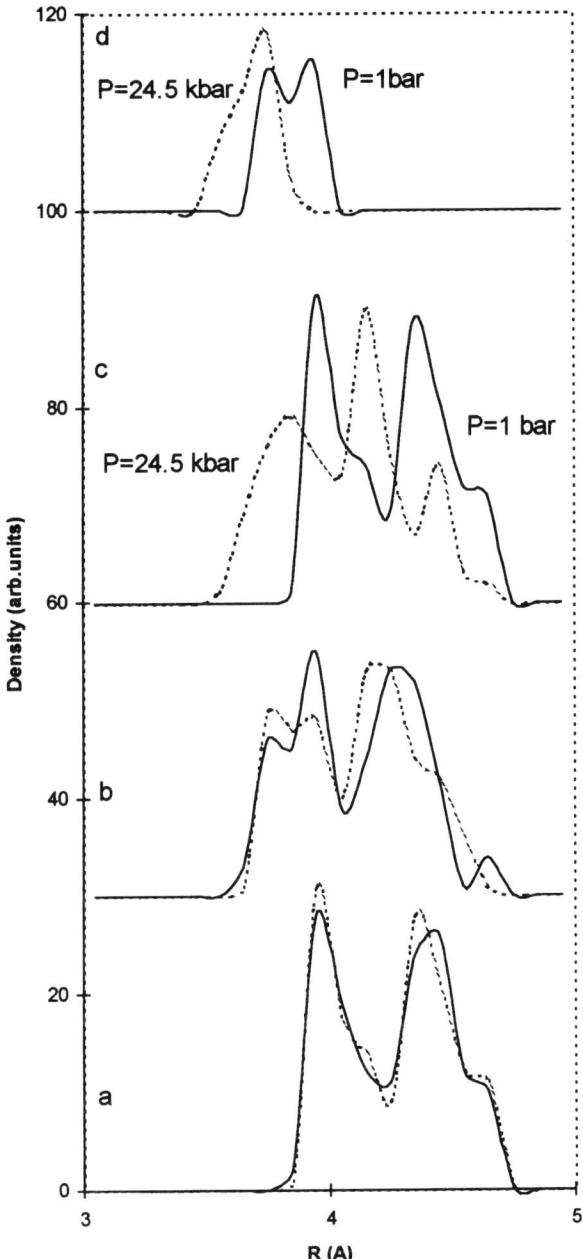

FIGURE 3. The averaged radial distributions of equilibrium oxygen positions around the centers of large and small cavities for the xenon hydrate at $T = 10$ K, comparison for (**a**) models A and B, large cavities at ambient pressure; (**b**) models A and B, large cavities at $P = 16$ kbar; (**c**) model B, large cavities at $P = 24.5$ kbar and ambient pressure; (**d**) model B, small cavities at $P = 24.5$ kbar and ambient pressure.

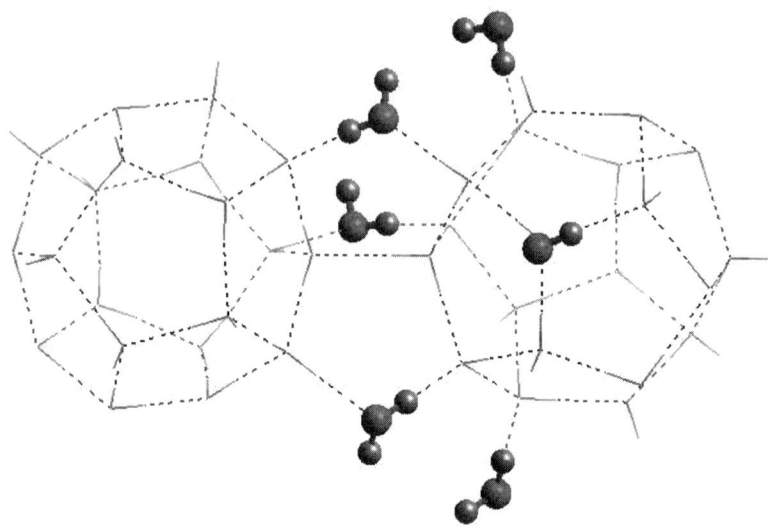

FIGURE 4. The unit cell of structure I clathrate hydrate.

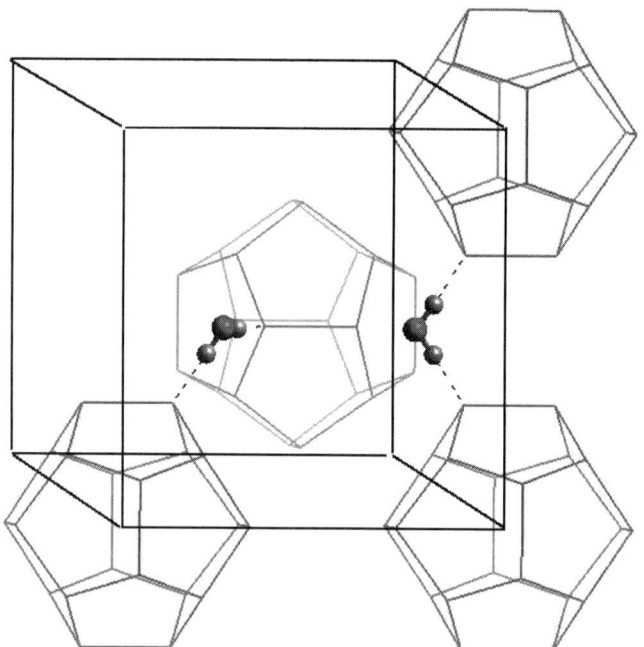

FIGURE 5. The illustration of two possible dodecahedra linked by water molecules for hydrate Structure I.

structural difference between the two structural models is that in model A, five out of six of the water molecules link dodecahedra from different sublattices. In model B, five out of six water molecules link dodecahedra from the same sublattice. In FIGURE 5 the two types of linkages for Structure I hydrate are shown. Under isotropic compression, in structural model A, the forces acting are mainly tangential to the dodecahedron surface. In the second structural model (B) these forces are mainly normal to the dodecahedron surface and, thus, effectively counteracted by guest–host repulsions. This effect is also related to the different distances between dodecahedra linked by the water molecules.

It is known that the proton configuration, which varies in thermodynamic equilibrium at high temperatures, can be frozen at low temperatures. At ambient pressure the proton configurations in Ih ice and some hydrates are frozen at $T = 105$ K and $T = 85$ K, respectively (a glassy state transition). The configuration lifetime in this case is much longer than the vibrational excitation lifetime. There is a good experimental approach for the annihilation of glass effects in the low temperature region that consists in doping of ice or hydrates[14] with a small fraction of KOH. Mobility of protons is found to be greatly enhanced in this case due to the creation of defects and the system quickly changes to some equilibrium ordered state. It might be expected that boundary of mechanical instability also changes after this transition.

SUMMARY

The mechanical stability of the methane and xenon hydrates has been investigated by lattice dynamics calculations. To solve the stability problem, the elastic constant calculation, optimization of molecular positions, and cell shape at different (P, T) points have been performed in quasiharmonic approximation. The predicted instability pressure is in the same pressure range as observed in experiments. However, different choices of proton configurations have significant effects on the structure behavior and the instability pressure. It can cause variation in the transition pressure for the glass state at low temperatures. To check these predictions more detailed experimental investigations of hydrates at the high pressure region are necessary.

REFERENCES

1. HANDA, Y.P., J.S. TSE, D.D. KLUG & E. WHALLEY. 1991. Pressure-induced phase transitions in clathrate hydrates. J. Chem. Phys. **94**: 623–627.
2. SHPAKOV, V.P., J.S. TSE, V.R. BELOSLUDOV & R.V. BELOSLUDOV. 1997. Elastic moduli and instability in molecular crystals. J. Phys. Cond. Mat. **9**: 5853–5865.
3. SHPAKOV, V.P., J.S. TSE, C. TULK et al. 1998. Elastic moduli calculation and instability in structure I methane clathrate hydrate. Chem. Phys. Lett. **282**: 107–114.
4. CALIFANO, S., V. SCHETTINO & N. NETO. 1961. Lattice Dynamics of Molecular Crystal. Springer, New York–London.
5. BELOSLUDOV, R.V. et al. 1994. Dynamical and thermodynamical properties of the acetylacetones of copper, aluminium, indium, and rhodium. Mol. Phys. **82**: 51–66.
6. LEIBFRIED, G. & W. LUDWIG. 1961. Theory of Anharmonic Effects in Crystals. Academic Press, New York–London.
7. GILL, P.E. et al. 1981. Practical Optimization. Academic Press, London.
8. WATSON, G.M. et al. 1997. In Computer Modeling in Inorganic Crystallography. C.R.A. Catlow, Ed. Academic, San Diego.

9. BORN, M. & H. HUANG. 1954. Dynamical Theory of Crystal Lattices. Clarendon Press, Oxford.
10. MCMULLAN, R.K. & G.A. JEFFREY. 1965. Polyhedral clathrate hydrates. IX structure of ethylene oxide hydrate. J. Chem. Phys. **42:** 2725–2731.
11. BERNAL, J.D. & R.H. FOWLER. 1933. The theory of water and ionic solution, with particular reference to hydrogen. J. Chem. Phys. **1:** 515–549.
12. FORRISDAHL, O.K. *et al.* 1996. Methane clathrate hydrates: melting, supercooling, and phase separation from molecular dynamics computer simulations. Mol. Phys. **89:** 819–830.
13. TANAKA, H. & K. KIYOHARA. 1993. The thermodynamic stability of clathrate hydrate. II. Simultaneous occupation of larger and smaller cages. J. Chem. Phys. **98:** 8110.
14. SUGA, H., T. MATSUO & O. YAMAMURO. 1992. Thermodynamic study of ice and clathrate hydrates. Pure Appl. Chem. **64:** 17–26.

Laboratory Measurements of Compressional and Shear Wave Speeds through Methane Hydrate

WILLIAM F. WAITE,[a,b] MICHAEL B. HELGERUD,[c]
AMOS NUR,[c] JOHN C. PINKSTON,[b] LAURA A. STERN,[b]
STEPHEN H. KIRBY,[b] AND WILLIAM B. DURHAM[d]

[b]*U.S. Geological Survey, 345 Middlefield Road, MS 977, Menlo Park, California 94025, USA*

[c]*Geophysics Department, Stanford University, Stanford, California 94305-2215, USA*

[d]*U.C. Lawrence Livermore National Laboratory, P.O. Box 808, Livermore, California 94550, USA*

ABSTRACT: Simultaneous measurements of compressional and shear wave speeds through polycrystalline methane hydrate have been made. Methane hydrate, grown directly in a wave speed measurement chamber, was uniaxially compacted to a final porosity below 2%. At 277 K, the compacted material had a compressional wave speed of 3,650 ± 50 m/s. The shear wave speed, measured simultaneously, was 1,890 ± 30 m/s. From these wave speed measurements, we derive V_p/V_s, Poisson's ratio, bulk, shear, and Young's moduli.

INTRODUCTION

Clathrate hydrates of natural gases are nonstoichiometric crystalline solids in which a hydrogen bonded water lattice is stabilized by individual guest molecules encaged in interstitial cavities. Of particular interest is methane (CH_4) hydrate, a Structure I hydrate with a unit cell composed of 46 water molecules, with eight cavities available for guest molecule occupation (ideally $CH_4 \cdot 5.75H_2O$). Current estimates of CH_4 hydrate distributions suggest vast reservoirs exist in the shallow geosphere,[1,2] and their high CH_4 content has lead to the promotion of hydrates as a potential energy resource. On a localized scale, drilling operations can destabilize hydrate-rich sediments, causing sediment collapse and well bore failures. Hydrate can also break down naturally on larger scales, triggering massive submarine landslides that can displace nearly 4,000 km^3 of material, jeopardizing waste-site integrity, cables and other submarine structures.[1,3–6] Additionally, the effectiveness of CH_4 as a greenhouse gas suggests that hydrate stability influences our global climate.[1,5,7]

---

[a]Address for correspondence: William F. Waite, U.S. Geological Survey, 345 Middlefield Road, MS 977, Menlo Park, California 94025, USA. Voice: 650-329-4803; fax: 650-329-5143.
  wwaite@usgs.gov

Hydrate studies focused on resource management, hazard mitigation, or climate change all require accurate physical property values, which have proven difficult to measure for CH_4 hydrate. There is no consensus in the few published measurements, and most CH_4 hydrate property estimates are based on the behavior of analog materials. Here we describe laboratory measurements of compressional and shear wave speeds (V_p and V_s) through well-characterized CH_4 hydrate, grown directly in a wave speed measurement chamber. From our simultaneous V_p and V_s measurements, and by assuming our samples are homogeneous and isotropic, we derive a suite of physical properties for dense, polycrystalline CH_4 hydrate.

EXPERIMENTAL METHOD

Sample Preparation

CH_4 hydrate samples were produced in a custom-built cylindrical pressure vessel (FIGURE 1A) by slowly heating granular H_2O ice in a pressurized CH_4 atmosphere, as described by Stern et al.[8] Ice used to seed this reaction was grown from triply distilled water, ground and sieved to obtain a 180–250 µm grain size distribution. The resultant CH_4 hydrate is polycrystalline, with random grain orientation, and approximately 28% porosity.

Following synthesis, samples are uniaxially compacted to reduce the porosity below 2%. We estimated the final hydrate porosity from the sample length measured during compaction and the known mass of ice used to seed the experiment. Hydrate extruded during compaction drives our calculated porosity lower than the actual porosity, but we cannot seal the sample chamber prior to compaction because CH_4 gas must be admitted for hydrate synthesis to occur. To balance our synthesis and compaction requirements, slots are cut in the Teflon sample jacket. These slots extend 5 mm past the compaction piston into the sample, allowing gas into the sample chamber during synthesis. Rapid piston displacement during the initial stage of compaction blocks these slots, minimizing hydrate extrusion during the remaining approximately 10 mm of compaction.

Wave Speed Measurement

Both pressure vessel pistons house a 1-MHz center-frequency piezo-electric transducer (either P- or S-wave) used for pulse-transmission wave speed measurements (FIG. 1B). The transducer remains at atmospheric pressure and supports none of the compressional loading during compaction. An HP Model 214A pulse generator drives the source transducer in the compaction piston, and an HP Model 465A amplifier boosts the signal detected by the transducer in the fixed piston. The signal is displayed on a Tektronix TDS-340 oscilloscope and recorded by a computer running National Instrument's LabView™ data acquisition and display software. Shear and/or compressional wave speed measurements, taken throughout the compaction process, are given by the ratio of sample length to the time of flight of the waveform through the sample.

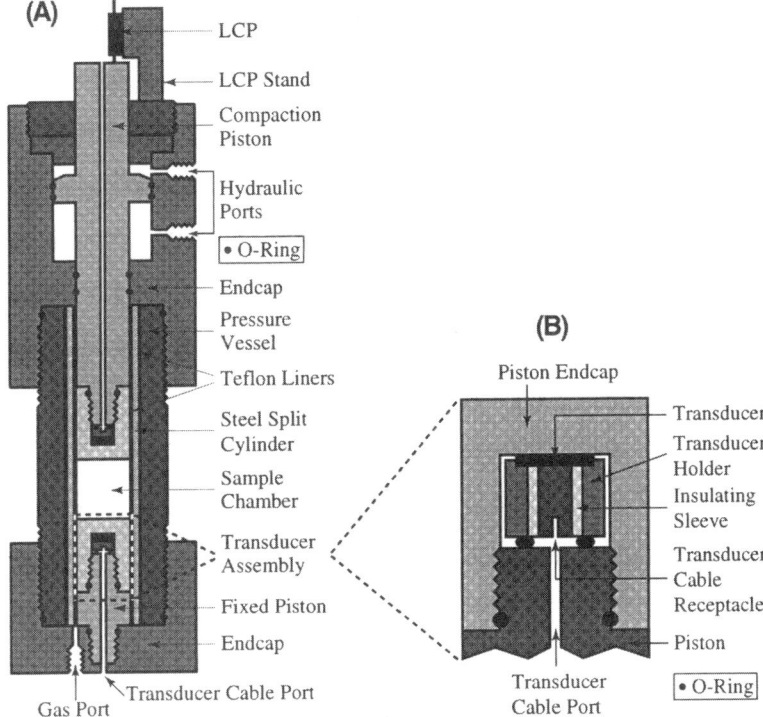

FIGURE 1. (A) Pressure vessel schematic. Polycrystalline methane hydrate is synthesized directly in the sample chamber, then uniaxially compacted *in situ*. Wave speed measurements are completed without handling the methane hydrate or otherwise removing it from the hydrate stability field. The sample length is monitored using a linear conductive plastic (LCP). (B) Transducer assembly schematic. Using a 1-MHz center-frequency S- or P-wave transducer, shear, and/or compressional wave speed measurements can be made throughout the compaction process.

The sample length is calculated from the known dimensions of the pressure vessel and measured position of the compaction piston relative to the pressure vessel. The LCP continuously monitors piston position changes, and periodic measurements of the absolute piston position are made with a depth micrometer to check the LCP results and verify the sample length. Differences between the LCP and depth micrometer results are less than 0.5% of the compacted sample length.

Four methods are used to measure the travel time of the signal through the sample. For compacted samples, cross correlation, Hilbert envelope, phase spectral analysis, and zero crossing pick results differ by less than 1.5%, our stated velocity uncertainty. Agreement between these different procedures that use different aspects of the measured waveform to estimate the signal travel time suggest our travel time estimates are independent of the theory from which they are obtained.

RESULTS

To test the validity of our measurement methods, we performed a control experiment on pure, polycrystalline H_2O ice compacted under vacuum at 260 K and uniaxial load of 40 MPa. The recovered H_2O ice sample was translucent, indicating that the sample was nearly fully dense, with a final porosity below 1%. Our method reproduces published wave speed results within the scatter of individual studies (TABLE 1).

Low noise levels in our ice and our hydrate experiments allow us to unambiguously pick arrival times for the precursor P-wave event generated by the S-wave transducer (see FIGURE 2). In a test using the precursor event, the calculated compressional wave speed through a compacted hydrate sample was indistinguishable from that observed using the dedicated P-wave transducers on a separate sample. This agreement between results obtained using different compressional wave sources on separate samples demonstrates the repeatability of our hydrate synthesis and compaction procedure and it shows that the S-wave transducer can be used to provide reliable P- and S-wave speed measurements simultaneously.

To draw meaningful conclusions from wave speed comparisons between our results and those already published, it is important to characterize our samples as completely as possible. When forming hydrate from small ice grains warmed in a pressurized methane atmosphere,[8] it is possible that a portion of the seed ice will melt rather than form hydrate. Fortunately, there are several indications of incomplete reaction that are measurable while the sample is in the synthesis chamber. Pressure and temperature (*PT*) effects are described in detail by Stern *et al.*[8,12] In the absence of observable *PT* effects, X-ray analysis of recovered samples synthesized according to their recipe show less than 3% ice,[8] some of which may have formed during the X-ray analysis.

No *PT* effects from incomplete reaction were observed during our reported experiments. Although we performed no X-ray analyses on compacted samples, the wave speed measurement itself provides a direct indication of unreacted material in the sample chamber. Prior to our reported wave speed measurements, our samples are held at 277 K for a minimum of 24 hours. If present, unreacted H_2O would be liquid, tending to lower our wave speed results relative to that expected for pure

TABLE 1. Comparison of compressional (V_p) and shear (V_s) wave speed measurements of polycrystalline H_2O ice (Ih)

Author	Measurement	V_p (m/s)	V_s (m/s)
This study	pulse-transmission	3900 ± 40	1970 ± 20
Gagnon, Ref. 9	Brillouin spectroscopy	3914	1995
Smith *et al.*, Ref. 10	pulse-echo	3940	1990
Shaw, Ref. 11	pulse-transmission	3890	1900

NOTE: The wave speeds reported in *this study* were measured at 260 K after releasing the uniaxial load. The excellent agreement of these values with the known properties of H_2O ice (Ih) demonstrates the reliability of our experimental apparatus and procedure.

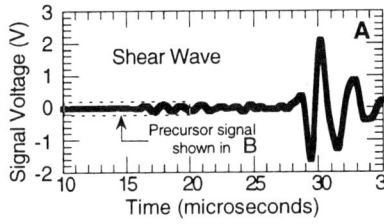
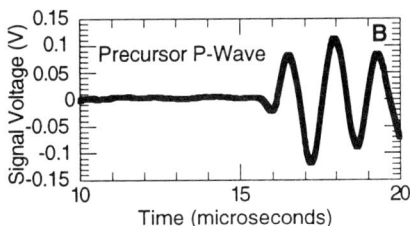

FIGURE 2. Measured waveforms produced by an S-wave transducer. (**A**) The shear wave signal, arriving near 30 μs, is preceded by a precursor P-wave arriving just after 15 μs. (**B**) Low system noise allows use of both waveforms to obtain simultaneous shear and compressional wave speeds.

hydrate. This unreacted water would transform to ice as the compacted sample cools from 277 K to 250 K following our experiment, causing a wave speed increase. No such increase was observed, meaning that if unreacted material was present in our experiments, our wave speed measurements were not sensitive enough to be affected. For these reasons, we believe our wave speed results to be representative of polycrystalline CH_4 hydrate.

There are very few published compressional wave speed measurements for CH_4 hydrate, and no shear wave experiments are available for comparison (TABLE 2). Briefly, Whalley[13] and Pearson et al.[14] derive compressional wave speeds for CH_4 hydrate relative to that of ice (Ih) from differences between several mechanical and thermodynamic parameters for the two materials. The Shpakov et al.[15] estimate is based on elastic moduli derived from lattice dynamics investigations of CH_4 hydrate.

TABLE 2. Comparison of V_p and V_s wave speed measurements through CH_4 hydrate

Author	Measurement	V_p (m/s)	V_s (M/S)
This study	pulse-transmission	3650 ± 50	1890 ± 30
Pearson et al.[14]	theory	3730[a]	—
Whalley[13]	theory	3660[b]	—
Mathews et al.[16]	DSDP Site 570 Log	3600	—
Whiffen et al.[17]; Kiefte et al.[18]	Brillouin spectroscopy	3400	—
Shpakov et al.[15]	theory	2500	—

NOTE: Sample compaction reported in *this study* was conducted at 277 K, with 10 MPa CH_4 pore pressure and a uniaxial load approaching 100 MPa. The final sample porosity was less than 2%. By assuming our samples are homogeneous and isotropic, we can use our simultaneous measurements of V_p and V_s to derive additional elastic parameters (see TABLE 3).

[a] V_p should be considered a lower bound for this reference. The reported velocity climbs from 3,730 m/s to 3,780 m/s as the cage occupancy drops from 100 to 80%. We believe our occupancy rate to be above 90%.

[b] Obtained from the Whalley[13] conclusion that V_p for CH_4 hydrate is 0.939 that of ice, taken from TABLE 1 to be 3,900 m/s.

On the experimental side, the Site 570 down-hole log result[16] comes from *in situ* measurements made on DSDP leg 84 after drilling through a three to four meter thick hydrate layer. A meter-long core section recovered from this interval was a solid hydrate mass, largely free of sediment. Brillouin spectroscopy measurements[17,18] look at the frequency of laser light scattered from thermally induced elastic waves in a clear sample. Producing a clear CH_4 hydrate sample is difficult. As Kiefte *et al.*[18] explain, although they successfully acquired scattered light spectra for several other structure I and II hydrates, only two weak spectra were obtained for CH_4 hydrate. They suggest their weak spectra may stem from the insufficient penetration of focused laser light into their sample.

DISCUSSION

By assuming our samples are homogeneous and isotropic, we can use our simultaneous V_p and V_s measurements to derive additional physical properties. Physical properties for both our ice and hydrate results are compared with published estimates for CH_4 hydrate in TABLE 3. To obtain our adiabatic moduli values, we used a density of 0.92 g/cc for ice,[11] and 0.90 g/cc for our CH_4 hydrate, which we assume to have a stoichiometry of $CH_4 \cdot 6H_2O$. To obtain our isothermal moduli, we used linear expansion coefficients of 52×10^{-6} K^{-1} for ice[19] and 88×10^{-6} K^{-1} for CH_4 hydrate.[18] For specific heat at constant pressure, we used 2.09 J/g·K for ice,[20] and 2.07 J/g·K for CH_4 hydrate.[21]

TABLE 3. Comparison of elastic property values of polycrystalline H_2O ice, polycrystalline CH_4 hydrate, and published estimates for CH_4 hydrate

Property	H_2O Ice this work	CH_4 Hydrate this work	Prior Estimates for CH_4 Hydrate
V_p/V_s	1.98 ± 0.02	1.93 ± 0.01	1.95[a]
Poisson's ratio	0.33 ± 0.01	0.317 ± 0.006	0.33[b]
Shear modulus (GPa)	3.6 ± 0.1	3.2 ± 0.1	2.4[a]
Adiabatic bulk modulus (GPa)	9.2 ± 0.2	7.7 ± 0.2	5.6[a]
Isothermal bulk modulus (GPa)	9.0 ± 0.3	7.1 ± 0.3	—
Adiabatic Young's modulus (GPa)	9.5 ± 0.2	8.5 ± 0.2	—
Isothermal Young's modulus (GPa)	9.1 ± 0.3	7.8 ± 0.3	8.4[b]

NOTE: None of the prior elastic property estimates for CH_4 hydrate were measured on Structure I hydrate.

[a]These values, from Pandit and King,[22] were measured on Structure II propane hydrate, but are cited elsewhere as estimates of Structure I hydrate.

[b]These estimates by Davidson[23] are based on the theoretical work of Whalley.[13]

CONCLUSION

Differences between our results, based on actual measurements of CH_4 hydrate, and published estimates underscore the importance of making physical property measurements directly on well characterized CH_4 hydrate. Accurate physical property values are essential for planning viable strategies to manage CH_4 hydrate as a global resource and address the challenges it presents. The synthesis procedure developed by Stern et al.[8] provides a promising foundation for extending the current description of CH_4 hydrate and should be considered in the implementation of future thermal and mechanical property measurements on pure CH_4 hydrate or mixtures of CH_4 hydrate with sediment.

ACKNOWLEDGMENTS

This work was supported under NSF grant OCE-97-10506, DOE grants DE-FG0386ER 13601 and DE-FG07-96ER 14723, GRI grant 5094-210-3235-1, USGS grant 1434-HG-97GR-03, NEDO, and an ODP Graduate Research Fellowship.

REFERENCES

1. KVENVOLDEN, K.A. 1993. Gas hydrates as a potential energy resource—a review of their methane content. *In* The Future of Energy Gases. USGS Prof. Paper 1570: 555–561.
2. HOLBROOK, W.S. et al. 1996. Methane hydrate and free gas on the Blake Ridge from vertical seismic profiling. Science **273**: 1840–1843.
3. CAMPBELL, K.J. 1999. Deepwater geohazards: how significant are they? The Leading Edge **18**: 514–519.
4. DILLON, W.P. et al. 1998. Evidence for faulting related to dissociation of gas hydrate and release of methane off the southeastern United States. *In* Gas Hydrates: Relevance to World Margin Stability and Climate Change. J.-P. Henriet & J. Mienert, Eds. **137**: 293–302. Special Publications, Geological Society, London.
5. HAQ, B. 1998. Natural gas hydrates: searching for the long-term climatic and slope stability records. *In* Gas Hydrates: Relevance to World Margin Stability and Climate Change. J.-P. Henriet & J. Mienert, Eds. **137**: 303–318. Special Publications, Geological Society, London.
6. MIENERT, J. et al. 1998. Gas hydrates along the northeastern Atlantic margin: possible gas hydrates bound margin instabilities and possible release of methane. *In* Gas Hydrates: Relevance to World Margin Stability and Climate Change. J.-P. Henriet & J. Mienert, Eds. **137**: 275–291. Special Publications, Geological Society, London.
7. THORPE, R.B. et al. 1998. What does the ice-core imply concerning the maximum climatic impact of possible gas hydrate release at Termination IA? *In* Gas Hydrates: Relevance to World Margin Stability and Climate Change. J.-P. Henriet & J. Mienert, Eds. **137**: 319–326. Special Publications, Geological Society, London.
8. STERN, L.A. et al. 1996. Peculiarities of methane clathrate hydrate formation and solid-state deformation, including possible superheating of water ice. Science **273**: 1843–1848.
9. GAGNON, R.E. et al. 1987. Elastic constants of ice Ih, up to 2.8 kbar, by Brillouin Spectroscopy. J. Phys. (Paris) **48**: 23–35.
10. SMITH, A.C. et al. 1986. Measurement of the speed of sound in ice. AIAA J. **24**: 1713–1715.
11. SHAW, G.H. 1986. Elastic properties and equation of state of high pressure ice. J. Chem. Phys. **84**: 5862–5868.

12. STERN, L.A. *et al.* 1998. Polycrystalline methane hydrate: Synthesis from superheated ice, and low temperature mechanical properties. Energy & Fuels **12:** 201–211.
13. WHALLEY, E. 1980. Speed of longitudinal sound in clathrate hydrates. J. Geophys. Res. **85:** 2539–2542.
14. PEARSON, C.F. 1983. Natural gas hydrate deposits: A review of *in situ* properties. J. Phys. Chem. **87:** 4180–4185.
15. SHPAKOV, V.P. *et al.* 1998. Elastic moduli calculation and instability in structure I methane clathrate hydrate. Chem. Phys. Lett. **282:** 107–114.
16. MATHEWS, M.A. *et al.* 1985. Site 570 methane hydrate zone. *In* Init. Repts. DSDP **84:** 773–790.
17. WHIFFEN, B.L. *et al.* 1982. Determination of acoustic velocities in xenon and methane hydrates by Brillouin spectroscopy. Geophys. Res. Lett. **9:** 645–648.
18. KIEFTE, H. *et al.* 1985. Determination of acoustic velocities of clathrate hydrates by Brillouin spectroscopy. J. Phys. Chem. **89:** 3103–3108.
19. WHALLEY, E. 1973. Lattice dynamics of ice. *In* Physics and Chemistry of Ice. S. Jones & L. Gold, Eds.: 73–81. Royal Society of Canada, Ottawa.
20. GIAUQUE, W.F. *et al.* 1936. The entropy of water and the third law of thermodynamics. J. Chem. Phys. **58:** 1144–1150.
21. HANDA, Y.P. 1986. Compositions, enthalpies of dissociation, and heat capacities in the range 85 to 270 K for clathrate hydrates of methane, ethane, and propane, and enthalpy of dissociation of isobutane hydrate, as determined by a heat-flow calorimeter. J. Chem. Thermodynam. **18:** 915–921.
22. PANDIT, B.I. & M.S. KING. 1982. Elastic wave propagation in propane gas hydrates. Proceedings of the 4th Canadian Permafrost Conference: 335–342.
23. DAVIDSON, D.W. 1983. Gas hydrates as clathrate ices. *In* Natural Gas Hydrates: Properties, Occurrence and Recovery. J. Cox, Ed.: 1–15. Butterworths, Boston.

Dissociation of Natural Gas Hydrates Observed by X-ray CT Scanner

JUN MIKAMI,[a] YOSHIHIRO MASUDA,[b] TAKASHI UCHIDA,[a] TOHRU SATOH,[a] AND HIDEAKI TAKEDA[c]

[a]*JAPEX Research Center, 1-2-1 Hamada, Mihama-ku, Chiba, 261-0025 Japan*

[b]*Department of Geosystem Engineering, University of Tokyo, 7-3-1 Hongou, Bunkyo-ku, Tokyo, 113-8656 Japan*

[c]*Teikoku Oil Company, 1-31-10 Hatagaya, Shibuya-ku, Tokyo, 151-0072 Japan*

ABSTRACT: Core samples containing pore-space gas hydrate within granular sands were collected from 913.76 m of the research well named JAPEX/JNOC/GSC Mallik 2L-38. X-ray CT images of the core were acquired while warming from −18 to 4°C, and subsequently during stepped decreases of 0.1 MPa in the chamber pressure below the methane hydrate equilibrium pressure. Discharged gas flows and sample temperatures were monitored continuously. Changes in CT values indicated that gas hydrate dissociated simultaneously both on the exposed surfaces and within the pore spaces of the sample in response to pressure changes. This suggested that pressure reductions were effectively transmitted through the sample most likely because the samples contained some amount of fluids. The result of gas flow measurements indicated that a larger pressure drawdown caused a higher dissociation rate.

INTRODUCTION

This research represents a contribution to the ongoing Japanese methane hydrate research program, with financial support from the Japan National Oil Corporation (JNOC), along with ten additional oil, gas, and electric companies in Japan. In February and March of 1998, the JAPEX/JNOC/GSC Mallik 2L-38 gas hydrate research well was drilled to the depth of 1150 m at the northern edge of the Mackenzie Delta, Northwest Territories, Canada. A drill site adjacent to Imperial Mallik L-38 (an industry exploration well drilled in 1972) was selected for the gas hydrate research well.[1–3] Gas-hydrate-bearing core samples, collected from beneath the permafrost interval (from 896 to 953 m), provided a unique opportunity to study the physical and chemical properties of natural gas hydrate samples. Core samples contained a variety of geologically existing forms of gas hydrate within sands and granular sands in the interval from 896 to 926 m. A crystal structure of gas hydrate was thought to be structure I from NMR and Raman spectra analyses.[4,5] This paper describes an

Telecommunication.
[a]Voice: 81-43-275-9311; fax: 81-43-275-9316. mikami@rc.japex.co.jp (Jun Mikami); uchida@rc.japex.co.jp (Takashi Uchida); sato@rc.japex.co.jp (Tohru Satoh)
[b]Voice: 81-3-5841-7063; fax: 81-3-3818-7492. masuda@geosys.t.u-tokyo.ac.jp
[c]Voice: 81-3-3466-1249; fax: 81-3-3468-3504. h_takeda@pop.teikokuoil.co.jp

experiment in which X-ray computerized tomography (CT) scanning techniques were used to observe the dissociation characteristics of Mallik 2L-38 core sample. Analyses of methane hydrate by X-ray CT scanner are useful for investigating location in sediments where gas hydrate exists and the relationship of gas hydrate to the host sediments, without significantly disturbing the samples. Mallik 2L-38 core samples containing gas hydrate were imaged using X-ray CT scanning techniques before being subjected to further analyses. X-ray CT images were obtained using the Toshiba X-ray CT scanner "X-force" at the JNOC Technology Research Center. Time series observations of CT responses during dissociation of gas hydrate within the sample are presented in this paper. Induced dissociation was accomplished through a controlled pressure reduction from the pressure–temperature (P-T) equilibrium threshold for gas hydrate stability. Results of this experiment should provide insight into gas production characteristics from natural gas hydrate reservoirs using pressure drawdown techniques.

SAMPLES AND METHODS

The sample used for the experiment was a gas-hydrate-bearing granular sand taken from the depth of 913.76 m. It was pressurized with nitrogen gas to 6 MPa in a pressure vessel immediately after taken at the well site, transported in an insulated box chilled with dry ice and kept in a freezer at about $-40°C$ before the experiment. The dimensions of the test sample were roughly 5 cm by 5 cm by 7 cm, with a bulk volume of approximately 115 cm^3 which was calculated from integration of CT slices. The average porosity of the sand was estimated to be about 30%[5,6] from the other porosity measurements of adjacent samples. The maximum volume of gas that could be incorporated into gas hydrate within the sample pore was about 4,800 cm^3 assuming 100% hydrate saturation and 80% cage occupancy of hydrate.

An objective of this experiment was to observe a dissociation process of natural gas hydrate in the sample. Additionally, dissociation rates were measured for different drawdown pressures from its equilibrium. An X-ray CT scanner was used to observe changes occurring in the interior of the sample interior during dissociation. During the experiment, scanning was executed at the final state of each temperature or pressure step. The interior of the sample is presented as cross-section images using CT values observed. TABLE 1 shows the scanning condition of the X-ray CT scanner.

TABLE 1. Scanning condition of X-ray CT scanner

X-ray tube voltage	130 kV
X-ray tube current	100 mA
Scanning time	1.5 sec
Width per slice	2.0 mm
Number of slices per scan	41
Total length of scan	80 mm

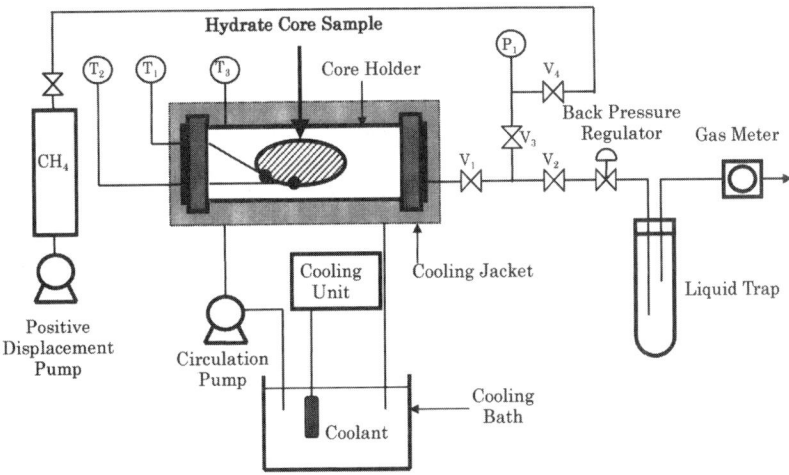

FIGURE 1. Schematic drawing of the experimental apparatus.

FIGURE 1 shows a schematic drawing of the apparatus used for the experiment. The core holder was made of an aluminum/fiberglass composite with the size of 117 mm inner diameter, 135 mm outer diameter and 360 mm length. It was used as a pressure vessel with a coiled tube jacket around itself for cooling. During the experiment it was pressurized by methane and was chilled by circulating coolant in the coiled tubes to prevent an abrupt dissociation of gas hydrate.

The dissociation experiment was conducted according to the following procedure:
1. The sample was fixed in the core holder that was previously chilled below −30°C. Glass beads of 1 mm diameter in the tray held the irregular shaped sample by their arbitrary surface contact. After setting the sample in the core holder, the system was pressurized by methane gas to 5.2 MPa.
2. The core holder was placed on the bed of the X-ray CT scanner. The temperature of the sample was raised gradually from −18 to 4°C. While warming the sample, the chamber pressure was continuously monitored.
3. After the temperature was stabilized to 4°C, the gas hydrate equilibrium pressure at 4°C was found to be 4.3 MPa from observations of sample by reducing the system pressure sequentially.
4. The core was repressurized with methane gas to about 5.7 MPa and maintained at constant temperature. The system pressure was again reduced to 4.3 MPa before the dissociation period.
5. Stepwise pressure drawdowns of 0.1 MPa, 0.2 MPa, 0.3 MPa, 0.5 MPa, and 1.0 MPa were imposed on the sample by bleeding gas from the apparatus, with the intent of inducing dissociation of gas hydrate within the core sample. The pressure drawdown is defined as the difference between the hydrate equilibrium pressure and each of the system pressures. Additional CT scans were acquired during each pressure drawdown stage.

RESULTS AND DISCUSSION

During the experiment, the pressure, temperature, and gas flow rate were recorded from the onset of system temperature increase until the end of the dissociation period (FIGURE 2). FIGURE 3 shows the data for each dissociation step. The origin on the time axis is the start of the dissociation period. Gas released from the dead volume initially caused a high apparent rate of gas production. Because the dead volume of the core holder was large compared to the sample volume and the control capacity of the back pressure regulator was limited, considerable time was required for gas flow rate stabilization. The gas expansion in the dead volume caused an initially high production rate for each dissociation step. The gas generated by hydrate dissociation was also produced with gas in the dead volume. Therefore, we could not calculate accurately the dissociation rate of gas hydrate from the measurement of gas flow rates. At the fourth step (pressure drawdown of 0.5 MPa) the gas flow rate was separated into two stages, due to the necessity for adjustment of the back pressure regulator in the middle of the step.

We have considered the effects of increasing temperature on gas hydrate dissociation. The core sample was initially stored in a freezer at the temperature about −40°C and, then, warmed from −18 to −5°C. Frequency histograms of the changes in CT values at −18°C, 0°C and 4°C are shown in FIGURE 4. No changes were observed in the sample during warming from −18 to −5°C, with CT values remaining almost constant during warming. However, upon further warming, CT values were found to increase slightly around 0°C. We attribute this slight increase in CT values to the melting of ice to water within the sample, because the CT value for freshwater ice is typically about −65 while fresh water is about zero.[7] It can be considered that free water initially coexisted with gas hydrate within the sediment pores.

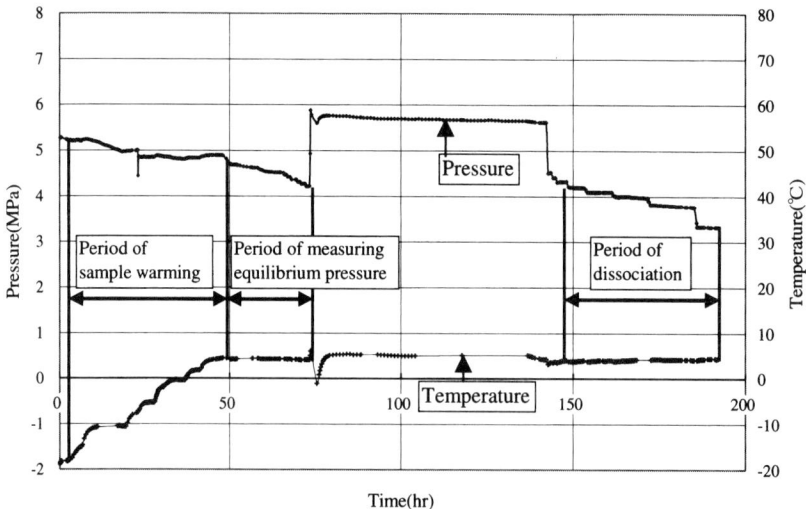

FIGURE 2. Pressure–temperature history of the experiment.

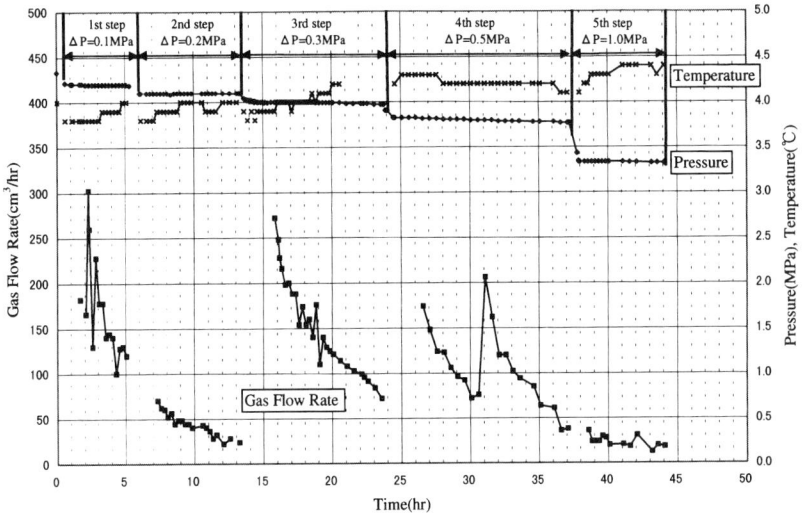

FIGURE 3. Pressure, temperature, and gas flow rate during the dissociation period.

This water was converted to ice during sample storage in the freezer. A slight decrease in CT values was observed at 4°C at about 5.7 MPa, and this may have been due to the formation of additional gas hydrate within the sample. This conclusion is supported by the slight decrease in pressure observed during this period. The formation of additional gas hydrate should result in lower net CT values, because the CT value for pure methane hydrate (−100 to −250) is lower than that for free water (around zero)[7,8] from which the gas hydrate formed.

CT images taken during the hydrate dissociation period indicate significant changes in the core during the pressure drawdown of 0.3 MPa. Changes in CT values were less apparent during subsequent pressure drawdowns. FIGURE 5 shows the differential CT values between an image taken prior to the start of dissociation, and three other images of the same slice taken after pressure drawdown of 0.1 MPa, 0.2 MPa, and 0.3 MPa from 4.3 MPa. This figure, therefore, indicates changes in CT values as dissociation progresses within the sample. In general, the darker parts within the sample indicate decreasing CT values between the two scans, whereas the brighter parts indicate no changes in CT values. The shift towards more negative CT values under these stepwise pressure reductions is interpreted as resulting from the presence of additional free gas (CT value near −1,000) within sample pore spaces as gas hydrate dissociation progressed. FIGURE 5A indicates very little change within the sample under a 0.1 MPa pressure drawdown. Under a 0.2 MPa pressure drawdown, changes in CT values are more apparent within the interior of the sample (FIG. 5B). The most extensive changes in CT values within the sample interior were observed under a pressure drawdown of 0.3 MPa (FIG. 5C). These results suggest that gas hydrate dissociation in the porous sand sample occurred both at the sample surface and also within the interior of the sample.

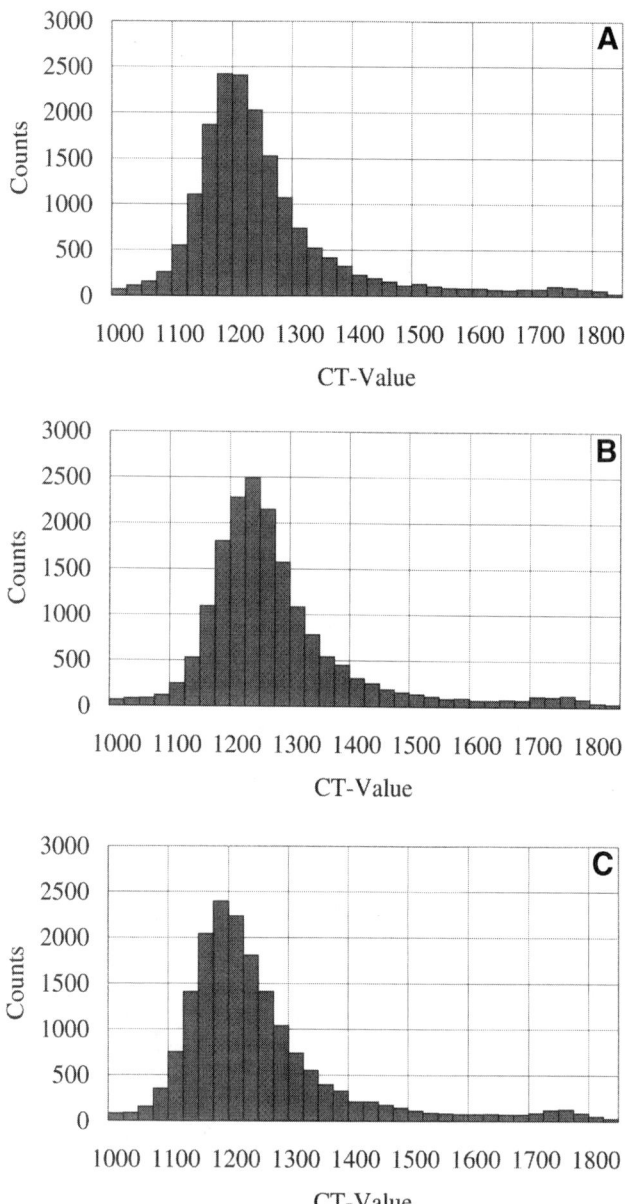

FIGURE 4. Frequency change in CT values. **A.** $T = -18°C$ ($P = 5.2$ MPa). **B.** $T = -0°C$ ($P = 4.8$ MPa). **C.** $T = 4°C$ ($P = 4.9$ MPa).

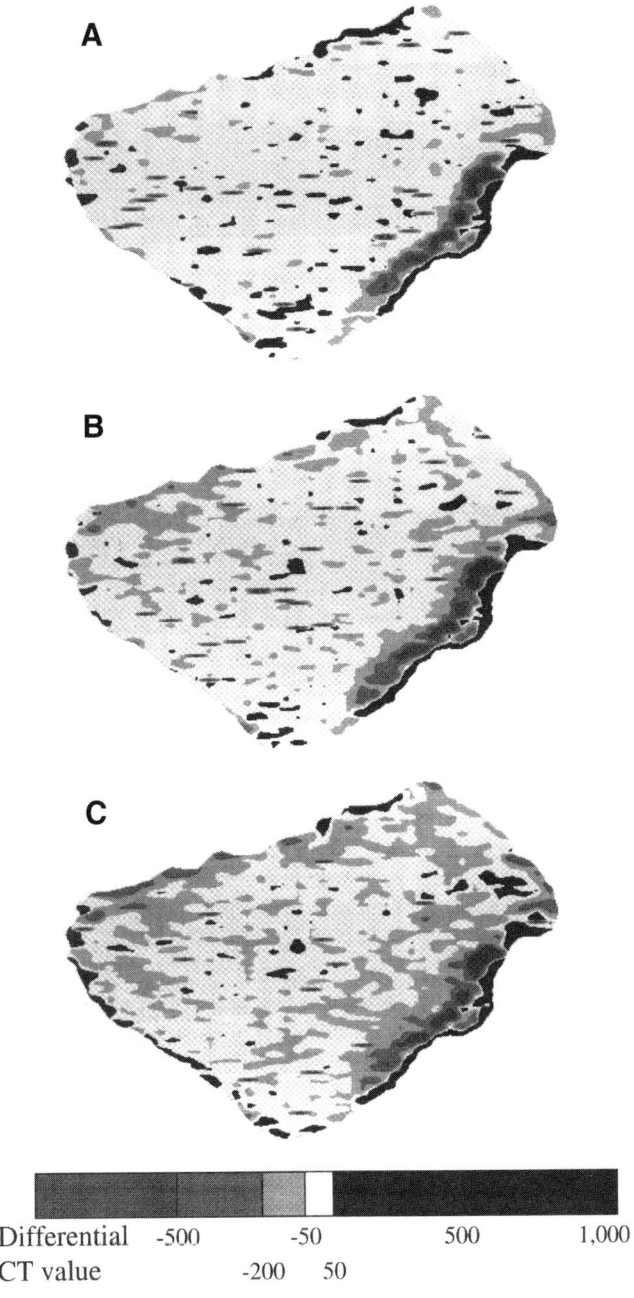

FIGURE 5. Differential CT values of sample at: **A.** $\Delta P = 0.1$ MPa; **B.** $\Delta P = 0.2$ MPa; and **C.** $\Delta P = 0.3$ MPa.

The following interpretations were made based on the observations:
1. Sediment pores were not completely filled with gas hydrate and therefore some pressure conduits existed in the sample. At temperatures above 0°C, free water can coexist along with gas hydrate in sediment pores. Free water can serve as a pressure conduit to transmit pressure changes into the interior of the sample, thereby initiating dissociation of gas hydrate under isothermal conditions. Since the sample was chilled to between −70 and −35°C for transportation and storage, any free water existing in sediment pores at the time of recovery would have been frozen prior to the start of the experiment. The slight increase in CT values as the sample warmed above 0°C is consistent with this interpretation.
2. The sediment fabric can be deformed by repeated fluctuations in pressure, particularly if the gas hydrate is serving as the primary bonding agent between sediment grains. Note that during the experiment the sample was exposed to gas pressure only, without any simulated overburden (confining) pressure. Therefore, flow paths could more readily be kept open once the bonding agents were disrupted.

We initially hypothesized that dissociation of gas hydrate should begin at the surface of the sample and advance inward. CT images suggest, however, that dissociation actually occurred simultaneously within the interior and on the surface of the sample.[9] Transportation and storage conditions may also have affected physical properties of the samples. Additional investigation is needed to clarify the affects of different transportation and storage methods on the properties of gas-hydrate–bearing sediments.

The rates of gas production at each pressure drawdown are plotted in FIGURE 6. Because the duration of the first dissociation step (0.1 MPa pressure drawdown) was not enough to stabilize the flow rate, these data were included in the second step

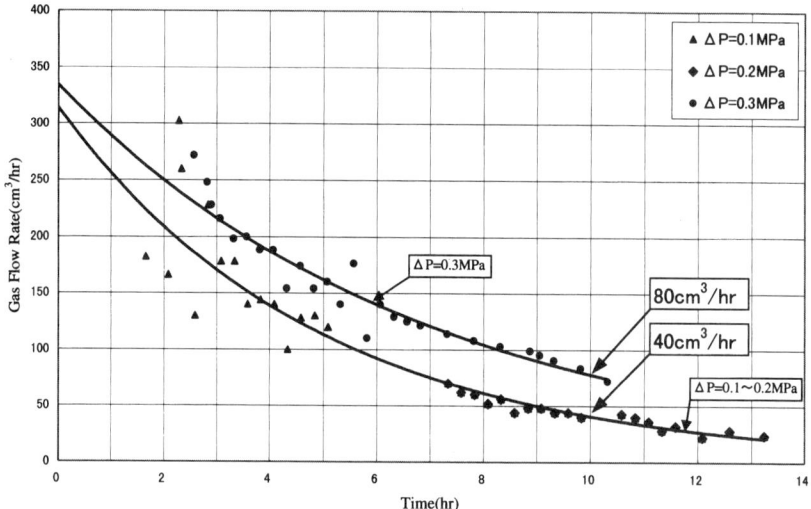

FIGURE 6. Gas flow rate versus time during pressure drawdown and gas hydrate dissociation.

(0.2 MPa pressure drawdown). Comparison of gas flow rates from the second and third step indicate very different gas production rates (40 cm^3/hr and 80 cm^3/hr respectively, after 10 hours). After the fourth step (0.5 MPa pressure drawdown), the gas production rate decreased and little change was observed in the CT images. These results indicated that almost all the gas hydrate in the sample was dissociated before the end of the third step (0.3 MPa pressure drawdown), and that only a small amount of gas hydrate remained in the sample after the fourth step. Gas production from gas hydrate dissociation under a 0.3 MPa pressure drawdown may be double the rate observed for 0.2 MPa drawdown under pseudo-stabilized conditions. This suggests that higher drawdown pressures may cause higher dissociation rates for gas hydrate in sediments.

SUMMARY AND CONCLUSIONS

The dissociation of gas hydrate in a sand sample from the Mallik 2L-38 well was investigated using X-ray CT scanning technique. During the experiment, time-series CT images of the interior of a gas-hydrate-bearing sand were successfully acquired. The conclusions are as follows:

1. The CT images obtained during sample warming indicate that an increase in CT values at 0°C is related to the melting of pore ice formed from free water that initially coexisted with gas hydrate in the sediment pores prior to freezing of the samples for transportation and storage. A decrease in CT values at about 4°C may be indicative of additional gas hydrate formation prior to initiation of the controlled dissociation stage.

2. Changes in CT values during forced dissociation indicate that gas hydrate dissociation was initiated nearly simultaneously within the interior of the sample and at the sample surface. It is believed that sample pores were not completely filled with gas hydrate, and free water therefore may act as conduits for the transmission of pressure reduction. Transportation and storage conditions may have also affected the nature of the sample.

3. The rate of gas production during the dissociation period can be calculated with limited accuracy only, due to a number of experimental uncertainties. Gas flow measurements suggest, however, that a larger pressure drawdown causes a higher rate of dissociation of gas hydrate within the sediment pore volume.

ACKNOWLEDGMENT

We are indebted to the Geological Survey of Canada (GSC) and the ongoing research program supported by JNOC and ten oil, gas, and electric companies. Special thanks to J-S. Vincent, F. Wright and S. Dallimore of GSC Ottawa, and A. Nakamura and M. Imazato of JNOC, and T. Ohara and O. Senoh of JAPEX who were responsible for carrying out the Mallik 2L-38 well project successfully as well as to JNOC Technology Research Center for their permission to use the X-ray CT scanner.

REFERENCES

1. BILY, C. & J.W.L. DICK. 1974. Natural occurring gas hydrate in the Mackenzie Delta, Northwest Territories. Bull. Canad. Petrol. Geol. **22:** 340–352.
2. COLLETT, T.S. & S.R. DALLIMORE. 1998. Quantitative assessment of gas hydrates in the Mallik L-38 Well, Mackenzie Delta, N.W.T. 7th International Conference on Permafrost, Yellowknife, Canada. 189–194.
3. DALLIMORE, S.R. & T.S. COLLETT. 1998. Gas hydrates associated with deep permafrost in the Mackenzie Delta, N.W.T., Canada. 7th International Conference on Permafrost, Yellowknife, Canada. 201–206.
4. TULK, C.A., C.I. RATCLIFFE & J.A. RIPMEESTER. 1999. Chemical and physical analysis of natural gas hydrate from the JAPEX/JNOC/GSC Mallik 2L-38 gas hydrate research well. *In* GSC Bulletin 544; Scientific Results from JAPEX/JNOC/GSC Mallik 2L-38 Gas Hydrate Research Well, Mackenzie Delta, Northwest Territories, Canada. S.R. Dallimore, T. Uchida & T.S. Collett, Eds.: 251–262.
5. UCHIDA, T. *et al.* 1999. Summary of physicochemical properties of natural gas hydrate and associated gas-hydrate-bearing sediments, JAPEX/JNOC/GSC Mallik 2L-38 gas hydrate research well, by the Japanese research consortium. GSC Bull. **544:** 205–228.
6. KATSUBE, T.J. *et al.* 1999. Petrophysical environment of sediments hosting gas hydrate, JAPEX/JNOC/GSC Mallik 2L-38 gas hydrate research well. GSC Bull. **544:** 109–124.
7. UCHIDA, T. *et al.* 1997. Methane hydrates in deep marine sediments—X-ray CT and NMR studies of ODP Leg 164 hydrates. Chishitsu News (J. Geolog. Survey Japan). **510:** 36–42. (In Japanese.)
8. MATSUMOTO, R. *et al.* 2000. Occurrence, structure and composition of natural gas hydrates recovered from the Blake Ridge, ODP Leg 164, Northwest Atlantic. Proceedings of the Ocean Drilling Program, Scientific Reports, College Station, Texas. 13–28.
9. UCHIDA, T. *et al.* 1999. Dissociation properties of natural gas hydrate from the JAPEX/JNOC/GSC Mallik 2L-38 gas hydrate research well by X-ray computerized tomography (CT) experiments. GSC Bulletin **544:** 269–280.

Occurrences of Natural Gas Hydrates beneath the Permafrost Zone in Mackenzie Delta

Visual and X-ray CT Imagery

TAKASHI UCHIDA,[a,b] SCOTT DALLIMORE,[d,e] AND JUN MIKAMI[a,c]

[a]*JAPEX Research Center, 1-2-1 Hamada, Mihama, Chiba, 261-0025 Japan*

[d]*Geological Survey of Canada, 601 Booth Street, Ottawa K1A 0E8, Canada*

ABSTRACT: The JAPEX/JNOC/GSC Mallik 2L-38 research well was drilled to a depth of 1,150 m beneath the permafrost zone in the Mackenzie Delta, N.W.T., Canada, early in 1998. A large amount of natural gas hydrates were successfully retrieved from a variety of sandy and gravel sediments. Over 110 m of gas hydrate-bearing sediments were found to be distributed between 897 m and 1,100 m deep. Approximately 37 meters of core were recovered in this interval with most of the recovered gas hydrates being less than 2 mm in size occurring mainly in intergranular porosity of silty to clean massive sand and conglomerate (granule to pebble). Typically, hydrate-bearing strata were between 10 cm and more than one meter thick with an estimated porosity of 25 to 35%. The largest form of hydrate was about 2 cm in diameter, occurring as clasts and intergranular porosity within granular sands. Occurrences of natural gas hydrate have been observed visually at the drill site and in core samples preserved in pressurized storage vessels utilizing an X-ray CT scanner technique. Quantitative assessments of gas hydrate concentrations in core samples have been made based on pressure response of dissociation vessels and direct volumetric measurements. Six types of gas hydrate have been recognized: (1) pore-space hydrate, (2) platy hydrate, (3) layered/massive hydrate, (4) disseminated hydrate, (5) nodule hydrate, and (6) vein/dyke hydrate. The X-ray CT images proved useful for characterizing macroscopic forms of gas hydrate. Finer grained occurrences were more difficult to study, however the distribution of gas hydrates and granular grains can be recognized. The occurrences of natural gas hydrates in the Mallik well are compared to the previous natural gas hydrate core samples obtained from ODP/DSDP programs and other field studies.

INTRODUCTION

This paper is a contribution to the ongoing research program on methane hydrates, with financial support from JNOC (Japan National Oil Corporation) and ten oil developing, gas supplying, and electricity developing companies in Japan. Gas hydrates are widespread in several Arctic sedimentary basins associated with

Telecommunication.
[b]Takashi Uchida: uchida@rc.japex.co.jp
[c]Jun Mikami: mikami@rc.japex.co.jp
[e]Scott Dallimore: SDallimo@NRCan.gc.ca

deep permafrost, and offshore along the continental margins. Although gas hydrates have been identified in these settings, relatively little is known about the geologic aspects that control their occurrence and distribution. Laboratory data generated for pure water and methane systems provide a general guide to methane hydrate stability conditions. In areas with thick permafrost, conventional pressure and temperature plots indicate that methane hydrate may exist at depths to 1,500 m depending on the geothermal gradient. Methane hydrate can form within ice-bonded sediments (intrapermafrost gas hydrate) and beneath the base of ice bonding (sub-permafrost gas hydrate).

In February and March of 1998, the JNOC/JAPEX/GSC Mallik 2L-38 research well was drilled to the depth of 1,150 m at the Northeastern edge of the Mackenzie Delta, Northwest Territories in Canada.[1] A drill site near the Mallik L-38, an industry exploratory well drilled in 1972 by Imperial Oil, was selected for the location of the gas hydrate research well. This site was chosen since it offered favorable logistics and has the thickest known gas hydrate occurrence in the region. In addition, detailed geologic, geophysical and engineering data were available from the original well from the archives of the National Energy Board and Imperial Oil Ltd. Cores collected from the Mallik 2L-38 well represent the first documented samples from beneath permafrost, where gas hydrate of various forms was observed in a variety of sediments.[2] The purposes of this paper are to describe the natural gas hydrates found within Mallik sediments and to review X-ray CT analyses of several samples. An additional goal is provide a comparison of gas hydrates obtained from the Mallik 2L-38 research well with samples collected from other sites from around the world.

GENERAL GEOLOGY

The Mallik 2L-38 well, which reached a target depth of 1,150 m, penetrated Iperk, Mackenzie Bay, and Kugmalit Sequences of the Oligocene to Holocene age.[3,4] The Iperk Sequence (0–346 m) includes Pliocene to Holocene sediments which lie on a marked basal unconformity. Iperk strata at the Mallik site are expected to be unconsolidated sediments with a dominance of fluvial deposits representing delta plain and coastal plain environments. The Mackenzie Bay Sequence (346–926.5 m) includes upper Oligocene to Miocene sediments. Strata at the Mallik site consist mainly of unconsolidated to weakly cemented sands and silts. The Kugmallit Sequence (926.5–1,150 m) consists of Oligocene sediments. Kugmallit strata at the Mallik site are expected to consist of interbedded, unconsolidated to weakly cemented sandstone with interbeds of siltstone and mudstone. Delta plain and delta front depositional environments are represented. A number of regional hydrocarbon fields occur within Kugmallit strata, especially in offshore areas of the Beaufort Sea.

Permafrost is ubiquitous beneath terrestrial areas of the Mackenzie Delta extending to a depth of 640 m at the Mallik 2L-38 well site, creating suitable pressure and temperature conditions for up to 1,200 m of methane hydrate. The occurrence of natural gas hydrates in the Mackenzie Delta–Beaufort Sea area has been known for more than 20 years. Indirect evidence, such as well log interpretation or gas shows, has been cited in 63% of the offshore wells and 17% of the onshore wells.[5] Prior to the Mallik 2L-38 the only actual gas hydrate samples collected from the area were from 92GSCTaglu, a 451-m deep research core hole completed by the Geological Survey of Canada in 1992.[6] Intrapermafrost methane hydrate veins and pore space hydrates were described within a variety of unconsolidated sediments.[7] It was noted that methane hydrate core samples displayed a metastable behavior at negative temperatures and atmospheric pressures. This phenomenon, referred to as the self-preservation behavior, was speculated to occur *in situ* at depths as shallow as 119 m.

The gas hydrate occurrences within Mallik L-38 are reported by Bily and Dick,[8] who interpreted a layered sequence of hydrates between 819–1,007 m primarily on the basis of electric resistivity and mud gas anomalies. Ten significant hydrate layers can be identified with a total thickness of approximately 112 m. Volumetric calculations based on well-log response by Collett and Dallimore,[9] suggested that these strata had porosities in the range of 33 to 40%, with very high pore-space hydrate concentrations ranging from 50% to more than 80%.

NATURAL GAS HYDRATE SAMPLES AND THEIR OCCURRENCES

In total, 37 meters of core were recovered from the Mallik 2L-38 between 886 and 952.6 m in the interval anticipated to contain gas hydrates. Well log interpretations,[10,11] and mud gas readings confirm that this interval contained high gas hydrate concentrations. In addition to the descriptions of gas hydrates within the core samples, gas hydrate dissociation tests conducted in the field confirmed gas hydrate concentrations from trace amounts to more than 40% of the pore space from 896 to 925m. Modeling by Wright *et al.*,[12] has confirmed that *in situ* gas hydrate concentrations are very high.

After initial descriptions and testing at the drill site, a limited number of core samples containing gas hydrate were stored either in pressure vessels or liquid nitrogen for specialized analyses. The samples in plastic bags were stored in 300 ml and 920 ml pressure vessels each charged with nitrogen gas to 7 MPa and transported to Japan. These samples were maintained at –30 to –70°C until analyses at various JNOC and JAPEX (Japan Petroleum Exploration Company) laboratories.

Within the cored interval between 886 m and 952 m, gas hydrates were observed to occur primarily in fine- to coarse-grained sands, to pebbly sands interbedded with hydrate-free or low hydrate content silts, and clays. Most of the gas hydrates recovered from Mallik 2L-38 are smaller than 2 mm in diameter, occurring mostly in the intergranular pore space. Typically gas-hydrate–bearing strata were between 10 cm and more than 1.5 m thick with an estimated porosity of 25 to 35%.[2] The largest visible form of pure gas hydrate was about 2 cm in diameter, occurring as small clasts, nodules, and fills of intergranular porosity within sandy gravel at 913.7 m.[2] It is noted that gas hydrates were almost never found in muddy sediments such as silts and clays.

The mode of gas hydrate dissociation was observed at the well site, and has been also recorded by video under laboratory conditions (about 20°C). In a few instances gas hydrates were found to dissociate violently with relatively rapid gas release. However, the dissociation in most cores was relatively calm with gradual effervescence of methane gas from intergranular porosity in sandy to pebbly sediments. Even with relatively rapid warming, the rate of dissociation was not high enough to generate *in situ* combustion. We attribute the relative stability of the gas hydrate to be caused in part by the self-preservation phenomena[13] described as an apparent metastable behavior of gas hydrate whereby the dissociation is slowed at atmospheric pressures by endothermic cooling causing an ice coating.

X-RAY CT OBSERVATIONS

The X-ray CT scanner is a useful nondestructive technique for recognizing occurrences of gas hydrate within sediments and their relationship to surrounding host sediments.[14] All of the core samples taken to Japan in pressure vessels were subjected to X-ray CT imagery before undertaking other analyses, led by the Japanese research consortium. The analyses were conducted with a Toshiba X-ray CT scanner *X-force* owned by JNOC Technology Research Center. This equipment, originally designed for medical use on human subjects, was slightly modified for observation of sediment/rock samples. The resolution of images acquired was 0.35 square mm per pixel and 1 mm in thickness due to the low output voltage of 130 kV. Because X-rays can penetrate aluminum, the system has the capability to observe hydrate-bearing sediments in an aluminum vessel pressurized by nitrogen or methane. However, CT images of the Mallik 2L-38 samples were directly measured without an aluminum container to obtain sharper images. The samples were placed on the measuring bed with dry ice to prevent gas hydrate dissociation. A major benefit of the method is that the images can be collected relatively rapidly with almost no gas hydrate dissociation (about 16 images/min).

Measured CT values are unitless, and depend mainly on the X-ray absorption coefficient of that related to the density of substance scanned. According to Uchida *et al.*,[14] and Matsumoto *et al.*,[15] who conducted an X-ray CT study of gas hydrate samples from Leg 164 of the Ocean Drilling Program, CT values of methane hydrates range from −100 to −250, whereas those of pure water, pure water ice, and dry ice are 0, −65, and 400, respectively (see FIGURE 1). CT values of sediment

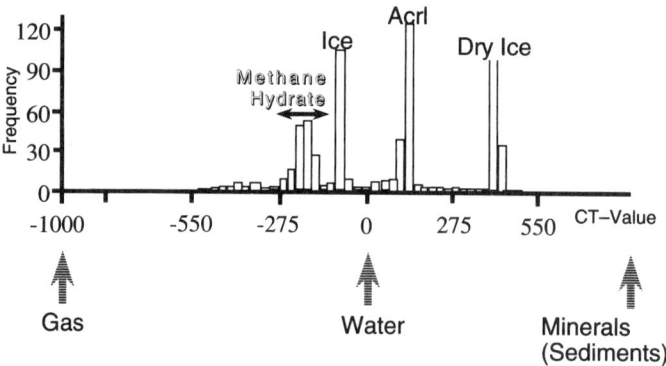

FIGURE 1. Frequency of CT values in the nodule hydrate from the ODP Leg 164 (Uchida *et al.*, 1997).

grains (minerals) usually exceed 1,000, so those of common gases diminish to −1,000, which reflect their densities.

RESULTS AND DISCUSSION

The X-ray CT imageries of gas hydrate–bearing sediments obtained from 913.7 and 903.0 m are shown in FIGURES 2 and 3 with associated frequency histograms of CT values shown in FIGURES 4 and 5, respectively. The CT value ranges are divided into eight regions: (1) CT value > 1,600, (2) 1,600 > CT value > 1,200, (3) 1,200 > CT value > 1,000, (4) 1,000 > CT value > 900, (5) 900 > CT value > 200, (6) 200 > CT value > 0, (7) 0 > CT value > −250, and (8) −250 > CT value.

The two CT images and their histograms show quite different characteristics, attributed to the sediment properties, as well as forms of gas hydrate present in the samples. Visual observations of the sediment sample of 913.7 m revealed that it consisted of granulars and pebbles (maximum 1 cm) with lesser amounts of sand and gas hydrate and/or ice. The coarse granules and pebbles fragments are represented by the areas whose CT values are larger than 1,600 and 1,200 indicating mineral components (FIGURES 2 and 4). The intergranular pore system of the sand, which is

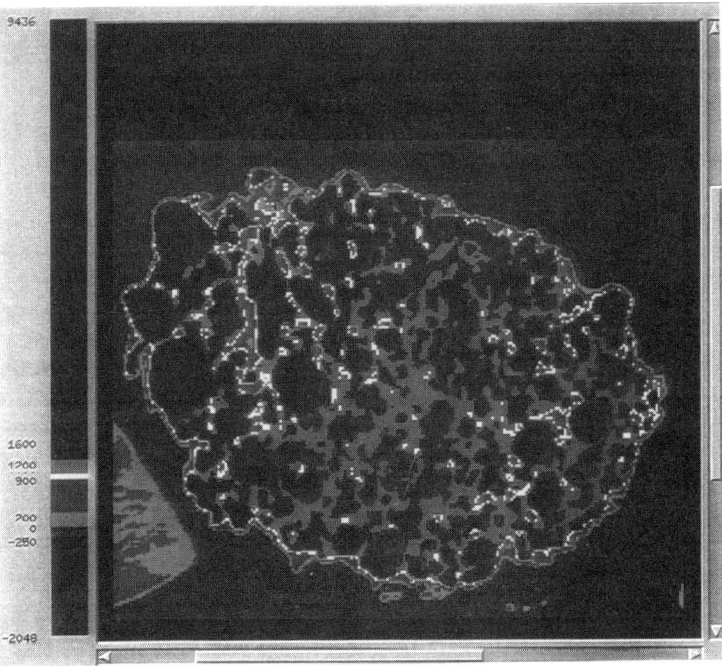

FIGURE 2. Computerized tomography image of the sand sample obtained from a depth of 913.7 m (GH91370) showing the biggest form of pore-space and nodule hydrates that fills the intergranular porosity of the granular sand.

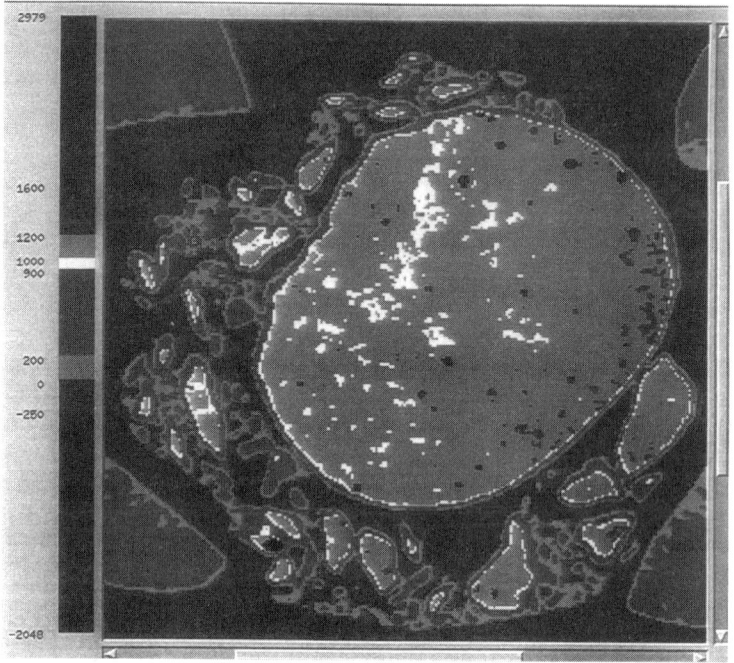

FIGURE 3. Computerized tomography image of the sand sample obtained from a depth of 903.0 m (GH90300) showing the pore-space hydrate that fills intergranular porosity of the unconsolidated sand.

highly saturated by gas hydrate, is considered to be represented by the areas whose CT values range from 900 to 1,200. The sample from 903.0 m showed a little different CT character (FIG. 3) with a more uniform CT values, similarly ranging from 900 to 1,200. Visual observations indicated that this sample consisted of a uniform, medium-grained sand with high pore-space gas hydrate concentrations. The histogram confirmed that the mineral fragments with high CT values (above 1,200), observed in the sample from 913.7 m, are absent (FIGURES 4 and 5).

The CT values of some substances depend on their densities, and are computed by means of the following equation:

$$\text{CT value}_i = \frac{\mu_i - \mu_w}{\mu_w} \times 1000,$$

where CT value$_i$ is a CT value of substance, μ_i is density of substance i, and μ_w is the density of water (1.0). The CT value of quartz is calculated to be 1,650 due to its density of 2.65, and the CT value of methane hydrate is considered to range from −100 to −250 (Uchida et al., 1997).[14] FIGURE 6 shows the relationships between the quartz grain content and methane hydrates, which should indicate the methane hydrate saturation within sediments, assuming the CT value of methane hydrate to be −200. When the CT values of some parts within the sediment sample range from

1,000 to 1,250, the methane hydrate content of those parts are considered to range from 23 to 36 volume percent, which is consistent with the porosities of sediments obtained from the Mallik 2L-38 well[2] and may indicate a high saturation of methane hydrate in the intergranular porosity of the sand sample.

Subsurface occurrences of gas hydrate within sediments are classified into six types (FIG. 2): (1) pore-space hydrate, (2) platy hydrate, (3) layered/massive hydrate, (4) disseminated hydrate, (5) nodule hydrate, and (6) vein/dyke hydrate. Each type may be characterized as follows;

1. A pore-space hydrate fills intergranular porosities of sands and sandstones, and is expected to be interconnected in their pore systems, which clearly contrasts with nodule and disseminated types. Most gas hydrates retrieved from the Mallik 2L-38 were pore-space types with a small amount of possible nodule type. A pore-space hydrate is small-sized and ranges up to 10 mm. However, it is considered to decompose continuously and effectively when produced.

2. A platy hydrate should be similar to thin layer of hydrate, whose thickness may range from 10 to 100 mm.

3. A layered hydrate and massive hydrate are extensively continuous horizontally and concordant to a strata, whose thickness should exceed 100 mm. The difference

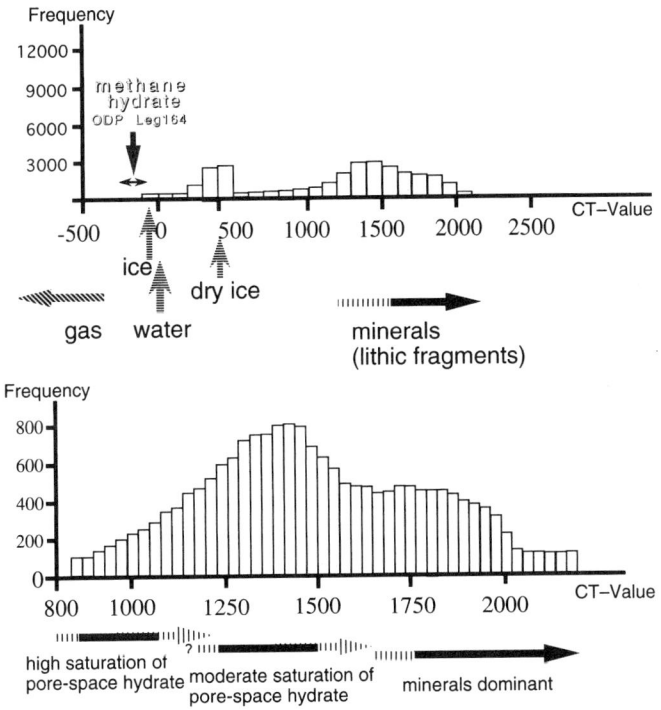

FIGURE 4. Frequency of CT values in GH91370. **Top:** a frequency in a wider range. **Bottom:** a part of the top histogram in a limited range.

between a layered hydrate and massive hydrate cannot be clearly described since no unquestionable occurrence of massive hydrate has yet been found.

4. Each of disseminated hydrates should range from approximately 1 to 50 mm in size, but may not be interconnected in pore system of sediments that is different from the interganular hydrate.

5. A nodule hydrate should be larger than approximately 50 mm in size, and distribute sporadically in sediment strata. This type of occurrence has been frequently documented from many fields such as ODP Leg 164.[14]

6. Vein hydrate and dyke hydrate are usually thin and may elongate discordantly to the strata. Those types of occurrence have been reported from ODP Leg 112,[16] Leg 164,[14,17] and Mackenzie Delta (the research well of 92GSCTAGLU [6,7]), and so on. The abundance of gas hydrate and the ideal reservoir characteristics of the host strata suggest that they represent a considerable natural gas resource in this environment.

For comparative purposes, a summary of natural gas hydrates from around the world are listed in TABLE 1. The physical nature of the gas hydrate samples are interpreted from descriptions of each paper according to the six occurrence types.

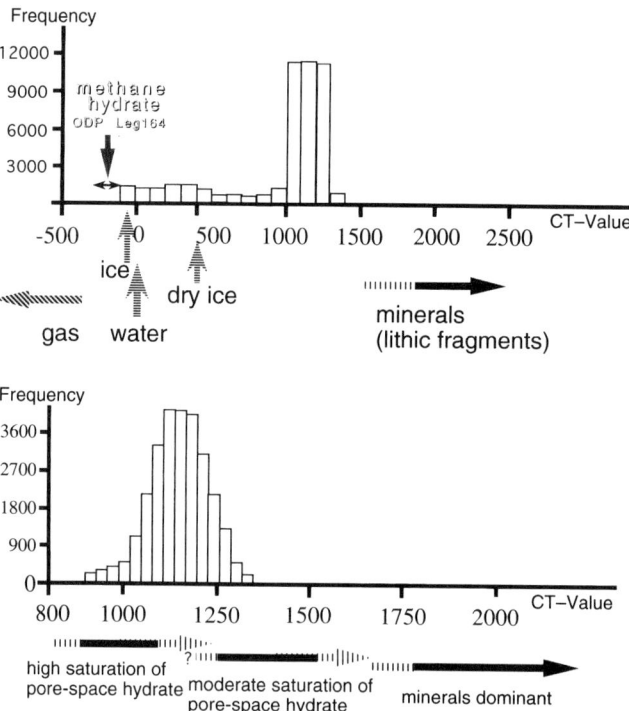

FIGURE 5. Frequency of CT values in GH90300. **Top:** a frequency in a wider range. **Bottom:** a part of the top histogram in a limited range.

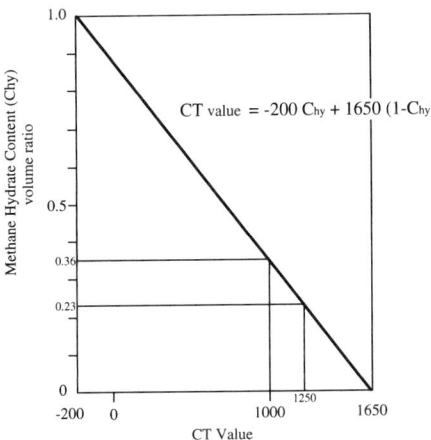

FIGURE 6. Relationships between the contents of quartz grains and methane hydrate.

Gas hydrates retrieved from the Mallik 2L-38 research well mostly consist of pore-space type possibly with a small amount of nodular type, whereas such a pore-space type has never been documented worldwide, except a disseminated type. In addition to these examples, other hydrate samples have been recovered from the Eel River Basin off California,[18] Black,[19] Caspian Sea,[20] Okhotsk Sea,[21] Baikal Lake,[21] North Slope,[22] and Messoyakha,[23] whose properties have not yet been clarified.

SUMMARY AND CONCLUSIONS

Subsurface occurrences of natural gas hydrate are sorted into six types: (1) pore-space hydrate, (2) platy hydrate, (3) layered/massive hydrate, (4) disseminated hydrate, (5) nodule hydrate, and (6) vein/dyke hydrate. Gas hydrates retrieved from Mallik 2L-38 mostly consist of pore-space type, possibly with a small amount of nodular type. Prior occurrences of natural gas hydrates that have been reported world-wide were investigated in comparison with the gas hydrates obtained from Mallik 2L-38. This is the first natural gas hydrate sample retrieved from beneath the permafrost zone.

Concerning X-ray CT observations of gas hydrate-bearing sands and granular sand, the CT value range separates into eight regions and relationships among granular lithic fragments, associated coarse mineral, and gas hydrates highly saturated the intergranular pore system of the sediment samples, and their shapes, occurrences, and textural relations to the surrounding host sediments can be recognized by X-ray CT imagery. Depending on the CT values of some parts of the sediment sample, the methane hydrate contents of those parts are estimated to range from 23 to 36 volume percent, which is consistent with the porosities of sediments obtained from

TABLE 1. Occurrence of natural gas hydrates[2]

Area	Water depth m	Subbottom depth m	Gas Hydrates					Reference
			Occurrences	Size	$C_1/(C_2+C_3)$	$\delta_{13}C(C_1)$ ‰	Gas/Water m³/ml	
Onshore								
Mackenzie Delta 1998	—	900+	pore space, vein	≤12 mm				this study
Mackenzie Delta 1992	—	336, 354	platy, pore space, vein	≤50 mm				Dallimore & Matthews (1997)
Offshore								
Okhots Sea 1991	710	1	layered, nodule	≤10 mm	18600	−64		Ginsberg et al. (1993)
Offshore Oregon 1996?	?	0.5	layered	≤10 mm				
Offshore Oregon								
Leg 146, Site 892	670	3.6	disseminated	4–5 mm	969	−66.8		Hovland et al. (1995)
Offshore Mexico								
Leg 66, Site 492	1935	141	disseminated	a few mm	1200	−66.8	−20	Shipley & Didyk (1982)
Offshore Guatemala								
Leg 67, Site 497	2347	368	disseminated?	2 cm	1808	−66.8	−20	Harrison & Curiale (1982)
Leg 67, Site 498	5478	310	disseminated?		416			Harrison & Curiale (1982)
Leg 84, Site 568	2010	404	nodule?	~ 10.4 cm³	530		30	Kvenvolden & McDonald (1985)
Leg 84, Site 570	1698	249	nodule, dyke/massive?	1.05 m×6cm	440	−44	29 ~ 82	Kvenvolden & McDonald (1985)

TABLE 1/continued. Occurrence of natural gas hydrates[2]

Area	Water depth m	Subbottom depth m	Gas Hydrates					Reference
			Occurrences	Size	$C_1/(C_2+C_3)$	$\delta_{13}C(C_1)$ ‰	Gas/Water m^3/ml	
Offshore Costa Rica								
Leg 84, Site 565	3099	285, 319	disseminated?	~1.5 cm^3	2000		133	Kvenvolden et al. (1985)
Offshore Peru								
Leg 112, Site 685	5070	99, 116	disseminated?	1.18 cm^3	14000	−65	100	Kvenvolden & Kastner (1990)
Leg 112, Site 688	3820	141	disseminated? vein	a few mm	42000	−60	13–26	Kvenvolden & Kastner (1990)
Blake Outer Ridge								
Leg 76, Site 533	3184	238	nodule?	4 cm φ	2900	−70	20	Kvenvolden & Barnard (1983)
Leg 164, Site 997	2800	330	nodule, vein	up to 35 cm	5163	−66.2	13.8	Uchida et al. (1997) Matsumoto et al. (1996)
Okushiri Ridge								
Leg 127, Site 796	2570	90	disseminated?	4 cm^3	2900			Shipboard Scientific Party (1983)
Gulf of Mexico								
Leg 96, Site 618	2400	27	disseminated?	1–4 mm	159	−71.3		Brooks et al. (1986)
Garden Banks-388	850	2.8	disseminated?	20 mm	829	−70.4	68	Brooks et al. (1986)
Green Canyon-257	880	4.2	disseminated?	10^3 mm	>10000	−69.2	35	Brooks et al. (1986)
Green Canyon-234	590	1.2, 2.8	nodule?	>150 mm	4.4	−43.2	177	Brooks et al. (1986)
Mississippi Canyon	1300	3.8	disseminated?	2 mm	37.4	−48.2		Brooks et al. (1986)

the Mallik 2L-38 well[2] and may indicate a high saturation of methane hydrate in the intergranular porosity of the sand sample.

ACKNOWLEDGMENTS

We are indebted to the Geological Survey of Canada and the ongoing research program supported by JNOC and ten oil developing companies/gas supplying companies/electric developing companies. Special thanks to Drs. J. Vincent and F. Wright of GSC Ottawa and Messrs. A. Nakamura and M. Imazato of JNOC and T. Ohara and O. Senoh of JAPEX who were responsible for carrying out the Mallik 2L-38 well project successfully as well as to JNOC Technology Research Center for permission to use the X-ray CT scanner.

REFERENCES

1. DALLIMORE, S.R., T. UCHIDA & T.S. COLLETT. 1998. JAPEX/JNOC/GSC Mallik 2L-38 gas hydrate research well: Overview of science program. Proceedings of the International Symposium on Methane Hydrates Resources in the Near Future? JNOC-TRC, Chiba City, Japan, October 20–22. 311–318.
2. UCHIDA, T., S. DALLIMORE, J. MIKAMI & M. NIXON. 1999. Occurrences and X-ray computerized tomography (CT) observations of natural gas hydrate, JAPEX/JNOC/GSC Mallik 2L-38 gas hydrate research well. In Scientific Results from JAPEX/JNOC/GSC Mallik 2L-38 Gas Hydrate Research Well, Mackenzie Delta, Northern Territories, Canada. S.R. Dallimore, T. Uchida, and T.S. Collett, Eds. Geological Survey of Canada, Bulletin 544. 197–204.
3. DIXSON, J., J.R. DIETRICH & D.H. MCNEIL. 1992. Upper Cretaceous to Pliocene sequence stratigraphy of the Beaufort–Mackenzie and Banks Island areas, northwest Canada. Geological Survey of Canada Bulletin 407, 90p.
4. JENNER, K.A., S.R. DALLIMORE, F.M. NIXON, W.J. WINTERS & T. UCHIDA. 1998. Sedimentology of methane hydrate host strata from JAPEX/JNOC/GSC Mallik 2L-38. Proceedings of the International Symposium on Methane Hydrates Resources in the Near Future? JNOC-TRC, Chiba City, Japan, October 20–22, 319–326.
5. SMITH, S.L. & A.S. JUDGE. 1993. Gas hydrate database for Canadian Arctic and selected east coast wells. Geological Survey of Canada Open File Report 2746, 7.
6. DALLIMORE, S.R. J.V. MATTHEWS. 1997. The Mackenzie Delta Borehole Project. Environmental Studies Research Funds Report No. 135, 1 CD Rom.
7. DALLIMORE, S.R. & T.S. COLLETT. 1995. Intrapermafrost gas hydrates from a deep core hole in the Mackenzie Delta, Northwest Territories, Canada. Geology 23: 527–530.
8. BILY, C. & J.W.L. DICK. 1974. Natural occurring gas hydrates in the Mackenzie Delta, Northwest Territories. Bulletin of Canadian Petroleum Geology 22: 340–352.
9. COLLETT, T.S. & S.R. DALLIMORE. 1998. Quantitative assessment of gas hydrates in the Mallik L-38 Well, Mackenzie Delta, N.W.T. Proceedings of the 7th International Conference on Permafrost, Yellowknife, Canada, June 1998.
10. COLLETT, T.S., T. UCHIDA, S.R. DALLIMORE & T.H. MROZ. 1998. Downhole well log evaluation of gas hydrates in the the Mallik 2L-38 well, Mackenzie Delta, N.W.T., Canada. Proceedings of the International Symposium on Methane Hydrates Resources in the Near Future? JNOC-TRC, Chiba City, Japan, October 20–22, 349–358.

11. MIYAIRI, M., K. AKIHISA, T. UCHIDA, T. COLLETT & S. DALLIMORE. 1999. Well log interpretation of natural gas hydrate-bearing formations in the Mallik 2L-38 Well. *In* Scientific Results from JAPEX/JNOC/GSC Mallik 2L-38 Gas Hydrate Research Well, Mackenzie Delta, Northern Territories, Canada. S.R. Dallimore, T. Uchida & T.S. Collett, Eds. Geological Survey of Canada, Bulletin **544**: 281–294.
12. WRIGHT, J.F. & A.E. TAYLOR. 1999. Estimating in-situ gas hydrate saturation from core temperature observations: Mallik 2L-38 well site, Mackenzie Delta, Canada. *In* Scientific Results from JAPEX/JNOC/GSC Mallik 2L-38 Gas Hydrate Research Well, Mackenzie Delta, Northern Territories, Canada. S.R. Dallimore, T. Uchida & T.S. Collett, Eds. Geological Survey of Canda, Bulletin **544**: 101–108.
13. ERSHOVE, E.D. & V.S. YAKUSHEV. 1992, Experimental research on gas hydrate decomposition in frozen rocks. Cold Regions Science and Technology **20**: 147–156.
14. UCHIDA, T., J. YAMAMOTO, S. OKADA, A. WASEDA, K. BABA, K. OKATSU, R. MATSUMOTO & SHIPBOARD SCIENTIFIC PARTY OF ODP LEG 164. 1997. Methane hydrates in deep marine sediments—X-ray CT and NMR studies of ODP Leg 164 hydrates. Chishitsu News (Journal of the Geological Survey of Japan) **510**: 36–42 (in Japanese).
15. MATSUMOTO, R., T. UCHIDA, A. WASEDA, T. UCHIDA, S. TAKEYA, T. HIRANO, K. YAMADA, Y. MAEDA & T. OKUI. 2000. Occurrence, structure and composition of natural gas hydrates recovered from the Blake Ridge, ODP Leg 164, Northwest Atlantic. ODP Proc. Sci. Results **164**: 13–28.
16. KVENVOLDEN, K.A. & M. KASTNER. 1990. Gas hydrates of the Peruvian outer continental margin. Proc. ODP Sci. Results **112**: 517–526.
17. MATSUMOTO, R., Y. WATANABE, M. SATOH, H. OKADA, Y. HIROKI, M. KAWASAKI & ODP LEG 164 SHIPBOARD SCIENTIFIC PARTY. 1996. Distribution and occurrence of marine gas hydrates—Preliminary results of ODP Leg 164-Blake Ridge Drilling. J. Geol. Soc. Japan **102**: 932–944.
18. KENNICUT, M.C., J.M. BROOKS & H.B. COX. 1993. The origin and distribution of gas hydrates in marine sediments. *In* M.H. Engel & S.A. Macko, Eds.: 535–544. Org. Geochem. Plenum Press, New York.
19. YEFREMOVA, A.G. & B.P. ZHIZHCHENKO. 1972. Doklady Nefti Gaza **2**: 1179.
20. YEFREMOVA, A.G. & N.D. GRITCHINA. 1981. Geologuva Nefti Gaza **2**: 32.
21. MAKOGON, Y.F. 1988. Natural gas hydrates: the state of study in the USSR and perspectives for its use. The Third Chemical Congress of North America, Toronto, Canada.
22. COLLETT, T.S. 1993. Natural gas hydrates of the Prudhoe Bay and Kuparuk River Area, North Slope, Alaska. A.A.P.G. Bull. **77**: 793–812.
23. MAKOGON, Y.F., F.A. TREBIN, A.A. TROFIMUK, V.P. TSAREV & N.W. CHERSKY. 1972. Detection of a pool of natural gas in a solid (hydrate gas) state. Doklady Academy of Sciences U.S.S.R., Earth Science Section. **196**: 197–200.

Hydrate Phase Composition for Multicomponent Gas Mixtures

HENG-JOO NG[a]

DB Robinson Research Ltd., 9419–20th Avenue, Edmonton, Alberta, Canada T6N 1E5

ABSTRACT: This investigation was carried out to determine the composition of guest components in the equilibrium hydrate phase. Measurements were made for six gas mixtures containing methane, ethane, propane, iso-butane, *n*-butane, and carbon dioxide. Gas hydrate was formed in the presence of water at two pressures, 2.07 and 6.89 MPa, and at temperatures 2°C below the corresponding equilibrium hydrate temperature. The results obtained showed that, with some exceptions, the van der Waals-Platteeuw thermodynamic model can be used for hydrate phase composition calculations.

INTRODUCTION

Hydrate formation has long been known to cause problems in natural gas and offshore untreated reservoir fluid production, processing, and transportation. Most research has been in the area of understanding the thermodynamics and kinetics of hydrate formation and inhibition. The discovery of *in situ* hydrate deposits creates interest in hydrate formation in porous media as well as production of hydrate from porous media. The phenomenon of hydrate formation can be used for separation technology because the composition of the hydrate formers in the hydrate crystal differs from that in the original mixture of the hydrocarbon (gas or liquid) phase. One can separate a hydrocarbon (gas or liquid) mixture by forming hydrates under conditions that will result in the desired fractionation. Sour gases such as hydrogen sulfide and carbon dioxide may be separated from natural gas by using hydrate technology. Happel *et al.*[1] presented a novel method separating nitrogen from methane.

A statistical thermodynamic method developed by van der Waals and Platteeuw[2] is the most widely used method for hydrate equilibrium calculations. The distribution of hydrate formers in the various cavities can be calculated from this method.[3,4] However, the accuracy of the composition of the hydrate phase for multicomponent systems, such as natural gas systems, calculated by the model has not been verified. Equilibrium composition measurements of the hydrate phase have typically been avoided due to experimental difficulties such as water occlusion, solid phase inhomogeneities, and measurements of solid phase concentrations. Hence, experimental data on composition distribution reported in the open literature is limited. Sum *et al.*[5] reported hydrate phase composition data for two methane/carbon dioxide binary mixtures using Raman spectroscopy.

[a]Telecommunication. Voice: 780-463-8638; fax: 780-450-1668.
hjng@dbra.com

In this study, a novel experimental method has been developed to determine the equilibrium hydrate phase composition. Experimental data for six gas mixtures are presented.

EXPERIMENTAL DETAILS

Experimental Equipment

The experimental measurements were carried out in a variable volume, blind, 316 stainless steel equilibrium cell, shown schematically in FIGURE 1. The equilibrium cell was immersed in a temperature controlled cooling bath. The contents of the cell were agitated with a retractable turbine blade impeller. The impeller was driven by a variable speed D.C. motor with a maximum speed of 1,000 rpm. The cell had a maximum working pressure of 20 MPa and was equipped with a floating piston. The maximum working volume of the cell was about 300 cm^3.

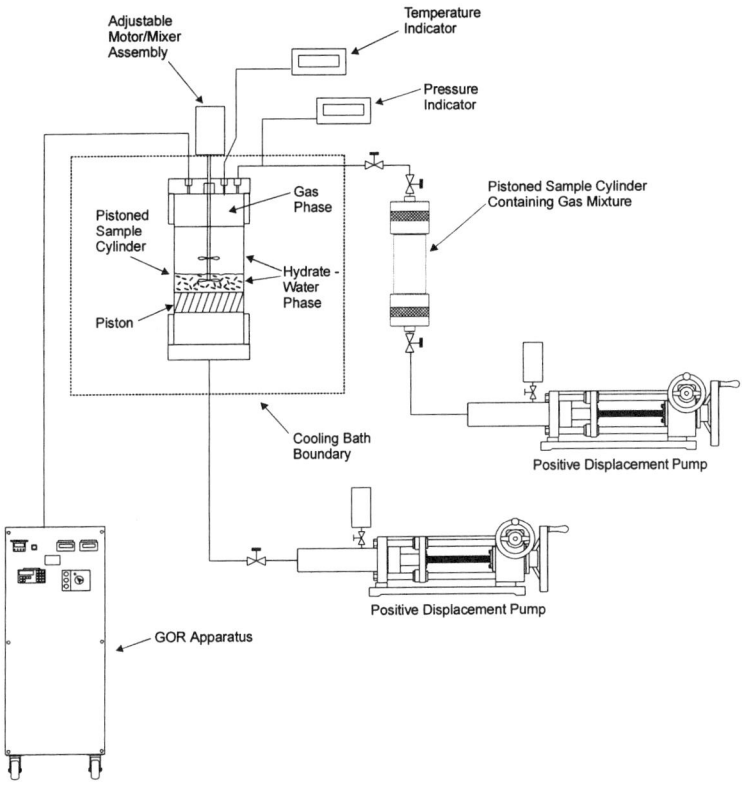

FIGURE 1. Schematic of the experimental apparatus.

Temperatures were measured with a resistance temperature device (RTD) and are known to within ±0.05°C. Pressures were measured using a 20 MPa calibrated Heise digital gauge. Pressures are known within ±7 kPa. An HP5890 chromatograph with a combination TCD and FID detectors was used for the compositional analyses. The concentration of major components determined by chromatography is known to within ±0.003 mole fraction.

Experimental Procedures and Methods

The experimental gas mixtures were prepared gravimetrically from pure components and stored at elevated pressures in the single-phase condition in stainless steel pipettes. In this study, each gas mixture was spiked with 0.5 mole percent of n-pentane (a non-hydrate former). Pentane in the gas mixture was used as the internal standard for hydrate phase composition determination for each run.

At the start of a run, the cell was connected to a displacement pump and evacuated. A known amount (approximately 30 cm^3) of water that had been melted from ice was then added to the cell, followed by about 150 cm^3 of the prepared gas mixture at the specified run pressure.

Following the charging process, the pressure and temperature of the cell were adjusted to the required values and equilibration commenced. The speed of the stirrer was controlled to about 300 rpm. The temperature of the cell was controlled at 2°C below the hydrate temperature for the specified pressure of 2.07 or 6.89 MPa. The hydrate formation temperature at the specified pressure was based on the value predicted by the EQUI-PHASE hydrate program.[6] In order to reduce the hydrate induction time, the pressure of the cell was increased by the injection of hydraulic fluid into the back of the piston. After initial hydrate formation, the system pressure was reduced to the original value and the conversion of water to hydrate would last for about 72 hours. During this period of time, the gas mixture in the cell was isobarically displaced and replaced with fresh charges of gas mixture. This step was repeated a number of times in a procedure to that ensure hydrates were formed in the presence of the specified gas mixture. Otherwise the composition of the hydrate phase would be different from that in the gas phase. For a typical run, a total of 125 volume of gas mixture (at run conditions) per volume of water was used.

At the end of the hydrate conversion process, the gas mixture in the cell was displaced and flashed with the original gas mixture. The gas phase of the cell was then displaced isobarically from the cell as much as possible. This was to minimize the amount of original gas mixture remaining in the cell. Following this, the temperature of the cell was increased and the hydrate in the cell was allowed to dissociate. The gas released from the hydrate was expanded and the volume of the gas released was measured by the gasometer at ambient conditions. The measured gas volume is known to within 1.0 cm^3. The gas in the gasometer was homogenized and a sample of the gas mixture was taken for compositional analysis. Based on the composition of n-pentane in the overall mixture, the amount of original gas mixture could be subtracted from the overall gas released. Hence the gas composition in the hydrate phase could be determined.

RESULTS AND DISCUSSIONS

Experimental measurements were made to determine the hydrate phase composition in equilibrium with six gas mixtures. The compositions of the gas mixtures are given in TABLE 1. For each gas mixture, the measurements were made in the hydrate–gas–liquid region at two pressures, 2.07 and 6.89 MPa. The temperature of the run was approximately 2°C below the corresponding incipient hydrate temperature.

A total of twelve experimental runs were made for the six gas mixtures. The data obtained are presented in TABLE 2.

TABLE 1. Composition of the gas mixtures

Mixture	Concentration, mole%					
	Methane	Ethane	Propane	i-Butane	n-Butane	CO_2
1	90	0	10	0	0	0
2	99	0	1	0	0	0
3	84	10	4	2	0	0
4	88	0	8	2	0	2
5	80	10	5	2	3	0
6	86.5	6	4	0.5	1	2

TABLE 2. Experimental compositions of gas mixtures in the hydrate phase

Pressure MPa	Temperature °C (Calc.)[a]	Feed Composition Mixture	Hydrate Composition, mole%					
			C_1	C_2	C_3	iC_4	nC_4	CO_2
2.07	11.4	1	58.6	—	41.4	—	—	—
6.89	20.1	1	65.7	—	34.3	—	—	—
2.07	3.7	2	72.6	—	27.4	—	—	—
6.89	13.2	2	77.1	—	22.9	—	—	—
2.07	10.9	3	60.6	8.7	18.2	12.4	—	—
6.89	19.4	3	59.0	8.7	20.0	12.3	—	—
2.07	12.9	4	60.0	—	29.7	9.2	—	1.02
6.89	21.6	4	63.0	—	28.9	7.8	—	1.04
2.07	11.3	5	58.1	7.4	20.6	10.7	3.1	—
6.89	19.6	5	61.7	8.9	18.1	8.7	2.7	—
2.07	9.8	6	58.4	7.6	26.5	4.4	1.8	1.1
6.89	18.7	6	62.1	7.7	23.8	3.9	1.1	1.0

[a] Calculated by using the EQUI-PHASE HYDRATE program.

The volume of gas released from the hydrate phase was also determined from the total gas volume measured by the gasometer and the n-pentane content in the overall gas composition. Based on the known amount of water charged and the volume of gas released from the hydrate phase, it was possible to determine the occupancy of the hydrate cavities assuming a total conversion of water to hydrate. Results of the determined molar ratios of water to gas of the hydrate phase are presented in TABLE 3. The molar ratios of water to gas determined from the experiments vary from 6.6 to 14. Since the six mixtures studied would form hydrate of structure II and occupy both small and large cavities, the ideal molar ratio is 5.67 water to 1 guest molecule. Results of the test indicate the cavities are not fully occupied. However, the experimental procedure could not ensure total water conversion, and hence the experimental molar ratios obtained should not be used for model development.

The composition of the hydrate phase can be calculated using the EQUI-PHASE HYDRATE program (Windows based application) employing the van der Waals and Platteeuw statistical thermodynamics model modified by Ng and Robinson.[7] FIGURE 2 shows the predicted hydrate phase compositions for the twelve experimental conditions. Except for ethane, the predicted compositions agree quite well with the experimental values for the system studied.

TABLE 3. Measured molar ratio of water to gas in the hydrate phase

Pressure MPa	Temperature C (Calc.)[a]	Number of moles			% of Gas Subtracted	Molar Ratio H_2O/Gas
		Water	Total Gas	Net Gas		
2.07	11.4	2.544	0.334	0.280	16.1	9.1
6.89	20.1	1.166	0.110	0.101	8.1	11.6
2.07	3.7	1.180	0.144	0.133	7.7	8.9
6.89	13.2	1.021	0.181	0.154	14.8	6.6
2.07	10.9	1.027	0.146	0.131	9.9	7.8
6.89	19.4	1.005	0.135	0.098	27.8	10.3
2.07	12.9	1.072	0.112	0.107	4.8	10.1
6.89	21.6	1.035	0.173	0.125	27.8	8.3
2.07	11.3	1.068	0.182	0.164	9.7	6.5
6.89	19.6	1.033	0.165	0.139	15.5	7.4
2.07	9.8	1.076	0.100	0.092	8.0	11.7
6.89	18.7	1.007	0.135	0.119	11.6	8.4

[a]Calculated by using the EQUI-PHASE HYDRATE program.
KEY: Water, number of moles of water charged into the cell; total gas, number of moles of gas released from the hydrate phase and dead volume of the cell; net gas, number of moles of gas released from hydrate based on internal standard-pentane analysis; % of gas subtracted, 100*(Total Gas − Net Gas)/Total Gas. Theoretical molar ratio for structure II is 5.67.

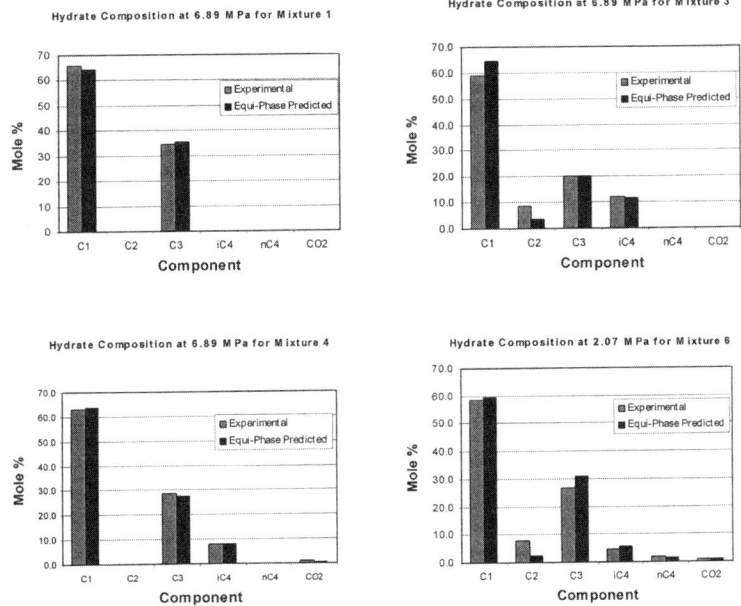

FIGURE 2. Comparison of experimental and predicted hydrate phase compositions.

ACKNOWLEDGMENTS

The author is thankful to Mr. Craig Borman for performing the experimental work and to the Gas Processors Association for their permission to release the experimental work.

REFERENCES

1. HAPPEL, J., M.A. HNATOW & H. MEYER. 1994. The study of separation of nitrogen from methane by hydrate formation using a novel apparatus. International Conference on Natural Gas Hydrates. Ann. N.Y. Acad. Sci. **715:** 412–424.
2. VAN DER WAALS, J.H. & J.C. PLATTEEUW. 1959. Clathrate solutions. Adv. Chem. Phys. **2:** 1–57.
3. SLOAN, E.D., JR. 1997. Clathrate Hydrates of Natural Gas. Marcel Dekker Inc., New York.
4. MAKOGON, Y.F. 1997. Hydrates of Hydrocarbons. PennWell Publishing Company, Tulsa.
5. SUM, A.K., R.C. BURRUSS & E.D. SLOAN, JR. 1996. Measurements of clathrate hydrates properties via Raman spectroscopy. Proceedings 2nd International Conference on Gas Hydrates. J.P. Monfort, Ed.: 51–58. Toulouse.
6. DB ROBINSON & ASSOCIATES. 1999. EQUI-PHASE HYDRATE PROGRAM (Windows based application). Proprietary software. DB Robinson & Associates Ltd., Edmonton, Alberta, Canada.
7. NG, H.-J. & D.B. ROBINSON. 1976. The measurement and prediction of hydrate formation in liquid hydrocarbon-water systems. Ind. Eng. Chem. Fundam. **15:** 293–298.

Index of Contributors

Ahmadi, G., 420–427
Akibayashi, S., 211–225
Allison, E., 437–440
Andersson, V., 322–329, 403–410, 851–858
Andreassen, K., 126–135, 200–210
Aoki, Y., 136–145
Argo, C.B., 355–365
Austvik, T., 294–303, 822–831
Aya, I., 254–260

Ballard, A.L., 702–712
Bartlett, D.B., 32–38
Baxter, T., 832–842
Bazin, N.M., 112–115
Belosludov, R.V., 993–1002
Belosludov, V.R., 993–1002
Bernard, B.B., 76–93
Bidle, K., 32–38
Bishnoi, P.R., 556–563, 576–582
Bloys, B., 350–354
Boissonnas, R., 159–166
Bollavaram, P., 533–543
Booth, J.S., 887–896
Borowski, W.S., 23–31
Borthne, G., 350–354
Brammer, L., 940–949
Brewer, P.G., 195–199
Brooks, J.M., 76–93
Bryant, W.R., 76–93
Bryn, P., 126–135
Burgass, R.W., 832–842, 924–931

Camargo, R., 906–916
Cameron, N.R., 76–93
Canselier, J.P., 753–765
Carson, B., 32–38
Carver, T.J., 658–668
Chapman, N.R., 65–75
Cherry, R.S., 623–632
Cingotti, B., 766–776

Circone, S., 544–555
Clarke, M., 556–563
Clennell, M.B., 887–896
Colgate, S.O., 983–992
Collett, T.S., 51–64, 94–100
Cournil, M., 564–575
Cranston, R.E., 94–100

Dallimore, S.R., 94–100, 1021–1033
Danesh, A., 392–402, 411–419, 832–842, 924–931
de Deugd, R.M., 502–514, 917–923
de Swaan Arons, J., 502–514, 917–923
Dec, S.F., 873–886
Devarakonda, S., 533–543
Dickens, G., 23–31
Drew, M.G.B., 658–668
Duchkov, A.D., 112–115
Durand, J.P., 766–776
Durham, W.B., 544–555, 1003–1010
Dvorkin, J., 116–125
Dyadin, Yu.A., 112–115, 967–972
Dybvik, L., 744–752

Ebinuma, T., 593–601
Edwards, R.N., 65–75, 146–158, 167–172
El Hafaia, T., 753–765
Englezos, P., 576–582

Fadnes, F.H., 722–734
Freer, E.M., 651–657

Geletiy, V.F., 112–115
Gering, K.L., 623–632
Gettrust, J., 65–75
Gjertsen, L.H., 294–303, 722–734, 822–831
Glenat, P., 906–916
Goland, A.N., 940–949
Goldberg, D., 159–166

Goodwin, S.P., 339–349, 355–365
Gudmundsson, J.S., 322–329, 403–410, 602–613, 851–858, 950–957

Haneda, H., 261–271
Hayashi, T., 136–145
Helgerud, M.B., 116–125, 1003–1010
Henderson, A., 355–365
Heng-Joo Ng, H.-J., 1034–1039
Henry, P., 887–896
Herri, J.-M., 564–575
Hirai, S., 246–253
Holder, G.D., *xiii–xiv*, 226–234, 614–622
Holditch, S.A., 777–796
Hondoh, T., 685–692, 973–982
Hori, A., 685–692
Hovland, M., 887–896
Hyndman, R.D., 65–75

Imai, N., 452–459, 932–939
Irvin, G., 515–526
Itoh, H., 693–701

Jaeger, J., 32–38
Jager, M.D., 917–923
Jannasch, H., 32–38
Javanmardi, J., 713–721
Jenner, K.A., 94–100
Ji, C., 420–427
John, V.T., 460–473, 515–526
Jøssang, A., 807–821
Jussaume, L., 753–765

Kalmychkov, G., 112–115
Karaaslan, U., 735–743
Kastner, M., 32–38
Katsube, J.T., 94–100
Kawamura, K., 693–701
Kawazoe, Y., 993–1002
Kelland, M.A., 281–293, 744–752
Khokhar, A.A., 950–957
Kini, R.A., 873–886
Kirby, S.H., 544–555, 1003–1010

Klooster, W.T., 940–949
Klug, D.D., 859–872
Knight, C.A., 441–451
Koetzle, T.F., 940–949
Komai, T., 261–271, 272–280
Kornai, T., 797–806
Krason, J., 173–188
Kuwano, K., 246–253
Kuzmin, M.I., 112–115
Kuznetsov, F.A., 101–111, 112–115
Kvamme, B., 496–501
Kvenvolden, K.A., 17–22, 23–31

Larionov, E.G., 112–115, 967–972
Larsen, R., 441–451, 897–905
Latychev, K., 146–158
Lee, M.W., 51–64
Lee, S.-Y., 614–622
Levik, O.I., 403–410, 602–613, 851–858
Li, S., 515–526
Li, X., 294–303, 822–831
Lorenson, T.D., 23–31, 189–194
Lorimer, S., 366–373
Lukas, D., 200–210
Lynn, R.J., 226–234

Maddox, R.N., 713–721
Mae, S., 593–601
Maekawa, T., 452–459, 932–939
Mahajan, D., 940–949
Mahov, G.M., 112–115
Makogon, T.Y., 777–796
Makogon, Y.F., 777–796
Manakov, A.Yu., 112–115, 967–972
Masuda, Y., 1011–1020
Matsubayashi, O, 167–172
Matsumoto, R., 39–50
Matthews, P.N., 330–338
Max, M.D., 460–473, 515–526
McGee, T.M., 527–532
McMullan, R.K., 940–949
McPherson, G., 515–526
Mehta, A., 366–373
Mienert, J., 126–135, 200–210

INDEX OF CONTRIBUTORS

Mihajlovic, G., 146–158
Mikami, J., 1011–1020, 1021–1033
Mironov, Yu.I., 112–115
Mitchell, G.F., 314–321
Mochizuki, T., 633–641, 642–650
Monfort, J.P., 753–765
Mooijer-van den Heuvel, M.M., 502–514
Mori, Y.H., 633–641, 642–650, 958–966
Mork, M., 403–410, 897–905
Moshfeghian, M., 713–721
Murakoshi, S., 678–684

Nagashima, K., 797–806
Nakayama, K., 136–145
Namba, T., 281–293
Namie, S., 254–260
Nariai, H., 254–260
Narita, H., 593–601
Nixon, F.M., 94–100
Notz, P.K., 330–338
Nur, A., 116–125, 1003–1010

Oelfke, R.H., 314–321
Ogawa, K., 246–253
Ohga, K., 272–280
Ohmura, R., 958–966
Okabe, R., 593–601
Okazaki, K., 246–253
Okuda, Y., 136–145
Østergaard, K.K., 411–419, 832–842, 924–931
Øvsthus, J., 281–293

Palermo, T., 339–349, 355–365, 766–776, 906–916
Parlaktuna, M., 735–743, 851–858
Paull, C.K., 23–31
Pellenbarg, R.E., 460–473, 515–526
Peters, C.J., 502–514, 917–923
Peters, D., 304–313
Pic, J.-S., 564–575
Pinkston, J.C., 544–555, 1003–1010
Posewang, J., 200–210
Prukop, G., 330–338

Reuvers, M., 502–514
Rider, K.T., 441–451
Ripmeester, J.A., 1–16, 859–872
Rodger, P.M., 474–482, 658–668, 669–677
Rogers, R.E., 843–850

Saito, S., 159–166
Sakai, A., 374–391
Sasaki, K., 211–225
Satoh, T., 1011–1020
Schei, G., 897–905
Selim, M.S., 304–313, 533–543
Servio, P., 576–582
Shigetomi, T., 958–966
Shimizu, S., 136–145
Shpakov, V.P., 993–1002
Simmons, B., 515–526
Singh, S.C., 126–135
Sinquin, A., 766–776, 906–916
Sloan, Jr., E.D., 304–313, 441–451, 533–543, 583–592, 651–657, 702–712, 873–886, 917–923, 950–957, 983–992
Smith, D.H., 420–427
Smoljakov, B.S., 112–115
Spence, G.D., 65–75
Stange, E., 807–821
Stern, L.A., 544–555, 1003–1010
Storr, M.T., 669–677
Subramanian, S., 583–592, 873–886, 983–992
Svartaas, T.M.., 281–293, 744–752

Tabatabaei, A.R., 392–402
Tabe, Y., 246–253
Takeda, H., 1011–1020
Takeya, S., 973–982
Talley, L.D., 314–321
Tanaka, T., 136–145
Teng, H, 235–245
Thieu, V., 983–992
Thomas, M., 887–896
Todd, A.C., 392–402, 411–419, 832–842, 924–931

Tohidi, B., 392–402, 411–419, 832–842, 924–931
Tse, J.S., 993–1002
Tsypkin, G.G., 428–436
Tulk, C.A., 859–872

Uchida, T., 94–100, 593–601, 973–982, 1011–1020, 1021–1033
Ussler III, W., 23–31

Vaessen, R.J.C., 483–496
van der Ham, F., 483–496

Waite, W.F., 1003–1010
Wakatsuki, M, 235–245
Wakisaka, A., 797–806
Walia, R., 65–75
Walsh, J., 366–373
Warzinski, R.P., 226–234

Weinberg, D.M., 623–632
Widener, M.W., 330–338
Willoughby, E.C., 146–158
Winters, W.J., 94–100, 887–896
Witkamp, G.J., 483–496
Woolsey, J.R., 527–532
Wright, J.F., 94–100

Yamada, K., 235–245
Yamamoto, Y., 261–271, 272–280, 797–806
Yamane, K., 254–260
Yamane, T., 136–145
Yamasaki, A, 235–245
Yanagisawa, Y, 235–245
Yasuoka, K., 678–684

Zhong, Yu., 843–850
Zhurko, F.V., 967–972